D1072165

Handbook of
# Reactive Chemical Hazards

AN INDEXED GUIDE TO PUBLISHED DATA

# Handbook of Reactive Chemical Hazards

L. BRETHERICK, BSc, CChem, FRIC

Project Leader, New
Technology Division,
Sunbury Research Centre,
British Petroleum Company, Ltd.

BUTTERWORTHS
LONDON - BOSTON
Sydney - Wellington - Durban - Toronto

The Butterworth Group

| | |
|---|---|
| United Kingdom London | Butterworth & Co (Publishers) Ltd 88 Kingsway, WC2B 6AB |
| Australia Sydney | Butterworths Pty Ltd 586 Pacific Highway, Chatswood, NSW 2067 Also at Melbourne, Brisbane, Adelaide and Perth |
| South Africa Durban | Butterworth & Co (South Africa) (Pty) Ltd 152–154 Gale Street |
| New Zealand Wellington | Butterworths of New Zealand Ltd T & W Young Building, 77–85 Customhouse Quay, 1 CPO Box 472 |
| Canada Toronto | Butterworth & Co (Canada) Ltd 2265 Midland Avenue Scarborough, Ontario, M1P 4S1 |
| USA Boston | Butterworth (Publishers) Inc 10 Tower Office Park, Woburn, Massachusetts 01801 |






First published 1975
Second edition 1979
Reprinted 1979

ISBN 0 408 70927 8



**British Library Cataloguing in Publication Data**

Bretherick, Leslie
Handbook of reactive chemical hazards. – 2nd ed.
1. Hazardous substances 2. Reactivity (Chemistry)
I. Title
541'.393 T55.3.H3 78-40367

ISBN 0–408–70927–8

Typeset by Scribe Design, Chatham, Kent
Printed in England by Billings & Sons Ltd, Guildford and London

# Foreword

During the two years in which I chaired the Enquiry Committee on Safety and Health of people at their place of work, what was certainly borne upon us was the tremendous speed of advance of modern technology and the importance of keeping up to date with the safety factors surrounding new developments.

This handbook, the product of Mr. L. Bretherick, certainly adds a new dimension to this whole problem.

It is very clear from its exposition that every single contingency that can at present be contemplated in the whole field of laboratory safety has been covered, and there is no doubt that from time to time there will require to be reviews in the light of chemical development. Sufficient to say, however, that within these pages lies an enormous amount of research work in this whole area of laboratory work in the chemical industry, that there has been collected a tremendous amount of information on hazards in laboratories and that Mr. Bretherick has become recognised as an expert in this field.

Safety Officers who are responsible for safety in industrial laboratories will find this book absolutely essential reading if they are to do their own work to the satisfaction of themselves and of the company which they serve.

I know of no previous book containing so much information as this one, and there is no doubt at all that if the book is diligently used there will be found within its pages answers to many of the problems in relation to safety in the use of chemicals.

I can only say, having spent so much time in this whole field, that the development of the modern technologies is the area in which safety research needs to be constantly updated. In the older industries already there is sufficient information to enable us to conduct our affairs in such a manner as to provide adequate safety and a proper healthy environment for the workers in those fields. It is in the modern fields that the problems arise, because with great frequency there are new discoveries that require fresh thought in relation to safety hazards.

Mr. Bretherick has undoubtedly produced for all those in this field a handbook that will enable the greatest possible safety precautions to accompany the developing sciences. That Mr. Bretherick should have been able to produce what is a monumental contribution to safety is a matter of great credit to him and the enormous amount of research that was required in order to produce this book.

I therefore recommend most strongly a careful survey of the entries in this book relevant to the particular activities of companies in this field, and feel very strongly that Safety Officers require this constantly at their elbow in order to provide the right kind of environment in which work in laboratories can be conducted with greater safety than is possible without the research benefits that have come from this book.

It is absolutely essential in the modern world that safety factors of those employed in research must be adequately safeguarded, and this means that the maximum amount of information must be available at all times, based upon present knowledge. It is because I recognise the value of the work that this book embraces in the whole difficult field of modern research and technology that I unhesitatingly in this foreword make the strongest recommendation for its close attention and use in laboratories.

In our travels as a Committee of Enquiry, we found that many countries in the world look to the United Kingdom for a lead in these matters. This book produces further evidence of the care and attention required in this highly critical area of laboratory research and, therefore, will be welcomed both in the United Kingdom and overseas.

I am certain that as a result of this work a great many people engaged in actual research, and the companies concerned, will find this of considerable help and assistance in safeguarding all those who are engaged in breaking through the frontiers of knowledge.

Robens

LORD ROBENS OF WOLDINGHAM

November, 1974

# Preface to the Second Edition

It has become apparent during the four years since preparation of the first edition that some changes in content would greatly increase the value of the Handbook, not only to practising chemists as originally envisaged, but also to those concerned with chemical safety but not so directly involved in handling chemical materials.

An introductory chapter has been added to try and give an overview of the complex subject and an understanding of the basic principles involved in minimising hazards arising from chemical reactivity.

For most of the 7000-odd entries a structural formula is now included to help to bring out the relation between structure and reactivity, both for individual compounds and for the collective group items.

A full alphabetical index to all the chemicals named in the text is also provided, and this should allow those not sufficiently familiar with the empirical formula-based arrangement of the main section to gain effective access to what they are seeking. Some synonyms of the IUPAC names, including some from the ASE-preferred selection, are given as cross-references in this index.

Some other minor changes in layout and content have been incorporated in an attempt to make this compilation more useful to a wider spectrum of users.

A considerable amount of new and revised information has become available or has been located in the 4 year period since completion of the manuscript of the first edition. The main new sources have been the major inorganic textbooks edited by Bailar and Pascal, the laboratory handbook by Sorbe, the continuing compilations of Leleu and the Institution of Chemical Engineers, and, above all, the one keyword 'Safety' used in *Chemical Abstracts* indexes since January 1974. A further source of much relevant information is the *Manual of Hazardous Chemical Reactions*.

The new material available up to September 1977 has been treated in the same general way as previously, and has been added into the revised text format. Of the 150-odd new alphabetical group entries, those bringing together information on the hazards of industrial unit process operations (nitration, polymerisation, etc.) and all the preparative hazards noted may prove to be the most useful.

I am again indebted to my employers, The British Petroleum Company, Ltd, for their continued generous support of my activities in this area, and to my colleagues, both at Sunbury and elsewhere in the Company, for assistance with information sources.

Other professional colleagues, both in the UK and overseas, have usually responded generously to my written requests for particular information. On the other hand, it is slightly disappointing to have to record that of the many thousands who have by now consulted the first edition, only two persons have been inspired to send information for inclusion in this second edition. However, the appearance of the new *Journal of Hazardous Materials* may be expected to give further stimulus to the dissemination of information vital to those making, handling and using chemicals.

I again record my thanks to the editorial staff of Butterworths, and to Dr R.S. Cahn and Dr E.R. Smith of the Chemical Society for assistance with matters of IUPAC nomenclature.

I am again greatly indebted to my wife and family for their continuing cheerful acceptance of my time-consuming occupation.

L.B.

# Preface to the First Edition

Although I had been aware during most of my career as a preparative chemist of a general lack of information relevant to the reactive hazards associated with the use of chemicals, the realisation that this book needed to be compiled came soon after my reading *Chemistry & Industry* for June 6th, 1964. This issue contained an account of an unexpected violent laboratory explosion involving chromium trioxide and acetic anhydride, a combination which I knew to be extremely hazardous from close personal experience 16 years previously.

This hazard had received wide publicity in the same journal in 1948, but during the intervening years had apparently lapsed into relative obscurity. It was then clear that currently existing arrangements for communicating 'well-known' reactive chemical hazards to practising chemists and students were largely inadequate. I resolved to try to meet this obvious need for a single source of information with a logically arranged compilation of available material. After a preliminary assessment of the over-all problems involved, work began in late 1964.

By late 1971, so much information had been uncovered but remained to be processed that it was apparent that the compilation would never be finished on the spare-time basis then being used. Fortunately I then gained the support of my employers, the British Petroleum Company, Ltd., and have now been able to complete this compilation as a supporting research objective since January, 1972.

The detailed form of presentation adopted has evolved steadily since 1964 to meet the dual needs for information on reactive chemical hazards in both specific and general terms, and the conflicting practical requirements of completeness and brevity. A comprehensive explanation of how this has been attempted, with suggestions on using this Handbook to best advantage, is given in the Introduction.

In an attempt to widen the scope of this work, unpublished information has been sought from many sources, both by published appeals and by correspondence. In this latter area, the contribution made by a friend, the late Mr. A. Kruk-Schuster, of Laboratory

Chemicals Disposal Company, Ltd., Billericay, has been outstanding. During 1965–1968 his literature work and global letter campaign to 2000 University chemistry departments and industrial institutions yielded some 300 contributions.

The coverage attempted in this Handbook is wide, but is certainly incomplete because of the difficulties in retrieving relevant information from original literature when it does not appear in the indices of either primary or abstract journals. Details of such new material known to users of this Handbook and within the scope given in the Introduction will be welcomed for inclusion in supplementary or revised editions of this work.

L.B.

# Acknowledgments

It is a pleasure to acknowledge the contributions of unpublished information on hazardous chemical reactions which have been made by both close colleagues and distant correspondents during the course of several years. Many of the earlier contributors must have despaired of ever seeing their help acknowledged in print (as private communications) in the text, but I hope that they will feel the long delay has been worthwhile. Thanks are also due to those whose contributions could not be incorporated.

I should like to record my special indebtedness to the following, who all provided substantial quantities of information from their own files or experience:

Mr R. N. Beer, BP Chemicals International Limited, Grangemouth
Dr A. F. Crowther, ICI Pharmaceuticals Limited, Macclesfield
Mr K. Everett, Safety Officer, University of Leeds
Dr C. S. Foote, University of California, Los Angeles
Dr H. Feuer, Ohio State University, Columbus
Mr. C. H. R. Gentry }
Mr H. Sutcliffe } Mullard Radio Valve Company, Mitcham
Dr P. R. Huddleston, Trent Polytechnic, Nottingham
Dr E. W. Jenkins, Centre for Studies in Science Education, University of Leeds
Mr A. Kruk-Schuster, Laboratory Chemicals Disposal Co. Ltd, Billericay
Dr E. Levens, Douglas Aircraft Co. Inc., Santa Monica
Dr G.D. Muir }
Mr W.G. Moss } BDH Chemicals Limited, Poole
Dr T. Poles, Engelhard Sales Limited, Cinderford
Dr C. H. Tilford, Warner-Lambert Research Institute, Morris Plains

Thanks are also due to the British Chemical Industries Safety Council, London, and the Manufacturing Chemists Association Inc.,

Washington, whose secretaries gave free use of their regularly published safety material.

I am indebted to colleagues, library and clerical staff at the BP Research Centre for much general encouragement and assistance, and to my employers, the British Petroleum Company Limited, without whose close support since 1971 this work would not have been completed. My former colleagues at May and Baker Limited, Dagenham, were largely responsible for the original development of my personal interest in laboratory safety matters.

The assistance rendered during preparation of this book by the Publisher's staff and their IUPAC nomenclature consultant, Dr R. S. Cahn, is also gratefully acknowledged.

Finally, I would record my heartfelt gratitude to my wife, who has cheerfully during several years converted miles of illegibility into a coherent typescript. To her I dedicate this work.

# Contents

# Introduction

**THIS SHOULD BE READ THROUGH CAREFULLY TO GAIN FULL BENEFIT FROM THE FOLLOWING SECTIONS.**

## Aims of the Handbook

This compilation has been prepared mainly to give the research student, practising chemist or safety officer access to a wide selection of documented information to allow him or her readily to assess the likely reaction hazard-potential associated with an existing or proposed chemical compound or reaction system.

A secondary, longer-term purpose is to present the information in a way which will, as far as possible, bring out the causes of, and interrelationships between, apparently disconnected facts and incidents. This should encourage an increased awareness of potential chemical reactivity hazards in school, college and university teaching laboratories and help to dispel the considerable ignorance of such matters which appears to be widespread in this important area of safety training during formative years.

Others involved in a more general way with the storage, handling, packing and transport of chemicals are likely to find information of relevance to their activities.

## Scope and Source Coverage

This Handbook includes all information which had become available to the author by September 1977 on the reactivity hazards of individual elements or compounds, either alone or in combination. Appropriate source references are included to give access to more expansive information than that compressed into the text entries.

A wide variety of possible sources of published information has been scanned to ensure maximum coverage. Primary sources have largely been restricted to journals known to favour or specialise in publication of safety matters, and the textbook series specialising in synthetic procedures.

Secondary sources have been a fairly wide variety of both specialised and general textbooks (notably those of Bailar, Pascal, Mellor and Sidgwick in the inorganic area), part of *Chemical Abstracts*, and various safety manuals, compilations, summaries, data sheets and case histories which cover the industrial area. Details of all the secondary sources and the primary textbooks will be found in Appendix 1. Textbook references are characterised by the absence of an author's initials.

Information on toxic hazards has been specifically excluded because this is collectively available elsewhere and its attempted inclusion would have delayed publication by several more years. For similar reasons, no attempt has been made to include details of all flammable materials capable of burning explosively when mixed with air and ignited, or any incidents related to this frequent cause of accidents.

However, to focus attention on the potential hazards always associated with the use of flammable substances, some 530 gases and liquids with flash points below 25°C and/or autoignition temperatures below 225°C have been included, their names prefixed with a dagger. The numerical values of the fire-related properties of flash point, autoignition temperature and explosive limits in air where known are given in the tabular Appendix 2. Those elements or compounds which ignite on exposure to air are also included in the text.

## General Arrangement

The information herein on reactive hazards is of two main types, specific or general, and these have been arranged differently in their respective separate sections. Specific information on instability of individual chemical compounds and on hazardous interactions of elements and/or compounds is contained in the main formula-based second section of the Handbook.

*See* ETHYL PERCHLORATE, $C_2H_5ClO_4$
NITRIC ACID, $HNO_3$ : Acetone

General information relating to types or groups of elements or compounds possessing similar structural or hazardous characteristics is contained in the smaller alphabetically based first section:

*See* ACYL NITRATES
PYROPHORIC METALS

Individual materials of variable composition and materials which cannot conveniently be formulated are also included in this section:

*See* BLEACHING POWDER
CELLULOSE NITRATE

Both theoretical and practical hazard topics, some indirectly related to the main theme of this book, are also included:

*See* DISPOSAL
EXPLOSIBILITY
GAS CYLINDERS
OXYGEN ENRICHMENT

## Specific Formula Entries

1. A single unstable compound of known composition is placed in the main (formula) section on the basis of its molecular formula expressed in the Hill system used by *Chemical Abstracts* (C and H if present, then all other elements alphabetically). The use of this indexing basis permits a compound to be located if its structure can be drawn, irrespective of whether it can be named or not. A representation of the structure of each compound, usually in abbreviated linear form, is given except where the empirical formula coincides exactly with the structural formula. The IUPAC name (and occasionally a synonym) corresponding to the indexing formulation appears in capitals to the left of the formula. References to the information source are given, followed by a statement of the observed hazard, with any relevant explanation. Cross-reference to similar compounds, preferably as a group entry, completes the item:

*See* TRIFLUOROACETYL NITRITE, $C_2F_3NO_3$

2a. Where two or more elements or compounds are involved in a reactive hazard, and an intermediate or product of reaction is identifiable as being responsible for the hazard, both reacting substances are normally cross-referred to the identified product. The well-known reaction of ammonia and iodine to give explosive nitrogen triiodide–ammonia is of this type. The entries

AMMONIA, $H_3N$: Halogens
IODINE, $I_2$ : Ammonia

are referred back to the main entry under the identified material

NITROGEN TRIIODIDE–AMMONIA, $I_3N{\cdot}H_3N$

No attempt has been made, however, to list all combinations of reactants which can lead to the formation of a particular unstable main entry compound.

2b. In a multi-reactant system where no identification of an unstable product was possible, one of the reactants had to be selected as primary reactant to prepare and index the main entry, with the other material(s) as secondary reactant(s). No strictly logical basis of choice for this is obvious.

However, it emerged during the compilation phase that most two-component reaction hazard systems of this type involve a fairly obvious oxidant material as one of the reactants. Where this situation was recognised, the oxidant has normally been selected as primary (indexing) reactant with its name in capitals, secondary reactant(s) following in lower case type:

*See* POTASSIUM PERMANGANATE, $KMnO_4$ : Acetic acid, etc.

In the markedly fewer cases where an obvious reducant has been involved as one reactant, this has normally been selected as primary reactant:

*See* LITHIUM TETRAHYDROALUMINATE, $AlH_4Li$: 3,5-Dibromocyclopentene

In the relatively few cases where neither (or none) of the reactants can be recognised as oxidant or reducant, the choice was made which appeared to give the more informative main entry text:

*See* CHLOROFORM, $CHCl_3$ : Acetone, etc.

Where some hazard has been noted during the preparation of a specific compound, but without it being possible to identify a specific cause, an entry for that compound states 'preparative hazard', and back-refers to the reactants involved in the preparation:

*See* SULPHUR DIOXIDE, $O_2S$

Occasionally, departures from these considerations have been made where such action appeared advantageous in bringing out a relationship between formally unrelated compounds or hazards. In all multi-component cases, however, the secondary reactants (except air and water) appear as formula entries back-referred to the main entry text, so the latter is accessible from either primary or secondary reactants:

*See* DIMETHYL SULPHOXIDE, $C_2H_6OS$: Acyl halides (main entry)
ACETYL CHLORIDE, $C_2H_3ClO$: Dimethyl sulphoxide (back reference)

## Grouping of Reactants

There are advantages to be gained in grouping together elements or compounds showing similar reactivities, because this tends to bring out the relationships between structure and reactivity more clearly than separate treatment. This course has been adopted widely for primary reactants (see next section), and for secondary reactants where one primary reactant has been separately involved in hazardous reactions with a large number of secondary materials. Where possible, the latter have been collected together under a suitable general group title indicative of the composition or characteristics of the collected materials, with a general formula where possible.

*See* CHLORINE, $Cl_2$ : Hydrocarbons
HYDROGEN PEROXIDE, $H_2O_2$ : Metals, Metal oxides, Metal salts
HYDROGEN SULPHIDE, $H_2S$: Oxidants

This arrangement means, however, that some practice will be necessary on the user's part in deciding into what group an individual secondary reactant falls before the longer-term advantages of the system become apparent. The group titles listed in Appendix 2 will be of use in this connection.

## General Group Entries

In some cases literature references relevant to well-defined hazard topics or groups of hazardous compounds have been found, and these are given, with a condensed version of appropriate information, at the beginning of the topic or group entry.

Cross-references to related group or sub-group entries are also included, with a group list of individually indexed main-section entries which lie within the structural or functional scope of the group entry, these lists being arranged in order of formulae. Compounds which are closely similar to, but not in strict conformity with, the group definition are indicated by an initial asterisk.

The group entries thus serve as sub-indexes for each structurally based group of hazardous compounds. Conversely, each individual compound entry is back-referred to the group entry, and thence to all its strict structural analogues and related congeners included in this Handbook.

These features should be useful in attempts to estimate the stability or reactivity of a compound or reaction system which does not appear in this Handbook. The effects on stability or reactivity of changes in the molecular structure to which the destabilising group(s) is attached are in some cases discussed in the group entry. Otherwise such information may be gained from comparison of the information available from the individual compound entries collectively listed in the group entry.

Care is, however, necessary in extrapolating from described hazardous compounds or systems to others in which the user of this Handbook may be interested. Due allowance must be made for changes in elemental reactivity up or down the columns of the Periodic Table, and for the effects of variation in chain length, branching and point of group-attachment in organic systems. Purity of materials, possible catalytic effects (positive or negative) of impurities, and scale of operations may all have a direct bearing upon a particular reaction rate. These and other related matters are dealt with in more detail in the following Introductory Chapter.

## Nomenclature

With the direct encouragement and assistance of the Publishers, an attempt has been made in this Handbook to use chemical names which conform to the most recent recommendations of IUPAC. While this has not been an essential part of the compilation, because each name has a corresponding molecular formula adjacent, it seems nonetheless desirable to minimise possible confusion by adopting the unambiguous system of nomenclature presented in the IUPAC publications.

Where the IUPAC name for a compound is very different from the previously used trivial name, the latter is included in parentheses (and in quotes where no longer an acceptable name). Generally, retained trivial names have not been used as main entry titles, but they have been used in the entry texts. Rarely, on the grounds of brevity, names not strictly conforming to IUPAC principles but recommended for chemicals used in industry in BS 2474: 1965 (amended 1970) have been used. The prefix *mixo*-, to represent the mixtures of isomers sometimes used as industrial materials, is a case in point.

No attempt has been made to use only one of the nomenclature systems acceptable to IUPAC, but no non-preferred or deprecated names are believed still to lurk in the text.

Some of the rigidly systematic names selected by the Association for Science Education in 1972 from the IUPAC possibilities are given as synonyms in the Index of Chemical Names (Appendix 5). This should assist those who will be coming into industry and research with a command of ASE nomenclature but who may be unfamiliar with the current variety of names used for chemicals.

In connection with the group titles adopted for the alphabetical section, it has been necessary in some cases to devise group names (particularly in the inorganic field) to indicate in a very general way the chemical structures involved in various classes, groups or sub-groups of compounds.

For this purpose, all elements have been considered either as METALS or NON-METALS, and, of the latter, HALOGENS, HYDROGEN, NITROGEN, OXYGEN and SULPHUR were selected as specially important. Group names have then been coined from suitable combinations of these, such as the simple

METAL OXIDES, NON-METAL SULPHIDES
*N*-HALOGEN COMPOUNDS, NON-METAL HALIDES
METAL NON-METALLIDES, COMPLEX HYDRIDES

or the more complex

METAL OXOHALOGENATES
AMMINECHROMIUM PEROXOCOMPLEXES
OXOSALTS OF NITROGENOUS BASES
METAL OXONON-METALLATES

Organic group entries are fairly conventional, such as

HALOALKENES
NITROSO COMPOUNDS

Where necessary, such group names are explained in the appropriate group entry, of which a full listing is given in Appendix 4.

## Cross-reference System

The cross-reference system adopted in this Handbook plays a large part in providing maximum access to, and use of, the rather heterogeneous collection of information herein. The significance of the four types of cross-reference which have been used is as follows.

*See* . . . . . . . . cross-refers to a directly related material or entry.
*See also*. . . . . cross-refers to an indirectly related material or entry.
*See other* . . . cross-refers to strict analogues of the compound or type.
*See related* . . cross-refers to related compounds (congeners) which are fairly closely but not strictly analogous structurally.

## Information Content of Individual Entries

A conscious effort has been made throughout this compilation to exclude all fringe information not directly relevant to the involvement of chemical reactivity in the various incidents, with just enough detail present to allow the reader to judge the relevance or otherwise of the quoted references to his or her particular reactivity problems.

It must be stressed that this book can do no more than serve as a guide to much more detailed information available via the quoted references. In all but a few cases it cannot relieve the student or chemist of their moral and now legal obligation to themselves and others to equip themselves with the fullest possible information from the technical literature resources which are available, before attempting any experimental work with materials known, or suspected, to be hazardous or potentially so.

THE ABSENCE OF A MATERIAL OR A COMBINATION OF MATERIALS FROM THIS HANDBOOK CANNOT BE TAKEN TO IMPLY THAT NO HAZARD EXISTS. LOOK THEN FOR ANALOGOUS MATERIALS USING THE GROUP ENTRY SYSTEM.

One aspect which, although it is excluded from all entry texts is nevertheless of vital importance, is that of the potential for damage, injury or death associated with the various materials and reaction systems dealt with in this Handbook.

Though some of the incidents have involved little or no damage (see CAN OF BEANS), others have involved personal injuries, often of unexpected severity (see SODIUM PRESS), and material damage is often immense (PERCHLORIC ACID, $ClHO_4$ : Cellulose derivatives, ref. 1, involved damage to 116 buildings and a loss approaching $3 M). The death-roll associated with reactive chemical hazards has ranged from 1 or 2 (see TETRAFLUOROETHYLENE, $C_2F_4$ : Iodine pentafluoride) to some 600 with 2000 injured in the incident at Oppau in 1923 (see AMMONIUM NITRATE, $H_4N_2O_3$, ref. 4).

This potential for destruction emphasises again the need to gain detailed knowledge before starting to use an unfamiliar chemical.

# Reactive Chemical Hazards

This introductory chapter seeks to present an overview of the complex subject of reactive chemical hazards, drawing attention to the underlying principles and to some practical aspects of minimising such hazards. It also serves in some measure to correlate some of the topic entries in the first, alphabetically arranged, section of the Handbook.

## Basics

All chemical reactions implicitly involve energy changes (energy of activation + energy of reaction), for these are the driving force. The majority of reactions liberate energy as heat (occasionally as light or sound) and are termed exothermic. In a minority of reactions, the reaction energy is absorbed into the products, when both the reaction and its products are described as endothermic.

All reactive hazards involve the release of energy in quantities or at rates too high to be absorbed by the immediate environment of the reacting system, and material damage occurs. The source of the energy may be an exothermic multi-component reaction, or the exothermic decomposition of a single unstable (often endothermic) compound.

All measures to minimise the occurrence of reactive chemical hazards are therefore directed at controlling the extent and rate of release of energy in a reacting system. Some of the factors which contribute to the possibility of excessive energy release, and appropriate means of their control, are now outlined briefly, with references to examples.

## Kinetic Factors

The rate of an exothermic chemical reaction determines the rate of energy release; so factors which affect reaction kinetics are important so far as possible hazards are concerned.

The effects of proportions and concentrations of reactants upon reaction rate are governed by the Law of Mass Action, and there are many examples where changes in proportion and/or concentration of reagents have converted an established uneventful procedure into a violent incident. For examples of the effect of increase in proportion,

*see* o-CHLORONITROBENZENE, $C_6H_4ClNO_2$ : Ammonia
SODIUM *p*-NITROPHENOLATE, $C_6H_4NNaO_3$

For the effect of increase in concentration upon reaction velocity,

*see* DIMETHYL SULPHATE, $C_2H_6O_4S$ : Ammonia
NITROBENZENE, $C_6H_5NO_2$ : Methanol, Potassium hydroxide

The effects of catalysts (which effectively reduce the energy of activation), either intentional or unsuspected, is also relevant in this context. Increase in the concentration of a catalyst (normally used at 1–2%) may have a dramatic effect on reaction velocity.

*See* TRIFLUOROMETHANESULPHONIC ACID, $CHF_3O_3S$ : Acyl chlorides, etc.
*o*-NITROANISOLE, $C_7H_7NO_3$ : Hydrogen
HYDROGENATION CATALYSTS

The presence of an unsuspected contaminant or catalytic impurity may affect the velocity or change the course of reaction.

*See* CATALYSIS BY IMPURITIES

In the same context, but in opposite sense, presence of inhibitors (negative catalysts, increasing energy of activation) may seriously interfere with the smooth progress of a reaction. An inhibitor may initiate an induction period which can lead to problems in establishing and controlling a desired reaction. For further details,

*see* INDUCTION PERIOD

Undoubtedly the most important factor affecting control of reaction rates is that of temperature. It follows from the Arrhenius equation that the rate of reaction will increase exponentially with temperature increase. Practically, it is found that an increase of 10°C in reaction temperature often doubles or trebles the reaction velocity.

Because most reactions are exothermic, they will tend to accelerate as reaction proceeds unless the rate of cooling is sufficient to prevent rise in temperature. Note that the exponential temperature effect accelerating the reaction will exceed the (usually) linear effect of falling reactant concentration in decelerating the reaction. Where the exotherm is large and cooling capacity is inadequate, the resulting accelerating reaction may proceed to the point of loss of control, and decomposition, fire or explosion may ensue. Reactions at high pressure may be exceptionally hazardous owing to the enhanced kinetic energy content of the system.

The great majority of incidents described in the text may be attributed to this primary cause of thermal runaway reactions. The scale of the damage produced is related directly to the size, and more particularly to the *rate*, of energy release.

Although detailed consideration of explosions is outside the scope of this Handbook, three levels of intensity of explosions (i.e. rates of fast energy release) can be discerned and roughly equated to the material damage potential.

Deflagration involves combustion of a material, usually in presence of air. In a normal liquid pool fire, combustion will normally proceed without explosion. Mixtures of gases or vapours with air within the explosive limits which are subsequently ignited will burn at normal flame velocity (a few m/s) to produce a 'soft' explosion, with minor material damage, often limited to scorching by the moving flame front. Injuries to personnel may well be more severe.

If the mixture (or a dust cloud) is confined, significant pressure effects can occur. Fuel–air mixtures near to stoicheiometric composition and closely confined will develop pressures of several bar within milliseconds, and material damage will be severe. Unconfined vapour explosions of large dimensions may involve higher flame velocities and generate significant pressure effects, as shown in the Flixborough disaster.

Detonation is an extreme form of explosion where the propagation velocity becomes supersonic in gaseous, liquid or solid states. The temperatures and particularly pressures associated with detonation are higher by orders of magnitude than in deflagration. Energy release occurs in a few microseconds and the resulting shattering effects are characteristic of detonation. Deflagration may accelerate to detonation if the burning material and geometry of confinement are appropriate (endothermic compounds, long narrow vessels or pipelines).

*See* ENDOTHERMIC COMPOUNDS
EXPLOSIONS
HIGH-PRESSURE TECHNIQUES
UNIT PROCESS INCIDENTS

Factors of importance in preventing such thermal runaway reactions are mainly related to the control of reaction velocity and temperature within suitable limits. These may involve such considerations as adequate heating and particularly cooling capacity in both liquid and vapour phases of a reaction system; proportions of reactants and rates of addition (allowing for an induction period); use of solvents as diluents and to reduce viscosity of the reaction medium; adequate agitation and mixing in the reaction vessel; control of reaction or distillation pressure; use of an inert atmosphere.

In some cases it is important not to overcool a reacting system, so that the energy of activation is maintained.

*See* ACETYLENE (ETHYNE), $C_2H_2$ : Halogens (reference 1)

## Adiabatic Systems

Because process heating is expensive, lagging is invariably applied to heated process vessels to minimise heat loss, particularly during long-term hot storage. Such adiabatic or near-adiabatic systems are potentially hazardous if materials of limited thermal stability, or which possess self-heating capability, are used in them. Insufficiently stabilised bulk-stored monomers come into the latter category.

*See* POLYMERISATION INCIDENTS
THERMAL STABILITY OF REACTION MIXTURES
VIOLENT POLYMERISATION

## Reactivity vs Composition and Structure

The ability to predict reactivity and stability of chemical compounds from their composition and structure is as yet limited, so the ability to accurately foresee potential hazards during preparation, handling and processing of chemicals and their mixtures is also restricted. Although some progress has been made in the use of computer programs to predict hazards, the best available approach still appears to be appraisal based on analogy with, or extrapolation from, data from existing compounds and processes, in conjunction with bench tests for thermal stability.

*See* ASSESSMENT OF REACTIVE CHEMICAL HAZARDS
COMPUTATION OF REACTIVE CHEMICAL HAZARDS
DIFFERENTIAL THERMAL ANALYSIS
REACTION SAFETY CALORIMETRY

It has long been recognised that instability in single compounds, or high reactivity in combinations of different materials, is often associated with particular groupings of atoms or other features of molecular structure, such as high proportions or local concentrations of oxygen or nitrogen. Full details of such features associated with explosive instability are collected under the heading EXPLOSIBILITY.

An approximate indication of likely instability in a compound may be gained from inspection of the empirical molecular formula to establish stoicheiometry.

*See* HIGH-NITROGEN COMPOUNDS
OXYGEN BALANCE

Endothermic compounds, formed as the energy-rich products of endothermic reactions, are thermodynamically unstable and may be liable to energetic decomposition.

*See* ENDOTHERMIC COMPOUNDS

## Reaction Mixtures

So far as reactivity between different compounds is concerned, some subdivisions can be made on the basis of the chemical types involved.

Oxidants are undoubtedly the most common chemical type to be involved in hazardous incidents, the other components functioning as fuels or other electron sources. Air (21% oxygen) is the most widely dispersed oxidant, and air-reactivity may lead to either short- or long-term hazards.

Where reactivity is very high, oxidation may proceed so fast that ignition occurs.

*See* PYROPHORIC MATERIALS

Slow reaction with air may lead to the longer-term hazard of peroxide formation.

*See* AUTOXIDATION
PEROXIDES IN SOLVENTS
PEROXIDISABLE COMPOUNDS

Oxidants more concentrated than air are of greater hazard potential, and the extent of involvement of the common chemicals

METAL CHLORATES
PERCHLORIC ACID, $ClHO_4$
CHLORINE, $Cl_2$
NITRIC ACID, $HNO_3$
HYDROGEN PEROXIDE, $H_2O_2$
SULPHURIC ACID, $H_2O_4S$

may be judged from the large number of incidents listed under each of those entries, and also under other OXIDANTS.

At the practical level, experimental oxidation reactions should be conducted to maintain in the reacting system a minimum oxygen balance consistent with other processing requirements. This may involve adding the oxidant slowly with appropriate mixing and cooling to the other reaction materials to maintain the minimum effective concentration of oxidant for the particular reaction. It will be essential to determine that the desired reaction has become established to prevent build-up of unused oxidant and a possible approach to the oxygen balance point.

Reducants in conjunction with reducible materials (or electron sinks) feature less frequently in hazardous incidents.

*See* REDUCANTS

Interaction of potent oxidants and reducants is invariably highly energetic and of high hazard potential.

*See* DIBENZOYL PEROXIDE, $C_{14}H_{10}O_4$ : Lithium tetrahydroaluminate
HYDRAZINE, $H_4N_2$ : Oxidants
ROCKET PROPELLANTS

Similar considerations apply to those compounds which contain both oxidising and reducing functions in the same molecular structure.

*See* REDOX COMPOUNDS

Water is, after air, one of the most common reagents likely to come into contact with reactive materials and several classes of compound will react violently, particularly with restricted amounts of water.

*See* WATER-REACTIVE COMPOUNDS

Most of the above has been written with deliberate processing conditions in mind, but it must not be forgotten that the same considerations will apply probably to a greater degree under the uncontrolled reaction conditions prevailing when accidental contact of reactive chemicals occurs in storage or transit.

Adequate planning is therefore necessary in storage arrangements to segregate oxidants from fuels and reducants, and fuels and combustible materials from compressed gases and water-reactive compounds. This will minimise the possibility of accidental contact and violent reaction arising from faulty containers or handling operations, and will prevent intractable problems in the event of fire in the storage areas.

Unexpected sources of ignition may lead to ignition of flammable materials during chemical processing operations.

*See* FIRE
IGNITION SOURCES

## Protective Measures

The need to provide protective measures will be directly related to the level of potential hazards which may be assessed from the procedures outlined above. Measures concerned with reaction control are frequently mentioned in the following text, but details of techniques and equipment for personal protection, though usually excluded from the scope of this work, are obviously of great importance.

Careful attention to such details is necessary as a second line of defence against the effects of reactive hazards. The level of protection considered necessary may range from the essential and absolute minimum of eye protection, via the safety screen, fume cupboard or enclosed reactor, up to the ultimate of a remotely controlled and blast-resistant isolation cell (usually for high-pressure operations). In the absence of facilities appropriate to the assessed level of hazard, operations must be deferred until such facilities are available.

# Class, Group and Topic Section

**(Entries arranged in alphabetical order)**

**EXPLANATORY NOTES**

The entries in this section give general information on the hazardous behaviour of some recognisably discrete classes or groups of the individual compounds of which details are given in Section 2. The group-lists of compounds serve as an index to analogues and homologues of a particular individual compound. Those compounds of generally similar, but not identical, structure to the majority in the group-list are prefixed*.

Where possible, a general structural formula is given to indicate the typical structure of the members of each class or group. In these formulae, the following symbols are used in addition to the usual symbols for the elements.

Ar = AROMATIC nucleus
E = NON-METALLIC element
M = METALLIC element
R = ORGANIC residue
X = HALOGEN
Z = non-halogen ANION species

Information on the derivation of the class and group names is given in the Introduction (page xx), and there is a full index to these names in Appendix 4.

Details of corrections of typographical or factual errors, or of further items for inclusion in the text, will be welcomed by the author.

## ACETYLENIC COMPOUNDS $C\equiv C$

1. Davidsohn, W.B. *et al.*, *Chem. Rev.*, 1967, **67**, 74
2. Mushii, R. Ya. *et al.*, *Chem. Abs.*, 1967, **67**, 92449
3. Dutton, G.G.S., *Chem. Age*, 1947, **56**(1436), 11–13

The presence of the endothermic acetylene group confers explosive instability on a wide range of acetylenic compounds (notably when halogen is also present) and derivatives of metals (especially of heavy metals) [1]. Explosive properties of butadiyne, buten-3-yne, hexatriyne, propyne and propadiene have been reviewed, with 74 references [2]. The tendency of higher acetylenes to explosive decomposition may be reduced by dilution with methanol [3]. The class includes the separately treated groups:

ACETYLENIC PEROXIDES
COMPLEX ACETYLIDES
ETHOXYETHYNYL ALCOHOLS
HALOACETYLENE DERIVATIVES
METAL ACETYLIDES

and the individually indexed compounds:

ACETYLENE, $C_2H_2$
PROPIOLOYL CHLORIDE, $C_3HClO$
PROPIOLALDEHYDE, $C_3H_2O$
PROPYNE, $C_3H_4$
METHOXYACETYLENE, $C_3H_4O$
2-PROPYN-1-OL (PROPARGYL ALCOHOL), $C_3H_4O$
2-PROPYNE-1-THIOL, $C_3H_4S$
1,3-BUTADIYNE, $C_4H_2$
1-BUTEN-3-YNE, $C_4H_4$
METHYL PROPIOLATE, $C_4H_4O_2$
ETHYNYL VINYL SELENIDE, $C_4H_4Se$
1- or 2-BUTYNE, $C_4H_6$
ETHOXYACETYLENE, $C_4H_6O$
3-METHOXYPROPYNE, $C_4H_6O$
2-BUTYNE-1, 4-DIOL, $C_4H_6O_2$
2-BUTYNE-1-THIOL, $C_4H_6S$
DIMETHYLAMINOACETYLENE, $C_4H_7N$
1,3-PENTADIYNE, $C_5H_4$
2-METHYL-1-BUTEN-3-YNE, $C_5H_6$
2-PENTEN-4-YN-3-OL, $C_5H_6O$
2-PROPYNYL VINYL SULPHIDE, $C_5H_6S$
1- or 2-PENTYNE, $C_5H_8$
1-ETHOXY-2-PROPYNE, $C_5H_8O$
2-METHYL-3-BUTYN-2-OL, $C_5H_8O$
1,3,5-HEXATRIYNE, $C_6H_2$
BUTADIYNE-1,4-DICARBOXYLIC ACID, $C_6H_2O_4$
TRIETHYNYLARSINE, $C_6H_3As$
TRIETHYNYLPHOSPHINE, $C_6H_3P$

1,3-HEXADIEN-5-YNE, $C_6H_6$
1,5-HEXADIEN-3-YNE (DIVINYLACETYLENE), $C_6H_6$
2,4-HEXADIYNE, $C_6H_6$
DI(2-PROPYN-1-YL) ETHER (DIPROPARGYL ETHER), $C_6H_6O$
4,5-HEXADIEN-2-YN-1-OL, $C_6H_6O$
DIMETHYL ACETYLENEDICARBOXYLATE, $C_6H_6O_4$
3-METHYL-2-PENTEN-4-YN-1-OL, $C_6H_8O$
*tert*-BUTYLNITROACETYLENE, $C_6H_9NO_2$
1-, 2- or 3-HEXYNE, $C_6H_{10}$
BUTOXYACETYLENE, $C_6H_{10}O$
HEPTA-1,3,5-TRIYNE, $C_7H_4$
1-HEPTEN-4,6-DIYNE, $C_7H_6$
2-HEPTYN-1-OL, $C_7H_{12}O$
1-ETHOXY-3-METHYL-1-BUTYN-3-OL, $C_7H_{12}O_2$
NITROPHENYLACETYLENE, $C_8H_5NO_2$
PHENOXYACETYLENE, $C_8H_6O$
1,5,7-OCTATRIEN-3-YNE, $C_8H_8$
DIETHYL ACETYLENEDICARBOXYLATE, $C_8H_{10}O_4$
1-DIETHYLAMINO-1-BUTEN-3-YNE, $C_8H_{13}N$
1-, 2-, 3- or 4-OCTYNE, $C_8H_{14}$
2-NONEN-4,6,8-TRIYN-1-AL, $C_9H_4O$
*o*-NITROPHENYLPROPIOLIC ACID, $C_9H_5NO_4$
BENZYLOXYACETYLENE, $C_9H_8O$
2-METHYL-3,5,7-OCTATRIYN-2-OL, $C_9H_8O$
OCTATETRAYNE-1,8-DICARBOXYLIC ACID, $C_{10}H_2O_4$
2,4-DIETHYNYLPHENOL, $C_{10}H_6O$
2,4-DIETHYNYL-5-METHYLPHENOL, $C_{11}H_8O$
3-BUTYN-1-YL *p*-TOLUENESULPHONATE, $C_{11}H_{12}O_3S$
1,3,5-TRIETHYNYLBENZENE, $C_{12}H_6$
1,2-DI(3-BUTEN-1-YNYL)CYCLOBUTANE, $C_{12}H_{12}$
1-BUTYLOXYETHYL 3-TRIMETHYLPLUMBYLPROPIOLATE, $C_{12}H_{22}O_3Pb$
DIBUTYL-3-METHYL-3-BUTEN-1-YNYLBORANE, $C_{13}H_{23}B$
BIS(DIBUTYLBORINO)ACETYLENE, $C_{18}H_{36}B_3$

*See also* PEROXIDISABLE COMPOUNDS

Metals
*Chemical Intermediates*, 1972 Catalogue, 158, Tamaqua (Pa), Air Products and Chemicals Inc., 1972
Acetylenic compounds with replaceable acetylenically bound hydrogen atoms must be kept out of contact with copper, silver, magnesium, mercury or alloys containing them, to avoid formation of explosive metal acetylides.
*See* METAL ACETYLIDES

**ACETYLENIC PEROXIDES** **C≡C–C–OOH**

Milas, N. A. *et al., Chem. Eng. News,* 1959, **37**(37), 66; *J. Amer. Chem. Soc.,* 1952, **75**, 1472; *J. Amer. Chem. Soc.,* 1953, **76**, 5970
The importance of strict temperature control to prevent explosion during preparation (second and third references) of acetylenic

peroxides is stressed. Use of inert solvent to prevent undue thickening and consequent poor temperature control is recommended.

## ACID ANHYDRIDES — $RCOOCOR$, $RSO_2OSO_2R$

Relatively few members of this reactive group have been involved in hazardous incidents.

*See* MALEIC ANHYDRIDE, $C_4H_2O_3$
ACETIC ANHYDRIDE, $C_4H_6O_3$
PHTHALIC ANHYDRIDE, $C_8H_4O_3$
BENZENESULPHONIC ANHYDRIDE, $C_{12}H_{10}O_5S_2$

## ACYL AZIDES — $RCON_3$


1. Smith, P. A. S., *Org. React.*, 1946, **3**, 373–375
2. Houben-Weyl, 1952, Vol. 8, 680
3. Lieber, E. *et al.*, *Chem. Rev.*, 1965, **65**, 377
4. Balabanov, G.P. *et al.*, *Chem. Abs.*, 1969, **70**, 59427
5. Renfrow, W.B. *et al.*, *J. Org. Chem.*, 1975, **40**, 1526


Azides of low molecular weight (more than 25% nitrogen content) should not be isolated from solution since the concentrated material is likely to be dangerously explosive [1]. The concentration of such solutions (prepared below 10°C) should be below 10% [2].

Carbonyl azides are explosive compounds, some exceptionally so; suitable handling precautions are necessary [3].

The sensitivity to friction, impact, and heat of benzenesulphonyl azide, its *p*- chloro-, methyl-, methoxy-, methoxycarbonylamino-, and nitro-derivatives, and *m*-benzenedisulphonyl diazide were studied [4]. Benzene- and *o*-toluene-sulphonyl azides may be smoothly thermolysed in benzene [5]. Individually indexed compounds are:

TRIFLUOROMETHANESULPHONYL AZIDE, $CF_3N_3O_2S$
AZIDODITHIOFORMIC ACID, $CHN_3S_2$
CARBONYL DIAZIDE, $CN_6O$
TRIFLUOROACETYL AZIDE, $C_2F_3N_3O$
DIAZOACETYL AZIDE, $C_2HN_5O$
1,2-DIAZIDOCARBONYLHYDRAZINE, $C_2H_2N_8O_2$
ACETYL AZIDE, $C_2H_3N_3O$
BIS(AZIDOTHIOCARBONYL) DISULPHIDE, $C_2N_6S_4$
4-AZIDOCARBONYL-1,2,3-THIADIAZOLE, $C_3HN_5OS$
ETHYL AZIDOFORMATE, $C_3H_5N_3O_2$
*N*-BUTYLAMIDOSULPHURYL AZIDE, $C_4H_{10}N_4O_2S$
2-FUROYL AZIDE, $C_5H_3N_3O_2$
GLUTARYL DIAZIDE, $C_5H_6N_6O_2$
PIVALOYL AZIDE, $C_5H_9N_3O$
*tert*-BUTYL AZIDOFORMATE, $C_5H_9N_3O_2$
*m*-BENZENEDISULPHONYL DIAZIDE, $C_6H_4N_6O_4S_2$
PHENYLPHOSPHONIC AZIDE CHLORIDE, $C_6H_5ClN_3OP$
PHENYLPHOSPHONIC DIAZIDE, $C_6H_5N_6OP$
PHENYLTHIOPHOSPHONIC DIAZIDE, $C_6H_5N_6PS$

*p*-BROMOBENZOYL AZIDE, $C_7H_4BrN_3O$
*p*-CHLOROBENZOYL AZIDE, $C_7H_4ClN_3O$
BENZOYL AZIDE, $C_7H_5N_3O$
4(?)-CARBOXYBENZENESULPHONYL AZIDE, $C_7H_5N_3O_4S$
*N*-AZIDOCARBONYLAZEPINE, $C_7H_6N_4O$
*p*-TOLUENESULPHONYL AZIDE, $C_7H_7N_3O_2S$
PHTHALOYL DIAZIDE, $C_8H_4N_6O_2$
3-PHENYLPROPIONYL AZIDE, $C_9H_9N_3O$
6-QUINOLINECARBONYL AZIDE, $C_{10}H_6N_4O$
1,3,5-TRIS(*p*-AZIDOSULPHONYLPHENYL)-1,3,5-TRIAZINETRIONE, $C_{21}H_{12}N_{12}O_9S_3$
SULPHURYL AZIDE CHLORIDE, $ClN_3O_2S$
FLUOROTHIOPHOSPHORYL DIAZIDE, $FN_6PS$
AMIDOSULPHURYL AZIDE, $H_2N_4O_2S$
POTASSIUM AZIDOSULPHATE, $KN_3O_3S$
POTASSIUM AZIDODISULPHATE, $KN_3O_6S_2$
SULPHURYL DIAZIDE, $N_6O_2S$
DISULPHURYL DIAZIDE, $N_6O_5S_2$

*See also* SULPHINYL AZIDES

## ACYL CHLORIDES — RCOCl

Aromatic hydrocarbons,
Trifluoromethanesulphonic acid

*See* TRIFLUOROMETHANESULPHONIC ACID, $CHF_3OS$: Acyl chlorides, etc.

## ACYL HALIDES — RCOX

This group tends to react violently with protic solvents, and with dimethylformamide. Individually indexed compounds are:

ACETYL BROMIDE, $C_2H_3BrO$
ACETYL CHLORIDE, $C_2H_3ClO$
* 2,4,6-TRICHLORO-1,3,5-TRIAZINE (CYANURIC CHLORIDE), $C_3Cl_3N_3$
TRIFLUOROACRYLOYL FLUORIDE, $C_3F_4O$
PERFLUOROPROPIONYL FLUORIDE, $C_3F_6O$
PROPIOLOYL CHLORIDE, $C_3HClO$
PROPIONYL CHLORIDE, $C_3H_5ClO$
FUROYL CHLORIDE, $C_5H_3ClO_2$
*o*-NITROBENZOYL CHLORIDE, $C_7H_4ClNO_3$
BENZOYL CHLORIDE, $C_7H_5ClO$
ISOPHTHALOYL DICHLORIDE, $C_8H_4Cl_2O_2$
4-METHOXY-3-NITROBENZOYL CHLORIDE, $C_8H_6ClNO_4$
SEBACOYL DICHLORIDE, $C_{10}H_{16}Cl_2O_2$
4,4-DIFERROCENYLPENTANOYL CHLORIDE, $C_{25}H_{27}ClFe_2O$

## ACYL NITRATES — $RCO_2NO_2$

A thermally unstable class, tending to violent decomposition or explosion on heating.

*See* ACETYL NITRATE, $C_2H_3NO_4$
* ACETYL PEROXONITRATE, $C_2H_3NO_5$

* PROPIONYL PEROXONITRATE, $C_3H_5NO_5$
BUTYRYL NITRATE, $C_4H_7NO_4$
*m*-NITROBENZOYL NITRATE, $C_7H_4N_2O_6$
BENZOYL NITRATE, $C_7H_5NO_4$

*See also* NITRIC ACID, $HNO_3$ : Phthalic anhydride
SODIUM NITRITE, $NNaO_2$ : Phthalic acid

## ACYL NITRITES — $RCO_2NO$

Ferrario, E., *Gazz. Chim. Ital.*, 1901, **40**(2), 98–99
Francesconi, L. *et al.*, *Gazz. Chim. Ital.*, 1895, **34**(1), 442
The stabilities of propionyl and butyryl nitrites are greater than that of acetyl nitrite, butyryl nitrite being the least explosive of these homologues. Benzyl nitrite is also unstable.

*See* ACETYL NITRITE, $C_2H_3NO_3$
TRIFLUOROACETYL NITRITE, $C_2F_3NO_3$
HEPTAFLUOROBUTYRYL NITRITE, $C_4F_7NO_3$

## ALKALI-METAL DERIVATIVES OF HYDROCARBONS — RM, ArM

Sidgwick, 1950, 68, 75
Alkali-metal derivatives of aliphatic or aromatic hydrocarbons, such as methyllithium, ethylsodium or phenylpotassium, are the most reactive towards moisture and air, immediately igniting in the latter. Derivatives of benzyl-type compounds, such as benzylsodium, are of slightly lower activity, usually, but not always, igniting in air. Derivatives of hydrocarbons with definitely acidic hydrogen atoms (acetylene, phenylacetylene, cyclopentadiene, fluorene), though readily oxidised, are usually relatively stable in ambient air. Sodium phenylacetylide if moist with ether ignites; derivatives of triphenylmethane also when dry.

## ALKALI METALS

1. *Handling and Uses of the Alkali Metals* (Advances in Chemistry Series No. 19), Washington, American Chemical Society, 1957
2. Markowitz, M.M., *J. Chem. Educ.*, 1963, **40**, 633–636
3. *Alkali Metal Dispersions*, Fatt, I. *et al.*, London, Van Nostrand, 1961

The collected papers of a symposium at Dallas, April 1956, covering the handling, use and hazards of lithium, sodium, potassium, their alloys, oxides and hydrides, in 19 chapters.

Interaction of all five alkali metals with water under various circumstances has been discussed comparatively [2]. In a monograph covering properties, preparation, handling and applications of the enhanced reactivity of metals dispersed finely in hydrocarbon diluents, the hazardous nature of potassium dispersions, and especially of rubidium and caesium dispersions, is stressed [3]. Alkaline earth metal dispersions are of relatively low hazard.

## ALKANETHIOLS — RSH

Ethylene oxide

*See* ETHYLENE OXIDE, $C_2H_4O$: Alkanethiols

## ALKENES — –C=C–

Oxides of nitrogen

*See* NITROGEN OXIDE, NO: Dienes, Oxygen

## ALKENYL NITRATES — –C=C–$ONO_2$

Tetrafluorohydrazine

*See* TETRAFLUOROHYDRAZINE, $F_4N_2$: Alkenyl nitrates

## ALKYLALUMINIUM ALKOXIDES AND HYDRIDES — $R_2AlOR$, $R_2AlH$

Though substitution of a hydrogen atom or an alkoxy group for one alkyl group in a trialkylaluminium tends to increase stability and reduce reactivity and the tendency to ignition, these compounds are still of high potential hazard, the hydrides being used industrially as powerful reducants.

*See* ALKYLALUMINIUM DERIVATIVES (references 1, 3, 6)

Individually indexed compounds of this relatively small sub-group of commercially available compounds are:

DIETHYLALUMINIUM HYDRIDE, $C_4H_{11}Al$
TETRAMETHYLDIALUMINIUM DIHYDRIDE, $C_4H_{14}Al_2$
DIPROPYLALUMINIUM HYDRIDE, $C_6H_{15}Al$
ETHOXYDIETHYLALUMINIUM, $C_6H_{15}AlO$
* TRIETHOXYDIALUMINIUM TRIBROMIDE, $C_6H_{15}Al_2Br_3O_3$
DIISOBUTYLALUMINIUM HYDRIDE, $C_8H_{19}Al$
* 4-ETHOXYBUTYLDIETHYLALUMINIUM, $C_{10}H_{23}AlO$
ETHOXYDIISOBUTYLALUMINIUM, $C_{10}H_{23}AlO$

## ALKYLALUMINIUM DERIVATIVES


1. Mirviss, S. B. *et al., Ind. Eng. Chem.*, 1961, **53**(1), 53A–56A
2. Heck, W. B. *et al., Ind. Eng. Chem.*, 1962, **54**(12), 35–38
3. Kirk-Othmer, 1963, Vol. 2, 38, 40
4. Houben-Weyl, 1970, Vol. 13(4), 19
5. 'Aluminum Alkyls', Brochure PB 3500/1 New 568, New York, Ethyl Corp., 1968

6. 'Aluminum Alkyls', New York, Texas Alkyls, 1971
7. Knap, J. E. *et al., Ind. Eng. Chem.*, 1957, **49**, 875

This main class of ALKYLALUMINIUM DERIVATIVES is divided into 3 groups for convenience:

TRIALKYLALUMINIUMS
ALKYLALUMINIUM ALKOXIDES AND HYDRIDES
ALKYLALUMINIUM HALIDES

Individual compounds are indexed under their appropriate group or sub-group headings, but since most of the available compounds are liquids with similar hazardous properties, these will be described collectively here. This aspect of the class has been extensively reviewed and documented [1–7].

Compounds with alkyl groups of $C_4$ and below all ignite immediately on exposure to air, unless diluted with a hydrocarbon solvent to 10–20% concentration [6]. Even these may ignite on prolonged exposure to air, owing to exothermic autoxidation, which becomes rapid if solutions are spilled [2, 6]. Compounds with $C_5$–$C_{14}$ alkyl groups (safe at 20–30% concentration) smoke in air but do not burn unless ignited externally or if the air is very moist. Contact with air containing more oxygen than normal (21%) will cause explosive oxidation.

Fires involving alkylaluminium compounds are difficult to control and must be treated appropriately to particular circumstances [1, 5, 6], usually with dry-powder extinguishers. Halocarbon fire extinguishants (carbon tetrachloride, chlorobromomethane, etc.), water or water-based foam must not be applied to alkylaluminium fires. Carbon dioxide is ineffective unless dilute solutions are involved [5, 6].

Alcohols

Alkylaluminium derivatives up to $C_4$ react explosively with methanol or ethanol, and triethylaluminium also with 2-propanol [1].

Halocarbons

With the exception of chlorobenzene and 1,2-dichloroethane, halocarbon solvents are unsuitable diluents, since carbon tetrachloride and chloroform may react violently with alkylaluminium derivatives [1].

Oxidants

In view of the generally powerful reducing properties of alkylaluminium derivatives, deliberate contact with known oxidants must be under careful control with appropriate precautions [3].

Water

Interaction of alkylaluminium derivatives up to $C_9$ chain length with liquid water is explosive [1] and violent shock effects have been noted [4].

Suitable handling and disposal techniques have been detailed for both laboratory [1, 2, 5–7] and manufacturing [5,6] scales of operation.

*See other* ALKYLMETAL HALIDES
ALKYLMETALS

**ALKYLALUMINIUM HALIDES** **$RAlX_2$, $R_2AlX$**

Three main structural sub-groups can be recognised: (1) dialkyl aluminium halides; (2) alkylaluminium dihalides; (3) trialkyl-dialuminium trihalides (equimolar complexes of an aluminium tri-halide and a trialkylaluminium).

While this is generally a very reactive group of compounds, similar in reactivity characteristics to trialkylaluminium compounds, increase in size of the alkyl groups and the number of halogens present tends to reduce pyrophoricity.

*See* ALKYLALUMINIUM DERIVATIVES (references 1, 2)

Individually indexed compounds of the ALKYLALUMINIUM HALIDE group, many of which are commercially available in bulk, are:

METHYLALUMINIUM DIIODIDE, $CH_3AlI_2$
ETHYLALUMINIUM BROMIDE IODIDE, $C_2H_5AlBrI$
ETHYLALUMINIUM DIBROMIDE, $C_2H_5AlBr_2$
ETHYLALUMINIUM DICHLORIDE, $C_2H_5AlCl_2$
DIMETHYLALUMINIUM BROMIDE, $C_2H_6AlBr$
DIMETHYLALUMINIUM CHLORIDE, $C_2H_6AlCl$
TRIMETHYLDIALUMINIUM TRICHLORIDE, $C_3H_9Al_2Cl_3$
DIETHYLALUMINIUM BROMIDE, $C_4H_{10}AlBr$
DIETHYLALUMINIUM CHLORIDE, $C_4H_{10}AlCl$
DIETHYLALUMINIUM THIOCYANATE: see the trimer, $C_{15}H_{30}Al_3N_3S_3$
TRIETHYLDIALUMINIUM TRICHLORIDE, $C_6H_{15}Al_2Cl_3$
DIISOBUTYLALUMINIUM CHLORIDE, $C_8H_{18}AlCl$
* HEXAETHYLTRIALUMINIUM TRITHIOCYANATE, $C_{15}H_{30}Al_3N_3S_3$

**ALKYLBORANES** **$R_3B$, $R_2BH$**

1. Sidgwick, 1951, 371
2. Mirviss, S. B. *et al., Ind. Eng. Chem.*, 1961, **53**(1), 53A

Trimethyl- and triethyl-borane ignite in air, and tributylborane ignites in a thinly diffuse layer, as when poured on cloth [1]. Generally, the pyrophoric tendency of trialkylboranes decreases with increasing branching on the 2- and 3-carbon atoms [2].

*See other* ALKYLNON-METAL HYDRIDES
ALKYLNON-METALS

**ALKYL CHLORITES** **ROClO**

*See* SILVER CHLORITE, $AgClO_2$: Alkyl iodides

## ALKYL HYDROPEROXIDES **ROOH**

Swern, 1970, Vol. 1, 19; 1971, Vol. 2, 1, 29

Most alkyl monohydroperoxides are liquid, the explosivity of the lower members (possibly due to traces of the dialkyl peroxides) decreasing with increasing chain length.

Transition metal complexes

Skibida, I.P., *Russ. Chem. Rev.*, 1975, 789–800

The kinetics and mechanism of decomposition of organic hydroperoxides in presence of transition metal compounds have been reviewed.

*See* METHYL HYDROPEROXIDE, $CH_4O_2$
ETHYL HYDROPEROXIDE, $C_2H_6O_2$
ALLYL HYDROPEROXIDE, $C_3H_6O_2$
ISOPROPYL HYDROPEROXIDE, $C_3H_8O_2$
*tert*-BUTYL HYDROPEROXIDE, $C_4H_{10}O_2$
2-CYCLOHEXENYL HYDROPEROXIDE, $C_6H_{10}O_2$
DI(2-HYDROPEROXYBUTYL-2) PEROXIDE, $C_8H_{18}O_6$
2-PHENYL-2-PROPYL HYDROPEROXIDE, $C_9H_{12}O_2$
1,2,3,4-TETRAHYDRO-1-NAPHTHYL HYDROPEROXIDE, $C_{10}H_{12}O_2$
2-METHYL-1-PHENYL-2-PROPYL HYDROPEROXIDE, $C_{10}H_{14}O_2$
1,2- or 1,4-BIS(2-HYDROPEROXY-2-PROPYL)BENZENE, $C_{12}H_{18}O_4$
* 1-ACETOXY-6-OXOCYCLODECYL HYDROPEROXIDE, $C_{12}H_{20}O_5$
* BIS(1-HYDROPEROXYCYCLOHEXYL) PEROXIDE, $C_{12}H_{22}O_6$
* DI(2-HYDROPEROXY-4-METHYL-2-PENTYL) PEROXIDE, $C_{12}H_{26}O_6$
α-BENZENEDIAZOBENZYL HYDROPEROXIDE, $C_{13}H_{12}N_2O_2$

*See also* HYDRAZONES

## ALKYLIMINIOFORMATE CHLORIDES ('IMINOESTER HYDROCHLORIDES') **$RC(=N^+H_2)OR'\ Cl^-$**

Preparative hazard

*See* HYDROGEN CHLORIDE, ClH: Alcohols, etc.

## ALKYLMETAL HALIDES **RMX**

This highly reactive class of organometallic compounds includes the group of

ALKYLALUMINIUM HALIDES

as well as the individually indexed compounds:

METHYLMAGNESIUM IODIDE, $CH_3IMg$
METHYLZINC IODIDE, $CH_3IZn$
* BISCHLOROMETHYLTHALLIUM CHLORIDE, $C_2H_4Cl_3Tl$
ETHYLMAGNESIUM IODIDE, $C_2H_5IMg$
DIMETHYLBISMUTH CHLORIDE, $C_2H_6BiCl$
DIMETHYLANTIMONY CHLORIDE, $C_2H_6ClSb$
DIETHYLBISMUTH CHLORIDE, $C_4H_{10}BiCl$

This reactive and usually pyrophoric class includes the groups

DIALKYLZINCS
DIPLUMBANES
TRIALKYLALUMINIUMS
TRIALKYLBISMUTHS

as well as the individually indexed compounds:

POLY(METHYLENEMAGNESIUM), $(CH_2Mg)_n$
METHYLSILVER, $CH_3Ag$
* BIS-METHYLSILVER–SILVER NITRATE, $(CH_3Ag)_2 \cdot AgNO_3$
* METHYLBISMUTH OXIDE, $CH_3BiO$
METHYLCOPPER, $CH_3Cu$
METHYLPOTASSIUM, $CH_3K$
METHYLLITHIUM, $CH_3Li$
METHYLSODIUM, $CH_3Na$
* METHYLSTIBINE, $CH_5Sb$
* VINYLLITHIUM, $C_2H_3Li$
ETHYLLITHIUM, $C_2H_5Li$
ETHYLSODIUM, $C_2H_5Na$
DIMETHYLBERYLLIUM, $C_2H_6Be$
* DIMETHYLBERYLLIUM–1,2-DIMETHOXYETHANE, $C_2H_6Be \cdot C_4H_{10}O_2$
DIMETHYLCADMIUM, $C_2H_6Cd$
DIMETHYLMERCURY, $C_2H_6Hg$
DIMETHYLMAGNESIUM, $C_2H_6Mg$
POLY(DIMETHYLMANGANESE), $(C_2H_6Mn)_n$
* DIMETHYLTIN DINITRATE, $C_2H_6N_2O_6Sn$
DIMETHYLZINC, $C_2H_6Zn$
PROPYLCOPPER(I), $C_3H_7Cu$
PROPYLSODIUM, $C_3H_7Na$
TRIMETHYLGALLIUM, $C_3H_9Ga$
TRIMETHYLINDIUM, $C_3H_9In$
TRIMETHYLANTIMONY, $C_3H_9Sb$
TRIMETHYLTHALLIUM, $C_3H_9Tl$
* DIVINYLMAGNESIUM, $C_4H_6Mg$
n- or *tert*-BUTYLLITHIUM, $C_4H_9Li$
DIETHYLBERYLLIUM, $C_4H_{10}Be$
DIETHYLCADMIUM, $C_4H_{10}Cd$
DIETHYLMAGNESIUM, $C_4H_{10}Mg$
* DIETHYLLEAD DINITRATE, $C_4H_{10}N_2O_6Pb$
DIETHYLZINC, $C_4H_{10}Zn$
* DIETHYLGALLIUM HYDRIDE, $C_4H_{11}Ga$
* DILITHIUM TETRAMETHYLCHROMATE(2–), $C_4H_{12}CrLi_2$
* BIS-DIMETHYLSTIBINYL OXIDE, $C_4H_{12}OSb_2$
TETRAMETHYLLEAD, $C_4H_{12}Pb$
TETRAMETHYLPLATINUM, $C_4H_{12}Pt$
TETRAMETHYLDISTIBANE, $C_4H_{12}Sb_2$
TETRAMETHYLTIN, $C_4H_{12}Sn$
DIMETHYL-1-PROPYNYLTHALLIUM, $C_5H_9Tl$
* 2-CHLOROVINYLTRIMETHYLLEAD, $C_5H_{11}ClPb$
* LITHIUM PENTAMETHYLTITANATE–BIS-2,2′-BIPYRIDINE, $C_5H_{15}LiTi \cdot C_{20}H_{16}N_4$
PENTAMETHYLTANTALUM, $C_5H_{15}Ta$
TRIVINYLBISMUTH, $C_6H_9Bi$
BIS(DIMETHYLTHALLIUM) ACETYLIDE, $C_6H_{12}Tl_2$

DIISOPROPYLBERYLLIUM, $C_6H_{14}Be$
TRIETHYLBISMUTH, $C_6H_{15}Bi$
TRIETHYLGALLIUM, $C_6H_{15}Ga$
TRIETHYLANTIMONY, $C_6H_{15}Sb$
* LITHIUM HEXAMETHYLCHROMATE(3–), $C_6H_{18}CrLi_3$
* BIS(DIMETHYLAMINO)DIMETHYLSTANNANE, $C_6H_{18}N_2Sn$
* (DIMETHYLSILYLMETHYL)TRIMETHYLLEAD, $C_6H_{18}PbSi$
HEXAMETHYLRHENIUM, $C_6H_{18}Re$
HEXAMETHYLTUNGSTEN, $C_6H_{18}W$
* TETRAVINYLLEAD, $C_8H_{12}Pb$
* DI-2-BUTENYLCADMIUM, $C_8H_{14}Cd$
OCTYLSODIUM, $C_8H_{17}Na$
TETRAETHYLLEAD, $C_8H_{20}Pb$
TETRAETHYLTIN, $C_8H_{20}Sn$
* TRIALLYLCHROMIUM, $C_9H_{15}Cr$
DIMETHYL-PHENYLETHYNYLTHALLIUM, $C_{10}H_{11}Tl$
* TETRAALLYLURANIUM, $C_{12}H_{20}U$
TRIBUTYLINDIUM, $C_{12}H_{27}In$
TETRAISOPROPYLCHROMIUM, $C_{12}H_{28}Cr$
* HEXACYCLOHEXYLDILEAD, $C_{36}H_{66}Pb_2$

*See also* ARYLMETALS
METAL ACETYLIDES

*See related* LITHIUM PERALKYLURANATES
TRIALKYLSILYLOXY ORGANOLEAD DERIVATIVES

## ALKYL NITRATES $RONO_2$

Alone,
or Lewis acids

1. Boschan, R. *et al.*, *Chem. Rev.*, 1955, **55**, 505
2. Slavinskaya, R. A., *J. Gen. Chem. (USSR)*, 1957, **27**, 844
3. Boschan, R. *et al.*, *J. Org. Chem.*, 1960, **25**, 2012
4. Smith, 1966, Vol. 2, 458
5. Rüst, 1948, 284

The class has been reviewed [1]. Ethyl [1], isopropyl, butyl, benzyl and triphenylmethyl nitrates [3] in contact with sulphuric acid [2], tin(IV) chloride [2, 3] or boron trifluoride [3] interact violently (after an induction period of up to several hours) with gas evolution. An autocatalytic mechanism is proposed.

Although pure alkyl nitrates are stable in storage, traces of oxides of nitrogen sensitise them to decomposition, and may cause explosion on heating or storage at ambient temperature [4].

The most important applications of alkyl nitrates are based on their explosive properties. Methyl and ethyl nitrates are too volatile for widespread use; propyl nitrate has been used as a liquid monopropellant. Ethylene dinitrate, glyceryl trinitrate and pentaerythrityl tetranitrate are widely used as explosives, the two former adsorbed on or blended with other compounds [1].

The nitrate esters of many polyhydroxy compounds (ethylene glycol, glycerol, 1-chloro-2,3-dihydroxypropane, erythritol, mannitol, sugars or cellulose) are all more or less shock-sensitive and will ignite or explode on heating [5].

*See* METHYL NITRATE, $CH_3NO_3$
ETHYLIDENE DINITRATE, $C_2H_4N_2O_6$
ETHYL NITRATE, $C_2H_5NO_3$
1-CHLORO-2,3-PROPYLENE DINITRATE, $C_3H_5ClN_2O_6$
ISOPROPYL NITRATE, $C_3H_7NO_3$
PROPYL NITRATE, $C_3H_7NO_3$
2,2'-OXYDI(ETHYL NITRATE), $C_4H_8N_2O_7$
* 2-DIETHYLAMMONIOETHYL NITRATE NITRATE, $C_6H_{15}N_3O_6$
BENZYL NITRATE, $C_7H_7NO_3$
2(*trans*-1-AZIDO-1,2-DIHYDROACENAPHTHYL) NITRATE, $C_{12}H_8N_4O_3$

*See also* LITHIUM AZIDE, $LiN_3$: Alkyl nitrates, etc.

## ALKYL NITRITES — RONO

1. Sorbe, 1968, 146
2. Rüst, 1948, 285

Many alkyl nitrites are thermally unstable and may easily decompose or explode on heating [1]. Methyl nitrite explodes more violently than ethyl nitrite [2]. Individually indexed compounds are:

METHYL NITRITE, $CH_3NO_2$
ETHYL NITRITE, $C_2H_5NO_2$
ISOPROPYL NITRITE, $C_3H_7NO_2$
BUTYL NITRITE, $C_4H_9NO_2$
n- and ISO-PENTYL NITRITE, $C_5H_{11}NO_2$

## ALKYLNON-METAL HALIDES — REX

This small class of flammable, air-sensitive and usually pyrophoric compounds includes the individual entries for:

DICHLORO(METHYL)ARSINE, $CH_3AsCl_2$
TRICHLORO(METHYL)SILANE, $CH_3Cl_3Si$
DICHLORO(METHYL)SILANE, $CH_4Cl_2Si$
TRICHLORO(VINYL)SILANE, $C_2H_3Cl_3Si$
TRICHLORO(ETHYL)SILANE, $C_2H_5Cl_3Si$
IODODIMETHYLARSINE, $C_2H_6AsI$
CHLORODIMETHYLPHOSPHINE, $C_2H_6ClP$
DICHLORODIMETHYLSILANE, $C_2H_6Cl_2Si$
DICHLORO(ETHYL)SILANE, $C_2H_6Cl_2Si$
BUTYLDICHLOROBORANE, $C_4H_9BCl_2$
CHLORODIETHYLBORANE, $C_4H_{10}BCl$
DICHLORODIETHYLSILANE, $C_4H_{10}Cl_2Si$
METHYLPHENYLCHLOROSILANE, $C_7H_9ClSi$

*See related* HALOSILANES

## ALKYLNON-METAL HYDRIDES — $R_2EH$, $REH_2$

Griffiths, S.T. *et al.*, *Combust. Flame*, 1958, **2**, 244–252

Measurements of the autoignition temperatures for several series of tetra-, tri-, di- and mono-alkylsilanes showed that the ease of oxidation decreases with increasing substitution.

Several of the partially lower-alkylated derivatives of non-metal hydrides are pyrophoric in air. Individually indexed compounds are:

METHYLPHOSPHINE, $CH_5P$
METHYLSILANE, $CH_6Si$
DIMETHYLARSINE, $C_2H_7As$
DIMETHYLPHOSPHINE, $C_2H_7P$
ETHYLPHOSPHINE, $C_2H_7P$
1,2-BISPHOSPHINOETHANE, $C_2H_8P_2$
1,1-DIMETHYLDIBORANE, $C_2H_{10}B_2$
1,2-DIMETHYLDIBORANE, $C_2H_{10}B_2$
DIETHYLARSINE, $C_4H_{11}As$
DIETHYLPHOSPHINE, $C_4H_{11}P$
* DIPHENYLPHOSPHINE, $C_{12}H_{11}P$

*See also* the fully alkylated class, ALKYLNON-METALS

## ALKYLNON-METALS $R_nE$

Several of the fully lower-alkylated non-metals are pyrophoric in air. Separately treated groups are:

SILANES
SILYLHYDRAZINES

in addition to the individually indexed compounds:

TRIMETHYLARSINE, $C_3H_9As$
TRIMETHYLBORANE, $C_3H_9B$
TRIMETHYLPHOSPHINE, $C_3H_9P$
* 2,4,6-TRIMETHYLPERHYDRO-2,4,6-TRIBORA-1,3,5-TRIAZINE, $C_3H_{12}B_3N_3$
ETHYLDIMETHYLPHOSPHINE, $C_4H_{11}P$
TETRAMETHYLDIARSANE, $C_4H_{12}As_2$
* BIS-DIMETHYLARSINYL OXIDE, $C_4H_{12}As_2O$
* BIS-DIMETHYLARSINYL SULPHIDE, $C_4H_{12}As_2S$
TETRAMETHYLDIPHOSPHANE, $C_4H_{12}P_2$
TETRAMETHYLSILANE, $C_4H_{12}Si$
DIETHYLMETHYLPHOSPHINE, $C_5H_{13}P$
DIMETHYL(TRIMETHYLSILYL)PHOSPHINE, $C_5H_{15}PSi$
TRIETHYLARSINE, $C_6H_{15}As$
* TRIETHYLPHOSPHINEGOLD(I) NITRATE, $C_6H_{15}AuNO_3$
TRIETHYLBORANE, $C_6H_{15}B$
* DIETHYL ETHANEPHOSPHONITE, $C_6H_{15}O_2P$
TRIETHYLPHOSPHINE, $C_6H_{15}P$
* DIETHOXYDIMETHYLSILANE, $C_6H_{16}O_2Si$
ETHYLENEBIS(DIMETHYLPHOSPHINE), $C_6H_{16}P$
* BIS-TRIMETHYLSILYL OXIDE, $C_6H_{18}OSi_2$
* (DIMETHYLSILYLMETHYL)TRIMETHYLLEAD, $C_6H_{18}PbSi$
TETRAETHYLDIARSANE, $C_8H_{20}As_2$
'TETRAETHYLDIBORANE', $C_8H_{22}B_2$
TRIISOPROPYLPHOSPHINE, $C_9H_{21}P$

ETHYLENEBIS(DIETHYLPHOSPHINE), $C_{10}H_{22}P_2$
*mixo*-TRIBUTYLBORANE, $C_{12}H_{27}B$
TRIBUTYLPHOSPHINE, $C_{12}H_{27}P$
'TETRAPROPYLDIBORANE', $C_{12}H_{30}B_2$
DIBUTYL-3-METHYL-3-BUTEN-1-YNYLBORANE, $C_{13}H_{23}B$
* BIS(DIPROPYLBORINO)ACETYLENE, $C_{14}H_{28}B_2$
BIS(DIBUTYLBORINO)ACETYLENE, $C_{18}H_{36}B_2$
TRIBENZYLARSINE, $C_{21}H_{21}As$
* HEXAPHENYLHEXAARSANE, $C_{36}H_{30}As_6$

*See related* TRIALKYLSILYLOXY ORGANOLEAD DERIVATIVES
*See also* the partially alkylated class, ALKYLNON-METAL HYDRIDES

## ALKYL PERCHLORATES $ROClO_3$

1. Hofmann, K. A. *et al., Ber.*, 1909, **42**, 4390; Meyer, J. *et al., Z. Anorg. Chem.*, 1936, **228**, 341; Schumacher, 1960, 214
2. Burton, H. *et al., Analyst*, 1955, **80**, 4
3. Hoffman, D. M., *J. Org. Chem.*, 1971, **36**, 1716

Methyl, ethyl and propyl perchlorates, readily formed from the alcohol and anhydrous perchloric acid, are highly explosive oils, sensitive to shock, heat and friction. Many of the explosions which have occurred on contact of hydroxylic compounds with concentrated perchloric acid or anhydrous metal perchlorates are attributable to the formation and decomposition of perchlorate esters [1, 2].

Safe procedures for preparation of solutions of 14 *sec*-alkyl perchlorates are described. Hot evaporation of solvent caused explosions in all cases [3].

*See* TRIFLUOROMETHYL PERCHLORATE, $CClF_3O_4$
TRICHLOROMETHYL PERCHLORATE, $CCl_4O_4$
METHYL PERCHLORATE, $CH_3ClO_4$
ETHYL PERCHLORATE, $C_2H_5ClO_4$
3-CHLORO-2-HYDROXYPROPYL PERCHLORATE, $C_3H_6Cl_2O_5$
PROPYL PERCHLORATE, $C_3H_7ClO_4$
2(2-HYDROXYETHOXY)ETHYL PERCHLORATE, $C_4H_9ClO_6$

## ALKYL TRIALKYLLEAD PEROXIDES $ROOPbR'_3$

Houben-Weyl, 1975, Vol. 13.3, 111

These unstable compounds may decompose very violently on heating.

*See other* ORGANOMINERAL PEROXIDES

## ALKYNES $R{-}C{\equiv}C{-}$

Oxides of nitrogen

*See* NITROGEN OXIDE, NO: Dienes, etc.

## ALLOYS (INTERMETALLIC COMPOUNDS)

The individually indexed alloys or intermetallic compounds are:

ALUMINIUM–COPPER–ZINC ALLOY, Al–Cu–Zn
ALUMINIUM AMALGAM, Al–Hg
ALUMINIUM–MAGNESIUM ALLOY, Al–Mg
ALUMINIUM–NICKEL ALLOYS, Al–Ni
ALUMINIUM–TITANIUM ALLOYS, Al–Ti
BISMUTH PLUTONIDE, BiPu
COPPER–ZINC COUPLE, Cu–Zn
SODIUM GERMANIDE, GeNa
ZINC AMALGAM, Hg–Zn
POTASSIUM–SODIUM ALLOY, K–Na
LITHIUM–TIN ALLOYS, Li–Sn
SODIUM–ANTIMONY ALLOY, Na–Sb
LEAD–ZIRCONIUM ALLOY, Pb–Zr

## ALLYL COMPOUNDS $CH_2{=}CHCH_2-$

Several allyl compounds are notable for their high flammability and reactivity, including:

1-BROMO-2-PROPENE, $C_3H_5Br$
1-CHLORO-2-PROPENE, $C_3H_5Cl$
1-IODO-2-PROPENE, $C_3H_5I$
2-PROPEN-1-OL, $C_3H_6O$
2-PROPEN-1-THIOL, $C_3H_6S$
ALLYLAMINE, $C_3H_7N$
3-CYANOPROPENE, $C_4H_5N$
ALLYL ISOTHIOCYANATE, $C_4H_5NS$
ALLYL FORMATE, $C_4H_6O_2$
ALLYL ETHYL ETHER, $C_5H_{10}O$
DI(2-PROPEN-1-YL) ETHER (DIALLYL ETHER), $C_6H_{10}O$
DIALLYL SULPHATE, $C_6H_{10}O_4S$
DIALLYL SULPHIDE, $C_6H_{10}S$
DIALLYLAMINE, $C_6H_{11}N$
DIALLYL PHOSPHITE, $C_6H_{11}O_3P$
DIALLYL PEROXYDICARBONATE, $C_8H_{10}O_6$
ALLYL BENZENESULPHONATE, $C_9H_{10}O_3S$
TRIALLYLCHROMIUM, $C_9H_{15}Cr$
TRIALLYL PHOSPHATE, $C_9H_{15}O_4P$
ALLYL *p*-TOLUENESULPHONATE, $C_{10}H_{12}O_3S$
TETRAALLYL-2-TETRAZENE, $C_{12}H_{20}N_4$
TETRAALLYLURANIUM, $C_{12}H_{20}U$

*See other* PEROXIDISABLE COMPOUNDS

## AMINATION INCIDENTS

For amination incidents involving reactions with either ammonia or amines, see the entries:

CELLULOSE NITRATE: Amines
GOLD(III) CHLORIDE, $AuCl_3$ : Ammonia
ACETALDEHYDE, $C_2H_4O$

ETHYLENE OXIDE, $C_2H_4O$: Ammonia
: Trimethylamine
DIMETHYL SULPHATE, $C_2H_6O_4S$: Ammonia
1-CHLORO-2-PROPYNE, $C_3H_3Cl$: Ammonia
ACRYLALDEHYDE, $C_3H_4O$: Acids, or Bases
1-CHLORO-2,3-EPOXYPROPANE, $C_3H_5ClO$: Isopropylamine
MALEIC ANHYDRIDE, $C_4H_2O_3$ : Cations, or Bases
1-PERCHLORYLPIPERIDINE, $C_5H_{10}ClNO_3$ : Piperidine
POTASSIUM HEXACYANOFERRATE(3–) ('FERRICYANIDE'), $C_6FeK_3N_6$ : Ammonia
TRIETHYNYLALUMINIUM, $C_6H_3Al$: Trimethylamine
1-CHLORO-2,4-DINITROBENZENE, $C_6H_3ClN_2O_4$ : Ammonia
*o*- and *p*-CHLORONITROBENZENE, $C_6H_4ClNO_2$ : Ammonia
BENZENEDIAZONIUM-2-CARBOXYLATE, $C_7H_4N_2O_2$ : Aniline
DIISOPROPYLPEROXYDICARBONATE, $C_8H_{14}O_6$ : Amines
CHLOROGERMANE, $ClGeH_3$ : Ammonia
CHLORINE AZIDE, $ClN_3$ : Ammonia
THIOTRITHIAZYL CHLORIDE, $ClN_3S_4$ : Ammonia
SULPHINYL CHLORIDE (THIONYL CHLORIDE), $Cl_2OS$: Ammonia
RHENIUM TETRACHLORIDE OXIDE, $Cl_4ORe$: Ammonia
TELLURIUM TETRACHLORIDE, $Cl_4Te$: Ammonia
AMMONIA, $H_3N$: Boron halides
: Halogens, or Interhalogens
: Silver compounds
STIBINE, $H_3Sb$: Ammonia
MERCURY, Hg: Ammonia

## 'AMINE PERCHROMATES'

*See* AMMINECHROMIUM PEROXOCOMPLEXES

## AMINIUM IODATES AND PERIODATES $RN^+H_3 IO_3^-$, $RN^+H_3 IO_4^-$

Fogel'zang, A.E. *et al.*, *Chem. Abs.*, 1975, **83**, 8849
Combustion rates of the periodate salts of amines exceed those of the corresponding iodate salts.

*See other* OXOSALTS OF NITROGENOUS BASES

## AMINIUM PERCHLORATES $RN^+H_3 ClO_4^-$

Datta, R.L. *et al.*, *J. Chem. Soc.*, 1919, **115**, 1006–1010
Many perchlorate salts of amines explode in the range 215–310°C, and some on impact at ambient temperature.

*See other* OXOSALTS OF NITROGENOUS BASES
*See also* POLY(AMINIUM PERCHLORATES)

## AMMINECHROMIUM PEROXOCOMPLEXES $H_3N{\rightarrow}Cr{-}OO{-}$

This class of compounds, previously described as 'amine perchromates', is characterised by the presence of basic nitrogen and peroxo ligands within the same coordination sphere. This creates a high tendency towards explosive decomposition, which sometimes apparently occurs spontaneously. Individually indexed compounds are:

- * DIMETHYLETHEROXODIPEROXOCHROMIUM(VI), $C_2H_6CrO_6$
- 1,2-DIAMINOETHANEAQUADIPEROXOCHROMIUM(IV), $C_2H_{10}CrN_2O_5$
- 1,2-DIAMINOETHANEAMMINEDIPEROXOCHROMIUM(IV), $C_2H_{11}CrN_3O_4$
- 1,2-DIAMINOPROPANEAQUADIPEROXOCHROMIUM(IV), $C_3H_{12}CrN_2O_5$
- 1,2-DIAMINO-2-METHYLPROPANEOXODIPEROXOCHROMIUM(VI), $C_4H_{12}CrN_2O_5$
- DIETHYLENETRIAMINEDIPEROXOCHROMIUM(IV), $C_4H_{13}CrN_3O_4$
- 1,2-DIAMINO-2-METHYLPROPANEAQUADIPEROXOCHROMIUM (IV), $C_4H_{14}CrN_2O_5$
- OXODIPEROXODIPYRIDINECHROMIUM(VI), $C_{10}H_{10}CrN_2O_5$
- OXODIPEROXODIPIPERIDINECHROMIUM(VI), $C_{10}H_{22}CrN_2O_5$
- DIANILINEOXODIPEROXOCHROMIUM(VI), $C_{12}H_{14}CrN_2O_5$
- TRIAMMINEDIPEROXOCHROMIUM(IV), $CrH_9N_3O_4$

*See related* AMMINEMETAL OXOSALTS

## AMMINECOBALT(III) AZIDES $H_3N{\rightarrow}Co{-}N_3$

Joyner, T. B., West States Sect. Combust. Inst., 1967, Paper WSS/CI-67-15 (*Chem. Abs.*, 1970, **72**, 113377h)

The explosive properties of a series of five amminecobalt(III) azides were examined in detail. Compounds are hexaammine-triazide; pentaammine azido-diazide; *cis*- and *trans*-tetraammine-diazido-azide; triammine-triazide.

*See related* METAL AZIDES

## AMMINEMETAL OXOSALTS $H_3N{\rightarrow}M^+\ EO_n^-$

1. Mellor, 1941, Vol. 2, 341–364, 404
2. Tomlinson, W. R. *et al., J. Amer. Chem. Soc.*, 1949, **71**, 375
3. Wendlandt, W. W. *et al., Thermal Properties of Transition Metal Complexes,* Barking, Elsevier, 1967
4. Bretherick, L., *J. Chem. Educ.*, 1970, **47**, A204
5. Ray, P., *Chem. Rev.*, 1961, **61**, 313
6. Joyner, T.B., *Can. J. Chem.*, 1969, **47**, 2729–2730
7. Hoppesch, C.W. *et al., Amer. Chem. Soc., Div. Fuel Chem. Preprints*, 1963, **7**(3), 235–241
8. Friederich, W. *et al., Chem. Abs.*, 1927, **21**, 1184
9. Anagnostopoulos, A. *et al., J. Inorg. Nucl. Chem.*, 1974, **36**, 2235–2238
10. Patil, K.C. *et al., Thermochim. Acta*, 1976, **15**, 257–260

Metal compounds containing both coordinated ammonia, hydrazine or similar nitrogenous donors, and coordinated or ionic perchlorate,

chlorate, nitrate, nitrite, nitro, permanganate or other oxidising groups, will decompose violently under various conditions of impact, friction or heat [1, 2].

From tabulated data for 17 such compounds of Co and Cr, it is considered that oxygenated *N*-coordination compounds cover a wide range of explosive types; many may explode powerfully with little provocation, and should be considered extremely dangerous, as some are sensitive enough to propagate explosion under water. The same considerations may be expected to apply to ammines of silver, gold, cadmium, lead and zinc which contain oxidising radicals [2]. The topic has been reviewed [3] and possible hazards in published student preparations were emphasised [4]. Some of the derivatives of metal biguanide and guanylurea complexes [5] are of this class.

Unexpected uniformities observed in the impact-sensitivities of a group of 22 amminecobalt oxosalts are related to kinetic factors during the initiation process [6]. A series of ammine derivatives of cadmium, cobalt, copper, mercury, nickel, platinum and zinc with (mainly) iodate anions was prepared and evaluated as explosives [7]. Earlier, ammine and hydrazine derivatives of some of the same metals (not mercury, platinum or zinc) with chlorate or perchlorate anions had been evaluated as detonators. Dihydrazinecopper(II) chlorate had exploded when dried at ambient temperature [8].

A series of pyrazole complexes which decompose explosively above 200°C is notable in that the anion is sulphate, rather than the more obvious oxidant species usually present. The compounds are manganese sulphate complexed with four molecules of pyrazole or its 3-methyl derivative; and cadmium or zinc sulphate complexed with four and three molecules of 3-methylpyrazole, respectively [9]. At about 100°C the hexaammine diperchlorates of copper and zinc decompose to the tetraammines, and those of cadmium, cobalt, manganese and nickel to the diammines. At about 220°C all these lower ammines decompose explosively to the metal oxides (or the chloride for cadmium) [10].

Individually indexed compounds in this large class are:

DIAMMINESILVER PERMANGANATE, $AgH_6MnN_2O_4$
PENTAAMMINETHIOCYANATOCOBALT(III) PERCHLORATE, $CH_{15}Cl_2CoN_6O_8S$
PENTAAMMINETHIOCYANATORUTHENIUM(III) PERCHLORATE, $CH_{15}Cl_2N_6O_8RuS$
TETRAAMMINEDITHIOCYANATOCOBALT(III) PERCHLORATE, $C_2H_{12}ClCoN_6O_4S_2$
*cis*- and *trans*-BIS(1,2-DIAMINOETHANE)DINITRITOCOBALT(III) PERCHLORATES, $C_4H_{16}ClCoN_6O_8$
TETRATHIOUREAMANGANESE(II) PERCHLORATE, $C_4H_{16}Cl_2MnN_8O_8S_4$
BIS-1,2-DIAMINOETHANEDICHLOROCOBALT(III) CHLORATE, $C_4H_{16}Cl_3CoN_4O_3$
BIS-1,2-DIAMINOETHANEDICHLOROCOBALT(III) PERCHLORATE, $C_4H_{16}Cl_3CoN_4O_4$

*cis*-BIS-1,2-DIAMINOETHANEDINITROCOBALT(III) IODATE, $C_4H_{16}CoIN_6O_7$

BIS(1,2-DIAMINOETHANE)HYDROXOOXORHENIUM(V) DIPERCHLORATE, $C_4H_{17}Cl_2N_4O_{10}Re$

PENTAAMMINEPYRAZINERUTHENIUM(II) DIPERCHLORATE, $C_4H_{19}Cl_2N_7O_8Ru$

BIS(1,2-DIAMINOETHANEAQUA)COBALT(III) PERCHLORATE, $C_4H_{20}Cl_3CoN_4O_{14}$

TETRAAMMINE-2,3-BUTANEDIIMINERUTHENIUM(III) PERCHLORATE, $C_4H_{20}Cl_3N_6O_{12}Ru$

PENTAAMMINEPYRIDINERUTHENIUM(II) DIPERCHLORATE, $C_5H_{20}Cl_2N_6O_8Ru$

BIS-1,2-DIAMINOPROPANE-*cis*-DICHLOROCHROMIUM(III) PERCHLORATE, $C_6H_{20}Cl_3CrN_4O_4$

HEXAUREAGALLIUM(III) PERCHLORATE, $C_6H_{24}Cl_3GaN_{12}O_{18}$

TRIS-1,2-DIAMINOETHANECOBALT(III) NITRATE, $C_6H_{24}CoN_9O_9$

HEXAUREACHROMIUM(III) NITRATE, $C_6H_{24}CrN_{15}O_{15}$

BIS-DIETHYLENETRIAMINECOBALT(III) PERCHLORATE, $C_8H_{26}Cl_3CoN_6O_{12}$

DIPYRIDINESILVER(I) PERCHLORATE, $C_{10}H_{10}AgClN_2O_4$

6,6′-DIHYDRAZINO-2,2′-BIPYRIDYLNICKEL(II) PERCHLORATE, $C_{10}H_{12}Cl_2N_6NiO_8$

1,4,8,11-TETRAAZACYCLOTETRADECANENICKEL(II) PERCHLORATE, $C_{10}H_{24}Cl_2N_4NiO_8$

TETRAACRYLONITRILECOPPER(I) PERCHLORATE, $C_{12}H_{12}ClCuN_4O_4$

TETRAACRYLONITRILECOPPER(II) PERCHLORATE, $C_{12}H_{12}Cl_2CuN_4O_8$

TETRAKIS(PYRAZOLE)MANGANESE(II) SULPHATE, $C_{12}H_{16}MnN_8O_4S$

TRIS(3-METHYLPYRAZOLE)ZINC SULPHATE, $C_{12}H_{18}N_6O_4SZn$

HEXA(DIMETHYL SULPHOXIDE)CHROMIUM(III) PERCHLORATE, $C_{12}H_{36}CrCl_3O_{18}S_6$

* $\mu$-PEROXO-BIS[AMMINE(2,2′,2″-TRIAMINOTRIETHYLAMINE)-COBALT(III)] (4+) PERCHLORATE, $C_{12}H_{42}Cl_4Co_2N_{10}O_{18}$

DI[TRIS-1,2-DIAMINOETHANECOBALT(III)] TRIPEROXODISULPHATE, $C_{12}H_{48}Co_2N_{12}O_{24}S_6$

DI[TRIS-1,2-DIAMINOETHANECHROMIUM(III)] TRIPEROXODISULPHATE, $C_{12}H_{48}Cr_2N_{12}O_{24}S_6$

TETRAKIS(3-METHYLPYRAZOLE)CADMIUM SULPHATE, $C_{16}H_{24}CdN_8O_4S$

TETRAKIS(3-METHYLPYRAZOLE)MANGANESE(II) SULPHATE, $C_{16}H_{24}MnN_8O_4S$

BROMO-5,7,7,12,14,14-HEXAMETHYL-1,4,8,11-TETRAAZA-4, 11-CYCLOTETRADECADIENEIRON(II) PERCHLORATE, $C_{16}H_{32}BrClFeN_4O_4$

IODO-5,7,7,12,14,14-HEXAMETHYL-1,4,8,11-TETRAAZA-4, 11-CYCLOTETRADECADIENEIRON(II) PERCHLORATE, $C_{16}H_{32}ClFeIN_4O_4$

CHLORO-5,7,7,12,14,14-HEXAMETHYL-1,4,8,11-TETRAAZA-4,11-CYCLOTETRADECADIENEIRON(II) PERCHLORATE, $C_{16}H_{32}Cl_2FeN_4O_4$

DICHLORO-5,7,7,12,14,14-HEXAMETHYL-1,4,8,11-TETRAAZA-4,11-CYCLOTETRADECADIENEIRON(III) PERCHLORATE, $C_{16}H_{32}Cl_3FeN_4O_4$

HEXAMETHYLENETETRAMMONIUM TETRAPEROXOCHROMATE (V) (?), $C_{18}H_{48}Cr_4N_{12}O_{32}$

DIACETONITRILE-5,7,7,12,14,14-HEXAMETHYL-1,4,8,11-TETRAAZA-4,11-CYCLOTETRADECADIENEIRON(II) PERCHLORATE, $C_{20}H_{38}Cl_2FeN_6O_8$

ACETONITRILEIMIDAZOLE-5,7,7,12,14,14-HEXAMETHYL-1,4,8,11-TETRAAZA-4,11-CYCLOTETRADECADIENEIRON(II) PERCHLORATE, $C_{21}H_{39}Cl_2FeN_7O_8$

4-[2-(4-HYDRAZINO-1-PHTHALAZINYL)HYDRAZINO]-4-METHYL-2-PENTANONE (4-HYDRAZINO-1-PHTHALAZINYL) HYDRAZONEDINICKEL(II) TETRAPERCHLORATE, $C_{22}H_{28}N_{12}Ni_2Cl_4O_{16}$

TRIS-2,2′-BIPYRIDINESILVER(II) PERCHLORATE, $C_{30}H_{24}AgCl_2N_6O_8$

TRIS-2,2′-BIPYRIDINECHROMIUM(II) PERCHLORATE, $C_{30}H_{24}Cl_2CrN_6O_8$

D-(+)-TRIS(*o*-PHENANTHROLINE)RUTHENIUM(II) PERCHLORATE, $C_{30}H_{24}Cl_2N_6O_8Ru$

* TRIS-2,2′-BIPYRIDINECHROMIUM (0), $C_{30}H_{24}CrN_6$

CARBONYL-BIS-TRIPHENYLPHOSPHINEIRIDIUM–SILVER DIPERCHLORATE, $C_{37}H_{30}AgCl_2IrO_9P_2$

HEXAPYRIDINEIRON(II) TRIDECACARBONYLTETRAFERRATE (2–), $C_{43}H_{30}Fe_5N_6O_{13}$

TETRAKIS(4-*N*-METHYLPYRIDINIO)PORPHINECOBALT(II)(4+) PERCHLORATE, $C_{44}H_{36}Cl_4CoN_8O_{16}$

OXYBIS (*N,N*-DIMETHYLACETAMIDETRIPHENYLSTIBONIUM) DIPERCHLORATE, $C_{44}H_{50}Cl_2N_2O_{11}Sb_2$

TETRAAMMINECADMIUM(II) PERMANGANATE, $CdH_{12}Mn_2N_4O_8$

DIHYDRAZINECOBALT(II) CHLORATE, $Cl_2CoH_8N_4O_6$

* HEXAAQUACOBALT(II) PERCHLORATE, $Cl_2CoH_{12}O_{14}$

PENTAAMMINEPHOSPHINATOCOBALT(III) PERCHLORATE, $Cl_2CoH_{17}N_5O_{10}P$

* BISHYDROXYLAMINEZINC(II) CHLORIDE, $Cl_2H_6N_2O_2Zn$

BISHYDRAZINENICKEL(II) PERCHLORATE, $Cl_2H_8N_4NiO_8$

* BISHYDRAZINETIN(II) CHLORIDE, $Cl_2H_8N_4Sn$

TETRAAMMINEBIS-DINITROGENOSMIUM DIPERCHLORATE, $Cl_2H_{12}N_8O_8Os$

PENTAAMMINECHLOROCOBALT(III) PERCHLORATE, $Cl_3CoH_{15}N_5O_8$

PENTAAMMINEAQUACOBALT(III) CHLORATE, $Cl_3CoH_{17}N_5O_{10}$

HEXAAMMINECOBALT(III) CHLORITE, $Cl_3CoH_{18}N_6O_6$

HEXAAMMINECOBALT(III) CHLORATE, $Cl_3CoH_{18}N_6O_9$

HEXAAMMINECOBALT(III) PERCHLORATE, $Cl_3CoH_{18}N_6O_{12}$

DIAMMINENITRATOCOBALT(II) NITRATE, $CoH_6N_4O_6$

* TRIAMMINETRINITROCOBALT(III), $CoH_9N_6O_6$

TRIHYDRAZINECOBALT(II) NITRATE, $CoH_{12}N_8O_6$

* AMMONIUM HEXANITROCOBALTATE (3–), $CoH_{12}N_9O_{12}$

PENTAAMMINENITRATOCOBALT(III) NITRITE, $CoH_{15}N_8O_7$

HEXAAMMINECOBALT(III) IODATE, $CoH_{18}I_3N_6O_9$

HEXAAMMINECOBALT(III) PERMANGANATE, $CoH_{18}Mn_3N_6O_{12}$

HEXAHYDROXYLAMINECOBALT(III) NITRATE, $CoH_{18}N_9O_{15}$

HEXAAMMINECOBALT (3+) HEXANITROCOBALTATE (3–), $Co_2H_{18}N_{12}O_{12}$

HEXAAMMINECHROMIUM(III) NITRATE, $CrH_{18}N_9O_9$

* CAESIUM AMIDOPENTAFLUOROTELLURATE, $CsF_5H_2NTe$
TETRAAMMINECOPPER(II) SULPHATE, $CuH_{12}N_4O_4S$
TETRAAMMINECOPPER(II) NITRITE, $CuH_{12}N_6O_4$
TETRAAMMINECOPPER(II) NITRATE, $CuH_{12}N_6O_6$
*cis*-DIAMMINEPLATINUM(II) NITRITE ('*cis*-DIAMMINEDINITRO-PLATINUM(II)'), $H_6N_4O_4Pt$
DIAMMINEPALLADIUM(II) NITRATE, $H_6N_4O_6Pd$
DIHYDRAZINEMANGANESE(II) NITRATE, $H_8MnN_6O_6$
TRIAMMINENITRATOPLATINUM(II) NITRATE, $H_9N_5O_6Pt$
TETRAAMMINEZINC PEROXODISULPHATE, $H_{12}N_4O_8S_2Zn$
TETRAAMMINEPALLADIUM(II) NITRATE, $H_{12}N_6O_6Pd$
TRIHYDRAZINENICKEL(II) NITRATE, $H_{12}N_8NiO_6$
TETRAAMMINEHYDROXONITRATOPLATINUM(IV) NITRATE, $H_{13}N_7O_{10}Pt$

*See related* AMMINECHROMIUM PEROXOCOMPLEXES
[14] DIENE-$N_4$ IRON COMPLEXES
HYDRAZINEMETAL NITRATES
URANYL MACROCYCLIC PERCHLORATES

## APROTIC SOLVENTS

1. Buckley, A., *J. Chem. Educ.*, 1965, **42**, 674
2. Banthorpe, D. V., *Nature*, 1967, **215**, 1296

Many aprotic (non-hydroxylic) solvents are not inert towards other reagents and care is necessary when using untried combinations of solvents and reagents for the first time.

A further potential hazard which should be considered is that some aprotic solvents, notably dimethyl sulphoxide [1] and *N,N*-dimethylformamide [2], may greatly promote the toxic properties of solutes because of their unique ability readily to penetrate synthetic rubber protective gloves and the skin.

The ether and cyclic ether solvents are also subject to peroxidation in storage.

*See* METALS: Halocarbons
FORMAMIDE, $CH_3NO$
DIMETHYL SULPHOXIDE, $C_2H_6OS$
*N,N*-DIMETHYLFORMAMIDE, $C_3H_7NO$
TETRAHYDROFURAN, $C_4H_8O$
*p*-DIOXANE, $C_4H_8O_2$
1,2-DIMETHOXYETHANE, $C_4H_{10}O_2$
BIS(2-METHOXYETHYL) ETHER, $C_6H_{14}O_2$

## AQUA REGIA

Fawcett, H. H. *et al.*, *Chem. Eng. News*, **33**, 897, 1406, 1622, 1844

Aqua regia (nitric and hydrochloric acids, 1:4), which had been used for cleaning purposes, was stored in screw-capped winchesters. Internal pressure developed overnight, one bottle being shattered. Aqua regia decomposes with evolution of gas and should not be stored in tightly closed bottles. Undersized stoppers are suggested. It is also a powerful oxidant.

## ARENEDIAZO ARYL SULPHIDES $ArN_2SAr$

*See* DIAZONIUM SULPHIDES AND DERIVATIVES

## ARENEDIAZOATES $ArN_2OR$

Houben-Weyl, 1965, Vol. 10 (3), 563–564

Alkyl and aryl arenediazoates ('diazoethers'), Ar–N=N–O–R, are generally unstable and even explosive compounds. They are produced by interaction of alcohols with (explosive) bis(arenendiazo) oxides, or of *p*-blocked phenols with diazonium salts. The thio analogues are similar.

*See* METHYL 4-BROMOBENZENEDIAZOATE, $C_7H_7BrN_2O$
METHYL 2-NITROBENZENEDIAZOATE, $C_7H_7N_3O_3$
METHYL BENZENEDIAZOATE, $C_7H_8N_2O$
4-NITROPHENYL 2-CARBOXYBENZENEDIAZOATE, $C_{13}H_9N_3O_5$

*See related* DIAZONIUM SULPHIDES

## ARENEDIAZONIUMOLATES $N_2^+ArO^-$

This group of internal diazonium salts (previously called diazooxides) contains those which are, like many other internal diazonium salts, explosively unstable and shock-sensitive materials.

Individually indexed compounds are:

4-CHLORO-2,5-DINITROBENZENEDIAZONIUM-6-OLATE, $C_6HClN_4O_5$
3,4-DIFLUORO-2-NITROBENZENEDIAZONIUM-6-OLATE, $C_6HF_2N_3O_3$
3,6-DIFLUORO-2-NITROBENZENEDIAZONIUM-4-OLATE, $C_6HF_2N_3O_3$
3,5-DINITROBENZENEDIAZONIUM-2-OLATE, $C_6H_2N_4O_5$
4,6-DINITROBENZENEDIAZONIUM-2-OLATE, $C_6H_2N_4O_5$
3-BROMO-2,7-DINITRO-5-BENZO[b]THIOPHENEDIAZONIUM-4-OLATE, $C_8HBrN_4O_5S$

*See also* DIAZONIUM CARBOXYLATES
5-DIAZONIOTETRAZOLIDE, $CN_6$
BENZENEDIAZONIUM-4-SULPHONATE, $C_6H_4N_2O_3S$

## ARYL CHLOROFORMATES $ArOCOCl$

Water

Muir, G.D., private comm., 1968

During preparation of aryl chloroformates, it is essential to keep the reaction mixture really cold during water washing to prevent vigorous decomposition. Phenyl and naphthyl chloroformates may be distilled, but benzyl chloroformate is considered too unstable.

## ARYLMETALS

ArM

This reactive group includes the individually indexed compounds:

CYCLOPENTADIENYLGOLD, $C_5H_5Au$
CYCLOPENTADIENYLSODIUM, $C_5H_5Na$
*m*-DILITHIOBENZENE, $C_6H_4Li_2$
*p*-DILITHIOBENZENE, $C_6H_4Li_2$
PHENYLSILVER, $C_6H_5Ag$
* PHENYLVANADIUM(V) DICHLORIDE OXIDE, $C_6H_5Cl_2OV$
PHENYLLITHIUM, $C_6H_5Li$
PHENYLSODIUM, $C_6H_5Na$
TOLYLCOPPER (*o*-, *m*-, *p*-), $C_7H_7Cu$
NAPHTHYLSODIUM, $C_{10}H_7Na$
DIPHENYLMERCURY, $C_{12}H_{10}Hg$
DIPHENYLMAGNESIUM, $C_{12}H_{10}Mg$
DIPHENYLTIN, $C_{12}H_{10}Sn$
* BIS($\eta$-CYCLOPENTADIENYL)PENTAFLUOROPHENYLZIRCONIUM HYDROXIDE, $C_{16}H_{11}F_5OZr$
TRIPHENYLALUMINIUM, $C_{18}H_{15}Al$
TRIPHENYLCHROMIUM TETRAHYDROFURANATE, $C_{18}H_{15}Cr{\cdot}3C_4H_8O$
* TRIPHENYLLEAD NITRATE, $C_{18}H_{15}NO_3Pb$
* TRIPHENYLMETHYLPOTASSIUM, $C_{19}H_{15}K$
* BIS-$\eta$-CYCLOPENTADIENYL-BIS-PENTAFLUOROPHENYLZIRCONIUM, $C_{22}H_{10}F_{10}Zr$
* TETRAKIS(PENTAFLUOROPHENYL)TITANIUM, $C_{24}F_{20}Ti$
TETRAPHENYLLEAD, $C_{24}H_{20}Pb$
* LITHIUM HEXAPHENYLTUNGSTATE(2–), $C_{36}H_{30}Li_2W$

## ARYLTHALLIC ACETATE PERCHLORATES

ArTl(OAc)$ClO_4$

*See* PERCHLORIC ACID, $ClHO_4$ : Ethylbenzene, etc.

## ASSESSMENT OF REACTIVE CHEMICAL HAZARDS

Bretherick, L., in *Chemical Process Hazards with Special Reference to Plant Design-V*, 1–15, Kneale, M. (Editor), Symp. Ser. No. 39a, London, Institution of Chemical Engineers, 1975

Literature sources of information available to early 1974 and related to assessment of reactive chemical hazards are listed, grouped and discussed in relation to the types of information made available and their application to various situations. Eighty references are given, and those considered suitable for the nucleus of a small safety library are indicated. Toxic hazards are not covered.

*See also* COMPUTATION OF REACTIVE CHEMICAL HAZARDS

## AUTOIGNITION TEMPERATURE

1. Hilado, C. J. *et al., Chem. Eng.*, 1972, **79**(19), 75–80
2. *Fire Hazard Properties of Flammable Liquids, Gases, Solids,* No. 325M, Boston, Nat. Fire Prot. Ass., 1969
3. Shimy, A.A., *Fire Technol.*, 1970, **6**, 135–139

Autoignition temperature (AIT) is the temperature at which a material in contact with air undergoes oxidation at a sufficiently high rate to initiate combustion without an external ignition source.

Though only those compounds with unusually low AITs (below 225°C) have been included in this Handbook, the reference above is to a compilation of data for over 300 organic compounds, which also includes the theoretical background and discussion of the effect of variations in test methods upon AIT values obtained [1]. Further AIT data are given in the tabulated publication [2].

Semi-empirical formulae, based only on molecular structure, have been developed which allow AITs to be calculated for hydrocarbons and alcohols, usually with a reasonable degree of accuracy. Flammability limits, flash points, boiling points and flame temperatures may also be calculated for these classes [3].

## AUTOXIDATION

1. Davies, 1961, 11
2. Ingold, K. U., *Chem. Rev.*, 1961, **61**, 563

Autoxidation (interaction of a substance with molecular oxygen at below 120°C without flame [1]) has often been involved in the generation of hazardous materials from reactive compounds exposed to air. Methods of inhibiting autoxidation of organic compounds in the liquid phase have been reviewed [2].

*See* PEROXIDISABLE COMPOUNDS

## AZIDE COMPLEXES OF COBALT(III)

1. Druding, L.F. *et al.*, *J. Coord. Chem.*, 1973, **3**, 105
2. Druding, L.F. *et al.*, *Anal. Chem.*, 1975, **47**(1), 176–177

A series of 12 complexes of Co(III) with both ionic and covalent azide groups was prepared and most were easily detonable as dry salts, especially at elevated temperatures [1]. Polarography is an accurate and safe method of analysis for azides [2].

*See other* METAL AZIDES

## AZIDES $-N_3$

Many compounds of both organic and inorganic derivation, which contain the azide function, are unstable or explosive under appropriate conditions of initiation. The large number of compounds has been subdivided for convenience on the basis of structure. The inorganic groups are:

ACYL AZIDES
AMMINECOBALT(III) AZIDES
AZIDE COMPLEXES OF COBALT(III)

METAL AZIDES
NON-METAL AZIDES

and the organic groups:

ACYL AZIDES
2-AZIDOCARBONYL COMPOUNDS
ORGANIC AZIDES

## 2-AZIDOCARBONYL COMPOUNDS $-CO-CN_3$

Boyer, J.H. *et al.*, *Chem. Rev.*, 1954, **54**, 33
Weyler, J. *et al.*, *J. Org. Chem.*, 1973, **38**, 3865
Certain 2-azidocarbonyl compounds and congeners have long been known as unstable substances.

Some members of a group of 2,5-dialkyl-3,6-diazido-1,4-benzoquinones decompose violently on melting.

* AZIDOACETONITRILE, $C_2H_2N_4$
AZIDOACETALDEHYDE, $C_2H_3N_3O$
AZIDOACETONE, $C_3H_5N_3O$
AZIDOACETONE OXIME, $C_3H_6N_4O$
* DIAZIDOMALONONITRILE, $C_3N_8$
TETRAAZIDO-*p*-BENZOQUINONE, $C_6N_{12}O_2$
ETHYL *a*-AZIDO-*N*-CYANOPHENYLACETIMIDATE, $C_{11}H_{11}N_5O$

## AZO COMPOUNDS $C-N=N-C$

The common structural characteristic of this group of unstable compounds is the $-N=N-$ group linked to two different carbon atoms. Individually indexed compounds are:

* METHYLDIAZENE, $CH_4N_2$
AZO-*N*-CHLOROFORMAMIDINE, $C_2H_4Cl_2N_6$
AZO-*N*-NITROFORMAMIDINE, $C_2H_4N_8O_4$
AZOMETHANE, $C_2H_6N_2$
DICYANODIAZENE (AZOCARBONITRILE), $C_2N_4$
* ISOPROPYLDIAZENE, $C_3H_8N_2$
DIMETHYL AZODIFORMATE, $C_4H_6N_2O_4$
AZO-*N*-METHYLFORMAMIDE, $C_4H_8N_4O_2$
DIETHYL AZODIFORMATE, $C_6H_{10}N_2O_4$
METHYL 3-METHOXYCARBONYLAZOCROTONATE, $C_7H_{10}N_2O_4$
AZOISOBUTYRONITRILE, $C_8H_{12}N_4$
3,4-DIMETHYL-4-(3,4-DIMETHYL-5-ISOXAZOLYLAZO)-ISOXAZOLIN-5-ONE, $C_{10}H_{12}N_4O_3$
* POTASSIUM AZODISULPHONATE (2−), $K_2N_2O_6S_2$

## *N*-AZOLIUM NITROIMIDATES

Katritzsky, A.R. *et al.*, *J. Chem. Soc., Perkin Trans. 1*, 1973, 2624–2626
Some of the internal salts containing the system hetero-$\overset{+}{N}-\overset{-}{N}-NO_2$ are dangerously explosive solids, sensitive to heat or impact and liable to violent spontaneous decomposition, even in solution. A related

'*N*-nitroimide' (hetero-N–NH–$NO_2$) was also explosive, but the potassium salts, N–NK–$NO_2$, were stable. The unstable compounds are:

4-NITROAMINO-1,2,4-TRIAZOLE, $C_2H_3N_5O_2$
BENZOTRIAZOLIUM 1- or 2-NITROIMIDATE, $C_6H_4N_5O_2$
BENZIMIDAZOLIUM 1-NITROIMIDATE, $C_7H_5N_4O_2$

*See other* HIGH-NITROGEN COMPOUNDS
*N*-NITRO COMPOUNDS

**BIS-ARENEDIAZO OXIDES** **$(ArN_2)_2O$**

1. Bamberger, E., *Ber.*, 1896, **29**, 451
2. Kaufmann, T. *et al.*, *Ann.*, 1960, **634**, 77

Action of alkalies on diazonium solutions, or of acids on alkali diazoates to give a final pH of 5–6, causes these compounds ('diazoanhydrides') to separate as oils or solids. Many of these are violently explosive (some exceeding nitrogen trichloride in effect), sensitive to friction and heat or contact with aromatic hydrocarbons.

*See* BIS-2,4,5-TRICHLOROBENZENEDIAZO OXIDE, $C_{12}H_4Cl_6N_4O$
BIS-*p*-CHLOROBENZENEDIAZO OXIDE, $C_{12}H_8Cl_2N_4O$
BIS-BENZENEDIAZO OXIDE, $C_{12}H_{10}N_4O$
BIS-TOLUENEDIAZO OXIDE, $C_{14}H_{14}N_4O$

*See related* BIS-ARENEDIAZO SULPHIDES

**BIS-ARENEDIAZO SULPHIDES** **$(ArN_2)_2S$**

Some of the products of interaction of diazonium salts with sulphides may have this structure.

*See* DIAZONIUM SULPHIDES

**BIS(SULPHUR DIIMIDES)** **Z(N=S=NR)$_2$**

Appel, R. *et al.*, *Chem. Ber.*, 1976, **109**, 2444

During the preparation of a series of the bis-imides, (Z = $(CH_2)_{2 \text{ or } 3}$; R = $CH_3$, $C_2H_5$), work-up operations must be at below 100°C to avoid violent decomposition.

**BITUMEN**

Agaev, A.S. *et al.*, *Chem. Abs.*, 1975, **82**, 88310

Factors leading to ignition or explosion during preparation of high-melting bitumens by air-blowing petroleum residues were identified as control of vapour temperature (often above AIT), presence of lower hydrocarbon vapours, and lack of control of free oxygen content. Measures for controlling these are discussed.

*See also* OXYGEN, $O_2$: Hydrocarbons

**BLEACHING POWDER**

1. Mellor, 1956, Vol. 2, Suppl. 1,564–567
2. *Accid. Bull.* No. 30, Washington, Amer. Rail Assoc. Bur. Explosives, 1921
3. Gill, A. H., *Ind. Eng. Chem.*, 1924, **16**, 577
4. 'Leaflet No. 6', London, Inst. Chem., 1941

Bleaching powder is effectively a mixture of calcium hypochlorite, calcium hydroxide and a non-hygroscopic form of calcium chloride [1] and may therefore be regarded as a less active form of oxidant than undiluted calcium hypochlorite. There is a long history of explosions, many apparently spontaneous, involving bleaching powder. On storage or heating, several modes of decomposition are possible, one involving formation of chlorate which may increase the hazard potential.

Of the three possible routes for thermal decomposition, that involving liberation of oxygen predominates as the water content decreases, and at 150°C the decomposition becomes explosive [1].

Material which has been stored for a long time is liable to explode on exposure to sunlight, or on overheating of tightly packed material in closed containers [2]. The spontaneous explosion of material packed in drums was attributed to catalytic liberation of oxygen by iron and manganese oxides present in the lime used for manufacture [3]. Traces of metallic cobalt, iron, magnesium or nickel may also catalyse explosive decomposition [1].

When the lever-lid of a 6-month-old tin of bleaching powder was being removed, it flew off with explosive violence, possibly due to rust-catalysed slow liberation of oxygen [4].

Bis (2-chloroethyl) sulphide

Mellor, 1956, Vol. 2, Suppl. 1, 567

Interaction is very exothermic and ignition may occur, particularly in presence of water.

Wood

*ABCM Quart. Safety Summ.*, 1933, **4**, 15

A mixture of sawdust and bleaching powder ignites when moistened.

*See other* OXIDANTS

**BORANES**

Only unsubstituted boranes and their complexes are grouped here; alkylated derivatives are grouped as

ALKYLNON-METAL HYDRIDES or
ALKYLNON-METALS

Individually indexed compounds are:

POLY[BORANE(1)], $(BH)_n$

* BORANE–TETRAHYDROFURAN, $BH_3 \cdot C_4H_8O$
* BORANE–PYRIDINE, $BH_3 \cdot C_5H_5N$
* BORANE–PHOSPHORUS TRIFLUORIDE, $BH_3 \cdot F_3P$
* BORANE–AMMONIA, $BH_3 \cdot H_3N$
HYDRAZINE–MONOBORANE, $BH_7N_2$
* CHLORODIBORANE, $B_2ClH_5$
DIBORANE, $B_2H_6$
HYDRAZINE–BISBORANE, $B_2H_{10}N_2$
TETRABORANE(10), $B_4H_{10}$
PENTABORANE(9), $B_5H_9$
PENTABORANE(11), $B_5H_{11}$
* DIAMMINEBORONIUM HEPTAHYDROTETRABORATE (PENTABORANE(9) DIAMMONIATE), $B_5H_{15}N_2$
HEXABORANE(12), $B_6H_{12}$
DECABORANE(14), $B_{10}H_{14}$

Carbon tetrachloride

Hermanek, S., *J. Chromatogr. Libr.*, 1975, **3**, 945–951 (*Chem. Abs.*, 1976, **84**, 38384)

Carbon tetrachloride is not recommended as an eluting solvent in the chromatographic separation of boranes, carbaboranes or their derivatives because of the danger of explosion.

*See other* NON-METAL HYDRIDES

## *tert*-BUTYL PEROXOPHOSPHATE DERIVATIVES $Me_3COOP(O)=$

Rieche, A. *et al.*, *Chem. Ber.*, 1962, **95**, 381–388

Although dialkyl *tert*-butyl peroxophosphate derivatives are relatively stable, the diaryl esters and bis-*tert*-butyl peroxo esters decompose violently on attempted isolation. Individually indexed compounds are:

*tert*-BUTYL PEROXOPHOSPHORYL DICHLORIDE, $C_4H_9Cl_2O_3P$
DI(*O–O-tert*-BUTYL) ETHYL DIPEROXOPHOSPHATE, $C_{10}H_{23}O_6P$
*O–O-tert*-BUTYL DIPHENYL MONOPEROXOPHOSPHATE, $C_{16}H_{19}O_5P$
*O–O-tert*-BUTYL DI(*p*-TOLYL) MONOPEROXOPHOSPHATE, $C_{18}H_{23}O_5P$

*See related* PEROXY ESTERS

## CAN OF BEANS

Foote, C. S., private comm., 1965

An unopened can of beans, placed in an oven initially at 110°C, later reset to 150°C, exploded, causing extensive damage. Comments were judged to be superfluous.

## CARBONACEOUS DUSTS

*Explosion and Ignition Hazards,* Report 6597, Washington, US Bureau of Mines, 1965

Hazards of 241 industrial dusts which may explode or burn because of their carbon content are defined, covering particle size and chemical composition in 10 categories.

*See* PETROLEUM COKE
CARBON, C

## CARBONYLMETALS $M(CO)_n$

Bailar, 1973, Vol. 1, 1227

The 'carbonylalkali-metals', previously formulated as monomeric compounds, are either dimeric acetylene derivatives of the general formula MOC≡COM (M = K, Li, Na), or are trimers of the latter, and formulated as salts of hexahydroxybenzene.

Many members of this reactive class are air-sensitive and pyrophoric, not always immediately, individually indexed compounds being:

'CARBONYLPOTASSIUM': see the hexamer, $C_6K_6O_6$
'CARBONYLLITHIUM': see the hexamer, $C_6Li_6O_6$
'CARBONYLSODIUM': see the hexamer, $C_6Na_6O_6$
* SODIUM TETRACARBONYLFERRATE(2–), $C_4FeNa_2O_4$
TETRACARBONYLNICKEL, $C_4NiO_4$
PENTACARBONYLIRON, $C_5FeO_5$
* SODIUM PENTACARBONYLRHENATE(2–), $C_5Na_2O_5Re$
HEXACARBONYLCHROMIUM, $C_6CrO_6$
* POTASSIUM BENZENEHEXOLATE, $C_6K_6O_6$
* LITHIUM BENZENEHEXOLATE ('CARBONYLLITHIUM'), $C_6Li_6O_6$
HEXACARBONYLMOLYBDENUM, $C_6MoO_6$
* SODIUM BENZENEHEXOLATE, $C_6Na_6O_6$
HEXACARBONYLVANADIUM, $C_6O_6V$
HEXACARBONYLTUNGSTEN, $C_6O_6W$
OCTACARBONYLDICOBALT, $C_8Co_2O_8$
* LITHIUM OCTACARBONYLTRINICCOLATE(2–), $C_8Li_2Ni_3O_8$
NONACARBONYLDIIRON, $C_9Fe_2O_9$
DECACARBONYLDIRHENIUM, $C_{10}O_{10}Re_2$
DODECACARBONYLTETRACOBALT, $C_{12}Co_4O_{12}$
DODECACARBONYLTRIIRON, $C_{12}Fe_3O_{12}$
* POTASSIUM TRICARBONYLTRIS(PROPYNL)MOLYBDATE(3–), $C_{12}H_9KMoO_3$
* BIS(DICARBONYL-$\eta$-CYCLOPENTADIENYLIRON)-BIS(TETRAHYDROFURAN)MAGNESIUM, $C_{22}H_{26}Fe_2MgO_6$
* TRIS(BIS-2-METHOXYETHYL ETHER)POTASSIUM HEXACARBONYLNIOBATE(1–), $C_{24}H_{42}KNbO_{15}$
* TETRAKIS(PYRIDINE)BIS(TETRACARBONYLCOBALT)MAGNESIUM, $C_{28}H_{20}Co_2MgN_4O_8$
* BIS[DICARBONYL-$\eta$-CYCLOPENTADIENYL(TRIBUTYLPHOSPHINE)MOLYBDENUM] TETRAKIS(TETRAHYDROFURAN)MAGNESIUM, $C_{54}H_{96}MgMo_2O_8P_2$

## CARBABORANES

Carbon tetrachloride

*See* BORANES: Carbon tetrachloride

## CATALYSIS BY IMPURITIES

For incidents where presence of impurities, often in trace (catalytic) amounts, has significantly decreased stability or increased reactivity,

*see* SILVER AZIDE, $AgN_3$
TRISILVER NITRIDE, $Ag_3N$
TETRAFLUOROAMMONIUM TETRAFLUOROBORATE, $BF_8N$
SODIUM PEROXOBORATE, $BNaO_3$
BROMINE TRIOXIDE, $BrO_3$
CARBON TETRACHLORIDE, $CCl_4$: Dimethylformamide
CHLOROFORM, $CHCl_3$: Acetone, Alkali
: Sodium hydroxide, Methanol
CHLOROMETHANE, $CH_3Cl$: Metals
SODIUM CARBONATE, $CNa_2O_3$: An aromatic amine, etc.
1,2-BIS(DIFLUOROAMINO)-*N*-NITROETHYLAMINE, $C_2H_4F_4N_4O_2$
AZIRIDINE (ETHYLENEIMINE), $C_2H_5N$: Acids
MERCURY(II) OXALATE, $C_2HgO_4$
ACRYLALDEHYDE, $C_3H_4O$
*N,N*-DIMETHYLFORMAMIDE, $C_3H_7NO$: Halogenated compounds, etc.
TRIMETHYL PHOSPHATE, $C_3H_9O_4P$
MALEIC ANHYDRIDE, $C_4H_2O_3$
DIKETENE, $C_4H_4O_2$
2-BUTYNE-1,4-DIOL, $C_4H_6O_2$
ACETIC ANHYDRIDE, $C_4H_6O_3$: Ethanol, etc.
BUTYRALDEHYDE OXIME, $C_4H_9NO$
*N,N*-DIMETHYLACETAMIDE, $C_4H_9NO$
DIETHYL PHOSPHOROCHLORIDATE, $C_4H_{10}ClOP$
ETHYL 2-FORMYLPROPIONATE OXIME, $C_6H_{11}NO_3$
BENZOYL AZIDE, $C_7H_5N_3O$
1-ETHOXY-3-METHYL-1-BUTYN-3-OL, $C_7H_{12}O_2$
*O*-MESITYLENESULPHONYLHYDROXYLAMINE, $C_9H_{13}NO_3S$
TRIALLYL PHOSPHATE, $C_9H_{15}O_4P$
AMMONIUM PERCHLORATE, $ClH_4NO_4$
POTASSIUM CHLORATE, $ClKO_3$: Manganese dioxide
CHLORINE, $Cl_2$: Hydrocarbons (reference 9)
: Methanol
SULPHINYL CHLORIDE (THIONYL CHLORIDE), $Cl_2OS$: Dimethylformamide
HYDROGEN PEROXIDE, $H_2O_2$
HYDRAZINE, $H_4N_2$: Metal catalysts
AMMONIUM NITRATE, $H_4N_2O_3$
NITROGEN OXIDE ('NITRIC OXIDE'), NO: Carbon, etc.
NICKEL, Ni: Hydrogen, Oxygen
OXYGEN (Liquid), $O_2$: Carbon, Iron(II) oxide

For the opposite effect of negative catalysis,

*see* INDUCTION PERIOD

## CELLULOSE

Calcium oxide

*See* CALCIUM OXIDE, CaO: Water

Oxidants

*See* BLEACHING POWDER: Wood

PERCHLORATES; Organic matter
PERCHLORIC ACID, $ClHO_4$ : Cellulose and derivatives
SODIUM CHLORATE, $ClNaO_3$ : Paper, etc; Wood
MAGNESIUM DIPERCHLORATE, $Cl_2MgO_8$ : Cellulose, etc.
FLUORINE, $F_2$ : Miscellaneous materials
NITRIC ACID, $HNO_3$ : Cellulose
ZINC PERMANGANATE, $Mn_2O_8Zn$: Cellulose
SODIUM NITRITE, $NNaO_2$ : Wood
SODIUM NITRATE, $NNaO_3$: Fibrous material
SODIUM PEROXIDE, $Na_2O_2$ : Fibrous materials

**CELLULOSE NITRATE**

1. Kirk-Othmer, 1964, Vol. 4, 627–628
2. Anon., *Accidents,* 1961 (46), 12
3. *ABCM Quart. Safety Summ.,* 1963, **34**, 13
4. *MCA Case History No. 1614*

Cellulose nitrate is very easily ignited and burns very rapidly or explosively, depending on the degree of confinement and state of subdivision. Unless very pure and stabilised, it deteriorates in storage and may ignite spontaneously [1].

Removal of the emulsion coating from celluloid film base gives unstable material which may spontaneously ignite on prolonged storage in an enclosed space [2].

Unused but aged centrifuge tubes, made of 'Nitrocellulose' plasticised with dibutyl phthalate, ignited and exploded while being steam-sterilised in an autoclave at 125°C. The violent decomposition was attributed to the age of the tubes, the high temperature and the presence of steam [3].

During hacksaw cutting of a pipe containing cellulose nitrate residues, a violent explosion occurred [4].

Amines

1. *ABCM Quart. Safety Summ.,* 1956, **27**, 2
2. Thurlow, G.K. *et al.,* private comm., 1973

Cellulose nitrate of high surface area (dry or alcohol-wet guncotton or scrap) spontaneously ignited in contact with various amines used as curing agents for epoxide resins. These included: 1,2-diaminoethane, *N*-hydroxyethyl-1,2-diaminoethane, diethylenetriamine, triethylenetetramine, *N*-2-hydroxyethyltriethylenetetramine, tetraethylenepentamine, 2-hydroxyethylamine, 2-hydroxyethyl-dimethylamine, 2-hydroxypropylamine, 3-dimethylaminopropylamine, 3-diethylaminopropylamine, 3-dibutylaminopropylamine, morpholine, diethylamine. Ethylamine and dibutylamine caused charring but not ignition [1].

Similar results were obtained during an investigation of the compatibility of cellulose nitrate with a range of amine and amide components used in paint manufacture. Preliminary small-scale (12g) tests in which ethyl acetate solutions of the nitrate and other components were mixed in a lagged boiling tube showed large exotherms (which boiled the solvent off) with 1,4-diazabicyclo-[2.2.2]octane, 2,4,6-tris(dimethylaminomethyl)phenol, morpholine and benzyldimethylamine. Smaller exotherms were shown by dodecanamine, dodecyldimethylamine (both fat-derived, containing homologues) and a polyamide resin, Versamid 140.

Subsequent tests in which small portions of these undiluted liquid amines and dried cellulose nitrate linters were contacted (with a little added butyl acetate for the solid phenol) under various conditions gave ignition with the first three amines, and exotherms to 110°C with foaming decomposition with the remaining four.

Other amine resin components showed slight or no exotherms in either test [2].

Iron red,
Plasticiser

Penczek, P. *et al.*, *Chem. Abs.*, 1976, **85**, 63924

During roller-blending to disperse iron red pigment into plasticised cellulose nitrate, the mixture became a gel after 15 passes and had a tendency to self-ignition.

*See other* PYROPHORIC MATERIALS

## CHLORINATED POLY(DIMETHYLSILOXANES)

Sosa, J.M., *Thermochim. Acta*, 1975, **13**, 100–104

Chlorinated silicone oil (DC200, with 15, 30 or 40% chlorine content) decomposed violently on heating. Thermal stability decreased with increasing chlorine content and was investigated by TGA and DSC techniques.

## CHLORINATED RUBBER

Metal oxides or hydroxides

1. Anon., *Chem. Trade J.*, 1962, **151**, 672
2. *Euro. Chem. News,* 1963 (May 24th), 29
3. 'Report GCS 27130', London, ICI, 1963
4. *ABCM Quart. Safety Summ.*, 1963, **34**, 12

Intimate mixtures of chlorinated rubber and zinc oxide or powdered zinc, with or without hydrocarbon or chlorinated solvents, react violently or explosively when heated at about 216°C. If in milling such mixtures local overheating occurs, a risk of a violent reaction exists. Such risks can be minimised by controlling milling temperatures, by cooling, or by using a mixture of maximum possible fluidity [1]. Similar reactions have been observed with antimony or lead oxides, or aluminium, barium or zinc hydroxides [2]. The full report [3] has been abstracted [4].

## CHLORINE-CONTAINING SYSTEMS

Mal'tseva, A.S. *et al.*, *Chem. Abs.*, 1975, **82**, 45981

Explosive limits and hazards of various binary and ternary systems containing chlorine or its compounds are reviewed and discussed.

## CHLORITE SALTS $ClO_2^-$

Mellor, 1941, Vol. 2, 284; 1956, Vol. 2, Suppl. 1, 573–575

Pascal, 1960, Vol. 16, 263

Many of the salts which have been prepared are explosive and sensitive to heat or impact. These include chlorites of copper (violent on impact), hydrazine (monochlorite, inflames when dry), lead, mercury (spontaneously explosive dry), nickel (explodes at 100°C but not on impact), silver (at 105°C or on impact), sodium, tetramethylammonium and thallium (which shows detonator properties). Several other chlorites not isolated and unstable in solution include ammonium chlorite and its mono-, di- and tri-methyl derivatives. The metal salts are powerful oxidants.

Chlorites are much less stable than the corresponding chlorates, and most will explode under shock or on heating to around 100°C. Individually indexed metal salts are:

SILVER CHLORITE, $AgClO_2$
BARIUM CHLORITE, $BaCl_2O_4$
POTASSIUM CHLORITE, $ClKO_2$
SODIUM CHLORITE, $ClNaO_2$
THALLIUM(I) CHLORITE, $ClO_2Tl$
MERCURY(II) CHLORITE, $Cl_2HgO_4$
MERCURY(I) CHLORITE, $Cl_2Hg_2O_4$
LEAD DICHLORITE, $Cl_2O_4Pb$

*See also* OXOSALTS OF NITROGENOUS BASES

## *N*-CHLORONITROAMINES $RN(Cl)NO_2$

Grakauskas, V. *et al.*, *J. Org. Chem.*, 1972, **37**, 334

*N*-Chloronitroamines and the derived *N*-chloro-*N*-nitrocarbamates are explosive compounds and decompose rapidly on storage.

*See also* *N*-FLUORO-*N*-NITROBUTYLAMINE, $C_4H_9FN_2O_2$
*See other* *N*-HALOGEN COMPOUNDS
*N*-NITRO COMPOUNDS

## CHLOROPHYLL

Ichimura, S. *et al.*, Japan Kokai, 74 86 512, 1974 (*Chem. Abs.*, 1976, **84**, 7220)

Chlorophyll adsorbed on clay, or powdered unicellular green algae (chlorella), can be caused to explode by focused irradiation from a powerful ruby laser.

## COMMERCIAL ORGANIC PEROXIDES — C–OO–

1. Boyars, C., *AD Rept. No. 742770*, USNTIS, 1972
2. Lee, P.R., *J. Appl. Chem.*, 1969, **19**, 345–351
3. Connor, J., *RARDE Memo.*, Waltham Abbey, 1974, **15**, 1–58

This group of compounds is widely used in industry as a radical source for initiation of polymerisation. They are available from several manufacturers in a very wide range of formulations in various diluents to minimise operational hazards. These have been classified into six hazard levels and of the many materials available, the few below (all dry and unformulated except for *, which is water-wetted) are included in the highest risk category. This specifies the material as being sensitive to friction or mechanical shock equivalent to the dissipation of 1kg-m or less of energy within the sample.

Existing and proposed methods of evaluating transportation hazards of organic peroxides exposed to impact, explosive shock or thermal surge stimuli are reviewed, and a hazard classification system is proposed [1]. Commercial 2-butanone peroxide ('MEK peroxide') as a 40% solution in dimethyl phthalate was previously thought to be safe in normal storage and transport situations, but several road- and rail-tanker explosion incidents showed evidence of detonation. Application of steady-state thermal explosion theory allowed the prediction of critical mass and induction-period in relation to temperature in bulk storage. The critical mass seems likely to have been attained in some of the incidents [2]. In a review of thermal decomposition of organic peroxides, hazards of self-accelerating decomposition are assessed [3].

*See* BIS(3-CARBOXYPROPIONYL) PEROXIDE, $C_8H_{10}O_8$
*O–O-tert*-BUTYL HYDROGENMONOPEROXYMALEATE, $C_8H_{12}O_5$
* ACETYL CYCLOHEXANESULPHONYL PEROXIDE, $C_8H_{14}O_5S$
DIISOPROPYL PEROXYDICARBONATE, $C_8H_{14}O_6$
BIS(1-HYDROPEROXYCYCLOHEXYL) PEROXIDE, $C_{12}H_{22}O_6$
BIS(2,4-DICHLOROBENZOYL) PEROXIDE, $C_{14}H_6Cl_4O_4$

## COMPLEX ACETYLIDES — $K_m[M(C{\equiv}CR)_n$

1. Bailar, 1973, Vol. 4, 810
2. Nast, R. *et al.*, *Chem. Ber.*, 1962, **95**, 1470–1483
3. Nast, R. *et al.*, *Z. Anorg. Allgem. Chem.*, 1955, **279**, 146–156

The salts $K_m[M(C{\equiv}CR)_n]$, where M is Cr(III), Co(II or III), Cu(0 or I), Au(I), Fe(II or III), Mn(II or III), Ni(0 or II), Pd(0), Pt(0 or II) or Ag(I), are frequently explosive [1]. A series of dialkynyl- palladates and platinates [2] and a tetraalkynylniccolate [3] are pyrophoric, while other tetraalkynylniccolates are explosive [3]. Most react violently with water, and individually indexed compounds are:

POTASSIUM* BIS(ETHYNYL)PALLADATE(2–), $C_4H_2K_2Pd$
POTASSIUM* BIS(ETHYNYL)PLATINATE(2–), $C_4H_2K_2Pt$
POTASSIUM* BIS(PROPYNYL)PALLADATE(2–), $C_6H_6K_2Pd$
POTASSIUM* BIS(PROPYNYL)PLATINATE(2–), $C_6H_6K_2Pt$
POTASSIUM TETRAKIS(ETHYNYL)NICCOLATE(2–), $C_8H_4K_2Ni$
POTASSIUM* TETRAKIS(ETHYNYL)NICCOLATE(4–), $C_8H_4K_4Ni$
POTASSIUM HEXAKIS(ETHYNYL)COBALTATE(4–), $C_{12}H_6CoK_4$
POTASSIUM HEXAKIS(ETHYNYL)MANGANATE(3–), $C_{12}H_6K_3Mn$
POTASSIUM TRICARBONYLTRIS(PROPYNYL)MOLYBDATE(3–), $C_{12}H_9KMoO_3$
POTASSIUM TETRAKIS(PROPYNYL)NICCOLATE(4–), $C_{12}H_{12}K_4Ni$
POTASSIUM* BIS(PHENYLETHYNYL)PALLADATE(2–), $C_{16}H_{10}K_2Pd$
POTASSIUM* BIS(PHENYLETHYNYL)PLATINATE(2–), $C_{16}H_{10}K_2Pt$
SODIUM HEXAKIS(PROPYNYL)FERRATE(4–), $C_{18}H_{18}FeNa_4$
POTASSIUM* TETRAKIS(PHENYLETHYNYL)NICCOLATE(4–), $C_{32}H_{20}K_4Ni$

*Sodium salts are similar

*See other* ACETYLENIC COMPOUNDS

## COMPLEX HYDRIDES $[MH_n]^-, [EH_n]^-$

Gaylord, 1956
Semenenko, K.N. *et al.*, *Russ. Chem. Rev.*, 1973, 1–13

This class of highly reactive compounds includes several which have found extensive use as reducants in preparative chemistry.
Properties and reactions of several covalent tetrahydroborates have been more recently reviewed. Individually indexed compounds are:

ALUMINIUM TETRAHYDROBORATE, $AlB_3H_{12}$
* ALUMINIUM DICHLORIDE HYDRIDE DIETHYLETHERATE, $AlCl_2H \cdot C_4H_{10}O$
CAESIUM HEXAHYDROALUMINATE (3–), $AlCs_3H_6$
COPPER(I) TETRAHYDROALUMINATE, $AlCuH_4$
LITHIUM TETRAHYDROALUMINATE, $AlH_4Li$
SODIUM TETRAHYDROALUMINATE, $AlH_4Na$
POTASSIUM HEXAHYDROALUMINATE (3–), $AlH_6K_3$
MAGNESIUM TETRAHYDROALUMINATE, $Al_2H_8Mn$
MANGANESE(II) TETRAHYDROALUMINATE, $Al_2H_8Mn$
CERIUM(III) TETRAHYDROALUMINATE, $Al_3CeH_{12}$
LITHIUM TETRAHYDROBORATE, $BH_4Li$
SODIUM TETRAHYDROBORATE, $BH_4Na$
HYDRAZINE–MONOBORANE, $BH_7N_2$
BERYLLIUM TETRAHYDROBORATE, $B_2BeH_8$
HYDRAZINE–BISBORANE, $B_2H_{10}N_2$
DIAMMINEBORONIUM TETRAHYDROBORATE, $B_2H_{12}N_2$
SODIUM OCTAHYDROTRIBORATE, $B_3H_8Na$
URANIUM(III) TETRAHYDROBORATE, $B_3H_{12}U$
HAFNIUM(IV) TETRAHYDROBORATE, $B_4H_{16}Hf$
ZIRCONIUM(IV) TETRAHYDROBORATE, $B_4H_{16}Zr$

* CAESIUM LITHIUM UNDECAHYDROTHIONONABORATE, $B_9CsH_{11}LiS$
CAESIUM LITHIUM TRIDECAHYDRONONABORATE, $B_9CsH_{13}Li$
BERYLLIUM TETRAHYDROBORATE-TRIMETHYLAMINE, $C_3H_{17}B_2BeN$
BIS(DIMETHYLAMINOBORANE)–ALUMINIUM TETRAHYDROBORATE, $C_4H_{22}AlB_3N_2$
HEPTAKIS(DIMETHYLAMINO)TRIALUMINIUM TRIBORON PENTAHYDRIDE, $C_{14}H_{47}Al_3B_3N_7$
LITHIUM DIHYDROCUPRATE, $CuH_2Li$
LITHIUM TETRAHYDROGALLATE, $GaH_4Li$
SODIUM TETRAHYDROGALLATE, $GaH_4Na$
POTASSIUM TRIHYDROMAGNESATE, $H_3KMg$
POTASSIUM TETRAHYDROZINCATE(2–), $H_4K_2Zn$

*See other* REDUCANTS
*See also* METAL HYDRIDES

## COMPUTATION OF REACTIVE CHEMICAL HAZARDS


1. Treweek, D.N. *et al., J. Haz. Mat.*, 1976, **1**, 173–189
2. Domalski, E.S. *et al., Proc. 4th Int. Symp. on Transp. of Haz. Cargoes*, 1975, USNTIS PB 254214
3. *CHETAH, Chemical Thermodynamic and Energy Release Evaluation Program*, Philadelphia, Amer. Soc. for Testing and Materials, 1975
4. Sherwood, R.M., *Chem. Brit.*, 1975, **11**, 417


During the past few years, several computational methods for predicting instability (or 'self-reactivity') of chemicals in bulk storage have been developed. These have been variously based on structural and calculated thermodynamic parameters, amplified in some cases with experimentally determined data, to produce hazard rating systems.

These methods have now been jointly evaluated for their ability to predict successfully instability in a range of compound types. It was concluded that the relationship between the parameters considered and chemical stability is too obtuse for conventional statistical analysis. However, application of pattern-recognition techniques to statistical analysis was more fruitful, and 13 of the more promising parameters were successfully evaluated. Total under- and over-estimation error was below 10% relative to generally accepted forms of consensus grading into the three stability categories of explosive, hazardous decomposition or polymerisation, and non-hazardous. The methods appear to have great potential utility in minimising hazards in storage, transfer and transportation of chemicals [1].

In an alternative assessment of the effectiveness of these computer programs, it is concluded that explosive power is over-emphasised in relation to the more practically important aspect of sensitivity to initiation, and many compounds are indicated as hazardous when they are not. There is also no provision for considering polymerisation as a hazardous possibility, and there are also very few quantitative data available on this. The parameter best correlating with material sensitivity is the bond-dissociation energy.

It is recommended that, at present, regulations specifying the handling and transport of chemicals should be based on the concept of known self-reactivity of functional groups present (nitro, nitroamine, peroxide, azide, etc.) [2].

One of the computer programs [3] is relatively simple to apply, and gives estimates of the maximum energy release possible for any covalent compound or mixture of compounds containing C, H, O, N and up to 18 other specified elements. Reactions may be specified if they are known, but the only input from the user is the structure of each compound involved. The result serves as a screening guide to permit decisions on which reaction systems need more detailed and/or experimental investigation [4].

*See* ASSESSMENT OF REACTIVE CHEMICAL HAZARDS
*See also* EXPLOSIBILITY (references 3–5)

## COTTON

Oils

Anon., *Textil-Rundschau*, 1957, **12**, 273

Cotton waste in contact with fatty oils (especially of unsaturated character) is much more subject to autoignition than animal fibres. Various factors affecting storage hazards of baled cotton, including the aggravating effect of moisture, are discussed. Mineral oils, though generally less subject to oxidative heating than vegetable oils, have also been involved in autoignition incidents.

## CRYOGENIC LIQUIDS

1. Zabetakis, M. G., *Safety with Cryogenic Fluids,* London, Heywood, 1967
2. *Cryogenics Safety Manual–A Guide to Good Practice,* London, Safety Panel, British Cryogenics Council, 1970
3. *Safety Problems in Handling Low-Temperature Industrial Fluids*, 1964, I.Ch.E. Sympos. papers, publ. in *Chem. Engr.*, 1965, **43**(185), CE7–10; (186), CE36–48
4. Bernstein, J.T., *Cryogenics*, 1973, **13**, 600–602
5. B.S. Code of Practice BS5429-1976

Two reference works are available which together cover this important technological field.

The first is a monograph concisely presenting principles of safety applicable to cryogenics in 6 chapters: Introduction; Physiological Hazards; Physical Hazards; Chemical Hazards; Laboratory Safety; Plant and Test Site Safety. There are 3 Appendices: Physical Constants and Conversion Factors; Safety Data Sheets (for air, argon, carbon dioxide, carbon monoxide, ethylene, fluorine, helium, hydrogen, krypton, methane, neon, nitrogen, oxygen, ozone, xenon); Disaster Investigation (Explosions).

The second was produced for the guidance of those concerned with operation and maintenance of plant for producing, storing and handling commercial gases which liquefy at relatively low temperatures. This illustrated book covers both general safety requirements and specific safety requirements for: Air Separation Plants Producing Oxygen, Nitrogen and Argon; Liquefied Natural Gas; Hydrogen Separation Plants; Ethylene and Ethane.

The symposium covered possible fire and explosion hazards in general terms, as well as in detail for liquid hydrogen, acetylene, natural gas, and a low-temperature nitrogen-washing process for ammonia synthesis gas [3]. Safety aspects of sampling and handling cryogenic liquids have been reviewed [4]. The new Code of Practice provides users of liquid oxygen, nitrogen, argon or natural gas with a basic appreciation of the problems and hazards associated with the small-scale use and storage of these materials [5].

*See* LIQUEFIED GASES: Water
ETHYLENE, $C_2H_4$: Steel-braced tyres
HYDROGEN (Liquid), $H_2$
NITROGEN (Liquid), $N_2$
OXYGEN (Liquid), $O_2$

## CRYSTALLINE HYDROGEN PEROXIDATES

Castrantas, 1965, 4
Emeléus, 1960, 432
Kirk-Othmer, 1966, Vol. 11, 395
Mellor, 1971, Vol. 8, Suppl. 3, 824

A few compounds will crystallise out with hydrogen peroxide in the crystal lattice, analogous to crystalline hydrates. These represent a form of concentrated hydrogen peroxide, which may react violently in close contact (grinding or heating) with oxidisable materials.

*See* SODIUM BORATE HYDROGEN PEROXIDATE, $BNaO_2 \cdot H_2O_2$
UREA HYDROGEN PEROXIDATE, $CH_4N_2O \cdot H_2O_2$
SODIUM CARBONATE HYDROGEN PEROXIDATE, $CNa_2O_3 \cdot 1.5H_2O_2$
TRIETHYLAMINE HYDROGEN PEROXIDATE, $C_6H_{15}N \cdot 4H_2O_2$
POTASSIUM FLUORIDE HYDROGEN PEROXIDATE, $FK \cdot H_2O_2$
SODIUM PYROPHOSPHATE HYDROGEN PEROXIDATE, $Na_4O_7P_2 \cdot 2H_2O_2$

*See also* HYDROGEN PEROXIDE, $H_2O_2$: Nitric acid, etc.

## CYANO COMPOUNDS —C≡N, $CN^-$

Metal cyanides are readily oxidised and those of some heavy metals show instability. Many organic nitriles are unusually reactive under appropriate circumstances, and *N*-cyano derivatives are reactive or unstable. The class includes the groups:

3-CYANOTRIAZENES
METAL CYANIDES AND CYANOCOMPLEXES

as well as the individually indexed compounds:

DISILVER CYANAMIDE, $CAg_2N_2$
CYANOGEN FLUORIDE, CFN
HYDROGEN CYANIDE, CHN
CYANAMIDE, $CH_2N_2$
NITROSYL CYANIDE, $CN_2O$
CYANOGEN AZIDE, $CN_4$
ACETONITRILE, $C_2H_3N$
* METHYL ISOCYANIDE, $C_2H_3N$
GLYCOLONITRILE, $C_2H_3NO$
CYANOGUANIDINE ('DICYANDIAMIDE'), $C_2H_4N_4$
* MERCURY(II) THIOCYANATE, $C_2HgN_2S_2$
CHLOROCYANOACETYLENE, $C_3ClN$
2-CHLOROACRYLONITRILE, $C_3H_2ClN$
MALONONITRILE, $C_3H_2N_2$
ACRYLONITRILE, $C_3H_3N$
* VINYL ISOCYANIDE, $C_3H_3N$
CYANOACETIC ACID, $C_3H_3NO_2$
5-CYANO-2-METHYLTETRAZOLE, $C_3H_3N_5$
2-CHLORO-1-CYANOETHANOL, $C_3H_4ClNO$
* ETHYL ISOCYANIDE, $C_3H_5N$
PROPIONITRILE, $C_3H_5N$
2-AMINOPROPIONITRILE, $C_3H_6N_2$
MESOXALONITRILE, $C_3N_2O$
PHOSPHORUS TRICYANIDE, $C_3N_3P$
DICYANODIAZOMETHANE, $C_3N_4$
DIAZIDOMALONONITRILE, $C_3N_8$
2-CYANO-1,2,3-TRIS(DIFLUOROAMINO)PROPANE, $C_4H_4F_6N_4$
1- and 3-CYANOPROPENE, $C_4H_5N$
1-CYANO-2-PROPEN-1-OL, $C_4H_5NO$
2-CYANO-2-PROPYL NITRATE, $C_4H_6N_2O_3$
ISOPROPYLISOCYANIDE DICHLORIDE–IRON(III) CHLORIDE, $C_4H_7Cl_2N \cdot Cl_3Fe$
BUTYRONITRILE, $C_4H_7N$
ISOBUTYRONITRILE, $C_4H_7N$
DIMETHYLAMINOACETONITRILE, $C_4H_8N_2$
DICYANOACETYLENE, $C_4N_2$
PIVALONITRILE, $C_5H_9N$
CYANODIETHYLGOLD: see the tetramer, $C_{20}H_{40}Au_4N_4$
3-DIMETHYLAMINOPROPIONITRILE, $C_5H_{10}N_2$
1,4-DICYANO-2-BUTENE, $C_6H_6N_2$
BIS(ACRYLONITRILE)NICKEL(0), $C_6H_6N_2Ni$
BIS(2-CYANOETHYL)AMINE, $C_6H_9N_3$
*o*-NITROBENZONITRILE, $C_7H_4N_2O_2$
*N*-CYANO-2-BROMOETHYLBUTYLAMINE, $C_7H_{13}BrN_2$
4-CYANO-3-NITROTOLUENE, $C_8H_6N_2O_2$
PHENYLACETONITRILE, $C_8H_7N$
AZOISOBUTYRONITRILE, $C_8H_{12}N_4$
* 2-ISOCYANOETHYL BENZENESULPHONATE, $C_9H_9NO_3S$
*N*-CYANO-2-BROMOETHYLCYCLOHEXYLAMINE, $C_9H_{15}BrN_2$
TETRAACRYLONITRILECOPPER(I) PERCHLORATE, $C_{12}H_{12}ClCuN_4O_4$
TETRAACRYLONITRILECOPPER(II) PERCHLORATE, $C_{12}H_{12}Cl_2CuN_4O_8$

3-ETHOXYCARBONYL-4,4,5,5-TETRACYANO-3-TRIMETHYLPLUMBYL-4,5-DIHYDRO-3*H*-PYRAZOLE, $C_{13}H_{14}N_6O_2Pb$
TETRACYANOOCTAETHYLTETRAGOLD, $C_{20}H_{40}Au_4N_4$

### 3-CYANOTRIAZENES

**N=N–N(CN)**

Bretschneider, H. *et al.*, *Monatsh.*, 1950, **81**, 981

Many aromatic 3-cyanotriazenes are shock-sensitive, explosive compounds.

*See other* HIGH-NITROGEN COMPOUNDS
TRIAZENES

### CYCLIC PEROXIDES

**COOC, (C–OO)$_n$**

Swern, 1970, Vol. 1, 37; 1972, Vol. 3, 67, 81

Generally produced *inter alia* during peroxidation of aldehydes or ketones, the lower members are often violently explosive. Dimeric and trimeric ketone peroxides are the most dangerous classes of organic peroxide, exploding on heating, touching or friction.

*See* 3,3,4,5-TETRACHLORO-3,6-DIHYDRO-1,2-DIOXIN, $C_4H_4Cl_4O_2$
3,6-DICHLORO-3,6-DIMETHYLTETRAOXANE, $C_4H_6Cl_2O_4$
*p*-DIOXENEDIOXETANE (2,3-EPIDIOXYDIOXANE), $C_4H_6O_4$
3,6-DIMETHYL-1,2,4,5-TETRAOXANE, $C_4H_8O_4$
1,6-DIAZA-3,4,8,9,12,13-HEXAOXABICYCLO[4.4.4] TETRADECANE ('HEXAMETHYLENETRIPEROXYDIAMINE'), $C_6H_{12}N_2O_6$
TETRAMETHYL-1,2-DIOXETANE, $C_6H_{12}O_2$
3,3,6,6-TETRAMETHYL-1,2,4,5-TETRAOXANE, $C_6H_{12}O_4$
1,4-EPIDIOXY-1,4-DIHYDRO-6,6-DIMETHYLFULVENE, $C_8H_{10}O_2$
3,6-DIETHYL-3,6-DIMETHYL-1,2,4,5-TETRAOXANE, $C_8H_{16}O_4$
3,3,6,6-TETRAKIS(BROMOMETHYL)-9,9-DIMETHYL-1,2,4,5,7,8-HEXAOXAONANE, $C_9H_{14}Br_4O_6$
3,3,6,6,9,9-HEXAMETHYL-1,2,4,5,7,8-HEXAOXAONANE, $C_9H_{18}O_6$
3,6,9-TRIETHYL-1,2,4,5,7,8-HEXAOXAONANE, $C_9H_{18}O_6$
1,4-EPIDIOXY-2-*p*-MENTHENE (ASCARIDOLE), $C_{10}H_{20}O_4$
3,6-DI(SPIROCYCLOHEXANE)TETRAOXANE, $C_{12}H_{20}O_4$
3,6,9-TRIETHYL-3,6,9-TRIMETHYL-1,2,4,5,7,8-HEXAOXAONANE, $C_{12}H_{24}O_6$
9,10-EPIDIOXYANTHRACENE, $C_{14}H_8O_2$
TRI(SPIROCYCLOPENTANE)-1,1,4,4,7,7-HEXAOXAOXONANE, $C_{15}H_{24}O_6$

*See also* DIOXETANES

### DEVARDA'S ALLOY

1. Cameron, W.G., *Chem. & Ind.*, 1948, 158
2. Chaudhuri, B.B., ibid., 462

The analytical use of the alloy to reduce nitrates is usually accompanied by the risk of a hydrogen explosion, particularly if heating is effected by flame. Use of a safety screen, and flameless heating, coupled with displacement of hydrogen by an inert gas, are recommended

precautions [1]. The explosion was later attributed to gas pressure in a restricted system [2].
*See other* ALLOYS (INTERMETALLIC COMPOUNDS)

**DIACYL PEROXIDES** **CO–OO–CO, $SO_2$–OO–$SO_2$**

Swern, 1970, Vol. 1, 70
Mageli, O.L. *et al.*, US Pat. 3 956 396, 1976
Most of the isolated diacyl (including sulphonyl) peroxides are solids with relatively low decomposition temperatures, and are explosive, and sensitive to shock, heat or friction. Several, particularly the lower members, will detonate on the slightest disturbance. Autocatalytic, self-accelerating decomposition, which is promoted by tertiary amines, is involved.

Solvents suitable for preparation of safe solutions of diacetyl, dipropionyl, diisobutyryl and di-2-phenylpropionyl peroxides are disclosed.

Individually indexed compounds are:

BISFLUOROFORMYL PEROXIDE, $C_2F_2O_4$
DIMETHANESULPHONYL PEROXIDE, $C_2H_6O_6S_2$
* *O*-TRIFLUOROACETYL-*S*-FLUOROFORMYL THIOPEROXIDE, $C_3F_4O_3S$
BISTRICHLOROACETYL PEROXIDE, $C_4Cl_6O_4$
BISTRIFLUOROACETYL PEROXIDE, $C_4F_6O_4$
* ACETYL 1,1-DICHLOROETHYL PEROXIDE, $C_4H_6Cl_2O_3$
DIACETYL PEROXIDE, $C_4H_6O_4$
POTASSIUM BENZENESULPHONYLPEROXOSULPHATE, $C_6H_5KO_7S_2$
DIPROPIONYL PEROXIDE, $C_6H_{10}O_4$
POTASSIUM *O–O*-BENZOYLPEROXOSULPHATE, $C_7H_5KO_6S$
PHTHALOYL PEROXIDE, $C_8H_4O_4$
DICROTONOYL PEROXIDE, $C_8H_{10}O_4$
BIS(3-CARBOXYPROPIONYL) PEROXIDE, $C_8H_{10}O_8$
DIISOBUTYRYL PEROXIDE, $C_8H_{14}O_4$
ACETYL CYCLOHEXANESULPHONYL PEROXIDE, $C_8H_{14}O_5S$
DI-2-FUROYL PEROXIDE, $C_{10}H_6O_6$
DI-2-METHYLBUTYRYL PEROXIDE, $C_{10}H_{18}O_4$
BIS-3,4-DICHLOROBENZENESULPHONYL PEROXIDE, $C_{12}H_6Cl_4O_6S_2$
BIS-*p*-BROMOBENZENESULPHONYL PEROXIDE, $C_{12}H_8Br_2O_6S_2$
BIS-*p*-CHLOROBENZENESULPHONYL PEROXIDE, $C_{12}H_8Cl_2O_6S_2$
DIBENZENESULPHONYL PEROXIDE, $C_{12}H_{10}O_6S_2$
DIHEXANOYL PEROXIDE, $C_{12}H_{22}O_4$
BIS-2,4-DICHLOROBENZOYL PEROXIDE, $C_{14}H_6Cl_4O_4$
BIS-*o*-AZIDOBENZOYL PEROXIDE, $C_{14}H_8N_6O_4$
2,2′-BIPHENYLDICARBONYL PEROXIDE, $C_{14}H_8O_4$
DIBENZOYL PEROXIDE, $C_{14}H_{10}O_4$
BIS-3-(2-FURYL)ACRYLOYL PEROXIDE, $C_{14}H_{10}O_6$
DI-*p*-TOLUENESULPHONYL PEROXIDE, $C_{14}H_{14}O_6S_2$
DICYCLOHEXYLCARBONYL PEROXIDE, $C_{14}H_{22}O_4$
DI-3-CAMPHOROYL PEROXIDE, $C_{20}H_{30}O_8$
DI-1-NAPHTHOYL PEROXIDE, $C_{22}H_{14}O_4$
DIDODECANOYL PEROXIDE (DILAUROYL PEROXIDE), $C_{24}H_{46}O_4$

*See* PEROXYCARBONATE ESTERS

### DIALKYL HYPONITRITES RON=NOR

Partington, J. R. *et al., J. Chem. Soc.*, 1932, **135**, 2593

The violence of the explosion when the ethyl ester was heated at 80°C was not so great as previously reported. The propyl and butyl esters explode if heated rapidly, but decompose smoothly if heated gradually.

*See related* AZO COMPOUNDS

### DIALKYLMAGNESIUMS $R_2Mg$

Sidgwick, 1950, 233

This series, either as the free alkyls or as their etherates, is extremely reactive, igniting in air or carbon dioxide and reacting violently or explosively with alcohols, ammonia or water.

*See* DIMETHYLMAGNESIUM, $C_2H_6Mg$
DIETHYLMAGNESIUM, $C_4H_{10}Mg$
DIPHENYLMAGNESIUM, $C_{12}H_{10}Mg$

*See other* ALKYLMETALS

### DIALKYL PEROXIDES ROOR

1. Castrantas, 1965, 12; Swern, 1970, Vol. 1, 38, 54
2. Davies, 1961, 75

The high and explosive instability of the lower dialkyl peroxides and 1,1-bis-peroxides decreases rapidly with increasing chain length and degree of branching, the di-*tert*-alkyl derivatives being amongst the most stable class of peroxides [1]. Though many 1,1-bisperoxides have been reported, few have been purified because of the higher explosion hazards compared with the monofunctional peroxides [2].

*See* DIMETHYL PEROXIDE, $C_2H_6O_2$
ETHYL METHYL PEROXIDE, $C_3H_8O_2$
DIETHYL PEROXIDE, $C_4H_{10}O_2$
DIPROPYL PEROXIDE, $C_6H_{14}O_2$
DI-*tert*-BUTYL PEROXIDE, $C_8H_{18}O_2$
2,2-DI(*tert*-BUTYLPEROXY)BUTANE, $C_{12}H_{26}O_4$

### DIALKYLZINCS $R_2Zn$

1. Sidgwick, 1950, 266
2. Noller, C.R., *J. Amer. Chem. Soc.*, 1929, **51**, 597
3. Houben-Weyl, 1973, Vol. 13.2a, 560, 576

The dialkylzincs up to the dibutyl derivative readily ignite and burn in air. The higher alkyls fume but do not always ignite [1]. During preparation of dialkylzincs, reaction of the copper–zinc alloy with mixed alkyl bromides and iodides must begin (exotherm, often after a long induction period) before too much halide mixture is added, or violent explosions may occur [2]. Reaction with water may be explosive [3].

Acyl halides

Houben-Weyl, 1973, Vol. 13.2a, 781

Too-fast addition of acyl halides during preparation of ketones may lead to explosive reactions.

Alkyl chlorides

Noller, C.R., *J. Amer. Chem. Soc.*, 1929, **51**, 599

During interaction to give hydrocarbons, too much chloride must not be added before reaction sets in (induction period), or explosions may occur.

Methanol

Houben-Weyl, 1973, Vol. 13.2a, 855

Contact of the neat liquids with uncooled methanol is explosively violent and leads to ignition. For analysis, ampouled samples of dialkylzincs must first be frozen in liquid nitrogen before being broken under methanol–heptane mixtures at –60°C.

Individually indexed compounds are:

DIMETHYLZINC, $C_2H_6Zn$
DIETHYLZINC, $C_4H_{10}Zn$

*See other* ALKYLMETALS

## DIAZIRINES

CN=N

Schmitz, E. *et al.*, *Org. Synth.*, 1965, **45**, 85

Liu, M.T.H., *Chem. Eng. News*, 1974, **52**(36), 3

Diazirine and several of its 3-substituted homologues, formally cyclic azo compounds, are explosive on heating or impact.

The shock-sensitivity of all diazirine compounds and the inadvisability of their handling in undiluted state have again been stressed.

*See* DIAZIRINE, $CH_2N_2$
3-CHLORO-3-TRICHLOROMETHYLDIAZIRINE, $C_2Cl_4N_2$
3-CHLORO-3-METHYLDIAZIRINE, $C_2H_3ClN_2$
3-METHYLDIAZIRINE, $C_2H_4N_2$
3-PROPYLDIAZIRINE, $C_4H_8N_2$
3,3-PENTAMETHYLENEDIAZIRINE, $C_6H_{10}N_2$
PHENYLCHLORODIAZIRINE, $C_7H_5ClN_2$

*See related* AZO COMPOUNDS

## DIAZO COMPOUNDS

$CN_2$

In this group of reactive and unstable compounds the common structural feature is two nitrogen atoms attached to the same carbon atom. Individually indexed compounds are:

DIDEUTERIODIAZOMETHANE, $CD_2N_2$
DIAZOMETHYLLITHIUM, $CHLiN_2$
DIAZOMETHYLSODIUM, $CHN_2Na$
DIAZOMETHANE, $CH_2N_2$

DINITRODIAZOMETHANE, $CN_4O_4$
DIAZOACETONITRILE, $C_2HN_3$
DIAZOACETYL AZIDE, $C_2HN_5O$
VINYLDIAZOMETHANE, $C_3H_4N_2$
METHYL DIAZOACETATE, $C_3H_4N_2O_2$
DICYANODIAZOMETHANE, $C_3N_4$
ETHYL DIAZOACETATE, $C_4H_6N_2O_2$
DIAZOCYCLOPENTADIENE, $C_5H_4N_2$
2-DIAZOCYCLOHEXANONE, $C_6H_8N_2O$
2-BUTEN-1-YL DIAZOACETATE, $C_6H_8N_2O_2$
*tert*-BUTYL DIAZOACETATE, $C_6H_{10}N_2O_2$
*o*-NITROPHENYLSULPHONYLDIAZOMETHANE, $C_7H_5N_3O_4S$
PHENYLDIAZOMETHANE, $C_7H_6N_2$
BIS(ETHOXYCARBONYLDIAZOMETHYL)MERCURY, $C_8H_{10}HgN_4O_4$
*tert*-BUTYL 2-DIAZOACETOACETATE, $C_8H_{12}N_2O_3$
1-DIAZOINDENE, $C_9H_6N_2$
3-METHOXY-2-NITROBENZOYLDIAZOMETHANE, $C_9H_7N_3O_4$
1,1-BENZOYLPHENYLDIAZOMETHANE, $C_{14}H_{10}N_2O$

## DIAZONIUM CARBOXYLATES $N_2^+ArCO_2^-$

Several of these internal salts, prepared by diazotisation of anthranilic acids, are explosive in the solid state, or react violently with various materials.

*See* 4-IODOBENZENEDIAZONIUM-2-CARBOXYLATE, $C_7H_3IN_2O_2$
BENZENEDIAZONIUM-2-, 3- or 4-CARBOXYLATE, $C_7H_4N_2O_2$
4-HYDROXYBENZENEDIAZONIUM-3-CARBOXYLATE, $C_7H_4N_2O_3$
3,6-DIMETHYLBENZENEDIAZONIUM-2-CARBOXYLATE, $C_9H_8N_2O_2$
4,6-DIMETHYLBENZENEDIAZONIUM-2-CARBOXYLATE, $C_9H_8N_2O_2$

## DIAZONIUM PERCHLORATES $ArN_2^+ClO_4^-$

Hofmann, K. A. *et al.*, *Ber.*, 1906, **39**, 3146; ibid., 1910, **43**, 2624
Vorländer, D., *Ber.*, 1906, **39**, 2713–2715
Burton, H. *et al.*, *Analyst*, 1955, **80**, 4

Extremely explosive, shock-sensitive materials when dry, some even when damp. The salt derived from diazotised *p*-phenylenediamine was considered to be more explosive than any other substance known in 1910.

*See* *m*-NITROBENZENEDIAZONIUM PERCHLORATE, $C_6H_4ClN_3O_6$
BENZENEDIAZONIUM PERCHLORATE, $C_6H_5ClN_2O_4$
*p*-AMINOBENZENEDIAZONIUM PERCHLORATE, $C_6H_6ClN_3O_4$
*o*-TOLUENEDIAZONIUM PERCHLORATE, $C_7H_7ClN_2O_4$
1- or 2-NAPHTHALENEDIAZONIUM PERCHLORATE, $C_{10}H_7ClN_2O_4$
4,4′-BIPHENYLENEBISDIAZONIUM PERCHLORATE, $C_{12}H_8Cl_2N_4O_8$
1-(2′-, 3′ or 4′-DIAZONIOPHENYL-2-METHYL-4,6-DIPHENYLPYRIDINIUM DIPERCHLORATES, $C_{24}H_{19}N_3O_8$
1-(2′-, 3′- or 4′-DIAZONIOPHENYL)-2,4,6-TRIPHENYLPYRIDINIUM DIPERCHLORATES, $C_{29}H_{21}Cl_2N_3O_8$

## DIAZONIUM SALTS

$ArN_2^+$

1. Houben-Weyl, 1965, Vol. 10(3), 32–38
2. Doyle, W. H., *Loss Prevention*, 1969, **3**, 14
3. Fogel'zang, A. E. *et al.*, *Chem. Abs.*. 1974, **81**, 155338

A few diazonium salts are unstable in solution, and many are in the solid state. Of these, the azides, chromates, nitrates, perchlorates (outstandingly), picrates, sulphides, triiodides and xanthates are noted as being explosive, and sensitive to friction, shock, heat and radiation. In view of their technical importance, diazonium salts are often isolated as their zinc chloride (or other) double salts, and although these are considerably more stable, some incidents involving explosive decomposition have been recorded.

During bottom discharge of an undefined diazonium chloride preparation, operation of a valve initiated explosion of the friction-sensitive chloride which had separated from solution. The latter did not occur with the corresponding sulphate [2].

The combustive and explosive behaviour of solid diazonium salts at various pressures was studied, including benzenediazonium chloride and nitrate, and *m*- and *p*-nitrobenzenediazonium chlorides [3].

Separately treated groups are:

ARENEDIAZONIUMOLATES
DIAZONIUM CARBOXYLATES
DIAZONIUM PERCHLORATES
DIAZONIUM SULPHATES
DIAZONIUM SULPHIDES AND DERIVATIVES
DIAZONIUM TETRAHALOBORATES
DIAZONIUM TRIIODIDES

Individually indexed compounds are:

* 1,10-BISDIAZONIODECABORAN(8)ATE, $B_{10}H_8N_4$
TETRAZOLE-5-DIAZONIUM CHLORIDE, $CHClN_6$
5-DIAZONIOTETRAZOLIDE, $CN_6$
3,4,5-TRIIODOBENZENEDIAZONIUM NITRATE, $C_6H_2I_3N_3O_3$
BENZENEDIAZONIUM-2-SULPHONATE, $C_6H_4N_2O_3S$
BENZENEDIAZONIUM-4-SULPHONATE, $C_6H_4N_2O_3S$
4-NITROBENZENEDIAZONIUM NITRATE, $C_6H_4N_4O_5$
4-NITROBENZENEDIAZONIUM AZIDE, $C_6H_4N_6O_2$
BENZENEDIAZONIUM TRIBROMIDE, $C_6H_5Br_3N_2$
BENZENEDIAZONIUM CHLORIDE, $C_6H_5ClN_2$
BENZENEDIAZONIUM NITRATE, $C_6H_5N_3O_3$
BENZENEDIAZONIUM HYDROGENSULPHATE, $C_6H_6N_2O_4S$
MERCURY 2-NAPHTHALENEDIAZONIUM TRICHLORIDE, $C_{10}H_7Cl_3HgN_2$
1-(4-DIAZONIOPHENYL)-1,2-DIHYDROPYRIDINE-2-IMINOSULPHINATE, $C_{11}H_8N_4O_2S$
DI(BENZENEDIAZONIUM) ZINC TETRACHLORIDE, $C_{12}H_{10}Cl_4N_4Zn$
BIS-5-CHLOROTOLUENEDIAZONIUM ZINC TETRACHLORIDE, $C_{14}H_{12}Cl_6N_4Zn$

## DIAZONIUM SULPHATES

$(ArN_2^+)_2\ SO_4^{2-}$

Bersier, P. *et al.*, *Chem. Ing. Tech.*, 1971, **43**(24), 1311–1315

During investigation after the violent explosion of a 6-chloro-2, 4-dinitrobenzenediazonium sulphate preparation made in nitrosylsulphuric acid, it was found that above certain minimum concentrations some diazonium sulphates prepared in sulphuric acid media could be brought to explosive decomposition by local application of thermal shock. Classed as dangerous were the diazonium derivatives of 6-chloro-2, 4-dinitroaniline (at 1.26 mmol/g, very dangerous at 1.98 mmol/g); 6-bromo-2,4-dinitroaniline (very dangerous at 1.76 mmol/g); 2,4-dinitroaniline (2.0 mmol/g). Classed as suspect were the diazonium derivatives above at lower concentrations, and those of 2-chloro-5-trifluoromethylaniline (at 1.84 mmol/g); 2,6-dichloro-4-nitroaniline (0.80 mmol/g); 2-methanesulphonyl-4-nitroaniline (0.80 mmol/g); 2-cyano-4-nitroaniline (1.04 mmol/g). A further 11 derivatives were not found to be unstable. Details of several stability testing methods are given.

These salts and other diazonium sulphates are:

6-BROMO-2,4-DINITROBENZENEDIAZONIUM HYDROGENSULPHATE, $C_6H_3BrN_4O_8S$
6-CHLORO-2,4-DINITROBENZENEDIAZONIUM HYDROGENSULPHATE, $C_6H_3ClN_4O_8S$
2,6-DICHLORO-4-NITROBENZENEDIAZONIUM HYDROGENSULPHATE, $C_6H_3Cl_2N_3O_6S$
* BENZENEDIAZONIUM-2- or 4-SULPHONATE, $C_6H_4N_2O_3S$
2,4-DINITROBENZENEDIAZONIUM HYDROGENSULPHATE, $C_6H_4N_4O_8S$
BENZENEDIAZONIUM HYDROGENSULPHATE, $C_6H_6N_2O_4S$
2-CHLORO-5-TRIFLUOROMETHYLBENZENEDIAZONIUM HYDROGENSULPHATE, $C_7H_4ClF_3N_2O_4S$
2-CYANO-4-NITROBENZENEDIAZONIUM HYDROGENSULPHATE, $C_7H_4N_4O_6S$
2-METHANESULPHONYL-4-NITROBENZENEDIAZONIUM HYDROGENSULPHATE, $C_7H_7N_3O_8S_2$

*See* THERMAL STABILITY OF REACTION MIXTURES
*See other* DIAZONIUM SALTS

## DIAZONIUM SULPHIDES AND DERIVATIVES $ArN_2^+S^-$

1. Graebe, C. *et al., Ber.*, 1882, **15**, 1683
2. Bamberger, E. *et al., Ber.*, 1896, **29**, 272
3. Nawiasky, P. *et al., Chem. Eng. News,* 1945, **23**, 1247
4. Hodgson, H. H., *Chem. & Ind.*, 1945, 362
5. Tomlinson, W. R., *Chem. Eng. News,* 1957, **29**, 5473
6. Hollingshead, R. G. W. *et al., Chem. & Ind.*, 1953, 1179
7. Anon., *Angew. Chem. (Nachr.),* 1962, **10**, 278
8. Parham, W. E. *et al., Org. Synth.*, 1967, **47**, 107
9. *BCISC Quart. Safety Summ.*, 1969, **40**, 17
10. Zemlyanskii, N.I. *et al.*, *Zh. Obsch. Khim.*, 1970, **40**, 1976–1978 (*Chem. Abs.*, 1971, **74**, 53204d)

There is a long history of the preparation of explosive solids or oils from interaction of diazonium salts with solutions of various sulphides and related derivatives. Such products have arisen from benzene- and toluene-diazonium salts with hydrogen, ammonium or sodium sulphides [1,5]; 2- or 3-chlorobenzene-, 4-chloro-2-methylbenzene-, 2- or 4-nitrobenzene- or 1- or 2-naphthalene-diazonium solutions with hydrogen sulphide, sodium hydrogensulphide or sodium mono-, di- or polysulphides [1–4, 7].

4-Bromobenzenediazonium solutions gave with hydrogen sulphide at –5°C a product which exploded under water at 0°C [2], and every addition of a drop of 3-chlorobenzenediazonium solution to sodium disulphide solution at 0°C caused a violent explosion [4]. In general, these compounds appear to be bis(arenediazo) sulphides or hydrogensulphides, since some of the corresponding disulphides are considerably more stable [2].

Interaction of 2-, 3- or 4-chlorobenzenediazonium salts with *O*-alkyldithiocarbonate ('xanthate') solutions [8] or thiophenolate solutions [9] produces explosive products, possibly arenediazo aryl sulphides. The intermediate diazonium *O*-ethyldithiocarbonate produced during the preparation of *m*-thiocresol can be dangerously explosive under the wrong conditions [8]. The product of interaction of 2-chlorobenzenediazonium chloride and sodium 2-chlorothiophenolate exploded violently on heating to 100°C, and the oil precipitated from interaction of potassium thiophenolate with 3-chlorobenzenediazonium chloride exploded during mixing of the solutions [9].

Interaction of substituted arenediazonium salts with potassium *O,O*-diphenyl dithiophosphate gave a series of solid diazonium salts which decomposed explosively when heated dry [10].

The unique failure of diazotised anthranilic acid solutions to produce any explosive sulphide derivatives under a variety of conditions has been investigated and discussed [6].

Individually detailed compounds of this type relevant to the above include:

*m*-THIOCRESOL, $C_7H_8S$
BIS(*p*-NITROBENZENEDIAZO) SULPHIDE, $C_{12}H_8N_6O_4S$
DI(BENZENEDIAZO) SULPHIDE, $C_{12}H_{10}N_4S$

*See* HYDROGEN TRISULPHIDE, $H_2S_3$: Benzenediazonium chloride

## DIAZONIUM TETRAHALOBORATES $ArN_2^+BX_4^-$

1. Olah, G. A. *et al., J. Org. Chem.,* 1961, **26**, 2053
2. Doak, G. O. *et al., Chem. Eng. News,* 1967, **45**(53), 8
3. Fletcher, T. L., *Chem. & Ind.,* 1972, 370
4. Pavlath, A. E. *et al., Aromatic Fluorine Compounds,* 15, New York, Reinhold, 1962

Solid diazonium tetrachloroborates decompose very vigorously, sometimes explosively, on heating in absence of solvent. Dry *o*-nitrobenzenediazonium tetrachloroborate is liable to explode spontaneously during storage at ambient temperature [1]. Hazards involved in drying off diazonium tetrafluoroborates have been discussed. Decomposition temperature of any new salt should be checked first on a small sample, and only if it is above 100°C should the bulk be dried off and stored. Salts which show signs of decomposition at or below ambient temperature must be kept moist and used immediately [2]. The need to use an inert solvent in any deliberate thermal decomposition is stressed in the later publication [3], which draws attention to an erroneous reference to use of tetrahydrofuran which could be hazardous. The presence of nitro substituent groups may greatly increase the decomposition temperature, so that preparative decomposition may become violent or even explosive [4].

*See also* 3-PYRIDINEDIAZONIUM TETRAFLUOROBORATE, $C_5H_4BF_4N_3$
*o*-AZIDOMETHYLBENZENEDIAZONIUM TETRAFLUOROBORATE, $C_7H_6BF_4N_5$
5-(*p*-DIAZONIOBENZENE SULPHONAMIDO)THIAZOLE TETRAFLUOROBORATE, $C_9H_7BF_4N_4S$

**DIAZONIUM TRIIODIDES** $ArN_2^+ I_3^-$

Carey, J. G. *et al., Chem. & Ind.*, 1960, 97

The products produced by interaction of diazonium salts and iodides are unstable and liable to be explosive in the solid state. They are usually the triiodides, but monoiodides have been isolated under specific conditions from diazotised aniline and *o*-toluidine. Products prepared from diazotised *o*-, *m*- or *p*-nitroanilines, *m*-chloro-, -methoxy- or -methylaniline are too unstable to isolate, decomposing below 0°C.

Isolated compounds were:

*p*-CHLOROBENZENEDIAZONIUM TRIIODIDE, $C_6H_4ClI_3N_2$
* BENZENEDIAZONIUM IODIDE, $C_6H_5IN_2$
BENZENEDIAZONIUM TRIIODIDE, $C_6H_5I_3N_2$
* *m*-TOLUENEDIAZONIUM IODIDE, $C_7H_7IN_2$
* *o*-TOLUENEDIAZONIUM IODIDE, $C_7H_7IN_2$
*p*-TOLUENEDIAZONIUM TRIIODIDE, $C_7H_7I_3N_2$
*o*-METHOXYBENZENEDIAZONIUM TRIIODIDE, $C_7H_7I_3N_2O$
*p*-METHOXYBENZENEDIAZONIUM TRIIODIDE, $C_7H_7I_3N_2O$
2,4-DIMETHYLBENZENEDIAZONIUM TRIIODIDE, $C_8H_9I_3N_2$

*See other* DIAZONIUM SALTS
IODINE COMPOUNDS

**DIAZOTISATION**

Houben-Weyl, 1965, Vol. 10.3, 1–112

In the extensive review of diazotisation frequent reference is made to the need for close temperature control during processing operations, and to the explosive nature of isolated diazonium salts. Most of the incidents are covered in the group entries

DIAZONIUM CARBOXYLATES
DIAZONIUM PERCHLORATES
DIAZONIUM SALTS
DIAZONIUM SULPHATES
DIAZONIUM SULPHIDES AND DERIVATIVES
DIAZONIUM TETRAHALOBORATES
DIAZONIUM TRIIODIDES

## (DIBENZOYLDIOXYIODO)BENZENES $(PhCO_2)_2IPh$

Plesnicar, B. *et al., Angew. Chem. (Intern. Ed.),* 1970, **9**, 797

Compounds of the general formula $XArI(O_2COArY)_2$, where X = H, *p*-Cl or *o*-$CH_3$ and Y = *m*-Cl or *p*-$NO_2$, are extremely powerful oxidants, unstable when dry, and will explode during manipulation at ambient temperature (particularly with metal spatulae), or on heating to 80–120°C. The group exceeds the oxidising power of organic peroxy acids.

*See* *p*-CHLORO(BIS-*p*-NITROBENZOYLDIOXYIODO)BENZENE, $C_{20}H_{12}ClIN_2O_8$
*p*-CHLORO(BIS-*m*-CHLOROBENZOYLDIOXYIODO)BENZENE, $C_{20}H_{12}Cl_3IO_4$
(BIS-*m*-CHLOROBENZOYLDIOXYIODO)BENZENE, $C_{20}H_{13}Cl_2IO_4$
(BIS-*p*-NITROBENZOYLDIOXYIODO)BENZENE, $C_{20}H_{13}IN_2O_8$
*o*-METHOXY(BIS-*p*-NITROBENZOYLDIOXYIODO)BENZENE, $C_{21}H_{15}IN_2O_9$

*See other* DIACYL PEROXIDES
IODINE COMPOUNDS
OXIDANTS

## DICHROMATE SALTS OF NITROGENOUS BASES $(N^+)_2\ Cr_2\ O_7^{2-}$

Gibson, G. M., *Chem. & Ind.*, 1966, 553

The dichromates of 1-phenylbiguanide, its *p*-chloro-*p*-methyl- and 1-naphthyl- analogues all decompose violently at *ca*. 130°C.

*See also* DIPYRIDINIUM DICHROMATE, $C_{10}H_{12}Cr_2N_2O_7$
DIANILINIUM DICHROMATE, $C_{12}H_{16}Cr_2N_2O_7$

*See other* OXOSALTS OF NITROGENOUS BASES

## [14] DIENE-$N_4$ IRON COMPLEXES formula on p. 190

Goedken, V. L. *et al., J. Amer. Chem. Soc.*, 1972, **94**, 3397

The macroheterocyclic 5,7,7,12,14,14-hexamethyl-1,4,8,11-tetraaza-4,11-cyclotetradecadiene (abbreviated to [14] diene-$N_4$) forms cationic complexes with iron (II) or (III), also containing acetonitrile, imidazole, phenanthroline or halogen ligands. When the anion is perchlorate, the products are explosive, sensitive to heat and impact, and some appear to decompose on storage (1 week) and become sensitive to slight disturbance.

Specific compounds include:

BROMO-5,7,7,12,14,14-HEXAMETHYL-1,4,8,11-TETRAAZA-4,11-CYCLOTETRADECADIENEIRON(II) PERCHLORATE, $C_{16}H_{32}BrClFeN_4O_4$
IODO-5,7,7,12,14,14-HEXAMETHYL-1,4,8,11-TETRAAZA-4,11-CYCLOTETRADECADIENEIRON(II) PERCHLORATE, $C_{16}H_{32}ClFeIN_4O_4$
CHLORO-5,7,7,12,14,14-HEXAMETHYL-1,4,8,11-TETRAAZA-4,11-CYCLOTETRADECADIENEIRON(II) PERCHLORATE, $C_{16}H_{32}Cl_2FeN_4O_4$
DICHLORO-5,7,7,12,14,14-HEXAMETHYL-1,4,8,11-TETRAAZA-4,11-CYCLOTETRADECADIENEIRON(III) PERCHLORATE, $C_{16}H_{32}Cl_3FeN_4O_4$
DIACETONITRILE-5,7,7,12,14,14-HEXAMETHYL-1,4,8,11-TETRAAZA-4,11-CYCLOTETRADECADIENEIRON(II) PERCHLORATE, $C_{20}H_{38}Cl_2FeN_6O_8$
ACETONITRILEIMIDAZOLE-5,7,7,12,14,14-HEXAMETHYL-1,4,8,11-TETRAAZACYCLODODECADIENEIRON(II) PERCHLORATE, $C_{21}H_{39}Cl_2FeN_7O_8$
5,7,7,12,14,14-HEXAMETHYL-1,4,8,11-TETRAAZA-4,11-CYCLOTETRADECADIENE-1,10-PHENANTHROLINEIRON(II) PERCHLORATE, $C_{28}H_{40}Cl_2FeN_6O_8$
BIS(5,7,7,12,14,14-HEXAMETHYL-1,4,8,11-TETRAAZA-4,11-CYCLOTETRADECADIENE)HYDROXODIIRON(II) TRIPERCHLORATE, $C_{32}H_{65}Cl_3Fe_2N_8O_{13}$
BIS[AQUA-5,7,7,12,14,14-HEXAMETHYL-1,4,8,11-TETRAAZA-4,11-CYCLOTETRADECADIENEIRON(II)] OXIDE TETRAPERCHLORATE, $C_{32}H_{68}Cl_4Fe_2N_8O_{19}$

*See other* AMMINEMETAL OXOSALTS

## DIENES C=C=C, C=C–C=C

The 1,2- and 1,3-dienes (vicinal and conjugated, respectively) are more reactive than the separated dienes. Individually indexed compounds are:

PROPADIENE, $C_3H_4$
PERFLUOROBUTADIENE, $C_4F_6$
2-CHLORO-1,3-BUTADIENE ('CHLOROPRENE'), $C_4H_5Cl$
1,2-BUTADIENE, $C_4H_6$
1,3-BUTADIENE, $C_4H_6$
*cis*-POLYBUTADIENE, $(C_4H_6)_n$
CYCLOPENTADIENE, $C_5H_6$
2-METHYL-1,3-BUTADIENE (ISOPRENE), $C_5H_8$
1,3-PENTADIENE, $C_5H_8$
1,4-PENTADIENE, $C_5H_8$
1(or 2),3,4,5,6-PENTAFLUOROBICYCLO[2.2.0]HEXA-2,5-DIENE, $C_6HF_5$
* 1,5-HEXADIEN-3-YNE (DIVINYLACETYLENE), $C_6H_6$
1,3-CYCLOHEXADIENE, $C_6H_8$
1,4-CYCLOHEXADIENE, $C_6H_8$
2,3-DIMETHYL-1,3-BUTADIENE, $C_6H_{10}$

1,4-HEXADIENE, $C_6H_{10}$
1,5-HEXADIENE, $C_6H_{10}$
2-METHYL-1,3-PENTADIENE, $C_6H_{10}$
4-METHYL-1,3-PENTADIENE, $C_6H_{10}$
* 1,3,5,7-CYCLOOCTATETRAENE, $C_8H_8$
* 1,5,7-OCTATRIEN-3-YNE, $C_8H_8$
* 1,3,5-CYCLOOCTATRIENE, $C_8H_{10}$
4-VINYLCYCLOHEXENE, $C_8H_{12}$
1,7-OCTADIENE, $C_8H_{14}$

*See other* PEROXIDISABLE COMPOUNDS

Oxides of nitrogen

*See* NITROGEN OXIDE ('NITRIC OXIDE'), NO: Dienes, Oxygen

**DIFFERENTIAL THERMAL ANALYSIS**


1. Corignan, Y. P. *et al., J. Org. Chem.*, 1967, **32**, 285
2. Prugh, R. W., *Chem. Eng. Prog.*, 1967, **63**(11), 53
3. Krien, G., *Differential Therm. Anal.*, 1972, **2**, 353–377
4. Duswalt, A.A., *Thermochim. Acta*, 1974, **8**(1–2), 57–68
5. Berthold, W. *et al., Chem. Ing. Tech.*, 1975, **47**, 368–373


Thermal stabilities of 40 explosive or potentially explosive *N*-nitroamines, aminium nitrates and guanidine derivatives were studied in relation to structure; 14 of the compounds decomposed violently when the exotherm occurred [1]. The value of the DTA technique in assessing reactive hazards of compounds or reaction mixtures is discussed [2].

DTA data on thermally unstable materials (peroxides) as well as explosives is reviewed [3]. The use of various DTA methods for defining processing, storage, and safe reaction conditions has been investigated, and experiences and conclusions are discussed [4].

Use of DTA and DSC methods for predictions in preventing thermal explosions by assessing the stable range of operations in large storage containers are discussed, with experimental details and exemplification using nonyl nitrate [5].

**DIFLUOROAMINOALKANOLS** **$F_2NC(RR')OH$**

Fokin, A.V. *et al., Chem. Abs.*, 1970, **72**, 78340

A series of 1-difluoroaminoalkanols prepared from difluoroamine and an aldehyde or ketone, together with their acetates or bis-ethers, were all shock- and friction-sensitive explosives.

*See other* DIFLUOROAMINO COMPOUNDS

## DIFLUOROAMINO COMPOUNDS $RNF_2$

Freeman, J. P., *Advan. Fluorine Chem.*, 1970, **6**, 321, 325

All organic compounds containing one or more difluoroamino groups should be treated as explosive oxidants and excluded from contact with strong reducing agents. If the ratio of $CH_2$ to $NF_2$ groups is below 5:1, the compound will be impact-sensitive. Direct combustion for elemental analysis is unsafe, but polarography is applicable.

*See other* *N*-HALOGEN COMPOUNDS

## DIFLUOROAMINOPOLYNITROAROMATIC COMPOUNDS $F_2N(NO_2)_mAr$

Lerom, M.W. *et al.*, *J. Chem. Eng. Data*, 1974, **19**, 389–392

Several difluoroaminopolynitro derivatives of stilbene, biphenyl, terphenyl and their precursors are explosives sensitive to initiation by impact, shock, friction and rapid heating.

*See other* DIFLUOROAMINO COMPOUNDS

## DIFLUOROPERCHLORYL SALTS $F_2ClO_3^+ X^-$

Organic materials,
or Water

Christe, K.O. *et al.*, *Inorg. Chem.*, 1973, **12**, 1358

Difluoroperchloryl tetrafluoroborate or hexafluoro-arsenate or -platinate all react violently with organic materials or water.

*See also* CHLORINE DIOXYGEN TRIFLUORIDE, $ClF_3O_2$

## 1,2-DIOLS –C(OH)C(OH)–

Preparative hazard

*See* SODIUM CHLORATE, $ClNaO_3$: Alkenes, Potassium osmate

## DIOXETANES $R_2\overline{COOC}R_2$

1. Mumford, C., *Chem. Brit.*, 1975, **11**, 402
2. Kopecky, K.R. *et al.*, *Can. J. Chem.*, 1975, **53**, 1103–1122

This group of cyclic peroxides, though thermodynamically unstable, contains some compounds of sufficient kinetic stability to exist as solids at ambient temperature [1]. However, several of the compounds are explosive in the solvent-free state [2].

*See* TETRAMETHYL-1,2-DIOXETANE, $C_6H_{12}O_2$

*See other* CYCLIC PEROXIDES

## DIOXYGENYL POLYFLUOROSALTS $O_2^-[MF_n]^+$ or $O_2^-[EF_n]^+$

1. Bailar, 1973, Vol. 2, 778–779

2. Griffiths, J.E. *et al.*, *Spectrochim. Acta*, 1975, **31A**, 1208

Dioxygenyl hexafluoro -antimonate, -arsenate, -phosphate, -platinate and -stannate, and tetrafluoroborate, all react very violently with water, liberating ozone and oxygen [1]. Irradiation for Raman spectroscopy of a range of the title compounds and their nitrosyl analogues in glass or quartz capillaries caused many of them to decompose explosively. A special rotating sapphire/Teflon cell overcame the problem. Compounds examined were dioxygenyl hexafluoro -arsenate, -bismuthate, -ruthenate, -rhodate, -platinate and -aurate; dioxygenyl undecafluorodi-antimonate, -niobate and -tantalate; and their nitrosyl analogues [2].

**DIPLUMBANES** **$(R_3Pb)_2$**

Sidgwick, 1950, 595

The higher homologues of the hexaalkyldiplumbane series may explosively disproportionate during distillation.

*See other* ALKYLMETALS

**1-(1,3-DISELENOLYLIDENE)PIPERIDINIUM PERCHLORATES** **formula on p. 190**

Anderson, J.R. *et al.*, *J. Org. Chem.*, 1975, **40**, 2016

Four substituted title salts (below), intermediates in the preparation of 1,3-diselenole-2-selenones, exploded on heating, ignition or shock. The tetrafluoroborate salts were stable, safe intermediates for large-scale application.

1-(4-METHYL-1,3-DISELENOLYLIDENE)PIPERIDINIUM PERCHLORATE, $C_9H_{14}ClNO_4Se_2$
1-(4,5-DIMETHYL-1,3-DISELENOLYLIDENE)PIPERIDINIUM PERCHLORATE, $C_{10}H_{16}ClNO_4Se_2$
2-(1,3-DISELENA-4,5,6,7-TETRAHYDROINDANYLIDENE)-PIPERIDINIUM PERCHLORATE, $C_{11}H_{16}ClNO_4Se_2$
1-(4-PHENYL-1,3-DISELENOLYLIDENE)PIPERIDINIUM PERCHLORATE, $C_{14}H_{16}ClNO_4Se_2$

*See other* OXOSALTS OF NITROGENOUS BASES

**DISPOSAL**

1. *Laboratory Waste Disposal Manual,* Washington, MCA, 2nd Ed., 1969 (revised, November 1972).
2. Gaston, P.J., *The Care, Handling and Disposal of Dangerous Chemicals,* Inst. Sci. Tech., Aberdeen, Northern Publishers, 1965
3. Voegelein, J. F., *J. Chem. Educ.,* 1966, **43**, A151–157
4. Teske, J. W., *J. Chem. Educ.*, 1970, **47**, A291–295

The problems of devising effective methods for the disposal of surplus reactive chemicals or hazardous chemical wastes or residues have been well covered in several publications. A comprehensive, classified and tabulated guide to disposal procedures [1] covers the same ground as, but in more detail than, an earlier publication [2], and includes details of facilities, reagents, protective clothing and equipment required. Further publications have described the practical solutions to disposal problems adopted by an explosives research laboratory [3] and an American university [4], the latter including details of the use of explosives to rupture corroded and unusable cylinders of compressed gases or liquids. Many of the techniques described, however, need a remote area for safe operations, which may not be accessible to laboratories in urban locations. In such cases, the services of a specialist chemical disposal contractor may be the most practical solution. It is possible to minimise the disposal problem by careful stock room procedures, adequately durable labelling arrangements for samples and materials in storage, and careful segregation of materials for disposal.

**DUST EXPLOSIONS**

1. Rep. RI 7132, Dorsett, H. G. *et al.*, Washington, US Bureau of Mines, 1968
2. Hartmann, I., *Ind. Eng. Chem.*, 1948, **40**, 752
3. Palmer, K. N., *Dust Explosions and Fires*, London, Chapman and Hall, 1973
4. Nagy, J. *et al.*, Rept. BM-RI 7208, Washington, USAEC, 1968
5. Yowell, R. L. *et al.*, *Loss Prevention*, 1968, **2**, 29–53
6. Strizhevskii, I. *et al.*, *Chem. Abs.*, 1975, **82**, 61407

Laboratory dust explosion data are presented for 73 chemical compounds, 29 drugs, 27 dyes and 46 pesticides, including ignition temperatures of clouds and layers, minimum ignition energy, explosion-limiting concentrations and pressures, and rates of rise at various dust concentrations. Explosibility indices are computed where possible and variation of explosibility parameters with chemical composition is discussed. General means of minimising dust explosion hazard are reviewed [1].

Of the 17 dusts investigated earlier, the four metals examined—aluminium, magnesium, titanium and zirconium—were among the most hazardous. Aluminium and magnesium showed the maximum rates of pressure rise and final pressures, magnesium having a low minimum explosive concentration. Ignition of zirconium dust often occurred spontaneously, apparently due to static electric discharges, and undispersed layers of the dust could be ignited by less than 1$\mu$J compared with 15$\mu$J for a dispersed dust. All except aluminium ignite in carbon dioxide atmospheres [2].

Recently a comprehensive account of practical and theoretical aspects of laboratory and plant-scale dust explosions and fires has appeared. Tabulated data for over 300 dusts of industrial significance are included in an Appendix [3].

Laboratory dust explosion data for 181 miscellaneous hazardous materials and 88 non-hazardous materials are presented [4]. Papers and symposium discussions on several aspects have been published [5]. Further data on dust-explosivity of several metallic and non-metallic industrial materials have been published [6].

*See* CARBONACEOUS DUSTS
METAL DUSTS
*See also* TITANIUM, Ti: Air
ZINC, Zn: Air

**DUSTS**

Taubkin, S.I. *et al.*, *Chem. Abs.*, 1976, **85**, 110611

A new method of classifying hazardous dusts into the three groups, highly explosible, explosible and fire-hazardous, is proposed, based on lower explosive limits and maximum rate of pressure increase. An extension of the system to classify workrooms taking account of their volumes is also suggested.

**ELECTRIC FIELDS**

Kabanov, A.A. *et al.*, *Russ. Chem. Rev.*, 1975, **44**, 538–551

Application of electric fields to various explosive heavy metal derivatives (silver oxalate, barium, copper, lead, silver or thallium azides, or silver acetylide) accelerates the rate of thermal decomposition. Possible mechanisms are discussed.

**ENDOTHERMIC COMPOUNDS**

1. Weast, 1976, D67–84
2. Mellor, 1941, Vol. 1, 706–707

Most chemical reactions are exothermic, but in the few endothermic reactions heat is absorbed into the reaction product(s) which are known as endothermic (or energy-rich) compounds. Such compounds are thermodynamically unstable, because heat would be released on decomposition to their elements. The majority of endothermic compounds possess a tendency towards instability and possibly explosive decomposition under various circumstances of initiation. Often, endothermic compounds possess features of multiple bonding ('unsaturation'), as in acetylene, hydrogen cyanide, silver fulminate, mercury azide or chlorine dioxide. Other, singly bonded, endothermic compounds are hydrazine, diborane, dichlorine monoxide.

Many, but not all, endothermic compounds have been involved in violent decompositions, reactions or explosions, and, in general, compounds with significantly positive values of standard heat of formation may be considered suspect on stability grounds. Values of thermodynamic constants for elements and compounds are conveniently tabulated [1], but it should also be noted that endothermicity may change to exothermicity with increase in temperature [2].

Examples of endothermic compounds will be found in the groups

ACETYLENIC COMPOUNDS
ALKYLMETALS
AZIDES
BORANES
CYANO COMPOUNDS
DIENES
HALOGEN OXIDES
METAL ACETYLIDES
METAL FULMINATES
OXIDES OF NITROGEN

Some of the more notable individually indexed compounds are:

HYDROGEN CYANIDE, CHN
TETRANITROMETHANE, $CN_4O_8$
ACETYLENE (ETHYNE), $C_2H_2$
ACETONITRILE, $C_2H_3N$
METHYL ISOCYANIDE, $C_2H_3N$
ETHYLENE, $C_2H_4$
MERCURY(II) THIOCYANATE, $C_2HgN_2S_2$
DICYANOGEN, $C_2N_2$
1-BROMO-2-PROPYNE, $C_3H_3Br$
1-CHLORO-2-PROPYNE, $C_3H_3Cl$
VINYL ISOCYANIDE, $C_3H_3N$
PROPYNE, $C_3H_4$
1,3-BUTADIYNE, $C_4H_2$
BUTEN-3-YNE, $C_4H_4$
DICYANOACETYLENE, $C_4N_2$
NITROSYL CHLORIDE, ClNO
HYDROGEN SELENIDE, $H_2Se$
TETRASULPHUR TETRANITRIDE, $N_4S_4$
OZONE, $O_3$

## EPOXIDATION

Swern, 1971, Vol. 2, 428–533

Preparation of epoxides (oxirans) on the commercial scale as resin or polymer components is widely practised. Careful control of conditions is necessary to avoid hazards, and the several factors involved are reviewed.

*See* HYDROGEN PEROXIDE, $H_2O_2$: Unsaturated compounds

## 1,2-EPOXIDES

C–C–O (ring)

The three lower members of this class are bulk industrial chemicals and their high reactivities have been involved in several large-scale incidents. Individually indexed compounds are:

ETHYLENE OXIDE, $C_2H_4O$
* THIIRAN (ETHYLENE SULPHIDE), $C_2H_4S$

1,CHLORO-2,3-EPOXYPROPANE, $C_3H_5ClO$
* 2,3-EPOXYPROPIONALDEHYDE OXIME, $C_3H_5NO_2$
PROPYLENE OXIDE, $C_3H_6O$
3,4-EPOXYBUTENE, $C_4H_6O$
1,2-EPOXYBUTANE, $C_4H_8O$
EPOXYETHYLBENZENE, $C_8H_8O$
* 2,3-EPOXYPROPIONALDEHYDE 2,4-DINITROPHENYLHYDRAZONE, $C_9H_8N_4O_5$
2-*tert*-BUTYL-3-PHENYLOXAZIRANE, $C_{11}H_{15}NO$

## ETHERS — ROR


1. Jackson, H. L. *et al., J. Chem. Educ.*, 1970, **47**, A175
2. Davies, A. G., *J.R. Inst. Chem.*, 1956, **80**, 386–389
3. Dasler, W. *et al., Ind. Eng. Chem. (Anal. Ed.)* 1946, **18**, 52
4. *Laboratory Waste Disposal Manual,* 145, Washington, MCA, 1969
5. Robertson, R., *Chem. & Ind.*, 1933, **52**, 274
6. Morgan, G.T. *et al.*, ibid., 1936, **55**, 421–422
7. Williams, E.C., ibid., 580–581
8. Karnojitsky, V.J., *Chem. et Ind.* (Paris), 1962, **88**, 233–238
9. Unpublished observations, 1973


There is a long history of laboratory and plant fires and explosions involving the high flammability and/or tendency to peroxide formation in these widely used solvents, diisopropyl ether being the most notorious. Methods of controlling peroxide hazards in the use of ethers have been reviewed [1] and information on storage, handling, purification [2,3] and disposal [4] have been detailed.

Three violent explosions of diisopropyl ether had been reported [5,6] and a general warning on the hazards of peroxidised ethers given by 1936 [7]. The latter reference mentions diethyl, ethyl *tert*-butyl, ethyl *tert*-amyl and diisopropyl ethers as very hazardous, while methyl *tert*-alkyl ethers, lacking non-methyl hydrogen atoms adjacent to the ether link, are relatively safe. Di(2-methyl-2-propen-1-yl) ether ('dimethallyl ether') is unusual in that while it forms peroxides with extreme rapidity, these are destroyed without danger during distillation [7]. The mechanism of peroxidation of ethers has been reviewed [8].

When solvents have been freed of peroxide by percolation through a column of activated alumina, the adsorbed peroxides must promptly be desorbed by treatment with methanol or water, which should be discarded safely. Small columns used to deperoxidise diethyl ether or tetrahydrofuran were left to dry out by evaporation. When moved several days later, the peroxide concentrated on the alumina at the top of the columns exploded mildly and cracked the glass columns [9].

See the individual entries:

TETRAHYDROFURAN, $C_4H_8O$
*m*- and *p*-DIOXAN, $C_4H_8O_2$
DIETHYL ETHER, $C_4H_{10}O$

1,1- and 1,2-DIMETHOXYETHANE, $C_4H_{10}O_2$
DIISOPROPYL ETHER, $C_6H_{14}O$
BIS(2-METHOXYETHYL) ETHER, $C_6H_{14}O_3$

*See other* PEROXIDISABLE COMPOUNDS

## ETHOXYETHYNYL ALCOHOLS **EtOC≡CCR₂OH**

1. Arens, J. F., *Adv. Org. Chem.*, 1960, **2**, 126
2. Brandsma, 1971, 12, 78

Vigorous decompositions or violent explosions have been observed on several occasions during careless handling (usually overheating) of ethoxyethynyl alcohols (structures not stated) [1]. The explosions noted [2] when magnesium sulphate was used to dry their ethereal solutions were attributed to the slight acidity of the salt causing exothermic rearrangement of the alcohols to acrylic esters and subsequent explosive reactions. Glassware used for distillation must be pretreated with ammonia gas to remove traces of acid [2].

*See* 1-ETHOXY-3-METHYL-1-BUTYN-3-OL, $C_7H_{12}O_2$
*See other* ACETYLENIC COMPOUNDS

## EXPLOSIBILITY

1. Lothrop, W. C. *et al.*, *Chem. Rev.*, 1949, **45**, 419–445
2. Tomlinson, W. R. *et al.*, *J. Chem. Educ.*, 1950, **27**, 606–609
3. Coffee, R. D., *J. Chem. Educ.*, 1972, **49**, A343–349
4. Van Dolah, R. W., *Ind. Eng. Chem.*, 1961, **53**(7), 50A–53A
5. Stull, D. R., 'Prediction of Real Chemical Hazards,' 65th AIChE Meeting, New York, 1973

This may be defined as the tendency of a chemical system (involving one or more compounds) to undergo violent or explosive decomposition under appropriate conditions of reaction or initiation. It is obviously of great practical interest to be able to predict which compound or reaction systems are likely to exhibit explosibility, and much work has been devoted to this end.

Early work [1] on the relationship between structure and performance of 176 organic explosives (mainly nitro or nitrate ester compounds) was summarised and extended in general terms to multi-component systems [2]. The contribution of various structural factors (bond-groupings) was discussed in terms of heats of decomposition and oxygen balance of the compound or compounds involved in the system. Materials or systems approaching zero oxygen balance are the most powerfully explosive (give maximum heat release). Bond groupings (below) known to confer explosibility were classed as 'plosophors', and explosibility-enhancing groups as 'auxoploses' by analogy with dyestuffs terminology.

The latter groups (ether, nitrile or oximino) tend to increase the proportion of nitrogen and/or oxygen in the molecule towards (or past) zero oxygen balance.

| *BOND GROUPINGS* *as in* | *CLASS ENTRY* |
|---|---|
| $-C{\equiv}C-$ | ACETYLENIC COMPOUNDS |
| $-C{\equiv}C-Metal$ | METAL ACETYLIDES |
| $-C{\equiv}C-X$ | HALOACETYLENE DERIVATIVES |
| $N{=}N$ bridged by $>C<$ (three-membered ring) | DIAZIRINES |
| $>CN_2$ | DIAZO COMPOUNDS |
| $\geq C-N{=}O$ | NITROSO COMPOUNDS |
| $\geq C-NO_2$ | NITROALKANES, *C*-NITRO and POLYNITROARYL COMPOUNDS |
| $>C(NO_2)_2$ | POLYNITROALKYL COMPOUNDS<br>TRINITROETHYL ORTHOESTERS |
| $\geq C-O-N{=}O$ | ACYL OR ALKYL NITRITES |
| $\geq C-O-NO_2$ | ACYL OR ALKYL NITRATES |
| $>C-C<$ bridged by $O$ (three-membered ring) | 1,2-EPOXIDES |
| $>C{=}N-O-Metal$ | METAL FULMINATES or *aci*-NITRO SALTS, OXIMATES |
| $NO_2-C(F)-NO_2$ | FLUORODINITROMETHYL COMPOUNDS |
| $>N-Metal$ | *N*-METAL DERIVATIVES |
| $-N{=}Hg^{+}{=}N-$ | POLY(DIMERCURYIMMONIUM SALTS) |
| $>N-N{=}O$ | NITROSO COMPOUNDS |
| $>N-NO_2$ | *N*-NITRO COMPOUNDS |
| $>N^{+}-N^{-}-NO_2$ | *N*-AZOLIUM NITROIMIDATES |
| $\geq C-N{=}N-C\leq$ | AZO COMPOUNDS |

| *BOND GROUPINGS* | *as in* | *CLASS ENTRY* |
|---|---|---|
| $>C-N=N-O-C<$ | | ARENEDIAZOATES |
| $>C-N=N-S-C<$ | | ARENEDIAZO ARYL SULPHIDES |
| $>C-N=N-O-N=N-C<$ | | BIS-ARENEDIAZO OXIDES |
| $>C-N=N-S-N=N-C<$ | | BIS-ARENEDIAZO SULPHIDES |
| $>C-N=N-N(R)-C<$ R(R = H, –CN, –OH, –NO) | | TRIAZENES |
| $-N=N-N=N-$ | | HIGH-NITROGEN COMPOUNDS<br>TETRAZOLES |
| $>C-O-O-H$ | | ALKYLHYDROPEROXIDES, PEROXYACIDS |
| $>C-O-O-C<$ | | PEROXIDES (CYCLIC, DIACYL, DIALKYL), PEROXYESTERS |
| –O–O–Metal | | METAL PEROXIDES, PEROXOACID SALTS |
| –O–O–Non-metal | | PEROXOACIDS, PEROXYESTERS |
| $N \rightarrow Cr-O_2$ | | AMMINECHROMIUM PEROXOCOMPLEXES |
| $-N_3$ | | AZIDES (ACYL, HALOGEN, NON-METAL, ORGANIC) |
| $C-N_2^+O^-$ (C and O linked) | | ARENEDIAZONIUMOLATES |
| $>C-N_2^+\ S^-$ | | DIAZONIUM SULPHIDES AND DERIVATIVES, 'XANTHATES' |
| $N^+-H\ Z^-$ | | HYDRAZINIUM SALTS,<br>OXOSALTS OF NITROGENOUS BASES |
| $-N^+-OH\ Z^-$ | | HYDROXYLAMMONIUM SALTS |
| $>C-N_2^+\ Z^-$ | | DIAZONIUM CARBOXYLATES or SALTS |
| $[N \rightarrow Metal]^+\ Z^-$ | | AMMINEMETAL OXOSALTS |
| Ar–Metal–X<br>X–Ar–Metal | | HALO-ARYLMETALS<br>HALOARENEMETAL π-COMPLEXES |
| N–X | | HALOGEN AZIDES<br>*N*-HALOGEN COMPOUNDS<br>*N*-HALOIMIDES |
| $-NF_2$ | | DIFLUOROAMINO COMPOUNDS<br>*N,N,N*-TRIFLUOROALKYLAMIDINES |

| *BOND GROUPINGS* | *as in* | *CLASS ENTRY* |
|---|---|---|
| –O–X | | ALKYL PERCHLORATES |
| | | AMINIUM PERCHLORATES |
| | | CHLORITE SALTS |
| | | HALOGEN OXIDES |
| | | HYPOHALITES |
| | | PERCHLORIC ACID (no class entry: *see* $ClHO_4$) |
| | | PERCHLORYL COMPOUNDS |
| | | *See also* ENDOTHERMIC COMPOUNDS |
| | | REDOX COMPOUNDS |
| | | STRAINED RING COMPOUNDS |

Although the semi-empirical approach outlined above is of some value in assessing potential explosibility hazards, more fundamental work has recently been done to institute a quantitative basis for such assessment.

A combination of thermodynamical calculations with laboratory thermal stability and impact sensitivity determinations has enabled a system to be developed which indicates the relative potential of a given compound or reaction system for sudden energy release, and the relative magnitude of the latter [3].

A similar treatment, specifically for compounds expected to be explosive, has also been separately developed [4].

A further computational technique which takes account of both thermodynamic and kinetic considerations has permitted of the development of a system which provides a numerical Reaction Hazard Index (RHI) for each compound, which is a real, rather than a potential, indication of hazard. The RHIs calculated for 80 compounds are in fairly close agreement with the relative hazard values (assessed on the basis of experience) assigned on the NFPA Reactivity Rating scale for these same compounds [5].

*See* COMPUTATION OF REACTIVE CHEMICAL HAZARDS
OXYGEN BALANCE

**EXPLOSIONS**

Stull, 1977

Fundamental factors which contribute to the occurrence of fire and explosions in chemical processing operations have been collected and reviewed in this 120-page book, which serves as an extremely informative guide to the whole and complex subject, subdivided into 12 chapters. These are: Introduction; Thermochemistry; Kinetochemistry; Ignition; Flames; Dust Explosions; Thermal Explosions; Gas Phase Detonations; Condensed Phase Detonations; Evaluating Reactivity Hazard Potential; Blast Effects, Fragments, Craters; Protection against Explosions. There are also three Appendices, 153 references and a Glossary of Technical Terms.

**EXPLOSIVE BOILING**

1. Vogel, 1957, 4
2. Weston, F. E., *Chem. News*, 1908, 27

The hazards associated with addition of nucleating agents (charcoal, porous pot, pumice, 'anti-bumping granules', etc.) to liquids at or above their boiling points have been adequately described [1]. The violent or near-explosive boiling which ensues is enhanced by gases adsorbed onto such solids of high surface area. Incidents involving sudden boiling of salt solutions being concentrated for crystallisation in vacuum desiccators have also been described. The heat liberated by rapid crystallisation probably caused violent local boiling to occur, and the crystallising dishes were fractured [2].

**EXPLOSIVE COMBUSTION**

Lafitte, P. *et al.*, *Hautes Temp. Leurs Util. Phys. Chim.*, 1973, **1**, 1–40

A comprehensive review includes autoignition of gas mixtures, explosions at low and high temperatures, properties of flames and combustion of metals.

**EXPLOSIVES**

1. Federoff, 1960
2. *Explosives, Propellants and Pyrotechnic Safety Covering Laboratory, Pilot Plant and Production Operations:* Manual AD272–424, Washington, US Naval Ordnance Laboratory, 1962
3. Kirk-Othmer, 1965, Vol. 8, 581
4. Urbanski, 1964–1967, Vols. 1, 2, 3

Explosive materials intended as such are outside the scope of this Handbook, but several specialist reference works on them contain much information relevant to safety practices for unstable materials [1–4].

**FIRE**

1. Williamson, J. J., *General Fire Hazards and Fire Protection*, London, Pitman, 6th edn, 1971
2. Mealing, P., Vol. 28A, Boreham Wood, Fire Res. Establ., 1977
3. *Chemical Fires and Chemicals at Fires*, Leicester, Inst. of Fire Engrs., 1962
4. Bahme, 1972
5. Meidl, 1972
6. *Fire and Related Properties of Industrial Chemicals*, London, Fire Protection Association, 4th edn, revised 1974
7. *Matrix of Electrical and Fire Hazard Properties and Classification of*

*Chemicals*, Elec. Haz. Panel, Washington, Nat. Res. Council, 1975, NTIS AD-A027 181/7GA

The book covers the CII syllabus for general fire hazards, which, together with means for their prevention, are presented simply and concisely. Constructional features, management considerations, heating, lighting, etc., are considered in detail. An alphabetical list of hazardous materials and an extensive bibliography are included [1].

An annual compilation of references to the scientific literature on fire is classified under headings: Occurrence of Fire; Fire Hazards; Initiation and Development of Combustion; Fire Precautions; Fire Resistance; Fire Fighting; Nuclear Energy; General. Name and Subject Index. The latest available volume is reference 2.

The special hazards associated with the involvement, usually on a large scale, of dangerous chemicals in fire situations are adequately described in the quoted references. Particular extinguishing problems and requirements are covered [3–5]. A new edition of a tabulated list of the physical and fire-hazardous properties of some 800 industrially significant chemicals, solvents and gases has been published [6]. A matrix relating fire-hazardous properties of 226 commercial chemicals to the classification groups of the US National Electrical Code has been made generally available [7].

## FIRE EXTINGUISHERS

1. *Fire Safety Data Sheets 6001–6003*, London, Fire Protection Association, 1973
2. Hirst, R., *Chem. Engr.*, 1974, 627–628, 636

The three illustrated data sheets cover the choice; siting, care and maintenance; and use of portable fire extinguishers [1]. A detailed account of the properties, mechanism of action and applications of modern fire extinguishing agents is available [2].

## FLAMMABILITY

1. Coward, H. F. *et al., Limits of Flammability of Gases and Vapors,* Bull. 503, Washington, US Bureau of Mines, 1952
2. Zabetakis, M. G., *Flammability Characteristics of Combustible Gases and Vapors,* Bull. 627, Washington, US Bureau of Mines, 1965
3. Shimy, A. A., *Fire Technol.*, 1970, **6**(2), 135–139
4. Hilado, C. J., *J. Fire Flammability*, 1975, **6**, 130–139
5. Li, C. C. *et al.*, ibid., 1977, **8**, 38–40

The hazards associated with flammability characteristics of combustible gases and vapours are excluded from detailed consideration in this Handbook, since the topic is adequately covered in standard reference works on combustion, including the two sources of much of the data on flammability limits [1,2].

However, to reinforce the constant need to consider flammability problems in laboratory and plant operations, the explosive limits have been included (where known) for those individual substances with flash point below 25°C. With the few noted exceptions, explosive limits quoted are those at ambient temperature and are expressed in % by volume in air in Appendix 2.

Semi-empirical formulae, based only on molecular structure, have been derived which allow flammability limits and flame temperatures to be calculated for hydrocarbons and alcohols. Flash points, auto-ignition temperatures and boiling points may also be calculated from molecular structures of these classes. Quoted examples indicate the methods to be reasonably accurate in most cases [3]. Equations are given for calculating upper and lower flammability limits of 102 organic compounds (hydrocarbons, alcohols, ethers, esters, aldehydes, ketones, epoxides, amines and halides) from structure and stoicheiometry in air [4]. An empirical method has been described to correlate relative flash points of organic compounds as a linear function of relative normal boiling point, with acetic acid as a reference base [5]. A more recently introduced concept of relevance in the context of flammability is that of the minimum oxygen concentration to support combustion. The Oxygen Index is a quantified measure of this, and is becoming of wide-spread use for non-volatile flammable solids of many types.

*See* OXYGEN INDEX

*See also* OXYGEN ENRICHMENT

## FLASH POINTS

1. *Flash Points,* Poole, BDH Ltd., 1962
2. 'Catalogue KL4', Colnbrook, Koch-Light Laboratories Ltd., 1973
3. *Flash Point Index of Trade Name Liquids,* 325A, Boston, Nat. Fire Prot. Assoc., 1972
4. *Properties of Flammable Liquids,* 325M, Boston, Nat. Fire Prot. Assoc., 1969
5. Prugh, R. W., *J. Chem. Educ.*, 1973, **50**, A85–89
6. Shimy, A. A., *Fire Technol.*, 1970, **6**(2), 135–139
7. Libman, B. *et al.*, *Chem. Abs.*, 1976, **85**, 145377
8. MacDermott, P. E., *Chem. Brit.*, 1974, **10**, 228

Flash point is defined as the minimum temperature at which a flammable liquid or volatile solid gives off sufficient vapour to form a flammable mixture with air.

There is usually a fair correlation between flash point and probability of involvement in fire if an ignition source is present in the vicinity of a source of the vapour; materials with low flash points being more likely to be so involved than those with higher flash points. While no attempt has been made to include in this Handbook details of all known combustible materials, it has been thought worthwhile to include substances with flash points below 25°C, a likely maximum ambient temperature in many laboratories in warm temperature zones. These materials have been included to draw attention to the high probability of fire if such flammable materials are handled with insufficient care to prevent contact of their vapours with an ignition source (stirrer motor, hot-plate, energy controller, flame, etc.).

The figures for flash points quoted in Appendix 2 are closed cup values except where indicated by (o) and most are reproduced by permission of the two companies concerned [1,2].

A comprehensive listing of flash points for commercial liquids and formulated mixtures is also available [3,4].

Recently a method for estimating approximate flash-point temperatures based upon the boiling point and molecular structure of a given compound has been published. After calculation of the stoicheiometric concentration in air, a nomograph is used to estimate the flash point to within 11°C [5].

An alternative semi-empirical method for calculating flash points, based only on molecular structure of hydrocarbons and alcohols, also exists. Accuracy appears to be reasonably good in most examples quoted [6]. Calculation of flash point of alcohols from their boiling points gives maximum errors of 9.8% [7]. The commonly accepted fallacy that liquids at temperature below their flash points are incapable of giving rise to flammable mixtures in air is dispelled with some examples of process operations with solvents at sub-atmospheric pressures. Under such conditions, flammable atmospheres may be generated at temperatures below the flash point. Thus, the real criterion of safety should be whether flammable mixtures can exist under given process conditions, rather than a flammable liquid being below its flash point [8].

## FLUORINATED COMPOUNDS — RF

*See* LITHIUM TETRAHYDROALUMINATE, $AlH_4Li$: Fluoroamides
SODIUM, Na: Fluorinated compounds
: Halocarbons (Ref. 6)

## FLUORINATED CYCLOPROPENYL METHYL ETHERS — $CFCF_2CHOCH_3$, etc.

Smart, B. E., *J. Org. Chem.*, 1976, **41**, 2377–2378

There are hazards involved during the preparation and after isolation of these materials. Addition of sodium methoxide powder to perfluoropropene in diglyme at –60°C led to ignition in some cases, and the products, 1,3,3-trifluoro-2-methoxycyclopropene (very volatile and flammable), or 3,3-difluoro-1,2-dimethoxycyclopropene, react violently with water or methanol, as does 1-chloro-3,3-difluoro-2-methoxycyclopropane.

## FLUORINATED PEROXIDES $F_5SOO-$

*See* PENTAFLUOROSULPHUR PEROXYACETATE, $C_2H_3F_5O_3S$
PENTAFLUOROSULPHUR PEROXYHYPOCHLORITE, $ClF_5O_2S$

## FLUORINATION

Grakauskas, V., *J. Org. Chem.*, 1970, **35**, 723; 1969, **34**, 2835
Sharts, C.M. *et al.*, *Org. React.*, 1974, **21**, 125–265
Safety precautions applicable to direct liquid-phase fluorination of aromatic compounds are discussed.

Attention is drawn to the hazards attached to the use of many newer fluorination reagents.

*For further references see* FLUORINE, $F_2$

## FLUORODINITRO COMPOUNDS $FC(NO_2)_2-$, $FCH_2C(NO_2)_2-$

1. Peters, H. M. *et al.*, *J. Chem. Eng. Data*, 1975, **20**(1), 113–117
2. *Rept. UCID-16141*, Snaeberger, D. F., Livermore Lab., Univ. of Cal., 1972 (*Chem. Abs.*, 1975, **83**, 12944n)

Several fluorodinitro derivatives of methane (1,1,1-) and ethane (1,2,2-) are described as explosive, sensitive to initiation by impact, shock, friction or other means [1]. Procedures for safe handling of fluorine and explosive fluoronitro compounds are detailed [2].

## FLUORODINITROMETHYL COMPOUNDS $-C(NO_2)_2F$

Kamlet, M. J. *et al.*, *J. Org. Chem.*, 1968, **33**, 3073
Witucki, E. F. *et al.*, *J. Org. Chem.*, 1972, **37**, 152
Adolph, H. J., ibid., 749
Gilligan, W. H., ibid., 3947
Coon, C. L. *et al.*, *Synthesis*, 1973, (10), 605–607
Several of this group are explosives of moderate-to-considerable sensitivity to impact or friction and need careful handling. Fluorodinitromethane and fluorodinitroethanol are also vesicant.

1-Fluoro-1,1-dinitro derivatives of ethane, butane, 2-butene and 2-phenylethane are explosive.

*See* FLUORODINITROMETHYL AZIDE, $CFN_5O_4$
FLUORODINITROMETHANE, $CHFN_2O_4$
2-FLUORO-2,2-DINITROETHANOL, $C_2H_3FN_2O_5$
2-FLUORO-2,2-DINITROETHYLAMINE, $C_2H_4FN_3O_4$
BIS(2-FLUORO-2,2-DINITROETHYL)AMINE, $C_4H_5F_2N_5O_8$

## FOAM RUBBER

1. *BRE CP 36/75*, Woolley, W. D. *et al.*, Garston, Building Res. Establ., 1975
2. Smith, E. E., *Consum. Prod. Flammability*, 1975, **2**, 58–69

A fire involving foamed rubber mattresses in a storeroom led to an unexpected and serious explosion. Subsequent investigation showed that an explosion risk may exist when the flammable smoke and vapour from smouldering of large quantities of foamed rubber are confined in an enclosed space. Suitable strict precautions are recommended [1]. Methods of quantitatively evaluating potential fire hazards from release rates of heat, smoke and toxic gases of various foamed rubber and plastics materials are developed and their application to real situations is discussed [2].

Oxygen
*See* OXYGEN (Gas), $O_2$ : Polymers

## FRICTIONAL IGNITION OF GASES

Powell, F., *Ind. Eng. Chem.*, 1969, **61**(12), 29
*Rept. Invest. RI 8005*, Desy, D. H. *et al.*, Washington, US Bur. Mines, 1975

The ignition of flammable gases and vapours by friction or impact is reviewed, with 82 references.

Ignition of methane by frictional impact of aluminium alloys and rusted steel has been investigated.

*See* THERMITE

## FULLER'S EARTH

Turpentine
*See* TURPENTINE: Diatomaceous earth

## GAS CYLINDERS

1. Whalley, E. W. F. (UKAEA), London, HMSO, 1965
2. *Properties of Gases,* Wall chart 57012, Poole, BDH Chemicals Ltd., 1973
3. Braker, W. *et al., Matheson Gas Data Book,* E. Rutherford, N.J., Matheson Gas Products, 5th Ed., 1971

In a report on safety in use of gas cylinders, the nature of hazards associated with gas cylinders and their contents and statutory requirements are discussed. Appendices outline model safeguards for storage, handling, use and transport of gas cylinders [1]. An inexpensive wall chart summarises the important properties of 116 gases and volatile liquids [2] and the *Gas Data Book* [3] gives comprehensive details of handling techniques and cylinder equipment necessary for safe use of 130 gases.

## GOLD CATALYSTS

Cusumano, J. A., *Nature*, 1974, **247**, 456

Supported metal catalysts containing gold should never be prepared by impregnation of a support with solutions containing both gold and ammonia. Dried catalysts so prepared contain extremely sensitive gold–nitrogen compounds which may explode at a light touch.

*See* below

## GOLD COMPOUNDS

Gold compounds exhibit a tendency to decompose violently with separation of the metal. Individually indexed gold compounds are:

GOLD(III) CHLORIDE, $AuCl_3$
GOLD(III) HYDROXIDE–AMMONIA, $2AuH_3O_3 \cdot 3H_3N$
SODIUM TRIAZIDOAURATE(?), $AuN_9Na$
BIS(DIHYDROXYGOLD) IMIDE, $Au_2H_5NO_4$
GOLD(III) OXIDE, $Au_2O_3$
GOLD(III) SULPHIDE, $Au_2S_3$
GOLD NITRIDE–AMMONIA, $Au_3N \cdot H_3N$
TRIGOLD DISODIUM HEXAAZIDE, $Au_3N_{18}Na_2$
GOLD(I) CYANIDE, CAuN
DIGOLD ACETYLIDE, $C_2Au_2$
DIMETHYLGOLD SELENOCYANATE, $C_3H_6AuNSe$
DIETHYLGOLD BROMIDE, $C_4H_{10}AuBr$
DIMETHYLGOLD(III) AZIDE: see the dimer $C_4H_{12}Au_2N_6$
TETRAMETHYLDIGOLD DIAZIDE, $C_4H_{12}Au_2N_6$
CYCLOPENTADIENYLGOLD(I), $C_5H_5Au$
CYANODIETHYLGOLD: see the tetramer $C_{20}H_{40}Au_4N_4$
DIMETHYL(TRIMETHYL)SILOXOGOLD: see the dimer $C_{10}H_{30}Au_2O_2Si_2$
TRIETHYLPHOSPHINEGOLD(I) NITRATE, $C_6H_{15}AuNO_3P$
1,2-DIAMINOETHANEBIS-TRIMETHYLGOLD, $C_8H_{26}Au_2N_2$
TETRAMETHYLBIS-(TRIMETHYLSILOXO)DIGOLD, $C_{10}H_{30}Au_2O_2Si_2$
TETRACYANOCTAETHYLTETRAGOLD, $C_{20}H_{40}Au_4N_4$

*See also* PLATINUM COMPOUNDS
*See other* HEAVY METAL DERIVATIVES
METAL AZIDES

## GRAPHITE OXIDE

Boehm, H. P. *et al., Z. Anorg. Chem.*, 1965, **335**, 74–79

The oxide is thermally unstable and on rapid heating will deflagrate at a temperature dependent on the method of preparation. This temperature is lowered by the presence of impurities, and dried samples of iron(III) chloride-impregnated oxide explode on heating.

## GRIGNARD REAGENTS **RMgX**

1. Kharasch and Reinmuth, 1954
2. Bondarenko, V. G. *et al., Chem. Abs.*, 1975, **82**, 89690

Preparation of Grignard reagents is frequently beset by practical difficulties in establishing the reaction [1]. Recently, improved equipment and control methods for safer preparations have been described [2].

*See* PHENYLMAGNESIUM BROMIDE, $C_6H_5BrMg$: Chlorine
*o*-, *m*- or *p*-TRIFLUOROMETHYLMAGNESIUM BROMIDE, $C_7H_4BrF_3Mg$

*See other* HALOARYLMETALS

## HALOACETYLENE DERIVATIVES —C≡CX


1. Whiting, M. C. *Chem. Eng. News*, 1972, **50**(23), 86
2. Brandsma, 1971, 99


The tendency towards explosive decomposition noted for dihalo-2,4-hexadiyne derivatives appears to be associated with the coexistence of halo- and acetylene functions in the same molecule, rather than with its being a polyacetylene. Haloacetylenes should be used with exceptional precautions [1]. Explosions may occur during distillation of bromoacetylenes when bath temperatures are too high, or if air is admitted to a hot vacuum-distillation residue [2].

*See* LITHIUM BROMOACETYLIDE, $C_2BrLi$
DIBROMOACETYLENE, $C_2Br_2$
LITHIUM CHLOROACETYLIDE, $C_2ClLi$
SODIUM CHLOROACETYLIDE, $C_2ClNa$
DICHLOROACETYLENE, $C_2Cl_2$
BROMOACETYLENE, $C_2HBr$
CHLOROACETYLENE, $C_2HCl$
FLUOROACETYLENE, $C_2HF$
IODOACETYLENE, $C_2HI$
DIIODOACETYLENE, $C_2I_2$
SILVER TRIFLUOROMETHYLACETYLIDE, $C_3AgF_3$
CHLOROCYANOACETYLENE, $C_3ClN$
LITHIUM TRIFLUOROMETHYLACETYLIDE, $C_3F_3Li$
* PROPIOLOYL CHLORIDE, $C_3HClO$
3,3,3-TRIFLUOROPROPYNE, $C_3HF_3$
1-BROMO-2-PROPYNE, $C_3H_3Br$
1-CHLORO-2-PROPYNE, $C_3H_3Cl$
1,4-DIBROMO-1,3-BUTADIYNE, $C_4Br_2$
1,4-DICHLORO-1,3-BUTADIYNE, $C_4Cl_2$
1-IODO-1,3-BUTADIYNE, $C_4HI$
1,4-DICHLORO-2-BUTYNE, $C_4H_4Cl_2$
1,4-DIIODO-1,3-BUTADIYNE, $C_4I_2$
1-IODO-3-PENTEN-1-YNE, $C_5H_5I$
1,6-DICHLORO-2,4-HEXADIYNE, $C_6H_4Cl_2$
2,4-HEXADIYNYLENE BISCHLOROSULPHITE, $C_6H_4Cl_2O_4S_2$
TETRA(CHLOROETHYNYL)SILANE, $C_8Cl_4Si$
2,4-HEXADIYNYLENE BISCHLOROFORMATE, $C_8H_4Cl_2O_4$
1-IODOOCTA-1,3-DIYNE, $C_8H_9I$
1-IODO-3-PHENYL-2-PROPYNE, $C_9H_7I$
1-BROMO-1,2-CYCLOTRIDECADIEN-4,8,10-TRIYNE, $C_{13}H_9Br$

## HALOALKENES $-C{=}CX$

Of the lower members of this reactive class, the more lightly substituted are of high flammability and many are classed as peroxidisable compounds. Individually indexed members are:

BROMOTRIFLUOROETHYLENE, $C_2BrF_3$
CHLOROTRIFLUOROETHYLENE, $C_2ClF_3$
TETRACHLOROETHYLENE, $C_2Cl_4$
TETRAFLUOROETHYLENE, $C_2F_4$
TRICHLOROETHYLENE, $C_2HCl_3$
1,1-DICHLOROETHYLENE, $C_2H_2Cl_2$
*cis*- or *trans*-1,2-DICHLOROETHYLENE, $C_2H_2Cl_2$
1,1-DIFLUOROETHYLENE, $C_2H_2F_2$
BROMOETHYLENE (VINYL BROMIDE), $C_2H_3Br$
CHLOROETHYLENE (VINYL CHLORIDE), $C_2H_3Cl$
FLUOROETHYLENE (VINYL FLUORIDE), $C_2H_3F$
HEXAFLUOROPROPENE, $C_3F_6$
3,3,3-TRIFLUOROPROPENE, $C_3H_3F_3$
* TRIFLUOROVINYL METHYL ETHER, $C_3H_3F_3O$
1,3-DICHLOROPROPENE, $C_3H_4Cl_2$
2,3-DICHLOROPROPENE, $C_3H_4Cl_2$
1-BROMO-2-PROPENE (ALLYL BROMIDE), $C_3H_5Br$
1-CHLORO-1-PROPENE, $C_3H_5Cl$
2-CHLOROPROPENE, $C_3H_5Cl$
1-CHLORO-2-PROPENE (ALLYL CHLORIDE), $C_3H_5Cl$
1-IODO-2-PROPENE (ALLYL IODIDE), $C_3H_5I$
1,1,4,4-TETRACHLOROBUTATRIENE, $C_4Cl_4$
1,1,4,4-TETRAFLUOROBUTATRIENE, $C_4F_4$
PERFLUOROBUTADIENE, $C_4F_6$
2-CHLORO-1,3-BUTADIENE, $C_4H_5Cl$
1-BROMO-2-BUTENE, $C_4H_7Br$
4-BROMO-1-BUTENE, $C_4H_7Br$
2-CHLORO-2-BUTENE, $C_4H_7Cl$
3-CHLORO-1-BUTENE, $C_4H_7Cl$
3-CHLORO-2-METHYL-1-PROPENE, $C_4H_7Cl$
1(or 2),3,4,5,6-PENTAFLUOROBICYCLO[2.2.0] HEXA-2,5-DIENE, $C_6HF_5$

*See other* PEROXIDISABLE COMPOUNDS

## HALOARENEMETAL π-COMPLEXES $[M(C_6F_6)_2]$, etc.


1. Klabunde, K. *et al., Inorg. Chem.*, 1175, **14**, 790–791
2. Klabunde, K. *et al., Angew. Chem. (Internat. Ed.)*, 1975, **14**, 288
3. Klabunde, K. *et al., J. Fluorine Chem.*, 1974, **4**, 115
4. Graves, V. *et al., Inorg. Chem.*, 1976, **15**, 578


The π-complexes formed between chromium(0), vanadium (0) or other transition metals, and mono- or poly-fluorobenzene show extreme sensitivity to heat and are explosive [1,2]. Hexafluorobenzenenickel(0) exploded at 70°C [3], and presence of two or more fluorine substituents leads to unstable, very explosive chromium(0) complexes [1]. Apparently, the aryl fluorine atoms are quite labile, and, on decomposition, M–F bonds are formed very exothermically. Laboratory workers should be wary of this possible behaviour in any haloarenemetal $\pi$-complex of this type [1]. However, in later work, no indications of explosivity, or indeed of any complex formation, were seen [4]. Individually indexed compounds are:

BIS(HEXAFLUOROBENZENE)COBALT(0), $C_{12}CoF_{12}$
BIS(HEXAFLUOROBENZENE)CHROMIUM(0), $C_{12}CrF_{12}$
BIS(HEXAFLUOROBENZENE)IRON(0), $C_{12}F_{12}Fe$
BIS(HEXAFLUOROBENZENE)NICKEL(0), $C_{12}F_{12}Ni$
BIS(HEXAFLUOROBENZENE)TITANIUM(0), $C_{12}F_{12}Ti$
BIS(HEXAFLUOROBENZENE)VANADIUM(0), $C_{12}F_{12}V$
BIS(1,4(?)-DIFLUOROBENZENE)CHROMIUM(0), $C_{12}H_8CrF_4$
BIS(FLUOROBENZENE)CHROMIUM(0), $C_{12}H_{10}CrF_2$
BIS(FLUOROBENZENE)VANADIUM(0), $C_{12}H_{10}F_2V$

*See other* HALO-ARYLMETALS

## HALOARYL COMPOUNDS — ArX

Though normally not very reactive, haloaryl compounds if sufficiently activated by other substituents may undergo violent reactions. Individually indexed compounds in this small class are:

1,2,4,5-TETRACHLOROBENZENE, $C_6H_2Cl_4$
1-CHLORO-2,4-DINITROBENZENE, $C_6H_3ClN_2O_4$
1-FLUORO-2,4-DINITROBENZENE, $C_6H_3FN_2O_4$
*p*-CHLORONITROBENZENE, $C_6H_4ClNO_2$
* 2,6-DINITROBENZYL BROMIDE, $C_7H_5BrN_2O_4$
2-CHLORO-4-NITROTOLUENE, $C_7H_6ClNO_2$
4-CHLORO-2-METHYLPHENOL, $C_7H_7ClO$
2-IODO-3,5-DINITROBIPHENYL, $C_{12}H_7IN_2O_4$

## HALO-ARYLMETALS — ArMX, XArM

The name adopted for this class of highly reactive (and in some circumstances self-reactive) compounds is intended to cover both arylmetal halides (halogen bonded to the metal) and haloaryl metals (halogen attached to the aryl nucleus). Individually indexed compounds are:

PENTAFLUOROPHENYLALUMINIUM DIBROMIDE, $C_6AlBr_2F_5$
PENTAFLUOROPHENYLLITHIUM, $C_6F_5Li$
*m*- or *p*-BROMOPHENYLLITHIUM, $C_6H_4BrLi$
*m*- or *p*-CHLOROPHENYLLITHIUM, $C_6H_4ClLi$
*p*-FLUOROPHENYLLITHIUM, $C_6H_4FLi$
*o*-, *m*- or *p*-TRIFLUOROMETHYLPHENYLMAGNESIUM BROMIDE, $C_7H_4BrF_3Mg$
*o*-, *m*- or *p*-TRIFLUOROMETHYLPHENYLLITHIUM, $C_7H_4F_3Li$
BIS(PENTAFLUOROPHENYL)ALUMINIUM BROMIDE, $C_{12}AlBrF_{10}$
BIS(CYCLOPENTADIENYL)BIS(PENTAFLUOROPHENYL)ZIRCONIUM, $C_{22}H_{10}F_{10}Zr$

*See also* GRIGNARD REAGENTS
HALOARENEMETAL $\pi$-COMPLEXES
ORGANOLITHIUM REAGENTS

## HALOCARBONS RX

This generic name has been adopted to designate a range of halogenated aliphatic hydrocarbons widely used in research and industrial operations, often as solvents or diluents. Few are completely inert chemically but, in general, reactivity decreases with increasing substitution of halogen atoms (particularly of fluorine) for hydrogen atoms in the saturated or unsaturated parent hydrocarbons. For other reactants which have been involved, see:

METALS: Halocarbons
PENTABORANE(9), $B_5H_9$ : Reactive solvents
CALCIUM DISILICIDE, $CaSi_2$ : Carbon tetrachloride
FLUORINE, $F_2$ : Halocarbons
DISILANE, $H_6Si_2$ : Non-metal halides
DINITROGEN TETRAOXIDE, $N_2O_4$: Halocarbons
OXYGEN (Liquid), $O_2$: Halocarbons

Unsaturated halocarbons are separately listed as:

HALOALKENES

and individually indexed saturated halocarbons are:

BROMOTRICHLOROMETHANE, $CBrCl_3$
BROMOTRIFLUOROMETHANE, $CBrF_3$
CARBON TETRABROMIDE, $CBr_4$
DICHLORODIFLUOROMETHANE, $CCl_2F_2$
TRICHLOROFLUOROMETHANE, $CCl_3F$
CARBON TETRACHLORIDE, $CCl_4$
POLY(CARBON MONOFLUORIDE), $(CF)_n$
CARBON TETRAFLUORIDE, $CF_4$
BROMOFORM, $CHBr_3$
CHLORODIFLUOROMETHANE, $CHClF_2$
CHLOROFORM, $CHCl_3$
IODOFORM, $CHI_3$
DIBROMOMETHANE, $CH_2Br_2$
DICHLOROMETHANE, $CH_2Cl_2$
DIIODOMETHANE, $CH_2I_2$
BROMOMETHANE, $CH_3Br$
CHLOROMETHANE, $CH_3Cl$
FLUOROMETHANE, $CH_3F$
IODOMETHANE, $CH_3I$
HEXABROMOETHANE, $C_2Br_6$
HEXACHLOROETHANE, $C_2Cl_6$
PENTACHLOROETHANE, $C_2HCl_5$
1,1,1,2-TETRACHLOROETHANE, $C_2H_2Cl_4$
1-CHLORO-1,1-DIFLUOROETHANE, $C_2H_3ClF_2$
1,1,1-TRICHLOROETHANE, $C_2H_3Cl_3$
1,1,2-TRICHLOROETHANE, $C_2H_3Cl_3$
1,2-DIBROMOETHANE, $C_2H_4Br_2$
1,1-DICHLOROETHANE, $C_2H_4Cl_2$
1,2-DICHLOROETHANE, $C_2H_4Cl_2$
1,1-DIFLUOROETHANE, $C_2H_4F_2$
BROMOETHANE, $C_2H_5Br$
CHLOROETHANE, $C_2H_5Cl$
FLUOROETHANE, $C_2H_5F$

IODOETHANE, $C_2H_5I$
3-BROMO-1,1,1-TRICHLOROPROPANE, $C_3H_4BrCl_3$
1-CHLORO-3,3,3-TRIFLUOROPROPANE, $C_3H_4ClF_3$
1,1-DICHLOROPROPANE, $C_3H_6Cl_2$
1,2-DICHLOROPROPANE, $C_3H_6Cl_2$
2,2-DICHLOROPROPANE, $C_3H_6Cl_2$
1-BROMOPROPANE, $C_3H_7Br$
2-BROMOPROPANE, $C_3H_7Br$
1-CHLOROPROPANE, $C_3H_7Cl$
2-CHLOROPROPANE, $C_3H_7Cl$
1-IODOPROPANE, $C_3H_7I$
2-IODOPROPANE, $C_3H_7I$
TETRACARBON MONOFLUORIDE, $C_4F$
*mixo*-DICHLOROBUTANE, $C_4H_8Cl_2$
1-BROMOBUTANE, $C_4H_9Br$
2-BROMOBUTANE, $C_4H_9Br$
1-BROMO-2-METHYLPROPANE, $C_4H_9Br$
2-BROMO-2-METHYLPROPANE, $C_4H_9Br$
1-CHLOROBUTANE, $C_4H_9Cl$
2-CHLOROBUTANE, $C_4H_9Cl$
1-CHLORO-2-METHYLPROPANE, $C_4H_9Cl$
2-CHLORO-2-METHYLPROPANE, $C_4H_9Cl$
2-IODOBUTANE, $C_4H_9I$
1-IODO-2-METHYLPROPANE, $C_4H_9I$
2-IODO-2-METHYLPROPANE, $C_4H_9I$
1-BROMO-3-METHYLBUTANE, $C_5H_{11}Br$
2-BROMOPENTANE, $C_5H_{11}Br$
1-CHLORO-3-METHYLBUTANE, $C_5H_{11}Cl$
2-CHLORO-2-METHYLBUTANE, $C_5H_{11}Cl$
1-CHLOROPENTANE, $C_5H_{11}Cl$
2-IODOPENTANE, $C_5H_{11}I$
PERFLUOROHEXYL IODIDE, $C_6F_{13}I$
BROMOBENZENE, $C_6H_5Br$
CHLOROBENZENE, $C_6H_5Cl$
IODOBENZENE, $C_6H_5I$
HEXACHLOROCYCLOHEXANE, $C_6H_6Cl_6$
2,2-DICHLORO-3,3-DIMETHYLBUTANE, $C_6H_{12}Cl_2$

## HALOGENATION INCIDENTS

For incidents involving halogenation reactions,

*see* *N*-HALOIMIDES
SILANES: Chloroform, etc.
SILVER DIFLUORIDE, $AgF_2$ : Dimethyl sulphoxide
: Hydrocarbons
BROMINE, $Br_2$ : Acetone
: Acrylonitrile
: *N,N*-Dimethylformamide
: Ethanol
: Non-metal halides
*N*-BROMOSUCCINIMIDE, $C_4H_4BrNO_2$ : Aniline, etc.
: Dibenzoyl peroxide, etc.
CHLORINE, $Cl_2$ : Antimony trichloride, Tetramethylsilane
: 1-Chloro-2-propyne

: Cobalt(III) chloride, Methanol
: Hydrocarbons, Lewis acids
: Synthetic rubber

COBALT TRIFLUORIDE, $CoF_3$

FLUORINE, $F_2$ : Caesium heptafluoropropoxide
: Cyanoguanidine
: Halocarbons
: Hydrocarbons
: Hydrogen fluoride, Seleninyl fluoride
: Sodium dicyanamide

IODINE PENTAFLUORIDE, $F_5I$: Benzene
: Dimethyl sulphoxide

IODINE, $I_2$ : Butadiene, etc.
: Ethanol, etc.
: Metals
: Non-metals

## HALOGEN AZIDES $XN_3$

Metals,

or Phosphorus

Dehnicke, K., *Angew. Chem. (Intern. Ed.),* 1967, **6**, 240

A comprehensive review covering stability relationships and reactions of these explosive compounds and their derivatives. Bromine, chlorine and iodine azides all explode violently in contact with magnesium, sodium, zinc or white phosphorus.

*See* BROMINE AZIDE, $BrN_3$
* CYANOGEN AZIDE, $CN_4$
CHLORINE AZIDE, $ClN_3$
FLUORINE AZIDE, $FN_3$
IODINE AZIDE, $IN_3$

*See other* AZIDES
*N*-HALOGEN COMPOUNDS

## *N*-HALOGEN COMPOUNDS —NX

1. Kovacic, P. *et al., Chem. Rev.,* 1970, **70**, 640
2. Petry, R. C. *et al., J. Org. Chem.,* 1967, **32**, 4034
3. Freeman, J. P. *et al., J. Amer. Chem. Soc.,* 1969, **91**, 4778

Many compounds containing one or more N—X bonds show unstable or explosive properties (and are also oxidants), and this topic has been reviewed [1]. Difluoroamino compounds, ranging from difluoroamine and tetrafluorohydrazine to polydifluoroamino compounds, are notably explosive and suitable precautions have been detailed [2, 3]. Within this class fall the separate groups:

DIFLUOROAMINO COMPOUNDS
HALOGEN AZIDES
*N*-HALOIMIDES

*N,N,N'*-TRIFLUOROALKYLAMIDINES

as well as the individually indexed compounds:

DIFLUOROAMMONIUM HEXAFLUOROARSENATE, $AsF_8H_2N$
TETRAFLUOROAMMONIUM TETRAFLUOROBORATE, $BF_8N$
BROMOAMINE, $BrH_2N$
NITROGEN TRIBROMIDE HEXAAMMONIATE, $Br_3N \cdot H_{18}N_6$
NITROSYL TRIBROMIDE, $Br_3NO$
TETRAFLUORODIAZIRIDINE, $CF_4N_2$
3-DIFLUOROAMINO-1,2,3-TRIFLUORODIAZIRIDINE, $CF_5N_3$
PENTAFLUOROGUANIDINE, $CF_5N_3$
BIS(DIFLUOROAMINO)DIFLUOROMETHANE, $CF_6N_2$
1-DICHLOROAMINOTETRAZOLE, $CHCl_2N_5$
1,1-DIFLUOROUREA, $CH_2F_2N_2O$
*N,N*-DIBROMOMETHYLAMINE, $CH_3Br_2N$
*N,N*-DICHLOROMETHYLAMINE, $CH_3Cl_2N$
*N*-FLUOROIMINODIFLUOROMETHANE, $CNF_3$
PERFLUORO-*N*-CYANODIAMINOMETHANE, $C_2F_5N_3$
PERFLUORO-1-AMINOMETHYLGUANIDINE, $C_2F_8N_4$
PERFLUORO-*N*-AMINOMETHYLTRIAMINOMETHANE, $C_2F_{10}N_4$
*N,N*-DICHLOROGLYCINE, $C_2H_3Cl_3NO_2$
1-BROMOAZIRIDINE, $C_2H_4BrN$
*N*-BROMOACETAMIDE, $C_2H_4BrNO$
1-CHLOROAZIRIDINE, $C_2H_4ClN$
*N*-CHLOROACETAMIDE, $C_2H_4ClNO$
AZO-*N*-CHLOROFORMAMIDINE, $C_2H_4Cl_2N_6$
1,2-BIS(DIFLUOROAMINO)ETHANOL, $C_2H_4F_4N_2O$
1,2-BIS(DIFLUOROAMINO)-*N*-NITROETHYLAMINE, $C_2H_4F_4N_4O_2$
DIMETHYL *N,N*-DICHLOROPHOSPHORAMIDATE, $C_2H_6Cl_2NO_3P$
2,4,6-TRIS(DICHLOROAMINO)-1,3,5-TRIAZINE (HEXACHLOROMELAMINE), $C_3Cl_6N_6$
2,4,6-TRIS(BROMOAMINO)-1,3,5-TRIAZINE, $C_3H_3Br_3N_6$
2,4,6-TRIS(CHLOROAMINO)-1,3,5-TRIAZINE, $C_3H_3Cl_3N_6$
*N,N*-DICHLORO-β-ALANINE, $C_3H_5Cl_2NO_2$
*N*-BROMOTRIMETHYLAMMONIUM BROMIDE(?), $C_3H_9Br_2N$
*N*-BROMOSUCCINIMIDE, $C_4H_4BrNO_2$
2-CYANO-1,2,3-TRIS(DIFLUOROAMINO)PROPANE, $C_4H_4F_6N_4$
*N*-CHLORO-4-METHYL-2-IMIDAZOLINONE, $C_4H_5ClN_2O$
*N*-CHLORO-5-METHYL-2-OXAZOLIDINONE, $C_4H_6ClNO_2$
*N*-CHLORO-3-MORPHOLINONE, $C_4H_6ClNO_2$
DI-1,2-BIS(DIFLUOROAMINO)ETHYL ETHER, $C_4H_6F_8N_4O$
4,4-BIS(DIFLUOROAMINO)-3-FLUOROIMINO-1-PENTENE, $C_5H_6F_5N_3$
TETRAKIS-(*N,N*-DICHLOROAMINOMETHYL)METHANE $C_5H_8Cl_8N_4$
*N*-CHLOROPIPERIDINE, $C_5H_{10}ClN$
*N*-BROMOTETRAMETHYLGUANIDINE, $C_5H_{12}BrN_3$
*N*-CHLOROTETRAMETHYLGUANIDINE, $C_5H_{12}ClN_3$
2,6-DIBROMOBENZOQUINONE-4-CHLOROIMINE, $C_6H_2Br_2ClNO$
2,6-DICHLOROBENZOQUINONE-4-CHLOROIMINE, $C_6H_2Cl_3NO$
BENZOQUINONE-1-CHLOROIMINE, $C_6H_4ClNO$
1-CHLOROBENZOTRIAZOLE, $C_6H_4ClN_3$
BENZOQUINONE-1,4-BIS(CHLOROIMINE), $C_6H_4Cl_2N_2$
*N*-CHLORO-4-NITROANILINE, $C_6H_5ClN_2O_2$
*N,N*-DICHLOROANILINE, $C_6H_5Cl_2N$

*N,N,N',N'*-TETRACHLOROADIPAMIDE, $C_6H_8Cl_4N_2O_2$
SODIUM *N*-CHLORO-*p*-TOLUENESULPHONAMIDE, $C_7H_7ClNNaO_2S$
*N*-CHLOROCINNAMALDIMINE, $C_9H_8ClN$
NITROGEN CHLORIDE DIFLUORIDE, $ClF_2N$
CHLOROAMINE, $ClH_2N$
NITROSYL CHLORIDE, ClNO
*N*-CHLOROSULPHINYLIMIDE, ClNOS
NITRYL CHLORIDE, $ClNO_2$
DICHLOROFLUOROAMINE, $Cl_2FN$
DICHLOROAMINE, $Cl_2HN$
NITROGEN TRICHLORIDE, $Cl_3N$
1,3,5-TRICHLOROTRITHIA-1,3,5-TRIAZINE (THIAZYL CHLORIDE), $Cl_3N_3S_3$
CHLORIMINOVANADIUM TRICHLORIDE, $Cl_4NV$
DIAMMINEDICHLOROAMINOTRICHLOROPLATINUM(IV), $Cl_5H_7N_3Pt$
FLUOROAMINE, $FH_2N$
NITROSYL FLUORIDE, FNO
NITRYL FLUORIDE, $FNO_2$
DIFLUOROAMINE, $F_2HN$
DIFLUORODIAZENE, $F_2N_2$
NITROGEN TRIFLUORIDE, $F_3N$
TETRAFLUOROHYDRAZINE, $F_4N_2$
DIIODOAMINE, $HI_2N$
NITROGEN TRIIODIDE–SILVER AMIDE, $I_3N \cdot AgH_2N$
NITROGEN TRIIODIDE–AMMONIA, $I_3N \cdot NH_3$

## HALOGEN OXIDES $XO_n$

The various compounds arising from union of oxygen with one or more halogens are a class of generally unstable but powerful oxidants, individually indexed compounds being:

BROMINE PERCHLORATE, $BrClO_4$
BROMYL FLUORIDE, $BrFO_2$
PERBROMYL FLUORIDE, $BrFO_3$
BROMINE DIOXIDE, $BrO_2$
BROMINE TRIOXIDE, $BrO_3$
* DICYANOGEN *N,N'*-DIOXIDE, $C_2N_2O_2$
CHLORYL HYPOFLUORITE, $ClFO_3$
PERCHLORYL FLUORIDE, $ClFO_3$
FLUORINE PERCHLORATE, $ClFO_4$
CHLORINE TRIFLUORIDE OXIDE, $ClF_3O$
CHLORINE DIOXYGEN TRIFLUORIDE, $ClF_3O_2$
CHLORINE DIOXIDE, $ClO_2$
CHLORINE TRIOXIDE, see dimeric $Cl_2O_6$
DICHLORINE OXIDE, $Cl_2O$
DICHLORINE TRIOXIDE, $Cl_2O_3$
CHLORINE PERCHLORATE, $Cl_2O_4$
CHLORYL PERCHLORATE, $Cl_2O_6$
PERCHLORYL PERCHLORATE (DICHLORINE HEPTOXIDE), $Cl_2O_7$
OXYGEN DIFLUORIDE, $F_2O$
DIOXYGEN DIFLUORIDE, $F_2O_2$

HEXAOXYGEN DIFLUORIDE, $F_2O_6$
IODINE DIOXYGEN TRIFLUORIDE, $F_3IO_2$
IODINE(V) OXIDE, $I_2O_5$
IODINE(VII) OXIDE, $I_2O_7$

### HALOGENS $X_2$

The reactivity hazard of this group of oxidants towards other materials decreases progressively from fluorine, which reacts violently with most materials under appropriate conditions (except for the metals on which resistant fluoride films form), through chlorine and bromine to iodine. Astatine may be expected to continue this trend.

### *N*-HALOIMIDES –CON(X)CO–

Alcohols,
or Amines,
or Diallyl sulphide,
or Hydrazine,
or Xylene

Martin, R. H., *Nature*, 1951, **168**, 32
Many of the reactions of several *N*-chloro- and *N*-bromoimides are extremely violent or explosive. Those observed include *N*-chlorosuccinimide with aliphatic alcohols or benzylamine or hydrazine hydrate; *N*-bromosuccinimide with aniline, diethyl sulphide or hydrazine hydrate; or 3-nitro-*N*-bromophthalimide with tetrahydrofurfuryl alcohol; 1,3-dichloro-5,5-dimethyl-2,4-imidazolidindione with xylene (violent explosion).

Other *N*-haloimides which may react similarly are:

POTASSIUM 1,3-DIBROMO-2,4-DIKETO-1,3,5-TRIAZINE-6-OLATE, $C_3Br_2KN_3O_3$
SODIUM 1,3-DICHLORO-2,4-DIKETO-1,3,5-TRIAZINE-6-OLATE, $C_3Cl_2N_3NaO_3$
1,3,5-TRICHLORO-1,3,5-TRIAZINETRIONE, $C_3Cl_3N_3O_3$

*N*-Haloamides may be expected to react similarly.
*See* *N*-BROMOACETAMIDE, $C_2H_4BrNO$
*N*-CHLOROACETAMIDE, $C_2H_4ClNO$
*See other* *N*-HALOGEN COMPOUNDS

### -HALOMETHYL-FURANS OR -THIOPHENES CH=C(CH₂X)OCH=CH, etc.

*Heterocyclic Compounds*, **1**, 207, Elderfield, R. C. (Editor), New York, Wiley, 1950
The great instability of 2-bromomethyl- and 2-chloromethyl-furans, often manifest as violent or explosive decomposition on attempted isolation or distillation, is reviewed. The furan nucleus is sensitive to traces of halogen acids, and decomposition becomes autocatalytic.

*See* 2-CHLORO-5-CHLOROMETHYLTHIOPHENE, $C_5H_4Cl_2S$
2-BROMOMETHYLFURAN, $C_5H_5BrO$
2-CHLOROMETHYLFURAN, $C_5H_5ClO$
2-CHLOROMETHYLTHIOPHENE, $C_5H_5ClS$
2,5-BIS(CHLOROMETHYL)THIOPHENE, $C_6H_6Cl_2S$
2-BROMOMETHYL-5-METHYLFURAN, $C_6H_7BrO$
2-CHLOROMETHYL-5-METHYLFURAN, $C_6H_7ClO$

## HALOSILANES $SiX_n$

Schumb, W. C. *et al., Inorg. Synth.*, 1939, **1**, 46

When heated, the vapours of the higher chlorosilanes (hexachlorodisilane to dodecachloropentasilane) ignite in air. Other halo- and alkylhalo-silanes ignite without heating or have low flash points.

*See* POLY(DIBROMOSILYLENE), $(Br_2Si)_n$
TRIBROMOSILANE, $Br_3HSi$
TRICHLORO(VINYL)SILANE, $C_2H_3Cl_3Si$
TRICHLORO(ETHYL)SILANE, $C_2H_5Cl_3Si$
CHLOROTRIMETHYLSILANE, $C_3H_9ClSi$
CHLORO(METHYL,PHENYL)SILANE, $C_7H_9ClSi$
TRIBROMOSILANE, $Br_3HSi$
TRICHLOROSILANE, $Cl_3HSi$
HEXACHLORODISILANE, $Cl_6Si_2$
OCTACHLOROTRISILANE, $Cl_8Si_3$
DECACHLOROTETRASILANE, $Cl_{10}Si_4$
DODECACHLOROPENTASILANE, $Cl_{12}Si_5$
POLY(DIFLUOROSILYLENE), $(F_2Si)_n$
TRIFLUOROSILANE, $F_3HSi$
IODODISILANE, $H_5ISi_2$

*See other* NON-METAL HALIDES

*See related* ALKYLNON-METAL HALIDES
NON-METAL HYDRIDES

## HAZARDOUS MATERIALS

Cloyd, D. R. *et al.*, *Handling Hazardous Materials*, NASA Technology Survey SP-5032, Washington, NASA, 1965

*Fire Protection Guide on Hazardous Materials*, Boston, National Fire Protection Association, 5th edn, 1973

A survey of hazards and safety procedures involved in handling rocket fuels and oxidisers, including liquid hydrogen, pentaborane, fluorine, chlorine trifluoride, ozone, dinitrogen tetraoxide, hydrazine, methylhydrazine and 1,1-dimethylhydrazine.

The later volume is a compendium of five NFPA publications:

325A, *Flashpoint Index of Trade Name Liquids* (8800 items), 1972

325M, *Fire Hazard Properties of Flammable Liquids, Gases and Volatile Solids* (1300), 1969

49, *Hazardous Chemicals Data* (388), 1973

491M, *Manual of Hazardous Chemical Reactions* (2350), 1971

704M, *Recommended System for the Identification of the Fire Hazard of Materials*, 1969

## HEAVY METAL DERIVATIVES

This class of compounds showing explosive instability deals with heavy metals bonded to elements other than nitrogen and contains the separately treated groups:

GOLD COMPOUNDS
LITHIUM PERALKYLURANATES
MERCURY COMPOUNDS
METAL ACETYLIDES
METAL FULMINATES
METAL OXALATES
PLATINUM COMPOUNDS
SILVER COMPOUNDS

as well as the individually indexed compounds:

THALLIUM BROMATE, $BrO_3Tl$
LEAD METHYLENEBISNITROAMIDE, $CH_2N_4O_4Pb$
DIGOLD(I) KETENDIIDE, $C_2Au_2O$
DICOPPER(I) KETENDIIDE, $C_2Cu_2O$
HYDROXYCOPPER(II) GLYOXIMATE, $C_2H_4CuN_2O_3$
LEAD DITHIOCYANATE, $C_2N_2PbS_2$
LEAD ACETATE–LEAD BROMATE, $C_4H_6O_4Pb{\cdot}Br_2O_6Pb$
CYCLOPENTADIENYLSILVER PERCHLORATE, $C_5H_5AgClO_4$
DIMETHYLTHALLIUM *N*-METHYLACETOHYDROXAMATE, $C_5H_{12}NO_2Tl$
LEAD 2,4,6-TRINITRORESORCINOLATE (LEAD STYPHNATE), $C_6HN_3O_8Pb$
COPPER DIPICRATE, $C_{12}H_4CuN_6O_{14}$
LEAD DIPICRATE, $C_{12}H_4N_6O_{14}Pb$
ZINC DIPICRATE, $C_{12}H_4N_6O_{14}Zn$
COPPER(II) 3,5-DINITROANTHRANILATE, $C_{14}H_8CuN_6O_{12}$
COPPER BIS(1-BENZENEAZOTHIOCARBONYL-2-PHENYLHYDRAZINE), $C_{26}H_{22}CuN_8S_2$
LEAD BIS(1-BENZENEAZOTHIOCARBONYL-2-PHENYLHYDRAZINE), $C_{26}H_{22}N_8PbS_2$
ZINC BIS(1-BENZENEAZOTHIOCARBONYL-2-PHENYLHYDRAZINE), $C_{26}H_{22}N_8S_2Zn$
LEAD OLEATE, $C_{36}H_{66}O_4Pb$

*See also* METAL AZIDES
METAL CYANIDES AND CYANOCOMPLEXES
*N*-METAL DERIVATIVES

## HEXAFLUOROCHLORONIUM SALTS $[ClF_6]^+Z^-$

1. Christe, K. O., *Inorg. Chem.*, 1973, **12**, 1582
2. Roberts, F. Q., *Inorg. Nucl. Chem. Lett.*, 1972, **8**, 737

They are very powerful oxidants and react explosively with organic materials or water [1]. They are not in themselves explosive, contrary to an earlier report [2].

*See other* OXIDANTS

## HIGH-NITROGEN COMPOUNDS

This class heading is intended to include not only those compounds containing a high total proportion of nitrogen (up to 87%) but also those containing high local concentrations in substituent groups (notably azide and diazonium) within the molecule.

Many organic molecular structures containing several chain-linked atoms of nitrogen are unstable or explosive and the tendency is exaggerated by attachment of azide or diazonium groups, or a high-nitrogen heterocyclic nucleus. Closely related but separately treated classes or groups include:

AZIDES (in several sections)
DIAZO COMPOUNDS
DIAZONIUM SALTS
HYDRAZINIUM SALTS
*N*-NITRO COMPOUNDS
TETRAZOLES
TRIAZENES

Individually indexed compounds in this class are:

5-AMINO-1,2,3,4-THIATRIAZOLE, $CH_2N_4S$
METHYLDIAZENE, $CH_4N_2$
1-AMINO-3-NITROGUANIDINE, $CH_5N_5O_2$
CARBONOHYDRAZIDE, $CH_6N_4O$
DINITRODIAZOMETHANE, $CN_4O_4$
DIAZOACETONITRILE, $C_2HN_3$
DIAZOACETYL AZIDE, $C_2HN_5O$
AZIDOACETONITRILE, $C_2H_2N_4$
1,2,4,5-TETRAZINE, $C_2H_2N_4$
1,2,3-TRIAZOLE, $C_2H_3N_3$
5-METHOXY-1,2,3,4-THIATRIAZOLE, $C_2H_3N_3OS$
CYANOGUANIDINE (‘DICYANDIAMIDE’), $C_2H_4N_4$
1,6-BIS(5-TETRAZOLYL)HEXAAZA-1,5-DIENE, $C_2H_4N_{14}$
1,2-DIMETHYLNITROSOHYDRAZINE, $C_2H_7N_3O$
DICYANODIAZENE (AZOCARBONITRILE), $C_2N_4$
DISODIUM DICYANODIAZENE, $C_2N_4Na_2$
DIAZIDOMETHYLENECYANAMIDE, $C_2N_8$
DIAZIDOMETHYLENEAZINE, $C_2N_{14}$
BIS(1,2,3,4-THIATRIAZOL-5-YLTHIO)METHANE, $C_3H_2N_6S_4$
DIAZIDOMALONONITRILE, $C_3N_8$
2,4,6-TRIAZIDO-1,3,5-TRIAZINE, $C_3N_{12}$
1,3-BIS(5-AMINO-1,3,4-TRIAZOL-2-YL)TRIAZENE, $C_4H_7N_{11}$
TETRAMETHYL-2-TETRAZENE, $C_4H_{12}N_4$
2-DIAZONIO-4,5-DICYANOIMIDAZOLATE (DIAZODICYANO-IMIDAZOLE), $C_5N_6$
POTASSIUM 3,5-DINITRO-2(1-TETRAZENYL)PHENOLATE, $C_6H_5KN_6O_5$
TETRAAZIDO-*p*-BENZOQUINONE, $C_6N_{12}O_2$

1,3-DIPHENYLTRIAZENE, $C_{12}H_{11}N_3$
1,5-DIPHENYL-1,4-PENTAAZADIENE, $C_{12}H_{11}N_5$
TETRAALLYL-2-TETRAZENE, $C_{12}H_{20}N_4$
1,3,5-TRIPHENYL-1,4-PENTAAZADIENE, $C_{18}H_{15}N_5$
1,3,6,8-TETRAPHENYLOCTAAZATRIENE, $C_{24}H_{20}N_8$
*trans*-TETRAAMMINEDIAZIDOCOBALT(III) *trans*-DIAMMINE-TETRAAZIDOCOBALTATE(1−), $Co_2H_{18}N_{22}$
PENTAZOLE, $HN_5$
TETRAIMIDE, $H_4N_4$
SODIUM TETRASULPHURPENTANITRIDATE, $N_5NaS_4$
TRIPHOSPHORUS PENTANITRIDE, $N_5P_3$
'TETRAPHOSPHORUS HEXANITRIDE', $N_6P_4$
'PHOSPHORUS PENTAAZIDE', $N_{15}P$

## HIGH-PRESSURE REACTION TECHNIQUES

1. Rooymans, C.J.M. in *Preparative Methods in Solid State Chemistry*, Hagenmuller, P. (Editor), New York, Academic Press, 1972
2. *High Pressure Safety Code*, London, High Pressure Association, 1975

The chapter which reviews high pressure techniques in liquid and gas systems for preparative purposes also includes safety considerations [1]. The Code deals mainly with mechanical hazards, but attention is drawn to the fact that application of high pressure to chemical systems may influence stability adversely [2].

The special equipment and conditions necessary for the use of high-pressure reaction techniques involve some potential hazards not encountered during processing at or near normal pressure. The large amount of kinetic energy contained in a high-pressure gas or vapour system is additive in effect to any potential hazard arising from reactivity considerations. The pressure vessel is of necessity of heavy construction and high thermal inertia and rapid heating, or more particularly cooling, cannot usually be achieved in the absence of special (internal) heat transfer arrangements. In heating such closed systems, proper allowance must be made for the expansion of liquid contents to avoid the possibility of hydraulic bursting of the vessel or the rupture disc.

## HYDRAZINEMETAL NITRATES $[M(N_2H_4)_m][NO_3]_n$

Medard, L. *et al.*, *Mém. Poudres*, 1952, **34**, 159–166

Hydrazine complexes of cadmium, cobalt, manganese, nickel and zinc nitrates were prepared as possible initiator materials. Dihydrazine-manganese(II) nitrate ignites at 150°C but is not shock-sensitive, while trihydrazinecobalt(II) nitrate, which decomposes violently at 206–211°C or in contact with conc. acids, and trihydrazinenickel(II) nitrate (deflagrates at 212–215°C) are fairly shock-sensitive. Trihydrazinecadmium nitrate (defl. 212–245°C) and trihydrazinezinc nitrate are of moderate and low sensitivity, respectively.

*See* TRIHYDRAZINENICKEL(II) NITRATE, $H_{12}N_8NiO_6$
*See related* AMMINEMETAL OXOSALTS

## HYDRAZINIUM SALTS $H_2N{-}NH_3^+$

Mellor, 1940, Vol. 8, 327; 1967, Vol. 8, Suppl. 2.2, 84–86
Salvadori, J., *Gazz. Chim. Ital.*, 1907, **37**(2), 32
Levi, G. R., *Gazz. Chim. Ital.*, 1923, [2], **53**, 105

Several salts are explosively unstable, including hydrazinium azide (explodes on rapid heating or on initiation by a detonator even when damp), chlorate (explodes at m.p. 80° C), chlorite (also highly flammable when dry), hydrogenselenate, hydrogensulphate (decomposes explosively when melted), nitrate, nitrite and the highly explosive perchlorate and diperchlorates used as propellants.

*See* PROPELLANTS

## HYDRAZONES $-C{=}NNH_2$

Air

1. Swern, 1971, Vol. 2, 19
2. Busch, M. *et al.*, *Ber.*, 1914, **47**, 3277

Alkyl- and aryl-hydrazones of aldehydes and ketones readily peroxidise in solution and rearrange to diazo hydroperoxides [1], some of which are explosively unstable [2].

*See* α-BENZENEDIAZOBENZYL HYDROPEROXIDE, $C_{13}H_{12}N_2O_2$

## HYDROGENATION CATALYSTS

1. Augustine, R.L., *Catalytic Hydrogenation*, 23, 28, London, Arnold, 1965
2. Anon., *Chem. Brit.*, 1974, **10**, 367
3. Poles, T., private comm., 1973
4. Freifelder, 1971, 74, 78, 81–83, 168, 175, 262, 263

Many hydrogenation catalysts are sufficiently active to effect rapid interaction of hydrogen and/or solvent vapour with air, causing ignition or explosion. This is particularly so where hydrogen is adsorbed on the catalyst either before a hydrogenation (Raney cobalt, nickel, etc.) or after a hydrogenation during separation of catalyst from the reaction mixture. Exposure to air of such a catalyst should be avoided until complete purging with an inert gas, such as nitrogen, has been effected.

With catalysts of high activity and readily reducible substrates, control of the exotherm may be required to prevent runaway reactions, particularly at high pressures [1].

A proprietary form of Raney nickel catalyst in which the finely divided metal particles are coated with a fatty amine is claimed to be free of pyrophoric hazards if it dries out [2].

Platinum-metal catalysts are preferably introduced to the reactor or hydrogenation system in the form of a water-wet paste or slurry. The latter is charged to the empty reactor: air is removed by purging with nitrogen or by several evacuations alternating with nitrogen filling: the reaction mixture is charged, after which hydrogen is admitted. The same procedure applies where it is mandatory to charge the catalyst in the dry state, but in this case the *complete* removal of air before introduction of the reaction mixture and/or hydrogen is of vital importance.

Platinum, palladium and rhodium catalysts are non-pyrophoric as normally manufactured. Iridium and, more particularly, ruthenium catalysts may exhibit pyrophoricity in their fully reduced form and for this reason are usually manufactured in the unreduced form and reduced *in situ.*

Spent catalysts should be purged from hydrogen and washed free from organics with water before storage in the water-wet condition. Under no circumstances should any attempt be made to dry a spent catalyst [3].

Specialist advice on safety and other problems in the use of catalysts and associated equipment is freely available from Engelhard Industries Ltd. at Cinderford, Gloucester, where a model high-pressure hydrogenation laboratory with full safety facilities is maintained.

In a volume devoted to hydrogenation techniques and applications, there are several references to safety aspects of catalytic hydrogenation. For noble metal and nickel catalysts, low-boiling solvents should be avoided to reduce risk of ignition when catalysts are added. Risks are highest with carbon-supported catalysts, which tend to float at the air interface of the solvent. The need for dilute solutions of nitro- and polynitro-aromatics or oximes, and for relatively low concentrations of catalysts to minimise the relatively large exotherms, are stressed [4].

*See* below

## HYDROGENATION INCIDENTS

*See* the entries:

SODIUM TETRAHYDROBORATE, $BH_4Na$: Ruthenium salts
FORMIC ACID, $CH_2O_2$: Palladium–carbon catalyst
1,1,1-TRIS(HYDROXYMETHYL)NITROMETHANE, $C_4H_9NO_5$: Hydrogen, etc.
1,1,1-TRIS(AZIDOMETHYL)ETHANE, $C_5H_9N_9$: Hydrogen, etc.
3-METHYL-2-PENTEN-4-YN-1-OL, $C_6H_8O$
*o*-NITROANISOLE, $C_7H_7NO_3$: Hydrogen
HYDROGEN, $H_2$: Acetylene, Ethylene

: Air, Catalysts
: Catalysts, Vegetable oils
IRIDIUM, Ir
NICKEL, Ni : Alone
: *p*-Dioxane
: Hydrogen
: Methanol
: Sulphur compounds
PALLADIUM, Pd: Carbon
PLATINUM, Pt: Ethanol
: Hydrogen, etc.
RHODIUM, Rh

**HYDROXOOXODIPEROXOCHROMATE SALTS** **$M[Cr(OH)O(O_2)_2]$**

Bailar, 1973, Vol. 3, 699

The ammonium, potassium and thallium salts are all violently explosive.

**HYDROXYLAMINIUM SALTS** **$N^+H_3OH\ Z^-$**

Anon., *Chem. Processing (Chicago),* 1963, **26**(24), 30

Some decompositions of salts of hydroxylamine are discussed, including violent decomposition of crude chloride solutions at 140°C, the exothermic decomposition of the pure chloride and explosive decomposition of the solid sulphate at 170°C. The phosphinate and nitrate decompose violently above 92 and 100°C, respectively.

*See other* OXOSALTS OF NITROGENOUS BASES

**3-HYDROXYTRIAZENES** **N=N–N(OH)**

Houben-Weyl, 1965, Vol. 10(3), 717

Many of the 3-hydroxytriazene derivatives produced by diazo-coupling on to *N*-alkyl or *N*-aryl hydroxylamines decompose explosively above their m.p.s. The heavy-metal derivatives are, however, stable and used in analytical chemistry.

*See other* TRIAZENES

**HYPOHALITES** **–OX**

1. Sidgwick, 1950, 1218; Chattaway, F.D., *J. Chem. Soc.,* 1923, **123**, 2999
2. Anbar, M. *et al., Chem. Rev.,* 1954, **54**, 927

This class of oxidant compounds, all containing O–X bonds, includes widely differing types and many compounds of limited stability. Alkyl hypochlorites, readily formed from alcohols and chlorinating agents, will explode on ignition, irradiation or contact with copper powder [1].

Of the many alkyl hypohalites described, only ethyl, *tert*-butyl and *tert*-pentyl are stable enough to isolate, purify and handle [2], though care is needed. Individually indexed compounds are:

* CHLOROPEROXYTRIFLUOROMETHANE, $CClF_3O_2$
  TRIFLUOROMETHYL HYPOFLUORITE, $CF_4O$
  DIFLUOROMETHYLENE DIHYPOFLUORITE, $CF_4O_2$
  METHYL HYPOCHLORITE, $CH_3ClO$
  TRIFLUOROACETYL HYPOFLUORITE, $C_2F_4O_2$
  ACETYL HYPOBROMITE, $C_2H_3BrO_2$
  ETHYL HYPOCHLORITE, $C_2H_5ClO$
  PERFLUOROISOPROPYL HYPOCHLORITE, $C_3ClF_7O$
  PENTAFLUOROPROPIONYL HYPOFLUORITE, $C_3F_6O_2$
  PERFLUOROPROPYL HYPOFLUORITE, $C_3F_8O$
  ISOPROPYL HYPOCHLORITE, $C_3H_7ClO$
  HEPTAFLUOROBUTYRYL HYPOFLUORITE, $C_4F_8O_2$
  *tert*-BUTYL HYPOCHLORITE, $C_4H_9ClO$
  CHLORYL HYPOFLUORITE, $ClFO_3$
* CHLORINE FLUOROSULPHATE, $ClFO_3S$
* PENTAFLUOROSULPHUR PEROXYHYPOCHLORITE, $ClF_5O_2S$
* NITRYL HYPOCHLORITE (FLUORINE NITRATE), $ClNO_3$
  NITRYL HYPOFLUORITE, $FNO_3$
* FLUORINE FLUOROSULPHATE, $F_2O_3S$
  PENTAFLUOROSULPHUR HYPOFLUORITE, $F_6OS$
  PENTAFLUOROSELENIUM HYPOFLUORITE, $F_6OSe$

*See also* BLEACHING POWDER
METAL OXOHALOGENATES

## IGNITION SOURCES

Enstad, G., A Reconsideration of the Concept of Minimum Ignition Energy, paper presented at meeting of Euro. Fed. Chem. Engrs. Working Party on 16th March, 1975

The main purpose of the paper was to explain why less energetic sparks would ignite a dust cloud, whereas more energetic sparks would not. The latter expel dust particles from the ignition zone, while the former do not, allowing ignition to occur.

Unexpected sources of ignition have led to many fires and explosions, usually in cases of leakage of spillage of flammable materials.

*See* FRICTIONAL IGNITION OF GASES
HYDROGENATION CATALYSTS
LIGHT ALLOYS
MILK POWDER
PYROPHORIC MATERIALS
STATIC INITIATION
ALUMINIUM–MAGNESIUM–ZINC ALLOYS, Al–Mg–Zn: Rusted steel
CARBON DIOXIDE, $CO_2$
ETHYLENE, $C_2H_4$: Steel-braced tyres
SULPHUR, S: Static discharges

## INDUCTION PERIOD

In the absence of anything to prevent it, a chemical reaction will begin when the components and any necessary energy of activation are present in the reaction system. If an inhibitor (negative catalyst or chain-breaker) is present in the system, it will prevent the onset of normal reaction until the concentration of the inhibitor has been reduced by decomposition or side reactions to a sufficiently low level for reaction to begin. This delay in onset of reaction is termed the induction period.

The existence of an induction period can significantly interfere with the course of a reaction where this involves control of the rate of reaction by regulating the rate of addition of one of the reagents or the reaction temperature. In the absence of reaction, the concentration of the reagent (or the temperature) may be increased to a level at which, once reaction does begin, it may accelerate out of control and become hazardous. It is therefore essential in the early stages of a reaction to ensure that the desired transformation has begun, particularly if large quantities of material are involved. This may be effected by watching for a change of colour or appearance in the reaction mixture, by increase in the rate of heat evolution (as judged from reflux rate, need for cooling, etc.), or if necessary by removing a sample for chemical testing.

Grignard reactions are notorious for the existence of induction periods, and an extensive account of methods for their elimination by various activation procedures is given at the beginning of the treatise on Grignard reactions by Kharasch and Reinmuth (1954). Another long-used method of promoting the onset of reaction in on-going processes is the addition to the reaction mixture of a small quantity of reaction liquor kept from a previous batch.

For some relevant examples,

*see* ALUMINIUM TETRAHYDROBORATE, $AlB_3H_{12}$: Alkenes, Oxygen
LITHIUM TETRAHYDROALUMINATE, $AlH_4Li$: 3,5-Dibromocyclopentene
DIBORANE, $B_2H_6$ : Air, etc. (reference 2)
CALCIUM ACETYLIDE (CARBIDE), $C_2Ca$: Methanol
TETRAFLUOROETHYLENE, $C_2F_4$ : Iodine pentafluoride, etc.
GLYCOLONITRILE, $C_2H_3NO$
ACRYLALDEHYDE, $C_3H_4O$ (reference 1)
TETRACARBONYLNICKEL, $C_4NiO_4$ : Oxygen, etc.
NITRIC ACID, $HNO_3$ : 2,4,6-Trimethyltrioxane

## INORGANIC AZIDES $EN_3$, $MN_3$

1. Mellor, 1967, Vol. 8, Suppl. 2,42
2. Evans, B. L. *et al.*, *Chem. Rev.*, 1959, **59**, 515
3 Deb, S. K. *et al. Proc. 8th Combust. Symp.*, 1960, 829

Relationships existing between structure, stability and thermal, photochemical and explosive decomposition (sometimes spontaneous) of the inorganic azides have been extensively investigated and reviewed [1,2]. The ignition characteristics of explosive inorganic azides, with or without added impurities under initiation by heat or light have been discussed [3]. For individually indexed compounds, see:

HALOGEN AZIDES
METAL AZIDE HALIDES
METAL AZIDES
NON-METAL AZIDES

## INORGANIC PEROXIDES $O_2^{2-}$

1. Castrantas, 1970
2. Castrantas, 1965

The guide to safe handling and storage of peroxides also contains a comprehensive bibliography of detailed information [1]. The earlier publication contains tabulated data on fire and explosion hazards of inorganic peroxides, with a comprehensive bibliography [2].

*See* HYDROGEN PEROXIDE, $H_2O_2$
METAL PEROXIDES
PEROXOACIDS
PEROXOACID SALTS

## INTERHALOGENS

Kirk-Othmer, 1966, Vol. 9, 585–598

The fluorine-containing members of this class are oxidants at least as powerful as fluorine itself. Individually indexed compounds are:

BROMINE FLUORIDE, BrF
BROMINE TRIFLUORIDE, $BrF_3$
BROMINE PENTAFLUORIDE, $BrF_5$
CHLORINE FLUORIDE, ClF
CHLORINE TRIFLUORIDE, $ClF_3$
CHLORINE PENTAFLUORIDE, $ClF_5$
IODINE CHLORIDE, ClI
IODINE TRICHLORIDE, $Cl_3I$
IODINE PENTAFLUORIDE, $F_5I$
IODINE HEPTAFLUORIDE, $F_7I$

## IODINE COMPOUNDS

Several iodine compounds are explosively unstable, individually indexed compounds being:

IODINE ISOCYANATE, CINO
2-IODOSYLVINYL CHLORIDE, $C_2H_2ClIO$
2-IODYLVINYL CHLORIDE, $C_2H_2ClIO_2$
DIIODOACETYLENE, $C_2I_2$

$o$-DIIODOBENZENE, $C_6H_4I_2$
1,6-DIIODO-2,4-HEXADIYNE, $C_6H_4I_2$
[(CHROMYLDIOXY)IODO] BENZENE, $C_6H_5CrIO_4$
IODOBENZENE, $C_6H_5I$
PHENYLIODINE DINITRATE, $C_6H_5IN_2O_6$
IODOSYLBENZENE, $C_6H_5IO$
IODYLBENZENE, $C_6H_5IO_2$
IODYLBENZENE PERCHLORATE, $C_6H_6ClIO_6$
IODINE TRIACETATE, $C_6H_9IO_6$
$p$-IODOTOLUENE, $C_7H_7I$
$p$-IODOSYLTOLUENE, $C_7H_7IO$
$p$-IODYLTOLUENE, $C_7H_7IO_2$
$p$-IODYLANISOLE, $C_7H_7IO_3$
3,5-DIMETHYL-4-[BIS(TRIFLUOROACETOXY)IODO] ISOXAZOLE, $C_9H_6F_6INO_5$
1-IODO-3-PHENYL-2-PROPYNE, $C_9H_7I$
2-IODO-3,5-DINITROBIPHENYL, $C_{12}H_7IN_2O_4$
CALCIUM BIS-$p$-IODYLBENZOATE, $C_{14}H_8CaI_2O_8$
9-PHENYL-9-IODAFLUORENE, $C_{18}H_{13}I$
IODINE(III) PERCHLORATE, $Cl_3IO_{12}$
AMMONIUM PERIODATE, $H_4INO_4$
IODINE AZIDE, $IN_3$
NITROGEN TRIIODIDE–AMMONIA, $I_3N \cdot H_3N$

## *vic*-IODO-ALKOXY OR -ACETOXY COMPOUNDS

**–C(I)–C(OR)– or –C(I)–C(OAc)–**

Dimmel, D. R., *Chem. Eng. News*, 1977, **55**(27), 38

Treatment of 3,7,7-trimethylbicyclo[4.1.0] heptane ('Δ3-carene') with iodine and copper acetate in methanol gave 3-iodo-4-methoxy-4,7,7-trimethylbicyclo[4.1.0]heptane. A 50 g sample exploded violently after standing at ambient temperature in a closed container for 10 days. This and the corresponding iodoacetoxy compound showed large exotherms at 90°C on DTA examination, the latter calculated at 4.12 MJ/mol. Similar products derived from methylcyclohexane also exhibited substantial exotherms from 60°C upwards. It is recommended that *vicinal* iodo-alkoxy or -acetoxy derivatives in the terpene area should be handled very cautiously.

## ION EXCHANGE RESINS

Anon., *Chem. Eng. News*, 1953, **31**, 5120; *Chem. Eng. News*, 1968, **26**, 1480

*MCA Case History No. 2155*

A three-year-old sample of ion exchange resin was soaked in dilute hydrochloric acid, and then charged into a 2.5 cm diameter glass column. After soaking in distilled water for 15 min, the tube exploded violently, presumably owing to swelling of the resin. Process resins as far as possible before charging into column. The earlier incident involved a column charged with dry resin which burst when wetted.

The case history involved dry resin which expanded and split a glass column when wetted with a salt solution.

*See* NITRIC ACID, $HNO_3$ : Ion exchange resins

## ISOCYANIDES

**RN=C:**

Acids

Sidgwick, 1950, 673

Acid-catalysed hydrolysis of isocyanides ('carbylamines'), to primary amine and formic acid is very rapid, sometimes explosively so.

## ISOPROPYLIDENE COMPOUNDS

**$Me_2C{=}C$**

Ozone

*See* OZONE, $O_3$ : Isopropylidene compounds

## KETONE PEROXIDES

Davies, 1961, 72

Kirk-Othmer, 1967, Vol. 14, 777

The variety of peroxides (monomeric, dimeric and polymeric) which can be produced from interaction of a given ketone with hydrogen peroxide is very wide (see group-types below). The proportions of the products in the reaction mixture depend on the reaction conditions used, as well as the structure of the ketone. Many of the products appear to coexist in equilibrium, and several types of structure are explosive and sensitive in varying degrees to heat or shock. Extreme caution is therefore required in handling ketone peroxides in high concentrations, particularly those derived from ketones of low molecular weight. Acetone is thus entirely unsuitable as a reaction or cleaning solvent whenever hydrogen peroxide is used.

*See* CYCLIC PEROXIDES
1-OXYPEROXY COMPOUNDS
POLYPEROXIDES
3,3,6,6-TETRAMETHYL-1,2,4,5-TETRAOXANE, $C_6H_{12}O_4$
3,6-DIETHYL-3,6-DIMETHYL-1,2,4,5-TETRAOXANE, $C_8H_{16}O_4$
3,3,6,6,9,9-HEXAMETHYL-1,2,4,5,7,8-HEXAOXAONANE, $C_9H_{18}O_6$
3,6-DI(SPIROCYCLOHEXANE) TETRAOXANE, $C_{12}H_{20}O_4$
3,6,9-TRIETHYL-3,6,9-TRIMETHYL-1,2,4,5,7,8-HEXAOXAONANE, $C_{12}H_{24}O_6$
TRI(SPIROCYCLOPENTANE)-1,1,4,4,7,7-HEXAOXAONANE, $C_{15}H_{24}O_6$

## KETOXIMINOSILANES

**$RSi(ON{=}CMeEt)_3$**

Tyler, L. J., *Chem. Eng. News*, 1974, **52**(35), 3

During manufacture of tris(ketoximino)silanes, two violent explosions attributed to acid-catalysed exothermic rearrangement/decomposition reactions occurred. Although these silane derivatives can be distilled under reduced pressure, the presence of acidic impurities (e.g. 2-butanone oxime hydrochloride, produced during silane preparation) drastically reduces thermal stability. Iron(III) chloride at 500 p.p.m. caused degradation to occur at 150°C, and at 2% concentration violent decomposition set in at 50°C.

*See* 2-BUTANONE OXIME HYDROCHLORIDE, $C_4H_{10}ClNO$

*See other* OXIMES

**KJELDAHL METHOD**

Beet, A. E., *J.R. Inst. Chem.*, 1955, **79**, 163, 299

Possible hazards introduced by variations in experimental technique in Kjeldahl nitrogen determination are discussed.

**LANTHANIDE METALS**

Oxidants

Bailar, 1973, Vol. 4, 70

While there are considerable variations in reactivity, several of the series ignite in halogens above 200°C or in air or oxygen above 150–180°C, or lower in the presence of moisture.

*See* EUROPIUM, Eu

*See other* METALS

**LASSAIGNE TEST**

Lance, R. C. *et al.*, *Microchem. J.*, 1975, **20**, 103–110

Use of sodium–lead alloy in place of sodium in the elemental fusion test is safer and controllable, especially with fluorinated compounds.

*See* SODIUM, Na: Fluorinated compounds
: Halocarbons

**LIGHT ALLOYS**

*Fire Prot. Assoc. J.*, 1959 (44), 28

Experiments to determine the probability of ignition of gas or vapour by incendive sparks arising from impact of aluminium-, magnesium- and zinc-containing alloys with rusty steel are described. The risk is greatest with magnesium alloys, where, the higher the magnesium content, the lower the impact energy necessary for incendive sparking. Wide ranges of ignitable gas concentrations also tend to promote ignition.

*See* DIIRON TRIOXIDE, $Fe_2O_3$ : Aluminium
: Aluminium, Propene
MAGNESIUM, Mg: Metal oxides

## LINSEED OIL

Watts, B. G., private comm., 1965
Taradoire, F., *Rev. Prod. Chim.*, 1925, **28**, 114–115 (*Chem. Abs.*, 1925, **19**, 1500)

Cloths used to apply linseed oil to laboratory benches were not burnt as directed, but dropped into a waste bin. A fire developed during a few hours and destroyed the laboratory. Tests showed that heating and ignition were rapid if a draught of warm air impinged on the oil-soaked cloth. Many other incidents involving ignition of peroxidisable materials dispersed on absorbent combustible fibrous materials have been recorded.

Practical tests on the spontaneous combustion of cotton waste soaked in linseed oil and other paint materials had been reported 40 years earlier.

## LIQUEFIED GASES

Water

1. Urano, Y. *et al.*, *Chem. Abs.*, 1976, **84**, 107979n; 1977, **86**, 108970f
2. Porteous, W. M. *et al.*, *Chem. Eng. Progr.*, 1976, **72**(5), 83–89

The explosive phenomenon produced by contact of liquefied gases with water was studied. Chlorodifluoromethane produced explosions when the liquid–water temperature differential exceeded 92°C, and propene did so at differentials of 96–109°C. Liquid propane did, but ethylene did not produce explosions under the conditions studied [1]. The previous literature on superheated vapour explosions has been critically reviewed, and new experimental work shows the phenomenon to be more widespread than had been thought previously. The explosions may be quite violent, and mixtures of liquefied gases may produce over-pressures above 7 bar [2].

*See* below

## LIQUEFIED NATURAL GAS

Organic liquids,
or Water

1. Katz, D.L. *et al.*, *Hydrocarbon Proc.*, 1971, **50**, 240; Anon., *Chem. Eng. News*, 1972, **50**(8), 57
2. Yang, K., *Nature*, 1973, **243**, 221–222

The quite loud 'explosions' (either immediate or delayed) which occur when LNG (containing unusually high contents of heavier materials) is spilled on to water are non-combustive and harmless. Superheating and shock-wave phenomena are involved [1]. There is a similar effect when LNG of normal composition (90% methane) is spilled on to some $C_5-C_8$ hydrocarbons or methanol, acetone or 2-butanone [2].

*See* above

## LIQUEFIED PETROLEUM GASES

1. *Storage and Handling of LPG,* London, Fire Prot. Assoc., 1964
2. *LPG Safety, Model Code,* part 9, London, Inst. Petr., 1967

The considerable fire hazards of propane and butane, the most commonly used LPGs, reside in their high volatility, high vapour density and tendency to collect in drains, hollows and basements, and a lower explosive limit of only 2% by volume. The booklet gives guidance on precautions necessary for bulk storage sites and for filling and storage of cylinders, and refers to existing Codes of Practice and Standards [1], of which the most recent is reference 2.

## LIQUID AIR

Liquid air, formerly used widely as a laboratory or industrial cryogenic liquid, has been involved in many violent incidents. Many of these have involved the increased content of residual liquid oxygen produced by fractional evaporation of liquid air during storage. However, liquid air is still a powerful oxidant in its own right. Liquid nitrogen, now widely available, is recommended as a safer coolant than liquid air.

Carbon disulphide
Biltz, W., *Chem. Ztg.*, 1925, **49**, 1001
A mixture prepared as a cooling bath exploded violently and apparently spontaneously.

Charcoal
Taylor, J., *J. Sci. Instrum.,* 1928, **5**, 24
Accidental contact via a cracked tube caused violent explosion. Nitrogen is a safer coolant.

Ether
Danckwort, P. W., *Angew. Chem.,* 1927, **40**, 1317
Addition of liquid air to ether in a dish caused a violent explosion after a short delay. Previous demonstrations had been uneventful, though it was known that such mixtures were impact- and friction-sensitive.

Hydrocarbons
McCartey, L. V. *et al.*, *Chem. Eng. News*, 1949, 27, 2612

All hydrocarbons (and most reducing agents) form explosive mixtures with liquid air.

*See* NITROGEN (Liquid), $N_2$
OXYGEN (Liquid), $O_2$

**LITHIUM PERALKYLURANATES** **Li[UR$_n$]**

Sigurdson, E. R. *et al.*, *J. Chem. Soc., Dalton Trans.*, 1977, 818

In a series of solvated lithium peralkyluranate(IV) and (V) complexes all were pyrophoric, and the former type may explode unpredictably at ambient temperature.

*See related* ALKYLMETALS
*See other* HEAVY METAL DERIVATIVES
PYROPHORIC MATERIALS

**MERCURY COMPOUNDS**

A number of mercury compounds show explosive instability or reactivity in various degrees.

*See* POLY(DIMERCURYIMMONIUM) COMPOUNDS

and the individually indexed compounds:

MERCURY(I) BROMATE, $Br_2Hg_2O_6$
MERCURY DIAZOCARBIDE, $(CHgN_2)_n$
MERCURY(I) CYANAMIDE, $CHg_2N_2$
CHLORATOMERCURIO(FORMYL)METHYLENEMERCURY(II), $C_2HClHg_2O_4$
MERCURY(II) METHYLNITROLATE, $C_2H_2HgN_4O_8$
1,2-BIS(HYDROXOMERCURIO)-1,1,2,2-BIS(OXYDIMERCURIO)-ETHANE ('ETHANE HEXAMERCARBIDE'), $C_2H_2Hg_6O_4$
MERCURY(II) FORMHYDROXAMATE, $C_2H_4HgN_2O_4$
2-HYDROXYETHYLMERCURY(II) NITRATE, $C_2H_5HgNO_4$
DIMETHYLMERCURY, $C_2H_6Hg$
MERCURY(II) ACETYLIDE, $C_2Hg$
MERCURY(II) CYANIDE, $C_2HgN_2$
DIMERCURY DICYANIDE OXIDE, $C_2Hg_2N_2O$
ALLYLMERCURY(II) IODIDE, $C_3H_5HgI$
MERCURY BIS(CHLOROACETYLIDE), $C_4Cl_2Hg$
2-METHYL-2-NITRATOMERCURIOPROPYL(NITRATO)DIMERCURY(II), $C_4H_8Hg_3N_2O_6$
$\mu$-1,2-BIS(CYANOMERCURIO)ETHANEDIYLIDENE-DIMERCURY(II), $C_4Hg_4N_2$
DIPROPYLMERCURY, $C_6H_{14}Hg$
BIS(ETHOXYCARBONYLDIAZOMETHYL)MERCURY, $C_8H_{10}HgN_4O_4$
DIISOPENTYLMERCURY, $C_{10}H_{22}Hg$
MERCURY DIPICRATE, $C_{12}H_4HgN_6O_{14}$
DIPHENYLMERCURY, $C_{12}H_{10}Hg$
MERCURY(II) PEROXYBENZOATE, $C_{14}H_{10}HgO_6$
MERCURY(II) CHLORITE, $Cl_2HgO_4$
MERCURY(I) CHLORITE, $Cl_2Hg_2O_4$

## METAL ABIETATES

formula on p. 190

*Fire Prot. Assoc. J.*, 1958, 255

Aluminium, zinc, calcium, sodium, cobalt, lead and manganese abietates ('resinates'), when finely divided, are subject to spontaneous heating and ignition. Store in sealed metal containers away from fire hazards.

## METAL ACETYLIDES

**C≡CM**

1. Brameld, V.F. *et al.*, *J. Soc. Chem. Ind.*, 1947, **66**, 346
2. Houben-Weyl, 1970, Vol. 13(1), 739
3. Rutledge, 1968, 85–86
4. Miller, 1965, Vol. 1, 486
5. Dolgopolskii, I. M. *et al.*, *Chem. Abs.*, 1947, **41**, 6721

Previous literature on formation of various types of copper acetylides is discussed and the mechanism of their formation is examined, with experimental detail. Whenever a copper or a copper-rich alloy is likely to come into contact with atmospheres containing (1) ammonia, water vapour and acetylene or (2) lime-sludge, water vapour and acetylene, or a combination of these two, there is the probability of acetylide formation with danger of explosion.

The action is aided by the presence of air, or air with carbon dioxide, and hindered by the presence of nitrogen. Explosive acetylides may be formed on copper and brasses containing more than 50% copper when these are exposed to acetylene atmospheres. The acetylides produced by action of acetylene on ammoniacal or alkaline solutions of copper(II) salts are more explosive than those from the corresponding copper(I) salts [1]. The hydrated forms are less explosive than the anhydrous material [2].

Catalytic forms of copper, mercury and silver acetylides, supported on alumina, carbon or silica and used for polymerisation of alkanes, are relatively stable [3].

In contact with acetylene, silver and mercury salt solutions will also give explosive acetylides, the mercury derivatives being complex [4]. Many of the metal acetylides react violently with oxidants.

Impact sensitivities of the dry copper derivatives of acetylene, buten-3-yne and 1,3-hexadien-5-yne under impact were determined as 2.4, 2.4 and 4.0 kg m, respectively. The copper derivative of a poly-acetylene mixture generated by low-temperature polymerisation of acetylene detonated under 1.2 kg m impact. Sensitivities were much lower for the moist compounds [5].

Individually indexed compounds in this often dangerously explosive class are:

DISILVER ACETYLIDE, $C_2Ag_2$

SILVER ACETYLIDE–SILVER NITRATE, $C_2Ag_2 \cdot AgNO_3$
DIGOLD(I) ACETYLIDE, $C_2Au_2$
BARIUM ACETYLIDE, $C_2Ba$
CALCIUM ACETYLIDE (CARBIDE), $C_2Ca$
DICAESIUM ACETYLIDE, $C_2Cs_2$
COPPER(II) ACETYLIDE, $C_2Cu$
DICOPPER(I) ACETYLIDE, $C_2Cu_2$
SILVER ACETYLIDE, $C_2HAg$
CAESIUM ACETYLIDE, $C_2HCs$
POTASSIUM ACETYLIDE, $C_2HK$
LITHIUM ACETYLIDE, $C_2HLi$
SODIUM ACETYLIDE, $C_2HNa$
RUBIDIUM ACETYLIDE, $C_2HRb$
LITHIUM ACETYLIDE–AMMONIA, $C_2HLi \cdot H_3N$
MERCURY(II) ACETYLIDE, $C_2Hg$
DIPOTASSIUM ACETYLIDE, $C_2K_2$
DILITHIUM ACETYLIDE, $C_2Li_2$
DISODIUM ACETYLIDE, $C_2Na_2$
DIRUBIDIUM ACETYLIDE, $C_2Rb_2$
STRONTIUM ACETYLIDE, $C_2Sr$
SILVER TRIFLUOROMETHYLACETYLIDE, $C_3AgF_3$
* PROPIOLIC ACID, $C_3H_2O_2$: Ammonia, etc.
SILVER 3-HYDROXYPROPYNIDE, $C_3H_3AgO$
SODIUM METHOXYACETYLIDE, $C_3H_3NaO$
* CERIUM DICARBIDE, $C_4Ce$
MERCURY BIS(CHLOROACETYLIDE), $C_4Cl_2Hg$
MANGANESE(II) BISACETYLIDE, $C_4H_2Mn$
SILVER BUTEN-3-YNIDE, $C_4H_3Ag$
SODIUM ETHOXYACETYLIDE, $C_4H_5NaO$
* LANTHANUM DICARBIDE, $C_4La$
1,3-PENTADIYN-1-YLSILVER, $C_5H_3Ag$
1,3-PENTADIYN-1-YLCOPPER, $C_5H_3Cu$
SILVER CYCLOPROPYLACETYLIDE, $C_5H_5Ag$
DIMETHYL-1-PROPYNYLTHALLIUM, $C_5H_9Tl$
DISILVER 1,3,5-HEXATRIYNIDE, $C_6Ag_2$
TRIETHYNYLALUMINIUM, $C_6H_3Al$
TRIETHYNYLANTIMONY, $C_6H_3Sb$
BIS(DIMETHYLTHALLIUM) ACETYLIDE, $C_6H_{12}Tl_2$
TETRAETHYNYLGERMANIUM, $C_8H_4Ge$
TETRAETHYNYLTIN, $C_8H_4Sn$
SODIUM PHENYLACETYLIDE, $C_8H_5Na$
COPPER 1,3,5-OCTATRIEN-7-YNIDE, $C_8H_7Cu$
3-BUTEN-1-YNYLDIETHYLALUMINIUM, $C_8H_{13}Al$
DIMETHYL-PHENYLETHYNYLTHALLIUM, $C_{10}H_{11}Tl$
3-BUTEN-1-YNYLTRIETHYLLEAD, $C_{10}H_{18}Pb$
3-METHYL-3-BUTEN-1-YNYLTRIETHYLLEAD, $C_{11}H_{20}Pb$
3-BUTEN-1-YNYLDIISOBUTYLALUMINIUM, $C_{12}H_{21}Al$
* 1-BUTYLOXYETHYL 3-TRIMETHYLPLUMBYLPROPIOLATE, $C_{12}H_{22}O_3Pb$
* ACETYLENEBIS-TRIETHYLLEAD, $C_{14}H_{30}Pb_2$
BIS(TRIETHYLTIN)ACETYLENE, $C_{14}H_{30}Sn_2$
* TRIBUTYL(PHENYLETHYNYL)LEAD, $C_{20}H_{32}Pb$

*See other* ACETYLENIC COMPOUNDS

## METAL ALKOXIDES $M(OR)_n$

Brochure, 'Alkali and Alkali-Earth Metal Alkoxides', Troisdorf-Oberlar, Dynamit Nobel AG, 1974

An Appendix is devoted to safe handling of the alkoxides indicated by a prefix dash in the list below, not all of which are represented as indexed entries. These materials are readily hydrolysed and the exotherm may ignite the solids in presence of moist air, the potassium derivatives being the most reactive. This tendency is increased by acidic conditions, and combustion of the solid spreads rapidly. Potassium *tert*-butoxide is especially reactive towards a range of liquids and solvent vapours.

*See* POTASSIUM *tert*-BUTOXIDE, $C_4H_9KO$: Acids, or Reactive solvents

Individually indexed compounds are:

-POTASSIUM METHOXIDE, $CH_3KO$
-SODIUM METHOXIDE, $CH_3NaO$
-POTASSIUM ETHOXIDE, $C_2H_5KO$
-SODIUM ETHOXIDE, $C_2H_5NaO$
-MAGNESIUM METHOXIDE, $C_2H_6MgO_2$
-SODIUM ISOPROPOXIDE, $C_3H_7NaO$
TITANIUM(III) METHOXIDE, $C_3H_9O_3Ti$
-POTASSIUM BUTOXIDE, $C_4H_9KO$
-POTASSIUM *tert*-BUTOXIDE, $C_4H_9KO$
-SODIUM BUTOXIDE, $C_4H_9NaO$
-MAGNESIUM ETHOXIDE, $C_4H_{10}MgO_2$
-SODIUM *tert*-PENTYLOXIDE, $C_5H_{11}NaO$
TITANIUM TETRAISOPROPOXIDE, $C_{12}H_{28}O_4Ti$

## METAL AMIDOSULPHATES $H_2NSO_2OM$

Metal nitrates,
or Nitrites

Heubel, J. *et al., Compt. Rend.* [3], 1963, **257**, 684

Heating mixtures of barium, potassium or sodium amidosulphates or amidosulphuric acid with sodium or potassium nitrates or nitrites, lead to reactions which may be explosive. Thermogravimetric plots are given.

## METAL AZIDE HALIDES $MXN_3$

Dehnicke, K., *Angew. Chem. (Intern. Ed.),* 1967, **6**, 243

Metal halides and halogen azides react to give a range of metal azide halides, many of which are explosive.

*See* SILVER AZIDE CHLORIDE, $AgClN_3$
TUNGSTEN AZIDE PENTABROMIDE, $Br_5N_3W$
CHROMYL AZIDE CHLORIDE, $ClCrN_3O_2$
VANADYL AZIDE DICHLORIDE, $Cl_2N_3OV$
TIN AZIDE TRICHLORIDE, $Cl_3N_3Sn$

TITANIUM AZIDE TRICHLORIDE, $Cl_3N_3Ti$
MOLYBDENUM DIAZIDE TETRACHLORIDE, $Cl_4MoN_6$
VANADIUM AZIDE TETRACHLORIDE, $Cl_4N_3V$
MOLYBDENUM AZIDE PENTACHLORIDE, $Cl_5MoN_3$
URANIUM AZIDE PENTACHLORIDE, $Cl_5N_3U$
TUNGSTEN AZIDE PENTACHLORIDE, $Cl_5N_3W$

## METAL AZIDES

**$MN_3$**

Mellor, 1940, Vol. 8, 344–355; Vol. 8, Suppl. 2, 16–54
This large and well-documented group of explosive compounds contains some which are widely used industrially. Related to this group are:

METAL AZIDE HALIDES

and, of the many known simple and complex explosive metal azides, those few selected for individual indexing here are:

SILVER AZIDE, $AgN_3$
LITHIUM TETRAAZIDOALUMINATE(1–), $AlLiN_{12}$
ALUMINIUM TRIAZIDE, $AlN_9$
SODIUM TRIAZIDOAURATE(?), $AuN_9Na$
TRIGOLD DISODIUM HEXAAZIDE, $Au_3N_{18}Na_2$
BARIUM DIAZIDE, $BaN_6$
* METHYLCADMIUM AZIDE, $CH_3CdN_3$
* TRIMETHYLPLATINUM(IV) AZIDE: see the tetramer, $C_{12}H_{36}N_{12}Pt_4$
DIMETHYLGOLD AZIDE: see the dimer, $C_4H_{12}Au_2N_6$
* TETRAMETHYLDIGOLD DIAZIDE, $C_4H_{12}Au_2N_6$
* BIS(2-AMINOETHYL)AMINECOBALT(III) AZIDE, $C_4H_{13}CoN_{12}$
* DIAZIDO-BIS(1,2-DIAMINOETHANE)RUTHENIUM(II) HEXAFLUOROPHOSPHATE, $C_4H_{16}F_6N_{10}PRu$
* LEAD(IV) TRIACETATE AZIDE, $C_6H_9N_3O_6Pb$
* DICARBONYL-π-CYCLOHEPTATRIENYLTUNGSTEN AZIDE, $C_9H_7N_3O_2W$
* BIS(π-CYCLOPENTADIENYL)TUNGSTEN DIAZIDE OXIDE, $C_{10}H_{10}N_6OW$
* TRIMETHYLPLATINUM(IV) AZIDE TETRAMER, $C_{12}H_{36}N_{12}Pt_4$
CALCIUM DIAZIDE, $CaN_6$
CADMIUM DIAZIDE, $CdN_6$
* PENTAAMMINEAZIDORUTHENIUM(III) CHLORIDE, $Cl_2H_{15}N_8Ru$
* POTASSIUM TRIAZIDOCOBALTATE(1–), $CoKN_9$
*trans*-TETRAAMMINEDIAZIDOCOBALT(III) *trans*-DIAMMINE-TETRAAZIDOCOBALTATE(1–), $Co_2H_{18}N_{24}$
* CHROMYL AZIDE, $CrN_6O_2$
* COPPER(II) AZIDE HYDROXIDE, $CuHN_3O$
TETRAAMMINECOPPER(II) AZIDE, $CuH_{12}N_{10}$
LITHIUM HEXAAZIDOCUPRATE(4–), $CuLi_4N_{18}$
COPPER(I) AZIDE, $CuN_3$
COPPER(II) AZIDE, $CuN_6$
* AZIDOGERMANE, $GeH_3N_3$
MERCURY(II) AZIDE, $HgN_6$
MERCURY(I) AZIDE, $Hg_2N_6$
POTASSIUM AZIDE, $KN_3$
LITHIUM AZIDE, $LiN_3$
SODIUM AZIDE, $N_3Na$

THALLIUM(I) AZIDE, $N_3Tl$
LEAD(II) AZIDE, $N_6Pb$
STRONTIUM AZIDE, $N_6Sr$
ZINC AZIDE, $N_6Zn$
LEAD(IV) AZIDE, $N_{12}Pb$
* THALLIUM(I) TETRAAZIDOTHALLATE, $N_{12}Tl_2$

*See related* AMMINECOBALT(III) AZIDES
AZIDE COMPLEXES OF COBALT(III)

*See also* RUBBER: Metal azides
ELECTRIC FIELDS

## METAL CHLORATES $M(ClO_3)_n$

Acids

1. Mellor, 1941, Vol. 2, 315
2. Stossel, E. *et al.*, US Pat. 2 338 268, 1944

Additionally to being oxidants in contact with strong acids, metal chlorates liberate explosive chlorine dioxide gas. With concentrated sulphuric acid, a violent explosion may occur unless effective cooling is used [1]. Heating a moist mixture of a metal chlorate and a dibasic organic acid (tartaric or citric acid) liberates chlorine dioxide mixed with carbon dioxide [2].

Ammonium salts

Mixtures are incompatible.

*See* AMMONIUM CHLORATE, $ClH_4NO_3$

Phosphorus,
or Sugar,
or Sulphur

1. Black, H. K., *School Sci. Rev.*, 1963, 44(53), 462
2. *59th Ann. Rep. HM Insp. Explosives* (Cmd. 4934), 5, London, HMSO, 1934
3. Berger, A., *Arbeits-Schutz.*, 1934, **2**, 20
4. Taradoire, F. *Chem. Abs.*, 1938, **32**, $1455_6$

The extremely hazardous nature of the mixtures of metal chlorates with phosphorus, sugar or sulphur, sometimes with addition of permanganates and metal powders, frequently prepared as amateur fireworks, is stressed. Apart from being powerfully explosive, such mixtures are dangerously sensitive to friction or shock, and spontaneous ignition sometimes occurs [1]. Chlorates containing 1–2% of bromates or sulphur as impurities are liable to spontaneous explosion [3]. The danger of mixtures of chlorates with sulphur or phosphorus is such that their preparation without licence was prohibited by Orders in Council many years ago [2].

Mixtures of sulphur with lead, silver, potassium and barium chlorates ignited at 63–67, 74, 160–162 and 108–111°C, respectively [4].

Sulphur,
Initiators

Taradoire, F., *Bull. Soc. Chim. Fr.*, 1942, **9**, 610–620

The effect of initiators (diluted sulphuric acid, chlorine dioxide, sulphur dioxide or disulphur dichloride) on ignition of mixtures of barium, lead or potassium chlorates with sulphur was examined.

Individually indexed metal chlorates are:

ALUMINIUM CHLORATE, $AlCl_3O_9$
POTASSIUM CHLORATE, $ClKO_3$
SODIUM CHLORATE, $ClNaO_3$
MAGNESIUM CHLORATE, $Cl_2MgO_6$
MANGANESE(II) CHLORATE, $Cl_2MnO_6$
LEAD(II) CHLORATE, $Cl_2O_6Pb$
ZINC DICHLORATE, $Cl_2O_6Zn$

*See other* METAL HALOGENATES

## METAL CYANIDES AND CYANOCOMPLEXES $M(CN)_n$, $[M(CN)_n]^-$

1. von Schwartz, 1918, 399, 327; Pieters, 1957, 30
2. *Res. Rep. No. 2,* New York, Nat. Board Fire Underwriters, 1950

Several members of this class which contain heavy metals tend to explosive instability, and most are capable of violent oxidation under appropriate circumstances. Fusion of mixtures of metal cyanides with metal chlorates, perchlorates, nitrates or nitrites causes a violent explosion [1]. Addition of one solid component (even as a residue in small amount) to another molten component is also highly dangerous [2]. Individually indexed compounds are:

SILVER CYANIDE, CAgN
GOLD(I) CYANIDE, CAuN
POTASSIUM CYANIDE, CKN
SODIUM CYANIDE, CNNa
CADMIUM DICYANIDE, $C_2CdN_2$
COPPER(II) CYANIDE, $C_2CuN_2$
MERCURY(II) CYANIDE, $C_2HgN_2$
* DIMERCURY DICYANIDE OXIDE, $C_2Hg_2N_2O$
NICKEL(II) CYANIDE, $C_2N_2Ni$
LEAD(II) CYANIDE, $C_2N_2Pb$
ZINC DICYANIDE, $C_2N_2Zn$
POTASSIUM TETRACYANOMERCURATE (2–), $C_4HgK_2N_4$
POTASSIUM TETRACYANOTITANATE(4–), $C_4K_4N_4Ti$
* SODIUM TETRACYANATOPALLADATE(2–), $C_4N_4Na_2O_4Pd$
SODIUM PENTACYANONITROSYLFERRATE (2–), $C_5FeN_6Na_2O$
POTASSIUM HEXACYANOFERRATE(3–) ('FERRICYANIDE'), $C_6FeK_3N_6$
POTASSIUM HEXACYANOFERRATE (4–) ('FERROCYANIDE'), $C_6FeK_4N_6$
* AMMONIUM HEXACYANOFERRATE(4–), $C_6H_{16}FeN_{10}$
POTASSIUM OCTACYANODICOBALTATE(8–), $C_8Co_2K_8N_8$

IRON(3+) HEXACYANOFERRATE(4–), $C_{18}Fe_7N_{18}$

*See also* MOLTEN SALT BATHS

## *N*-METAL DERIVATIVES N–M

Many metal derivatives of nitrogenous systems containing one or more bonds linking nitrogen to a metal (usually but not exclusively a heavy metal) show explosive instability. Individually indexed compounds are:

SILVER PERCHLORYLAMIDE, $AgClHNO_3$
SILVER AMIDE, $AgH_2N$
SILVER *N*-NITROSULPHURIC DIAMIDATE, $AgH_2N_3O_4S$
DISILVER IMIDE, $Ag_2HN$
NITROGEN TRIODIDE–SILVER AMIDE, $I_3N \cdot AgH_2N$
TRISILVER NITRIDE, $Ag_3N$
SILVER 2,4,6-TRIS(DIOXOSELENA)PERHYDROTRIAZINE-1,3,5-TRIIDE ('SILVER TRISELENIMIDATE'), $Ag_3N_3O_6Se_3$
TETRASILVER DIIMIDOTRIPHOSPHATE, $Ag_4H_3N_2O_8P_3$
SILVER DIIMIDODIOXOSULPHATE(4–), $Ag_4N_2O_2S$
PENTASILVER DIIMIDOTRIPHOSPHATE, $Ag_5H_2N_2O_8P_3$
PENTASILVER DIAMIDOPHOSPHATE, $Ag_5N_2O_2P$
BIS(DIHYDROXYGOLD)IMIDE, $Au_2H_5NO_4$
GOLD NITRIDE–AMMONIA, $Au_3N \cdot H_3N$
BISMUTH AMIDE OXIDE, $BiH_2NO$
BISMUTH NITRIDE, BiN
*N,N*-BIS(BROMOMERCURIO)HYDRAZINE, $Br_2H_2Hg_2N_2$
DISILVER CYANAMIDE, $CAg_2N_2$
DISILVER DIAZOMETHANEDIIDE, $CAg_2N_2$
CALCIUM CYANAMIDE, $CCaN_2$
SILVER TETRAZOLIDE, $CHAgN_4$
SILVER NITROUREIDE, $CH_2AgN_3O_3$
SILVER 5-AMINOTETRAZOLIDE, $CH_2AgN_5$
SILVER NITROGUANIDIDE, $CH_4AgN_3$
POTASSIUM METHYLAMIDE, $CH_4KN$
MERCURY(I) CYANAMIDE, $CHg_2N_2$
5-HYDROXY-1(*N*-SODIO-5-TETRAZOLYLAZO)TETRAZOLE, $C_2HN_{10}NaO$
SILVER 1,2,3-TRIAZOLIDE, $C_2H_2AgN_3$
SILVER 1,3-DI(5-TETRAZOLYL)TRIAZENE, $C_2H_2AgN_{11}$
*N,N'*-DISODIUM *N,N'*-DIMETHOXYSULPHONYLDIAMIDE, $C_2H_6N_2Na_2O_{14}S$
COPPER(II)1,3-DI(5-TETRAZOLYL)TRIAZENIDE, $C_4H_4CuN_{22}$
* SELENINYL BIS(DIMETHYLAMIDE), $C_4H_{12}N_2OSe$
LITHIUM 1,1-DIMETHYL(TRIMETHYLSILYL)HYDRAZIDE, $C_5H_{15}LiN_2Si$
*N,N-p*-TRILITHIOANILINE, $C_6H_4Li_3N$
LITHIUM TRIETHYLSILYLAMIDE, $C_6H_{16}LiNSi$
DILITHIUM 1,1-BIS(TRIMETHYLSILYL)HYDRAZIDE, $C_6H_{18}Li_2N_2Si_2$
BIS(DIMETHYLAMINO)DIMETHYLSTANNANE, $C_6H_{18}N_2Sn$
TRIS(DIMETHYLAMINO)ANTIMONY, $C_6H_{18}N_3Sb$
3-SODIO-(5-NITRO-2-FURFURYLIDENEAMINO)IMIDAZOLIDIN-2,4-DIONE, $C_8H_5N_4NaO_5$

TETRAKIS(DIMETHYLAMINO)TITANIUM, $C_8H_{24}N_4Ti$
SILVER HEXANITRODIPHENYLAMIDE, $C_{12}H_4AgN_7O_{12}$
TRICALCIUM DINITRIDE, $Ca_3N_2$
CADMIUM DIAMIDE, $CdH_4N_2$
TRICADMIUM DINITRIDE, $Cd_3N_2$
CERIUM NITRIDE, CeN
MONOPOTASSIUM PERCHLORYLAMIDE, $ClHKNO_3$
MERCURY(II) AMIDE CHLORIDE, $ClH_2HgN$
*N,N'*-BIS(CHLOROMERCURIO)HYDRAZINE, $Cl_2H_2Hg_2N_2$
COBALT(III) AMIDE, $CoH_6N_3$
COBALT NITRIDE, CoN
CHROMIUM NITRIDE, CrN
CAESIUM AMIDE, $CsH_2N$
TRICAESIUM NITRIDE, $Cs_3N$
COPPER(I) NITRIDE, $Cu_3N$
GERMANIUM(II) IMIDE, GeHN
POLY(DIMERCURYIMMONIUM HYDROXIDE), $(HHg_2NO)_n$
LEAD IMIDE, HNPb
POTASSIUM AMIDE, $H_2KN$
SODIUM AMIDE, $H_2NNa$
SODIUM HYDRAZIDE, $H_3N_2Na$
ZINC DIHYDRAZIDE, $H_6N_4Zn$
TRIMERCURY DINITRIDE, $Hg_3N_2$
POTASSIUM NITRIDOOSMATE (1–), $KNO_3Os$
POTASSIUM SULPHURDIIMIDE (2–), $K_2N_2S$
POTASSIUM THALLIUM(I) AMIDE (2–) AMMONIATE, $K_2NTl{-}H_3N$
POTASSIUM NITRIDE (3–), $K_3N$
POTASSIUM 2,4,6-TRIS(DIOXOSELENA)PERHYDROTRIAZINE-1,3,5-TRIIDE ('POTASSIUM TRISELENIMIDATE'), $K_3N_3O_6Se_3$
LITHIUM SODIUM NITROXYLATE, $LiNNaO_2$
LITHIUM NITRIDE, $Li_3N$
SODIUM NITROXYLATE (2–), $NNa_2O_2$
SODIUM NITRIDE (3–), $NNa_3$
PLUTONIUM NITRIDE, NPu
RUBIDIUM NITRIDE (3–), $NRb_3$
ANTIMONY(III)NITRIDE, NSb
THALLIUM(I) NITRIDE, $NTl_3$
URANIUM(III) NITRIDE, NU
TRILEAD DINITRIDE, $N_2Pb_3$
THALLIUM 2,4,6-TRIS(DIOXOSELENA)PERHYDROTRIAZINE-1,3,5-TRIIDE ('THALLIUM TRISELENIMIDATE'), $N_3O_6Se_3Tl_3$
TRIS(THIONITROSYL)THALLIUM, $N_3S_3Tl$
* TETRASELENIUM TETRANITRIDE, $N_4Se_4$
* TRITELLURIUM TETRANITRIDE, $N_4Te_3$
TRITHORIUM TETRANITRIDE, $N_4Th_3$

*See* PERCHLORYLAMIDE SALTS

**METAL DUSTS**


1. Jacobson, M. *et al.*, *Rep. Invest.*, 6516(9) Washington, US Bureau of Mines, 1961
2. Brown, H. R., *Chem. Eng. News,* 1956, **34,** 87
3. Nedin, V. V. *et al., Chem. Abs.*, 1972, 77, 167636–167641

4. Martin, R., *Powder Metall.*, 1976, **19**, 70–73
5. Alekseev, A. G. *et al.*, *Chem. Abs.*, 1977, **86**, 175311e

Of the 313 samples examined, the dust explosion hazards of finely divided aluminium, aluminium–magnesium alloys, magnesium, thorium, titanium and uranium, and the hydrides of thorium and uranium, are rated highest [1]. The need to exercise caution when handling dusts of some recently introduced reactive metals is briefly discussed. Some will form explosive mixtures, not only with air or oxygen, but also with nitrogen and carbon dioxide, reacting to give the carbonate or nitride. Beryllium, cerium, germanium, hafnium, lithium, niobium, potassium, sodium, thorium, titanium, uranium and zirconium are discussed [2].

The pyrophoric capabilities of titanium and titanium–aluminium powders were studied, the effect of particle size on flammability and explosion parameters of aluminium and magnesium powders dispersed in air was determined, and explosion hazards of aluminium, magnesium and aluminium-based alloy powders and use of inert gas media as preventives were assessed. In a study to determine explosivity of ferrochromium, ferromanganese, ferrosilicon, ferrotitanium, manganese and calcium–silicon powders, the latter proved to be the most active [3]. Metal dust and powder explosion risks have been reviewed recently [4], including those during spraying operations in powder production [5].

*See* the individual metals

**METAL FULMINATES** **MC≡N→O**

Urbanski, 1967, Vol. 3, 157

Hackspill, L. *et al.*, *Chem. Abs.*, 1938, **32**, 4377a

The metal salts are all powerfully explosive. Of several salts examined, those of cadmium, copper and silver were more powerful detonators than mercury fulminate, while thallium fulminate was much more sensitive to heating and impact. Formally related salts are also explosive.

Sodium, potassium, rubidium and caesium fulminates are all easily detonated by feeble friction or heat. They all form double salts with mercury(II) fulminate which also explode readily, that of the rubidium salt at 45°C.

Individually indexed compounds are:

SILVER FULMINATE, CAgNO
SODIUM FULMINATE, CNNaO
THALLIUM FULMINATE, CNOTl
CADMIUM FULMINATE, $C_2CdN_2O_2$
COPPER(II) FULMINATE, $C_2CuN_2O_2$
* MERCURY(II) METHYLNITROLATE, $C_2H_2HgN_4O_8$
* MERCURY(II) FORMHYDROXAMATE, $C_2H_4HgN_2O_4$
MERCURY(II) FULMINATE, $C_2HgN_2O_2$
DIMETHYLTHALLIUM FULMINATE, $C_3H_6NOTl$
* DIMETHYLTHALLIUM *N*-METHYLACETOHYDROXAMATE, $C_5H_{12}NO_2Tl$

* TRIPROPYLLEAD FULMINATE, $C_{10}H_{21}NOPb$

## METAL HALIDES

**$MX_n$**

Members of this class have usually featured as secondary reagents in hazardous combination of chemicals.

Individually indexed compounds are:

SILVER CHLORIDE, AgCl
SILVER FLUORIDE, AgF
SILVER DIFLUORIDE, $AgF_2$
ALUMINIUM TRIBROMIDE, $AlBr_3$
ALUMINIUM TRICHLORIDE, $AlCl_3$
ALUMINIUM TRIIODIDE, $AlI_3$
AMERICIUM CHLORIDE, $AmCl_3$
GOLD(III) CHLORIDE, $AuCl_3$
BERYLLIUM FLUORIDE, $BeF_2$
BISMUTH PENTAFLUORIDE, $BiF_5$
CALCIUM BROMIDE, $Br_2Ca$
COBALT(II) BROMIDE, $Br_2Co$
IRON(II) BROMIDE, $Br_2Fe$
MERCURY(II) BROMIDE, $Br_2Hg$
TITANIUM DIBROMIDE, $Br_2Ti$
ZIRCONIUM DIBROMIDE, $Br_2Zr$
IRON(III) BROMIDE, $Br_3Fe$
* VANADIUM TRIBROMIDE OXIDE, $Br_3OV$
TELLURIUM TETRABROMIDE, $Br_4Te$
CALCIUM CHLORIDE, $CaCl_2$
* AMMONIUM CHLORIDE, $ClH_4N$
* MANGANESE CHLORIDE TRIOXIDE, $ClMnO_3$
SODIUM CHLORIDE, ClNa
THALLIUM(I) CHLORIDE, ClTl
COBALT(II) CHLORIDE, $Cl_2Co$
CHROMYL CHLORIDE, $Cl_2CrO_2$
MAGNESIUM CHLORIDE, $Cl_2Mg$
MANGANESE(II) CHLORIDE, $Cl_2Mn$
* MANGANESE DICHLORIDE DIOXIDE, $Cl_2MnO_2$
LEAD DICHLORIDE, $Cl_2Pb$
TIN DICHLORIDE, $Cl_2Sn$
TITANIUM DICHLORIDE, $Cl_2Ti$
ZIRCONIUM DICHLORIDE, $Cl_2Zr$
COBALT(III) CHLORIDE, $Cl_3Co$
CHROMIUM(III) CHLORIDE, $Cl_3Cr$
IRON(III) CHLORIDE, $Cl_3Fe$
* HEXAAMMINETITANIUM(III) CHLORIDE, $Cl_3H_{18}N_6Ti$
* MANGANESE TRICHLORIDE OXIDE, $Cl_3MnO$
* ANTIMONY TRICHLORIDE OXIDE, $Cl_3OSb$
* VANADIUM TRICHLORIDE OXIDE, $Cl_3OV$
RHODIUM(III) CHLORIDE, $Cl_3Rh$
RUTHENIUM(III) CHLORIDE, $Cl_3Ru$
ANTIMONY TRICHLORIDE, $Cl_3Sb$
TITANIUM TRICHLORIDE, $Cl_3Ti$
VANADIUM TRICHLORIDE, $Cl_3V$

ZIRCONIUM TRICHLORIDE, $Cl_3Zr$
GERMANIUM TETRACHLORIDE, $Cl_4Ge$
* RHENIUM TETRACHLORIDE OXIDE, $Cl_4ORe$
LEAD TETRACHLORIDE, $Cl_4Pb$
TIN TETRACHLORIDE, $Cl_4Sn$
TITANIUM TETRACHLORIDE, $Cl_4Ti$
ZIRCONIUM TETRACHLORIDE, $Cl_4Zr$
ANTIMONY PENTACHLORIDE, $Cl_5Sb$
URANIUM HEXACHLORIDE, $Cl_6U$
COBALT TRIFLUORIDE, $CoF_3$
CHROMIUM PENTAFLUORIDE, $CrF_5$
CAESIUM FLUORIDE, CsF
LEAD DIFLUORIDE, $F_2Pb$
MANGANESE TRIFLUORIDE, $F_3Mn$
PALLADIUM TRIFLUORIDE, $F_3Pd$
MANGANESE TETRAFLUORIDE, $F_4Mn$
PLATINUM TETRAFLUORIDE, $F_4Pt$
RHODIUM TETRAFLUORIDE, $F_4Rh$
IRIDIUM HEXAFLUORIDE, $F_6Ir$
NEPTUNIUM HEXAFLUORIDE, $F_6Np$
OSMIUM HEXAFLUORIDE, $F_6Os$
PLATINUM HEXAFLUORIDE, $F_6Pt$
PLUTONIUM HEXAFLUORIDE, $F_6Pu$
RHENIUM HEXAFLUORIDE, $F_6Re$
URANIUM HEXAFLUORIDE, $F_6U$
IRON(II) IODIDE, $FeI_2$
* AMMONIUM IODIDE, $H_4IN$
MERCURY(II) IODIDE, $HgI_2$
POTASSIUM IODIDE, IK
SODIUM IODIDE, INa
TITANIUM DIIODIDE, $I_2Ti$
ZINC IODIDE, $I_2Zn$
ZIRCONIUM TETRAIODIDE, $I_4Zr$

## METAL HALOGENATES $M(XO_3)_n$

Metals and oxidisable derivatives,
or Non-metals,
or Oxidisable materials

Mellor, 1946, Vol. 2, 310; 1956, Vol. 2, Suppl. 1, 583–584; 1941, Vol. 3, 651
von Schwartz, 1918, 323
Intimate mixtures of chlorates, bromates or iodates of barium, cadmium, calcium, magnesium, potassium, sodium or zinc, with finely divided aluminium, arsenic, copper; carbon, phosphorus, sulphur; hydrides of alkali and alkaline earth metals; sulphides of antimony, arsenic, copper or tin, metal cyanides, thiocyanates or impure manganese dioxide may react violently or explosively, either spontaneously (especially in presence of moisture) or on initiation by heat, friction, impact, sparks or addition of sulphuric acid.

Mixtures of sodium or potassium chlorate with sulphur or phosphorus are rated as being exceptionally dangerous on frictional initiation.

*See* METAL CHLORATES
*See other* METAL OXOHALOGENATES

**METAL HYDRIDES** **$MH_n$, $[MH_n]^-$**

1. Banus, M. D., *Chem. Eng. News,* 1954, **32**, 2424–2427
2. Mackay, 1966, 66
3. *Metal Hydrides*, Mueller, W. M. *et al.*, New York, Academic Press, 1968

Precautions necessary for safe handling of three main groups of hydrides of commercial significance are discussed. The first group (sodium hydride; lithium or sodium tetrahydroaluminates) ignite or explode in contact with liquid water or high humidity, while the second group (lithium, calcium, strontium, barium hydrides; sodium or potassium tetrahydroborates) do not. Burning sodium hydride is reactive enough to explode with the combined water in concrete. The third group ('alloy' or non-stoicheiometric hydrides of titanium, zirconium, thorium, uranium, vanadium, tantalum and palladium) are produced commercially in very finely divided form. Though less pyrophoric than the corresponding powdered metals, once burning is established, they are difficult to extinguish and water-, carbon dioxide- or halocarbon-based extinguishers caused violent explosions. Powdered dolomite is usually effective in smothering such fires [1]. The trihydrides of the lanthanoids (rare earth metals) are pyrophoric in air and the dihydrides, though less reactive, must be handled under inert atmosphere [2].

A later reference states, however, that finely divided metal hydrides of the second group (lithium, calcium, barium and strontium hydrides) will ignite in air or react violently, sometimes explosively, with water or air of high humidity [3].

Individually indexed members of this class of active reducants are:

ALUMINIUM HYDRIDE, $AlH_3$
ALUMINIUM HYDRIDE–TRIMETHYLAMINE, $AlH_3 \cdot C_3H_9N$
ALUMINIUM HYDRIDE–DIETHYL ETHERATE, $AlH_3 \cdot C_4H_{10}O$
BARIUM HYDRIDE, $BaH_2$
BERYLLIUM HYDRIDE, $BeH_2$
CALCIUM DIHYDRIDE, $CaH_2$
CADMIUM HYDRIDE, $CdH_2$
CERIUM DIHYDRIDE, $CeH_2$
CERIUM TRIHYDRIDE, $CeH_3$
* CHLOROGERMANE, $ClGeH_3$
CAESIUM HYDRIDE, CsH
COPPER(I) HYDRIDE, CuH
POLY(GERMANIUM MONOHYDRIDE), $(GeH)_n$
POLY(GERMANIUM DIHYDRIDE), $(GeH_2)_n$

GERMANE, $GeH_4$
DIGERMANE, $Ge_2H_6$
TRIGERMANE, $Ge_3H_8$
POTASSIUM HYDRIDE, HK
LIT'IIUM HYDRIDE, HLi
SOLIUM HYDRIDE, HNa
RUBIDIUM HYDRIDE, HRb
MAGNESIUM HYDRIDE, $H_2Mg$
THORIUM DIHYDRIDE, $H_2Th$
TITANIUM DIHYDRIDE, $H_2Ti$
ZINC HYDRIDE, $H_2Zn$
'ZIRCONIUM HYDRIDE', $H_2Zr$
LANTHANUM TRIHYDRIDE, $H_3La$
PLUTONIUM(III) HYDRIDE, $H_3Pu$
URANIUM(III) HYDRIDE, $H_3U$
THORIUM HYDRIDE, $H_4Th$
URANIUM(IV) HYDRIDE, $H_4U$

*See other* PYROPHORIC MATERIALS
*See also* COMPLEX HYDRIDES

## METAL HYPOCHLORITES $M(OCl)_n$

A widely used group of industrial oxidants which has been involved in numerous incidents, some with nitrogenous materials leading to formation of nitrogen trichloride.

*See* BLEACHING POWDER
CALCIUM HYPOCHLORITE, $CaCl_2O_2$
SODIUM HYPOCHLORITE, ClNaO
*See also* CHLORINE, $Cl_2$ : Nitrogen compounds

Amines

Kirk-Othmer, 1963, Vol. 2, 104

Primary or secondary amines react with sodium or calcium hypochlorites to give *N*-chloroamines, some of which are explosive when isolated. Application of other chlorinating agents to amines or their precursors may also produce the same result under appropriate conditions.

*See related* HYPOHALITES
*See other* METAL OXOHALOGENATES

## METAL NITRATES $M(NO_3)_n$

Aluminium

*See* ALUMINIUM, Al: Metal nitrates, etc.

Citric acid

Shannon, I. R., *Chem. & Ind.*, 1970, 149

During vacuum evaporation of an aqueous mixture of unspecified mixed metal nitrates and citric acid, the amorphous solid exploded when nearly dry. This was attributed to oxidation of the organic residue by the nitrates present, possibly catalysed by one of the oxides expected to be produced.

Esters,
or Phosphorus,
or Tin(II) chloride

Pieters, 1957, 30

Mixtures of metal nitrates with alkyl esters may explode, owing to formation of alkyl nitrates. Mixtures of a nitrate with phosphorus, tin(II) chloride or other reducing agents may react explosively.

Metal phosphinates

1. Mellor, 1940, Vol. 8, 881
2. Costa, R.L., *Chem. Eng. News,* 1947, **25**, 3177

Mixtures of metal nitrates and phosphinates, previously proposed as explosives [1], explode on heating.

Organic matter

Bowen, H. J. M., *Anal. Chem.*, 1968, **40**, 969; private comm.
Grewelling, T., *Anal. Chem.*, 1969, **41**, 540–541

When organic matter is destroyed for residue analysis by heating with equimolar potassium–sodium nitrate mixture to 390° C, a twentyfold excess of nitrate must be used. If over 10% of organic matter is present, pyrotechnic reactions occur which could be explosive.

Subsequent to an explosion during ashing of a sodium citrate–nitric acid residue (i.e. heating citric acid with sodium nitrate) at below 500°C, experiments on the effect of heating various organic materials with metal nitrates showed the tendency for explosion to increase from magnesium through calcium to sodium nitrate. This is in order of m.p. of the nitrates, and explosion may occur when the nitrates melt and make intimate contact with the organic matter. Pretreatment with nitric acid may reduce the explosion risk.

*See* CITRIC ACID, above

Potassium hexanitrocobaltate (3–)

*See* POTASSIUM HEXANITROCOBALTATE (3–), $CoK_3N_6O_{12}$

Individually indexed metal nitrates are:

SILVER NITRATE, $AgNO_3$
BARIUM NITRATE, $BaN_2O_6$
CALCIUM NITRATE, $CaN_2O_6$
COBALT(II) NITRATE, $CoN_2O_6$
COPPER(II) NITRATE, $CuN_2O_6$

IRON(III) NITRATE, $FeN_3O_9$
MERCURY(II) NITRATE, $HgN_2O_6$
MERCURY(I) NITRATE, $Hg_2N_2O_6$
POTASSIUM NITRATE, $KNO_3$
LITHIUM NITRATE, $LiNO_3$
MAGNESIUM NITRATE, $MgN_2O_6$
MANGANESE(II) NITRATE, $MnN_2O_6$
SODIUM NITRATE, $NNaO_3$
LEAD(II) NITRATE, $N_2O_6Pb$
TIN(II) NITRATE, $N_2O_6Sn$
ZINC NITRATE, $N_2O_6Zn$
* TIN(II) NITRATE OXIDE, $N_2O_7Sn_2$
* URANYL NITRATE, $N_2O_8U$
* VANADIUM TRINITRATE OXIDE, $N_3O_{10}V$
PLUTONIUM(IV) NITRATE, $N_4O_8Pu$

*See* MOLTEN SALT BATHS
*See other* METAL OXONON-METALLATES

**METAL NITRITES** $M(NO_2)_n$

Metal cyanides

*See* SODIUM NITRITE, $NNaO_2$ : Metal cyanides

Nitrogenous bases

Metal nitrites react with salts of nitrogenous bases to give the corresponding nitrite salts, many of which are unstable.

*See* NITRITE SALTS OF NITROGENOUS BASES

Potassium hexanitrocobaltate (3–)

*See* POTASSIUM HEXANITROCOBALTATE (3–), $CoK_3N_6O_{12}$
*See* MOLTEN SALT BATHS
*See other* METAL OXONON-METALLATES

**METAL NON-METALLIDES** M–E

This class includes the products of combination of metals and non-metals except C (as acetylene), H, N, O and S, which are separately treated in the groups:

METAL ACETYLIDES
*N*-METAL DERIVATIVES
METAL HYDRIDES
METAL OXIDES
METAL SULPHIDES

Individually indexed compounds are:

ALUMINIUM PHOSPHIDE, AlP
DIALUMINIUM OCTAVANADIUM TRIDECASILICIDE, $Al_2Si_{13}V_8$
PLATINUM DIARSENIDE, $As_2Pt$

MAGNESIUM BORIDE, $B_2Mg_3$
TRIBARIUM TETRANITRIDE, $Ba_3N_4$
IRON CARBIDE, $CFe_3$
TITANIUM CARBIDE, CTi
TUNGSTEN CARBIDE, CW
DITUNGSTEN CARBIDE, $CW_2$
THORIUM DICARBIDE, $C_2Th$
URANIUM DICARBIDE, $C_2U$
ZIRCONIUM DICARBIDE, $C_2Zr$
TETRAALUMINIUM TRICARBIDE, $C_3Al_4$
POTASSIUM GRAPHITE, $C_8K$
RUBIDIUM GRAPHITE, $C_8Rb$
CALCIUM SILICIDE, CaSi
CALCIUM DISILICIDE, $CaSi_2$
TRICALCIUM DIPHOSPHIDE, $Ca_3P_2$
CADMIUM SELENIDE, CdSe
TRICADMIUM DIPHOSPHIDE, $Cd_3P_2$
COPPER MONOPHOSPHIDE, CuP
COPPER DIPHOSPHIDE, $Cu_2P$
TRICOPPER DIPHOSPHIDE, $Cu_3P_2$
IRON–SILICON, Fe–Si
TRIS(IODOMERCURI)PHOSPHINE, $Hg_3I_3P$
TRIMERCURY TETRAPHOSPHIDE, $Hg_3P_4$
POTASSIUM SILICIDE, KSi
HEXALITHIUM DISILICIDE, $Li_6Si_2$
MAGNESIUM SILICIDE, $Mg_2Si$
TRIMAGNESIUM DIPHOSPHIDE, $Mg_3P_2$
MANGANESE(II) TELLURIDE, MnTe
TRIMANGANESE DIPHOSPHIDE, $Mn_3P_2$
SODIUM TETRASULPHURPENTANITRIDATE, $N_5NaS_4$
SODIUM SILICIDE, NaSi
SODIUM PHOSPHIDE (3–), $Na_3P$
TRIZINC DIPHOSPHIDE, $P_2Zn_3$
LEAD PENTAPHOSPHIDE, $P_5Pb$

## METAL OXALATES $(CO_2M)_2$

Mellor, 1941, Vol. 1, 706
The tendency for explosive decomposition of heavy metal oxalates is related to the value of the heat of decomposition.

*See* SILVER OXALATE, $C_2Ag_2O_4$
COPPER(I) OXALATE, $C_2Cu_2O_4$
MERCURY(II) OXALATE, $C_2HgO_4$
IRON(III) OXALATE, $C_6Fe_2O_{12}$
*See other* HEAVY METAL DERIVATIVES

## METAL OXIDES $MO_n$

Malinin, G. V. *et al., Russ. Chem. Rev.*, 1975, **44**, 392–397
Thermal decomposition of metal oxides is reviewed. Some oxides (tricobalt tetraoxide, copper(II) oxide, trilead tetraoxide, uranium

dioxide, triuranium octaoxide) liberate quite a high proportion of atomic oxygen, with a correspondingly higher potential for oxidation of fuels than molecular oxygen.

This large class covers a wide range of types of reactivity and there is a separate class entry for

METAL PEROXIDES

Individually indexed compounds are:

SILVER(II) OXIDE, AgO
SILVER(I) OXIDE, $Ag_2O$
DISILVER PENTATIN UNDECAOXIDE, $Ag_2Sn_5O_{11}$
DIALUMINIUM TRIOXIDE, $Al_2O_3$
GOLD(III) OXIDE, $Au_2O_3$
BARIUM OXIDE, BaO
BERYLLIUM OXIDE, BeO
DIBISMUTH TRIOXIDE, $Bi_2O_3$
CALCIUM OXIDE, CaO
CADMIUM OXIDE, CdO
* MANGANESE CHLORIDE TRIOXIDE, $ClMnO_3$
COBALT(II) OXIDE, CoO
DICOBALT TRIOXIDE, $Co_2O_3$
CHROMIUM(II) OXIDE, CrO
CHROMIUM TRIOXIDE, $CrO_3$
DICHROMIUM TRIOXIDE, $Cr_2O_3$
CAESIUM TRIOXIDE ('OZONATE'), $CsO_3$
DICAESIUM OXIDE, $Cs_2O$
COPPER(II) OXIDE, CuO
COPPER(I) OXIDE, $Cu_2O$
* MANGANESE FLUORIDE TRIOXIDE, $FMnO_3$
IRON(II) OXIDE, FeO
DIIRON TRIOXIDE, $Fe_2O_3$
TRIIRON TETRAOXIDE, $Fe_3O_4$
DIGALLIUM OXIDE, $Ga_2O$
MERCURY(II) OXIDE, HgO
'MERCURY(I) OXIDE', $Hg_2O$
INDIUM(II) OXIDE, InO
IRIDIUM(IV) OXIDE, $IrO_2$
POTASSIUM DIOXIDE (SUPEROXIDE), $KO_2$
* POTASSIUM TRIOXIDE (POTASSIUM OZONATE), $KO_3$
DILANTHANUM TRIOXIDE, $La_2O_3$
MAGNESIUM OXIDE, MgO
MANGANESE(II) OXIDE, MnO
MANGANESE(IV) OXIDE, $MnO_2$
MANGANESE(VII) OXIDE, $Mn_2O_7$
MOLYBDENUM(IV) OXIDE, $MoO_2$
MOLYBDENUM(VI) OXIDE, $MoO_3$
SODIUM OXIDE, $Na_2O$
NIOBIUM(V) OXIDE, $Nb_2O_5$
NICKEL(II) OXIDE, NiO
NICKEL(IV) OXIDE, $NiO_2$
DINICKEL TRIOXIDE, $Ni_2O_3$
LEAD(II) OXIDE, OPb
PALLADIUM(II) OXIDE, OPd
TIN(II) OXIDE, OSn

ZINC OXIDE, $OZn$
OSMIUM(IV) OXIDE, $O_2Os$
LEAD(IV) OXIDE, $O_2Pb$
PALLADIUM OXIDE, $O_2Pd$
PLATINUM(IV) OXIDE, $O_2Pt$
TIN(IV) OXIDE, $O_2Sn$
TITANIUM(IV) OXIDE, $O_2Ti$
URANIUM(IV) OXIDE, $O_2U$
TUNGSTEN(IV) OXIDE, $O_2W$
ZINC PEROXIDE, $O_2Zn$
DIPALLADIUM TRIOXIDE, $O_3Pd_2$
ANTIMONY(III) OXIDE, $O_3Sb_2$
THALLIUM(III) OXIDE, $O_3Tl_2$
VANADIUM(III) OXIDE, $O_3V_2$
TUNGSTEN(VI) OXIDE, $O_3W$
OSMIUM(VIII) OXIDE, $O_4Os$
TRILEAD TETRAOXIDE, $O_4Pb_3$
RUTHENIUM(VIII) OXIDE, $O_4Ru$
TANTALUM(V) OXIDE, $O_5Ta_2$
VANADIUM(V) OXIDE, $O_5V_2$
TRIURANIUM OCTAOXIDE, $O_8U_3$

## METAL OXOHALOGENATES

**$MXO_n$**

Solymosi, F., *Acta Phys. Chem.*, 1976, 22, 75–115

This group covers the four levels of oxidation represented in the series hypochlorite, chlorite, chlorate and perchlorate, and, as expected, the oxidising power of the anion is roughly proportional to the oxygen content.

The group has been subdivided under the headings:

METAL HYPOCHLORITES
CHLORITE SALTS
METAL HALOGENATES
METAL PERCHLORATES

for which separate entries exist. Individually indexed compounds for all four sub-groups are:

SILVER BROMATE, $AgBrO_3$
SILVER CHLORITE, $AgClO_2$
SILVER PERCHLORATE, $AgClO_4$
SILVER IODATE, $AgIO_3$
ALUMINIUM CHLORATE, $AlCl_3O_9$
BARIUM BROMATE, $BaBr_2O_6$
BARIUM CHLORITE, $BaCl_2O_4$
BARIUM PERCHLORATE, $BaCl_2O_8$
BERYLLIUM PERCHLORATE, $BeCl_2O_8$
POTASSIUM BROMATE, $BrKO_3$
MERCURY(I) BROMATE, $Br_2Hg_2O_6$
LEAD BROMATE, $Br_2O_6Pb$
ZINC BROMATE, $Br_2O_6Zn$
* METHYLMERCURY PERCHLORATE, $CH_3ClHgO_4$
* LEAD ACETATE–LEAD BROMATE, $C_4H_6O_4Pb \cdot Br_2O_6Pb$

CALCIUM HYPOCHLORITE, $CaCl_2O_2$
CALCIUM CHLORITE, $CaCl_2O_4$
CADMIUM CHLORATE, $CdCl_2O_6$
POTASSIUM CHLORATE, $ClKO_3$
POTASSIUM PERCHLORATE, $ClKO_4$
LITHIUM PERCHLORATE, $ClLiO_4$
SODIUM HYPOCHLORITE, ClNaO
SODIUM CHLORITE, $ClNaO_2$
SODIUM CHLORATE, $ClNaO_3$
SODIUM PERCHLORATE, $ClNaO_4$
THALLIUM(I) CHLORITE, $ClO_2Tl$
MERCURY(II) CHLORITE, $Cl_2HgO_4$
MERCURY(I) CHLORITE, $Cl_2Hg_2O_4$
MAGNESIUM HYPOCHLORITE, $Cl_2MgO_2$
MAGNESIUM CHLORATE, $Cl_2MgO_6$
MAGNESIUM PERCHLORATE, $Cl_2MgO_8$
MANGANESE DIPERCHLORATE, $Cl_2MnO_8$
NICKEL DIPERCHLORATE, $Cl_2NiO_8$
LEAD DICHLORITE, $Cl_2O_4Pb$
ZINC DICHLORATE, $Cl_2O_6Zn$
LEAD DIPERCHLORATE, $Cl_2O_8Pb$
URANYL DIPERCHLORATE, $Cl_2O_{10}U$
GALLIUM TRIPERCHLORATE, $Cl_3GaO_{12}$
VANADYL TRIPERCHLORATE, $Cl_3O_{13}V$
TITANIUM TETRAPERCHLORATE, $Cl_4O_{16}Ti$
* TETRAZIRCONIUM TETRAOXIDE HYDROGEN NONAPERCHLORATE, $Cl_9HO_{40}Zr_4$
* AMMONIUM IODATE, $H_4INO_3$
* AMMONIUM PERIODATE, $H_4INO_4$
POTASSIUM IODATE, $IKO_3$
POTASSIUM PERIODATE, $IKO_4$
SODIUM IODATE, $INaO_3$

## METAL OXOMETALLATES $M^+ MO_n^-$

Salts with oxometallate anions function as oxidants, those with oxygen present as peroxogroups being naturally the more powerful and separately grouped under

PEROXOACID SALTS

Individually indexed oxometallate salts are:

DIBISMUTH DICHROMIUM NONAOXIDE ('BISMUTH CHROMATE'), $Bi_2Cr_2O_9$
CALCIUM PERMANGANATE, $CaMn_2O_8$
COPPER CHROMATE, $CrCuO_4$
LEAD CHROMATE, $CrO_4Pb$
POTASSIUM DICHROMATE, $Cr_2K_2O_7$
SODIUM DICHROMATE, $Cr_2Na_2O_7$
* AMMONIUM PERMANGANATE, $H_4MnNO_4$
POTASSIUM PERMANGANATE, $KMnO_4$
* POTASSIUM NITRIDOOSMATE(1−), $KNO_3Os$
SODIUM PERMANGANATE, $MnNaO_4$
ZINC PERMANGANATE, $Mn_2O_8Zn$

## METAL OXONON-METALLATES

$M^+ EO_n^-$

This large and commonly used class of salts covers a wide range of oxidising potential. Among the most powerful oxidants are

METAL OXOHALOGENATES and
PEROXOACID SALTS

which are dealt with separately. There is a rough gradation down the sub-groups

METAL NITRATES
METAL NITRITES
METAL SULPHATES
METAL AMIDOSULPHATES

for which separate entries emphasise the individual features. Less highly oxidised anions function as reducants. Individually indexed entries are:

SILVER NITRATE, $AgNO_3$
SILVER HYPONITRITE, $Ag_2N_2O_2$
* HEPTASILVER NITRATE OCTAOXIDE, $Ag_7NO_{11}$
BARIUM NITRATE, $BaN_2O_6$
BARIUM SULPHATE, $BaO_4S$
CALCIUM CARBONATE, $CCaO_3$
POTASSIUM CARBONATE, $CK_2O_3$
LITHIUM CARBONATE, $CLi_2O_3$
MAGNESIUM CARBONATE, $CMgO_3$
SODIUM CARBONATE, $CNa_2O_3$
LEAD CARBONATE, $CO_3Pb$
SODIUM ACETATE, $C_2H_3NaO_2$
CALCIUM SULPHATE, $CaO_4S$
* CHROMYL FLUOROSULPHATE, $CrF_2O_8S_2$
CHROMIUM(II) SULPHATE, $CrO_4S$
COPPER(II) PHOSPHINATE, $CuH_4O_4P_2$
COPPER(II) NITRATE, $CuN_2O_6$
COPPER(II) SULPHATE, $CuO_4S$
IRON(II) SULPHATE, $FeO_4S$
* SODIUM HYDROGENSULPHATE, $HNaO_4S$
POTASSIUM AMIDOSULPHATE, $H_2KNO_3S$
POTASSIUM PHOSPHINATE ('HYPOPHOSPHITE'), $H_2KO_2P$
SODIUM AMIDOSULPHATE, $H_2NNaO_3S$
SODIUM PHOSPHINATE ('HYPOPHOSPHITE'), $H_2NaO_2P$
* AMMONIUM NITRITE, $H_4N_2O_2$
* AMMONIUM NITRATE, $H_4N_2O_3$
* HYDROXYLAMMONIUM PHOSPHINATE, $H_6NO_3P$
* AMMONIUM AMIDOSULPHATE ('SULPHAMATE'), $H_6N_2O_3S$
* AMMONIUM SULPHATE, $H_8N_2O_4S$
* HYDROXYLAMMONIUM SULPHATE, $H_8N_2O_6S$
MERCURY(I) NITRATE, $HgNO_3$
MERCURY(II) NITRATE, $HgN_2O_6$
POTASSIUM NITRITE, $KNO_2$
POTASSIUM NITROSODISULPHATE (2—), $K_2NO_7S_2$
POTASSIUM DINITROSOSULPHITE, $K_2N_2O_5S$
LITHIUM SODIUM NITROXYLATE, $LiNNaO_2$
LITHIUM NITRATE, $LiNO_3$
MAGNESIUM NITRATE, $MgN_2O_6$

MAGNESIUM SULPHATE, $MgO_4S$
SODIUM NITRITE, $NNaO_2$
SODIUM NITRATE, $NNaO_3$
SODIUM HYPONITRITE, $N_2Na_2O_2$
SODIUM TRIOXODINITRATE (2–), $N_2Na_2O_3$
SODIUM TETRAOXODINITRATE (2–), $N_2Na_2O_4$
SODIUM PENTAOXODINITRATE (2–), $N_2Na_2O_5$
SODIUM HEXAOXODINITRATE (2–), $N_2Na_2O_6$
LEAD HYPONITRITE, $N_2O_2Pb$
LEAD(II) NITRATE, $N_2O_6Pb$
ZINC NITRATE, $N_2O_6Zn$
* TIN(II) NITRATE OXIDE, $N_2O_7Sn_2$
* URANYL NITRATE, $N_2O_8U$
* VANADIUM TRINITRATE OXIDE, $N_3O_{10}V$
PLUTONIUM(IV) NITRATE, $N_4O_8Pu$
SODIUM THIOSULPHATE, $Na_2O_3S_2$
SODIUM METASILICATE, $Na_2O_3Si$
SODIUM SULPHATE, $Na_2O_4S$
SODIUM DITHIONITE ('HYDROSULPHITE'), $Na_2O_4S_2$
SODIUM DISULPHITE (METABISULPHITE), $Na_2O_5S_2$
LEAD SULPHATE, $O_4PbS$

*See also* DIMETHYL SULPHOXIDE, $C_2H_6OS$: Metal oxosalts

## METAL PERCHLORATES $M(ClO_4)_n$

Burton, M., *Chem. Eng. News*, 1970, **48**(51), 55

Though metal periodates and perbromates are known, the perchlorates have most frequently been involved in hazardous incidents over a long period. These highly stable salts are powerful oxidants and contact with combustible materials or reducants must be under controlled conditions. A severe restriction on the use of metal perchlorates in laboratory work has been recommended.

*See* SILVER PERCHLORATE, $AgClO_4$
BARIUM PERCHLORATE, $BaCl_2O_8$
DIMETHYL SULPHOXIDE, $C_2H_6OS$: Magnesium perchlorate
: Metal oxosalts
* DIETHYLTHALLIUM PERCHLORATE, $C_4H_{10}ClO_4Tl$
2,2-DIMETHOXYPROPANE, $C_5H_{12}O_2$: Metal perchlorates
AMMONIUM PERCHLORATE, $ClH_4NO_4$
INDIUM(I) PERCHLORATE, $ClInO_4$
POTASSIUM PERCHLORATE, $ClKO_4$
LITHIUM PERCHLORATE, $ClLiO_4$
SODIUM PERCHLORATE, $ClNaO_4$
IRON(II) PERCHLORATE, $Cl_2FeO_8$
MAGNESIUM PERCHLORATE, $Cl_2MgO_8$
TIN(II) PERCHLORATE, $Cl_2O_8Sn$
DIDYMIUM PERCHLORATE, $Cl_3DyO_{12}$
ERBIUM PERCHLORATE, $Cl_3ErO_{12}$
IRON(III) PERCHLORATE, $Cl_3FeO_{12}$
HYDRAZINIUM SALTS

Calcium hydride

Mellor, 1941, Vol. 3, 651

Rubbing a mixture of calcium (or strontium) hydride with a metal perchlorate in a mortar causes a violent explosion.

Sulphuric acid
Pieters, 1957, 30
Schumacher, 1960, 190
Metal perchlorates with highly concentrated or anhydrous acid form the explosively unstable anhydrous perchloric acid.
*See* PERCHLORIC ACID, $ClHO_4$ : Dehydrating agents
*See other* METAL OXOHALOGENATES

**METAL PEROXIDES** $M(O_2)_n$

Castrantas, 1965, 1,4
Bailar, 1973, Vol. 2, 784
This group contains many powerful oxidants, the most common being sodium peroxide. Undoubtedly the most hazardous is potassium superoxide, readily formed on the metal exposed to air (but as the monovalent $O_2^-$ ion it is not a true peroxide). Many transition metal peroxides are dangerously explosive.
*See* BARIUM PEROXIDE, $BaO_2$
* DIMETHYLETHEROXODIPEROXOCHROMIUM(VI), $C_2H_6CrO_6$
CALCIUM PEROXIDE, $CaO_2$
MERCURY PEROXIDE, $HgO_2$
* POTASSIUM DIOXIDE (SUPEROXIDE), $KO_2$
POTASSIUM PEROXIDE, $K_2O_2$
* OXODIPEROXOMOLYBDENUM–HEXAMETHYLPHOSPHORAMIDE, $MoO_5 \cdot C_6H_{18}N_3OP$
SODIUM PEROXIDE, $Na_2O_2$
STRONTIUM PEROXIDE, $O_2Sr$
ZINC PEROXIDE, $O_2Zn$

**METAL PEROXOMOLYBDATES** $M_2^+[Mo(O_2)_4]^{2-}$

Sidgwick, 1950, 1045
Many of the red metal peroxomolybdates are explosive.
*See* POTASSIUM TETRAPEROXOMOLYBDATE (2–), $K_2MoO_8$
SODIUM TETRAPEROXOMOLYBDATE (2–), $MoNa_2O_8$

**METAL POLYHALOHALOGENATES** $M^+[XX'_4]^-$

Organic solvents,
or Water
1. Whitney, E. D. *et al., J. Amer. Chem. Soc.*, 1964, **86**, 2583
2. Sharpe, A. G. *et al., J. Chem. Soc.*, 1948, 2135
Potassium, rubidium and caesium tetrafluorochlorates and hexafluorobromates react violently with water, and explosively with common

organic solvents, analogously to the parent halogen fluorides [1]. Silver and barium tetrafluorobromates ignite in contact with ether, acetone, dioxan and petrol [2].

## METAL PYRUVATE NITROPHENYLHYDRAZONES

$MO_2CC(Me){=}NNHC_6H_4NO_2$

Ragno, M., *Gazz. Chim. Ital.*, 1945, **75**, 186–192

A wide range of the title salts of mono-, di- and trivalent metals, with an *o*-, *m*- or *p*-nitro group present showed unstable or explosive behaviour on heating. The lead salt exploded violently at 240°C, while the aluminium, beryllium and silver salts are only feebly explosive.

## METALS

Individually indexed metals are:

SILVER, Ag
ALUMINIUM, Al
GOLD, Au
BARIUM, Ba
BERYLLIUM, Be
BISMUTH, Bi
CALCIUM, Ca
CADMIUM, Cd
CERIUM, Ce
COBALT, Co
CHROMIUM, Cr
CAESIUM, Cs
COPPER, Cu
EUROPIUM, Eu
IRON, Fe
GALLIUM, Ga
GERMANIUM, Ge
HAFNIUM, Hf
MERCURY, Hg
INDIUM, In
IRIDIUM, Ir
POTASSIUM, K
LANTHANUM, La
LITHIUM, Li
MAGNESIUM, Mg
MANGANESE, Mn
MOLYBDENUM, Mo
SODIUM, Na
NIOBIUM, Nb
NICKEL, Ni
OSMIUM, Os
LEAD, Pb
PALLADIUM, Pd
PLATINUM, Pt
PLUTONIUM, Pu
RUBIDIUM, Rb
RHENIUM, Re
RHODIUM, Rh
RUTHENIUM, Ru
ANTIMONY, Sb
TIN, Sn
STRONTIUM, Sr
TANTALUM, Ta
TECHNETIUM, Tc
THORIUM, Th
TITANIUM, Ti
THALLIUM, Tl
URANIUM, U
VANADIUM, V
TUNGSTEN, W
ZINC, Zn
ZIRCONIUM, Zr

Halocarbons

Lenze, F. *et al.*, *Z. Ges. Schiess- u. Sprengstoffw.*, 1932, **27**, 255, 293, 337, 373; *Chem. Ztg.*, 1932, **56**, 921–923

Various combinations of alkali and alkaline earth-metals with halocarbons were found to be highly heat- or impact-sensitive explosives.

*See* ALUMINIUM, Al: Halocarbons

BARIUM, Ba: Halocarbons
BERYLLIUM, Be: Halocarbons
POTASSIUM, K: Halocarbons
POTASSIUM–SODIUM ALLOY, K–Na: Halocarbons
LITHIUM, Li: Halocarbons
MAGNESIUM, Mg: Halocarbons
SODIUM, Na: Halocarbons
PLUTONIUM, Pu: Carbon tetrachloride
SAMARIUM, Sm: 1,1,2-Trichlorotrifluoroethane
TITANIUM, Ti: Halocarbons
ZINC, Zn: Halocarbons
ZIRCONIUM, Zr: Carbon tetrachloride

## METAL SALICYLATES

**formula on p. 190**

Nitric acid

*See* NITRIC ACID, $HNO_3$: Metal salicylates

## METAL SALTS

By far the largest class of compound in this Handbook, the metal (and ammonium) salts have been allocated into two sub-classes dependent on the presence or absence of oxygen in the anion.

The main groupings adopted for the non-oxygenated salts are:

METAL ACETYLIDES
METAL AZIDES
METAL CYANIDES AND CYANOCOMPLEXES
*N*-METAL DERIVATIVES
METAL FULMINATES
METAL HALIDES
METAL POLYHALOHALOGENATES
METAL PYRUVATE NITROPHENYLHYDRAZONES

and for the oxosalts:

METAL OXALATES
METAL OXHALOGENATES (anion an oxoderivative of a halogen)
METAL OXOMETALLATES (anion an oxoderivative of a metal)
METAL OXONON-METALLATES (anion an oxoderivative of a non-metal)
PEROXOACID SALTS (anion a peroxoderivative of a metal or non-metal)

There is a separate entry for

OXOSALTS OF NITROGENOUS BASES

In some cases it has been convenient to subdivide the latter groups into smaller sub-groups and such sub-division is indicated under the appropriate group heading.

## METAL SULPHATES

**$M(SO_4)_n$**

Aluminium

*See* ALUMINIUM, Al: Metal oxides, etc.

Magnesium
*See* MAGNESIUM, Mg: Metal oxosalts

## METAL SULPHIDES $M_mS_n$

Some metal sulphides are so readily oxidised as to be pyrophoric in air. Individually indexed compounds are:

SILVER SULPHIDE, $Ag_2S$
DIGOLD TRISULPHIDE, $Au_2S_3$
BARIUM SULPHIDE, BaS
DIBISMUTH TRISULPHIDE, $Bi_2S_3$
CALCIUM SULPHIDE, CaS
CALCIUM POLYSULPHIDE, $CaS_x$
DICERIUM TRISULPHIDE, $Ce_2S_3$
COBALT(II) SULPHIDE, CoS
CHROMIUM(II) SULPHIDE, CrS
* DICAESIUM SELENIDE, $Cs_2Se$
COPPER IRON(II) SULPHIDE, $CuFeS_2$
COPPER(II) SULPHIDE, CuS
EUROPIUM(II) SULPHIDE, EuS
IRON(II) SULPHIDE, FeS
IRON DISULPHIDE, $FeS_2$
DIIRON TRISULPHIDE, $Fe_2S_3$
GERMANIUM(II) SULPHIDE, GeS
* AMMONIUM SULPHIDE, $H_8N_2S$
MERCURY(II) SULPHIDE, HgS
POTASSIUM SULPHIDE, $K_2S$
MANGANESE(II) SULPHIDE, MnS
MANGANESE(IV) SULPHIDE, $MnS_2$
MOLYBDENUM(IV) SULPHIDE, $MoS_2$
SODIUM SULPHIDE, $Na_2S$
SODIUM DISULPHIDE, $Na_2S_2$
SODIUM POLYSULPHIDE, $Na_2S_x$
* THORIUM OXIDE SULPHIDE, OSTh
* ZIRCONIUM OXIDE SULPHIDE, OSZr
RHENIUM(VII) SULPHIDE, $Re_2S_7$
SAMARIUM SULPHIDE, SSm
TIN(II) SULPHIDE, SSn
TIN(IV) SULPHIDE, $S_2Sn$
TITANIUM(IV) SULPHIDE, $S_2Ti$
URANIUM(IV) SULPHIDE, $S_2U$
DIANTIMONY TRISULPHIDE, $S_3Sb_2$

## METAL THIOCYANATES $M(SC{\equiv}N)_n$

Oxidants

1. von Schwartz, 1918, 299–300
2. *MCA Case History No. 853*

Metal thiocyanates are oxidised explosively by chlorates or nitrates when fused, or if intimately mixed, at 400°C or on spark or flame ignition [1]. Nitric acid violently oxidised an aqueous thiocyanate solution [2].

*See* NITRIC ACID, $HNO_3$: Metal thiocyanate

**MILD STEEL**

*MCA Case History No. 947*

A small mild steel cylinder suitable for high pressure was two-thirds filled with liquid ammonia by connecting it to a large ammonia cylinder and cooling it to −70°C by immersion in dry ice and acetone. Some hours after filling, the small cylinder burst, splitting cleanly along its length. This was attributed to cryogenic embrittlement and weakening of the mild steel cylinder.

**MILK POWDER**

Buma, T. J. *et al., Chem. Abs.*, 1977, **87**, 83300

Lumps of powdered milk formed near the hot-air inlet of spray driers may spontaneously ignite. The thermochemical mechanism has been investigated.

*See other* IGNITION SOURCES

**MOLECULAR SIEVE**

Ethylene

Doyle, W. H., *Loss Prevention*, 1969, **3**, 15

5A molecular sieve, not previously soaked in dilute ethylene, was used to dry compressed ethylene gas in a flow system. An exothermic reaction attained red heat and caused explosive failure of the dryer. The smaller-pored 3A sieve is not catalytically active towards ethylene.

Triaryl phosphates

Schmitt, C. R., *J. Fire Flamm.*, 1973, (4), 113–131

A molecular sieve bed was used to continuously purify the fire-resistant hydraulic fluid (a triaryl phosphate) in a large hydraulic press. Periodically the sieve bed was regenerated by treatment with steam, and then purified air at 205°C. After 9 years of uneventful operation, the bed ignited on admission of hot air. The fuel source was traced to the accumulation of organic residues (phenols or cresols?) on the sieve and which were not removed by the steam/air treatment.

**MOLTEN METALS**

Water

The mechanism of the violent explosions which occur when molten metals other than water-reactive metals contact liquid water have been investigated.

*See* ALUMINIUM, Al: Water
COPPER, Cu: Water
TIN, Sn: Water

**MOLTEN SALT BATHS**


1. 'Precautions in the Use of Nitrate Salt Baths', Min. of Labour, SHW booklet, London, HMSO, 1964
2. Pieters, 1957, 30
3. *Potential Hazards in the Use of Salt Baths for Heat Treatment of Metals,* NBFU Res. Rep. No. 2, New York, 1946
4. Beck, W., *Aluminium*, 1935, **17**, 3–6


The booklet covers hazards attendant upon the use of molten nitrate salt baths for heat treatment of metals, including storage and disposal of salts, starting up, electrical heating, and emptying of salt baths. Readily oxidisable materials must be rigorously excluded from the vicinity of nitrate baths [1].

Earlier it had been reported that aluminium and its alloys if contaminated with organic matter may explode in nitrate–nitrite fused salt heating baths [2].

Uses, composition and precautions in the use of molten salt baths are discussed. Most common causes of accidents are: steam explosions, trapping of air, explosive reactions with metals (magnesium) and organic matter or cyanides from other heat-treatment processes [3].

Explosions involving use of aluminium in nitrate baths have also been attributed to accelerated corrosive failure of the iron container, rather than to direct interaction with aluminium [4].

**MOLTEN SALTS**

Zhuchkov, P.A. *et al.*, *Chem. Abs.*, 1974, **80**, 28651

The causes of furnace blasts occurring in soda-regeneration plants when water is spilt into molten sodium carbonate, chloride, hydroxide, sulphate or sulphide are discussed, together with methods of prevention.

*See also* SMELT: Water

**NITRATING AGENTS**

Dubar, J. *et al., Compt. Rend., Ser. C,* 1968, **266**, 1114

The potentially explosive character of various nitration mixtures (2-cyano-2-propyl nitrate in acetonitrile; solutions of dinitrogen tetraoxide in esters, ethers or hydrocarbons; dinitrogen pentaoxide in methylene chloride; nitronium tetrafluoroborate in sulpholane) are mentioned.

## NITRATION

1. Biasutti, G. S. *et al.*, *Loss Prevention*, 1974, **8**, 123–125
2. Albright, Hanson, 1976
3. Obnovlenskii, P. A. *et al.*, *Chem. Abs.*, 1975, **83**, 168135
4. Raczynski, S., *Chem. Abs.*, 1963, **59**, 15114e
5. Rüst, 1948, 317–319

Accident statistics reveal nitration as the most widespread and powerfully destructive industrial unit process operation. This is because nitric acid can, under certain conditions, effect complete and highly exothermal conversion of organic molecules to gases, the reactions being capable of acceleration to deflagration or detonation. Case histories are described and safety aspects of continuous nitration processes are discussed in detail [1]. Of the 25 chapters of the book [2], each a paper presented at the symposium on 'Advances in Industrial and Laboratory Nitrations' at Philadelphia in April 1975, three deal with safety aspects of nitration:

Ch. 8, Hanson, C. *et al.*, *Side Reactions during Aromatic Nitration*

Ch. 22, Biasutti, G. S., *Safe Manufacture and Handling of Liquid Nitric Esters*

Ch. 23, Brunberg, B., *Safe Manufacture and Handling of Liquid Nitric Esters: the Injector Nitration Process*

Technological difficulties associated with methods of reducing explosion hazards in aromatic nitration have been discussed [4], and several case histories of violent incidents during nitration operations have been collected [5].

## NITRATION INCIDENTS

*See*

2-METHYL-5-NITROIMIDAZOLE, $C_4H_5N_3O_2$
ACETIC ANHYDRIDE, $C_4H_6O_3$ : Metal nitrates
2-FORMAMIDO-1-PHENYL-1,3-PROPANEDIOL, $C_{10}H_{13}NO_3$
NITRIC ACID, $HNO_3$ : Alcohols
: Chlorobenzene
: 2-Methylbenzimidazole, etc.
: Methylthiophene
: Nitrobenzene, Sulphuric acid

*See also* NITRATING AGENTS

## NITRITE SALTS OF NITROGENOUS BASES $N^+ NO_2^-$

1. Mellor, 1940, Vol. 8, 289, 470–472
2. Ray, P. C. *et al.*, *J. Chem. Soc.*, 1911, **99**, 1470; *J. Chem. Soc.*, 1912 **101**, 141, 216

Ammonium and substituted-ammonium salts exhibit a range of

instability, and reaction mixtures which may be expected to yield these products should be handled with care. Ammonium nitrite will decompose explosively either as the solid, or in concentrated aqueous solution when heated to 60—70°C. Presence of traces of acid lowers the decomposition temperature markedly. Hydroxylammonium nitrite appears to be so unstable that it decomposes immediately in solution. Hydrazinium(1+) nitrite is a solid which explodes violently on percussion, or less vigorously if heated rapidly, and hydrogen azide may be a product of decomposition [1]. Mono- and dialkylammonium nitrites decompose at temperatures below 60—70°C, but usually without violence [2].

**NITRO-ACYL HALIDES** **$o$-$NO_2ArCOCl$, etc.**

Aromatic acyl halides containing a nitro group adjacent to the halide function show a tendency towards violent thermal decomposition. The few individually indexed compounds are:

2-NITROTHIOPHENE-4-SULPHONYL CHLORIDE, $C_4H_2ClNO_4S_2$
2,4-DINITROBENZENESULPHENYL CHLORIDE, $C_6H_3ClN_2O_4S$
*o*-NITROBENZOYL CHLORIDE, $C_7H_4ClNO_3$
2,4-DINITROPHENYLACETYL CHLORIDE, $C_8H_5ClN_2O_5$
3-METHYL-2-NITROBENZOYL CHLORIDE, $C_8H_6ClNO_3$
*o*-NITROPHENYLACETYL CHLORIDE, $C_8H_6ClNO_3$
4-METHOXY-3-NITROBENZOYL CHLORIDE, $C_8H_6ClNO_4$

**NITROALKANES** **$RNO_2$**

1. 'Nitroparaffins', TDS1, New York, Commercial Solvents Corp., 1968
2. Hass, H.B. *et al., Chem. Rev.,* 1943, **32**, 388
3. Noble, P. *et al., Chem. Rev.,* 1964, **64**, 20

The nitroalkanes are mild oxidants under ordinary conditions, but precautions should be taken when they are subjected to high temperatures and pressures, since violent reactions may occur [1]. The polynitroalkanes, being more in oxygen balance than the mono-derivatives, tend to explode more easily [2], and caution is urged, particularly during distillation [3].

*See also* POLYNITROALKYL COMPOUNDS

Alkali metals,
or Inorganic bases

Watts, C. E., *Chem. Eng. News,* 1952, **30**, 2344

Contact of nitroalkanes with inorganic bases must be effected under conditions which will avoid isolation in a dry state of the explosive metal salts of the isomeric *aci*-nitroparaffins (or nitronic acids).

Metal oxides

Hermoni, A. *et al., Chem. & Ind.,* 1960, 1265

Hermoni, A. *et al.*, *Proc. 8th Combust. Symp.*, 1960, 1084–1088
Contact with metal oxides increases the sensitivity of nitromethane, nitroethane and 1-nitropropane to heat (and of nitromethane to detonation). Twenty-four oxides were examined in a simple quantitative test, and a mechanism is proposed. Cobalt, nickel, chromium, lead and silver oxides were most effective in lowering ignition temperatures.

At 39 bar initial pressure, the catalytic decomposition by chromium or iron oxides becomes explosive at above 245°C.

*See* *aci*-NITRO SALTS

## NITROALKYL PEROXONITRATES $NO_2ROONO_2$

*See* DINITROGEN TETRAOXIDE (NITROGEN DIOXIDE), $N_2O_4$: Cycloalkenes, etc.

## NITROAROMATIC–ALKALI HAZARDS

1. Merz, V. *et al.*, *Ber.*, 1871, **4**, 981–982
2. Uhlmann, P. W., *Chem. Ztg.*, 1914, **38**, 389–390
3. *MCA Data Sheets*; *Haz. Chem. Data*, 1975

It is widely stated in the usual reference texts that nitroaromatic compounds and more particularly polynitroaromatic compounds may present a severe explosion risk if subjected to shock, or if heated rapidly and uncontrollably, such as in fire situations. However, the same reference texts make no mention of the fact that there is also a risk of violent decomposition or explosion when nitroaromatic compounds are heated more moderately with caustic alkalies, even when water or organic solvents are present. It was known more than 100 years ago that mononitroaromatics (nitro-benzene, -toluene, -naphthalene) would react violently on heating with caustic alkalies 'with generation of 1½ foot flames', and that dinitro compounds were almost completely carbonised [1]. By 1914 the potential hazards involved in heating di- or tri-nitroaryl compounds with alkalies or ammonia were sufficiently well-recognised for a general warning on the possibilities of violent or explosive reactions in such systems to have been published [2]. Knowledge of these potential hazards apparently has faded to the point where they are not mentioned in standard sources of information [3]. Several industrial explosions have occurred during the past 40 years which appear attributable to this cause, but there has been little recognition of this, or of the common features in many of the incidents. Too little investigational work in this area has been reported to allow any valid conclusions to be drawn as to the detailed course of the observed reactions. However, it may be more than coincidence that in all the incidents reported, the structures of the nitroaromatic compounds involved were such that *aci*-nitroquinonoid salt species could have been formed under the reaction conditions. Many of these salts are of very limited thermal stability. For the individual incidents of this type,

*See* SODIUM 2,4-DINITROPHENOLATE, $C_6H_3N_2NaO_5$
1,3,5-TRINITROBENZENE, $C_6H_3N_3O_6$ : Methanol, Potassium hydroxide
*p*-CHLORONITROBENZENE, $C_6H_4ClNO_2$ : Sodium methoxide
SODIUM *o*-NITROTHIOPHENOLATE, $C_6H_4NNaO_2S$
SODIUM *p*-NITROPHENOLATE, $C_6H_4NNaO_3$
2,4-DINITROPHENOL, $C_6H_4N_2O_5$ : Bases
NITROBENZENE, $C_6H_5NO_2$ : Alkali
2,4,6-TRINITROTOLUENE, $C_7H_5N_3O_6$ : Potassium hydroxide
2-CHLORO-4-NITROTOLUENE, $C_7H_6ClNO_2$ : Sodium hydroxide
2,4-DINITROTOLUENE, $C_7H_6N_2O_4$ : Alkali
: Sodium oxide
*o*- or *p*-NITROTOLUENE, $C_7H_7NO_2$ : Sodium hydroxide
4-METHYL-2-NITROPHENOL, $C_7H_7NO_3$ : Sodium hydroxide, etc.
*o*-NITROANISOLE, $C_7H_7NO_3$ : Sodium hydroxide, etc.
POLYNITROARYL COMPOUNDS: Bases, etc.

*See also* *aci*-NITROQUINONOID COMPOUNDS

## *C*-NITRO COMPOUNDS $C—NO_2$

This group contains compounds with a single nitro group (attached to either an aliphatic or an aromatic nucleus) which have been involved in hazardous incidents. Poly-substitution is covered in the separate groups:

FLUORODINITROMETHYL COMPOUNDS
POLYNITROALKYL COMPOUNDS
POLYNITROARYL COMPOUNDS
TRINITROETHANOL ORTHOESTERS

and individually indexed mononitro compounds are:

TRIBROMONITROMETHANE, $CBr_3NO_2$
TRICHLORONITROMETHANE (CHLOROPICRIN), $CCl_3NO_2$
5-NITROTETRAZOLE, $CHN_5O_2$
CHLORONITROMETHANE, $CH_2ClNO_2$
NITROOXIMINOMETHANE (METHYLNITROLIC ACID), $CH_2N_2O_3$
NITROMETHANE, $CH_3NO_2$
POTASSIUM 1-NITROETHOXIDE, $C_2H_4KNO_3$
2-NITROACETALDEHYDE OXIME, $C_2H_4N_2O_3$
1-NITRO-1-OXIMINOETHANE (ETHYLNITROLIC ACID), $C_2H_4N_2O_3$
NITROETHANE, $C_2H_5NO_2$
2-NITROETHANOL, $C_2H_5NO_3$
2-CARBAMOYL-2-NITROACETONITRILE ('FULMINURIC ACID'), $C_3H_3N_3O_3$
1-NITROPROPANE, $C_3H_7NO_2$
2-METHYL-5-NITROIMIDAZOLE, $C_4H_5N_3O_2$

1-NITRO-3-BUTENE, $C_4H_7NO_2$
*tert*-NITROBUTANE, $C_4H_9NO_2$
ETHYL 2-NITROETHYL ETHER, $C_4H_9NO_3$
1,1,1-TRIS(HYDROXYMETHYL)NITROMETHANE, $C_4H_9NO_5$
METHYLNITROTHIOPHENE, $C_5H_5NO_2S$
*o*- or *p*-CHLORONITROBENZENE, $C_6H_4ClNO_2$
4-HYDROXY-3-NITROBENZENESULPHONYL CHLORIDE, $C_6H_4ClNO_5S$
NITROBENZENE, $C_6H_5NO_2$
*o*-NITROPHENOL, $C_6H_5NO_3$
*o*-NITROANILINE, $C_6H_6N_2O_2$
*m*-NITROBENZALDEHYDE, $C_7H_5NO_3$
2-CHLORO-4-NITROTOLUENE, $C_7H_6ClNO_2$
*o*- or *p*-NITROTOLUENE, $C_7H_7NO_2$
3-METHYL-4-NITROPHENOL, $C_7H_7NO_3$
4-METHYL-2-NITROPHENOL, $C_7H_7NO_3$
*o*-NITROANISOLE, $C_7H_7NO_3$
2-METHYL-5-NITROBENZENESULPHONIC ACID, $C_7H_7NO_5S$
2-METHOXY-5-NITROANILINE, $C_7H_8N_2O_3$
3-METHYL-4-NITRO-1-BUTEN-3-YL ACETATE, $C_7H_{11}NO_4$
3-METHYL-4-NITRO-2-BUTEN-1-YL ACETATE, $C_7H_{11}NO_4$
*o*-NITROACETOPHENONE, $C_8H_7NO_3$
2-METHYL-5-NITROBENZIMIDAZOLE, $C_8H_7N_3O_2$
2-(2-AMINOETHYLAMINO)-5-CHLORONITROBENZENE, $C_8H_{10}ClN_3O_2$
2-(2-AMINOETHYLAMINO)-NITROBENZENE, $C_8H_{11}N_3O_2$
NITROINDANE, $C_9H_9NO_2$
2-(2-AMINOETHYLAMINO)-5-METHOXYNITROBENZENE, $C_9H_{13}N_3O_3$
1-NITRONAPHTHALENE, $C_{10}H_7NO_2$
2,6-DI-*tert*-BUTYL-4-NITROPHENOL, $C_{14}H_{21}NO_3$

*See also* NITROALKANES

## *N*-NITRO COMPOUNDS

**N–NO₂**

Romburgh, P. van, *Chem. Weekblad*, 1934, **31**, 732–733

Many *N*-nitro derivatives show explosive instability, and the explosive properties of *N*-alkyl-*N*-nitroarylamines have been discussed briefly. Individually indexed compounds are:

NITROUREA, $CH_3N_3O_3$
*N*-NITROMETHYLAMINE, $CH_4N_2O_2$
NITROGUANIDINE, $CH_4N_4O_2$
METHYLENEBISNITROAMINE, $CH_4N_4O_4$
1-AMINO-3-NITROGUANIDINE, $CH_5N_5O_2$
1,2-BIS(DIFLUOROAMINO)*N*-NITROETHYLAMINE, $C_2H_4F_4N_4O_2$
AZO-*N*-NITROFORMAMIDINE, $C_2H_4N_8O_2$
1-METHYL-3-NITRO-1-NITROSOGUANIDINE, $C_2H_5N_5O_3$
*N,N'* DINITRO-1,2-DIAMINOETHANE, $C_2H_6N_4O_4$
1-METHYL-3-NITROGUANIDIUM PERCHLORATE, $C_2H_7ClN_4O_6$
1-METHYL-3-NITROGUANIDIUM NITRATE, $C_2H_7N_5O_5$
2-HYDROXY-4,6-BIS(*N*-NITROAMINO)-1,3,5-TRIAZINE, $C_3H_3N_7O_5$
*N,N'*-DINITRO-*N*-METHYL-1,2-DIAMINOETHANE, $C_3H_8N_4O_4$

$N,N'$-DIMETHYL-$N,N'$-DINITROOXAMIDE, $C_4H_6N_4O_6$
1,3,5,7-TETRANITROPERHYDRO-1,3,5,7-TETRAZOCINE, $C_4H_8N_8O_8$
$N$,2,3,5-TETRANITROANILINE, $C_6H_3N_5O_8$
$N$,2,4,6-TETRANITROANILINE, $C_6H_3N_5O_8$
$N,N'$-DIACETYL-$N,N'$-DINITRO-1,2-DIAMINOETHANE, $C_6H_{10}N_4O_6$
1-NITRO-3(2,4-DINITROPHENYL)UREA, $C_7H_5N_5O_7$
$N$,2,4,6-TETRANITRO-$N$-METHYLANILINE (TETRYL), $C_7H_5N_4O_8$
NITRIC AMIDE (NITRAMIDE), $H_2N_2O_2$

*See also* $N$-AZOLIUM NITROIMIDATES

## *aci*-NITROQUINONOID COMPOUNDS **formula on p. 190**

Generally, aromatic nitro compounds cannot form *aci*-nitro salts with bases unless there is an *o*- or *p*-substituent present (or is introduced by the action of the base) bearing a labile hydrogen atom. Then, isomerisation to produce *o*- or *p*-quinonoid *aci*-nitro species, then the salt, may be possible. Many salts of this type are unstable or explosive, and such species may have been involved in various incidents with nitroaromatics and bases. Individually indexed salts of this type are:

SODIUM 3-HYDROXYMERCURIO-2,6-DINITRO-4-*aci*-NITRO-2,5-CYCLOHEXADIENONE, $C_6H_2HgN_3NaO_8$
SODIUM 2-HYDROXYMERCURIO-6-NITRO-4-*aci*-NITRO-2,5-CYCLOHEXADIENONE, $C_6H_3HgN_2NaO_6$
POTASSIUM 6-*aci*-NITRO-2,4-DINITRO-2,4-CYCLOHEXADIENIMINE, $C_6H_3KN_4O_6$
* SODIUM 2,4-DINITROPHENOLATE, $C_6H_3N_2NaO_5$
* SILVER *p*-NITROPHENOLATE, $C_6H_4AgNO_3$
SODIUM 2-HYDROXYMERCURIO-4-*aci*-NITRO-2,5-CYCLOHEXADIENONE, $C_6H_4HgNO_4$
SODIUM 1,4-BIS-*aci*-NITRO-2,5-CYCLOHEXADIENE, $C_6H_4N_2Na_2O_4$
SODIUM 6-*aci*-NITRO-4-NITRO-2,4-CYCLOHEXADIENIMINE, $C_6H_4N_3NaO_4$
SODIUM 1,3-BIS-*aci*-NITROCYCLOHEXEN-2,4-DIIMINE, $C_6H_4N_4Na_2O_4$
2,4,6-TRINITROTOLUENE, $C_7H_5N_3O_6$ : Potassium hydroxide
POTASSIUM 4-METHOXY-1-*aci*-NITRO-3,5-DINITRO-2,5-CYCLOHEXADIENE, $C_7H_6KN_3O_7$
SODIUM 4,4-DIMETHOXY-1-*aci*-NITRO-3,5-DINITRO-2,5-CYCLOHEXADIENE, $C_8H_8N_3NaO_8$
* 2-(2-AMINOETHYLAMINO)-5-CHLORONITROBENZENE, $C_8H_{10}ClN_3O_2$
* 2-(2-AMINOETHYLAMINO)NITROBENZENE, $C_8H_{11}N_3O_2$
* 2-(2-AMINOETHYLAMINO)-5-METHOXYNITROBENZENE, $C_9H_{13}N_3O_3$
POTASSIUM 6-*aci*-NITRO-2,4-DINITRO-1-PHENYLIMINO-2,4-CYCLOHEXADIENE, $C_{12}H_7KN_4O_6$

*See* NITROAROMATIC–ALKALI HAZARDS

## *aci*-NITRO SALTS **–C=N(O)OM**

Many *aci*-nitro salts derived from action of bases on nitroalkanes are explosive in the dry state. Individually indexed compounds are:

SODIUM *aci*-NITROMETHANE, $CH_2NNaO_2$
AMMONIUM *aci*-NITROMETHANE, $CH_6N_2O_2$
DIPOTASSIUM *aci*-NITROACETATE, $C_2HK_2NO_4$
MERCURY(II) METHYLNITROLATE, $C_2H_2HgN_4O_8$
MONOPOTASSIUM *aci*-1,1-DINITROETHANE, $C_2H_3KN_2O_4$
DIAMMONIUM *N,N'*-DINITRO-1,2-DIAMINOETHANE, $C_2H_{12}N_6O_4$
DIPOTASSIUM BIS-*aci*-TETRANITROETHANE, $C_2K_2N_4O_8$
DILITHIUM BIS-*aci*-TETRANITROETHANE, $C_2Li_2N_4O_8$
DISODIUM BIS-*aci*-TETRANITROETHANE, $C_2N_4Na_2O_8$
SODIUM NITROMALONALDEHYDE, $C_3H_2NNaO_4$
POTASSIUM *aci*-1,1-DINITROPROPANE, $C_3H_5KN_2O_4$
MONOPOTASSIUM *aci*-2,5-DINITROCYCLOPENTANONE, $C_5H_5KN_2O_5$
* POTASSIUM 4,6-DINITROBENZOFUROXAN HYDROXIDE COMPLEX, $C_6H_3KN_4O_7$
* POTASSIUM *p*-NITROPHENOLATE, $C_6H_4KNO_3$
* SODIUM *o*-NITROTHIOPHENOLATE, $C_6H_4NNaO_2S$
POTASSIUM *aci*-1-NITRO-1-PHENYLNITROMETHANE, $C_7H_5KN_2O_4$
THALLIUM *aci*-PHENYLNITROMETHANE, $C_7H_6NO_2Tl$
DISODIUM 1,3-DIHYDROXY-1,3-BIS(*aci*-NITROMETHYL)-2,2,4,4-TETRAMETHYLCYCLOBUTANE, $C_{10}H_{16}N_2Na_2O_6$

*See* NITROALKANES: Alkali metals or Inorganic bases

## NITROSATED NYLON —CON(NO)—

*ABCM Quart. Safety Summ.*, 1963, **34**, 20

Nylon, nitrosated with dinitrogen trioxide according to Belg. Patent 606 944 and stored cold, exploded on being allowed to warm to ambient temperature. The *N*-nitroso nylon would be similar to *N*-nitroso-*N*-alkylamides, some of which are unstable. Nylon components should therefore be excluded from contact with nitrosating agents.

*See other* NITROSO COMPOUNDS

## NITROSO COMPOUNDS C—N=O, N—N=O

A number of compounds containing nitroso or coordinated nitrosyl groups exhibit instability under appropriate conditions. Individually indexed compounds are:

NITROSOTRIFLUOROMETHANE, $CF_3NO$
NITROSOGUANIDINE, $CH_4N_4O$
NITROSYL CYANIDE, $CN_2O$
*N*-METHYL-*N*-NITROSOUREA, $C_2H_5N_3O_2$
1-METHYL-3-NITRO-1-NITROSOGUANIDINE, $C_2H_5N_5O_3$
1,2-DIMETHYLNITROSOHYDRAZINE, $C_2H_7N_3O$
1,3,5-TRINITROSOHEXAHYDRO-1,3,5-TRIAZINE, $C_3H_6N_6O_3$
PERFLUORO-*tert*-NITROSOBUTANE, $C_4F_9NO$
*N,N'*-DIMETHYL-*N,N'*-DINITROSOOXAMIDE, $C_4H_6N_4O_4$
ETHYL *N*-METHYL-*N*-NITROSOCARBAMATE, $C_4H_8N_2O_3$
*N*-NITROSOETHYL(2-HYDROXYETHYL)AMINE, $C_4H_{10}N_2O_2$
3,7-DINITROSO-1,3,5,7-TETRAAZABICYCLO[3.3.1] NONANE, $C_5H_{10}N_6O_2$

TRINITROSOPHLOROGLUCINOL, $C_6H_3N_3O_6$
SODIUM *p*-NITROSOPHENOLATE, $C_6H_4NNaO_2$
*o*- or *p*-NITROSOPHENOL, $C_6H_5NO_2$
*N*-NITROSO-6-HEXANELACTAM, $C_6H_{10}N_2O_2$
*N*-NITROSOACETANILIDE, $C_8H_8N_2O_2$
$\eta$-CYCLOPENTADIENYL(METHYL)-BIS(*N*-METHYL-*N*-NITROSOHYDROXYLAMINO)TITANIUM, $C_8H_{14}N_4O_4Ti$
2-CHLORO-1-NITROSO-2-PHENYLPROPANE, $C_9H_{10}ClO$
LEAD(II) TRINITROSOPHLOROGLUCINOLATE, $C_{12}N_6O_{12}Pb_3$
NITROSYLRUTHENIUM TRICHLORIDE, $Cl_3NORu$
NITROSYLSULPHURIC ACID, $HNO_5S$
POTASSIUM NITROSODISULPHATE(2–), $K_2NO_7S_2$
POTASSIUM DINITROSOSULPHITE(2–), $K_2N_2O_5S$
DINITROSYLNICKEL, $N_2NiO_2$
NITRITONITROSONICKEL, $N_2NiO_3$
* TRIS(THIONITROSYL)THALLIUM, $N_3S_3Tl$

*See also* NITROSATED NYLON
3-NITROSOTRIAZENES

## 3-NITROSOTRIAZENES N=N–N(NO)

Müller, E. *et al.*, *Ber.*, 1962, **95**, 1255

A very unstable series of compounds, many decomposing at well below 0°C. The products formed from sodium triazenes and nitrosyl chloride explode violently on being disturbed with a wooden spatula, and are much more sensitive than those derived from silver triazenes. These exploded under a hammer blow, or on friction from a metal spatula.

*See other* TRIAZENES

## NON-METAL AZIDES $E(N_3)_n$

This group contains compounds with azide groups linked to non-oxygenated non-metals, individually indexed compounds being:

ALUMINIUM TRIS(TETRAAZIDOBORATE), $AlB_3N_{36}$
BORON AZIDE DICHLORIDE, $BCl_2N_3$
TRIAZIDOBORANE, $BN_9$
BROMINE AZIDE, $BrN_3$
BIS(TRIFLUOROMETHYL)PHOSPHORUS(III) AZIDE, $C_2F_6N_3P$
AZIDODIMETHYLBORANE, $C_2H_6BN_3$
DIAZIDODIMETHYLSILANE, $C_2H_6N_6Si$
DIPHENYLPHOSPHORUS(III) AZIDE, $C_{12}H_{10}N_3P$
* TRIAZIDOCHLOROSILANE, $ClN_9Si$
PHOSPHORUS AZIDE DIFLUORIDE, $F_2N_3P$
PHOSPHORUS AZIDE DIFLUORIDE–BORANE, $F_2N_3P \cdot BH_3$
HYDROGEN AZIDE (HYDRAZOIC ACID), $HN_3$
AZIDOSILANE, $H_3N_3Si$
* AMMONIUM AZIDE, $H_4N_4$
HYDRAZINIUM AZIDE, $H_5N_5$
* NITROSYL AZIDE, $N_4O$
* PHOSPHORUS TRIAZIDE OXIDE, $N_9OP$

PHOSPHORUS TRIAZIDE, $N_9P$
SILICON TETRAAZIDE, $N_{12}Si$
'PHOSPHORUS PENTAAZIDE', '$N_{15}P$'
* SODIUM HEXAAZIDOPHOSPHATE(1–), $N_{18}NaP$
1,1,3,3,5,5-HEXAAZIDO-2,4.6-TRIAZA-1,3,5-TRIPHOSPHORINE, $N_{21}P_3$

*See related* ACYL AZIDES
HALOGEN AZIDES

## NON-METAL HALIDES AND THEIR OXIDES

**$EX_n$, $EOX_n$**

This highly reactive class includes the separately treated groups

*N*-HALOGEN COMPOUNDS
HALOSILANES

as well as the individually indexed compounds:

ARSENIC TRICHLORIDE, $AsCl_3$
ARSENIC PENTAFLUORIDE, $AsF_5$
ARSINE–BORON TRIBROMIDE, $AsH_3 \cdot BBr_3$
BORON BROMIDE DIIODIDE, $BBrI_2$
BORON DIBROMIDE IODIDE, $BBr_2I$
BORON TRIBROMIDE, $BBr_3$
* DICHLOROBORANE, $BCl_2H$
BORON TRICHLORIDE, $BCl_3$
BORON TRIFLUORIDE, $BF_3$
BORON DIIODOPHOSPHIDE, $BI_2P$
BORON TRIIODIDE, $BI_3$
DIBORON TETRACHLORIDE, $B_2Cl_4$
DIBORON TETRAFLUORIDE, $B_2F_4$
1,3,5-TRICHLORO-2,4,6-TRIFLUOROBORAZINE, $B_3Cl_3F_3N_3$
(*B*)1,3,5-TRICHLOROBORAZINE, $B_3Cl_3H_3N_3$
TRIBORON PENTAFLUORIDE, $B_3F_5$
TETRABORON TETRACHLORIDE, $B_4Cl_4$
SELENINYL BROMIDE, $Br_2OSe$
SULPHUR DIBROMIDE, $Br_2S$
* SILICON DIBROMIDE SULPHIDE, $Br_2SSi$
POLY(DIBROMOSILYLENE), $(Br_2Si)_n$
PHOSPHORUS TRIBROMIDE, $Br_3P$
* BIS(DIFLUOROBORYL)METHANE, $CH_2B_2F_4$
TETRACARBON MONOFLUORIDE, $C_4F$
* PHENYLPHOSPHORYL DICHLORIDE, $C_6H_5Cl_2OP$
PHOSPHORUS CHLORIDE DIFLUORIDE, $ClF_2P$
* THIOPHOSPHORYL CHLORIDE DIFLUORIDE, $ClF_2PS$
SULPHINYL CHLORIDE (THIONYL CHLORIDE), $Cl_2OS$
SELENINYL CHLORIDE, $Cl_2OSe$
SULPHONYL DICHLORIDE (SULPHURYL CHLORIDE), $Cl_2O_2S$
DISULPHURYL DICHLORIDE, $Cl_2O_5S_2$
SULPHUR DICHLORIDE, $Cl_2S$
DISULPHUR DICHLORIDE, $Cl_2S_2$
* DISELENIUM DICHLORIDE, $Cl_2Se_2$
* TRICHLOROSILANE, $Cl_3HSi$
PHOSPHORYL CHLORIDE, $Cl_3OP$

PHOSPHORUS TRICHLORIDE, $Cl_3P$
* THIOPHOSPHORYL CHLORIDE, $Cl_3PS$
TETRACHLORODIPHOSPHANE, $Cl_4P_2$
TETRACHLOROSILANE, $Cl_4Si$
TELLURIUM TETRACHLORIDE, $Cl_4Te$
PHOSPHORUS PENTACHLORIDE, $Cl_5P$
SULPHINYL FLUORIDE, $F_2OS$
SELENIUM DIFLUORIDE DIOXIDE, $F_2O_2Se$
DISULPHURYL DIFLUORIDE, $F_2O_5S_2$
POLY(DIFLUOROSILYLENE), $(F_2Si)_n$
XENON DIFLUORIDE, $F_2Xe$
* TRIFLUOROSILANE, $F_3HSi$
NITROGEN TRIFLUORIDE, $F_3N$
PHOSPHORUS TRIFLUORIDE, $F_3P$
* THIOPHOSPHORYL FLUORIDE, $F_3PS$
SULPHUR TETRAFLUORIDE, $F_4S$
SELENIUM TETRAFLUORIDE, $F_4Se$
SILICON TETRAFLUORIDE, $F_4Si$
SULPHUR HEXAFLUORIDE, $F_6S$
* BIS(*S,S*-DIFLUORO-*N*-SULPHIMIDO)SULPHUR TETRAFLUORIDE, $F_8N_2S_3$
* PHOSPHORUS DIIODIDE TRISELENIDE, $I_2P_4Se_3$
PHOSPHORUS TRIIODIDE, $I_3P$
TETRAIODODIPHOSPHANE, $I_4P_2$

*See also* PERFLUOROSILANES

## NON-METAL HYDRIDES $EH_n$

There is a separate group entry for

BORANES

Most members of this readily oxidised class ignite in air, individually indexed compounds being:

ARSINE, $AsH_3$
ARSINE–BORON TRIBROMIDE, $AsH_3 \cdot BBr_3$
* *B*-CHLORO-*N,N*-DIMETHYLAMINODIBORANE, $C_2H_{10}B_2ClN$
'SOLID PHOSPHORUS HYDRIDE', $HP_2$
SILICON MONOHYDRIDE, $(HSi)_n$
'UNSATURATED' SILICON HYDRIDE, $(H_{1.5}Si)_n$
* POTASSIUM DIHYDROGENPHOSPHIDE, $H_2KP$
* SODIUM DIHYDROGENPHOSPHIDE, $H_2NaP$
* OXOSILANE, $H_2OSi$
HYDROGEN SULPHIDE, $H_2S$
HYDROGEN SELENIDE, $H_2Se$
POLYSILYLENE, $(H_2Si)_n$
AMMONIA, $H_3N$
PHOSPHINE, $H_3P$
STIBINE, $H_3Sb$
HYDRAZINE, $H_4N_2$
* OXODISILANE, $H_4OSi_2$
DIPHOSPHANE, $H_4P_2$
SILANE, $H_4Si$
* DISILYL OXIDE, $H_6OSi_2$
2,4,6-TRISILATRIOXANE ('TRIPROSILOXANE'), $H_6O_3Si_3$

* DISILYL SULPHIDE, $H_6SSi_2$
DISILANE, $H_6Si_2$
TRISILANE, $H_8Si_3$
* TRISILYLAMINE, $H_9NSi_3$
TRISILYLPHOSPHINE, $H_9PSi_3$
TETRASILANE, $H_{10}Si_4$
* TETRASILYLHYDRAZINE, $H_{12}N_2Si_4$
* TETRAAMMINELITHIUM DIHYDROGENPHOSPHIDE, $H_{14}LiN_4P$

*See also* SILANES
*See related* ALKYLNON-METAL HYDRIDES

## NON-METAL OXIDES $E_mO_n$

The generally acidic materials in this class may function as oxidants, some quite powerful, under appropriate conditions. Individually indexed compounds are:

DIARSENIC TRIOXIDE, $As_2O_3$
DIARSENIC PENTAOXIDE, $As_2O_5$
DIBORON DIOXIDE, $B_2O_2$
DIBORON TRIOXIDE, $B_2O_3$
CARBON MONOXIDE, CO
CARBON DIOXIDE, $CO_2$
* POLY(DIHYDROXODIOXODISILANE), $(H_2O_4Si_2)_n$
* TETRAHYDROXODIOXOTRISILANE ('TRISILICIC ACID'), $H_4O_6Si_3$
NITROGEN OXIDE ('NITRIC OXIDE'), NO
DINITROGEN OXIDE ('NITROUS OXIDE'), $N_2O$
DINITROGEN TRIOXIDE, $N_2O_3$
DINITROGEN TETRAOXIDE (NITROGEN DIOXIDE), $N_2O_4$
DINITROGEN PENTAOXIDE, $N_2O_5$
SILICON OXIDE, OSi
SULPHUR DIOXIDE, $O_2S$
SELENIUM DIOXIDE, $O_2Se$
SILICON DIOXIDE, $O_2Si$
PHOSPHORUS(III) OXIDE, $O_3P_2$
SULPHUR TRIOXIDE, $O_3S$
ANTIMONY(III) OXIDE, $O_3Sb_2$
SELENIUM TRIOXIDE, $O_3Se$
TELLURIUM TRIOXIDE, $O_3Te$
* TETRAPHOSPHORUS TETRAOXIDE TRISULPHIDE, $O_4P_4S_3$
PHOSPHORUS(V) OXIDE, $O_5P_2$
* TETRAPHOSPHORUS HEXAOXIDE–BIS(BORANE), $O_6P_4 \cdot B_2H_6$
* TETRAPHOSPHORUS HEXAOXIDE TETRASULPHIDE, $O_6P_4S_4$
DISULPHUR HEPTAOXIDE, $O_7S_2$

*See also* HALOGEN OXIDES

## NON-METAL PERCHLORATES $E(ClO_4)_n$

1. Solymosi, F., *Chem. Abs.*, 1972, **77**, 42498
2. Sunderlin, K. G. R., *Chem. Eng. News*, 1974, **52**(31), 3

The thermal stability, structures and physical properties of various non-metal perchlorates (nitrosyl, nitronium, hydrazinium, ammonium, etc.)

have been reviewed [1]. Many organic perchlorates were examined for sensitivity to impact. The salts of cations containing only C, H and S exploded quite consistently, while those with C, H and N, or C, H, N and S did not explode, except 1,2-bis(ethylammonio)ethane diperchlorate [2].

Several perchlorate derivatives of non-metallic elements (including some non-nitrogenous organic compounds) are noted for explosive instability.

*See* 1,3-DITHIOLIUM PERCHLORATE, $C_3H_4ClO_4S_2$
TROPYLIUM PERCHLORATE, $C_7H_7ClO_4$
2,4,6-TRIMETHYLPYRILIUM PERCHLORATE, $C_8H_{11}ClO_5$
TRIS(ETHYLTHIO)CYCLOPROPENYLIUM PERCHLORATE, $C_9H_{15}ClO_4S_3$
THIANTHRENIUM PERCHLORATE, $C_{12}H_8ClO_4S_2$
4,4′-DIPHENYL-2,2′-BI(1,3-DITHIOL)-2′-YL-2-YLIUM PERCHLORATE, $C_{18}H_{12}ClO_4S_4$
FLUORONIUM PERCHLORATE, $ClFH_2O_4$
PHOSPHONIUM PERCHLORATE, $ClH_4O_4P$
* NITROSYL PERCHLORATE, $ClNO_5$
IODINE(III) PERCHLORATE, $Cl_3IO_{12}$
CAESIUM TETRAPERCHLORATOIODATE, $Cl_4CsIO_{16}$
SILICON TETRAPERCHLORATE, $Cl_4O_{16}Si$
BIS-TRIPERCHLORATOSILICON OXIDE, $Cl_6O_{25}Si_2$

*See* OXOSALTS OF NITROGENOUS BASES
*See other* PERCHLORATES
*See related* ORGANOSILICON PERCHLORATES

## NON-METALS

Most members of this group of elements are readily oxidised with more or less violence dependent upon the oxidant and conditions involved. Individually indexed elements are:

ARSENIC, As
BORON, B
CARBON, C
HYDROGEN, $H_2$
NITROGEN, $N_2$
OXYGEN, $O_2$
PHOSPHORUS, P
SULPHUR, S
SELENIUM, Se
SILICON, Si
TELLURIUM, Te

## NON-METAL SULPHIDES $E_mS_n$

In this group of readily oxidisable materials, individually indexed compounds are:

ARSENIC DISULPHIDE, $AsS_2$
DIARSENIC DISULPHIDE, $As_2S_2$
DIBORON TRISULPHIDE, $B_2S_3$
* CARBONYL SULPHIDE, COS
CARBON SULPHIDE, CS
CARBON DISULPHIDE, $CS_2$
HYDROGEN SULPHIDE, $H_2S$
HYDROGEN DISULPHIDE, $H_2S_2$

HYDROGEN TRISULPHIDE, $H_2S_3$
DISULPHUR DINITRIDE, $N_2S_2$
TETRASULPHUR DINITRIDE, $N_2S_4$
TETRASULPHUR TETRANITRIDE, $N_4S_4$
* SODIUM TETRASULPHURPENTANITRIDATE, $N_5NaS_4$
* TETRAPHOSPHORUS TETRAOXIDE TRISULPHIDE, $O_4P_4S_3$
PHOSPHORUS(V) SULPHIDE, $P_2S_5$
TETRAPHOSPHORUS TRISULPHIDE, $P_4S_3$
SILICON MONOSULPHIDE, SSi

## OIL–LAGGING FIRES

1. Macdermott, P. E., *Chem. Brit.*, 1976, **12**, 69
2. Markham, H., ibid., 205
3. Carrettes, G. V., ibid., 204
4. Hilado, C. J., *J. Fire Flamm.*, 1974, **5**, 326–333
5. *Fires in Oil-Soaked Lagging, CP35/74*, Bowes, P. C., Garston, Building Res. Establ., 1974

Fires frequently occur when combustible liquids leak into lagging material on hot surfaces, and the temperature of ignition is often well below the conventionally determined AIT. Experiments showed an ignition temperature some 100–200°C below the AIT depending on the nature of the oils and of the lagging materials into which they were soaked [1]. The experimental technique used was criticised on the grounds that excess air is supplied to the sample, whereas in practice access of air is often severely limited. An alternative procedure to simulate practice more closely is described, and the results for leakage of heat-transfer oils are given, ignition generally occurring at 160–200°C. The need for care to avoid fire when removing oil-contaminated lagging is stressed [2].

Attention is drawn to an existing test for the self-heating properties of textile oils, and the co-oxidant effect of iron or copper in reducing the auto-ignition temperature of the oils [3]. The presence of combustibles incorporated into lagging materials during manufacture can give rise to significant self-heating effects in uncontaminated lagging [4]. The theoretical background and current experimental progress have been reviewed [5].

## ORGANIC AZIDES $RN_3$

Boyer, J. H. *et al., J. Chem. Eng. Data,* 1964, **9**, 480; *Chem. Eng. News,* 1964, **42**(31),6

The need for careful and small-scale handling of organic azides, which are usually heat- or shock-sensitive compounds of varying degrees of stability, has been discussed. The presence of more than one azido group, particularly if on the same atom (C or N) greatly reduces the stability.

This class contains the separately treated groups:

ACYL AZIDES
2-AZIDOCARBONYL COMPOUNDS

as well as the individually indexed compounds:

TRIAZIDOMETHYLIUM HEXACHLOROANTIMONATE, $CCl_6N_9Sb$
FLUORODINITROMETHYL AZIDE, $CFN_5O_4$
5-AZIDOTETRAZOLE, $CHN_7$
METHYL AZIDE, $CH_3N_3$
AZIDOACETONITRILE, $C_2H_2N_4$
3-AZIDO-1,2,4-TRIAZOLE, $C_2H_2N_6$
VINYL AZIDE, $C_2H_3N_3$
AZIDOACETIC ACID, $C_2H_3N_3O_2$
1,1-DIAZIDOETHANE, $C_2H_4N_6$
1,2-DIAZIDOETHANE, $C_2H_4N_6$
ETHYL AZIDE, $C_2H_5N_3$
* *N*-AZIDODIMETHYLAMINE, $C_2H_6N_4$
THIOCARBONYL AZIDE THIOCYANATE, $C_2N_4S_2$
DIAZIDOMETHYLENECYANAMIDE, $C_2N_8$
DIAZIDOMETHYLENEAZINE, $C_2N_{14}$
1,3-DIAZIDOPROPENE, $C_3H_4N_6$
DIAZIDOMALONONITRILE, $C_3N_8$
2,4,6-TRIAZIDO-1,3,5-TRIAZINE, $C_3N_{12}$
PICRYL AZIDE, $C_6H_2N_6O_6$
1,3- or 1,4-DIAZIDOBENZENE, $C_6H_4N_6$
PHENYL AZIDE, $C_6H_5N_3$
BENZYL AZIDE, $C_7H_7N_3$
TRIPHENYLMETHYL AZIDE, $C_{19}H_{15}N_3$

*See related* NON-METAL AZIDES

## ORGANIC PEROXIDES　　C–OO


1. Castrantas, 1970
2. Varjarvandi, J. *et al., J. Chem. Educ.*, 1971, **48**, A451
3. 'Code of Practice for Storage of Organic Peroxides', London, Laporte Chemicals Ltd., 1970
4. Jackson, H. L. *et al., J. Chem. Educ.*, 1970, **47**, A175
5. Castrantas, 1965
6. Davies, 1961
7. Swern, 1970, Vol. 1, 1–104
8. Houben-Weyl, 1952, Vol. 8(3), 1
9. Swern, 1972, Vol. 3, 341–364
10. Stevens, H. C., US Pat. 2 415 971, 1947
11. Cookson, P. G. *et al., J. Chem. Soc., Chem. Comm.*, 1976, 1022–1023


Of two general guides to the safe handling and use of peroxides, the second includes details of hazard evaluation tests, and the first has a comprehensive bibliography [1,2]. Storage aspects are rather specific [1,3]. Procedures for the safe handling of peroxidisable compounds have also been described [4]. Tabulated data on fire and explosion hazards of classes of organic peroxides with an extensive bibliography are available [5]. Theoretical aspects have been considered [5–7]. The hazards involved in synthesis of organic peroxides have been detailed [8], and a further review on the evaluation and management of peroxide hazards has appeared recently [9].

The use of iodine to stabilise liquid organic peroxides against explosive or incendiary decomposition has been claimed [10]. A new safe general method for rapid preparation of primary, secondary or tertiary alkyl hydroperoxides and peroxides has been described [11].

Index entries have been assigned to the structurally based sub-groups:

ALKYL HYDROPEROXIDES
*tert*-BUTYL PEROXOPHOSPHATE DERIVATIVES
CYCLIC PEROXIDES
DIACYL PEROXIDES
DIALKYL PEROXIDES
DIOXETANES
KETONE PEROXIDES
1-OXYPEROXY COMPOUNDS
OZONIDES
PEROXYACIDS
PEROXYCARBONATE ESTERS
PEROXYESTERS
POLYPEROXIDES

*See also* COMMERCIAL ORGANIC PEROXIDES
ORGANOMINERAL PEROXIDES

## ORGANOLITHIUM REAGENTS

**XArLi**


1. Bretherick, L., *Chem. & Ind.*, 1971, 1017
2. Gilman, H., private comm., 1971
3. 'Benzotrifluorides Catalogue 6/15', West Chester, Pa., Marshallton Res. Labs., 1971
4. Coates, 1960, 19
5. *MCA Case History No. 1834*


Several halo-aryllithium compounds are explosive in the solid state in absence or near-absence of solvents or diluents, and operations with them should be designed to avoid their separation from solution. Such compounds include *m*- and *p*-bromo-, *m*-chloro-, *p*-fluoro-, *m*- and *p*-trifluoromethyl-phenyllithiums [1] and 3:4-dichloro-2,5-dilithiothiophene [2], but *m*-bromophenyl- and *o*-trifluoromethyl-phenyllithium appear to be explosive in presence of solvent also [1,3]. *m*- and *p*-Dilithiobenzene are also explosively unstable under appropriate conditions.

Most organolithium compounds are pyrophoric when pure (especially those of high lithium content) and are usually handled in solution and under inert atmosphere [4]. A completed preparation of *o*-trifluoromethylphenyllithium refluxing in ether under nitrogen suddenly exploded violently [5].

*See* PENTAFLUOROPHENYLLITHIUM, $C_6F_5Li$
*See other* HALO-ARYLMETALS

## ORGANOMETALLICS

This miscellaneous group of organometallic compounds contains the individually indexed compounds:

POLY(η-CYCLOPENTADIENYLTITANIUM(III) DICHLORIDE), $(C_5H_5Cl_2Ti)_n$
CYCLOPENTADIENYLSODIUM, $C_5H_5Na$
3-KALIOBENZOCYCLOBUTENE, $C_8H_5K$
DIPOTASSIUM CYCLOOCTATETRAENE, $C_8H_8K_2$
η-CYCLOPENTADIENYL(METHYL)-BIS(*N*-METHYL-*N*-NITROSO-HYDROXYLAMINO)TITANIUM, $C_8H_{14}N_4O_4Ti$
η-CYCLOPENTADIENYLTRIMETHYLTITANIUM, $C_8H_{14}Ti$
* DICARBONYL-π-CYCLOHEPTATRIENYLTUNGSTEN AZIDE, $C_9H_7N_3O_2W$
BIS(η-CYCLOPENTADIENYLDINITROSYLCHROMIUM), $C_{10}H_{10}Cr_2N_4O_4$
BIS(η-CYCLOPENTADIENYL)MAGNESIUM, $C_{10}H_{10}Mg$
BIS-η-CYCLOPENTADIENYLLEAD, $C_{10}H_{10}Pb$
BIS-η-CYCLOPENTADIENYLTITANIUM, $C_{10}H_{10}Ti$
BIS-η-CYCLOPENTADIENYLZIRCONIUM, $C_{10}H_{10}Zr$
BIS(η-CYCLOPENTADIENYL)TETRAHYDROBORATONIOBIUM(III), $C_{10}H_{14}BNb$
BIS(2,4-PENTANEDIONATO)CHROMIUM, $C_{10}H_{14}CrO_4$
η-BENZENE-η-CYCLOPENTADIENYLIRON(II) PERCHLORATE, $C_{11}H_{11}ClFeO_4$
POTASSIUM HEXAETHYNYLCOBALTATE(4–), $C_{12}H_6CoK_4$
(2,2-DICHLORO-1-FLUOROVINYL)FERROCENE, $C_{12}H_7ClFFe$
BIS(η-BENZENE)CHROMIUM(0), $C_{12}H_{12}Cr$
BIS(η-BENZENE)IRON(0), $C_{12}H_{12}Fe$
2-(DIMETHYLAMINOMETHYL)FLUOROFERROCENE, $C_{13}H_{16}FFeN$
DIMETHYLAMINOMETHYLFERROCENE, $C_{13}H_{17}FeN$
TRIS-η-CYCLOPENTADIENYLPLUTONIUM, $C_{15}H_{15}Pu$
TRIS-η-CYCLOPENTADIENYLURANIUM, $C_{15}H_{15}U$
TRIS(2,4-PENTANEDIONATO)MOLYBDENUM(III), $C_{15}H_{21}MoO_6$
BIS(η-CYCLOPENTADIENYL)PENTAFLUOROPHENYLZIRCONIUM HYDROXIDE, $C_{16}H_{11}F_5OZr$
BIS-η-CYCLOPENTADIENYL-BIS(PENTAFLUOROPHENYL)-ZIRCONIUM, $C_{22}H_{10}F_{10}Zr$
BIS(DICARBONYL-η-CYCLOPENTADIENYLIRON)-BIS(TETRA-HYDROFURAN)MAGNESIUM, $C_{22}H_{26}Fe_2MgO_6$
1,3-BIS(DI-η-CYCLOPENTADIENYLIRON)-2-PROPEN-1-ONE, $C_{23}H_{20}Fe_2O$
BIS(DI-η-BENZENECHROMIUM) DICHROMATE, $C_{24}H_{24}Cr_4O_7$
TRIS-2,2′-BIPYRIDINECHROMIUM(0), $C_{30}H_{24}CrN_6$
BIS[DICARBONYL-η-CYCLOPENTADIENYL(TRIBUTYLPHOSPHINE) MOLYBDENUM] TETRAKIS(TETRAHYDROFURAN)-MAGNESIUM, $C_{54}H_{96}MgMo_2O_8P_2$

*See related* ALKYLMETAL HALIDES
ALKYLMETALS
HALO-ARYMETALS
ORGANOLITHIUM REAGENTS

## ORGANOMINERAL PEROXIDES $RM{-}OO$, $RE{-}OO$

Castrantas, 1965, 18

Swern, 1970, Vol. 1, 13

Sosnovsky, G. *et al., Chem. Rev.*, 1966, **66**, 529

Available information suggests that both hydroperoxides and peroxides in this extensive class are in many cases stable to heat at temperatures rather below 100°C, but may decompose explosively at higher temperatures. There are, however, exceptions.

*See* TRIMETHYLSILYL HYDROPEROXIDE, $C_3H_{10}O_2Si$
DIETHYLHYDROXYTIN HYDROPEROXIDE, $C_4H_{12}O_3Sn$
TRIETHYLTIN HYDROPEROXIDE, $C_6H_{16}O_2Sn$
BIS(TRIETHYLTIN) PEROXIDE, $C_{12}H_{30}O_2Sn_2$
TRIPHENYLTIN HYDROPEROXIDE, $C_{18}H_{16}O_2Sn$

*See also* ALKYL TRIALKYLLEAD PEROXIDES

## ORGANOSILICON PERCHLORATES $R_3SiClO_4$

Wannagat, U. *et al., Z. Anorg. Chem.*, 1950, **302**, 185–198

Several trialkyl- or triaryl-silicon perchlorates explode on heating, including:

TRIMETHYLSILICON PERCHLORATE, $C_3H_9ClO_4Si$
TRIETHYLSILICON PERCHLORATE, $C_6H_{15}ClO_4Si$
TRIPROPYLSILICON PERCHLORATE, $C_9H_{21}ClO_4Si$
TRIPHENYLSILICON PERCHLORATE, $C_{18}H_{15}ClO_4Si$
TRI-*p*-TOLYLSILICON PERCHLORATE, $C_{21}H_{21}ClO_4Si$

*See related* NON-METAL PERCHLORATES

## OXIDANTS

*Inorganic High-Energy Oxidisers: Synthesis, Structure and Properties*, Lawless, E. W. and Smith, I. C., London, Arnold, 1968

Developments in inorganic oxidants, mainly derivatives of fluorine, are covered in this reference.

Members of this class of materials have been involved in the majority of the two-component reactive systems included in this Handbook, and the whole class is very large. Most oxidants have been collectively treated in the structurally based entries:

ALKYL HYDROPEROXIDES
ALKYL TRIALKYLLEAD PEROXIDES
AMINIUM IODATES AND PERIODATES
* BLEACHING POWDER
CHLORITE SALTS
* COMMERCIAL ORGANIC PEROXIDES
CYCLIC PEROXIDES
DIACYL PEROXIDES
DIALKYL PEROXIDES
(DIBENZOYLDIOXYIODO)BENZENES
DIFLUOROAMINO COMPOUNDS

DIFLUOROPERCHLORYL SALTS
DIOXETANES
DIOXYGENYL POLYFLUOROSALTS
FLUORINATED PEROXIDES
FLUORODINITRO COMPOUNDS
*N*-HALOGEN COMPOUNDS
HALOGEN OXIDES
HALOGENS
*N*-HALOIMIDES
HEXAFLUOROCHLORONIUM SALTS
HYDROXOOXODIPEROXOCHROMATE SALTS
HYPOHALITES
INTERHALOGENS
KETONE PEROXIDES
* LIQUID AIR
METAL CHLORATES
METAL HALOGENATES
METAL HYPOCHLORITES
METAL NITRATES
METAL NITRITES
METAL OXOHALOGENATES
METAL OXOMETALLATES
METAL OXONON-METALLATES
METAL OXIDES
METAL PERCHLORATES
METAL PEROXIDES
METAL PEROXOMOLYBDATES
METAL POLYHALOHALOGENATES
* MOLTEN SALT BATHS
NITROALKANES
NON-METAL PERCHLORATES
ORGANOMINERAL PEROXIDES
OXIDES OF NITROGEN
OXOHALOGEN ACIDS
* OXYGEN ENRICHMENT
OXYGEN FLUORIDES
1-OXYPEROXY COMPOUNDS
OZONIDES
PERCHLORYL COMPOUNDS
PEROXOACIDS
PEROXOACID SALTS
PEROXYACIDS
PEROXYCARBONATE ESTERS
POLYNITROALKYL COMPOUNDS
POLYPEROXIDES
XENON COMPOUNDS

Other individually indexed oxidants (not covered in the above) are:

DIOXYGENYL TETRAFLUOROBORATE, $BF_4O_2$
TRIFLUOROMETHYL HYPOFLUORITE, $CF_4O$
NITROMETHANE, $CH_3NO_2$
NITROSYL TETRAFLUOROCHLORATE, $ClF_4NO$
CHLOROSULPHURIC ACID, $ClHO_3S$
NITRYL CHLORIDE, $ClNO_2$
NITRYL HYPOCHLORITE (CHLORINE NITRATE), $ClNO_3$

NITROSYL PERCHLORATE, $ClNO_5$
NITRYL PERCHLORATE, $ClNO_6$
CHROMYL CHLORIDE, $Cl_2CrO_2$
CHROMIUM PENTAFLUORIDE, $CrF_5$
CHROMIC ACID, $CrH_2O_4$
FLUOROSELENIC ACID, $FHO_3Se$
MANGANESE FLUORIDE TRIOXIDE, $FMnO_3$
NITROSYL FLUORIDE, FNO
NITRYL HYPOFLUORITE ('FLUORINE NITRATE'), $FNO_3$
PEROXODISULPHURYL DIFLUORIDE, $F_2O_6S_2$
OSMIUM HEXAFLUORIDE, $F_6Os$
PLATINUM HEXAFLUORIDE, $F_6Pt$
PERMANGANIC ACID, $HMnO_4$
NITROUS ACID, $HNO_2$
NITRIC ACID, $HNO_3$
HYDROGEN PEROXIDE, $H_2O_2$
SULPHURIC ACID, $H_2O_4S$
DIOXONIUM HEXAMANGANATO(VII)MANGANATE, $H_6Mn_7O_{26}$
OXYGEN, $O_2$ (Gas or liquid)
OZONE, $O_3$

## OXIDANTS AS HERBICIDES

Cook, W. H., *Can. J. Res.*, 1933, **8**, 509

The effect of humidity upon combustibility of various mixtures of organic matter and sodium chlorate was studied. Addition of a proportion of hygroscopic material (calcium or magnesium chlorides) effectively reduces the hazard. Similar effects were found for sodium dichromate and barium chlorate.

*See* SODIUM CHLORATE, $ClNaO_3$ : Organic matter

## *N*-OXIDES

**N→O**

Baumgarten, H. E. *et al.*, *J. Amer. Chem. Soc.*, 1957, **79**, 3145

A procedure for preparing *N*-oxides is described which avoids formation of peracetic acid. After prolonged treatment of the amine at 35–40°C with excess 30% hydrogen peroxide, excess of the latter is catalytically decomposed with platinum oxide.

## OXIDES OF NITROGEN

**$NO_n$**

The oxides of nitrogen collectively are oxidants with power increasing with the level of oxygen content. Dinitrogen oxide will often support violent combustion, since its oxygen content (36.5%) approaches double that of atmospheric air.

*See* NITROGEN OXIDE ('NITRIC OXIDE'), NO
DINITROGEN OXIDE ('NITROUS OXIDE'), $N_2O$
DINITROGEN TRIOXIDE, $N_2O_3$
DINITROGEN TETRAOXIDE (NITROGEN DIOXIDE), $N_2O_4$
DINITROGEN PENTAOXIDE, $N_2O_5$

Glyptal resin
*ABCM Quart. Safety Summ.*, 1937, **8**, 31
A new wooden fume cupboard was varnished with glyptal (glyceryl phthalate) resin. After a few weeks' use with 'nitrous fumes', the resin spontaneously and violently ignited. This was attributed to formation of glyceryl trinitrate.

'Oxides of nitrogen', usually as mixtures of unspecified composition, are thought to have been involved in several violent incidents.

*See* HYDROGEN, $H_2$: Liquid nitrogen
NITROGEN OXIDE ('NITRIC OXIDE'), NO: Dienes, Oxygen
NITROGEN DIOXIDE, $NO_2$: Alkenes
DINITROGEN TETRAOXIDE, $N_2O_4$: Unsaturated hydrocarbons

## OXIMES RC=NOH

1. Horner, L. in *Autoxidation and Antioxidants*, Lundberg, W. O. (Editor), Vol. 1, 184–186, 197–202, New York, Interscience, 1961
2. Tyler, L. J., *Chem. Eng. News*, 1974, **52**(35), 3

Several explosions or violent decompositions during distillation of aldoximes may be attributable to presence of peroxides arising from autoxidation. The peroxide may form on the –CH=NOH system (both aldehydes and hydroxylamines peroxidise [1]) or perhaps from unreacted aldehyde.

Attention has been drawn to an explosion hazard inherent to ketoximes and many of their derivatives. The hazard is attributed to inadvertent occurrence of acidic conditions leading to highly exothermic Beckmann rearrangement reactions accompanied by potentially catastrophic gas evolution. Presence of acidic salts (iron(III) chloride, or the ketoxime hydrochloride) markedly lowers decomposition temperatures [2].

Individually indexed compounds are:

* NITROOXIMINOMETHANE (METHYLNITROLIC ACID), $CH_2N_2O_3$
CHLOROACETALDEHYDE OXIME, $C_2H_4ClNO$
* HYDROXYCOPPER(II) GLYOXIMATE, $C_2H_4CuNO$
2-NITROACETALDEHYDE OXIME, $C_2H_4N_2O_3$
* 1-NITRO-1-OXIMINOETHANE(ETHYLNITROLIC ACID), $C_2H_4N_2O_3$
2,3-EPOXYPROPIONALDEHYDE OXIME, $C_3H_5NO_2$
BROMOACETONE OXIME, $C_3H_6BrNO$
AZIDOACETONE OXIME, $C_3H_6N_4O$
1-HYDROXYIMIDAZOLE-2-CARBOXALDOXIME 3-OXIDE, $C_4H_5N_3O_3$
2-METHYLACRYLALDEHYDE OXIME, $C_4H_7NO$
2,3-BUTANEDIONE MONOXIME, $C_4H_7NO_2$
2-BUTANONE OXIME, $C_4H_9NO$
BUTYRALDEHYDE OXIME, $C_4H_9NO$
2-BUTANONE OXIME HYDROCHLORIDE, $C_4H_{10}ClNO$
2-ETHYLACRYLALDEHYDE OXIME, $C_5H_9NO$
1,2,3- or 1,3,5-CYCLOHEXANETRIONE TRIOXIME, $C_6H_9N_3O_3$
2-ISOPROPYLACRYLALDEHYDE OXIME, $C_6H_{11}NO$
ETHYL 2-FORMYLPROPIONATE OXIME, $C_6H_{11}NO_3$

* TRIPOTASSIUM CYCLOHEXANEHEXONE-1,3,5-TRIOXIMATE, $C_6K_3N_3O_6$
* 2,2′-OXYDI[(IMINOMETHYL)FURAN] MONO-*N*-OXIDE ('DEHYDROFURFURAL OXIME'), $C_{10}H_8N_2O_4$

*See* KETOXIMINOSILANES
SULPHURIC ACID, $H_2O_4S$, Cyclopentanone oxime
HYDROXYLAMINE, $H_3NO$: Carbonyl compound, etc.

## OXOHALOGEN ACIDS

**$HOXO_n$**

The oxidising power of the group of oxohalogen acids increases directly with oxygen content, though the high stability of the perchlorate ion at ambient temperature must be taken into account. The corresponding 'anhydrides' (halogen oxides) are also powerful oxidants, several being explosively unstable.

*See* BROMIC ACID, $BrHO_3$
HYPOCHLOROUS ACID, ClHO
CHLORIC ACID, $ClHO_3$
PERCHLORIC ACID, $ClHO_4$
PERIODIC ACID, $HIO_4$
ORTHOPERIODIC ACID, $H_5IO_6$

*See also* HALOGEN OXIDES

## OXOSALTS OF NITROGENOUS BASES

**$N^+EO_n^-$**

1. Fogel'zang, A. G. *et al.*, *Chem. Abs.*, 1971, **75**, 142412h
2. Mikhailova, T. A. *et al.*, *Chem. Abs.*, 1976, **85**, 201766

Burning rates of ammonium salts were investigated at various constant pressures. Ammonium permanganate burns faster than ammonium bromate, chlorate, dichromate, iodate, nitrite, perchlorate, periodate or triperchromate [1]. When submitting explosive nitrogenous bases or their salts to elemental nitrogen determination, it is advantageous to pre-treat the samples with phosphoric acid which allows the more stable pyro- and meta-phosphates (formed at 260° and 300°C, respectively) to be fully oxidised without explosion in the Dumas nitrogen procedure [2].

Many of the salts of nitrogenous bases (particularly of high nitrogen content) with oxoacids are unstable or explosive.

There are separate group entries for:

AMINIUM IODATES AND PERIODATES
AMINIUM PERCHLORATES
DIAZONIUM PERCHLORATES
DICHROMATE SALTS OF NITROGENOUS BASES
1-(1,3-DISELENOLYLIDENE)PIPERIDINIUM PERCHLORATES
HYDRAZINIUM SALTS
HYDROXYLAMINIUM SALTS
POLY(AMINIUM PERCHLORATES)

NITRITE SALTS OF NITROGENOUS BASES

and individually indexed compounds are:

AMMONIUM BROMATE, $BrH_4NO_3$
CHLOROFORMAMIDINIUM NITRATE, $CH_4ClN_3O_3$
CHLOROFORMAMIDINIUM PERCHLORATE, $CH_4Cl_2N_2O_4$
URONIUM PERCHLORATE (UREA PERCHLORATE), $CH_5ClN_2O_5$
URONIUM NITRATE (UREA NITRATE), $CH_5N_3O_4$
METHYLAMMONIUM CHLORITE, $CH_6ClNO_2$
METHYLAMMONIUM PERCHLORATE, $CH_6ClNO_4$
GUANIDINIUM PERCHLORATE, $CH_6ClN_3O_4$
GUANIDINIUM NITRATE, $CH_6N_4O_3$
AMINOGUANIDINIUM NITRATE, $CH_7N_5O_3$
DIAMINOGUANIDINIUM NITRATE, $CH_8N_6O_3$
TRIAMINOGUANIDINIUM PERCHLORATE, $CH_9ClN_6O_4$
TRIAMINOGUANIDINIUM NITRATE, $CH_9N_7O_3$
2-AZA-1,3-DIOXOLANIUM PERCHLORATE (ETHYLENE-DIOXYAMMONIUM PERCHLORATE), $C_2H_6ClNO_6$
1-METHYL-3-NITROGUANIDINIUM PERCHLORATE, $C_2H_7ClN_4O_6$
1-METHYL-3-NITROGUANIDINIUM NITRATE, $C_2H_7N_5O_5$
DIMETHYLAMMONIUM PERCHLORATE, $C_2H_8ClNO_4$
2-AMINOETHYLAMMONIUM PERCHLORATE, $C_2H_9ClNO_4$
1,2-ETHYLENEBIS-AMMONIUM PERCHLORATE, $C_2H_{10}Cl_2N_2O_8$
GUANIDINIUM DICHROMATE, $C_2H_{12}Cr_2N_6O_7$
TRIMETHYLAMINE *N*-OXIDE PERCHLORATE, $C_3H_{10}ClNO_5$
4-CHLORO-1-METHYLIMIDAZOLIUM NITRATE, $C_4H_5ClN_3O_3$
TETRAMETHYLAMMONIUM CHLORITE, $C_4H_{12}ClNO_2$
PYRIDINIUM PERCHLORATE, $C_5H_6ClO_4$
2,4-DINITROPHENYLHYDRAZINIUM PERCHLORATE, $C_6H_7ClN_4O_8$
* 2-DIETHYLAMMONIOETHYLNITRATE NITRATE, $C_6H_{15}N_3O_6$
TRIETHYLAMMONIUM NITRATE, $C_6H_{16}NO_3$
1,2-BIS(ETHYLAMMONIO)ETHANE PERCHLORATE, $C_6H_{18}Cl_2N_2O_8$
2-METHOXYANILINIUM NITRATE, $C_7H_{10}N_2O_4$
1-*p*-CHLOROPHENYLBIGUANIDIUM HYDROGENDICHROMATE, $C_8H_{12}ClCr_2N_5O_7$
1-PHENYLBIGUANIDIUM HYDROGENDICHROMATE, $C_8H_{13}Cr_2N_5O_7$
3-AZONIABICYCLO[3.2.2]NONANE NITRATE, $C_8H_{16}N_2O_3$
1,3,6,8-TETRAAZONIATRICYCLO [6.2.1.1$^{3,6}$] DODECANE TETRANITRATE, $C_8H_{20}N_8O_{12}$
TETRAMETHYLAMMONIUM PENTAPEROXODICHROMATE, $C_8H_{24}Cr_2N_2O_{12}$
1,2,3,4-TETRAHYDROISOQUINOLINIUM NITRATE, $C_9H_{12}N_2O_3$
1,2,3,4-TETRAHYDROQUINOLINIUM NITRATE, $C_9H_{12}N_2O_3$
*p*-TOLYLBIGUANIDIUM HYDROGENDICHROMATE, $C_9H_{15}Cr_2N_5O_7$
HEXAMETHYLENETETRAMMONIUM TETRAPEROXOCHROMATE(V) (3–) (?), $C_{18}H_{43}Cr_4N_{12}O_{32}$
TRI-*p*-TOLYLAMMONIUM PERCHLORATE, $C_{21}H_{22}ClNO_4$
AMMONIUM CHLORATE, $ClH_4NO_3$
AMMONIUM PERCHLORATE, $ClH_4NO_4$
HYDROXYLAMINIUM PERCHLORATE, $ClH_4NO_5$
PHOSPHONIUM PERCHLORATE, $ClH_4O_4P$
HYDRAZINIUM CHLORITE, $ClH_5N_2O_2$
HYDRAZINIUM CHLORATE, $ClH_5N_2O_3$
HYDRAZINIUM PERCHLORATE, $ClH_5N_2O_4$

* AMMONIUM FLUOROCHROMATE, $CrFH_4NO_3$
AMMONIUM DICHROMATE, $Cr_2H_8N_2O_7$
HYDRAZINIUM DIPERCHLORATE, $Cl_2H_6N_2O_8$
AMMONIUM IODATE, $H_4INO_3$
AMMONIUM PERIODATE, $H_4INO_4$
AMMONIUM PERMANGANATE, $H_4MnNO_4$
AMMONIUM NITRITE, $H_4N_2O_2$
AMMONIUM NITRATE, $H_4N_2O_3$
HYDROXYLAMINIUM NITRATE, $H_4N_2O_4$
HYDRAZINIUM NITRITE, $H_5N_3O_2$
HYDRAZINIUM NITRATE, $H_5N_3O_3$
HYDROXYLAMINIUM PHOSPHINATE, $H_6NO_3P$
AMMONIUM AMIDOSULPHATE ('SULPHAMATE'), $H_6N_2O_3S$
* AMMONIUM AMIDOSELENATE, $H_6N_2O_3Se$
HYDRAZINIUM HYDROGENSELENATE, $H_6N_2O_4Se$
HYDRAZINIUM DINITRATE, $H_6N_4O_6$
* AMMONIUM THIOSULPHATE, $H_8N_2O_3S_2$
HYDROXYLAMINIUM SULPHATE, $H_8N_2O_6S$
AMMONIUM PEROXODISULPHATE, $H_8N_2O_8S_2$

*See also* CHLORITE SALTS
DICHROMATE SALTS OF NITROGENOUS BASES

## OXYGEN BALANCE

1. Kirk-Othmer, 1965, Vol. 8, 581
2. Slack, R., private comm., 1957

Oxygen balance is the difference between the oxygen content of a chemical compound and that required fully to oxidise the carbon, hydrogen and other oxidisable elements present to carbon dioxide, water, etc. The concept is of particular importance in the design of explosive compounds or compositions, since the explosive power is maximal at equivalence, or zero oxygen balance. If there is a deficiency of oxygen present, the balance is negative, while an excess of oxygen gives a positive balance, and such compounds can function as oxidants. The balance is usually expressed as a percentage. The nitrogen content of a compound is not considered as oxidisable, as it is usually liberated as the gaseous element in explosive decomposition [1].

While it is, then, possible to recognise highly explosive materials by consideration of their oxygen balance (e.g. ETHYLENE DINITRATE, $C_2H_4N_2O_6$ is zero-balanced; 3,4,-BIS(1,2,3,4-THIATRIAZOL-5-YLTHIO)-MALEIMIDE, $C_4HN_7O_2S_4$, has a positive balance), the tendency to instability becomes apparent well below the zero-balance point. The empirical statement that the stability of any organic compound is doubtful when the oxygen or sulphur content approaches that necessary to convert the other elements present to their lowest state of oxidation (one sulphur atom equalling two oxygen atoms here) forms a useful guide [2].

## OXYGEN ENRICHMENT

1. 'Oxygen Enrichment of Confined Areas', Information Sheet, London, Inst. of Welding, 1966

2. Wilk, I. J., *J. Chem. Educ.*, 1968, **45**, A547–551
3. Johnson, J. E. *et al.*, *NRL Rep 6470*, Washington, Nav. Res. Lab., 1966
4. Woods, F. J. *et al.*, *NRL Rep. 6606*, Washington, Nav. Res. Lab., 1967
5. Denison, D. M. *et al.*, *Nature*, 1968, **218**, 1111–1113

With the widening industrial use of oxygen, accidents caused by atmospheric enrichment are increasing. Most materials, especially clothing, burn fiercely in an atmosphere containing more than the usual 21% of oxygen. In presence of petroleum products, fire and explosion can be spontaneous. Equipment which may emit or leak oxygen should be used sparingly, and never stored, in confined spaces [1].

Fourteen case histories of accidents caused by oxygen enrichment of the atmosphere are discussed and safety precautions described [2].

The flammability of textiles and other solids was studied under the unusual atmospheric conditions which occur in deep diving operations. The greatest effect upon ease of ignition and linear burning rate was caused by oxygen enrichment; increase in pressure had a similar effect [3].

Ignition and flame spread of fabrics and paper were measured at pressures from 21 bar down to the limiting pressure for ignition to occur. Increase in oxygen concentration above 21% in mixtures with nitrogen caused rapid decrease of minimum pressure for ignition.

In general, but not invariably, materials ignite less readily but burn faster in helium mixtures than in nitrogen mixtures. Nature of material has a marked influence on effect of variables on rate of burning. At oxygen concentrations of 41% all materials examined would burn except for glass and polytetrafluoroethylene, which resisted ignition attempts in pure oxygen. Flame retardants become ineffective on cotton in atmospheres containing above 32% oxygen [4].

A brief summary of known hazards and information in this general area is available [5].

## OXYGEN FLUORIDES $O_2F$, FOF, FOOF, $O_2FOF$, $(O_2F)_2$

Streng, A. G., *Chem. Rev.*, 1963, **63**, 607

In the series oxygen difluoride, dioxygen difluoride, trioxygen difluoride and tetraoxygen difluoride, as the oxygen content increases, the stability decreases and the oxidising power increases, tetraoxygen difluoride, even at −200°C, being one of the most potent oxidants known. Applications to both chemical reaction and rocket propulsion systems are covered in some detail.

*See* OXYGEN DIFLUORIDE, $F_2O$
DIOXYGEN DIFLUORIDE, $F_2O_2$
TRIOXYGEN DIFLUORIDE, $F_2O_3$
*See other* HALOGEN OXIDES

## OXYGEN INDEX

1. *Oxygen Index Test*, ASTM D2863, 1970
2. Isaacs, J. S., *J. Fire Flamm.*, 1970, **1**, 36–47
3. Kamp, A. C. F. *et al.*, *Proc. 1st Euro. Sympos. Therm. Anal.*, 1976, 440–443

The flammability properties of volatile materials are readily gauged from the values of the figures for flash point and limits of flammability in air. For involatile solid materials, a range of empirical tests provides a measure of flammability properties under various circumstances. One of the tests, originally developed to measure the minimum concentration of oxygen in which a sample of a plastics material will continue to burn candle-like, is simple, accurate and reproducible [1]. Results are expressed as an Oxygen Index (O.I.), which is the minimum proportion of oxygen in a mixture with nitrogen to just sustain combustion of a sample of standard size in a chimney-type apparatus. The method has also been applied to textile materials [2], and may well be applicable to a much wider range of materials.

Determination of the Oxygen Index over a wide temperature range gives a better understanding of flammability parameters, and the Temperature Index (T.I., the temperature at which O.I. is 20.8) will rank flammability of materials. The O.I. at ambient temperature indicates potential hazard at the primary ignition stage of a fire, while the T.I. and the O.I.–temperature relation is related to practical fire situations [3].

## 1-OXYPEROXY COMPOUNDS — C(OH)OO–

Swern, 1970, Vol. 1, 29, 33

This group of compounds includes those monomers with one or more carbon atoms carrying a hydroperoxy or peroxy group and also singly bonded to an oxygen atom present as hydroxyl, ether or cyclic ether functions. While the group of compounds, in general, is moderately stable, the lower 1-hydroxy- and 1,1′-dihydroxy-alkyl peroxides or hydroperoxides are explosive.

*See* HYDROXYMETHYL HYDROPEROXIDE, $CH_4O_3$
HYDROXYMETHYL METHYL PEROXIDE, $C_2H_6O_3$
BIS-HYDROXYMETHYL PEROXIDE, $C_2H_6O_4$
1-HYDROXY-3-BUTYL HYDROPEROXIDE, $C_4H_{10}O_3$
BIS(1-HYDROXYCYCLOHEXYL) PEROXIDE, $C_{12}H_{22}O_4$
1(1′-HYDROPEROXY-1′-CYCLOHEXYLPEROXY)-CYCLOHEXANOL, $C_{12}H_{22}O_5$
BIS(1-HYDROPEROXYCYCLOHEXYL) PEROXIDE, $C_{12}H_{22}O_6$
1-ACETOXY-1-HYDROPEROXY-6-CYCLODODECANONE, $C_{14}H_{24}O_5$

## OZONIDES — formula on p. 190

1. Rieche, A., *Angew. Chem.*, 1958, **70**, 251
2. Swern, 1970, Vol. 1, 39; Bailey, P. S., *Chem. Rev.*, 1958, **58**, 928

3. Greenwood, F. L. *et al., J. Org. Chem.*, 1967, **32**, 3373
4. Rieche, A. *et al., Ann.*, 1942, **553**, 187, 224

The preparation, properties and uses of ozonides have been comprehensively reviewed [1]. Many pure ozonides are generally stable to storage; some may be distilled under reduced pressure. The presence of peroxidic impurities is thought to cause the violently explosive decomposition often observed in this group. Use of ozone is not essential for their formation, as they are also produced by dehydration of $\alpha$, $\alpha'$-dihydroxy peroxides [2].

Polymeric alkene ozonides are shock-sensitive; that of *trans*-2-butene exploded when exposed to friction in a ground glass joint. The use of GLC to analyse crude ozonisation products is questionable because of the heat-sensitivity of some constituents [3]. Ozonides are decomposed, sometimes explosively, by finely divided palladium, platinum or silver, or by iron(II) salts [4].

Individually indexed compounds are:

ETHYLENE OZONIDE, $C_2H_4O_3$
PROPENE OZONIDE, $C_3H_6O_3$
MALEIC ANHYDRIDE OZONIDE, $C_4H_2O_6$
VINYL ACETATE OZONIDE, $C_4H_6O_5$
*trans*-2-BUTENE OZONIDE, $C_4H_8O_3$
ISOPRENE DIOZONIDE, $C_5H_8O_6$
*trans*-2-PENTENE OZONIDE, $C_5H_{10}O_3$
BENZENE TRIOZONIDE, $C_6H_6O_9$
*trans*-2-HEXENE OZONIDE, $C_6H_{12}O_3$
1,2-DIMETHYLCYCLOPENTENE OZONIDE, $C_7H_{12}O_3$
4-HYDROXY-4-METHYL-1,6-HEPTADIENE DIOZONIDE, $C_8H_{14}O_7$
2,6-DIMETHYL-2,5-HEPTADIEN-4-ONE DIOZONIDE, $C_9H_{14}O_7$
1,3-DIPHENYL-1,3-EPIDIOXY-1,3-DIHYDROISOBENZOFURAN, $C_{20}H_{14}O_3$

## PAPER TOWELS

Unpublished observations, 1970

The increasing use of disposable paper towels in chemical laboratories accentuates the fire hazard potentially created by disposal of solid oxidising agents or reactive residues into a bin containing such towels. The partially wet paper, necessarily of high surface area and absorbency, presents favourable conditions for fire to be initiated and spread. Separate bins for paper towels and chemical residues seem advisable.

*See* SODA-LIME

## PERCHLORATES —$OClO_3$

1. Schumacher, 1960
2. Burton, H. *et al.*, *Analyst*, 1955, **80**, 4

All perchlorates have some potential for hazard when in contact with other reactive materials, while many are intrinsically hazardous, owing to the high oxygen content.

Existing knowledge on perchloric acid and its salts was extensively reviewed in 1960 in a monograph including the chapters: Perchloric Acid; Alkali Metal, Ammonium and Alkaline Earth Perchlorates; Metal Perchlorates; Miscellaneous Perchlorates; Manufacture of Perchloric Acid and Perchlorates; Analytical Chemistry of Perchlorates; Perchlorates in Explosives and Propellants; Miscellaneous Uses of Perchlorates; Safety Considerations in Handling Perchlorates [1].

There is a shorter earlier review, with a detailed treatment of the potentially catastrophic acetic anhydride–acetic acid–perchloric acid system. The violently explosive properties of methyl, ethyl and lower alkyl perchlorates, and the likelihood of their formation in alcohol–perchloric acid systems, are stressed. The instability of diazonium perchlorates, some when damp, is mentioned [2].

The group has been divided into the separately treated sub-groups:

ALKYL PERCHLORATES
AMMINEMETAL OXOSALTS
AMMONIUM PERCHLORATES
DIAZONIUM PERCHLORATES
METAL PERCHLORATES
NON-METAL PERCHLORATES

Glycol,
Polymer

*MCA Case History No. 464*

A mixture of an inorganic perchlorate, a glycol and a polymer exploded violently after heating at 265–270°C. It was stated that the glycol may have become oxidised, but formation of a perchlorate ester seems a more likely cause.

Organic matter

Schumacher, 1960, 188

Mixtures with finely divided or fibrous organic material are likely to be explosive. Porous or fibrous materials exposed to aqueous solutions and then dried are rendered explosively flammable and are easily ignited.

Reducants

Mellor, 1941, Vol. 2, 387; Vol. 3, 651
Schumacher, 1960, 188

Perchlorate salts react explosively when rubbed in a mortar with calcium hydride or with sulphur and charcoal; when melted with reducants; or on contact with glowing charcoal. Mixtures with finely divided aluminium, magnesium, zinc or other metals are explosives.

**PERCHLORYLAMIDE SALTS** **$NHClO_3^-$**

'Perchloryl Fluoride', Booklet DC-1819, Philadelphia, Pennsalt Chem. Corp., 1957

Ammonium perchlorylamide and the corresponding silver and barium salts are shock-sensitive when dry and may detonate. Extreme care is required when handling such salts.
*See other* *N*-METAL DERIVATIVES

## PERCHLORYL COMPOUNDS $RClO_3$

1. *Friedel–Crafts and Related Reactions*, Olah, G. A. (Editor), Vol. 3.2, 1507–1516, New York, Interscience, 1964
2. Baum, K. *et al.*, *J. Amer. Chem. Soc.*, 1974, **96**, 3233–3237
3. Beard, C. D. *et al.*, ibid., 3237–3239

The need for great care to avoid the possibility of detonation of perchloryl compounds by exposure to shock, overheating or sparks is stressed. The compounds are generally more sensitive to impact than mercury fulminate or 1,3,5-trinitrohexahydro-1,3,5-triazine (RDX), and are of comparable sensitivity to lead azide [1]. A range of highly explosive alkyl perchlorates [2] and perchlorylamines [3] has been prepared by interaction of dichlorine heptaoxide with alcohols or amines in carbon tetrachloride solution. The solutions of the products were not sensitive to mechanical shock and could be used directly for further reactions. Adequate warning of the explosion hazards involved in isolating the perchloryl compounds is given.

Organic compounds containing the perchloryl substituent are inherently shock-sensitive and powerful explosives. The few available examples are:

1-PERCHLORYLPIPERDINE, $C_5H_{10}ClNO_3$
2,6-DINITRO-4-PERCHLORYLPHENOL, $C_6H_3ClN_2O_8$
NITROPERCHLORYLBENZENE, $C_6H_4ClNO_5$
PERCHLORYLBENZENE, $C_6H_5ClO_3$
2,6-DIPERCHLORYL-4,4′-DIPHENOQUINONE, $C_{12}H_6Cl_2O_8$
PERCHLORYL FLUORIDE, $ClFO_3$

*See also* PERCHLORYLAMIDE SALTS

## PERFLUOROSILANES $Si_mF_n$

Bailar, 1973, Vol. 1, 1385
The compounds in this class ($Si_2F_6$ to $Si_4F_{10}$ are mentioned later) ignite in air and react violently with water.
*See other* NON-METAL HALIDES

## PEROXIDES –OO–

This group name probably covers the largest group of hazardous compounds and there are three main divisions:

INORGANIC PEROXIDES
ORGANIC PEROXIDES
ORGANOMINERAL PEROXIDES

## PEROXIDES IN SOLVENTS

Many laboratory accidents have been ascribed to presence of peroxides in solvents, usually, but not exclusively, ethers.

When peroxides are removed from solvents by adsorption on alumina columns, the concentrated band at the top of the column may become hazardous if the solvent evaporates.

*See* ETHERS
PEROXIDISABLE COMPOUNDS
4-METHYL-2-PENTANONE, $C_6H_{12}O$
HYDROGEN PEROXIDE, $H_2O_2$: Acetone, etc.
PEROXOMONOSULPHURIC ACID, $H_2O_5S$: Acetone

## PEROXIDISABLE COMPOUNDS

1. Jackson, H.L. *et al.*, *J. Chem. Educ.*, 1970, **47**, A175
2. Brandsma, 1971, 13
3. *MCA Case History No. 1693*
4. *Recognition and Handling of Peroxidisable Compounds*, Data sheet 655, Chicago, National Safety Council, 1976

An account of a Du Pont safety study of the control of peroxidisable compounds covers structure examples, handling procedures, distillation of peroxidisable compounds, and detection and elimination of peroxides [1].

Essential organic structural features for a peroxidisable hydrogen atom are recognised as:

| | |
|---|---|
| $>\underset{\text{H}}{\underset{\mid}{\text{C}}}-\text{O}-$ | as in acetals, ethers, oxygen heterocycles |
| $\begin{matrix}-\text{CH}_2\\-\text{CH}_2\end{matrix}>\underset{\text{H}}{\underset{\mid}{\text{C}}}-$ | as in isopropyl compounds, decahydronaphthalenes |
| $>\text{C}=\text{C}-\underset{\text{H}}{\underset{\mid}{\text{C}}}-$ | as in allyl compounds |
| $>\text{C}=\underset{\text{H}}{\underset{\mid}{\text{C}}}\diagup\text{X}$ | as in haloalkenes |
| $>\text{C}=\underset{\text{H}}{\underset{\mid}{\text{C}}}\diagup$ | as in other vinyl compounds (monomeric esters, ethers, etc.) |
| $>\text{C}=\underset{\text{H}}{\underset{\mid}{\text{C}}}-\underset{\text{H}}{\underset{\mid}{\text{C}}}=\text{C}<$ | as in dienes |

$$>C=C(H)-C\equiv C-$$ as in vinylacetylenes

$$-C-C(H)-Ar$$ as in cumenes, tetrahydronaphthalenes, styrenes

$$-C(H)=O$$ as in aldehydes

$$-C(=O)-N-C(H)<$$ as in *N*-alkyl-amides or -ureas, lactams

While the two latter types readily peroxidise, the products are readily degraded and do not accumulate to a hazardous level.

Inorganic compounds which readily peroxidise are listed as potassium and higher alkali metals, alkali metal alkoxides and amides, and organometallic compounds.

Three lists of specific compounds or compound types indicate different types of potential hazard, and appropriate storage, handling and disposal procedures are detailed.

List A, giving examples of compounds which form explosive peroxides in storage only, include diisopropyl ether, divinylacetylene, vinylidene chloride, potassium and sodium amide. Review of stocks and testing for peroxide content by given tested procedures at three-monthly intervals is recommended, together with safe disposal of any peroxidic samples.

List B, giving examples of liquids where a degree of concentration is necessary before hazardous levels of peroxide will develop includes several common solvents containing one ether function (diethyl ether, tetrahydrofuran, ethyl vinyl ether) or two ether functions (*p*-dioxan, 1,1-diethoxyethane, the dimethyl ethers of ethylene glycol or 'diethylene glycol') as well as the susceptible hydrocarbons propyne, butadiyne, dicyclopentadiene, cyclohexene, and tetra- and decahydronaphthalenes. Checking stocks at 12-monthly intervals, with peroxidic samples being discarded or repurified, is recommended here [1].

A simple method of effectively preventing accumulation of dangerously high concentrations of peroxidic species in distillation residues is that detailed in an outstanding practical textbook of preparative acetylene chemistry [2]. The material to be distilled is mixed with an equal volume of non-volatile mineral oil. This remains after distillation as an inert diluent for polymeric peroxidic materials.

List C contains peroxidisable monomers where the presence of peroxide may initiate exothermic polymerisation of the bulk of material. Precautions and procedures for storage and use of monomers with or without the presence of inhibitors are discussed in detail. Examples cited are acrylic acid, acrylonitirile, butadiene, 2-chlorobutadiene, chlorotrifluoroethylene, methyl methacrylate, styrene, tetrafluoroethylene, vinyl acetate, vinylacetylene, vinyl chloride, vinylidene chloride and vinylpyridine [1].

In general terms, the presence of two or more of the structural features indicated above in the same compound will tend to increase the likelihood of hazard. The selection of compound classes and of individually indexed compounds below includes compounds known to have been involved or those with a multiplicity of such structural features which would be expected to be especially susceptible to peroxide formation.

Subsequent to incidents involving peroxidation of stored bottles of vinylidene chloride, a labelling procedure and list of peroxidisable compounds has been prepared [3]. Of the 108 compounds listed, 35 are noted as forming peroxides with ease, and these are marked + in the list below, although many have no corresponding text entry. A new data sheet exists [4].

Separately treated groups are:

ACETYLENIC COMPOUNDS
ALLYL COMPOUNDS
DIENES
HALOALKENES
TETRAHYDROPYRANYL ETHER DERIVATIVES

and individually indexed compounds:

+ 1,1-DICHLOROETHYLENE, $C_2H_2Cl_2$
ACETALDEHYDE, $C_2H_4O$
2-CHLOROACRYLONITRILE, $C_3H_2ClN$
ACRYLALDEHYDE, $C_3H_4O$
2-PROPYNE-1-THIOL, $C_3H_4S$
1,1,2,3-TETRACHLORO-1,3-BUTADIENE, $C_4H_2Cl_4$
BUTEN-3-YNE, $C_4H_4$
+ 2-CHLORO-1,3-BUTADIENE ('CHLOROPRENE'), $C_4H_5Cl$
1,2- or 1,3-BUTADIENE, $C_4H_6$
CROTONALDEHYDE, $C_4H_6O$
DIVINYL ETHER, $C_4H_6O$
VINYL ACETATE, $C_4H_6O_2$
ETHYL VINYL ETHER, $C_4H_8O$
+ TETRAHYDROFURAN, $C_4H_8O$
*m*-DIOXANE, $C_4H_8O_2$
*p*-DIOXANE, $C_4H_8O_2$
+ DIETHYL ETHER, $C_4H_{10}O$
+ 1,1- or 1,2-DIMETHOXYETHANE, $C_4H_{10}O_2$
2-PENTEN-4-YN-3-OL, $C_5H_6O$
ALLYL VINYL ETHER, $C_5H_8O$
2,3-DIHYDROPYRAN, $C_5H_8O$
1-ETHOXY-2-PROPYNE, $C_5H_8O$
ALLYL ETHYL ETHER, $C_5H_{10}O$
ISOPROPYL VINYL ETHER, $C_5H_{10}O$
TETRAHYDROPYRAN, $C_5H_{10}O$
3,3-DIMETHOXYPROPENE, $C_5H_{10}O_2$
2,2-DIMETHYL-1,3-DIOXOLAN, $C_5H_{10}O_2$
+ 1,3-DIOXEPANE, $C_5H_{10}O_2$
2-METHOXYETHYL VINYL ETHER, $C_5H_{10}O_2$
4-METHYL-1,3-DIOXAN, $C_5H_{10}O_2$
ETHYL ISOPROPYL ETHER, $C_5H_{12}O$

+ DIETHOXYMETHANE, $C_5H_{12}O_2$
+ 1,1-DIMETHOXYPROPANE, $C_5H_{12}O_2$
+ 2,2-DIMETHOXYPROPANE, $C_5H_{12}O_2$
DI(2-PROPYN-1-YL) ETHER (DIPROPARGYL ETHER), $C_6H_6O$
4,5-HEXADIEN-2-YN-1-OL, $C_6H_6O$
+ 3-ISOPROPOXYPROPIONITRILE, $C_6H_7NO$
+ CYCLOHEXENE, $C_6H_{10}$
+ DIALLYL ETHER (DI(2-PROPEN-1-YL) ETHER), $C_6H_{10}O$
DIETHYLKETENE, $C_6H_{10}O$
ISOBUTYL VINYL ETHER, $C_6H_{12}O$
2,6-DIMETHYL-1,4-DIOXAN, $C_6H_{12}O_2$
+ 1,2-EPOXY-3-ISOPROPOXYPROPANE, $C_6H_{12}O_2$
+ 1,3,3-TRIMETHOXYPROPENE, $C_6H_{12}O_3$
+ 1-CHLORO-2,2-DIETHOXYETHANE, $C_6H_{13}ClO_2$
+ DIISOPROPYL ETHER, $C_6H_{14}O$
+ ISOPROPYL PROPYL ETHER, $C_6H_{14}O$
+ 1,1-DIETHOXYETHANE (DIETHYLACETAL), $C_6H_{14}O_2$
1,2-DIETHOXYETHANE, $C_6H_{14}O_2$
3,3-DIETHOXYPROPENE, $C_7H_{14}O_2$
+ DIPROPOXYMETHANE, $C_7H_{16}O_2$
4-VINYLCYCLOHEXENE, $C_8H_{12}$
+ DIETHYL FUMARATE, $C_8H_{12}O_4$
+ CYCLOOCTENE, $C_8H_{14}$
DIBUTYL ETHER, $C_8H_{18}O$
CINNAMALDEHYDE, $C_9H_8O$
+ ALLYL PHENYL ETHER, $C_9H_{10}O$
+ BENZYL ETHYL ETHER, $C_9H_{12}O$
+ 3,3,5-TRIMETHYL-2-CYCLOHEXEN-1-ONE (ISOPHORONE), $C_9H_{14}O$
+ 1,2,3,4-TETRAHYDRONAPHTHALENE, $C_{10}H_{12}$
+ DECAHYDRONAPHTHALENE, $C_{10}H_{18}$
2-ETHYLHEXYL VINYL ETHER, $C_{10}H_{20}O$
+ BENZYL BUTYL ETHER, $C_{12}H_{18}O$
+ DIBENZYL ETHER, $C_{14}H_{14}O$
*a*-PENTYLCINNAMALDEHYDE, $C_{14}H_{18}O$
+ 1,2-DIBENZYLOXYETHANE, $C_{16}H_{18}O_2$
+ BENZYL 1-NAPHTHYL ETHER, $C_{17}H_{14}O$
+ *p*-DIBENZYLOXYBENZENE, $C_{20}H_{18}O_2$
POTASSIUM AMIDE, $H_2KN$
SODIUM AMIDE, $H_2NNa$
POTASSIUM, K

## PEROXOACIDS — EOOH

Inorganic acids with a peroxide function are given the IUPAC group name above, which distinguishes them from the organic PEROXYACIDS. Collectively they are a group of very powerful oxidants.

*See* PEROXONITRIC ACID, $HNO_4$
PEROXOMONOSULPHURIC ACID, $H_2O_5S$
PEROXODISULPHURIC ACID, $H_2O_8S_2$
PEROXOMONOPHOSPHORIC ACID, $H_3O_5P$

## PEROXOACID SALTS — $EOO^-$, $MOO^-$

Many of the salts of peroxoacids are unstable or explosive, are capable of initiation by heat, friction or impact, and all are powerful oxidants. Individually indexed compounds are:

SILVER PEROXOCHROMATE, $AgCrO_5$
AMMONIUM PEROXOBORATE, $BH_4NO_3$
SODIUM PEROXOBORATE, $BNaO_3$
THALLIUM(I) PEROXOBORATE, $B_2O_7Tl_2 \cdot H_2O$
SODIUM PEROXYACETATE, $C_2H_3NaO_3$
POTASSIUM TRICYANODIPEROXOCHROMATE(3–), $C_3CrK_3N_3O_4$
POTASSIUM *O–O*-BENZOYLMONOPEROXOSULPHATE, $C_7H_5KO_6S$
TETRAMETHYLAMMONIUM PENTAPEROXODICHROMATE(2–), $C_8H_{24}Cr_2N_2O_{12}$
MERCURY PEROXYBENZOATE, $C_{14}H_{10}HgO_6$
HEXAMETHYLENETETRAMMONIUM TETRAPEROXOCHROMATE(V) (3–) (?), $C_{18}H_{48}Cr_4N_{12}O_{32}$
CALCIUM PEROXODISULPHATE, $CaO_8S_2$
CALCIUM PEROXOCHROMATE (3–), $Ca_3Cr_2O_{12}$
POTASSIUM HYDROXOOXODIPEROXOCHROMATE, $CrHKO_6$
THALLIUM HYDROXOOXODIPEROXOCHROMATE, $CrHO_6Tl$
AMMONIUM HYDROXOOXODIPEROXOCHROMATE, $CrH_5NO_6$
AMMONIUM TETRAPEROXOCHROMATE (3–), $CrH_{12}N_2O_8$
POTASSIUM TETRAPEROXOCHROMATE (3–), $CrK_3O_8$
SODIUM TETRAPEROXOCHROMATE(3–), $CrNa_3O_8$
AMMONIUM PENTAPEROXODICHROMATE(2–), $Cr_2H_8N_2O_{12}$
POTASSIUM PENTAPEROXODICHROMATE(2–), $Cr_2K_2O_{12}$
POTASSIUM PEROXOFERRATE (2–), $FeK_2O_5$
POTASSIUM PEROXOMONOSULPHATE, $HKO_5S$
AMMONIUM PEROXODISULPHATE, $H_8N_2O_8S_2$
TETRAAMMINEZINC PEROXODISULPHATE, $H_{12}N_4O_8S_2Zn$
POTASSIUM DIPEROXOMOLYBDATE(2–), $K_2MoO_6$
POTASSIUM TETRAPEROXOMOLYBDATE (2–), $K_2MoO_8$
POTASSIUM DIPEROXOORTHOVANADATE(2–), $K_2O_6V$
POTASSIUM PEROXODISULPHATE(2–), $K_2O_8S_2$
POTASSIUM TETRAPEROXOTUNGSTATE(2–), $K_2O_8W$
SODIUM TETRAPEROXOMOLYBDATE (2–), $MoNa_2O_8$
SODIUM TETRAPEROXOTUNGSTATE (2–), $Na_2O_8W_2$

*See other* OXIDANTS

## PEROXYACIDS

**RCO–OOH**


1. Castrantas, 1965, 12; Swern, 1970, Vol. 1, 59, 337
2. Isard, A. *et al.*, Ger. Offen., 1 643 158, 1968
3. Isard, A. *et al.*, *Chemical Tech.*, 1974, **4**, 380
4. Swern, D., *Chem. Rev.*, 1945, **45**, 3–16


The peroxyacids were until recently the most powerful oxidants of all organic peroxides, and it is often unnecessary to isolate them from the mixture of acid and hydrogen peroxide used to generate them. The pure lower aliphatic members are explosive (performic, particularly) at high, but not low, concentrations, being sensitive to heat but not usually to shock. Dipicolinic acid, or phosphates, have been used to stabilise these solutions. The detonable limits of peroxyacid solutions can be plotted by extrapolation from known data. Aromatic peroxyacids are generally more stable, particularly if ring substituents are present [1].

The patented preparation of peroxyacids [2] by interaction of carboxylic acids with hydrogen peroxide in presence of metaboric acid needs appropriate safeguards to prevent accidental separation of the concentrated peroxyacids [3]. Much descriptive data on stabilities of a wide selection of peroxyacids has been summarised [4]. Salts of peroxy acids are listed with peroxoacid salts above.

*See* PEROXYFORMIC ACID, $CH_2O_3$
PEROXYTRIFLUOROACETIC ACID, $C_2HF_3O_3$
PEROXYACETIC ACID, $C_2H_4O_3$
PEROXYPROPIONIC ACID, $C_3H_6O_3$
PEROXYCROTONIC ACID, $C_4H_6O_3$
MONOPEROXYSUCCINIC ACID, $C_4H_6O_5$
PEROXYFUROIC ACID, $C_5H_4O_4$
BENZENEPEROXYSULPHONIC ACID, $C_6H_6O_4S$
PEROXYHEXANOIC ACID, $C_6H_{12}O_3$
PEROXYBENZOIC ACID, $C_7H_6O_3$
DIPEROXYTEREPHTHALIC ACID, $C_8H_6O_6$
3-PEROXYCAMPHORIC ACID (3-PEROXY-1,2,2-TRIMETHYL-1,3-CYCLOPENTANEDICARBOXYLIC ACID), $C_{10}H_{16}O_5$

*See* (DIBENZOYLDIOXYIODO)BENZENES
POLYMERIC PEROXYACIDS

## PEROXYCARBONATE ESTERS —$OCO_2OR$, —$OOCO_2OR$

Strain, F. *et al., J. Amer. Chem. Soc.*, 1950, **72**, 1254
Kirk-Othmer, 1967, Vol. 14, 803

Of the three possible types of peroxycarbonate esters—dialkyl monoperoxycarbonates, dialkyl diperoxycarbonates and dialkyl peroxydicarbonates—the latter are by far the least stable class. Several of the 16 alkyl and substituted alkyl esters prepared decomposed violently or explosively at temperatures only slightly above the temperature (0–10°C) of preparation, owing to self-accelerating exothermic decomposition. Several were also explosive on exposure to heat, friction or shock.

Amines and certain metals cause accelerated decomposition of dialkyl peroxydicarbonates by a true catalytic mechanism

*See* DIMETHYL PEROXYDICARBONATE, $C_4H_6O_6$
DIETHYL PEROXYDICARBONATE, $C_6H_{10}O_6$
DIALLYL PEROXYDICARBONATE, $C_8H_{10}O_6$
DIISOPROPYL PEROXYDICARBONATE, $C_8H_{14}O_6$
DIPROPYL PEROXYDICARBONATE, $C_8H_{14}O_6$
DI(2-METHOXYETHYL) PEROXYDICARBONATE, $C_8H_{14}O_8$
DI-*tert*-BUTYL PEROXYDICARBONATE, $C_9H_{18}O_5$
DIDODECYL PEROXYDICARBONATE, $C_{26}H_{50}O_6$

*See other* PEROXIDES

## PEROXY COMPOUNDS C—OO—

Castrantas, 1965

Detonation theory is used to clarify the explosive characteristics of peroxy compounds. Some typical accidents are described. Hazards involved in use of a large number of peroxy compounds (including all those commercially available) are tabulated. 134 References.

## PEROXYESTERS

**$-CO_2OR$**

Castrantas, 1965, 13

Swern, 1970, Vol. 1, 79

Though as a group they are noted for instability, there is a fairly wide variation in stability between particular sub-groups and compounds.

See the group:

PEROXYCARBONATE ESTERS

and the individually indexed compounds:

TRIFLUOROMETHYL PEROXONITRATE, $CF_3NO_4$
* ACETYL PEROXONITRATE, $C_2H_3NO_5$
TRIFLUOROMETHYL PEROXYACETATE, $C_3H_3F_3O_3$
* PROPIONYL PEROXONITRATE, $C_3H_5NO_5$
* BUTYRYL PEROXONITRATE, $C_4H_7NO_5$
1-HYDROXYETHYL PEROXYACETATE, $C_4H_8O_4$
*tert*-BUTYL CHLOROPEROXYFORMATE, $C_5H_9ClO_3$
TRIFLUOROMETHYL 3-FLUOROFORMYLHEXAFLUOROPEROXY-BUTYRATE, $C_6F_{10}O_4$
*tert*-BUTYL PEROXYACETATE, $C_6H_{12}O_3$
ISOBUTYL PEROXYACETATE, $C_6H_{12}O_3$
BIS-TRIMETHYLSILYL PEROXOMONOSULPHATE, $C_6H_{18}O_5SSi_2$
*O,O-tert*-BUTYL HYDROGEN MONOPEROXOMALEATE, $C_8H_{12}O_5$
DI-*tert*-BUTYL DIPEROXYOXALATE, $C_{10}H_{13}O_6$
*tert*-BUTYL *p*-NITROPEROXYBENZOATE, $C_{11}H_{13}NO_5$
*tert*BUTYL PEROXYBENZOATE, $C_{11}H_{14}O_3$
*tert*-BUTYL 1-ADAMANTANEPEROXYCARBOXYLATE, $C_{15}H_{24}O_3$
Di-*tert*-BUTYL DIPEROXYPHTHALATE, $C_{16}H_{22}O_6$
1,1-BIS(*p*-NITROBENZOYLPEROXY)CYCLOHEXANE, $C_{20}H_{18}N_2O_{10}$
1,1-BIS(BENZOYLPEROXY)CYCLOHEXANE, $C_{20}H_{20}O_6$
1,1,6,6-TETRAKIS(ACETYLPEROXY)CYCLODODECANE, $C_{20}H_{32}O_{12}$

*See related* *tert*-BUTYL PEROXOPHOSPHATE DERIVATIVES
(DIBENZOYLDIOXYIODO)BENZENES

## PETROLEUM COKE

Hulisz, S., *Chem. Abs.*, 1977, **86**, 19269s

The flammability and explosivity of high-sulphur petroleum coke dust (particle size $< 75 \mu m$) were examined. Air-dried powder was non-explosive but fire-prone above 400°C. A 5 mm layer became incandescent at 420–470° and a dust cloud ignited at 520–660°C.

*See also* CARBON, C
CARBONACEOUS DUSTS

## PHOSPHINE DERIVATIVES $RPH_2$, $R_2PH$

Halogens

Van Wazer, 1958, Vol. 1, 196

Organic derivatives of phosphine react very vigorously with halogens.

*See* ALKYLNON-METALS, etc.

## PICRATES $(NO_2)_3C_6H_2O^-$

1. Anon., *Angew. Chem. (Nachr.)*, 1954, **2**, 21
2. Hopper, J. D., *J. Franklin Inst.*, 1938, **225**, 219–225
3. Kast, H., *Chem. Abs.*, 1911, **5**, 2178

While the melting point of a picrate was being determined in a silicone oil bath approaching 250°C, an explosion occurred, scattering hot oil. It is recommended that picrates, styphnates and similar derivatives should not be heated above 210°C in a liquid-containing m.p. apparatus.

In an investigation of the sensitivity to impact of a range of hydrated and anhydrous metal picrates, anhydrous nickel picrate was found to be particularly sensitive [2]. The explosive characteristics of a range of 13 mono- to trivalent metal picrates and methylpicrates were determined [3].

*See* LEAD DIPICRATE, $C_{12}H_4N_6O_{14}Pb$
NITRIC ACID, $HNO_3$: Metal salicylates

## PLATINUM COMPOUNDS

Cotton, F. A., *Chem. Rev.*, 1955, **55**, 577

Several platinum compounds, including trimethylplatinum derivatives, are explosively unstable. Individually indexed compounds are:

POTASSIUM DINITROOXALATOPLATINATE(2–), $C_2K_2N_2O_8Pt$
TRIMETHYLPLATINUM HYDROXIDE, $C_3H_{10}OPt$
DIACETATOPLATINUM(II), $C_4H_6O_4Pt^{2+}$
TETRAMETHYLPLATINUM, $C_4H_{12}Pt$
HEXAMETHYLDIPLATINUM, $C_6H_{18}Pt_2$
TRIMETHYLPLATINUM(IV) AZIDE TETRAMER, $C_{12}H_{36}N_{12}Pt_4$
DIAMMINEDICHLOROAMINOTRICHLOROPLATINUM(IV), $Cl_5H_7N_3Pt$
AMMONIUM HEXACHLOROPLATINATE(2–), $Cl_6H_8N_2Pt$
POTASSIUM HEXACHLOROPLATINATE(2–), $Cl_6K_2Pt$
*cis*-DIAMMINEPLATINUM(II) NITRITE ('*cis*-DIAMMINEDINITROPLATINUM(II)'), $H_6N_4O_4Pt$
SODIUM HEXAHYDROXOPLATINATE(2–), $H_6Na_2O_6Pt$
AMMINEPENTAHYDROXOPLATINUM, $H_8NO_5Pt$
AMMONIUM TETRANITROPLATINATE(II), $H_8N_6O_8Pt$
TRIAMMINENITRATOPLATINUM(II) NITRATE, $H_9N_5O_6Pt$
AMMINEDECAHYDROXODIPLATINUM, $H_{13}NO_{10}Pt_2$
TETRAAMMINEHYDROXONITRATOPLATINUM(IV) NITRATE, $H_{13}N_7O_{10}Pt$

* PLATINUM(IV) OXIDE, $O_2Pt$

*See also* GOLD COMPOUNDS
*See other* HEAVY METAL DERIVATIVES

**POLY(AMINIUM PERCHLORATES)** $\{[CH_2]_nN^+H_2\ ClO_4^-\}_n$

Thomas, T. J. *et al., Amer. Inst. Aero. Astron. J.*, 1976, 14, 1334–1335

Ignition temperatures were determined by DTA for the perchlorate salts of ethylamine, isopropylamine, 4-ethylpyridine, poly(ethyleneimine), poly(propyleneimine) and poly(2- or 4-vinyl pyridine). In contrast to the low ignition temperatures (175–250°C) of the polymeric salts, mixtures of the polymeric bases with ammonium perchlorate decompose above 300°C.

*See* AMINIUM PERCHLORATES
*See other* OXOSALTS OF NITROGENOUS BASES

**POLY(DIMERCURYIMMONIUM) COMPOUNDS** $(Hg{=}\overset{+}{N}{=}Hg\ Z^-)_n$

1. Ciusa, W., *Chem. Abs.*, 1943, **37**, $3271_9$; 1944, **38**, $4133_4$
2. Sorbe, 1968, 97
3. Bailar, 1973, Vol. 3, 313

Several explosive salts including the acetylide, azide, borate, bromate, chlorate, chromate, iodate (and ammonium iodate double salt), nitrite, perchlorate (and ammonium perchlorate double salt), periodate, permanganate, picrate and trinitrobenzoate were prepared. The three latter salts and the acetylide, azide and bromate are impact-sensitive detonators [1].

It appears probable that many of the explosively unstable compounds [2], formed in various ways from interaction of mercury or its compounds and ammonia or its salts, may have the common polymeric structure now recognised for Millon's base [3]. This is a silica-like network of $N^+$ and Hg in 4- and 2-coordination, respectively, with $OH^-$ and water in the interstitial spaces.

This group includes the compounds:

POLY(DIMERCURYIMMONIUM BROMATE), $(BrHg_2NO_3)_n$
POLY(DIMERCURYIMMONIUM ACETYLIDE), $(C_2HHg_2N)_n$
POLY(DIMERCURYIMMONIUM PERCHLORATE), $(ClHg_2NO_4)_n$
POLY(DIMERCURYIMMONIUM HYDROXIDE) ('MILLON'S BASE ANHYDRIDE'), $(HHg_2NO)_n$
POLY(DIMERCURYIMMONIUM IODIDE HYDRATE), $(H_2Hg_2INO)_n$
POLY(DIMERCURYIMMONIUM PERMANGANATE), $(Hg_2MnNO_4)_n$
POLY(DIMERCURYIMMONIUM AZIDE), $(Hg_2N_4)_n$

*See other* MERCURY COMPOUNDS
*N*-METAL DERIVATIVES

**POLY(DIMETHYLSILYL) CHROMATE** $[Si(Me)_2OCrO_3]_n$

*See* BIS-TRIMETHYLSILYL CHROMATE, $C_6H_{18}CrO_4Si_2$

## POLYMERIC PEROXYACIDS

1. Takagi, T. J., *Polymer Sci. (B), Polymer Lett.*, 1967, **5**, 1031–1035
2. Harrison, C. R. *et al.*, *J. Chem. Soc., Chem. Comm.*, 1974, 1009

An ion exchange resin based on poly-acrylic or -methacrylic acids and containing aliphatic peroxyacid groups [1] readily explodes on impact, but a polystyrene resin containing aromatic peroxyacid groups on *ca.* 70% of the phenyl residues could not be caused to explode on impact [2].

*See other* PEROXYACIDS

## POLYMERISATION INCIDENTS

There are many records of violent or explosive incidents involving runaway exothermic polymerisation reactions.

*See* DIENES
POLYPEROXIDES
ALUMINIUM TRICHLORIDE, $AlCl_3$ : Alkenes
: Nitrobenzene
TRIBORON PENTAFLUORIDE, $B_3F_5$ : Tetrafluoroethylene
CHLOROPEROXYTRIFLUOROMETHANE, $CClF_3O_2$ : Tetrafluoroethylene
CYANAMIDE, $CH_2N_2$
FORMALDEHYDE, $CH_2O$: Phenol
TETRAFLUOROETHYLENE, $C_2F_4$
ACETYLENE (ETHYNE), $C_2H_2$ : Cobalt
1,1-DICHLOROETHYLENE, $C_2H_2Cl_2$
ACETONITRILE, $C_2H_3N$: Sulphuric acid, etc.
GLYCOLONITRILE, $C_2H_3NO$
ETHYLENE, $C_2H_4$
ACETALDEHYDE, $C_2H_4O$: Acetic acid
: Metals
: Sulphuric acid
ETHYLENE OXIDE, $C_2H_4O$: Ammonia
: Contaminants
: Trimethylamine
THIIRAN (ETHYLENE SULPHIDE), $C_2H_4S$
AZIRIDINE (ETHYLENEIMINE), $C_2H_5N$
DIMETHYL SULPHOXIDE, $C_2H_6OS$: Acyl halides, etc.
MALONODINITRILE, $C_3H_2N_2$
PROPIOLALDEHYDE, $C_3H_2O$
METHYL TRICHLOROACETATE, $C_3H_3Cl_3O_2$
ACRYLONITRILE, $C_3H_3N$: Initiators
ACRYLALDEHYDE, $C_3H_4O$
ACRYLIC ACID, $C_3H_4O_2$
1-CHLORO-2-PROPENE (ALLYL CHLORIDE), $C_3H_5Cl$
CHLOROACETONE, $C_3H_5ClO$
ACRYLAMIDE, $C_3H_5NO$
2,3-EPOXYPROPIONALDEHYDE OXIME, $C_3H_5NO_2$
PROPENE, $C_3H_6$ : Lithium nitrate, etc.
2-AMINOPROPIONITRILE, $C_3H_6N_2$
METHYL VINYL ETHER, $C_3H_6O$
2-PROPEN-1-OL (ALLYL ALCOHOL), $C_3H_6O$: Sulphuric acid

PROPYLENE OXIDE, $C_3H_6O$: Sodium hydroxide
2-METHYLAZIRIDINE (PROPYLENEIMINE), $C_3H_7N$: Acids
DIKETENE, $C_4H_4O_2$
1-CYANO-2-PROPEN-1-OL, $C_4H_5NO$
1,3-BUTADIENE, $C_4H_6$
1-BUTEN-3-ONE, $C_4H_6O$
METHYL ACRYLATE, $C_4H_6O_2$
VINYL ACETATE, $C_4H_6O_2$
ETHYL VINYL ETHER, $C_4H_8O$
CYCLOPENTADIENE, $C_5H_6$
FURFURYL ALCOHOL, $C_5H_6O_2$ : Acids
METHYL METHACRYLATE, $C_5H_8O_2$
1,4-DICYANOBUTENE, $C_6H_6N_2$
DIALLYL SULPHATE, $C_6H_{10}O_4S$
DIALLYL PHOSPHITE, $C_6H_{11}O_3P$
2(?)-VINYLPYRIDINE, $C_7H_7N$
1,5,7-OCTATRIEN-3-YNE, $C_8H_8$
STYRENE, $C_8H_8$ : Butyllithium
: Dibenzoyl peroxide
: Initiators
DIETHYL ACETYLENEDICARBOXYLATE, $C_8H_{10}O_4$ : 1,3,5-Cyclooctatriene
DIALLYL PEROXYDICARBONATE, $C_8H_{10}O_6$
4-HYDROXY-*trans*-CINNAMIC ACID, $C_9H_8O_3$
ALLYL BENZENESULPHONATE, $C_9H_{10}O_3S$
TRIALLYL PHOSPHATE, $C_9H_{15}O_4P$
ALLYL *p*-TOLUENESULPHONATE, $C_{10}H_{12}O_3S$
2-BUTEN-1-YL BENZENESULPHONATE, $C_{10}H_{12}O_3S$
1,3,5-TRIETHYNYLBENZENE, $C_{12}H_6$
1,2-DI(3-BUTEN-1-YL)CYCLOBUTANE, $C_{12}H_{12}$
OCTAKIS(TRIFLUOROPHOSPHINE)DIRHODIUM, $F_{24}P_8Rh_2$

**POLYNITROALKYL COMPOUNDS** **$C(NO_2)_n$, $C(NO_2)–CNO_2$**

Hammond, G. S. *et al.*, *Tetrahedron*, 1963, **19**, Suppl. 1, 177, 188

Trinitromethane ('nitroform'), dinitroacetonitrile, their salts, and polynitroalkanes are all potentially dangerous, and must be carefully handled as explosive compounds. Individually indexed compounds are:

DICHLORODINITROMETHANE, $CCl_2N_2O_4$
FLUOROTRINITROMETHANE, $CFN_3O_6$
TRINITROMETHANE ('NITROFORM'), $CHN_3O_6$
DINITROMETHANE, $CH_2N_2O_4$
POTASSIUM TRINITROMETHANIDE ('NITROFORM' SALT), $CKN_3O_6$
DINITRODIAZOMETHANE, $CN_4O_4$
TETRANITROMETHANE, $CN_4O_8$
SILVER CYANODINITROMETHANIDE, $C_2AgN_3O_4$
DINITROACETONITRILE, $C_2HN_3O_4$
SODIUM 5-DINITROMETHYLTETRAZOLIDE, $C_2HN_6NaO_4$
SILVER DINITROACETAMIDE, $C_2H_2AgN_3O_5$
POTASSIUM DINITROACETAMIDE, $C_2H_2KN_3O_5$
SODIUM DINITROACETAMIDE, $C_2H_2N_3NaO_5$
2(?)-FLUORO-1,1-DINITROETHANE, $C_2H_3FN_2O_4$
2,2,2-TRINITROETHANOL, $C_2H_3N_3O_7$

TRINITROACETONITRILE, $C_2N_4O_6$
HEXANITROETHANE, $C_2N_6O_{12}$
1,1-DINITRO-3-BUTENE, $C_4H_6N_2O_4$
2,3-DINITRO-2-BUTENE, $C_4H_6N_2O_4$

*See also* FLUORODINITROMETHYL COMPOUNDS
NITROALKANES
TRINITROETHYL ORTHOESTERS

*See related* *aci*-NITRO SALTS

## POLYNITROARYL COMPOUNDS $Ar(NO_2)_n$

1. Urbanski, 1964, Vol. 1
2. Shipp, K. G. *et al.*, *J. Org. Chem.*, 1972, **37**, 1966
3. Shipp, K. G. *et al.*, US Pat. 3 941 853, 1976

Polynitro derivatives of monocyclic aromatic systems (trinitrobenzene, trinitrotoluene, tetranitro-*N*-methylaniline, trinitrophenol, etc.) have long been used as explosives [1]. It has recently been found that a series of polynitro derivatives of biphenyl, diphenylmethane and 1,2-diphenylethylene (stilbene) are explosives liable to detonate on grinding or impact [2]. The same may be true of other polynitro derivatives of polycyclic systems not normally used as explosives, e.g. polynitro-fluorenones, -carbazoles, etc.

Penta- and hexa-nitrobenzophenones are also high-energy explosives [3].

The presence of two or more nitro groups (each with two oxygen atoms) on an aromatic nucleus often increases the reactivity of other substituents and the tendency towards explosive instability as oxygen balance is approached.

Bases
or Salts

Uhlmann, P. W., *Chem. Ztg.*, 1914, **38**, 389–390

In view of previous violent or explosive reactions, heating of di- and tri-nitroaryl compounds with alkalies, ammonia or *O*-ethylsulphuric acid salts in autoclaves should be avoided.

*See* DIETHYL SULPHATE, $C_4H_{10}O_4S$: 2,7-Dinitro-9-phenylphenanthridine
NITROAROMATIC–ALKALI HAZARDS

Individually indexed compounds are:

2-HYDROXY-3,5-DINITROPYRIDINE, $C_5H_3N_3O_5$
1,3-DICHLORO-4,6-DINITROBENZENE, $C_6H_2Cl_2N_2O_4$
1,3-DIFLUORO-4,6-DINITROBENZENE, $C_6H_2F_2N_2O_4$
SODIUM PICRATE, $C_6H_2N_3NaO_7$
2,3,4,6-TETRANITROPHENOL, $C_6H_2N_4O_9$
1-CHLORO-2,4-DINITROBENZENE, $C_6H_3ClN_2O_4$
2,4-DINITROBENZENESULPHENYL CHLORIDE, $C_6H_3ClN_2O_4S$
* 2,6-DINITRO-4-PERCHLORYLPHENOL, $C_6H_3ClN_2O_8$
1-FLUORO-2,4-DINITROBENZENE, $C_6H_3FN_2O_4$

POTASSIUM 4,6-DINITROBENZOFUROXAN HYDROXIDE COMPLEX, $C_6H_3KN_4O_7$
SODIUM 2,4-DINITROPHENOLATE, $C_6H_3N_2NaO_5$
PICRIC ACID, $C_6H_3N_3O_7$
TRINITRORESORCINOL (STYPHNIC ACID), $C_6H_3N_3O_8$
1,3,5-TRIHYDROXYTRINITROBENZENE (TRINITROPHLOROGLUCINOL), $C_6H_3N_3O_9$
*N*,2,3,5- or *N*,2,4,6-TETRANITROANILINE, $C_6H_3N_5O_8$
4-CHLORO-2,6-DINITROANILINE, $C_6H_4ClN_3O_4$
6-CHLORO-2,4-DINITROANILINE, $C_6H_4ClN_3O_4$
2,4-DINITROPHENOL, $C_6H_4N_2O_5$
POTASSIUM 3,5-DINITRO-2(1-TETRAZENYL)PHENOLATE, $C_6H_5KN_6O_5$
2,4-DINITROANILINE, $C_6H_5N_3O_4$
2-AMINO-4,6-DINITROPHENOL, $C_6H_5N_3O_5$
TRIAMINOTRINITROBENZENE, $C_6H_6N_6O_6$
2,4-DINITROPHENYLHYDRAZINIUM PERCHLORATE, $C_6H_7ClN_4O_8$
2,4,6-TRINITROBENZOIC ACID, $C_7H_3N_3O_8$
SILVER 3,5-DINITROANTHRANILATE, $C_7H_4AgN_3O_6$
* *o*-NITROBENZOYL CHLORIDE, $C_7H_4ClNO_3$
2,6-DINITROBENZYL BROMIDE, $C_7H_5BrN_2O_4$
2,4,6-TRINITROTOLUENE, $C_7H_5N_3O_6$
3-METHYL-2,4,6-TRINITROPHENOL, $C_7H_5N_3O_7$
1-NITRO-3(2,4-DINITROPHENYL)UREA, $C_7H_5N_5O_7$
*N*,2,4,6-TETRANITRO-*N*-METHYLANILINE (TETRYL), $C_7H_5N_5O_8$
2,4-DINITROTOLUENE, $C_7H_6N_2O_4$
* *mixo*-NITROTOLUENE, $C_7H_7NO_2$
2,4-DINITROPHENYLACETYL CHLORIDE, $C_8H_5ClN_2O_5$
3,5-DINITRO-2-TOLUAMIDE, $C_8H_7N_3O_5$
TETRANITRONAPHTHALENE, $C_{10}H_4N_4O_8$
DINITRO-1-PICRYLBENZOTRIAZOLES, $C_{12}H_4N_8O_{10}$
1,3,6,8-TETRANITROCARBAZOLE, $C_{12}H_5N_5O_8$
1,8-DIHYDROXY-2,4,5,7-TETRANITROANTHRAQUINONE, $C_{14}H_4N_4O_{12}$
COPPER(II) 3,5-DINITROANTHRANILATE, $C_{14}H_8CuN_6O_{12}$
*mixo*-DIMETHOXYDINITROANTHRAQUINONE, $C_{16}H_{10}N_2O_8$
2,2,4-TRIMETHYLDECAHYDROQUINOLINE PICRATE, $C_{18}H_{26}N_4O_7$
2,7-DINITRO-9-PHENYLPHEANTHRIDINE, $C_{19}H_{12}N_3O_4$

*See also* *C*-NITRO COMPOUNDS

## POLYNITROAZOPYRIDINES $(NO_2)_2C_5H_2NN{=}NC_5H_2N(NO_2)_2$

Coburn, M. D., *J. Heterocyclic Chem.*, 1974, **11**, 1099–1100
Most of a series of azodinitropyridines and their hydrazine precursors were explosive with a sensitivity comparable to that of RDX. Individual compounds are:

2,2′-AZO-3,5-DINITROPYRIDINE, $C_{10}H_4N_8O_8$
3,5-DINITRO-2-(PICRYLAZO)PYRIDINE, $C_{11}H_4N_8O_{10}$
1-(3,5-DINITRO-2-PYRIDYL)-2-PICRYLHYDRAZINE, $C_{11}H_6N_8O_{10}$
2,6-BIS(2-PICRYLAZO)-3,5-DINITROPYRIDINE, $C_{17}H_5N_{13}O_{16}$

2,6-BIS(2-PICRYLHYDRAZINO)-3,5-DINITROPYRIDINE, $C_{17}H_9N_{13}O_{16}$

*See other* HIGH-NITROGEN COMPOUNDS
POLYNITROARYL COMPOUNDS

## POLYPEROXIDES

This group covers polymeric peroxides of indeterminate structure rather than polyfunctional molecules of known structure. Polymeric peroxide species described as hazardous include those derived from: butadiene (highly explosive); isoprene, dimethylbutadiene (both strongly explosive); 1,5-*p*-menthadiene, 1,3-cyclohexadiene (both explode at 110°C); methyl methacrylate, vinyl acetate, styrene (all explode above 40°C); diethyl ether (extremely explosive even below 100°C); and 1,1-diphenylethylene, cyclopentadiene (both explode on heating).

*See* TETRAFLUOROETHYLENE, $C_2F_4$: Air
1,1-DICHLOROETHYLENE, $C_2H_2Cl_2$: Air
CHLOROETHYLENE (VINYL CHLORIDE), $C_2H_3Cl$: Air
POLY(ETHYLIDENE PEROXIDE), $(C_2H_4O_2)_n$
HEXAFLUOROPROPENE, $C_3F_6$: Air
1,3-BUTADIENE, $C_4H_6$: Air
DIMETHYLKETENE, $C_4H_6O$: Air
VINYL ACETATE, $C_4H_6O_2$: Oxygen
POLY(1,3-BUTADIENE PEROXIDE), $(C_4H_6O_2)_n$
POLY(DIMETHYLKETENE PEROXIDE), $(C_4H_6O_3)_n$
POLY(VINYL ACETATE PEROXIDE), $(C_4H_6O_4)_n$
TETRAHYDROFURAN, $C_4H_8O$
DIETHYL ETHER, $C_4H_{10}O$: Air
CYCLOPENTADIENE, $C_5H_6$: Oxygen
2-METHYL-1,3-BUTADIENE (ISOPRENE), $C_5H_8$: Air
METHYL METHACRYLATE, $C_5H_8O_2$: Air
POLY(METHYL METHACRYLATE PEROXIDE), $(C_5H_8O_4)_n$
1,3-CYCLOHEXADIENE, $C_6H_8$: Air
POLY(CYCLOHEXADIENE PEROXIDE), $(C_6H_8O_2)_n$
2,3-DIMETHYL-1,3-BUTADIENE, $C_6H_{10}$: Air
* PHTHALOYL PEROXIDE, $C_8H_4O_4$
STYRENE, $C_8H_8$: Oxygen
POLY(STYRENE PEROXIDE), $(C_8H_8O_2)_n$
6,6-DIMETHYLFULVENE, $C_8H_{10}$: Air
1,5-*p*-MENTHADIENE, $C_{10}H_{16}$: Air
1,1-DIPHENYLETHYLENE, $C_{14}H_{12}$: Oxygen

*See also* HYDROGEN PEROXIDE, $H_2O_2$: Ketones, Nitric acid

## PREPARATIVE HAZARDS

Preparative hazards have been noted for the following compounds:

SILVER TETRAFLUOROBORATE, $AgBF_4$
ALUMINIUM COPPER(I) SULPHIDE, $AlCuS_2$
PLATINUM DIARSENIDE, $As_2Pt$

DIBORANE, $B_2H_6$
BROMOGERMANE, $BrGeH_3$
HYDROGEN BROMIDE, BrH
DIBROMOGERMANE, $Br_2GeH_2$
TRIFLUOROMETHYL PERCHLORATE, $CClF_3O_4$
CYANOGEN CHLORIDE, CClN
PHOSPHORYL DICHLORIDE ISOCYANATE, $CCl_2NO_2P$
POLY(CARBON MONOFLUORIDE), $(CF)_n$
FLUOROTRINITROMETHANE, $CFN_3O_6$
UREA, $CH_4N_2O$
BISFLUOROFORMYL PEROXIDE, $C_2F_2O_4$
SODIUM 2,2,2-TRIFLUOROETHOXIDE, $C_2H_2F_3NaO$
ACETYL CHLORIDE, $C_2H_3ClO$
PEROXYACETIC ACID, $C_2H_4O_3$
IODOETHANE, $C_2H_5I$
METHYL IMINIOFORMATE CHLORIDE, $C_2H_6ClNO$
DIMETHYLAMMONIUM PERCHLORATE, $C_2H_8ClNO_4$
2-AMINO-5-NITROTHIAZOLE, $C_3H_3N_3O_2S$
3-BROMO-1,1,1-TRICHLOROPROPANE, $C_3H_4BrCl_3$
PROPIONYL CHLORIDE, $C_3H_5ClO$
HYDROXYACETONE, $C_3H_6O_2$
3-HYDROXYTHIETANE-1,1-DIOXIDE, $C_3H_6O_3S$
DIAZOMETHYLDIMETHYLARSINE, $C_3H_7AsN_2$
ETHYL IMINIOFORMATE CHLORIDE, $C_3H_8ClN$
CHLOROTRIMETHYLSILANE, $C_3H_9ClSi$
PERFLUORO-2,5-DIAZAHEXANE 2,5-DIOXYL, $C_4F_{10}N_2O_2$
PERFLUORO-*tert*-BUTANOL, $C_4HF_9O$
2-METHYL-5-NITROIMIDAZOLE, $C_4H_5N_3O_2$
CHROMYL ACETATE, $C_4H_6CrO_6$
1,1,1-TRIS(BROMOMETHYL)METHANE, $C_4H_7Br_3$
*tert*-NITROBUTANE, $C_4H_9NO_2$
CHLORODIETHYLBORANE, $C_4H_{10}BCl$
DIETHYLAMINOSULPHINYL CHLORIDE, $C_4H_{10}ClNOS$
TRIMETHYL ORTHOFORMATE, $C_4H_{10}O_3$
ZINC ETHYLSULPHINATE, $C_4H_{10}O_4S_2Zn$
1-AMINO-2,2,2-TRIS(HYDROXYMETHYL)METHANE, $C_4H_{11}NO_3$
TETRAMETHYLAMMONIUM OZONIDE, $C_4H_{12}NO_3$
METHYLNITROTHIOPHENE, $C_5H_5NO_2S$
4-BROMOCYCLOPENTENE, $C_5H_7Br$
2,2-DIMETHYLPROPANE (NEOPENTANE), $C_5H_{12}$
1,1,1-TRIS(AMINOMETHYL)ETHANE, $C_5H_{15}N_3$
1,3-DIFLUORO-4,6-DINITROBENZENE, $C_6H_2F_2N_2O_4$
3,5-DINITROBENZENEDIAZONIUM-2-OLATE, $C_6H_2N_4O_5$
2,4,5-TRICHLOROPHENOL, $C_6H_3Cl_3O$
2-CHLORO-5-NITROBENZENESULPHONIC ACID, $C_6H_4ClNO_5S$
6-CHLORO-2,4-DINITROANILINE, $C_6H_4ClN_3O_4$
1,4,5,8-TETRAHYDRO-1,4,5,8-TETRATHIAFULVALENE, $C_6H_4S_4$
PHENYLLITHIUM, $C_6H_5Li$
2,4- or 2,5-DINITROANILINE, $C_6H_5N_3O_4$
PHENYLSODIUM, $C_6H_5Na$
*o*-NITROANILINE, $C_6H_6N_2O_2$
*N*-PHENYLHYDROXYLAMINE, $C_6H_7NO$
4-OXIMINO-4,5,6,7-TETRAHYDROBENZOFURAZAN, $C_6H_7N_3O_3$
1,2-CYCLOHEXANEDIONE, $C_6H_8O_2$
TRIACETYL BORATE, $C_6H_9BO_6$

3,3-DIMETHYL-1-BUTYNE, $C_6H_{10}$
*N*-NITROSO-6-HEXANELACTAM, $C_6H_{10}N_2O_2$
2-HYDROXY-2-METHYLGLUTARIC ACID, $C_6H_{10}O_3$
TETRAMETHOXYETHYLENE, $C_6H_{12}O_4$
2,4,6-TRINITROBENZOIC ACID, $C_7H_3N_3O_8$
*p*-CHLOROPHENYL ISOCYANATE, $C_7H_4ClNO$
*o*-, *m*-, or *p*-TRIFLUOROMETHYLPHENYLLITHIUM, $C_7H_4F_3Li$
2-CHLORO-5-METHYLANILINE, $C_7H_5ClN$
2,5-PYRIDINEDICARBOXYLIC ACID, $C_7H_5NO_4$
2-METHYL-5-NITROBENZENESULPHONIC ACID, $C_7H_7NO_5S$
2-METHOXY-5-NITROANILINE, $C_7H_8N_2O_3$
*m*-THIOCRESOL, $C_7H_8S$
PHTHALIC ANHYDRIDE, $C_8H_4O_3$
3-NITROPHTHALIC ACID, $C_8H_5NO_6$
NITROTEREPHTHALIC ACID, $C_8H_5NO_6$
4-METHOXY-3-NITROBENZOYL CHLORIDE, $C_8H_6ClNO_4$
4-CYANO-3-NITROTOLUENE, $C_8H_6N_2O_2$
*p*-(BROMOMETHYL)BENZOIC ACID, $C_8H_7BrO_2$
TETRAMETHYLSUCCINODINITRILE, $C_8H_{12}N_2$
6-AMINOPENICILLANIC ACID *S*-OXIDE, $C_8H_{12}N_2O_4S$
DI-*tert*-BUTYL PEROXIDE, $C_8H_{18}O_2$
3,5-DIMETHYL-4-[BIS(TRIFLUOROACETOXY)IODO]ISOXAZOLE, $C_9H_6F_6INO_5$
1-DIAZOINDENE, $C_9H_6N_2$
NITROINDANE, $C_9H_9NO_2$
3,5-DIMETHYLBENZOIC ACID, $C_9H_{10}O_2$
NITROMESITYLENE, $C_9H_{11}NO_2$
2,4-DIETHYNYLPHENOL, $C_{10}H_6O$
BIS-$\eta$-CYCLOPENTADIENYLLEAD, $C_{10}H_{10}Pb$
3-*p*-CHLOROPHENYLBUTANOIC ACID, $C_{10}H_{11}ClO_2$
BUTYLBENZENE, $C_{10}H_{14}$
2-METHYL-1-PHENYL-2-PROPYL HYDROPEROXIDE, $C_{10}H_{14}O_2$
3,7-DIMETHYL-2,6-OCTADIENAL (CITRAL), $C_{10}H_{16}O$
*cis*-CYCLODECENE, $C_{10}H_{18}$
2,4-DI-*tert*-BUTYL-2,2,4,4-TETRAFLUORO-1,3-DIMETHYL-1,3,2,4-DIAZAPHOSPHETIDINE, $C_{10}H_{24}F_4N_2P_2$
2,4-DIETYHYNYL-5-METHYLPHENOL, $C_{11}H_8O$
2,4,6-TRI(ALLYLOXY)-1,3,5-TRIAZINE ('TRIALLYL CYANURATE'), $C_{12}H_{15}N_3O_3$
1,4,7,10,13,16-HEXAOXACYCLOOCTADECANE ('18-CROWN-6'), $C_{12}H_{24}O_6$
2-(DIMETHYLAMINOMETHYL)FLUOROFERROCENE, $C_{13}H_{16}FFeN$
POLY(1,1-DIPHENYLETHYLENE PEROXIDE), $(C_{14}H_{12}O_2)_n$
*N*-METHYL-*p*-NITROANILINIUM 2(*N*-METHYL-*N*-*p*-NITROPHENYL-AMINOSULPHONYL)ETHYLSULPHATE, $C_{16}H_{20}N_4O_{10}S_2$
1,4-OCTADECANOLACTONE, $C_{18}H_{34}O_2$
*cis*-DICHLOROBIS(2,2′-BIPYRIDYL)COBALT(III) CHLORIDE, $C_{20}H_{16}Cl_3CoN_4$
1,5-DIBENZOYLNAPHTHALENE, $C_{24}H_{16}O_2$
CADMIUM SELENIDE, CdSe
CHLOROTETRAFLUOROPHOSPHORANE, $ClF_4P$
HYDROGEN CHLORIDE, ClH
CHLOROAMINE, $ClH_2N$
XENON TETRAFLUORIDE OXIDE, $F_4OXe$
DIPOTASSIUM HEXAFLUOROMANGANATE(IV), $F_6K_2Mn$

NITROSYLSULPHURIC ACID, $HNO_5S$
AZIDOSILANE, $H_3N_3Si$
PEROXOMONOPHOSPHORIC ACID, $H_3O_5P$
PENTAAMMINEDINITROGENRUTHENIUM(II) SALTS, $H_{15}N_7Ru^{2+}$
TETRAIODODIPHOSPHANE, $I_4P_2$
DINITROGEN OXIDE ('NITROUS OXIDE'), $N_2O$
OSMIUM(IV) OXIDE, $O_2Os$
TETRAPHOSPHORUS HEXAOXIDE TETRASULPHIDE, $O_6P_4S_4$

*See* 1,2-DIOLS

## PROPELLANTS

Gould, R. F., *Advanced Propellant Chemistry* (ACS 54), Washington, Amer. Chem. Soc., 1966
This deals, in 26 chapters in 5 sections, with theoretical and practical aspects of the use and safe handling of powerful oxidisers, and their complementary reactive fuels.

Materials include: nitrogen pentaoxide, perfluoroammonium ion and salts, nitronium tetrafluoroborate, hydrazinium mono- and diperchlorates, nitronium perchlorate, tricyanomethyl compounds, difluoroamine and its alkyl derivatives, oxygen difluoride, chlorine trifluoride, dinitrogen tetraoxide, bromine trifluoride, nitrogen fluorides, liquid ozone–fluorine system.

*See also* ROCKET PROPELLANTS

## PYROPHORIC ALLOYS

Schmitt, C. R., *J. Fire Flamm.*, 1971, **2**, 163–164
Alloys of reactive metals are often more pyrophoric than the parent metals. Examples are alloys of titanium with zirconium; thorium with copper, silver or gold; uranium with tin, lead or gold; magnesium with aluminium; hafnium with iron.

*See* CERIUM, Ce: Alone, or Metals

*See other* PYROPHORIC MATERIALS

## PYROPHORIC CATALYSTS

'Laboratory Handling of Metal Catalysts', *Chem. Safety,* 1949, (2), 5
*Catalyst Handbook*, 180–181, London, Wolfe, 1970
Proposed Code of Practice for laboratory handling of possibly pyrophoric catalysts includes: storage in tightly closed containers; extreme care in transfer operations, with provision for immediate cleaning up of spills and copious water flushing; avoidance of air-drying during filtration, and storage of residues under water; use of water-flush in event of ignition.

The later reference details precautions to prevent fires in catalysts discharged from industrial reactors.

*See* HYDROGENATION CATALYSTS
DODECACARBONYL TETRACOBALT, $C_{12}Co_4O_{12}$
CHROMIUM–COPPER CATALYST, Cr–Cu
NICKEL, Ni: Magnesium silicate

## PYROPHORIC IRON–SULPHUR COMPOUNDS

1. Schultze, R. *et al.*, *Arbeitsschutz*, 1964, 194–196
2. Dodonov, Ya. Ya. *et al.*, *Chem. Abs.*, 1964, **60**, 5058h–5059a
3. Anon., *Loss. Prev. Bull.*, 1977, (012), 1–6

Iron(III) salts and thiols in alcoholic solution interact to produce highly pyrophoric mixtures containing iron thiolates, oxide, hydrates, sulphides and sulphur. Effects of variation in reaction conditions and structure of thiols upon pyrophoricity were examined. Treatment of the pyrophoric mixtures with nitrogen oxide to form nitrosyl complexes effectively deactivates them [1]. Laboratory treatment of hydrated iron oxides with hydrogen sulphide simulated the production of pyrophoric iron sulphides which frequently cause fires in petroleum refining operations. Presence of gasoline during sulphide preparation gave pyrophoric materials which retained their activity longer than when gasoline was absent [2]. Several petroleum refinery fires and incidents are detailed [3].

*See* PHTHALIC ANHYDRIDE, $C_8H_4O_3$
IRON(II) SULPHIDE, FeS
DIIRON TRISULPHIDE, $Fe_2S_3$

## PYROPHORIC MATERIALS

*Spontaneously Combustible Solids – Literature Survey, Rept. 75-159*, Kayser, E. G. *et al.*, White Oaks(Md.), US Naval Surface Weapons Centre, 1975 (USNTIS AD-A019919)

Existing information on solids spontaneously combustible in contact with air or water has been reviewed, with 145 references. Data relevant to the causes and prevention of spontaneous ignition are included as well as the application of mathematical treatments to the problem, and available testing methods for assessing relevant factors in natural and manufactured products are discussed. The relevant groups:

ALKYLMETALS
ARYLMETALS
BORANES
CARBONYLMETALS
COMPLEX ACETYLIDES
COMPLEX HYDRIDES
METAL HYDRIDES
NON-METAL HYDRIDES
ORGANOMETALLICS
PYROPHORIC ALLOYS

PYROPHORIC CATALYSTS
PYROPHORIC IRON–SULPHUR COMPOUNDS
PYROPHORIC METALS

are treated separately. Other individually indexed compounds are:

CELLULOSE NITRATE: Iron red, etc.
BORON, B: Dichromates, etc.
DICHLOROBORANE, $BCl_2H$
TETRABORON TETRACHLORIDE, $B_4Cl_4$
BISMUTH PLUTONIDE, BiPu
TITANIUM DIBROMIDE, $Br_2Ti$
ZIRCONIUM DIBROMIDE, $Br_2Zr$
URANIUM MONOCARBIDE, CU
1-METHYL-3-NITRO-1-NITROSOGUANIDINE, $C_2H_5N_5O_3$
TITANIUM(III) METHOXIDE, $C_3H_9O_3Ti$
SODIUM TETRACARBONYLFERRATE(2–), $C_4FeNa_2O_4$
CHROMIUM DIACETATE, $C_4H_6CrO_4$
POTASSIUM *tert*-BUTOXIDE, $C_4H_9KO$
DIMETHYL ETHANEPHOSPHONITE, $C_4H_{11}O_2P$
2-FURALDEHYDE, $C_5H_4O_2$: Sodium hydrogencarbonate
*N,N,p*-TRILITHIOANILINE, $C_6H_4Li_3N$
BIS(ACRYLONITRILE)NICKEL(0), $C_6H_6N_2Ni$
POTASSIUM OCTACYANODICOBALTATE(8–), $C_8Co_2K_8N_8$
* SILVER ISOPHTHALATE, $C_8H_4Ag_2O_4$
PHTHALIC ANHYDRIDE, $C_8H_4O_3$: (Preparative hazard)
TETRAKIS(ETHYLTHIO)URANIUM, $C_8H_{20}S_4U$
POTASSIUM DINITROGENTRIS(TRIMETHYLPHOSPHINE)-
COBALTATE(1–), $C_9H_{27}CoKN_2P_3$
*N,N,N'*-TRIS(TRIMETHYLSILYL)DIAMINOPHOSPHANE,
$C_9H_{29}N_2PSi_3$
MANGANESE(II) *N,N*-DIETHYLDITHIOCARBAMATE, $C_{10}H_{20}MnN_2S_4$
4-ETHOXYBUTYLDIETHYLALUMINIUM, $C_{10}H_{23}AlO$
IRON(III) MALEATE, $C_{12}H_6Fe_2O_{12}$: Iron(III) hydroxide
POTASSIUM TRICARBONYLTRIS(PROPYNYL)MOLYBDATE,
$C_{12}H_9KMoO_3$
DIPHENYLPHOSPHINE, $C_{12}H_{11}P$
IRON(II) CHELATE OF BIS-*N,N'*-(2-PENTANON-4-YLIDENE)-1,3-
DIAMINO-2-HYDROXYPROPANE, $C_{13}H_{20}FeN_2O_3$
TRIS-$\eta$-CYCLOPENTADIENYLPLUTONIUM, $C_{15}H_{15}Pu$
TRIS-$\eta$-CYCLOPENTADIENYLURANIUM, $C_{15}H_{15}U$
TETRAKIS(BUTYLTHIO)URANIUM, $C_{16}H_{36}S_4U$
* TRIPHENYLCHROMIUM TETRAHYDROFURANATE, $C_{18}H_{15}Cr \cdot 3C_4H_8O$
TRIS(BIS-2-METHOXYETHYL ETHER)POTASSIUM HEXACARBONYL-
NIOBATE(1–), $C_{24}H_{42}KNbO_{15}$
COBALT TRIS(DIHYDROGENPHOSPHIDE), $CoH_6P_3$
COBALT NITRIDE, CoN
COBALT(II) SULPHIDE, CoS
CHROMIUM(II) OXIDE, CrO
COPPER IRON(II) SULPHIDE, $CuFeS_2$
POLY(DIFLUOROSILYLENE), $(F_2Si)_n$
IRON(II) HYDROXIDE, $FeH_2O_2$
IRON(II) OXIDE, FeO
MAGNESIUM HYDRIDE, $H_2Mg$
ZINC HYDRAZIDE, $H_2N_2Zn$
TITANIUM DIIODIDE, $I_2Ti$

INDIUM(II) OXIDE, InO
TRIPOTASSIUM ANTIMONIDE, $K_3Sb$
MOLYBDENUM(IV) OXIDE, $MoO_2$
URANIUM(III) NITRIDE, NU
ZIRCONIUM OXIDE SULPHIDE, OSZr
LEAD PENTAPHOSPHIDE, $P_5Pb$
ZINC, Zn: Acetic acid

*See also* METAL DUSTS (reference 3)
THORIUM FURNACE RESIDUES

## PYROPHORIC METALS

1. Feitknecht, W., *Conference on Finely Divided Solids,* Commis à l'Énergie Atom. Saclay, 27–29 Sept., 1967
2. Peer, L. H. *et al., Mill & Factory,* 1959, **65**(2), 79
3. Anon., *Chem. Eng. News,* 1952, **30**, 3210
4. Breidenfeld, J., *Metall.*, 1954, **8**, 94–97
5. Koelman, B. *et al., Metal Progress,* 1953, **63**(2), 77–79
6. Popov, E. L. *et al., Chem. Abs.*, 1975, **83**, 135768
7. Evans, J. D. *et al., Powder Metall.*, 1976, **19**, 17–21
8. Schmitt, C. R., *J. Fire Flamm.*, 1971, **2**, 157–172
9. Nedin, V. V.; Kostina, E. S., *Chem. Abs.*, 1977, **86**, 157758, 157759

Finely divided metal powders develop pyrophoricity when a critical specific surface area is exceeded; this is ascribed to high heat of oxide formation on exposure to air. Safe handling is possible in relatively low concentrations of oxygen in an inert gas [1].

Safe handling, storage, disposal and fire fighting techniques for hafnium, titanium, uranium, thorium and hazards of machining the two latter metals are discussed [2].

Dry, finely divided tantalum, thorium, titanium, zirconium metals or titanium–nickel, zirconium–copper alloys are not normally shock sensitive. However, if they are enclosed in glass bottles which break on impact, ignition will occur. Storage of these materials moist and in metal containers is recommended [3].

Heat of combustion, thermal conductivity, surface area and other factors influencing pyrophoricity of aluminium, cobalt, iron, magnesium and nickel powders are discussed [4]. The relationship between heat of formation of the metal oxide and particle size of metals in pyrophoric powders is discussed for several metals and alloys, including copper [5]. Further work on the relationship of surface area and ignition temperature for copper, manganese and silicon [6], and for iron and titanium [7] is reported. The latter also includes a simple calorimetric test to determine ignition temperature. In a literature review with 115 references, factors influencing the pyrophoricity of metals are identified as particle size, presence of moisture, nature of the surface of the particle, heat of formation of the oxide or nitride, mass, hydrogen content, stress, purity and presence of oxide, among others. Static charge hazards, fire and explosion incidents, handling procedures and transport considerations are also discussed. References are given to reviews of incidents involving barium, beryllium, magnesium, paper, plutonium, polypropylene, thorium, titanium, zinc-rich coatings and zirconium [8]. Equations to calculate the lower ignition limits for explosive suspensions in air of aluminium, iron, magnesium, manganese, tantalum, tin and titanium powders have been derived. Results for the latter agree well with experimental findings [9].

Individually indexed pyrophoric metals are:

BARIUM, Ba
CALCIUM, Ca
CADMIUM, Cd
CERIUM, Ce
COBALT, Co
CHROMIUM, Cr
CAESIUM, Cs
EUROPIUM, Eu
IRON, Fe
HAFNIUM, Hf
IRIDIUM, Ir
POTASSIUM, K
LITHIUM, Li
MANGANESE, Mn
SODIUM, Na
NICKEL, Ni
LEAD, Pb
PALLADIUM, Pd
PLATINUM, Pt
PLUTONIUM, Pu
RUBIDIUM, Rb
STRONTIUM, Sr
TANTALUM, Ta
THORIUM, Th
TITANIUM, Ti
URANIUM, U
ZIRCONIUM, Zr

*See also* ALUMINIUM AMALGAM, Al–Hg
BISMUTH PLUTONIDE, BiPu

## QUALITATIVE ANALYSIS

*See* LEAD DIPICRATE, $C_{12}H_4N_6O_{14}Pb$
NITRIC ACID, $HNO_3$: Metal salicylates

## REACTION SAFETY CALORIMETRY

1. *Proc. Int. Sympos. Prevent. Risks Chem. Ind.* (Frankfurt, 1976), Heidelberg, Int. Soc. Sec. Assoc., 1976
2. Hub, L., *Proc. Chem. Process Haz. Sympos.* (Manchester, 1977), No. 49, 39–46, Rugby, IChE, 1977

One of the sessions of the Symposium was largely devoted to presentations and discussion on the use of various experimental calorimetric methods for use in assessing possible hazards in chemical processing operations. The methods described cover a wide range of sample sizes and degrees of complexity:

Grewer, T. Adiabatic small-scale reaction test in Dewar, simple to operate.

Janin, R. Measurements of heat release and pressure development in sealed micro-capsules by DSC. Results are data-processed.

Lemke, D. Heat accumulation tests of medium-scale samples of thermally unstable technical materials under adiabatic storage.

Schleicher, K. Survey of general hazard testing methods.

Eigenmann, K. Use of micromethods in DTA, DSC.

Hub, L. Medium-small-scale safety calorimeter, usable under isothermal, quasi-isothermal or adiabatic conditions.

Regenass, W. Medium-scale heat flow calorimeter for measurement of heat release and cooling requirements under realistic reaction conditions.

Schofield, F. Use of a range of tests to determine detonation capability, localised thermal decomposition, thermal stability of reaction masses and effects of prolonged storage; translation of these results to industrial-scale processing operations.

Berthold, W. Use of adiabatic autoclave to simulate possibility of thermal explosion in large containers of reactive materials.

Hub, L. (reference 2). Use of medium-scale heat flow calorimeter for separate measurement of reaction heat removed via reaction vessel walls and reflux condenser system under fully realistic process conditions. Results are data-processed.

*See* ASSESSMENT OF REACTIVE CHEMICAL HAZARDS
DIFFERENTIAL THERMAL ANALYSIS

**REACTIVE METALS**

Stout, E. L., *Los Alamos Scientific Lab. Rep.*, Washington, USAEC, 1957; *Chem. Eng. News*, 1958, **36**(8), 64–65

Safety considerations in handling plutonium, uranium, thorium, alkali metals, titanium, magnesium and calcium are discussed.

**REDOX COMPOUNDS**

Compounds which contain oxidant and reducant functions in close proximity on a molecular basis tend towards explosive instability, and usually with a low energy of activation. Relevant types are salts of reducant bases with oxidant acids, and metal oxosalts with coordinated nitrogenous reducants.

*See* AMMINEMETAL OXOSALTS
OXOSALTS OF NITROGENOUS BASES

and the individually indexed compounds:

POTASSIUM CYANIDE–POTASSIUM NITRITE, $CKN \cdot KNO_2$
POTASSIUM TRICYANODIPEROXOCHROMATE(3–), $C_3CrK_3N_3O_4$
POTASSIUM PENTACYANODIPEROXOCHROMATE(5–), $C_5CrK_5N_5O_4$
* TRIPHENYLPHOSPHINE OXIDE–HYDROGEN PEROXIDE, $C_{18}H_{15}OP \cdot H_2O_2$
INDIUM(I) PERCHLORATE, $ClInO_4$
TIN(II) PERCHLORATE, $Cl_2O_8Sn$
LEAD(II) NITRATE PHOSPHINATE, $H_2NO_5PPb$
TETRAHYDROXOTIN(2+) NITRATE, $H_4N_2O_{10}Sn$

HYDRAZINIUM NITRATE, $H_5N_3O_3$
HYDROXYLAMINIUM PHOSPHINATE, $H_6NO_3P$
HYDRAZINIUM DINITRATE, $H_6N_4O_6$
* SODIUM NITRATE, $NNaO_3$: Diarsenic trioxide, Iron(II) sulphate
TIN(II) NITRATE, $N_2O_6Sn$
* URANYL NITRATE, $N_2O_8U$: Diethyl ether

## REDUCANTS

Most of the compounds showing powerful reducing action have been separately collected under the group headings

COMPLEX HYDRIDES
METAL ACETYLIDES
METAL HYDRIDES

The remaining individually indexed compounds are:

POTASSIUM HYPOBORATE, $BH_3KO$
SODIUM HYPOBORATE, $BH_3NaO$
BARIUM PHOSPHINATE ('HYPOPHOSPHITE'), $BaH_4O_4P_2$
1,1-DIMETHYLHYDRAZINE, $C_2H_8N_2$
CALCIUM PHOSPHINATE ('HYPOPHOSPHITE'), $CaH_4O_4P_2$
DIGALLIUM OXIDE, $Ga_2O$
POTASSIUM PHOSPHINATE ('HYPOPHOSPHITE'), $H_2KO_2P$
HYPONITROUS ACID, $H_2N_2O_2$
SODIUM PHOSPHINATE ('HYPOPHOSPHITE'), $H_2NaO_2P$
HYDROXYLAMINE, $H_3NO$
PHOSPHINIC ('HYPOPHOSPHOROUS') ACID, $H_3O_2P$
PHOSPHONIUM IODIDE, $H_4IP$
HYDRAZINE, $H_4N_2$
HYDROXYLAMMONIUM PHOSPHINATE, $H_6NO_3P$
AMMONIUM THIOSULPHATE, $H_8N_2O_3S_2$
MAGNESIUM, Mg
POTASSIUM, K
SODIUM, Na
SODIUM THIOSULPHATE, $Na_2O_3S_2$
SODIUM DITHIONITE ('HYDROSULPHITE'), $Na_2O_4S_2$
SODIUM DISULPHITE ('METABISULPHITE'), $Na_2O_5S_2$

## ROCKET PROPELLANTS

1. Kirk-Othmer, 1965, Vol. 8, 659
2. Urbanski, 1967, Vol. 3, 291
3. *ACS 88*, 1969

All of the theoretically possible high-energy (and potentially hazardous) oxidant–fuel systems have been considered for use, and many have been evaluated, in rocket propulsion systems (with apparently the exception of the most potent combination, liquid ozone–liquid acetylene). Some of the materials which have been examined are listed below, and it is apparent that any preparative reactions deliberately involving oxidant–fuel pairs must be conducted under controlled conditions with appropriate precautions.

| *OXIDANTS* | *FUELS* |
| --- | --- |
| CHLORINE TRIFLUORIDE | ALCOHOLS |
| DINITROGEN TETRAOXIDE | AMINES |
| FLUORINE | AMMONIA |
| FLUORINE OXIDES | BERYLLIUM ALKYLS |
| HALOGEN FLUORIDES | BORANES |
| METHYL NITRATE | DICYANOGEN |
| NITRIC ACID | HYDRAZINES |
| NITROGEN TRIFLUORIDE | HYDROCARBONS |
| OXYGEN | HYDROGEN |
| OXYGEN FLUORIDES | NITROALKANES |
| OZONE | POWDERED METALS |
| PERCHLORIC ACID | SILANES |
| PERCHLORYL FLUORIDE | THIOLS |
| TETRAFLUOROHYDRAZINE | |
| TETRANITROMETHANE | |

Many of the above combinations are hypergolic (ignite on contact) or can be made so with additives.

A few single compounds have been examined as monopropellants, (alkyl nitrates, ethylene oxide, hydrazine, hydrogen peroxide), the two latter being catalytically decomposed in this application.

Solid propellant mixtures, which are of necessity storage-stable, often contain ammonium or hydrazinium perchlorates as oxidants.

The hazardous aspects of rocket propellant technology has been surveyed [3].

*See also* PROPELLANTS

## RUBBER

Air,
Cotton

*Res. Rept. 137*, Jones, S., Buxton, Safety in Mines Res. Establ., 1956

If compressed air leaks from a rubberised cotton hose and causes squealing vibration to occur, enough heat may be generated to cause ignition. Fires have been started by squealing poorly patched hoses, and ignition sources have been caused by the use of rubberised cotton gaskets (cut from used conveyor belting, etc.) in plain flanged joints.

Sodium chlorate

*See* SODIUM CHLORATE, $ClNaO_3$: Aluminium, Rubber

Metal azides

Tanaka, J. *et al.*, *Chem. Abs.*, 1952, **46**, 11743h

During the preparation of cellular rubber by thermal decomposition of calcium, strontium or barium azides, various additives are necessary to prevent explosive decomposition of the azide in the blended mixture.

## RUST

Rust (a hydrated basic iron carbonate) is one of the most common contaminants in non-stainless chemical processing plant, and has been involved in several incidents.

*See* FRICTIONAL IGNITION OF GASES
CARBON DISULPHIDE, $CS_2$: Air, Rust
HYDRAZINE, $H_2N_4$: Rust
HYDROGEN SULPHIDE, $H_2S$: Rust
MAGNESIUM, Mg: Metal oxides (reference 2)

## SHOCK-SENSITIVE MATERIALS

*Recommended Safe Practices and Procedures: Storage and Handling of Shock- and Impact-sensitive Materials,* Pamphlet SE-7, Washington, MCA, 1961

Materials classified as explosives are excluded.

## SILANES

**$H(SiH_2)_nH$, $RSiH_3$, etc.**

1. Stock, A. *et al., Ber.,* 1922, **55**, 3961
2. Kirk-Othmer, 1969, Vol. 18, 177

All the lower silanes are extremely sensitive to oxygen and ignite in air. The liberated hydrogen often ignites explosively [1]. Only under certain critical experimental conditions can they be mixed with oxygen without igniting [2].

Chloroform,
or Carbon tetrachloride,
Oxygen

Stock, A. *et al., Ber.,* 1923, **56,** 1087

The chlorination of the lower silanes proceeds explosively in presence of oxygen, but catalytic presence of aluminium chloride controls the reaction.

Halogens

Stock, A. *et al., Ber.,* 1919, **52**, 695

Reaction of silanes with chlorine or bromine is violent.

*See other* NON-METAL HYDRIDES

## SILICONE GREASE

Bromine trifluoride

*See* BROMINE TRIFLUORIDE, $BrF_3$: Silicone grease

## SILVER COMPOUNDS

Many silver compounds are explosively unstable.

*See* METAL ACETYLIDES

and the individually indexed compounds:

SILVER PERCHLORYLAMIDE, $AgClHNO_3$
SILVER AZIDE CHLORIDE, $AgClN_3$
SILVER CHLORITE, $AgClO_2$
SILVER CHLORATE, $AgClO_3$
SILVER PERCHLORATE, $AgClO_4$
SILVER PEROXOCHROMATE, $AgCrO_5$
SILVER AMIDE, $AgH_2N$
SILVER *N*-NITROSULPHURIC DIAMIDATE, $AgH_2N_3O_4S$
SILVER AZIDE, $AgN_3$
SILVER HEXAHYDROHEXABORATE(2–), $Ag_2B_6H_6$
DISILVER IMIDE, $Ag_2HN$
SILVER HYPONITRITE, $Ag_2N_2O_2$
SILVER OSMATE, $Ag_2O_4Os$
DISILVER PENTATIN UNDECAOXIDE, $Ag_2Sn_5O_{11}$
TRISILVER NITRIDE, $Ag_3N$
SILVER 2,4,6-TRIS(DIOXOSELENA)PERHYDROTRIAZINE-1,3,5-TRIIDE, $Ag_3N_3O_6Se_3$
TETRASILVER DIIMIDOTRIPHOSPHATE, $Ag_4H_3N_2O_8P_3$
SILVER DIIMIDODIOXOSULPHATE(4–), $Ag_4N_2O_2S$
PENTASILVER DIIMIDOTRIPHOSPHATE, $Ag_5H_2N_2O_8P_3$
PENTASILVER DIAMIDOPHOSPHATE, $Ag_5N_2O_2P$
SILVER CYANIDE, CAgN
SILVER CYANATE, CAgNO
SILVER FULMINATE, CAgNO
SILVER TRINITROMETHANIDE, $CAgN_3O_6$
SILVER AZIDODITHIOFORMATE, $CAgN_3S_2$
SILVER TRICHLOROMETHANEPHOSPHONATE(2–), $CAg_2Cl_3O_3P$
DISILVER CYANAMIDE, $CAg_2N_2$
DISILVER DIAZOMETHANEDIIDE, $CAg_2N_2$
SILVER TETRAZOLIDE, $CHAgN_4$
SILVER NITROUREIDE, $CH_2AgN_3O_3$
SILVER 5-AMINOTETRAZOLIDE, $CH_2AgN_5$
METHYLSILVER, $CH_3Ag$
SILVER NITROGUANIDIDE, $CH_4AgN_3$
SILVER CYANODINITROMETHIDE, $C_2AgN_3O_4$
DISILVER KETENDIIDE, $C_2Ag_2O$
DISILVER KETENDIIDE–SILVER NITRATE, $C_2Ag_2O \cdot AgNO_3$
SILVER OXALATE, $C_2Ag_2O_4$
SILVER ACETYLIDE, $C_2HAg$
SILVER 1,2,3-TRIAZOLIDE, $C_2H_2AgN_3$
SILVER DINITROACETAMIDE, $C_2H_2AgN_3O_5$
SILVER 1,3-DI(5-TETRAZOLYL)TRIAZINE, $C_2H_2AgN_{11}$
SILVER TRIFLUOROMETHYLACETYLIDE, $C_3AgF_3$
SILVER MALONATE, $C_3H_2Ag_2O_4$
SILVER 3-HYDROXYPROPYNIDE, $C_3H_3AgO$
SILVER BUTEN-3-YNIDE, $C_4H_3Ag$
1,3-PENTADIYN-1-YLSILVER, $C_5H_3Ag$
CYCLOPENTADIENYLSILVER PERCHLORATE, $C_5H_5AgClO_4$
DISILVER 1,3,5-HEXATRIYNIDE, $C_6Ag_2$

SILVER *p*-NITROPHENOLATE, $C_6H_4AgNO_3$
PHENYLSILVER, $C_6H_5Ag$
SILVER PHENOLATE, $C_6H_5AgO$
SILVER PHENYLSELENONATE, $C_6H_5AgO_3Se$
SILVER 3,5-DINITROANTHRANILATE, $C_7H_4AgN_3O_6$
SILVER ISOPHTHALATE, $C_8H_4Ag_2O_4$
SILVER 1-BENZENEAZOTHIOCARBONYL-2-PHENYLHYDRAZINE, $C_{13}H_{11}AgN_4S$
TRIS-2,2'-BIPYRIDINESILVER(II) PERCHLORATE, $C_{30}H_{24}AgCl_2N_6O_8$

*See other* HEAVY METAL DERIVATIVES

## SILVER-CONTAINING EXPLOSIVES

Luchs, J. A., *Photog. Sci. Eng.*, 1966, **10**, 334

Silver solutions used in photography can become explosive under a variety of conditions. Ammoniacal silver nitrate solutions, on storage, heating or evaporation, eventually deposit silver nitride ('fulminating silver'). Silver nitrate and ethanol may give silver fulminate, and, in contact with azides or hydrazine, silver azide. These are all dangerously sensitive explosives and detonators.

*See also* SILVERING SOLUTIONS
TOLLENS' REAGENT

## SILVERING SOLUTIONS

1. Smith, I. C. P., *Chem. & Ind.*, 1965, 1070; *J. Brit. Soc. Glassblowers*, 1964, 45
2. Ermes, M., *Diamant*, 1929, **51**, 62, 587
3. Lohmann, E., ibid., 526; Mylius, W., ibid., 42
4. Sivertz, C. *et al.*, Ger. Pat. 2 162 263, 1972

Brashear's silvering solution (alkaline ammoniacal silver oxide containing glucose) or residues therefrom should not be kept for more than two hours after preparation, since an explosive precipitate forms on standing [1]. The danger of explosion may be avoided by working with dilute silver solutions (0.35M) in the Brashear process, when formation of $Ag(NH_3)_2OH$ (and explosive $AgNH_2$ and $Ag_3N$ therefrom) is minimised. The use of Rochelle salt, rather than caustic, and shielding of solutions from direct sunlight, are also recommended safeguards [2,3].

Addition of sodium gluconate or tartrate to ammoniacal silver salt–base mixtures inhibits the formation of fulminating silver [4].

*See* TOLLENS' REAGENT

## SILYLHYDRAZINES $(R_3Si)_2N{-}NHSiR_3$, etc.

Oxidants

1. Wannagat, U. *et al.*, *Z. Anorg. Allgem. Chem.*, 1959, **299**, 341–349
2. Wannagat, U. *et al.*, *Monats.*, 1966, **97**, 1157–1162

During analysis of a series of tris(organosilyl)hydrazines, treatment with 1:1 mixtures of nitric and sulphuric acids had caused explosive reactions [1]. Subsequently the hypergolic behaviour of a range of 20 silylhydrazines and congeners in contact with fuming nitric acid was examined. All di- or tri-silyl derivatives showed ignition delays of 10 ms or less, several also exploding after ignition. All the derivatives ignited on dropping into gaseous fluorine, and into concentrated liquid ozone–oxygen mixtures, most also exploding in the latter [2].

The most reactive compounds are:

LITHIUM 1,1-DIMETHYL(TRIMETHYLSILYL)HYDRAZIDE, $C_5H_{15}LiN_2Si$
1,2-DIMETHYL-2-(TRIMETHYLSILYL)HYDRAZINE, $C_5H_{16}N_2Si$
LITHIUM TRIETHYLSILYLAMIDE, $C_6H_{16}LiNSi$
DILITHIUM 1,1-BIS(TRIMETHYLSILYL)HYDRAZIDE, $C_6H_{18}Li_2N_2Si_2$
1,2-BIS(TRIMETHYLSILYL)HYDRAZINE, $C_6H_{20}N_2Si_2$
TRIS(TRIMETHYLSILYL)HYDRAZINE, $C_9H_{28}N_2Si_3$
1,2-BIS(TRIETHYLSILYL)HYDRAZINE, $C_{12}H_{32}N_2Si_2$
TRIETHYLSILYL-1,2-BIS(TRIMETHYLSILYL)HYDRAZINE, $C_{12}H_{34}N_2Si_3$
1,2-BIS(TRIETHYLSILYL)TRIMETHYLSILYLHYDRAZINE, $C_{15}H_{40}N_2Si_3$
1,2-BIS(TRIPROPYLSILYL)HYDRAZINE, $C_{18}H_{44}N_2Si_2$
3,3,6,6-TETRAPHENYLPERHYDRO-3,6-DISILATETRAZINE, $C_{24}H_{24}N_4Si_2$

*See other* *N*-METAL DERIVATIVES

## SLAG WOOL

Potassium permanganate

*See* POTASSIUM PERMANGANATE, $KMnO_4$: Slag wool

## SMELT


1. Nelson, W., *Chem. Abs.*, 1973, **78**, 126458j
2. Morgan, H. W. *et al.*, *Chem. Abs.*, 1967, **66**, 56874t
3. Anon., ibid., 56876v
4. Duda, Z., *Chem. Abs.*, 1976, **84**, 32851


'Smelt', the residue left from burning of spent sulphite liquor from wood-pulp treatment and consisting largely of sodium sulphide and sodium carbonate, explodes violently on contact with water while still hot. The mechanism is discussed [1], and operational procedures are recommended for avoiding the possibility of explosions [2,3]. A further possible cause of explosions is the generation of hydrogen and/or hydrogen sulphide at the high temperatures involved [4].

## SOAP POWDERS

Anon., *Chem. Abs.*, 1935, **29**, 6759[7]

General factors affecting spontaneous ignition of soap powders are discussed.

## SODA–LIME

**$NaOH \cdot Ca(OH)_2$**

Hydrogen sulphide

Bretherick, L., *Chem. & Ind.*, 1971, 1042

Soda-lime, after absorbing hydrogen sulphide, exhibits a considerable exotherm (100°C) when exposed simultaneously to moisture and air, particularly with carbon dioxide enrichment, and has caused fires in laboratory waste bins containing moist paper wipes. Saturation with water and disposal in sealed containers is recommended.

## SODIUM PRESS

Blau, K., private comm., 1965

The jet of a sodium press became blocked during use, and the ram was tightened to free it. It suddenly cleared and a piece of sodium wire was extruded, piercing a finger, which had to be amputated later. Sodium in a blocked die should be dissolved out in a dry alcohol.

*See* POTASSIUM, K: Alcohols

## SOLVENTS

Davies, A. G., *J. R. Inst. Chem.*, 1956, **80**, 386

A short, detailed account, with references, of explosion hazards of autoxidised solvents, including: the autoxidation reaction; solvent and peroxide content; inhibition of peroxide formation; detection and estimation of peroxides; removal of peroxides.

## SPILLAGES

'How to Deal with Spillages of Hazardous Chemicals', Poole, BDH Chemicals Ltd., 1970

A revised wall chart, with standardised disposal procedures for some 330 toxic and hazardous chemicals.

## SPONTANEOUS IGNITION

Virtala, V. *et al.*, *Chem. Abs.*, 1952, **44**, 7770i–7771b

Methods for assessing the potential for spontaneous (oxidative) ignition for a range of 25 organic liquids and solids are described. Case histories are included for ignition of castor oil on peat, mineral oil on iron turnings and wood shavings round steam pipes at 100°C, with 56 references.

*See* SOAP POWDERS

## STARCH

Calcium hypochlorite,
Sodium hydrogensulphate

*See* CALCIUM HYPOCHLORITE, $CaCl_2O_2$: Sodium hydrogensulphate, etc.

## STATIC INITIATION

Initiation of explosive decomposition by sparks derived from static electricity is thought to have been involved in the incidents involving the compounds:

METHYLMERCURY PERCHLORATE, $CH_3ClHgO_4$
TITANIUM CARBIDE, CTi
CHLOROETHYLENE (VINYL CHLORIDE), $C_2H_3Cl$
DICYANODIAZOMETHANE, $C_3N_4$
BIS-*o*-AZIDOBENZOYL PEROXIDE, $C_{14}H_8N_6O_4$
POTASSIUM PERCHLORATE, $ClKO_4$: Metal powders
SODIUM CHLORATE, $ClNaO_3$: Organic matter
: Paper

## STEEL WOOL

Anon., *Fire Prot. Assoc. J.*, 1953, **21**, 53

Ignition can occur if steel wool short-circuits the contacts of even a small dry-cell torch battery.

## STRAINED-RING COMPOUNDS

Some molecules with small distorted rings (of high strain-energy) are explosively unstable.

*See* BICYCLO[2.1.0]PENT-2-ENE, $C_5H_6$
1(or 2),3,4,5,6-PENTAFLUOROBICYCLO[2,2,2]HEXA-2,5-DIENE, $C_6HF_5$
3,6,9-TRIAZATETRACYCLO[$6.1.0.0^{2,4}.0^{5,7}$]NONANE, $C_6H_3N_3$
BENZVALENE, $C_6H_6$
PRISMANE, $C_6H_6$
1,4-DIHYDRODICYCLOPROPA[*b,g*]NAPHTHALENE, $C_{12}H_8$
TETRASULPHUR DINITRIDE, $N_2S_4$

## SULPHINYL AZIDES $ArSON_3$

Maricich, T. J. *et al.*, *J. Amer. Chem. Soc.*, 1974, **96**, 7771, 7776

Benzene-, *p*-nitrobenzene- and *p*-toluene-sulphinyl azides are thermally unstable, decomposing explosively when warmed alone or as concentrated solutions. They may be prepared safely and handled as solutions at −20°C or below. The isolated solids may be stored at −80°C, but benzene sulphinyl azide explodes at 11°C, and *p*-toluenesulphinyl azide at 8°C.

*See other* ACYL AZIDES

## SULPHONATION INCIDENTS

*See* the entries:

* 4-HYDROXY-3-NITROBENZENESULPHONYL CHLORIDE, $C_6H_4ClNO_5S$

*p*-NITROTOLUENE, $C_7H_7NO_2$ : Sulphuric acid, etc.

* 2-METHOXYANILINIUM NITRATE, $C_7H_{10}N_2O_4$ : Sulphuric acid

SULPHURIC ACID, $H_2O_4S$: *p*-Chloronitrobenzene, Sulphur trioxide
: Nitroaryl bases
: Tetramethylbenzenes

## SULPHUR BLACK

Anon., *Ind. Eng. Chem.*, 1919, **11**, 892

Twenty-four hours after several barrels of the dyestuff were bulked, blended and repacked, spontaneous heating occurred. This was attributed to the aerial oxidation of excess sodium polysulphide used during manufacture.

*See* SODIUM SULPHIDE, $Na_2S$

## TETRAHYDROPYRANYL ETHER DERIVATIVES $\overline{O(CH_2)_3CHO}$–

Meyers, A.I. *et al.*, *Tetrahedron Lett.*, 1976, 2417–2418

The tetrahydropyranyl group, commonly used in synthetic procedures to protect hydroxyl groups, appears not to be safe when peroxidising reagents are used with tetrahydropyranyl ether derivatives, because explosive peroxides, not destroyed by the usual reagents, are produced.

The hydroboration product of 2-methyl-2-propenyl tetrahydropyranyl ether was routinely oxidised with alkaline hydrogen peroxide, then treated with sodium sulphite solution during work-up. The product gave a negative test for peroxides but exploded violently during attempted distillation at 0.06 mbar from a vessel at 120°C. Epoxidation of 3-methyl-3-butenyl tetrahydropyranyl ether with peroxyacetic acid gave, after sulphite treatment, an apparently peroxide-free product. However, after distillation at 1 mbar, the fore-run (b.p. 40–70°C) exploded violently when the flask was disturbed. The main fractions were subsequently found to give strong positive indications of the presence of peroxides, which were only removed after prolonged treatment with sodium thiosulphate.

Other acetal-type protecting groups (tetrahydrofurfuryl ethers, methoxymethyl ethers, 1,3-dioxolanes) are also considered to be incompatible with peroxidising reagents.

*See other* PEROXIDISABLE COMPOUNDS

## TETRA(*N*-METHYLPYRIDYL)PORPHINE PERCHLORATES

Reid, J. B. *et al., Inorg. Chem.*, 1977, **16**, 968

Variously metallated derivatives of the porphine nitrate were converted to the perchlorate salts, but several exploded during drying. The need for great caution in attempting to prepare the anhydrous perchlorates is stressed.

*See* TETRAKIS(4-*N*-METHYLPYRIDINIO)PORPHINECOBALT(II)($4^+$) PERCHLORATE, $C_{44}H_{36}Cl_4CoN_8O_{16}$

*See other* AMMINEMETAL OXOSALTS

## TETRAZOLES

N=NNHN=CH (ring)

1. Benson, F. R., *Chem. Rev.*, 1947, **41**, 4–5
2. Morisson, H., *Util. Elem. Pyrotech. Explos. Syst. Spatiaux, Colloq. Int.*, 1968 (publ. 1969), 111–120
3. Schroeder, M. A., *Rept. AD-A018652*, Richmond (Va.), NTIS, 1975

There is a wide variation in thermal stability in derivatives of this high-nitrogen nucleus and several show explosive properties [1]. The characteristics of explosive tetrazole salts have been summarised [2], and the relationship between structure and reactivity of isomeric 1- and 2-tetrazole derivatives has been reviewed in a ballistics context [3].

1-DICHLOROAMINOTETRAZOLE, $CHCl_2N_5$
5-NITROTETRAZOLE, $CHN_5O_2$
5-AZIDOTETRAZOLE, $CHN_7$
TETRAZOLE, $CH_2N_4$
5-*N*-NITROAMINOTETRAZOLE, $CH_2N_6O_2$
5-AMINOTETRAZOLE, $CH_3N_5$
5-DIAZONIOTETRAZOLIDE, $CN_6$
SODIUM 5-DINITROMETHYLTETRAZOLIDE, $C_2HN_6NaO_4$
5-HYDROXY-1(*N*-SODIO-5-TETRAZOLYLAZO)TETRAZOLE, $C_2HN_{10}NaO$
1,3-DI(5-TETRAZOLYL)TRIAZENE, $C_2H_3N_{11}$
2-METHYLTETRAZOLE, $C_2H_4N_4$
1,2-DI(5-TETRAZOLYL)HYDRAZINE, $C_2H_4N_{10}$
1,6-BIS(5-TETRAZOLYL)HEXAAZA-1,5-DIENE, $C_2H_4N_{14}$
POTASSIUM 1-TETRAZOLEACETATE, $C_3H_3KN_4O_2$
SODIUM 1-TETRAZOLEACETATE, $C_3H_3N_4NaO_2$
5-CYANO-2-METHYLTETRAZOLE, $C_3H_3N_5$
2- or 5-ETHYLTETRAZOLE, $C_3H_6N_4$
5-AMINO-2-ETHYLTETRAZOLE, $C_3H_7N_5$
2-METHYL-5-VINYLTETRAZOLE, $C_4H_6N_4$
1,2-DIHYDROPYRIDO[2,1,*e*] TETRAZOLE, $C_5H_5N_4$
5-PHENYLTETRAZOLE, $C_7H_6N_4$

3-PHENYL-1-TETRAZOLYL-1-TETRAZENE, $C_7H_8N_8$
5(4-DIMETHYLAMINOBENZENEAZO)TETRAZOLE, $C_9H_{11}N_7$
1(2-NAPHTHYL)-3(5-TETRAZOLYL)TRIAZENE, $C_{11}H_9N_7$

*See other* HIGH-NITROGEN COMPOUNDS

## THERMAL STABILITY OF REACTION MIXTURES

1. Grewer, T., *Chem. Ing. Tech.*, 1975, **47**, 230–235
2. Hartgerink, J. W., *Proc. 3rd Symp. Chem. Probl. Connected Stab. Explos.*, 1973 (publ. 1974), 220–230 (*Chem. Abs.*, 1975, **82**, 173005b)

The importance of gaining knowledge on the reaction parameters of exothermic reaction systems to assess potential processing hazards is discussed in detail. The roles of differential thermal analysis, adiabatic thermal storage tests and adiabatic reaction tests are discussed and suitable techniques described with reference to practical examples of thermally unstable systems [1]. Two storage tests (adiabatic storage and isothermal heat generation) are described which give information on the induction period of instability. An exothermal decomposition meter, 100 times more sensitive than a DTA apparatus, is described [2].

*See* DIAZONIUM SULPHATES
DIFFERENTIAL THERMAL ANALYSIS
BENZENEDIAZONIUM CHLORIDE, $C_6H_5ClN_2$ (second reference)
2-CYANO-4-NITROBENZENEDIAZONIUM HYDROGENSULPHATE, $C_7H_4N_4O_6S$: Sulphuric acid
SULPHURIC ACID, $H_2O_4S$: *p*-Chloronitrobenzene, etc.

## THERMITE REACTIONS

*See* LIGHT ALLOYS
ALUMINIUM, Al: Metal oxides, etc.
: Sodium sulphate
ALUMINIUM–MAGNESIUM ALLOY, Al–Mg: Iron(III) oxide
ALUMINIUM–MAGNESIUM–ZINC ALLOYS, Al–Mg–Zn: Rusted steel
DIIRON TRIOXIDE, $Fe_2O_3$: Calcium disilicide
MAGNESIUM, Mg: Metal oxides

## THIOPHENOLATES

**$ArS^-$**

Diazonium salts

*See* DIAZONIUM SULPHIDES AND DERIVATIVES

## THORIUM FURNACE RESIDUES

Schmitt, C. R., *J. Fire Flamm.*, 1971, **2**, 163

Many furnace residues (fine powders and salts) deposited in the upper parts of furnaces used for thorium melting operations are highly pyrophoric and often ignite as the furnace is opened. Such residues may be rendered safe by storage under water for 60–90 days. If the water is drained off early, ignition may occur.

*See other* PYROPHORIC MATERIALS

## TOLLENS' REAGENT

1. Green, E., *Chem. & Ind.*, 1965, 943
2. Waldman, H., *Chimia*, 1959, **13**, 297–298
3. Coltoff, W., *Chem. Weekblad*, 1932, **29**, 737

This mixture of ammoniacal silver oxide and sodium hydroxide solution is potentially dangerous, since if kept for a few hours, it deposits a highly explosive precipitate. This danger was described by Tollens in 1882 but it is not generally known now. Prepare the reagent in small amounts just before use, in the tube to be used for the test, and discard immediately after use, NOT into a container for silver residues [1]. Several earlier references to hazards of storing the reagent before or after use are discussed [2]. On one occasion a violent explosion of the reagent occurred 1 hour after preparation and before a precipitate had formed [3].

*See* SILVERING SOLUTIONS

## TOXIC HAZARDS

While toxic hazards have been specifically excluded from consideration in this Handbook, such hazards are at least as important as reactive ones, particularly on a long-term basis. Due account of toxic hazards must therefore be taken in planning and executing laboratory work, particularly if unfamiliar materials are being brought into use.

It is perhaps appropriate to point out that many of the elements or compounds listed in this Handbook are here because of a high degree of reactivity towards other materials. It may therefore be broadly anticipated that under suitable circumstances of contact with animal organisms, a high degree of interaction will ensue, with possible subsequent onset of toxic or other deleterious effects.

*See* APROTIC SOLVENTS

## TRIALKYLALUMINIUMS $R_3Al$

A highly reactive group of compounds, of which the lower members are extremely pyrophoric, with very short ignition delays of use in rocket- or jet-fuel systems. Storage stability is generally high (decomposition with alkene and hydrogen evolution begins above about 170–180°C), but branched alkylaluminiums (notably triisobutylaluminium) decompose above 50°C.

*See* ALKYLALUMINIUM DERIVATIVES (references 1–6)

Individually indexed compounds of the TRIALKYLALUMINIUM group, many of which are commercially available in bulk, are:

TRIMETHYLALUMINIUM, $C_3H_9Al$
TRIETHYLALUMINIUM, $C_6H_{15}Al$
3-BUTEN-1-YNYLDIETHYLALUMINIUM, $C_8H_{13}Al$
TRIISOPROPYLALUMINIUM, $C_9H_{21}Al$
TRIPROPYLALUMINIUM, $C_9H_{21}Al$
3-BUTEN-1-YNYLDIISOBUTYLALUMINIUM, $C_{12}H_{21}Al$
TRIISOBUTYLALUMINIUM, $C_{12}H_{27}Al$

## TRIALKYLBISMUTHS

**$R_3Bi$**

Oxidants

Gilman, H. *et al.*, *Chem. Rev.*, 1942, **30**, 291

The lower alkylbismuths ignite in air, and explode in contact with oxygen, concentrated nitric or sulphuric acids.

*See* TRIMETHYLBISMUTH, $C_3H_9Bi$
TRIVINYLBISMUTH, $C_6H_9Bi$
TRIETHYLBISMUTH, $C_6H_{15}Bi$
TRIBUTYLBISMUTH, $C_{12}H_{27}Bi$

*See other* ALKYLMETALS

## TRIALKYLSILYLOXY ORGANOLEAD DERIVATIVES

**$R_3SiOPbR'_3$**

Houben-Weyl, 1975, Vol. 13.7, 118

Compounds containing Si–O–Pb bonds may interact explosively with oxygen at about 140°C, or with aluminium chloride, acyl halides or anhydrides.

*See related* ALKYLMETALS
ALKYLNON-METALS

## TRIAZENES

**N=N–N**

Houben-Weyl, 1965, Vol. 10(3), 700, 717, 722, 731

A number of triazene derivatives bearing –H, –CN, –OH or –NO on the terminal nitrogen of the chain are explosively unstable, mainly to heat.

*See* 3-CYANOTRIAZENES
3-HYDROXYTRIAZENES
3-NITROSOTRIAZENES
1,3-DI(5-TETRAZOLYL)TRIAZENE, $C_2H_3N_{11}$
1,3-DIMETHYLTRIAZENE ('DIAZOAMINOMETHANE'), $C_2H_7N_3$
3,3-DIMETHYL-1-PHENYLTRIAZENE, $C_8H_{11}N_3$
3,3-DIMETHYL-1-(3-QUINOLYL)TRIAZENE, $C_{11}H_{12}N_4$
1,3-DIPHENYLTRIAZENE, $C_{12}H_{11}N_3$
1,3-BIS(PHENYLTRIAZENO)BENZENE, $C_{18}H_{16}N_6$

*See other* HIGH-NITROGEN COMPOUNDS

## $N,N,N'$-TRIFLUOROALKYLAMIDINES $-C(NF)NF_2$

Ross, D. L. *et al.*, *J. Org. Chem.*, 1970, **35**, 3093

All the N–F compounds involved in the synthesis of a group of $N,N,N'$-trifluoroalkylamidines ($C_3$–$C_7$) were shock-sensitive, explosive compounds in varying degrees. Several were only stable in solution, and others exploded during analytical combustion.

*See other* $N$-HALOGEN COMPOUNDS

## TRINITROETHYL ORTHOESTERS $HC[OCH_2C(NO_2)_3]_3, C[OCH_2C(NO_2)_3]_4$

Shimio, K. *et al.*, *Chem. Abs.*, 1976, **85**, 194924c

Tris(2,2,2-trinitroethyl) orthoformate and tetrakis(2,2,2-trinitroethyl) orthocarbonate form powerfully explosive solutions in nitromethane. The oxygen balances of the esters are +154 and +182%, respectively.

*See* OXYGEN BALANCE

*See other* POLYNITROALKYL COMPOUNDS

## TURPENTINE

Diatomaceous earth

Anon., *Ind. Eng. Chem.*, 1950, **42**(7), 77A

A large quantity of discoloured (and peroxidised) turpentine was heated with fuller's earth to decolourise it, and subsequently exploded. Fuller's earth causes exothermic catalytic decomposition of peroxides and rearrangement of the terpene molecule.

Halogens,
or Oxidants,
or Tin(IV) chloride

Mellor, 1941, Vol. 2, 11, 90; 1941, Vol. 7, 446: 1943, Vol. 11, 395

Turpentine ignites in contact with fluorine (at −210°C), chlorine, iodine, chromium trioxide and chromyl chloride, and usually with tin(IV) chloride. Other highly unsaturated liquid hydrocarbons may be expected to react similarly.

## UNIT PROCESS INCIDENTS

Incidents have been grouped for each of the unit process headings:

AMINATION
DIAZOTISATION
EPOXIDATION
GRIGNARD REAGENT FORMATION
HALOGENATION
HYDROGENATION
NITRATION

POLYMERISATION
SULPHONATION

## UNSATURATED OILS

*See* LINSEED OIL
CARBON, C: Unsaturated oils

## URANYL MACROCYCLIC PERCHLORATE LIGANDS

Vidali, M. *et al., J. Inorg. Nucl. Chem.*, 1975, **37**, 1715–1719

A series of uranyl complexes of macrocyclic azomethines were used as ligands for transition metal ions, with perchlorate anions. Raman spectra of the uranyl–metal complexes could not be recorded because the samples exploded during attempted measurements.

*See other* AMMINEMETAL OXOSALTS

## VEGETABLE OILS

Catalysts,
Hydrogen

*See* HYDROGEN, $H_2$ : Catalysts, Vegetable oils

## VIOLENT POLYMERISATION

*A Review of Violent Monomer Polymerisation*, Harmon, M. *et al.*, *Rept. AD-A017443*, Richmond (Va.), USNTIS, 1974

Literature related to the possibility of violent polymerisation of the ten monomers most significant industrially has been classified and reviewed, and 209 annotated references are given. The compounds covered are acrylic acid, acrylonitrile, 1,3-butadiene, ethylene, ethylene oxide, methyl acrylate, methyl methacrylate, propiolactone, styrene, vinyl acetate, vinyl chloride and vinylidene chloride. All except ethylene, methyl acrylate and propiolactone have been involved in explosive polymerisation incidents. For each compound, data and selected references on physical properties, reactivity, inhibition and handling procedures are given.

*See also* POLYMERISATION INCIDENTS

## WATER-REACTIVE COMPOUNDS

A large number of individual compounds react exothermically and violently with water, particularly with restricted quantities. Many such compounds come within the groups:

ACID ANHYDRIDES

ACYL HALIDES
ALKALI METALS
ALKYLALUMINIUM DERIVATIVES
ALKYLNON-METAL HALIDES
COMPLEX HYDRIDES
METAL HALIDES
METAL HYDRIDES
METAL OXIDES
NON-METAL HALIDES AND THEIR OXIDES
NON-METAL OXIDES

**WAX FIRE**

Carbon tetrachloride

Gilmont, R., *Chem. Eng. News,* 1947, **25**, 2853

Use of carbon tetrachloride to extinguish a wax fire caused an explosion. This was attributed to a violent reaction between unsaturated wax components and carbon tetrachloride initiated by radicals from decomposing peroxides.

*See* DIBENZOYL PEROXIDE, $C_{14}H_{10}O_4$: Carbon tetrachloride, etc.

**WOOL**

Lee, P. R., *J. Appl. Chem.*, 1969, **19**, 345–351

The self-heating and ignition of baled or loose wool in bulk storage is discussed and analysed, and steady-state thermal explosion theory is applied to the prediction of critical masses and induction periods for storage and transportation situations in relation to ambient temperature. Results obtained were consistent with current safety practices.

**'XANTHATES'** **EtOC(S)S$^-$**

Sorbe, 1968, 74

Derivatives of alkyldithiocarbonates ('xanthates') are hazardous as dusts, forming explosive suspensions in air. The lower-alkyl salts are explosive in the solid state when dry.

*See* POTASSIUM *O*-ETHYLDITHIOCARBONATE (XANTHATE), $C_3H_5KOS_2$

**XENON COMPOUNDS** **Xe–F, Xe–O**

Jha, N. K., *RIC Rev.*, 1971, **4**, 167–168

Several references to hazards associated with xenon compounds have been collected. Individually indexed compounds are:

CAESIUM BROMOXENATE, $BrCsO_3Xe$
XENON(II) FLUORIDE TRIFLUOROMETHANESULPHONATE, $CF_4O_3SXe$
XENON(II) FLUORIDE TRIFLUOROACETATE, $C_2F_4O_2Xe$
XENON(II) TRIFLUOROACETATE, $C_4F_6O_4Xe$
XENON(II) FLUORIDE PERCHLORATE, $ClFO_4Xe$
CAESIUM HYDROGENXENATE, $CsHO_4Xe$
XENON DIFLUORIDE OXIDE, $F_2OXe$
XENON DIFLUORIDE, $F_2Xe$
XENON TETRAFLUORIDE OXIDE, $F_4OXe$
XENON TETRAFLUORIDE, $F_4Xe$
XENON HEXAFLUORIDE, $F_6Xe$
XENON(II) PENTAFLUOROORTHOTELLURATE, $F_{10}O_2Te_2Xe$
XENON(II) PENTAFLUOROORTHOSELENATE, $F_{10}O_2Se_2Xe$
XENON TETRAHYDROXIDE, $H_4O_4Xe$
POTASSIUM HEXAOXOXENONATE(4−)–XENON TRIOXIDE, $K_4O_6Xe \cdot 2O_3Xe$
XENON TRIOXIDE, $O_3Xe$
XENON TETRAOXIDE, $O_4Xe$

# CYCLIC FORMULAE FOR PART 1

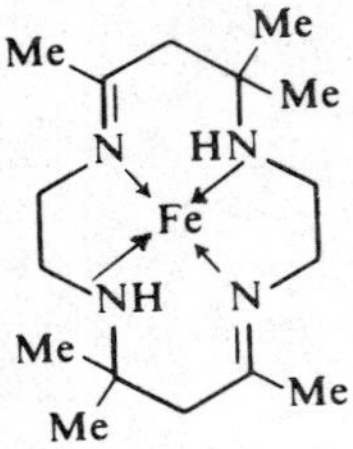

[14] Diene-$N_4$ iron complexes (p. 51)

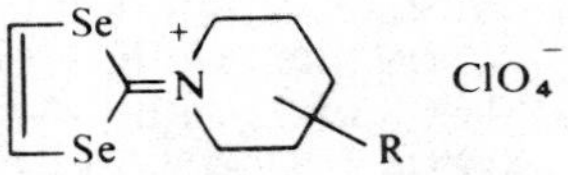

1-(1,3-Diselenonylidene)-piperidinium perchlorates (p. 55)

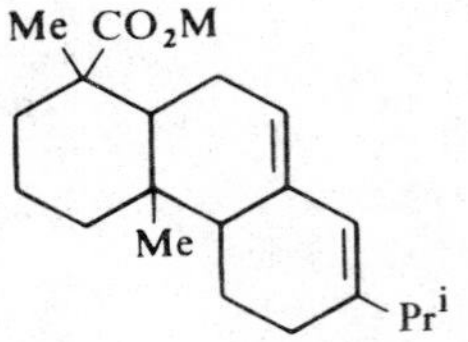

Metal abietates (p. 96)

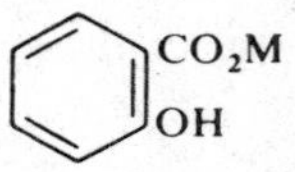

Metal salicylates (p. 119)

N(O)OM

O=⟨ ⟩=N(O)OM, etc.

*aci*-Nitroquinonoid compounds (p. 128)

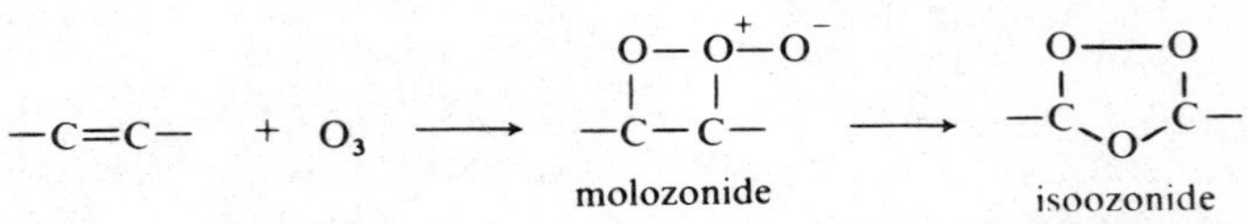

$-C{=}C- \; + \; O_3 \longrightarrow$ molozonide $\longrightarrow$ isoozonide (more stable)

Ozonides (p. 147)

# Specific Chemical Section

**(Elements and Compounds arranged in formula order)**

## EXPLANATORY NOTES

This section gives detailed information on the hazardous properties of individual chemicals, either alone or in combination with other compounds. The items are arranged in order of the empirical formula (at right of bold capital title line) corresponding to the chemical name used as the title of each item. An abbreviated (usually linear) representation of the structure is given at the head of each item entry, except where the empirical and structural formulae are identical.
A † prefixed to the title name indicates information on fire-related properties in Appendix 2.

Chemicals involved with the title compound in a reactive incident follow under the title. Where these secondary chemicals are described in group terms (e.g. metal halogenates), reference to the group entries in Section 1 may suggest other analogous possibilities of hazards.

References to original literature then follow, and sufficient of the relevant information is given to allow a general picture of the extent and degree of hazard to be seen. All temperatures are expressed in degrees Celsius; pressures in bars or mbars; volumes in $m^3$, litres or ml; and energy as joules, kJ or MJ. Where appropriate, attention is drawn to closely similar or related materials or events by *See* or *See also* cross-references.

Finally, if a compound is a member of one of the general classes or groups in Section 1, it is related to those by a *See other* cross-reference. If the compound is not strictly classifiable, a *See related* reference establishes a less direct link to the compound index lists in Section 1.

An alphabetical index of the IUPAC chemical names used as titles in Section 2, together with synonyms, is given in Appendix 5.

Details of corrections of typographical or factual errors, or of further items for inclusion in the text, will be welcomed by the author.

## SILVER

**Ag**

Acetylenic compounds

*See* ACETYLENIC COMPOUNDS: Metals

Aziridine

*See* AZIRIDINE (ETHYLENEIMINE), $C_2H_5N$: Silver

Bromine azide

*See* BROMINE AZIDE, $BrN_3$

1-Bromo-2-propyne

*See* 1-BROMO-2-PROPYNE, $C_3H_3Br$: Metals

Carboxylic acids

Koffolt, J. H., private comm., 1965.

Silver is incompatible with oxalic or tartaric acids, since the silver salts decompose on heating. Silver oxalate explodes at 140°C, and silver tartarate loses carbon dioxide.

*See also* METAL OXALATES

Chlorine trifluoride

*See* CHLORINE TRIFLUORIDE, $ClF_3$: Metals

Electrolytes,
Zinc

Britz, W. K. *et al.*, *Power Sources Symp. Proc.*, 1974, **26**, 162–165 (*Chem. Abs.*, 1975, **83**, 150293)

Causes of spontaneous combustion and other hazards of silver–zinc batteries were investigated.

Ethanol,
Nitric acid

Luchs, J. K., *Photog. Sci. Eng.*, 1966, **10**, 334

Action of silver on nitric acid in presence of ethanol may form the readily detonable silver fulminate.

*See* NITRIC ACID, $HNO_3$: Alcohols
*See also* SILVER-CONTAINING EXPLOSIVES

Ethylene oxide

*See* ETHYLENE OXIDE, $C_2H_4O$: Silver

Ethyl hydroperoxide

*See* ETHYL HYDROPEROXIDE, $C_2H_6O_2$: Silver

Hydrogen peroxide

*See* HYDROGEN PEROXIDE, $H_2O_2$: Metals

Ozonides

*See* OZONIDES

Peroxomonosulphuric acid

*See* PEROXOMONOSULPHURIC ACID, $H_2O_5S$: Catalysts

Peroxyformic acid

*See* PEROXYFORMIC ACID, $CH_2O_3$: Metals

*See other* METALS

**SILVERED COPPER** **Ag–Cu**

Ethylene glycol

*See* ETHYLENE GLYCOL, $C_2H_6O_2$: Silvered copper wire

**SILVER TETRAFLUOROBORATE** **$AgBF_4$**

Preparative hazard

1. Meerwein, H. *et al.*, *Arch. Pharm.*, 1958, **291**, 541–554
2. Lemal, D.M. *et al.*, *Tetrahedron Lett.*, 1961, 776–777
3. Olah, G. A. *et al.*, *J. Inorg. Nucl. Chem.*, 1960, **14**, 295–296

Experimental directions must be followed exactly to prevent violent spontaneous explosions during preparation of the salt from silver oxide and boron trifluoride etherate in nitromethane, according to the earlier method [1]. The later method [3] is generally safer than the earlier [2].

**SILVER TETRAFLUOROBROMATE** **$Ag[BrF_4]$** **$AgBrF_4$**

*See* METAL POLYHALOHALOGENATES

**SILVER BROMATE** **$AgBrO_3$**

Sulphur compounds

1. Taradoire, F., *Bull. Soc. Chim. Fr.*, 1945, **12**, 94–95
2. Pascal, 1960, Vol. 13.1, 1004

The bromate is a powerful oxidant, and unstable mixtures with sulphur ignite at 73–75°C, and with disulphur dibromide on contact [1]. Hydrogen sulphide ignites on contact with the bromate [2].

*See other* METAL OXOHALOGENATES

## SILVER CHLORIDE

**AgCl**

Aluminium

*See* ALUMINIUM, Al: Silver chloride

Ammonia

1. Mellor, 1941, Vol. 3, 382
2. Kauffmann, G. B., *J. Chem. Educ.*, 1977, **54**, 132
3. Ranganathan, S. *et al.*, ibid., 1976, **53**, 347

Exposure of ammoniacal silver chloride solutions to air or heat produces a black crystalline deposit of 'fulminating silver', mainly silver nitride, with disilver imide and silver amide also possibly present. Attention is drawn [2] to the potential explosion hazard in a method of recovering silver from the chloride by passing an ammoniacal solution of the chloride through an ion exchange column to separate the $Ag(NH_3)^+$ ion, prior to elution as the nitrate [3]. It is essential to avoid letting the ammoniacal solution stand for several hours, either alone or on the column [2].

*See* TRISILVER NITRIDE, $Ag_3N$

*See other* METAL HALIDES

## SILVER PERCHLORYLAMIDE

**$AgNHClO_3$** **$AgClHNO_3$**

*See* PERCHLORYLAMIDE SALTS

## SILVER AZIDE CHLORIDE

**$N_3AgCl$** **$AgClN_3$**

Frierson, W. J. *et al., J. Amer. Chem. Soc.*, 1943, **65**, 1698

It is shock-sensitive when dry.

*See other* METAL AZIDE HALIDES

## SILVER CHLORITE

**AgOClO** **$AgClO_2$**

Alkyl iodides

Levi, G. R., *Gazz. Chim. Ital [2]*, 1923, **53**, 40

Attempts to react the chlorite with methyl or ethyl iodides caused explosions, immediate in the absence of solvents, or delayed in presence of solvents. Silver chlorite itself is impact-sensitive, cannot be ground finely and explodes at 105°C.

*See other* CHLORITE SALTS

Hydrochloric acid,
or Sulphur

Mellor, 1941, Vol. 2, 284

It explodes in contact with hydrochloric acid or on rubbing with sulphur.

Non-metals

Pascal, 1960, Vol. 16, 264

Finely divided carbon, sulphur or red phosphorus are oxidised violently by the chlorite.

*See other* CHLORITE SALTS
METAL OXOHALOGENATES

**SILVER CHLORATE** **$AgClO_3$**

Sorbe, 1968, 126

An explosive salt.

*See other* METAL OXOHALOGENATES

**SILVER PERCHLORATE** **$AgClO_4$**

Alkynes,
Mercury

Comyns, A. E. *et al., J. Amer. Chem. Soc.*, 1957, **79**, 4342

Concentrated solutions of the perchlorate in 2-pentyne or 3-hexyne (complexes are formed) explode on contact with mercury.

*See* METAL ACETYLIDES

Aromatic compounds

1. Sidgwick, 1950, 1234
2. Brinkley, S. R., *J. Amer. Chem. Soc.,* 1940, **62**, 3524
3. Peone, J. *et al., Inorg. Synth.*, 1974, **15**, 69

Silver perchlorate forms solid complexes with aniline, pyridine, toluene, benzene and many other aromatic hydrocarbons [1]. A sample of the benzene complex exploded violently on crushing in a mortar. The ethanol complex also exploded similarly, and unspecified perchlorates dissolved in organic solvents were observed to explode [2]. Solutions of the perchlorate in benzene are said to be dangerously explosive [3], but this may be in error for the solid benzene complex.

Diethyl ether

1. Heim, F., *Angew. Chem.*, 1957, **69**, 274
2. Anon., *Angew. Chem. (Nachr.)*, 1962, **10**, 2

After crystallisation from ether, the material exploded violently on crushing in a mortar. It has been considered stable previously, since it melts without decomposition [1]. However, there was a similar incident with silver perchlorate not previously in contact with organic materials [2].

*See other* METAL OXOHALOGENATES

**SILVER PEROXOCHROMATE** $Ag[(Cr(O_2)O_3]$ $AgCrO_5$

Sulphuric acid

Riesenfeld, E. H., *et al.*, *Ber.*, 1914, **47**, 548

In attempts to prepare 'perchromic acid', a mixture of silver (or barium) peroxochromate and 50% sulphuric acid prepared at −80°C reacted explosively on slow warming to about −30°C.

*See other* PEROXOACID SALTS

**SILVER FLUORIDE** $AgF$

Calcium hydride

*See* CALCIUM DIHYDRIDE, $CaH_2$ : Silver halides

Non-metals

Mellor, 1941, Vol. 3, 389

Boron reacts explosively when ground with silver fluoride; silicon reacts violently.

Titanium

Mellor, 1941, Vol. 7, 20

Interaction at 320°C is incandescent.

*See other* METAL HALIDES

**SILVER DIFLUORIDE** $AgF_2$

Boron,
Water

Tulis, A. J. *et al.*, *Proc. 7th Symp. Explos. Pyrotechnics*, 1971, **3**(4), 1–12

Mixtures of boron and silver difluoride function as detonators when contacted with water

Dimethyl sulphoxide

*See* IODINE PENTAFLUORIDE, $F_5I$ : Dimethyl sulphoxide

Hydrocarbons,
or Water

Priest, H. F., *Inorg. Synth.*, 1950, **3**, 176

It reacts even more vigorously with most substances than does cobalt fluoride.

*See other* METAL HALIDES

**SILVER AMIDE** **$AgNH_2$** **$AgH_2N$**

Brauer, 1965, Vol. 2, 1045

Extraordinarily explosive when dry.

*See* NITROGEN TRIIODIDE–SILVER AMIDE, $I_3N{\cdot}AgH_2N$

*See other* *N*-METAL DERIVATIVES

**SILVER *N*-NITROSULPHURIC DIAMIDATE** **$AgH_2N_3O_4S$**

**$AgN(NO_2)SO_2NH_2$**

Sorbe, 1968, 120

This silver salt of the nitro-amide is explosive.

*See other* *N*-METAL DERIVATIVES

**DIAMMINESILVER PERMANGANATE** **$AgH_6MnN_2O_4$**

Pascal, 1960, Vol. 16, 1062 **$[Ag(NH_3)_2]MnO_4$**

It may explode on impact or shock.

*See other* AMMINEMETAL OXOSALTS

**SILVER IODATE** **$AgIO_3$**

Metals

*See* POTASSIUM, K: Oxidants
SODIUM, Na: Iodates

Tellurium

Pascal, 1960, Vol. 13.2, 1961

Interaction is violent.

*See other* METAL OXOHALOGENATES

**SILVER NITRATE** **$AgNO_3$**

Acetylene and derivatives

Mellor, 1946, Vol. 5, 854

Silver nitrate (or other soluble salt) reacts with acetylene in presence of ammonia to form silver acetylide, a sensitive, powerful detonator when dry. In the absence of ammonia, or when calcium acetylide is added to silver nitrate solution, explosive double salts of silver acetylide and silver nitrate are produced. Mercurous acetylide precipitates silver acetylide from the aqueous nitrate.

*See other* METAL ACETYLIDES

Acrylonitrile

*See* ACRYLONITRILE, $C_3H_3N$ : Silver nitrate

Ammonia

1. *MCA Case History No. 2116*
2. *CISHC Chem. Safety Summ.*, 1976, **47**, 31

A bottle containing Gomari tissue-staining solution (ammoniacal silver nitrate) prepared 2 weeks previously exploded when disturbed. The solution must be prepared freshly each day, and discarded immediately after use with appropriate precautions [1]. A large quantity of ammoniacal silver nitrate solution exploded violently when disturbed by removing a glass rod [2].

Ammonia,
Ethanol

*MCA Case History No. 1733*

A silvering solution exploded when disturbed. This is a particularly dangerous mixture, because both silver nitride and silver fulminate would be formed.

*See* Ethanol, below

Ammonia,
Sodium carbonate

Vasbinder, H., *Pharm. Weekblad*, 1952, **87**, 861–865

A mixture of the components in gum arabic solution (marking ink) exploded when warmed.

Ammonia,
Sodium hydroxide

1. Milligan, T. W. *et al.*, *J. Org. Chem.*, 1962, **27**, 4663
2. *MCA Case History No. 1554*
3. Morse, J. R., *School Sci. Rev.*, 1955, **37**(131), 147
4. Baldwin, J., *School Sci. Rev.*, 1967, **48**(165), 586

During preparation of an oxidising agent on a larger scale than described [1], addition of warm sodium hydroxide solution to warm ammoniacal silver nitrate with stirring caused immediate precipitation of black silver nitride which exploded [2]. Similar incidents had been reported previously [3], including one where explosion appeared to be initiated by addition of Devarda's alloy (Al–Cu–Zn [4]).

*See* TRISILVER NITRIDE, $Ag_3N$

*See also* SILVERING SOLUTIONS
TOLLENS' REAGENT

Arsenic

Mellor, 1941, Vol. 3, 470

A finely divided mixture with excess nitrate ignited when shaken out on to paper.

Chlorine trifluoride

*See* CHLORINE TRIFLUORIDE, $ClF_3$ : Metals, etc.

Chlorosulphuric acid

Mellor, 1941, Vol. 3, 470

Interaction is violent, nitrosulphuric acid being formed.

Disilver acetylide

*See* DISILVER ACETYLIDE–SILVER NITRATE, $C_2Ag_2 \cdot AgNO_3$

Disilver ketendiide

*See* DISILVER DIKETENDIIDE–SILVER NITRATE, $C_2Ag_2O \cdot AgNO_3$

Ethanol

1. Tully, J. P., *Ind. Eng. Chem. (News Ed.)* 1941, **19**, 250
2. Luchs, J. K., *Photog. Sci. Eng.*, 1966, **10**, 334
3. Garin, D.L. *et al.*, *J. Chem. Educ.*, 1970, **47**, 741

Reclaimed silver nitrate crystals, damp with the alcohol used for washing, exploded violently when touched with a spatula, generating a strong smell of ethyl nitrate [1]. The explosion was attributed to formation of silver fulminate (which is produced on addition of ethanol to silver nitrate solutions). Ethyl nitrate may also have been involved. Alternatives to avoid ethanol washing of recovered silver nitrate are discussed [2], including use of 2-propanol [3].

*See* SILVER FULMINATE, CAgNO

Magnesium,
Water

Marsden, F., private comm., 1973

Lyness, D. J. *et al.*, *School. Sci. Rev.*, 1953, **35**(125), 139

An intimate mixture of dry powdered magnesium and silver nitrate may ignite explosively on contact with a drop of water.

Non-metals

Mellor, 1941, Vol. 3, 469–473

Under a hammer blow, a mixture with charcoal ignites, while mixtures with phosphorus and sulphur explode, the former violently.

Phosphine

Mellor, 1941, Vol. 3, 471

Rapid passage of gas into the concentrated nitrate solution caused an explosion, or ignition with a slower gas stream. The explosion may have been due to rapid oxidation of the precipitated silver phosphide derivative by the co-produced nitric acid or dinitrogen tetraoxide.

Phosphonium iodide

*See* PHOSPHONIUM IODIDE, $H_4IP$ : Oxidants
*See other* METAL OXONON-METALLATES

**SILVER AZIDE** **$AgN_3$**

1. Mellor, 1940, Vol. 8, 349; 1967, Vol. 8 Suppl. 2, 47
2. Gray, P. *et al.*, *Chem. & Ind.*, 1955, 1255
3. Kabanov, A. A. *et al.*, *Russ. Chem. Rev.*, 1975, **44**, 538

While pure silver azide explodes at 340°C [1], the presence of impurities may cause explosion at 270°C. It is also impact-sensitive and explosions are usually violent [2]. Its use as a detonator has been proposed.

Application of an electric field to crystals of the azide will detonate them, at down to −100°C [3].

Chlorine azide

*See* CHLORINE AZIDE, $ClN_3$ : Ammonia, etc.

Halogens

Mellor, 1940, Vol 8, 336

Silver azide, itself a sensitive compound, is converted by ethereal iodine into the less stable and explosive compound, iodine azide. Similarly, contact with nitrogen-diluted bromine vapour gives bromine azide, often causing explosions.

*See* SILVER(II)AZIDE CHLORIDE, $AgClN_3$

Oxide or sulphide impurities

Kurochkin, E. S. *et al.*, *Chem. Abs.*, 1975, **83**, 201390q

Pure silver azide explodes at 340°C, but presence of below 10% of copper(I) or (II) oxides or sulphides, copper(I) selenide or bismuth(III) sulphide reduces the detonation temperature to 235°C. Concentrations of 10% of copper(II) oxide, copper(I) selenide or sulphide further reduced it to 200, 190 and 170°C, respectively.

Photosensitising dyes

Aleksandrov, E. *et al.*, *Chem. Abs.*, 1974, **81**, 31755b

In a study of thermolysis of dye-sensitised silver azide, it was found that many dyes caused explosions in the initial stages.

*See other* METAL AZIDES

## (UNKNOWN STRUCTURE) $AgN_5S_3$

*See* 1,3,5-TRICHLOROTRITHIA-1,3,5-TRIAZINE (THIAZYL CHLORIDE), $Cl_3N_3S_3$ : Ammonia, etc.

## SILVER(II) OXIDE $AgO$

Hydrogen sulphide

*See* HYDROGEN SULPHIDE, $H_2S$: Metal oxides

## SILVER HEXAHYDROHEXABORATE(2–) $Ag_2B_6H_6$

Bailar, 1973, Vol. 1, 808 $Ag_2[B_6H_6]$

It is a detonable salt.

*See other* HEAVY METAL DERIVATIVES

## DISILVER IMIDE $Ag_2NH$ $Ag_2HN$

Bailar, 1973, Vol. 3, 101

It explodes very violently when dry.

*See other* *N*-METAL DERIVATIVES

## SILVER HYPONITRITE AgON=NOAg $Ag_2N_2O_2$

*See* HYPONITROUS ACID, $H_2N_2O_2$

*See other* SILVER COMPOUNDS

## SILVER(I) OXIDE $Ag_2O$

Aluminium

*See* COPPER(II) OXIDE, CuO : Metals

Ammonia

Vasbinder, H., *Pharm. Weekblad*, 1952, **87**, 861–865

The clear solution, obtained by centrifuging a solution of the oxide in aqueous ammonia which had been treated with silver nitrate until precipitation started, exploded on two occasions after 10–14 days' storage in closed bottles in the dark. This was ascribed to slow precipitation of amorphous disilver imide, which is very explosive even when wet.

*See* DISILVER IMIDE, $Ag_2HN$

Ammonia or hydrazine,
Ethanol

Silver oxide and ammonia or hydrazine slowly form explosive silver nitride and, in presence of alcohol, silver fulminate may also be produced.

*See* SILVER-CONTAINING EXPLOSIVES
SILVERING SOLUTIONS

Boron trifluoride etherate,
Nitromethane

*See* SILVER TETRAFLUOROBORATE, $AgBF_4$

Carbon monoxide

Mellor, 1941, Vol. 3, 377

Carbon monoxide is exothermically oxidised over silver oxide, and the temperature may attain 300°C.

Chlorine,
Ethylene

*See* ETHYLENE, $C_2H_4$ : Chlorine

Dichloro(methyl)silane

*See* DICHLORO(METHYL)SILANE, $CH_4Cl_2Si$: Oxidants

Hydrogen sulphide

*See* HYDROGEN SULPHIDE, $H_2S$ : Metal oxides

Magnesium

Mellor, 1941, Vol. 3, 378

Oxidation of magnesium proceeds explosively when warmed with silver oxide in a sealed tube.

Metal sulphides

Mellor, 1941, Vol. 3, 376

Mixtures with auric, antimony or mercuric sulphides ignite on grinding.

Nitroalkanes

*See* NITROALKANES : Metal oxides

Non-metals

Mellor, 1941, Vol. 3, 376–377

Selenium, sulphur or phosphorus ignite on grinding with the oxide.

Potassium–sodium alloy

*See* POTASSIUM–SODIUM ALLOY, K–Na: Metal oxides

Seleninyl chloride

*See* SELENINYL CHLORIDE, $Cl_2OSe$ : Metal oxides

Selenium disulphide

Mellor, 1941, Vol. 3, 377

A mixture may ignite under impact.

*See other* METAL OXIDES

## SILVER PEROXIDE $Ag_2O_2$

Polyisobutene

Mellinger, T., *Arbeitsschutz*, 1972, 248

Mixtures of silver peroxide with 1% of polyisobutene exploded on three separate occasions. Use of a halogenated polymer to replace the polyisobutene was safe.

*See other* METAL PEROXIDES

## SILVER OSMATE $Ag_2OsO_4$ $Ag_2O_4Os$

Sorbe, 1968, 126

Explodes on impact or heating.

*See other* HEAVY METAL DERIVATIVES

## SILVER SULPHIDE $Ag_2S$

Potassium chlorate

*See* POTASSIUM CHLORATE, $ClKO_3$ : Metal sulphides

## DISILVER PENTATIN UNDECAOXIDE $Ag_2Sn_5O_{11}$

**(unknown structure)**

Mellor, 1941, Vol. 7, 418

The compound 'silver beta-stannate' is formed by long contact between solutions of silver and stannous nitrates, and loses water on heating and decomposes explosively.

*See other* METAL OXIDES

## TRISILVER NITRIDE $Ag_3N$

Hahn, H., *et al., Z. Anorg. Chem.*, 1949, **258**, 77

Very sensitive to contact with hard objects, exploding even when moist. An extremely sensitive explosive when dry, initiable by friction, impact or heating. The impure product produced by allowing ammoniacal silver oxide solution to stand seems even more sensitive, often exploding spontaneously in suspension.

*See* SILVER CHLORIDE, AgCl : Ammonia
SILVER-CONTAINING EXPLOSIVES

SILVERING SOLUTIONS
TOLLENS' REAGENT
*See other* *N*-METAL DERIVATIVES

## SILVER 2,4,6-TRIS(DIOXOSELENA)-PERHYDROTRIAZINE-1,3,5-TRIIDE ('SILVER TRISELENIMIDATE')

$Ag_3N_3O_6Se_6$

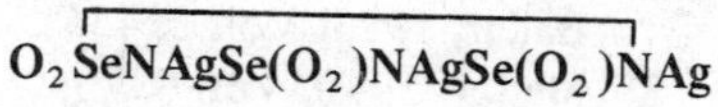

$O_2SeNAgSe(O_2)NAgSe(O_2)NAg$ (cyclic)

*See* SELENIUM DIFLUORIDE DIOXIDE, $F_2O_2Se$: Ammonia
*See other* *N*-METAL DERIVATIVES

## TETRASILVER DIIMIDOTRIPHOSPHATE

$Ag_4H_3N_2O_8P_3$

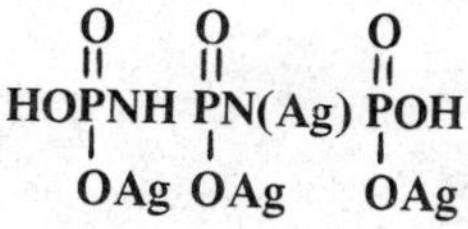

HOP(O)(OAg)NH P(O)(OAg)N(Ag) P(O)(OAg)OH

Alone,
or Sulphuric acid

Mellor, 1940, Vol. 8, 705; 1971, Vol. 8, Suppl. 3, 787
The dry material explodes on heating, and ignites in contact with sulphuric acid. The molecule contains one N–Ag bond.
*See other* *N*-METAL DERIVATIVES

## SILVER DIIMIDODIOXOSULPHATE(4–)

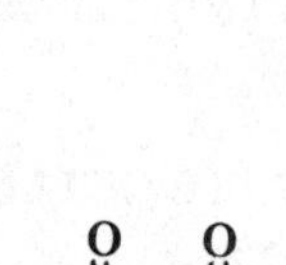

$Ag_4N_2O_2S$

(AgN=)₂S(OAg)₂

Nachbaur, E. *et al., Angew. Chem. (Intern. Ed.),* 1973, **12**, 339
The dry salt explodes on friction or impact.
*See other* *N*-METAL DERIVATIVES

## PENTASILVER DIIMIDOTRIPHOSPHATE

$Ag_5H_2N_2O_8P_3$

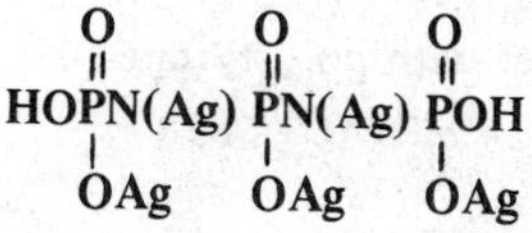

HOP(O)(OAg)N(Ag) P(O)(OAg)N(Ag) P(O)(OAg)OH

Alone,
or Sulphuric acid

Mellor, 1940, Vol. 8, 705; 1971, Vol. 8, Suppl. 3, 787
The salt is explosive and may readily be initiated by friction, heat or contact with sulphuric acid. The molecule contains two N–Ag bonds.
*See other* N-METAL DERIVATIVES

**PENTASILVER DIAMIDOPHOSPHATE** **$AgOPO(NAg_2)_2$** **$Ag_5N_2O_2P$**

1. Bailar, 1973, Vol. 2, 455
2. Mellor, 1940, Vol. 8, 705

The salt contains 85% silver and four N–Ag bonds [1], and detonates readily on friction, heating or contact with sulphuric acid [2].
*See other* N-METAL DERIVATIVES

**HEPTASILVER NITRATE OCTAOXIDE** **$(Ag_3O_4)_2 \cdot AgNO_3$** **$Ag_7NO_{11}$**

Alone,
or Sulphides,
or Non-metals
Mellor, 1941, Vol. 3, 483–485
The crystalline product produced by electrolytic oxidation of silver nitrate (possibly $(Ag_3O_4)_2 \cdot AgNO_3$) detonates feebly at 110°C. Mixtures with phosphorus or sulphur explode on impact, hydrogen sulphide ignites on contact, and antimony trisulphide ignites when ground with the salt.
*See other* METAL OXONON-METALLATES

**ALUMINIUM** **Al**

1. *Haz. Chem. Data*, 1973, 44
2. *Dust Explosion Prevention: Aluminium Powder*, NFPA Standard Code 651, 1967
3. Popov, E. I. *et al., Chem. Abs.*, 1975, **82**, 61411z
4. Iida, K. *et al.*, Japan Kokai 75 13 233, 1975

Finely divided aluminium powder or dust forms highly explosive dispersions in air [1]. All aspects of prevention of aluminium dust explosions are covered in a US standard code [2]. The effects on ignition properties of impurities introduced by recycled metal used to prepare dust were studied [3]. Pyrophoricity is eliminated by surface-coating aluminium powder with polystyrene [4].
*See* METAL DUSTS

Aluminium halides,
Carbon oxides
Guntz, A. *et al., Compt. Rend.*, 1897, **124**, 187–190

Aluminium powder burns when heated in carbon dioxide, and presence of aluminium chloride or iodide vapour in carbon monoxide or dioxide accelerates the reaction to incandescence.

Ammonium nitrate
Mellor, 1946, Vol. 5, 219
Mixtures with the powdered metal are used as an explosive, sometimes with addition of carbon, hydrocarbons and other oxidising agents.
*See* AMMONIUM NITRATE, $H_4N_2O_3$ : Metals

Ammonium peroxodisulphate
*See* AMMONIUM PEROXODISULPHATE, $H_8N_2O_8S_2$ : Aluminium

Antimony,
or Arsenic
Matignon, C., *Compt. Rend.*, 1900, **130**, 1393
Powdered aluminium reacts violently on heating with antimony or arsenic.

Antimony trichloride
Matignon, C., *Compt. Rend.*, 1900, **130**, 1393
Powdered aluminium ignites in antimony trichloride vapour.

Arsenic trioxide,
Sodium arsenate,
Sodium hydroxide
*MCA Case History No. 1832*
An aluminium ladder was used (instead of the usual wooden one) to gain access to a tank containing the alkaline arsenical mixture. Hydrogen produced by alkaline attack on the ladder generated arsine, which poisoned the three workers involved.

Barium peroxide
*See* BARIUM PEROXIDE, $BaO_2$ : Metals

Bismuth
Mellor, 1947, Vol. 9, 626
The finely divided mixture of metals produced by hydrogen reduction of co-precipitated bismuth and aluminium hydroxides is pyrophoric.

Butanol
Luberoff, B. J., private comm., 1964
Butanol, used as a solvent in an autoclave preparation at *ca.* 100°C, severely attacked the aluminium gasket, liberating hydrogen which

caused a sharp rise in pressure. Other alcohols would behave similarly, forming the aluminium alkoxide.

Carbon,
Chlorine trifluoride
*See* CHLORINE TRIFLUORIDE, $ClF_3$ : Metals, etc.

Carbon dioxide
*See* CARBON DIOXIDE, $CO_2$ : Metals

Carbon dioxide,
Sodium peroxide
*See* SODIUM PEROXIDE, $Na_2O_2$ : Metals, etc.

Carbon disulphide
Matignon, C., *Compt. Rend.*, 1900, **130**, 1391
Aluminium powder ignites in carbon disulphide vapour.

Chloroformamidinium nitrate
*See* CHLOROFORMAMIDINIUM NITRATE, $CH_4ClN_3O_3$ : Alone, or Metals

Copper,
Sulphur
Donohue, P. C., US Pat. 3 932 291, 1973 (*Chem. Abs.*, 1976, **84**, 98532g)
During preparation of aluminium copper(I) disulphide from the elements in an air-free silica tube at 900–1250°C, initial heating must be slow to prevent explosion of the tube by internal pressure of unreacted sulphur vapour.

Copper oxide
Anon., *Chem. Age* , 1932, **27**, 23
A mixture of aluminium powder and hot copper oxide exploded violently during mixing with a steel shovel on an iron plate. The frictional mixing initiated the thermite-like mixture.
*See* Metal oxides, etc., below

Diborane
*See* DIBORANE, $B_2H_6$ : Metals

Disodium acetylide
*See* DISODIUM ACETYLIDE, $C_2Na_2$ : Metals

Disulphur dibromide
*See* DISULPHUR DIBROMIDE, $Br_2S_2$ : Metals

Explosives

1. Stettbacher, A., *Chem. Abs.*, 1944, **38**, $4445_4$
2. Muraour, H., ibid., $4445_7$

The addition of substantial amounts (up to 32%) of aluminium powder to conventional explosives enhances the energy release by up to 100% [1], involving high-temperature reduction of liberated carbon dioxide and water by the aluminium [2].

Formic acid

Matignon, C., *Compt. Rend.*, 1900, **130**, 1392

The metal reduces the acid with incandescence.

Halocarbons

1. Anon., *Angew. Chem.*, 1950, **62**, 584
2. Anon., *Chem. Age,* 1950, **63**, 155
3. Anon., *Chem. Eng. News,* 1954, **32**, 258
4. *Pot. Incid. Rep., ASESB,* 1968, **39**
5. Anon., *Chem. Eng. News*, 1955, **33**, 942
6. Eiseman, B. J., *J. Amer. Soc. Htg. Refr. Air Condg. Eng.*, 1963, **5**, 63
7. Eiseman, B. J., *Chem. Eng. News,* 1961, **39**(27), 44
8. Laccabue, J. R., *Fluorolube-Aluminum Detonation Point* : Report 7E.1500, San Diego, Gen. Dynamics, 1958
9. Atwell, V. J., *Chem. Eng. News,* 1954, **32**, 1824
10. ICI Mond Div., private comm., 1968
11. Anon., *Ind. Acc. Prev. Bull. RoSPA*, 1953, **21**, 60
12. Wendon, W. G., private comm., 1973
13. Coffee, R.D., *Loss Prevention*, 1971, **5**, 113
14. Schwab, R.F., ibid., 113
15. Corley, R., ibid., 114
16. Heinrich, H. J., *Arbeitsschutz*, 1966, 156–157
17. Lamoroux, A. *et al.*, *Mém. Poudres*, 1957, **39**, 435–445
18. Hamstead, A. C. *et al.*, *Corrosion*, 1958, **14**, 189t–190t
19. Hartmann, I., *Ind. Eng. Chem.*, 1948, **40**, 756
20. *MCA Case History No. 2160*
21. Arias, A., *Ind. Eng. Chem., Prod. Res. Dev.*, 1976, **15**, 150
22. Wendon, G. W., ibid., 1977, 16, 112
23. Arias, A., ibid., 112

Heating aluminium powder with carbon tetrachloride, chloromethane or carbon tetrachloride–chloroform mixtures in closed systems to 152°C may cause an explosion, particularly if traces of aluminium chloride are present [1]. A mixture of carbon tetrachloride and aluminium powder exploded during ball-milling [2], and it was later shown that heavy impact would detonate the mixture [3]. Mixtures with monofluorotrichlorethane and trichlorotrifluoroethane will flash or spark on heavy impact [4].

A virtually unvented aluminium tank containing a mixture of

*o*-dichlorobenzene, 1, 2-dichloroethane and 1, 2-dichloropropane exploded violently seven days after filling. This was attributed to formation of aluminium chloride which catalysed further accelerating attack on the aluminium tank [5].

In a dichlorodifluoromethane system, frictional wear exposed fresh metal surfaces on an aluminium compressor impellor, causing an exothermic reaction with the halocarbon which melted much of the impellor. Later tests showed similar results decreasing in order of intensity, with: tetrafluoromethane; chlorodifluoromethane; bromotrifluoromethane; dichlorodifluoromethane; 1,2-dichlorotetrafluoroethane; 1,1,2-trichlorotrifluoroethane [6].

In similar tests molten aluminium dropped into liquid dichlorodifluoromethane burned incandescently below the liquid [6].

Aluminium bearing surfaces under load react explosively with polychlorotrifluoroethylene greases or oils. The inactive oxide film will be removed from the metal by friction, and hot spots will initiate reaction [8].

An attempt to scale up the methylation of 2-methylpropane with chloromethane in presence of aluminium chloride and aluminium went out of control and detonated, destroying the autoclave. The preparation had been done on a smaller scale on 20 previous occasions without incident [9].

*See* BROMOMETHANE, $CH_3Br$ : Metals

Violent decomposition, with evolution of hydrogen chloride, may occur when 1,1,1-trichloroethane comes into contact with aluminium or its alloys with magnesium [10].

Aluminium-dusty, greasy overalls were cleaned by immersion in trichloroethylene. During subsequent drying, violent ignition occurred. This was attributed to presence of free hydrogen chloride in the solvent, which reacted to produce aluminium chloride [10]. This is known to catalyse polymerisation of trichloroethylene, producing more hydrogen chloride and heat. The reaction is self-accelerating and can develop a temperature of 1350°C [11].

Trichloroethylene cleaning baths must be kept neutral with sodium carbonate, and free of aluminium dust.

Halocarbons now available with added stabilisers (probably amines) show a reduced tendency to react with aluminium powder [12].

Aluminium powder undergoes an exothermic and uncontrollable reaction with dichloromethane above 95°C under appropriate pressure [13]. Several cases of violent reaction between aluminium and trichloroethylene or tetrachloroethylene in vapour degreasers have been noted [14]. Chloromethane in liquefied storage diffused 70 m along a nitrogen inerting line into the pressure regulator. Interaction with aluminium components of the regulator formed alkylaluminium compounds which ignited when the regulator was dismantled [15]. An explosion in an aluminium degreasing plant using tetrachloroethylene was attributed to overheating of residues on the heating coils. Subsequent tests showed that simultaneous presence of water and aluminium chloride in an aluminium powder–tetrachloroethylene mixture lowered the initiation temperature to below 250°C. Presence of cutting oils reduced it below 150°C, and a temperature of 300°C was reached within 100 min [16].

Reaction of aluminium powder with hexachloroethane in alcohol is not initially violent, but may become so [17]. An aluminium transfer pipe failed after a few hours' service carrying refined 1,2-dichloropropane in warm weather. The corrosive attack was simulated and studied under laboratory conditions [18]. In dichlorodifluoromethane vapour, aluminium dust ignited at 580°C, and suspensions of the dust in the vapour on sparking gave strong explosions [19]. A fire occurred at a liquid outlet from a 40 $m^3$ mild steel tanker of chloromethane. This was traced to the presence of trimethylaluminium produced by interaction of chloromethane and (unsuspected) aluminium baffle plates in the tanker [20]. A proposal to prepare pure aluminium chloride by ball-milling aluminium in carbon tetrachloride [21] is criticised [22] as potentially hazardous. Possible modifications (use of inert solvents, continuously fed mills, etc.) to improve the procedure are suggested [23].

*See other* METALS: Halocarbons

Halogens

1. Mellor, 1946, Vol. 2, 92, 135; Vol. 5, 209
2. Azmathulla, S. *et al.*, *J. Chem. Educ.*, 1955, **32**, 447; *School Sci. Rev.*, 1956, **38** (134), 107
3. Hammerton, C. M., *School Sci. Rev.*, 1957, **38**(136), 459

Aluminium powder ignites in chlorine without heating, and foil reacts vigorously with liquid bromine at 15°C, and incandesces on warming in the vapour [1]. The metal and iodine react violently in the presence of water, either as liquid, vapour or that in hydrated salts [2]. Moistening a powdered mixture causes incandescence and will initiate a thermite mixture [3].

Hydrochloric acid,
or Hydrofluoric acid

Kirk-Othmer, 1963, Vol. 1, 952

The metal is attacked violently by the aqueous acids.

Hydrogen chloride

Batty, G. F., private comm., 1972

Erroneous use of aluminium instead of alumina in a hydrogen chloride purification reactor caused a vigorous exothermic reaction which distorted the steel reactor shell.

Interhalogens

*See* BROMINE PENTAFLUORIDE, $BrF_5$ : Acids, etc.

CHLORINE FLUORIDE, ClF : Aluminium
IODINE CHLORIDE, ClI : Metals
IODINE PENTAFLUORIDE, $F_5I$ : Metals
IODINE HEPTAFLUORIDE, $F_7I$ : Metals

Iron,
Water
Chen, W. Y. *et al.*, *Ind. Eng. Chem.*, 1955, **47**(7), 32A
A sludge of aluminium dust (containing iron and sand) removed from castings in water was found to undergo sudden exotherms (in summer weather) to 95°C with hydrogen evolution. Similar effects with aluminium-sprayed steel plates exposed to water were attributed to electrolytic action, as addition of iron filings to an aluminium dust slurry in water caused hydrogen evolution to occur without application of heat.
*See* Water, below

Mercury(II) salts
Woelfel, W. C., *J. Chem. Educ.*, 1967, **44**, 484
Aluminium foil is unsuitable as a packing material in contact with mercury(II) salts in presence of moisture, when vigorous amalgamation ensues.
*See* ALUMINIUM AMALGAM, Al–Hg

Metal nitrates,
Potassium perchlorate,
Water
Johansson, S. R. *et al.*, *Chem. Abs.*, 1973, **78**, 18435r
A pyrotechnic mixture of aluminium powder with potassium perchlorate, barium nitrate, potassium nitrate and water exploded after 24 h storage under water. Tests revealed the exothermic interaction of finely divided aluminium with nitrate and water to produce ammonia and aluminium hydroxide. Under the conditions prevailing in the stored mixture, the reaction would be expected to accelerate, finally involving the perchlorate as oxidant and causing ignition of the mixture.

Metal nitrates,
Sulphur,
Water
Anon., *Chem. Eng. News,* 1954, **32**, 258
*Berufsgenossenschaft,* 1954, 184
Aluminium powder, barium and potassium nitrates, sulphur and vegetable adhesives, mixed to a paste with water, exploded on two occasions. Laboratory investigation showed initial interaction of water and aluminium to produce hydrogen. It is supposed that nascent hydrogen reduced the nitrates present, increasing the

alkalinity and hence the rate of attack on aluminium, the reaction becoming self-accelerating. Cause of ignition was unknown. Other examples of interaction of aluminium with water are known.

*See* Iron, Water, above
Water, below

Metal oxides,
or Oxosalts,
or Sulphides

1. Mellor, 1946, Vol. 5, 217
2. Price, D. J. *et al.*, *Chem. Met. Eng.*, 1923, **29**, 878

Many metal oxo-compounds (nitrates, oxides and particularly sulphates) and sulphides are reduced violently or explosively (i.e. undergo 'thermite' reaction) on heating an intimate mixture with aluminium powder to a suitably high temperature. Contact of massive aluminium with molten salts may give explosions [1]. Application of sodium carbonate to molten (red-hot) aluminium caused an explosion [2].

*See* MOLTEN SALT BATHS
DIIRON TRIOXIDE, $Fe_2O_3$ : Aluminium

Methanol

*See* MAGNESIUM, Mg: Methanol

Nitro compounds,
Water

Hajek, V. *et al.*, *Research*, 1951, **4**, 186–191

Dry mixtures of picric acid and aluminium powder are inert, but addition of water causes ignition after a delay dependent upon quantity added. Other nitro compounds and nitrates are discussed in this context.

Non-metal halides

1. Matignon, C., *Compt. Rend.*, 1900, **130**, 1393
2. Lenher, V. *et al.*, *J. Amer. Chem. Soc.*, 1926, **48**, 1553

Powdered aluminium ignites in the vapour of arsenic trichloride or sulphur dichloride, and incandesces in phosphorus trichloride vapour [1]. Above 80°C, aluminium reacts incandescently with diselenium dichloride [2].

Non-metals

Matignon, C., *Compt. Rend.*, 1900, **130**, 1393–1394

Powdered aluminium reacts violently with phosphorus, sulphur or selenium, and a mixture of powdered metal with red phosphorus exploded when violently shocked.

Oleic acid

de Ment, J., *J. Chem. Educ.*, 1959, **36**, 308

Shortly after mixing the two, an explosion occurred. This could not be repeated. The acid may have been peroxidised.

Oxidants

1. Kirshenbaum, 1956, 4, 13
2. Mellor, 1947, Vol. 2, 310
3. Annikov, V. E. *et al.*, *Chem. Abs.*, 1976, **85**, 145389y

Mixtures of aluminium powder with liquid chlorine, dinitrogen tetraoxide or tetranitromethane are detonable explosives, but not as powerful as aluminium–liquid oxygen mixtures, some of which exceed TNT in effect by a factor of 3 to 4 [1]. Mixtures of the powdered metal and various bromates may explode on impact, heating or friction. Iodates and chlorates act similarly [2].

Detonation properties of gelled slurries of aluminium powder in aqueous nitrate or perchlorate salt solutions have been studied [3].

Other combinations are:

Halogens, above
POTASSIUM CHLORATE, $ClKO_3$ : Metals
POTASSIUM PERCHLORATE, $ClKO_4$ : Aluminium, etc.
: Metal powders
SODIUM CHLORATE, $ClNaO_3$: Aluminium, Rubber
NITRYL FLUORIDE, $FNO_2$ : Metals
AMMONIUM PEROXODISULPHATE, $H_8N_2O_8S_2$ : Aluminium, etc.
SODIUM NITRATE, $NNaO_3$ : Aluminium
SODIUM PEROXIDE, $Na_2O_2$ : Aluminium, etc.
: Metals
ZINC PEROXIDE, $O_2Zn$ : Metals

Paint

*See* ZINC, Zn; Paint primer base

Phosphorus pentachloride

Berger, E. *Compt. Rend.*, 1920, **170**, 29

Aluminium powder ignites in contact with the pentachloride.

2-Propanol

Wilds, A. L., *Org. React.*, 1944, **2**, 198

Muir, G. D., private comm., 1968

Dissolution of aluminium in 2-propanol to give the isopropoxide is rather exothermic, but often subject to an induction period similar to that in preparation of Grignard reagents. Only small amounts of aluminium should be present until reaction begins.

*See also* MAGNESIUM, Mg: Methanol

Silver chloride

Anon., *Chem. Eng. News,* 1954, **32**, 258

An intimate mixture of the two powders may lead to reaction of explosive violence, unless excess aluminium is present.

Sodium hydroxide

1. *MCA Case History No. 1115*
2. *MCA Case History No. 1888*

In the first of two incidents involving corrosive attack of aluminium by sodium hydroxide solutions, the vigorous evolution of hydrogen was noticed before a tank trailer (supposed to be mild steel) had perforated. In the second, corrosion caused failure of an aluminium coupling between a pressure gauge and a pump, causing personal contamination.

Sodium sulphate

Kohlmeyer, E. J., *Aluminium*, 1942, **24**, 361–362

The violent explosion experienced when an 8:3 molar mixture of metal powder and salt was heated to 800°C was attributed to thermal dissociation (at up to 3000°C) of the metal sulphide(s) formed as primary reactants.

*See other* THERMITE REACTIONS

Steel

*MCA Case History Nos. 2161, 2184* (same incident)

A joint between a mild steel valve screwed onto an aluminium pipe was leaking a resin–solvent mixture, and when the joint was tightened with a wrench, a flash fire occurred. This was attributed to generation of sparks by a thermite reaction between the (rusted) steel and aluminium metal when the joint was tightened.

Sulphur

Read, C. W. W., *School Sci. Rev.*, 1940, **21**(83), 977

The violent interaction of aluminium powder and sulphur on heating is considered too dangerous for a school experiment.

Water

1. *MCA Case History No. 462*
2. Shidlovskii, A. A., *Chem. Abs.*, 1947, **41**, 1105d
3. Bamberger, M. *et al.*, *Z. Angew. Chem.*, 1913, **26**, 353–355
4. Gibson, 1969, 2

Cans of aluminium paint contaminated with water contained a considerable pressure of hydrogen from interaction of finely divided metal and moisture [1]. Mixtures of powdered aluminium and water can be caused to explode powerfully by initiation with a boosted detonator [2]. During granulation of aluminium by pouring the molten metal through a sieve into water, a violent explosion occurred. This was attributed to steam trapped in the cooling metal [3]. Moist finely divided aluminium may ignite in air [4].

Zinc

*MCA Case History No. 1722*

Ball-milling aluminium–zinc (not stated if alloy or mixture) under inadequate inerting arrangements led to fires during operation or discharge of the mill.

*See other* METALS

**ALUMINIUM–COPPER–ZINC ALLOY** **Al–Cu–Zn**

*See* DEVARDA'S ALLOY
SILVER NITRATE, $AgNO_3$ : Ammonia, etc.

*See other* ALLOYS (INTERMETALLIC COMPOUNDS)

**ALUMINIUM AMALGAM** **Al–Hg**

1. Neely, T. A. *et al.*, *Org. Synth.*, 1965, **45**, 109
2. Calder, A. *et al.*, *Org. Synth.*, 1972, 52, 78

The amalgamated aluminium wool remaining from preparation of triphenylaluminium will rapidly oxidise and become hot on exposure to air. Careful disposal is necessary [1]. Amalgamated aluminium foil may be pyrophoric and should be kept moist and used immediately [2].

*See other* ALLOYS (INTERMETALLIC COMPOUNDS)

**ALUMINIUM–MAGNESIUM ALLOY** **Al–Mg**

Barium nitrate

*See* BARIUM NITRATE, $BaN_2O_6$ : Aluminium–magnesium alloy

Iron(III) oxide,
Water

Maischak, K. D. *et al.*, *Neue Huette*, 1970, **15**, 662–665

Accidental contact of the molten alloy (26% Al) with a wet rusty iron surface caused violent explosions with brilliant light emission. Initial evolution of steam, causing fine dispersion of the alloy, then interaction of the fine metals with rust in a 'thermite' reaction, were postulated as likely stages. Direct interaction of the magnesium (74%) with steam may also have been involved.

*See other* THERMITE REACTIONS

**ALUMINIUM–MAGNESIUM–ZINC ALLOYS** **Al–Mg–Zn**

Rusted steel

Yoshino, H. *et al.*, *Chem. Abs.*, 1966, **64**, 14017j

Impact of an alloy containing 6% aluminium and 3% zinc with rusted steel caused incendive sparks which ignited liquefied petroleum gas–air mixtures in 11 out of 20 attempts.

*See other* IGNITION SOURCES
THERMITE REACTIONS

**ALUMINIUM–NICKEL ALLOYS** **Al–Ni**

Water

Anon., *Angew. Chem. (Nachr.)*, 1968, **16**, 2

Heating moist Raney nickel alloy containing 20% of aluminium in an autoclave under hydrogen caused the aluminium and water to interact explosively, generating 1000 bar pressure of hydrogen.

*See also* HYDROGENATION CATALYSTS

*See other* ALLOYS (INTERMETALLIC COMPOUNDS)

**ALUMINIUM–TITANIUM ALLOYS** **Al–Ti**

Oxidants

Mellor, 1941, Vol. 7, 20–21

Alloys ranging from $Al_3Ti_2$ to $Al_4Ti$ have been described, which ignite or incandesce on heating in chlorine; or bromine or iodine vapour; or hydrogen chloride; or oxygen.

*See other* ALLOYS (INTERMETALLIC COMPOUNDS)

**ALUMINIUM TETRAHYDROBORATE** $Al[BH_4]_3$ $AlB_3H_{12}$

1. Schlessinger, H. I. *et al.*, *J. Amer. Chem. Soc.*, 1940, **62**, 3421
2. Badin, F. J. *et al.*, *J. Amer. Chem. Soc.*, 1949, **71**, 2950

The vapour is spontaneously inflammable in air [1], and explodes in oxygen, but only in presence of traces of moisture [2].

Alkenes,
Oxygen

Gaylord, 1956, 26

The tetrahydroborate reacts with alkenes and, in presence of oxygen, combustion is initiated, even in absence of moisture. Butene explodes after an induction period, while butadiene explodes immediately.

Water

Semenenko, K. N. *et al.*, *Russ. Chem. Rev.*, 1973, 4

Interaction at ambient temperature is explosive.

*See other* COMPLEX HYDRIDES

## ALUMINIUM TRIS(TETRAAZIDOBORATE) $AlB_3N_{36}$

Mellor, 1967, Vol. 8, Suppl. 2.2, 2 $Al[B(N_3)_4]_3$

A very shock-sensitive explosive, containing nearly 90% wt. of nitrogen.

*See other* HIGH-NITROGEN COMPOUNDS
NON-METAL AZIDES

## ALUMINIUM TRIBROMIDE $AlBr_3$

Water

Nicholson, D. G. *et al., Inorg. Synth.*, 1950, **3**, 35

The anhydrous bromide should be destroyed by melting and pouring slowly into running water. Hydrolysis is very violent and may destroy the container if water is added to the container.

*See other* METAL HALIDES

## ALUMINIUM DICHLORIDE HYDRIDE DIETHYL ETHERATE $AlCl_2H \cdot C_4H_{10}O$

Dibenzyl ether

Marconi, W. *et al., Ann. Chim.*, 1965, **55**, 897

During attempted reductive cleavage of the ether with the etherate an explosion occurred. Peroxides may have been present.

*See related* COMPLEX HYDRIDES

## ALUMINIUM TRICHLORIDE $AlCl_3$

1. Popov, P. V., *Savodskaya Lab.*, 1946, **13**, 127 *(Chem. Abs.*, 1947, **41**, 6723d); Kitching, A. F., *School Sci. Rev.*, 1930, **12**(45), 79
2. *MCA SD-62,* 1956
3. Bailar, 1973, Vol. 1, 1019

Long storage of the anydrous salt in closed containers caused spontaneous decomposition and occasional explosion on opening [1]. General handling precautions are detailed [2].

The need is emphasised for extreme care in experiments in which the chloride is heated in sealed tubes. High internal pressure may be generated, not only by the vapour pressure and pressure of desorbed hydrogen chloride, but also by the near doubling in volume which occurs when the chloride melts [3].

Alkenes

Jenkins, P. A., unpublished information, 1975

Mixtures of $C_4$ alkene isomers (largely isobutene) are polymerised commercially in contact with low levels of aluminium chloride (or other Lewis acid) catalysts. The highly exothermic runaway reactions occasionally experienced in practice are caused by events leading to the production of high local levels of catalyst. Rapid increases in temperature and pressure of 160°C and 18 bar, respectively, have been observed experimentally when alkenes are brought into contact with excess solid aluminium chloride. The runaway reaction appears to be more severe in the vapour phase, and a considerable amount of catalytic degradation contributes to the overall large exotherm.

*See* POLYMERISATION INCIDENTS

Aluminium,
Carbon oxides

*See* ALUMINIUM, Al: Aluminium halides, Carbon oxides

Aluminium,
Sodium peroxide

*See* SODIUM PEROXIDE, $Na_2O_2$ : Aluminium, etc.

Benzoyl chloride,
Naphthalene

Clar, E. *et al.*, *Tetrahedron*, 1974, **30**, 3296

During preparation of 1,5-dibenzoylnaphthalene, addition of aluminium chloride to a mixture of benzoyl chloride and naphthalene must be effected above the m.p. of the mixture to avoid a violent reaction.

Ethylene oxide

*See* ETHYLENE OXIDE, $C_2H_4O$ : Contaminants

Nitrobenzene

Riethman, J. *et al.*, *Chem. Ing. Tech.*, 1974, **48**, 729

Mixtures of nitrobenzene and aluminium chloride are thermally unstable and may lead to explosive decomposition. Subsequent to an incident involving rupture of a 4000 litre vessel, the decomposition reaction was investigated and a three-stage mechanism involving formation and subsequent polymerisation of *o*- and *p*-chloronitrosobenzene is proposed, with supporting evidence.

*See other* POLYMERISATION INCIDENTS

Nitrobenzene,
Phenol

Anon., *Chem. Eng. News*, 1953, **31**, 4915

Addition of aluminium chloride to a large volume of recovered nitrobenzene containing 5% of phenol caused a violent explosion. Experiment showed that mixtures containing all three components reacted violently at 120°C.

Nitromethane

*See* ALUMINIUM TRICHLORIDE–NITROMETHANE, $AlCl_3 \cdot CH_3NO_2$

Oxygen difluoride

*See* OXYGEN DIFLUORIDE, $F_2O$ : Halogens, etc.

Phenyl azide

*See* PHENYL AZIDE, $C_6H_5N_3$ : Lewis acids

Perchlorylbenzene

*See* PERCHLORYLBENZENE, $C_6H_5ClO_3$ : Aluminium trichloride

Sodium tetrahydroborate

*See* SODIUM TETRAHYDROBORATE, $BH_4Na$: Aluminium chloride, etc.

Water

Anon., *Ind. Eng. Chem. (News Ed.)*, 1934, **12**, 194

An unopened bottle of anhydrous aluminium chloride erupted when the rubber bung with which it was sealed was removed. The accumulation of pressure was attributed to absorption of moisture by the anhydrous chloride before packing. The presence of an adsorbed layer of moisture in the bottle used for packing may have contributed. Reaction with liquid water is violently exothermic.

*See other* METAL HALIDES

## ALUMINIUM TRICHLORIDE–NITROMETHANE $AlCl_3 \cdot CH_3NO_2$

An alkene

Cowen, F.M. *et.al.*, *Chem. Eng. News,* 1948, **26**, 2257

A gaseous alkene was passed into a cooled autoclave containing the complex, initially with agitation, and later without. Later, when the alkene was admitted to a pressure of 5.6 bar at 2°C, a slight exotherm occurred, followed by an explosion. The autoclave contents were completely carbonised. Mixtures of ethylene, aluminium chloride and nitromethane had exploded previously, but at 75°C.

*See* ETHYLENE, $C_2H_4$ : Aluminium trichloride

Carbon monoxide,
Phenol

Webb, H. F., *Chem. Eng. News*, 1977, **55**(12), 4

An attempt to formylate phenol by heating a mixture with nitromethane and aluminium chloride in an autoclave under carbon monoxide at 100 bar pressure at 110°C led to a high-energy explosion after 30 min.

*See* above

**ALUMINIUM CHLORATE** $Al(ClO_3)_3$ $AlCl_3O_9$

Sidgwick, 1950, 428

During evaporation, its aqueous solution evolves chlorine dioxide, and eventually explodes.

*See other* METAL OXOHALOGENATES

**ALUMINIUM PERCHLORATE** $Al(ClO_4)_3$ $AlCl_3O_{12}$

Dimethyl sulphoxide

*See* DIMETHYL SULPHOXIDE, $C_2H_6OS$: Metal oxosalts

**CAESIUM HEXAHYDROALUMINATE (3–)** $AlCs_3H_6$

$Cs_3[AlH_6]$

*See* POTASSIUM HEXAHYDROALUMINATE (3–), $AlH_6K_3$
*See other* COMPLEX HYDRIDES

**COPPER(I) TETRAHYDROALUMINATE** $Cu[AlH_4]$ $AlCuH_4$

Aubry, J. *et al.*, *Compt. Rend.*, 1954, **238**, 2535

The unstable hydride decomposed at $-70°C$, and ignited on contact with air.

*See other* COMPLEX HYDRIDES

**ALUMINIUM COPPER(I) SULPHIDE** $AlCuS_2$

Preparative hazard

*See* ALUMINIUM, Al: Copper, Sulphur

**LITHIUM TETRADEUTEROALUMINATE** $AlD_4Li$

Leleu, *Cahiers*, 1977, (86), 99 $Li[AlD_4]$

It may ignite in moist air.

*See related* COMPLEX HYDRIDES

**ALUMINIUM HYDRIDE** $AlH_3$

Mirviss, S. B. *et al.*, *Ind. Eng. Chem.*, 1961, **53**(1), 54A

Gibson, 1969, 66

It is very unstable and has been known to decompose spontaneously at ambient temperature with explosive violence. Its complexes (particularly the diethyl etherate) are considerably more stable. The hydride ignites in air with or without oxygen enrichment [2].

*See also* ALUMINIUM HYDRIDE–TRIMETHYLAMINE, $AlH_3 \cdot C_3H_9N$
ALUMINIUM HYDRIDE–DIETHYL ETHERATE, $AlH_3 \cdot C_4H_{10}O$

Carbon dioxide,
Methyl ethers

Barbaras, G. *et al.*, *J. Amer. Chem. Soc.*, 1948, **70**, 877

Presence of carbon dioxide in solutions of the hydride in dimethyl or di(2-methoxyethyl) ethers can cause a violent decomposition on warming the residue from evaporation. Presence of aluminium chloride tends to increase the vigour of decomposition to explosion. Lithium tetrahydroaluminate may behave similarly, but is generally more stable.

Tetrazole derivatives

Fetter, N. R. *et al.*, US Pat. 3 396 170, 1968 (*Chem. Abs.*, 1968, **69**, 87170b)

The 1:1 complexes arising from interaction of the hydride (as a complex with ether or trimethylamine) and various tetrazole derivatives are explosive. Tetrazoles mentioned are 2-methyl-, 2-ethyl-, 5-ethyl-, 2-methyl-5-vinyl-, 5-amino-2-ethyl-, 1-alkyl-5-amino- and 5-cyano-2-methyl-tetrazole.

*See* TETRAZOLES
*See other* METAL HYDRIDES

## ALUMINIUM HYDRIDE–TRIMETHYLAMINE $AlH_3 \cdot C_3H_9N$

Water

Ruff, J. K., *Inorg. Synth.*, 1967, **9**, 34

It ignites in moist air and is explosively hydrolysed by water.

*See other* METAL HYDRIDES

## ALUMINIUM HYDRIDE–DIETHYL ETHERATE $AlH_3 \cdot C_4H_{10}O$

Water

Schmidt, D. L. *et al.*, *Inorg. Synth.*, 1973, **14**, 51

Interaction of the solid with water or moist air is violent and may be explosive.

*See* RELATED METAL HYDRIDES

## ALUMINIUM TRIHYDROXIDE $Al(OH)_3$ $AlH_3O_3$

Chlorinated rubber

*See* CHLORINATED RUBBER : Metal oxides or hydroxides

## LITHIUM TETRAHYDROALUMINATE $Li[AlH_4]$ $AlH_4Li$

1. Augustine, 1968, 12

2. Gaylord, 1956, 37
3. Sakaliene, A. *et al.*, *Chem. Abs.*, 1970, **73**, 19115v
4. *MCA Case History No. 1832*
5. Walker, E. R. H., *Chem. Soc. Rev.*, 1976, **5**, 36

Care is necessary in handling this powerful reducant, which may ignite if lumps are pulverised with a pestle and mortar, even in a dry box [1]. A rubber mallet is recommended for the purpose [2].

The explosive thermal decomposition of the aluminate at 150–170°C is due to its interaction with partially hydrolysed decomposition products [3]. A spilled mixture with ether ignited after the ether evaporated [4]. Sodium bis-(2-methoxyethoxy)dihydroaluminate, which is of similar reducing capability to lithium tetrahydroaluminate, is safer in that it does not ignite in moist air or oxygen and is stable at 200°C [5].

Alkyl benzoates
1. Tados, W. *et al.*, *J. Chem. Soc.*, 1954, 2353
2. Field, B. O. *et al.*, ibid., 1955, 1111

Application of a method of reducing benzaldehydes to the corresponding alcohols with a fourfold excess of tetrahydroaluminate [1] to alkyl benzoates proved to be difficult to control and frequently dangerous [2].

Bis(2-methoxyethyl) ether
1. Watson, A. R., *Chem. & Ind.*, 1964, 665
2. Adams, R. M., *Chem. Eng. News,* 1953, **31**, 2334
3. Barbaras, G. *et al., J. Amer. Chem. Soc.,* 1948, **70**, 877
4. *MCA Case History No. 1494*

The peroxide-free ether, being dried by distillation at 162°C under inert atmosphere at ambient pressure, exploded violently when the heating bath temperature had been raised to 200°C towards the end of distillation. This was attributed to local overheating of an insulating crust of hydride in contact with oxygen-containing organic material [1]. Two previous explosions were attributed to peroxides [2] and the high solubility of carbon dioxide in such ethers [3]. Stirring during distillation would probably prevent crust formation. Alternatively, drying could be effected with a column of molecular sieve or activated alumina. During distillation of the solvent from the aluminate at 100°C at atmospheric pressure, the flask broke and its contents ignited explosively. The aluminate decomposes at 125–135°C [4].

*See* ALUMINIUM TRIHYDRIDE, $AlH_3$ : Carbon dioxide

Boron trifluoride diethyl etherate
1. Scott, R. B., *Chem. Eng. News,* 1967, **45**(28), 7; ibid., **45**(21), 51
2. Shapiro, I. *et al.*, *J. Amer. Chem. Soc.*, 1952, **74**, 90

Use of lumps of the solid aluminate, rather than its ethereal solution, and of peroxide-containing etherate [1], rather than the peroxide-free material specified [2], caused an explosion during the attempted preparation of diborane.

Dibenzoyl peroxide

*See* DIBENZOYL PEROXIDE, $C_{14}H_{10}O_4$ : Lithium tetrahydroaluminate

3,5-Dibromocyclopentene

Johnson, C. R. *et al.*, *Tetrahedron Lett.*, 1964, **45**, 3327

Preparation of the 4-bromo compound by partial debromination of crude 3,5-dibromocyclopentene by addition of its ethereal solution to the aluminate in ice-cold ether is hazardous. Explosions have occurred on two occasions about an hour after addition of dibromide.

1,2-Dimethoxyethane

1. *MCA Case History No. 1182*
2. Hoffmann, K. A. *et al.*, *Org. Synth.*, 1968, **48**, 62

The finely powdered aluminate was charged through a funnel into a nitrogen-purged flask. When the solvent was added through the same funnel, ignition occurred, possibly due to local absence of purge gas in the funnel caused by turbulence [1]. Distillation of the solvent from the solid must not be taken to dryness, to avoid explosive decomposition of the residual aluminate [2].

Dioxane

Anon., private comm., 1976

Dioxane was purified by distillation from the complex hydride in a glass still, and when the residue was cooling down, a severe explosion and fire occurred. This seems likely to have been caused by ingress of air into the cooling dioxane vapour (flammability limits are 2–22%) and subsequent heating of finely divided hydride on the upper parts of the still on contact with air to above the rather low autoignition temperature of dioxane (180°C). Nitrogen purging will render the operation safe.

*See* Bis(2-methoxyethyl) ether, above (reference 4)
1,2-Dimethoxyethane, above (reference 1)

Ethyl acetate

1. Bessant, K. H. C., *Chem. & Ind.*, 1957, 432
2. Yardley, J. T., ibid., 433

Following a reductive dechlorination in ether, a violent explosion occurred when ethyl acetate was added to decompose excess aluminate [1]. Ignition was attributed to the strongly exothermic reaction occurring when undiluted (and reducible) ethyl acetate contacts the solid aluminate. Addition of a solution of ethyl acetate in inert solvent or of a moist unreactive solvent to destroy excess reagent is preferable [2].

Fluoroamides



1. Karo, W., *Chem. Eng. News*, 1955, **33**, 1368
2. Reid, T. S. *et al.*, *Chem. Eng. News*, 1951, **29**, 3042

The reduction of amides of fluorocarboxylic acids with the tetrahydroaluminate appears generally hazardous at all stages. During reduction of *N*-ethylheptafluorobutyramide in ether, violent and prolonged gas evolution caused a fire. Towards the end of reduction of trifluoroacetamide in ether, solid separated and stopped the stirrer. Attempts to restart the stirrer by hand caused a violent explosion [1]. During decomposition by water of the reaction complex formed by interaction with tetrafluorosuccinamide in ether, a violent explosion occurred. Experiment showed this was due to the low stability of the complex, particularly in absence of ether, when detonation at room temperature occurred. Reaction complexes similarly obtained from trifluoroacetic acid, heptafluorobutyramide and octafluoroadipamide also showed instability, decomposing when heated. General barricading of all reductions of fluoro compounds with lithium tetrahydroaluminate is recommended [2].

Hydrogen peroxide

*See* HYDROGEN PEROXIDE, $H_2O_2$: Lithium tetrahydroaluminate

Pyridine

Augustine, 1968, 22–23

Addition of the aluminate (0.5 g) to pyridine (50 ml) must be effected very slowly with cooling. Addition of 1 g portions may cause a highly exothermic reaction.

Tetrahydrofuran

Moffett, R. B., *Chem. Eng. News*, 1954, **32**, 4328

The solvent had been dried over the aluminate and then stored over calcium hydride for 2 years 'to prevent peroxide formation'. Subsequent addition of more aluminate caused a strong exotherm and ignition of liberated hydrogen. Calcium hydride does not prevent peroxide formation in solvents.

1,2,3,4-Tetrahydro-1-naphthyl hydroperoxide

*See* 1,2,3,4-TETRAHYDRO-1-NAPHTHYL HYDROPEROXIDE, $C_{10}H_{12}O_2$: Lithium tetrahydroaluminate

Water

Leleu, *Cahiers*, 1977, (86), 100

Interaction is very vigorous, attaining incandescence and ignition of the evolved hydrogen.

*See other* COMPLEX HYDRIDES

## SODIUM TETRAHYDROALUMINATE $Na[AlH_4]$ $AlH_4Na$

Tetrahydrofuran

Del Giudia, F. P. *et al.*, *Chem. Eng. News*, 1961, **39**(40), 57

During synthesis from its elements in tetrahydrofuran, a violent explosion occurred when absorption of hydrogen had stopped. This was attributed to deposition of solid above the liquid level, overheating and reaction with solvent to give butoxyaluminohydrides. Vigorous stirring and avoiding local overheating are essential.

*See other* COMPLEX HYDRIDES

## POTASSIUM HEXAHYDROALUMINATE (3–) $AlH_6K_3$

$K_3[AlH_6]$

Ashby, E. C., *Chem. Eng. News*, 1969, **47**(1), 9

A 20 g sample, prepared and stored dry in a dry box for several months, developed a thin crust of oxidation/hydrolysis products. When the crust was disturbed, a violent explosion occurred (later confirmed as equivalent to 230 g TNT). A weaker explosion was observed with potassium tetrahydroaluminate. The effect was attributed to superoxidation of traces of metallic potassium, and subsequent interaction of the hexahydroaluminate and superoxide after frictional initiation. Precautions advised include use of freshly prepared material, minimal storage in a dry diluent under an inert atmosphere and destruction of solid residues.

Potassium hydrides and caesium hexahydroaluminate may behave similarly, as caesium also superoxidises in air.

*See other* COMPLEX HYDRIDES

## ALUMINIUM TRIIODIDE $AlI_3$

Bailar, 1973, Vol. 1, 1023

It reacts violently with water and on heating produces flammable vapour, which may explode if mixed with air and ignited.

Aluminium,
Carbon oxides

*See* ALUMINIUM, Al: Aluminium halides, Carbon oxides

*See other* METAL HALIDES

## LITHIUM TETRAAZIDOALUMINATE(1–) $AlLiN_{12}$

$Li[Al(N_3)_4]$

Mellor, 1967, Vol. 8, Suppl. 2.2, 2

A shock-sensitive explosive.

*See other* METAL AZIDES

## ALUMINIUM TRIAZIDE $Al(N_3)_3$ $AlN_9$

Brauer, 1963, Vol. 1, 829
May be detonated by shock.
*See other* METAL AZIDES

## ALUMINIUM PHOSPHIDE $AlP$

Mineral acids

1. Wang, C. C. *et al., J. Inorg. Nucl. Chem.*, 1963, **25**, 327
2. Mellor, 1971, Vol. 8, Suppl. 3, 306

Evolution of phosphine is slow in contact with water or alkali, but explosively violent with dilute mineral acids [1]. However, reports of violent interaction with concentrated or dilute hydrochloric acid, and of explosive reaction with 1:1 aqua regia, have been questioned [2].

## MAGNESIUM TETRAHYDROALUMINATE $Al_2H_8Mg$

$Mg[AlH_4]_2$

Gaylord, 1956, 25
It is similar to the lithium salt.
*See other* COMPLEX HYDRIDES

## MANGANESE(II) TETRAHYDROALUMINATE $Al_2H_8Mn$

$Mn[AlH_4]_2$

Aubry, J. *et al., Compt. Rend.*, 1954, **238**, 2535
The unstable hydride decomposed at $-80°C$ and ignited in contact with air.
*See other* COMPLEX HYDRIDES

## DIALUMINIUM TRIOXIDE $Al_2O_3$

Chlorine trifluoride
*See* CHLORINE TRIFLUORIDE, $ClF_3$: Metals, etc.

Ethylene oxide
*See* ETHYLENE OXIDE, $C_2H_4O$: Contaminants

Halocarbons,
Heavy metals
Burbidge, B. W., unpublished information, 1976
It is known that alumina is chlorinated exothermically at above 200°C by contact with halocarbon vapours, and hydrogen chloride, phosgene, etc., are produced. It has now been found that a Co/Mo–alumina catalyst will generate a substantial exotherm in contact with vapour of carbon tetrachloride or 1,1,1-trichloroethane at ambient temperature in presence

of air. In absence of air the effect is less intense. Two successive phases appear to be involved: first, adsorption raises the temperature of the alumina; then reaction, presumably metal-catalysed, sets in, with a further exotherm.

Oxygen difluoride

*See* OXYGEN DIFLUORIDE, $F_2O$: Adsorbents

Sodium nitrate

*See* SODIUM NITRATE, $NNaO_3$: Aluminium, etc.

Vinyl acetate

*See* VINYL ACETATE, $C_4H_6O_2$: Desiccants

**DIALUMINIUM OCTAVANADIUM TRIDECASILICIDE** **$Al_2Si_{13}V_8$**

**(complex structure)**

Hydrofluoric acid

Sidgwick, 1950, 833

The silicide reacts violently with aqueous hydrofluoric acid.

*See other* METAL NON-METALLIDES

**CERIUM(III) TETRAHYDROALUMINATE** **$Al_3CeH_{12}$**

**$Ce[AlH_4]_3$**

Aubry, J. *et al.*, *Compt. Rend.*, 1954, **238**, 2535

The unstable hydride decomposed at $-80^\circ C$, and ignited in contact with air.

*See other* COMPLEX HYDRIDES

**AMERICIUM TRICHLORIDE** **$AmCl_3$**

*MCA Case History No. 1105*

A multi-wall shipping container, containing 400 $cm^3$ of a solution of americium chloride in a polythene bottle and sealed for $3\frac{1}{2}$ months, exploded. The reason could have been a slow pressure build-up of radiolysis products. Venting and other precautions are recommended.

*See other* METAL HALIDES

**ARSENIC** **As**

Muller, W. J., *Z. Angew. Chem.*, 1914, **27**, 338

An explosive variety (or compound) of arsenic was produced as a surface layer on the exposed iron surfaces of a corroded lead-lined vessel which contained 35% sulphuric acid with a high arsenic content. It exploded on friction or ignition, and contained no hydrogen, but variable small amounts of iron and lead. It may have been analogous to explosive antimony.

Bromine azide

*see* BROMINE AZIDE, $BrN_3$

Dirubidium acetylide

*See* DIRUBIDIUM ACETYLIDE, $C_2Rb_2$ : Non-metals

Halogens or Interhalogens

1. Mellor, 1946, Vol. 2, 92
2. Mellor, 1956, Vol. 2, Suppl. 1, 379

The finely powdered element inflames in gaseous chlorine or liquid chlorine at –33°C [1]. The latter is doubtful [2].

*See* BROMINE TRIFLUORIDE, $BrF_3$ : Halogens, etc.
BROMINE PENTAFLUORIDE, $BrF_5$ : Acids, etc.
CHLORINE TRIFLUORIDE, $ClF_3$ : Metals, etc.
IODINE PENTAFLUORIDE, $F_5I$: Metals, etc.

Metals

Mellor, 1940, Vol. 4, 485–486; 1942, Vol. 15, 629; 1937, Vol. 16, 161

Palladium or zinc and arsenic react on heating with evolution of light and heat, and platinum with vivid incandescence.

*See* ALUMINIUM, Al: Antimony, etc.

Nitrogen trichloride

*See* NITROGEN TRICHLORIDE, $Cl_3N$: Initiators

Oxidants

*See* SILVER NITRATE, $AgNO_3$: Arsenic
DICHLORINE OXIDE, $Cl_2O$: Oxidisable materials
CHROMIUM TRIOXIDE, $CrO_3$ : Arsenic
NITROSYL FLUORIDE, FNO: Metals, etc.
POTASSIUM PERMANGANATE, $KMnO_4$ : Antimony, etc.
POTASSIUM DIOXIDE (SUPEROXIDE), $KO_2$ : Metals
SODIUM PEROXIDE, $Na_2O_2$ : Non-metals

*See other* NON-METALS

## DIFLUOROPERCHLORYL HEXAFLUOROARSENATE $AsClF_8O_2$

$[F_2ClO_2] [AsF_6]$

*See* DIFLUOROPERCHLORYL SALTS

## ARSENIC TRICHLORIDE $AsCl_3$

Aluminium

*See* ALUMINIUM, Al: Non-metal halides

Hexafluoroisopropylideneaminolithium
*See* HEXAFLUOROISOPROPYLIDENEAMINOLITHIUM, $C_3F_6LiN$: Non-metal halides

## ARSENIC PENTAFLUORIDE $AsF_5$

Benzene,
Potassium methoxide
Kolditz, L. *et al., Z. Anorg. Allgem. Chem.*, 1965, **341**, 88–92
Interaction of the pentafluoride and methoxide proceeded smoothly in trichlorotrifluoroethane at 30–40°C, whereas in benzene as solvent repeated explosions occurred.
*See other* NON-METAL HALIDES

## DIFLUOROAMMONIUM HEXAFLUOROARSENATE $AsF_8H_2N$

$F_2NH_2[AsF_6]$

Christe, K. O., *Inorg. Chem.*, 1975, **14**, 2821–2824
Solutions of this and of the hexafluoroantimonate salt in hydrogen fluoride, kept for extended periods between –50 and +25°C, burst the Kel-F or Teflon FEP containers. This was attributed to excess pressure of hydrogen fluoride and nitrogen arising from decomposition of the salts. The variable rates of decomposition indicated catalysis by trace impurities. The solids also decompose exothermally after a short period at ambient temperature.
*See other* *N*-HALOGEN COMPOUNDS

## TRIFLUOROSELENIUM HEXAFLUOROARSENATE $AsF_9Se$

$SeF_3[AsF_6]$

Water
Bartlett, N. *et al.*, *J. Chem. Soc.*, 1956, 3423
Violent interaction.

## † ARSINE $AsH_3$

Rüst, 1948, 301
Arsine is strongly endothermic, and can be detonated by suitably powerful initiation.

Chlorine
*See* CHLORINE, $Cl_2$ : Non-metal hydrides

Nitric acid
*See* NITRIC ACID, $HNO_3$ : Non-metal hydrides
*See other* NON-METAL HYDRIDES

## ARSINE–BORON TRIBROMIDE $AsH_3 \cdot BBr_3$

Oxidants

1. Stock, A., *Ber.*, 1901, **34**, 949
2. Mellor, 1939, Vol. 9, 57

Unlike arsine, the complex ignites on exposure to air or oxygen, even at below 0°C [1]. It is violently oxidized by nitric acid [2].

*See* NON-METAL HYDRIDES

## ARSENIC DISULPHIDE $AsS_2$

Potassium nitrate

*See* POTASSIUM NITRATE, $KNO_3$: Metal sulphides

## DIARSENIC TRIOXIDE $As_2O_3$

*MCA SD-60,* 1956

*See* CHLORINE TRIFLUORIDE, $ClF_3$: Metals, etc.
HYDROGEN FLUORIDE, FH: Oxides
SODIUM NITRATE, $NNaO_3$: Diarsenic trioxide, etc.
ZINC, Zn: Arsenic trioxide

## DIARSENIC PENTAOXIDE $As_2O_5$

Bromine pentafluoride

*See* BROMINE PENTAFLUORIDE, $BrF_5$ : Acids, etc.

## PLATINUM DIARSENIDE $As_2Pt$

Preparative hazard

*See* PLATINUM, Pt: Arsenic

## DIARSENIC DISULPHIDE $As_2S_2$

Chlorine

*See* CHLORINE, $Cl_2$: Sulphides

## DIARSENIC TRISULPHIDE $As_2S_3$

Potassium chlorate

*See* POTASSIUM CHLORATE, $ClKO_3$: Diarsenic trisulphide

## GOLD $Au$

Hydrogen peroxide

*See* HYDROGEN PEROXIDE, $H_2O_2$ : Metals

## GOLD(III) CHLORIDE $AuCl_3$

Ammonia and derivatives

Mellor, 1941, Vol.3, 582–583

Sidgwick, 1950, 178

Action of ammonia or ammonium salts on gold chloride, oxide or other salts under a wide variety of conditions gives explosive or 'fulminating' gold. Of uncertain composition but containing Au–N bonds, this is a heat-, friction- and impact-sensitive explosive when dry, similar to the related mercury and silver compounds.

*See other* METAL HALIDES

## GOLD(III) HYDROXIDE–AMMONIA $2AuH_3O_3 \cdot 3H_3N$

**$2Au(OH)_3 \cdot 3NH_3$**

Sorbe, 1958, 63

Explosive gold, formed from the hydroxide and ammonia, is formulated as above.

## SODIUM TRIAZIDOAURATE (?) $Na[Au(N_3)_3]$ $AuN_9Na$

Rodgers, G. T., *J. Inorg. Nucl. Chem.*, 1958, **5**, 339–340

The material (of unknown structure, analysing as $Au_{1.5}N_9Na$ and possibly impure $Na^+AuN_9^-$) explodes at 130°C.

*See other* GOLD COMPOUNDS
METAL AZIDES

## BIS(DIHYDROXYGOLD)IMIDE $Au_2H_5NO_4$

**$HN[Au(OH)_2]_2$**

Mellor, 1940, Vol. 8, 259

An explosive compound.

*See other* GOLD COMPOUNDS
*N*-METAL DERIVATIVES

## GOLD(III) OXIDE $Au_2O_3$

Ammonium salts

*See* GOLD(III) CHLORIDE, $AuCl_3$ : Ammonia, etc.

## GOLD(III) SULPHIDE

$Au_2S_3$

Silver oxide

*See* SILVER OXIDE, $Ag_2O$: Metal sulphides

## GOLD NITRIDE–AMMONIA

$Au_3N \cdot H_3N$

Raschig, F., *Ann.*, 1886, **235**, 349

An explosive compound, probably present in explosive gold, produced from action of ammonia on gold(I) oxide.

*See other* *N*-METAL DERIVATIVES

## TRIGOLD DISODIUM HEXAAZIDE

$Au_3N_{18}Na_2$

Mellor, 1940, Vol. 8, 349; 1967, Vol. 8, Suppl. 2, 30

Of uncertain constitution, both the dry solid and its aqueous solutions may explode.

*See other* METAL AZIDES

## BORON

B

Bailar, 1973, Vol. 1, 692

Many of the previously described violent reactions of boron with a variety of reagents are ascribed to the use of impure or uncharacterised 'boron'. The general impression of the reactivity of pure boron, even when finely divided, is one of extreme inertness, except to highly oxidising reagents at high temperatures.

Ammonia

Mellor, 1940, Vol. 8, 109

Boron incandesces when heated in dry ammonia, hydrogen being evolved.

Dichromates,
Silicon

Howlett, S. *et al.*, *Thermochimica Acta*, 1974, **9**, 213–216

The mechanisms of ignition and combustion of pyrotechnic mixtures of boron with potassium and/or sodium dichromate in presence or absence of silicon are discussed.

Dirubidium acetylide

*See* DIRUBIDIUM ACETYLIDE, $C_2Rb_2$: Non-metals

Halogens or Interhalogens

Mellor, 1946, Vol. 2, 92

Bailar, 1973, Vol. 1, 690–691

Boron ignites in gaseous chlorine or fluorine at ambient temperature, attaining incandescence in fluorine.

Powdered boron reacts spontaneously with the halogens from fluorine to iodine at 20, 400, 600 and 700°C, respectively.

*See* BROMINE TRIFLUORIDE, $BrF_3$: Halogens, etc.
BROMINE PENTAFLUORIDE, $BrF_5$: Acids, etc.
CHLORINE TRIFLUORIDE, $ClF_3$: Boron-containing materials
IODINE PENTAFLUORIDE, $F_5I$: Metals, etc.

*See* Oxidants, below

Metal fluorides

Mellor 1946, Vol. 3, 389; Vol. 5, 15

Explosive interaction when boron and lead or silver fluoride are ground together at ambient temperature.

Oxidants

Mellor, 1946, Vol. 5, 16

The presence of traces of nitrites in fused metal nitrates increases the violence of interaction.

*See* Halogens or Interhalogens, above
SILVER DIFLUORIDE, $AgF_2$: Boron, etc.
NITROSYL FLUORIDE, FNO: Metals, etc.
NITRYL FLUORIDE, $FNO_2$: Non-metals
OXYGEN DIFLUORIDE, $F_2O$ : Non-metals
NITRIC ACID, $HNO_3$: Non-metals
POTASSIUM NITRITE, $KNO_2$ : Boron
POTASSIUM NITRATE, $KNO_3$ : Non-metals
NITROGEN OXIDE ('NITRIC OXIDE'), NO: Non-metals
DINITROGEN OXIDE ('NITROUS OXIDE'), $N_2O$ : Boron
SODIUM PEROXIDE, $Na_2O_2$: Non-metals
LEAD(II) OXIDE, OPb: Non-metals
LEAD(IV) OXIDE, $O_2Pb$: Non-metals

Water

1. Bailar, 1973, Vol. 1, 691
2. Shidlovskii, A. A., *Chem. Abs.*, 1963, **59**, 11178d

Interaction of powdered boron and steam may become violent at red-heat [1]. The highly exothermic reactions of water with boron or silicon might become combustive or explosive processes at sufficiently high temperatures and pressures [2].

*See other* NON-METALS

## BORON BROMIDE DIIODIDE — $BBrI_2$

Water

Mellor, 1946, Vol. 5, 136

Interaction is violent, as for the tribromide or triiodide.

*See other* NON-METAL HALIDES

## BORON DIBROMIDE IODIDE $BBr_2I$

Water
Mellor, 1946, Vol, 5, 136
Interaction is violent, as for the tribromide or triiodide.
*See other* NON-METAL HALIDES

## BORON TRIBROMIDE $BBr_3$

Sodium
*See* SODIUM, Na: Non-metal halides

Water
1. *BCISC Quart. Safety Summ.*, 1966, **37**, 22
2. Anon., *Lab. Pract.*, 1966, **15**, 797

Boron halides react violently with water, and, particularly if there is a deficiency of water, a violent explosion may result. It is therefore highly dangerous to wash glass ampoules of boron tribromide in running water, or to surround a container of boron tribromide with water under any circumstances. Experiment showed that an ampoule of boron tribromide, when deliberately broken under water, caused a violent explosion, possibly a detonation. Dry non-polar solvents should be used for cleaning or cooling purposes [1]. Small quantities of boron tribromide may be destroyed by cautious addition to a large volume of water, or water containing ice [2].
*See other* NON-METAL HALIDES

## DIFLUOROPERCHLORYL TETRAFLUOROBORATE $BClF_6O_2$

$[F_2ClO_2][BF_4]$

*See* DIFLUOROPERCHLORYL SALTS

## DICHLOROBORANE $Cl_2BH$ $BCl_2H$

Bailar, 1973, Vol. 1, 742
Ignites in air.
*See related* NON-METAL HALIDES

## BORON AZIDE DICHLORIDE $B(N_3)Cl_2$ $BCl_2N_3$

1. Anon., *Angew. Chem. (Nachr.)*, 1970, **18**, 27
2. Paetzold, P. I., *Z. Anorg. Chem.*, 1963, **326**, 47

A hard crust of sublimed material exploded when crushed with a spatula [1]. Previous explosions on sublimation or during solvent removal were known [2].
*See other* NON-METAL AZIDES

## BORON TRICHLORIDE $BCl_3$

Aniline
Jones, R. G., *J. Amer. Chem. Soc.*, 1939, **61**, 1378
In absence of cooling or a diluent, interaction is violent.

Hexafluorisopropylideneaminolithium
*See* HEXAFLUORISOPROPYLIDENEAMINOLITHIUM, $C_3F_6LiN$: Non-metal halides
*See other* NON-METAL HALIDES

## BORON TRIFLUORIDE $BF_3$

Alkali metals,
or Alkaline earth metals (not magnesium)
*Merck Index,* 1968, 162
Interaction hot causes incandescence.

Alkyl nitrates
*See* ALKYL NITRATES: Lewis acids
*See other* NON-METAL HALIDES

## TETRAFLUOROBORIC ACID $HBF_4$ $BF_4H$

Acetic anhydride
*See* ACETIC ANHYDRIDE, $C_4H_6O_3$: Tetrafluoroboric acid

## NITRONIUM TETRAFLUOROBORATE $BF_4NO_2$

$NO_2{}^+[BF_4]^-$

Tetrahydrothiophene-1,1-dioxide
*See* NITRATING AGENTS

## DIOXYGENYL TETRAFLUOROBORATE $BF_4O_2$

$O_2^+[BF_4]^-$

Organic materials
Goetschel, C. T. *et al.*, *J. Amer. Chem Soc.*, 1969, **91**, 4706
It is a very powerful oxidant, addition of a small particle to small samples of benzene or 2-propanol at ambient temperature causing ignition. A mixture prepared at −196°C with either methane or ethane exploded when the temperature was raised to −78°C.
*See other* OXIDANTS

## TETRAFLUOROAMMONIUM TETRAFLUOROBORATE $BF_8N$

$[NF_4]^+[BF_4]^-$

2-Propanol

Goetschel, C. T. *et al., Inorg. Chem.*, 1972, **11**, 1700

When the fluorine used during synthesis contained traces of oxygen, the solid behaved as a powerful oxidant (causing 2-propanol to ignite on contact) and it also exploded on impact. Material prepared from oxygen-free fluorine did not show these properties, which were ascribed to the presence of traces of dioxygenyl tetrafluoroborate. borate.

*See* DIOXYGENYL TETRAFLUOROBORATE, $BF_4O_2$

*See other* *N*-HALOGEN COMPOUNDS

## POLY[BORANE(1)] $(BH)_n$

Bailar, 1973, Vol. 1, 740

Ignites in air.

*See other* BORANES

## BORANE–TETRAHYDROFURAN $BH_3 \cdot C_4H_8O$

1. Bruce, M. I., *Chem. Eng. News*, 1974, **52**(41), 3
2. Hopps, H., ibid.
3. Gaines, D. F. *et al., Inorg. Chem.*, 1963, **2**, 526
4. Kollonitsch, J., *Chem. Eng. News*, 1974, **52**(47), 3

A glass bottle containing a 1M solution of the complex in tetrahydrofuran exploded after 2 weeks' undisturbed laboratory storage out of direct sunlight at 15°C [1]. The problem of pressure build-up during storage of such commercial solutions (which are stabilised with 5 mol% of sodium tetrahydroborate) at above 0°C had been noted previously, and was attributed to presence of moisture in the original containers [2].

However, by analogy with the known generation of hydrogen in tetrahydroborate–diborane–bis(2-methoxyethyl) ether systems [3], it is postulated that the tetrahydroborate content may destabilise the borane–tetrahydrofuran reagent, with generation of hydrogen pressure in the closed bottle [4]. Storage at 0°C and opening bottles behind a screen are recommended [2].

*See related* BORANES

## BORANE–PYRIDINE $BH_3 \cdot C_5H_5N$

1. Brown, H. C. *et al., J. Amer. Chem. Soc.*, 1956, **78**, 5385
2. Baldwin, R. A. *et al., J. Org. Chem.*, 1961, **26**, 3550

Decomposition was rapid at 120°C/7.5 mbar [1], and sometimes violent on attempted distillation at reduced pressure [2].

*See related* BORANES

**BORANE–PHOSPHORUS TRIFLUORIDE** $BH_3 \cdot F_3P$

$BH_3 \cdot PF_3$

Mellor, 1971, Vol. 8, Suppl. 3, 442

The unstable gas ignites in air.

*See related* BORANES

**BORANE–AMMONIA** $BH_3 \cdot NH_3$ $BH_3 \cdot H_3N$

Sorbe, 1968, 56

It may explode on rapid heating.

*See related* BORANES

**POTASSIUM HYPOBORATE** $KOBH_3$ $BH_3KO$

Mellor, 1946, Vol. 5, 38

As a reducant stronger than the phosphinate, it would be expected to interact more vigorously with oxidants.

*See* POTASSIUM PHOSPHINATE ('HYPOPHOSPHITE'), $H_2KO_2P$: Nitric acid

*See other* REDUCANTS

**SODIUM HYPOBORATE** $NaOBH_3$ $BH_3NaO$

Mellor, 1946, Vol. 5, 38

As a reducant stronger than the phosphinate, it would be expected to interact more vigorously with oxidants.

*See* SODIUM PHOSPHINATE, $H_2NaO_2P$: Oxidants

*See other* REDUCANTS

**BORIC ACID** $B(OH)_3$ $BH_3O_3$

Acetic anhydride

*See* ACETIC ANHYDRIDE, $C_4H_6O_3$: Boric acid

Potassium

*See* POTASSIUM, K: Oxidants

**LITHIUM TETRAHYDROBORATE** $Li[BH_4]$ $BH_4Li$

Water

Gaylord, 1965, 22

Contact with limited amounts of water, either as liquid or present as moisture in cellulose fibres, may cause ignition after a delay.

*See other* COMPLEX HYDRIDES

## AMMONIUM PEROXOBORATE $NH_4OBO_2$ $BH_4NO_3$

Menzel, H. *et al.*, *Oesterr. Chem. Zeit.*, 1925, 28, 162

Explosive decomposition under vacuum.

*See other* PEROXOACID SALTS

## SODIUM TETRAHYDROBORATE $Na[BH_4]$ $BH_4Na$

Acids

Bailar, 1973, Vol. 1, 768

Interaction of sodium and other tetrahydroborates with anhydrous acids (fluorophosphoric, phosphoric or sulphuric) to generate diborane is very exothermic, and may be dangerously violent with rapid mixing. Safer methods of making diborane are given.

Alkali

Anon., *Angew. Chem. (Nachr.)*, 1960, 8, 238

Mikheeva, V. I. *et al.*, *Chem. Abs.*, 1969, **71**, 85064n

A large volume of alkaline tetrahydroborate solution spontaneously heated and decomposed, liberating large volumes of hydrogen which burst the container. Decomposition is rapid when pH is below 10.5.

Dry mixtures with sodium hydroxide containing 15–40% of tetrahydroborate liberate hydrogen explosively at 230–270°C.

Aluminium chloride,
Bis(2-methoxyethyl) ether

1. de Jongh, H. A. P., *Chem. Eng. News*, 1977, **55**(31), 31
2. Brown, H. C., ibid., (35), 5

Addition of a 4% solution of hydroborate in diglyme containing 0.09% of water to a 27% solution of aluminium chloride in the same solvent led to a violent explosion, attributed to formation and ignition of hydrogen. The ignition source arose from contact of the hydroborate solution with the solid chloride, as demonstrated experimentally. Nitrogen purging is essential for all hydride reductions [1], and also for hydroboration, organoborane, Grignard and organometallic reactions generally [2]. Previous work had shown that clear solutions of the sodium tetrahydroborate–aluminium chloride reagent did not ignite in dry air, but the solid-containing reagent could lead to ignition [2].

Diborane,
Bis(2-methoxyethyl) ether

*See* BORANE–TETRAHYDROFURAN, $BH_3 \cdot C_4H_8O$ (reference 4)

Palladium

*See* PALLADIUM, Pd : Sodium tetrahydroborate

Ruthenium salts

Cusumana, J. A., *Nature*, 1974, **247**, 456

Use of borohydride solutions to reduce ruthenium salt solutions to the metal or an alloy gave solid products (possibly hydrides) which when dry exploded violently on contact with water or when disturbed by a spatula. Hydrazine appears to be a safe reducant for ruthenium salt solutions.

Sulphuric acid

Pascal, 1961, Vol. 6, 337

Ignition may occur if the mixture is not cooled.

*See other* COMPLEX HYDRIDES

**HYDRAZINE–MONOBORANE** $N_2H_4 \cdot BH_3$ $BH_7N_2$

Gunderloy, F. C., *Inorg. Synth.*, 1967, **9**, 13

It is shock-sensitive and highly flammable, like the bis-compound.

*See* COMPLEX HYDRIDES
NON-METAL HYDRIDES

**BORON DIIODOPHOSPHIDE** $BI_2P$

Chlorine

*See* CHLORINE, $Cl_2$ : Phosphorus compounds

Metals

Mellor, 1947, Vol. 8, 845

It ignites in contact with mercury vapour or magnesium powder.

*See other* NON-METAL HALIDES

**BORON TRIIODIDE** $BI_3$

Ammonia

Mellor, 1945, Vol. 5, 136

Strong exotherm on contact.

Phosphorus

Mellor, 1946, Vol. 5, 136

Warm red or white phosphorus reacts incandescently.

Water

1. Moissan, H., *Compt. Rend.*, 1892, **115**, 204
2. Unpublished information

Violent reaction [1], particularly with limited amounts of water [2].

*See other* NON-METAL HALIDES

## BORON NITRIDE $BN$

Sodium peroxide

*See* SODIUM PEROXIDE, $Na_2O_2$: Boron nitride

## TRIAZIDOBORANE $B(N_3)_3$ $BN_9$

Anon., *Angew. Chem. (Nachr.)*, 1970, **18**, 27

A sample of the vacuum-distilled pyridine complex exploded in a heated capillary sampling tube.

*See other* NON-METAL AZIDES

## SODIUM BORATE HYDROGEN PEROXIDATE $BNaO_2 \cdot H_2O_2$

*See* CRYSTALLINE HYDROGEN PEROXIDATES

*See also* SODIUM PEROXOBORATE, $BNaO_3$

## SODIUM PEROXOBORATE $NaOBO_2$ $BNaO_3$

1. Anon., *Angew. Chem.*, 1963, **65**, 41
2. Castrantas, 1965, 5

The true peroxoborate has been reported to detonate on light friction [1]. The common 'tetrahydrate' is not a peroxoborate, but $NaBO_2 \cdot H_2O_2 \cdot 3H_2O$, and while subject to catalytic decomposition by heavy metals and their salts, or easily oxidisable foreign matter, it is relatively stable under mild grinding with other substances [2].

*See* CRYSTALLINE HYDROGEN PEROXIDATES

*See other* PEROXOACID SALTS

## BORON PHOSPHIDE $BP$

Oxidants

*See* NITRIC ACID, $HNO_3$: Non-metals

SODIUM NITRATE, $NNaO_3$: Boron phosphide

## BERYLLIUM TETRAHYDROBORATE $Be[BH_4]_2$ $B_2BeH_8$

Mackay, 1966, 169

Semenenko, K. N. *et al.*, *Russ. Chem. Rev.*, 1973, 4

It vigorously ignites and often explodes in air. It reacts explosively with cold water.

*See other* COMPLEX HYDRIDES

## CHLORODIBORANE $ClHBH_2BH_2$ $B_2ClH_5$

Bailar, 1973, Vol. 1, 778

A gas above $-11°C$, it ignites in air.

*See related* BORANES

## DIBORON TETRACHLORIDE $Cl_2B{-}BCl_2$ $B_2Cl_4$

Air

Wartik, T. *et al., Inorg. Synth.*, 1967, **10**, 125

Sudden exposure to air may cause explosion.

Dimethylmercury

Wartik, T. *et al., Inorg. Chem.*, 1971, **10**, 650

The reaction, starting at $-63°C$ under vacuum, exploded violently on two occasions, after 23 uneventful runs. Investigation pending.

*See other* NON-METAL HALIDES

## DIBORON TETRAFLUORIDE $F_2B{-}BF_2$ $B_2F_4$

Metal oxides

Holliday, A. K. *et al., J. Chem. Soc.*, 1964, 2732

Mixtures with mercury(II) oxide and manganese dioxide prepared at $-80°C$ ignited at $20°C$ and reacted violently at $15°C$, respectively. Copper(II) oxide reacted vigorously at $25°C$ without ignition.

Oxygen

Trefonas, L. *et al., J. Chem. Phys.*, 1958, **28**, 54

The gas is extremely explosive in presence of oxygen.

*See other* NON-METAL HALIDES

## † DIBORANE $H_2B(\mu\text{-}H)_2BH_2$ $B_2H_6$

The autoignition range is lowered by moisture.

*MCA SD-84,* 1961

Preparative hazard

*See* SODIUM TETRAHYDROBORATE, $BH_4Na$: Acids

Air,
Benzene,
or Moisture

1. Mellor, 1946, Vol. 5, 36
2. Schlessinger, H. I. *et al., Chem Rev.*, 1942, **31**, 8

3. Simons, H. P. *et al.*, *Ind. Eng. Chem.*, 1958, **50**, 1665, 1669
4. *Hydrides of Boron and Silicon*, Stock, A. E., Ithaca, Cornell Univ. Press, 1933

Usually ignites in air unless dry and free of impurities [1]. Ignition delays of 3–5 days, followed by violent explosions, have been experienced [2]. Effects of presence of moisture or benzene vapour in air on spontaneously explosive reaction have been assessed quantitatively [3].

Explosion followed spillage of liquid diborane [4].

Ammonia

*See* DIAMMINEBORONIUM TETRAHYDROBORATE, $B_2H_{12}N_2$

Chlorine

*See* CHLORINE, $Cl_2$ : Non-metal hydrides

Dimethyl sulphoxide

*See* DIMETHYL SULPHOXIDE, $C_2H_6OS$: Boron compounds

Halocarbons

*Haz. Chem. Data,* 1971, 88

Diborane reacts violently with halocarbon liquids used as vaporising fire-extinguishants.

Metals

*Haz. Chem. Data*, 1973, 108

Interaction with aluminium or lithium gives complex hydrides which may ignite in air.

Octanal oxime,
Sodium hydroxide

Augustine, 1968, 78

Addition of sodium hydroxide solution during work-up of a reaction mixture of oxime and diborane in THF is very exothermic, a mild explosion being noted on one occasion.

Oxygen

Whatley, A. T. *et al.*, *J. Amer. Chem. Soc.*, 1954, **76**, 1997–1999

Mixtures at 105–165°C exploded spontaneously after an induction period dependent on temperature and composition.

*See* Air, etc., above

Tetravinyllead

Houben-Weyl, 1975, Vol. 13.3, 253

Interaction is explosively violent at ambient temperature.

*See other* NON-METAL HYDRIDES

**HYDRAZINE–BISBORANE** $N_2H_4 \cdot 2BH_3$ $B_2H_{10}N_2$

Gunderloy, F. C., *Inorg. Synth.*, 1967, **9**, 14

Explodes on impact or at over 100°C and is extremely flammable.

*See other* COMPLEX HYDRIDES
NON-METAL HYDRIDES

**DIAMMINEBORONIUM TETRAHYDROBORATE** $B_2H_{12}N_2$

$[BH_2(NH_3)_2]^+[BH_4]^-$

Sidgwick, 1950, 354; Bailar, 1973, Vol. 1, 924

The diammine complex of diborane (formulated as above), though less reactive than diborane, ignites on heating in air.

*See other* COMPLEX HYDRIDES

**MAGNESIUM BORIDE** $Mg_3B_2$ $B_2Mg_3$

Acids

Mellor, 1946, Vol. 5, 25

The crude product containing some silicide evolves, in contact with hydrochloric or sulphuric acid, boron and silicon hydrides, which may ignite.

*See other* METAL NON-METALLIDES

**DIBORON DIOXIDE** **OBBO** $B_2O_2$

Water

Halliday, A. K. *et al., Chem. Rev.*, 1962, **62**, 316

At 400°C, traces of water react causing a violent eruption and incandescence.

*See other* NON-METAL OXIDES

**DIBORON TRIOXIDE** **OBOBO** $B_2O_3$

Bromine pentafluoride

*See* BROMINE PENTAFLUORIDE, $BrF_5$ : Acids, etc.

**THALLIUM(I) PEROXOBORATE** $Tl_2B_2O_7 \cdot H_2O$ $B_2O_7Tl_2 \cdot H_2O$

Bailar, 1973, Vol. 1, 1154

It liberates oxygen at 18°C and explodes on further warming.

*See other* PEROXOACID SALTS

**DIBORON TRISULPHIDE** **SBSBS** $B_2S_3$

Chlorine

*See* CHLORINE, $Cl_2$ : Sulphides

**1,3,5-TRICHLORO-2,4,6-TRIFLUOROBORAZINE** $B_3Cl_3F_3N_3$

**ClNBFNClBFNClBF**

Water

Elter, G. *et al.*, *Angew. Chem. (Intern. Ed.)*, 1975, **14**, 709

Interaction of this *N*-chloro-*B*-fluoro compound with water is explosively violent.

*See other* NON-METAL HALIDES

***B*-1,3,5-TRICHLOROBORAZINE** $B_3Cl_3H_3N_3$

**ClBNHBClNHBClNH**

Water

Niedenzu, K. *et al.*, *Inorg. Synth.*, 1967, **10**, 141

Violent interaction.

**TRIBORON PENTAFLUORIDE** $B_3F_5$

Timms, P. L., *J. Amer. Chem. Soc.*, 1967, **89**, 1631

It reacts explosively with air or water.

Tetrafluoroethylene

Timms, P. L., *J. Amer. Chem. Soc.*, 1967, **89**, 1631

The pentafluoride catalyses polymerisation of the tetrafluoride smoothly below –100°C, but explosively above that temperature.

*See other* POLYMERISATION INCIDENTS
NON-METAL HALIDES

**BORAZINE** **HBNHBHNHBHNH** $B_3H_6N_3$

Niedenzu, K. *et al.*, *Inorg. Synth.*, 1967, **10**, 144

Samples sealed into ampoules exploded when stored in daylight, but not in the dark.

*See other* NON-METAL HYDRIDES

**SODIUM OCTAHYDROTRIBORATE** $NaB_3H_8$ $B_3H_8Na$

Dewkett, W. J. *et al.*, *Inorg. Synth.*, 1974, **15**, 116

Air should not be drawn through solutions of the compound in ether, or through its solid complex with *p*-dioxan, because such materials have occasionally ignited in air.

*See other* COMPLEX HYDRIDES

**URANIUM(III) TETRAHYDROBORATE** $U[BH_4]_3$ $B_3H_{12}U$

Semenenko, K. N. *et al.*, *Russ. Chem. Rev.*, 1973, 4

It ignites in air and explodes on heating, unlike the U(VI) analogue.

*See other* COMPLEX HYDRIDES

**TETRABORON TETRACHLORIDE** $B_4Cl_4$

Urry, G. *et al.*, *Inorg. Chem.*, 1963, 2, 398

It ignites in air.

*See other* NON-METAL HALIDES
PYROPHORIC MATERIALS

**TETRABORANE(10)** $B_4H_{10}$

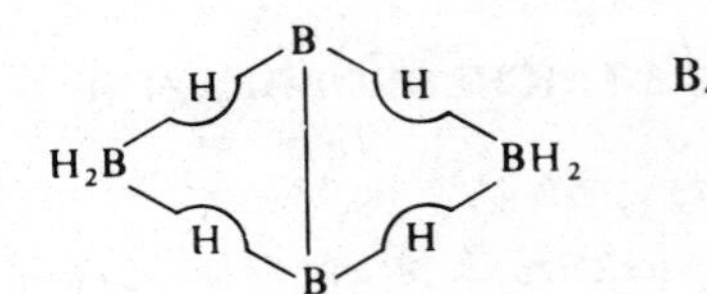


Oxidants

Mellor, 1946, Vol. 5, 36

Bailar, 1973, Vol. 1, 790

Ignites in air or oxygen, and explodes with concentrated nitric acid. The later reference states that the pure compound does not ignite in air.

*See also* NITRIC ACID, $HNO_3$: Non-metal hydrides

*See other* BORANES

**HAFNIUM(IV) TETRAHYDROBORATE** $B_4H_{16}Hf$

$Hf[BH_4]_4$

Gaylord, 1956, 58

Violent ignition on exposure to air.

*See other* COMPLEX HYDRIDES

**ZIRCONIUM(IV) TETRAHYDROBORATE** $B_4H_{16}Zr$

$Zr[BH_4]_4$

Gaylord, 1956, 58

Violent ignition on exposure to air.

*See other* COMPLEX HYDRIDES

**SODIUM TETRABORATE (2–)** $B_4Na_2O_7$

$Na_2[B_4O_5(OH)_4] \cdot 3H_2O$

Zirconium

*See* ZIRCONIUM, Zr: Oxygen-containing compounds

## † PENTABORANE(9)

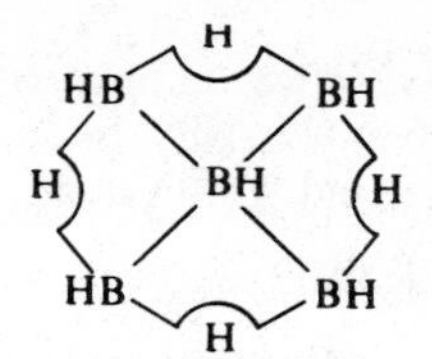


$B_5H_9$

It ignites spontaneously if impure.
*MCA SD-84*, 1961

Ammonia
*See* DIAMMINEBORONIUM HEPTAHYDROTETRABORATE (PENTABORANE(9)DIAMMONIATE), $B_5H_{15}N_2$

Oxygen
*Pentaborane*, Tech. Bull. LF202, Alton (Ill.), Olin Corp. Energy Div., 1960
Reaction of pentaborane with oxygen is often violently explosive.

Reactive solvents
Cloyd, 1965, 35
Miller, V. R., *et al.*, *Inorg. Synth.*, 1974, **15**, 118–122
Pentaborane is stable in inert hydrocarbon solvents but forms shock-sensitive solutions in most other solvents containing carbonyl, ether or ester functional groups and/or halogen substitutents.

The later reference gives detailed directions for preparation and handling of this exceptionally reactive compound, including a list of 26 solvents and compounds rated as potentially dangerous in the presence of pentaborane. When large quantities are stored at low temperature in glass, a phase change involving expansion of the solid borane may rupture the container.

*See* DIMETHYL SULPHOXIDE, $C_2H_6OS$: Boron compounds
*See other* BORANES

## PENTABORANE(11)

$B_5H_{11}$

Kit and Evered, 1960, 69
Ignites in air.
*See other* NON-METAL HYDRIDES

## DIAMMINEBORONIUM HEPTAHYDROTETRABORATE (PENTABORANE(9) DIAMMONIATE)

$B_5H_{15}N_2$

$[H_2B\ 2NH_3][B_4H_7]$

Kodama, G., *J. Amer. Chem. Soc.*, 1970, **92**, 3482
The diammoniate of pentaborane(9) decomposes spectacularly on standing at ambient temperature.
*See related* BORANES

**HEXABORANE(12)** **$B_6H_{12}$**

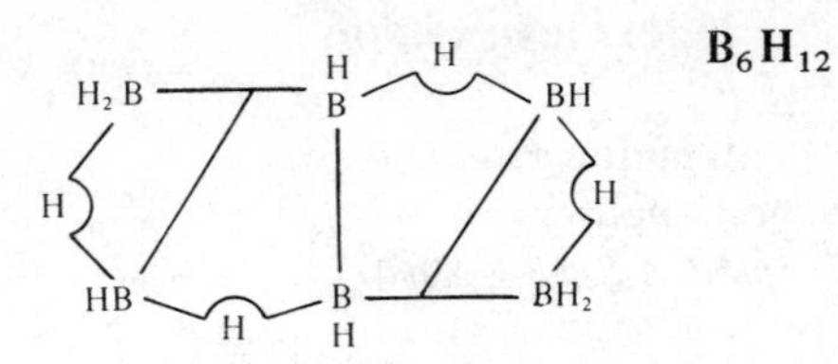


Mellor, 1946, Vol. 5, 36

It is unstable, and ignites in air.

*See other* BORANES

**CAESIUM LITHIUM UNDECAHYDROTHIONONABORATE** **$B_9CsH_{11}LiS$**

**$CsLiB_9H_{11}S$**

Siedle, A. R. *et al.*, *Inorg. Chem.*, 1974, **13**, 2737

It ignites in air.

*See related* COMPLEX HYDRIDES

**CAESIUM LITHIUM TRIDECAHYDRONONABORATE** **$B_9CsH_{13}Li$**

**$CsLiB_9H_{13}$**

Siedle, A. R. *et al.*, *Inorg. Chem.*, 1974, **13**, 2737

It ignites in air.

*See other* COMPLEX HYDRIDES

**1,10-BISDIAZONIODECABORAN(8)ATE** **$B_{10}H_8(N_2)_2$** **$B_{10}H_8N_4$**

The precursor is explosive.

*See* AMMONIUM DECAHYDRODECABORATE(2–), $B_{10}H_{18}N_2$: Nitrous acid

*See related* DIAZONIUM SALTS

**DECABORANE(14)** **$B_{10}H_{14}$**

Ethers,
or Halocarbons,
or Oxygen

*MCA SD-84*, 1961

Hawthorne, M. F., *Inorg. Synth.*, 1967, **10**, 93–94

It forms impact-sensitive mixtures with ethers (dioxan, etc.) and halocarbons (carbon tetrachloride) and ignites in oxygen at 100°C.

*See* PENTABORANE(9), $B_5H_9$: Reactive solvents
DIMETHYL SULPHOXIDE, $C_2H_6OS$: Boron compounds

*See other* NON-METAL HYDRIDES

**AMMONIUM DECAHYDRODECABORATE(2–)** **$B_{10}H_{18}N_2$**

**$(NH_4)_2B_{10}H_{10}$**

Nitrous acid

Knoth, W. H., *J. Amer. Chem. Soc.*, 1964, **86**, 115

Interaction of the $B_{10}H_{10}(2-)$ anion with excess nitrous acid gives an inner diazonium salt (of unknown structure, possibly containing a nitronium ion) which is highly explosive in the dry state. It is readily reduced wet to the non-explosive 1,10-bisdiazoniodecaboran(8)ate inner salt.

**BARIUM** **Ba**

Halocarbons

1. *Serious Acc. Ser.*, 1952, **23** and Suppl., Washington, USAEC
2. Anon., *Ind. Res.*, 1968, (9), 15
3. *Pot. Incid. Rep.*, 1968, **39**

A violent reaction occurred when cleaning lump metal under carbon tetrachloride [1]. Finely divided barium, slurried with trichlorotrifluoroethane, exploded during transfer owing to frictional initiation [2]. Granular barium in contact with fluorotrichloromethane, carbon tetrachloride, 1,1,2-trichlorotrifluoroethane, tetrachloroethylene or trichloroethylene is susceptible to detonation [3].

*See other* METALS: Halocarbons

Interhalogens

*See* BROMINE PENTAFLUORIDE, $BrF_5$: Acids, etc.
IODINE HEPTAFLUORIDE, $F_7I$: Metals

Oxidising gases

Kirk-Othmer, 1964, Vol. 3, 78

The finely divided metal may ignite or explode in air or other oxidising gases.

Water

Sidgwick, 1950, 844

Interaction is more violent than with calcium or strontium, but less so than with sodium.

*See other* METALS

**BARIUM TETRAFLUOROBROMATE** **$Ba[BrF_4]_2$** **$BaBr_2F_8$**

*See* METAL POLYHALOHALOGENATES

**BARIUM BROMATE** **$Ba(BrO_3)_2$** **$BaBr_2O_6$**

Hackspill, L. *et al.*, *Compt. Rend.*, 1930, **191**, 663

Thermal decomposition with evolution of oxygen is almost explosive at 300°C.

Disulphur dibromide

*See* DISULPHUR DIBROMIDE, $Br_2S_2$ : Oxidants

Sulphur

Taradoire, F., *Bull. Soc. Chim. Fr.*, 1945, **12**, 94, 447

Mixtures are unstable, and may ignite 2–11 days after preparation at 22–25°C, or immediately at 91–93°C. Presence of moisture (as water of crystallisation) accelerates ignition.

*See other* METAL OXOHALOGENATES

**BARIUM PERCHLORYLAMIDE** **$Ba(NHClO_3)_2$** **$BaCl_2H_2N_2O_6$**

*See* PERCHLORYLAMIDE SALTS

**BARIUM CHLORITE** **$Ba(ClO_2)_2$** **$BaCl_2O_4$**

Solymosi, F. *et al.*, *Chem. Abs.*, 1968, **68**, 51465s

When heated rapidly, barium chlorite decomposes explosively at 190°C, and the lead salt at 112°C.

Dimethyl sulphate

Pascal, 1960, Vol. 16.1, 264

The sulphate ignites in contact with the unheated chlorite, presumably owing to formation of very unstable methyl chlorite.

*See other* CHLORITE SALTS

**BARIUM PERCHLORATE** **$Ba(ClO_4)_2$** **$BaCl_2O_8$**

Alcohols

Kirk-Othmer, 1964, Vol. 5, 75

Distillation of mixtures with $C_1$–$C_3$ alcohols gives the highly explosive alkyl perchlorates. Extreme shock-sensitivity is still shown by *n*-octyl perchlorate.

*See* ALKYL PERCHLORATES

*See other* METAL OXOHALOGENATES

**BARIUM HYDRIDE** **$BaH_2$**

1. Mellor, 1941, Vol. 3, 650
2. Gibson, 1969, 74

The heated hydride ignites in oxygen [1], or in air cold if finely divided [2].

Metal halogenates

*See* METAL HALOGENATES: Metals, etc.

## BARIUM HYDROXIDE $BaH_2O_2$

Chlorinated rubber

*See* CHLORINATED RUBBER: Metal oxides or hydroxides

## BARIUM AMIDOSULPHATE $Ba(H_2NSO_3)_2$ $BaH_4N_2O_6S_2$

Metal nitrates or nitrites

*See* METAL AMIDOSULPHATES

## BARIUM PHOSPHINATE ('HYPOPHOSPHITE') $BaH_4O_4P_2$

$Ba(H_2PO_2)_2$

Potassium chlorate

*See* POTASSIUM CHLORATE, $ClKO_3$ : Reducants

## BARIUM NITRATE $BaN_2O_6$

Aluminium,
Potassium nitrate,
Potassium perchlorate,
Water

*See* ALUMINIUM, Al: Metal nitrates, etc.

Aluminium–magnesium alloy

Tomlinson, W. R. *et al.*, *J. Chem. Educ.*, 1950, **27**, 606

An intimate mixture of the finely divided components, widely used as a photoflash composition, is readily ignitable and extremely sensitive to friction or impact.

*See other* METAL OXONON-METALLATES

## BARIUM DIAZIDE $Ba(N_3)_2$ $BaN_6$

1. Fagan, C. P., *J. and Proc. R. Inst. Chem.*, 1947, 126
2. Ficheroulle, H. *et al.*, *Mem. Poudres,* 1956, **33**, 7
3. Gyunter, P. L. *et al.*, *Chem. Abs.*, 1943, **37**, $1270_9$–$1271_2$
4. Verneker, V. R-P. *et al.*, *J. Phys. Chem.*, 1968, **72**, 778–783

This material is impact-sensitive when dry and is supplied and stored damp with alcohol. It is used as a saturated solution and it is important to prevent total evaporation of such solution, or the slow growth of large crystals which may become dried and shock-sensitive. Lead drains must not be used, to avoid formation of the detonator, lead azide. Exposure to acid conditions may generate explosive hydrazoic acid [1]. More recently it has been stated that barium azide is relatively insensitive to impact but highly sensitive to friction [2]. Strontium and, particularly, calcium azides show much more marked explosive properties than barium azide.

The explosive properties appear to be associated with the method of formation of the azide [3]. Factors affecting the sensitivity of the azide include surface area, solvent used and ageing. Presence of barium metal, sodium or iron ions as impurities increases the sensitivity [4].

*See other* METAL AZIDES

## BARIUM OXIDE — BaO

Dinitrogen tetraoxide
*See* DINITROGEN TETRAOXIDE (NITROGEN DIOXIDE), $N_2O_4$: Barium oxide

Hydroxylamine
*See* HYDROXYLAMINE, $H_3NO$: Oxidants

Sulphur trioxide
*See* SULPHUR TRIOXIDE, $O_3S$: Metal oxides

Triuranium octaoxide
*See* TRIURANIUM OCTAOXIDE, $O_8U_3$: Barium oxide

## BARIUM PEROXIDE — $BaO_2$

Acetic anhydride
Rüst, 1948, 337
The peroxide was substituted for (unavailable) potassium permanganate in a process for purifying the crude anhydride in an open vessel. After several operations, when only minor explosions occurred, a violent explosion and fire ensued. Acetyl peroxide would be produced.

Delay compositions
Stupp, J., *Chem. Abs.*, 1975, **82**, 113746j
The spontaneous ignition of the peroxide in (unspecified) tracer-ignition delay compositions is described.

Hydrogen sulphide
*See* HYDROGEN SULPHIDE, $H_2S$: Metal oxides

Hydroxylamine
*See* HYDROXYLAMINE, $H_3NO$: Oxidants

Metals

Pascal, 1958, Vol. 4, 775

Powdered aluminium or magnesium ignites in intimate contact with the peroxide.

Non-metal oxides

Pascal, 1958, Vol. 4, 773–774

The heated oxide will attain incandescence in a rapid stream of carbon dioxide or sulphur dioxide.

Organic materials,
Water

Koffolt., J. H., private comm., 1966

Contact of barium peroxide and water will readily produce a temperature and a local oxygen concentration high enough to ignite many organic compounds.

Peroxyformic acid

*See* PEROXYFORMIC ACID, $CH_2O_3$ : Metals, etc.

Propane

Hoffmann, A. B. *et al.*, *J. Chem. Educ.*, 1974, **51**, 419, note 7

Heating barium peroxide under gaseous propane at ambient pressure caused a violent exothermic reaction which deformed the glass container.

Selenium

Johnson, L. B., *Ind. Eng. Chem.*, 1960, **52**, 241–244

Powdered mixtures ignite at 265°C.

Wood

Dupré, A., *J. Soc. Chem. Ind.*, 1897, **16**, 492

Friction of the peroxide between wooden surfaces ignited the latter.

*See other* METAL PEROXIDES

**BARIUM SULPHATE** **$BaSO_4$** **$BaO_4S$**

Aluminium

*See* ALUMINIUM, Al: Metal oxosalts

Phosphorus

*See* PHOSPHORUS, P: Metal sulphates

**BARIUM SULPHIDE** **BaS**

Dichlorine oxide

*See* DICHLORINE OXIDE, $Cl_2O$: Oxidisable materials

Oxidants

Mellor, 1941, Vol. 3, 745

Barium sulphide explodes weakly on heating with lead dioxide or potassium chlorate, and strongly with potassium nitrate. Calcium and strontium sulphides are similar.

Phosphorus(V) oxide

Pascal, 1958, Vol. 4, 832

Interaction is violent, attaining incandescence.

*See other* METAL SULPHIDES

## TRIBARIUM TETRANITRIDE $Ba_3N_4$

Sorbe, 1968, 34

Reacts violently with air or water.

*See other* METAL NON-METALLIDES

## BERYLLIUM Be

Carbon dioxide,
Nitrogen

*See* CARBON DIOXIDE, $CO_2$: Metals, Nitrogen

Halocarbons

*Pot. Incid. Rep.*, 1968, **39**

Mixtures of powdered beryllium with carbon tetrachloride or trichloroethylene will flash on heavy impact.

*See other* METALS: Halocarbons

Halogens

Pascal, 1958, Vol. 4, 22

Warm beryllium incandesces in fluorine or chlorine.

Phosphorus

*See* PHOSPHORUS, P: Metals

*See other* METALS

## BERYLLIUM PERCHLORATE $Be(ClO_4)_2$ $BeCl_2O_8$

Laran, R. J., US Pat. 3 157 464, 1964

A powerful oxidant, insensitive to heat or shock and useful in propellant and igniter systems.

*See other* METAL OXOHALOGENATES

## BERYLLIUM FLUORIDE $BeF_2$

Magnesium

*See* MAGNESIUM, Mg: Beryllium fluoride

## BERYLLIUM HYDRIDE $BeH_2$

Methanol,
or Water

Barbaras, G. D., *J. Amer. Chem. Soc.*, 1951, **73**, 48

Reaction of the ether-containing hydride with methanol or water is violent, even at −196°C.

*See other* METAL HYDRIDES

## BERYLLIUM OXIDE BeO

Magnesium

*See* MAGNESIUM, Mg: Metal oxides

## BISMUTH Bi

Aluminium

*See* ALUMINIUM, Al: Bismuth

Oxidants

*See* BROMINE PENTAFLUORIDE, $BrF_5$: Acids, etc.
CHLORIC ACID, $ClHO_3$: Metals, etc.
PERCHLORIC ACID, $ClHO_4$: Bismuth
NITROSYL FLUORIDE, FNO: Metals
IODINE PENTAFLUORIDE, $F_5I$: Metals
NITRIC ACID, $HNO_3$: Metals
AMMONIUM NITRATE, $H_4N_2O_3$: Metals

## BISMUTH PENTAFLUORIDE $BiF_5$

von Wartenberg, H., *Z. Anorg. Chem.*, 1940, **224**, 344

It reacts vigorously with water, sometimes with ignition.

*See other* METAL HALIDES

## BISMUTHIC ACID $HOBiO_2$ $BiHO_3$

Hydrofluoric acid

Mellor, 1939, Vol. 9, 657

Interaction of the solid acid with 40% hydrofluoric acid solution is violent, ozonised oxygen being evolved.

**BISMUTH AMIDE OXIDE** $H_2NBiO$ $BiH_2NO$

Watt, G. W. *et al.*, *J. Amer. Chem. Soc.*, 1939, **61**, 1693

The solid, prepared in liquid ammonia, explodes when free of ammonia and exposed to air.

*See other* *N*-METAL DERIVATIVES

**BISMUTH NITRIDE** BiN

Alone,

or Water

1. Fischer F. *et al.*, *Ber.*, 1910, **43**, 1471
2. Franklin, E. C., *J. Amer. Chem. Soc.*, 1905, **27**, 847

Very unstable, exploded on shaking [1] or heating, or in contact with water or dilute acids [2].

*See other* *N*-METAL DERIVATIVES

**BISMUTH PLUTONIDE** BiPu

Williamson, G. K., *Chem. & Ind.*, 1960, 1384

Extremely pyrophoric.

*See other* ALLOYS (INTERMETALLIC COMPOUNDS)

**DIBISMUTH DICHROMIUM NONAOXIDE ('BISMUTH CHROMATE')** $Bi_2O_3 \cdot 2CrO_3$ $Bi_2Cr_2O_9$

Hydrogen sulphide

Pascal, 1960, Vol. 13.1, 1025

The gas may ignite on contact with the 'chromate'.

*See other* METAL OXOMETALLATES

**DIBISMUTH TRIOXIDE** $Bi_2O_3$

Chlorine trifluoride

*See* CHLORINE TRIFLUORIDE, $ClF_3$: Metals, etc.

Potassium

*See* POTASSIUM, K: Metal oxides

Sodium

*See* SODIUM, Na: Metal oxides

## DIBISMUTH TRISULPHIDE — $Bi_2S_3$

Glatz, A. C. *et al.*, *J. Electrochem. Soc.*, 1963, **110**, 1231

Possible causes of explosions in direct synthesis are discussed.

*See also* SULPHUR,S: Metals

*See other* METAL SULPHIDES

## BROMINE PERCHLORATE — $BrClO_4$

Schack, C. J. *et al.*, *Inorg. Chem.*, 1971, **10**, 1078

Shock-sensitive.

Perfluorobutadiene

Schack, C. J. *et al.*, *Inorg. Chem.*, 1975, **14**, 151, footnote 8

During adduct formation, the perchlorate must be present in excess to prevent formation of a mono-adduct, which may well be explosive.

*See* CHLORINE PERCHLORATE, $Cl_2O_4$ : Chlorotrifluoroethylene

*See other* HALOGEN OXIDES

## CAESIUM HEXAFLUOROBROMATE(1−) — $Cs[BrF_6]$ — $BrCsF_6$

*See* METAL POLYHALOHALOGENATES

## CAESIUM BROMOXENATE — $Cs[XeO_3Br]$ — $BrCsO_3Xe$

Jaselskis, B. *et al.*, *J. Amer. Chem. Soc.*, 1969, **91**, 1875

Aqueous solutions are extremely unstable and caution is required if isolation of the compound is contemplated.

*See other* XENON COMPOUNDS

## BROMINE FLUORIDE — $BrF$

Sidgwick, 1950, 1149

Chemically it behaves like the other bromine fluorides, but is more reactive.

Hydrogen

Pascal, 1960, Vol. 16.1, 412

Hydrogen ignites in the fluoride at ambient temperature.

*See other* INTERHALOGENS

## BROMYL FLUORIDE — $BrO_2F$ — $BrFO_2$

Water

Bailar, 1973, Vol. 2, 1388

Hydrolysis may proceed explosively.

*See other* HALOGEN OXIDES

Polymers

Johnson, G. K. *et al.*, *Inorg. Chem.*, 1972, **11**, 800

It is considerably more reactive than perchloryl fluoride, and attacks glass and the usually inert polytetrafluoroethylene and polychlorotrifluoroethylene.

*See other* HALOGEN OXIDES

## BROMINE TRIFLUORIDE $BrF_3$

Davis, R. A. *et al.*, *J. Org. Chem.*, 1967, **32**, 3478

Musgrove, W. K. R., *Advan. Fluorine Chem.*, 1964, **1**, 12

The hazards and precautions involved in the use of this very reactive fluorinating agent are outlined. Contact with rubber, plastics or other organic materials may be explosively violent, and reaction with water is very vigorous.

Ammonium halides

Sharpe, A. G. *et al.*, *J. Chem. Soc.*, 1948, 2137

Explosive reaction.

Antimony trichloride oxide

Mellor, 1956, Vol. 2, Suppl. 1, 166

Interaction is violent, even more so than with diantimony trioxide.

Carbon monoxide

Mellor, 1956, Vol. 2, Suppl. 1, 166

At temperatures rather above 30°C, explosions occurred.

Carbon tetrachloride

Dixon, K. R. *et al.*, *Inorg. Synth.*, 1970, **12**, 233

Excess bromine trifluoride may be destroyed conveniently in a hood by slow addition to a large volume of the solvent, interaction being vigorous but not dangerous.

*See* Solvents, below

Halogens
or Metals,
or Non-metals,
or Organic materials

Mellor, 1941, Vol. 2, 113; 1956, Vol. 2, Suppl. 1, 164–167

'Chlorine Trifluoride Technical Bull.', Morristown, Baker and Adamson, 1970

Incandescence is caused by contact with bromine, iodine, arsenic, antimony (even at −10°C); powdered molybdenum, niobium, tantalum, titanium, vanadium; boron, carbon, phosphorus or sulphur. Carbon tetraiodide, chloromethane, benzene, ether ignite or explode on contact, as do organic materials generally. Silicon also ignites.

*See* Uranium, below

2-Pentanone

Stevens, T. E., *J. Org. Chem.*, 1961, **26**, 1629, footnote 11

During evaporation of solvent hydrogen fluoride, an exothermic reaction between residual ketone and bromine trifluoride set in and accelerated to explosion.

Potassium hexachloroplatinate(2–)

Dixon, K. R. *et al.*, *Inorg. Synth.*, 1970, **12**, 233–237

Interaction of the reagents in bromine as diluent to produce the trichlorotrifluoro- and then hexafluoro-platinates is so vigorous that increase in scale above 1g of salt is not recommended.

Pyridine

Kirk-Othmer, 1966, Vol. 9, 592

The solid, produced by action of bromine trifluoride on pyridine in carbon tetrachloride, ignites when dry. 2-Fluoropyridine reacts similarly.

Silicone grease

Sharpe, A. G. *et al.*, *J. Chem. Soc.*, 1948, 2136

As it reacts explosively in bulk, the amount of silicone grease used on ground joints must be minimal.

Solvents

1. Sharpe, A. G. *et al.*, *J. Chem. Soc.*, 1948, 2135
2. Simons, J. H., *Inorg. Synth.*, 1950, **3**, 185

Bromine trifluoride explodes on contact with acetone or ether [1], and the frozen solid at –80°C reacts violently with toluene at –80°C.

*See* Halogens, etc., above

Tin dichloride

Mellor, 1956, Vol. 2, Suppl. 1, 165

Contact causes ignition.

Uranium,
Uranium hexafluoride

Johnson, R. *et al.*, *6th Nucl. Eng. Sci. Conf.*, *New York, 1960*, Reprint Paper No. 23

Uranium may ignite or explode during dissolution in bromine

trifluoride, particularly when high concentrations of the hexafluoride are present. Causative factors are identified.

*See* Halogens, etc., above

Water

Mellor, 1941, Vol. 2, 113

Interaction is violent, oxygen being evolved.

*See other* INTERHALOGENS

**BROMINE PENTAFLUORIDE** **$BrF_5$**

Acids,
or Halogens,
or Metal halides,
or Metals,
or Non-metals,
or Oxides

Mellor, 1956, Vol. 2, Suppl. 1, 172

Sidgwick, 1950, 1158

Contact with the following at ambient or slightly elevated temperatures is violent, ignition often occurring: strong nitric and sulphuric acids; chlorine (explodes on heating), iodine; ammonium chloride, potassium iodide; antimony, arsenic, boron powder, selenium, tellurium; aluminium powder, barium, bismuth, cobalt powder, chromium, iridium powder, iron powder, lithium powder, manganese, molybdenum, nickel powder, rhodium powder, tungsten, zinc; charcoal, red phosphorus, sulphur; arsenic pentoxide, boron trioxide, calcium oxide, carbon monoxide, chromium trioxide, iodine pentoxide, magnesium oxide, molybdenum trioxide, phosphorus pentoxide, sulphur dioxide, tungsten trioxide.

Hydrogen-containing materials

1. Mellor, 1956, Vol. 2, Suppl. 1., 172
2. Braker, 1971, 49

Contact with the following materials, containing combined hydrogen, is likely to cause fire or explosion : acetic acid, ammonia, benzene, ethanol, hydrogen, hydrogen sulphide, methane; cork, grease, paper, wax, etc. The carbon content further contributes to the reactivity observed [1] . Chloromethane reacts with explosive violence [2] .

Perchloryl perchlorate

*See* PERCHLORYL PERCHLORATE (DICHLORINE HEPTAOXIDE), $Cl_2O_7$: Bromine pentafluoride

Water

Sidgwick, 1950, 1158

Contact with water causes a violent reaction or explosion, oxygen being evolved.

*See other* INTERHALOGENS

**POTASSIUM HEXAFLUOROBROMATE(1–)** $K[BrF_6]$ $BrF_6K$

*See* METAL POLYHALOHALOGENATES

**RUBIDIUM HEXAFLUOROBROMATE(1–)** $Rb[BrF_6]$ $BrF_6Rb$

*See* METAL POLYHALOHALOGENATES

**BROMOGERMANE** $BrGeH_3$

Preparative hazard

*See* BROMINE, $Br_2$ : Germane

**HYDROGEN BROMIDE** HBr BrH

Preparative hazard

*See* BROMINE, $Br_2$ : Phosphorus

1,2-Diaminoethaneamminediperoxochromium(IV)

*See* 1,2-DIAMINOETHANEAMMINEDIPEROXOCHROMIUM(IV), $C_2H_{11}CrN_3O_4$ : Hydrogen bromide

Fluorine

*See* FLUORINE, $F_2$ : Hydrogen halides

Ozone

*See* OZONE, $O_3$ : Hydrogen bromide

**BROMIC ACID** $HBrO_3$ $BrHO_3$

In contact with oxidisable materials, reactions are similar to those of the metal bromates.

*See* METAL HALOGENATES

**BROMOAMINE** $BrNH_2$ $BrH_2N$

Jander, J. *et al.*, *Z. Anorg. Chem.*, 1958, **296**, 117

The isolated material decomposes violently at –70°C, while an ethereal solution is stable for a few hours at that temperature.

*See other* *N*-HALOGEN COMPOUNDS

## BROMOSILANE $BrSiH_3$ $BrH_3Si$

Ward, L. G. L., *Inorg. Synth.*, 1968, **11**, 161

Ignites in air (gas above 2°C).

*See other* HALOSILANES

## AMMONIUM BROMIDE $NH_4Br$ $BrH_4N$

Bromine trifluoride

*See* BROMINE TRIFLUORIDE, $BrF_3$: Ammonium halides

## AMMONIUM BROMATE $NH_4BrO_3$ $BrH_4NO_3$

1. Sorbe, 1968, 129
2. Shidlovskii, A. A., *et al.*, *Chem. Abs.*, 1968, **69**, 78870c

It is a combustible and explosive salt which is very friction-sensitive [1], and may explode spontaneously [2].

*See also* LEAD BROMATE, $Br_2O_6Pb$

*See other* OXOSALTS OF NITROGENOUS BASES

## POLY(DIMERCURYIMMONIUM BROMATE) $(BrHg_2NO_3)_n$

$(Hg{=}N^+{=}Hg\ BrO_3^-)_n$

Sorbe, 1968, 97

Highly explosive.

*See other* POLY(DIMERCURYIMMONIUM) COMPOUNDS

## POTASSIUM BROMATE $KBrO_3$ $BrKO_3$

Disulphur dibromide

*See* DISULPHUR DIBROMIDE, $Br_2S_2$: Oxidants

Non-metals

1. Taradoire, F., *Bull. Soc. Chim. Fr.*, 1945, **12**, 94, 466
2. Pascal, 1961, Vol. 6, 440

Mixtures with sulphur are unstable, and may ignite some hours after preparation, depending on the state of subdivision and atmospheric humidity [1]. Selenium reacts violently witb aqueous solutions of the oxidant.

*See other* METAL OXOHALOGENATES

## BROMINE AZIDE

$BrN_3$

Alone,
or Arsenic,
or Metals,
or Phosphorus

Mellor, 1940, Vol. 8, 336

The solid, liquid and vapour are all very shock-sensitive, and the liquid explodes on contact with arsenic; sodium, silver foil; or phosphorus. Concentrated solutions in organic solvents may explode on shaking.

*See other* HALOGEN AZIDES

## SODIUM BROMATE

$NaBrO_3$ $BrNaO_3$

Fluorine

*See* FLUORINE, $F_2$ : Sodium bromate

Grease

*MCA Case History No. 874*

A bearing assembly from a sodium bromate crusher had been degreased at 120°C, and, while still hot, the sleeve was hammered to free it. The assembly exploded violently probably because of the presence of a hot mixture of sodium bromate and a grease component (possibly a sulphurised derivative). It is known that mixtures of bromates and organic or sulphurous matter are heat- and friction-sensitive.

*See other* METAL OXOHALOGENATES

## BROMINE DIOXIDE

$BrO_2$

Brauer, 1963, Vol. 1, 306

Unstable unless stored at low temperatures, it may explode if heated rapidly.

*See other* HALOGEN OXIDES

## BROMINE TRIOXIDE

$BrO_3$

1. Lewis, B. *et al.*, *Z. Elektrochem.*, 1929, **35**, 648–652
2. Pflugmacher, A. *et al.*, *Z. Anorg. Chem.*, 1955, **279**, 313

The solid produced at –5°C by interaction of bromine and ozone is only stable at –80°C or in presence of ozone, and decomposition may be violently explosive in presence of trace impurities [1]. The structure may be the dimeric bromyl perbromate, analogous to $Cl_2O_6$ [2].

*See other* HALOGEN OXIDES

## THALLIUM BROMATE $BrO_3Tl$

Pascal, 1961, Vol. 6, 950

It decomposes explosively around 140°C.

*See other* HEAVY METAL DERIVATIVES
METAL OXOHALOGENATES

## BROMINE $Br_2$

*MCA SD-49,* 1968

Acetone

Levene, P. A., *Org. Synth.,* 1943, Coll. Vol. 2, 89

During bromination of acetone to bromoacetone, presence of a large excess of bromine must be avoided to prevent a sudden very violent reaction.

Acrylonitrile

*See* ACRYLONITRILE, $C_3H_3N$:Bromine

Ammonia

Mellor, 1967, Vol. 8, Suppl. 2, 417

Interaction at normal or elevated temperatures, followed by cooling to –95°C, gives an explosive red oil.

*See* NITROGEN TRIBROMIDE HEXAAMMONIATE, $Br_3N{\cdot}H_{18}N_6$

Boron

*See* BORON, B: Halogens

Bromine trifluoride

*See* BROMINE TRIFLUORIDE, $BrF_3$ : Halogens, etc.

Carbonyl compounds

*MCA SD-49*, 1968

Organic compounds containing active hydrogen atoms adjacent to a carbonyl group (aldehydes, ketones, carboxylic acids) may react violently in unmoderated contact with bromine.

Copper(I) hydride

*See* COPPER(I) HYDRIDE, CuH: Halogens

Diethyl ether

Tucker, H., private comm., 1972

Shortly after adding bromine to ether the solution erupted violently (or exploded softly).

*See also* CHLORINE, $Cl_2$ : Diethyl ether

Diethylzinc
Houben-Weyl, 1973, Vol. 13.2a, 757
Interaction without diluents may produce dangerous explosions.
Even with diluents (ether), interaction of dialkylzincs and halogens is initially violent at 0 to –20°C.

*N,N*-Dimethylformamide
Tayim, H. A. *et al.*, *Chem. & Ind.*, 1973, 347
Interaction is extremely exothermic, and under confinement in an autoclave the internal temperature and pressure exceeded 100°C and 135 bar, causing failure of the bursting disc. The product of interaction is hydroxymethylenedimethylammonium bromide, and the explosive decomposition may have involved formation of *N*-bromodimethylamine, carbon monoxide and hydrogen bromide.
*See other* *N*-HALOGEN COMPOUNDS

Ethanol,
Phosphorus
Read, C. W. W., *School Sci. Rev.*, 1940, **21**(83), 967
Interaction of ethanol, phosphorus and bromine to give bromoethane is considered too dangerous for a school experiment.

Fluorine
*See* FLUORINE, $F_2$ : Halogens

Germane
Swiniarski, M. F. *et al.*, *Inorg. Synth.*, 1974, **15**, 157–160
During the preparation of mono- or di-bromogermane, either the scale of operation or the rate of bromine addition must be closely controlled to prevent explosive reaction occurring.
*See* Non-metal hydrides, below

Hydrogen
Mellor, 1956, Vol. 2, Suppl. 1, 707
Combination is explosive under appropriate pressure and temperature conditions.

Metal acetylides and carbides
Several of the mono- and di-alkali metal acetylides and copper acetylides ignite at ambient temperature or on slight warming, with either liquid or vapour. The alkaline earth, iron, uranium and zirconium carbides ignite in the vapour on heating.
*See* TRIIRON CARBIDE, $CFe_3$ : Halogens
DICAESIUM ACETYLIDE, $C_2Cs_2$ : Halogens
COPPER ACETYLIDE, $C_2Cu$: Halogens
DICOPPER(I) ACETYLIDE, $C_2Cu_2$ : Halogens

CAESIUM ACETYLIDE, $C_2HCs$
RUBIDIUM ACETYLIDE, $C_2HRb$
DILITHIUM ACETYLIDE, $C_2Li_2$ : Halogens
DIRUBIDIUM ACETYLIDE, $C_2Rb_2$ : Halogens
STRONTIUM ACETYLIDE, $C_2Sr$: Halogens
URANIUM DICARBIDE, $C_2U$: Halogens
ZIRCONIUM DICARBIDE, $C_2Zr$: Halogens

Metal azides
Mellor, 1940, Vol. 8, 336
Nitrogen-diluted bromine vapour passed over silver or sodium azide formed bromine azide, and often caused explosions.

Metals
1. Staudinger, H., *Z. Elektrochem.*, 1925, **31**, 549
2. Mellor, 1941, Vol. 2, 469; 1963, Vol. 2, Suppl. 2.2, 1563, 2174
3. *MCA SD-49*, 1968
4. Mellor, 1941, Vol. 7, 260
5. Mellor, 1939, Vol. 3, 379

Lithium is stable in contact with dry bromine, but heavy impact will initiate explosion, while sodium in contact with bromine needs only moderate impact for initiation [1]. Potassium ignites in bromine vapour and explodes violently in contact with liquid bromine, and rubidium ignites in bromine vapour [2]. Aluminium, mercury and titanium react violently with dry bromine [3].

Warm germanium ignites in bromine vapour [4], and antimony ignites in bromine vapour and reacts explosively with the liquid halogen [5].

*See* GALLIUM, Ga: Halogens

Methanol
1. Muir, G. D., *Chem. Brit.*, 1972, **8**, 136
2. Bush, E. L., private comm., 1968

Reaction may be vigorously exothermic. A mixture of bromine (9 ml) and methanol (15 ml) boiled in 2 min and in a previous incident such a mixture had erupted from a measuring cylinder [1]. The exotherm with industrial ethanol (containing 5% methanol) is much greater, and addition of 10 ml of bromine to 40 ml of IMS rapidly causes violent boiling [2].

Non-metal hydrides
1. Stock, A. *et al.*, *Ber.*, 1917, **50**, 1739
2. Sujishi, S. *et al.*, *J. Amer. Chem. Soc.*, 1954, **76**, 4631
3. Geisler, T. C. *et al.*, *Inorg. Chem.*, 1972, **11**, 1710
4. Merck, 1968, 823

Interaction of silane and its homologues with bromine at ambient temperature is explosively violent [1] and temperatures of below

$-30°C$ are necessary to avoid ignition of the reactants [2]. Ignition of disilane at $-95°C$ and of germane at $-112°C$ emphasises the need for good mixing to dissipate the large exotherm [3].

Phosphine reacts violently with bromine at ambient temperature [4].

*See also* ETHYLPHOSPHINE, $C_2H_7P$: Halogens
PHOSPHINE, $H_3P$: Halogens

Oxygen difluoride

*See* OXYGEN DIFLUORIDE, $F_2O$: Halogens

Ozone

*See* OZONE, $O_3$: Bromine

Phosphorus

1. Bandar, L. S. *et al., Zh. Prikl. Khim.*, 1966, **39**, 2304
2. 'Leaflet No. 2', Inst. of Chem., London, 1939

During preparation of hydrogen bromide by addition of bromine to a suspension of red phosphorus in water, the latter must be freshly prepared to avoid possibility of explosion. This is due to formation of peroxides in the suspension on standing and subsequent thermal decomposition [1]. In the earlier description of such an explosion, action of bromine on boiling tetralin was preferred to generate hydrogen bromide [2], which is now also available in cylinders.

*See also* PHOSPHORUS, P: Halogens

Rubber

Pascal, 1960, Vol. 16.1, 371

Bromine reacts violently in contact with natural rubber, but more slowly with some synthetic rubbers.

Sodium hydroxide

*MCA Case History No. 1636*

A bucket containing 25% sodium hydroxide solution was used to catch and neutralise bromine dripping from a leak. Lack of stirring allowed a layer of unreacted bromine to form below the alkali. Many hours later, a violent eruption occurred when the layers were disturbed during disposal operations. Continuous stirring is essential to prevent stratification of slowly reacting mutually insoluble liquids.

Tetracarbonylnickel

*See* TETRACARBONYLNICKEL, $C_4NiO_4$: Bromine

Tetraselenium tetranitride

*See* TETRASELENIUM TETRANITRIDE, $N_4Se_4$

Trialkylboranes

Coates, 1967, Vol. 1, 199

The lower homologues tend to ignite in bromine or chlorine.

Trimethylamine

Böhme, H. *et al.*, *Chem. Ber.*, 1951, **84**, 170–181

The 1:1 adduct (presumably *N*-bromotrimethylammonium bromide) decomposes explosively when heated in a sealed tube.

Trioxygen difluoride

*See* TRIOXYGEN DIFLUORIDE, $F_2O_3$

Tungsten,

Tungsten trioxide

Tillack, J., *Inorg. Synth.*, 1973, **14**, 116–120

During preparation of tungsten(IV) dibromide oxide, appropriate proportions of reactants are heated in an evacuated sealed glass ampoule to 400–450°C. Initially only one end should be heated to prevent excessive pressure bursting the ampoule.

*See other* HALOGENS

## CALCIUM BROMIDE $CaBr_2$ $Br_2Ca$

Potassium

*See* POTASSIUM, K: Metal halides

## COBALT(II) BROMIDE $CoBr_2$ $Br_2Co$

Sodium

*See* SODIUM, Na: Metal halides

## COPPER(I) BROMIDE $Cu_2Br_2$ $Br_2Cu_2$

*tert*-Butyl peroxybenzoate,

Limonene

*See* *tert*-BUTYL PEROXYBENZOATE, $C_{11}H_{14}O_3$: Copper(I) bromide, etc.

## IRON(II) BROMIDE $FeBr_2$ $Br_2Fe$

Potassium

*See* POTASSIUM, K: Metal halides

Sodium

*See* SODIUM, Na: Metal halides

**DIBROMOGERMANE** **$Br_2GeH_2$**

Preparative hazard

*See* BROMINE, $Br_2$: Germane

***N,N*-BIS(BROMOMERCURIO)HYDRAZINE** **$Br_2H_2Hg_2N_2$**

**$(BrHg)_2NNH_2$**

Hofmann, K. A. *et al.*, *Ann.*, 1899, **305**, 217

An explosive compound.

*See other* *N*-METAL DERIVATIVES

**MERCURY(II) BROMIDE** **$HgBr_2$** **$Br_2Hg$**

Indium

Clark, R. J. *et al.*, *Inorg. Synth.*, 1963, **7**, 19–20

Interaction at 350°C is so vigorous that it is unsafe to increase the scale of this preparation of indium bromide.

*See other* METAL HALIDES

**MERCURY(I) BROMATE** **$Hg_2(BrO_3)_2$** **$Br_2Hg_2O_6$**

Hydrogen sulphide

Pascal, 1960, Vol. 13.1, 1004

Contact of the gas with the solid oxidant causes ignition.

*See other* METAL OXOHALOGENATES

**SELENINYL BROMIDE** **$Se(O)Br_2$** **$Br_2OSe$**

Metals

Mellor, 1947, Vol. 10, 912

Sodium and potassium react explosively (the latter more violently), and zinc dust ignites, in contact with the liquid bromide.

*See* SODIUM, Na: Non-metal halides

Phosphorus

Mellor, 1947, Vol. 10, 912

Red phosphorus ignites, and white phosphorus explodes in contact with the liquid bromide.
*See other* NON-METAL HALIDES

**LEAD BROMATE** **$Pb(BrO_3)_2$** **$Br_2O_6Pb$**

Sidgwick, 1950, 1227
An explosive salt.
*See also* AMMONIUM BROMATE, $BrH_4NO_3$
LEAD ACETATE–LEAD BROMATE, $C_4H_6BrO_4Pb \cdot Br_2O_6Pb$
*See other* METAL OXOHALOGENATES

**ZINC BROMATE** **$Zn(BrO_3)_2$** **$Br_2O_6Zn$**

*See* METAL HALOGENATES

**SULPHUR DIBROMIDE** **$SBr_2$** **$Br_2S$**

Nitric acid
*See* NITRIC ACID, $HNO_3$: Sulphur halides

Sodium
*See* SODIUM, Na: Non-metal halides

**SILICON DIBROMIDE SULPHIDE** **$SiBr_2S$** **$Br_2SSi$**

Water
Bailar, 1973, Vol. 1, 1415
Hydrolysis of the sulphide is explosive.
*See related* NON-METAL HALIDES

**DISULPHUR DIBROMIDE** **BrSSBr** **$Br_2S_2$**

Metals
Mellor, 1947, Vol. 10, 652
Pascal, 1960, Vol. 13.2, 1162
Thin sections of potassium or sodium usually ignite and incandesce in the liquid bromide. Iron at about 650°C ignites and incandesces in the vapour.

Interaction with finely divided aluminium or antimony is violent.

Oxidants

Taradoire, F., *Bull. Soc. Chim. Fr.*, 1945, **12**, 95

Interaction with moist barium bromate is very violent, and mixtures with potassium bromate and water (3–4%) ignite at 20°C. In absence of water, ignition occurs at 125°C. Silver bromate also leads to deflagration.

*See* SILVER BROMATE, $AgBrO_3$ : Sulphur compounds

*See also* NITRIC ACID, $HNO_3$ : Non-metal halides

Phosphorus

*See* PHOSPHORUS, P: Non-metal halides

*See other* NON-METAL HALIDES

**POLY(DIBROMOSILYLENE)** **$(Br_2Si)_n$**

Oxidants

Brauer, 1963, Vol. 1, 688

Ignites in air at 120°C and reacts explosively with oxidants (e.g. nitric acid).

*See* HALOSILANES

*See other* NON-METAL HALIDES

**TITANIUM DIBROMIDE** **$TiBr_2$** **$Br_2Ti$**

Gibson, 1969, 60–61

It may ignite in moist air.

*See other* PYROPHORIC MATERIALS

**ZIRCONIUM DIBROMIDE** **$ZrBr_2$** **$Br_2Zr$**

Pascal, 1963, Vol. 9, 558

It ignites in air, reacts violently with water and incandesces in steam.

*See other* PYROPHORIC MATERIALS

**IRON(III) BROMIDE** **$FeBr_3$** **$Br_3Fe$**

Potassium

*See* POTASSIUM, K: Metal halides

Sodium

*See* SODIUM, Na: Metal halides

**TRIBROMOSILANE** **$Br_3SiH$** **$Br_3HSi$**

Schumb, W. C., *Inorg. Synth.*, 1939, **1**, 42

It usually ignites when poured in air.
*See* HALOSILANES

## NITROGEN TRIBROMIDE HEXAAMMONIATE $Br_3N \cdot H_{18}N_6$

$NBr_3 \cdot 6NH_3$

Mellor, 1967, Vol. 8, Suppl. 2, 417; 1940, Vol. 8, 605
The compound ($NBr_3 \cdot 6NH_3$) prepared by condensation of its vapour at −95°C explodes suddenly at −67°C. Prepared in another way, it is stable under water but explodes violently in contact with phosphorus or arsenic.
*See other N*-HALOGEN COMPOUNDS

## NITROSYL TRIBROMIDE $N(O)Br_3$ $Br_3NO$

Sodium–antimony alloy
Mellor, 1940, Vol. 8, 621
Powdered sodium antimonide ignites when dropped into the vapour.
*See other N*-HALOGEN COMPOUNDS

## VANADIUM TRIBROMIDE OXIDE $VOBr_3$ $Br_3OV$

Bailar, 1973, Vol. 3, 508
The bromide (and corresponding chloride) is violently hygroscopic.
*See related* METAL HALIDES

## PHOSPHORUS TRIBROMIDE $PBr_3$ $Br_3P$

Calcium hydroxide,
Sodium carbonate
Seager, J. F., *Chem. Brit.*, 1976, **12**, 105
During disposal of the tribromide by a recommended method involving slow addition to a mixture of soda ash and dry slaked lime, a violent reaction, accompanied by flame, occurred a few seconds after the first drop. Cautious addition of the bromide to a large volume of ice water is suggested.

Oxygen
*See* OXYGEN (Gas), $O_2$ : Phosphorus tribromide

Phenylpropanol
Taylor, D. A. H., *Chem. Brit.*, 1974, **10**, 101–102
During dropwise addition of the bromide to the liquid alcohol, the mechanical stirrer stopped, presumably allowing a layer of the dense

tribromide to accumulate below the alcohol. Later manual shaking caused an explosion, probably owing to sudden release of gas on mixing.

Potassium

*See* POTASSIUM, K: Non-metal halides

Sodium,
Water

*See* SODIUM, Na: Non-metal halides

1,1,1-Trishydroxymethylmethane

1. Derfer, J. M. *et al.*, *J. Amer. Chem. Soc.*, 1949, **71**, 175
2. Farber, S. *et al.*, *Synth. Comm.*, 1974, **4**, 243

Interaction to form the corresponding tribromide is extremely hazardous, even using previously specified precautions [1]. Several fires occurred in the effluent gases, and reaction residues ignited on exposure to air [2], doubtless owing to phosphine derivatives.

Water

Mellor, 1940, Vol. 8, 1032

Interaction with warm water is very rapid and may be violent with limited quantities.

*See other* NON-METAL HALIDES

## TELLURIUM TETRABROMIDE $TeBr_4$ $Br_4Te$

Ammonia

Sorbe, 1968, 154

Interaction gives a mixture of tritellurium tetranitride and tellurium bromide nitride, which explodes on heating.

*See* TRITELLURIUM TETRANITRIDE, $N_4Te_3$

*See other* METAL HALIDES

## TUNGSTEN AZIDE PENTABROMIDE $W(N_3)Br_5$ $Br_5N_3W$

Extremely explosive.

*See other* METAL AZIDE HALIDES

## CARBON C

Air

1. *Fire Prot. Assoc. J.*, 1964, 337
2. Cameron, A. *et al.*, *J. Appl. Chem.*, 1972, **22**, 1007
3. Zav'yalov, A. N. *et al.*, *Chem. Abs.*, 1976, **84**, 166521e

Activated carbon is a potential fire hazard because of its very high surface area and adsorptive capacity. Freshly prepared material may heat spontaneously in air, and presence of water accelerates this. Spontaneous heating and ignition may occur if contamination by drying oils or oxidising agents occurs [1]. The spontaneous heating effect has now been related to the composition and method of preparation of activated carbon, and the relative hazards may be readily assessed [2].

Free radicals present in charcoal are responsible for auto-ignition effects, and charcoal may be stabilised for storage and transport without moistening by treatment with hot air at 50°C [3]. For generation of pyrophoric carbon,

*See* BARIUM ACETYLIDE, $C_2Ba$
*See also* PETROLEUM COKE

Alkali metals

Bailar, 1973, Vol. 1, 443

Graphite in contact with liquid potassium, rubidium or caesium at 300°C gives intercalation compounds, $C_8M$, which ignite in air and may react explosively with water.

*See* POTASSIUM, K: Carbon
SODIUM, Na: Non-metals

Iron(II) oxide,
Oxygen (liquid)

*See* OXYGEN (Liquid), $O_2$: Carbon, Iron(II) oxide

Metal salts

*MCA Case History No. 1094*

Dry metal-impregnated charcoal catalyst was being added from a polythene bag to an aqueous solution under nitrogen. Static so generated ignited the charcoal dust and caused a flash fire. The risk was eliminated by adding a slurry of catalyst from a metal container.

*See* COBALT(II) NITRATE, $CoN_2O_6$: Carbon

Nitrogen oxide,
Potassium hydrogentartrate

*See* NITROGEN OXIDE ('NITRIC OXIDE'), NO: Carbon, Potassium hydrogentartrate

Oxidants

Carbon has been frequently involved in hazardous reactions, particularly finely divided or high-porosity forms exhibiting a high ratio of surface area to mass (up to 2000 $m^2/g$). It then functions as an unusually active fuel which possesses adsorptive and catalytic properties to accelerate the rate of energy release involved in combustion reactions with virtually any oxidant.

Less active forms of carbon will ignite or explode on suitably intimate contact with oxygen, oxides, peroxides, oxosalts, halogens, interhalogens and other oxidising species. Individual combinations are:

SILVER NITRATE, $AgNO_3$ : Non-metals
BROMINE TRIFLUORIDE, $BrF_3$ : Halogens, etc.
BROMINE PENTAFLUORIDE, $BrF_5$ : Acids, etc.
PEROXYFORMIC ACID, $CH_2O_3$ : Non-metals
CHLORINE TRIFLUORIDE, $ClF_3$ : Metals, etc.
AMMONIUM PERCHLORATE, $ClH_4NO_4$ : Carbon
DICHLORINE OXIDE, $Cl_2O$: Carbon
: Oxidisable materials
COBALT(II) NITRATE, $CoN_2O_6$ : Carbon
FLUORINE, $F_2$ : Non-metals
OXYGEN DIFLUORIDE, $F_2O$: Non-metals
TRIOXYGEN DIFLUORIDE, $F_2O_3$
NITROGEN TRIFLUORIDE, $F_3N$: Charcoal
IODINE HEPTAFLUORIDE, $F_7I$: Carbon
IODINE(V) OXIDE, $I_2O_5$ : Non-metals
POTASSIUM PERMANGANATE, $KMnO_4$ : Non-metals
POTASSIUM NITRATE, $KNO_3$ : Non-metals
POTASSIUM DIOXIDE, $KO_2$ : Carbon
SODIUM NITRATE, $NNaO_3$ : Non-metals
NITROGEN OXIDE ('NITRIC OXIDE'), NO: Non-metals
ZINC NITRATE, $N_2O_6Zn$: Carbon
SODIUM PEROXIDE, $Na_2O_2$ : Non-metals
SODIUM SULPHIDE, $Na_2S$: Carbon
OXYGEN (Liquid), $O_2$ : Carbon

Unsaturated oils

Bahme, C. W., *NFPA Quart.*, 1952, **45**, 431
von Schwartz, 1918, 326

Unsaturated (drying) oils, like linseed oil, etc., will rapidly heat and ignite when distributed on active carbon, owing to enormous increase in surface area of the oil exposed to air. A similar, but slower, effect occurs on fibrous material such as cotton waste.

*See other* NON-METALS

## SILVER CYANIDE AgCN CAgN

Phosphorus tricyanide

*See* PHOSPHORUS TRICYANIDE, $C_3N_3P$

Fluorine

*See* FLUORINE, $F_2$ : Metal salts

**SILVER CYANATE** **AgOCN** **CAgNO**

Sorbe, 1968, 125

Explodes on heating

*See other* HEAVY METAL DERIVATIVES

**SILVER FULMINATE** **AgC≡N→O** **CAgNO**

Urbanski, 1967, Vol. 3, 157

Dimeric silver fulminate, readily formed from silver or its salts, nitric acid and ethanol, is a much more sensitive and powerful detonator than mercuric fulminate.

*See other* METAL FULMINATES

Hydrogen sulphide

Boettger, A., *J. Prakt. Chem.*, 1868, **103**, 309

Contact with hydrogen sulphide at ambient temperature initiates violent explosion of the fulminate.

*See other* HEAVY METAL DERIVATIVES

**SILVER TRINITROMETHANIDE** **$AgC(NO_2)_3$** **$CAgN_3O_6$**

Witucki, E. F. *et al.*, *J. Org. Chem.*, 1972, **37**, 152

The explosive silver salt may be replaced with advantage by the potassium salt in preparation of 1,1,1-trinitroalkanes.

*See other* HEAVY METAL DERIVATIVES

**SILVER AZIDODITHIOFORMATE** **$AgSC(S)N_3$** **$CAgN_3S_2$**

Sorbe, 1968, 126

The tetrahydrated salt explodes on the slightest friction.

*See* AZIDODITHIOFORMIC ACID, $CHN_3S_2$

*See other* HEAVY METAL DERIVATIVES

**SILVER TRICHLOROMETHANEPHOSPHONATE(2–)** **$CAg_2Cl_3O_3P$**

**$(AgO)_2P(O)CCl_3$**

Yakubovich, A. Ya. *et al.*, *Chem. Abs.*, 1953, **47**, 2685i

It explodes on heating.

*See other* HEAVY METAL DERIVATIVES

**DISILVER CYANAMIDE** **$Ag_2NCN$** **$CAg_2N_2$**

Chrétien, A. *et al.*, *Compt. Rend.*, 1951, **232**, 1114

Deb. S. K. *et al.*, *Trans. Farad. Soc.*, 1959, **55**, 106–113

Cradock, S., *Inorg. Synth* , 1974, **15**, 167

During pyrolysis to silver (via silver dicyanamide), initial heating must be slow to avoid explosion.

High-intensity illumination will also cause explosive decomposition of a confined sample. Safety precautions for preparation and subsequent use of the explosive salt are detailed in the third reference.

*See other* *N*-METAL DERIVATIVES

**DISILVER DIAZOMETHANEDIIDE** **$Ag_2CN_2$** **$CAg_2N_2$**

Blues, E. T. *et al.*, *J. Chem. Soc., Chem. Comm.*, 1974, 466–467

Both the disilver derivative and its precursory dipyridine complex are highly explosive and extremely shock-sensitive when dry.

*See other* HEAVY METAL DERIVATIVES

**GOLD(I) CYANIDE** **AuCN** **CAuN**

Magnesium

*See* MAGNESIUM, Mg: Metal cyanides

**CARBON TETRABORIDE** **$CB_4$**

Chlorine trifluoride

*See* CHLORINE TRIFLUORIDE, $ClF_3$: Boron-containing materials

**BROMOTRICHLOROMETHANE** **$BrCCl_3$** **$CBrCl_3$**

Ethylene

*See* ETHYLENE, $C_2H_4$: Bromotrichloromethane

**BROMOTRIFLUOROMETHANE** **$BrCF_3$** **$CBrF_3$**

Aluminium

*See* ALUMINIUM, Al: Halocarbons

**TRIBROMONITROMETHANE** **$Br_3CNO_2$** **$CBr_3NO_2$**

Sorbe, 1968, 40

It is used as an explosive.

*See other* *C*-NITRO COMPOUNDS

## CARBON TETRABROMIDE $CBr_4$

Hexacyclohexyldilead

*See* HEXACYCLOHEXYLDILEAD, $C_{36}H_{66}Pb_2$ : Halocarbons

Lithium

*See* LITHIUM, Li: Halocarbons

## CALCIUM CYANAMIDE CaNCN $CCaN_2$

Water

Pieri, M., *Chem. Abs.*, 1952, **46**, 8335i

Absorption of water during handling or storage of technical calcium cyanamide may cause explosions, owing to liberation of acetylene from the calcium carbide content (up to 2%). Precautions are discussed.

*See other N*-METAL DERIVATIVES

## CALCIUM CARBONATE $CaCO_3$ $CCaO_3$

Fluorine

*See* FLUORINE, $F_2$ : Metal salts

## CHLOROPEROXYTRIFLUOROMETHANE $CClF_3O_2$

$ClOOCF_3$

Tetrafluoroethylene

Ratcliffe, C. T. *et al.*, *J. Amer. Chem. Soc.*, 1971, **93**, 3887–3888

The peroxy compound initiated explosive polymerisation of tetrafluoroethylene when a mixture prepared at $-196°$ warmed to $-110°C$.

*See related* HYPOHALITES

## TRIFLUOROMETHYL PERCHLORATE $F_3CClO_4$ $CClF_3O_4$

1. Schack, C. J. *et al.*, *Inorg. Chem.*, 1974, **13**, 2375
2. Schack, C. J. *et al.*, *Inorg. Nucl. Chem. Lett.*, 1974, **10**, 449

Though apparently not explosively unstable [1], its synthesis [2] was occasionally accompanied by deflagrations.

*See other* ALKYL PERCHLORATES

## TRIFLUOROMETHANESULPHENYL CHLORIDE $CClF_3S$

$CF_3SCl$

Chlorine fluorides

*See* BIS(TRIFLUOROMETHYL)SULPHUR DIFLUORIDE, $C_2F_8S$: Chlorine fluorides

Hexafluoroisopropylideneaminolithium

*See* HEXAFLUOROISOPROPYLIDENEAMINOLITHIUM, $C_3F_6LiN$: Non-metal halides

**CYANOGEN CHLORIDE** **CNCl** **CClN**

Preparative hazard

Alfenaar, M. *et al.*, Neth. Appl. 73 07 035, 1974 (*Chem. Abs.*, 1976, **84**, 10497r)

Cyanogen halides may be prepared by electrolysis of hydrogen cyanide or its salts mixed with halide salts. If ammonium chloride is used as the halide salt, precautions to prevent formation of explosive nitrogen trichloride are necessary.

**CHLOROSULPHONYLISOCYANATE** **$ClSO_2NCO$** **$CClNO_3S$**

Water

Graf, R., *Org. Synth.*, 1966, **46**, 25

Interaction is violent.

**DICHLORODIFLUOROMETHANE** **$Cl_2CF_2$** **$CCl_2F_2$**

Aluminium

*See* ALUMINIUM, Al: Halocarbons

Magnesium

*See* MAGNESIUM, Mg: Halocarbons

Water

*See* LIQUEFIED GASES: Water

**PHOSPHORYL DICHLORIDE ISOCYANATE** **$Cl_2P(O)NCO$** **$CCl_2NO_2P$**

Preparative hazard

*See* *N*-CARBOMETHOXYIMINOPHOSPHORYL CHLORIDE, $C_2H_3Cl_3NO_2P$

**DICHLORODINITROMETHANE** **$Cl_2C(NO_2)_2$** **$CCl_2N_2O_4$**

Hocking, M. B. *et al.*, *Chem. & Ind.*, 1976, 952

It exploded during attempted distillation at atmospheric pressure, but was distilled uneventfully at 31°C/13 mbar.

*See other* POLYNITROALKYL COMPOUNDS

## CARBONYL DICHLORIDE (PHOSGENE) $COCl_2$ $CCl_2O$

*MCA SD-95*, 1967

Hexafluoroisopropylideneaminolithium

*See* HEXAFLUOROISOPROPYLIDENEAMINOLITHIUM, $C_3F_6LiN$: Non-metal halides

Potassium

*See* POTASSIUM, K: Non-metal halides

## TRICHLOROFLUOROMETHANE $CCl_3F$

Metals

*See* ALUMINIUM, Al: Halocarbons
BARIUM, Ba: Halocarbons
LITHIUM, Li: Halocarbons

*See other* METALS: Halocarbons

## TRICHLORONITROMETHANE (CHLOROPICRIN) $CCl_3NO_2$

Anon., *Chem. Eng. News.*, 1972, **50**(38), 13

Tests have shown that, above a critical volume, bulk containers of chloropicrin can be shocked into detonation. Containers below 700 kg content will now be the maximum size as against rail-tanks previously used.

Aniline

Jackson, K. E., *Chem. Rev.*, 1934, **14**, 269

Reaction at 145°C with excess aniline is violent.

1-Bromo-2-propyne

*BCISC Quart. Safety Summ.*, 1968, **39**, 12

An insecticidal mixture in a rail tanker exploded with great violence during pump-transfer operations, possibly owing to the pump running dry and overheating. Both components of the mixture are in fact explosive and the mixture was also shock- and heat-sensitive.

*See other* ACETYLENIC COMPOUNDS

Sodium hydroxide

Scholz, S., *Explosivstoffe*, 1963, **11**, 159, 181

During destruction of chemical warfare ammunition, pierced shells containing chloropicrin reacted violently with alcoholic sodium hydroxide.

Sodium methoxide

Ramsey, B. G. *et al.*, *J. Amer. Chem. Soc.*, 1966, 88, 3059

During addition of the nitro compound in methanol to sodium methoxide solution, the temperature must not be allowed to fall much below 50°C. If this happens, excess nitro compound will accumulate and cause a violent and dangerous exotherm.

*See other* *C*-NITRO COMPOUNDS

## CARBON TETRACHLORIDE $CCl_4$

*MCA SD-3,* 1963

Aluminium trichloride,
Triethylaluminium

*See* TRIETHYLDIALUMINIUM TRICHLORIDE, $C_6H_{15}Al_2Cl_3$: Carbon tetrachloride

Boranes

*See* BORANES: Carbon tetrachloride

Calcium disilicide

*See* CALCIUM DISILICIDE, $CaSi_2$ : Carbon tetrachloride

Chlorine trifluoride

*See* CHLORINE TRIFLUORIDE, $ClF_3$ : Carbon tetrachloride

Decaborane(14)

*See* DECABORANE(14), $B_{10}H_{14}$: Ethers, etc.

Dibenzoyl peroxide,
Ethylene

*See* ETHYLENE, $C_2H_4$ : Carbon tetrachloride
DIBENZOYL PEROXIDE, $C_{14}H_{10}O_4$ : Carbon tetrachloride, Ethylene

*See also* WAX FIRE

*N,N*-Dimethylformamide

'DMF Brochure', Billingham, ICI, 1965

There is a potentially dangerous reaction of carbon tetrachloride with dimethylformamide in presence of iron. The same occurs with 1,2,3,4,5,6-hexachlorocyclohexane, but not with dichloromethane or 1,2-dichloroethane under the same conditions.

*See* *N,N*-DIMETHYLACETAMIDE, $C_4H_9NO$: Halogenated compounds

Dinitrogen tetraoxide

*See* DINITROGEN TETRAOXIDE (NITROGEN DIOXIDE), $N_2O_4$ : Halocarbons

Fluorine

*See* FLUORINE, $F_2$ : Halocarbons

Metals

*See* ALUMINIUM, Al: Halocarbons
BARIUM, Ba: Halocarbons
BERYLLIUM, Be : Halocarbons
POTASSIUM, K: Halocarbons
POTASSIUM–SODIUM ALLOY, K–Na: Halocarbons
SODIUM, Na: Halocarbons
ZINC, Zn: Halocarbons

Potassium *tert*-butoxide

*See* POTASSIUM *tert*-BUTOXIDE, $C_4H_9KO$: Acids, etc.

1,11-Diamino-3,6,9-triazundecane ('tetraethylenepentamine')

1. Hudson, F. L., private comm., 1973
2. Collins, R. F., *Chem. & Ind.*, 1957, 704

A mixture erupted violently one hour after preparation [1]. Interaction (not vigorous) of amines and halocarbons at ambient temperature had been previously recorded [2]. The presence of five basic centres in the amine would be expected to exaggerate exothermic effects.

*See other* HALOCARBONS

## TRICHLOROMETHYL PERCHLORATE $CCl_3OClO_3$ $CCl_4O_4$

Sidgwick, 1950, 1236

An extremely explosive liquid, only capable of preparation in minute amounts.

*See other* ALKYL PERCHLORATES

## TRIAZIDOMETHYLIUM HEXACHLOROANTIMONATE $CCl_6N_9Sb$

$[C(N_3)_3]^+[SbCl_6]^-$

Müller, U. *et al.*, *Angew. Chem.*, 1966, **78**, 825

The salt containing the $C(N_3)_3$ cation is sensitive to shock or rapid heating.

*See other* ORGANIC AZIDES

## DIDEUTERIODIAZOMETHANE $CD_2N_2$

Gassman, P. G. *et al.*, *Org. Synth.*, 1973, **53**, 38–43

The explosive properties will be similar to those of diazomethane, for which precautions are extensively summarised.

*See other* DIAZO COMPOUNDS

**POLY(CARBON MONOFLUORIDE)** **$(CF)_n$**

Preparative hazard

*See* FLUORINE, $F_2$: Graphite

Hydrogen

Bailar, 1973, Vol. 1, 1269

Above 400°C in hydrogen, deflagration and flaming of the polymer occurs, the vigour depending on the fluorine content. Rapid heating to 500°C in an inert atmosphere causes explosive deflagration.

*See other* HALOCARBONS

**CYANOGEN FLUORIDE** **CNF** **CFN**

Hydrogen fluoride

Bailar, 1973, Vol. 1, 1246

Polymerisation of cyanogen fluoride is rapid at ambient temperature and explosive in presence of hydrogen fluoride.

*See other* CYANO COMPOUNDS

**FLUOROTRINITROMETHANE** **$FC(NO_2)_3$** **$CFN_3O_6$**

Preparative hazard

*See* FLUORINE, $F_2$: Trinitromethane

**FLUORODINITROMETHYL AZIDE** **$F(NO_2)_2CN_3$** **$CFN_5O_4$**

Unstable at ambient temperature.

*See other* FLUORODINITROMETHYL COMPOUNDS
ORGANIC AZIDES

**CARBONYL DIFLUORIDE** **$COF_2$** **$CF_2O$**

Hexafluoroisopropylideneaminolithium

*See* HEXAFLUOROISOPROPYLIDENEAMINOLITHIUM, $C_3F_6LiN$: Non-metal halides

**NITROSOTRIFLUOROMETHANE** **$NOCF_3$** **$CF_3NO$**

1. Spaziante, P. M., *Intern. Rev. Sci.: Inorg. Chem. Ser. 1*, 1972, **3**, 141
2. Banks, R. E. *et al.*, *J. Chem. Soc., Perkin Trans., 1*, 1974, 2534–2535

Suggestions of untoward hazards inherent in the preparation of nitrosotrifluoromethane by pyrolysis of trifluoroacetyl nitrite [1] are discounted in the later reference, which gives full details of equipment and procedure that had been used uneventfully during the preceding decade [2].

*See other* NITROSO COMPOUNDS

**TRIFLUOROMETHYL PEROXONITRATE** **$CF_3NO_4$**

**$NO_2OOCF_3$**

Hohorst, F. A. *et al.*, *Inorg. Chem.*, 1974, **13**, 715

A small sample exploded under a hammer blow.

*See other* PEROXYESTERS

**TRIFLUOROMETHANESULPHONYL AZIDE** **$CF_3N_3O_2S$**

**$CF_3SO_2N_3$**

Cavender, C. J. *et al.*, *J. Org. Chem.*, 1972, **37**, 3568

An explosion occurred when the azide separated during its preparation in absence of solvent.

*See other* ACYL AZIDES

**CARBON TETRAFLUORIDE** **$CF_4$**

Aluminium

*See* ALUMINIUM, Al: Halocarbons

**TETRAFLUORODIAZIRIDINE** **$FNCF_2NF$** (ring) **$CF_4N_2$**

Firth, W. C., *J. Org. Chem.*, 1968, **33**, 3489

Explodes on evaporation or condensation at low temperatures. Treat as a powerful explosive and handle on small scale.

*See other N*-HALOGEN COMPOUNDS

**PERFLUOROUREA** **$CO(NF_2)_2$** **$CF_4N_2O$**

Acetonitrile

Fraser, G. W. *et al.*, *Chem. Comm.*, 1966, 532

The solution of the perfluoro compound in acetonitrile prepared at –40°C must not be kept at room temperature, since difluorodiazene is formed.

*See* ACETONITRILE, $C_2H_3N$: *N*-Fluoro compounds

## TRIFLUOROMETHYL HYPOFLUORITE $CF_4O$

$CF_3OF$

Hydrocarbons

Allison, J. A. C. *et al.*, *J. Amer. Chem. Soc.*, 1959, **81**, 1089

In absence of nitrogen as diluent, interaction with acetylene, cyclopropane or ethylene is explosive on mixing.

Hydrogen-containing solvents

1. *Catalogue G-7*, 10, Gainesville (Fl.), Peninsular Chem. Research, 1973
2. Barton, D. H. R. *et al.*, *Chem. Comm.*, 1968, 804

Contact of the extremely reactive compound with hydrogen-containing solvents or conventional plastics tubing, even at –80°C, is undesirable [1]. Fully halogenated solvents are preferred, and some general precautions are described [2].

Lithium

Porter, R. S. *et al.*, *J. Amer. Chem. Soc.*, 1957, **79**, 5625

Interaction set in at about 170°C with a sufficient exotherm to melt the glass container.

Polymers

Barton, D. H. R. *et al.*, *J. Org. Chem.*, 1972, **37**, 329

It is a powerful oxidant and only all-glass apparatus, free of polythene, PVC, rubber or similar elastomers should be used. Appreciable concentrations of the gas in oxidisable materials should be avoided.

Pyridine

Barton, D. H. R. *et al.*, *Chem. Comm.*, 1968, 804 (footnote)

Use of pyridine as acid-acceptor in reactions involving trifluoromethyl hypofluorite is discouraged, as a highly explosive by-product is formed.

*See other* HYPOHALITES

## TRIFLUOROMETHANESULPHINYL FLUORIDE $CF_4OS$

$CF_3S(O)F$

Hexafluoroisopropylideneaminolithium

*See* HEXAFLUOROISOPROPYLIDENEAMINOLITHIUM, $C_3F_6LiN$: Non-metal halides

## DIFLUOROMETHYLENE DIHYPOFLUORITE $CF_4O_2$

$CF_2(OF)_2$

Haloalkenes

Hohorst, F. A. *et al.*, *Inorg. Chem.*, 1968, 7, 624

Attempts to react the fluoro compound with *trans*-dichloroethylene

or tetrafluoroethylene at room temperature in absence of diluent caused violent explosions. The title compound should not be allowed to contact organic or easily oxidised material without adequate precautions.
*See other* HYPOHALITES

**XENON(II) FLUORIDE TRIFLUOROMETHANESULPHONATE** **$FXeOSO_2CF_3$** **$CF_4O_3SXe$**

Wechsberg, M. *et al., Inorg. Chem.*, 1972, **11**, 3066
Unless a deficiency of xenon difluoride was used in the preparation at 0°C or below, the product exploded violently on warming to ambient temperature.
*See other* XENON COMPOUNDS

**3-DIFLUOROAMINO-1,2,3-TRIFLUORODIAZIRIDINE** **$CF_5N_3$**

**$FNC(NF_2)FNF$** (ring)

Firth, W. C., *J. Org. Chem.*, 1968, **33**, 3489
Explodes on evaporation or condensation at low temperatures. Treat as a powerful explosive and handle on small scale.
*See other* *N*-HALOGEN COMPOUNDS

**PENTAFLUOROGUANIDINE** **$FN{=}C(NF_2)_2$** **$CF_5N_3$**

Zollinger, J. L. *et al., J. Org. Chem.*, 1973, **38**, 1070–1071
This, and several of its adducts with alcohols, are shatteringly explosive compounds, frequently exploding during phase changes at low temperatures, or on friction or impact.

Liquid fuels
Scurlock, A. C. *et al.*, US Pat. 3 326 732, 1967
This compound with multiple N–F bonding is useful as an oxidant in propellant technology
*See other* *N*-HALOGEN COMPOUNDS

**BIS(DIFLUOROAMINO)DIFLUOROMETHANE** **$CF_6N_2$**

**$(NF_2)_2CF_2$**

Koshar, R. J. *et al., J. Org. Chem.*, 1966, **31**, 4233
Explosions occurred during handling of this material, especially during phase transitions. Use of protective equipment is recommended for preparation, handling and storage, even on microscale.
*See other* *N*-HALOGEN COMPOUNDS

**IRON CARBIDE** **$Fe_3C$** **$CFe_3$**

Halogens

Mellor, 1946, Vol. 5, 898

Incandesces in chlorine below 100°C, and in bromine at 100°C.

*See other* METAL NON-METALLIDES

**SILVER TETRAZOLIDE** **$\overline{N{=}NNAgN{=}CH}$** **$CHAgN_4$**

Thiele, J., *Ann.*, 1892, **270**, 59

It explodes on heating.

*See other N*-METAL DERIVATIVES

**BROMOFORM** **$CHBr_3$**

Acetone,
Potassium hydroxide

1. Willgerodt, C., *Ber.*, 1881, **14**, 2451
2. Weizmann, C. *et al.*, *J. Amer. Chem. Soc.*, 1948, **70**, 1189

Interaction in presence of powdered potassium hydroxide (or other bases) is violently exothermic, even in presence of diluting solvents [1, 2].

*See also* CHLOROFORM, $CHCl_3$: Acetone, etc.

Metals

*See* POTASSIUM, K: Halocarbons
LITHIUM, Li: Halocarbons

**CHLORODIFLUOROMETHANE** **$CHClF_2$**

Aluminium

*See* ALUMINIUM, Al: Halocarbons

**TETRAZOLE-5-DIAZONIUM CHLORIDE** **$\overline{N{=}NNHN{=}C}N_2^+\ Cl^-$** **$CHClN_6$**

Shevlin, P. B. *et al.*, *J. Amer. Chem. Soc.*, 1977, **99**, 2628

The crystalline diazonium salt will detonate at the touch of a spatula. An ethereal solution exploded violently after 1 h at −78°C, presumably owing to separation of the solid salt.

*See* 5-AMINOTETRAZOLE, $CH_3N_5$

*See other* DIAZONIUM SALTS

## 1-DICHLOROAMINOTETRAZOLE $Cl_2NN{-}N{=}NN{=}CH$ $CHCl_2N_5$

Karrer, 1950, 804

1-Dichloroaminotetrazole and its 5-derivatives are extremely explosive, as expected in an *N,N*-dichloro derivative of a high-nitrogen nucleus.

*See other N*-HALOGEN COMPOUNDS

## CHLOROFORM $CHCl_3$

*MCA SD-89*, 1962

Acetone,
Alkali

1. Willgerodt, C., *Ber.*, 1881, **14**, 258
2. King, H. K., *Chem. & Ind.*, 1970, 185
3. Ekely, J. B. *et al., J. Amer. Chem. Soc.*, 1924, **46**, 1253
4. Grant, D. H., *Chem. & Ind.*, 1970, 919

Chloroform and acetone interact vigorously and exothermally in presence of potassium hydroxide or calcium hydroxide to form 1,1,1-trichloro-2-hydroxy-2-methylpropane [1], and a laboratory solvent-residues explosion was attributed to this cause [2]. No reaction occurs in absence of base [1], and other haloforms and ketones also react similarly in presence of base [3]. A minor explosion (or sudden boiling) of a chloroform–acetone mixture in new glassware may have been caused by surface alkali [4].

*See also* BROMOFORM, $CHBr_3$: Acetone, etc.

Bis(dimethylamino)dimethylstannane

*See* BIS(DIMETHYLAMINO)DIMETHYLSTANNANE, $C_6H_{18}N_2Sn$: Chloroform

Dinitrogen tetraoxide

*See* DINITROGEN TETRAOXIDE (NITROGEN DIOXIDE), $N_2O_4$: Halocarbons

Fluorine

*See* FLUORINE, $F_2$: Halocarbons

Metals

Davis, T. L. *et al., J. Amer. Chem. Soc.*, 1938, **60**, 720–722

The mechanism of the explosive interaction on impact of chloroform with sodium or potassium has been studied.

*See* METALS: Halocarbons

Nitromethane

*See* NITROMETHANE, $CH_3NO_2$: Haloforms

Potassium *tert*-butoxide

*See* POTASSIUM *tert*-BUTOXIDE, $C_4H_9KO$: Acids, etc.

*See also* Sodium methoxide, below

Sodium,
Methanol

Unpublished information, 1948

During attempted preparation of trimethyl orthoformate, addition of sodium to an inadequately cooled chloroform–methanol mixture caused a violent explosion.

*See* Sodium methoxide, below

Sodium hydroxide,
Methanol

*MCA Case History No. 498*

*MCA Case History No. 1913*

A chloroform–methanol mixture was put into a drum contaminated with sodium hydroxide. A vigorous reaction set in, and the drum exploded. Chloroform normally reacts slowly with sodium hydroxide owing to the insolubility of the latter. The presence of methanol (or other solubiliser) increases the rate of reaction by increasing the degree of contact between chloroform and alkali.

In the later incident addition of chloroform to a 4:1 mixture of methanol and 50 w/v% sodium hydroxide solution caused explosion of the drum.

Sodium methoxide

1. *MCA Case History No. 693*
2. Kaufmann, W. E. *et al.*, *Org. Synth.*, 1944, Coll. Vol. 1, 258

For the preparation of methyl orthoformate, solid sodium methoxide, methanol and chloroform were mixed together. The mixture boiled violently and then exploded [1]. The analogous preparation of ethyl orthoformate [2] involves the slow addition of sodium or sodium ethoxide solution to a chloroform–ethanol mixture. The explosion was caused by the addition of the sodium methoxide in one portion.

*See* Sodium, above

Triisopropylphosphine

*See* TRIISOPROPYLPHOSPHINE, $C_9H_{21}P$: Chloroform

*See other* HALOCARBONS

**FLUORODINITROMETHANE** $FCH(NO_2)_2$ $CHFN_2O_4$

Potentially explosive.

*See* FLUORODINITROMETHYL COMPOUNDS

## TRIFLUOROMETHANESULPHONIC ACID $CHF_3O_3S$

$CF_3SO_3H$

Acyl chlorides,
Aromatic hydrocarbons

Effenberger, F. *et al., Angew. Chem. (Intern. Ed.)*, 1972, **11**, 300

Addition of catalytic amounts (1%) of trifluoromethanesulphonic acid to mixtures of acyl chlorides and aromatic hydrocarbons causes more or less violent evolution of hydrogen chloride, depending on reactivity of the components.

## IODOFORM $CHI_3$

Acetone

*See* CHLOROFORM, $CHCl_3$ : Acetone, etc.

Hexamethylenetetramine

Sorbe, 1968, 137

The 1:1 addition complex exploded at 178°C.

*See related* *N*-HALOGEN COMPOUNDS
*See other* HALOCARBONS

## DIAZOMETHYLLITHIUM $LiCHN_2$ $CHLiN_2$

Müller, E. *et al., Chem. Ber.*, 1954, **87**, 1887

Alkali metal salts of diazomethane are very explosive when exposed to air in the dry state, and should be handled, preferably wet with solvent, under an inert atmosphere.

*See other* DIAZO COMPOUNDS

## † HYDROGEN CYANIDE HCN CHN

*MCA SD-67,* 1961

Gause, E. H. *et al., J. Chem. Eng. Data,* 1960, **5**, 351

Wöhler, L. *et al., Chem. Ztg.*, 1926, **50**, 761, 781

Anhydrous hydrogen cyanide is stable at or below room temperature if inhibited with acid (e.g. 0.1% sulphuric acid). In absence of inhibitor, exothermic polymerisation occurs, and if temperature attains 184°C, explosively rapid polymerisation occurs.

Presence of alkali favours explosive polymerisation.

*See also* MERCURY(II) CYANIDE, $C_2HgN_2$: Hydrogen cyanide

Alcohols,
Hydrogen chloride

*See* HYDROGEN CHLORIDE, ClH: Alcohols, etc.

Ammonium chloride

*See* CYANOGEN CHLORIDE, CClN: Preparative hazard

*See other* CYANO COMPOUNDS

**FULMINIC ACID** HC≡N→O **CHNO**

Sorbe, 1968, 72

It is fairly stable as an ethereal solution, but the isolated acid is explosively unstable, and sensitive to heat, shock or friction.

*See* METAL FULMINATES

**DIAZOMETHYLSODIUM** $NaCHN_2$ **$CHN_2Na$**

*See* DIAZOMETHYLLITHIUM, $CHLiN_2$

**TRINITROMETHANE ('NITROFORM')** $CH(NO_2)_3$ **$CHN_3O_6$**

Marans, N. S. *et al., J. Amer. Chem. Soc.*, 1950, 72, 5329

Explosions occurred during distillation of this polynitro compound.

Divinyl ketone

Graff, M. *et al., J. Org. Chem.*, 1968, **33**, 1247

One attempted reaction of trinitromethane with impure ketone caused an explosion at refrigerator temperature.

2-Propanol

*MCA Case History No. 1010*

Frozen mixtures of trinitromethane–2-propanol (9:1) exploded during thawing. The former dissolves exothermally in the alcohol, the heat effect increasing directly with the concentration above 50% w/w. Traces of nitric acid may also have been present.

*See other* POLYNITROALKYL COMPOUNDS

**AZIDODITHIOFORMIC ACID** $N_3C(S)SH$ **$CHN_3S_2$**

1. Mellor, 1947, Vol. 8, 338
2. Smith, G. B. L., *Inorg. Synth.*, 1939, **1**, 81

The isolated solid or its salts are shock- and heat-sensitive explosives [1]. Safe preparative procedures have been detailed. The heavy metal salts, though powerful detonators, are too sensitive for practical use [2].

*See* CARBON DISULPHIDE, $CS_2$: Metal azides

*See also* BIS(AZIDOTHIOCARBONYL)DISULPHIDE, $C_2N_6S_4$

*See other* ACYL AZIDES

**5-NITROTETRAZOLE** $\overline{N=NNHN=C}NO_2$ $CHN_5O_2$

Jenkins, J. M., *Chem. Brit.*, 1970, **6**, 401

An acidified solution of the sodium salt was allowed to evaporate during 3 days and spontaneously exploded 2 weeks later. Nature of the explosive species, possibly the tetrazolic acid, is being sought.

*See other* HIGH-NITROGEN COMPOUNDS
*C*-NITRO COMPOUNDS

**5-AZIDOTETRAZOLE** $\overline{N=NNHN=C}N_3$ $CHN_7$

Alone,
or Acetic acid,
or Alkali

1. Thiele, J. *et al.*, *Ann.*, 1895, **287**, 238
2. Lieber, E. *et al.*, *J. Amer. Chem. Soc.*, 1951, **73**, 1313

Though explosive, it (and its ammonium salt) is much less sensitive to impact or friction than its sodium or potassium salts [1]. A small sample of the latter exploded violently during suction filtration. The parent compound explodes spontaneously even in acetone (but not in ethanol or aqueous) solution if traces of acetic acid are present [2]. The salts are readily formed from diaminoguanidine salts and alkali nitrites. The ammonium salt explodes on heating, and the silver salt is violently explosive, even when wet [1]. The sodium salt is also readily formed from cyanogen azide.

*See other* HIGH-NITROGEN COMPOUNDS
*N*-METAL DERIVATIVES
ORGANIC AZIDES

**SODIUM HYDROGENCARBONATE** $NaOCO_2H$ $CHNaO_3$

2-Furaldehyde

*See* 2-FURALDEHYDE (FURFURAL), $C_5H_4O$

**SILVER NITROUREIDE** $AgN(NO_2)CONH_2$ $CH_2AgN_3O_3$

*See* NITROUREA, $CH_3N_3O_3$

**SILVER-5-AMINOTETRAZOLIDE** $\overline{N=NNAgN=C}NH_2$ $CH_2AgN_5$

Thiele, J., *Ann.*, 1892, **270**, 59

Explodes on heating, similarly to silver tetrazolide.

*See other* *N*-METAL DERIVATIVES

**BIS(DIFLUOROBORYL)METHANE** $(F_2B)_2CH_2$ $CH_2B_2F_4$

Maraschin, N. J. *et al.*, *Inorg. Chem.*, 1975, **14**, 1856

Highly reactive, it explodes on exposure to air or water.

*See related* NON-METAL HALIDES

**DIBROMOMETHANE** $CH_2Br_2$

Potassium

*See* POTASSIUM, K: Halocarbons

**CHLORONITROMETHANE** $ClCH_2NO_2$ $CH_2ClNO_2$

1. Seigle, L. W. *et al.*, *J. Org. Chem.*, 1940, **5**, 100
2. Libman, D. D., private comm., 1968

Chlorination of nitromethane following the published general method [1] gave a product which decomposed explosively during distillation at 95 mbar [2]. A b.p. of 122°C/1 bar is quoted in the literature.

*See other C*-NITRO COMPOUNDS

**DICHLOROMETHANE** $CH_2Cl_2$

*MCA SD-86,* 1962

Normally non-flammable, but will form ignitable mixtures of wide limits in oxygen-enriched air or at elevated temperatures.

Air,
Methanol

Coffee, R. D. *et al.*, *J. Chem. Eng. Data*, 1972, **17**, 89–93

Dichloromethane, previously considered to be non-flammable except in oxygen, becomes flammable in air at 102°C/1 bar, at 27°C/1.7 bar or at 27°C/1 bar in presence of less than 0.5 vol% of methanol. Other data are also given.

Dinitrogen pentaoxide

*See* NITRATING AGENTS

Dinitrogen tetraoxide

*See* DINITROGEN TETRAOXIDE, $N_2O_4$: Halocarbons

Metals

*See* ALUMINIUM, Al: Halocarbons
LITHIUM, Li: Halocarbons
SODIUM, Na: Halocarbons

Nitric acid

*See* NITRIC ACID, $HNO_3$: Dichloromethane

Potassium *tert*-butoxide

*See* POTASSIUM *tert*-BUTOXIDE, $C_4H_9KO$: Acids, etc.
*See other* HALOCARBONS

**1,1-DIFLUOROUREA** $F_2NCONH_2$ $CH_2F_2N_2O$

Parker, C. O. *et al.*, *Inorg. Synth.*, 1970, **12**, 309
Concentrated aqueous solutions of difluorourea decompose above −20°C with evolution of tetrafluorohydrazine and difluoroamine, both explosive gases.

*See other* *N*-HALOGEN COMPOUNDS

**DIIODOMETHANE** $CH_2I_2$

Alkenes,
Diethylzinc

*See* DIETHYLZINC, $C_4H_{10}Zn$: Alkenes, diiodomethane

Metals

*See* COPPER–ZINC COUPLE, Cu–Zn: Diiodomethane, etc.
POTASSIUM, K: Halocarbons
LITHIUM, Li: Halocarbons

**POLY(METHYLENEMAGNESIUM)** $(CH_2Mg)_n$

Ziegler, K. *et al.*, *Z. Anorg. Chem.*, 1955, **282**, 345
The polymer ignites in air.

*See other* ALKYLMETALS

**SODIUM *aci*-NITROMETHANE** $H_2C{=}N(O)ONa$ $CH_2NNaO_2$

Zelinsky, N., *Ber.*, 1894, **27**, 3407
Salts of *aci*-nitromethane are easily detonated and powerful explosives.

Mercury(II) chloride,
Acids

Nef, J. U., *Ann.*, 1894, **280**, 263, 305
Interaction gives mercury nitromethane, which is converted by acids to mercury fulminate.

*See* MERCURY(II) FULMINATE, $C_2HgN_2O_2$

1,1,3,3-Tetramethyl-2,4-cyclobutanedione

*See* DISODIUM BIS-*aci*-1,3-DIHYDROXY-1,3-DI(NITROMETHYL)-2,2,4,4-TETRAMETHYLCYCLOBUTANE, $C_{10}H_{16}N_2Na_2O_6$

Water

Nef, J. U., *Ann.*, 1894, **280**, 273

The *aci*-sodium salt, normally crystallising with one molecule of ethanol and stable, will explode if moistened with water. This is due to liberation of heat and conversion to sodium fulminate.

*See* DISODIUM 1,3-DIHYDROXY-1,3-BIS(*aci*-NITROMETHYL)-2,2,4,4-TETRAMETHYLCYCLOBUTANE, $C_{10}H_{16}N_2Na_2O_6$

*See other aci*-NITRO SALTS

**CYANAMIDE** **$CNNH_2$** **$CH_2N_2$**

Anon., *Fire Prot. Assoc. J.*, 1966, 243

Anon., *Sichere Chemiearbeit*, 1976, **28**, 63

Cyanamide is thermally unstable and needs storage under controlled conditions. Contact with moisture, acids or alkalies accelerates the rate of decomposition, and at temperatures above 49°C thermal decomposition is rapid and may become violent. A maximum storage temperature of 27°C is recommended.

Commercial cyanamide is stabilised with boric acid or sodium dihydrogenphosphate but vacuum distillation produces a neutral unstabilised distillate, which immediately may decompose spontaneously. Small-scale storage tests showed that unstabilised cyanamide was 47% decomposed after 18 days at 20°C and 75% decomposed after 29 days at 30°C, whereas stabilised material showed only 1% decomposition under each of these conditions. Larger-scale tests with 1–2 kg samples led to sudden and violently exothermic polymerisation after storage for 14 days at ambient temperature.

If small samples of unstabilised cyanamide are required, they are best prepared by freezing out from aqueous solutions of the stabilised material. Such small samples should be used immediately or stored under refrigeration.

*o*-Phenylenediamine salts

Sawatari, K. *et al.*, Japan Kokai, 76 16 669, 1976 (*Chem. Abs.*, 1976, **85**, 63069e)

During the preparation of 2-aminobenzimidazoles, reaction conditions are maintained at below 90°C to prevent explosive polymerisation of cyanamide.

Water

Pinck, L. A. *et al.*, *Inorg. Synth*, 1950, **3**, 41

Evaporation of aqueous solutions to dryness is hazardous owing to possibility of explosive polymerisation in concentrated solution.

*See other* CYANO COMPOUNDS

**DIAZIRINE** $\overline{CH_2N{=}N}$ $CH_2N_2$

1. Graham, W. H., *J. Org. Chem.*, 1965, **30**, 2108
2. Schmitz, E. *et al.*, *Chem. Ber.*, 1962, **95**, 800

This cyclic isomer of diazomethane is also a gas (b.p., –14°C) which explodes on heating. Several homologues are also thermally unstable [1,2].

*See* DIAZIRINES

**DIAZOMETHANE** **(linear resonance structures)** $CH_2N_2$

1. Eistert, B., in *Newer Methods of Preparative Organic Chemistry*, 517–518, New York, Interscience, 1948
2. de Boer, H. J. *et al.*, *Org. Synth.*, 1963, Coll. Vol. 4, 250
3. Gutsche, C. D., *Org. React.*, 1954, **8**, 392–393
4. Zollinger, H., *Azo and Diazo Chemistry*, 22, London, Interscience, 1961
5. Fieser, L. *et al.*, *Reagents for Organic Synthesis*, Vol. 1, 191, New York, Wiley, 1967
6. Horàk, V. *et al.*, *Chem. & Ind.*, 1961, 472

Diazomethane boils at –23°C and the undiluted liquid or concentrated solutions may explode if impurities or solids are present [1], including freshly crystallised products [2]. Gaseous diazomethane, even when diluted with nitrogen, may explode at elevated temperatures (100°C or above), or under high-intensity lighting, or if rough surfaces are present [1]. Ground glass apparatus or glass-sleeved stirrers are therefore undesirable when working with diazomethane. Explosive intermediates may also be formed during its use as a reagent, for which cold dilute solutions have been frequently used uneventfully [1]. Further safety precautions have been detailed [2–4]. Many precursors for diazomethane are available [5], including the stable water-soluble intermediate *N*-nitroso-3-methylaminosulpholane [6]. Many of the explosions observed are attributed to unsuitable conditions of contact between concentrated alkali and undiluted nitroso precursors [1].

Alkali metals

Eistert, B., in *Newer Methods of Preparative Organic Chemistry*, 518, New York, Interscience, 1948

Contact of diazomethane with alkali metals causes explosions.

*See* DIAZOMETHYLLITHIUM, $CHLiN_2$

Calcium sulphate

Gutsche, C. D., *Org. React.*, 1954, **8**, 392

Calcium sulphate is an unsuitable desiccant for drying-tubes in diazomethane systems. Contact of diazomethane vapour and the sulphate causes an exotherm which may lead to detonation. Potassium hydroxide is a suitable desiccant.

Dimethylaminodimethylarsine,
Trimethyltin chloride

Krommes, P. *et al.*, *J. Organomet. Chem.*, 1976, **110**, 195–200

Interaction in ether to produce diazomethyldimethylarsine is accompanied by violent foaming, and eye protection is essential.

*See other* DIAZO COMPOUNDS

**ISODIAZOMETHANE** $H_2NN{=}C{:}$ $CH_2N_2$

Müller, E. *et al.*, *Chem. Ber.*, 1954, **87**, 1887; *Ann.*, 1968, **713**, 87

This unstable liquid begins to decompose at 15°C and explodes exothermically at 35–40°C. It is *N*-aminoisonitrile.

*See other* DIAZO COMPOUNDS

**NITROOXIMINOMETHANE (METHYLNITROLIC ACID)** $CH_2N_2O_3$

$NO_2CH{=}NOH$

Sorbe, 1968, 147

An unstable and explosive crystalline solid, formally a nitro-oxime.

*See related* OXIMES

**DINITROMETHANE** $(NO_2)_2CH_2$ $CH_2N_2O_4$

Sorbe, 1968, 148

It explodes at 100°C.

*See other* POLYNITROALKYL COMPOUNDS

**TETRAZOLE** $\overline{N{=}NNHN{=}CH}$ (ring) $CH_2N_4$

Benson, F.R., *Chem. Rev.*, 1947, **41**, 5

It explodes above its m.p., 155°C.

*See other* TETRAZOLES

**LEAD METHYLENEBISNITROAMIDE** $Pb(NNO_2)_2CH_2$ $CH_2N_4O_4Pb$

*See* METHYLENEBISNITROAMINE, $CH_4N_4O_4$

*See other* HEAVY METAL DERIVATIVES

**5-AMINO-1,2,3,4-THIATRIAZOLE** $\overline{SN{=}NN{=}C}NH_2$ (ring) $CH_2N_4S$

Lieber, E. *et al.*, *Inorg. Synth.*, 1960, **6**, 44

It decomposes with a slight explosion in a capillary tube at 136°C.

*See other* HIGH-NITROGEN COMPOUNDS

### 5-*N*-NITROAMINOTETRAZOLE $N{=}NNHN{=}CNHNO_2$ $CH_2N_6O_2$

1. Lieber, E. *et al.*, *J. Amer. Chem. Soc.*, 1951, **73**, 2328
2. O'Connor, T. E. *et al.*, *J. Soc. Chem. Ind.*, 1949, **68**, 309

It and its monopotassium salt explode at 140°C and the diammonium salt explodes at 220°C after melting [1]. The disodium salt explodes at 207°C [2].

*See other* TETRAZOLES

### † FORMALDEHYDE HCHO $CH_2O$

*MCA SD-1*, 1960

Hydrogen peroxide

*See* HYDROGEN PEROXIDE, $H_2O_2$: Oxygenated compounds

Magnesium carbonate

*BCISC Quart. Safety Summ.*, 1965, (143), 44

During neutralisation of the formic acid present in formaldehyde solution by shaking with magnesium carbonate in a screw-capped bottle, the latter exploded owing to pressure of carbon dioxide. Periodical release of pressure should avoid this.

Nitromethane

*See* NITROMETHANE, $CH_3NO_2$: Formaldehyde

Peroxyformic acid

*See* PEROXYFORMIC ACID, $CH_2O_3$: Organic materials

Phenol

Taylor, H. D. *et al.*, *Major Loss Prevention in the Process Industries*, 1971, I.Ch.E. Sympos. Series No. 34, 46

At least nine cases of catalysed plant-scale preparations of phenol–formaldehyde resin which ran away with sudden pressure development and failure of bursting discs or reactors are briefly mentioned. No details of process conditions are given.

*See other* POLYMERISATION INCIDENTS

### FORMIC ACID $HCO_2H$ $CH_2O_2$

1. *BCISC Quart. Safety Summ.*, 1973, **44**, 18
2. Falconer, J. *et al.*, *Proc. 2nd Int. Conf. Solid Surf.*, 1974, 525

The slow decomposition in storage of 98–100% formic acid with liberation of carbon monoxide led to rupture of the sealed glass containers. In absence of gas leakage, a full 2.5 l bottle would develop a pressure of over 7 bar during 1 year at 25°C [1]. Explosive decomposition of formic acid on a clean nickel (1.1.0) surface was studied, using deuteroformic acid [2].

Aluminium

*See* ALUMINIUM, Al: Formic acid

Hydrogen peroxide

*See* HYDROGEN PEROXIDE, $H_2O_2$ : Oxygenated compounds

Nitromethane

*See* NITROMETHANE, $CH_3NO_2$ : Acids

Palladium–carbon catalyst

Freifelder, 1971, 188

Addition of dry catalyst to 98% formic acid used as a hydrogenation solvent can be extremely hazardous, because hydrogen is released from decomposition of the acid. Addition of acid to the water-wetted catalyst is safer.

*See other* HYDROGENATION INCIDENTS

Phosphorus pentaoxide

Muir, G. D., private comm., 1968

Attempted dehydration of 95% acid to anhydrous formic acid caused rapid evolution of carbon monoxide.

**PEROXYFORMIC ACID** **$HCO_2OH$** **$CH_2O_3$**

1. Greenspan, F. P., *J. Amer. Chem. Soc.*, 1946, **68**, 907
2. D'Ans, J. *et al.*, *Ber.*, 1915, **48**, 1136
3. Weingartshofer, A. *et al.*, *Chem. Eng. News*, 1952, **30**, 3041
4. Swern, 1970, Vol. 1, 337
5. Isard, A. *et al.*, *Chemical Tech.*, 1974, **4**, 380

Peroxyformic acid solutions are unstable and undergo a self-accelerating exothermic decomposition at ambient temperature [1]. An 80% solution exploded at 80–85°C [2]. A small sample of the pure vacuum-distilled material cooled to below −10°C exploded violently when the flask was moved [3]. Though the acid has occasionally been distilled, this is an extremely dangerous operation [4]. During preparation of peroxyformic acid by a patented procedure involving interaction of formic acid with hydrogen peroxide in the presence of metaboric acid, an explosion occurred which was attributed to spontaneous separation of virtually pure peroxyformic acid [5].

Metals,
or Oxides

D'Ans, J. *et al.*, *Ber.*, 1915, **48**, 1136

Violence of reaction depends on concentration of acid and scale and proportion of reactants. The following observations were made with additions to 2–3 drops of *ca.* 90% acid. Nickel powder, becomes violent;

mercury, colloidal silver and thallium powder readily cause explosions. Zinc powder causes a violent explosion immediately. Iron powder (and silicon) are ineffective alone, but a trace of maganese dioxide promotes deflagration. Barium peroxide, copper(I) oxide, impure chromium trioxide, iridium dioxide, lead dioxide, manganese dioxide and vanadium pentoxide all cause violent decomposition, sometimes accelerating to explosion. Lead oxide, trilead tetraoxide and sodium peroxide all cause an immediate violent explosion.

Non-metals

D'Ans, J. *et al., Ber.*, 1915, **48**, 1136

Impure carbon and red phosphorus are oxidised violently, and silicon, promoted by traces of manganese dioxide, is oxidised with ignition.

Organic materials

1. D'Ans, J. *et al., Ber.*, 1915, **48**, 1136
2. Anon., *Chem. Eng. News*, 1950, **28**, 418
3. Shanley, E. S., ibid., 3067

Formaldehyde, benzaldehyde and aniline react violently with 90% performic acid [1]. An unspecified organic compound was added to the acid (preformed from formic acid and 90% hydrogen peroxide), and soon after the initial vigorous reaction had subsided, the mixture exploded violently [2]. Reaction with alkenes is vigorously exothermic, and adequate cooling is necessary [3]. Reactions with performic acid can be more safely accomplished by slow addition of hydrogen peroxide to a solution of the compound in formic acid. Adequate safety screens should be used with all peracid preparations [3].

Sodium nitrate

D'Ans, J. *et al., Ber.*, 1915, **48**, 1139

The salt may lead to explosive decomposition of the peroxy acid.

*See other* PEROXYACIDS

**METHYLSILVER** **$CH_3Ag$**

Thiele, H., *Z. Electrochem.*, 1943, **49**, 426

Prepared at −80°C, the addition compound with silver nitrate decomposes explosively on warming to −20°C.

*See other* ALKYLMETALS

**SILVER NITROGUANIDIDE** **$CH_3AgN_4O_2$**

**$H_2NC(NH)NAgNO_2$**

*See* NITROGUANIDINE, $CH_4N_4O_2$

*See other* *N*-METAL DERIVATIVES

## METHYLALUMINIUM DIIODIDE $CH_3AlI_2$

Nitromethane

*See* NITROMETHANE, $CH_3NO_2$: Alkylmetal halides
*See also* ALKYLALUMINIUM HALIDES

## DICHLORO(METHYL)ARSINE $CH_3AsCl_2$

Chlorine

*See* CHLORINE, $Cl_2$: Dichloro(methyl)arsine

## METHYLBISMUTH OXIDE $CH_3BiO$

Marquardt, A., *Ber.*, 1887, **20**, 1522
Ignites on warming in air.
*See related* ALKYLMETALS

## BROMOMETHANE $CH_3Br$

No Fl.P.; but E.L., 10–15% (5–25% at 10 bar)

1. *MCA SD-35,* 1968
2. *MCA Case History No. 746*

Though bromomethane is used as a fire extinguishant it does form explosive mixtures with air within narrow limits (13.5–14.5% also recorded) at atmospheric pressure, but considerably wider at higher pressure [1, 2].

Dimethyl sulphoxide

*See* TRIMETHYLSULPHOXONIUM BROMIDE, $C_3H_9BrOS$

Ethylene oxide

*See* ETHYLENE OXIDE, $C_2H_4O$: Air, Bromomethane

Metals

*MCA Case History No. 746* and addendum
Metallic components of zinc, aluminium and magnesium are unsuitable for service with bromomethane because of the formation of pyrophoric Grignard-type compounds. The Case History attributes a severe explosion to ignition of a bromomethane–air mixture by pyrophoric methylaluminium bromides produced by corrosion of an aluminium component.
*See other* METALS: Halocarbons

**_N,N_-DIBROMOMETHYLAMINE** $CH_3NBr_2$ $CH_3Br_2N$

Cooper, J. C. *et al.*, *Explosivstoffe*, 1969, **17**(6), 129–130

Like the dichloro analogue, it appears to be more sensitive to impact or shock than *N*-chloromethylamine.

*See other* *N*-HALOGEN COMPOUNDS

**METHYLCADMIUM AZIDE** $CH_3CdN_3$

Dehnicke, K. *et al.*, *J. Organomet. Chem.*, 1966, **6**, 298

Surprisingly it is thermally stable to 300°C. It is hygroscopic and very readily hydrolysed to explosive hydrogen azide.

*See related* METAL AZIDES

**† CHLOROMETHANE** $CH_3Cl$

*MCA SD-40,* 1970

Aluminium

*See* ALUMINIUM, Al: Halocarbons

Aluminium trichloride,
Ethylene

*See* ETHYLENE, $C_2H_4$: Aluminium trichloride

Interhalogens

*See* BROMINE TRIFLUORIDE, $BrF_3$: Halogens, etc.
BROMINE PENTAFLUORIDE, $BrF_5$: Hydrogen-containing materials

Metals

*MCA SD-40,* 1970

In presence of catalytic amounts of aluminium trichloride, powdered aluminium and chloromethane interact to form pyrophoric trimethylaluminium. Chloromethane may react explosively with magnesium, or potassium, sodium or their alloys. Zinc probably reacts similarly to magnesium.

*See* ALUMINIUM, Al: Halocarbons
SODIUM, Na: Halocarbons

*See other* HALOCARBONS

**METHYLMERCURY PERCHLORATE** $CH_3HgClO_4$ $CH_3ClHgO_4$

Anon., *Angew. Chem. (Nachr.)*, 1970, **18**, 214

On rubbing with a glass rod, a sample exploded violently. As the explosion could not be reproduced using a metal rod, initiation by static electricity was suspected.

*See* STATIC INITIATION

**METHYL HYPOCHLORITE** $CH_3OCl$ $CH_3ClO$

Sandmeyer, T., *Ber.*, 1886, **19**, 859

The liquid could be very gently distilled (12°C) but the superheated vapour readily and violently explodes, as does the liquid on ignition.

*See other* HYPOHALITES

**METHYL PERCHLORATE** $CH_3OClO_3$ $CH_3ClO_4$

Sidgwick, 1950, 1236

The high explosive instability is due in part to the covalent character of the alkyl perchlorates, and also to the excess of oxygen in the molecule over that required to completely combust the other elements present (i.e. negative oxygen balance).

*See other* ALKYL PERCHORATES

***N,N*-DICHLOROMETHYLAMINE** $CH_3NCl_2$ $CH_3Cl_2N$

1. Bamberger, E. *et al., Ber.*, 1895, **28**, 1683
2. Okon, K. *et al., Chem. Abs.*, 1960, **54**, 17887b

A mixture with water exploded violently on warming [1]. Contact with solid sodium sulphide or distillation over calcium hypochlorite also caused explosions [2].

*See other N*-HALOGEN COMPOUNDS

**† TRICHLORO(METHYL)SILANE** $Cl_3SiCH_3$ $CH_3Cl_3Si$

**METHYLCOPPER** $CH_3Cu$

Coates, 1960, 348

Ikariya, T. *et al., J. Organometal. Chem.*, 1974, **72**, 146

The dry solid is very impact-sensitive and may explode spontaneously on being allowed to dry out at room temperature.

Methylcopper decomposes explosively at ambient temperature, and violently in presence of a little air.

*See other* ALKYLMETALS

**† FLUOROMETHANE** $CH_3F$

**XENON(II) FLUORIDE METHANESULPHONATE** $CH_3FO_3SXe$

$FXeOSO_2CH_3$

Wechsberg, M. *et al., Inorg. Chem.*, 1972, **11**, 3066

The solid explodes on warming from 0°C to ambient temperature.

*See other* XENON COMPOUNDS

## IODOMETHANE $CH_3I$

Silver chlorite

*See* SILVER CHLORITE, $AgClO_2$: Alkyl iodides

Sodium

1. Anon., *J. Chem. Educ.*, 1966, **43**, A236
2. Braidech, M. M., *J. Chem. Educ.*, 1967, **44**, A324

The first stage of a reaction involved the addition of sodium dispersed in toluene to a solution of adipic ester in toluene. The subsequent addition of iodomethane (b.p., 42°C) was too fast and vigorous boiling ejected some of the flask contents. Exposure of sodium particles to air caused ignition, and a violent toluene-vapour explosion followed [1]. When a reagent as volatile and reactive as iodomethane is added to a reaction mixture, controlled addition, and one or more wide-bore reflux condensers, are essential. A similar incident but involving benzene solvent was also reported [2].

*See also* SODIUM, Na: Halocarbons

## METHYLMAGNESIUM IODIDE $CH_3MgI$ $CH_3IMg$

Vanadium trichloride

*See* VANADIUM TRICHLORIDE, $Cl_3V$: Methylmagnesium iodide

## METHYLZINC IODIDE $CH_3ZnI$ $CH_3IZn$

Nitromethane

*See* NITROMETHANE, $CH_3NO_2$: Alkylmetal halides

## METHYLPOTASSIUM $CH_3K$

Weiss E. *et al.*, *Angew. Chem. (Intern. Ed.)*, 1968, 7, 133

The dry material is highly pyrophoric.

*See other* ALKYLMETALS

## POTASSIUM METHANEDIAZOATE $CH_3N{=}NOK$ $CH_3KN_2O$

Hantzsch, A. *et al.*, *Ber.*, 1902, **35**, 901

Interaction with water is explosively violent.

*See related* ARENEDIAZOATES

**POTASSIUM METHOXIDE** $KOCH_3$ $CH_3KO$

Arsenic pentafluoride,
Benzene

*See* ARSENIC PENTAFLUORIDE, $AsF_5$: Benzene, etc.

**POTASSIUM METHYLSELENIDE** $KSeCH_3$ $CH_3KSe$

*o*-Nitroacetophenone

*See* *o*-NITROACETOPHENONE, $C_8H_7NO_3$: Potassium methylselenide

**METHYLLITHIUM** $CH_3Li$

Sidgwick, 1950, 71
Ignites and burns brilliantly in air.

*See also* ALKALI-METAL DERIVATIVES OF HYDROCARBONS
*See other* ALKYLMETALS

**FORMAMIDE** $HCONH_2$ $CH_3NO$

Iodine,
Pyridine,
Sulphur trioxide

Anon., *J. Chem. Educ.*, 1973, **50**, A293
Bottles containing a modified Karl Fischer reagent with formamide replacing methanol developed gas pressure during several months and exploded. No reason is yet apparent.

† **METHYL NITRITE** $CH_3ONO$ $CH_3NO_2$

**NITROMETHANE** $CH_3NO_2$

1. McKitterick, D. S. *et al.*, *Ind. Eng. Chem. (Anal. Ed.)*, 1938, **10**, 630
2. Makovky, A. *et al.*, *Chem. Rev.*, 1958, **58**, 627
3. Travis, J. R., *Los Alamos Rept. DC 6994*, Washington, USAEC, 1965
4. Sorbe, 1968, 148

Conditions under which it may explode by detonation, heat or shock were determined. It was concluded that it is potentially very explosive and precautions are necessary to prevent its exposure to severe shock or high temperatures in use [1]. Later work, following two rail tank explosions, showed that shock caused by sudden application of gas pressure, or by forced high-velocity flow through restrictions, could detonate the liquid. The stability and decomposition of nitromethane relevant to its use as a rocket fuel is also reviewed [2].

The role of discontinuities in the initiation of shock-compressed nitromethane has been experimentally evaluated [3]. It explodes at about 230°C [4].

Acids,
or Bases
Makovky, A. *et al.*, *Chem. Rev.*, 1958, **58**, 631
Addition of bases or acids to nitromethane renders it susceptible to initiation by a detonator. These include aniline, diaminoethane, morpholine, methylamine, ammonium hydroxide, potassium hydroxide, sodium carbonate, and formic, nitric, sulphuric and phosphoric acids.
*See also* 1,2-Diaminoethane, etc., below

Alkylmetal halides
Traverse, G., US Pat. 2 775 863, 1957
Contact with $R_m MX_n$ (R is methyl, ethyl; M is aluminium, zinc; X is Br, I) causes ignition. Diethylaluminium bromide, dimethylaluminium bromide, ethylaluminium bromide iodide, methylzinc iodide and methylaluminium diiodide are claimed as specially effective.

Aluminium trichloride
*See* ALUMINIUM TRICHLORIDE–NITROMETHANE, $AlCl_3 \cdot CH_3NO_2$

Aluminium trichloride,
Ethylene
*See* ETHYLENE, $C_2H_4$: Aluminium trichloride

Aminium salts,
Organic solvents
Runge, W. F. *et al.*, US Pat. 3 915 768, 1975 (*Chem. Abs.*, 1976, **84**, 76547p)
Presence of, e.g., 5% of methylammonium acetate and 5% of methanol sensitises nitromethane to shock-initiation.

Bis-(2-aminoethyl)amine
Runge, W. F. *et al.*, US Pat. 3 798 902, 1974
Explosive solutions of nitromethane in dichloromethane, sensitised by addition of 2–12% of the amine, retained their sensitivity at –40°C, and with 10–12% of amine, to below –50°C

Boron trifluoride etherate,
Silver oxide
*See* SILVER TETRAFLUOROBORATE, $AgBF_4$

Calcium hypochlorite
*See* CALCIUM HYPOCHLORITE, $CaCl_2O_2$: Nitromethane

1,2-Diaminoethane,
*N*,2,4,6-tetranitro-*N*-methylaniline
*MCA Case History No. 1564*
During preparations to initiate the explosion of nitromethane sensitised by addition of 20% of the diamine, accidental contact of the liquid mixture with the solid 'tetryl' detonator caused ignition of the latter.
*See also* Acids, or Bases, above

Formaldehyde
Noland, W. E., *Org. Synth.*, 1961, **41**, 69
Interaction of nitromethane and formaldehyde in presence of alkali gives not only 2-nitroethanol but also di- and tri-condensation products. After removal of the 2-nitroethanol by vacuum distillation the residue must be cooled before admitting air into the system to prevent a flash explosion or violent fume-off.

Haloforms
Presles, H. N. *et al., Acta Astronaut.*, 1976, **3**, 531–540
Mixtures with chloroform or bromoform are detonable.

Hydrazine,
Methanol
Forshey, D. R. *et al., Explosivstoffe*, 1969, **17**(6), 125–129
Addition of hydrazine strongly sensitises to detonation nitromethane and its mixtures with methanol.

Hydrocarbons
1. Watts, C. E., *Chem. Eng. News,* 1952, **30**, 2344
2. Makovky, A. *et al., Chem. Rev.*, 1958, **58**, 631

Nitromethane may act as a mild oxidant and should not be heated with hydrocarbons or readily oxidisable materials under confinement [1]. Explosions may occur during cooling of such materials heated to high temperatures and pressures [2]. Mixtures of nitromethane and solvents which are to be heated above the boiling point of nitromethane should be first subjected to small-scale explosive tests [2].

Lithium perchlorate
1. Titus, J. A., *Chem. Eng. News,* 1971, **49**(23), 6
2. 'Nitroparaffin Data Sheet TDS 1', New York, Commercial Solvents Corp., 1965
3. Egly, R. S., *Chem. Eng. News,* 1973, **51**(6), 30

Explosions which occurred at the auxiliary electrode during electro-oxidation reactions in nitromethane-lithium perchlorate electrolytes, may have been caused by lithium fulminate. This could have been produced by formation of the lithium salt of nitromethane and subsequent dehydration to the fulminate [1], analogous to the known formation of mercuric fulminate [2]. This explanation is not considered tenable, however [3].

Metal oxides
*See* NITROALKANES: Metal oxides

Nitric acid
*See* NITRIC ACID, $HNO_3$: Nitromethane
*See other* C-NITRO COMPOUNDS

**METHYL NITRATE** **$CH_3ONO_2$** **$CH_3NO_3$**

1. Kit and Evered, 1960, 268
2. Black, A. P. *et al.*, *Org. Synth.*, 1943, Coll. Vol. 2, 412
3. Goodman, H. *et al.*, *Comb. and Flame*, 1972, **19**, 157

It has high shock-sensitivity and thermal sensitivity, exploding at 65°C. It is too sensitive for use as a rocket mono-propellant [1]. Conditions during preparation of the ester from methanol and mixed nitric–sulphuric acids are fairly critical, and explosions may occur if the ester is suddenly heated, or distilled in presence of acid [2]. Spontaneous ignition or explosion of the vapour at 250–316°C in presence of gaseous diluents has been studied [3].
*See other* ALKYL NITRATES

**METHYL AZIDE** **$CH_3N_3$**

1. *MCA Case History No. 887*
2. Boyer, J. H. *et al.*, *Chem. Rev.*, 1954, **54**, 32

The product, prepared by interaction of sodium azide with dimethyl sulphate and sodium hydroxide, exploded during concurrent vacuum distillation. The explosion was attributed to formation and codistillation with the product, of hydrogen azide, due to variation in pH below 5 during the reaction. Free hydrogen azide itself is explosive, and it may also have reacted with mercury in a manometer to form the detonator, mercuric azide [1]. Methyl azide is stable at ambient temperature but may detonate on rapid heating [2].

Mercury
Currie, C. L. *et al.*, *Can. J. Chem.*, 1963, **41**, 1048
Presence of mercury in methyl azide markedly reduces the stability towards shock or electric discharge.

Methanol

Grundmann, C. *et al., Angew. Chem.*, 1950, **62**, 410

In spite of extensive cooling and precautions, a mixture of methyl azide, methanol and dimethyl malonate exploded violently while being sealed into a Carius tube. The vapour of the azide is very easily initiated by heat, even at low concentrations.

*See other* ORGANIC AZIDES

## NITROUREA $NO_2NHCONH_2$ $CH_3N_3O_3$

Urbanski, 1967, Vol. 3, 34

A rather unstable explosive material which gives mercuric and silver salts. These are much more impact-sensitive.

*See other* *N*-METAL DERIVATIVES
*N*-NITRO COMPOUNDS

## 5-AMINOTETRAZOLE N=NNHN=CNH$_2$ $CH_3N_5$

Nitrous acid

1. Elmore, D. T., *Chem. Brit.*, 1966, **2**, 414
2. Thiele, J., *Ann.*, 1892, **270**, 59
3. Gray, E. J. *et al., J. Chem. Soc., Perkin Trans. 1*, 1976, 1503

Diazotised 5-aminotetrazole is unstable under the conditions recommended for its use as a biochemical reagent. While the pH of the diazotised material (the cation of which contains 87% of nitrogen) at 0°C was being reduced to 5 by addition of potassium hydroxide, a violent explosion occurred [1]. This may have been caused by a local excess of alkali causing the formation of 5-diazoniotetrazolide, which will explode in concentrated solution at 0°C [2]. The diazonium chloride is also very unstable in concentrated solution at 0°C.

Small-scale diazotisation (2 g of amine) and subsequent coupling at pH 3 with ethyl cyanoacetate to prepare ethyl 2-cyano-(1*H*-tetrazol-5-ylhydrazono)acetate proceeded uneventfully, but on double the scale a violent explosion occurred [3].

*See other* DIAZONIUM SALTS
HIGH-NITROGEN COMPOUNDS
TETRAZOLES

## METHYLSODIUM $CH_3Na$

Schlenk, W. *et al., Ber.*, 1917, **50**, 262

Ignites immediately in air. Tendency to ignition decreases with ascent of the homologous series of alkylsodiums.

*See other* ALKYLMETALS

## SODIUM METHOXIDE $CH_3ONa$ $CH_3NaO$

Chloroform,
Methanol

*See* CHLOROFORM, $CHCl_3$ : Sodium methoxide

*p*-Chloronitrobenzene

*See* *p*-CHLORONITROBENZENE, $C_6H_4ClNO_2$ : Sodium methoxide

Perfluorocyclopropene

*See* FLUORINATED CYCLOPROPENYL METHYL ETHERS

## 'SODIUM PERCARBONATE' '$CH_3Na_2O_6$'

*See* SODIUM CARBONATE HYDROGEN PEROXIDATE, $CNa_2O_3 \cdot 1.5H_2O_2$

## † METHANE $CH_4$

Halogens or Interhalogens

*See* BROMINE PENTAFLUORIDE, $BrF_5$ : Hydrogen-containing materials
CHLORINE TRIFLUORIDE, $ClF_3$ : Methane
CHLORINE, $Cl_2$ : Hydrocarbons
FLUORINE, $F_2$ : Hydrocarbons
IODINE HEPTAFLUORIDE, $F_7I$: Carbon, etc.

Oxidants

*See* DIOXYGENYL TETRAFLUOROBORATE, $BF_4O_2$ : Organic materials
DIOXYGEN DIFLUORIDE, $F_2O_2$
TRIOXYGEN DIFLUORIDE, $F_2O_3$
OXYGEN (Liquid), $O_2$ : Hydrocarbons
: Liquefied gases

## CHLOROFORMAMIDINIUM NITRATE $CH_4ClN_3O_3$

$ClC(NH)NH_3NO_3$

Alone,
or Amines,
or Metals

1. Sauermilch, W., *Explosivstoffe*, 1961, **9**, 71–74 (*Chem. Abs.*, 1961, **55**, 21589f)
2. Sorbe, 1968, 144

It is powerfully explosive, and also an oxidant which reacts violently with ammonia or amines, and causes explosive ignition of wet magnesium powder [1], or powdered aluminium or iron [2].

*See* CHLOROFORMAMIDINIUM CHLORIDE, $CH_4Cl_2N_2$ : Oxyacids, etc.

*See other* OXOSALTS OF NITROGENOUS BASES

## CHLOROFORMAMIDINIUM CHLORIDE $CH_4Cl_2N_2$

$ClC(NH)NH_3Cl$

Oxyacids and salts

Sauermilch, W., *Explosivstoffe*, 1961, **9**, 71–74, 256 (*Chem. Abs.*, 1961, **55**, 21589f; 1962, **57**, 8434d)

The title compound ('cyanamide dihydrochloride') reacts with perchloric acid to form the perchlorate salt, and with ammonium nitrate, the nitrate salt, both being highly explosive.

## CHLOROFORMAMIDINIUM PERCHLORATE $CH_4Cl_2N_2O_4$

$ClC(NH)NH_3ClO_4$

*See* CHLOROFORMAMIDINIUM CHLORIDE, $CH_4Cl_2N_2$: Oxyacids, etc.

*See other* OXOSALTS OF NITROGENOUS BASES

## · DICHLORO(METHYL)SILANE $CH_3SiCl_2H$ $CH_4Cl_2Si$

Oxidants

1. Mueller, R. *et al., J. Prakt. Chem.*, 1966, **31**, 1–6
2. Sorbe, 1968, 128

The pure material is not ignited by impact, but it is in presence of potassium permanganate or trilead tetraoxide [1], or by copper or silver oxides, even under an inert gas [2].

*See also* TRICHLOROSILANE, $Cl_3HSi$

*See other* ALKYLNON-METAL HALIDES

## POTASSIUM METHYLAMIDE $KNHCH_3$ $CH_4KN$

Makhija, R. C. *et al., Can. J. Chem.*, 1971, **49**, 807

Extremely hygroscopic and pyrophoric; may explode on contact with air.

*See other N*-METAL DERIVATIVES

## METHYLDIAZENE $CH_3N{=}NH$ $CH_4N_2$

Oxygen

Ackermann, M. N. *et al., Inorg. Chem.*, 1972, **11**, 3077

Interaction on warming a mixture from $-196°C$ rapidly to ambient temperature is explosive.

*See other* AZO COMPOUNDS

## UREA $CO(NH_2)_2$ $CH_4N_2O$

Preparative hazard for $^{15}N$-labelled compound

*See* OXYGEN (Liquid), $O_2$: Ammonia, etc.

Chromyl chloride

*See* CHROMYL CHLORIDE, $Cl_2CrO_2$: Urea

Nitrosyl perchlorate

*See* NITROSYL PERCHLORATE, $ClNO_5$: Organic materials

Phosphorus pentachloride

*See* PHOSPHORUS PENTACHLORIDE, $Cl_5P$: Urea

Sodium nitrite

*See* SODIUM NITRITE, $NNaO_2$: Urea

Titanium tetrachloride

*See* TITANIUM TETRACHLORIDE, $Cl_4Ti$: Urea

## UREA HYDROGEN PEROXIDATE $CH_4N_2O \cdot H_2O_2$

$CO(NH_2)_2 \cdot H_2O_2$

*MCA Case History No. 719*

The contents of a screw-capped brown glass bottle spontaneously erupted after 4 years' storage at ambient temperature. All peroxides should be kept in special storage and periodically checked.

*See other* CRYSTALLINE HYDROGEN PEROXIDATES

## *N*-NITROMETHYLAMINE $NO_2NHCH_3$ $CH_4N_2O_2$

Sulphuric acid

Urbanski, 1967, Vol. 3, 16

The nitroamine is decomposed explosively by concentrated sulphuric acid.

*See other N*-NITRO COMPOUNDS

## AMMONIUM THIOCYANATE $NH_4SCN$ $CH_4N_2S$

Potassium chlorate

*See* POTASSIUM CHLORATE, $ClKO_3$: Thiocyanates

**THIOUREA** $CS(NH_2)_2$ $CH_4N_2S$

Acrylaldehyde

*See* ACRYLALDEHYDE, $C_3H_4O$: Acids, etc.

Hydrogen peroxide,
Nitric acid

*See* HYDROGEN PEROXIDE, $H_2O_2$: Nitric acid, etc.

**NITROSOGUANIDINE** $NONHC(NH)NH_2$ $CH_4N_4O$

Water

Henry, R. A. *et al.*, *Ind. Eng. Chem.*, 1949, **41**, 846–849
The compound is normally stored and transported as a water-wet paste which slowly decomposes at elevated ambient temperatures, evolving nitrogen. Precautions are described to prevent pressure build-up in sealed containers.

*See other* NITROSO COMPOUNDS

**NITROGUANIDINE** $NO_2NHC(NH)NH_2$ $CH_4N_4O_2$

1. Urbanski, 1967, Vol. 3, 31
2. McKay, A. F., *Chem. Rev.*, 1952, **51**, 301

Nitroguanidine is difficult to detonate, but its mercury and silver complex salts are much more impact-sensitive [1]. Many nitroguanidine derivatives have been considered as explosives [2].

*See other* *N*-METAL DERIVATIVES
*N*-NITRO COMPOUNDS

**METHYLENEBISNITROAMINE** $NO_2NHCH_2NHNO_2$ $CH_4N_4O_4$

Glowiak, B., *Chem. Abs.*, 1960, **54**, 21761e
It is a powerful and sensitive explosive, the explosion temperature being 217°C (lead salt, 195°C).

*See other* *N*-NITRO COMPOUNDS

**† METHANOL** $CH_3OH$ $CH_4O$

*MCA SD-22,* 1970

*See* Water, below

Acetyl bromide

*See* ACETYL BROMIDE, $C_2H_3BrO$: Hydroxylic compounds

Air

Ferris, T. V., *Loss Prevention*, 1974, **8**, 15–19

Explosive behaviour of methanol–air mixtures at 1.81 bar and 120°C was determined, with or without addition of oxygen and water.

Alkylaluminium solution

*MCA Case History No. 1778*

Accidental use of methanol in place of hexane to rinse out a hypodermic syringe used for a dilute alkylaluminium solution caused a violent reaction which blew the plunger out of the barrel.

*See* ALKYLALUMINIUM DERIVATIVES: Alcohols

Beryllium dihydride

*See* BERYLLIUM DIHYDRIDE, $BeH_2$: Methanol

Chloroform,
Sodium

*See* CHLOROFORM, $CHCl_3$: Sodium, Methanol

Chloroform,
Sodium hydroxide

*See* CHLOROFORM, $CHCl_3$: Sodium hydroxide, Methanol

Cyanuric chloride

*See* 2,4,6-TRICHLORO-1,3,5-TRIAZINE (CYANURIC CHLORIDE), $C_3Cl_3N_3$: Methanol

Dichloromethane

*See* DICHLOROMETHANE, $CH_2Cl_2$: Air, etc.

Diethylzinc

*See* DIALKYLZINCS: Methanol
DIETHYLZINC, $C_4H_{10}Zn$: Methanol

Metals

*See* POTASSIUM, K: Air (slow oxidation)
MAGNESIUM, Mg: Methanol

Oxidants

*See* *N*-HALOIMIDES: Alcohols
BARIUM PERCHLORATE, $BaCl_2O_8$: Alcohols
BROMINE, $Br_2$: Methanol
SODIUM HYPOCHLORITE, ClNaO: Methanol
CHLORINE, $Cl_2$: Methanol
LEAD DIPERCHLORATE, $Cl_2O_8Pb$: Methanol
CHROMIUM TRIOXIDE, $CrO_3$: Alcohols
NITRIC ACID, $HNO_3$: Alcohols
HYDROGEN PEROXIDE, $H_2O_2$: Oxygenated compounds

Phosphorus(III) oxide

*See* PHOSPHORUS(III) OXIDE, $O_3P_2$: Organic liquids

Potassium *tert*-butoxide

*See* POTASSIUM *tert*-BUTOXIDE, $C_4H_9KO$: Acids, etc.

Water

*MCA Case Histories Nos. 1822, 2085*

Static discharge ignited the contents of a polythene bottle being filled with a 40:60 mixture of methanol and water at 30°C, and a later similar incident in a fibre-lined tank involved a 30:70 mixture.

**METHYL HYDROPEROXIDE** $CH_3OOH$ $CH_4O_2$

Rieche, A. *et al., Ber.*, 1929, **62**, 2458

Violently explosive, shock-sensitive, especially on warming; great care is necessary in handling. The barium salt is dangerously explosive when dry.

Phosphorus(V) oxide

Rieche, A. *et al., Ber.*, 1929, **62**, 2460

The peroxide explodes violently on contact with the desiccant.

Platinum

Rieche, A. *et al., Ber.*, 1929, **62**, 2460

A 50% aqueous solution of the peroxide decomposed explosively on warming with spongy platinum.

*See other* ALKYL HYDROPEROXIDES

**HYDROXYMETHYL HYDROPEROXIDE** $HOCH_2OOH$ $CH_4O_3$

Rieche, A., *Ber.*, 1931, **64**, 2328; ibid., 1935, **68**, 1465

Explodes on heating, but is friction-insensitive. Higher homologues are not explosive.

*See other* 1-OXYPEROXY COMPOUNDS

**METHANESULPHONIC ACID** $CH_3SO_3H$ $CH_4O_3S$

Ethyl vinyl ether

*See* ETHYL VINYL ETHER, $C_4H_3O$: Methanesulphonic acid

**† METHANETHIOL** $CH_3SH$ $CH_4S$

Mercury(II) oxide

*See* MERCURY(II) OXIDE, HgO: Methanethiol

## URONIUM PERCHLORATE (UREA PERCHLORATE) $CH_5ClN_2O_5$

$NH_2CONH_3ClO_4$

Aromatic nitro compounds

1. Shimio, K. *et al.*, Jap. Kokai 75 25 720, 1975 (*Chem. Abs.*, 1975, **83**, 134504q)
2. Fujiwara, S. *et al.*, Jap. Kokai 74 134 812, 1974

Concentrated aqueous solutions of the urea salt will dissolve solid or liquid aromatic nitro compounds (e.g. picric acid [1] or nitrobenzene [2]) to give liquid high-velocity explosives.

*See other* OXOSALTS OF NITROGENOUS BASES

## † METHYLAMINE $CH_3NH_2$ $CH_5N$

*MCA SD-57*, 1955

Nitromethane

*See* NITROMETHANE, $CH_3NO_2$: Acids, etc.

## URONIUM NITRATE (UREA NITRATE) $NH_2CONH_3NO_3$ $CH_5N_3O_4$

1. Kirk-Othmer, 1970, Vol. 21, 38
2. Markalous, F. *et al.*, Czech Pat. 152 080, 1974
3. Lazarov, S. *et al.*, *Chem. Abs.*, 1973, **79**, 94169t

The nitrate decomposes explosively when heated [1]. Prepared in the presence of phosphates, the salt is much more stable, even when dry [2]. The manufacture and explosive properties of urea nitrate and its mixtures with other explosives are discussed in detail [3].

*See other* OXOSALTS OF NITROGENOUS BASES

## 1-AMINO-3-NITROGUANIDINE $NH_2NHC(NH)NHNO_2$ $CH_5N_5O_2$

Lieber, E. *et al.*, *J. Amer. Chem. Soc.*, 1951, **73**, 2328

Explodes at the m.p., 190°C

*See other* HIGH-NITROGEN COMPOUNDS

## METHYLPHOSPHINE $CH_3PH_2$ $CH_5P$

Houben-Weyl, 1963, Vol. 12.1, 69

Primary lower-alkylphosphines readily ignite in air.

*See other* ALKYLNON-METALS

## METHYLSTIBINE $CH_3SbH_2$ $CH_5Sb$

Sorbe, 1968, 28

Alkylstibines decompose explosively on heating or shock.

*See related* ALKYLMETALS

## METHYLAMMONIUM CHLORITE $CH_3NH_3^+ClO_2^-$ $CH_6ClNO_2$

Levi, G. R., *Gazz. Chim. Ital.*, 1922, [1], **52**, 207

A concentrated solution caused a slight explosion when poured on to a cold iron plate.

*See other* CHLORITE SALTS
OXOSALTS OF NITROGENOUS BASES

## METHYLAMMONIUM PERCHLORATE $CH_3NH_3^+ClO_4^-$ $CH_6ClNO_4$

Kasper, F., *Z. Chem.*, 1969, **9**, 34

The semi-crystalline mass exploded when stirred after standing overnight. The preparation was based on a published method used uneventfully for preparation of ammonium, dimethylammonium and piperidinium perchlorates.

*See other* AMINIUM PERCHLORATES
OXOSALTS OF NITROGENOUS BASES

## GUANIDINIUM PERCHLORATE $NH_2C(NH)NH_3^+ClO_4^-$ $CH_6ClN_3O_4$

Davis, 1943, 121; Schumacher, 1960, 213

Unusually sensitive to initiation, and of high explosive power, it decomposes violently at 350°C.

*See* DIFFERENTIAL THERMAL ANALYSIS

Diiron trioxide

Isaev, R. N. *et al.*, *Chem. Abs.*, 1970, **73**, 132626a

Addition of 10% of iron(III) oxide reduces the thermal stability of the salt.

*See other* OXOSALTS OF NITROGENOUS BASES

## † METHYLHYDRAZINE $CH_3NHNH_2$ $CH_6N_2$

Dicyanofurazan

*See* DICYANOFURAZAN, $C_4N_4O$: Nitrogenous bases

Oxidants

Kirk-Othmer, 1966, Vol. 11, 186

A powerful reducing agent and fuel, hypergolic with many oxidants such as dinitrogen tetraoxide or hydrogen peroxide.

*See* ROCKET PROPELLANTS

## AMMONIUM *aci*-NITROMETHANE $NH_4^+CH_2{=}N(O)O^-$ $CH_6N_2O_2$

Watts, C. E., *Chem. Eng. News*, 1952, **30**, 2344

The isolated salt is a friction-sensitive explosive.

*See other aci*-NITRO SALTS

**'UREA PEROXIDE'** '$CH_6N_2O_3$'

*See* UREA HYDROGEN PEROXIDATE, $CH_4N_2O \cdot H_2O_2$

**AMINOGUANIDINE** $NH_2NHC(NH)NH_2$ $CH_6N_4$

Kurzer, F. *et al.*, *Chem. & Ind.*, 1962, 1585

All the oxoacid salts are potentially explosive, including the wet nitrate.

*See* AMINOGUANIDINIUM NITRATE, $CH_7N_5O_3$

**CARBONOHYDRAZIDE** $CO(NHNH_2)_2$ $CH_6N_4O$

Sorbe, 1968, 74

It explodes on heating.

Nitrous acid

Curtius, J. *et al.*, *Ber.*, 1894, 27, 55

Interaction forms the highly explosive carbonyl diazide.

*See* CARBONYL DIAZIDE, $CN_6O$

*See other* HIGH-NITROGEN COMPOUNDS

**GUANIDINIUM NITRATE** $NH_2C(NH)NH_3^+NO_3^-$ $CH_6N_4O_3$

1. Davis, T. L., *Org. Synth.*, 1941, Coll. Vol. 1, 302
2. Smith, G. B. L. and Schmidt, M. T., *Inorg. Synth.*, 1939, **1**, 96, 97

According to an *O. S.* amendment sheet, the procedure as described is dangerous because the reaction mixture [1] (dicyanodiamide and ammonium nitrate) is similar in composition to commercial blasting explosives. This probably also applies to similar earlier preparations [2].

*See other* OXOSALTS OF NITROGENOUS BASES

**METHYLSILANE** $CH_3SiH_3$ $CH_6Si$

Mercury,

Oxygen

Stock, A. *et al.*, *Ber.*, 1919, **52**, 706

Does not ignite in air, but explodes if shaken with mercury in oxygen.

*See other* ALKYLNON-METAL HYDRIDES

**AMINOGUANIDINIUM NITRATE** $CH_7N_5O_3$

$NH_2NHC(NH)NH_3^+NO_3^-$

Koopman, H., *Chem. Weekbl.*, 1957, **53**, 97

An aqueous solution exploded violently during evaporation on a steam bath. Nitrate salts of many organic bases are unstable and should be avoided.

*See* AMINOGUANIDINE, $CH_6N_4$
*See other* HIGH-NITROGEN COMPOUNDS
OXOSALTS OF NITROGENOUS BASES

**DIAMINOGUANIDINIUM NITRATE** $CH_8N_6O_3$

$(NH_2NH)_2C{=}NH_2^+\ NO_3^-$

Violent decomposition occurred at 260°C.

*See* DIFFERENTIAL THERMAL ANALYSIS
*See other* OXOSALTS OF NITROGENOUS BASES

**TRIAMINOGUANIDINIUM PERCHLORATE** $CH_9ClN_6O_4$

$(NH_2NH)_2C{=}NNH_3^+\ ClO_4^-$

Violent decomposition occurred at 317°C.

*See* DIFFERENTIAL THERMAL ANALYSIS
*See other* OXOSALTS OF NITROGENOUS BASES

**TRIAMINOGUANIDINIUM NITRATE** $CH_9N_7O_3$

$(NH_2NH)_2C{=}NNH_3^+\ NO_3^-$

Violent decomposition occurred at 230°C.

*See* DIFFERENTIAL THERMAL ANALYSIS
*See other* OXOSALTS OF NITROGENOUS BASES

**PENTAAMMINETHIOCYANATOCOBALT(III) PERCHLORATE**

$CH_{15}Cl_2CoN_6O_8S$

$[Co(NH_3)_5SCN][ClO_4]_2$

Explodes at 325°C; medium impact-sensitivity.

*See* AMMINEMETAL OXOSALTS

**PENTAAMMINETHIOCYANATORUTHENIUM(III) PERCHLORATE**

$CH_{15}Cl_2N_6O_8RuS$

$[Ru(NH_3)_5SCN][ClO_4]_2$

Armor, J. N., private comm., 1969

After washing with ether, 0.1 g of the complex exploded violently when touched with a spatula.

*See other* AMMINEMETAL OXOSALTS

**MERCURY DIAZOCARBIDE** $[HgC(N_2)]_n$ $(CHgN_2)_n$

Houben-Weyl, 1974, Vol. 13.2b, 19

Interaction of diazomethane with bis(trimethylsilylamino)mercury gives a polymeric explosive solid.

*See other* HEAVY METAL DERIVATIVES
MERCURY COMPOUNDS

**MERCURY(I) CYANAMIDE** **$Hg_2NCN$** **$CHg_2N_2$**

Deb, S.K. *et al., Trans. Faraday Soc.*, 1959, **55**, 106–113
Relatively large particles explode on rapidly heating to 325°C, or under high-intensity illumination when confined.

*See other* HEAVY METAL DERIVATIVES
*N*-METAL DERIVATIVES

**IODINE ISOCYANATE** **IN=C=O** **CINO**

Rosen, S. *et al., Anal. Chem.*, 1966, **38**, 1394
On storage, solutions of iodine isocyanate gradually deposit a touch-sensitive, mildly explosive solid (possibly cyanogen peroxide).

*See other* IODINE COMPOUNDS

**CARBON TETRAIODIDE** **$CI_4$**

Bromine trifluoride
*See* BROMINE TRIFLUORIDE, $BrF_3$ : Halogens, etc.

Lithium
*See* LITHIUM, Li: Halocarbons

**POTASSIUM CYANIDE** **KC≡N** **CKN**

Mercury(II) nitrate
*See* MERCURY(II) NITRATE, $HgN_2O_6$ : Potassium cyanide

Nitrogen trichloride
*See* NITROGEN TRICHLORIDE, $Cl_3N$: Initiators

Perchloryl fluoride
*See* PERCHLORYL FLUORIDE, $ClFO_3$ : Calcium acetylide, etc.

Sodium nitrite
*See* SODIUM NITRITE, $NNaO_2$ : Metal cyanides

## POTASSIUM CYANIDE–POTASSIUM NITRITE $CKN \cdot KNO_2$

$KCN \cdot KNO_2$

Sorbe, 1968, 68

The double salt (a redox compound) is explosive.

*See other* REDOX COMPOUNDS

## POTASSIUM THIOCYANATE $KSC{\equiv}N$ CKNS

Calcium chlorite

*See* CALCIUM CHLORITE, $CaCl_2O_4$ : Potassium thiocyanate

Perchloryl fluoride

*See* PERCHLORYL FLUORIDE, $ClFO_3$ : Calcium acetylide, etc.

## POTASSIUM TRINITROMETHANIDE ('NITROFORM' SALT) $CKN_3O_6$

$KC(NO_2)_3$

1. Sandler, S. R. *et al., Organic Functional Group Preparations*, 433, New York, Academic Press, 1968
2. Shulgin, A. T., private comm., 1968

This intermediate, produced by action of alkali on tetranitromethane, must be kept damp and used as soon as possible with great care, as it may be explosive [1]. Material produced as a by-product in a nitration reaction using tetranitromethane was washed with acetone. It exploded very violently after several months' storage [2].

*See also* TETRANITROMETHANE, $CN_4O_8$

*See other* POLYNITROALKYL COMPOUNDS

## 'CARBONYLPOTASSIUM' 'CKO'

*See* the hexameric POTASSIUM BENZENEHEXOLATE, $C_6K_6O_6$

*See also* CARBONYLMETALS

## POTASSIUM CARBONATE $K_2CO_3$ $CK_2O_3$

Carbon

Druce, J. G. F., *School Sci. Rev.*, 1926, 7(28), 261

Potassium metal prepared by the old process of distilling an intimate mixture of the carbonate and carbon contained some carbonylpotassium and several explosions with old samples of potassium may have involved this compound.

*See* POTASSIUM, K

Chlorine trifluoride

*See* CHLORINE TRIFLUORIDE, $ClF_3$ : Metals, etc.

Magnesium

*See* MAGNESIUM, Mg: Potassium carbonate
*See other* METAL OXONON-METALLATES

## 'CARBONYLLITHIUM' 'CLiO'

*See* the hexameric LITHIUM BENZENEHEXOLATE ('CARBONYLLITHIUM'), $C_6Li_6O_6$
*See also* CARBONYLMETALS

## LITHIUM CARBONATE $Li_2CO_3$ $CLi_2O_3$

Fluorine

*See* FLUORINE, $F_2$: Metal salts

## MAGNESIUM CARBONATE $MgCO_3$ $CMgO_3$

Formaldehyde

*See* FORMALDEHYDE, $CH_2O$: Magnesium carbonate

## *N*-FLUOROIMINODIFLUOROMETHANE $CNF_3$

$FN{=}CF_2$

Ginsberg, V. A. *et al., Zh. Obsch. Khim.*, 1967, **37**, 1413
It boils at −60°C and explodes on warming.
*See other* *N*-HALOGEN COMPOUNDS

## SODIUM CYANIDE $NaC{\equiv}N$ CNNa

It is oxidised violently or explosively in contact with hot oxidants.
*See* METAL CYANIDES AND CYANOCOMPLEXES

## SODIUM FULMINATE $NaC{\equiv}N{\rightarrow}O$ CNNaO

Smith, 1966, Vol. 2, 99
Even sodium fulminate detonates when touched lightly with a glass rod.
*See other* METAL FULMINATES

## SODIUM THIOCYANATE $NaSC{\equiv}N$ CNNaS

Sodium nitrite

*See* SODIUM NITRITE, $NNaO_2$: Sodium thiocyanate

**THALLIUM FULMINATE** $TlC{\equiv}N{\rightarrow}O$ **CNOTl**

Boddington, T. *et al.*, *Trans. Faraday Soc.*, 1969, **65**, 509

Explosive, even more shock- and heat-sensitive than mercury(II) fulminate.

*See* METAL FULMINATES

**NITROSYL CYANIDE** $O{=}N{-}C{\equiv}N$ **$CN_2O$**

(Nitrogen oxide)

Kirby, G. W., *Chem. Soc. Rev.*, 1977, 9

Direct preparation of the gas is potentially hazardous, and explosive decomposition of the impure gas in the condensed state (below −20°C) has occurred. A safe procedure involving isolation of the 1:1 adduct with 9,10-dimethylanthracene is preferred. The impure gas contains NO and it is known that nitrosyl cyanide will react with the latter to form an explosive compound.

*See other* CYANO COMPOUNDS
NITROSO COMPOUNDS

**THALLIUM(I) AZIDODITHIOCARBONATE** **$CN_3S_2Tl$**

$TlSC(S)N_3$

Sulphuric acid

Bailar, 1973, Vol. 1, 1155

The highly unstable explosive salt is initiated by contact with sulphuric acid.

*See also* AZIDODITHIOCARBONIC ACID, $CHN_3S_2$

**CYANOGEN AZIDE** $N{\equiv}CN_3$ **$CN_4$**

Marsh, F. D., *J. Amer. Chem. Soc.*, 1964, **86**, 4506; *J. Org. Chem.*, 1972, **37**, 2966

Coppolino, A. P., *Chem. Eng. News*, 1974, **52**(25), 3

It detonates with great violence when subjected to mild mechanical, thermal or electrical shock. Special precautions are necessary to prevent separation of the azide from solution and to decontaminate equipment and materials.

The azide is not safe in storage at −20°C as previously stated. A sample exploded violently immediately after removal from refrigerated storage.

*See also* DIAZIDOMETHYLENECYANAMIDE, $C_2N_8$

Sodium hydroxide

Marsh, F. D., *J. Org. Chem.*, 1972, **37**, 2966

Interaction with 10% alkali forms sodium 5-azidotetrazolide, which is violently explosive if isolated.

*See* 5-AZIDOTETRAZOLE, $CHN_7$

*See related* HALOGEN AZIDES

**DINITRODIAZOMETHANE** $(NO_2)_2CN_2$ $CN_4O_4$

Schallkopf, V. *et al., Angew. Chem. (Intern. Ed.),* 1969, 8, 612

It explodes on impact, rapid heating or contact with concentrated sulphuric acid.

*See other* DIAZO COMPOUNDS

POLYNITROALKYL COMPOUNDS

**TETRANITROMETHANE** $C(NO_2)_4$ $CN_4O_8$

Liang, P., *Org. Synth.,* 1955, Coll. Vol. 3, 804

During its preparation from fuming nitric acid and acetic anhydride, strict temperature control and rate of addition of anhydride are essential to prevent a runaway violent reaction. Tetranitromethane should not be distilled, since explosive decomposition may occur.

*See* NITRIC ACID, $HNO_3$: Acetic anhydride

Aluminium

*See* ALUMINIUM, Al: Oxidants

Amines

Gol'binder, A. A., *Chem. Abs.*, 1963, **59**, 9730b

Mixtures of e.g. aniline with tetranitromethane (19.5%) ignite in 35–55 s, and will proceed to detonation if the depth of liquid is above a critical value.

Aromatic nitrocompounds

Urbanski, 1964, Vol. 1, 592

Mixtures of nitrobenzene, *o*- or *p*-nitrotoluene, *m*-dinitrobenzene or 1-nitronaphthalene with tetranitromethane were found to be high explosives of high sensitivity and detonation velocities.

Hydrocarbons

1. Stettbacher, A., *Z. Ges. Schiess. und Sprengstoffw*., 1930, **25**, 439
2. Hager, K. F., *Ind. Eng. Chem.*, 1949, **41**, 2168
3. Tschinkel, J. G., *Ind. Eng. Chem.* 1956, **48**, 732
4. Stettbacher, A., *Tech. Ind. Schweiz. Chem. Ztg*., 1941, **24**, 265–271
5. Anon., *Chem. Ztg*., 1920, **44**, 497 (*Chem. Abs*., 1920, **14**, 2988)

When mixed with hydrocarbons in approximately stoicheiometric proportions, a sensitive, highly explosive mixture is produced, which needs careful handling [1,2]. The use of such mixtures as rocket propellants has also been investigated [3].

Explosion of only 10 g of a mixture with toluene caused 10 deaths and 20 severe injuries [4]. The mixture contained excess toluene in error [5].

Sodium ethoxide

Macbeth, A. K., *Ber.*, 1913, **46**, 2537–2538

Addition of the last of several small portions of the ethoxide solution caused the violent explosion of 30 g of tetranitromethane.

Toluene,
Cotton

Winderlich, R., *J. Chem. Educ.*, 1950, **27**, 669

A demonstration mixture, to show combustion of cotton by combined oxygen, exploded with great violence soon after ignition.

*See* Hydrocarbons, above

*See other* POLYNITROALKYL COMPOUNDS

**5-DIAZONIOTETRAZOLIDE** $N{=}NN^{-}N{=}CN_2^+$ (ring) $CN_6$

Thiele, J., *Ann.*, 1892, **270**, 60

Concentrated aqueous solutions of this internal diazonium salt explode at 0°C.

*See other* DIAZONIUM SALTS
TETRAZOLES

**CARBONYL DIAZIDE** $CO(N_3)_2$ $CN_6O$

1. Chapman, L. E. *et al.*, *Chem. & Ind.*, 1966, 1266
2. Kesting, W., *Ber.*, 1924, **57**, 1321

It is a violently explosive solid, which should be used only in solution, and on a small scale [1]. It exploded violently even under ice-water [2].

*See other* ACYL AZIDES

**'CARBONYLSODIUM'** 'CNaO'

*See* the hexameric SODIUM BENZENEHEXOLATE, $C_6Na_6O_6$

*See also* CARBONYLMETALS

**SODIUM CARBONATE** $Na_2CO_3$ $CNa_2O_3$

Ammonia,
Silver nitrate

*See* SILVER NITRATE, $AgNO_3$: Ammonia, etc.

An aromatic amine,
A chloronitro compound

*MCA Case History No. 1964*

An unspecified process had been operated for 20 years using synthetic sodium carbonate powder (soda-ash) to neutralise the hydrogen chloride as it was formed by interaction of the amine and chloro-compound in a non-aqueous (and probably flammable) solvent in a steel reactor. Substitution of the powdered sodium carbonate by the crystalline sodium carbonate–sodium hydrogencarbonate double salt ('trona, natural soda') caused a reduction in the rate of neutralisation, the reaction mixture became acid, and attack of the steel vessel led to contamination by iron. These changed conditions initiated exothermic side reactions, which eventually ran out of control and caused failure of the reactor.

Subsequent laboratory work confirmed this sequence and showed that presence of dissolved iron(III) was necessary to catalyse the side reactions.

*See* CATALYSIS BY IMPURITIES

2,4-Dinitrotoluene

*See* 2,4-DINITROTOLUENE, $C_7H_6N_2O_4$: Alkali

Fluorine

*See* FLUORINE, $F_2$: Metal salts

Lithium

*See* LITHIUM, Li: Sodium carbonate

Phosphorus pentaoxide

*See* PHOSPHORUS(V) OXIDE, $O_5P_2$: Inorganic bases

Sodium sulphide,
Water

*See* SMELT: Water

Sulphuric acid

*MCA Case History No. 888*

Lack of any mixing arrangements caused stratification of strong sulphuric acid and (probably) sodium carbonate solutions in the same tank. When gas evolution caused intermixture of the layers, a violent eruption of the tank contents occurred.

2,4,6-Trinitrotoluene

*See* 2,4,6-TRINITROTOLUENE, $C_7H_5N_3O_6$: Added impurities

*See other* METAL OXONON-METALLATES

**SODIUM CARBONATE HYDROGEN PEROXIDATE** **$CNa_2O_3 \cdot 1.5H_2O_2$**

**$Na_2CO_3 \cdot 1.5H_2O_2$**

*See* CRYSTALLINE HYDROGEN PEROXIDATES

**† CARBON MONOXIDE** **CO**

Aluminium,
Aluminium halides

*See* ALUMINIUM, Al: Aluminium halides, Carbon oxides

Dinitrogen oxide

*See* DINITROGEN OXIDE, $N_2O$: Carbon monoxide

Fluorine,
Oxygen

*See* BISFLUOROFORMYL PEROXIDE, $C_2F_2O_4$

Interhalogens

*See* BROMINE TRIFLUORIDE, $BrF_3$: Carbon monoxide
BROMINE PENTAFLUORIDE, $BrF_5$: Acids, etc.
IODINE HEPTAFLUORIDE, $F_7I$: Carbon, etc.

Metal oxides

*See* SILVER(I) OXIDE, $Ag_2O$: Carbon monoxide
DICAESIUM OXIDE, $Cs_2O$: Halogens, etc.
DIIRON TRIOXIDE, $Fe_2O_3$: Carbon monoxide

Metals

*See* POTASSIUM, K: Non-metal oxides
SODIUM, Na: Non-metal oxides

Oxidants

*See* CHLORINE DIOXIDE, $ClO_2$: Carbon monoxide
PEROXODISULPHURYL DIFLUORIDE, $F_2O_6S_2$: Carbon monoxide
OXYGEN (Liquid), $O_2$: Liquefied gases

**· CARBONYL SULPHIDE** **COS**

**CARBON DIOXIDE** **$CO_2$**

Acrylaldehyde

*See* ACRYLALDEHYDE, $C_3H_4O$: Acids

Aluminium,
Aluminium halides

*See* ALUMINIUM, Al: Aluminium halides, Carbon oxides

Aziridine

*See* AZIRIDINE (ETHYLENEIMINE), $C_2H_5N$: Acids

Barium peroxide

*See* BARIUM PEROXIDE, $BaO_2$ : Non-metal oxides

Dicaesium oxide

*See* DICAESIUM OXIDE, $Cs_2O$: Halogens, etc.

Flammable materials

1. *BCISC Quart. Safety Summ*., 1973, **44**(174), 10
2. Leonard, J. T. *et al*., *Inst. Phys. Conf. Ser*., 1975, **27**(Static Electrif.), 301–310 (*Chem. Abs*., 1976, **84**, 107987p)
3. Anon., *Chem. Eng. News*, 1976, **54**(30), 18

Dangers attached to the use of carbon dioxide in fire prevention and extinguishing systems in confined volumes of air and flammable vapours are discussed. The main hazard arises from the production of very high electrostatic charges, due to the presence of small solid particles of carbon dioxide generated during discharge of the compressed gas. Highly incendive sparks (5–15 mJ at 10–20 kV) readily may be produced. Nitrogen is preferred as an inerting gas for enclosed volumes of air and flammable gases or vapours [1].

The electrostatic hazard in discharge of carbon dioxide in tank fire-extinguishing systems is associated with the plastics distribution horns, and their removal eliminated the hazardous potential [2]. Alternatively, a metal sleeve may be fitted inside the plastics horn [3].

Metal hydrides

*See* ALUMINIUM TRIHYDRIDE, $AlH_3$ : Carbon dioxide
LITHIUM TETRAHYDROALUMINATE, $AlH_4Li$: Bis(2-methoxyethyl) ether

Metal acetylides

*See* POTASSIUM ACETYLIDE, $C_2HK$: Non-metal oxides
LITHIUM ACETYLIDE–AMMONIA, $C_2Li \cdot H_3N$: Gases
DISODIUM ACETYLIDE, $C_2Na_2$ : Non-metal oxides
DIRUBIDIUM ACETYLIDE, $C_2Rb_2$ : Non-metal oxides

Metals

1. Hartmann, I., *Ind. Eng. Chem*., 1948, **40**, 756
2. Rhein, R. A., *Rept. No. CR-60125*, Washington, NASA, 1964

Dusts of magnesium, zirconium, titanium and some magnesium–aluminium alloys [1], and (when heated) of aluminium, chromium and manganese [2], when suspended in carbon dioxide atmospheres are ignitable and explosive, and several bulk metals will burn in the gas.

*See* Metals, Nitrogen below

ALUMINIUM, Al: Aluminium halides, Carbon oxides
POTASSIUM, K: Non-metal oxides
POTASSIUM–SODIUM ALLOY, K–Na: Carbon dioxide
LITHIUM, Li: Atmospheric gases
: Non-metal oxides
MAGNESIUM, Mg: Carbon dioxide, etc.
SODIUM, Na: Non-metal oxides
TITANIUM, Ti: Carbon dioxide
URANIUM, U: Carbon dioxide

Metals,
Nitrogen

Rhein, R. A., *Rept. No. CR-60125*, Washington, NASA, 1964
Powdered beryllium, calcium, cerium (and alloys), thorium, titanium, uranium and zirconium ignite on heating in mixtures of carbon dioxide and nitrogen; ignition temperatures were determined. Aluminium, chromium, magnesium and manganese ignite only in absence of nitrogen.

Metals,
Sodium peroxide

*See* SODIUM PEROXIDE, $Na_2O_2$: Metals, Carbon dioxide
*See other* NON-METAL OXIDES

## LEAD CARBONATE $PbCO_3$ $CO_3Pb$

Fluorine

*See* FLUORINE, $F_2$: Metal salts

## CARBON SULPHIDE $CS^{\cdot}$ CS

Pearson, T. G., *School. Sci. Rev.*, 1938, **20**(78), 189
The gaseous radical readily polymerises explosively.

## † CARBON DISULPHIDE $CS_2$

*MCA SD-12,* 1967

Air,
Rust

1. Mee, A. J., *School Sci. Rev.*, 1940, **22**(86), 95
2. Dickens, G. A., *School Sci. Rev.*, 1950, **31**(114), 264
3. Anon., *J.R. Inst. Chem.*, 1956, **80**, 664

Disposal of 2 litres of the solvent into a rusted iron sewer caused an explosion. Initiation of the solvent–air mixture by the rust was suspected [1]. A hot gauze falling from a tripod into a laboratory sink containing some carbon disulphide initiated two explosions [2]. It is a hazardous solvent because of its high volatility and flammability. The vapour or liquid has been known to ignite on contact with steam pipes, particularly if rusted [3].

Halogens

*See* CHLORINE, $Cl_2$: Carbon disulphide
FLUORINE, $F_2$: Sulphides

Metal azides

Mellor, 1947, Vol. 8, 338

Carbon disulphide and aqueous solutions of metal azides interact to produce metal azidodithioformates most of which are explosive, with varying degrees of power and sensitivity to shock or heat.

*See also* AZIDODITHIOFORMIC ACID, $CHN_3S_2$

Metals

*See* ALUMINIUM, Al: Carbon disulphide
POTASSIUM–SODIUM ALLOY, K–Na: Carbon dioxide, etc.
SODIUM, Na: Sulphides
ZINC, Zn: Carbon disulphide

Oxidants

*See* Halogens, above
PERMANGANIC ACID, $HMnO_4$: Organic materials
NITROGEN OXIDE ('NITRIC OXIDE'), NO: Carbon disulphide
DINITROGEN TETRAOXIDE (NITROGEN DIOXIDE), $N_2O_4$: Carbon disulphide

**TITANIUM CARBIDE** **TiC** **CTi**

*MCA Case History No. 618*

A violent, and apparently spontaneous, dust explosion occurred while the finely ground carbide was being removed from a ball-mill. Static initiation seems a likely possibility.

*See also* DUST EXPLOSIONS
*See other* METAL NON-METALLIDES

**URANIUM MONOCARBIDE** **UC** **CU**

Schmitt, C. R., *J. Fire Flammability*, 1971, 2, 163

The finely divided carbide is pyrophoric.

*See other* PYROPHORIC MATERIALS

**TUNGSTEN CARBIDE** **WC** **CW**

Fluorine

*See* FLUORINE, $F_2$: Metal acetylides and carbides

Nitrogen oxides

Mellor, 1946, Vol. 5, 890

At about 600°C, the carbide ignites and incandesces in dinitrogen mono- or tetra-oxides.

**DITUNGSTEN CARBIDE** **$W_2C$** **$CW_2$**

Oxidants

Mellor, 1946, Vol. 5, 890

The carbide burns incandescently at red heat in contact with dinitrogen mono- or tetra-oxide.

*See also* FLUORINE, $F_2$: Metal acetylides and carbides

*See other* METAL NON-METALLIDES

**SILVER CYANODINITROMETHIDE** **$AgC(CN)(NO_2)_2$** **$C_2AgN_3O_4$**

Parker, C. O. *et al.*, *Tetrahedron*, 1962, **17**, 86

A sample in a m.p. capillary exploded at 196°C, shattering the apparatus.

*See other* HEAVY METAL DERIVATIVES
POLYNITROALKYL COMPOUNDS

**DISILVER ACETYLIDE** **AgC≡CAg** **$C_2Ag_2$**

Miller, 1965, Vol. 1, 486

Bailar, Vol. 3, 102

Reynolds, R. J. *et al.*, *Analyst*, 1971, **96**, 319

Silver acetylide is a more powerful detonator than the copper derivative, but both will initiate explosive acetylene-containing gas mixtures.

It decomposes violently when heated to 120–140°C. Formation of a deposit of this explosive material was observed when silver-containing solutions were aspirated into an acetylene-fuelled atomic absorption spectrometer. Precautions to prevent formation are discussed.

**DISILVER ACETYLIDE–SILVER NITRATE** **$C_2Ag_2 \cdot AgNO_3$**

**$AgC{\equiv}CAg \cdot AgNO_3$**

Baker, W. E. *et al.*, *Chem. Eng. News*, 1965, **43**(49), 46

*OTS Rept. AD 419 625*, Hogan, V. D. *et al.*, Washington, US Dept. Comm., 1960

The dry complex is exploded by high-intensity light pulses, or by heat or sparks. As a slurry in acetone, it is stable for a week if kept dark.

The complex explodes violently at 245°C/1 bar and at 195°C/1.3 mbar.

*See related* METAL ACETYLIDES

**DISILVER KETENDIIDE** **$Ag_2C{=}C{=}O$** **$C_2Ag_2O$**

Blues, E. T. *et al.*, *Chem. Comm.*, 1970, 699

This and its pyridine complex explode violently if heated or struck.

*See other* HEAVY METAL DERIVATIVES

**DISILVER KETENDIIDE–SILVER NITRATE** **$C_2Ag_2O \cdot AgNO_3$**
**$Ag_2C{=}C{=}O \cdot AgNO_3$**

1. Bryce-Smith, D., *Chem. & Ind.*, 1975, 154
2. Blues, E. T. *et al.*, *J. Chem. Soc., Chem. Comm.*, 1970, 701

The red complex is more dangerously explosive than disilver ketendiide itself, now described as mildly explosive in the dry state when heated strongly or struck [1]. It was previously formulated as a 2:1 complex [2].

*See other* HEAVY METAL DERIVATIVES

**SILVER OXALATE** **$AgO_2CCO_2Ag$** **$C_2Ag_2O_4$**

1. Sidgwick, 1950, 126
2. Anon., *BCISC Quart. Safety Summ.*, 1973, **44**, 19
3. Kabanov, A. A. *et al.*, *Russ. Chem. Rev.*, 1975, **44**, 538
4. MacDonald, J. Y. *et al.*, *J. Chem. Soc.*, 1925, **127**, 2675

Above 140°C its exothermic decomposition to metal and carbon dioxide readily becomes explosive [1]. A 1 kg batch which had been thoroughly dried at 50°C exploded violently when mechanical grinding in an end-runner mill was attempted [2]. It is a compound of zero oxygen balance. The explosion temperature of silver oxalate is lowered appreciably (from 143 to 122°C) by application of an electric field [3].

*See* ELECTRIC FIELDS

The salt prepared from silver nitrate with excess of sodium oxalate is much less stable than that from excess nitrate [4].

*See other* HEAVY METAL DERIVATIVES
METAL OXALATES

**DIGOLD ACETYLIDE** **$AuC{\equiv}CAu$** **$C_2Au_2$**

1. Mellor, 1946, Vol. 5, 855

2. Matthews, J. A. *et al.*, *J. Amer. Chem. Soc.*, 1900, 22, 110
Produced by action of acetylene on sodium gold thiosulphate (or other gold salts), the acetylide is explosive and readily initiated by light impact, friction or rapid heating to 83°C [1]. This unstable detonator is noted for high brisance [2].
*See other* METAL ACETYLIDES

**DIGOLD(I) KETENDIIDE** **$Au_2C{=}C{=}O$** **$C_2Au_2O$**

Blues, E. T. *et al.*, *J. Chem. Soc., Chem. Comm.*, 1974, 513–514
The ketenide is shock-sensitive when dry, and it or its complexes with tertiary heterocyclic bases (pyridine, methylpyridines, 2,6-dimethylpyridine, quinoline) explode when heated above 100°C.
*See other* HEAVY METAL DERIVATIVES

**BARIUM ACETYLIDE** **$baC{\equiv}Cba$** **$C_2Ba$**

Masdupay, E. *et al.*, *Compt. Rend.*, 1951, **232**, 1837–1839
It is much more reactive than the diacetylide, and ignites in contact with water or ethanol in air. It may incandesce on heating to 150°C under vacuum or hydrogen, the product from the latter treatment being very pyrophoric owing to the presence of pyrophoric carbon.

Halogens
Mellor, 1946, Vol. 5, 862
Barium acetylide incandesces with chlorine, bromine and iodine at 140 130, 122°C, respectively.

Selenium
Mellor, 1946, Vol. 5, 862
A mixture incandesces when heated to 150°C.
*See other* METAL ACETYLIDES

**BARIUM THIOCYANATE** **$Ba(SCN)_2$** **$C_2BaN_2S_2$**

*See* POTASSIUM CHLORATE, $ClKO_3$ : Metal thiocyanates

Sodium nitrate
*See* SODIUM NITRATE, $NNaO_3$ : Barium thiocyanate

**† BROMOTRIFLUOROETHYLENE** **$BrFC{=}CF_2$** **$C_2BrF_3$**

Oxygen
*See* OXYGEN (Gas), $O_2$ : Halocarbons
*See other* HALOALKENES

**LITHIUM BROMOACETYLIDE** **LiC≡CBr** **$C_2BrLi$**

*See* LITHIUM CHLOROACETYLIDE, $C_2ClLi$

**DIBROMOACETYLENE** **BrC≡CBr** **$C_2Br_2$**

Rodd, 1951, Vol. 1A, 284

It ignites in air and explodes on heating.

*See other* HALOACETYLENE DERIVATIVES

**OXALYL DIBROMIDE** **BrCOCOBr** **$C_2Br_2O_2$**

Potassium

*See* POTASSIUM, K: Oxayl dihalides

**HEXABROMOETHANE** **$C_2Br_6$**

Hexacyclohexyldilead

*See* HEXACYCLOHEXYLDILEAD, $C_{36}H_{66}Pb_2$

**CALCIUM ACETYLIDE (CARBIDE)** **caC≡Cca** **$C_2Ca$**

1. Anon., *Ind. Safety Bull.*, 1940, **8**, 41
2. Déribère, M., *Chem. Abs.*, 1936, **30**, $8619_8$
3. Orlov, V. N., *Chem. Abs.*, 1975, **83**, 168138p

Use of a steel chisel to open a drum of carbide caused an incendive spark which ignited traces of acetylene in the drum. The non-ferrous tools normally used for this purpose should be kept free from embedded ferrous particles [1]. If calcium carbide is warm when filled into drums which are then hermetically sealed, absorption of the nitrogen from the trapped air may enrich the oxygen content up to 28%. In this case, 3% of acetylene (liberated by moisture) is enough to form an explosive mixture, which may be initiated on opening the sealed drum. Other precautions are detailed [2]. Use of carbon dioxide to purge carbide drums, and of brass or bronze non-sparking tools to open them are advocated [3].

Halogens

Mellor, 1946, Vol. 5, 862

The acetylide incandesces with chlorine, bromine and iodine at 245, 350 and 305°C, respectively. Strontium and barium acetylides are more reactive.

Hydrogen chloride

Mellor, 1946, Vol. 5, 862

Incandescence on warming; strontium and barium acetylides are similar.

Iron(III) chloride,
Iron(III) oxide
Partington, 1967, 372
The carbide is an energetic reducant. A powdered mixture with iron oxide and chloride burns violently when ignited, producing molten iron

Lead difluoride
Mellor, 1946, Vol. 5, 864
Incandescence on contact at ambient temperature.

Magnesium
Mellor, 1940, Vol. 4, 271
A mixture incandesces when heated in air.

Methanol
Unpublished observations, 1951
Interaction of calcium acetylide with boiling methanol to give calcium methoxide is very vigorous, but subject to an induction period of variable length. Once reaction starts, evolution of acetylene gas is very fast, and a wide-bore condenser and adequate ventilation are necessary.

Perchloryl fluoride
*See* PERCHLORYL FLUORIDE, $ClFO_3$: Calcium acetylide

Selenium
*See* SELENIUM, Se: Metal acetylides

Silver nitrate
Luchs, J. K., *Photog. Sci. Eng.*, 1966, **10**, 334
Addition of calcium acetylide to silver nitrate solution precipitates silver acetylide, a highly sensitive explosive. Copper salt solutions would behave similarly.
*See other* HEAVY METAL DERIVATIVES

Sodium peroxide
*See* SODIUM PEROXIDE, $Na_2O_2$ Calcium acetylide

Sulphur
*See* SULPHUR, S: Metal acetylides

Tin dichloride
Mellor, 1941, Vol. 7, 430

A mixture can be ignited with a match, and reduction to metallic tin proceeds with incandescence.

Water

Jones, G. W. *et al., Explosions in Med. Press. Acetylene Generators,* Invest. Report 3755, Washington, US Bureau of Mines, 1944

Moll, H., *Chem. Weekbl.*, 1933, **30**, 108

At the end of a generation run, maximum temperatures and high moisture content of acetylene may cause the finely divided acetylide to overheat and initiate explosion of compressed gas.

During analysis of technical carbide by addition of water, the explosive mixture formed in the unpurged reaction vessel exploded, ignited either by excessive local temperature or possibly by formation of diphosphane.

*See other* METAL ACETYLIDES

**CADMIUM DICYANIDE** $Cd(CN)_2$ $C_2CdN_2$

Magnesium

*See* MAGNESIUM, Mg: Metal cyanides

**CADMIUM FULMINATE** $Cd(C{\equiv}N{\rightarrow}O)_2$ $C_2CdN_2O_2$

*See* METAL FULMINATES

† **CHLOROTRIFLUOROETHYLENE** $ClFC{=}CF_2$ $C_2ClF_3$

Chlorine perchlorate

*See* CHLORINE PERCHLORATE, $Cl_2O_4$: Chlorotrifluoroethylene

1,1-Dichloroethylene

*See* 1,1-DICHLOROETHYLENE, $C_2H_2Cl_2$ : Chlorotrifluoroethylene

Ethylene

*See* ETHYLENE, $C_2H_4$ : Chlorotrifluoroethylene

Oxygen

*See* OXYGEN (Gas), $O_2$ : Halocarbons

**POLY(CHLOROTRIFLUOROETHYLENE)** $(C_2ClF_3)_n$

$(CF_2{-}CFCl)_n$

*See* ALUMINIUM, Al: Halocarbons

## LITHIUM CHLOROACETYLIDE $LiC{\equiv}CCl$ $C_2ClLi$

Cadiot, P., *BCISC Quart. Safety Summ.*, 1965, (142), 27

During preparation of a chloroethynyl compound with lithium chloroacetylide in liquid ammonia, some of the salt separated as a crust due to evaporation of solvent, and exploded violently. Such salts are stable in solution, but dangerous in the solid state. Evaporation of ammonia must be prevented or made good until unreacted salt has been decomposed by addition of ammonium chloride. Lithium trifluoromethylacetylide behaves similarly.

*See other* HALOACETYLENE DERIVATIVES

## SODIUM CHLOROACETYLIDE $NaC{\equiv}CCl$ $C_2ClNa$

Viehe, H. G., *Chem. Ber.*, 1959, **92**, 1271

Though stable and usable in solution, the sodium salt, like the lithium and calcium salts, is dangerously explosive in the solid state.

*See other* HALOACETYLENE DERIVATIVES

## DICHLOROACETYLENE $ClC{\equiv}CCl$ $C_2Cl_2$

1. Wotiz, J. H. *et al., J. Org. Chem.*, 1961, **26**, 1626
2. Ott, E., *Ber.*, 1942, **75**, 1517
3. Kirk-Othmer, 1964, Vol. 5, 203–205
4. Siegel, J. *et al., J. Org. Chem.*, 1970, **35**, 3199
5. Riemschneider, R. *et al., Ann.*, 1961, **640**, 14
6. *MCA Case History No. 1989*

Dichloroacetylene is a heat-sensitive explosive gas which ignites in contact with air. However, its azeotrope with diethyl ether (55.4% dichloroacetylene) is not explosive and is stable to air [1, 2]. It is formed on catalysed contact between acetylene and chlorine, or sodium hypochlorite at low temperature; or by the action of alkali upon polychloro-ethane and -ethylene derivatives, notably trichloro ethylene [3]. A safe synthesis has been described [4]. Ignition of 58% mol solution in ether on exposure to air of high humidity and violent explosion of a concentrated solution in carbon tetrachloride shortly after exposure to air have been reported. Stirring the ethereal solution with tap-water usually caused ignition and explosion [5].

Dichloroacetylene had been collected without incident on six previous occasions as a dilute solution. When the cooling system was modified by lowering the temperature in the water-separator, liquid chloroacetylene separated there and exploded when a stopcock was turned [6].

*See* TETRACHLOROETHYLENE, $C_2Cl_4$ : Sodium hydroxide
TRICHLOROETHYLENE, $C_2HCl_3$ : Alkali
: Epoxides

*See also* CHLOROCYANOACETYLENE, $C_3ClN$
*See other* HALOACETYLENE DERIVATIVES

**1,2-DICHLOROTETRAFLUOROETHANE** $C_2Cl_2F_4$

**$CClF_2CClF_2$**

Aluminium
*See* ALUMINIUM, Al: Halocarbons

**OXALYL DICHLORIDE** **ClCOCOCl** $C_2Cl_2O_2$

Potassium
*See* POTASSIUM, K: Oxalyl dihalides

**1,1,2-TRICHLOROTRIFLUOROETHANE** $C_2Cl_3F_3$

**$CCl_2FCClF_2$**

Metals
*See* ALUMINIUM, Al: Halocarbons
BARIUM, Ba: Halocarbons
LITHIUM, Li: Halocarbons
SAMARIUM, Sm: 1,1,2-Trichlorotrifluoroethane
TITANIUM, Ti: Halocarbons

**SODIUM TRICHLOROACETATE** **$Cl_3CCO_2Na$** $C_2Cl_3NaO_2$

Doyle, W. H., *Loss Prevention*, 1969, **3**, 15
Bags of the salt ignited in storage (cause unknown).

**TETRACHLOROETHYLENE** **$Cl_2C{=}CCl_2$** $C_2Cl_4$

Aluminium
*See* ALUMINIUM, Al: Halocarbons

Dinitrogen tetraoxide
*See* DINITROGEN TETRAOXIDE (NITROGEN DIOXIDE), $N_2O_4$: Halocarbons

Metals
*See* BARIUM, Ba: Halocarbons
LITHIUM, Li: Halocarbons

Sodium hydroxide
Mitchell, P. R., private comm., 1973

The presence of 0.5% of trichloroethylene as impurity in tetrachloroethylene during unheated drying over solid sodium hydroxide caused the generation of dichloroacetylene. After subsequent fractional distillation, the volatile fore-run exploded.

*See* DICHLOROACETYLENE, $C_2Cl_2$
TRICHLOROETHYLENE, $C_2HCl_3$ : Alkali

*See other* HALOALKENES

### 3-CHLORO-3-TRICHLOROMETHYLDIAZIRINE $C_2Cl_4N_2$

$Cl_3C\overline{C(Cl)N{=}N}$

Liu, M. T. H., *Chem. Eng. News*, 1974, **52**(36), 3
An ampoule of the compound exploded when scored with a file, indicating high shock-sensitivity.

*See other* DIAZIRINES

### HEXACHLOROETHANE $C_2Cl_6$

Metals

*See* ALUMINIUM, Al: Halocarbons
ZINC, Zn: Halocarbons

### CHROMYL ISOTHIOCYANATE $CrO_2(N{=}C{=}S)_2$ $C_2CrN_2O_2S_2$

Forbes, G. S. *et al.*, *J. Amer. Chem. Soc.*, 1943, **65**, 2273
The salt, prepared in carbon tetrachloride solution, exploded feebly several times as the exothermic reaction proceeded. Oxidation of thiocyanate by chromium(VI) is postulated.

### CHROMYL ISOCYANATE $CrO_2(N{=}C{=}O)_2$ $C_2CrN_2O_4$

Forbes, G. S. *et al.*, *J. Amer. Chem. Soc.*, 1943, **65**, 2273
The product is stable during unheated vacuum evaporation of its solutions in carbon tetrachloride. Evaporation with heat at atmospheric pressure led to a weak explosion of the salt.

### DICAESIUM ACETYLIDE $CsC{\equiv}CCs$ $C_2Cs_2$

Acids
Mellor, 1946, Vol. 5, 848

It ignites in contact with gaseous hydrogen chloride or its concentrated aqueous solution, and explodes with nitric acid.

Diiron trioxide
Mellor, 1946, Vol. 5, 849
Incandescence on warming.

Halogens
Mellor, 1946, Vol. 5, 848
It burns in all four halogens at or near ambient temperature.

Lead dioxide
*See* LEAD(IV) OXIDE, $O_2Pb$: Metal acetylides

Non-metals
Mellor, 1946, Vol. 5, 848
Mixtures of boron or silicon with dicaesium acetylide react vigorously on heating.
*See other* METAL ACETYLIDES

**COPPER(II) ACETYLIDE** **$CuC_2$** **$C_2Cu$**

Urbanski, 1967, Vol. 3, 228
Copper(II) acetylide (black or brown) is much more sensitive to impact, friction and heat than copper(I) acetylide (red-brown), which is used in electric fuses or detonators.
*See other* METAL ACETYLIDES

**COPPER(II) CYANIDE** **$Cu(CN)_2$** **$C_2CuN_2$**

Magnesium
*See* MAGNESIUM, Mg: Metal cyanides

**COPPER(II) FULMINATE** **$Cu(C{\equiv}N{\rightarrow}O)_2$** **$C_2CuN_2O_2$**

*See* METAL FULMINATES

**DICOPPER(I) ACETYLIDE** **$CuC{\equiv}CCu$** **$C_2Cu_2$**

1. Mellor, 1946, Vol. 5, 851, 852
2. Rutledge, 1968, 84–85
3. Morita, S., *J. Soc. High Press. Gas Ind.*, 1955, **19**, 167–176

Readily formed from copper or its compounds and acetylene, it detonates on impact or heating above 100°C. If warmed in air or oxygen, it explodes on subsequent contact with acetylene [1].

Explosivity of the precipitate increases with acidity of the salt solutions, while the stability increases in the presence of reducing agents (formaldehyde, hydrazine, or hydroxylamine). The form with a metallic lustre was the most explosive acetylide made. Catalysts with the acetylide supported on a porous solid are fairly stable [2].

The ignition temperature of the pure red acetylide is 260–270°C. On exposure to air or oxygen, it is converted to the black copper(II) acetylide, which ignites and explodes at 100°C [3].

Halogens

Mellor, 1946, Vol. 5, 852

Ignition occurs on contact with chlorine, bromine vapour or finely divided iodine.

Silver nitrate

Mellor, 1946, Vol. 5, 853

Contact with silver nitrate solution transforms copper(I) acetylide into a sensitive and explosive mixture of silver acetylide and silver.

*See also* SILVER NITRATE, $AgNO_3$ : Acetylene, etc.

Talc

Chambionnat, A., *Chem. Abs.*, 1951, **45**, 7791e

The effect upon the sensitivity of the acetylide of adding talc as an inert desensitiser has been studied.

*See other* METAL ACETYLIDES

**DICOPPER(I) KETENDIIDE** **$Cu_2C{=}C{=}O$** **$C_2Cu_2O$**

Blues, E. T. *et al.*, *J. Chem. Soc., Chem. Comm.*, 1973, 921

The dry compound is mildly explosive.

*See other* HEAVY METAL DERIVATIVES

**COPPER(I) OXALATE** **$(CO_2Cu)_2$** $C_2Cu_2O_4$

Sidgwick, 1950, 126

Explodes feebly on heating.

*See other* METAL OXALATES

**BISFLUOROFORMYL PEROXIDE** **FCO–OO–COF** $C_2F_2O_4$

1. Talbot, R. L., *J. Org. Chem.*, 1968, **33**, 2095
2. Czerepinski, R. *et al.*, *Inorg. Chem.*, 1968, **7**, 109

Several explosions occurred during the preparation, which involves charging carbon monoxide into a mixture of fluorine and oxygen [1]. It has been known to decompose or explode at elevated temperatures,

and all samples should be maintained below 30°C and well shielded [2].
*See other* DIACYL PEROXIDES

## TRIFLUOROACETYL NITRITE $CF_3COONO$ $C_2F_3NO_3$

1. Taylor, C. W. *et al., J. Org. Chem.*, 1962, **27**, 1064
2. Gibbs, R. *et al., J. Chem. Soc. Perkin II*, 1972, 1340
3. Banks, R. E. *et al., J. Chem. Soc. (C)*, 1956, 1350–1353
4. Banks, R. E. *et al., J. Chem. Soc., Perkin 1*, 1974, 2535

Though much more stable than acetyl nitrite even at 100°C, the vapour of trifluoroacetyl nitrite will explode at 160–200°C unless diluted with inert gas to below about 50% vol. concentration. Higher perfluoro-homologues are more stable [1]. A detailed examination of the explosion parameters has been made [2].

This and higher polyfluoroacyl nitrites tend to explode above 140°C at atmospheric pressure, and handling of large quantities should be avoided [3], especially during pyrolysis [4].
*See other* ACYL NITRITES

## TRIFLUOROACETYL AZIDE $F_3CCON_3$ $C_2F_3N_3O$

Sprenger, G. H. *et al., Inorg. Chem.*, 1973, **12**, 2891
It explodes on exposure to mechanical or thermal shock. Care is necessary during preparation to eliminate hydrogen chloride from the precursory acid chloride, to prevent formation of hydrogen azide.
*See other* ACYL AZIDES

## † TETRAFLUOROETHYLENE $F_2C{=}CF_2$ $C_2F_4$

Graham, D. P., *J. Org. Chem.*, 1966, **31**, 956
A terpene inhibitor is usually added to the monomer to prevent spontaneous polymerisation, and in its absence the monomer will spontaneously explode at pressures above 2.7 bar. The inhibited monomer will explode if ignited.

Air
Muller, R. *et al., Paste und Kaut.*, 1967, **14**(12), 903
Liquid tetrafluoroethylene, being collected in an open liquid nitrogen-cooled trap, formed a peroxidic polymer which exploded.

Air,
Hexafluoropropene
Dixon, G. D., private comm., 1968
A mixture of the two monomers and air sealed in an ampoule formed a gummy peroxide during several weeks. The residue left after opening the tube exploded violently on warming.
*See other* POLYPEROXIDES

Chloroperoxytrifluoromethane

*See* CHLOROPEROXYTRIFLUOROMETHANE, $CClF_3O_2$: Tetrafluoroethylene

Difluoromethylene dihypofluorite

*See* DIFLUOROMETHYLENE DIHYPOFLUORITE, $CF_4O_2$: Haloalkenes

Dioxygen difluoride

*See* DIOXYGEN DIFLUORIDE, $F_2O_2$

Iodine pentafluoride,
Limonene

*MCA Case History No. 1520*

Accidental contamination of a tetrafluoroethylene gas supply system with iodine pentafluoride caused a violent explosion in the cylinders. Exothermic reaction of the contaminant with the inhibitor (limonene) present in the gas cylinders may have initiated the explosion.

Metal alkoxides

*See* TRIFLUOROVINYL METHYL ETHER, $C_3H_3F_3O$

Oxygen

*See* OXYGEN (Gas), $O_2$: Tetrafluoroethylene

Triboron pentafluoride

*See* TRIBORON PENTAFLUORIDE, $B_3F_5$: Tetrafluoroethylene

*See other* HALOALKENES

## POLYTETRAFLUOROETHYLENE $(CF_2CF_2)_n$ $(C_2F_4)_n$

Fluorine

*See* FLUORINE, $F_2$: Polymeric materials

## TRIFLUOROACETYL HYPOFLUORITE $C_2F_4O_2$

$CF_3COOF$

Alone,
or Potassium iodide,
Water

Cady, G. H. *et al., J. Amer. Chem. Soc.*, 1953, **75**, 2501–2502

The gas explodes on sparking and often during preparation or

distillation. Unless much diluted with nitrogen, it explodes on contact with aqueous potassium iodide.
*See other* HYPOHALITES

**XENON(II) FLUORIDE TRIFLUOROACETATE** **$C_2F_4O_2Xe$**

**$CF_3CO_2XeF$**

Jha, N. K., *RIC Rev.*, 1971, **4**, 157
Explodes on thermal or mechanical shock.
*See other* XENON COMPOUNDS

**PERFLUORO-*N*-CYANODIAMINOMETHANE** **$C_2F_5N_3$**

**$NF_2CF_2NFCN$**

*See* FLUORINE, $F_2$: Sodium dicyanamide

**BISTRIFLUOROMETHYL NITROXIDE** **$C_2F_6NO$**

**$(F_3C)_2NO^{\cdot}$**

Platinum hexafluoride
*See* PLATINUM HEXAFLUORIDE, $F_6Pt$: Bistrifluoromethyl nitroxide

**BIS(TRIFLUOROMETHYL)PHOSPHORUS(III) AZIDE** **$C_2F_6N_3P$**

**$(CF_3)_2PN_3$**

1. Allcock, H. R., *Chem. Eng. News*, 1968, **46**(18), 70
2. Tesi, G. *et al.*, *Proc. Chem. Soc.*, 1960, 219

Explosive, but stable if stored cold [1]. Previously it was found to be unpredictably unstable, violent explosions having occurred even at −196°C [2].
*See other* NON-METAL AZIDES

**PERFLUORO-1-AMINOMETHYLGUANIDINE** **$C_2F_8N_4$**

**$NF_2CF_2NFC(NF)NF_2$**

*See* FLUORINE, $F_2$: Cyanoguanidine

**BIS(TRIFLUOROMETHYL)SULPHUR DIFLUORIDE** **$C_2F_8S$**

**$(F_3C)_2SF_2$**

Chlorine fluorides
Sprenger, G. H. *et al.*, *J. Fluorine Chem.*, 1974, **7**, 335
Explosions have occurred when chlorine mono- or tri-fluorides were treated with 2–3 mmol quantities of the sulphide or trifluoromethylsulphenyl chloride in absence of a solvent. Because chlorine trifluoride

is known to deprotonate organic solvents with formation of explo-
carbene species, only fully halogenated solvents are suitable as diluents.

**PERFLUORO-*N*-AMINOMETHYLTRIAMINOMETHANE** $C_2F_{10}N_4$

$NF_2CF_2NFCF(NF_2)_2$

*See* FLUORINE, $F_2$: Cyanoguanidine

**SILVER ACETYLIDE** AgC≡CH $C_2HAg$

Anon., *Chem. Trade J.*, 1966, **158**, 153

A poorly stoppered dropping bottle of silver nitrate solution absorbed sufficient acetylene from the atmosphere in an acetylene plant laboratory to block the dropping tube. A violent detonation occurred on moving the dropping tube.

*See also* METAL ACETYLIDES

**BROMOACETYLENE** BrC≡CH $C_2HBr$

1. Tanaka, R. *et al.*, *J. Org. Chem.*, 1971, **36**, 3856
2. Hucknall, D. J. *et al.*, *Chem. & Ind.*, 1972, 116

It is dangerous and may burn or explode in contact with air, even when solid at −196°C [1]. Procedures for safe generation, transfer and storage, all under nitrogen, are described [2].

*See other* HALOACETYLENE DERIVATIVES

**CHLOROACETYLENE** ClC≡CH $C_2HCl$

Tanaka, R. *et al.*, *J. Org. Chem.*, 1971, **36**, 3856

Chloroacetylene may burn or explode in contact with air, and its greater volatility makes it more dangerous than bromoacetylene. Procedures for safe generation, handling and storage, all under nitrogen, are described.

*See also* TRICHLOROETHYLENE, $C_2HCl_3$: Alkali
*See other* HALOACETYLENE DERIVATIVES

**CHLORATOMERCURIO(FORMYL)METHYLENEMERCURY(II)**

$C_2HClHg_2O_4$

$ClO_3HgC(CHO)=Hg$

Whitmore, 1921, 154

An extremely sensitive crystalline solid, which explodes on shaking with its crystallisation liquor, or when dry upon gentle mixing with copper oxide.

*See other* MERCURY COMPOUNDS

## † TRICHLOROETHYLENE $Cl_2C{=}CHCl$ $C_2HCl_3$

*MCA SD-14,* 1956 considers it to be non-flammable at normal ambient temperatures, but note its very wide limits.

Alkali

1. Fabian, F., private comm., 1960
2. *ABCM Quart. Safety Summ.*, 1956, **27**, 17

An emulsion, formed during extraction of a strongly alkaline liquor with trichloroethylene, decomposed with the evolution of the spontaneously flammable gas, dichloroacetylene [1]. This reaction could also occur if alkaline metal-stripping preparations were used in conjunction with trichloroethylene de-greasing preparations, some of which also contain amine inhibitors which could cause the same reaction [2].

*See also* TETRACHLOROETHYLENE, $C_2Cl_4$ : Sodium hydroxide

Aluminium

*See* ALUMINIUM, Al: Halocarbons

Epoxides

Dobinson, B. *et al., Chem. & Ind.,* 1972, 214

1-Chloro-2,3-epoxypropane, the mono- and di-2,3-expoypropyl ethers of 1,4-butanediol, and 2,2-bis [4(2′,3′-epoxpropoxy) phenyl] -propane can, in presence of catalytic quantities of halide ions, cause dehydrochlorination of trichloroethylene to dichloro acetylene, which causes minor explosions when the mixture is boiled under reflux. A mechanism is discussed.

*See also* Alkali, above

Metals

*See* ALUMINIUM, Al: Halocarbons
BARIUM, Ba: Halocarbons
BERYLLIUM, Be: Halocarbons
LITHIUM, Li: Halocarbons
MAGNESIUM, Mg: Halocarbons
TITANIUM, Ti: Halocarbons

Oxidants

*See* PERCHLORIC ACID, $ClHO_4$ : Trichloroethylene
DINITROGEN TETRAOXIDE (NITROGEN DIOXIDE), $N_2O_4$ : Halocarbons
OXYGEN (Gas), $O_2$ : Halocarbons
OXYGEN (Liquid), $O_2$ : Halocarbons

*See other* HALOALKENES

## PENTACHLOROETHANE $CCl_3CHCl_2$ $C_2HCl_5$

Potassium

*See* POTASSIUM, K: Halocarbons

**CAESIUM ACETYLIDE** **CsC≡CH** $C_2HCs$

Mellor, 1946, Vol. 5, 849–850

It reacts similarly to the dicaesium compound.

*See* DICAESIUM ACETYLIDE, $C_2Cs_2$

*See other* METAL ACETYLIDES

**FLUOROACETYLENE** **FC≡CH** $C_2HF$

Alone,
or Bromine

Middleton, W. J. *et al.*, *J. Amer. Chem. Soc.*, 1959, **81**, 803–804

Liquid fluoroacetylene is treacherously explosive close to its b.p., –80°C. The gas does not ignite in air and is not explosive. Ignition occurred in contact with a solution of bromine in carbon tetrachloride. Mercury and silver salts were both stable to impact, but the latter exploded on heating, whereas the former decomposed violently.

*See other* HALOACETYLENE DERIVATIVES

† **TRIFLUOROETHYLENE** **$F_2C{=}CHF$** $C_2HF_3$

**TRIFLUOROACETIC ACID** **$CF_3CO_2H$** $C_2HF_3O_2$

Lithium tetrahydroaluminate

*See* LITHIUM TETRAHYDROALUMINATE, $AlH_4Li$: Fluoroamides

**PEROXYTRIFLUOROACETIC ACID** **$CF_3CO_2OH$** $C_2HF_3O_3$

Sundberg, R. J. *et al.*, *J. Org. Chem.*, 1968, **33**, 4098

This extremely powerful oxidising agent must be handled and used with great care.

4-Iodo-3,5-dimethylisoxazole

Plepys, R. A. *et al.*, US Pat. 3 896 140, 1975

Interaction to produce 3,5-dimethyl-4-bis(trifluoroacetoxy)iodoisoxazole yields a detonable by-product, believed to be iodine pentaoxide contaminated with organic material.

*See* IODINE(V) OXIDE, $I_2O_5$: Non-metals

*See other* PEROXYACIDS

**POLY(DIMERCURYIMMONIUM ACETYLIDE)** $(C_2HHg_2N)_n$

**$(Hg{=}N^+{=}Hg\ C_2H^-)_n$**

Sorbe, 1968, 97

Highly explosive.

*See other* POLY(DIMERCURYIMMONIUM) COMPOUNDS

**IODOACETYLENE** **IC≡CH** **$C_2HI$**

Sorbe, 1968, 65

It explodes above 85°C.

*See other* HALOACETYLENE COMPOUNDS

**POTASSIUM ACETYLIDE** **KC≡CH** **$C_2HK$**

Chlorine

Mellor, 1946, Vol. 5, 849

It ignites in chlorine.

Non-metal oxides

Mellor, 1946, Vol. 5, 849

Interaction with sulphur dioxide at ambient temperature, or with carbon dioxide on warming, causes incandescence.

*See other* METAL ACETYLIDES

**DIPOTASSIUM *aci*-NITROACETATE** **$C_2HK_2NO_4$**

**KOCOCH=N(O)OK**

Water

1. Lyttle, D. A., *Chem. Eng. News,* 1949, 27, 1473
2. Whitmore, F. C., *Org. Synth.,* 1941, Coll. Vol. 1, 401

An inhomogeneous mixture of the dry salt with a little water exploded violently after 30 min [1]. This was probably due to exothermic decarboxylation generating free nitromethane in an alkaline system. The decomposition of sodium nitroacetate proceeds exothermically above 80°C [2].

*See* NITROMETHANE, $CH_3NO_2$: Acids, etc.

*See other aci*-NITRO SALTS

**LITHIUM ACETYLIDE** **LiC≡CH** **$C_2HLi$**

Mellor, 1946, Vol. 5, 849

It reacts similarly to the dilithium compound.

*See* DILITHIUM ACETYLIDE, $C_2Li_2$

## LITHIUM ACETYLIDE–AMMONIA $LiC{\equiv}CH \cdot NH_3$ $C_2HLi \cdot H_3N$

Gases,
or Water

Mellor, 1946, Vol. 5, 8

The complex ignites in contact with carbon dioxide, sulphur dioxide, chlorine or water.

*See other* METAL ACETYLIDES

## DIAZOACETONITRILE $N_2CHCN$ $C_2HN_3$

Phillips, D. D. *et al., J. Amer. Chem. Soc.,* 1956, **78**, 5452

*MCA Case History No. 2169*

Removal of a rubber stopper from a flask of a concentrated solution in methylene chloride initiated an explosion, probably through friction on solvent-free material between the flask neck and bung. Handling only in dilute solution is recommended.

An explosion also occurred during operation of the glass stopcock of a dropping funnel while 20 ml of the apparently undiluted nitrile was being added to a reaction mixture.

*See other* DIAZO COMPOUNDS
HIGH-NITROGEN COMPOUNDS

## DINITROACETONITRILE $CH(NO_2)_2CN$ $C_2HN_3O_4$

*See* POLYNITROALKYL COMPOUNDS

## DIAZOACETYL AZIDE $N_2CHCON_3$ $C_2HN_5O$

Neunhöffer, G. *et al., Ann.,* 1968, **713**, 97–98

The material may be distilled cautiously at 20–21°C/0.27 mbar but is preferably used in solution as prepared. Either the solid (below 7°C) or the liquid explodes with great violence on impact or friction.

*See other* ACYL AZIDES
DIAZO COMPOUNDS

## SODIUM 5-DINITROMETHYLTETRAZOLIDE $C_2HN_6NaO_4$

```
N——CC (NO2)2
‖    ‖
N    N      Na+
 \  /
  N
  H
```

Einberg, F., *J. Org. Chem.,* 1964, **29**, 2021

It explodes violently at m.p., 160°C, and may be an *aci*-nitro salt.

*See other* POLYNITROALKYL COMPOUNDS
TETRAZOLES

## 5-HYDROXY-1(*N*-SODIO-5-TETRAZOLYLAZO)TETRAZOLE $C_2HN_{10}NaO$

```
N——C — N=N
‖     ‖     N     N
N     N     ‖     ‖
 \ N /      N——COH
  Na
```

Thiele, J. *et al., Ann.,* 1893, **273**, 150

The salt explodes very violently on heating.

*See other* *N*-METAL DERIVATIVES
TETRAZOLES

## SODIUM ACETYLIDE NaC≡CH $C_2HNa$

1. Mellor, 1946, Vol. 5, 849
2. Greenlee, K. W., *Inorg. Synth.,* 1947, **2**, 81
3. Houben-Weyl, 1970, Vol. 13.1, 280

It reacts similarly to the disodium compound [1]. If heated to 150°C, it decomposes extensively, evolving gas which ignites in air owing to presence of pyrophoric carbon. The residual carbon is also highly reactive [2].

The dry powder (from solution in liquid ammonia) may ignite if exposed to air as an extended layer, e.g. on filter-paper [3].

*See* DISODIUM ACETYLIDE, $C_2Na_2$
*See other* METAL ACETYLIDES

## RUBIDIUM ACETYLIDE RbC≡CH $C_2HRb$

Mellor, 1946, Vol. 5, 849

It reacts similarly to the dirubidium compound.

*See* DIRUBIDIUM ACETYLIDE, $C_2Rb_2$
*See other* METAL ACETYLIDES

## † ACETYLENE (ETHYNE) HC≡CH $C_2H_2$

The endothermic gas may explosively decompose without air.

*MCA SD*-7, 1957

1. Mayes, H. A., *Chem. Engr.,* 1965, (185), 25
2. Nedwick, J. J., *Ind. Eng. Chem. Proc. Des. Dev.,* 1962, **1**, 137; Tedeschi, R. J. *et al., ibid.,* 1968, **7**, 303
3. Miller, 1965, Vol. 1, 506
4. Kirk-Othmer, 1963, Vol. 1, 195–202
5. Rimarski, W., *Angew. Chem.,* 1929, **42**, 933
6. Foote, C. S., private comm., 1965
7. Sutherland, M. E. *et al., Chem. Eng. Progr.,* 1973, **69**(4), 48–51
8. Williams, A. *et al., Chem. Rev.,* 1970, **70**, 270–271
9. Yantovskii, S. A., *Chem. Abs.,* 1964, **61**, 13117f

10. Landesman, Ya. M. *et al., Chem. Abs.*, 1974, **81**, 151352q
11. Glikin, M. A. *et al., Chem. Abs.*, 1975, **83**, 100289h
12. Bityutskii, V. K. *et al., Chem. Abs.*, 1975, **83**, 65042y
13. Rüst, 1948, 297

Addition of up to 30% of miscible diluent did not sufficiently desensitise (explosive) liquid acetylene to permit of its transportation. The solid has similar properties [1]. However, safe techniques for the use of liquid acetylene in reaction systems at pressures up to 270 bar have been described [2].

Solid acetylene in presence of liquid nitrogen at −181°C is sensitised by presence of grit (carborundum) and may readily explode on impact [3]. The hazards of preparing and using acetylene have been adequately described [3–5]. A flask which had contained acetylene was left open to the air for a week, and then briefly purged with air. It still contained enough acetylene to form an explosive mixture [6].

The explosive decomposition of acetylene in absence of air readily escalates to detonation in tubular vessels. This type of explosive decomposition has been experienced in a 7 mile acetylene pipeline system [7].

In a review on combustion and oxidation of acetylene, the factors affecting spontaneous ignition at low temperatures are discussed [8]. The explosive decomposition of acetylene and conditions for spontaneous ignition to occur have been reviewed [9–11]. 'Decomposition region and localisation of explosive decomposition of acetylene' (title only translated) [12]. Explosion of a vessel filled with liquid acetylene will generate a pressure of 5464 bar and a temperature of 3016°C [13].

*See also* MAGNESIUM, Mg: Barium carbonate, Water

Air,
Butane,
or Propane

Konovalov, E. N. *et al.*, Russ. Pat., 449 905, 1974 (*Chem. Abs.*, 1975, **82**, 111548e)

Addition of 0.3–3.5 wt.% of propane or butane reduces the explosion hazards of acetylene–air mixtures.

Bleaching powder

*See* CALCIUM HYPOCHLORITE, $CaCl_2O_2$: Acetylene

Cobalt

Mellor, 1942, Vol. 14, 513

Finely divided (pyrophoric) cobalt decomposes and polymerises acetylene on contact, becoming incandescent.

Copper

Anon., *ABCM Quart. Safety Summ.*, 1946, **17**, 24

Rubber-covered electric cable, used as a makeshift handle in the effluent pit of an acetylene plant, formed copper acetylide with residual acetylene and the former detonated when disturbed and initiated explosion of the latter. All heavy metals must be rigorously excluded from locations where acetylene may be present.

*See* METAL ACETYLIDES

Copper (I) acetylide

*See* COPPER (I) ACETYLIDE, $C_2Cu_2$

Ethylene,
Hydrogen

*See* HYDROGEN (Gas), $H_2$ : Acetylene, Ethylene

Halogens

1. Muir, G. D., private comm., 1968
2. von Schwartz, 1918, 142, 321
3. *MCA SD-7,* 1957
4. Humphreys, V., *Chem. & Ind.*, 1971, 681–682
5. Sokolova, E. I. *et al., Chem. Abs.*, 1970, **72**, 71120m
6. Davenport, A. P., *School Sci. Rev.*, 1973, **55**(191), 332

Tetrabromoethane is made by passing acetylene into bromine in carbon tetrachloride at reflux. The rate of reaction falls off rapidly below reflux temperature, and if the rate of addition of acetylene is insufficient to maintain the temperature, high concentrations of unreacted acetylene build up, with the possibility of a violent delayed reaction [1]. In absence of a diluent, interaction may be explosive [2]. Mixtures of acetylene and chlorine may explode upon initiation by sunlight or other UV source, or high temperature, sometimes very violently [2]. Interaction with fluorine is very violent [3] and with iodine possibly explosive [2].

Dilution of equimolar mixtures of chlorine and acetylene with 55 mol% of nitrogen or 70% of air prevented spontaneous explosion. At higher dilutions sparking did not initiate explosion [4]. Explosive interaction of acetylene and chlorine in the dark is initiated by presence of oxygen between 0.1 and about 40 vol.%. The reaction is inhibited by inert gases or oxygen at higher concentrations [5]. A safe technique for demonstrating explosive combination of acetylene and chlorine is described [6].

Heavy metals and salts

*See* METAL ACETYLIDES

Mercury(II) salts,
Nitric acid,
(Sulphuric acid)

*See* NITRIC ACID, $HNO_3$: Acetylene, Mercury(II) salts

Oxides of nitrogen

*See* NITROGEN OXIDE ('NITRIC OXIDE'), NO: Dienes, Oxygen

Oxygen

1. Fowles, G. *et al.*, *School Sci. Rev.*, 1940, **22**(85), 6; 1962, **44**(152), 161; 1963, **44**(154), 706; 1964, **45**(156), 459
2. Kiyama, R. *et al.*, *Rev. Phys. Chem., Japan*, 1953, **23**, 43–48
3. Moye, A., *Chem. Ztg.*, 1922, **46**, 69
4. Ivanov, U., *Chem. Abs.*, 1976, **84**, 57968k

The explosion of acetylene–oxygen mixtures in open vessels is a very dangerous experiment (stoicheiometric mixtures detonate with great violence, completely shattering the container). Full precautions are essential for safety [1]. When a mixture of acetylene and oxygen (54:46) at 270°C/10.9 bar was compressed in 0.7s to 56.1 bar, the resulting explosion attained a pressure of several kbar. In other tests rapid compression of acetylene or its mixtures with air caused no explosions [2]. Previously, passage of acetylene into liquid air to deliberately generate a paste of solid acetylene and liquid oxygen, 'by far the most powerful of all explosives', had been proposed [3]. Acetylene had been collected for teaching purposes over water in a pneumatic trough. Later, oxygen was collected in the same way without changing the water, and the sample exploded violently when exposed to a glowing splint. Acetylene remaining dissolved in the water had apparently been displaced by the oxygen stream, the lower explosive limit for acetylene being only 2.5% in air and less in oxygen [4].

Potassium

Berthelot, M., *Bull. Soc. Chim. Fr.* [2], 1866, **5**, 188

Molten potassium ignites in acetylene, then explodes.

Trifluoromethyl hypofluorite

*See* TRIFLUOROMETHYL HYPOFLUORITE, $CF_4O$: Hydrocarbons

*See other* ACETYLENIC COMPOUNDS

**SILVER 1,2,3-TRIAZOLIDE** **AgNN=NCH=CH** **$C_2H_2AgN_3$**

*See* 1,2,3-TRIAZOLE, $C_2H_3N_3$

## SILVER DINITROACETAMIDE $C_2H_2AgN_3O_5$

$AgC(NO_2)_2CONH_2$

Parker, C. O. *et al., Tetrahedron*, 1962, **17**, 108
The sodium salt explodes at 170°C and the potassium and silver salts at 159 and 130°C, respectively.
*See other* POLYNITROALKYL COMPOUNDS

## SILVER 1,3-DI(5-TETRAZOLYL)TRIAZENE $C_2H_2AgN_{11}$

N=NNHN=CN=NNHC=NNAgN=N

*See* 1,3-DI(5-TETRAZOLYL)TRIAZENE, $C_2H_3N_{11}$

## 2-IODOSYLVINYL CHLORIDE ClCH=CHIO $C_2H_2ClIO$

Alone,
or Water
Thiele, J. *et al., Ann.*, 1909, **369**, 131
It explodes at 63°C and contact with water disproportionates it to the more explosive iodyl compound.

## 2-IODYLVINYL CHLORIDE $ClCH{=}CHIO_2$ $C_2H_2ClIO_2$

Thiele, J. *et al., Ann.*, 1909, **369**, 131
Explodes violently on impact, friction or heating to 135°C.
*See other* IODINE COMPOUNDS

## † 1,1-DICHLOROETHYLENE $Cl_2C{=}CH_2$ $C_2H_2Cl_2$

Air
1. Reinhardt, R. C., *Chem. Eng. News*, 1947, **25**, 2136
2. *MCA Case History No. 1172*
3. *MCA Case History No. 1693*
4. Harmon, 1974, iv, 2.21

When stored between −40° and +25°C in the absence of inhibitor and presence of air, vinylidene chloride rapidly absorbs oxygen with formation of a violently explosive peroxide. The latter initiates polymerisation, producing an insoluble polymer which adsorbs the peroxide. Separation of this polymer in a dry state must be avoided, since if more than 15% of peroxide is present, the polymer may be detonable by slight shock or heat. Hindered phenols are suitable inhibitors to prevent peroxidation [1]. The Case History describes an explosion during handling of a pipe used to transfer vinylidene chloride [2].

Two further cases of formation of the polymeric peroxide in bottles of vinylidene chloride under refrigerated storage, and subsequent explosions during handling or disposal, are recorded [3]. The monomer is normally handled at −10°C in absence of light or water, which tend to promote self-polymerisation [4].

*See* VIOLENT POLYMERISATION
*See other* POLYMERISATION INCIDENTS
POLYPEROXIDES

Chlorotrifluoroethylene
Raasch, M. S. *et al., Org. Synth.*, 1962, **42**, 46
Condensation of the reactants at 180°C under pressure to give 1,1,2-trichloro-2,3,3-trifluorocyclobutane was effected smoothly several times in a 1 litre autoclave. Scaling up to a 3 litre autoclave led to uncontrolled polymerisation which distorted the autoclave.

Ozone
'Vinylidene Chloride Monomer', 9, Midland (Mich.), Dow Chemical Co., 1966
The reaction products formed with ozone are particularly dangerous.

Perchloryl fluoride
*See* PERCHLORYL FLUORIDE, $ClFO_3$: Hydrocarbons, etc.
*See other* HALOALKENES

**† *cis*-1,2-DICHLOROETHYLENE** **CHCl=CHCl** $C_2H_2Cl_2$

**† *trans*-1,2-DICHLOROETHYLENE** **CClH=CHCl** $C_2H_2Cl_2$

*BCISC Quart. Safety Summ.*, 1964, **35**, 37
*Sichere Chemiearbeit*, 1964, **16**, 35
Under appropriate conditions, dichlorethylene, previously thought to be non-flammable, can cause a fire hazard. Addition of a hot liquid to the cold solvent caused sudden emission of sufficient vapour to cause a flame to flash back 12 m from a fire. Although the bulk of the solvent did not ignite, various items of paper and wood in the room were ignited by the transient flame.

Alkalies
*Fire Accident Prev.*, 1956, **42**, 28
Thron, H., *Chem. Ztg.*, 1924, **48**, 142
1,2-Dichloroethylene in contact with solid caustic alkalies or their concentrated solutions will form chloroacetylene which ignites in air.

Distillation of a mixture of ethanol containing 0.25% of the halocarbon with aqueous sodium hydroxide gave a distillate which ignited in air.

*See also* TRICHLOROETHYLENE, $C_2HCl_3$ : Alkalies
1,1,2,2-TETRACHLOROETHANE, $C_2H_2Cl_4$ : Alkalies

Difluoromethylene dihypofluorite
*See* DIFLUOROMETHYLENE DIHYPOFLUORITE, $CF_4O_2$ : Haloalkenes

Dinitrogen tetraoxide
*See* DINITROGEN TETRAOXIDE (NITROGEN DIOXIDE), $N_2O_4$ : Halocarbons
*See other* HALOALKENES

## 1,1,1,2-TETRACHLOROETHANE $CCl_3CH_2Cl$ $C_2H_2Cl_4$

Dinitrogen tetraoxide
*See* DINITROGEN TETRAOXIDE (NITROGEN DIOXIDE), $N_2O_4$ : Halocarbons

## 1,1,2,2-TETRACHLOROETHANE $CHCl_2CHCl_2$ $C_2H_2Cl_4$

Alkalies
*MCA SD-34*, 1949
It is not an inert solvent, and on heating with solid potassium hydroxide or other base, hydrogen chloride is eliminated and chloro- or dichloroacetylene, which ignite in air, are formed.
*See also* TRICHLOROETHYLENE, $C_2HCl_3$ : Alkalies
1,2-DICHLOROETHYLENE, $C_2H_2Cl_2$ : Alkalies

Metals
*See* POTASSIUM, K: Halocarbons
SODIUM, Na: Halocarbons
*See other* HALOCARBONS

## † 1,1-DIFLUOROETHYLENE $CF_2{=}CH_2$ $C_2H_2F_2$

## TRIFLUOROACETAMIDE $CF_3CONH_2$ $C_2H_2F_3NO$

Lithium tetrahydroaluminate
*See* LITHIUM TETRAHYDROALUMINATE, $AlH_4Li$: Fluoroamides

## SODIUM 2,2,2-TRIFLUOROETHOXIDE $NaOCH_2CF_3$ $C_2H_2F_3NaO$

Schmutz, J. L. *et al., Inorg. Chem.*, 1975, **14**, 2437

During preparation from sodium and excess alcohol in ether, attempts to remove last traces of alcohol by warming under vacuum led to explosions.

**MERCURY(II) METHYLNITROLATE** $C_2H_2HgN_4O_8$

$[NO_2CH{=}N(O)O]_2Hg$

Urbanski, 1967, Vol. 3, 158

This salt of *aci*-dinitromethane also shows detonator properties.

*See related* METAL FULMINATES

**1,2-BIS(HYDROXOMERCURIO)-1,1,2,2-BIS(OXYDIMERCURIO)-ETHANE ('ETHANE HEXAMERCARBIDE')** $C_2H_2Hg_6O_4$

$OHg_2C(HgOH)C(HgOH)Hg_2O$

Hofmann, K. A., *Ber.*, 1898, **31**, 1904

This compound, formulated as 1,2-bis(hydroxomercurio)-1,1: 2,2-bis(oxydimercuri)ethane, explodes very violently at 230°C.

*See other* HEAVY METAL DERIVATIVES

**POTASSIUM DINITROACETAMIDE** $C_2H_2KN_3O_5$

$KC(NO_2)_2CONH_2$

*See* SILVER DINITROACETAMIDE, $C_2H_2AgN_3O_5$

**DIAZOACETALDEHYDE** $N_2CHCHO$ $C_2H_2N_2O$

Arnold, Z., *Chem. Comm.*, 1967, 299

It may be distilled out continuously as formed at 40°C/0.013 bar, but readily detonates with great violence if overheated

*See other* DIAZOCOMPOUNDS

**SODIUM DINITROACETAMIDE** $C_2H_2N_3NaO_5$

$NaC(NO_2)_2CONH_2$

*See* SILVER DINITROACETAMIDE, $C_2H_2AgN_3O_5$

**AZIDOACETONITRILE** $N_3CH_2CN$ $C_2H_2N_4$

Freudenberg, K. *et al.*, *Ber.*, 1932, **65**, 1188

With over 51% nitrogen content, it is, as expected, explosive, sensitive to impact or heating to 250°C.

*See other* HIGH-NITROGEN COMPOUNDS
ORGANIC AZIDES

### 1,2,4,5-TETRAZINE $C_2H_2N_4$

CH=N–N=CH–N=N (ring)

Sulphuric acid

Sorbe, 1968, 139

The solid base decomposes violently in contact with the concentrated acid.

*See other* HIGH-NITROGEN COMPOUNDS

### 3-AZIDO-1,2,4-TRIAZOLE $N_3C{=}NNHCH{=}N$ (ring) $C_2H_2N_6$

Denault, G. C. *et al., J. Chem. Eng. Data,* 1968, **13**, 514

Samples exploded during analytical combustion.

*See other* ORGANIC AZIDES

### 1,2-DIAZIDOCARBONYLHYDRAZINE $C_2H_2N_8O_2$

$N_3CONHNHCON_3$

Kesting, W., *Ber.*, 1924, **57**, 1321

Similar in properties to lead or silver azide, it explodes on heating or impact.

*See other* ACYL AZIDES
HIGH-NITROGEN COMPOUNDS

### KETENE $H_2C{=}C{=}O$ $C_2H_2O$

Hydrogen peroxide

*See* HYDROGEN PEROXIDE, $H_2O_2$ : Ketene

### † GLYOXAL CHOCHO $C_2H_2O_2$

Preparative hazard

*See* NITRIC ACID, $HNO_3$ : 2,4,6-Trimethyltrioxane

### OXALIC ACID $CO_2HCO_2H$ $C_2H_2O_4$

Silver

*See* SILVER, Ag: Carboxylic acids

Sodium chlorite

*See* SODIUM CHLORITE, $ClNaO_2$ : Oxalic acid

## TRILEAD DICARBONATE DIHYDROXIDE $C_2H_2O_8Pb_3$

$Pb_3(CO_3)_2(OH)_2$

Fluorine

*See* FLUORINE, $F_2$ : Metal salts

## † BROMOETHYLENE (VINYL BROMIDE) $C_2H_3Br$

$BrCH{=}CH_2$

Non-flammable except at 6–15% with arc, hot flame or other high-energy source.

*See other* HALOALKENES

## ACETYL BROMIDE $CH_3COBr$ $C_2H_3BrO$

Hydroxylic compounds

Merck, 1976, 11

Interaction with water, methanol or ethanol is violent, hydrogen bromide being evolved.

*See other* ACYL HALIDES

## ACETYL HYPOBROMITE $CH_3COOBr$ $C_2H_3BrO_2$

Reilly, J. J. *et al., J. Org. Chem.*, 1974, **39**, 3292

The instability of the hypobromite is noted, and protective shielding is recommended for work with quantities greater than 5g.

*See other* HYPOHALITES

## † CHLOROETHYLENE (VINYL CHLORIDE) $C_2H_3Cl$

$ClCH{=}CH_2$

*MCA SD-56*, 1972

*MCA Case History No. 625*

Harmon, 1974, 2.20, 4.74–4.77

Discharge of a spray of vapour and liquid under pressure from a cylinder into a fume hood caused ignition of the vapour, due to static electricity. Discharge of the gas only did not cause static build-up.

Vinyl chloride tends to self-polymerise explosively if peroxidation occurs, and several industrial explosions have been recorded.

*See* VIOLENT POLYMERISATION

*See other* HALOALKENES
POLYMERISATION INCIDENTS

Air

*MCA Case History No. 1551*

Accidental exposure of the recovered monomer to atmospheric oxygen for a long period caused formation of an unstable polyperoxide which initiated an explosion. Suitable precautions are discussed, including use of aqueous 20–30% sodium hydroxide solution to destroy the peroxide.

*See also* 1,1-DICHLOROETHYLENE, $C_2H_2Cl_2$: Air

*See other* POLYPEROXIDES

**† 1-CHLORO-1,1-DIFLUOROETHANE** **$CClF_2CH_3$** **$C_2H_3ClF_2$**

**3-CHLORO-3-METHYLDIAZIRINE** **$\overline{ClCMeN{=}N}$** **$C_2H_3ClN_2$**

1. Liu, M. T. H., *Chem. Eng. News*, 1974, **52**(36), 10
2. Jones, W. E. *et al.*, *J. Photochem.*, 1976, **5**, 233–239

Extremely shock-sensitive and violently explosive; initiation has been caused by prolonged freezing at −198°C, or by sawing a stopcock off a metal trap containing trace amounts [1]. It may be stored safely at −80°C [2].

*See other* DIAZIRINES

**† ACETYL CHLORIDE** **$CH_3COCl$** **$C_2H_3ClO$**

Preparative hazard

*See* PHOSPHORUS TRICHLORIDE, $Cl_3P$: Acetic acid

Dimethyl sulphoxide

*See* DIMETHYL SULPHOXIDE, $C_2H_6OS$: Acyl halides

Water

*Haz. Chem. Data,* 1971, 24

Violent interaction

**† METHYL CHLOROFORMATE** **$ClCO_2Me$** **$C_2H_3ClO_2$**

***N,N*-DICHLOROGLYCINE** **$C_2H_3Cl_2NO_2$**

**$Cl_2NCH_2CO_2H$**

Vit, J. *et al.*, *Synth. Commun.*, 1976, **6**(1), 1–4

It explodes at 65°C, so solutions must be evaporated at below 40°C.

*See other* *N*-HALOGEN COMPOUNDS

## 1,1,1-TRICHLOROETHANE $CCl_3CH_3$ $C_2H_3Cl_3$

*MCA SD-90*, 1965

Dinitrogen tetraoxide

*See* DINITROGEN TETRAOXIDE (NITROGEN DIOXIDE), $N_2O_4$: Halocarbons

Metals

*See* ALUMINIUM, Al: Halocarbons
POTASSIUM, K: Halocarbons
POTASSIUM–SODIUM ALLOY, K–Na: Halocarbons
MAGNESIUM, Mg: Halocarbons

Oxygen

*See* OXYGEN (Gas), $O_2$: Halocarbons
OXYGEN (Liquid), $O_2$: Halocarbons

## 1,1,2-TRICHLOROETHANE $CHCl_2CH_2Cl$ $C_2H_3Cl_3$

Potassium

*See* POTASSIUM, K: Halocarbons

## *N*-CARBOMETHOXYIMINOPHOSPHORYL CHLORIDE $C_2H_3Cl_3NO_2P$

$Cl_3P{=}NCO_2Me$

Mellor, 1971, Vol. 8, Suppl. 3, 589

This intermediate (or its ethyl homologue), produced during the preparation of phosphoryl dichloride isocyanate from interaction of phosphorus pentachloride and methyl (or ethyl) carbamate, is unstable. Its decomposition to the required product may be violent or explosive unless moderated by presence of a halogenated solvent.

## 1,1,1-TRICHLOROETHANOL $CCl_3CH_2OH$ $C_2H_3Cl_3O$

Sodium hydroxide

*MCA Case History No. 1574*

Accidental contact of 50% sodium hydroxide solution with residual trichloroethanol in a pump caused an explosion. This was confirmed in laboratory experiments. Chlorohydroxyacetylene, the isomeric chloroketene or chlorooxirene, may have been formed by elimination of hydrogen chloride.

**† TRICHLORO(VINYL)SILANE** $Cl_3SiCH{=}CH_2$ $C_2H_3Cl_3Si$

It may ignite in air.

*See other* ALKYLNON-METAL HALIDES

**† FLUOROETHYLENE (VINYL FLUORIDE)** $C_2H_3F$

$FCH{=}CH_2$

**1-FLUORO-1,1-DINITROETHANE** $FC(NO_2)_2CH_3$ $C_2H_3FN_2O_4$

*See* FLUORODINITROMETHYL COMPOUNDS

**2(?)-FLUORO-1,1-DINITROETHANE** $C_2H_3FN_2O_4$

$(NO_2)_2CHCH_2F$

*MCA Case History No. 784*

During a prolonged fractionation of the crude material at 75°C/0.05 bar, an exothermic decomposition began. As a remedial measure, air was admitted to the hot, decomposing residue, causing a violent explosion. Admission of nitrogen, or cooling of hot residues before admitting air, might have avoided the incident.

*See other* FLUORODINITRO COMPOUNDS

**2-FLUORO-2,2-DINITROETHANOL** $C_2H_3FN_2O_5$

$CF(NO_2)_2CH_2OH$

Cochoy, R. E. *et al., J. Org. Chem.*, 1972, **37**, 3041

It is a potentially explosive vesicant, from which a series of esters, expected to be explosive, was prepared.

*See other* FLUORODINTROMETHYL COMPOUNDS
POLYNITROALKYL COMPOUNDS

**† 1,1,1-TRIFLUOROETHANE** $F_3CCH_3$ $C_2H_3F_3$

**PENTAFLUOROSULPHUR PEROXYACETATE** $C_2H_3F_5O_3S$

$CH_3CO_2OSF_5$

Hopkinson, M. J. *et al., J. Fluorine Chem.*, 1976, **7**, 505

The peroxyester and its homologues, like other fluoroperoxy compounds, are potentially explosive and may detonate on thermal or mechanical shock.

*See related* PEROXYESTERS
*See other* FLUORINATED PEROXIDES

**MONOPOTASSIUM *aci*-1,1-DINITROETHANE** $C_2H_3KN_2O_3$

$CH_3NO_2C{=}N(O)OK$

Rodd, 1965, Vol. 1B, 98

Isolated salts of nitrolic acids are explosive.

*See other aci*-NITRO SALTS

**VINYLLITHIUM** $CH_2$=CHLi $C_2H_3Li$

Juenge, E. C. *et al., J. Org. Chem.,* 1961, **26**, 564

When freshly prepared, it is violently pyrophoric but on storage it becomes less reactive and slow to ignite in air, possibly owing to polymerisation.

*See related* ALKYLMETALS

**† ACETONITRILE** $CH_3CN$ $C_2H_3N$

2-Cyano-2-propyl nitrate

*See* NITRATING AGENTS

Dinitrogen tetraoxide,
Indium

*See* DINITROGEN TETRAOXIDE (NITROGEN DIOXIDE), $N_2O_4$: Acetonitrile, Indium

*N*-Fluoro compounds

Fraser, G. W., *et al., Chem. Comm.,* 1966, 532

Nitrogen–fluorine compounds are potentially explosive in contact with acetonitrile.

*See* PERFLUOROUREA, $CF_4N_2O$: Acetonitrile

Iron(III) perchlorate

*See* IRON(III) PERCHLORATE, $Cl_3FeO_{12}$

Nitric acid

*See* NITRIC ACID, $HNO_3$: Acetonitrile

Perchloric acid

*See* PERCHLORIC ACID, $ClHO_4$: Acetonitrile

Sulphuric acid,
Sulphur trioxide

Lee, S.A., private comm., 1972

A mixture of acetonitrile and sulphuric acid on heating (or self-heating) to 53°C underwent an uncontrollable exotherm to 160°C in a few seconds. The presence of 28 mol % of sulphur trioxide reduces the initiation temperature to about 15°C. Polymerisation of acetonitrile is suspected.

*See other* CYANO COMPOUNDS

## METHYL ISOCYANIDE $CH_3N{=}C$: $C_2H_3N$

Lemoult, M. D., *Compt. Rend.*, 1906, **143**, 902

Stein, A. R., *Chem. Eng. News*, 1968, **46**(45), 7–8

It exploded when heated in a sealed ampoule: during redistillation of the liquid at 59°C/1 bar, a drop fell back into the dry boiler-flask and exploded violently.

*See* ETHYL ISOCYANIDE, $C_3H_5N$

*See related* CYANO COMPOUNDS

## GLYCOLONITRILE $HOCH_2CN$ $C_2H_3NO$

1. *BCISC Quart. Safety Summ.*, 1964, **35**, 2
2. Gaudry, R., *Org. Synth.*, 1955, Coll. Vol. 3, 436
3. Anon., *Chem. Eng. News*, 1966, **44**, (49), 50

A year-old bottled sample, containing syrupy phosphoric acid as stabiliser, and which already showed signs of tar formation, exploded in storage. The pressure explosion appeared to be due to polymerisation, after occlusion of inhibitor in tar, in a container in which the stopper had become cemented by polymer [1]. A similar pressure explosion occurred when dry, redistilled nitrile, stabilised with ethanol [2], polymerised after 13 days [3].

Alkali

Sorbe, 1968, 146

Traces of alkali promote violent polymerisation.

*See other* CYANO COMPOUNDS

## † METHYL ISOCYANATE $CH_3N{=}C{=}O$ $C_2H_3NO$

## ACETYL NITRITE $CH_3COONO$ $C_2H_3NO_3$

Francesconi, L. *et al.*, *Gazz. Chim. Ital.*, 1895, [1], **34**, 439

An unstable liquid, decomposed by light, of which the vapour is violently explosive on mild heating.

*See other* ACYL NITRITES

## ACETYL NITRATE $CH_3COONO_2$ $C_2H_3NO_4$

1. Pictet, A. *et al.*, *Ber.*, 1907, **40**, 1164
2. Bordwell, F. G. *et al.*, *J. Amer. Chem. Soc.*, 1960, **82**, 3588
3. Konig, W., *Angew. Chem.*, 1955, **67**, 517
4. Wibaut, J. P., *Chem. Weekbl.*, 1942, **39**, 534

Acetyl nitrate, readily formed above 0°C from acetic anhydride and concentrated nitric acid, is thermally unstable, and its solutions may decompose violently above 60°C (forming tetranitromethane, a powerful oxidant). The pure nitrate explodes violently on rapid heating to above 100°C [1]. Isolation before use as a nitrating agent at −10°C is not necessary, but the mixture must be preformed at 20–25°C before cooling to −10°C to avoid violent reactions [2]. Spontaneous explosions of pure isolated material a few days old had been reported previously [3].

Spontaneous detonation of freshly distilled material has also been recorded, as well as explosion on touching with a glass rod [4].

*See also* TETRANITROMETHANE, $CN_4O_8$

Mercury(II) oxide

Chrétien, A. *et al., Compt. Rend.*, 1945, **220**, 823

Acetyl nitrate explodes when mixed with red mercury oxide, or other 'active' oxides.

*See other* ACYL NITRATES

**PEROXYACETYL NITRATE** $CH_3COOONO_2$ $C_2H_3NO_5$

1. Stephens, E. R. *et al., J. Air Poll. Control Ass.*, 1969, **19**, 261–264
2. Stephens, E. R., *Anal. Chem.*, 1964, **36**, 928
3. Louw, R. *et al., J. Amer. Chem. Soc.*, 1975, **97**, 4396

Accidental production of the liquid nitrate during overcooled storage at 0°C of a rich mixture of the vapour and helium is thought to have caused a violent explosion [1]. Mixtures of up to 0.1 vol.% at 7 bar stored at 10°C or above are quite safe. During dilution with nitrogen of vapour samples in evacuated bulbs, slow pressurisation is necessary to avoid explosion [2]. It is extremely explosive and may only be handled in high dilution with air or nitrogen. The propionyl analogue is similarly explosive, but higher homologues less so [3].

**1,2,3-TRIAZOLE** HNN=NCH=CH $C_2H_3N_3$

Baltzer, O. *et al., Ann.*, 1891, **262**, 320, 322

The vapour readily explodes if superheated (above 200°C), and the silver derivative explodes on heating.

*See other* HIGH-NITROGEN COMPOUNDS

**VINYL AZIDE** $CH_2$=$CHN_3$ $C_2H_3N_3$

Wiley, R. H. *et al., J. Org. Chem.*, 1957, **22**, 995

A sample contained in a flask detonated when the ground-joint was rotated. Literature statements that it is surprisingly stable are erroneous.

*See other* ORGANIC AZIDES

**ACETYL AZIDE** $CH_3CON_3$ $C_2H_3N_3O$

Smith, 1966, Vol. 2, 214

It is treacherously explosive.

*See other* ACYL AZIDES

**AZIDOACETALDEHYDE** $N_3CH_2CHO$ $C_2H_3N_3O$

Forster, M. O., *et al., J. Chem. Soc.*, 1908, **93**, 1870

It decomposed with vigorous gas evolution below 80°C at 5 mbar. A reaction mixture of chloroacetaldehyde hydrate and sodium azide had previously exploded mildly on heating in absence of added water.

*See other* 2-AZIDOCARBONYL COMPOUNDS

**5-METHOXY-1,2,3,4-THIATRIAZOLE** $\overline{SN{=}NN{=}C}OMe$ $C_2H_3N_3OS$

Jensen, K. A., *et al., Acta. Chem. Scand.*, 1964, **18**, 825

It explodes at ambient temperature, and its higher homologues are unstable.

*See other* HIGH-NITROGEN COMPOUNDS

**AZIDOACETIC ACID** $N_3CH_2CO_2H$ $C_2H_3N_3O_2$

Borowski, S. J., *Chem. Eng. News*, 1976, **54**(44), 5

The pure acid is insensitive to shock at up to 175°C, and not explosively unstable at up to 250°C. In contrast, the presence of iron or its salts leads to rapid exothermic decomposition of the acid at 25°C, and explosion at 90°C or lower under strong illumination with visible light.

*See* CATALYSIS BY IMPURITIES

*See other* ORGANIC AZIDES

**2,2,2-TRINITROETHANOL** $C(NO_2)_3CH_2OH$ $C_2H_3N_3O_7$

1. Marans, N. S. *et al., J. Amer. Chem. Soc.*, 1950, **72**, 5329
2. Cochoy, R. E. *et al., J. Org. Chem.*, 1972, **37**, 3041

It is a moderately shock-sensitive explosive which has exploded during distillation [1]. A series of its esters, expected to be explosive, was prepared [2].

*See other* POLYNITROALKYL COMPOUNDS

**4-NITROAMINO-1,2,4-TRIAZOLE** $C_2H_3N_5O_2$

$NO_2NH\overline{NCH{=}NN{=}CH}$

Explodes at m.p., 172°C.

*See* *N*-AZOLIUM NITROIMIDATES

### 1,3-DI(5-TETRAZOLYL)TRIAZENE $C_2H_3N_{11}$

N=NNHN=CN=NNHC=NNHN=N

Hofmann, K. A. *et al.*, *Ber.*, 1910, **43**, 1869–1870
The barium salt explodes weakly on heating, and the copper and silver salts strongly on heating or friction.
*See other* TETRAZOLES
TRIAZENES

### SODIUM ACETATE $CH_3CO_2Na$ $C_2H_3NaO_2$

Diketene
*See* DIKETENE, $C_4H_4O_2$: Acids, etc.

Potassium nitrate
*See* POTASSIUM NITRATE, $KNO_3$: Sodium acetate

### SODIUM PEROXYACETATE $CH_3CO_2ONa$ $C_2H_3NaO_3$

Humber, L. G., *J. Org. Chem.*, 1959, **24**, 1789
A sample of the dry salt exploded at room temperature.
*See other* PEROXOACID SALTS

### † ETHYLENE (ETHENE) $H_2C{=}CH_2$ $C_2H_4$

1. Waterman, H. I. *et al.*, *J. Inst. Petr. Tech.*, 1931, **17**, 506–510
2. Conrad, D. *et al.*, *Chem. Ing. Tech.*, 1975, **47**, 265 (summary only)
3. Lawrence, W. W. *et al.*, *Loss Prevention*, 1967, **1**, 10–12
4. McKay, F. F. *et al.*, *Hydrocarbon Proc.*, 1977, **56** (11), 487–494

Explosive decomposition occurred at 350°C under a pressure of 170 bar [1]. The limiting pressures and temperatures for explosive decomposition of ethylene with electric initiation were determined in the ranges 100–250 bar and 120–250°C. Limiting conditions are much lower for high-energy (exploding wire) initiation [2]. Previously, reaction parameters including effect of pipe size and presence of nitrogen as inert diluent on the propagation of explosive decomposition had been studied, using thermite initiation. 10 vol.% of nitrogen markedly interferes with propagation [3].

Sudden pressurising with ethylene of part of an air-containing pipeline system from 1 to 88.5 bar led to adiabatic compressive heating and autoignition of the ethylene, slow propagation of the decomposition flame upstream into the main 30 cm pipeline and subsequent rupture of the latter at an estimated wall temperature of 700–800°C. The AIT for ethylene in air at 1 bar is 492°C, but this falls with increase in pressure and could lie between 204 and 371°C at 68–102 bar, and the ethylene decomposition reaction may start at 315–371°C within these pressure limits. Detonation was not, however, observed. Precautions against such incidents in ethylene installations are discussed [4].

Acetylene,
Hydrogen

*See* HYDROGEN (Gas), $H_2$: Acetylene, etc.

Air

Tanimoto, S., in *Safety in Polyethylene Plants*, 1974, **2**, 14–21 (a CEP Manual), Amer. Inst. Chem. Engrs.

In a published symposium mainly devoted to engineering aspects of large-scale polyethylene manufacture, this paper describes an extensive large-scale field experiment to determine the blast and destructive effects when various ethylene–air mixtures were detonated to simulate a compressor-house explosion.

Air,
Chlorine

Fujiwara, T. *et al.*, Jap. Pat. 7 432 841, 1974 (*Chem. Abs*, 1975, **82**, 111554w)

During oxychlorination of ethylene to 1,2-dichloroethane, excess hydrogen chloride is used to maintain the reaction mixture outside the explosive limits.

Aluminium trichloride

Waterman, H. I. *et al., J. Inst. Pet.*, 1947, **33**, 254

Mixtures of ethylene and aluminium chloride, initially at 30–60 bar, rapidly heat and explode in presence of supported nickel catalyst, methyl chloride or nitromethane.

Aluminium trichloride,
Nitromethane

*See* ALUMINIUM TRICHLORIDE–NITROMETHANE, $AlCl_3 \cdot CH_3NO_2$: Alkene

Bromotrichloromethane

Elsner, H. *et al., Angew. Chem.*, 1962, **74**, 253

Following a literature method for preparation of 1-bromo-3,3,3-trichloropropane, the reagents were being heated at 120°C/51 bar. During the fourth preparation a violent explosion occurred.

*See* Carbon tetrachloride, below

Carbon tetrachloride

Zakaznov, V. F. *et al., Khim. Prom.*, 1968, 8, 584

Mixtures of ethylene and carbon tetrachloride can be initiated to explode at temperatures between 25 and 105°C and pressures of 30–80 bar, causing a sixfold pressure increase. At 100°C and 61 bar explosion initiated in the gas phase propagated into the liquid phase. Increase of carbon tetrachloride concentrations in the gas phase decreased the limiting decomposition pressure.

*See* Bromotrichloromethane, above
DIBENZOYL PEROXIDE, $C_{14}H_{10}O_4$ : Carbon tetrachloride, etc.

Chlorine

*See* CHLORINE, $Cl_2$ : Hydrocarbons

Chlorotrifluoroethylene

Colombo, P. *et al.*, *J. Polymer Sci.*, 1963, **B1**(8), 435–436

Mixtures containing ratios of about 20:1 and 12:1 of ethylene: haloalkene undergoing polymerisation under gamma irradiation at 308 krad/h exploded violently after a total dose of 50 krad. Dose rate and haloalkene concentration were both involved in the initiation process.

Copper

Dunstan, A. E. *et al.*, *J. Soc. Chem. Ind.*, 1932, **51**, 132T

Polymerisation of ethylene in presence of metallic copper becomes violent above a pressure of 54 bar at ~400°C, much carbon being deposited.

Lithium

*See* LITHIUM, Li: Ethylene

Molecular sieve

*See* MOLECULAR SIEVE: Ethylene

Oxides of nitrogen

*See* NITROGEN OXIDE ('NITRIC OXIDE'), NO: Dienes, Oxygen

Ozone

*See* OZONE, $O_3$ : Combustible gases
: Ethylene

Steel-braced tyres

*CISHC Chem. Safety Summ.*, 1974–5, **45–46**, 2–3

Two hours after a road tanker had crashed, causing the load of liquid ethylene to leak, one of the tyres of the tanker burst and ignited the spill, eventually causing the whole tanker to explode. The tyre failed because it froze and became embrittled, and it is known that such failure of steel-braced tyres gives off showers of sparks. This could therefore be a common ignition source in cryogenic spillage incidents.

*See* IGNITION SOURCES

Tetrafluoroethylene

Coffman, D. D. *et al., J. Amer. Chem. Soc.*, 1949, **71**, 492

A violent explosion occurred when a mixture of tetrafluoroethylene and excess ethylene was heated at 160°C and 480 bar. Traces of oxygen must be rigorously excluded. Other olefins reacted smoothly.

*See* TETRAFLUOROETHYLENE, $C_2F_4$: Air

Trifluoromethyl hypofluorite

*See* TRIFLUOROMETHYL HYPOFLUORITE, $CF_4O$: Hydrocarbons

**1-BROMOAZIRIDINE** $\overline{CH_2NBrCH_2}$ $C_2H_4BrN$

Graefe, A. F., *J. Amer. Chem. Soc.*, 1958, **80**, 3940

The compound is very unstable and always decomposes, sometimes explosively during or shortly after distillation.

*See other N*-HALOGEN COMPOUNDS

***N*-BROMOACETAMIDE** $CH_3CONHBr$ $C_2H_4BrNO$

'Organic Positive Bromine Compounds', Brochure, Boulder, Arapahoe Chemicals Inc., 1962

It tends to decompose rapidly at elevated temperatures in presence of moisture and light.

*See also N*-HALOIMIDES

*See other N*-HALOGEN COMPOUNDS

**1,2-DIBROMOETHANE** $BrCH_2CH_2Br$ $C_2H_4Br_2$

Magnesium

*See* MAGNESIUM, Mg: Halocarbons

**1-CHLOROAZIRIDINE** $\overline{CH_2NClCH_2}$ $C_2H_4ClN$

1. Davies, C. S., *Chem. Eng. News*, 1964, **42**(8), 41
2. Graefe, A. F., *Chem. Eng. News*, 1958, **36**(43), 52

A sample of redistilled material exploded after keeping in an amber bottle at ambient temperature for 3 months [1]. A similar sample had exploded very violently when dropped [2].

*See other N*-HALOGEN COMPOUNDS

**CHLOROACETALDEHYDE OXIME** $ClCH_2CH{=}NOH$ $C_2H_4ClNO$

Brintzinger, H. *et al., Chem. Ber.*, 1952, **85**, 345

Vacuum distillation of the product at 61°C/27 mbar must be interrupted when a solid separates from the residue to avoid an explosion.
*See also* BROMOACETONE OXIME, $C_3H_6BrNO$

**N-CHLOROACETAMIDE** **$CH_3CONHCl$** **$C_2H_4ClNO$**

Muir, G. D., private comm., 1968
It has exploded during desiccation of the solid or during concentration of its chloroform solution. It may be purified safely by pouring a solution in acetone into water and air-drying the product.
*See also* *N*-HALOIMIDES
*See other* *N*-HALOGEN COMPOUNDS

**† 1,1-DICHLOROETHANE** **$CHCl_2CH_3$** **$C_2H_4Cl_2$**

**† 1,2-DICHLOROETHANE** **$CH_2ClCH_2Cl$** **$C_2H_4Cl_2$**

*MCA SD-18*, 1971

Dinitrogen tetraoxide
*See* DINITROGEN TETRAOXIDE (NITROGEN DIOXIDE), $N_2O_4$: Halocarbons

Metals
*See* ALUMINIUM, Al: Halocarbons
POTASSIUM, K: Halocarbons

Nitric acid
*See* NITRIC ACID, $HNO_3$: 1,2-Dichloroethane

**AZO-N-CHLOROFORMAMIDINE** **$C_2H_4Cl_2N_6$**
**$NH_2C(NCl)N{=}NC(NCl)NH_2$**

Braz, G. I. *et al., Zh. Prikl. Khim. USSR*, 1944, **17**, 565
Decomposes explosively at 155°C.
*See other* AZO COMPOUNDS
*N*-HALOGEN COMPOUNDS

**† BIS-CHLOROMETHYL ETHER** **$CH_2ClOCH_2Cl$** **$C_2H_4Cl_2O$**

**ETHYLENE DIPERCHLORATE** **$(CH_2OClO_3)_2$** **$C_2H_4Cl_2O_8$**

Schumacher, 1960, 214

A highly sensitive, violently explosive material, capable of initiation by addition of a few drops of water.
*See other* ALKYL PERCHLORATES

**BISCHLOROMETHYLTHALLIUM CHLORIDE** $C_2H_4Cl_3Tl$

$(ClCH_2)_2\ TlCl$

Yakubovich, A. Ya. *et al., Dokl. Akad. Nauk SSSR*, 1950, **3**, 957
An explosive solid of low stability.
*See related* ALKYLMETAL HALIDES

**HYDROXYCOPPER(II) GLYOXIMATE** $C_2H_4CuN_2O_3$

HOCuON=CHCHNOH

Morpurgo, G. O. *et al., J. Chem. Soc., Dalton Trans.*, 1977, 746
The complex loses weight up to 140–150°C, then explodes.
*See related* OXIMES
*See other* HEAVY METAL DERIVATIVES

**2-FLUORO-2,2-DINITROETHYLAMINE** $C_2H_4FN_3O_4$

$CF(NO_2)_2CH_2NH_2$

Adolph, H. G. *et al., J. Org. Chem.,* 1969, **34**, 47
Outstandingly explosive amongst fluorodinitromethyl compounds, samples stored neat at ambient temperature regularly exploded within a few hours. Occasionally concentrated solutions in dichloromethane have decomposed violently after long storage.
*See other* FLUORODINITROMETHYL COMPOUNDS

† **1,1-DIFLUOROETHANE** $CHF_2CH_3$ $C_2H_4F_2$

**1,2-BIS(DIFLUOROAMINO)ETHANOL** $C_2H_4F_4N_2O$

$NF_2CH_2CHOHNF_2$

Reed, S. F., *J. Org. Chem.*, 1967, **32**, 2894
It is slightly more impact-sensitive than glyceryl trinitrate.
*See* DIFLUOROAMINO COMPOUNDS
*See other* *N*-HALOGEN COMPOUNDS

**1,2-BIS(DIFLUOROAMINO)-*N*-NITROETHYLAMINE** $C_2H_4F_4N_4O_2$

$NF_2CH_2CH(NF_2)NHNO_2$

Tyler, W. E., US Pat. 3 344 167, 1967

The crude product tends to explode spontaneously on storage, though the triple-distilled material appears stable on prolonged storage. Generally, such nitroamines are unstable and explode at 75°C or above.

*See* DIFLUOROAMINO COMPOUNDS
*See other N*-NITRO COMPOUNDS

**MERCURY(II) FORMHYDROXAMATE** **$C_2H_4HgN_2O_4$**

**$Hg(ONHCHO)_2$**

Urbanski, 1967, Vol. 3, 158
It also possesses detonator properties.
*See related* METAL FULMINATES

**POTASSIUM 1-NITROETHOXIDE** **$KOCH(NO_2)CH_3$** **$C_2H_4KNO_3$**

Meyer, V. *et al., Ber.*, 1872, **5**, 1032
The solid exploded on heating, and an aqueous solution sealed into a Carius tube exploded violently when the tube was cracked open with a red-hot glass rod.
*See other C*-NITRO COMPOUNDS

**3-METHYLDIAZIRINE** **$MeCHN{=}N$** (ring) **$C_2H_4N_2$**

Schmitz, E. *et al., Chem. Ber.*, 1962, **95**, 795
The gas explodes on heating.
*See* DIAZIRINES

**2-NITROACETALDEHYDE OXIME** **$C_2H_4N_2O_3$**

**$NO_2CH_2CH{=}NOH$**

Sorbe, 1968, 149
The oxime (an isomer of ethylnitrolic acid) decomposes gradually at ambient temperature but explosively above 110°C.
*See other* OXIMES

**1-NITRO-1-OXIMINOETHANE (ETHYLNITROLIC ACID)** **$C_2H_4N_2O_3$**

**$CH_3C(NO_2){=}NOH$**

Sorbe, 1968, 147
An explosive solid, formally a nitro-oxime.
*See related* OXIMES

**ETHYLIDENE DINITRATE** **$CH_3CH(ONO_2)_2$** **$C_2H_4N_2O_6$**

Kacmarek, A. J. *et al., J. Org. Chem.*, 1975, **40**, 1853

An analytical sample exploded in the hot zone of the combustion apparatus.

*See* DINITROGEN PENTAOXIDE, $N_2O_5$: Acetaldehyde

*See other* ALKYL NITRATES

**CYANOGUANIDINE ('DICYANDIAMIDE')** $C_2H_4N_4$

**CNNHC(NH)NH$_2$**

Oxidants

Baumann, J., *Chem. Ztg.*, 1920, **44**, 474

Mixtures of cyanoguanidine with ammonium nitrate or potassium chlorate, etc., were formerly proposed for use as powerful explosives.

*See other* CYANO COMPOUNDS
HIGH-NITROGEN COMPOUNDS

**2-METHYLTETRAZOLE** **MeNN=NCH=N** (ring) $C_2H_4N_4$

Aluminium hydride

*See* ALUMINIUM HYDRIDE, $AlH_3$: Tetrazole derivatives

*See other* TETRAZOLES

**1,1-DIAZIDOETHANE** $(N_3)_2CHCH_3$ $C_2H_4N_6$

Forster, M.O. *et al.*, *J. Chem. Soc.*, 1908, **93**, 1070

The extreme instability and explosive behaviour of this diazide caused work on other *gem*-diazides to be abandoned.

**1,2-DIAZIDOETHANE** $N_3CH_2CH_2N_3$ $C_2H_4N_6$

Alone,
or Sulphuric acid

Forster, M. O., *et al.*, *J. Chem. Soc.*, 1908, **93**, 1070

Though less unstable than the 1,1-isomer (above), it explodes on heating, and unreproducibly in contact with sulphuric acid.

*See other* ORGANIC AZIDES

**AZIDOCARBONYLGUANIDINE** $C_2H_4N_6O$

$N_3CONHC(NH)NH_2$

Thiele, J. *et al.*, *Ann.*, 1898, **303**, 93

It explodes violently on rapid heating.

*See other* ACYL AZIDES

**AZO-N-NITROFORMAMIDINE** $C_2H_4N_8O_4$

$NO_2NHC(NH)N=NC(NH)NHNO_2$

Wright, G. F., *Can. J. Chem.*, 1952, **30**, 64
Explosive decomposition at 165°C.
*See other* AZO COMPOUNDS
*N*-NITRO COMPOUNDS

**1,2-DI(5-TETRAZOLYL)HYDRAZINE** $(N=NNHN=CNH)_2$ $C_2H_4N_{10}$

Thiele, J., *Ann.*, 1898, **303**, 66
Explodes without melting when heated.
*See other* TETRAZOLES

**1,6-BIS(5-TETRAZOLYL)HEXAAZA-1,5-DIENE** $C_2H_4N_{14}$

$(N=NNHN=CN=NNH)_2$

Hofmann, K. A. *et al., Ber.*, 1911, **44**, 2953
This very high-nitrogen compound (87.5%) explodes violently on pressing with a glass rod, or on heating to 90°C.
*See other* HIGH-NITROGEN COMPOUNDS
TETRAZOLES

**ACETALDEHYDE** $CH_3CHO$ $C_2H_4O$

*MCA SD-43*, 1952
Vervalin, 1973, 90–91
The MCA Data Sheet describes acetaldehyde as extremely or violently reactive with: acid anhydrides, alcohols, halogens, ketones, phenols, amines, ammonia, hydrogen cyanide or hydrogen sulphide.

Acetaldehyde vapour leaking into a building equipped only with flameproof electrical equipment nevertheless ignited, possibly on contact with rusted steel, corroded aluminium or hot steam lines.

Acetic acid
*MCA Case History No. 1764*
A drum contaminated with acetic acid was filled with acetaldehyde. The ensuing exothermic polymerisation reaction caused a mild eruption lasting for several hours.

Air
White, A. G. *et al., J. Soc. Chem. Ind.*, 1950, **69**, 206
Mixtures of 30–60% of acetaldehyde vapour with air or 60–80% with oxygen may ignite on surfaces at 176 and 105°C, respectively,

owing to formation and subsequent violent decomposition of peracetic acid.

Cobalt acetate,
Oxygen

Phillips, B. *et al., J. Amer. Chem. Soc.,* 1957, **79**, 5982

Bloomfield, G. F. *et al., J. Soc. Chem. Ind.,* 1935, **54**, 129T

Oxygenation of acetaldehyde in presence of cobalt acetate at −20°C caused precipitation of 1-hydroxyethyl peracetate (acetaldehyde hemiperacetate), which exploded violently on stirring. Ozone or UV light also catalyses the autoxidation.

Desiccants,
Hydrogen peroxide

*See* HYDROGEN PEROXIDE, $H_2O_2$ : Acetaldehyde, Desiccants

Dinitrogen pentaoxide

*See* DINITROGEN PENTAOXIDE, $N_2O_5$ : Acetaldehyde

Hydrogen peroxide

*See* HYDROGEN PEROXIDE, $H_2O_2$ : Oxygenated compounds

Mercury(II) oxosalts

Sorbe, 1968, 97

Some of the products of interaction of acetaldehyde and mercury(II) salts (chlorate or perchlorate) are highly explosive and extremely shock-sensitive.

*See* CHLORATOMERCURIO(FORMYL)METHYLENEMERCURY(II), $C_2HClHg_2O_4$

Metals

Sorbe, 1968, 103

Impure material will readily polymerise in presence of trace metals (iron) or acids.

*See* Sulphuric acid, below

Oxygen

*MCA Case History No. 117*

Oxygen leaked into a free space in an acetaldehyde storage tank normally purged with nitrogen. Accelerating exothermic oxidation led to detonation.

*See* Air, above

Sulphuric acid

Sorbe, 1968, 103

Acetaldehyde is polymerised violently by the concentrated acid.

*See* Metals, above
*See other* POLYMERISATION INCIDENTS
PEROXIDISABLE COMPOUNDS

## † ETHYLENE OXIDE

$\overline{CH_2OCH_2}$ $C_2H_4O$

1. Hess, L. G. *et al.*, *Ind. Eng. Chem.*, 1950, **42**, 1251
2. *MCA SD-38*, 1971
3. *MCA Case History No. 1666*
4. Marston, 1974, 2.9

Liquid ethylene oxide is not detonable, but the vapour may be readily initiated into explosive decomposition. Recommendations for storage and handling are discussed in detail [1]. Metal fittings containing copper, silver, mercury or magnesium should not be used in ethylene oxide service, since traces of acetylene could produce explosive acetylides capable of detonating ethylene oxide vapour [2].

Ethylene oxide exposed to heating and subsequently cooled (e.g. by exposure to fire conditions) may continue to polymerise exothermically, leading to container pressurisation and explosion. A mechanism consistent with several observed incidents is proposed [3]. Presence of hot spots in processing plant is identified as a particular hazard [4].

*See* Contaminants, below
VIOLENT POLYMERISATION
*See other* POLYMERISATION INCIDENTS

Air,
Bromomethane

Baratov, A. N. *et al.*, p.24 of British Lending Library translation (628. 74, issued 1966) of Russian book on *Fire Prevention and Firefighting Symposium*. The addition of bromomethane to ethylene oxide (used for germicidal sterilising) to reduce the risk of explosion is relatively ineffective, the inhibiting concentration being 31.2, as against 5.8 for hexane and 13.5% vol. for hydrogen.

Alkanethiols,
or An alcohol

Meigs, D. P., *Chem. Eng. News*, 1942, **20**, 1318
Autoclave reactions involving ethylene oxide with unspecified alkanethiols or an alcohol went out of control and exploded violently. Similar previous reactions had been uneventful.

Ammonia

*MCA Case History No. 792*

Accidental contamination of an ethylene oxide feed tank by ammonia caused violently explosive polymerisation.

*See* Contaminants, below
Trimethylamine, below

Contaminants

Gupta, A. K., *J. Soc. Chem. Ind.*, 1949, **68**, 179

Precautions designed to prevent explosive polymerisation of ethylene oxide are discussed, including rigid exclusion of acids, covalent halides such as aluminium, iron(III) and tin(IV) chlorides, basic materials like alkali hydroxides, ammonia, amines, metallic potassium, and catalytically active solids such as aluminium or iron oxides or rust.

*See* Ammonia, above
Trimethylamine, below

Lagging materials

Hilado, C. J., *J. Fire Flamm.*, 1974, **5**, 321–326

The self-heating temperature (effectively the AIT) of a 50:50 mixture of ethylene oxide and air is reduced from 456°C on passage through various thermal insulation (lagging) materials to 251–416°C, depending on the particular material (of which 13 were tested).

*See also* OIL–LAGGING FIRES

*m*-Nitroaniline

Anon., *Angew. Chem. (Nachr.)*, 1958, **70**, 150

Interaction of the two compounds in an autoclave at 150–160°C is described as safe in Swiss Pat. 171 721. During careful repetition of the reaction with stepwise heating, an autoclave exploded at 130°C.

Nitrogen

Grose-Wortmann, H., *Chem. Ing. Tech.*, 1970, **42**, 85–86

Concentration boundaries for explosive decomposition of the oxide and its mixtures with nitrogen at elevated temperatures and pressures are presented graphically.

Polyhydric alcohol,
Propylene oxide

Vervalin, 1973, 82

A polyether-alcohol, prepared by co-condensation of ethylene and propylene oxides with a polyhydric alcohol, was stored at above 100°C and exposed to air via a vent line. After 10–15 h, violent decomposition occurred, rupturing the vessel. It was subsequently found that exothermic oxidation of the product occurred above 100°C, and that at 300°C a rapid exothermic reaction set in, accompanied by vigorous gas evolution.

Trimethylamine

1. *BCISC Quart. Safety Summ.*, 1966, **37**, 44
2. Anon., *Chem. Trade J.*, 1956, **138**, 1376

Accidental contamination of a large ethylene oxide feed-cylinder by reaction liquor containing trimethylamine caused the cylinder to explode 18 h later. Contamination was possible because of a faulty pressure gauge and suck-back of froth above the liquid level [1]. A similar incident occurred previously [2].

*See* Ammonia, above
Contaminants, above

*See other* 1,2-EPOXIDES

**THIOACETIC *S*-ACID** **$CH_3COSH$** **$C_2H_4OS$**

**ACETIC ACID** **$CH_3CO_2H$** **$C_2H_4O_2$**

*MCA SD-41*, 1951

Acetaldehyde

*See* ACETALDEHYDE, $C_2H_4O$: Acetic acid

Acetic anhydride,
Water

*See* ACETIC ANHYDRIDE, $C_4H_6O_3$: Acetic acid, Water

5-Azidotetrazole

*See* 5-AZIDOTETRAZOLE, $CHN_7$: Acetic acid

Oxidants

*See* BROMINE PENTAFLUORIDE, $BrF_5$: Hydrogen-containing materials
CHROMIUM TRIOXIDE, $CrO_3$: Acetic acid
HYDROGEN PEROXIDE, $H_2O_2$: Acetic acid
POTASSIUM PERMANGANATE, $KMnO_4$: Acetic acid
SODIUM PEROXIDE, $Na_2O_2$: Acetic acid

Potassium *tert*-butoxide

*See* POTASSIUM *tert*-BUTOXIDE, $C_4H_9KO$: Acids

**† METHYL FORMATE** **$HCO_2Me$** **$C_2H_4O_2$**

**POLY(ETHYLIDENE PEROXIDE)** **$(-CHMeOO-)_n$** **$(C_2H_4O_2)_n$**

Rieche, A. *et al.*, *Angew. Chem.*, 1936, **49**, 101

The highly explosive polyperoxide is present in peroxidised diethyl ether and has been responsible for many accidents during distillation of the solvent.

*See other* POLYPEROXIDES

## ETHYLENE OZONIDE

$CH_2CH_2O{-}O^+{-}O^-$ $C_2H_4O_3$

1. Harries, C. *et al., Ber.,* 1909, **42**, 3305
2. Briner, E. *et al., Helv. Chim. Acta,* 1921, **12**, 154
3. Criegee, R., *Angew. Chem.*, 1958, **65**, 398–399

It explodes very violently on heating, friction or shock [1]. Stable at 0°C but often decomposes explosively at ambient temperature [2].

Pouring the liquid from one vessel to another initiated a violent explosion [3].

*See other* OZONIDES

## PEROXYACETIC ACID

$CH_3CO_2OH$ $C_2H_4O_3$

1. Swern, D., *Chem. Rev.,* 1949, **45**, 7
2. Davies, 1961, 56
3. Phillips, B. *et al., J. Org. Chem.,* 1959, **23**, 1823
4. Anon., *Angew. Chem. (Nachr.),* 1957, **5**, 178
5. Smith, I.C.P., private comm., 1973
6. *MCA Case History No. 1804*

It is insensitive to impact but explodes violently at 110°C [1]. The solid acid has exploded at −20°C [2]. Safe procedures (on basis of detonability experiments) for preparation of anhydrous peracetic acid solutions in chloroform or ester solvents have been described [3]. However, a case of explosion on impact has been recorded [4]. During vacuum distillation, turning a ground glass stopcock in contact with the liquid initiated a violent explosion, but the grease may have been involved as well as friction [5].

During a pilot-scale preparation of a solution of the acid by adding acetic anhydride slowly to 90% hydrogen peroxide in dichloromethane, an explosion of great violence occurred. Subsequent investigation revealed that during the early stages of the addition a two-phase system existed, which was difficult to detect because of closely similar densities and refractive indices. The peroxide-rich phase became extremely shock-sensitive when between 10 and 30% of the anhydride had been added [6].

*See* HYDROGEN PEROXIDE, $H_2O_2$: Acetic acid
: Acetic anhydride
: Vinyl acetate

Acetic anhydride
*MCA Case History No. 1795*

During an attempt to prepare an anhydrous 25% solution of peroxyacetic acid in acetic acid by dehydrating a water-containing solution with acetic anhydride, a violent explosion occurred. Mistakes in the operational procedure allowed heated evaporation to begin before the anhydride had been hydrolysed. Acetyl peroxide could have been formed from the anhydride and peroxyacid, and the latter may have detonated and/or catalysed violent hydrolysis of the anhydride.

*See* ACETIC ANHYDRIDE, $C_4H_6O_3$ : Water

5-*p*-Chlorophenyl-2,2-dimethyl-3-hexanone

Gillespie, J. S. *et al.*, *Tetrahedron*, 1975, **31**, 5

During the attempted preparation of 3-*p*-chlorophenylbutanoic acid by addition of the ketone (2 g-mol.) to peracetic acid (50%) in acetic acid at 65–70°C, a serious explosion occurred.

3-Methyl-3-buten-1-yl tetrahydropyranyl ether

*See* TETRAHYDROPYRANYL ETHER DERIVATIVES

*See other* PEROXYACIDS

## 2,4-DITHIA-1,3-DIOXANE-2,2,4,4-TETRAOXIDE ('CARBYL SULPHATE')

**$C_2H_4O_6S_2$**

**$\overline{CH_2CH_2OSO_2OSO_2}$**

*N*-Methyl-*p*-nitroaniline

1. Deucker, W. *et al.*, *Chem. Ing. Tech.*, 1973, **45**, 1040–1041
2. Wooton, D. L. *et al.*, *J. Org. Chem.*, 1974, **39**, 2112

The title cyclic ester and substituted aniline (>2 mol) condense when heated in nitrobenzene solution, initially at 45°C and eventually at 75°C under vacuum in a sealed vessel, to give the ester-salt, *N*-methyl-*p*-nitroanilinium 2(*N*-methyl-*N*-*p*-nitrophenylaminosulphonyl)ethylsulphate. The reaction technique had been used uneventfully during 8 years, but a 3 $m^3$ containing vessel was violently ruptured when the reaction ran away and developed internal pressure exceeding 50–70 bar.

Subsequent DTA investigation showed that an exothermic reaction set in above 75°C after an induction period depending on the initial temperature and concentration of reactants, which attained nearly 300°C, well above the decomposition point of the cyclic ester component (170°C). The reaction technique used could have permitted local over-concentration and overheating effects to occur owing to slow dissolution of the clumped solid ester and aniline in the nitrobenzene solvent [1]. Crude 'carbyl sulphate' contains excess sulphur trioxide [2].

*See* DIFFERENTIAL THERMAL ANALYSIS

**THIIRAN (ETHYLENE SULPHIDE)** $\overline{CH_2CH_2S}$ $C_2H_4S$

Acids

Sorbe, 1968, 122

The sulphide may polymerise violently in presence of acids, especially when warm.

*See related* 1,2-EPOXIDES

**ETHYLALUMINIUM BROMIDE IODIDE** $C_2H_5AlBrI$

Nitromethane

*See* NITROMETHANE, $CH_3NO_2$: Alkylmetal halides

*See other* ALKYLALUMINIUM HALIDES

**ETHYLALUMINIUM DIBROMIDE** $C_2H_5AlBr_2$

*See* ALKYLALUMINIUM HALIDES

**ETHYLALUMINIUM DICHLORIDE** $C_2H_5AlCl_2$

*See* ALKYLALUMINIUM HALIDES

† **BROMOETHANE** $C_2H_5Br$

Preparative hazard

*See* BROMINE, $Br_2$: Ethanol, etc.

† **CHLOROETHANE** $C_2H_5Cl$

*MCA SD-50,* 1953

Potassium

*See* POTASSIUM, K: Halocarbons

† **CHLOROMETHYL METHYL ETHER** $ClCH_2OCH_3$ $C_2H_5ClO$

**ETHYL HYPOCHLORITE** EtOCl $C_2H_5ClO$

Alone,

or Copper

Sandmeyer, T., *Ber.*, 1885, **18**, 1768

Though distillable slowly (36°C), ignition or rapid heating of the vapour causes explosion, as does contact of copper powder with the cold liquid.

*See other* HYPOHALITES

**ETHYL PERCHLORATE** $EtOClO_3$ $C_2H_5ClO_4$

Sidgwick, 1950, 1235

Reputedly the most explosive substance known, it is very sensitive to impact, friction and heat.

*See other* ALKYL PERCHLORATES

**2,2-DICHLOROETHYLAMINE** $Cl_2CHCH_2NH_2$ $C_2H_5Cl_2N$

Roedig, A. *et al., Chem. Ber.*, 1966, **99**, 121

A violent explosion occurred during evaporation of an ethereal solution at 260 mbar from a bath at 80–90°C. No explosion occurred when the bath temperature was limited to 40–45°, or during subsequent distillation at 60–64°/76 mbar. Aziridine derivatives may have been formed.

**† TRICHLORO(ETHYL)SILANE** $Cl_3SiEt$ $C_2H_5Cl_3Si$

**† FLUOROETHANE** $C_2H_5F$

**2-HYDROXYETHYLMERCURY(II) NITRATE** $C_2H_5HgNO_4$

$HOC_2H_4HgNO_3$

Whitmore, 1921, 110

It decomposes with a slight explosion on heating.

*See other* MERCURY COMPOUNDS

**IODOETHANE** $C_2H_5I$

Preparative hazard

*See* IODINE, $I_2$ : Ethanol, etc.

Silver chlorite

*See* SILVER CHLORITE, $AgClO_2$ : Alkyl iodides

## ETHYLMAGNESIUM IODIDE  EtMgI  $C_2H_5IMg$

Ethoxyacetylene

Jordan, C. F., *Chem, Eng. News,* 1966, **44**(8), 40

A stirred mixture of ethoxyacetylene with methylmagnesium iodide in ether exploded when the agitator was turned off. The corresponding bromide had been used, without incident, in earlier attempts on smaller scale.

*See other* ALKYLMETAL HALIDES

## ETHYLLITHIUM  $C_2H_5Li$

Sorbe, 1968, 82

It ignites in air.

*See other* ALKYLMETALS

## † AZIRIDINE (ETHYLENEIMINE)  $\overline{CH_2NHCH_2}$  $C_2H_5N$

Preparative hazard

1. Wystrach, V. P. *et al., J. Amer. Chem. Soc.,* 1955, **77**, 5915
2. Wystrach, V. P. *et al., Chem. Eng. News,* 1956, **34**, 1274

The procedure described previously [1] for the preparation is erroneous. 2-Chloroethylamine hydrochloride must be added with stirring as a 33% solution in water to strong sodium hydroxide solution. Addition of the solid hydrochloride to the alkali caused separation in bulk of 2-chloroethylamine, which polymerised explosively. Adequate dilution and stirring, and a temperature below 50°C are all essential [2].

Acids

'Ethyleneimine', Brochure 125-521-65, Midland, Mich., Dow Chemical Co., 1965

It is very reactive chemically and subject to aqueous acid-catalysed exothermic polymerisation, which may be violent if uncontrolled by dilution, slow addition or cooling. Ethyleneimine is normally stored over solid caustic alkali, to minimise polymerisation catalysed by presence of carbon dioxide.

Chlorinating agents

Graefe, A. F. *et al., J. Amer. Chem. Soc.,* 1958, **80**, 3939

It gives the explosive 1-chloroaziridine on treatment with sodium hypochlorite solution.

*See* 1-CHLOROAZIRIDINE, $C_2H_4ClN$

Silver

'Ethyleneimine', Brochure 125-521-65, Midland, Mich., Dow Chemical Co., 1965

Explosive silver derivatives may be formed in contact with silver or its alloys, including silver solder, which is therefore unsuitable in handling equipment.

*See other N*-METAL DERIVATIVES

**VINYLAMINE** **$H_2C{=}CHNH_2$** **$C_2H_5N$**

Isoprene

Seher, A., *Ann.*, 1952, **575**, 153–161

Attempts at interaction led to explosion.

**† ACETALDEHYDE OXIME** **$CH_3CH{=}NOH$** **$C_2H_5NO$**

**† *N*-METHYLFORMAMIDE** **$HCONHCH_3$** **$C_2H_5NO$**

**† ETHYL NITRITE** **EtONO** **$C_2H_5NO_2$**

**METHYL CARBAMATE** **$H_2NCO_2Me$** **$C_2H_5NO_2$**

Phosphorus pentachloride

*See* *N*-CARBOMETHOXYIMINOPHOSPHORYL CHLORIDE, $C_2H_3Cl_3NO_2P$

**NITROETHANE** **$C_2H_5NO_2$**

Metal oxides

*See* NITROALKANES: Metal oxides

**† ETHYL NITRATE** **$EtONO_2$** **$C_2H_5NO_3$**

Lewis acids

*See* ALKYL NITRATES

**2-NITROETHANOL** **$NO_2CH_2CH_2OH$** **$C_2H_5NO_3$**

*ABCM Quart. Safety Summ.*, 1956, 27, 24

An explosion occurred towards the end of vacuum distillation of a relatively small quantity of nitroethanol. This was attributed to the presence of peroxides, but the presence of traces of alkali seems a possible alternative cause.
*See other* C-NITRO COMPOUNDS

**ETHYL AZIDE** **$EtN_3$** **$C_2H_5N_3$**

1. Boyer, J. H. *et al.*, *Chem. Rev.*, 1954, **54**, 32
2. Koch, E., *Angew. Chem. (Nachr.)*, 1970, **18**, 26

Though stable at room temperature, it may detonate on rapid heating [1]. A sample stored at −55°C exploded after a few minutes' exposure at laboratory temperature, possibly owing to development of internal pressure in the stoppered vessel. It will also explode if dropped from 1 m in a small flask on to a stone floor [2].
*See other* ORGANIC AZIDES

**BIURET** **$C_2H_5N_3O_2$**

**$NH_2CONHCONH_2$**

Chlorine
*See* CHLORINE, $Cl_2$: Nitrogen compounds

***N*-METHYL-*N*-NITROSOUREA** **$NH_2CON(NO)Me$** **$C_2H_5N_3O_2$**

1. Arndt, F., *Org. Synth.*, 1943, Coll. Vol. 2, 462
2. Anon., *Angew. Chem. (Nachr.)*, 1957, **5**, 198

This must be stored under refrigeration to avoid sudden decomposition after storage at room temperature or slightly above (30°C) [1]. More stable materials for generation of diazomethane are now available. Material stored at 20°C exploded after 6 months [2].
*See* DIAZOMETHANE, $CH_2N_2$
*See other* NITROSO COMPOUNDS

**1-METHYL-3-NITRO-1-NITROSOGUANIDINE** **$C_2H_5N_5O_3$**

**$MeN(NO)C(NH)NHNO_2$**

Eisendrath, J. N., *Chem. Eng. News*, 1953, **31**, 3016
Aldrich advertisement, *J. Org. Chem.*, 1974, **39**(7), cover 4

Formerly used as a diazomethane precursor, this material will detonate under high impact, and a sample exploded when melted in a sealed capillary tube.

Although the crude product from the aqueous nitrosation is pyrophoric, recrystallised material is a stable but powerfully mutagenic compound.

*See other* *N*-NITRO COMPOUNDS
NITROSO COMPOUNDS

**ETHYLSODIUM** $C_2H_5Na$

Houben-Weyl, 1970, Vol. 13.1, 384
The powder ignites in air.
*See other* ALKYLMETALS

**SODIUM DIMETHYLSULPHINATE** $Na^+Me_2SO^-$ $C_2H_5NaOS$

*p*-Chlorotrifluoromethylbenzene
*See* *p*-CHLOROTRIFLUOROMETHYLBENZENE, $C_7H_4ClF_3$ : Sodium dimethylsulphinate
*See also* DIMETHYL SULPHOXIDE, $C_2H_6OS$: Sodium hydride

**SODIUM 2-HYDROXYETHOXIDE** $NaOC_2H_4OH$ $C_2H_5NaO_2$

(Polychlorobenzenes)
Milnes, H. H., *Nature*, 1971, **232**, 395–396
It was found that the monosodium salt of ethylene glycol decomposed at about 230°C, during investigation of an explosion in the preparation of 2,3,5-trichlorophenol by hydrolysis of 1,2,4,5-tetrachlorobenzene with sodium hydroxide in ethylene glycol at 180°C. Decomposition in the presence of polychlorobenzenes was exothermic and proceeded rapidly and uncontrollably, raising the temperature to 410°C, and liberating large volumes of white vapours, one constituent of which was identified as ethylene oxide. The decomposition proceeded identically in presence of the tetrachlorobenzene or the three isomeric trichlorobenzenes (and seems also likely to occur in the absence of these or the chlorophenols produced by hydrolysis).
*See* 1,2,4,5-TETRACHLOROBENZENE, $C_6H_2Cl_4$ : Sodium hydroxide

**† ETHANE** $C_2H_6$

Chlorine
*See* CHLORINE, $Cl_2$ : Hydrocarbons

Dioxygenyl tetrafluoroborate
*See* DIOXYGENYL TETRAFLUOROBORATE, $BF_4O_2$ : Organic materials

**DIMETHYLALUMINIUM BROMIDE** **$Me_2AlBr$** **$C_2H_6AlBr$**

Nitromethane

*See* NITROMETHANE, $CH_3NO_2$ : Alkylmetal halides

*See other* ALKYLALUMINIUM HALIDES

**DIMETHYLALUMINIUM CHLORIDE** **$Me_2AlCl$** **$C_2H_6AlCl$**

Gaines, D. F. *et al., Inorg. Chem.*, 1974, **15**, 203–204

It ignites in air and reacts violently with water.

*See other* ALKYLALUMINIUM HALIDES

**IODODIMETHYLARSINE** **$IAsMe_2$** **$C_2H_6AsI$**

Millar, I. T. *et al., Inorg. Synth.*, 1960, **6**, 117

It ignites when heated in air.

*See other* ALKYLNON-METAL HALIDES

**DIMETHYLGOLD(III) AZIDE** **$Me_2AuN_3$** **$C_2H_6AuN_3$**

*See* the dimer, $C_4H_{12}Au_2N_6$

**AZIDODIMETHYLBORANE** **$N_3BMe_2$** **$C_2H_6BN_3$**

Anon., *Angew. Chem. (Nachr.)*, 1970, **18**, 27

A sample exploded on contact with a warm sampling capillary.

*See other* NON-METAL AZIDES

**BARIUM METHYL PEROXIDE** **$Ba(OOMe)_2$** **$C_2H_6BaO_4$**

*See* METHYL HYDROPEROXIDE, $CH_4O_2$

**DIMETHYLBERYLLIUM** **$Me_2Be$** **$C_2H_6Be$**

Coates, 1967, Vol. 1, 106

Ignites in moist air or in carbon dioxide, and reacts explosively with water.

*See other* ALKYLMETALS

**DIMETHYLBERYLLIUM–1,2-DIMETHOXYETHANE** **$C_2H_6Be \cdot C_4H_{10}O_2$**

**$Me_2Be \cdot MeOC_2H_4OMe$**

Houben-Weyl, 1973, Vol. 13.2a, 38

It ignites in air.

*See related* ALKYLMETALS

**DIMETHYLBISMUTH CHLORIDE** **$Me_2BiCl$** **$C_2H_6BiCl$**

Sidgwick, 1950, 781

Ignites when warm in air.

*See other* ALKYLMETAL HALIDES

**DIMETHYLCADMIUM** **$Me_2Cd$** **$C_2H_6Cd$**

1. Egerton, A. *et al.*, *Proc. R. Soc.*, 1954, **A225**, 429; Davies A. G., *Chem. & Ind.*, 1958, 1177
2. Sidgwick, 1950, 268

On exposure to air, dimethylcadmium peroxide is formed as a crust which explodes on friction [1]. Ignition of dimethylcadmium may occur if a large area : volume ratio is involved, as when it is dropped on to filter paper [2].

*See other* ALKYLMETALS

**2-CHLOROETHYLAMINE** **$ClCH_2CH_2NH_2$** **$C_2H_6ClN$**

May polymerise explosively.

*See* AZIRIDINE (ETHYLENEIMINE), $C_2H_5N$: Preparative hazard

**METHYL IMINIOFORMATE CHLORIDE** **$C_2H_6ClNO$**

**$HC(N^+H_2)OMe\ Cl^-$**

Preparative hazard for this 'imidoester hydrochloride'.

*See* HYDROGEN CHLORIDE, ClH: Alcohols, Hydrogen chloride

**2-AZA-1,3-DIOXOLANIUM PERCHLORATE (ETHYLENEDIOXYAMMONIUM PERCHLORATE)** **$C_2H_6ClNO_6$**

$\overline{O[CH_2]_2ON^+}H_2\ ClO_4^-$

*MCA Case History No. 1622*

A batch of this sensitive compound exploded violently, probably during recrystallisation.

*See other* OXOSALTS OF NITROGENOUS BASES

**CHLORODIMETHYLPHOSPHINE** **$ClPMe_2$** **$C_2H_6ClP$**

1. Staendeke, H. *et al.*, *Angew. Chem. (Intern. Ed.)*, 1973, **12**, 877
2. Parshall, G. W., *Inorg. Synth.*, 1974, **15**, 192–193

It ignites on contact with air [1], and handling is detailed [2].

*See other* ALKYLNON-METAL HALIDES

**DIMETHYLANTIMONY CHLORIDE** **$Me_2SbCl$** **$C_2H_6ClSb$**

Sidgwick, 1950, 777

Ignites at 40°C in air.

*See other* ALKYLMETAL HALIDES

**DIMETHYL *N,N*-DICHLOROPHOSPHORAMIDATE** **$C_2H_6Cl_2NO_3P$**

**$(MeO)_2P(O)NCl_2$**

Block, H. D. *et al., Angew. Chem. (Intern. Ed.),* 1971, **10**, 491

During preparation on 1.5g mol scale, a violent explosion occurred during stirring after reaction. This did not occur on 0.5g mol scale, and longer reaction time may have led to liberation of explosive nitrogen--chlorine compounds. Suggested precautions include a working scale below 0.2g mol and short reaction time.

*See other N*-HALOGEN COMPOUNDS

† **DICHLORODIMETHYLSILANE** **$Cl_2Si(Me)_2$** **$C_2H_6Cl_2Si$**

† **DICHLORO(ETHYL)SILANE** **$Cl_2HSiEt$** **$C_2H_6Cl_2Si$**

**DIMETHYL ETHEROXODIPEROXOCHROMIUM(VI)** **$C_2H_6CrO_6$**

**$[CrMe_2O(O_2)_2O]$**

Schwarz, R. *et al., Ber.,* 1936, **69**, 575

The blue solid explodes powerfully at –30°C.

*See related* AMMINECHROMIUM PEROXOCOMPLEXES

**2,2,2,4,4,4-HEXAFLUORO-1,3-DIMETHYL-1,3,2,4-DIAZAPHOSPHETIDINE** **$C_2H_6F_6N_2P_2$**

**$\overline{MeNPF_3NMePF_3}$**

*tert*-Butyllithium

*See* *tert*-BUTYLLITHIUM, $C_4H_9Li$: 2,2,2,4,4,4-Hexafluoro-1,3-dimethyl-1,3,2,4-diazaphosphetidine

**DIMETHYLMERCURY** **$Me_2Hg$** **$C_2H_6Hg$**

Diboron tetrachloride

*See* DIBORON TETRACHLORIDE, $B_2Cl_4$: Dimethylmercury

**DIMETHYLMAGNESIUM** $Me_2Mg$ $C_2H_6Mg$

Gilman, H. *et al., J. Amer. Chem. Soc.*, 1930, **52**, 5049

Contact with moist air usually caused ignition of the dry powder, and water always ignited the solid or its ethereal solution.

*See other* ALKYLMETALS

**POLY(DIMETHYLMANGANESE)** $(Me_2Mn)_n$ $(C_2H_6Mn)_n$

Bailar, 1973, Vol. 3, 851; Vol. 4, 792

It is a readily explosive powder, and probably polymeric.

*See other* ALKYLMETALS

**AZOMETHANE** $CH_3N{=}NCH_3$ $C_2H_6N_2$

Allen, A.O. *et al., J. Amer. Chem. Soc.*, 1935, **57**, 310–317

The slow thermal decomposition of gaseous azomethane becomes explosive above 341°C/250 mbar and 386°C/24 mbar.

*See other* AZO COMPOUNDS

***N,N'*-DISODIUM *N,N'*-DIMETHOXYSULPHONYLDIAMIDE** $C_2H_6N_2Na_2O_4S$
$SO_2[N(Na)OMe]_2$

Goehring, 1957, 87

The dry solid readily explodes, as do many *N*-metal hydroxyiamides.

*See other N*-METAL DERIVATIVES

**DIMETHYL HYPONITRITE** $MeON{=}NOMe$ $C_2H_6N_2O_2$

Mendenhall, G. D. *et al., J. Org. Chem.*, 1975, **40**, 1646

Samples of the ester exploded violently during low-temperature distillation, and on freezing by liquid nitrogen. The undiluted liquid is exceptionally unpredictable, but addition of mineral oil before vaporisation improved safety aspects.

*See* DIALKYL HYPONITRITES

**DIMETHYLTIN DINITRATE** $Me_2Sn(NO_3)_2$ $C_2H_6N_2O_6Sn$

Sorbe, 1968, 159

It decomposes explosively on heating.

*See related* ALKYLMETALS

**$N$-AZIDODIMETHYLAMINE** $N_3NMe_2$ $C_2H_6N_4$

Bock, H. *et al., Angew. Chem.,* 1962, **74**, 327

Rather explosive.

*See* ORGANIC AZIDES

**$N,N'$-DINITRO-1,2-DIAMINOETHANE** $(NO_2NHCH_2)_2$ $C_2H_6N_4O_4$

Urbanski, 1967, Vol, 3, 20

This powerful but relatively insensitive explosive decomposes violently at 202°C, and gives lead and silver salts which are highly impact-sensitive.

*See also* DIFFERENTIAL THERMAL ANALYSIS
*See other* $N$-METAL DERIVATIVES
$N$-NITRO COMPOUNDS

**DIAZIDODIMETHYLSILANE** $(N_3)_2Si(Me)_2$ $C_2H_6N_6Si$

Anon., *Angew. Chem. (Nachr.),* 1970, **18**, 26–27

A three-year-old sample exploded violently on removing the ground stopper.

*See other* NON-METAL AZIDES

**† DIMETHYL ETHER** $CH_3OCH_3$ $C_2H_6O$

**† ETHANOL** $C_2H_5OH$ $C_2H_6O$

Acetic anhydride,
Sodium hydrogensulphate

*See* ACETIC ANHYDRIDE, $C_4H_6O_3$: Ethanol, etc.

Acetyl bromide

*See* ACETYL BROMIDE, $C_2H_3BrO$: Hydroxylic compounds

Ammonia,
Silver nitrate

*See* SILVER NITRATE, $AgNO_3$: Ammonia, Ethanol

Disulphuric acid,
Nitric acid

*See* NITRIC ACID, $HNO_3$: Alcohols, Disulphuric acid

Disulphuryl difluoride

*See* DISULPHURYL DIFLUORIDE, $F_2O_5S_2$: Ethanol

Nitric acid,
Silver

*See* SILVER, Ag : Ethanol, Nitric acid

Oxidants

*See* *N*-HALOMIDES: Alcohols
SILVER PERCHLORATE, $AgClO_4$ : Aromatic compounds
BARIUM PERCHLORATE, $BaCl_2O_8$ : Alcohols
BROMINE PENTAFLUORIDE, $BrF_5$ : Hydrogen-containing materials
POTASSIUM PERCHLORATE, $ClKO_4$
NITROSYL PERCHLORATE, $ClNO_5$ : Organic materials
CHROMYL CHLORIDE, $Cl_2CrO_2$ : Organic solvents
CHLORYL PERCHLORATE, $Cl_2O_{10}$ : Ethanol
URANYL DIPERCHLORATE, $Cl_2O_{10}U$: Ethanol
CHROMIUM TRIOXIDE, $CrO_3$ : Alcohols
FLUORINE NITRATE, $FNO_3$ : Organic materials
DIOXYGEN DIFLUORIDE, $F_2O_2$
URANIUM HEXAFLUORIDE, $F_6U$: Aromatics, etc.
IODINE HEPTAFLUORIDE, $F_7I$: Organic solvents
PERMANGANIC ACID, $HMnO_4$ : Organic materials
NITRIC ACID, $HNO_3$ : Alcohols
HYDROGEN PEROXIDE, $H_2O_2$ : Alcohols
: Oxygenated compounds
PEROXODISULPHURIC ACID, $H_2O_8S_2$ : Organic liquids
POTASSIUM DIOXIDE (SUPEROXIDE), $KO_2$ : Ethanol
SODIUM PEROXIDE, $Na_2O_2$ : Hydroxy compounds
RUTHENIUM(VIII) OXIDE, $O_4Ru$: Organic materials

Phosphorus(III) oxide

*See* PHOSPHORUS(III) OXIDE, $O_3P_2$ : Organic liquids

Platinum

*See* PLATINUM, Pt: Ethanol

Potassium

*See* POTASSIUM, K : Air (slow oxidation)

Potassium *tert*-butoxide

*See* POTASSIUM *tert*-BUTOXIDE, $C_4H_9KO$ : Acids, etc.

Silver nitrate

*See* SILVER NITRATE, $AgNO_3$ : Ethanol

Silver oxide

*See* SILVER(I) OXIDE, $Ag_2O$: Ammonia, etc.

Sodium

*See* SODIUM, Na: Ethanol

Acid anhydrides

*See* Trifluoroacetic anhydride, below

Acyl halides,
or Non-metal halides

1. Buckley, A., *J. Chem. Educ.*, 1965, **42**, 674
2. Allan, G. G. *et al.*, *Chem. & Ind.*, 1967, 1706
3. Santosusso, T. M. *et al.*, *Tetrahedron Lett.*, 1974, 4255–4258

In absence of diluent or other effective control of reaction rate, the sulphoxide reacts violently or explosively with the following: acetyl chloride, benzenesulphonyl chloride, cyanuric chloride, phosphorus trichloride, phosphoryl chloride, silicon tetrachloride, sulphur dichloride, disulphur dichloride, sulphuryl chloride or thionyl chloride [1]. These violent reactions are explained in terms of exothermic polymerisation of formaldehyde produced under a variety of conditions by interaction of the sulphoxide with reactive halides, acidic or basic reagents [2].

The thermolytic degradation of the sulphoxide to give acidic products which catalyse further decomposition has been discussed [3].

*See* Dinitrogen tetraoxide, below
Sodium hydride, below

*See also* PERCHLORIC ACID, $ClHO_4$: Sulphoxides

Boron compounds

Shriver, 1969, 209

Dimethyl sulphoxide forms an explosive mixture with $B_9H_9{}^{2-}$ and with diborane. It is probable that other boron hydrides and hydroborates behave similarly.

Bromomethane

*See* TRIMETHYLSULPHOXONIUM BROMIDE, $C_3H_9BrOS$

Dinitrogen tetraoxide

Buckley, A., *J. Chem. Educ.*, 1965, **42**, 674

Interaction may be violent or explosive.

*See* Acyl halides, above

Iodine pentafluoride

Lawless, E. M., *Chem. Eng. News*, 1969, **47**(13), 8

Interaction is explosive, after a delay, in either tetrahydrothiophene-1,1-dioxide or trichlorofluoromethane as solvent, on 0.15g mol scale, though not on one-tenth this scale. Reaction of dimethyl sulphoxide with silver difluoride is also violent.

Magnesium perchlorate

*See* MAGNESIUM PERCHLORATE, $Cl_2MgO_8$ : Dimethyl sulphoxide

Metal alkoxides

*MCA Case History No. 1718*

Addition of potassium *tert*-butoxide or of sodium isopropoxide to the solvent led to ignition of the latter. This was attributed to presence of free metals in the alkoxides, but a more likely explanation seems to be that of direct reaction between the powerful bases and the solvent.

*See* Acyl halides, etc., above (reference 2)
POTASSIUM *tert*-BUTOXIDE, $C_4H_9KO$: Acids, etc.

Metal oxosalts

Martin, 1971, 435

Dehn, H., Brit. Pat. 1 129 777, 1968

Mixtures of metal salts of oxoacids with the sulphoxide are powerful explosives. Examples are aluminium and sodium perchlorates and ferric nitrate.

The water in hydrated oxosalts (aluminium perchlorate, iron(III) nitrate or perchlorate) may be partially or totally replaced by dimethyl (or other) sulphoxide to give solvated salts useful as explosives.

*See also* PERCHLORIC ACID, $ClHO_4$ : Sulphoxides
MAGNESIUM PERCHLORATE, $Cl_2MgO_8$ : Dimethyl sulphoxide

Non-metal halides

*See* Acyl halides, above

Perchloric acid

*See* PERCHLORIC ACID, $ClHO_4$ : Sulphoxides

Periodic acid

*See* PERIODIC ACID, $HIO_4$ : Dimethyl sulphoxide

Phosphorus(III) oxide

*See* PHOSPHORUS(III) OXIDE, $O_3P_2$ : Organic liquids

Potassium

Houben-Weyl, 1970, Vol. 13.1, 295

Interaction of potassium 'sand' and the sulphoxide is violent in absence of a diluent and leads to partial decomposition of the potassium methylsulphinate. Tetrahydrofuran is a suitable diluent.

*See also* Metal alkoxides, above

Silver difluoride

*See* Iodine pentafluoride, above

Sodium hydride

1. French, F. A., *Chem. Eng. News*, 1966, **44**(15), 48
2. Olson, G. L., *Chem. Eng. News*, 1966, **44**(24), 7
3. Russell, G. A. *et al.*, *J. Org. Chem.*, 1966, **31**, 248
4. Batchelor, J. F., private comm., 1976

Two violent pressure-explosions occurred during preparations of dimethylsulphinyl anion on 3–4g mol scale by reaction of sodium hydride with excess solvent. In each case the explosion occurred soon after the separation of a solid. The first reaction involved addition of 4.5g mol of hydride to 18.4g mol of sulphoxide, heated to 70 °C [1]. and the second 3.27 and 19.5g mol, respectively, heated to 50 °C [2]. A smaller-scale reaction, at the original lower hydride concentration [3], did not explode but methylation was incomplete. For explanation, see Acyl halides, above.

Explosion and fire occurred when the hydride–solvent mixture was overheated (above 70°C) [4].

*See also* SODIUM DIMETHYLSULPHINATE, $C_2H_5NaOS$

Sulphur trioxide

*See* SULPHUR TRIOXIDE, $O_3S$: Dimethyl sulphoxide

Trifluoroacetic anhydride

Sharma, A. K. *et al.*, *Tetrahedron Lett.*, 1974, 1503–1506

Interaction of the sulphoxide with some acid anhydrides or halides may be explosive. The highly exothermic reaction with trifluoroacetic anhydride was adequately controlled in dichloromethane solution at below –40°C.

*See* Acyl halides, above
*See other* APROTIC SOLVENTS

## DIMETHYL PEROXIDE $CH_3OOCH_3$ $C_2H_6O_2$

1. Rieche, A. *et al.*, *Ber.*, 1928, **61**, 951
2. Baker, G. *et al.*, *Chem. & Ind.*, 1964, 1988

Extremely explosive, heat- and shock-sensitive as liquid or vapour [1]. During determination of the impact sensitivity of the confined material, rough handling of the container caused ignition. The material should only be handled in small quantity and with great care.

*See other* DIALKYL PEROXIDES

## ETHYLENE GLYCOL $HOCH_2CH_2OH$ $C_2H_6O_2$

Perchloric acid

*See* PERCHLORIC ACID, $ClHO_4$: Glycols and their ethers

Phosphorus pentasulphide

*See* PHOSPHORUS(V) SULPHIDE, $P_2S_5$ : Alcohols

Silvered copper wire

1. Downs, W. R., *TN D-4327*, Washington, NASA Tech. Note, 1968
2. Stevens, H. D., *Chem. Abs.*, 1975, **83**, 134485j

Contact of aqueous ethylene glycol solutions with d.c.-energised silvered copper wires causes ignition of the latter to occur. Bare copper or nickel- or tin-plated wires were inert and silver-plated wire can be made so by adding benzotriazole as a metal deactivator to the coolant solution [1]. This problem of electrical connector fires in aircraft has been studied in detail to identify the significant factors [2].

Sodium hydroxide

*See* SODIUM 2-HYDROXYETHOXIDE, $C_2H_5NaO_2$

**ETHYL HYDROPEROXIDE** **EtOOH** **$C_2H_6O_2$**

Baeyer, A. *et al., Ber.*, 1901, 34, 738

It explodes violently on superheating; the barium salt is heat- and impact-sensitive.

Hydriodic acid

Sidgwick, 1950, 873

The concentrated acid is oxidised explosively.

Silver

Sidgwick, 1950, 873

Finely divided 'molecular' silver decomposes the hydroperoxide, sometimes explosively.

*See other* ALKYL HYDROPEROXIDES

**HYDROXYMETHYL METHYL PEROXIDE** **$C_2H_6O_3$**

**$HOCH_2OOCH_3$**

Rieche, A. *et al., Ber.*, 1929, 62, 2458

Violently explosive, impact-sensitive when heated.

*See other* 1-OXYPEROXY COMPOUNDS

**DIMETHYL SULPHITE** **MeOS(O)OMe** **$C_2H_6O_3S$**

*See* PHOSPHORUS(III) OXIDE, $O_3P_2$ : Organic liquids

**BISHYDROXYMETHYL PEROXIDE** **$(HOCH_2O)_2$** **$C_2H_6O_4$**

Wieland, H. *et al., Ber.*, 1930, **63**, 66

Highly explosive, very friction-sensitive. Higher homologues are more stable.

*See other* 1-OXYPEROXY COMPOUNDS

**† DIMETHYL SULPHATE** **$(MeO)_2SO_2$** **$C_2H_6O_4S$**

*MCA Case History No. 1786*

The product of methylating an unnamed material at 110°C was allowed to remain in the reactor of a pilot-plant. After 80 min the reactor exploded. This was ascribed to exothermic decomposition of the mixture above 100°C, and subsequent acceleration and boiling decomposition at 150°C.

Ammonia

1. Lindlar, H., *Angew. Chem.*, 1963, **75**, 297
2. Claesson, P. *et al.*, *Ber.*, 1880, **13**, 1700

A violent reaction occurred which shattered the flask when litre quantities of dimethyl sulphate and conc. aqueous ammonia were accidentally mixed. Use dilute ammonia in small quantities to destroy dimethyl sulphate [1]. Similar incidents had been noted previously with ammonia and other volatile bases [2].

Barium chlorite

*See* BARIUM CHLORITE, $BaCl_2O_4$ : Dimethyl sulphate

Tertiary bases

Sorbe, 1968, 123

In absence of diluent, quaternation of some tertiary organic bases may proceed explosively.

*See* Ammonia, above

**DIMETHYL SELENATE** **$(MeO)_2SeO_2$** **$C_2H_6O_4Se$**

Sidgwick, 1950, 977

Dimethyl selenate, like its ethyl and propyl homologues, can be distilled under reduced pressure, but explodes at *ca* 150°C under ambient pressure, though less violently than the lower alkyl nitrates.

**DIMETHANESULPHONYL PEROXIDE** **$C_2H_6O_6S_2$**

**$CH_3SO_2OOSO_2CH_3$**

Haszeldine, R. N. *et al.*, *J. Chem. Soc.*, 1964, 4903

A sample heated in a sealed tube exploded violently at 70°C. Unconfined, the peroxide decomposed (sometimes explosively) immediately after melting at 79°C.

*See other* DIACYL PEROXIDES

## † DIMETHYL SULPHIDE $Me_2S$ $C_2H_6S$

Dibenzoyl peroxide

*See* DIBENZOYL PEROXIDE, $C_{14}H_{10}O_4$ : Dimethyl sulphide

*p*-Dioxane,
Nitric acid

*See* NITRIC ACID, $HNO_3$ : Dimethyl sulphide, *p*-Dioxane

Oxygen

*See* OXYGEN (Gas), $O_2$ : Dimethyl sulphide

## † ETHANETHIOL EtSH $C_2H_6S$

## † DIMETHYL DISULPHIDE $(MeS)_2$ $C_2H_6S_2$

## DIMETHYLZINC $Me_2Zn$ $C_2H_6Zn$

1. Egerton, A. *et al., Proc. R. Soc.*, 1954, **A225**, 429
2. Frankland, E., *Phil. Trans. R. Soc.*, 1852, 417

Ignites in air (owing to peroxide formation [1]), and explodes in oxygen [2].

2,2-Dichloropropane

Houben-Weyl, 1973, Vol. 13.2a, 767

Uncontrolled reaction is explosive.

*See* DIALKYLZINCS: Alkyl chlorides

Ozone

*See* OZONE, $O_3$ : Alkylmetals

*See other* ALKYLMETALS

## DIMETHYLALUMINIUM HYDRIDE $Me_2AlH$ $C_2H_7Al$

Houben-Weyl, 1970, Vol. 13.4, 58

Slight contact with air or moisture causes ignition.

*See other* ALKYLALUMINIUM ALKOXIDES OR HYDRIDES

## DIMETHYLARSINE $Me_2AsH$ $C_2H_7As$

von Schwartz, 1918, 322; Sidgwick, 1950, 762

Inflames in air, even at 0°C.

*See other* ALKYLNON-METAL HYDRIDES

## 1-METHYL-3-NITROGUANIDINIUM PERCHLORATE $C_2H_7ClN_4O_6$

$MeN^+H_2C(NH)NHNO_2\ ClO_4^-$

McKay, A.F. *et al.*, *J. Amer. Chem. Soc.*, 1947, **69**, 3029

The salt is sensitive to impact, exploding violently.

*See other* *N*-NITRO COMPOUNDS
OXOSALTS OF NITROGENOUS BASES

## † DIMETHYLAMINE $Me_2NH$ $C_2H_7N$

*MCA SD-57*, 1955

Acrylaldehyde

*See* ACRYLALDEHYDE, $C_3H_4O$: Acids, etc.

Fluorine

*See* FLUORINE, $F_2$: Nitrogenous bases

Maleic anhydride

*See* MALEIC ANHYDRIDE, $C_4H_2O_3$: Cations, etc.

## † ETHYLAMINE $EtNH_2$ $C_2H_7N$

Cellulose nitrate

*See* CELLULOSE NITRATE : Amines

## 2-HYDROXYETHYLAMINE (ETHANOLAMINE) $C_2H_7NO$

$HOCH_2CH_2NH_2$

Cellulose nitrate

*See* CELLULOSE NITRATE: Amines

## *O*-(2-HYDROXYETHYL)HYDROXYLAMINE $C_2H_7NO_2$

$HOC_2H_4ONH_2$

Sulphuric acid

Campbell, H. F., *Chem. Eng. News*, 1975, **53**(49), 5

After 30 min at 120°C under vacuum, an equimolar mixture with the concentrated acid exploded violently. Salts of unsubstituted hydroxylamine are thermally unstable.

*See* HYDROXYLAMINIUM SALTS

### 1,3-DIMETHYLTRIAZENE ('DIAZOAMINOMETHANE') $C_2H_7N_3$

**MeN=NNHMe**

Sorbe, 1968, 141

It explodes on heating.

*See other* TRIAZENES

### 1,2-DIMETHYLNITROSOHYDRAZINE **MeN(NO)NHMe** $C_2H_7N_3O$

Smith, 1966, Vol. 2, 459

The liquid deflagrates on heating.

*See other* NITROSO COMPOUNDS

### 1-METHYL-3-NITROGUANIDINIUM NITRATE $C_2H_7N_5O_5$

$MeN^+H_2C(NH)NHNO_2\ NO_3^-$

McKay, A. F. *et al., J. Amer. Chem. Soc.*, 1947, **69**, 3029

The salt could be exploded by impact between steel surfaces.

*See other* *N*-NITRO COMPOUNDS
OXOSALTS OF NITROGENOUS BASES

### DIMETHYLPHOSPHINE $Me_2PH$ $C_2H_7P$

Houben-Weyl, 1963, Vol. 12(1), 69

Parshall, G. W., *Inorg. Synth.*, 1968, **11**, 158

Secondary lower-alkylphosphines readily ignite in air.

*See other* ALKYLNON-METAL HYDRIDES

### ETHYLPHOSPHINE $EtPH_2$ $C_2H_7P$

Houben-Weyl, 1963, Vol. 12(1), 69

Primary lower-alkylphosphines readily ignite in air.

Halogens,
or Nitric acid

von Schwartz, 1918, 324–325

It explodes on contact with chlorine, bromine or fuming nitric acid, inflames with concentrated acid.

*See other* ALKYLNON-METAL HYDRIDES

### DIMETHYLAMMONIUM PERCHLORATE $C_2H_8ClNO_4$

$Me_2NH_2ClO_4$

Preparative hazard

1. Gore, P.H., *Chem. Brit.*, 1976, **12**, 205

2. Menzer, M. *et al.*, *Z. Chem.*, 1977, **17**, 344

A violent explosion occurred during vacuum evaporation of an aqueous mixture of excess dimethylamine and perchloric acid [1], and a similar incident occurred when the moist solid from a like preparation was moved [2].

*See other* AMINIUM PERCHLORATES
OXOSALTS OF NITROGENOUS BASES

**1,2-DIAMINOETHANE** **$(NH_2CH_2)_2$** **$C_2H_8N_2$**

Cellulose nitrate

*See* CELLULOSE NITRATE: Amines

Diisopropyl peroxydicarbonate

*See* DIISOPROPYL PEROXYDICARBONATE, $C_8H_{14}O_6$

Nitromethane

*See* NITROMETHANE, $CH_3NO_2$ : Acids, etc.
: 1,2-Diaminoethane, etc.

**† 1,1-DIMETHYLHYDRAZINE** **$Me_2NNH_2$** **$C_2H_8N_2$**

Dicyanofurazan

*See* DICYANOFURAZAN, $C_4N_4O$: Nitrogenous bases

Oxidants

Kirk-Othmer, 1966, Vol. 11, 186

Wannagat, U. *et al.*, *Monats.*, 1969, **97**, 1157–1162

A powerful reducing agent and fuel, hypergolic with many oxidants, such as dinitrogen tetraoxide, hydrogen peroxide, nitric acid.

The ignition delay with fuming nitric acid was determined as 8 ms, explosion also occurring.

*See* ROCKET PROPELLANTS
HEXANITROETHANE, $C_2N_6O_{12}$: 1,1-Dimethylhydrazine

*See other* REDUCANTS

**† 1,2-DIMETHYLHYDRAZINE** **$(MeNH)_2$** **$C_2H_8N_2$**

**1,2-BISPHOSPHINOETHANE** **$H_2PC_2H_4PH_2$** **$C_2H_8P_2$**

Taylor, R. C. *et al.*, *Inorg. Chem.*, 1973, **14**, 10–11

It ignites in air.

*See other* ALKYLNON-METAL HALIDES

## 2-AMINOETHYLAMMONIUM PERCHLORATE $C_2H_9ClN_2O_4$

$H_2NC_2H_4NH_3ClO_4$

Hay, R. W. *et al., J. Chem. Soc., Dalton Trans.*, 1975, 1467

Solutions of the perchlorate salts of 1,2-diaminoethane and other amines must be evaporated without heating to avoid the risk of violent explosions.

*See other* OXOSALTS OF NITROGENOUS BASES

## † 1,1-DIMETHYLDIBORANE $C_2H_{10}B_2$

$Me_2B(\mu\text{-}H)_2BH_2$

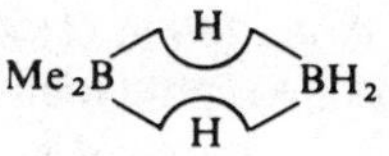

## † 1,2-DIMETHYLDIBORANE $C_2H_{10}B_2$

$MeHB(\mu\text{-}H)_2BHMe$

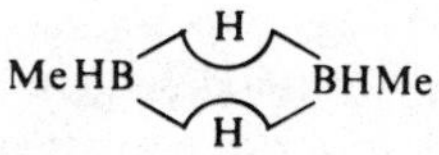

*See other* ALKYLNON-METAL HYDRIDES

## *B*-CHLORO-*N,N*-DIMETHYLAMINODIBORANE $C_2H_{10}B_2ClN$

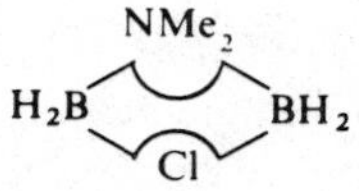

$H_2B(\mu\text{-}NMe_2)(\mu\text{-}Cl)BH_2$

Burg, A. B. *et al., J. Amer. Chem. Soc.*, 1949, **71**, 3454

Ignites in air.

*See other* NON-METAL HYDRIDES

## 1,2-ETHYLENEBIS-AMMONIUM PERCHLORATE $C_2H_{10}Cl_2N_2O_8$

$(CH_2N^+H_3)_2(ClO_4^-)_2$

Lothrop, W. C. *et al., Chem. Rev.*, 1949, **44**, 432

An explosive which appreciably exceeds the power and brisance of TNT.

*See* AMMONIUM PERCHLORATES

*See other* OXOSALTS OF NITROGENOUS BASES

## 1,2-DIAMINOETHANEAQUADIPEROXOCHROMIUM(IV) $C_2H_{10}CrN_2O_5$

$[Cr(C_2H_8N_2)H_2O(O_2)_2]$

Childers, R. F. *et al., Inorg. Chem.*, 1968, **7**, 749

House, D. A. *et al., Inorg. Chem.*, 1966, **5**, 840

The monohydrate is light-sensitive and explodes at 96–97°C if heated at 2°C/min. It effervesces vigorously on dissolution in perchloric acid.

*See other* AMMINECHROMIUM PEROXOCOMPLEXES

**1,2-DIAMINOETHANEAMMINEDIPEROXOCHROMIUM(IV)**

**$C_2H_{11}CrN_3O_4$**

**$[Cr(C_2H_8N_2)NH_3(O_2)_2]$**

House, D. A. *et al., Inorg. Chem.*, 1967, **6**, 1077

The monohydrate is potentially explosive at 25°C and decomposes or explodes at 115°C during slow or moderate heating.

Hydrogen bromide

Hughes, R. G. *et al., Inorg. Chem.*, 1968, **7**, 74

Interaction must be slow with cooling to prevent explosion.

*See other* AMMINECHROMIUM PEROXOCOMPLEXES

**TETRAAMMINEDITHIOCYANATOCOBALT(III) PERCHLORATE**

**$C_2H_{12}ClCoN_6O_4S_2$**

**$[Co(SCN)_2(NH_3)_4]ClO_4$**

Tomlinson, W. R. *et al., J. Amer. Chem. Soc.*, 1949, **71**, 375

Explodes at 335°C, medium impact-sensitivity.

*See other* AMMINEMETAL OXOSALTS

**GUANIDINIUM DICHROMATE** **$C_2H_{12}Cr_2N_6O_7$**

**$(H_2NC(N^+H_2)NH_2)_2Cr_2O_7^-$**

Ma, C., *J. Amer. Chem. Soc.*, 1951, **73**, 1333–1335

Heating causes orderly decomposition but, under confinement, a violent explosion.

*See other* DICHROMATE SALTS OF NITROGENOUS BASES

**DIAMMONIUM *N,N'*-DINITRO-1,2-DIAMINOETHANE** **$C_2H_{12}N_6O_4$**

**$(NH_4)_2(CH_2NNO_2)_2$**

Violent decomposition occurred at 191°C.

*See* DIFFERENTIAL THERMAL ANALYSIS

*See other aci*-NITROSALTS

**MERCURY(II) ACETYLIDE** **hgC≡Chg** **$C_2Hg$**

Bailar, 1973, Vol. 3, 314–315

Possibly polymeric, it explodes when heated or shocked.

*See other* METAL ACETYLIDES

## MERCURY(II) CYANIDE $Hg(CN)_2$ $C_2HgN_2$

Fluorine

*See* FLUORINE, $F_2$: Metal salts

Hydrogen cyanide

Wöhler, L. *et al.*, *Chem. Ztg.*, 1926, **50**, 761

The cyanide is a friction- and impact-sensitive explosive and may initiate detonation of liquid hydrogen cyanide. Other heavy metal cyanides are similar.

Magnesium

*See* MAGNESIUM, Mg: Metal cyanides

Sodium nitrite

*See* SODIUM NITRITE, $NNaO_2$: Metal cyanides

*See other* METAL CYANIDES

## MERCURY(II) FULMINATE $Hg(C{\equiv}N{\rightarrow}O)_2$ $C_2HgN_2O_2$

Alone,

or Sulphuric acid

Urbanski, 1967, Vol. 3, 135, 140

Carl, L. R., *J. Franklin Inst.*, 1945, **240**, 149 (*Chem. Abs.*, 1946, **40**, $209_4$)

Mercury fulminate, readily formed by interaction of mercury nitrate, nitric acid and ethanol, is a very widely used detonator. It may be initiated when dry by flame, heat, impact, friction or intense radiation. Contact with sulphuric acid causes explosion.

The effects of impurities on the preparation and decomposition of the salt have been described.

*See other* METAL FULMINATES

## MERCURY(II) THIOCYANATE $Hg(SCN)_2$ $C_2HgN_2S_2$

Unpublished information, 1950

A large batch of the damp salt became overheated in a faulty drying oven and decomposed vigorously, producing an enormous 'Pharaoh's serpent'.

*See related* METAL CYANIDES

## MERCURY(II) OXALATE $HgC_2O_4$ $C_2HgO_4$

1. *ABCM Quart. Safety Summ.*, 1953, **24**, 30, 45
2. Muir, G. D., private comm., 1968

When dry, it explodes readily on percussion, grinding or heating to

105°C. This instability is attributed to presence of impurities (nitrate, oxide or basic oxalate) in the product [1]. It is so thermally unstable that storage is inadvisable [2].
*See other* METAL OXALATES

**DIMERCURY DICYANIDE OXIDE** $Hg(CN)_2 \cdot HgO$ $C_2Hg_2N_2O$

1. *Merck Index, 1968*, 660
2. May & Baker Ltd, catalogue note, 1971
3. Kast, H. *et al., Chem. Abs.*, 1922, **16**, 4065 and cited references

This addition compound, $Hg(CN)_2 \cdot HgO$, when pure is explosive, sensitive to impact or heat [1]. It is stabilised for commerce by the presence of excess mercury(II) cyanide [2]. Several explosive incidents have been described [3].
*See related* METAL CYANIDES

**DIIODOACETYLENE** $IC{\equiv}CI$ $C_2I_2$

1. Anon., *Chemiearbeit,* 1955, **7**, 55
2. Vaughn, T. H. *et al., J. Amer. Chem. Soc.,* 1932, **54**, 789

Pure, recrystallised material exploded while being crushed manually in a mortar. The decomposition temperature is 125°C, and this may have been reached locally during crushing [1]. Explosion on impact, on heating to 84°C, and during attempted distillation at 98°C/5 mbar had been reported previously [2].
*See other* HALOACETYLENE DERIVATIVES

**DIPOTASSIUM ACETYLIDE** $KC{\equiv}CK$ $C_2K_2$

Water
Bahme, 1972, 80
Contact with limited amounts of water may cause ignition and explosion of evolved acetylene.
*See other* METAL ACETYLIDES

**POTASSIUM DINITROOXALATOPLATINATE (2−)** $C_2K_2N_2O_8Pt$
$K_2[Pt(NO_2)_2C_2O_4]$

Vèzes, M., *Compt. Rend.*, 1897, **125**, 525
The salt decomposes violently at 240°C.
*See other* PLATINUM COMPOUNDS

**DIPOTASSIUM BIS-*aci*-TETRANITROETHANE** $C_2K_2N_4O_8$
$KON(O){=}C(NO_2)C(NO_2){=}N(O)OK$

Borgardt, F. G. *et al., J. Org. Chem.*, 1966, **31**, 2806

This anhydrous salt, and the mono- and dihydrates of the analogous lithium and sodium salts, are all very impact-sensitive.
*See other aci*-NITRO SALTS
POLYNITROALKYL COMPOUNDS

**DILITHIUM ACETYLIDE** **LiC≡CLi** **$C_2Li_2$**

Halogens
Mellor, 1946, Vol. 5, 848
It burns brilliantly when cold in fluorine or chlorine but must be warm before ignition occurs in bromine or iodine vapours.

Lead oxide
*See* LEAD(II) OXIDE, OPb: Metal acetylides

Non-metals
Mellor, 1946, Vol. 5, 848
It burns vigorously in phosphorus, sulphur or selenium vapours.
*See other* METAL ACETYLIDES

**DILITHIUM BIS-*aci*-TETRANITROETHANE** **$C_2Li_2N_4O_8$**
**$LiON(O)=C(NO_2)C(NO_2)=N(O)OLi$**
*See* DIPOTASSIUM BIS-*aci*-TETRANITROETHANE, $C_2K_2N_4O_8$

**† DICYANOGEN** **N≡CC≡N** **$C_2N_2$**

Oxidants
The potential energy of mixtures of (endothermic) dicyanogen and powerful oxidants may be released explosively under appropriate circumstances.
*See* ROCKET PROPELLANTS
DICHLORINE OXIDE, $Cl_2O$: Dicyanogen
FLUORINE, $F_2$ : Halogens
OXYGEN (Liquid), $O_2$ : Liquefied gases
OZONE, $O_3$ : Dicyanogen

**NICKEL(II) CYANIDE** **$Ni(CN)_2$** **$C_2N_2Ni$**

Magnesium
*See* MAGNESIUM, Mg: Metal cyanides

**DICYANOGEN *N,N'*-DIOXIDE** **O←N≡CC≡N→O** **$C_2N_2O_2$**
Grundmann, C., *Angew. Chem.*, 1963, **75**, 450

*Ann.*, 1965, **687**, 194
Solid decomposes at –45°C under vacuum, emitting a brilliant light before exploding.
*See related* HALOGEN OXIDES

**LEAD(II) CYANIDE** **$Pb(CN)_2$** **$C_2N_2Pb$**

Magnesium
*See* MAGNESIUM, Mg : Metal cyanides

**LEAD DITHIOCYANATE** **$Pb(SCN)_2$** **$C_2N_2PbS_2$**
Urbanski, 1967, Vol. 3, 230
The explosive properties of lead dithiocyanate have found limited use.
*See other* HEAVY METAL DERIVATIVES

**THIOCYANOGEN** **N≡CSSC≡N** **$C_2N_2S_2$**
Söderbäck, E., *Ann.*, 1919, **419**, 217
Low-temperature storage is necessary, as it polymerises explosively above its m.p., –2°C (–7°C is also recorded).

**ZINC DICYANIDE** **$Zn(CN)_2$** **$C_2N_2Zn$**

Magnesium
*See* MAGNESIUM, Mg: Metal cyanides

**DICYANODIAZENE (AZOCARBONITRILE)** **$C_2N_4$**
**NCN=NCN**

1. Marsh, F. D. *et al.*, *J. Amer. Chem. Soc.*, 1965, **87**, 1819
2. Ittel, S. D. *et al.*, *Inorg. Chem.*, 1975, **14**, 1183

The solid explodes when mechanically shocked or heated in a closed vessel [1]. Preparative methods are hazardous because of the need to heat the explosive precursor, cyanogen azide [1,2].
*See other* AZO COMPOUNDS
CYANO COMPOUNDS

**DISODIUM DICYANODIAZENE** **$Na_2$[NCN=NCN]** **$C_2N_4Na_2$**
Marsh, F. D. *et al.*, *J. Amer. Chem. Soc.*, 1965, 87, 1820
This compound (a radical anion salt) is an explosive powder.
*See other* HIGH-NITROGEN COMPOUNDS

**DISODIUM BIS-*aci*-TETRANITROETHANE** $C_2N_4Na_2O_8$

$NaON(O)=C(NO_2)C(NO_2)=N(O)ONa$

*See* DIPOTASSIUM BIS-*aci*-TETRANITROETHANE, $C_2K_2N_4O_8$

**TRINITROACETONITRILE** $(NO_2)_3CCN$ $C_2N_4O_6$

Schischkow, A., *Ann. Chim.* [3], 1857, **49**, 310

Parker, C. O. *et al.*, *Tetrahedron*, 1962, **17**, 79, 84

It explodes if heated quickly to 220°C.

It is also a friction- and impact-sensitive explosive, which may be used conveniently in carbon tetrachloride solution to minimise handling problems.

*See other* POLYNITROALKYL COMPOUNDS

**THIOCARBONYL AZIDE THIOCYANATE** $C_2N_4S_2$

$N_3C(S)SCN$

Ammonia,
or Hydrazine

Audrieth, L. F. *et al.*, *J. Amer. Chem. Soc.*, 1930, **52**, 2799–2805

The unstable compound reacts explosively with ammonia gas, and violently with concentrated hydrazine solutions.

*See other* ORGANIC AZIDES

**HEXANITROETHANE** $(NO_2)_3CC(NO_2)_3$ $C_2N_6O_{12}$

1. Loewenschuss, A. *et al.*, *Spectrochim. Acta*, 1974, **30A**, 371–378
2. Sorbe, 1968, 149

Grinding the solid to record its IR spectrum was precluded on safety grounds [1]. It decomposes explosively above 140°C [2].

1,1-Dimethylhydrazine

Noble, P. *et al.*, *Am. Inst. Aeron. Astronaut. J.*, 1963, **1**, 395–397

It is a powerful oxidant and hypergolic with dimethylhydrazine or other strong organic bases.

Organic compounds

Will, M., *Ber.*, 1914, **47**, 961–965

Though relatively insensitive to friction, impact or shock, it can be detonated. With hydrogen-containing organic compounds, this oxygen-rich compound (+300% oxygen-balanced) forms powerfully explosive mixtures. The addition compound with *o*-nitroaniline ($C_8H_6N_8O_{14}$, −27% balance) is extremely explosive.

*See other* POLYNITROALKYL COMPOUNDS

**BIS(AZIDOTHIOCARBONYL) DISULPHIDE** $C_2N_6S_4$

$N_3C(S)SSC(S)N_3$

Smith, G. B. L., *Inorg. Synth.*, 1939, **1**, 81

This compound, readily formed by iodine oxidation of solutions of azidodithioformic acid or its salts, is a powerful explosive. It is sensitive to mechanical impact or heating to 40°C, and slow decomposition during storage increases the sensitivity. Preparative precautions are detailed.

*See* AZIDODITHIOFORMIC ACID, $CHN_3S_2$

*See other* ACYL AZIDES

**DIAZIDOMETHYLENECYANAMIDE** $(N_3)_2C{=}NCN$ $C_2N_8$

Marsh, F. D., *J. Org. Chem.*, 1972, **37**, 2967

This explosive solid may be produced during preparation of cyanogen azide, $CN_4$.

*See other* HIGH-NITROGEN COMPOUNDS

ORGANIC AZIDES

**DIAZIDOMETHYLENEAZINE** $(N_3)_2C{=}NN{=}C(N_3)_2$ $C_2N_{14}$

Houben-Weyl, 1965, Vol. 10(3), 793

This very explosive bis-*gem*-diazide contains over 89% of nitrogen.

*See other* HIGH-NITROGEN COMPOUNDS

ORGANIC AZIDES

**DISODIUM ACETYLIDE** $NaC{\equiv}CNa$ $C_2Na_2$

Opolsky, S., *Bull. Acad. Cracow,* 1905, 548

A brown explosive form is produced if excess sodium is used in preparation of thiophene homologues – possibly because of sulphur compounds.

Halogens

Mellor, 1946, Vol. 5, 848

Disodium acetylide burns in chlorine and (though not stated) probably also in fluorine, and in contact with bromine and iodine on warming.

Metals

Mellor, 1946, Vol. 5, 848

Trituration in a mortar with finely divided lead, aluminium, iron or mercury may be violent, carbon being liberated.

Metal salts

Mellor, 1946, Vol. 5, 848

Rubbing in a mortar with some chlorides or iodides may cause incandescence or explosion. Sulphates are reduced, and nitrates would be expected to behave similarly.

Non-metal oxides

von Schwartz, 1918, 328

Disodium acetylide incandesces in carbon dioxide or sulphur dioxide.

Oxidants

Mellor, 1946, Vol. 5, 848

Ignites on warming in oxygen, and incandesces at 150°C in dinitrogen pentaoxide.

*See* Halogens above

Phosphorus

*See* PHOSPHORUS, P: Metal acetylides

Water

1. Mellor, 1946, Vol. 5, 848
2. Davidsohn, W. E., *Chem. Rev.*, 1967, **67**, 74

Excess water is necessary to avoid explosion [1]; the need for care in handling is stressed [2].

*See other* METAL ACETYLIDES

**DIRUBIDIUM ACETYLIDE** **RbC≡CRb** $\mathbf{C_2Rb_2}$

Acids

Mellor, 1946, Vol. 5, 848

With concentrated hydrochloric acid ignition occurs, and contact with nitric acid causes explosion.

Halogens

Mellor, 1946, Vol. 5, 848

It burns in all four halogens.

Metal oxides

Mellor, 1946, Vol. 5, 848–850

Iron(III) and chromium(III) oxides react exothermically, and lead oxide explosively. Copper oxide and manganese dioxide react at 350°C incandescently.

Non-metal oxides

Mellor, 1946, Vol. 5, 848

Warming in carbon dioxide, nitrogen oxide or sulphur dioxide causes ignition.

Non-metals

Mellor, 1946, Vol. 5, 848

It reacts vigorously with boron and silicon on warming, ignites with arsenic, and burns in sulphur or selenium vapours.

*See other* METAL ACETYLIDES

## STRONTIUM ACETYLIDE $SrC_2$ $C_2Sr$

Halogens

Mellor, 1946, Vol. 5, 862

Strontium acetylide incandesces with chlorine, bromine and iodine at 197, 174 and 182°C, respectively.

*See other* METAL ACETYLIDES

## THORIUM DICARBIDE $ThC_2$ $C_2Th$

Non-metals,
or Oxidants

Mellor, 1946, Vol. 5, 885–886

Contact with selenium or sulphur vapour causes the heated carbide to incandesce. Contact of the carbide with molten potassium chlorate, potassium nitrate or even potassium hydroxide causes incandescence.

*See other* METAL NON-METALLIDES

## URANIUM DICARBIDE $UC_2$ $C_2U$

Air

Mellor, 1946, Vol. 5, 890

Sidgwick, 1950, 1071

Uranium dicarbide emits brilliant sparks on impact and ignites on grinding in a mortar or on heating in air to 400°C.

Halogens

Mellor, 1946, Vol. 5, 891

Incandescence occurs in warm fluorine, in chlorine at 300°C and weakly in bromine at 390°C.

Hydrogen chloride

*See* HYDROGEN CHLORIDE, ClH: Metal acetylides or carbides

Nitrogen oxide

*See* NITROGEN OXIDE ('NITRIC OXIDE'), NO: Metal acetylides or carbides

Water
Sidgwick, 1950, 1071
Mellor, 1946, Vol. 5, 890–891
Interaction with hot water is violent, and the carbide ignites in steam at dull red heat.
*See other* METAL NON-METALLIDES

**ZIRCONIUM DICARBIDE** **$ZrC_2$** **$C_2Zr$**

Halogens
Mellor, 1946, Vol. 5, 885
Ignites in cold fluorine, in chlorine at 250°C, bromine at 300°C and iodine at 400°C.
*See other* METAL NON-METALLIDES

**SILVER TRIFLUOROMETHYLACETYLIDE** **$C_3AgF_3$**

**$AgC{\equiv}CCF_3$**

Henne, A. L. *et al., J. Amer. Chem. Soc.,* 1951, **73**, 1042
Explosive decomposition on heating.
*See* HALOACETYLENE DERIVATIVES
*See other* HEAVY METAL DERIVATIVES

**TETRAALUMINIUM TRICARBIDE** **$Al_4C_3$** **$C_3Al_4$**

Oxidants
Mellor, 1946, Vol. 5, 872
Incandescence on warming with lead peroxide or potassium permanganate.
*See other* METAL NON-METALLIDES

**POTASSIUM 1,3-DIBROMO-2,4-DIKETO-1,3,5-TRIAZINE-6-OLATE**

**$C_3Br_2KN_3O_3$**

**N=C(OK)NBrCONBrCO** (ring: N to terminal CO)

It may be expected to show similar properties to the chloro-analogue.
*See* SODIUM 1,3-DICHLORO-2,4-DIKETO-1,3,5-TRIAZINE-6-OLATE, $C_3Cl_2N_3NaO_3$
*See other* *N*-HALOIMIDES

**PERFLUOROISOPROPYL HYPOCHLORITE** $C_3ClF_7O$

$(F_3C)_2CFOCl$

Schack, C. J. *et al.*, *J. Amer. Chem. Soc.*, 1969, **91**, 2904
Material condensed at −95°C may suddenly decompose completely and vaporise.

*See other* HYPOHALITES

**CHLOROCYANOACETYLENE** ClC≡CCN $C_3ClN$

Hashimoto, N. *et al.*, *J. Org. Chem.*, 1970, **35**, 675
Avoid contact with air at elevated temperature because of its low ignition temperature. Burns moderately in the open, but may explode in a nearly closed vessel. Presence of mono- and di-chloroacetylenes as impurities increases flammability hazard, which may be reduced by addition of 1% of ethyl ether.

*See other* HALOACETYLENE DERIVATIVES

**SODIUM 1,3-DICHLORO-2,4-DIKETO-1,3,5-TRIAZINE-6-OLATE**

$C_3Cl_2N_3NaO_3$

N=C(ONa)NClCONClCO (ring)

'FI-CLOR 60S', Brochure NH/FS/67.4, Loughborough, Fisons, 1967
This compound, used in chlorination of swimming pools, is a powerful oxidant, and indiscriminate contact with combustible materials must be avoided. Ammonium salts and other nitrogenous materials are incompatible in formulated products.

*See other* *N*-HALOIMIDES

**2,4,6-TRICHLORO-1,3,5-TRIAZINE (CYANURIC CHLORIDE)**

$C_3Cl_3N_3$

N=CClN=CClN=CCl (ring)

Allyl alcohol,
Alkali

Anon., *Loss Prev. Bull.*, 1974, (001), 11
When aqueous sodium hydroxide was added to a mixture of the chloride and alcohol at 28°C instead of the normal 5°C, a rapidly accelerating reaction led to rupture of the bursting disc and a gasket, and subsequently to a flash-fire and explosion.

*See* Methanol, below
Water, below

*N,N*-Dimethylformamide

*BCISC Quart. Safety Summ.*, 1964, **35**, 24

Cyanuric chloride reacts vigorously and exothermically with dimethylformamide after a deceptively long induction period. The 1:1 adduct initially formed decomposes above 60°C with evolution of carbon dioxide and formation of a dimeric unsaturated quaternary ammonium salt. Dimethylformamide is appreciably basic and is not a suitable solvent for acyl halides.

Dimethyl sulphoxide

*See* DIMETHYL SULPHOXIDE, $C_2H_6OS$ : Acyl halides

Methanol

*ABCM Quart. Safety Summ.*, 1960, **31**, 40

Cyanuric chloride dissolved in methanol reacted violently and uncontrollably with the solvent. This was attributed to the absence of an acid acceptor to prevent the initially acid-catalysed (and later auto-catalysed) exothermic reaction of all three chlorine atoms simultaneously.

Water

*MCA Case History No. 1869*

A reaction mixture containing the chloride and water, held in abeyance before processing, developed a high internal pressure in the containing vessel. Hydrolysis (or alcoholysis) of the chloride becomes rapidly exothermic above 30°C.

*See* Allyl alcohol, above
Methanol, above

*See related* ACYL HALIDES

**1,3,5-TRICHLORO-1,3,5-TRIAZINETRIONE** $C_3Cl_3N_3O_3$

NClCONClCONClCO

'FI-CLOR 91' Brochure, Loughborough, Fisons, 1967

This compound, used in chlorination of swimming pools, is a powerful oxidant, and indiscriminate contact with combustible materials must be avoided.

*See other N*-HALOIMIDES

**2,4,6-TRIS(DICHLOROAMINO)-1,3,5-TRIAZINE (HEXACHLOROMELAMINE)** $C_3Cl_6N_6$

$N{=}C(NCl_2)N{=}C(NCl_2)N{=}CNCl_2$ (ring closed between first N and last C)

As a trifunctional *N, N*-dichloro-compound, it is probably more reactive and less stable than the trichloro- derivative.

*See* 2,4,6-TRIS(CHLOROAMINO)1,3,5-TRIAZINE, $C_3H_3Cl_3N_6$

**POTASSIUM TRICYANODIPEROXOCHROMATE(3–)** $C_3CrK_3N_3O_4$

$K_3[Cr(CN)_3(O_2)_2]$

Bailar, 1973, Vol. 4, 167

A highly explosive material, with internal redox features.

*See other* PEROXOACID SALTS
REDOX COMPOUNDS

**LITHIUM TRIFLUOROMETHYLACETYLIDE** $C_3F_3Li$

$LiC{\equiv}CCF_3$

*See* LITHIUM CHLOROACETYLIDE, $C_2ClLi$
*See other* HALOACETYLENE DERIVATIVES

**TRIFLUOROACRYLOYL FLUORIDE** $F_2C{=}CFCOF$ $C_3F_4O$

Sodium azide

Middleton, W. J., *J. Org. Chem.*, 1973, **38**, 3294

The product of interaction was an unidentified highly explosive solid.

*See other* ACYL HALIDES

***O*-TRIFLUOROACETYL-*S*-FLUOROFORMYL THIOPEROXIDE** $C_3F_4O_3S$

$CF_3COOSCOF$

Anon., *Angew. Chem. (Nachr.)*, 1970, **18**, 378

It exploded spontaneously in a glass bomb closed with a PTFE-lined valve. No previous indications of instability had been noted during distillation, pyrolysis or irradiation.

*See related* DIACYL PEROXIDES

**† HEXAFLUOROPROPENE** $F_2C{=}CFCF_3$ $C_3F_6$

Air,
Tetrafluoroethylene

*See* TETRAFLUOROETHYLENE, $C_2F_4$ : Air, Hexafluoropropene

## HEXAFLUOROISOPROPYLIDENEAMINOLITHIUM $C_3F_6LiN$

$(CF_3)_2C{=}NLi$

Non-metal halides

Swindell, R.F., *Inorg. Chem.*, 1972, **11**, 242

Interaction of the lithium derivative with a range of chloro- and fluoro- derivatives of arsenic, boron, phosphorus, silicon and sulphur during warming to 25°C tended to be violently exothermic in absence of solvent. Thionyl chloride reacted with explosion.

*See* HEXAFLUOROISOPROPYLIDENEAMINE, $C_3HF_6N$ : Butyllithium

## PERFLUOROPROPIONYL FLUORIDE $C_2F_5COF$ $C_3F_6O$

Fluorinated catalysts

Sorbe, 1968, 62

The acid fluoride may decompose explosively in contact with fluorinated catalysts.

*See other* ACYL HALIDES

## PENTAFLUOROPROPIONYL HYPOFLUORITE $C_3F_6O_2$

$C_2F_5COOF$

Menefee, A. *et al., J. Amer. Chem. Soc.*, 1954, **76**, 2020

Less stable than its lower homologue, the hypofluorite explodes on sparking, or on distillation at atmospheric pressure (b.p., 2°C), though not at below 0.13 bar.

*See* FLUORINE, $F_2$ : Caesium heptafluoropropoxide

*See other* HYPOHALITES

## PERFLUOROPROPYL HYPOFLUORITE $C_3F_8O$

$C_3F_7OF$

Sorbe, 1968, 62

An explosive compound.

*See other* HYPOHALITES

## PROPIOLOYL CHLORIDE $HC{\equiv}CCOCl$ $C_3HClO$

Balfour, W. J. *et al., J. Org. Chem.*, 1974, **39**, 726

The chloride purified by distillation at 58–60°C/1 bar usually ignites spontaneously in air owing to the presence of chloroacetylene, but vacuum distillation at cryogenic temperatures prevents formation of the impurity.

*See other* ACYL HALIDES

*See related* HALOACETYLENE DERIVATIVES

**3,3,3-TRIFLUOROPROPYNE** $HC{\equiv}CCF_3$ $C_3HF_3$

Haszeldine, R. N., *J. Chem. Soc.*, 1951, 590

It tends to explode during analytical combustion and the cuprous and silver derivatives decomposed violently (with occasional explosion) on rapid heating.

*See other* HALOACETYLENE DERIVATIVES

**HEXAFLUOROISOPROPYLIDENEAMINE** $C_3HF_6N$

$(CF_3)_2C{=}NH$

Butyllithium

Swindell, R. F. *et al., Inorg. Chem.*, 1972, **11**, 242

The exothermic reaction which set in on warming the reagents in hexane to 0°C sometimes exploded if concentrated solutions of butyllithium (above 2.5 M) were used, but not if diluted (to about 1.2 M) with pentane

**4-AZIDOCARBONYL-1,2,3-THIADIAZOLE** $C_3HN_5OS$

$N{=}NSCH{=}CCON_3$ (ring: N to C)

Pain, D. L. *et al., J. Chem. Soc.*, 1965, 5167

The azide is extremely explosive in the dry state.

*See other* ACYL AZIDES

**SILVER MALONATE** $CH_2(CO_2Ag)_2$ $C_3H_2Ag_2O_4$

Sorbe, 1968, 126

Explodes on heating.

*See other* SILVER COMPOUNDS

† **2-CHLOROACRYLONITRILE** $CH_2{=}CClCN$ $C_3H_2ClN$

**SODIUM NITROMALONALDEHYDE** $C_3H_2NNaO_4$

$Na^+NO_2C^-(CHO)_2$

Fanta, P. E., *Org. Synth.*, 1962, Coll. Vol. 4, 844

The monohydrate, possibly an *aci*-nitro salt, is an impact-sensitive solid and must be carefully handled with precautions.

*See other aci*-NITRO SALTS

## MALONONITRILE $CH_2(CN)_2$ $C_3H_2N_2$

Alone,
or Bases

1. Personal experience
2. 'Malononitrile' Brochure, p. 11, Basle, Lonza Ltd, 1974

It may polymerise violently on prolonged heating at 130°C, or in contact with strong bases at lower temperatures [1]. The stability of the molten nitrile decreases with increasing temperature and decreasing purity, but no violent decomposition at below 100°C has been recorded [2].

*See other* CYANO COMPOUNDS

## BIS(1,2,3,4-THIATRIAZOL-5-YLTHIO)METHANE $C_3H_2N_6S_4$

$(\overline{SN{=}NN{=}C}SCH_2)_2$

Pilgram, K. *et al., Angew. Chem.*, 1965, 77, 348

This compound, and its three longer-chain homologues, explodes loudly with a flash on impact or on heating to the m.p.

*See other* HIGH-NITROGEN COMPOUNDS

## PROPIOLALDEHYDE $HC{\equiv}CCHO$ $C_3H_2O$

Bases

Sauer, J. C., *Org. Synth.*, 1963, Coll. Vol. 4, 814

This acetylenic aldehyde undergoes vigorous polymerisation in presence of alkalies and, with pyridine, the reaction is almost explosive.

*See also* ACRYLALDEHYDE, $C_3H_4O$
*See other* ACETYLENIC COMPOUNDS

## PROPIOLIC ACID $HC{\equiv}CCO_2H$ $C_3H_2O_2$

Ammonia,
Heavy metal salts

Baudrowski, E., *Ber.*, 1882, **15**, 2701

Interaction of the acid with ammoniacal solutions of copper(I) or silver salts gives precipitates which explode on warming or impact. The structure is not given, but may be an amminemetal acetylide salt.

*See* METAL ACETYLIDES

## SILVER 3-HYDROXYPROPYNIDE $AgC{\equiv}CCH_2OH$ $C_3H_3AgO$

*Organic Chemistry*, 111, Karrer, P., London, Elsevier, 4th Engl. Edn., 1950

The silver salt is explosive.

*See other* METAL ACETYLIDES
SILVER COMPOUNDS

## ALUMINIUM TRIFORMATE $Al(HCO_2)_3$ $C_3H_3AlO_6$

*ABCM Quart. Safety Summ.*, 1939, **10**, 1

An aqueous solution of aluminium formate was being evaporated over a low flame. When the surface crust was disturbed, an explosion occurred. This seems likely to have been due to decomposition, liberation of carbon monoxide and ignition of the latter admixed with air.

## † 1-BROMO-2-PROPYNE $BrCH_2C{\equiv}CH$ $C_3H_3Br$

1. Coffee, R. D. *et al., Loss Prevention,* 1967, **1**, 6–9
2. Driedger, P. E. *et al., Chem. Eng. News,* 1972, **50**(12), 51
3. Forshey, D. R., *Fire Technol.*, 1969, **5**, 100–111

This liquid acetylenic compound may be decomposed by mild shock, and when heated under confinement, it decomposes with explosive violence and may detonate. Addition of 20–30% wt. of toluene makes the bromide insensitive to laboratory impact and confinement tests [1]. More recently, it was classed as extremely shock-sensitive [2].

It can be ignited by impact derived from the 'liquid-hammer' effect of accidental pressurisation of the aerated liquid, and will then undergo sustained (monopropellant) burning decomposition. The chloro analogue is similar, but less readily ignited [3].

Metals

*Dangerous Substances,* 1972, Sect. 1, 27

There is a danger of explosion in contact with copper, high-copper alloys, mercury or silver.

*See* METAL ACETYLIDES

Trichloronitromethane

*See* TRICHLORONITROMETHANE (CHLOROPICRIN), $CCl_3NO_2$: 1-Bromo-2-propyne

*See other* HALOACETYLENE DERIVATIVES

## 2,4,6-TRIS(BROMOAMINO)-1,3,5-TRIAZINE $C_3H_3Br_3N_6$

N=C(NHBr)N=C(NHBr)N=CNHBr (ring)

Vona, J.A., *et al., Chem. Eng. News,* 1952, **30**, 1916

Bromination with this and similar *N*-halo-compounds may become violent or explosive after an induction period as long as 15 min. Small-scale preliminary experiments, designed to avoid the initial presence of excess brominating agent, are recommended.

Allyl alcohol

Vona, J. A. *et al., Chem. Eng. News,* 1952, **30**, 1916

The components reacted violently 15 min after mixing at ambient temperature. This seems likely to have been a radical-initiated polymerisation of the alcohol (possibly peroxidised) in absence of diluent.

*See other N-*HALOGEN COMPOUNDS

**† 1-CHLORO-2-PROPYNE** **$ClCH_2C{\equiv}CH$** **$C_3H_3Cl$**

Doyle, W. H., *Loss Prevention,* 1969, **3**, 15

Pumping the liquid against a closed valve caused the pump to explode, which detonated the reservoir tank-car.

Ammonia

Anon., *Chemiearbeit,* 1956, 8(6), 45

Interaction of 1-chloro-2-propyne and liquid ammonia under pressure in a steel bomb had been used several times to prepare the amine. On one occasion the usual slow exothermic reaction did not occur, and the bomb was shaken mechanically. A rapid exothermic reaction, followed by an explosion, occurred. Other cases of instability in propyne derivatives are known.

*See* 2-PROPYNE-1-OL (PROPARGYL ALCOHOL), $C_3H_4O$
2-PROPYNE-1-THIOL, $C_3H_4S$

Chlorine

*See* CHLORINE, $Cl_2$ : 1-Chloro-2-propyne

*See other* HALOACETYLENE DERIVATIVES

**2,4,6-TRIS(CHLOROAMINO)-1,3,5-TRIAZINE** **$C_3H_3Cl_3N_6$**

**N=C(NHCl)N=C(NHCl)N=CNHCl** (ring: first N bonded to last C)

*See* 2,4,6-TRIS(BROMOAMINO)-1,3,5-TRIAZINE, $C_3H_3Br_3N_6$

*See other N-*HALOGEN COMPOUNDS

**METHYL TRICHLOROACETATE** **$CCl_3CO_2Me$** **$C_3H_3Cl_3O_2$**

Trimethylamine

Anon., *Angew. Chem. (Nachr.),* 1962, **10**, 197

A stirred, uncooled mixture in an autoclave reacted violently, the pressure developed exceeding 400 bar (6000 lb $in^{-2}$). Polymerisation of a reactive species formed by dehydrochlorination of the ester seems a possibility.

*See* SODIUM HYDRIDE, HNa: Ethyl 2,2,3-trifluoropropionate

**† 3,3,3-TRIFLUOROPROPENE** $H_2C{=}CHCF_3$ $C_3H_3F_3$

**† 1,1,1-TRIFLUOROACETONE** $CF_3COCH_3$ $C_3H_3F_3O$

**TRIFLUOROVINYL METHYL ETHER** $F_2C{=}CFOMe$ $C_3H_3F_3O$

1. Dixon, S., US Pat. 2 917 548, 1959
2. Anderson, A. W., *Chem. Eng. News*, 1976, **54**(16), 5

Trifluorovinyl methyl ether, b.p. 10.5–12.5°C, prepared from interaction of tetrafluoroethylene and sodium methoxide [1], has considerable explosive potential. Upon ignition it decomposes more violently than acetylene and should be treated with extreme caution [2]. Other trifluorovinyl ethers are similarly available from higher alkoxides [1], and although not tested for instability, should be handled carefully.

*See related* HALOALKENES

**TRIFLUOROMETHYL PEROXYACETATE** $C_3H_3F_3O_3$

$CH_3CO_2OCF_3$

Bernstein, P. A. *et al.*, *J. Amer. Chem. Soc.*, 1971, **93**, 3885

A 1 g sample cooled to –196°C exploded violently when warmed in a bath at 22°C.

*See other* PEROXY ESTERS

**POTASSIUM 1-TETRAZOLEACETATE** $C_3H_3KN_4O_2$

$\overline{N{=}NN{=}CHN}CH_2CO_2K$

Eizember, R. F. *et al.*, *J. Org. Chem.*, 1974, **39**, 1792–1793

During oven-drying, kilogram quantities of the salt exploded violently. Investigation showed that self-propagating and extremely rapid decomposition of a cold sample can be initiated by local heating to over 200°C by a flint spark, prolonged static spark or flame. The sodium salt could only be initiated by flame, and the free acid is much less sensitive.

*See other* TETRAZOLES

## † ACRYLONITRILE $CH_2{=}CHCN$ $C_3H_3N$

*MCA SD-31*, 1964
Harmon, 1974, 2.3
The monomer is sensitive to light, and even when inhibited, it will polymerise at above 200°C. It must never be stored uninhibited, or adjacent to basic materials.
*See* VIOLENT POLYMERISATION

### Acids

1. 'Acrylonitrile', London, British Hydrocarbon Chemicals Ltd., 1965
2. Kaszuba, F. J., *J. Amer. Chem. Soc.*, 1945, **67**, 1227
3. Shirley, D. A., *Preparation of Organic Intermediates*, 3, New York, Wiley, 1951
4. Kaszuba, F. J., *Chem. Eng. News*, 1952, **30**, 824

Contact of strong acids (sulphuric or nitric) with acrylonitrile may lead to vigorous reactions. Even small amounts of acid are potentially dangerous, as these may neutralise the aqueous ammonia present as polymerisation inhibitor and leave the nitrile unstabilised [1]. Precautions necessary in the hydrolysis of acrylonitrile [2] are omitted in the later reference [3]. It is essential to use well-chilled ingredients (acrylonitrile, diluted sulphuric acid, hydroquinone, copper powder) to avoid eruption and carbonisation. A really wide bore condenser is necessary to cope with vigorous boiling of unhydrolysed acrylonitrile.

### Bases

*MCA SD-31*, 1964
Acrylonitrile polymerises violently in contact with bases. In the absence of inhibitor, the pure nitrile will also polymerise.

### Bromine

*MCA Case History No. 1214*
Bromine was being added in portions to acrylonitrile with ice cooling, with intermediate warming to 20°C between portions. After half the bromine was added, the temperature increased to 70°C; then the flask exploded. This was attributed either to an accumulation of unreacted bromine (which would be obvious) or to violent polymerisation. The latter seems more likely, catalysed by hydrogen bromide formed by substitutive bromination.
*See* Acids, above

### Initiators

1. Zhulin, V. M. *et al.*, *Dokl. Akad. Nauk SSSR*, 1966, **170**, 1360–1363
2. Biesenberger, J. A. *et al.*, *Polymer Eng. Sci.*, 1976, **16**, 101–116

At pressures above 6000 bar free radical-initiated polymerisation sometimes proceeded explosively [1]. The parameters were determined in a batch reactor for thermal runaway polymerisation of acrylonitrile initiated by azoisobutyronitrile, dibenzoyl peroxide or di-*tert*-butyl peroxide.

*See* VIOLENT POLYMERISATION
*See other* POLYMERISATION INCIDENTS

Silver nitrate

*ABCM Quart. Safety Summ.*, 1962, **33**, 24

Acrylonitrile containing undissolved solid silver nitrate is liable, on long standing, to polymerise explosively and ignite. This is attributed to the slow deposition of a thermally insulating layer of polymer on the solid nitrate, which gradually gets hotter and catalyses rapid polymerisation.

Tetrahydrocarbazole,
Benzyltrimethylammonium hydroxide

*BCISC Quart. Safety Summ.*, 1968, **39**, 36

Cyanoethylation of 1,2,3,4-tetrahydrocarbazole initiated by the hydroxide had been effected smoothly on twice a published scale of working. During a further fourfold increase in scale, the initiator was added at 0°C, and shortly after cooling had been stopped and heating begun, the mixture exploded. A smaller proportion of initiator and very slow warming to effect reaction are recommended (to avoid rapid polymerisation of the nitrile by the base).

*See* Bases, above
*See other* CYANO COMPOUNDS

## VINYL ISOCYANIDE $H_2C{=}CHNC{:}$ $C_3H_3N$

Matteson, D. S. *et al.*, *J. Amer. Chem. Soc.*, 1968, **90**, 3765

The molar heat of formation of this endothermic compound (+230–250 kJ) is comparable with that of buten-3-yne (vinylacetylene). While no explosive decomposition of vinyl isocyanide has been reported, the possibility should be borne in mind.

*See related* CYANO COMPOUNDS
*See other* ENDOTHERMIC COMPOUNDS

## 2-THIONO-4-THIAZOLIDINONE (RHODANINE) $C_3H_3NOS_2$

$\overline{SC(S)NHC(O)CH_2}$

Merck, 1968, 916

It may explode on rapid heating.

**CYANOACETIC ACID** $CNCH_2CO_2H$ $C_3H_3NO_2$

*See* FURFURYL ALCOHOL, $C_5H_6O_2$ : Acids

**2-CARBAMOYL-2-NITROACETONITRILE ('FULMINURIC ACID')** $C_3H_3N_3O_3$

$H_2NCOCHNO_2CN$

Sorbe, 1968, 74

It explodes on heating.

*See other* *C*-NITRO COMPOUNDS

**2-AMINO-5-NITROTHIAZOLE** $C_3H_3N_3O_2S$

$SC(NH_2)=NCH=CNO_2$ (ring)

Preparative hazard

*See* NITRIC ACID, $HNO_3$ : 2-Aminothiazole, Sulphuric acid

**2,4,6-TRIHYDROXY-1,3,5-TRIAZINE (CYANURIC ACID)** $C_3H_3N_3O_3$

$N=COHN=COHN=COH$ (ring)

Chlorine

*See* CHLORINE, $Cl_2$ : Nitrogen compounds

**SODIUM 1-TETRAZOLEACETATE** $C_3H_3N_4NaO_2$

$N=NN=CHNCH_2CO_2Na$ (ring)

*See* POTASSIUM 1-TETRAZOLEACETATE, $C_3H_3KN_4O_2$

**5-CYANO-2-METHYLTETRAZOLE** $N=NNMeN=CCN$ (ring) $C_3H_3N_5$

Aluminium hydride

*See* ALUMINIUM HYDRIDE, $AlH_3$ : Tetrazole derivatives

**2-HYDROXY-4,6-BIS(*N*-NITROAMINO)-1,3,5-TRIAZINE** $C_3H_3N_7O_5$

**HOC=NC(NHNO$_2$)=NC(NHNO$_2$)=N**

Atkinson, E. R., *J. Amer. Chem. Soc.*, 1951, **73**, 4443–4444
The explosive nitration product of 'melamine' was identified as the title compound, which is easily detonated on impact.
*See other* *N*-NITRO COMPOUNDS

**SODIUM METHOXYACETYLIDE** **NaC≡COMe** $C_3H_3NaO$

Houben-Weyl, 1970, Vol. 13.1, 649
It may ignite in air.

Brine
Jones, E. R. H. *et al.*, *Org. Synth.*, 1963, Coll. Vol. 4, 406
During addition of saturated brine at −20°C to the sodium derivative at −70°C, minor explosions occur. These may have been due to particles of sodium igniting the liberated methoxyacetylene.
*See other* METAL ACETYLIDES

**† PROPADIENE (ALLENE)** **$CH_2$=C=$CH_2$** $C_3H_4$

Bondor, A. M. *et al.*, *Khim. Prom.*, 1965, **41**, 923
Forshey, D. R., *Fire Technol.*, 1969, **5**, 100–111
The pure diene can decompose explosively under a pressure of 2 bar, but this is also given as the upper limiting pressure for flame propagation during sustained (monopropellant) burning at 25°C.

Oxides of nitrogen
*See* NITROGEN OXIDE ('NITRIC OXIDE'), NO: Dienes, Oxygen
*See other* DIENES

**† PROPYNE** **HC≡$CCH_3$** $C_3H_4$

1. *MCA Case History No. 632*
2. Fitzgerald, F., *Nature*, 1960, **186**, 386–387
3. Hurden, D., *J. Inst. Fuel*, 1963, **36**, 50–54

The liquid material (which contains *ca.* 30% of propadiene) in cylinders is not shock-sensitive, but a temperature of 95°C (even very localised) accompanied by pressures of *ca.* 3.5 bar, will cause a detonation to propagate [1]. Induced decomposition of the endothermic hydrocarbon leads to flame propagation in absence of air above minimum pressures of 3.4 and 2.1 bar at 20 and 120°C, respectively [2]. Application as a monopropellant and possible hazards therefrom (including formation of explosive copper propynide) have been discussed [3].

*See other* ACETYLENIC COMPOUNDS

**3-BROMO-1,1,1-TRICHLOROPROPANE** $CCl_3CH_2CH_2Br$ $C_3H_4BrCl_3$

Preparative hazard

*See* ETHYLENE, $C_2H_4$ : Bromotrichloromethane

**† 1-CHLORO-3,3,3-TRIFLUOROPROPANE** $C_3H_4ClF_3$

$ClC_2H_4CF_3$

**2-CHLORO-1-CYANOETHANOL** $ClCH_2CH(CN)OH$ $C_3H_4ClNO$

Scotti, F. *et al.*, *J. Org. Chem.*, 1964, **29**, 1800

Distillation (at 110°C/4 mbar) is hazardous, since slight overheating may cause explosive decomposition to 2-chloroacetaldehyde and hydrogen cyanide.

*See other* CYANO COMPOUNDS

**1,3-DITHIOLIUM PERCHLORATE** $C_3H_4ClO_4S_2$

$(\overline{SCH_2SCH{=}CH})^+ \ ClO_4^-$

1. Ferraris, J. P. *et al.*, *Chem. Eng. News*, 1974, **52**(37), 3
2. Klingsberg, E., *J. Amer. Chem. Soc.*, 1964, **86**, 5292
3. Leaver, D. *et al.*, *J. Chem. Soc.*, 1962, 5109
4. Wudl, F. *et al.*, *J. Org. Chem.*, 1974, **39**, 3608–3609
5. Melby, L. R. *et al.*, ibid., 2456

The salt (an intermediate in the preparation of 1,4,5,8-tetrahydro-1,4,5,8-tetrathiafulvalene) exploded violently during removal from a sintered glass filter with a Teflon-clad spatula [1]. Previous references to the salt exploding at 250°C [2] or melting at 264°C had been made. Use of a salt alternative to the perchlorate is urged [1]. Safer methods suitable for small-scale [4] and large-scale [5] preparations have now been described.

*See other* NON-METAL PERCHLORATES

**† 1,3-DICHLOROPROPENE** $ClCH{=}CHCH_2Cl$ $C_3H_4Cl_2$

**† 2,3-DICHLOROPROPENE** $CH_2{=}CClCH_2Cl$ $C_3H_4Cl_2$

**2,2,3,3-TETRAFLUOROPROPANOL** **$CHF_2CF_2CH_2OH$** **$C_3H_4F_4O$**

Potassium hydroxide,
or Sodium

Bagnall, R. D., private comm., 1972

Attempted formation of sodium tetrafluoropropoxide by adding the alcohol to sodium (40 g) caused ignition and a fierce fire which melted the flask. This was attributed to alkoxide-induced elimination of hydrogen fluoride, and subsequent exothermic polymerisation. In an alternative preparation of the potassium alkoxide by adding alcohol to solid potassium hydroxide a vigorous exotherm occurred. This was not observed when the base was added slowly to the alcohol.

**VINYLDIAZOMETHANE** **$H_2C{=}CHCHN_2$** **$C_3H_4N_2$**

Salomon, R. G. *et al.*, *J. Org. Chem.*, 1975, **40**, 758

Potentially explosive, it should be stored in solution at 0°C and shielded from light.

*See other* DIAZO COMPOUNDS

**METHYL DIAZOACETATE** **$N_2CHCO_2Me$** **$C_3H_4N_2O_2$**

Searle, N. E., *Org. Synth.*, 1963, Coll. Vol. 4, 426

This ester must be handled with particular caution as it explodes with extreme violence on heating.

*See other* DIAZO COMPOUNDS

**2-AMINOTHIAZOLE** **$C_3H_4N_2S$**

**$\overline{SC(NH_2)=NCH=CH}$** (ring)

*MCA Case History No. 1587*

Drying 2-aminothiazole in an oven without forced air circulation caused development of hot spots and eventual ignition. It has a low auto-ignition temperature and will ignite after 3.5 h at 100°C.

Nitric acid,
Sulphuric acid

*See* NITRIC ACID, $HNO_3$: 2-Aminothiazole, Sulphuric acid

**IMIDAZOLIDINE-2,4-DITHIONE ('DITHIOHYDANTOIN')** **$C_3H_4N_2S_2$**

**$\overline{HNC(S)NHC(S)CH_2}$** (ring)

Pouwels, H., *Chem. Eng. News*, 1975, **53**(49), 5

A 70 g sample, sealed into a brown glass ampoule, exploded after storage at ambient temperature for 17 years. This was attributed to slow decomposition and gas generation.

*See also* PYRIMIDINE-2,4,5,6-TETRAONE ('ALLOXAN'), $C_4H_2N_2O_4$

**1,3-DINITRO-2-IMIDAZOLIDONE** $C_3H_4N_4O_5$

$\overline{N(NO_2)CON(NO_2)CH_2CH_2}$

Violent decomposition occurred at 238°C.

*See* DIFFERENTIAL THERMAL ANALYSIS

**1,3-DIAZIDOPROPENE** $N_3CH{=}CHCH_2N_3$ $C_3H_4N_6$

Forster, M. O. *et al., J. Chem. Soc.*, 1912, **101**, 489

A sample exploded while being weighed.

*See other* ORGANIC AZIDES

**† ACRYLALDEHYDE** $CH_2{=}CHCHO$ $C_3H_4O$

*MCA SD-85,* 1961

Acids,
or Bases

1. *MCA SD-85,* 1961
2. Hearsey, C. J., private comm., 1973
3. Catalogue note, Hopkin and Williams, 1973

Acrylaldehyde is very reactive and will polymerise rapidly, accelerating to violence, in contact with strong acid or basic catalysts. Normally an induction period, shortened by increase in contamination, water content or initial temperature, precedes the onset of polymerisation. Uncatalysed polymerisation sets in at 200°C in the pure material [1]. Exposure to weakly acid condition (nitrous fumes, sulphur dioxide, carbon dioxide), some hydrolysable salts, or thiourea will also cause exothermic and violent polymerisation. A two-year-old sample stored in a refrigerator close to a bottle of dimethylamine exploded violently, presumably after absorbing enough volatile amine (which penetrates plastics closures) to initiate polymerisation [2]. The stabilising effect of the added hydroquinone may cease after a comparatively short storage time. Such unstabilised material could polymerise explosively [3].

*See other* PEROXIDISABLE COMPOUNDS

## † METHOXYACETYLENE $CH_3OC{\equiv}CH$ $C_3H_4O$

*See* SODIUM METHOXYACETYLIDE, $C_3H_3NaO$: Brine
*See other* ACETYLENIC COMPOUNDS

## 2-PROPYN-1-OL (PROPARGYL ALCOHOL) $C_3H_4O$

$HC{\equiv}CCH_2OH$

Alkalies

**Anon.**, *Angew. Chem.(Nachr.)*, 1954, 2, 209

If propargyl alcohol and similar acetylenic compounds are dried with alkali before distillation, the residue may explode (probably owing to salt formation). Sodium sulphate is recommended as a suitable desiccant.

Mercury(II) sulphate,
Sulphuric acid

Nettleton, J., private comm., 1972

Reppe, W. *et al.*, *Ann.*, 1955, **596**, 38

Following the published procedure, hydroxyacetone was being prepared on half the scale by treating propargyl alcohol as a 30% wt. aqueous solution with mercury sulphate and sulphuric acid (6 g and 0.6 g per mol of alcohol, respectively). On stirring and warming the mixture to 70°C a violent exothermic eruption occurred. Quartering the scale of operations to 1g mol and reducing the amount of acid to 0.37g/mol gave a controllable reaction at 70°C. Adding the alcohol to the other reactants maintained at 70°C is an alternative possibility to avoid the suspected protonation and polymerisation of the propargyl alcohol.

*See other* ACETYLENIC COMPOUNDS

## ACRYLIC ACID $H_2C{=}CHCO_2H$ $C_3H_4O_2$

1. *Haz. Chem. Data*, 1975, 34
2. Anon., *RoSPA OS&H Bull.*, 1976(10), 3

It is normally supplied as the inhibited monomer, but because of its relatively high freezing point (14°C) it often partly solidifies, and the solid phase (and the vapour) will be free of inhibitor. Even the uninhibited acid may be stored safely below the m.p., but such material will polymerise exothermically at ambient temperature, and may accelerate to a violent or explosive state if confined. Narrow vents may become blocked by polymerisation of the uninhibited vapour [1].
A 17 $m^3$ tank trailer of glacial acrylic acid was being warmed by internal coils containing water supplied from an unmonitored steam and water mixer to prevent the acid freezing in subzero temperatures. The extremely violent explosion which occurred later probably involved explosive

polymerisation accelerated by both the latent heat liberated when the acid froze and the uncontrolled deliberate heating [2].

*See* VIOLENT POLYMERISATION

*See other* POLYMERISATION INCIDENTS

**† VINYL FORMATE** $HCO_2CH{=}CH_2$ $C_3H_4O_2$

**2-PROPYN-1-THIOL** $HC{\equiv}CCH_2SH$ $C_3H_4S$

1. Sato, K. *et al., Chem. Abs.*, 1956, **53**, 5112b
2. Brandsma, 1971, 179

When distilled at atmospheric pressure, it polymerised explosively. It distils smoothly under reduced pressure at 33–35°C/127 mbar [1]. The polymer produced on exposure to air may explode on heating. Presence of a stabiliser is essential during handling or storage under nitrogen at −20°C [2].

*See other* ACETYLENIC COMPOUNDS

**† 1-BROMO-2-PROPENE (ALLYL BROMIDE)** $C_3H_5Br$

$CH_2{=}CHCH_2Br$

**† 1-BROMO-2,3-EXPOXYPROPANE** $C_3H_5BrO$

$BrCH_2\overline{CHOCH_2}$

**† 1-CHLORO-1-PROPENE** $CHCl{=}CHCH_3$ $C_3H_5Cl$

**† 2-CHLOROPROPENE** $CH_2{=}CHClCH_3$ $C_3H_5Cl$

**† 1-CHLORO-2-PROPENE (ALLYL CHLORIDE)** $C_3H_5Cl$

$CH_2{=}CHCH_2Cl$

Lewis acids,
Metals

*MCA SD-99,* 1973

Contact with aluminium chloride, boron trifluoride, sulphuric acid, etc., may cause violently exothermic polymerisation. Products of contact with aluminium, magnesium, zinc (or galvanised metal) may produce similar results.

*See other* ALLYL COMPOUNDS
HALOALKENES

### 1-CHLORO-2,3-PROPYLENE DINITRATE $C_3H_5ClN_2O_6$

$ClCH_2CH(ONO_2)CH_2ONO_2$

Sorbe, 1968, 54

An explosive syrup.

*See other* ALKYL NITRATES

### CHLOROACETONE $CH_2ClCOCH_3$ $C_3H_5ClO$

1. Allen, C. F. H. *et al., Ind. Eng. Chem. (News Ed.),* 1931, 9, 184
2. Ewe, G. E., ibid., 229

Two separate incidents involved explosive polymerisation of chloroacetone stored in glass bottles under ambient conditions for extended periods.

*See also* BROMOACETONE OXIME, $C_3H_6BrNO$

### † 1-CHLORO-2,3-EPOXYPROPANE $ClCH_2\overline{CHOCH_2}$ $C_3H_5ClO$

Isopropylamine

Barton, N. *et al., Chem. & Ind.*, 1971, 994

With slow mixing and adequate cooling, smooth condensation to 1-chloro-3-isopropylamino-2-propanol occurs. With rapid mixing and poor cooling, a variable induction period with slow warming precedes a rapid, violent exotherm (to 350°C in 6s). Other primary and secondary amines behave similarly. Moderating effect of water and nature of the products are discussed.

*See* *N*-Substituted aniline, below

Potassium *tert*-butoxide

*See* POTASSIUM *tert*-BUTOXIDE, $C_4H_9KO$: Acids, etc.

*N*-Substituted aniline (unspecified)

Schierwater, F.-W., *Major Loss Prevention*, 1971, 47

Interaction is exothermic and the mixture was normally maintained at 60°C by stirring and cooling. Malfunction caused a temperature increase to 70°C and cooling capacity was insufficient to regain control. The temperature steadily increased to 120°C, when explosive decomposition occurred. This was attributed to thermal instability of the system and to inadequate pressure-release arrangements.

*See* Isopropylamine, above

Sulphuric acid

Leleu, *Cahiers*, 1974(75), 276

Interaction is violent.

Trichloroethylene

*See* TRICHLOROETHYLENE, $C_2HCl_3$: Epoxides

*See other* 1,2-EPOXIDES

**† PROPIONYL CHLORIDE** **EtCOCl** **$C_3H_5ClO$**

Preparative hazard

*See* PHOSPHORUS TRICHLORIDE, $Cl_3P$: Carboxylic acids

**† ETHYL CHLOROFORMATE** **$ClCO_2Et$** **$C_3H_5ClO_2$**

***N,N*-DICHLORO-β-ALANINE** **$C_3H_5Cl_2NO_2$**

**$Cl_2NCHMeCO_2H$**

Vit, J. *et al.*, *Synth. Commun.*, 1976, **6**(1), 1–4

It is thermally unstable above 95°C.

*See other* *N*-HALOGEN COMPOUNDS

***N*-CARBOETHOXYIMINOPHOSPHORYL CHLORIDE** **$C_3H_5Cl_3NO_2P$**

**$Cl_3P{=}NCO_2Et$**

*See* *N*-CARBOMETHOXYIMINOPHOSPHORYL CHLORIDE, $C_2H_3Cl_3NO_2P$

**ALLYLMERCURY(II) IODIDE** **$H_2C{=}CHCH_2HgI$** **$C_3H_5HgI$**

Potassium cyanide

Whitmore, 1921, 122

Among the products of interaction, a minor one is an explosive liquid.

*See other* MERCURY COMPOUNDS

**† 1-IODO-2-PROPENE (ALLYL IODIDE)** **$CH_2{=}CHCH_2I$** **$C_3H_5I$**

**POTASSIUM *aci*-1,1-DINITROPROPANE** **$C_3H_5KN_2O_4$**

**$EtC(NO_2){=}N(O)OK$**

Bisgrove, D. E. *et al.*, *Org. Synth.*, 1963, Coll. Vol. 4, 373

The potassium salt of 1,1-dinitropropane, isolated as a by-product during preparation of 3,4-dinitro-3-hexene, is a hazardous explosive.

*See* NITROALKANES: Alkali metals
*See other aci*-NITRO SALTS

**POTASSIUM *O*-ETHYL DITHIOCARBONATE (XANTHATE)** **KSC(S)OEt** **$C_3H_5KOS_2$**

Diazonium salts
*See* DIAZONIUM SULPHIDES AND DERIVATIVES

**ETHYL ISOCYANIDE** **EtNC:** **$C_3H_5N$**

Lemoult, M. P., *Compt. Rend.*, 1906, **143**, 903
This showed a strong tendency to explode while being sealed into glass ampoules.
*See* METHYL ISOCYANIDE, $C_2H_3N$
*See related* CYANO COMPOUNDS

**† PROPIONITRILE** **EtCN** **$C_3H_5N$**

**ACRYLAMIDE** **$H_2C{=}CHCONH_2$** **$C_3H_5NO$**

Muir, 1977, 125
It may polymerise with violence on melting at 85°C.
*See other* POLYMERISATION INCIDENTS

**2,3-EPOXYPROPIONALDEHYDE OXIME** **$C_3H_5NO_2$**

**$\overline{OCH_2CH}CH{=}NOH$**

Payne, G. B., *J. Amer. Chem. Soc.*, 1959, **81**, 4903
The residue from distillation at 48–49°C/1.3 mbar polymerised violently, and the distilled material polymerised explosively after 1–2 h at ambient temperature.
*See other* OXIMES
POLYMERISATION INCIDENTS
*See also* 2,3-EPOXYPROPIONALDEHYDE 2,4-DINITROPHENYLHYDRAZONE, $C_9H_8N_4O_5$

**PROPIONYL NITRITE** **EtCOONO** **$C_3H_5NO_3$**

*See* ACYL NITRITES

**PROPIONYL PEROXONITRATE** $C_2H_5CO_2ONO_2$ $C_3H_5NO_5$

1. Stephens, E. R., *Anal. Chem.*, 1964, **36**, 928–929
2. Louw, R. *et al.*, *J. Amer. Chem. Soc.*, 1975, **97**, 4396

Like the lower homologue, it is extremely explosive [1], and may only be handled in high dilution with air or nitrogen. Higher homologues are less explosive [2].

*See related* ACYL NITRATES
PEROXY ESTERS

**AZIDOACETONE** $N_3CH_2COCH_3$ $C_3H_5N_3O$

1. Spauschus, H. O. *et al.*, *J. Amer. Chem. Soc.*, 1951, **73**, 209
2. Forster, M. O. *et al.*, *J. Chem. Soc.*, 1908, **93**, 72

A small sample exploded after 6 months' storage in the dark [1]. The freshly prepared material explodes when dropped on to a hotplate and burns brilliantly [2].

*See other* 2-AZIDOCARBONYL COMPOUNDS

**5-AMINO-3-METHYLTHIO-1,2,4-OXADIAZOLE** $C_3H_5N_3OS$

$\overline{ON{=}C(SMe)N{=}C}NH_2$

Wittenbrook, L. S., *J. Heterocycl. Chem.*, 1975, **12**, 38

It undergoes rapid and moderately violent decomposition at the m.p., 97–99°C.

**ETHYL AZIDOFORMATE** $N_3CO_2Et$ $C_3H_5N_3O_2$

Forster, M. O. *et al.*, *J. Chem. Soc.*, 1908, **93**, 81

It is liable to explode if boiled at atmospheric pressure (at 114°C).

*See other* ACYL AZIDES

**† CYCLOPROPANE** $\overline{CH_2CH_2CH_2}$ $C_3H_6$

**† PROPENE** $CH_2{=}CHCH_3$ $C_3H_6$

*See* Acrylaldehyde, etc., below

*MCA SD-59*, 1956

Russell, F. R. *et al.*, *Chem. Eng. News*, 1952, **30**, 1239

Propene at 955 bar and 327°C was being subjected to further rapid compression. At 4860 bar explosive decomposition occurred, causing a pressure surge to 10 000 bar or above. Decomposition to carbon, hydrogen and methane must have occurred to account for this pressure. Ethylene behaves similarly at much lower pressure and cyclopentadiene, cyclohexadiene, acetylene and a few aromatic hydrocarbons have been explosively decomposed.

Acrylaldehyde,
Air

Inoue, H. *et al., J. Chem. Soc. Japan, Ind. Chem. Sect.*, 1968, **71**, 615–618

In a study of the explosive limits of propylene and/or acrylaldehyde in air (in presence or absence of steam), the narrower limits of 1.4–7.1% were determined for propene–air mixtures at room temperature.

Lithium nitrate,
Sulphur dioxide

Pitkethly, R.C., private comm., 1973

A mixture under confinement in a glass pressure bottle at 20°C polymerised explosively, the polymerisation probably being initiated by access of light through the clear glass container. Such alkene–sulphur dioxide copolymerisations will not occur above a ceiling temperature, different for each alkene.

Oxides of nitrogen

*See* NITROGEN OXIDE ('NITRIC OXIDE'), NO: Dienes, Oxygen

Trifluoromethyl hypofluorite

*See* TRIFLUOROMETHYL HYPOFLUORITE, $CF_4O$: Hydrocarbons

Water

*See* LIQUEFIED GASES: Water

**DIMETHYLGOLD SELENOCYANATE** **$Me_2AuSeCN$** **$C_3H_6AuNSe$**

Stocco, F. *et al., Inorg. Chem.*, 1971, **10**, 2640

Very shock-sensitive and explodes readily when precipitated from aqueous solution. Crystals obtained by slow evaporation of a carbon tetrachloride extract were less sensitive.

*See other* GOLD COMPOUNDS

**BROMOACETONE OXIME** **$C_3H_6BrNO$**

**$BrCH_2C(Me){=}NOH$**

Forster, M. O. *et al., J. Chem. Soc.*, 1908, **93**, 84

It decomposes explosively during distillation.

*See also* CHLOROACETALDEHYDE OXIME, $C_2H_4ClNO$
CHLOROACETONE, $C_3H_5ClO$

**† 1,1-DICHLOROPROPANE** **$CHCl_2Et$** **$C_3H_6Cl_2$**

**† 1,2-DICHLOROPROPANE** **$CH_2ClCHClCH_3$** **$C_3H_6Cl_2$**

Aluminium

*See* ALUMINIUM, Al: Halocarbons (reference 18)

Aluminium,
*o*-Dichlorobenzene,
1,2-Dichloroethane

*See* ALUMINIUM, Al: Halocarbons (reference 5)

*See other* HALOCARBONS

**2,2-DICHLOROPROPANE** **$CH_3CCl_2CH_3$** **$C_3H_6Cl_2$**

Dimethylzinc

*See* DIMETHYLZINC, $C_2H_6Zn$: 2,2-Dichloropropane

**3-CHLORO-2-HYDROXYPROPYL PERCHLORATE** **$C_3H_6Cl_2O_5$**

**$CH_2ClCHOHCH_2OClO_3$**

Hofmann, K. A. *et al., Ber.*, 1909, **42**, 4390

Explodes violently on heating in a capillary.

*See other* ALKYL PERCHLORATES

**DICHLOROMETHYLENEDIMETHYLAMMONIUM CHLORIDE** **$C_3H_6Cl_3N$**

**$CCl_2{=}\overset{+}{N}(Me)_2\ Cl^-$**

Bader, A. R., private comm., 1971

Senning, A., *Chem. Rev.*, 1965, **65**, 388

The compound does not explode on heating; the published reference is in error.

**DIMETHYLTHALLIUM FULMINATE** **$Me_2Tl(C{\equiv}N{\rightarrow}O)$** **$C_3H_6NOTl$**

Beck, W. *et al., J. Organomet. Chem.*, 1965, **3**, 55

Highly explosive, unlike the diphenyl analogue.

*See related* METAL FULMINATES

**2-AMINOPROPIONITRILE** **$CH_3CHNH_2CN$** **$C_3H_6N_2$**

Shuer, V. F., *Zh. Prikl. Khim.*, 1966, **39**, 2386

Vacuum-distilled material kept dark for 6 months was found to have largely polymerised to a yellow solid, and it exploded 15 days later.
*See other* CYANO COMPOUNDS

**2- or 5-ETHYLTETRAZOLE** $C_3H_6N_4$

**CH=NNEtN=N or HNN=NN=CEt**

Aluminium hydride
*See* ALUMINIUM HYDRIDE, $AlH_3$ : Tetrazole derivatives

**AZIDOACETONE OXIME** $C_3H_6N_4O$

**$N_3CH_2C(Me)$=NOH**

Forster, M. O. *et al., J. Chem. Soc.,* 1908, **93**, 83
During distillation at 84°C/2.6 mbar, the large residue darkened and finally exploded violently.
*See also* BROMOACETONE OXIME, $C_3H_6BrNO$
*See other* 2-AZIDOCARBONYL COMPOUNDS

**1,3,5-TRINITROSOHEXAHYDRO-1,3,5-TRIAZINE** $C_3H_6N_6O_3$

**$CH_2N(NO)CH_2N(NO)CH_2NNO$**

Sulphuric acid
Urbanski, 1967, Vol. 3, 122
Concentrated sulphuric acid causes explosive decomposition.
*See other* NITROSO COMPOUNDS

**† ACETONE** **$CH_3COCH_3$** $C_3H_6O$
*MCA SD-87*, 1962

Bromoform
*See* BROMOFORM, $CHBr_3$ : Acetone, etc.

Carbon,
Air
Boiston, D. A., *Brit. Chem. Eng.,* 1968, **13**, 85
Fires in plant to recover acetone from air with active carbon are due to the bulk surface effect of oxidative heating when air flow is too low to cool effectively.

Chloroform
*See* CHLOROFORM, $CHCl_3$ : Acetone, etc.

2-Methyl-1,3-butadiene

*See* 2-METHYL-1,3-BUTADIENE (ISOPRENE), $C_5H_8$ : Acetone

Nitric acid,
Sulphuric acid,

Fawcett, H.H., *Ind. Eng. Chem.*, 1959, **51**(4), 89A

Acetone will decompose with explosive violence if brought into contact with mixed nitric and sulphuric acids, particularly under confinement.

*See* Oxidants, below

Oxidants

*See* BROMINE TRIFLUORIDE, $BrF_3$ : Solvents
BROMINE, $Br_2$ : Acetone
NITROSYL CHLORIDE, ClNO: Acetone, etc.
NITROSYL PERCHLORATE, $ClNO_5$ : Organic materials
NITRYL PERCHLORATE, $ClNO_6$ : Organic solvents
CHROMYL CHLORIDE, $Cl_2CrO_2$ : Organic solvents
CHROMIUM TRIOXIDE, $CrO_3$ : Acetone
DIOXYGEN DIFLUORIDE, $F_2O_2$
NITRIC ACID, $HNO_3$ : Acetone, etc.
HYDROGEN PEROXIDE, $H_2O_2$ : Acetone, etc.
: Ketones
: Oxygenated compounds
PEROXOMONOSULPHURIC ACID, $H_2O_5S$ : Acetone
OXYGEN (Gas), $O_2$ : Acetone, Acetylene

*See* Nitric acid, Sulphuric acid, above

Potassium *tert*-butoxide

*See* POTASSIUM *tert*-BUTOXIDE, $C_4H_9KO$: Acids, etc.

Sulphur dichloride

*See* SULPHUR DICHLORIDE, $Cl_2S$: Acetone

Thiotrithiazyl perchlorate

*See* THIOTRITHIAZYL PERCHLORATE, $ClN_3O_4S_4$ : Organic solvents

## † 2-PROPEN-1-OL (ALLYL ALCOHOL) $CH_2{=}CHCH_2OH$ $C_3H_6O$

Alkali,
2,4,6-Trichloro-1,3,5-triazine (Cyanuric chloride)

*See* 2,4,6-TRICHLORO-1,3,5-TRIAZINE (CYANURIC CHLORIDE), $C_3Cl_3N_3$ : Allyl alcohol, etc.

Sulphuric acid

Senderens, J. B., *Compt. Rend.*, 1925, **181**, 698–700

During preparation of diallyl ether by dehydration of the alcohol with sulphuric acid, a violent explosion may occur (possibly due to polymerisation).

2,4,6-Tris(bromoamino)-1,3,5-triazine

*See* 2,4,6-TRIS(BROMOAMINO)-1,3,5-TRIAZINE, $C_3H_3Br_3N_6$:
Allyl alcohol

*See other* ALLYL COMPOUNDS

## † METHYL VINYL ETHER MeOCH=$CH_2$ $C_3H_6O$

Acids

1. Braker, 1971, 382
2. MVE Brochure, Billingham, ICI, 1962

Methyl vinyl ether is rapidly hydrolysed by contact with dilute acids to produce acetaldehyde, which is more reactive and has wider flammability limits than the ether [1]. Presence of base is essential during storage or distillation of the ether to prevent rapid acid-catalysed homopolymerisation, which is not prevented by antoxidants. Even mildly acidic solids (calcium chloride or some ceramics) will initiate exothermic polymerisation [2].

*See* ACETALDEHYDE, $C_2H_4O$

Halogens,
or Hydrogen halides

Braker, 1971, 382

Addition reactions with bromine, chlorine, hydrogen bromide or chloride are very vigorous and may be explosive if uncontrolled.

*See other* PEROXIDISABLE COMPOUNDS

## † PROPIONALDEHYDE EtCHO $C_3H_6O$

## † PROPYLENE OXIDE MeCHOCH$_2$ $C_3H_6O$

1. Kulik, M. K., *Science*, 1967, **155**, 400
2. Smith, R. S., *Science*, 1967, **156**, 12

Use of propylene oxide as a biological sterilant is hazardous because of ready formation of explosive mixtures with air (2.8–37%). Commercially available mixtures with carbon dioxide, though non-explosive, may be asphyxiant and vesicant [1]. Such mixtures may be ineffective, but neat propylene oxide may be safely used, provided that it is removed by evacuation using a water-jet pump [2].

Ethylene oxide,
Polyhydric alcohol

*See* ETHYLENE OXIDE, $C_2H_4O$: Polyhydric alcohol, etc.

Sodium hydroxide

*MCA Case History No. 31*

A drum of a crude product containing unreacted propylene oxide and sodium hydroxide catalyst exploded and ignited, probably owing to base-catalysed polymerisation of the oxide.

*See also* ETHYLENE OXIDE, $C_2H_4O$: Contaminants

*See other* 1,2-EPOXIDES

**ALLYL HYDROPEROXIDE** $CH_2=CHCH_2OOH$ $C_3H_6O_2$

Dykstra, S. *et al.*, *J. Amer. Chem. Soc.*, 1957, **79**, 3474

Seyfarth, H. E. *et al.*, *Angew. Chem.*, 1965, 77, 1078

When impure, the material is unstable towards heat and light and decomposes to give an explosive residue. The pure material is more stable to light, but detonates on heating or in contact with solid alkalies.

Preparation by action of oxygen on diallylzinc gives improved yields, but there is a risk of explosion. The peroxide is also impact-sensitive if sand is admixed.

*See other* ALKYL HYDROPEROXIDES

**† 1,3-DIOXOLANE** $\overline{CH_2OCH_2OCH_2}$ $C_3H_6O_2$

**† ETHYL FORMATE** $HCO_2Et$ $C_3H_6O_2$

**HYDROXYACETONE** $HOCH_2COMe$ $C_3H_6O_2$

Preparative hazard

*See* 2-PROPYNE-1-OL, $C_3H_4O$ : Mercury(II) sulphate, etc.

**† METHYL ACETATE** $MeCO_2Me$ $C_3H_6O_2$

**† DIMETHYL CARBONATE** $CO(OMe)_2$ $C_3H_6O_3$

Potassium *tert*-butoxide

*See* POTASSIUM *tert*-BUTOXIDE, $C_4H_9KO$: Acids, etc.

**LACTIC ACID** $MeCHOHCO_2H$ $C_3H_6O_3$

Hydrofluoric acid,
Nitric acid
*See* NITRIC ACID, $HNO_3$ : Lactic acid, etc.

**PEROXYPROPIONIC ACID** $EtCO_2OH$ $C_3H_6O_3$

1. Swern, D., *Chem. Rev.*, 1949, **45**, 9
2. Phillips, B. *et al.*, *J. Org. Chem.*, 1959, **23**, 1823

More stable than its lower homologues, it merely deflagrates on heating. Higher homologues appear to be more stable [1]. Safe procedures (on the basis of detonability experiments) for preparation of anhydrous solutions of peroxypropionic acid in chloroform or ethyl propionate have been described [2].
*See other* PEROXYACIDS

**PROPENE OZONIDE** $C_3H_6O_3$

$MeCHCH_2O{-}O^+{-}O^-$ (ring: $CH_2$ to $O$)

Briner, E. *et al.*, *Helv. Chim. Acta*, 1929, **12**, 181
Stable at 0°C but often explosively decomposes at ambient temperature.
*See other* OZONIDES

**1,3,5-TRIOXANE** $CH_2OCH_2OCH_2O$ (ring) $C_3H_6O_3$

*MCA Case History No. 1129*
Use of an axe to break up a large lump of trioxane caused ignition and a vigorous fire. Peroxides may have been involved.

Hydrogen peroxide,
Lead
*See* HYDROGEN PEROXIDE, $H_2O_2$: Lead, Trioxane

Oxygen (liquid)
*See* OXYGEN (Liquid), $O_2$: 1,3,5-Trioxane
*See other* PEROXIDISABLE COMPOUNDS

**3-HYDROXYTHIETANE-1,1-DIOXIDE** $O_2SCH_2CHOHCH_2$ (ring) $C_3H_6O_3S$

Dittmer, D. C. *et al.*, *J. Org. Chem.*, 1971, **36**, 1324
Evaporation of the residue after treatment of the thietane with hydrogen peroxide is liable to explode, and must be done in an open dish. This is

probably because of formation of a 2- or 3-hydroperoxide derivative.

*See* HYDROGEN PEROXIDE, $H_2O_2$: Acetic acid, 3-Hydroxythietane

**† 2-PROPEN-1-THIOL** **$CH_2=CHCH_2SH$** **$C_3H_6S$**

**DIAZOMETHYLDIMETHYLARSINE** **$N_2CHAsMe_2$** **$C_3H_7AsN_2$**

Preparative hazard

*See* DIAZOMETHANE, $CH_2N_2$: Dimethylaminodimethylarsine, etc.

**† 1-BROMOPROPANE** **$BrCH_2Et$** **$C_3H_7Br$**

**† 2-BROMOPROPANE** **MeCHBrMe** **$C_3H_7Br$**

**† 1-CHLOROPROPANE** **$ClCH_2Et$** **$C_3H_7Cl$**

**† 2-CHLOROPROPANE** **MeCHClMe** **$C_3H_7Cl$**

**† CHLOROMETHYL ETHYL ETHER** **$ClCH_2OEt$** **$C_3H_7ClO$**

**ISOPROPYL HYPOCHLORITE** **$Me_2CHOCl$** **$C_3H_7ClO$**

Chattaway, F. D. *et al., J. Chem. Soc.,* 1923, **123**, 3001

Of extremely low stability; explosions occurred during its preparation if cooling was inadequate.

*See other* HYPOHALITES

**1-CHLORO-2,3-PROPANEDIOL** **$C_3H_7ClO_2$**

**$CH_2ClCHOHCH_2OH$**

Perchloric acid

*See* PERCHLORIC ACID, $ClHO_4$, Glycols, etc.

**PROPYL PERCHLORATE** **$PrOClO_3$** **$C_3H_7ClO_4$**

*See* ALKYL PERCHLORATES

**PROPYLCOPPER(I)** **PrCu** $C_3H_7Cu$

Houben-Weyl, 1970, **13**(1), 737

Small quantities only should be handled because of the danger of explosion.

*See other* ALKYLMETALS

† **1-IODOPROPANE** $CH_2IEt$ $C_3H_7I$

† **2-IODOPROPANE** **MeCHIMe** $C_3H_7I$

† **ALLYLAMINE** $CH_2{=}CHCH_2NH_2$ $C_3H_7N$

† **CYCLOPROPYLAMINE** $\overline{CH_2CHNH_2CH_2}$ $C_3H_7N$

† **2-METHYLAZIRIDINE (PROPYLENEIMINE)** $C_3H_7N$

$CH_3\overline{CHNHCH_2}$

Acids

Inlow, R. O. *et al., J. Inorg. Nucl. Chem.*, 1975, **37**, 2353

Like the lower homologue ethyleneimine, it may polymerise explosively if exposed to acids or acidic fumes, so must always be stored over solid alkali.

*See other* POLYMERISATION INCIDENTS

***N,N*-DIMETHYLFORMAMIDE** $HCONMe_2$ $C_3H_7NO$

This powerful solvent is not inert and reacts vigorously or violently with a range of materials.

*See* BROMINE, $Br_2$: *N,N*-Dimethylformamide
CARBON TETRACHLORIDE, $CCl_4$: *N,N*-Dimethylformamide
2,4,6-TRICHLORO-1,3,5-TRIAZINE, $C_3Cl_3N_3$: *N,N*-Dimethylformamide
1,2,3,4,5,6-HEXACHLOROCYCLOHEXANE, $C_6H_6Cl_6$: *N,N*-Dimethylformamide
2,5-BIS-*endo*-DICHLORO-7-THIABICYCLO[2.2.1]HEPTANE, $C_6H_8Cl_2S$: *N,N*-Dimethylformamide
TRIETHYLALUMINIUM, $C_6H_{15}Al$: *N,N*-Dimethylformamide
SULPHINYL CHLORIDE, $Cl_2OS$: *N,N*-Dimethylformamide
CHROMIUM TRIOXIDE, $CrO_3$: *N,N*-Dimethylformamide
LITHIUM AZIDE, $LiN_3$: Alkyl nitrates, etc.
MAGNESIUM NITRATE, $MgN_2O_6$: *N,N*-Dimethylformamide
SODIUM, Na: *N,N*-Dimethylformamide
PHOSPHORUS(V) OXIDE, $O_5P_2$: Organic liquids

Halogenated compounds,
Iron

*DMF*, Brochure A-65510-2, Wilmington (Del.), Du Pont, 1972
The tertiary amide catalyses dehydrohalogenation reactions, which for some highly halogenated compounds (carbon tetrachloride, hexachlorocyclohexane) at elevated temperatures in the presence of iron may become violent.

*See* CATALYSIS BY IMPURITIES
*N,N*-DIMETHYLACETAMIDE, $C_4H_9NO$: Halogenated compounds, etc.
*See other* APROTIC SOLVENTS

**ETHYL CARBAMATE** **$H_2NCO_2Et$** **$C_3H_7NO_2$**

Phosphorus pentachloride

*See* *N*-CARBOMETHOXYIMINOPHOSPHORYL CHLORIDE, $C_2H_3Cl_3NO_2P$

**† ISOPROPYL NITRITE** **$Me_2CHONO$** **$C_3H_7NO_2$**

**1-NITROPROPANE** **$NO_2CH_2Et$** **$C_3H_7NO_2$**

Metal oxides

*See* NITROALKANES: Metal oxides

**† ISOPROPYL NITRATE** **$Me_2CHONO_2$** **$C_3H_7NO_3$**

*Tech. Bull. No. 2,* London, ICI Ltd., 1964
Brochet, C., *Astronaut. Acta*, 1970, **15**, 419–425
It covers storage and handling precautions for this fuel of low oxygen balance. The self-sustaining exothermic decomposition renders it suitable as a rocket monopropellant.

Conditions for detonation of the nitrate, an explosive of low sensitivity, have been investigated.

Lewis acids

*See* ALKYL NITRATES

**† PROPYL NITRATE** **$PrONO_2$** **$C_3H_7NO_3$**

*See* ALKYL NITRATES

## 5-AMINO-2-ETHYLTETRAZOLE $EtNN{=}NC(NH_2){=}N$ $C_3H_7N_5$

Aluminium hydride

*See* ALUMINIUM HYDRIDE, $AlH_3$: Tetrazole derivatives

## PROPYLSODIUM $C_3H_7Na$

Houben-Weyl, 1970, Vol. 13.1, 384

The powder ignites in air.

*See other* ALKYLMETALS

## SODIUM ISOPROPOXIDE $NaOCHMe_2$ $C_3H_7NaO$

Dimethyl sulphoxide

*See* DIMETHYL SULPHOXIDE, $C_2H_6OS$: Metal alkoxides

## † PROPANE $CH_3CH_2CH_3$ $C_3H_8$

Barium peroxide

*See* BARIUM PEROXIDE, $BaO_2$: Propane

## ETHYL IMINIOFORMATE CHLORIDE $C_3H_8ClNO$

$HC(N^+H_2)OEt\ Cl^-$

Preparative hazard for this 'imidoester hydrochloride'.

*See* HYDROGEN CHLORIDE, ClH: Alcohols, Hydrogen cyanide

## ISOPROPYLDIAZENE $Me_2CHN{=}NH$ $C_3H_8N_2$

Chlorine

Abendroth, H. J., *Angew. Chem.*, 1959, **71**, 340

The crude chlorination product of 'isoacetone hydrazone' (presumably an *N*-chloro- compound) exploded violently during drying at 0°C.

*See* *N*-HALOGEN COMPOUNDS

*See other* AZO COMPOUNDS

## *N,N'*-DINITRO-*N*-METHYL-1,2-DIAMINOETHANE $C_3H_8N_4O_4$

$CH_3N(NO_2)C_2H_4NHNO_2$

Violent decomposition occurred at 210°C.

*See* DIFFERENTIAL THERMAL ANALYSIS

*See other* *N*-NITRO COMPOUNDS

† **ETHYL METHYL ETHER** **EtOMe** **$C_3H_8O$**

† **PROPANOL** **PrOH** **$C_3H_8O$**

Potassium *tert*-butoxide

*See* POTASSIUM *tert*-BUTOXIDE, $C_4H_9KO$: Acids, etc.

† **2-PROPANOL** **MeCHOHMe** **$C_3H_8O$**

*MCA SD-98*, 1972

Aluminium

*See* ALUMINIUM, Al: 2-Propanol

Aluminium triisopropoxide,
Crotonaldehyde

Wagner-Jauregg, T., *Angew. Chem.*, 1939, **52**, 710

During distillation of 2-propanol recovered from the reduction of crotonaldehyde with aluminium isopropoxide, a violent explosion occurred. This was attributed either to peroxidised diisopropyl ether (a possible by-product) or to peroxidised crotonaldehyde.

*See* DIISOPROPYL ETHER, $C_6H_{14}O$

2-Butanone

1. Bathie, H. M., *Chem. Brit.*, 1974, **10**, 143
2. Unpublished observations, 1974
3. Pitt, M. J., *Chem. Brit.*, 1974, **10**, 312

Distillation to small volume of a small sample of a 4-year-old mixture of the alcohol with 0.5% of the ketone led to a violent explosion, and the presence of peroxides was subsequently confirmed [1]. Pure alcohols which can form stable radicals (secondary branched structures) may slowly peroxidise to a limited extent under normal storage conditions (isopropanol to 0.0015 M in brown bottle, subdued light during 6 months; to 0.0009 M in dark during 5 years) [2]. The presence of ketones markedly increases the possibility of peroxidation by sensitising photochemical oxidation of the alcohol. Acetone (produced during autoxidation of isopropanol) is not a good sensitiser, but the presence of 2-butanone in isopropanol would be expected to accelerate markedly peroxidation of the latter. Treatment of any mixture of a secondary alcohol with tin(II) chloride and then lime before distillation is recommended [3].

*See* 2-BUTANOL, $C_4H_{10}O$

Oxidants

*See* DIOXYGENYL TETRAFLUOROBORATE, $BF_4O_2$: Organic materials

BARIUM PERCHLORATE, $BaCl_2O_8$ : Alcohols
TRINITROMETHANE, $CHN_3O_6$ : 2-Propanol
CHROMIUM TRIOXIDE, $CrO_3$ : Alcohols
HYDROGEN PEROXIDE, $H_2O_2$ : Oxygenated compounds
OXYGEN (Gas), $O_2$ : Alcohols

Potassium *tert*-butoxide
*See* POTASSIUM *tert*-BUTOXIDE, $C_4H_9KO$: Acids, etc.

**† DIMETHOXYMETHANE** **$CH_2(OMe)_2$** $C_3H_8O_2$

Oxygen
Molem, M. J. *et al.*, *Chem. Abs.*, 1975, **83**, 45495v
The nature of the reaction and the products in methylal–oxygen mixtures during cool-flame or explosive oxidation were studied.
*See other* PEROXIDISABLE COMPOUNDS

**ETHYL METHYL PEROXIDE** **EtOOMe** $C_3H_8O_2$
Rieche, A., *Ber.*, 1929, **62**, 218
Shock-sensitive as liquid or vapour, it explodes violently on superheating.
*See other* DIALKYL PEROXIDES

**ISOPROPYL HYDROPEROXIDE** **$Me_2CHOOH$** $C_3H_8O_2$
Medvedev, S., *et al.*, *Ber.*, 1932, **65**, 133
Explodes just above b.p., 107–109°C.
*See other* ALKYL HYDROPEROXIDES

**GLYCEROL** $C_3H_8O_3$
**$CH_2OHCHOHCH_2OH$**

Oxidants
The violent or explosive reactions exhibited by glycerol in contact with many solid oxidants are due to its unique properties of having three hydroxylic points of attack in the same molecule, of being a liquid which ensures good contact, and of high boiling point and viscosity, which prevents dissipation of oxidative heat. The difunctional, less viscous liquid glycols show similar but less violent behaviour.
*See* CALCIUM HYPOCHLORITE, $CaCl_2O_2$ : Hydroxy compounds
CHLORINE, $Cl_2$ : Glycerol
CHROMIUM TRIOXIDE, $CrO_3$ : Glycerol
NITRIC ACID, $HNO_3$ : Glycerol, Hydrofluoric acid
: Glycerol, Sulphuric acid

POTASSIUM PERMANGANATE, $KMnO_4$ : Glycerol
SODIUM PEROXIDE, $Na_2O_2$ : Hydroxy compounds

Phosphorus triiodide
*See* PHOSPHORUS TRIIODIDE, $I_3P$: Hydroxylic compounds, etc.

Sodium hydride
*See* SODIUM HYDRIDE, HNa : Glycerol

**† ETHYL METHYL SULPHIDE** **EtSMe** **$C_3H_8S$**

**† PROPANETHIOL** **PrSH** **$C_3H_8S$**

**† 2-PROPANETHIOL** **$Me_2CHSH$** **$C_3H_8S$**

**TRIMETHYLALUMINIUM** **$Me_3Al$** **$C_3H_9Al$**

Kirk-Othmer, 1963, Vol. 2, 40

Extremely pyrophoric, ignition delay is 13 ms in air at 232°C/0.16 bar.

*See* TRIALKYLALUMINIUMS

**TRIMETHYLDIALUMINIUM TRICHLORIDE (ALUMINIUM TRICHLORIDE–TRIETHYLALUMINIUM COMPLEX)** **$C_3H_9Al_2Cl_3$**
**$AlCl_3 \cdot Et_3Al$**

*See* ALKYLALUMINIUM HALIDES

**TRIMETHYLARSINE** **$Me_3As$** **$C_3H_9As$**

Air,
or Halogens

Sidgwick, 1950, 762, 769

Inflames in air, and interaction with halogens is violent.

*See other* ALKYLNON-METALS

## TRIMETHYLBORANE $Me_3B$ $C_3H_9B$

Stock, A. *et al.*, *Ber.*, 1921, **54**, 535

The gas ignites in air.

*See also* CHLORINE, $Cl_2$: Trialkylboranes

*See other* ALKYLNON-METALS

## † TRIMETHYL BORATE $(MeO)_3B$ $C_3H_9BO_3$

## TRIMETHYLBISMUTH $Me_3Bi$ $C_3H_9Bi$

Coates, 1967, Vol. 1, 536

Sorbe, 1968, 157

It ignites in air, but is unreactive with water; it explodes at 110°C.

*See* TRIALKYLBISMUTHS

## TRIMETHYLSULPHOXONIUM BROMIDE $C_3H_9BrOS$

$Me_3S^+O\ Br^-$

Scaros, M. G. *et al.*, *Chem. Brit.*, 1973, **9**, 523; private comm., 1974

Preparation of the title compound by interaction of dimethyl sulphoxide and bromomethane, sealed into a resin-coated glass bottle and heated at 65°C, led to an explosion after 120 h. The isolated salt thermally decomposes above 180°C to produce formaldehyde and a residue of methanesulphonic acid. In solution in dimethyl sulphoxide, the salt begins to decompose after several hours at 74–80°C (and after exposure to light), the exothermic reaction accelerating with vigorous evolution of vapour (including formaldehyde and dimethyl sulphide), the residue of methanesulphonic acid finally attaining 132°C. Some white solid (probably poly-formaldehyde) is also produced. The explosion seems likely to have been a pressure-burst of the container under excessive internal pressure from the decomposition products.

A safe procedure has been developed in which maximum reaction temperatures and times of 65°C and 55 h are used. Once prepared, the salt should not be redissolved in dimethyl sulphoxide. In contrast, trimethylsulphoxonium iodide appears simply to dissolve on heating rather than decomposing like the bromide.

## *N*-BROMOTRIMETHYLAMMONIUM BROMIDE(?) $C_3H_9Br_2N$

$BrN^+Me_3\ Br^-$

*See* BROMINE, $Br_2$: Trimethylamine

## TRIMETHYLSILICON PERCHLORATE    $Me_3SiClO_4$    $C_3H_9ClO_4Si$

*See* ORGANOSILICON PERCHLORATES

## † CHLOROTRIMETHYLSILANE    $ClMe_3Si$    $C_3H_9ClSi$

Preparative hazard

*See* CHLORINE, $Cl_2$: Antimony trichloride, Tetramethylsilane

Hexafluoroisopropylideneaminolithium

*See* HEXAFLUOROISOPROPYLIDENEAMINOLITHIUM, $C_3F_6LiN$: Non-metal halides

*See other* HALOSILANES

## TRIMETHYLGALLIUM    $Me_3Ga$    $C_3H_9Ga$

Sidgwick, 1950, 461

It ignites in air, even at −76°C, and reacts violently with water.

*See other* ALKYLMETALS

## TRIMETHYLINDIUM    $Me_3In$    $C_3H_9In$

1. Bailar, 1973, Vol. 1, 1116
2. Houben-Weyl, 1970, Vol. 13.4, 351

Trimethylindium and other lower alkyl-derivatives ignite in air [1], including bis-trimethylindium diethyl etherate [2].

*See other* ALKYLMETALS

## † ISOPROPYLAMINE    $Me_2CHNH_2$    $C_3H_9N$

1-Chloro-2,3-epoxypropane

*See* 1-CHLORO-2,3-EPOXYPROPANE, $C_3H_5ClO$: Isopropylamine

Perchloryl fluoride

*See* PERCHLORYL FLUORIDE, $ClFO_3$: Nitrogenous bases

## † PROPYLAMINE    $PrNH_2$    $C_3H_9N$

Triethynylaluminium

*See* TRIETHYNYLALUMINIUM, $C_6H_3Al$ : Dioxan, etc.

## † TRIMETHYLAMINE    $Me_3N$    $C_3H_9N$

*MCA SD-57,* 1955

Bromine
*See* BROMINE, $Br_2$ : Trimethylamine

Ethylene oxide
*See* ETHYLENE OXIDE, $C_2H_4O$ : Trimethylamine

**1-AMINO-2-PROPANOL** **$NH_2CH_2CHOHCH_3$** **$C_3H_9NO$**

Cellulose nitrate
*See* CELLULOSE NITRATE : Amines

**TRIMETHYLAMINE OXIDE** **$Me_3N{\rightarrow}O$** **$C_3H_9NO$**

Ringel, C., *Z. Chem.*, 1969, **9**, 188
A preparation, allowed to stand for a week rather than the specified day, exploded during concentration.

**TRIMETHYLPLATINUM(IV) AZIDE** **$Me_3PtN_3$** **$C_3H_9N_3Pt$**

*See* the tetramer, $C_{12}H_{36}N_{12}Pt_4$

**TRIMETHYL PHOSPHITE** **$(MeO)_3P$** **$C_3H_9O_3P$**

Magnesium diperchlorate
*See* MAGNESIUM DIPERCHLORATE, $Cl_2MgO_8$ : Trimethyl phosphite

Trimethylplatinum(IV) azide tetramer
*See* TRIMETHYLPLATINUM(IV) AZIDE TETRAMER, $C_{12}H_{36}N_{12}Pt_4$ : Trimethyl phosphite

**TITANIUM(III) METHOXIDE** **$Ti(OMe)_3$** **$C_3H_9O_3Ti$**

Bailar, 1973, Vol. 3, 384
A pyrophoric solid.
*See other* METAL ALKOXIDES

**TRIMETHYL PHOSPHATE** **$(MeO)_3P{\rightarrow}O$** **$C_3H_9O_4P$**

*ABCM Quart. Safety Summ.*, 1953, **25**, 3
The residue from a large atmospheric distillation of trimethyl phosphate exploded violently. This was attributed to rapid decomposition of the ester, catalysed by acidic degradation products, with evolution of gaseous hydrocarbons. It is recommended that only small batches of alkylphosphate should be vacuum distilled and in presence of magnesium oxide to neutralise any acid by-products.

**TRIMETHYLPHOSPHINE** $Me_3P$ $C_3H_9P$

Personal experience

May ignite in air.

*See other* ALKYLNON-METALS

**TRIMETHYLANTIMONY** $Me_3Sb$ $C_3H_9Sb$

Air,

or Halogens

1. von Schwartz, 1918, 322
2. Sidgwick, 1950, 777

Inflames in air [1] and reacts violently with halogens [2].

*See other* ALKYLMETALS

**TRIMETHYLTHALLIUM** $Me_3Tl$ $C_3H_9Tl$

Sidgwick, 1950, 463

Liable to explode violently above 90°C, and ignites in air.

*See other* ALKYLMETALS

**TRIMETHYLAMINE *N*-OXIDE PERCHLORATE** $C_3H_{10}ClNO_5$

$Me_3\overset{+}{N}(O)H\ ClO_4^-$

Hofmann, K. A. *et al.*, *Ber.*, 1910, **43**, 2624

Explodes on heating, or under a hammer blow.

*See related* OXOSALTS OF NITROGENOUS BASES

**† 1,2-DIAMINOPROPANE** $C_3H_{10}N_2$

$NH_2CH_2CHNH_2CH_3$

**† 1,3-DIAMINOPROPANE** $C_3H_{10}N_2$

$NH_2[CH_2]_3NH_2$

**TRIMETHYLPLATINUM HYDROXIDE** $C_3H_{10}OPt$

$Me_3PtOH$

Hoechstetter, M. N. *et al.*, *Inorg. Chem.*, 1969, 8, 400

The compound (originally reported as tetramethyl platinum), and usually obtained in admixture with an alkoxy derivative, detonates on heating.

*See other* PLATINUM COMPOUNDS

## TRIMETHYLSILYL HYDROPEROXIDE $C_3H_{10}O_2Si$

$Me_3SiOOH$

Buncel, E. *et al.*, *J. Chem. Soc.*, 1958, 1551
It decomposes rapidly above 35°C and may have been involved in an explosion during distillation of the corresponding peroxide, which is stable to 135°C.
*See* ORGANOMINERAL PEROXIDES

## TRIMETHYL(PHOSPHINO)GERMANE $Me_3GePH_2$ $C_3H_{11}GeP$

Oxygen
Dahl, A. R. *et al.*, *Inorg. Chem.*, 1975, **14**, 1093
A solution in chloroform at –45°C reacts rapidly with oxygen, and at oxygen pressures above 25 mbar combustion occurs. Dimethyldiphosphinogermane behaves similarly.
*See related* PHOSPHINE DERIVATIVES

## 2,4,6-TRIMETHYLPERHYDRO-2,4,6-TRIBORA-1,3,5-TRIAZINE (*B*-TRIMETHYLBORAZINE) $C_3H_{12}B_3N_3$

HNBMeNHBMeNHBMe (ring)

Nitryl chloride
*See* NITRYL CHLORIDE, $ClNO_2$ : *B*-Trimethylborazine

## 1,2-DIAMINOPROPANEAQUADIPEROXO-CHROMIUM(IV) $C_3H_{12}CrN_2O_5$

$[Cr(C_3H_{10}N_2)H_2O(O_2)_2]$

House, D. A. *et al.*, *Inorg. Chem.*, 1966, **5**, 840
Several preparations of the dihydrate exploded spontaneously at 20–25°C, and it explodes at 88–90°C during slow heating.
*See other* AMMINECHROMIUM PEROXOCOMPLEXES

## BERYLLIUM TETRAHYDROBORATE–TRIMETHYLAMINE $C_3H_{17}B_2BeN$

$Be[BH_4]_2 \cdot Me_3N$

Burg, A. B. *et al.*, *J. Amer. Chem. Soc.*, 1940, **62**, 3427
The complex ignites in contact with air or water.
*See other* COMPLEX HYDRIDES

## MESOXALONITRILE $CO(CN)_2$ $C_3N_2O$

Water

Martin, E. L., *Org. Synth.*, 1971, **51**, 70

The nitrile reacts explosively with water.

*See other* CYANO COMPOUNDS

## PHOSPHORUS TRICYANIDE $P(CN)_3$ $C_3N_3P$

1. Absalom, R. *et al.*, *Chem. & Ind.*, 1967, 1593
2. Smith, T. D. *et al.*, ibid., 1969
3. Mellor, 1971, Vol. 8, Suppl. 3, 583

During vacuum sublimation at *ca.* 100°C, explosions occurred on three occasions. These were attributed to slight leakage of air into the sublimer. The use of joints remote from the heating bath, a high melting grease, and a cooling liquid other than water (to avoid rapid evolution of hydrogen cyanide in the event of breakage) are recommended, as well as working in a fume cupboard [1].

However, the later publication suggests that the silver cyanide used in the preparation could be the cause of the explosions. On long keeping, it discolours and may then contain the nitride (fulminating silver). Silver cyanide precipitated from slightly acidic solution and stored in a dark bottle is suitable [2]. It ignites in air if touched with a warm glass rod, and reacts violently with water [3].

*See other* CYANO COMPOUNDS

## DICYANODIAZOMETHANE $(CN)_2CN_2$ $C_3N_4$

Ciganek, E., *J. Org. Chem.*, 1965, **30**, 4200

Small (mg) quantities melt at 75°C, but larger amounts explode. It also has borderline sensitivity towards static electricity and must be handled with full precautions.

*See other* CYANO COMPOUNDS
DIAZO COMPOUNDS

## DIAZIDOMALONONITRILE $(N_3)_2C(CN)_2$ $C_3N_8$

*MCA Case History No. 820*

An ethereal solution of *ca.* 100 g of the crude nitrile was allowed to spontaneously evaporate and crystallise. The crystalline slurry so produced exploded violently without warning. Previously such material had been found not to be shock-sensitive to hammer blows, but dry, recrystallised material was very shock-sensitive. Traces of free hydrogen azide could have been present, and a metal spatula had been used to stir the slurry.

*See other* CYANO COMPOUNDS
ORGANIC AZIDES

**2,4,6-TRIAZIDO-1,3,5-TRIAZINE** $C_3N_{12}$

$N{=}C(N_3)N{=}C(N_3)N{=}CN_3$ (cyclic)

Ott, E. *et al.*, *Ber.*, 1921, **54**, 183
Explodes on impact, shock or rapid heating to 170–180°C.
*See other* HIGH-NITROGEN COMPOUNDS
ORGANIC AZIDES

**† PROPADIEN-1,3-DIONE ('CARBON SUBOXIDE')** $C_3O_2$

**O=C=C=C=O**

**1,4-DIBROMO-1,3-BUTADIYNE** **BrC≡CC≡CBr** $C_4Br_2$

Pettersen, R. C. *et al.*, *Chem. Eng. News*, 1977, **55**(48), 36
An explosion temperature of 67°C has been published, but a solvent-free sample standing unheated under nitrogen exploded after an hour.
*See other* HALOACETYLENE DERIVATIVES

**CERIUM DICARBIDE** $CeC_4$ $C_4Ce$

*See* LANTHANUM DICARBIDE, $C_4La$

**1,4-DICHLORO-1,3-BUTADIYNE** **ClC≡CC≡CCl** $C_4Cl_2$

Sorbe, 1968, 50
It explodes above 70°C.
*See other* HALOACETYLENE DERIVATIVES

**MERCURY BIS(CHLOROACETYLIDE)** $Hg(C{\equiv}CCl)_2$ $C_4Cl_2Hg$

Whitmore, 1921, 119
It explodes fairly violently above the m.p., 185°C.
*See other* MERCURY COMPOUNDS
METAL ACETYLIDES

### 3,4-DICHLORO-2,5-DILITHIOTHIOPHENE $C_4Cl_2Li_2S$

$ClC{=}C(Li)SC(Li){=}CCl$

Gilman, H., private comm., 1971
The dry solid is explosive, though relatively insensitive to shock.
*See other* ORGANOLITHIUM REAGENTS

### 1,1,4,4-TETRACHLOROBUTATRIENE $Cl_2C{=}C{=}C{=}CCl_2$ $C_4Cl_4$

Heinrich, B. *et al.*, *Angew. Chem. (Intern. Ed.)*, 1968, 7, 375
One of the by-products formed by heating this compound is an explosive polymer.
*See other* HALOALKENES

### BISTRICHLOROACETYL PEROXIDE $Cl_3COOOCOCl_3$ $C_4Cl_6O_4$

1. Swern, 1971, Vol. 2, 815
2. Miller, W. T., *Chem. Abs.*, 1952, **46**, 7889e

Pure material explodes on standing at room temperature [1], but the very shock-sensitive solid may be stored safely in trichlorofluoromethane solution at –20°C [2].
*See other* DIACYL PEROXIDES

### TETRACARBON MONOFLUORIDE $C_4F$

Bailar, 1973, Vol. 1, 1271
Though generally inert, rapid heating causes deflagration.
*See other* HALOCARBONS
NON-METAL HALIDES

### 1,1,4,4-TETRAFLUOROBUTATRIENE $F_2C{=}C{=}C{=}CF_2$ $C_4F_4$

Martin, E. L. *et al.*, *J. Amer. Chem. Soc.*, 1959, **81**, 5256
In the liquid state, it explodes above –5°C.
*See other* HALOALKENES

### PERFLUOROBUTADIENE $F_2C{=}CFCF{=}CF_2$ $C_4F_6$

Bromine perchlorate
*See* BROMINE PERCHLORATE, $BrClO_4$: Perfluorobutadiene

### TRIFLUOROACETIC ANHYDRIDE $F_3CCOOCOCF_3$ $C_4F_6O_3$

Dimethyl sulphoxide
*See* DIMETHYL SULPHOXIDE, $C_2H_6OS$: Trifluoroacetic anhydride

## BISTRIFLUOROACETYL PEROXIDE $CF_3COOOCOCF_3$ $C_4F_6O_4$

Swern, 1971, Vol. 2, 815

Pure material explodes on standing at room temperature.

*See other* DIACYL PEROXIDES

## HEPTAFLUOROBUTYRYL NITRITE $C_4F_7NO_3$

$C_3F_7COONO$

Banks, R. E. *et al., J. Chem. Soc.* (*C*), 1966, 1351

This may be explosive like its lower homologues, and suitable precautions are necessary during heating or distillation.

*See other* ACYL NITRITES

## HEPTAFLUOROBUTYRYL HYPOFLUORITE $C_4F_8O_2$

$C_3F_7COOF$

Cady, G. H. *et al., J. Amer. Chem. Soc.*, 1954, **76**, 2020

The vapour slowly decomposes at ambient temperature, but will decompose explosively if initiated by a spark.

*See other* HYPOHALITES

## PERFLUORO-*tert*-NITROSOBUTANE $(F_3C)_3CNO$ $C_4F_9NO$

Nitrogen oxides

Sterlin, S. R. *et al.*, Russ. Pat. 482 432, 1975 (*Chem. Abs.*, 1975, **83**, 147118e)

The danger of explosion during oxidation of the nitroso compound to perfluoro-*tert*-butanol with nitrogen oxides, and subsequent hydrolysis, was reduced by working in a flow system at 160–210°C with 8–10% of nitrogen oxides in air, and using concentrated sulphuric acid for hydrolysis.

*See other* NITROSO COMPOUNDS

## PERFLUORO-2,5-DIAZAHEXANE 2,5-DIOXYL $C_4F_{10}N_2O_2$

$F_3C\dot{N}(O)C_2F_4\dot{N}OCF_3$

Preparative hazard

Banks, R. E. *et al., J. Chem. Soc., Perkin Trans. 1*, 1974, 2535–2536

During an increased-scale preparation of the dioxyl by permanganate oxidation of the hydrolysate of a nitrosotrifluoromethane–tetrafluoroethylene–phosphorus trichloride adduct, an impurity in the dioxyl, trapped out at –96°C (– 196°C?) and <2.5 mbar, caused a violent explosion to occur when the trap content was allowed to warm up. A procedure to eliminate the hazard is detailed.

**DI-(BISTRIFLUOROMETHYLPHOSPHIDO)-MERCURY** $[(CF_3)_2P]_2Hg$ $C_4F_{12}HgP_2$

Grobe, J. *et al., Angew. Chem. (Intern. Ed.),* 1972, **11**, 1098
It ignites in air.

**SODIUM TETRACARBONYLFERRATE(2–)** $C_4FeNa_2O_4$
$Na_2[Fe(CO)_4]$

Collman, J. P., *Acc. Chem. Res.*, 1975, **8**, 343
It is extremely oxygen-sensitive and ignites in air.
*See related* CARBONYL METALS
*See other* PYROPHORIC MATERIALS

**GERMANIUM ISOCYANATE** $Ge(NCO)_4$ $C_4GeN_4O_4$

Water
Johnson, D. H., *Chem. Rev.,* 1951, **48**, 287
The rate of (exothermic) hydrolysis becomes dangerously fast above 80°C.

**PERFLUORO-*tert*-BUTANOL** $(F_3C)_3COH$ $C_4HF_9O$

Preparative hazard
*See* PERFLUORO-*tert*-NITROSOBUTANE, $C_4F_9NO$: Nitrogen oxides

**1-IODO-1,3-BUTADIYNE** IC≡CC≡CH $C_4HI$

1. Schluhbach, H. H. *et al., Ann.,* 1951, **573**, 118, 120
2. Kloster-Jensen, E., *Tetrahedron,* 1966, **22**, 969

Crude material exploded violently at 35°C during attempted vacuum distillation, and temperatures below 30° are essential for safe handling [1]. A sample of pure material exploded on scratching under illumination [2].
*See other* HALOACETYLENE DERIVATIVES

† **1,3-BUTADIYNE** HC≡CC≡CH $C_4H_2$

1. Armitage, J. B., *et al. J. Chem. Soc.,* 1951, 44
2. Mushii, R. Y. *et al. Khim, Prom.,* 1963, 109
3. Moshkovich, F. B. *et al., Khim. Prom.*, 1965, **41**, 137–139
4. Anon., *Chem. Age*, 1951, **64**(1667), 955–958
5. Schilling, H. *et al.*, Ger. Pat. 860 212 (*Chem. Abs.*, 1956, **50**, 4510a)

Potentially very explosive, it may be handled and transferred by low-temperature distillation. It should be stored at –25°C to prevent decomposition and formation of explosive polymers [1]. The critical pressure for explosion is 0.04 bar, but presence of 15–40% of diluents (acetylene, ammonia, carbon dioxide or nitrogen) will raise the critical pressure to 0.92 bar [2]. Further data on attenuation by inert diluents of the explosive decomposition of the diyne are available [3].

During investigation of the cause of a violent explosion in a plant for separation of higher acetylenes, the most important finding was to keep the concentration of 1,3-butadiyne below 12% in its mixtures. Methanol is a practical diluent [4]. The use of butane (at 70 mol%) or other diluents to prevent explosion of 1,3-butadiyne when heated under pressure has been claimed [5]. It polymerises rapidly above 0°C.

*See other* ACETYLENIC COMPOUNDS
PEROXIDISABLE COMPOUNDS

## 1,2-DIBROMO-1,2-DIISOCYANATOETHANE POLYMERS

$(C_4H_2Br_2N_2O_2)_{2\text{ or }3}$

$(ONCCHBrCHBrCNO)_{2\text{ or }3}$

2-Phenyl-2-propyl hydroperoxide

*See* 2-PHENYL-2-PROPYL HYDROPEROXIDE, $C_9H_{12}O_2$: 1,2-Dibromo-1,2-diisocyanatoethane polymers

## 2-NITROTHIOPHENE-4-SULPHONYL CHLORIDE $C_4H_2ClNO_4S_2$

$\overline{CH=C(NO_2)SCH=C}SO_2Cl$

Libman, D. D., private comm., 1968

After distilling the chloride up to 147°C/6 mbar, the residue decomposed vigorously.

*See other* NITRO-ACYL HALIDES

## 1,1,2,3-TETRACHLORO-1,3-BUTADIENE $C_4H_2Cl_4$

$Cl_2C=CClCCl=CH_2$

Autoxidises to an unstable peroxide.

*See* 3,3,4,5-TETRACHLORO-3,6-DIHYDRO-1,2-DIOXIN, $C_4H_2Cl_4O_2$

## HEPTAFLUOROBUTYRAMIDE $C_3F_7CONH_2$ $C_4H_2F_7NO$

*See* LITHIUM TETRAHYDROALUMINATE, $AlH_4Li$:Fluoroamides

## IRON(II) MALEATE $C_4H_2FeO_4$

$O_2^-CCH=CHCO_2^-\ Fe^{2+}$

Schwab, R. F. *et al.*, *Chem. Eng. Prog.*, 1970, **66**(9), 53

The finely divided maleate, a by-product of phthalic anhydride manufacture, is subject to rapid aerial oxidation above 150°C, and has been involved in plant fires.

**POTASSIUM BIS(ETHYNYL)PALLADATE(2–)** $C_4H_2K_2Pd$

$K_2[Pd(C{\equiv}CH)_2]$

Immediately pyrophoric in air, explosive decomposition with aqueous reagents; the sodium salt is similar.

*See* COMPLEX ACETYLIDES

**POTASSIUM BIS(ETHYNYL)PLATINATE(2–)** $C_4H_2K_2Pt$

$K_2[Pt(C{\equiv}CH)_2]$

Pyrophoric in air, explosive decomposition with water; the sodium salt is similar.

*See* COMPLEX ACETYLIDES

**MANGANESE(II) BISACETYLIDE** $Mn(C{\equiv}CH)_2$ $C_4H_2Mn$

Bailar, 1973, Vol. 3, 855

A highly explosive compound.

*See other* METAL ACETYLIDES

**PYRIMIDINE-2,4,5,6-TETRAONE ('ALLOXAN')** $C_4H_2N_2O_4$

HNCONHCOCOCO (ring)


1. Wheeler, A. S. and Bogert, M. T., *J. Amer. Chem. Soc.*, 1910, **32**, 809
2. Gartner, R. A., ibid., 1911, **33**, 85
3. Franklin, E. C., ibid., 1901, **23**, 1362
4. Sorbe, 1968, 140


The tetraone slowly decomposes during storage at ambient temperature with generation of carbon dioxide. Two incidents involving bursting of bottles [1] and of pressure generation [2,3] were reported. It also explodes at temperatures above 170°C [4].

*See also* IMIDAZOLIDINE-2,4-DITHIONE ('DITHIOHYDANTOIN'), $C_3H_4N_2S_2$

**MALEIC ANHYDRIDE** OCCH=CHCOO (ring) $C_4H_2O_3$

*MCA SD-88*, 1962

Cations,

or Bases


1. Vogler, C. E. *et al., J. Chem. Eng. Data,* 1963, **8**, 620

2. Davie, W. R., *Chem. Eng. News.*, 1964, **42**(8), 41
3. *MCA Case History No. 622*
4. *MCA Case History No. 2032*

Maleic anhydride decomposes exothermically, evolving carbon dioxide, in the presence of alkali- or alkaline earth-metal or ammonium ions, dimethylamine, triethylamine, pyridine or quinoline, at temperatures above 150°C [1]. Sodium ions and pyridine are particularly effective, even at concentrations below 0.1%, and decomposition is rapid [2]. The Case History describes gas-rupture of a large insulated tank of semi-solid maleic anhydride which had been contaminated by sodium hydroxide. Use of additives to reduce sensitivity of the anhydride has been described [3].

Accidental transfer of an aqueous solution of sodium 2-benzothiazolethiolate ('sodium mercaptobenzthiazole') into a bulk storage tank of the anhydride led to eventual explosive destruction of the tank [4].

### MALEIC ANHYDRIDE OZONIDE $C_4H_2O_6$

OCOCOCHCHO–O$^+$–O$^-$

Briner, E. *et al.*, *Helv. Chim. Acta*, 1937, **20**, 1211

Explodes on warming to –40°C.

*See other* OZONIDES

### SILVER BUTEN-3-YNIDE AgC≡CCH=$CH_2$ $C_4H_3Ag$

Nitric acid

Willstatter, R. *et al.*, *Ber.*, 1913, **46**, 535–538

The silver salt, which deflagrates on heating, explodes if moistened with fuming nitric acid.

*See other* METAL ACETYLIDES

### 1-CHLORO-3,3-DIFLUORO-2-METHOXYCYCLOPROPENE $C_4H_3ClF_2O$

ClC=C(OMe)$CF_2$

*See* FLUORINATED CYCLOPROPENYL ETHERS

### 1,3,3-TRIFLUORO-2-METHOXYCYCLOPROPENE $C_4H_3F_3O$

FC=C(OMe)$CF_2$

*See* FLUORINATED CYCLOPROPENYL ETHERS

## POTASSIUM 4-METHYLFUROXAN-5-CARBOXYLATE $C_4H_3KN_2O_4$

$\overline{ON{=}C(CO_2K)C(Me){=}N}{\rightarrow}O$

Gasco, A. *et al.*, *Tetrahedron Lett.*, 1974, 630, footnote 2
The dried salt explodes violently on heating, impact or friction.

## † BUTEN-3-YNE $H_2C{=}CHC{\equiv}CH$ $C_4H_4$

1. Hennion, G. F. *et al.*, *Org. Synth.*, 1963, Coll. Vol. 4, 684
2. Rutledge, 1968, 28
3. Strizhevskii, I. I. *et al.*, *Chem. Abs.*, 1973, **78**, 98670b
4. Vervalin, 1973, 38–43

It must be stored out of contact with air to avoid formation of explosive compounds [1]. The sodium salt is a safe source of butenyne for synthetic use [2].

Explosive properties of the liquid and gaseous hydrocarbon alone, or diluted with decahydronaphthalene, were determined [3]. Mechanical failure in a compressor circulating the gaseous hydrocarbon through a reaction system led to local overheating and explosive decomposition, which propagated throughout the large plant [4].

1,3-Butadiene

1. Jarvis, H. C., *Chem. Eng. Progr.*, 1971, **67**(6), 41–44
2. Freeman, R. H. *et al.*, ibid., 45–51
3. Keister, R. G. *et al.*, *Loss Prevention*, 1971, **5**, 67–75
4. Carver, F. W. S. *et al.*, *Proc. 5th Symp. Chem. Process Haz.*, No. 39a, London, I.Ch.E., 1975

Unusual conditions in the main fractionation column separating product butadiene from by-product buten-3-yne (vinylacetylene, thought to be safe at below 50 mol% concentration) caused the concentration of the latter to approach 60% in part of the column as fractionation proceeded. Explosive decomposition, possibly initiated by an overheated unstable organic material derived from sodium nitrite, destroyed the column and adjacent plant [1,2]. Subsequent investigation showed that all mixtures of buten-3-yne and 1,3-butadiene can reproducibly be caused to react exothermically and then decompose explosively at appropriately high heating rates under pressure. Butadiene alone will behave similarly at higher heating rates and pressures [3]. Another study on the detailed mechanism and type of decomposition of $C_4$ hydrocarbons containing buten-3-yne showed that both were dependent on energy of initiation and time of application, as well as on the composition and phases present [4].

*See* SODIUM NITRITE, $NNaO_2$ : 1,3-Butadiene

Oxygen

Dolgopolskii, I. M. *et al.*, *Chem. Abs.*, 1958, **52**, 19904g

The rate of absorption of oxygen by liquid butenyne increased with time, and eventually a yellow liquid phase separated. After evaporation of excess hydrocarbon, the yellow peroxidic liquid was explosive. Presence of 5% of chloroprene increased the rate of absorption 5–6 times, and of 2% of water decreased it by 50%, but residues were explosive in each case.

*See other* ACETYLENIC COMPOUNDS
PEROXIDISABLE COMPOUNDS

## BARIUM 1,3-DI(5-TETRAZOLYL)TRIAZENIDE — $C_4H_4BaN_{22}$

$Ba(\overline{N{=}NNHN{=}C}N{=}NN\overline{C{=}NNHN{=}N})_2$

*See* 1,3-DI(5-TETRAZOLYL)TRIAZENE, $C_2H_3N_{11}$
*See other* HEAVY METAL DERIVATIVES

## *N*-BROMOSUCCINIMIDE — $C_4H_4BrNO_2$

$\overline{CH_2CONBrCOCH_2}$

Aniline,
or Diallyl sulphide,
or Hydrazine hydrate

*See* *N*-HALOIMIDES: Alcohols, etc.

Dibenzoyl peroxide,
*p*-Toluic acid

1. Tcheou, F. K. *et al.*, *J. Chinese Chem. Soc.*, 1950, **17**, 150–153
2. Mould, R. W., private comm., 1975

Thirtyfold increase in scale of a published method [1] for radical-initiated side-chain bromination of the acid in carbon tetrachloride led to violent reflux and eruption of flask contents through the condenser [2].

*See other* HALOGENATION INCIDENTS
*N*-HALOGEN COMPOUNDS

## *N*-CHLOROSUCCINIMIDE — $C_4H_4ClNO_2$

$\overline{CH_2CONClCOCH_2}$

Alcohols,
or Benzylamine

*See* *N*-HALOIMIDES: Alcohols, etc.

Dust

Boscott, R. J., private comm., 1968
Smouldering in a stored drum of the chloroimide was attributed to dust contamination.

**1,4-DICHLORO-2-BUTYNE** $ClCH_2C{\equiv}CCH_2Cl$ $C_4H_4Cl_2$

Ford, M. *et al., Chem. Eng. News,* 1972, **50**(3), 67
Herberz, T., *Chem. Ber.*, 1952, **85**, 475–482
Preparation by conversion of the diol in pyridine with neat thionyl chloride is difficult to control and hazardous on a large scale. Use of dichloromethane as diluent and operation at –30°C renders the preparation reproducible and safer.

During distillation at up to 110°C/7–8 mbar, slight overheating of the residue to 120°C caused explosive decomposition.
*See other* HALOACETYLENE DERIVATIVES

**3,3,4,5-TETRACHLORO-3,6-DIHYDRO-1,2-DIOXIN** $C_4H_4Cl_4O_2$

$OOCCl_2CHClCHClCH_2$ (ring)

Akopyan, A. N. *et al., Chem. Abs.*, 1975, **82**, 111537t
This autoxidation product of 1,1,2,3-tetrachloro-1,3-butadiene exploded during attempted vacuum distillation.
*See other* CYCLIC PEROXIDES

**COPPER(II)1,3-DI(5-TETRAZOLYL)TRIAZENIDE** $C_4H_4CuN_{22}$

$Cu(N{=}NNHN{=}CN{=}NNC{=}NNHN{=}N)_2$ (rings)

*See* 1,3-DI(5-TETRAZOLYL)TRIAZENE, $C_2H_3N_{11}$
*See other* HEAVY METAL DERIVATIVES
*N*-METAL DERIVATIVES

**TETRAFLUOROSUCCINAMIDE** $(CF_2CONH_2)_2$ $C_4H_4F_4N_2O_2$

*See* LITHIUM TETRAHYDROALUMINATE, $AlH_4Li$: Fluoroamides

**POLY[BIS(TRIFLUOROETHOXY)PHOSPHAZENE]** $(C_4H_4F_6NO_2P)_n$

$[(F_3CCH_2O)_2P{=}N]_n$

Allcock, H. R. *et al., Macromolecules*, 1974, 7, 284–290
Sealed tubes containing the linear P=N polymer exploded after heating

at 300°C for 24–30 h (or after shorter times at higher temperatures), owing to pressure build-up from formation of cyclic tri- and tetra-mers.

**2-CYANO-1,2,3-TRIS(DIFLUOROAMINO)PROPANE** $C_4H_4F_6N_4$

$NF_2CCN(CH_2NF_2)_2$

Reed, S. F., *J. Org. Chem.*, 1968, **33**, 1861

The 95% pure product is shock-sensitive.

*See other* DIFLUOROAMINO COMPOUNDS

**SUCCINOYL DIAZIDE** $(CH_2CON_3)_2$ $C_4H_4N_6O_2$

1. Maclaren, J. A., *Chem. & Ind.*, 1971, 395
2. France, A. D. G. *et al.*, *Chem. & Ind.*, 1962, 2065

This intermediate for preparation of ethylene diisocyanate exploded violently during isolation [1]. Previously the almost dry solid had exploded violently on stirring with a spatula [2].

*See other* ACYL AZIDES

**† FURAN** $\overline{CH{=}CHOCH{=}CH}$ $C_4H_4O$

**DIKETENE** $CH_2{=}\overline{CCH_2COO}$ $C_4H_4O_2$

1. Vervalin, 1973, 86
2. Zdenek, F. *et al.*, Czech. Pat. 156 584, 1975 (*Chem. Abs.*, 1976, **85**, 46390k)

Diketene residues awaiting incineration in a tank trailer decomposed violently on standing, and blew off the dome cover and ignited [1]. The risk of autoignition during the exothermic dimerisation of diketene to dehydroacetic acid is eliminated by operating in a non-flammable solvent [2].

*See other* POLYMERISATION INCIDENTS

Acids,
or Bases,
or Sodium acetate

1. *Haz. Chem. Data,* 1969, 99
2. Laboratory Chemical Disposal Co. Ltd., confidential information, 1968

Presence of mineral, or Lewis acids, or bases including amines, will catalyse violent polymerisation of this very reactive dimer, accompanied by gas evolution [1]. Sodium acetate is sufficiently basic to cause violent polymerisation at 0.1% concentration when added to diketene at 60°C [2].

## METHYL PROPIOLATE    $HC{\equiv}CCO_2Me$    $C_4H_4O_2$

Octakis(trifluorophosphine)rhodium

*See* OCTAKIS(TRIFLUOROPHOSPHINE)DIRHODIUM, $F_{24}P_8Rh_2$: Acetylenic esters

## DIHYDROXYMALEIC ACID    $C_4H_4O_6$

$HO_2CC(OH){=}C(OH)CO_2H$

Axelrod, B., *Chem. Eng. News,* 1955, **33**, 3024

A glass ampoule of the acid exploded during storage at ambient temperature. It is unstable and slowly loses carbon dioxide, even at 4°C.

## † THIOPHENE    $\overline{CH{=}CHSCH{=}CH}$    $C_4H_4S$

Nitric acid

*See* NITRIC ACID, $HNO_3$: Thiophene

*N*-Nitrosoacetanilide

*See* *N*-NITROSOACETANILIDE, $C_8H_8N_2O_2$: Thiophene

## ETHYNYL VINYL SELENIDE    $HC{\equiv}CSeCH{=}CH_2$    $C_4H_4Se$

Brandsma, L. *et al., Rec. Trav. Chim.,* 1962, **81**, 539

Distillable at 30–40°C/80 mbar, it explosively decomposes on heating at 1 bar.

*See other* ACETYLENIC COMPOUNDS

## † 2-CHLORO-1,3-BUTADIENE ('CHLOROPRENE')    $C_4H_5Cl$

$H_2C{=}CClCH{=}CH_2$

Air

1. Bailey, H. C., *Oxidation of Organic Compounds-I, ACS No. 75* (Gould, R. F., Editor), 138–149, Washington, American Chemical Society, 1968
2. Bailey, H. C., private comm., 1974

'Chloroprene' will autoxidise very rapidly, and even at 0°C it produces an unstable peroxide (a mixed 1,2- and 1,4-addition copolymer with oxygen), which effectively will catalyse exothermic polymerisation of the monomer. The kinetics of autoxidation have been studied [1]. It forms 'popcorn polymer' at a greater rate than does butadiene [2].

*See other* DIENES
HALOALKENES

## *N*-CHLORO-4-METHYL-2-IMIDAZOLINONE $C_4H_5ClN_2O$

ClNCONHCMe=CH

Walles, W. E., US Pat. 3 850 920, 1974

It exploded after several hours at ambient temperature.

*See other* *N*-HALOGEN COMPOUNDS

## 4-CHLORO-1-METHYLIMIDAZOLIUM NITRATE $C_4H_5ClN_3O_3$

$ClC{=}CHN^+(Me)CH{=}N\ NO_3^-$

Personal experience

Towards the end of evaporation of a dilute aqueous solution, the residue decomposed violently. Nitrate salts of many organic bases are unstable and should be avoided.

*See other* OXOSALTS OF NITROGENOUS BASES

## 1-CHLORO-1-BUTEN-3-ONE $ClCH{=}CHCOCH_3$ $C_4H_5ClO$

Pohland, A. E. *et al., Chem. Rev.*, 1966, **66**, 164

If prepared from vinyl chloride, this unstable ketone will decompose almost explosively within one day, apparently owing to the presence of some *cis*-isomer.

## 1-FLUORO-1,1-DINITRO-2-BUTENE $C_4H_5FN_2O_4$

$FC(NO_2)_2CH{=}CHMe$

*See* FLUORODINITROCOMPOUNDS (reference 4)

## BIS(2-FLUORO-2,2-DINITROETHYL)AMINE $C_4H_5F_2N_5O_8$

$(FC(NO_2)_2CH_2)_2NH$

Gilligan, W. H., *J. Org. Chem.*, 1972, **37**, 3947

The amine and several derived amides are explosives and need appropriate care in handling.

*See other* FLUORODINITROMETHYL COMPOUNDS

## † ETHYL TRIFLUOROACETATE $CF_3CO_2Et$ $C_4H_5F_3O_2$

**POTASSIUM HYDROGENTARTRATE** $C_4H_5KO_6$

$KO_2CCHOHCHOHCO_2H$

Carbon,
Nitrogen oxide

*See* NITROGEN OXIDE ('NITRIC OXIDE'), NO: Carbon, etc.

**† 1-CYANOPROPENE** $CNCH{=}CHCH_3$ $C_4H_5N$

**† 3-CYANOPROPENE** $CH_2{=}CHCH_2CN$ $C_4H_5N$

**1-CYANO-2-PROPEN-1-OL** $CNCHOHCH{=}CH_2$ $C_4H_5NO$

Unpublished information
It shows a strong tendency to exothermic polymerisation of explosive violence in presence of light and air at temperatures above 25–30°C. Presence of an inhibitor and inert atmosphere are essential for stable storage.
*See other* CYANO COMPOUNDS

**ALLYL ISOTHIOCYANATE** $CH_2{=}CHCH_2NCS$ $C_4H_5NS$

Anon., *Ind. Eng. Chem.(News Ed.)*, 1941, **19**, 1408
A routine preparation by interaction of allyl chloride and sodium thiocyanate in an autoclave at 5.5 bar exploded violently at the end of the reaction. Peroxides were not present or involved and no other cause could be found, but extensive decomposition occurred when allyl isothiocyanate was heated to 250°C in glass ampoules.
*See other* ALLYL COMPOUNDS

**2-METHYL-5-NITROIMIDAZOLE** $C_4H_5N_3O_2$

$\overline{HNCMe{=}NCH{=}C}NO_2$

Preparative hazard
Zmojdzin, A. *et al.*, Brit. Pat. 1 418 538, 1974
2-Methylimidazole is difficult to nitrate, and use of conventional reagents under forcing conditions (excess sulphuric/nitric acids, high temperature) involves a high risk of cleavage and violent runaway oxidation. A new and safe process involves the use of nitration liquor from a previous batch (and optional use of excess nitric acid), both of which moderate the nitration reaction, in conjunction with balancing amounts of sulphuric acid, which tends to accelerate the nitration of

2-methylimidazole. Accurate control of the highly exothermic reaction is readily effected.

*See other* NITRATION INCIDENTS

Nitric acid,
Sulphuric acid

Zmojdzin, A. *et al.*, Fr. Demande 2 220 523, 1974 (*Chem. Abs.*, 1975, **82**, 156312g)

The danger of explosion during further nitration with nitrating acid is eliminated by addition of excess nitric acid as the reaction proceeds.

*See* NITRATION

**1-HYDROXYIMIDAZOLE-2-CARBOXALDOXIME 3-OXIDE** $C_4H_5N_3O_3$

**HONC(CH=NOH)=N(O)CH=CH** (ring)

Hayes, K., *J. Heterocycl. Chem.*, 1974, **11**, 615

It is an explosive solid, containing three types of N–O bonds.

*See other* OXIMES

**SODIUM ETHOXYACETYLIDE** **NaC≡COEt** $C_4H_5NaO$

1. Jones, E. R. H. *et al., Org. Synth,* 1963, Coll. Vol. 4, 405
2. Brandsma. 1971, 120

This sodium derivative is extremely pyrophoric, apparently even at −70°C [1]. The solid may explode after prolonged contact with air [2].

*See other* METAL ACETYLIDES

**† 1,2-BUTADIENE** $CH_2{=}C{=}CHCH_3$ $C_4H_6$

**† 1,3-BUTADIENE** $CH_2{=}CHCH{=}CH_2$ $C_4H_6$

*MCA SD-55,* 1954

Alone,
or Air

1. Scott, D. A., *Chem. Eng. News,* 1940, **18**, 404
2. Hendry, D. G. *et al., Ind. Eng. Chem. Prod. Res. Dev.*, 1968, **7**, 136, 145
3. Mayo, F. R. *et al., Prog. Rep.* No. 40, Sept. 1971, Stamford Res. Inst. Project PRC-6217
4. Keister, R. G. *et al., Loss Prevention*, 1971, **5**, 69
5. Penkina, O. M. *et al., Chem. Abs.*, 1975, **83**, 29457d

6. Bailey, H. C., private comm., 1974
7. Miller, G. H. *et al.*, *J. Polymer Sci. C*, 1964, 1109–1115
8. Vervalin, 1964, 335–338, 358–359
9. Harmon, 1974, 2-5–2-6

When heated under pressure, butadiene may violently decompose thermally.

Solid butadiene at below −113°C will absorb enough oxygen at subatmospheric pressure to make it explode violently when allowed to melt. The peroxides formed on long contact with air are explosives sensitive to heat or shock, but may also initiate polymerisation [1].

The hazards associated with peroxidation of butadiene are closely related to the fact that the polyperoxide is insoluble in butadiene and progressively separates. If local concentrations build up, self-heating from the initially slow, spontaneous decomposition begins, and when a large enough mass of peroxide accumulates, explosion occurs. Critical mass at 27°C is a 9 cm sphere; size decreases rapidly with increasing temperature [2].

Although isoprene and styrene also readily peroxidise, the peroxides are soluble in the monomers, and the degree of hazard is correspondingly less [3].

*See also* POLY(BUTADIENE PEROXIDE, $(C_4H_6O_2)_n$: Butadiene
2-METHYL-1,3-BUTADIENE (ISOPRENE), $C_5H_8$: Air

Butadiene will decompose explosively if heated under pressure at 30–40°C/min to exceed critical temperatures of 200–340°C and pressures of 1–1.2 kbar simultaneously [4]. A more recent study of the explosion properties of butadiene at elevated temperatures and pressures has been published (title only translated) [5]. Phenolic antioxidants (e.g. *tert*-butylcatechol at 0.02% wt.) are effective in stabilising butadiene against autoxidation in clean storage at moderate ambient temperatures. The presence of rust and/or water in steel storage cylinders rapidly consumes the antioxidant [6]. In prolonged storage, butadiene (even when very pure and in sealed glass containers) will produce 'popcorn' polymer [7]. Growth of this involves continued diffusion of monomer into an existing polymeric matrix which continues to increase in bulk. The expansion may eventually rupture the storage container [6].

*See* POLY(1,3-BUTADIENE PEROXIDE), $(C_4H_6O_2)_n$

Case histories of two industrial explosions involving peroxide formation with air have been detailed [8]. The literature relating to violent polymerisation has been reviewed [9].

*See* VIOLENT POLYMERISATION

Aluminium tris-tetrahydroborate

*See* ALUMINIUM TRIS-TETRAHYDROBORATE, $AlB_3H_{12}$: Alkenes, etc.

Boron trifluoride etherate,
Phenol

*MCA Case History No. 790*

The hydrocarbon–phenol reaction, catalysed by the etherate, was being run in petroleum ether solution in a sealed pressure bottle. The bottle burst, possibly owing to exothermic polymerisation of the diene.

*See* BUTADIENE PEROXIDE, $C_4H_6O_2$: Butadiene

Buten-3-yne

*See* BUTEN-3-YNE, $C_4H_4$: 1,3-Butadiene

Cobalt

Miller, G. H. *et al.*, *J. Polymer Sci. C*, 1964, 1109–1115

Butadiene vapour in contact with cobalt metal will initiate 'popcorn' polymerisation of the diene.

*See* Text to reference 7 above

Crotonaldehyde

*See* CROTONALDEHYDE, $C_4H_6O$: Butadiene

Ethanol,
Iodine,
Mercury oxide

*See* 2-ETHOXY-1-IODO-3-BUTENE, $C_6H_{11}IO$

Oxides of nitrogen,
Oxygen

Vervalin, 1973, 63–65

An explosion and fire occurred in the pipework of a vessel in which dilute butadiene was stored under 'inert' atmosphere, generated by combustion of fuel gas in a limited supply of air. The 'inert' gas, which contained up to 1.8% oxygen and traces of oxides of nitrogen, reacted over an extended period in the vapour phase to produce concentrations of gummy material containing up to 64% of butadiene peroxide and 4.2% of a butadiene–nitrogen oxide complex. The deposits eventually decomposed explosively.

*See* NITROGEN OXIDE ('NITRIC OXIDE'), NO: Dienes, Oxygen

Oxygen

*MCA Case History No. 303*

A leaking valve allowed butadiene to accumulate in a pipeline exposed to an inerting gas containing up to 2% of oxygen. Peroxide formed and initiated 'popcorn' polymerisation which burst the pipeline.

*See* Reference 6, above

Sodium nitrite

*See* SODIUM NITRITE, $NNaO_2$ : Butadiene
*See other* DIENES

**† 1-BUTYNE** $HC{\equiv}CC_2H_5$ $C_4H_6$

**† 2-BUTYNE** $CH_3C{\equiv}CCH_3$ $C_4H_6$

**† CYCLOBUTENE** $\overline{CH_2CH{=}CHCH_2}$ $C_4H_6$

***cis*-POLYBUTADIENE** $(CH_2CH{=}CHCH_2)_n$ $(C_4H_6)_n$

Sedov, V. V. *et al.*, *Chem. Abs.*, 1976, **85**, 109747e

Stereoregular *cis*-polybutadiene compositions may explode when heated at 337–427°C/1–0.01μbar, presumably owing to cyclisation.

*See other* DIENES

***N*-CHLORO-5-METHYL-2-OXAZOLIDINONE** $C_4H_6ClNO_2$

$\overline{OCONClCH_2CHMe}$

Walles, W. E., US Pat. 3 850 920, 1974

It exploded at 160°C.

*See other* *N*-HALOGEN COMPOUNDS

***N*-CHLORO-3-MORPHOLINONE** $C_4H_6ClNO_2$

$Cl\overline{NCH_2COOCH_2CH_2}$

Walles, W. E., US Pat. 3 850 920, 1974

It exploded at 115°C.

*See other* *N*-HALOGEN COMPOUNDS

**ACETYL 1,1-DICHLOROETHYL PEROXIDE** $C_4H_6Cl_2O_3$

$AcOOCCl_2Me$

Griesbaum, K. *et al.*, *J. Amer. Chem. Soc.*, 1976, **98**, 2880

A friction- and heat-sensitive viscous liquid, it exploded violently during attempted injection into a gas chromatograph.

*See related* DIACYL PEROXIDES

**3,6-DICHLORO-3,6-DIMETHYLTETRAOXANE** $C_4H_6Cl_2O_4$

$ClMe\overline{COOCClMeOO}$

Griesbaum, K. *et al.*, *J. Amer. Chem. Soc.*, 1976, **98**, 2880

Extremely sensitive to shock, heat or minor friction, it has exploded very violently when touched with a glass rod or spatula.

*See other* CYCLIC PEROXIDES

**CHROMIUM DIACETATE** $Cr(OCOCH_3)_2$ $C_4H_6CrO_4$

Ocone, L. R. *et al., Inorg. Synth,* 1966, 8, 129–131

The anhydrous salt is pyrophoric in air, or slowly chars in lower oxygen concentrations.

*See also* BIS(2,4-PENTANEDIONATO) CHROMIUM, $C_{10}H_{14}CrO_4$

**CHROMYL ACETATE** $CrO_2(OAc)_2$ $C_4H_6CrO_6$

Preparative hazard

*See* CHROMIUM TRIOXIDE, $CrO_3$: Acetic anhydride

**1,2-BIS(DIFLUOROAMINO)ETHYL VINYL ETHER** $C_4H_6F_4N_2O$

$NF_2CH_2CH(NF_2)OCH=CH_2$

Reed, S. F., *J. Org. Chem.,* 1967, **32**, 2894

It is slightly less impact-sensitive than glyceryl nitrate.

*See other* DIFLUOROAMINO COMPOUNDS

**DI-1,2-BIS(DIFLUOROAMINO)ETHYL ETHER** $C_4H_6F_8N_4O$

$(NF_2CH_2CHNF_2)_2O$

Reed, S. F., *J. Org. Chem.,* 1967, **32**, 2894

It is slightly more impact-sensitive than glyceryl nitrate.

*See other N*-HALOGEN COMPOUNDS

**DIVINYLMAGNESIUM** $(H_2C=CH)_2Mg$ $C_4H_6Mg$

Houben-Weyl, Vol. 13.2a, 203

The solid may ignite in air.

*See related* ALKYLMETALS

**ETHYL DIAZOACETATE** $N_2CHCO_2Et$ $C_4H_6N_2O_2$

Searle, N. E., *Org. Synth.,* 1963, Coll. Vol. 4, 426

It is explosive, and distillation, even under reduced pressure as described, may be dangerous.

*See other* DIAZO COMPOUNDS

**3-ETHYL-4-HYDROXY-1,2,5-OXADIAZOLE** $C_4H_6N_2O_2$

EtC=NON=COH

Sodium hydroxide

Barker, M. D., *Chem. & Ind.,* 1971, 1234

The sodium salt obtained by vacuum evaporation at 50°C of an aqueous alcoholic mixture of the above ingredients exploded violently when disturbed.

**2-CYANO-2-PROPYL NITRATE** **$Me_2C(CN)ONO_2$** $C_4H_6N_2O_3$

Freeman, J. P., *Org. Synth.*, 1963, **43**, 85

It is a moderately explosive material of moderate impact-sensitivity.

*See* NITRATING AGENTS

*See other* CYANO COMPOUNDS

**DIMETHYL AZODIFORMATE** $C_4H_6N_2O_4$

**$MeCO_2N{=}NCO_2Me$**

1. US Pat. 3 347 845, 1967
2. Kauer, J. C., *Org. Synth.*, 1963, Coll. Vol. 4, 412

Shock-sensitive, burns explosively.

*See* DIETHYL AZODIFORMATE, $C_6H_{10}N_2O_4$

*See other* AZO COMPOUNDS

**1,1-DINITRO-3-BUTENE** $C_4H_6N_2O_4$

**$(NO_2)_2CHCH_2CH{=}CH_2$**

Witucki, E. F. *et al., J. Org. Chem.*, 1972, **37**, 152

A fume-off during distillation of this compound (b.p., 63–65°C/2 mbar) illustrates the inherent instability of this type of compound.

*See also* 2,3-DINITRO-2-BUTENE, $C_4H_6N_2O_4$
1-NITRO-3-BUTENE, $C_4H_7NO_2$

**2,3-DINITRO-2-BUTENE** $C_4H_6N_2O_4$

**$CH_3CH(NO_2){=}CHNO_2CH_3$**

Bisgrove, D. E. *et al., Org. Synth*, 1963, Coll. Vol. 4, 374

Only one explosion has been recorded during vacuum distillation at 135°C/14 mbar.

*See other* POLYNITROALKYL COMPOUNDS

**DIACETATOPLATINUM(II) NITRATE** $C_4H_6N_2O_{10}Pt$

**$[Pt(OAc)_2][NO_3]_2$**

Preparative hazard

*See* NITRIC ACID, $HNO_3$ : Acetic acid, etc.

## 2-METHYL-5-VINYLTETRAZOLE $C_4H_6N_4$

$MeNN{=}NC(C_2H_3){=}N$ (ring)

Aluminium hydride

*See* ALUMINIUM HYDRIDE, $AlH_3$: Tetrazole derivatives

*See other* TETRAZOLES

## *N,N'*-DIMETHYL-*N,N'*-DINITROSOOXAMIDE $C_4H_6N_4O_4$

$[MeN(NO)CO]_2$

Preussmann, R., *Angew. Chem.*, 1963, **75**, 642

Removal of the solvent carbon tetrachloride (in which nitrosation had been effected) at atmospheric, rather than reduced, pressure caused a violent explosion at the end of distillation. Lowest possible temperatures should be maintained in the preparation. *N*-Nitroso-*p*-toluenesulphonmethylamide seems preferable as a source of diazomethane.

*See* DIAZOMETHANE, $CH_2N_2$

*See other* NITROSO COMPOUNDS

## *N,N'*-DIMETHYL-*N,N'*-DINITROOXAMIDE $C_4H_6N_4O_6$

$MeN(NO_2)COCON(NO_2)Me$

Allenby, O. C. W. *et al.*, *Can. J. Res.*, 1947, **25B**, 295–300

It is an explosive.

*See other* *N*-NITRO COMPOUNDS

## † 1-BUTEN-3-ONE $H_2C{=}CHCOCH_3$ $C_4H_6O$

*Haz. Chem. Data*, 1975, 207

The uninhibited monomer polymerises on exposure to heat or sunlight. The inhibited monomer may also polymerise if heated sufficiently (by exposure to fire) and lead to rupture of the containing vessel.

*See other* POLYMERISATION INCIDENTS

## † CROTONALDEHYDE $CH_3CH{=}CHCHO$ $C_4H_6O$

Butadiene

1. Greenlee, K. W., *Chem. Eng. News*, 1948, **26**, 1985
2. Hanson, E. S., ibid., 2551
3. Watson, K. M., *Ind. Eng. Chem.*, 1943, **35**, 398

An autoclave without a bursting disc and containing the two poorly mixed reactants was wrecked by a violent explosion which occurred on heating the autoclave to 180°C [1]. This was attributed to not allowing sufficient free space for liquid expansion to occur [2]. The need to calculate separate reactant volumes under reaction conditions for autoclave preparations is stressed [3].

*See* 1,3-BUTADIENE, $C_4H_6$: Alone, etc.

Nitric acid

*See* NITRIC ACID, $HNO_3$: Crotonaldehyde

*See other* PEROXIDISABLE COMPOUNDS

**DIMETHYLKETENE** **$Me_2C{=}C{=}O$** **$C_4H_6O$**

Air

1. Smith, G. W. *et al., Org. Synth.*, 1964, Coll. Vol. 4, 348
2. Staudinger, H., *Ber.*, 1925, **58**, 1079

Dimethylketene rapidly forms an extremely explosive peroxide when exposed to air at ambient temperatures. Drops of solution allowed to evaporate may explode. Inert atmosphere should be maintained above the monomer [1]. The peroxide is polymeric and very sensitive, exploding on friction at –80°C. Higher homologues are very unstable and unisolable [2].

*See other* POLYPEROXIDES

**† DIVINYL ETHER** **$(CH_2{=}CH)_2O$** **$C_4H_6O$**

Anon., *Chemist & Druggist*, 1947, **157**, 258

The presence of *N*-phenyl-1-naphthylamine as inhibitor considerably reduces the development of peroxide in the ether.

Nitric acid

*See* NITRIC ACID, $HNO_3$: Divinyl ether

*See other* PEROXIDISABLE COMPOUNDS

**† 3,4-EPOXYBUTENE** **$CH_2{=}CH\overline{CHOCH_2}$** **$C_4H_6O$**

**† ETHOXYACETYLENE** **$EtOC{\equiv}CH$** **$C_4H_6O$**

Jacobs, T. L. *et al., J. Amer. Chem. Soc.*, 1942, **64**, 223

Small samples rapidly heated in sealed tubes to around 100°C exploded.

Ethylmagnesium iodide

*See* ETHYLMAGNESIUM IODIDE, $C_2H_5IMg$: Ethoxyacetylene

*See other* ACETYLENIC COMPOUNDS

**† METHACRYLALDEHYDE** $CH_2=C(Me)CHO$ $C_4H_6O$

**3-METHOXYPROPYNE** $HC\equiv CCH_2OMe$ $C_4H_6O$

Brandsma, 1971, 13

It explodes on distillation at its atmospheric b.p., 61°C.

*See other* ACETYLENIC COMPOUNDS

**† ALLYL FORMATE** $HCO_2CH_2CH=CH_2$ $C_4H_6O_2$

**† BUTANE-2,3-DIONE** $CH_3COCOCH_3$ $C_4H_6O_2$

**2-BUTYNE-1,4-DIOL** $HOCH_2C\equiv CCH_2OH$ $C_4H_6O_2$

Alkalies,
or Halide salts,
or Mercury salts

Kirk-Othmer, 1960, Suppl. Vol. 2 (to 1st Ed.), 45

The pure diol may be distilled unchanged, but traces of alkali or alkaline earth hydroxides or halides may cause explosive decomposition during distillation. In presence of strong acids, mercury salts may cause violent decomposition of the diol.

*See other* ACETYLENIC COMPOUNDS

**BUTYROLACTONE** $\overline{O[CH_2]_3CO}$ $C_4H_6O_2$

Butanol,
2,4-Dichlorophenol,
Sodium hydroxide

*CISHC Chem. Safety Summ.*, 1977, **48**, 3

In an altered process to prepare 2,4-dichlorophenoxybutyric acid, the lactone was added to the other components, and soon after, the reaction temperature reached 165°C, higher than the usual 160°C. Application of cooling failed to check the thermal runaway, and soon after reaching 180°C the vessel began to fail and an explosion and fire occurred.

**† METHYL ACRYLATE** $CH_2=CHCO_2Me$ $C_4H_6O_2$

*MCA SD-79*, 1960

1. *MCA Case History No. 2033*
2. Harmon, 1974, 2.11

A 4 l glass bottle of the ester (sealed and partly polymerised old stock) exploded several hours after being brought from storage into a laboratory. An inhibitor had originally been present, but could have been consumed during prolonged storage, and peroxides may have initiated exothermic polymerisation of the remaining monomer [1]. The monomer is normally stored and handled inhibited and at below 10°C, but not under an inert atmosphere, because oxygen is essential to the inhibition process [2].

*See* VIOLENT POLYMERISATION

*See other* POLYMERISATION INCIDENTS

**† VINYL ACETATE** **$CH_3CO_2CH{=}CH_2$** **$C_4H_6O_2$**

*MCA SD-75*, 1970

Harmon, 1974, 2.19

The monomer is volatile and tends to self-polymerise, and is therefore stored and handled cool and inhibited, with storage limited to below 6 months. Several industrial explosions have been reported.

Air,
Water

*MCA SD-75*, 1970

Vinyl acetate is normally inhibited with hydroquinone to prevent polymerisation. A combination of too low a level of inhibitor and warm, moist storage conditions may lead to spontaneous polymerisation. This process involves autoxidation of acetaldehyde (a normal impurity produced by hydrolysis of the monomer) to a peroxide, which initiates exothermic polymerisation as it decomposes. In bulk, this may accelerate to a dangerous extent. Other peroxides or radical sources will initiate the exothermic polymerisation.

Desiccants

*MCA SD-75*, 1970

Vinyl acetate vapour may react vigorously in contact with silica gel or alumina.

Dibenzoyl peroxide,
Ethyl acetate

Vervalin, 1973, 81

Polymerisation of the ester with dibenzoyl peroxide in ethyl acetate accelerated out of control and led to discharge of a large volume of vapour which ignited and exploded.

*See other* POLYMERISATION INCIDENTS

Hydrogen peroxide

*See* HYDROGEN PEROXIDE, $H_2O_2$: Vinyl acetate

Oxygen

Barnes, C. E. *et al., J. Amer. Chem. Soc.*, 1950, 72, 210

The unstabilised polymer exposed to oxygen at 50°C generated an ester–oxygen interpolymeric peroxide which, when isolated, exploded vigorously on gentle heating.

*See other* POLYPEROXIDES

Ozone

*See* VINYL ACETATE OZONIDE, $C_4H_6O_5$

Toluene

*MCA Case History No. 2087*

The initial exotherm of solution polymerisation of the ester in boiling toluene in a 10 m$^3$ reactor was too great for the cooling and vent systems, and the reaction began to accelerate out of control. Failure of a gasket released a quantity of the flammable reaction mixture which became ignited and destroyed the containing building.

*See other* POLYMERISATION INCIDENTS
PEROXIDISABLE COMPOUNDS

## POLY(1,3-BUTADIENE PEROXIDE) $(C_4H_6O_2)_n$

**$(HOOCH{=}CHCH{=}CH_2)_n$, etc.**

Butadiene

1. Alexander, D. S., *Ind. Eng. Chem.*, 1959, **51**, 733
2. Hendry, D. G. *et al., Ind. Eng. Chem., Prod. Res. Dev.*, 1968, 7, 136, 145
3. *ASTM D1022-64*
4. Bailey, H. C., private comm., 1974

A violent explosion in a partially full butadiene storage sphere was traced to butadiene peroxide. The latter had been formed by contact with air over a long period, and had eventually initiated the exothermic polymerisation of the sphere contents. A tank monitoring and purging system was introduced to prevent recurrence [1].

*See* BUTADIENE, $C_4H_6$ : Air

The polyperoxide (formed even at 0°C) is a mixed 1,2- and 1,4-addition copolymer of butadiene with oxygen [2], effectively a dialkyl peroxide. Its concentration in butadiene is very seriously underestimated by a standard analytical method applicable to hydroperoxides [3], which indicates only some 5% of the true value. Alternative methods of greater reliability are likely to form the basis of a revised standard method currently under consideration [4].

*See other* POLYPEROXIDES

**POLY(DIMETHYLKETENE PEROXIDE)** $(C_4H_6O_3)_n$

$(Me_2C{=}C(O_2)O)_n$

*See* DIMETHYLKETENE, $C_4H_6O$: Air

*See other* POLYPEROXIDES

**ACETIC ANHYDRIDE** $Ac_2O$ $C_4H_6O_3$

*MCA SD-15*, 1962

Acetic acid,
Water

*MCA Case History No. 1865*

Erroneous addition of aqueous acetic acid into a tank of the anhydride caused violent exothermic hydrolysis of the latter.

*See* Water, below

Barium peroxide

*See* BARIUM PEROXIDE, $BaO_2$ : Acetic anhydride

Boric acid

1. Lerner, L. M., *Chem. Eng. News*, 1973, **51**(34), 42
2. *Experiments in Organic Chemistry*, 281, Fieser, L. F., Boston, Heath, 1955

Attempted preparation of acetyl borate by slowly heating a stirred mixture of the anhydride and solid acid led to an eruptive explosion at 60°C [1]. The republished procedure [2] omitted the reference to a violent reaction mentioned in the German original. Modifying the procedure by adding portions of boric acid to the hot stirred anhydride should give a smoother reaction.

Chromic acid

1. Dawkins, A. E., *Chem. & Ind.*, 1956, 196
2. Baker, W., ibid., 280

Addition of acetic anhydride to a solution of chromium trioxide in water caused violent boiling [1], due to the exothermic acid-catalysed hydrolysis of the anhydride [2].

Chromium trioxide

*See* CHROMIUM TRIOXIDE, $CrO_3$ : Acetic anhydride

1,3-Diphenyltriazene

*See* 1,3-DIPHENYLTRIAZENE, $C_{12}H_{11}N_3$ : Acetic anhydride

Ethanol,
Sodium hydrogensulphate

Staudinger, H., *Angew. Chem.*, 1922, **35**, 657

Accidental presence of the acid salt vigorously catalysed a large-scale preparation of ethyl acetate, causing violent boiling and emission of vapour which became ignited and exploded.

*See* CATALYSIS BY IMPURITIES

Glycerol,
Phosphoryl chloride

Bellis, M. P., *Hexagon Alpha Chi Sigma (Indianapolis)*, 1949, **40**(10), 40

Violent acylation occurs in catalytic presence of phosphoryl chloride. The high viscosity of the mixture in absence of solvent prevents mixing and dissipation of heat of reaction.

Hydrochloric acid,
Water

1. Vogel, 1957, 572–573
2. Batchelor, J. F., private comm., 1976

Crude dimethylaniline was being freed of impurities by treatment with acetic anhydride, according to a published procedure [1]. However, three times the recommended proportion of anhydride was used, and the reaction mixture was ice-cooled before addition of diluted hydrochloric acid to hydrolyse the excess anhydride. Hydrolysis then proceeded with explosive violence [2].

Hydrogen peroxide

*See* HYDROGEN PEROXIDE, $H_2O_2$: Acetic anhydride

Hypochlorous acid

*See* HYPOCHLOROUS ACID, ClHO: Acetic anhydride

Metal nitrates

1. Davey, W. *et al.*, *Chem. & Ind.*, 1948, 814
2. Collman, J. P. *et al.*, *Inorg. Synth.*, 1963, **7**, 205–207
3. James, B. D., *J. Chem. Educ.*, 1974, **51**(9), 568

Use of mixtures of nitrates with acetic anhydride as a nitrating agent may be hazardous, depending on the proportion of reactants and the cation; copper and sodium nitrates usually cause violent reaction [1]. An improved procedure for the use of the anhydride–copper(II) nitration mixture [2] has been further modified [3] to improve safety aspects.

Nitric acid

*See* NITRIC ACID, $HNO_3$: Acetic anhydride

Perchloric acid,
Water

1. *ABCM Quart. Safety Summ.*, 1954, **25**, 3

2. Turner, H. S. *et al.*, *Chem. & Ind.*, 1965, 1933
3. McG. Tegart, W. J., *The Electrolytic and Chemical Polishing of Metals in Research and Industry*, London, Pergamon, 2nd Ed., 1959

Anhydrous solutions of perchloric acid in acetic acid are prepared by using acetic anhydride to remove the water introduced as diluent of perchloric acid. It is essential that the acetic anhydride be added slowly to the aqueous perchloric–acetic acid mixture under conditions where it will readily react with the water, i.e. at about 10°C. Use of anhydride cooled in a freezing mixture caused delayed and violent boiling to occur. Full directions for the preparation are given [1]. A violent explosion occurred during the preparation of an electro-polishing solution by addition of perchloric acid solution to a mixture of water and acetic anhydride. The cause of the explosion, the vigorously exothermic acid-catalysed hydrolysis of acetic anhydride, is avoided if water is added last to the mixture produced by adding perchloric acid solution to acetic anhydride [2]. The published directions [3] are erroneous and insufficiently detailed for safe working.

*See also* PERCHLORIC ACID, $ClHO_4$ : Acetic anhydride
: Dehydrating agents

Peroxyacetic acid

*See* PEROXYACETIC ACID, $C_2H_4O_3$ : Acetic anhydride

Potassium permanganate

*See* POTASSIUM PERMANGANATE, $KMnO_4$ : Acetic acid, etc.

Tetrafluoroboric acid

Lichtenberg, D. W. *et al.*, *J. Organomet. Chem.*, 1975, **94**, 319

Dehydration of the aqueous 48% acid by addition to the anhydride is rather exothermic, and caution is advised.

*See* Chromic acid; Perchloric acid, etc., above
*p*-Toluenesulphonic acid, etc., below

*p*-Toluenesulphonic acid,
Water

1. Jones, B. J. *et al.*, *Clin. Chem.*, 1955, **1**, 305
2. Pearson, S. *et al.*, *Anal. Chem.*, 1953, **25**, 813–814

Caution is advised [1] to avoid explosions when using an analytical method involving sequential addition of acetic acid, aqueous *p*-toluene-sulphonic acid and acetic anhydride to serum. It is difficult to see why this should happen, unless the anhydride were being added before the sulphonic acid solution.

*See* Water, below

Water

1. Leigh, W. R. D. *et al., Chem. & Ind.*, 1962, 778
2. Benson, G., *Chem. Eng. News,* 1947, **25**, 3458

Accidental slow addition of water to a mixture of the anhydride (85%) and acetic acid (15%) led to a violent, large-scale explosion. This was simulated closely in the laboratory again in absence of mineral-acid catalyst [1]. If unmoderated, the rate of acid-catalysed hydrolysis of (water-insoluble) acetic anhydride can accelerate to explosive boiling [2].

*See other* ACID ANHYDRIDES

**PEROXYCROTONIC ACID** **MeCH=CHCO$_2$OH** **$C_4H_6O_3$**

Vasilina, T. U. *et al., Chem. Abs.*, 1974, **81**, 151446y

The decomposition of the acid has been mentioned in a safety context, but details were not translated.

*See other* PEROXY ACIDS

**DIACETYL PEROXIDE** **AcOOAc** **$C_4H_6O_4$**

1. Shanley, E. S., *Chem. Eng. News,* 1949, **27**, 175
2. Kuhn, L. P., *Chem. Eng. News,* 1948, **26**, 3197
3. Kharasch, M. S. *et al., J. Amer. Chem. Soc.*, 1948, **70**, 1269
4. Moore, C. G., *J. Chem. Soc.*, 1951, 236

Acetyl peroxide may be readily prepared and used in ethereal solution. It is essential to prevent separation of the crystalline peroxide even in traces, since, when dry, it is shock-sensitive and a high explosion risk [1]. Crystalline material, separated and dried deliberately, detonated violently [2]. The commercial material, supplied as a 30% solution in dimethyl phthalate, is free of the tendency to crystallise and is relatively safe. It is, however, a powerful oxidant [1].

Precautions necessary for the preparation and thermolysis of the peroxide have been detailed [3,4].

*See* FLUORINE, $F_2$: Sodium acetate

*See other* ACYL PEROXIDES

***p*-DIOXENEDIOXETANE (2,3-EPIDIOXYDIOXANE)** **$C_4H_6O_4$**

**OOCHCHOC$_2$H$_4$O**

Wilson, T. *et al., J. Amer. Chem. Soc.*, 1976, **98**, 1087

A small sample of the solid peroxide exploded on warming to ambient temperature. Storage at −20°C in solution appears safe.

*See other* CYCLIC PEROXIDES

## POLY(VINYL ACETATE PEROXIDE)

$(C_4H_6O_4)_n$

$(CH_3CO_2\dot{C}HCH_2OO-)_n$

*See* VINYL ACETATE, $C_4H_6O_2$ : Oxygen

*See other* POLYPEROXIDES

## LEAD ACETATE–LEAD BROMATE

$C_4H_6O_4Pb \cdot Br_2O_6Pb$

$Pb(OAc)_2 \cdot Pb(BrO_3)_2$


1. Berger, A., *Arbeits-Schutz*., 1934, **2**, 20
2. Leymann, –, *Chem. Fabrik*, 1929, 360–361


The compound (formulated as the double salt, rather than the mixed salt) may be formed during the preparation of lead bromate from lead acetate and potassium bromate in acetic acid, and is explosive and very sensitive to friction. Although lead bromate is stable up to 180°C, it is an explosive salt [1]. Further details of the incident are available [2].

*See other* HEAVY METAL DERIVATIVES

## DIACETATOPLATINUM(II)

$[Pt(OAc)_2]^{2+}$

$C_4H_6O_4Pt$

Preparative hazard

*See* NITRIC ACID, $HNO_3$ : Acetic acid, etc.

## MONOPEROXYSUCCINIC ACID

$C_4H_6O_5$

$HCO_2CH_2CH_2CO_2OH$


1. Lombard, R. *et al.*, *Bull. Soc. Chim. Fr.*, 1963, **12**, 2800
2. Castrantas, 1965, 16


It explodes in contact with flame [1] and is weakly shock-sensitive [2].

*See other* PEROXYACIDS

## VINYL ACETATE OZONIDE

$C_4H_6O_5$

$MeCO_2CHCH_2O{-}O^+{-}O^-$

Kirk-Othmer, 1970, Vol. 21, 320

The ozonide formed by vinyl acetate is explosive when dry.

*See other* OZONIDES

## DIMETHYL PEROXYDICARBONATE

$MeO_2COOCO_2Me$

$C_4H_6O_6$

Explodes on heating to 55–60°C, or readily under a hammer blow, and more powerfully than dibenzoyl peroxide.

*See* PEROXYCARBONATE ESTERS

**TARTARIC ACID** **$(CHOHCO_2H)_2$** **$C_4H_6O_6$**

Silver

*See* SILVER, Ag: Carboxylic acids

**2-BUTYNE-1-THIOL** **EtC≡CSH** **$C_4H_6S$**

Brandsma, 1971, 180

Presence of a stabiliser is essential during handling or storage at –20°C under nitrogen. Exposure to air leads to formation of a polymer which may explode on heating.

*See other* ACETYLENIC COMPOUNDS

**† 1-BROMO-2-BUTENE** **$BrCH_2CH{=}CHCH_3$** **$C_4H_7Br$**

**† 4-BROMO-1-BUTENE** **$CH_2{=}CHCH_2CH_2Br$** **$C_4H_7Br$**

Chloroplatinic acid,
Methylphenylchlorosilane

*See* METHYLPHENYLCHLOROSILANE, $C_7H_9ClSi$: 4-Bromobutene, etc.

**1,1,1-TRIS(BROMOMETHYL)METHANE** **$C_4H_7Br_3$**
**$(BrCH_2)_3CH$**

Preparative hazard

*See* PHOSPHORUS TRIBROMIDE, $Br_3P$: 1,1,1-Trishydroxymethylmethane

**† 2-CHLORO-2-BUTENE** **$CH_3CCl{=}CHCH_3$** **$C_4H_7Cl$**

**† 3-CHLORO-1-BUTENE** **$CH_2{=}CHCHClCH_3$** **$C_4H_7Cl$**

**† 3-CHLORO-2-METHYL-1-PROPENE** **$CH_2{=}CMeCH_2Cl$** **$C_4H_7Cl$**

**† BUTYRYL CHLORIDE** **PrCOCl** **$C_4H_7ClO$**

**1-CHLORO-2-BUTANONE** **$ClCH_2COEt$** **$C_4H_7ClO$**

Tilford, C. H., private comm., 1965

An amber bottle of stabilised material spontaneously exploded.

*See* CHLOROACETONE, $C_3H_5ClO$

**† 2-CHLOROETHYL VINYL ETHER** $C_4H_7ClO$

$ClCH_2CH_2OCH{=}CH_2$

**† ISOBUTYRYL CHLORIDE** $Me_2CHCOCl$ $C_4H_7ClO$

**† ISOPROPYL CHLOROFORMATE** $ClCO_2CHMe_2$ $C_4H_7ClO_2$

**ISOPROPYLISOCYANIDE DICHLORIDE–IRON(III) CHLORIDE** $C_4H_7Cl_2N \cdot Cl_3Fe$

$Me_2CHN{=}CCl_2 \cdot FeCl_3$

Metal oxides,
or Water

Fuks, R. *et al., Tetrahedron*, 1973, **29**, 297–298

Interaction of the 3:2 complex with calcium, mercury or silver oxides was usually too violent and uncontrollable for preparative purposes, but zinc oxide was satisfactory. Reaction with water was violent.

*See related* CYANO COMPOUNDS

**1-FLUORO-1,1-DINITROBUTANE** $FC(NO_2)_2Pr$ $C_4H_7FN_2O_4$

*See* FLUORODINITRO COMPOUNDS

**† BUTYRONITRILE** PrCN $C_4H_7N$

**DIMETHYLAMINOACETYLENE** $Me_2NC{\equiv}CH$ $C_4H_7N$

Brandsma, 1971, 139

The amine reacts extremely vigorously with neutral water.

*See other* ACETYLENIC COMPOUNDS

**† ISOBUTYRONITRILE** $Me_2CHCN$ $C_4H_7N$

**2-CYANO-2-PROPANOL** $Me_2C(CN)OH$ $C_4H_7NO$

Sulphuric acid

*See* SULPHURIC ACID, $H_2O_4S$ : 2-Cyano-2-propanol

## 3,6-DIHYDRO-1,2,2*H*-OXAZINE $C_4H_7NO$

$\overline{\text{ONHCH}_2\text{CH=CHCH}_2}$

Nitric acid

*See* NITRIC ACID, $HNO_3$ : 3,6-Dihydro-1,2,2*H*-oxazine

## 2-METHYLACRYLALDEHYDE OXIME $C_4H_7NO$

$H_2C{=}CMeCH{=}NOH$

Mowry, D. T. *et al., J. Amer. Chem. Soc.*, 1947, **69**, 1831

The viscous distillation residue may decompose explosively if overheated. A polymeric peroxide may have been involved.

*See other* OXIMES

## 2,3-BUTANEDIONE MONOXIME $C_4H_7NO_2$

$CH_3COC(NOH)CH_3$

Muir, G. D., private comm., 1968

Distillation from a heating mantle at 6.5 mbar has caused explosions on several occasions. Distillation at below 1.3 mbar from a steam bath seems safe.

## 1-NITRO-3-BUTENE $C_4H_7NO_2$

$NO_2CH_2CH_2CH{=}CH_2$

Witucki, E. F. *et al., J. Org. Chem.*, 1972, **37**, 152

A fume-off during distillation of one sample of this compound (b.p., 55°C/25 mbar) illustrates the inherent instability of this type of compound.

*See* also 1,1-DINITRO-3-BUTENE, $C_4H_6N_2O_4$
2,3-DINITRO-2-BUTENE, $C_4H_6N_2O_4$

## BUTYRYL NITRATE $C_3H_7COONO_2$ $C_4H_7NO_4$

Francis, F. E., *Ber.*, 1906, **39**, 3798

It detonates on heating.

*See other* ACYL NITRATES

## BUTYRYL PEROXONITRATE $PrCO_2ONO_2$ $C_4H_7NO_5$

Stephens, E. R., *Anal. Chem.*, 1964, **36**, 928

It is highly explosive, like its lower homologues.

*See related* PEROXYESTERS

**ETHYL DIAZOACETATE** $N_2CH_2CO_2Et$ $C_4H_7N_2O_2$

Tris(dimethylamino) antimony

Krommes, P. *et al.*, *J. Organomet. Chem.*, 1975, **97**, 63, 65

Interaction at room temperature, either initially or on warming from −20°C or below, becomes explosively violent.

*See other* DIAZO COMPOUNDS

**1,3-BIS(5-AMINO-1,3,4-TRIAZOL-2-YL)TRIAZENE** $C_4H_7N_{11}$

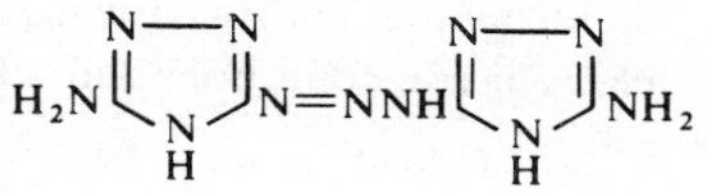

$H_2N$–(triazolyl, N–N, NH)–$N{=}NNH$–(triazolyl, N–N, NH)–$NH_2$

Hauser, M., *J. Org. Chem.*, 1964, **29**, 3449

It explodes at the m.p., 187°C.

*See other* HIGH-NITROGEN COMPOUNDS

**† 1-BUTENE** $CH_2{=}CHEt$ $C_4H_8$

Aluminium tris-tetrahydroborate

*See* ALUMINIUM TRIS-TETRAHYDROBORATE, $AlB_3H_{12}$: Alkenes, etc.

**† *cis*-2-BUTENE** $MeCH{=}CHMe$ $C_4H_8$

**† *trans*-2-BUTENE** $MeCH{=}CHMe$ $C_4H_8$

**† CYCLOBUTANE** $\overline{CH_2[CH_2]_2CH_2}$ $C_4H_8$

**† METHYLCYCLOPROPANE** $\overline{CH_2CHMeCH_2}$ $C_4H_8$

**† 2-METHYLPROPENE** $CH_2{=}CMeCH_3$ $C_4H_8$

**POLYISOBUTENE** $(Me_2\dot{C}CH_2-)_n$ $(C_4H_8)_n$

Silver peroxide

*See* SILVER PEROXIDE, $Ag_2O_2$: Polyisobutene

**4-MORPHOLINESULPHENYL CHLORIDE** $C_4H_8ClNOS$

$\overline{OC_2H_4N(SCl)CH_2CH_2}$

*MCA Case History No. 1806*
The product from chlorination of 4,4′-dithiodimorpholine exploded violently after vacuum stripping of the solvent carbon tetrachloride. As it was a published process that had been previously operated uneventfully, no reason was apparent. It seems remotely possible that an unstable *N*-chloro derivative could have been produced if chlorination conditions had differed from those previously employed.

**† *mixo*-DICHLOROBUTANE** $CH_2ClCHClEt$ + $CH_2ClCH_2CHClMe$ $C_4H_8Cl_2$

**BIS(2-CHLOROETHYL) SULPHIDE** $(ClCH_2CH_2)_2S$ $C_4H_8Cl_2S$

Bleaching powder
*See* BLEACHING POWDER: Bis(2-chloroethyl) sulphide

**BIS(1-CHLOROETHYLTHALLIUM CHLORIDE) OXIDE** $C_4H_8Cl_4OTl_2$
$(MeCHClTl_2)O$

Yakubovich, A. Ya. *et al., Dokl. Akad. Nauk SSSR,* 1950, **73**, 957–958
An explosive solid of low stability.

**2-METHYL-2-NITRATOMERCURIOPROPYL(NITRATO)DIMERCURY(II)**
$C_4H_8Hg_3N_2O_6$
$NO_3HgC(Me)_2CH_2HgHgNO_3$

Whitmore, 1921, 116
The compound, prepared from isobutene and mercury(II) nitrate, explodes on impact at 80°C.
*See other* MERCURY COMPOUNDS

**† DIMETHYLAMINOACETONITRILE** $Me_2NCH_2CN$ $C_4H_8N_2$

**3-PROPYLDIAZIRINE** $\overline{PrCHN{=}N}$ $C_4H_8N_2$
Schmitz, E., *et al., Ber.*, 1962, **95**, 800
It exploded on attempted distillation from calcium chloride at about 75°C.
*See other* DIAZIRINES

## ETHYL *N*-METHYL-*N*-NITROSOCARBAMATE $C_4H_8N_2O_3$

**MeN(NO)$CO_2$Et**

1. Hartman, W. W., *Org. Synth.*, 1943, Coll. Vol. 2, 465
2. Druckrey, H. *et al.*, *Nature*, 1962, 195, 1111

The material is unstable and explodes if distilled at atmospheric pressure [1] or may become explosive if stored above 15°C [2]. Its use for preparation of diazomethane has been superseded by stable intermediates.

*See* DIAZOMETHANE, $CH_2N_2$

*See other* NITROSO COMPOUNDS

## 2,2′-OXYDI(ETHYL NITRATE) $C_4H_8N_2O_7$

**$O(CH_2CH_2ONO_2)_2$**

Kit and Evered, 1960, 268

A powerful explosive, sensitive to vibration and mechanical shock, too heat-sensitive for a rocket propellant.

*See other* ALKYL NITRATES

## AZO-*N*-METHYLFORMAMIDE $C_4H_8N_4O_2$

**MeNHCON=NCONHMe**

Smith, R. F. *et al.*, *J. Org. Chem.*, 1975, **40**, 1855

Thermolysis is vigorously exothermic, the temperature increasing from 176 to 259°C during decomposition.

*See other* AZO COMPOUNDS

## 1,3,5,7-TETRANITROPERHYDRO-1,3,5,7-TETRAZOCINE $C_4H_8N_8O_8$

**$NO_2\overline{N[CH_2N(NO_2)]_3CH_2}$**

Violent decomposition occurred at 279°C.

*See* DIFFERENTIAL THERMAL ANALYSIS

*See other* *N*-NITRO COMPOUNDS

## † 2-BUTANONE $CH_3COC_2H_5$ $C_4H_8O$

*MCA SD-83*, 1961

Chloroform

*See* CHLOROFORM, $CHCl_3$: Acetone, etc.

Hydrogen peroxide,
Nitric acid

*See* HYDROGEN PEROXIDE, $H_2O_2$: Ketones, etc.

Isopropanol

*See* ISOPROPANOL $C_3H_8O$: 2-Butanone

Potassium *tert*-butoxide

*See* POTASSIUM *tert*-BUTOXIDE, $C_4H_9KO$; Acids, etc.

**† BUTYRALDEHYDE** **$C_3H_7CHO$** **$C_4H_8O$**

*MCA SD-78*, 1960

**† CYCLOPROPYL METHYL ETHER** **$\overline{CH_2CH_2CH}OMe$** **$C_4H_8O$**

**† 1,2-EPOXYBUTANE** **$\overline{CH_2OCH}Et$** **$C_4H_8O$**

**† ETHYL VINYL ETHER** **$EtOCH{=}CH_2$** **$C_4H_8O$**

Methanesulphonic acid

Eaton, P. E. *et al., J. Org. Chem.*, 1972, **37**, 1947

Methanesulphonic acid is too powerful a catalyst for *O*-alkylation with the vinyl ether, causing explosive polymerisation of the latter on the multimol scale. Dichloroacetic acid is a satisfactory catalyst on the 3g mol scale.

*See other* PEROXIDISABLE COMPOUNDS

**† ISOBUTYRALDEHYDE** **$Me_2CHCHO$** **$C_4H_8O$**

*MCA SD-78*, 1960

**† TETRAHYDROFURAN** **$\overline{O[CH_2]_3CH_2}$** **$C_4H_8O$**

1. 'Recommended Handling Procedures', THF Brochure FC3-664, Wilmington, Du Pont, 1964
2. Anon., *Org. Synth.*, 1966, **46**, 105
3. Schurz, J. *et al., Angew. Chem.*, 1956, **68**, 182
4. Davies, A. G., *J. R. Inst. Chem.*, 1956, **80**, 386

Like many other ethers and cyclic ethers, in absence of inhibitors tetrahydrofuran is subject to autoxidation on exposure to air, when initially the 2-hydroperoxide forms. This tends to decompose smoothly when heated, but if allowed to accumulate for a considerable period, it becomes transformed to other peroxidic species which will decompose violently [1].

Commercial material is supplied stabilised with a phenolic anti-oxidant which is effective under normal closed storage conditions in preventing the formation and accumulation of peroxide. Procedures for testing for the presence of peroxides and also for their removal are detailed [1,2,4]. The latest publication recommends the use of cuprous chloride for removal of trace quantities of peroxides. If more than trace quantities are present, the peroxidised solvent should be discarded by dilution and flushing away with water [2]. An attempt to remove peroxides by shaking with solid ferrous sulphate before distillation did not prevent explosion of the distillation residue [3]. Alkali treatment to destroy peroxides [1] appears not to be safe [2]. (See Caustic alkalies, below). Distillation or alkali treatment of stabilised tetrahydrofuran removes the involatile antioxidant and the solvent must be restabilised or stored under nitrogen to prevent peroxide formation during storage, which should not exceed a few days' duration in the absence of stabiliser [2]. The use of lithium tetrahydroaluminate is only recommended for drying tetrahydrofuran which is peroxide-free and not grossly wet [2].

*See* ETHERS (reference 9)
LITHIUM TETRAHYDROALUMINATE, $AlH_4Li$: Tetrahydrofuran
2-TETRAHYDROFURYL HYDROPEROXIDE, $C_4H_8O_3$

*See other* PEROXIDISABLE COMPOUNDS

Borane

*See* BORANE–TETRAHYDROFURAN, $BH_3 \cdot C_4H_8O$

Caustic alkalies

1. *NSC Newsletter, Chem. Section,* 1964(10); 1967(3)
2. Anon., *Org. Synth.,* 1966, **46**, 105

It is not safe to store quantities of tetrahydrofuran which have been purified (i.e. freed of the normally added inhibitors), since dangerous concentrations of peroxide may build up during storage. Peroxidised material should not be dried with sodium or potassium hydroxide, as explosions may occur [1,2].

Lithium tetrahydroaluminate

*See* LITHIUM TETRAHYDROALUMINATE, $AlH_4Li$: Tetrahydrofuran

Sodium tetrahydroaluminate

*See* SODIUM TETRAHYDROALUMINATE, $AlH_4Na$: Tetrahydrofuran

## BUTYRIC ACID $C_3H_7CO_2H$ $C_4H_8O_2$

Chromium trioxide

*See* CHROMIUM TRIOXIDE, $CrO_3$: Butyric acid

## † *m*-DIOXANE $\overline{O[CH_2]_3OCH_2}$ $C_4H_8O_2$

## † *p*-DIOXANE — $\overline{OC_2H_4OCH_2CH_2}$ — $C_4H_8O_2$

Air

Dasler, W. *et al., Ind. Eng. Chem. (Anal. Ed.),* 1946, 18, 52

Like all other ethers, dioxane is susceptible to autoxidation with formation of peroxides which may be hazardous if distillation (causing concentration) is attempted. Because it is water-miscible, treatment by shaking with aqueous reducants (iron(II) sulphate, sodium sulphide, etc.) is impracticable. Peroxides may be removed, however, under anhydrous conditions by passing dioxane (or any other ether) down a column of activated alumina. The peroxides (and any water) are removed by adsorption on to the alumina, which must then be washed with water or methanol to remove them before being discarded.

*See other* PEROXIDISABLE COMPOUNDS

Decaborane(14)

*See* DECABORANE(14), $B_{10}H_{14}$: Ethers

Nickel

Mozingo, R., *Org. Synth.,* 1955, Coll. Vol. 3, 182

Dioxane reacts with Raney nickel catalyst almost explosively above 210°C.

Sulphur trioxide

*See* SULPHUR TRIOXIDE, $O_3S$: Dioxane

Triethynylaluminium

*See* TRIETHYNYLALUMINIUM, $C_6H_3Al$: Dioxane

## † ETHYL ACETATE — $MeCO_2Et$ — $C_4H_8O_2$

*MCA SD-51,* 1953

Lithium tetrahydroaluminate

*See* LITHIUM TETRAHYDROALUMINATE, $AlH_4Li$ : Ethyl acetate

Potassium *tert*-butoxide

*See* POTASSIUM *tert*-BUTOXIDE, $C_4H_9KO$: Acids, etc.

## † ISOPROPYL FORMATE — $HCO_2CHMe_2$ — $C_4H_8O_2$

## † METHYL PROPIONATE — $EtCO_2Me$ — $C_4H_8O_2$

† **PROPYL FORMATE** **$HCO_2Pr$** **$C_4H_8O_2$**

Potassium *tert*-butoxide

*See* POTASSIUM *tert*-BUTOXIDE, $C_4H_9KO$; Acids, etc.

**TETRAHYDROTHIOPHENE-1,1-DIOXIDE ('SULPHOLANE')** **$C_4H_8O_2S$**

**$O_2S[CH_2]_3CH_2$**

Nitronium tetrafluoroborate

*See* NITRATING AGENTS

***trans*-2-BUTENE OZONIDE** **$C_4H_8O_3$**

**$MeCHCH(Me)O{-}O^+{-}O^-$**

*See* 2-HEXENE OZONIDE, $C_6H_{12}O_3$

*See other* OZONIDES

**2-TETRAHYDROFURYL HYDROPEROXIDE** **$C_4H_8O_3$**

**$O[CH_2]_3CHOOH$**

Rein, H. and Criegee, R., *Angew. Chem.*, 1950, **62**, 120

This occurs as the first product of the ready autoxidation of tetrahydrofuran, and is relatively stable. It readily changes, however, to a highly explosive polyalkylidene peroxide, which is responsible for the numerous explosions observed on distillation of peroxidised tetrahydrofuran.

*See* TETRAHYDROFURAN, $C_4H_8O$

*See other* ALKYL HYDROPEROXIDES

**3,6-DIMETHYL-1,2,4,5-TETRAOXANE** **MeCHOOCHMeOO** **$C_4H_8O_4$**

Rieche, A. *et al.*, *Ber.*, 1939, **72**, 1933

This dimeric 'ethylidene peroxide' is an extremely shock-sensitive solid which explodes violently on being touched. Extreme caution in handling is required.

*See other* CYCLIC PEROXIDES

**1-HYDROXYETHYL PEROXYACETATE** **$C_4H_8O_4$**

**$MeCO_2OCHOHMe$**

An explosive low-melting solid, readily formed during autoxidation of acetaldehyde.

*See* ACETALDEHYDE, $C_2H_4O$ : Cobalt acetate, oxygen

*See other* PEROXYESTERS

**† TETRAHYDROTHIOPHENE** **$\overline{S[CH_2]_3CH_2}$** **$C_4H_8S$**

Hydrogen peroxide

*See* HYDROGEN PEROXIDE, $H_2O_2$ : Tetrahydrothiophene

**BUTYLDICHLOROBORANE** **$BuBCl_2$** **$C_4H_9BCl_2$**

Air,

or Water

Niedenzu, K. *et al., Inorg. Synth.*, 1967, **10**, 126

Ignites on prolonged exposure to air; hydrolysis may be explosive.

*See other* ALKYLNON-METAL HALIDES

**† 1-BROMOBUTANE** **$BrCH_2Pr$** **$C_4H_9Br$**

Bromobenzene,

Sodium

*See* SODIUM, Na: Halocarbons (reference 7)

**† 2-BROMOBUTANE** **MeCHBrEt** **$C_4H_9Br$**

**† 1-BROMO-2-METHYLPROPANE** **$BrCH_2CHMe_2$** **$C_4H_9Br$**

**† 2-BROMO-2-METHYLPROPANE** **$BrC(Me)_3$** **$C_4H_9Br$**

**† 2-BROMOETHYL ETHYL ETHER** **$BrCH_2CH_2OEt$** **$C_4H_9BrO$**

**† 1-CHLOROBUTANE** **$ClCH_2Pr$** **$C_4H_9Cl$**

**† 2-CHLOROBUTANE** **MeCHClEt** **$C_4H_9Cl$**

**† 1-CHLORO-2-METHYLPROPANE** $ClCH_2CH(Me)_2$ $C_4H_9Cl$

**† 2-CHLORO-2-METHYLPROPANE** $ClC(Me)_3$ $C_4H_9Cl$

***tert*-BUTYL HYPOCHLORITE** $Me_3COCl$ $C_4H_9ClO$

1. Lewis, J. C., *Chem. Eng. News*, 1962, **40**(43), 62; Teeter, H.M. *et al.*, *Org. Synth.*, 1962, Coll. Vol. 4, 125
2. Mintz, M. J. *et al.*, *Org. Synth.*, 1973, Coll. Vol. 5, 185

This material, normally in sealed ampoules and used as a paper chromatography spray reagent, is photo-sensitive. Exposure to UV light causes exothermic decomposition to acetone and chloromethane. Ampoules have burst because of pressure build-up after exposure to fluorescent or direct daylight. Store cool and dark, and open ampoules with personal protection. The material also reacts violently with rubber [1]. It should not be heated to above its boiling point [2]. There is also a preparative hazard.

*See* CHLORINE, $Cl_2$ : *tert*-Butanol
*See other* HYPOHALITES

**2(2-HYDROXYETHOXY)ETHYL PERCHLORATE** $C_4H_9ClO_6$
$HOC_2H_4OC_2H_4OClO_3$

Hofman, K. A. *et al.*, *Ber.*, 1909, **42**, 4390

Explodes violently on heating in a capillary.

*See other* ALKYL PERCHLORATES

***tert*-BUTYL PEROXOPHOSPHORYL DICHLORIDE** $C_4H_9Cl_2O_3P$
$Me_3COOP(O)Cl_2$

The product decomposed violently after isolation.

*See other* *tert*-BUTYL PEROXOPHOSPHATE DERIVATIVES

***N*-FLUORO-*N*-NITROBUTYLAMINE** $BuNFNO_2$ $C_4H_9FN_2O_2$

Grakauskas, V. *et al.*, *J. Org. Chem.*, 1972, **37**, 334

A sample exploded on vacuum distillation at 60°C, though not at 40°C/33 mbar.

**† 2-IODOBUTANE** MeCHIEt $C_4H_9I$

**† 1-IODO-2-METHYLPROPANE** $ICH_2CH(Me)_2$ $C_4H_9I$

### † 2-IODO-2-METHYLPROPANE $IC(Me)_3$ $C_4H_9I$

### POTASSIUM *tert*-BUTOXIDE $KOCMe_3$ $C_4H_9KO$

Acids,
or Reactive solvents

Manwaring, R., *et al., Chem. & Ind.*, 1973, 172

Contact of 1.5 g portions of the solid butoxide with drops of the liquid (1) or with vapours (v) of the reagents below caused ignition after the indicated period (min).

Acetic acid,v,3; sulphuric acid, l,0.5
Methanol, l,2; ethanol, v,7; propanol, l,1; isopropanol, l,1
Ethyl acetate, v,2; butyl acetate, v,2; propyl formate, v,4; dimethyl carbonate, l,1; diethyl sulphate, l,1
Acetone, v,4, l,2; 2-butanone, v,1, l,0.5; 4-methyl-2-butanone, v,3
Dichloromethane, l,2; chloroform, v,2, l,0; carbon tetrachloride, l,1; 1-chloro-2,3-epoxypropane, l,1

The potentially dangerous reactivity with water, acids or halocarbons was already known, but that arising from contact with alcohols, esters and ketones was unexpected. Under normal reaction conditions, little significant danger should exist where excess of solvent will dissipate the heat, but accidental spillage of the solid butoxide could be hazardous.

Air,
or Oxygen

Fieser, 1967, Vol. 1, 911

This extremely powerful base may ignite if exposed to air or oxygen at elevated temperatures.

*See other* PYROPHORIC MATERIALS

Dimethyl sulphoxide

*See* DIMETHYL SULPHOXIDE, $C_2H_6OS$: Metal alkoxides
*See other* METAL ALKOXIDES

### BUTYLLITHIUM BuLi $C_4H_9Li$

Air,
or Carbon dioxide,
or Water

*MCA SD-91,* 1966

Dezmelyk, E. W. *et al., Ind. Eng. Chem.*, 1961, **53**(6), 56A

This reactive liquid is now normally supplied commercially as a solution (up to 25% wt.) in pentane, hexane or heptane, because it reacts with ether and such solutions must be stored under refrigeration. Reaction with atmospheric oxygen or water vapour is highly exothermic and solutions of 20% concentration will ignite on exposure to air at ambient temperature and about 70% relative humidity. Solutions of above 25% concentration ignite in air at any humidity. Contact with liquid water will cause immediate ignition of solutions in these highly flammable solvents.

Though contact with carbon dioxide causes an exothermic reaction, it is less than with atmospheric oxygen, so carbon dioxide extinguishers are effective on butyllithium fires.

Full handling, operating and disposal procedures are detailed.

Styrene

*See* STYRENE, $C_8H_8$: Butyllithium

*See other* ALKYLMETALS

**_tert_-BUTYLLITHIUM** **$Me_3CLi$** **$C_4H_9Li$**

Turbitt, T. D. *et al.*, *J. Chem. Soc., Perkin Trans. 2*, 1974, 183

The 2 M solution in heptane ignites in air.

2,2,2,4,4,4-Hexafluoro-1,3-dimethyl-1,3,2,4-diazaphosphetidine

Harris, R.,*et al., J. Chem. Soc., Dalton Trans.*, 1976, 23

Interaction in hexane to produce 2,4-di-*tert*-butyl-2,2,4,4-tetrafluoro-1,3,2,4-diazaphosphetidine often starts only after an induction period and may proceed very violently. Careful temperature control is imperative.

*See other* ALKYLMETALS

**† PYRROLIDINE** **$HN[CH_2]_3CH_2$** **$C_4H_9N$**

**2-BUTANONE OXIME** **$CH_3C(NOH)C_2H_5$** **$C_4H_9NO$**

Tyler, L. J., *Chem. Eng. News*, 1974, **52**(35), 3

Investigation following two violent explosions involving the oxime or its derivatives showed that it may be distilled at 152°C under atmospheric pressure only if highly purified. Presence of impurities, especially acidic impurities (e.g. the oxime hydrochloride) drastically lowers the temperature at which degradation occurs.

*See* 2-BUTANONE OXIME HYDROCHLORIDE, $C_4H_{10}ClNO$

*See other* OXIMES

**BUTYRALDEHYDE OXIME** $C_3H_7CH{=}NOH$ $C_4H_9NO$

Anon., *Chemiearbeit*, 1966, (3), 20

A large batch exploded violently (without flame) during vacuum distillation at 90–120°C/20–25 mbar. Since the distilled product contained up to 12% of butyronitrile, it was assumed that the oxime had rearranged to butyramide, and subsequently dehydrated to butyronitrile. The release of water into a system at 120°C would generate excessive pressure which the vessel could not withstand. The rearrangement may have been catalysed by metallic impurities.

*See also* ETHYL 2-FORMYLPROPIONATE OXIME, $C_6H_{11}NO_3$: Hydrogen chloride

SULPHURIC ACID, $H_2O_4S$ : Cyclopentanone oxime

*See other* OXIMES

***N,N*-DIMETHYLACETAMIDE** $Me_2NAc$ $C_4H_9NO$

Halogenated compounds

'DMAC Brochure A-79931', Wilmington, Du Pont, 1969

The tertiary amide acts as a dehydrohalogenating agent, and reaction with some highly halogenated compounds (carbon tetrachloride, hexachlorocyclohexane) is very exothermic and may become violent, particularly in the presence of iron.

*See* CATALYSIS BY IMPURITIES

*N,N*-DIMETHYLFORMAMIDE, $C_3H_7NO$: Halogenated compounds, etc.

*See other* APROTIC SOLVENTS

**MORPHOLINE** $HNC_2H_4OCH_2CH_2$ $C_4H_9NO$

Cellulose nitrate

*See* CELLULOSE NITRATE: Amines

Nitromethane

*See* NITROMETHANE, $CH_3NO_2$: Acids, etc.

† **BUTYL NITRITE** BuONO $C_4H_9NO_2$

***tert*-NITROBUTANE** $NO_2C(Me)_3$ $C_4H_9NO_2$

Preparative hazard

*See* POTASSIUM PERMANGANATE, $KMnO_4$ : Acetone, *tert*-Butylamine

Mee, A. J., *School Sci. Rev.*, 1940, **22**(85), 96

A sample exploded during distillation.

*See other* *C*-NITRO COMPOUNDS

## BUTYL NITRATE $BuONO_2$ $C_4H_9NO_3$

Lewis acids

*See* ALKYL NITRATES: Lewis acids

## ETHYL 2-NITROETHYL ETHER $EtOC_2H_4NO_2$ $C_4H_9NO_3$

Cann, P. F. *et al., J. Chem. Soc., Perkin Trans. 2*, 1974, 817–819

Addition of diphenyl ether as an inert diluent to the crude ether before distillation at 0.13 mbar is essential to prevent violent explosion of the residue after it has cooled.

*See other* *C*-NITRO COMPOUNDS

## 1,1,1-TRIS(HYDROXYMETHYL)NITROMETHANE $C_4H_9NO_5$

## $(HOCH_2)_3CNO_2$

Hydrogen,
Nickel

Ostis, E. K. *et al., Chem. Abs.*, 1975, **82**, 36578b

Catalytic hydrogenation of the title compound to the amine with hydrogen at 5–10 bar is described as hazardous, and an electrochemical process is recommended.

*See other* HYDROGENATION INCIDENTS
*C*-NITRO COMPOUNDS

## † BUTANE $MeC_2H_4Me$ $C_4H_{10}$

## † ISOBUTANE $Me_2CHCH_3$ $C_4H_{10}$

## DIETHYLALUMINIUM BROMIDE $Et_2AlBr$ $C_4H_{10}AlBr$

Nitromethane

*See* NITROMETHANE, $CH_3NO_2$: Alkylmetal halides
*See* ALKYLALUMINIUM HALIDES

## DIETHYLALUMINIUM CHLORIDE $Et_2AlCl$ $C_4H_{10}AlCl$

Chlorine azide

Houben-Weyl, 1970, Vol. 13.4, 76

Interaction produces distillable ethylaluminium azide chloride, but the residue is explosive.

*See* ALKYLALUMINIUM HALIDES

**DIETHYLGOLD BROMIDE** $Et_2AuBr$ $C_4H_{10}AuBr$

Sorbe, 1968, 63

Explodes at 70°C.

*See other* GOLD COMPOUNDS

**CHLORODIETHYLBORANE** $ClBEt_2$ $C_4H_{10}BCl$

Köster, R. *et al., Inorg. Synth.*, 1974, **15**, 149–152

It ignites in air: preparative hazard also described.

*See other* NON-METAL HALIDES

**† BORON TRIFLUORIDE DIETHYL ETHERATE** $C_4H_{10}BF_3O$

$Et_2O{\rightarrow}BF_3$

Lithium tetrahydroaluminate

*See* LITHIUM TETRAHYDROALUMINATE, $AlH_4Li$: Boron trifluoride diethyl etherate

**DIETHYLBERYLLIUM** $Et_2Be$ $C_4H_{10}Be$

Coates, 1967, Vol. 1, 106

Ignites in air, even when containing ether, and reacts explosively with water.

*See other* ALKYLMETALS

**DIETHYLBISMUTH CHLORIDE** $Et_2BiCl$ $C_4H_{10}BiCl$

Gilman, H. *et al., J. Org. Chem.*, 1939, **4**, 167

Ignites in air.

*See other* ALKYLMETAL HALIDES

**DIETHYLCADMIUM** $Et_2Cd$ $C_4H_{10}Cd$

Krause, E., *Ber.*, 1918, **50**, 1813

Houben-Weyl, 1973, Vol. 13.2a, 903, footnote 3; 908

The vapour decomposes explosively at 180°C. Exposure to ambient air produces white fumes which turn brown and then explode violently.

Moderate quantities decompose explosively when heated rapidly to 130°C, not 180°C as stated previously. This, apart from allyl-cadmium derivatives, is the only pyrophoric dialkylcadmium.

*See other* ALKYLMETALS

**2-BUTANONE OXIME HYDROCHLORIDE** $C_4H_{10}ClNO$

$MeC(N^+HOH)C_2H_5\ Cl^-$

Tyler, L. J., *Chem. Eng. News*, 1974, **52**(35), 3

The salt undergoes violent degradation at 50–70°C, and its presence in trace quantities may promote degradation of the oxime.

*See* OXIMES

**DIETHYLAMINOSULPHINYL CHLORIDE** **$C_4H_{10}ClNOS$**

**$Et_2NSOCl$**

Preparative hazard

*See* SULPHINYL CHLORIDE (THIONYL CHLORIDE), $Cl_2OS$: Bis(dimethylamino)sulphoxide

**DIETHYL PHOSPHOROCHLORIDATE** **$(EtO)_2P(O)Cl$** **$C_4H_{10}ClOP$**

Silbert, L. A., *J. Org. Chem.*, 1971, **36**, 2162

Presence of hydrogen chloride as impurity causes an uncontrollable exothermic reaction during preparation of diethyl phosphate from the title compound.

**DIETHYLTHALLIUM PERCHLORATE** **$Et_2TlClO_4$** **$C_4H_{10}ClO_4Tl$**

Cook, J. R. *et al., J. Inorg. Nucl. Chem.*, 1964, **26**, 1249

It explodes at the m.p., 250°C.

*See related* METAL PERCHLORATES

† **DICHLORODIETHYLSILANE** **$Cl_2SiEt_2$** **$C_4H_{10}Cl_2Si$**

**DIETHYLMAGNESIUM** **$Et_2Mg$** **$C_4H_{10}Mg$**

Gilman, H. *et al., J. Amer. Chem. Soc.*, 1930, **52**, 5048

Contact with moist air usually caused ignition of the dry powder, and water always ignited the solid or its ethereal solution.

*See other* ALKYLMETALS

**PIPERAZINE** **$C_4H_{10}N_2$**

**$HNC_2H_4NHCH_2CH_2$** (ring)

Dicyanofurazan

*See* DICYANOFURAZAN, $C_4N_4O$: Nitrogenous bases

**DIETHYL HYPONITRITE** **EtON=NOEt** **$C_4H_{10}N_2O_2$**

*See* DIALKYL HYPONITRITES

## *N*-NITROSOETHYL(2-HYDROXYETHYL)AMINE $C_4H_{10}N_2O_2$

**$EtN(NO)C_2H_4OH$**

Preussman, R., *Chem. Ber.*, 1962, 95, 1572

It decomposed explosively during attempted distillation at 18 mbar from a bath at 170°C, but was distilled with slight decomposition at 103–105°C/0.4 mbar.

*See other* NITROSO COMPOUNDS

## DIETHYLLEAD DINITRATE $Et_2Pb(NO_3)_2$ $C_4H_{10}N_2O_6Pb$

Potts, D. *et al.*, *Can. J. Chem.*, 1969, **47**, 1621

The salt is unstable above 0°C and explodes on heating.

*See related* ALKYLMETALS

## *N*-BUTYLAMIDOSULPHURYL AZIDE $BuNHSO_2N_3$ $C_4H_{10}N_4O_2S$

Shozda, R. J. *et al.*, *J. Org. Chem.*, 1967, 32, 2876

It exploded during analytical combustion.

*See other* ACYL AZIDES

## BUTANOL BuOH $C_4H_{10}O$

Aluminium

*See* ALUMINIUM, Al: Butanol

Chromium trioxide

*See* CHROMIUM TRIOXIDE, $CrO_3$: Alcohols

## † 2-BUTANOL $CH_3CHOHC_2H_5$ $C_4H_{10}O$

Podsnev, V. S. *et al.*, *Chem. Abs.*, 1977, **86**, 194347n

A powerful explosion which occurred during distillation of a 10-year-old sample of the alcohol was attributed to presence of peroxy compounds formed by autoxidation, possibly involving 2-butanone as an effective photochemical sensitiser.

*See* 2-PROPANOL, $C_3H_8O$: 2-Butanone

## † *tert*-BUTANOL $Me_3COH$ $C_4H_{10}O$

## † DIETHYL ETHER $Et_2O$ $C_4H_{10}O$

*MCA SD-29*, 1965

Air

1. Criegee, R. *et al., Angew. Chem.*, 1936, **49**, 101; ibid., 1958, **70**, 261
2. Mallinckrodt, E., *et al. Chem. Eng. News*, 1955, **33**, 3194
3. Ray, A., *J. Appl. Chem.*, 1955, 188
4. Anon., *Chemist & Druggist*, 1947, **157**, 258
5. Feinstein, R. N., *J. Org. Chem.*, 1959, **24**, 1172
6. Davies, A. G., *J.R. Inst. Chem.*, 1956, **80**, 386

The hydroperoxide initially formed by autoxidation of ether is not particularly explosive, but on standing and evaporation, polymeric 1-oxyperoxides are formed which are dangerously explosive, even below 100°C. Numerous laboratory explosions have been caused by evaporation of peroxidised ether [1]. Formation of peroxides in stored ether may be prevented by presence of sodium diethyldithiocarbamate (0.05 p.p.m; reference 2), which probably deactivates traces of metals which catalyse peroxidation [1]; of pyrogallol (1 p.p.m.; reference 3); or by larger proportions (0.005–0.02% of other inhibitors; reference 4). Once present in ether, peroxides may be detected by the iodine–starch test and removed by percolation through anion exchange resin [5], or activated alumina [6], which leaves the ether dry, or by shaking with ferrous sulphate or sodium sulphite solution. Many other methods have been described [6].

*See other* 1-OXYPEROXY COMPOUNDS
PEROXIDISABLE COMPOUNDS
POLYPEROXIDES

Halogens or Interhalogens

*See* BROMINE TRIFLUORIDE, $BrF_3$ : Halogens, etc.
: Solvents
BROMINE PENTAFLUORIDE, $BrF_5$ : Hydrogen-containing materials
BROMINE, $Br_2$ : Diethyl ether
CHLORINE, $Cl_2$ : Diethyl ether
IODINE HEPTAFLUORIDE, $F_7I$

Oxidants

*See* Halogens and Interhalogens, above
LIQUID AIR: Diethyl ether
SILVER PERCHLORATE, $AgClO_4$ : Diethyl ether
PERCHLORIC ACID, $ClHO_4$ : Diethyl ether
NITROSYL PERCHLORATE, $ClNO_5$ : Organic materials
NITRYL PERCHLORATE, $ClNO_6$ : Organic solvents
CHROMYL CHLORIDE, $Cl_2CrO_2$ : Organic solvents
FLUORINE NITRATE, $FNO_3$ : Organic materials
PERMANGANIC ACID, $HMnO_4$ : Organic materials
NITRIC ACID, $HNO_3$ : Diethyl ether
PEROXODISULPHURIC ACID, $H_2O_8S_2$ : Organic liquids

IODINE(VII) OXIDE, $I_2O_7$: Diethyl ether
SODIUM PEROXIDE, $Na_2O_2$: Organic liquids, etc.
OZONE, $O_3$: Diethyl ether

Peat soils

Walkley, A., *Austral. Chem. Inst. J.*, 1939, **6**, 310

Explosions occurred during extraction of fats and waxes with ether from the soils, as well as when heating the extract at 100°C. Although the latter is scarcely surprising (the ether contained 0.023% of peroxide), the former is unusual.

Sulphur and compounds

*See* THIOTRITHIAZYL PERCHLORATE, $ClN_3O_4S_4$: Organic solvents
SULPHONYL DICHLORIDE, $Cl_2O_2S$: Diethyl ether
SULPHUR, S: Diethyl ether

Uranyl nitrate

*See* URANYL NITRATE, $N_2O_8U$: Diethyl ether

Wood pulp extracts

Durso, D. F., *Chem. Eng. News*, 1957, **35**(34), 91, 115

Ethereal extracts of pulp exploded during or after concentration by evaporation. Although the ether used for extraction had been freed previously of peroxides with cerium(III) hydroxide, the ethereal extracts had been stored for 3 weeks before concentration was effected. During this time the ether and/or extracted terpenes would again form peroxides, but no attempt seems to have been made to test for, or to remove, them before distillation was begun.

*See other* PEROXIDISABLE COMPOUNDS

**† METHYL PROPYL ETHER** **MeOPr** **$C_4H_{10}O$**

*tert*-**BUTYL HYDROPEROXIDE** **$Me_3COOH$** **$C_4H_{10}O_2$**

1. Milas, N. A. *et al., J. Amer. Chem. Soc.*, 1946, **68**, 205
2. Castrantas, 1965, 15

Though relatively stable, explosions have been caused by distillation to dryness [1], or attempted distillation at atmospheric pressure [2].

*See other* ALKYL HYDROPEROXIDES

**DIETHYL PEROXIDE** **EtOOEt** **$C_4H_{10}O_2$**

1. Baeyer, A. *et al., Ber.*, 1900, **33**, 3387
2. Castrantas, 1965, 15; Baker, G. *et al., Chem. & Ind.*, 1964, 1988
3. Gray, P. *et al., Proc. R. Soc.*, 1971, **A325**, 175

While it is acknowledged as rather explosive, the stated lack of shock-sensitivity at room temperature [1] is countered by its alternative description as shock-sensitive and detonable [2]. The

vapour explodes above a certain critical pressure, at temperatures above 190°C [3].
*See other* DIALKYL PEROXIDES

**†1.1-DIMETHOXYETHANE** $(MeO)_2CHMe$ $C_4H_{10}O_2$

**†1.2-DIMETHOXYETHANE** $MeOC_2H_4OMe$ $C_4H_{10}O_2$

Lithium tetrahydroaluminate
*See* LITHIUM TETRAHYDROALUMINATE, $AlH_4Li$: 1,2-Dimethoxyethane

**1-HYDROXYBUTYL-3-HYDROPEROXIDE** $C_4H_{10}O_3$
$HOC_2H_4CH(Me)OOH$

Rieche, A., *Ber.*, 1930, **63**, 2642
Explodes on heating.
*See other* 1-OXYPEROXY COMPOUNDS

**†TRIMETHYL ORTHOFORMATE** $CH(OMe)_3$ $C_4H_{10}O_3$

Preparative hazard
*See* CHLOROFORM, $CHCl_3$ : Sodium, Methanol
: Sodium methoxide

**DIETHYL SULPHATE** $SO_2(OEt)_2$ $C_4H_{10}O_4S$

2,7-Dinitro-9-phenylphenanthridine,
Water
Hodgson, J. F., *Chem. & Ind.*, 1968, 1399
Accidental ingress of water to the heated mixture liberated sulphuric acid or its half ester which caused a violent reaction with generation of a very large volume of solid black foam.
*See* 4-NITROANILINE-2-SULPHONIC ACID, $C_6H_6N_2O_5S$
SULPHURIC ACID, $H_2O_4S$: Nitroaryl bases, etc.

Iron,
Water
Siebeneicher, K., *Angew. Chem.*, 1934, **47**, 105
Moisture in a sealed iron container caused hydrolysis to sulphuric acid, leading to corrosion of the metal and development of a high internal pressure of hydrogen, which ruptured the container.

Potassium *tert*-butoxide

*See* POTASSIUM *tert*-BUTOXIDE, $C_4H_9KO$: Acids, etc.

**ZINC ETHYLSULPHINATE** **$Zn(OS(O)Et)_2$** **$C_4H_{10}O_4S_2Zn$**

Preparative hazard

*See* DIETHYLZINC, $C_4H_{10}Zn$: Sulphur dioxide

**† 1-BUTANETHIOL** **BuSH** **$C_4H_{10}S$**

Nitric acid

*See* NITRIC ACID, $HNO_3$: Butanethiol

**† 2-BUTANETHIOL** **MeEtCHSH** **$C_4H_{10}S$**

**† DIETHYL SULPHIDE** **EtSEt** **$C_4H_{10}S$**

**† 2-METHYLPROPANETHIOL** **$Me_2CHCH_2SH$** **$C_4H_{10}S$**

**† 2-METHYL-2-PROPANETHIOL** **$Me_3CSH$** **$C_4H_{10}S$**

**DIETHYLZINC** **$Et_2Zn$** **$C_4H_{10}Zn$**

*Aluminum Alkyls and other Organometallics*, 5, Ethyl Corp., New York, 1967

Immediately pyrophoric in air, reacts violently with water.

Alkenes,
Diiodomethane

Houben-Weyl, 1973, Vol. 13.2a, 845

During preparation of cyclopropane derivatives, it is important to add the diiodomethane slowly to a solution of diethylzinc in the alkene. Addition of diethylzinc to an alkene–diiodomethane mixture may be explosively violent.

Halogens

*See* BROMINE, $Br_2$: Diethylzinc
CHLORINE, $Cl_2$: Diethylzinc

Methanol

Houben-Weyl, 1973, Vol. 13.2a, 855

Interaction is explosively violent and ignition ensues.

Ozone

*See* OZONE, $O_3$ : Alkylmetals

Sulphur dioxide

Houben-Weyl, 1973, Vol. 13.2a, 709

During preparation of zinc ethylsulphinate, addition of diethylzinc to liquid sulphur dioxide at −15°C leads to an explosively violent reaction. Condensation of the dioxide into cold diethylzinc leads to a controllable reaction on warming.

*See* ALKYLZINCS

**DIETHYLALUMINIUM HYDRIDE** **$Et_2AlH$** **$C_4H_{11}Al$**

*See* ALKYLALUMINIUM ALKOXIDES AND HYDRIDES

**DIETHYLARSINE** **$Et_2AsH$** **$C_4H_{11}As$**

Von Schwartz, 1918, 322,

Sidgwick, 1950, 762

Inflames in air, even at 0°C

*See other* ALKYLNON-METAL HYDRIDES

**DIETHYLGALLIUM HYDRIDE** **$Et_2GaH$** **$C_4H_{11}Ga$**

Eisch, J. J., *J. Amer. Chem. Soc.*, 1962, **84**, 3835

It ignites in air and reacts violently with water.

*See other* ALKYLMETAL HYDRIDES

† **BUTYLAMINE** **$BuNH_2$** **$C_4H_{11}N$**

Perchloryl fluoride

*See* PERCHLORYL FLUORIDE, $ClFO_3$ : Nitrogenous bases

† **2-BUTYLAMINE** **$MeEtCHNH_2$** **$C_4H_{11}N$**

† ***tert*-BUTYLAMINE** **$Me_3CNH_2$** **$C_4H_{11}N$**

† **DIETHYLAMINE** **$Et_2NH$** **$C_4H_{11}N$**

*MCA SD-97,* 1971

Cellulose nitrate

*See* CELLULOSE NITRATE: Amines

Dicyanofurazan

*See* DICYANOFURAZAN, $C_4N_4O$: Nitrogenous bases

**† ETHYLDIMETHYLAMINE** **$EtNMe_2$** **$C_4H_{11}N$**

**† ISOBUTYLAMINE** **$Me_2CHCH_2NH_2$** **$C_4H_{11}N$**

**$N$-2-HYDROXYETHYLDIMETHYLAMINE** **$C_4H_{11}NO$**

**$HOC_2H_4NMe_2$**

Cellulose nitrate

*See* CELLULOSE NITRATE: Amines

**1-AMINO-2,2,2-TRIS(HYDROXYMETHYL)METHANE** **$C_4H_{11}NO_3$**

**$H_2NC(CH_2OH)_3$**

Preparative hazard

*See* 1,1,1-TRIS(HYDROXYMETHYL)NITROMETHANE, $C_4H_9NO_5$: Hydrogen, etc.

**DIMETHYL ETHANEPHOSPHONITE** . **$C_2H_5P(OMe)_2$** **$C_4H_{11}O_2P$**

Arbuzov, B. A. *et al., Chem. Abs.*, 1953, **47**, 3226a

Absorbed onto filter paper, it ignites in air.

*See other* PYROPHORIC MATERIALS

**DIETHYLPHOSPHINE** **$Et_2PH$** **$C_4H_{11}P$**

Houben-Weyl, 1963, Vol. 12(1), 69

von Schwartz, 1918, 323

Secondary lower-alkylphosphines readily ignite in air.

*See other* ALKYLNON-METAL HYDRIDES

**ETHYLDIMETHYLPHOSPHINE** **$EtPMe_2$** **$C_4H_{11}P$**

Smith, J. F., private comm., 1970

May ignite in air.

*See other* ALKYLNON-METALS

**TETRAMETHYLDIARSANE** $Me_2AsAsMe_2$ $C_4H_{12}As_2$

Sidgwick, 1950, 770

Inflames in air.

*See other* ALKYLNON-METALS

**BIS-DIMETHYLARSINYL OXIDE** $(Me_2As)_2O$ $C_4H_{12}As_2O$

von Schwartz, 1918, 322

Inflames in air.

*See related* ALKYLNON-METALS

**BIS-DIMETHYLARSINYL SULPHIDE** $(Me_2As)_2S$ $C_4H_{12}As_2S$

von Schwartz, 1918, 322

Inflames in air.

*See related* ALKYLNON-METALS

**TETRAMETHYLDIGOLD DIAZIDE** $C_4H_{12}Au_2N_6$

$Me_2Au(N_3)_2AuMe_2$ (Me₂Au bridged by two N₃ groups to AuMe₂)

Beck, W. *et al.*, *Inorg. Chim. Acta,* 1968, 2, 468

The dimeric azide is extremely sensitive and may explode under water if touched.

*See other* GOLD COMPOUNDS
METAL AZIDES

**TETRAMETHYLAMMONIUM CHLORITE** $C_4H_{12}ClNO_2$

$NMe_4ClO_2$

Levi, G. R., *Gazz. Chim. Ital.*, 1922[1], 52, 207

The dry solid explodes on impact.

*See other* CHLORITE SALTS
OXOSALTS OF NITROGENOUS BASES

**DILITHIUM TETRAMETHYLCHROMATE** $C_4H_{12}CrLi_2$

$Li_2[CrMe_4]$

Kurras, E. *et al., J. Organomet. Chem.*, 1965, 4, 114–118

Isolated as a dioxane complex, it ignites in air.

*See related* ALKYLMETALS

**1,2-DIAMINO-2-METHYLPROPANEOXODIPEROXO-CHROMIUM (VI)** $C_4H_{12}CrN_2O_5$

$[Cr(C_4H_{12}N_2)O(O_2)_2]$

House, D. A. *et al.*, *Inorg. Chem.*, 1967, **6**, 1078, footnote 6

This blue precipitate is formed intermediately during preparation of 1,2-diamino-2-methylpropaneaquadiperoxochromium(IV) monohydrate, and its analogues are dangerously explosive and should not be isolated without precautions.

*See other* AMMINECHROMIUM PEROXOCOMPLEXES

**TETRAMETHYLAMMONIUM OZONIDE** $C_4H_{12}NO_3$

$Me_4NO_3$

Preparative hazard

*See* OZONE, $O_3$ : Tetramethylammonium hydroxide

† **2-DIMETHYLAMINOETHYLAMINE** $Me_2NC_2H_4NH_2$ $C_4H_{12}N_2$

***N*-HYDROXYETHYL-1,2-DIAMINOETHANE** $C_4H_{12}N_2O$

$HOC_2H_4NHC_2H_4NH_2$

Cellulose nitrate

*See* CELLULOSE NITRATE: Amines

**BIS(DIMETHYLAMINO) SULPHOXIDE** $C_4H_{12}N_2OS$

$(Me_2N)_2SO$

Sulphinyl chloride

*See* SULPHINYL CHLORIDE (THIONYL CHLORIDE), $Cl_2OS$: Bis(dimethylamino) sulphoxide

**SELENINYL BIS(DIMETHYLAMIDE)** $SeO(NMe_2)_2$ $C_4H_{12}N_2OSe$

Bailar, 1973, Vol. 2, 975

Vacuum distillation at temperatures around 50–60°C leads to explosive decomposition.

*See related* *N*-METAL DERIVATIVES

**TETRAMETHYL-2-TETRAZENE** $Me_2NN{=}NNMe_2$ $C_4H_{12}N_4$

Houben-Weyl, 1967, Vol. 10.2, 830

It explodes above its b.p., 130°C/1 bar.

*See other* HIGH-NITROGEN COMPOUNDS

**BIS-DIMETHYLSTIBINYL OXIDE** **$(Me_2Sb)_2O$** **$C_4H_{12}OSb_2$**

Sidgwick, 1950, 777

Ignites in air.

*See related* ALKYLMETALS

**DIETHYLHYDROXYTIN HYDROPEROXIDE** **$C_4H_{12}O_3Sn$**

**$Et_2HOSnOOH$**

Bailar, 1973, Vol. 2, 68

It is explosive.

*See other* ORGANOMINERAL PEROXIDES

**TETRAMETHYLDIPHOSPHANE** **$Me_2PPMe_2$** **$C_4H_{12}P_2$**

1. Kardosky, G. *et al., Inorg. Synth.*, 1973, **14**, 15
2. Goldsberry, R. *et al.*, ibid., 1972, **13**, 28

It ignites in air [1] and unreacted tetramethyldiphosphane residues or its solutions in diethyl ether may also ignite [2].

*See other* ALKYL NON-METALS

**TETRAMETHYLLEAD** **$Me_4Pb$** **$C_4H_{12}Pb$**

Sidgwick, 1950, 463

Liable to explode violently above 90°C.

*See other* ALKYLMETALS

**TETRAMETHYLPLATINUM** **$Me_4Pt$** **$C_4H_{12}Pt$**

Gilman, H. *et al., J. Amer. Chem. Soc.*, 1953, **75**, 2065

It explodes weakly on heating.

*See* TRIMETHYLPLATINUM HYDROXIDE, $C_3H_{10}OPt$

*See other* ALKYLMETALS

PLATINUM COMPOUNDS

**TETRAMETHYLDISTIBANE** **$Me_2SbSbMe_2$** **$C_4H_{12}Sb_2$**

Sidgwick, 1950, 779

Ignites in air.

*See other* ALKYLMETALS

† **TETRAMETHYLSILANE** **$Me_4Si$** **$C_4H_{12}Si$**

Chlorine

*See* CHLORINE, $Cl_2$: Antimony trichloride, etc

## † TETRAMETHYLTIN $Me_4Sn$ $C_4H_{12}Sn$

Dinitrogen tetraoxide

*See* DINITROGEN TETRAOXIDE (NITROGEN DIOXIDE), $N_2O_4$: Tetramethyltin

*See other* ALKYLMETALS

## BIS(2-AMINOETHYL)AMINESILVER NITRATE $C_4H_{13}AgN_4O_3$

$[AgC_4H_{13}N_3]NO_3$

Tennhouse, G. J. *et al., J. Inorg. Nucl. Chem.*, 1966, **28**, 682–684

The complex explodes at 200°C.

*See* AMMINEMETAL OXOSALTS

## BIS(2-AMINOETHYL)AMINECOBALT(III) AZIDE $C_4H_{13}CoN_{12}$

$[Co(H_2NC_2H_4)_2NH][N_3]_3$

Druding, L. F. *et al., Acta Cryst.*, 1974, **B30**, 2386

The shock-sensitivity and dangerously explosive nature is stressed.

*See* AMMINECOBALT(III) AZIDES

*See related* METAL AZIDES

## DIETHYLENETRIAMINEDIPEROXOCHROMIUM(IV) $C_4H_{13}CrN_3O_4$

$[Cr(C_4H_{13}N_3)(O_2)_2]$

House, D. A. *et al., Inorg. Chem.*, 1966, **5**, 840

The monohydrate explodes at 109–110°C during slow heating.

*See other* AMMINECHROMIUM PEROXOCOMPLEXES

## BIS(2-AMINOETHYL)AMINE (DIETHYLENETRIAMINE) $C_4H_{13}N_3$

$(H_2NC_2H_4)_2NH$

Cellulose nitrate

*See* CELLULOSE NITRATE : Amines

Nitromethane

*See* NITROMETHANE, $CH_3NO_2$ : Bis(2-aminoethyl)amine

## TETRAMETHYLDIALUMINIUM DIHYDRIDE $C_4H_{14}Al_2$

$Me_3Al \cdot MeAlH_2$

Wiberg, W. E. *et al., Angew. Chem.*, 1939, **52**, 372

It ignites and burns explosively in air.

*See other* ALKYLALUMINIUM ALKOXIDES, etc.

## 1,2-DIAMINO-2-METHYLPROPANEAQUADIPEROXO-CHROMIUM(IV)

$C_4H_{14}CrN_2O_5$

$[Cr(C_4H_{12}N_2)H_2O(O_2)_2]$

House, D. A. *et al.*, *Inorg. Chem.*, 1967, **6**, 1078

The monohydrate exploded at 83–84°C during slow heating, and is potentially explosive at 20–25°C. There is also a preparative hazard.

*See* 1,2-DIAMINO-2-METHYLPROPANEOXODIPEROXOCHROMIUM(VI), $C_4H_{12}CrN_2O_5$

*See other* AMMINECHROMIUM PEROXOCOMPLEXES

## TETRAMETHYLAMMONIUM AMIDE $NMe_4NH_2$ $C_4H_{14}N_2$

Ammonia

Musker, W. K., *J. Org. Chem.*, 1967, **32**, 3189

During the preparation, the liquid ammonia used as solvent must be completely removed at –45°C. The compound decomposes explosively at ambient temperature in presence of ammonia.

*See related N*-METAL DERIVATIVES

## *cis-* or *trans*-BIS(1,2-DIAMINOETHANE)DINITROCOBALT(III) PERCHLORATES

$C_4H_{16}ClCoN_6O_8$

$[Co(C_2H_8N_2)_2(NO_2)_2]ClO_4$

Seel, F. *et al.*, *Z. Anorg. Allgem. Chem.*, 1974, **408**, 281

They are dangerously explosive compounds.

*See other* AMMINEMETAL OXOSALTS

## TETRATHIOUREAMANGANESE(II) PERCHLORATE $C_4H_{16}Cl_2MnN_8O_8S_4$

$[Mn(CH_4N_2S)_4][ClO_4]_2$

Karnaukhov, A. S. *et al.*, *Chem. Abs.*, 1973, **79**, 140195y

The complex decomposes explosively at 257°C.

*See other* AMMINEMETAL OXOSALTS

## BIS-1,2-DIAMINOETHANEDICHLOROCOBALT(III) CHLORATE

$C_4H_{16}Cl_3CoN_4O_3$

$[Co(C_2H_8N_2)_2Cl_2]ClO_3$

It explodes at 320°C.

*See other* AMMINEMETAL OXOSALTS

**BIS-1,2-DIAMINOETHANEDICHLOROCOBALT(III) PERCHLORATE** $C_4H_{16}Cl_3CoN_4O_4$

$[Co(C_2H_8N_2)_2Cl_2]ClO_4$

It explodes at 300°C; low impact sensitivity.

*See other* AMMINEMETAL OXOSALTS

***cis*-BIS-1,2-DIAMINOETHANEDINITROCOBALT(III) IODATE** $C_4H_{16}CoIN_6O_7$

$[Co(C_2H_8N_2)_2(NO_2)_2]IO_3$

Lobanov, N. I., *Zh. Neorg. Khim.*, 1959, **4**, 151

It dissociates explosively on heating.

*See other* AMMINEMETAL OXOSALTS

**DIAZIDO-BIS(1,2-DIAMINOETHANE)RUTHENIUM(II) HEXAFLUOROPHOSPHATE** $C_4H_{16}F_6N_{10}PRu$

$[Ru(C_2H_8N_2)_2(N_3)_2][PF_6]$

Kane-Maguire, L. A. P. *et al.*, *Inorg. Synth.*, 1970, **12**, 24

Small explosions were observed on scratching the crystals with a metal spatula.

*See related* METAL AZIDES

**BIS(1,2-DIAMINOETHANE)HYDROXOOXORHENIUM(V) DIPERCHLORATE** $C_4H_{17}Cl_2N_4O_{10}Re$

$[Re(C_2H_8N_2)_2O(OH)][ClO_4]_2$

Murmann, R. K. *et al.*, *Inorg. Synth.*, 1966, **8**, 174–175

It explodes violently when dried at above room temperature and is also shock-sensitive. Several explosions occurred during analytical combustion.

*See other* AMMINEMETAL OXOSALTS

**PENTAAMMINEPYRAZINERUTHENIUM(II) DIPERCHLORATE** $C_4H_{19}Cl_2N_7O_8Ru$

$[Ru(NH_3)_5C_4H_4N_2][ClO_4]_2$

Creutz, C. A., private comm., 1969

After ether washing, 30 mg of the salt exploded violently when disturbed.

*See other* AMMINEMETAL OXOSALTS

## BIS(1,2-DIAMINOETHANEAQUA)COBALT(III) PERCHLORATE

$C_4H_{20}Cl_3CoN_4O_{14}$

$[Co(C_2H_8N_2)_2(H_2O)_2][ClO_4]_3$

Seel, F. *et al., Z. Anorg. Allgem. Chem.*, 1974, **408**, 281
During evaporation of a solution of the complex salt, a very violent explosion occurred.
*See other* AMMINEMETAL OXOSALTS

## TETRAAMMINE-2,3-BUTANEDIIMINERUTHENIUM(III) PERCHLORATE

$C_4H_{20}Cl_3N_6O_{12}Ru$

$[Ru(NH_3)_4C_4H_8N_2][ClO_4]_3$

Evans, I. P. *et al., J. Amer. Chem. Soc.*, 1977, **98**, 8042
It is explosive.
*See other* AMMINEMETAL OXOSALTS

## BIS(DIMETHYLAMINOBORANE)–ALUMINIUM TETRAHYDROBORATE

$C_4H_{22}AlB_3N_2$

$[Me_2NBH_3]_2[AlBH_4]$

Burg., A. B. *et al., J. Amer. Chem. Soc.*, 1951, **73**, 957
The impure oily product ignites in air and reacts violently with water.
*See other* COMPLEX HYDRIDES

## POTASSIUM TETRACYANOMERCURATE (2–)

$C_4HgK_2N_4$

$K_2[Hg(CN)_4]$

Ammonia
Pieters, 1957, 30
Contact may be explosive.
*See also* POTASSIUM HEXACYANOFERRATE (3–), $C_6FeK_3N_6$: Ammonia
*See other* METAL CYANIDES

## μ-1,2-BIS(CYANOMERCURIO)ETHANEDIYLIDENE-DIMERCURY(II)

$C_4Hg_4N_2$

CNHgC(=Hg)C(=Hg)HgCN

Whitmore, 1921, 128
It explodes slightly on heating.
*See other* MERCURY COMPOUNDS

**1,4-DIIODO-1,3-BUTADIYNE** **IC≡CC≡CI** **$C_4I_2$**

Sorbe, 1968, 66

It explodes at about 100°C.

*See other* HALOACETYLENE DERIVATIVES

**POTASSIUM TETRACYANOTITANATE(4–)** **$C_4K_4N_4Ti$**

**$K_4[Ti(CN)_4]$**

Water

Nicholls, D. *et al., J. Chem. Soc., Chem. Comm.*, 1974, 635–636

Interaction is violent.

*See other* METAL CYANIDES AND CYANOCOMPLEXES

**LANTHANUM DICARBIDE** **$La(C_2)_2$** **$C_4La$**

Borlas, R. A. *et al., Chem. Abs.*, 1976, **84**, 182199m

'On the flammability and explosiveness of lanthanum dicarbide and cerium dicarbide powders' – title only translated.

*See related* METAL ACETYLIDES

**DICYANOACETYLENE** **N≡CC≡CC≡N** **$C_4N_2$**

Ciganek, E. *et al., J. Org. Chem.*, 1968, **33**, 542

Highly endothermic, it is potentially explosive in the pure state or in concentrated solutions, but fairly stable in dilute solution.

*See related* HALOACETYLENE DERIVATIVES

*See other* CYANO COMPOUNDS

**SODIUM TETRACYANATOPALLADATE(2–)** **$C_4N_4Na_2O_4Pd$**

**$Na_2[Pd(CNO)_4]$**

Bailar, 1973, Vol. 3, 1288

It explodes on heating or impact.

*See related* METAL CYANIDES AND CYANOCOMPLEXES

**DICYANOFURAZAN** **NCC=NON=CCN** **$C_4N_4O$**

Homewood, R. H. *et al.*, US Pat. 3 832 249, 1974

It is a stable and relatively insensitive but powerful explosive.

Nitrogenous bases

Denson, D.B. *et al.*, US Pat. 3 740 947, 1973 (*Chem. Abs.*, 1973, **79**, 94195p)

Contact of dicyanofurazan, or its *N*-oxide (dicyanofuroxan), with hydrazine, mono- or di-methylhydrazine, piperidine, piperazine, diethylamine or their mixtures is instantaneously explosive.

## DICYANOFUROXAN (3,4-DICYANOFURAZAN *N*-OXIDE) $C_4N_4O_2$

NCC=NON(O)=CCN

Nitrogenous bases

*See* DICYANOFURAZAN, $C_4N_4O$: Nitrogenous bases

## † TETRACARBONYLNICKEL Ni(CO)$_4$ $C_4NiO_4$

Bromine

Blanchard, A. A. *et al., J. Amer. Chem. Soc.,* 1926, **48**, 872

The two interact explosively in the liquid state, but smoothly as vapour.

Dinitrogen tetraoxide

*See* DINITROGEN TETRAOXIDE (NITROGEN DIOXIDE), $N_2O_4$: Tetracarbonylnickel

Mercury,
Oxygen

Mellor, 1946, Vol. 5, 955

A mixture of the dry carbonyl and oxygen will explode on vigorous shaking with mercury.

Oxygen,
Butane

1. Egerton, A. *et al., Proc. R. Soc.,* 1954, **A225**, 427
2. Badin, E. J. *et al., J. Amer. Chem. Soc.,* 1948, **70**, 2055

The carbonyl on exposure to air produces a deposit which becomes peroxidised and may ignite. Mixtures with air or oxygen at low partial and total pressures explode after a variable induction period [1]. Addition of the carbonyl to a butane–oxygen mixture at 20–40°C caused explosive reaction in some cases [2].

*See other* CARBONYLMETALS

## HEXACHLOROCYCLOPENTADIENE ClC=CClCCl$_2$CCl=CCl $C_5Cl_6$

Sodium

*See* SODIUM Na: Halocarbons

**POTASSIUM PENTACYANODIPEROXOCHROMATE(5–)** $C_5CrK_5N_5O_4$

$K_5[Cr(CN)_5(O_2)_2]$

Bailar, 1973, Vol. 4, 167

A highly explosive material, with internal redox features.

*See other* REDOX COMPOUNDS

**SODIUM PENTACYANONITROSYLFERRATE (2–)** $C_5FeN_6Na_2O$

$Na_2[Fe(CN)_5NO]$

Sodium nitrite

*See* SODIUM NITRITE, $NNaO_2$: Metal cyanides

**† PENTACARBONYLIRON** $Fe(CO)_5$ $C_5FeO_5$

It may ignite in air.

Acetic acid,
Water

Braye, E. H. *et al., Inorg. Synth.*, 1966, **8**, 179

A brown pyrophoric powder is produced if the carbonyl is dissolved in acetic acid containing above 5% of water.

Nitrogen oxide

*See* NITROGEN OXIDE, NO: Pentacarbonyliron

Transition metal halides,
Zinc

Lawrenson, M. J., private comm., 1970

The preparation of metal carbonyls by treating a transition metal halide either with carbon monoxide and zinc or with iron pentacarbonyl is well-known and smooth. However, a violent eruptive reaction occurs if a methanolic solution of cobalt, rhodium or ruthenium halide is treated with both zinc and iron pentacarbonyl.

*See other* CARBONYLMETALS

**1,3-PENTADIYN-1-YLSILVER** $AgC{\equiv}CC{\equiv}CCH_3$ $C_5H_3Ag$

Schluhbach, H. H. *et al., Ann.*, 1950, **568**, 155

Very sensitive to impact or friction and explodes when moistened with sulphuric acid.

## FUROYL CHLORIDE

$C_5H_3ClO_2$

$\overline{HC{=}CHOC(COCl){=}CH}$

Preparative hazard

*See* PHOSPHORUS TRICHLORIDE, $Cl_3P$: Carboxylic acids

## 1,3-PENTADIYN-1-YLCOPPER

$CuC{\equiv}CC{\equiv}CCH_3$ $C_5H_3Cu$

Schluhbach, H. H. *et al., Ann.*, 1950, **568**, 155

Explodes on impact or friction.

*See other* METAL ACETYLIDES

## 2-FUROYL AZIDE

$C_5H_3N_3O_2$

$\overline{CH{=}CHOC(CON_3){=}CH}$

Dunlop, 1953, 544

The azide explodes violently on heating in absence of a solvent or diluent.

*See other* ACYL AZIDES

## 2-HYDROXY-3,5-DINITROPYRIDINE

$C_5H_3N_3O_5$

$\overline{N{=}C(OH)C(NO_2){=}CHC(NO_2){=}CH}$

Glowiak, B., *Chem. Abs.*, 1963, **58**, 498h

Various heavy metal salts show explosive properties, and the lead salt might be useful as an initiating explosive.

*See other* POLYNITROARYL COMPOUNDS

## 1,3-PENTADIYNE

$HC{\equiv}CC{\equiv}CCH_3$ $C_5H_4$

Brandsma, 1971, 7, 36

It explodes on distillation at atmospheric pressure.

*See other* ACETYLENIC COMPOUNDS

## 3-PYRIDINEDIAZONIUM TETRAFLUOROBORATE

$C_5H_4BF_4N_3$

$\overline{CH{=}NCH{=}CHCH{=}C}N_2^+\ BF_4^-$

1. Johnson, E. P. *et al., Chem. Eng. News,* 1967, **45**(44), 44
2. Roe, A. *et al., J. Amer. Chem. Soc.,* 1947, **69**, 2443

A sample, air-dried on aluminium foil, exploded spontaneously, while another sample exploded on heating to 47°C [1]. The earlier reference describes the instability of the salt above 15°C if freed of solvent, and both the 2- and 4-isomeric salts were found to be very unstable and incapable of isolation [2].

*See other* DIAZONIUM TETRAHALOBORATES

**2-CHLORO-5-CHLOROMETHYLTHIOPHENE** $C_5H_4Cl_2S$

$\overline{CH=C(Cl)SC(CH_2Cl)=CH}$

Rosenthal, N. A., *J. Amer. Chem. Soc.*, 1951, **73**, 590

Storage at ambient temperature may lead to explosively violent decomposition.

*See* 2-HALOMETHYL-FURANS OR -THIOPHENES

**2-FLUOROPYRIDINE** $C_5H_4FN$

$F\overline{C=NCH=CHCH=CH}$

*See* BROMINE TRIFLUORIDE, $BrF_3$: Pyridine

**DIAZOCYCLOPENTADIENE** $\overline{CH=CHCH=CHC}N_2$ $C_5H_4N_2$

1. Wedd, A. G., *Chem. & Ind.*, 1970, 109
2. Aarons, L. J. *et al., J. Chem. Soc., Faraday Trans. 2*, 1974, **70**, 1108
3. DeMore, W. B. *et al., J. Amer. Chem. Soc.*, 1959, **81**, 5875
4. Weil, T., *J. Org. Chem.*, 1963, **28**, 2472
5. Regitz, M. *et al., Tetrahedron*, 1967, **23**, 2706

The material exploded violently during distillation at 48–53°C/66 mbar. Use of solutions of undistilled material is recommended [1]. The compound exploded on one occasion in the solid state after condensation at –196°C, probably owing to fortuitous tribomechanical shock [2]. A similar explosion on chilling a sample in liquid nitrogen had been noted previously [3]. Of two simplified methods of isolating the material without distillation [4,5], the former appears to give higher yields.

*See other* DIAZO COMPOUNDS

**2-FURALDEHYDE (FURFURAL)** $\overline{OCH=CHCH=C}CHO$ $C_5H_4O_2$

Sodium hydrogencarbonate (bicarbonate)

Unpublished information, 1977

Several cases of spontaneous ignition after exposure to air of fine coke removed from filter strainers on a petroleum refinery furfural extraction unit have been noted. This has been associated with the use of sodium bicarbonate injected into the plant for pH control, which produced a pH of 10.5 locally. This would lead to a Cannizzaro reaction and conversion of the aldehyde to furfuryl alcohol with subsequent resinification catalysed by acids produced by autoxidation of the aldehyde. Pyrolysis GLC showed the presence of a significant proportion of furfuryl alcohol-derived resins in the coke. The latter is now discarded into drums of water, immediately after discharge from the strainers, to prevent further incidents.

*See other* PYROPHORIC MATERIALS

**PEROXYFUROIC ACID** $C_5H_4O_4$

$\overline{OCH{=}CHCH{=}C}CO_2OH$

Alone,
or Metal salts,
or Organic materials

Milas, N. A. *et al., J. Amer. Chem. Soc.*, 1934, **56**, 1221

Thermal decomposition of the peracid becomes violently explosive at 40°C. Intimate contact with finely divided metal salts (mainly halides), ergosterol, pyrogallol or animal charcoal led to explosive decomposition (often violent) in 18 out of 28 experiments.

*See other* PEROXYACIDS

**SILVER CYCLOPROPYLACETYLIDE** $AgC{\equiv}C\overline{CHCH_2CH_2}$ $C_5H_5Ag$

Slobodin, Ya. M. *et al., Chem. Abs.*, 1952, **46**, 10112e

It explodes on heating.

*See other* METAL ACETYLIDES

**CYCLOPENTADIENYLSILVER PERCHLORATE** $C_5H_5AgClO_4$

Ulbricht, T. L. V., *Chem. & Ind.*, 1961, 1570

The complex explodes on heating.

*See other* SILVER COMPOUNDS

**CYCLOPENTADIENYLGOLD(I)** $C_5H_5Au$

Huttel, R. *et al., Angew. Chem. (Intern. Ed.)*, 1967, **6**, 862

It is sensitive to friction and heat, often deflagrating on gentle warming.

*See other* GOLD COMPOUNDS

## 2-BROMOMETHYLFURAN

**$C_5H_5BrO$**

**$\overline{OCH{=}CHCH{=}C}CH_2Br$**

Dunlop, 1953, 231

It is very unstable, and the liberated hydrogen bromide accelerates further decomposition to explosive violence.

*See* 2-HALOMETHYLFURANS

## 2-CHLOROMETHYLFURAN

**$C_5H_5ClO$**

**$\overline{OCH{=}CHCH{=}C}CH_2Cl$**

*ABCM Quart. Safety Summ.*, 1962, **33**, 2

A small sample of freshly prepared and distilled material, when stored over a weekend, exploded violently owing to polymerisation or decomposition. Material should be prepared and used immediately, or if brief storage is inevitable, refrigeration is necessary.

*See* 2-HALOMETHYL-FURANS AND -THIOPHENES

## 2-CHLOROMETHYLTHIOPHENE

**$C_5H_5ClS$**

**$\overline{SCH{=}CHCH{=}C}CH_2Cl$**

1. Bergeim, F. H., *Chem. Eng. News,* 1952, **30**, 2546; Meyer, F.C., ibid., 3352
2. Wiberg, K. B., *Org. Synth.,* 1955, Coll. Vol. 3, 197

The material is unstable and gradually decomposes, even when kept cold and dark, with liberation of hydrogen chloride, which accelerates the decomposition. If kept in closed containers, pressure increase may cause an explosion [1]. Amines stabilise the material, which can then be kept cold in a vented container for several months [2].

*See* 2-HALOMETHYL-FURANS AND -THIOPHENES

## η-CYCLOPENTADIENYLTITANIUM(III) DICHLORIDE POLYMER

**$(C_5H_5Cl_2Ti)_n$**

**$(C_5H_5TiCl_2)_n$**

Lucas, C. R. *et al., Inorg. Synth.*, 1976, **16**, 238

Pyrophoric in air.

*See other* ORGANOMETALLICS

### 1-IODO-3-PENTEN-1-YNE $IC{\equiv}CCH{=}CHCH_3$ $C_5H_5I$

Vaughn, J. A. *et al., J. Amer. Chem. Soc.*, 1934, **56**, 1208
After distillation at 72°C/62 mbar, the residue always exploded if heating was continued.
*See other* HALOACETYLENE DERIVATIVES

### MONOPOTASSIUM *aci*-2,5-DINITROCYCLOPENTANONE $C_5H_5KN_2O_5$

$\overline{COCHNO_2[CH_2]_2C}{=}N(O)OK$

Wieland, H. *et al., Ann.*, 1928, **461**, 304
The monosalt explodes at 154–158°C and is less stable than the di-potassium salt (241–245°C).
*See other aci*-NITRO SALTS

### † PYRIDINE $C_5H_5N$

$\overline{N{=}CHCH{=}CHCH{=}CH}$

Formamide,
Iodine,
Sulphur trioxide
*See* FORMAMIDE, $CH_3NO$: Iodine, etc.

Maleic anhydride
*See* MALEIC ANHYDRIDE, $C_4H_2O_3$: Cations, etc.

Oxidants
*See* BROMINE TRIFLUORIDE, $BrF_3$: Pyridine
TRIFLUOROMETHYL HYPOFLUORITE, $CF_4O$: Pyridine
CHROMIUM TRIOXIDE, $CrO_3$: Pyridine
FLUORINE, $F_2$: Nitrogenous bases
DINITROGEN TETRAOXIDE, $N_2O_4$: Heterocyclic bases

### 1,2-DIHYDROPYRIDO[2,1-*e*]TETRAZOLE $C_5H_5N_4$

CH=CHCH=CHNC=NN=CH (ring structure)

Fargher, R. G. *et al., J. Chem. Soc.*, 1915, **107**, 695
Explodes on touching with a hot rod.
*See other* TETRAZOLES

**METHYLNITROTHIOPHENE** **$MeC_4H_2SNO_2$** **$C_5H_5NO_2S$**

Preparative hazard

*See* NITRIC ACID, $HNO_3$ : Methylthiophene

**CYCLOPENTADIENYLSODIUM** **$C_5H_5Na$**

King, R. B. *et al., Inorg. Synth.*, 1963, **7**, 101

The solid obtained by evaporation of the air-sensitive solution is pyrophoric in air.

Lead(II) nitrate

*See* LEAD(II) NITRATE, $N_2O_6Pb$: Cyclopentadienylsodium

*See* ALKALI-METAL DERIVATIVES OF HYDROCARBONS

*See other* ORGANOMETALLICS

**BICYCLO[2.1.0]PENT-2-ENE** **$C_5H_6$**

Andrist, A. H. *et al., Org. Synth.*, 1976, **55**, 18

A purified undiluted sample was reported to explode.

*See other* STRAINED RING COMPOUNDS

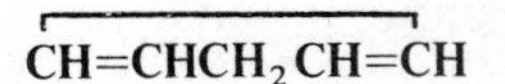

† **CYCLOPENTADIENE** **CH=CHCH$_2$CH=CH** **$C_5H_6$**

Kirk-Othmer, 1965, Vol. 6, 690

Raistrick, B. *et al., J. Chem. Soc.*, 1939, 1761–1773

Dimerisation is highly exothermic, the rate increasing rapidly with temperature, and may cause rupture of a closed uncooled container. The monomer may largely be prevented from dimerising by storage at $-80°C$ or below.

The polymerisation of the undiluted diene may become explosive within the range 0–40°C and at pressures up to 340 bar. The effect of diluents was also studied.

*See* PROPENE, $C_3H_6$

*See other* POLYMERISATION INCIDENTS

Nitric acid

*See* NITRIC ACID, $HNO_3$ : Hydrocarbons

Oxides of nitrogen

*See* NITROGEN OXIDE, NO: Dienes, Oxygen
DINITROGEN TETRAOXIDE, $N_2O_4$ : Hydrocarbons

Oxygen

Hock, H. *et al., Chem. Ber.*, 1951, **84**, 349

Exposure of the diene to oxygen gives peroxidic products containing

some monomeric, but largely polymeric, peroxides, which explode strongly on contact with a flame.
*See other* POLYPEROXIDES

Sulphuric acid
*See* SULPHURIC ACID, $H_2O_4S$: Cyclopentadiene
*See other* DIENES

**† 2-METHYL-1-BUTEN-3-YNE** **$CH_2{=}CMeC{\equiv}CH$** **$C_5H_6$**

**3,5-DIBROMOCYCLOPENTENE** **$C_5H_6Br_2$**

**$\overline{CH{=}CHCHBrCH_2CHBr}$**

Lithium tetrahydroaluminate
*See* LITHIUM TETRAHYDROALUMINATE, $AlH_4Li$: 3,5-Dibromocyclopentene

**PYRIDINIUM PERCHLORATE** **$C_5H_6ClNO_4$**

**$\overline{CH{=}CHCH{=}CHCH{=}N^+}H\ ClO_4^-$**

1. Kuhn, R. *et al.*, *Chem. Ztg.*, 1950, **74**, 139
2. Anon., *Chemiearbeit,* 1963, **15**(3), 19
3. Arndt, F. *et al.*, *Chem. Ztg.*, 1950, **74**, 140
4. Schumacher, 1960, 213
5. Zacherl, M. K., *Mikrochemie*, 1948, **33**, 387–388

It can be detonated on impact, but is normally considered a stable intermediate (m.p., 288°C), suitable for purification of pyridine [1]. Occasionally explosions have occurred when the salt was disturbed [2], which have been variously attributed to presence of ethyl perchlorate, ammonium perchlorate or chlorates. A safer preparative modification is described [3]. It explodes on heating to above 335°C, or at a lower temperature if ammonium perchlorate is present [4].

A violent explosion which occurred during the final distillation according to the preferred method [3] was recorded [5].
*See other* OXOSALTS OF NITROGENOUS BASES

**1,3-DICHLORO-5,5-DIMETHYL-2,4-IMIDAZOLIDINDIONE** **$C_5H_6Cl_2N_2O_2$**

**$ClN\overline{CONClCOC}Me_2$**

Xylene
Anon., *Chem. Trade J.*, 1951, **129**, 136

An attempt to chlorinate xylene with the 'dichlorohydantoin' caused a violent explosion.
*See other N*-HALOIMIDES

**3,3-DIFLUORO-1,2-DIMETHOXYCYCLOPROPENE** $C_5H_6F_2O_2$

MeOC=C(OMe)$CF_2$

*See* FLUORINATED CYCLOPROPENYL ETHERS

**4,4-BIS(DIFLUOROAMINO)-3-FLUOROIMINO-1-PENTENE** $C_5H_6F_5N_3$

$CH_2$=CHC(NF)C$(NF_2)_2CH_3$

Parker, C. O. *et al., J. Org. Chem.*, 1972, **37**, 922
Samples of this and related poly-difluoroamino compounds exploded during analytical combustion.
*See other N*-HALOGEN COMPOUNDS

**4-IODO-3,5-DIMETHYLISOXAZOLE** ON=CMeC(I)=CMe $C_5H_6INO$

Peroxytrifluoroacetic acid
*See* PEROXYTRIFLUOROACETIC ACID, $C_2HF_3O_3$: 4-Iodo-3,5-dimethylisoxazole

**GLUTARYL DIAZIDE** $N_3CO[CH_2]_3CON_3$ $C_5H_6N_6O_2$

Curtiss, T. *et al., J. Prakt. Chem.*, 1900, **62**, 196
A very small sample exploded sharply on heating.
*See other* ACYL AZIDES

**DIVINYL KETONE** $CH_2$=CHCOCH=$CH_2$ $C_5H_6O$

Trinitromethane
*See* TRINITROMETHANE, $CHN_3O_6$: Divinyl ketone

† **2-METHYLFURAN** OCH=CHCH=CMe $C_5H_6O$

**2-PENTEN-4-YN-3-OL** $CH_3CH$=C(OH)C≡CH $C_5H_6O$

Brandsma, 1971, 73
The residue from distillation at 20 mbar exploded vigorously at above 90°C.
*See other* ACETYLENIC COMPOUNDS

**FURFURYL ALCOHOL** $C_5H_6O_2$

$\overline{OCH=CHCH=C}CH_2OH$

Acids

1. Tobie, W. C., *Chem. Eng. News,* 1940, **18**, 72
2. *MCA Case History No. 858*
3. Dunlop 1953, 214, 221, 783
4. Golosova, L. V., Russ. Pat. 478 024, 1975 (*Chem. Abs.*, 1975, **83**, 194098v)

A mixture of the alcohol with formic acid rapidly self-heated, then reacted violently [1]. A stirred mixture with cyanoacetic acid exploded violently after application of heat [2]. Contact with acids or acidic materials causes self-condensation of the alcohol, which may be explosively violent under unsuitable physical conditions. The general mechanism has been discussed [3].

The explosion hazards associated with the use of acidic catalysts to polymerise furfuryl alcohol may be avoided by using as catalyst the condensation product of *m*-phenylenediamine and 1-chloro-2,3-epoxypropane [4].

*See* NITRIC ACID, $HNO_3$: Furfuryl alcohol

Hydrogen peroxide

*See* HYDROGEN PEROXIDE, $H_2O_2$: Alcohols

**† 2-METHYLTHIOPHENE** $\overline{SCH=CHCH=C}Me$ $C_5H_6S$

**2-PROPYNYL VINYL SULPHIDE** $HC\equiv CCH_2SCH=CH_2$ $C_5H_6S$

Brandsma, 1971, 7, 182

It decomposes explosively above 85°C.

*See other* ACETYLENIC COMPOUNDS

**4-BROMOCYCLOPENTENE** $\overline{CH=CHCH_2CHBrC}H_2$ $C_5H_7Br$

Preparative hazard

*See* LITHIUM TETRAHYDROALUMINATE, $AlH_4Li$: 3,5-Dibromocyclopentene

**ETHYL 2,2,3-TRIFLUOROPROPIONATE** $C_5H_7F_3O_2$

$FCH_2CF_2CO_2Et$

Sodium hydride

*See* SODIUM HYDRIDE, HNa: Ethyl 2,2,3-trifluoropropionate

**† 1-METHYLPYRROLE** $\overline{\text{MeNCH=CHCH=CH}}$ $C_5H_7N$

**† CYCLOPENTENE** $\overline{\text{CH=CH[CH}_2\text{]}_2\text{CH}_2}$ $C_5H_8$

**† 2-METHYL-1,3-BUTADIENE (ISOPRENE)** $C_5H_8$

$CH_2=CMeCH=CH_2$

Acetone

Lokhmacheva, I. K. *et al., Chem. Abs.*, 1975, **82**, 63856x

Prevention of peroxidation of isoprene–acetone mixtures, and other hazards involved in the industrial preparation of synthetic citral, are discussed.

Air

Kirk-Othmer, 1967, Vol. 12, 79

In absence of inhibitors, isoprene absorbs air with formation of peroxides, which do not separate from solution. Although the solution is not detonable, the gummy peroxide polymer obtained by evaporation can be detonated on impact under standard conditions.

*See also* BUTADIENE, $C_4H_6$: Alone, or Air

Ozone

Loveland, J. W., *Chem. Eng. News,* 1956, **34**(3), 292

Isoprene (1 g) dissolved in heptane was ozonised at −78°C. Soon after cooling was stopped, a violent explosion, followed by a lighter one, occurred. This was attributed to high concentrations of peroxides and ozonides building up at the low temperature employed. Operation at a higher temperature would permit the ozonides and peroxides to decompose, avoiding high concentrations in the reaction mixture.

*See other* OZONIDES

Vinylamine

*See* VINYLAMINE, $C_2H_5N$: Isoprene

*See other* DIENES

**† 1,3-PENTADIENE** $CH_2=CHCH=CHCH_3$ $C_5H_8$

**† 1,4-PENTADIENE** $CH_2=CHCH_2CH=CH_2$ $C_5H_8$

**† 1-PENTYNE** $HC{\equiv}CPr$ $C_5H_8$

**2-PENTYNE** $MeC{\equiv}CEt$ $C_5H_8$

Mercury,
Silver perchlorate
*See* SILVER PERCHLORATE, $AgClO_4$ : Alkynes, etc.

**TETRAKIS-(*N,N*-DICHLOROAMINOMETHYL)METHANE** $C_5H_8Cl_8N_4$
$(Cl_2NCH_2)_4C$

Lévy, R. S. *et al., Mem. Poudres,* 1958, **40**, 109
Shock- and heat-sensitive, it is a brisant more powerful than mercury fulminate.
*See other N*-HALOGEN COMPOUNDS

**† ALLYL VINYL ETHER** $C_5H_8O$
$CH_2{=}CHCH_2OCH{=}CH_2$

**CYCLOPENTANOTE** $\overline{CO[CH_2]_3CH_2}$ $C_5H_8O$

Hydrogen peroxide,
Nitric acid
*See* HYDROGEN PEROXIDE, $H_2O_2$ : Ketones, etc.

**† CYCLOPROPYL METHYL KETONE** $\overline{CH_2CH_2CH}COCH_3$ $C_5H_8O$

**† 2,3-DIHYDROPYRAN** $\overline{O[CH_2]_3CH{=}CH}$ $C_5H_8O$

**1-ETHOXY-2-PROPYNE** $EtOCH_2C{\equiv}CH$ $C_5H_8O$

Brandsma, 1971, 172
Distillation of a 1 kg quantity at 80°C/1 bar led to a violent explosion. As the compound had not been stored under nitrogen during the 3 weeks since preparation, peroxides were suspected.
*See other* ACETYLENIC COMPOUNDS

**† 2-METHYL-3-BUTYN-2-OL** $CH_3C(Me)OHC{\equiv}CH$ $C_5H_8O$

Sulphur tetrafluoride
*See* SULPHUR TETRAFLUORIDE, $F_4S$: 2-Methyl-3-butyn-2-ol

**† METHYL ISOPROPENYL KETONE** $CH_3COCMe{=}CH_2$ $C_5H_8O$

† **ALLYL ACETATE** $C_5H_8O_2$

$CH_3CO_2CH_2CH{=}CH_2$

† **ETHYL ACRYLATE** $CH_2{=}CHCO_2Et$ $C_5H_8O_2$

*MCA Case History No. 1759*

Inhibited monomer was transferred from a steel drum into a 4 l clear glass bottle exposed to sunlight in a laboratory in which the temperature was, for a time, higher than usual. Exothermic polymerisation set in and caused the bottle to burst. Precautions recommended included increase in inhibitor concentration tenfold (to 200 p.p.m.) for laboratory-stored samples, and use of metal or brown glass containers.

† **METHYL CROTONATE** $CH_3CH{=}CHCO_2Me$ $C_5H_8O_2$

† **METHYL METHACRYLATE** $CH_2{=}CMeCO_2Me$ $C_5H_8O_2$

Harmon, 1974, 2.13

The monomer tends to self-polymerise and this may become explosive. It must be stored inhibited.

*See* Initiation, below
VIOLENT POLYMERISATION

Air

Barnes, C. E. *et al., J. Amer. Chem. Soc.*, 1950, **72**, 210

Exposure of the purified (unstabilised) monomer to air at room temperature for 2 months generated an ester–oxygen interpolymer, which exploded on evaporation of the surplus monomer at 60°C (but not 40°C).

*See other* POLYPEROXIDES

Dibenzoyl peroxide

*See* DIBENZOYL PEROXIDE, $C_{14}H_{10}O_4$: Methyl methacrylate

Initiators

Biesenberg, J. S. *et al., Polym. Eng. Sci.*, 1976, **16**, 101–116

The parameters were determined in a batch reactor for thermal runaway polymerisation of methyl methacrylate, initiated by azoisobutyronitrile, dibenzoyl peroxide or di-*tert*-butyl peroxide.

*See other* POLYMERISATION INCIDENTS

† **ISOPROPENYL ACETATE** $CH_3CO_2CMe{=}CH_2$ $C_5H_8O_2$

† **VINYL PROPIONATE** $EtCO_2CH{=}CH_2$ $C_5H_8O_2$

**POLY(METHYL METHACRYLATE PEROXIDE)** $(C_5H_8O_4)_n$

$(HOOCH{=}C(Me)CO_2Me)_n$

*See* METHYL METHACRYLATE, $C_5H_8O_2$: Air

*See other* POLYPEROXIDES

**ISOPRENE DIOZONIDE** $C_5H_8O_6$

$O^-{-}O^+OCH_2C(Me)CHCH_2O{-}O^+{-}O^-$

*See* 2-METHYL-1,3-PENTADIENE, $C_6H_{10}$: Ozone

**† CHLOROCYCLOPENTANE** $ClCH[CH_2]_3CH_2$ $C_5H_9Cl$

***N*-CHLOROCARBONYLOXYTRIMETHYLUREA** $C_5H_9ClN_2O_3$

$ClCO_2NMeCONMe_2$

Groebner, P. *et al.*, *Eur. J. Med. Chem., Chim. Ther.*, 1974, **9**, 32–34

A small sample decomposed explosively during vacuum distillation.

**† PIVALOYL CHLORIDE** $Me_3CCOCl$ $C_5H_9ClO$

***tert*-BUTYL CHLOROPEROXYFORMATE** $C_5H_9ClO_3$

$ClCO_2OCMe_3$

1. Bartlett, P. D. *et al.*, *J. Amer. Chem. Soc.*, 1963, **85**, 1858
2. Hanson, P., *Chem. Brit.*, 1975, **11**, 418

The peroxyester is stable in storage at –25°C, and samples did not explode on friction, impact or heating. However, a 10 g sample stored at ambient temperature for 1 h heated spontaneously, exploded and ignited [1]. A sample stored at –30°C exploded during manipulation at 0–5°C [2]

*See other* PEROXYESTERS

**† PIVALONITRILE** $Me_3CCN$ $C_5H_9N$

**† 1,2,5,6-TETRAHYDROPYRIDINE** $HN[CH_2]_3CH{=}CH$ $C_5H_9N$

**CYCLOPENTANONE OXIME** $CH_2[CH_2]_3C{=}NOH$ $C_5H_9NO$

Sulphuric acid

*See* SULPHURIC ACID, $H_2O_4S$: Cyclopentanone oxime

**2-ETHYLACRYLALDEHYDE OXIME** $C_5H_9NO$

$H_2C{=}C(Et)CH{=}NOH$

Marvel, C. S. *et al., J. Amer. Chem. Soc.*, 1950, **72**, 5408
An explosion during distillation was recorded, possibly attributable to peroxide formation.

*See other* OXIMES

**PIVALOYL AZIDE** $Me_3CCON_3$ $C_5H_9N_3O$

Bühler, A. *et al., Helv. Chim. Acta,* 1943, **26**, 2123
The azide exploded very violently on warming, and on one occasion, on standing.

*See other* ACYL AZIDES

***tert*-BUTYL AZIDOFORMATE** $N_3CO_2CMe_3$ $C_5H_9N_3O_2$

1. Carpino, L. A. *et al., Org. Synth.*, 1964, **44**, 15
2. Sakai, K. *et al., J. Org. Chem.*, 1971, **36**, 2387
3. Insalaco, M. A. *et al., Org. Synth.*, 1970, **50**, 12
4. Fenlon, W. J., *Chem. Eng. News*, 1976, **54**(22), 3
5. Koppel, H. C., ibid., (39), 5
6. Aldrich advertisement, *J. Chem. Soc., Chem. Comm.*, 1976, (23), i
7. Feyen, P., *Angew. Chem. (Intern. Edn.)*, 1977, **16**, 115

Explosion during distillation at 74°C/92 mbar has been recorded on only one out of several hundred occasions [1]. An alternative preparation avoiding distillation has been described [2].

Although distillation at 57–61°C/53 mbar has been accomplished on many occasions, the compound's structure suggests it should be considered as potentially explosive [3]. The undiluted compound is a shock-sensitive explosive, which is thermally unstable between 100 and 135°C and autoignites at 143°C [4]. A safe and more effective reagent for *tert*-butoxycarbonylation of amino groups, 2-(*tert*-butoxycarbonyl-oxyimino)-2-phenylacetonitrile, has now superseded the azide [5,6]. An explosion during evaporation of an ethereal solution of the azide at 50°C has also been reported [7].

*See other* ACYL AZIDES

**1,1,1-TRIS(AZIDOMETHYL)ETHANE** $(N_3CH_2)_3CCH_3$ $C_5H_9N_9$

Hydrogen,
Palladium catalyst

Zompa, L. J. *et al., Org. Prep. Proced. Int.*, 1974, **6**, 103–106
There is an explosion hazard during the palladium-catalysed hydrogenation of the tris-azide in ethanol at 2 bar to the tris-amine.

**DIMETHYL-1-PROPYNYLTHALLIUM** **$Me_2TlC{\equiv}CCH_3$** **$C_5H_9Tl$**

Nast, R. *et al.*, *J. Organomet. Chem.*, 1966, **6**, 461

Explodes on heating and is sensitive to stirring or impact.

*See other* ALKYLMETALS
METAL ACETYLIDES

**† CYCLOPENTANE** **$CH_2[CH_2]_3CH_2$** **$C_5H_{10}$**

**† 1,1-DIMETHYLCYCLOPROPANE** **$Me_2CCH_2CH_2$** **$C_5H_{10}$**

**† ETHYLCYCLOPROPANE** **$EtCHCH_2CH_2$** **$C_5H_{10}$**

**† 2-METHYL-1-BUTENE** **$CH_2{=}CMeEt$** **$C_5H_{10}$**

**† 2-METHYL-2-BUTENE** **$CH_3CMe{=}CHMe$** **$C_5H_{10}$**

**† 3-METHYL-1-BUTENE** **$CH_2{=}CHCHMe$** **$C_5H_{10}$**

**† METHYLCYCLOBUTANE** **$MeCHCH_2CH_2CH_2$** **$C_5H_{10}$**

**† 1-PENTENE** **$CH_2{=}CHPr$** **$C_5H_{10}$**

**† 2-PENTENE** **$CH_3CH{=}CHEt$** **$C_5H_{10}$**

**CYANODIETHYLGOLD** **$C_5H_{10}AuN$**

*See the tetramer,* TETRACYANOOCTAETHYLTETRAGOLD, $C_{20}H_{40}Au_4N_4$

***N*-CHLOROPIPERIDINE** **$ClN[CH_2]_4CH_2$** **$C_5H_{10}ClN$**

Claxton, G. P. *et al., Org. Synth.*, 1977, **56**, 120

To avoid rapid spontaneous decomposition of the compound, solvent ether must only be partially distilled out of the extracted product from a water-bath maintained at below 60°C.

*See other* *N*-HALOGEN COMPOUNDS

### 1-PERCHLORYLPIPERIDINE $C_5H_{10}ClNO_3$

$O_3ClN[CH_2]_4CH_2$

Alone,
or Piperidine

Gardner, D. L. *et al., J. Org. Chem.*, 1964, **29**, 3738–3739

It is a dangerously sensitive oil which has exploded violently on storage, heating or contact with piperidine. Absorption in alumina was necessary to desensitise it to allow non-explosive analytical combustion.

*See other* PERCHLORYL COMPOUNDS

### DICHLOROMETHYLENEDIETHYLAMMONIUM CHLORIDE

$Cl_2C{=}N^+Et_2\ Cl^-$ $C_5H_{10}Cl_3N$

Bader, A. R., private comm., 1971

Viehe, H. G. *et al., Angew. Chem. (Intern. Ed.)*, 1971, **10**, 573

The compound does not explode on heating; the published statement is in error.

### † 3-DIMETHYLAMINOPROPIONITRILE $Me_2NC_2H_4CN$ $C_5H_{10}N_2$

### 3,7-DINITROSO-1,3,5,7-TETRAAZABICYCLO[3.3.1]NONANE

$C_5H_{10}N_6O_2$

$CH_2N(NO)CH_2NCH_2NCH_2N(NO)CH_2$

*MCA Case History No. 841*

A cardboard drum of this blowing agent (as a 40% dispersion in fine silica) ignited when roughly handled in storage.

*See other* NITROSO COMPOUNDS

### † ALLYL ETHYL ETHER $CH_2{=}CHCH_2OEt$ $C_5H_{10}O$

*ABCM Quart. Safety Summ.*, 1963, **34**, 7

A commercial sample was distilled without previously being tested for peroxide. An explosion occurred towards the end of the distillation, and the residue was later shown to contain peroxide.

*See other* PEROXIDISABLE COMPOUNDS

### † ETHYL PROPENYL ETHER $EtOCH{=}CHCH_3$ $C_5H_{10}O$

† ISOPROPYL VINYL ETHER $Me_2CHOCH{=}CH_2$ $C_5H_{10}O$

† ISOVALERALDEHYDE $Me_2CHCH_2CHO$ $C_5H_{10}O$

† 3-METHYL-2-BUTANONE $CH_3COCHMe_2$ $C_5H_{10}O$

† 3-METHYL-1-BUTEN-3-OL $C_5H_{10}O$

$CH_2{=}CHC(Me)OHCH_3$

† 2-METHYLTETRAHYDROFURAN $\overline{O[CH_2]_3CHMe}$ $C_5H_{10}O$

† 2-PENTANONE $CH_3COPr$ $C_5H_{10}O$

Bromine trifluoride

*See* BROMINE TRIFLUORIDE, $BrF_3$: 2-Pentanone

† 3-PENTANONE $Et_2CO$ $C_5H_{10}O$

Hydrogen peroxide,
Nitric acid

*See* HYDROGEN PEROXIDE, $H_2O_2$: Ketones, etc.

† 4-PENTEN-1-OL $HO[CH_2]_3CH{=}CH_2$ $C_5H_{10}O$

† TETRAHYDROPYRAN $\overline{O[CH_2]_4CH_2}$ $C_5H_{10}O$

† VALERALDEHYDE BuCHO $C_5H_{10}O$

† BUTYL FORMATE $HCO_2Bu$ $C_5H_{10}O_2$

† 3,3-DIMETHOXYPROPENE $CH_2{=}CHCH(OMe)_2$ $C_5H_{10}O_2$

† 2,2-DIMETHYL-1,3-DIOXOLANE $\overline{OCMe_2OCH_2CH_2}$ $C_5H_{10}O_2$

† ETHYL PROPIONATE — $EtCO_2Et$ — $C_5H_{10}O_2$

† ISOBUTYL FORMATE — $HCO_2CH_2CHMe_2$ — $C_5H_{10}O_2$

† ISOPROPYL ACETATE — $CH_3CO_2CHMe_2$ — $C_5H_{10}O_2$

† 2-METHOXYETHYL VINYL ETHER — $C_5H_{10}O_2$

$MeOCH_2CH_2OCH{=}CH_2$

† METHYL BUTYRATE — $PrCO_2Me$ — $C_5H_{10}O_2$

† 4-METHYL-1,3-DIOXANE — $C_5H_{10}O_2$

$\overline{OCH_2OCHMeCH_2CH_2}$

† METHYL ISOBUTYRATE — $Me_2CHCO_2Me$ — $C_5H_{10}O_2$

† PROPYL ACETATE — $CH_3CO_2Pr$ — $C_5H_{10}O_2$

TETRAHYDROFURFURYL ALCOHOL — $C_5H_{10}O_2$

$\overline{O[CH_2]_3C}H_2OH$

3-Nitro-*N*-bromophthalimide
*See* *N*-HALOIMIDES: Alcohols

TELLURANE-1,1-DIOXIDE — $\overline{TeO_2[CH_2]_4CH_2}$ — $C_5H_{10}O_2Te$

Morgan, G. T. *et al., J. Chem. Soc.*, 1928, 321
It exploded on rapid heating, and decomposed violently in contact with nitric or sulphuric acids.

† DIETHYL CARBONATE — $CO(OEt)_2$ — $C_5H_{10}O_3$

*trans*-2-PENTENE OZONIDE $C_5H_{10}O_3$

MeCHCH(Et)O–$O^+$–$O^-$

*See* 2-HEXENE OZONIDE, $C_6H_{12}O_3$
*See other* OZONIDES

† **1-BROMO-3-METHYLBUTANE** $BrCH_2CHMe_2$ $C_5H_{11}Br$

† **2-BROMOPENTANE** MeCHBrPr $C_5H_{11}Br$

† **1-CHLORO-3-METHYLBUTANE** $ClCH_2CHMe_2$ $C_5H_{11}Cl$

† **2-CHLORO-2-METHYLBUTANE** $EtCClMe_2$ $C_5H_{11}Cl$

† **1-CHLOROPENTANE** $ClCH_2Bu$ $C_5H_{11}Cl$

**2-CHLOROVINYLTRIMETHYLLEAD** $ClCH{=}CHPbMe_3$ $C_5H_{11}ClPb$

Steingross, W. *et al., J. Organomet. Chem.*, 1966, **6**, 109

It may explode violently during vacuum distillation at 102°C/62.5 mbar.

*See related* ALKYLMETALS

† **2-IODOPENTANE** MeCHIPr $C_5H_{11}I$

† **CYCLOPENTYLAMINE** $CH_2[CH_2]_3CHNH_2$ $C_5H_{11}N$

† **1-METHYLPYRROLIDINE** $MeN[CH_2]_3CH_2$ $C_5H_{11}N$

† **PIPERIDINE** $HN[CH_2]_4CH_2$ $C_5H_{11}N$

Dicyanofurazan

*See* DICYANOFURAZAN, $C_4N_4O$: Nitrogenous bases

*N*-Nitrosoacetanilide

*See* *N*-NITROSOACETANILIDE, $C_8H_8N_2O_2$ : Piperidine

1-Perchlorylpiperidine

*See* 1-PERCHLORYLPIPERIDINE, $C_5H_{10}ClNO_3$ : Piperidine

**† 4-METHYLMORPHOLINE** $O(CH_2CH_2)_2NMe$ $C_5H_{11}NO$

**† ISOPENTYL NITRITE** $Me_2CHC_2H_4ONO$ $C_5H_{11}NO_2$

**† PENTYL NITRITE** $BuCH_2ONO$ $C_5H_{11}NO_2$

Sorbe, 1968, 146
It explodes on heating to above 250°C.
*See other* ALKYL NITRITES

**† 2,2-DIMETHYLPROPANE (NEOPENTANE)** $C_5H_{12}$
$CMe_4$

Preparative hazard
*See* ALUMINIUM, Al: Halocarbons

**† 2-METHYLBUTANE (ISOPENTANE)** $EtCHMe_2$ $C_5H_{12}$

**† PENTANE** $Me[CH_2]_3Me$ $C_5H_{12}$

***N*-BROMOTETRAMETHYLGUANIDINE** $C_5H_{12}BrN_3$
$Me_2NC(=NBr)NMe_2$

Papa, A. J., *J. Org. Chem.*, 1966, **31**, 1426
The material is unstable even at 0°C, and explodes if heated above 50°C at atmospheric pressure.
*See other* *N*-HALOGEN COMPOUNDS

***N,N,N',N'*-TETRAMETHYLDEUTEROFORMAMIDINIUM PERCHLORATE** $C_5H_{12}ClDN_2O_4$
$Me_2NCD{=}N^+Me_2\ ClO_4^-$

Menzer, M. *et al., Z. Chem.*, 1977, **17**, 344
The undeuterated salt (150 mg) was refluxed with deuterium oxide to effect deuterium exchange, then the excess was evaporated off under vacuum. Towards the end of the evaporation the moist salt exploded violently.
*See* AMINIUM PERCHLORATES

***N*-CHLOROTETRAMETHYLGUANIDINE** $C_5H_{12}ClN_3$
$Me_2NC(=NCl)NMe_2$

Papa, A. J., *J. Org. Chem.*, 1966, **31**, 1426

The material is unstable even at 0°C, and explodes if heated above 50°C at atmospheric pressure.
*See other N*-HALOGEN DERIVATIVES

**DIMETHYLTHALLIUM *N*-METHYLACETOHYDROXAMATE** $C_5H_{12}NO_2Tl$
**$Me_2TlON(Me)COMe$**

Schwering, H. U. *et al., J. Organomet. Chem.*, 1975, **99**, 22
It exploded below 160°C, unlike the other dialkylmetal derivatives, which showed high thermal stability.
*See other* HEAVY METAL DERIVATIVES

**† BUTYL METHYL ETHER** **BuOMe** $C_5H_{12}O$

**† ETHYL ISOPROPYL ETHER** **$EtOCHMe_2$** $C_5H_{12}O$

**† ETHYL PROPYL ETHER** **EtOPr** $C_5H_{12}O$

**† ISOPENTANOL** **$Me_2CHC_2H_4OH$** $C_5H_{12}O$

Hydrogen trisulphide
*See* HYDROGEN TRISULPHIDE, $H_2S_3$ : Pentyl alcohol

**† 2-PENTANOL** **MeCHOHPr** $C_5H_{12}O$

**† *tert*-PENTANOL** **$Me_2C(OH)Et$** $C_5H_{12}O$

**† DIETHOXYMETHANE** **$(EtO)_2CH_2$** $C_5H_{12}O_2$

**† 1,1-DIMETHOXYPROPANE** **$(MeO)_2CHEt$** $C_5H_{12}O_2$

**† 2,2-DIMETHOXYPROPANE** **$(MeO)_2C(Me)_2$** $C_5H_{12}O_2$

Metal perchlorates
1. Dickinson, R. C. *et al., Chem. Eng. News,* 1970, **48**(28), 6; Cramer, R., ibid., **48**(45), 7
2. Mikulski, S. M., *Chem. Eng. News,* 1971, **49**(4), 8

During dehydration of the hydrated manganese and nickel salts with dimethoxypropane, heating above 65°C caused violent explosions, probably involving oxidation by the anion [1]. Triethyl orthoformate is recommended as a safer dehydrating agent [2].

**2(2-METHOXYETHOXY)ETHANOL** $C_5H_{12}O_3$

$MeOC_2H_4OC_2H_4OH$

Calcium hypochlorite
*See* CALCIUM HYPOCHLORITE, $CaCl_2O_2$: Hydroxy compounds

**PENTAERYTHRITOL** $C(CH_2OH)_4$ $C_5H_{12}O_4$

Thiophosphoryl chloride
*See* THIOPHOSPHORYL CHLORIDE, $Cl_3PS$ : Pentaerythritol

**† 2-METHYLBUTANE-2-THIOL** $Me_2C(Et)SH$ $C_5H_{12}S$

**† 3-METHYLBUTANETHIOL** BuSH $C_5H_{12}S$

**† PENTANETHIOL** $BuCH_2SH$ $C_5H_{12}S$

**† ISOPENTYLAMINE** $Me_2CHC_2H_4NH_2$ $C_5H_{13}N$

**† *N*-METHYLBUTYLAMINE** BuNHMe $C_5H_{13}N$

**† 1-PENTYLAMINE** $BuCH_2NH_2$ $C_5H_{13}N$

**DIETHYLMETHYLPHOSPHINE** $Et_2PMe$ $C_5H_{13}P$

Personal experience
May ignite in air with long exposure.
*See other* ALKYLNON-METALS

**† 2-DIMETHYLAMINO-*N*-METHYLETHYLAMINE** $C_5H_{14}N_2$

$Me_2NC_2H_4NHMe$

**3-DIMETHYLAMINOPROPYLAMINE** **$Me_2N[CH_2]_3NH_2$** **$C_5H_{14}N_2$**

Cellulose nitrate

*See* CELLULOSE NITRATE: Amines

**DIMETHYL(TRIMETHYL)SILOXOGOLD** **$C_5H_{15}AuOSi$**

*See the dimer*, TETRAMETHYLBIS (TRIMETHYLSILOXO)DIGOLD, $C_{10}H_{30}Au_2O_2Si_2$

**LITHIUM 1,1-DIMETHYL(TRIMETHYLSILYL)HYDRAZIDE** **$C_5H_{15}LiN_2Si$**

**$LiN(SiMe_3)NMe_2$**

Oxidants

Interaction with fluorine, nitric acid or ozone is hypergolic.

*See* SILYLHYDRAZINES

**LITHIUM PENTAMETHYLTITANATE–BIS-2,2′-BIPYRIDINE**

**$C_5H_{15}LiTi \cdot C_{20}H_{16}N_4$**

**$Li[TiMe_5] \cdot 2C_{10}H_8N_2$**

Houben-Weyl, 1975, Vol. 13.3, 288

Friction-sensitive, it decomposes explosively.

*See related* ALKYLMETALS

**DIMETHYLAMINOTRIMETHYLSILANE** **$C_5H_{15}NSi$**

**$Me_2NSiMe_3$**

Xenon difluoride

*See* XENON DIFLUORIDE, $F_2Xe$: Silicon–nitrogen compounds

**1,1,1-TRIS(AMINOMETHYL)ETHANE** **$(H_2NCH_2)_3CCH_3$** **$C_5H_{15}N_3$**

Preparative hazard

*See* 1,1,1-TRIS(AZIDOMETHYL)ETHANE, $C_5H_9N_9$: Hydrogen, etc.

**DIMETHYL(TRIMETHYLSILYL)PHOSPHINE** **$C_5H_{15}PSi$**

**$Me_2PSiMe_3$**

Goldsberry, R. *et al., Inorg. Chem.*, 1972, **13**, 29–32

It ignites in air and is hydrolysed to dimethylphosphine, also spontaneously flammable.

*See other* ALKYLNON-METALS

**PENTAMETHYLTANTALUM** **$Me_5Ta$** **$C_5H_{15}Ta$**

Mertis, K. *et al., J. Organomet. Chem.*, 1975, **97**, C65

It should be handled with extreme caution, even in absence of air, because dangerous explosions have occurred on warming frozen samples or during transfer operations.

*See other* ALKYLMETALS

**1,2-DIMETHYL-2-(TRIMETHYLSILYL)HYDRAZINE** **$C_5H_{16}N_2Si$**

**$MeNHNMeSiMe_3$**

Oxidants

Interaction with fluorine, nitric acid or ozone is hypergolic.

*See* SILYLHYDRAZINES

**PENTAAMMINEPYRIDINERUTHENIUM(II) DIPERCHLORATE**

**$C_5H_{20}Cl_2N_6O_8Ru$**

**$[Ru(NH_3)_5C_5H_5N][ClO_4]_2$**

Creutz, C. A., private comm., 1969

The dry salt exploded on touching.

*See other* AMMINEMETAL OXOSALTS

**2-DIAZONIO-4,5-DICYANOIMIDAZOLATE (DIAZODICYANOIMIDAZOLE)**

**$C_5N_6$**

Sheppard, W. A. *et al., J. Amer. Chem. Soc.*, 1973, **95**, 2696

It is highly shock-sensitive when dry and explodes at 150°C.

*See other* DIAZONIUM SALTS
HIGH-NITROGEN COMPOUNDS

**SODIUM PENTACARBONYLRHENATE(2–)** **$C_5Na_2O_5Re$**

**$Na_2[Re(CO)_5]$**

Bailar, 1973, Vol. 3, 955

This and other alkali salts are pyrophoric.

*See related* CARBONYLMETALS

**DISILVER 1,3,5-HEXATRIYNIDE** **$Ag[C{\equiv}C]_3Ag$** **$C_6Ag_2$**

Hunsman, W., *Chem. Ber.*, 1950, **83**, 216

The silver salt may be handled moist, but when dry it explodes violently on touching with a glass rod.

*See other* METAL ACETYLIDES
SILVER COMPOUNDS

**PENTAFLUOROPHENYLALUMINIUM DIBROMIDE** **$C_6AlBr_2F_5$**

**$C_6F_5AlBr_2$**

Chambers, R. D. *et al., J. Chem. Soc. (C)* 1967, 2185; *Tetrahedron Lett.*, 1965, 2389

Ignites in air; explodes violently on rapid heating to 195°C or during uncontrolled hydrolysis.

*See other* HALO-ARYLMETALS

**HEXACARBONYLCHROMIUM** **$(CO)_6Cr$** **$C_6CrO_6$**

Weast, 1972, B-83

It explodes at 210°C.

*See other* CARBONYLMETALS

**PENTAFLUOROPHENYLLITHIUM** **$C_6F_5Li$**

Deuterium oxide

Kinsella, E. *et al., Chem. & Ind.*, 1971, 1017

During addition of an ethereal solution of deuterium oxide (containing some peroxide) to a suspension of the organolithium reagent in pentane, a violent explosion occurred. This may have been initiated by the peroxide present.

*See other* HALO-ARYLMETALS

**† HEXAFLUOROBENZENE** **$C_6F_6$**

Metals

*See* HALOARENEMETAL π-COMPLEXES

**TRIFLUOROMETHYL 3-FLUOROFORMYLHEXAFLUOROPEROXYBUTYRATE** **$C_6F_{10}O_4$**

**$FCO[CF_2]_3CO_2OCF_3$**

Bernstein, P. A. *et al., J. Amer. Chem. Soc.*, 1971, **93**, 3885

A small sample exploded at 70°C.

*See other* PEROXYESTERS

## PERFLUOROHEXYL IODIDE $C_6F_{13}I$

Sodium

*See* SODIUM, Na : Halocarbons

## POTASSIUM HEXACYANOFERRATE (3–) ('FERRICYANIDE') $C_6FeK_3N_6$

$K_3[Fe(CN)_6]$

Ammonia

Pieters, 1957, 30

Sidgwick, 1950, 1359

Contact may be explosive, possibly owing to rapid oxidation of ammonia by alkaline 'ferricyanide'.

*See also* POTASSIUM TETRACYANOMERCURATE(2–), $C_4HgK_2N_4$

Chromium trioxide

*See* CHROMIUM TRIOXIDE, $CrO_3$ : Potassium hexacyanoferrate(3–)

Sodium nitrite

*See* SODIUM NITRITE, $NNaO_2$ : Metal cyanides

*See other* METAL CYANIDES

## POTASSIUM HEXACYANOFERRATE(4–) ('FERROCYANIDE') $C_6FeK_4N_6$

$K_4[Fe(CN)_6]$

Copper(II) nitrate

*See* COPPER(II) NITRATE, $CuN_2O_6$ : Potassium hexacyanoferrate(4–)

Sodium nitrite

*See* SODIUM NITRITE, $NNaO_2$ : Metal cyanides

*See other* METAL CYANIDES

## IRON(III) OXALATE $Fe_2(C_2O_4)_3$ $C_6Fe_2O_{12}$

Weinland, R. *et al., Z. Anorg. Chem.,* 1929, 178, 219

The salt, probably complex, decomposes at 100°C.

*See other* METAL OXALATES

## 4-CHLORO-2,5-DINITROBENZENEDIAZONIUM-6-OLATE $C_6HClN_4O_5$

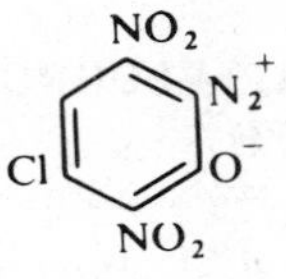

*See* NITRIC ACID, $HNO_3$ : 4-Chloro-2-nitroaniline

*See other* ARENEDIAZONIUMOLATES

**3,4-DIFLUORO-2-NITROBENZENEDIAZONIUM-6-OLATE**
**3,6-DIFLUORO-2-NITROBENZENEDIAZONIUM-4-OLATE** $C_6HF_2N_3O_3$

Finger, G. C. *et al., J. Amer. Chem. Soc.*, 1951, **73**, 148
Both isomers, produced under different conditions of diazotisation of 3,4,6-trifluoro-2-nitroaniline, exploded on heating, ignition or impact.
*See other* ARENEDIAZONIUMOLATES

**1(or 2),3,4,5,6-PENTAFLUOROBICYCLO[2.2.0]HEXA-2,5-DIENE** $C_6HF_5$

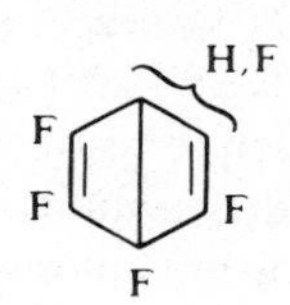


Rotajezak, E., *Roczn. Chem.*, 1970, **44**, 447–449
Both isomers were stored as frozen solids, for as liquids they are more explosive than the hexafluoro 'Dewar' benzene.
*See other* HALOALKENES
STRAINED-RING COMPOUNDS

**LEAD 2,4,6-TRINITRORESORCINOLATE (LEAD STYPHNATE)** $C_6HN_3O_8Pb$

1. *MCA Case History No. 957*
2. Okazaki, K. *et al., Chem. Abs.*, 1977, **86**, 57638u

Three beakers of lead styphnate were being heated in a laboratory oven to dry the explosive salt. When one of the beakers was moved, all three detonated. Other heavy metal salts of polynitrophenols are dangerously explosive when dry [1]. The desensitising effect of presence of water upon the friction-sensitivity of this priming explosive was studied. There is no desensitisation up to 2% content and little at 5%. Even at 20% water content it is still as sensitive as dry pentaerythritol tetranitrate [2].
*See other* HEAVY METAL DERIVATIVES
POLYNITROARYL COMPOUNDS

## 3,4-BIS(1,2,3,4-THIATRIAZOL-5-YLTHIO)MALEIMIDE $C_6HN_7O_2S_4$

Pilgram, K. *et al., Angew. Chem.*, 1965, **77**, 348
It explodes loudly on impact or at the m.p.
*See other* HIGH-NITROGEN COMPOUNDS

## 1,3,5-HEXATRIYNE $H[C{\equiv}C]_3H$ $C_6H_2$

1. Hunsman, W., *Chem. Ber.*, 1950, **83**, 213–217
2. Armitage, J. B. *et al., J. Chem. Soc.*, 1952, 2013
3. Bjornov, E. *et al., Spectrochim. Acta*, 1974, **30A**, 1255–1266

It polymerises slowly at –20°C, but rapidly at ambient temperature in air to give a friction-sensitive explosive solid, probably a peroxide [1]. Although stable under vacuum at –5°C for a limited period, exposure at 0°C to air led to violent explosions soon afterwards [2]. Explosive hazards have again been stressed [3].

*See other* ACETYLENIC COMPOUNDS

## 2,6-DIBROMOBENZOQUINONE-4-CHLOROIMINE $C_6H_2Br_2ClNO$

$O{=}C_6H_2Br_2{=}NCl$

1. Hartmann, W. W. *et al., Org. Synth.*, 1943, Coll. Vol. 2, 177
2. Horber, D. F. *et al., Chem. & Ind.*, 1967, 1551
3. Taranto, B. J., *Chem. Eng. News*, 1967, **45**(52), 54

The chloroimine decomposed violently on a drying tray at 60°C [1], and a bottle accidentally heated to 50°C exploded [2]. Decomposition of the heated solid occurs after a temperature-dependent induction period. Unheated material eventually exploded after storage at ambient temperature [3].

*See also* 2,6-DICHLOROBENZOQUINONE-4-CHLOROIMINE, $C_6H_2Cl_3NO$
*See other* *N*-HALOGEN COMPOUNDS

## 1,3-DICHLORO-4,6-DINITROBENZENE $C_6H_2Cl_2N_2O_4$

Cl
Cl
$NO_2$
$NO_2$

Dimethyl sulphoxide,
Potassium fluoride
Koch-Light Laboratories Ltd, private comm., 1976

A mixture of the dichloro compound with potassium fluoride in the solvent was heated to reflux to effect replacement of chlorine by fluorine. The reaction accelerated out of control and exploded, leaving much carbonised residue. Analogous replacement reactions had been done uneventfully on many previous occasions.

*See other* POLYNITROARYL COMPOUNDS

## 2,6-DICHLOROBENZOQUINONE-4-CHLOROIMINE $C_6H_2Cl_3NO$

$O{=}C_6H_2Cl_2{=}NCl$

Taranto, B. J., *Chem. Eng. News*, 1967, **45**(52), 54

The heated, solid material decomposed violently after an induction period dependent upon temperature. Unheated material exploded after storage at ambient temperature.

*See also* 2,6-DIBROMOBENZOQUINONE-4-CHLOROIMINE., $C_6H_2Br_2ClNO$

*See other* *N*-HALOGEN COMPOUNDS

## 1,2,4,5-TETRACHLOROBENZENE $C_6H_2Cl_4$

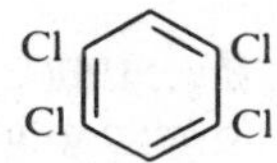


Sodium hydroxide

1. *MCA Case History No. 620*
2. *ABCM Quart. Safety Summ.*, 1952, **23**, 5
3. Calnan, A. B., private comm., 1972
4. Rawls, R. L. *et al.*, *Chem. Eng. News*, 1976, **54**(35), 27

Several serious incidents have been reported about the commercial preparation of 2,4,5-trichlorophenol by alkaline hydrolysis of tetrachlorobenzene. On two occasions the hydrolysis under pressure with methanolic alkali at 125°C went out of control, one reaching 400°C [1]. After hydrolysis in ethylene glycol solution, the residue from vacuum stripping exploded, probably owing to overheating. Electric heating was used without knowledge of the liquid temperature [2].

In another incident in 1968, a violent explosion occurred during hydrolysis in ethylene glycol at atmospheric pressure, which was regarded as a safe process. Serious after-effects were caused by the widespread distribution of the most powerful dermatitic compound known, di(4,5-dichlorobenzo)dioxin, which was produced by cyclisation during the explosion. Traces remaining on contaminated plant after 3 years' weathering still caused dermatitis at a concentration of 10 p.p.m. [3].

The 1976 Seveso incident also involved the ethylene glycol non-pressurised process [4].

*See* SODIUM 2-HYDROXYETHOXIDE, $C_2H_5NaO_2$: (Polychlorobenzenes)

*See other* HALOARYL COMPOUNDS

## 1,3-DIFLUORO-4,6-DINITROBENZENE $C_6H_2F_2N_2O_4$

$FC{=}CHCF{=}CNO_2CH{=}CNO_2$

Preparative hazard

*See* 1,3-DICHLORO-4,6-DINITROBENZENE, $C_6H_2Cl_2N_2O_4$:
Dimethyl sulphoxide, etc.

## † 1,2,4,5-TETRAFLUOROBENZENE $C_6H_2F_4$

## SODIUM 3-HYDROXYMERCURIO-2,6-DINITRO-4-*aci*-NITRO-2,5-CYCLOHEXADIENONE $C_6H_2HgN_3NaO_8$

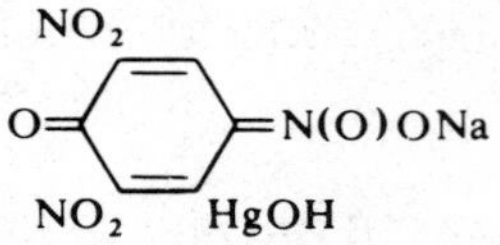


Hantzsch, A. *et al.*, *Ber.*, 1906, **39**, 1111
This salt of the mono-*aci p*-quinonoid form of 3-hydroxymercurio-2,4,6-trinitrophenol explodes on rapid heating.

*See* *aci*-NITROQUINONOID COMPOUNDS

## 3,4,5-TRIIODOBENZENEDIAZONIUM NITRATE $C_6H_2I_3N_3O_3$

$I_3C_6H_2N_2^+NO_3^-$

Kalb, L. *et al.*, *Ber.*, 1926, **59**, 1867
Unstable on warming, explodes on heating in a flame.

*See other* DIAZONIUM SALTS

## SODIUM PICRATE $C_6H_2N_3NaO_7$

$(NO_2)_3C_6H_2ONa$

Coningham, H., *H.M. Insp. Expl. Rept. 236*, London, 1919
A large-scale explosion involving initiation of wet sodium picrate by impact was investigated.

*See other* POLYNITROARYL COMPOUNDS

## 3,5-DINITROBENZENEDIAZONIUM-2-OLATE $C_6H_2N_4O_5$

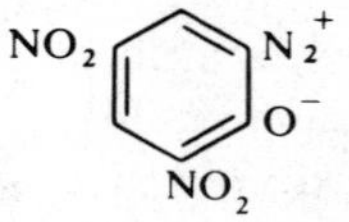


Preparative hazard

*See* *N*,2,3,5-TETRANITROANILINE, $C_6H_3N_5O_8$

**4,6-DINITROBENZENEDIAZONIUM-2-OLATE** $C_6H_2N_4O_5$

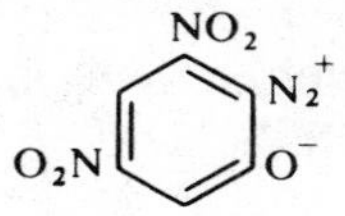

Urbanski, 1967, Vol. 3, 204

This priming explosive, as sensitive as mercuric fulminate, is much more powerful than metal-containing initiators.

*See other* AREN EDIAZONIUMOLATES

**2,3,4,6-TETRANITROPHENOL** $C_6H_2N_4O_9$

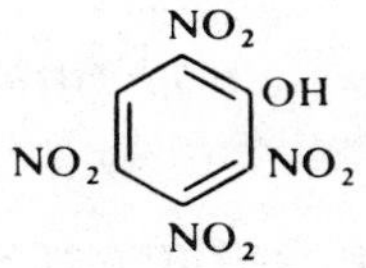

Sorbe, 1968, 152

Exceptionally explosive.

*See other* POLYNITROARYL COMPOUNDS

**PICRYL AZIDE** $(NO_2)_3C_6H_2N_3$ $C_6H_2N_6O_6$

Schrader, E., *Ber.*, 1917, **50**, 778

Explodes weakly under impact, but not on heating.

*See other* ORGANIC AZIDES

**BUTADIYNE-1,4-DICARBOXYLIC ACID** $C_6H_2O_4$

$HO_2C[C{\equiv}C]_2CO_2H$

Sorbe, 1968, 109

Explodes on heating.

*See other* ACETYLENIC COMPOUNDS

**TRIETHYNYLALUMINIUM** $(HC{\equiv}C)_3Al$ $C_6H_3Al$

Dioxane,

or Trimethylamine

Chini, P. *et al.*, *Chim. Ind. (Milan)*, 1962, **44**, 1220

Houben-Weyl, 1970, Vol. 13.4, 159

The residue from sublimation of the complex with dioxane is explosive and the complex should not be dried by heating. The trimethylamine complex may also explode on sublimation.

Triethynylaluminium or its complex with diethyl ether may decompose explosively on heating. Sublimation is not therefore advised as a purification method.

*See other* METAL ACETYLIDES

**TRIETHYNYLARSINE** $(HC{\equiv}C)_3As$ $C_6H_3As$

Voskuil, W. *et al., Rec. Trav. Chim.*, 1964, 83, 1301

Explodes on strong friction.

*See also* METAL ACETYLIDES

**6-BROMO-2,4-DINITROBENZENEDIAZONIUM HYDROGENSULPHATE** $C_6H_3BrN_4O_8S$

NO$_2$; Br; NO$_2$; $N_2^+$ $HSO_4^-$

*See* DIAZONIUM SULPHATES

**1-CHLORO-2,4-DINITROBENZENE** $ClC_6H_3(NO_2)_2$ $C_6H_3ClN_2O_4$

Sorbe, 1968, 56

It has been used as an explosive.

*See* 1-FLUORO-2,4-DINITROBENZENE, $C_6H_3FN_2O_4$

Ammonia

1. Uhlmann, P. W., *Chem. Ztg.*, 1914, **38**, 389–390
2. Wells, F. B. *et al., Org. Synth.*, 1943, Coll. Vol. 2, 221–222

During the preparation of 2,4-dinitroaniline by a well-established procedure involving heating the reactants in a direct-fired autoclave (170°C and 40 bar were typical conditions), a sudden increase in temperature and pressure preceded a violent explosion [1]. An alternative process avoiding the use of a sealed vessel is now used [2].

*See* POLYNITROARYL COMPOUNDS: Bases, or Salts
*p*-CHLORONITROBENZENE, $C_6H_4ClNO_2$ : Ammonia

*See other* AMINATION INCIDENTS
POLYNITROARYL COMPOUNDS

**2,4-DINITROBENZENESULPHENYL CHLORIDE** $C_6H_3ClN_2O_4S$

$(NO_2)_2C_6H_3SCl$

Kharasch, N. *et al., Org. Synth.*, 1964, **44**, 48

During vacuum removal of solvent, the residual chloride must not be overheated, as it may explode.

*See other* NITRO-ACYL HALIDES

**2,6-DINITRO-4-PERCHLORYLPHENOL** $C_6H_3ClN_2O_8$

$HO(NO_2)_2C_6H_2ClO_3$

Gardner, D. M. *et al., J. Org. Chem.*, 1963, **28**, 2652

This analogue of picric acid is dangerously explosive and very shock-sensitive.

*See other* PERCHLORYL COMPOUNDS

**6-CHLORO-2,4-DINITROBENZENEDIAZONIUM HYDROGENSULPHATE** $C_6H_3ClN_4O_8S$

$NO_2$

Cl $NO_2$

$N_2^+$ $HSO_4^-$

*See* DIAZONIUM SULPHATES

**2,6-DICHLORO-4-NITROBENZENEDIAZONIUM HYDROGENSULPHATE** $C_6H_3Cl_2N_3O_6S$

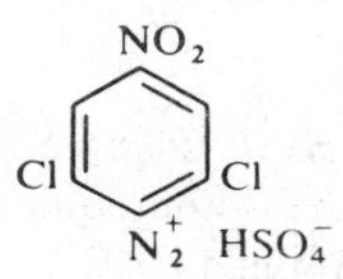


*See* DIAZONIUM SULPHATES

**2,4,5-TRICHLOROPHENOL** $Cl_3C_6H_2OH$ $C_6H_3Cl_3O$

Preparative hazard

*See* 1,2,4,5-TETRACHLOROBENZENE, $C_6H_2Cl_4$ : Sodium hydroxide

**1-FLUORO-2,4-DINITROBENZENE** $(NO_2)_2C_6H_3F$ $C_6H_3FN_2O_4$

1. Halpern, B. D., *Chem. Eng. News*, 1951, 29, 2666
2. Muir, G. D., private comm., 1969

The residue left from conversion of the chloro to the fluoro compound exploded during distillation at 1.3 mbar [1]. Reheating the residue from distillation caused violent decomposition [2].

Ether peroxides

Shafer, P. R., private comm., 1967

When air was admitted after vacuum evaporation of a (peroxidic) ether solution of the dinitro compound (5 g), a violent explosion occurred. Exploding ether peroxide may have initiated the dinitro compound.

*See other* HALOARYL COMPOUNDS

† **1,2,4-TRIFLUOROBENZENE** $C_6H_3F_3$

## SODIUM 2-HYDROXYMERCURIO-6-NITRO-4-*aci*-NITRO-2,5-CYCLOHEXADIENONE

$C_6H_3HgN_2NaO_6$

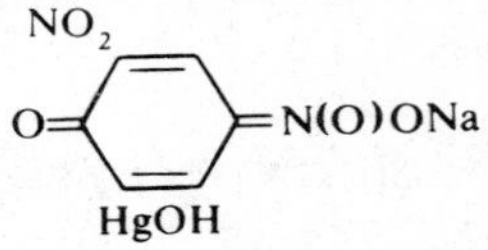

Hantzsch, A. *et al.*, *Ber.*, 1906, **39**, 1113

This salt of the mono-*aci-p*-quinonoid form of 2-hydroxymercurio-4,6-dinitrophenol explodes violently on strong heating.

*See* *aci*-NITROQUINONOID COMPOUNDS

## POTASSIUM 6-*aci*-NITRO-2,4-DINITRO-2,4-CYCLOHEXADIENIMINE

$C_6H_3KN_4O_6$

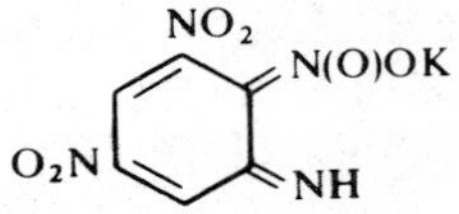

Green, A. G. *et al.*, *J. Chem. Soc.*, 1913, **103**, 508–513

This salt of the *aci-o*-iminoquinonoid form of 2,4,6-trinitroaniline explodes violently at 110°C.

*See* *aci*-NITROQUINONOID COMPOUNDS

## POTASSIUM 4,6-DINITROBENZOFUROXAN HYDROXIDE COMPLEX

$C_6H_3KN_4O_7$

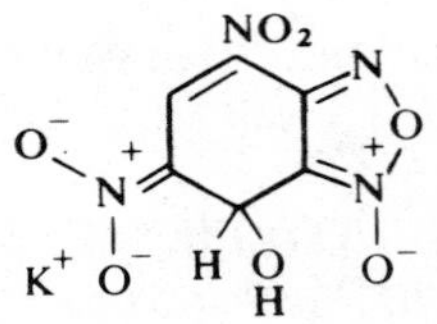

Boulton, A. J. *et al.*, *J. Chem. Soc.*, 1965, 5414

This explosive complex, similar to, but not formally an *aci*-nitro salt, has been evaluated as an explosive. It has a cyclised *aci-o*-iminoquinonoid structure.

*See related* *aci*-NITROQUINONOID COMPOUNDS

## SODIUM 2,4-DINITROPHENOLATE

$C_6H_3N_2NaO_5$

$NaOC_6H_3(NO_2)_2$

1. Hickson, B., *Chem. Age*, 1926, **14**, 522
2. Uhlmann, P. W., *Chem. Ztg.*, 1914, **38**, 388–390

A violent explosion occurred during centrifugal drying of an aqueous paste of the sodium salt [1], which was known to be explosive [2]. The instability may be associated with the possible presence of the *aci-o-* or *p*-quinonoid salt.

*See other* POLYNITROARYL COMPOUNDS

## 3,6,9-TRIAZATETRACYCLO[6.1.0.0$^{2,4}$.0$^{5,7}$]NONANE (*cis*-BENZENE TRIIMINE)

$C_6H_3N_3$

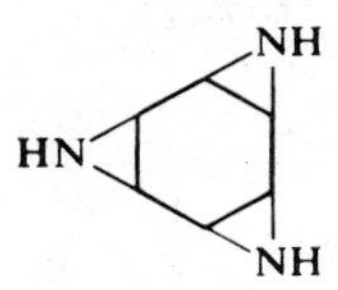


Schwesinger, R. *et al., Angew. Chem. (Intern. Ed.)*, 1973, **12**, 989

It explodes if rapidly heated to *ca.* 200°C.

*See other* STRAINED-RING COMPOUNDS

## 1,3,5-TRINITROBENZENE

$(NO_2)_3C_6H_3$ $C_6H_3N_3O_6$

Methanol,
Potassium hydroxide

*See* POTASSIUM 4-METHOXY-1-*aci*-NITRO-3,5-DINITRO-2,5-CYCLOHEXADIENE, $C_7H_6KN_3O_7$

## TRINITROSOPHLOROGLUCINOL

$C_6H_3N_3O_6$

NO
HO OH
ON NO
OH

Heavy metal salts may be hazardous.

*See* TRILEAD TRINITROSOPHLOROGLUCINOLATE, $C_{12}N_6O_{12}Pb_3$

*See other* NITROSO COMPOUNDS

## PICRIC ACID

$(NO_2)_3C_6H_2OH$ $C_6H_3N_3O_7$

Kirk-Othmer, 1965, Vol. 8, 617

Urbanski, 1964, Vol. 1, 518

Picric acid, in common with several other poly-nitrophenols, is an explosive material in its own right and is usually stored as a water-wet paste. It forms salts with many metals, some of which (lead, mercury, copper or zinc) are rather sensitive to heat, friction or impact. The salts with ammonia and amines and molecular complexes with hydrocarbons, etc. are, in general, not so sensitive. Contact of picric acid with concrete floors may form the friction-sensitive calcium salt.

Aluminium,
Water

*See* ALUMINIUM, Al: Nitro compounds, etc.

*See other* POLYNITROARYL COMPOUNDS

**TRINITRORESORCINOL (STYPHNIC ACID)** $C_6H_3N_3O_8$

$O_2N$ $NO_2$
HO OH
$NO_2$

Sorbe, 1968, 152

The acid and its salts are very explosive.

*See other* POLYNITROARYL COMPOUNDS

**1,3,5-TRIHYDROXYTRINITROBENZENE (TRINITROPHLOROGLUCINOL)** $C_6H_3N_3O_9$

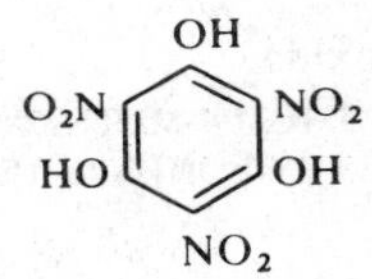


Sorbe, 1968, 153

It explodes on heating.

*See other* POLYNITROARYL COMPOUNDS

***N*,2,3,5-TETRANITROANILINE** $C_6H_3N_5O_8$

$(NO_2)_3C_6H_2NHNO_2$

Mudge, P. R. *et al., J. Chem. Soc., Chem. Comm.*, 1975, 569

Thermal rearrangement to the internal salt, 3,5-dinitrobenzenediazonium-2-olate proceeds explosively in absence of solvent.

*See other* *N*-NITRO COMPOUNDS

***N*,2,4,6-TETRANITROANILINE** $C_6H_3N_5O_8$

$(NO_2)_3C_6H_2NHNO_2$

Olsen, R. E. *et al., ACS 54*, 1966, 50

Impure tetryl must be recrystallised with care as it may deflagrate at only 50°C.

*See other* *N*-NITRO COMPOUNDS
POLYNITROARYL COMPOUNDS

**TRIETHYNYLPHOSPHINE** $(HC{\equiv}C)_3P$ $C_6H_3P$

Voskuil, W. *et al.*, *Rec. Trav. Chim.*, 1964, **83**, 1301

Explodes on strong friction and, on standing, it decomposes and may explode spontaneously.

*See other* ACETYLENIC COMPOUNDS

**TRIETHYNYLANTIMONY** $(HC{\equiv}C)_3Sb$ $C_6H_3Sb$

Voskuil, W., *et al.*, *Rec. Trav. Chim.*, 1964, **83**, 1301

Explodes on strong friction.

*See other* METAL ACETYLIDES

**SILVER *p*-NITROPHENOLATE** $AgOC_6H_4NO_2$ $C_6H_4AgNO_3$

Spiegel, L. *et al.*, *Ber.*, 1906, **39**, 2639

A sample decomposed explosively after intensive drying at 110°C. The *aci*-nitro-*p*-quinonoid salt may have been involved.

*See related* *aci*-NITROQUINONOID SALTS

*See other* SILVER COMPOUNDS

**SILVER BENZO-1,2,3-TRIAZOLE-1-OLATE** $C_6H_4AgN_3O$

$\overline{C_6H_4N(OAg)N{=}N}$

Deorha, D. S. *et al.*, *J. Indian Chem. Soc.*, 1964, **41**, 793

It is unsuitable as a gravimetric precipitate, as it explodes on heating rather than giving weighable silver.

*See other* SILVER COMPOUNDS

***o*-NITROBENZENEDIAZONIUM TETRACHLOROBORATE** $C_6H_4BCl_4N_3O_2$

$NO_2C_6H_4N_2^+\ BCl_4^-$

*See* DIAZONIUM TETRAHALOBORATES

***m*-BROMOPHENYLLITHIUM** $BrC_6H_4Li$ $C_6H_4BrLi$

*See* ORGANOLITHIUM REAGENTS

***p*-BROMOPHENYLLITHIUM** $BrC_6H_4Li$ $C_6H_4BrLi$

Anon., *Angew. Chem. (Nachr.)*, 1962, **10**, 65

It explodes if traces of oxygen are present in the inert atmosphere needed for the preparation.

*See* ORGANOLITHIUM REAGENTS

*See other* HALO-ARYLMETALS

## $p$-BROMOBENZENEDIAZONIUM SALTS $C_6H_4BrN_2^+Z^-$

$BrC_6H_4N_2^+Z^-$

Hydrogen sulphide

*See* DIAZONIUM SULPHIDES

## † $o$-, $m$- or $p$-CHLOROFLUOROBENZENE $C_6H_4ClF$

$ClC_6H_4F$

## $p$-CHLOROBENZENEDIAZONIUM TRIIODIDE $C_6H_4ClI_3N_2$

$ClC_6H_4N_2^+I_3^-$

*See* DIAZONIUM TRIIODIDES

## $m$-CHLOROPHENYLLITHIUM $ClC_6H_4Li$ $C_6H_4ClLi$

*See* ORGANOLITHIUM REAGENTS

## $p$-CHLOROPHENYLLITHIUM $ClC_6H_4Li$ $C_6H_4ClLi$

Anon., *Angew. Chem., (Nachr.),* 1962, **10**, 65

It explodes if traces of oxygen are present in the inert atmosphere needed for the preparation.

*See* ORGANOLITHIUM REAGENTS

*See other* HALO-ARYLMETALS

## BENZOQUINONE-1-CHLOROIMINE $C_6H_4ClNO$

O=⟨ring⟩=NCl

Sorbe, 1968, 112

It explodes on heating, like the bis-chloroimine.

*See other* *N*-HALOGEN COMPOUNDS

## $o$-CHLORONITROBENZENE $ClC_6H_4NO_2$ $C_6H_4ClNO_2$

Ammonia

*MCA Case History No. 1657*

A pressure reactor for preparing $o$-nitroaniline was accidentally overcharged with the chloro compound, and to avoid overfilling the reactor, the operator undercharged the aqueous ammonia, but did not tell the supervisor. During the subsequent heating cycle, the temperature could not be controlled, and the reaction ran away, finally bursting the vessel.

*See other* AMINATION INCIDENTS

## *p*-CHLORONITROBENZENE $ClC_6H_4NO_2$ $C_6H_4ClNO_2$

Doyle, W. H., *Loss Prevention*, 1969, **3**, 16

A steam-heated still used to top the crude material exploded after heating on total reflux for an hour.

Ammonia

Vincent, G. C., *Loss Prevention*, 1971, **5**, 46–52

During the large-scale preparation of *p*-nitroaniline at 160–180°C/30–40 bar in a jacketed autoclave, several concurrent abnormalities (excess chloro compound, too little ammonia solution, failure to apply cooling or to vent the autoclave, and non-failure of a rupture-disc) led to a runaway reaction and pressure-explosion of the vessel. Full analysis of the incident is given.

It seems remotely possible that the ammonium salt of the *aci*-quinonoid form of *p*-nitroaniline could have been formed under the extreme conditions.

*See other* AMINATION INCIDENTS

Sodium methoxide

*ABCM Quart. Safety Summ.*, 1944, **15**, 15

Addition of the chloro compound to a solution of sodium methoxide in methanol caused an unusually exothermic reaction to occur. The lid of the 450 litre vessel was blown off, and a fire and explosion followed. No cause for the unusual vigour of the reaction was found.

*See other* HALOARYL COMPOUNDS
*C*-NITRO COMPOUNDS

## NITROPERCHLORYLBENZENE $NO_2C_6H_4ClO_3$ $C_6H_4ClNO_5$

McCoy, G., *Chem. Eng. News*, 1960, 38(4), 62

The nitration product of perchlorylbenzene is explosive, comparable in shock sensitivity with lead azide, with a very high propagation rate.

*See other* PERCHLORYL COMPOUNDS

## 2-CHLORO-5-NITROBENZENESULPHONIC ACID $C_6H_4ClNO_5S$

$NO_2$

$SO_3H$

Cl

Preparative hazard

*See* SULPHURIC ACID, $H_2O_4S$: *p*-Chloronitrobenzene, etc.

## 4-HYDROXY-3-NITROBENZENESULPHONYL CHLORIDE $C_6H_4ClNO_5S$

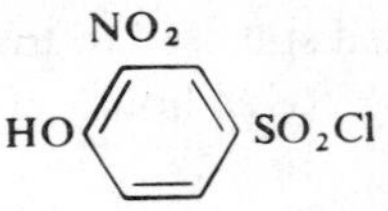

NO$_2$

HO–⟨benzene ring⟩–SO$_2$Cl

Vervalin, C. H., *Hydrocarbon Process.*, 1976, **55**(9), 321–322
The chloride, produced by interaction of *o*-nitrophenol and chlorosulphuric acid at 4°C, decomposed violently during discharge from the 2000 l vessel, leaving a glowing residue. It was subsequently found that accelerating exothermic decomposition sets in at 24–27°C.

*See related* SULPHONATION INCIDENTS
*See other* *C*-NITRO COMPOUNDS

## *m*-CHLOROBENZENEDIAZONIUM SALTS $C_6H_4ClN_2^+Z^-$

$ClC_6H_4N_2^+Z^-$

Potassium thiophenolate,
or Sodium polysulphide
*See* DIAZONIUM SULPHIDES

## *o*-CHLOROBENZENEDIAZONIUM SALTS $C_6H_4ClN_2^+Z^-$

$ClC_6H_4N_2^+Z^-$

Potassium 2-chlorothiophenolate
*See* DIAZONIUM SULPHIDES

## 1-CHLOROBENZOTRIAZOLE $C_6H_4NClN{=}N$ (ring) $C_6H_4ClN_3$

Hopps, H. B., *Chem. Eng. News*, 1971, **49**(30), 3
Spontaneous ignition during packaging.
*See other* *N*-HALOGEN COMPOUNDS

## 4-CHLORO-2,6-DINITROANILINE $C_6H_4ClN_3O_4$

$Cl(NO_2)_2C_6H_2NH_2$

Preparative hazard
*See* NITRIC ACID, $HNO_3$: 4-Chloro-2-nitroaniline

Nitrosylsulphuric acid
Anon., *Angew. Chem. (Nachr.)*, 1970, **18**, 62
During large-scale diazotisation of the amine, severe local overheating

is thought to have caused the explosion observed. The effect could not be reproduced in the laboratory.
*See other* POLYNITROARYL COMPOUNDS

**6-CHLORO-2,4-DINITROANILINE** $C_6H_4ClN_3O_4$

Cl
$O_2N$ $NH_2$
$NO_2$

Nitrosylsulphuric acid
Bersier, P. *et al., Chem. Ing. Tech.*, 1971, **43**, 1311–1315
The solid amine can be caused to detonate by heating, or by a powerful initiating charge. The third batch of a new process involving diazotisation of 6-chloro-2,4-dinitroaniline (at a higher than usual concentration in 40% nitrosylsulphuric acid at 30–40°C) exploded violently soon after the temperature had been increased to 50°C. Subsequent DSC work showed that the temperature at which thermal decomposition of the diazonium sulphate solution or suspension sets in is inversely proportional to the concentration of amine, falling from 160°C at 0.3 mmol/g to 80°C at 2 mmol/g. Thermal stability of 17 other diazonium derivatives was similarly investigated.
*See* DIAZONIUM SULPHATES
*See other* POLYNITROARYL COMPOUNDS

***m*-NITROBENZENEDIAZONIUM PERCHLORATE** $C_6H_4ClN_3O_6$

$NO_2C_6H_4N_2^+ ClO_4^-$

Schumacher, 1960, 205
Explosive, very sensitive to heat and shock.
*See other* DIAZONIUM PERCHLORATES

***o*-DICHLOROBENZENE** $C_6H_4Cl_2$

Aluminium
*See* ALUMINIUM, Al: Halocarbons

**1,6-DICHLORO-2,4-HEXADIYNE** $C_6H_4Cl_2$

$ClCH_2C{\equiv}CC{\equiv}CCH_2Cl$

1. Driedger, P. E. *et al., Chem. Eng. News,* 1972, **50**(12), 51
2. Whiting, M. C., *Chem. Eng. News,* 1972, **50**(23), 86
3. Ford, M. *et al., Chem. Eng. News,* 1972, **50**(30), 67

It is extremely shock-sensitive, a 4.0 kg cm shock causing detonation

in 50% of test runs (cf. 3.5 kg cm for propargyl bromide; 2.0 kg cm for glyceryl nitrate). The intermediate bis-chlorosulphite involved in the preparation needs low temperatures to prevent vigorous decomposition. The corresponding diiodo derivative was expected to be similarly hazardous [1], and this has been confirmed [2]. Improvements in preparative techniques (use of dichloromethane solvent at −30°C) to avoid violent reaction have also been described [3].
*See other* HALOACETYLENE DERIVATIVES

**BENZOQUINONE-1,4-BIS(CHLOROIMINE)** $C_6H_4Cl_2N_2$

ClN=⟨ring⟩=NCl

Sorbe, 1968, 112
It explodes on heating, like the mono-chloroimine.
*See other* *N*-HALOGEN COMPOUNDS

**DICHLOROPHENOLS** $Cl_2C_6H_3OH$ $C_6H_4Cl_2O$

Laboratory Chemical Disposal Co. Ltd., confidential information
During vacuum fractionation of the mixed dichlorophenols produced by hydrolysis of trichlorobenzene, admission of air caused the column contents to be forced down into the boiler at 210°C, when a violent explosion ensued.
*See also* 1,2,4,5-TETRACHLOROBENZENE, $C_6H_2Cl_4$: Sodium hydroxide

**2,4-HEXADIYNYLENE BISCHLOROSULPHITE** $C_6H_4Cl_2O_4S_2$
$ClSO_2OCH_2C{\equiv}CC{\equiv}CCH_2OSO_2Cl$

Driedger, P. E. *et al.*, *Chem. Eng. News*, 1972, **50**(12), 51
This is probably an intermediate in the preparation of the 1,6-dichloro compound from 2,4-hexadiyne-1,6-diol and thionyl chloride in dimethylformamide. The reaction mixture must be kept at a low temperature to avoid vigorous decomposition and charring. This may be due to interaction of the intermediate with the solvent.
*See also* HALOACETYLENE DERIVATIVES
DIMETHYLFORMAMIDE, $C_3H_7NO$
2,4-HEXADIYNYLENE BISCHLOROFORMATE, $C_8H_4Cl_2O_4$

***p*-FLUOROPHENYLLITHIUM** $FC_6H_4Li$ $C_6H_4FLi$

*See* ORGANOLITHIUM REAGENTS

† ***m*-DIFLUOROBENZENE** $FC_6H_4F$ $C_6H_4F_2$

**† *p*-DIFLUOROBENZENE** **$FC_6H_4F$** **$C_6H_4F_2$**

**OCTAFLUOROADIPAMIDE** **$C_6H_4F_8N_2O_2$**

**$H_2NCO[CF_2]_4CONH_2$**

Lithium tetrahydroaluminate

*See* LITHIUM TETRAHYDROALUMINATE, $AlH_4Li$: Fluoroamides

**SODIUM 2-HYDROXYMERCURIO-4-*aci*-NITRO-2,5-CYCLOHEXADIENONE** **$C_6H_4HgNO_4$**

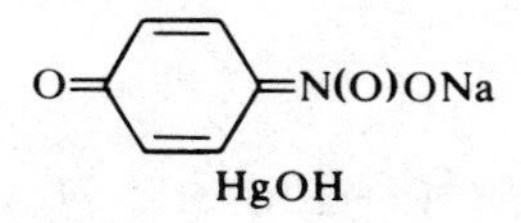


Hantzsch, A. *et al.*, *Ber.*, 1906, **39**, 1115

This salt of the *aci-p*-quinonoid form of 2-hydroxymercurio-4-nitrophenol explodes on heating.

*See* *aci*-NITROQUINONOID COMPOUNDS

***o*-DIIODOBENZENE** **$IC_6H_4I$** **$C_6H_4I_2$**

Rüst, 1948, 302

A small sample in a melting point capillary exploded violently at 181°C, breaking the surrounding oil-bath.

*See other* IODINE COMPOUNDS

**1,6-DIIODO-2,4-HEXADIYNE** **$ICH_2C{\equiv}CC{\equiv}CCH_2I$** **$C_6H_4I_2$**

*See* 1,6 DICHLORO-2,4-HEXADIYNE, $C_6H_4Cl_2$

**POTASSIUM *p*-NITROPHENOLATE** **$C_6H_4KNO_3$**

KO–$C_6H_4$–$NO_2$ ⇌ O=$C_6H_4$=N(O)OK

Personal experience

A sample of this potassium salt exploded after long storage in a non-evacuated desiccator. This may have been caused by the presence of the potassium salt of the tautomeric *aci-p*-quinones.

*See* *aci*-NITRO SALTS

***m*- or *p*-DILITHIOBENZENE** **$LiC_6H_4Li$** **$C_6H_4Li_2$**

*See* ORGANOLITHIUM REAGENTS

**N,N-p-TRILITHIOANILINE** **$LiC_6H_4NLi_2$** **$C_6H_4Li_3N$**

Gilman, H. *et al., J. Amer. Chem. Soc.*, 1949, **71**, 2934

Highly explosive in contact with air.

*See other* N-METAL DERIVATIVES
PYROPHORIC MATERIALS

**SODIUM p-NITROSOPHENOLATE** **$NaOC_6H_4NO$** **$C_6H_4NNaO_2$**

Sorbe, 1968, 86

The dry solid tends to ignite

*See other* NITROSO COMPOUNDS

**SODIUM o-NITROTHIOPHENOLATE** **$C_6H_4NNaO_2S$**

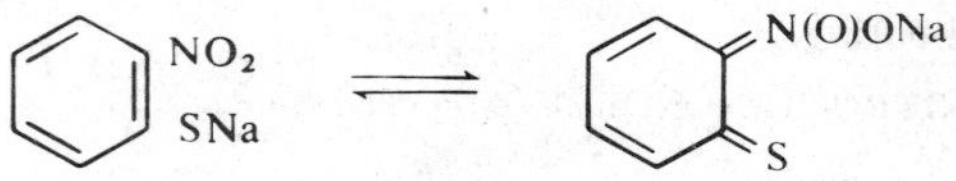

Davies, H. J., *Chem. & Ind.*, 1966, 257

The material exploded when the temperature of evaporation of the slurry was increased by adding a higher-boiling solvent. This may have been caused by the presence of some sodium salt of the tautomeric *aci-o*-thioquinone imine. Such salts are shock- or thermally unstable.

*See* NITROAROMATIC–ALKALI HAZARDS

**SODIUM p-NITROPHENOLATE** **$NaOC_6H_4NO_2$** **$C_6H_4NNaO_3$**

1. Semiganowski, N., *Z. Anal. Chem.*, 1927, **72**, 29
2. Baltzer, R., private comm., 1976

An ammoniacal solution of *p*-nitroaniline containing ammonium chloride was being treated with 50% sodium hydroxide solution to displace ammonia. In error, double the quantity required to give the usual 10% of free alkali was added, and during the subsequent degassing operation (heating to 130°C under pressure, followed by depressuring to vent ammonia), complete conversion to sodium *p*-nitrophenolate occurred unwittingly [1]. This solid product, separated by centrifuging, was then heated to dry it, when it decomposed violently and was ejected through the vessel opening like a rocket [2].

*See* NITROAROMATIC–ALKALI HAZARDS

## DISODIUM p-NITROPHENYL ORTHOPHOSPHATE $C_6H_4NNa_2O_6P$

$NO_2C_6H_4OP(O)(ONa)_2$

Preparative hazard

*See* NITRIC ACID, $HNO_3$: Disodium phenyl orthophosphate

## SODIUM 1,4-(BIS-*aci*-NITRO)-2,5-CYCLOHEXADIENE $C_6H_4N_2Na_2O_4$

NaO(O)N=⟨cyclohexadiene ring⟩=N(O)ONa

Meisenheimer, J. *et al., Ber.*, 1906, **39**, 2529

The sodium and potassium salts of the bis-*aci-p*-quinonoid form of *p*-dinitrobenzene deflagrate on heating.

*See* *aci*-NITROQUINONOID COMPOUNDS

## BENZO-1,2,3-THIADIAZOLE-1,1-DIOXIDE $C_6H_4N_2O_2S$

$C_6H_4SO_2N{=}N$ (ring)

Wittig, G. *et al., Chem. Ber.*, 1962, **95**, 2718; *Org. Synth.*, 1967, **47**, 6

The solid explodes at 60°C, on impact or friction, or sometimes spontaneously.

## BENZENEDIAZONIUM-2-SULPHONATE $C_6H_4N_2O_3S$

$N_2^+C_6H_4SO_3^-$

Franklin, E. C., *Amer. Chem. J.*, 1899, **20**, 459

The internal diazonium salt explodes on contact with flame or on percussion.

*See related* DIAZONIUM CARBOXYLATES

*See other* DIAZONIUM SALTS

## BENZENEDIAZONIUM-4-SULPHONATE $C_6H_4N_2O_3S$

$N_2^+C_6H_4SO_3^-$

1. Wichelhaus, H., *Ber.*, 1901, **34**, 11
2. *Chemiearbeit*, 1959, **11**(9), 59

This internal salt of diazotised sulphanilic acid (Pauly's reagent) exploded violently on touching when thoroughly dry [1]. Use of a metal spatula to remove a portion of a refrigerated sample of the solid diazonium compound caused a violent explosion. All solid diazo compounds must be stored in small quantities under refrigeration in loosely plugged containers. Handle gently with non-metallic spatulae using personal protection [2].

*See related* DIAZONIUM CARBOXYLATES

*See other* DIAZONIUM SALTS

**o-, m- or p-DINITROBENZENE** $NO_2C_6H_4NO_2$ $C_6H_4N_2O_4$

Nitric acid

*See* NITRIC ACID, $HNO_3$: Nitroaromatics

*See other* POLYNITROARYL COMPOUNDS

**m-DINITROBENZENE** $NO_2C_6H_4NO_2$ $C_6H_4N_2O_4$

Tetranitromethane

*See* TETRANITROMETHANE, $CN_4O_8$: Aromatic nitrocompounds

**2,4-DINITROPHENOL** $(NO_2)_2C_6H_3OH$ $C_6H_4N_2O_5$

Bases

Uhlmann, P. W., *Chem. Ztg.*, 1914, **38**, 389–390

Dinitrophenol forms explosive (*aci*-quinonoid?) salts with alkalies or ammonia, and should not be heated with them in closed vessels. Dinitrophenol was later classified as an explosive in the UK, and is normally available from laboratory suppliers wetted with 15% water, as is picric acid.

*See* *aci*-NITROQUINONOID COMPOUNDS

SODIUM 2,4-DINITROPHENOLATE, $C_6H_3N_2NaO_5$

*See other* POLYNITROARYL COMPOUNDS

**SODIUM 6-*aci*-NITRO-4-NITRO-2,4-CYCLOHEXADIENIMINE**

$C_6H_4N_3NaO_4$

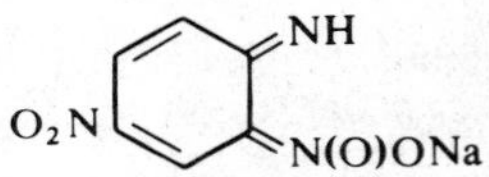

NH

$O_2N$

N(O)ONa

Green, A. G. *et al.*, *J. Chem. Soc.*, 1913, **103**, 508–513

The sodium and potassium salts of the mono-*aci*-*o*-iminoquinonoid form of 2,4-dinitroaniline readily deflagrate on heating.

*See* *aci*-NITROQUINONOID COMPOUNDS

**$o$-NITROBENZENEDIAZONIUM SALTS** **$C_6H_4N_3O_2^+\ Z^-$**

**$NO_2C_6H_4N_2^+\ Z^-$**

Disodium disulphide,
or Disodium polysulphide
*See* DIAZONIUM SULPHIDES

**$p$-NITROBENZENEDIAZONIUM SALTS** **$C_6H_4N_3O_2^+\ Z^-$**

**$NO_2C_6H_4N_2^+\ Z^-$**

Disodium sulphide,
or Hydrogen sulphide,
or Potassium *O,O*-diphenyl dithiophosphate
*See* DIAZONIUM SULPHIDES

**SODIUM 1,3-BIS-*aci*-NITROCYCLOHEXEN-2,4-DIIMINE** **$C_6H_4N_4Na_2O_4$**

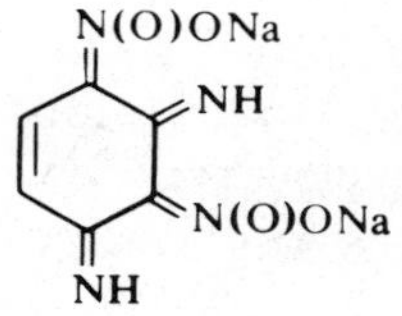

N(O)ONa
NH
N(O)ONa
NH

Meisenheimer, J. *et al., Ber.*, 1906, **39**, 2538
The disodium salt of the bis-*o*-quinonoid form of 2,4-dinitro-*m*-phenylenediamine deflagrates violently in contact with heat or moisture.
*See other* *aci*-NITROQUINONOID COMPOUNDS

**$p$-NITROBENZENESULPHINYL AZIDE** **$C_6H_4N_4O_3S$**

**$NO_2C_6H_4SON_3$**

*See* SULPHINYL AZIDES

**4-NITROBENZENEDIAZONIUM NITRATE** **$C_6H_4N_4O_5$**

**$NO_2C_6H_4N_2^+\ NO_3^-$**

Bamberger, E., *Ber.*, 1895, **28**, 239
The isolated dry salt explodes on heating, but not on friction.
*See other* DIAZONIUM SALTS

**2,4-DINITROBENZENEDIAZONIUM HYDROGENSULPHATE** **$C_6H_4N_4O_8S$**

**$(NO_2)_2C_6H_3N_2^+\ HSO_4^-$**

*See* DIAZONIUM SULPHATES

## BENZOTRIAZOLIUM 1- or 2-NITROIMIDATE $C_6H_4N_5O_2$

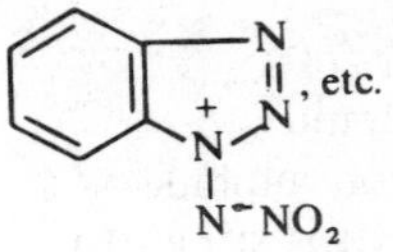

(structure: benzotriazolium ring with $\overset{+}{N}$–$N^{-}NO_2$), etc.

Dangerous solids, which explode at their m.p.s, 74.5 and 81°C, respectively. The 1-isomer is also impact-sensitive.

*See* N-AZOLIUM NITROIMIDATES

## 1,3-DIAZIDOBENZENE $(N_3)_2C_6H_4$ $C_6H_4N_6$

Sulphuric acid

Forster, M. O. *et al., J. Chem. Soc.*, 1907, **91**, 1953

The azide ignites and explodes mildly with concentrated acid.

## 1,4-DIAZIDOBENZENE $(N_3)_2C_6H_4$ $C_6H_4N_6$

Griess, P., *Ber.*, 1888, **21**, 1561

The diazide explodes very violently on heating.

*See other* ORGANIC AZIDES

## 4-NITROBENZENEDIAZONIUM AZIDE $C_6H_4N_6O_2$

$NO_2C_6H_4N_2^+ N_3^-$

Hantzsch, A., *Ber.*, 1903, **36**, 2058

The dry azide exploded violently with a brilliant flash.

*See other* DIAZONIUM SALTS

## *m*-BENZENEDISULPHONYL DIAZIDE $C_6H_4N_6O_4S_2$

$C_6H_4(SO_2N_3)_2$

Balabanov, G. P. *et al., Chem. Abs.*, 1969, **70**, 59427t

The title compound was the most explosive of a series of seven substituted benzenesulphonyl azides.

*See other* ACYL AZIDES

## 2-ETHYNYLFURAN $\overline{OCH{=}CHCH{=}C}C{\equiv}CH$ $C_6H_4O$

Dunlop, 1953, 77

It explodes on heating or in contact with concentrated nitric acid.

*See other* ACETYLENIC COMPOUNDS

**1,4,5,8-TETRAHYDRO-1,4,5,8-TETRATHIAFULVALENE** $C_6H_4S_4$

Preparative hazard

*See* 1,3-DITHIOLIUM PERCHLORATE, $C_3H_4ClO_4S_2$

**PHENYLSILVER** $C_6H_5Ag$

1. Houben-Weyl, 1970, Vol. 13.1, 774
2. Krause, E. *et al.*, *Ber.*, 1923, **56**, 2064

The dry solid explodes on warming to room temperature, or on light friction or stirring [1]. It is more stable wet with ether [2].

*See other* ARYLMETALS

**SILVER PHENOLATE** **AgOPh** $C_6H_5AgO$

Macomber, R. S. *et al.*, *Synth. React. Inorg. Met.-Org. Chem.*, 1977, 7, 111–122

The dried solid product from interaction of lithium phenolate and silver perchlorate in benzene (and probably largely silver phenolate) exploded on gentle heating. Other silver alkoxide derivatives were unstable.

*See* SILVER PERCHLORATE, $AgClO_4$

*See other* SILVER COMPOUNDS

**SILVER PHENYLSELENONATE** $C_6H_5AgO_3Se$

$AgOSe(O_2)C_6H_5$

*See* PHENYLSELENONIC ACID, $C_6H_6O_3Sc$

*See other* SILVER COMPOUNDS

**4-HYDROXY-3,5-DINITROBENZENEARSONIC ACID** $C_6H_5AsN_2O_8$

$HO(NO_2)_2C_6H_2AsO(OH)_2$

Phillips, M. A., *Chem. & Ind.*, 1947, 61

A heated, unstirred, water-wet sludge of this analogue of picric acid exploded violently, apparently owing to local overheating in the containing flask.

*See other* POLYNITROARYL COMPOUNDS

**BENZENEDIAZONIUM TETRAFLUOROBORATE** $C_6H_5BF_4N_2$

$C_6H_5N_2^+ BF_4^-$

Caesium fluoride,
Difluoroamine

*See* DIFLUOROAMINE, $F_2HN$: Benzenediazonium tetrafluoroborate, etc.

**BROMOBENZENE** **$BrC_6H_5$** **$C_6H_5Br$**

1-Bromobutane,
Sodium
*See* SODIUM, Na: Halocarbons (reference 7)

**PHENYLMAGNESIUM BROMIDE** **PhMgBr** **$C_6H_5BrMg$**

Chlorine
*See* CHLORINE, $Cl_2$: Phenylmagnesium bromide

**BENZENEDIAZONIUM TRIBROMIDE** **$PhN_2^+ Br_3^-$** **$C_6H_5Br_3N_2$**

Sorbe, 1968, 40
An explosive compound.
*See other* DIAZONIUM SALTS

**CHLOROBENZENE** **$ClC_6H_5$** **$C_6H_5Cl$**

Sodium
*See* SODIUM, Na: Halocarbons (reference 8)

**BENZENEDIAZONIUM CHLORIDE** **$C_6H_5N_2^+ Cl^-$** **$C_6H_5ClN_2$**

Hantzsch, A. *et al., Ber.*, 1901, **34**, 3338
Grewer, T., *Chem. Ing. Tech.*, 1975, **47**, 233–234
The dry salt was more or less explosive, depending on the method of preparation. DTA examination of a 15% solution of the diazonium salt in hydrochloric acid showed an exothermic decomposition reaction, with the exotherm peak at 65°C. Adiabatic decomposition of the solution from 20°C took 80 min to attain the maximum temperature of 80°C.
*See* THERMAL STABILITY OF REACTION MIXTURES
*See also* BIS-BENZENEDIAZONIUM ZINC TETRACHLORIDE, $C_{12}H_{10}Cl_4N_4Zn$

Potassium *O*-methyldithiocarbonate
*See* THIOPHENOL, $C_6H_6S$
DIAZONIUM SULPHIDES
*See other* DIAZONIUM SALTS

**_N_-CHLORO-4-NITROANILINE** **$ClNHC_6H_4NO_2$** **$C_6H_5ClN_2O_2$**

1. Goldschmidt, S. *et al., Ber.*, 1922, **55**, 2450

2. Pyetlewski, L. L., *Rept. AD A028841*, 24, Richmond (Va.), NTIS, 1976

Though stable at $-70°C$, it soon explodes at ambient temperature [1], and violently on heating slightly [2].

*See other* N-HALOGEN COMPOUNDS

**BENZENEDIAZONIUM PERCHLORATE** $C_6H_5ClN_2O_4$

$PhN_2^+ ClO_4^-$

1. Vorlander, D., *Ber.*, 1906, **39**, 2713–2715
2. Hofmann, K. A. *et al.*, ibid., 3146–3148

The salt explodes on friction, impact or contact with sulphuric acid when still wet with liquor [1], and is frightfully explosive when dry [2].

*See other* DIAZONIUM PERCHLORATES

**PHENYLPHOSPHONIC AZIDE CHLORIDE** $C_6H_5ClN_3OP$

$PhPON_3Cl$

Utvary, K., *Inorg. Nucl. Chem. Lett.*, 1965, **1**, 77–78

A violent explosion occurred during distillation at 80–88°C/0.46 mbar of mixtures containing the azide chloride and the diazide.

*See* PHENYLPHOSPHONIC DIAZIDE, $C_6H_5N_6OP$

*See other* ACYL AZIDES

**BENZENESULPHINYL CHLORIDE** $C_6H_5S(O)Cl$ $C_6H_5ClOS$

Anon., *Chem. Eng. News.* 1957, **35**, 57

A glass bottle, undisturbed for several months, exploded, probably from gas pressure caused by photolytic or hydrolytic decomposition.

**BENZENESULPHONYL CHLORIDE** $C_6H_5SO_2Cl$ $C_6H_5ClO_2S$

Dimethyl sulphoxide

*See* DIMETHYL SULPHOXIDE, $C_2H_6OS$: Acyl halides

**PERCHLORYLBENZENE** $ClO_3C_6H_5$ $C_6H_5ClO_3$

Aluminium trichloride

Bruce, W. F., *Chem. Eng. News*, 1960, **38**(4), 63

A mixture is quiescent for some time, then suddenly explodes. Many perchloryl aromatics appear to be inherently shock-sensitive.

*See other* PERCHLORYL COMPOUNDS

**N,N-DICHLOROANILINE** $C_6H_5NCl_2$ $C_6H_5Cl_2N$

Goldschmidt, S., *Ber.*, 1913, **46**, 2732

The oil prepared at −20°C explodes during warming to ambient temperature.

*See other* N-HALOGEN COMPOUNDS

**PHENYLPHOSPHORYL DICHLORIDE** $PhPOCl_2$ $C_6H_5Cl_2OP$

Koch-Light Ltd, private comm., 1976

A bottle of the dichloride exploded violently while on the bench. No cause was established, but diffusive ingress of moisture, or an intramolecular Friedel–Crafts polymerisation reaction, both of which would form hydrogen chloride, seem possible contributory factors.

*See related* NON-METAL HALIDES

**PHENYLVANADIUM(V) DICHLORIDE OXIDE** $C_6H_5Cl_2OV$

$PhVOCl_2$

Houben-Weyl, 1975, Vol. 13.3, 365

Thermal decomposition may be explosive.

*See related* ARYL METALS

**[(CHROMYLDIOXY)IODO] BENZENE** $\overline{OCrO_2OI}C_6H_5$ $C_6H_5CrIO_4$

Sidgwick, 1950, 1250

It explodes at 66°C.

*See other* IODINE COMPOUNDS

† **FLUOROBENZENE** $FC_6H_5$ $C_6H_5F$

**IODOBENZENE** $IC_6H_5$ $C_6H_5I$

Sorbe, 1968, 66

It explodes on heating above 200°C.

*See other* IODINE COMPOUNDS

**BENZENEDIAZONIUM IODIDE** $C_6H_5N_2^+ I^-$ $C_6H_5IN_2$

*See* DIAZONIUM TRIIODIDES

**PHENYLIODINE DINITRATE** $PhI(NO_3)_2$ $C_6H_5IN_2O_6$

Sorbe, 1968, 66

It explodes above 100°C.

*See other* IODINE COMPOUNDS

**IODOSYLBENZENE** $OIC_6H_5$ $C_6H_5IO$

1. Saltzman, A. *et al., Org. Synth.*, 1963, **43**, 60
2. Banks, D. F., *Chem. Rev.*, 1966, **66**, 255

It explodes at 210°C [1] and homologues generally explode on melting [2].

*See other* IODINE COMPOUNDS

**IODYLBENZENE** $O_2IC_6H_5$ $C_6H_5IO_2$

1. Sharefkin, J. G. *et al., Org. Synth.*, 1963, **43**, 65
2. Banks, D. F., *Chem. Rev.*, 1966, **66**, 259

It explodes at 230°C [1]. Extreme care should be used in heating, compressing or grinding iodyl compounds, as heat or impact may cause detonation of homologues [2].

*See other* IODINE COMPOUNDS

**BENZENEDIAZONIUM TRIIODIDE** $C_6H_5N_2^+ I_3^-$ $C_6H_5I_3N_2$

*See* DIAZONIUM TRIIODIDES

**POTASSIUM 3,5-DINITRO-2(1-TETRAZENYL)PHENOLATE** $C_6H_5KN_6O_5$

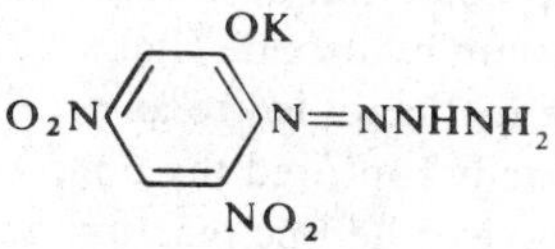

OK

$O_2N$ [ring] $N{=}NNHNH_2$

$NO_2$

Kenney, J. F., US Pat. 2 728 760, 1952

It is explosive.

*See other* HIGH-NITROGEN COMPOUNDS

**POTASSIUM BENZENESULPHONYLPEROXOSULPHATE** $C_6H_5KO_7S_2$

$C_6H_5SO_2OOSO_2OK$

Davies, 1961, 65

It explodes on friction or warming.

*See other* DIACYL PEROXIDES

**PHENYLLITHIUM** $C_6H_5Li$

Preparative hazard

*See* LITHIUM, Li: Bromobenzene

Titanium tetraethoxide

Gilman, H. *et al., J. Org. Chem.*, 1945, **10**, 507

The product of interaction at 0°C (of unknown composition) ignited in air and reacted violently with water.
*See other* ARYLMETALS

**p-BENZOQUINONE MONOIMINE** $O=C_6H_4=NH$ $C_6H_5NO$

Willstätter, R. *et al.*, *Ber.*, 1906, **37**, 4607
The solid decomposes with near-explosive violence.

**NITROBENZENE** $C_6H_5NO_2$

*MCA SD-21*, 1967

Alkali

1. Hatton, J. P., private comm., 1976
2. Anon., *Sich. Chemiearb.*, 1975, **27**, 89; Bilow, W., private comm., 1976
3. Wohl, A., *Ber.*, 1899, **32**, 3846; 1901, **34**, 2444
4. *ABCM Quart. Safety Summ.*, 1953, **24**, 42
5. Bretherick, L., *Chem. & Ind.*, 1976, 576

Heating a mixture of nitrobenzene, flake sodium hydroxide and a little water in an autoclave led to an explosion [1]. During the technical-scale preparation of a warm solution of nitrobenzene in methanolic potassium hydroxide (flake 90% material), accidental omission of most of the methanol led to an accelerating exothermic reaction which eventually ruptured the 6 $m^3$ vessel. Laboratory investigation showed that no exothermic reaction occurred between potassium hydroxide and nitrobenzene, either alone or with the full amount (3.4 vol.) of methanol, but that it did if only a little solvent was present. The residue was largely a mixture of azo- and azoxy-benzene [2].

The violent conversion of nitrobenzene to (mainly) *o*-nitrophenol by heating with finely powdered and anhydrous potassium hydroxide had been described [3]. Accidental substitution of nitrobenzene for aniline as diluent during large-scale fusion of benzanthrone with potassium hydroxide caused a violent explosion [4]. Related incidents are also discussed [5].

*See* NITROAROMATIC–ALKALI HAZARDS

Aluminium trichloride

*See* ALUMINIUM TRICHLORIDE, $AlCl_3$ : Nitrobenzene
: Nitrobenzene, etc.

Aniline,
Gycerol,
Sulphuric acid

*See* QUINOLINE, $C_9H_7N$

Oxidants

*See* TETRANITROMETHANE, $CN_4O_8$ : Aromatic nitrocompounds
SODIUM CHLORATE, $ClNaO_3$ : Nitrobenzene
NITRIC ACID, $HNO_3$ : Nitroaromatics
: Nitrobenzene
PEROXODISULPHURIC ACID, $H_2O_8S_2$ : Organic liquids
DINITROGEN TETRAOXIDE, $N_2O_4$ : Nitrobenzene

Phosphorus pentachloride

*See* PHOSPHORUS PENTACHLORIDE, $Cl_5P$: Nitrobenzene

Potassium

*See* POTASSIUM, K: Nitrogen-containing explosives

Sulphuric acid

*MCA Case History No. 678*

Nitrobenzene was washed with dilute (5%) sulphuric acid to remove amines, and became contaminated with some acid emulsion which had formed. After distillation, the hot, acid, tarry residue attacked the iron vessel, evolving hydrogen, and eventually exploded. It was later found that addition of the nitrobenzene to the diluted acid did not give emulsions, while the reversed addition did. A final wash with sodium carbonate solution was added to the process.

Uronium perchlorate (Urea perchlorate)

*See* URONIUM PERCHLORATE (UREA PERCHLORATE), $CH_5ClN_2O_5$ : Aromatic nitro compounds

*See other* C-NITRO COMPOUNDS

## *o*-NITROSOPHENOL $NOC_6H_4OH$ $C_6H_5NO_2$

Alone,
or Acids

Baeyer, A. *et al., Ber.*, 1902, **35**, 3037

It explodes on heating or in contact with concentrated acids.

*See other* NITROSO COMPOUNDS

## *p*-NITROSOPHENOL $NOC_6H_4OH$ $C_6H_5NO_2$

Milne, W. D., *Ind. Eng. Chem.*, 1919, **11**, 489

Kuznetsov, V., *Chem. Abs.*, 1940, **34**, 3498i

Stored barrels heated spontaneously and caused a fire. Contamination of the bulk material by acid or alkali may cause ignition.

The tendency to spontaneous ignition was found to be associated with the presence of nitrates in the sodium nitrite used in preparation. A diagnostic test for this tendency is to add sulphuric acid, when impure material will effervesce or ignite. The sodium salt is more suitable for storage purposes. (But see SODIUM *p*-NITROSOPHENOLATE, $C_6H_4NNaO_2$.)

*See other* NITROSO COMPOUNDS

**o-NITROPHENOL** **$NO_2C_6H_4OH$** **$C_6H_5NO_3$**

Chlorosulphuric acid

*See* 4-HYDROXY-3-NITROBENZENESULPHONYL CHLORIDE, $C_6H_4ClNO_5S$

**m-NITROBENZENESULPHONIC ACID** **$C_6H_5NO_5S$**

**$NO_2C_6H_4SO_3H$**

Sulphuric acid

*MCA Case Histories Nos. 1482, 944*

A 270 litre batch of a solution in sulphuric acid exploded violently after storage at ~150°C for several hours. An exotherm develops at 145°C, and the acid is known to decompose at ~200°C. The earlier Case History describes a similar incident when water, leaking from a cooling coil into the fuming sulphuric acid medium, caused an exotherm to over 150°C, and subsequent violent decomposition.

**BENZENEDIAZONIUM SALTS** **$C_6H_5N_2^+ Z^-$**

Alone

*See* THE CHLORIDE, $C_6H_5ClN_2$
THE NITRATE, $C_6H_5N_3O_3$

Ammonium sulphide,
or Hydrogen sulphide,
or Disodium sulphide

*See* DIAZONIUM SULPHIDES

**BENZOTRIAZOLE** **$\overline{C_6H_4NHN=N}$** **$C_6H_5N_3$**

Anon., *Chem. Eng. News*, 1956, **34**, 2450

A 1 tonne quantity, during distillation at 160°C/2.5 mbar, exothermally decomposed, then detonated. No cause was found, and similar batches had previously distilled satisfactorily. The multiple N–N bonding may tend to cause instability in the molecule, particularly if heavy metals are present. They were absent in this case.

*See other* HIGH-NITROGEN COMPOUNDS
*N*-METAL DERIVATIVES

**PHENYL AZIDE** $C_6H_5N_3$

Lindsay, R. O. *et al., Org. Synth.*, 1955, Coll. Vol. 3, 710

Though distillable at considerably reduced pressure, phenyl azide explodes when heated at atmospheric pressure, and occasionally at lower pressures.

Lewis acids

Boyer, J. H. *et al., Chem. Rev.*, 1954, **54**, 29

The ready decomposition of most aryl azides with sulphuric acid and Lewis acids may be vigorous or violent, depending on structure and conditions. In absence of a diluent (carbon disulphide), phenyl azide and aluminium trichloride exploded violently.

*See other* ORGANIC AZIDES

**BENZENESULPHINYL AZIDE** **$PhSON_3$** $C_6H_5N_3OS$

*See* SULPHINYL AZIDES

**BENZENESULPHONYL AZIDE** $C_6H_5SO_2N_3$ $C_6H_5N_3O_2S$

Dermer, O. C. *et al., J. Amer. Chem. Soc.*, 1955, **77**, 71

Whereas the pure azide decomposes rapidly but smoothly at 105°C, the crude material explodes violently on heating.

*See other* ACYL AZIDES

**BENZENEDIAZONIUM NITRATE** $C_6H_5N_2^+ \ NO_3^-$ $C_6H_5N_3O_3$

Urbanski, 1967, Vol. 3, 201

Although not a practical explosive, the isolated salt is highly sensitive to friction and impact, and explodes at 90°C.

*See other* DIAZONIUM SALTS

**2,4-DINITROANILINE** $(NO_2)_2C_6H_3NH_2$ $C_6H_5N_3O_4$

Preparative hazard

*See* 1-CHLORO-2,4-DINITROBENZENE, $C_6H_3ClN_2O_4$ : Ammonia

Sorbe, 1968, 151

It is a compound with high fire and explosion hazards.

*See other* POLYNITROARYL COMPOUNDS

## 2-AMINO-4,6-DINITROPHENOL $C_6H_5N_3O_5$

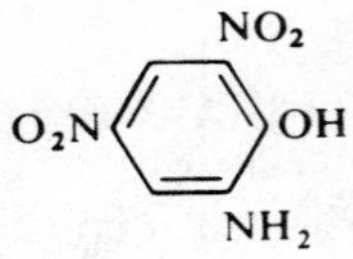

$O_2N$–($C_6H_2$)(–$NO_2$)(–OH)(–$NH_2$)

Sorbe, 1968, 152

This partially reduced derivative of picric acid, 'picramic acid', explodes very powerfully when dry.

*See other* POLYNITROARYL COMPOUNDS

## PHENYLPHOSPHONIC DIAZIDE $C_6H_5P(O)(N_3)_2$ $C_6H_5N_6OP$

Baldwin, R. A., *J. Org. Chem.*, 1965, **30**, 3866

Although a small sample had been distilled at 72–74°C/0.13 mbar without incident, it decomposed vigorously on exposure to flame, and the liquid was impact-sensitive, exploding vigorously. The material can be prepared and handled safely in pyridine solution.

*See other* ACYL AZIDES

## PHENYLTHIOPHOSPHONIC DIAZIDE $C_6H_5P(S)(N_3)_2$ $C_6H_5N_6PS$

Baldwin, R. A., *J. Org. Chem.*, 1965, **30**, 3866

During an attempt to distil crude material at 80°C/0.13 mbar, a violent explosion occurred. The material can be prepared and handled safely in pyridine solution.

*See other* ACYL AZIDES

## PHENYLSODIUM PhNa $C_6H_5Na$

Preparative hazard

*See* SODIUM, Na: Halocarbons (reference 8)

Sorbe, 1968, 86

Suspensions containing phenylsodium may ignite in moist air.

*See other* ARYLMETALS

## † BENZENE $C_6H_6$

*MCA SD-2,* 1960

Arsenic pentafluoride,
Potassium methoxide

*See* ARSENIC PENTAFLUORIDE, $AsF_5$: Benzene, etc.

Diborane

*See* DIBORANE, $B_2H_6$: Air, etc.

Interhalogens

*See* BROMINE TRIFLUORIDE, $BrF_3$ : Halogens, etc.
BROMINE PENTAFLUORIDE, $BrF_5$ : Hydrogen-containing materials
IODINE PENTAFLUORIDE, $F_5I$ : Benzene
IODINE HEPTAFLUORIDE, $F_7I$ : Organic solvents
*See* Uranium hexafluoride, below

Oxidants

*See* SILVER PERCHLORATE, $AgClO_4$ : Aromatic compounds
DIOXYGENYL TETRAFLUOROBORATE, $BF_4O_2$ : Organic materials
NITRYL PERCHLORATE, $ClNO_6$ : Organic solvents
DIOXYGEN DIFLUORIDE, $F_2O_2$
PERMANGANIC ACID, $HMnO_4$ : Organic materials
NITRIC ACID, $HNO_3$ : Hydrocarbons
PEROXOMONOSULPHURIC ACID, $H_2O_5S$: Aromatics
PEROXODISULPHURIC ACID, $H_2O_8S_2$ : Organic liquids
SODIUM PEROXIDE, $Na_2O_2$ : Organic liquids, etc.
OXYGEN (Liquid), $O_2$ : Hydrocarbons
OZONE, $O_3$ : Aromatic compounds
: Benzene, etc.

Uranium hexafluoride

*See* URANIUM HEXAFLUORIDE, $F_6U$: Aromatic hydrocarbons

**BENZVALENE** $C_6H_6$

Katz, T. J. *et al., J. Amer. Chem. Soc.*, 1971, **93**, 3782
When isolated, this highly strained valence tautomer of benzene exploded violently when scratched. It may be handled safely in solution in ether.
*See other* STRAINED RING COMPOUNDS

**1,3-HEXADIEN-5-YNE** **$H_2C{=}CHCH{=}CHC{\equiv}CH$** $C_6H_6$

Brandsma, 1971, 132
Distillation of this rather unstable material at normal pressure involves risk of an explosion.
*See other* ACETYLENIC COMPOUNDS

**† 1,5-HEXADIEN-3-YNE (DIVINYLACETYLENE)** $C_6H_6$

**$H_2C{=}CHC{\equiv}CCH{=}CH_2$**

1. Nieuwland, J. A. *et al., J. Amer. Chem. Soc.*, 1931, **53**, 4202
2. Cupery, M. E. *et al.*, ibid., 1934, **56**, 1167–1169

The extreme hazards involved in handling this highly reactive material are stressed. Freshly distilled material rapidly polymerises at ambient temperature to produce a gel and then a hard resin. These products can neither be distilled nor manipulated without explosions ranging from rapid decomposition to violent detonation. The hydrocarbon should be stored in the mixture with catalyst used to prepare it, and distilled out as required [1]. The dangerously explosive gel is a peroxidic species not formed in absence of air [2].

*See* 1,2-DI(3-BUTEN-1-YNYL)CYCLOBUTANE, $C_{12}H_{12}$

Air

Handy, C. T. *et al., J. Org. Chem.*, 1962, **27**, 41

The dienyne reacts readily with atmospheric oxygen, forming an explosively unstable polymeric peroxide. Equipment used with it should be rinsed with a dilute solution of a polymerisation inhibitor to prevent formation of unstable residual films. Adequate shielding of operations is essential.

*See other* ACETYLENIC COMPOUNDS
POLYPEROXIDES

**2,4-HEXADIYNE** **MeC≡CC≡CMe** $C_6H_6$

Bohlman, F., *Chem. Ber.*, 1951, **84**, 785–794

Filtration through alumina will prevent explosion during subsequent distillation at 55°C/13 μbar.

*See other* ACETYLENIC COMPOUNDS

**PRISMANE** $C_6H_6$

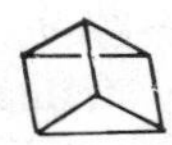

Katz, T. J. *et al., J. Amer. Chem. Soc.*, 1973, **95**, 2739

This pentacyclic isomer of benzene is an explosive liquid.

*See other* STRAINED RING COMPOUNDS

***p*-BROMOANILINE** $BrC_6H_4NH_2$ $C_6H_6BrN$

*See* *p*-BROMOBENZENEDIAZONIUM SALTS, $C_6H_4BrN_2^+X^-$

**IODYLBENZENE PERCHLORATE** $C_6H_5IO_2^+\ ClO_4^-$ $C_6H_6ClIO_6$

Masson, I., *J. Chem. Soc.*, 1935, 1674

A small sample exploded very violently while still damp.

*See other* IODINE COMPOUNDS

***o*-CHLOROANILINE** $ClC_6H_4NH_2$ $C_6H_6ClN$

*See* 2-CHLOROBENZENEDIAZONIUM SALTS $C_6H_4ClN_2^+X^-$

## *m*-CHLOROANILINE $ClC_6H_4NH_2$ $C_6H_6ClN$

*See* *m*-CHLOROBENZENEDIAZONIUM SALTS, $C_6H_4ClN_2^+X^-$

## *p*-AMINOBENZENEDIAZONIUM PERCHLORATE $C_6H_6ClN_3O_4$

$H_2NC_6H_4N_2^+\ ClO_4^-$

Hofmann, K. A. *et al.*, *Ber.*, 1910, **43**, 2624
Extremely explosive.
*See* DIAZONIUM PERCHLORATES

## 2,5-BIS(CHLOROMETHYL)THIOPHENE $C_6H_6Cl_2S$

$ClCH_2$–(thiophene, S)–$CH_2Cl$

Griffing, J. M. *et al.*, *J. Amer. Chem. Soc.*, 1948, **70**, 3417
It polymerises at ambient temperatures and must therefore be stored cold.
*See other* 2-HALOMETHYL-FURANS or -THIOPHENES

## HEXACHLOROCYCLOHEXANE $C_6H_6Cl_6$

*N,N*-Dimethylformamide
'DMF Brochure', Billingham, ICI, 1965
There is a potentially dangerous reaction of hexachlorocyclohexane with DMF in presence of iron. The same occurs with carbon tetrachloride, but not with dichloromethane or 1,2-dichloroethane under the same conditions.
*See also* *N,N*-DIMETHYLACETAMIDE, $C_4H_9NO$: Halogenated compounds

## *N*-ETHYLHEPTAFLUOROBUTYRAMIDE $C_6H_6F_7NO$

$C_3F_7CONHEt$

*See* LITHIUM TETRAHYDROALUMINATE, $AlH_4Li$: Fluoroamides

## POTASSIUM BIS(PROPYNYL)PALLADATE(2–) $C_6H_6K_2Pd$

$K_2[Pd(C{\equiv}CMe)_2]$

Immediately pyrophoric in air, and explosive decomposition with aqueous reagents. The sodium salt is similar.
*See* COMPLEX ACETYLIDES

## POTASSIUM BIS(PROPYNYL)PLATINATE(2–) $C_6H_6K_2Pt$

$K_2[Pt(C{\equiv}CMe)_2]$

Pyrophoric in air, explosive decomposition with water, and the sodium salt is similar.

*See* COMPLEX ACETYLIDES

## TRIPOTASSIUM CYCLOHEXANEHEXONE-1,3,5-TRIOXIMATE $C_6K_3N_3O_6$ *

NOK
OC CO
KON C NOK
O

Alone,
or Acids

Benedikt, R., *Ber.*, 1878, **11**, 1375

The salt explodes on heating above 130°C or if moistened with sulphuric or nitric acids. The lead salt also explodes violently on heating.

*See related* OXIMES

## *p*-BENZOQUINONE DIIMINE $HN{=}C_6H_4{=}NH$ $C_6H_6N_2$

Acids

Willstätter, R. *et al., Ber.*, 1906, **37**, 4607

The unstable solid decomposes explosively in contact with concentrated hydrochloric or sulphuric acids.

## 1,4-DICYANO-2-BUTENE $C_6H_6N_2$

$NCCH_2CH{=}CHCH_2CN$

*MCA Case History No. 1747*

Overheating in a vacuum evaporator initiated accelerating polymerisation-decomposition of dicyanobutene, and the rapid gas evolution eventually caused pressure-failure of the process equipment.

*See other* CYANO COMPOUNDS

## BIS(ACRYLONITRILE)NICKEL(0) $[Ni(CH_2{=}CHCN)_2]$ $C_6H_6N_2Ni$

Schrauzer, G. N., *J. Amer. Chem. Soc.*, 1959, **81**, 5310

Pyrophoric in air.

*See other* CYANO COMPOUNDS

*Out of sequence.

## *o*-NITROANILINE $NO_2C_6H_4NH_2$ $C_6H_6N_2O_2$

Preparative hazard

*See* *o*-CHLORONITROBENZENE, $C_6H_4ClNO_2$: Ammonia

Hexanitroethane

*See* HEXANITROETHANE, $C_2N_6O_{12}$: Organic compounds

Sulphuric acid

*See* *o*-NITROBENZENEDIAZONIUM SALTS, $C_6H_4N_3O_2^+X^-$
SULPHURIC ACID, $H_2O_4S$: Nitroaryl bases

## *m*-NITROANILINE $NO_2C_6H_4NH_2$ $C_6H_6N_2O_2$

Ethylene oxide

*See* ETHYLENE OXIDE, $C_2H_4O$: *m*-Nitroaniline

## *p*-NITROANILINE $NO_2C_6H_4NH_2$ $C_6H_6N_2O_2$

*See* *p*-NITROBENZENEDIAZONIUM SALTS, $C_6H_4N_3O_2^+X^-$

Sodium hydroxide

*See* SODIUM *p*-NITROPHENOLATE, $C_6H_4NNaO_3$

Sulphuric acid

*See* SULPHURIC ACID, $H_2O_4S$: Nitroaryl bases, etc.

## BENZENEDIAZONIUM HYDROGENSULPHATE $C_6H_6N_2O_4S$

$PhN_2^+ HSO_4^-$

Sorbe, 1968, 122

The salt explodes at about 100°C, but is not initiated by impact.

*See other* DIAZONIUM SULPHATES

## 4-NITROANILINE-2-SULPHONIC ACID $C_6H_6N_2O_5S$

$NO_2(SO_3H)C_6H_3NH_2$

*See* SULPHURIC ACID, $H_2O_4S$: Nitroaryl bases

## AMMONIUM PICRATE $(NO_3)_3C_6H_2O^-NH_4^+$ $C_6H_6N_4O_7$

Unpublished observations

A small sample, isolated incidentally during decomposition of a picrate with ammonia, exploded during analytical combustion.

Presence of traces of metallic picrates (arising from metal contact) increases temperature sensitivity. The salt may also explode on impact.

*See* PICRIC ACID, $C_6H_3N_3O_7$

**TRIAMINOTRINITROBENZENE** $(H_2N)_3C_6(NO_2)_3$ $C_6H_6N_6O_6$

Hydroxylaminium perchlorate

*See* HYDROXYLAMINIUM PERCHLORATE, $ClH_4NO_5$: Triaminotrinitrobenzene

**DI(2-PROPYN-1-YL) ETHER (DIPROPARGYL ETHER)** $C_6H_6O$

$(HC{\equiv}CCH_2)_2O$

*Guide for Safety*, 1972, 302

Distillation of the ether in a 230 l still led to an explosion; attributable to peroxidation.

*See other* ACETYLENIC COMPOUNDS
PEROXIDISABLE COMPOUNDS

**4,5-HEXADIEN-2-YN-1-OL** $C_6H_6O$

$H_2C{=}C{=}CHC{\equiv}CCH_2OH$

Brandsma, 1971, 8, 54

Dilution with white mineral oil before distillation is recommended to prevent explosion of the concentrated distillation residue.

*See other* ACETYLENIC COMPOUNDS
PEROXIDISABLE COMPOUNDS

**PHENOL** $C_6H_5OH$ $C_6H_6O$

*MCA SD-4*, 1964

Aluminium trichloride,
Nitrobenzene

*See* ALUMINIUM TRICHLORIDE, $AlCl_3$: Nitrobenzene, etc.

Aluminium trichloride–nitromethane complex,
Carbon monoxide

*See* ALUMINIUM TRICHLORIDE–NITROMETHANE, $AlCl_3 \cdot CH_3NO_2$

Formaldehyde

*See* FORMALDEHYDE, $CH_2O$: Phenol

Peroxodisulphuric acid

*See* PEROXODISULPHURIC ACID, $H_2O_8S_2$: Organic liquids

Peroxomonosulphuric acid

*See* PEROXOMONOSULPHURIC ACID, $H_2O_5S$: Aromatics

Sodium nitrate,
Trifluoroacetic acid

*See* SODIUM NITRATE, $NNaO_3$: Phenol, Trifluoroacetic acid

Sodium nitrite

*See* SODIUM NITRITE, $NNaO_2$: Phenol

**1,2-DIHYDROXYBENZENE (PYROCATECHOL)** **$C_6H_6O_2$**

**$(OH)_2C_6H_4$**

Nitric acid

*See* NITRIC ACID, $HNO_3$: Pyrocatechol

**PHENYLSELENONIC ACID** **$C_6H_5Se(O_2)OH$** **$C_6H_6O_3Se$**

Stoecker, M. *et al.*, *Ber.*, 1906, **39**, 2197

It explodes feebly at 180°C but the silver salt is more explosive.

**DIMETHYL ACETYLENEDICARBOXYLATE** **$C_6H_6O_4$**

**$MeO_2CC{\equiv}CCO_2Me$**

Octakis(trifluorophosphine)dirhodium

*See* OCTAKIS(TRIFLUOROPHOSPHINE)DIRHODIUM, $F_{24}P_8Rh_2$: Acetylenic esters

**BENZENEPEROXYSULPHONIC ACID** **$PhSO_2OOH$** **$C_6H_6O_4S$**

*See* HYDROGEN PEROXIDE, $H_2O_2$: Benzenesulphonic anhydride

**BENZENE TRIOZONIDE** **$C_6H_6O_9$**

Harries, C. *et al.*, *Ber.*, 1904, **37**, 3431

Extremely explosive at the slightest touch.

*See* OZONE, $O_3$: Aromatic compounds

*See other* OZONIDES

## † BENZENETHIOL (THIOPHENOL) $C_6H_5SH$ $C_6H_6S$

Preparative hazard

1. Leuckart, R., *J. Prakt. Chem.*, 1890, **41**, 179
2. Graesser, R., private comm., 1968

During preparation of thiophenol by addition of a cold solution of potassium *O*-methyldithiocarbonate to a cold solution of benzenediazonium chloride, a violent explosion, accompanied by an orange flash, occurred. This was attributed to the formation and decomposition of bis-benzenediazodisulphide. A preparation in which the diazonium salt was added to the 'xanthate' solution proceeded smoothly.

*See also* DIAZONIUM SULPHIDES

## 2-BROMOMETHYL-5-METHYLFURAN $C_6H_7BrO$

Me–(furan ring, O)–$CH_2Br$

Dunlop, 1953, 261

It distils with violent decomposition at 70°C/33 mbar.

*See other* 2-HALOMETHYL-FURANS

## 2,4-DINITROPHENYLHYDRAZINIUM PERCHLORATE $C_6H_7ClN_4O_8$

$(NO_2)_2C_6H_3NHN^+H_3\ ClO_4^-$

Anon., *Angew. Chem. (Nachr.)*, 1967, **15**, 78

Although solutions of the perchlorate are stable as prepared, explosive decomposition may occur during concentration by evaporation.

*See also* HYDRAZINIUM SALTS
*See other* OXOSALTS OF NITROGENOUS BASES

## 2-CHLOROMETHYL-5-METHYLFURAN $C_6H_7ClO$

Me–(furan ring, O)–$CH_2Cl$

Dunlop, 1953, 261

It is even more unstable than its lower homologue, 2-chloromethylfuran.

*See other* 2-HALOMETHYL-FURANS

## ANILINE $C_6H_5NH_2$ $C_6H_7N$

*MCA SD-17*, 1963

*See* BENZENEDIAZONIUM SALTS, $C_6H_5N_2^+Z^-$

Anilinium chloride

Anon., *Chem. Met. Eng.*, 1922, **27**, 1044

Heating a mixture of the components in an autoclave at 240–260°C/7.6 bar to produce diphenylamine led to a violent explosion, following a sudden increase in pressure to 17 bar. The equipment and process had been used previously and uneventfully.

Benzenediazonium-2-carboxylate
*See* BENZENEDIAZONIUM-2-CARBOXYLATE, $C_7H_4N_2O_2$

Boron trichloride
*See* BORON TRICHLORIDE, $BCl_3$ : Aniline

Dibenzoyl peroxide
*See* DIBENZOYL PEROXIDE, $C_{14}H_{10}O_4$ : Aniline

Nitromethane
*See* NITROMETHANE, $CH_3NO_2$ : Acids, etc.

Oxidants
*See* *N*-HALOIMIDES: Alcohols, Amines
PEROXYFORMIC ACID, $CH_2O_3$ : Organic materials
DIISOPROPYL PEROXYDICARBONATE, $C_8H_{14}O_6$
PERCHLORYL FLUORIDE, $ClFO_3$ : Nitrogenous bases
PERCHLORIC ACID, $ClHO_4$ : Aniline, etc.
FLUORINE NITRATE, $FNO_3$ : Organic materials
NITROSYL PERCHLORATE, $ClNO_5$ : Organic materials
FLUORINE, $F_2$ : Nitrogenous bases
NITRIC ACID, $HNO_3$ : Aromatic amines
PEROXOMONOSULPHURIC ACID, $H_2O_5S$: Aromatics
PEROXODISULPHURIC ACID, $H_2O_8S_2$ : Organic liquids
SODIUM PEROXIDE, $Na_2O_2$ : Organic liquids, etc.
OZONE, $O_3$ : Aromatic compounds

Tetranitromethane
*See* TETRANITROMETHANE, $CN_4O_8$ : Amines

Trichloronitromethane
*See* TRICHLORONITROMETHANE, $CCl_3NO_2$ : Aniline

## 2-METHYLPYRIDINE $C_6H_7N$

N=CHCH=CHCH=CMe (ring: N bonded to CMe)

Hydrogen peroxide,
Iron(II) sulphate,
Sulphuric acid
*See* HYDROGEN PEROXIDE, $H_2O_2$ : Iron(II) sulphate, etc.

## *N*-PHENYLHYDROXYLAMINE $C_6H_5NHOH$ $C_6H_7NO$

Preparative hazard

*See* ZINC, Zn: Nitrobenzene

## 4-OXIMINO-4,5,6,7-TETRAHYDROBENZOFURAZAN $C_6H_7N_3O_3$

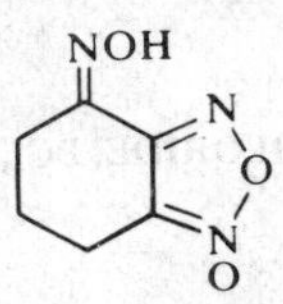


Preparative hazard

*See* SULPHINYL CHLORIDE (THIONYL CHLORIDE), $Cl_2OS$: 1,2,3-Cyclohexanetrione trioxime, etc.

## ETHYL 2-CYANO-2-(1-*H*-TETRAZOL-5-YLHYDRAZONO)ACETATE $C_6H_7N_7O_2$

$CHN_4NHN{=}C(CN)CO_2Et$

Preparative hazard

*See* 5-AMINOTETRAZOLE, $CH_3N_5$: Nitrous acid (reference 3)

## † 1,3-CYCLOHEXADIENE $CH_2[CH{=}CH]_2CH_2$ $C_6H_8$

*See also* PROPENE, $C_3H_6$

Air

1. Bodendorf, K., *Arch. Pharm.*, 1933, **271**, 11
2. Hock, H. *et al.*, *Chem. Ber.*, 1951, **84**, 349

Cyclohexadiene autoxidises slowly in air, but the residual (largely polymeric) peroxide explodes very violently on ignition [1]. The monomeric peroxide has also been isolated [2].

*See other* DIENES
POLYPEROXIDES

## † 1,4-CYCLOHEXADIENE $CH{=}CHCH_2CH{=}CHCH_2$ $C_6H_8$

*See also* PROPENE, $C_3H_6$
*See other* DIENES

## ANILINIUM CHLORIDE $PhN^+H_3Cl^-$ $C_6H_8ClN$

Aniline

*See* ANILINE, $C_6H_7N$: Anilinium chloride

**2,5-BIS-*endo*-DICHLORO-7-THIABICYCLO[2.2.1] HEPTANE** $C_6H_8Cl_2S$

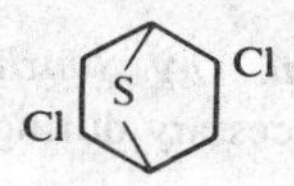


*N,N*-Dimethylformamide

1. Corey, E. J. *et al., J. Org. Chem.*, 1969, **34**, 1234 (footnote 5)
2. Paquette, L. A., private comm., 1975
3. Brown, H. C. *et al., J. Org. Chem.*, 1962, **27**, 1928–1929

To effect reduction to the parent heterocycle, a solution of the dichloride in DMF was being added to a hot solution of sodium tetrahydroborate in the same solvent, when a violent explosion occurred [1,2]. This seems likely to have arisen from interaction of the dichloride with the solvent. Use of aqueous diglyme as an alternative solvent [3] would only be applicable to this and other hydrolytically stable halides [2].

***N,N,N',N'*-TETRACHLOROADIPAMIDE** $C_6H_8Cl_4N_2O_2$

$Cl_2NCO[CH_2]_4CONCl_2$

Water

*See* DICHLOROAMINE, $Cl_2HN$

**PHENYLHYDRAZINE** $C_6H_5NHNH_2$ $C_6H_8N_2$

Lead(IV) oxide

*See* LEAD(IV) OXIDE, $O_2Pb$: Nitrogen compounds

Perchloryl fluoride

*See* PERCHLORYL FLUORIDE, $ClFO_3$: Nitrogenous bases

2-Phenylamino-3-phenyloxazirane

*See* 2-PHENYLAMINO-3-PHENYLOXAZIRANE (BENZALDEHYDE PHENYLHYDRAZONE OXIDE), $C_{13}H_{12}N_2O$: Phenylhydrazine

**2-DIAZOCYCLOHEXANONE** $\overline{CO[CH_2]_4CN_2}$ $C_6H_8N_2O$

Regitz, M. *et al., Org. Synth.*, 1971, **51**, 86

It may explode on being heated, and a relatively large residue should be left during vacuum distillation to prevent overheating.

*See other* DIAZO COMPOUNDS

**2-BUTEN-1-YL DIAZOACETATE** $C_6H_8N_2O_2$

$N_2CHCO_2CH_2CH{=}CHMe$

Blankley, C. J. *et al., Org. Synth.*, 1969, **49**, 25
Precautions are necessary during vacuum distillation of this potentially explosive material.
*See other* DIAZO COMPOUNDS

**ANILINIUM NITRATE** $C_6H_5N^+H_3\ NO_3^-$ $C_6H_8N_2O_3$

Nitric acid
*See* NITRIC ACID, $HNO_3$: Anilinium nitrate

† **2,5-DIMETHYLFURAN** $\overline{OC(Me){=}CHCH{=}CMe}$ $C_6H_8O$

**3-METHYL-2-PENTEN-4-YN-1-OL** $C_6H_8O$

$HC{\equiv}CCMe{=}CHCH_2OH$

Lorentz, F., *Loss Prevention,* 1967, **1**, 1–5
Fesenko, G. V. *et al., Chem. Abs.*, 1975, **83**, 47605e
Temperature control during pressure hydrogenation of *cis*- or *trans*-isomers is essential, since at 155°C violent decomposition to carbon, hydrogen and carbon monoxide with development of over 1000 bar pressure will occur. The material should not be heated above 100°C, particularly if acid or base is present, to avoid exothermic polymerisation.

The *cis*-isomer is readily cyclised to 2,3-dimethylfuran, which promotes fire and explosion hazards. These were measured for the *cis*- and *trans*-isomers, and for *trans*-3-methyl-1-penten-4-yn-3-ol.

Sodium hydroxide
1. *MCA Case History No. 363*
2. Silver, L., *Chem. Eng. Progr.*, 1967, **63**(8), 43
Presence of traces of sodium hydroxide probably caused formation of the acetylenic sodium salt, which exploded during high-vacuum distillation in a metal still [1]. A laboratory investigation which duplicated the explosion, without revealing the precise cause, was also reported [2].
*See other* ACETYLENIC COMPOUNDS

**1,2-CYCLOHEXANEDIONE** $\overline{CO[CH_2]_4CO}$ $C_6H_8O_2$

Preparative hazard
*See* NITRIC ACID, $HNO_3$: Cyclohexanol

## POLY(CYCLOHEXADIENE PEROXIDE) $(C_6H_8O_2)_n$

$$\left(\text{C}_6\text{H}_8\text{-OOH}\right)_n$$

*See* 1,3-CYCLOHEXADIENE, $C_6H_8$ : Air
*See other* POLYPEROXIDES

## CITRIC ACID $C_6H_8O_7$

$HO_2CC(OH)(CH_2CO_2H)_2$

Metal nitrates
*See* METAL NITRATES: Citric acid

## TRIACETYL BORATE $(AcO)_3B$ $C_6H_9BO_6$

Preparative hazard
*See* ACETIC ANHYDRIDE, $C_4H_6O_3$ : Boric acid

## TRIVINYLBISMUTH $(CH_2{=}CH)_3Bi$ $C_6H_9Bi$

Coates, 1967, Vol. 1, 538
Ignites in air.
*See related* ALKYLMETALS

## IODINE TRIACETATE $I(OAc)_3$ $C_6H_9IO_6$

1. Schutzenberger, P., *Compt. Rend.*, 1861, **52**, 135
2. Oldham, F. W. H. *et al.*, *J. Chem. Soc.*, 1941, 368

The acetate explodes at 140°C [1], and higher homologues decompose at about 120°C, the violence decreasing with ascent of the homologous series [2].
*See other* IODINE COMPOUNDS

## *tert*-BUTYLNITROACETYLENE $(Me)_3CC{\equiv}CNO_2$ $C_6H_9NO_2$

Amines
Jäger, V. *et al.*, *Angew. Chem. (Intern. Ed.)*, 1969, **8**, 273
Reaction with primary, secondary or tertiary amines proceeds explosively with ignition in absence of a solvent.
*See other* ACETYLENIC COMPOUNDS

**BIS(2-CYANOETHYL)AMINE** $(CNC_2H_4)_2NH$ $C_6H_9N_3$

*BCISC Quart. Safety Summ.*, 1967, **38**, 42

Some 18-month old bottles of iminodipropionitrile exploded, probably owing to slow hydrolysis and release of ammonia.

*See other* CYANO COMPOUNDS

***N*-NITROSOPHENYLHYDROXYLAMINE *O*-AMMONIUM SALT ('CUPFERRON')** $PhN(NO)ONH_4$ $C_6H_9N_3O_2$

Thorium salts

Pitwell, L. R., *J.R. Inst. Chem.*, 1956, **80**, 173

Solutions of this reagent are destabilised by the presence of thorium ions. If a working temperature of 10–15°C is much exceeded, the risk of decomposition, not slowed by cooling, and accelerating to explosion, exists. Titanium and zirconium salts also cause slight destabilisation, but decomposition temperatures are 35 and 40°C, respectively.

**1,2,3-CYCLOHEXANETRIONE TRIOXIME** $C_6H_9N_3O_3$

NOH
=NOH
NOH

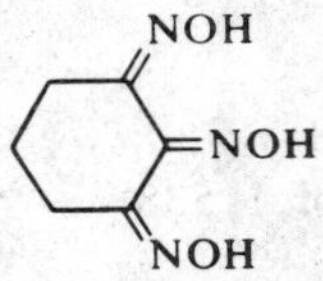

Sulphinyl chloride

*See* SULPHINYL CHLORIDE (THIONYL CHLORIDE), $Cl_2OS$: 1,2,3-Cyclohexanetrione trioxime, etc.

**1,3,5-CYCLOHEXANETRIONE TRIOXIME** $C_6H_9N_3O_3$

HON NOH
NOH

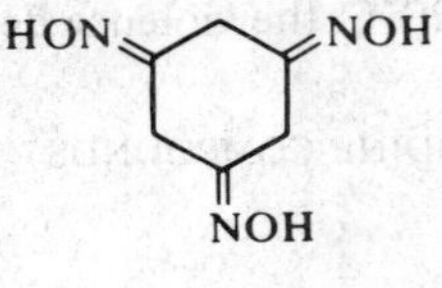

Baeyer, A., *Ber.*, 1886, **19**, 160

The oxime of phloroglucinol explodes rather violently at 155°C.

*See other* OXIMES

**LEAD(IV) TRIACETATE AZIDE** $Pb(OAc)_3N_3$ $C_6H_9N_3O_6Pb$

*See* LEAD(IV) AZIDE, $N_{12}Pb$

**SODIUM ETHYL ACETOACETATE** **$AcCHNaCO_2Et$** **$C_6H_9NaO_3$**

2-Iodo-3,5-dinitrobiphenyl

*See* 2-IODO-3,5-DINITROBIPHENYL, $C_{12}H_7IN_2O_4$ : Sodium, etc.

**† CYCLOHEXENE** **$\overline{CH=CH[CH_2]_3CH_2}$** **$C_6H_{10}$**

**2,3-DIMETHYL-1,3-BUTADIENE** **$CH_2=CMeCMe=CH_2$** **$C_6H_{10}$**

Air

Bodendorf, K., *Arch. Pharm.*, 1933, **271**, 33

A few mg of the autoxidation residue, a largely polymeric peroxide, exploded violently on ignition.

*See* POLYPEROXIDES

*See other* DIENES

**3,3-DIMETHYL-1-BUTYNE** **$C_6H_{10}$**

See p. 599

**† 1,4-HEXADIENE** **$C_6H_{10}$**

**$CH_2=CHCH_2CH=CHMe$**

**† 1,5-HEXADIENE** **$C_6H_{10}$**

**$CH_2=CHCH_2CH_2CH=CH_2$**

**† 1-HEXYNE** **$HC\equiv CBu$** **$C_6H_{10}$**

**† 2-HEXYNE** **$MeC\equiv CPr$** **$C_6H_{10}$**

**† 3-HEXYNE** **$EtC\equiv CEt$** **$C_6H_{10}$**

Mercury,

Silver perchlorate

*See* SILVER PERCHLORATE, $AgClO_4$ : Alkynes, etc.

**† 2-METHYL-1,3-PENTADIENE** **$CH_2=CMeCH=CHMe$** **$C_6H_{10}$**

**† 4-METHYL-1,3-PENTADIENE** **$CH_2=CHCH=C(Me)_2$** **$C_6H_{10}$**

**CADMIUM PROPIONATE** $Cd(O_2CEt)_2$ $C_6H_{10}CdO_4$

Anon., *Chem. Age,* 1957, 77, 794

The salt exploded during drying at 60–100°C in an electric oven, presumably because of overheating, and production and ignition of 3-pentanone vapour.

**3,3-PENTAMETHYLENEDIAZIRINE** $N{=}NC[CH_2]_4CH_2$ $C_6H_{10}N_2$

Schmitz, E. *et al., Org. Synth*., 1965, **45**, 85; *Ber*., 1961, **94**, 2168

Small-scale preparation is recommended, in view of a previous explosion during superheating of the liquid during distillation at 109°C.

*See* DIAZIRINES

***tert*-BUTYL DIAZOACETATE** $N_2CHCO_2C(Me)_3$ $C_6H_{10}N_2O_2$

Regitz, M. *et al., Org. Synth*., 1968, **48**, 36

Vacuum distillation of the product is potentially hazardous.

*See other* DIAZO COMPOUNDS

***N*-NITROSO-6-HEXANELACTAM** $ONN[CH_2]_5CO$ $C_6H_{10}N_2O_2$

Preparative hazard

*See* 6-HEXANELACTAM, $C_6H_{11}NO$: Acetic acid, etc.

**DIETHYL AZODIFORMATE** $C_6H_{10}N_2O_4$

$EtO_2CN{=}NCO_2Et$

U.S. Pat. 3 347 845, 1967

Kauer, J. C., *Org. Synth.,* 1963, Coll. Vol. 4, 412

Shock-sensitive, it burns explosively, like its lower homologue.

*See* DIMETHYL AZODIFORMATE, $C_4H_6N_2O_4$

*See other* AZO COMPOUNDS

***N,N'*-DIACETYL-*N,N'*-DINITRO-1,2-DIAMINOETHANE** $C_6H_{10}N_4O_6$

$AcN(NO_2)C_2H_4N(NO_2)Ac$

Violent decomposition occurred at 142°C.

*See* DIFFERENTIAL THERMAL ANALYSIS

*See other* *N*-NITRO COMPOUNDS

**BUTOXYACETYLENE** $BuOC{\equiv}CH$ $C_6H_{10}O$

Jacobs, T. L. *et al., J. Amer. Chem. Soc*., 1942, **64**, 223

Small samples rapidly heated in sealed tubes to around 100°C exploded.
*See other* ACETYLENIC COMPOUNDS

**CYCLOHEXANONE** $\overline{CO[CH_2]_4CH_2}$ $C_6H_{10}O$

Hydrogen peroxide,
Nitric acid
*See* HYDROGEN PEROXIDE, $H_2O_2$: Ketones, etc.

Nitric acid
*See* NITRIC ACID, $HNO_3$: 4-Methylcyclohexanone

**DIETHYLKETENE** $Et_2C{=}C{=}O$ $C_6H_{10}O$

Sorbe, 1968, 118
Like the lower homologue, it readily forms explosive peroxides with air at ambient temperature.

**† DIALLYL ETHER [DI(2-PROPENYL) ETHER]** $C_6H_{10}O$
$(H_2C{=}CHCH_2)_2O$

Preparative hazard
*See* 2-PROPEN-1-OL (ALLYL ALCOHOL), $C_3H_6O$: Sulphuric acid

*MCA Case History No. 412*
The ether was left exposed to air and sunlight for 2 weeks before distillation, and became peroxidised. During distillation to small bulk, a violent explosion occurred.
*See other* ALLYL COMPOUNDS

**HEXENAL** $C_6H_{10}O$
$H_2C{=}CH[CH_2]_3CHO$

Nitric acid
*See* NITRIC ACID, $HNO_3$: Hexenal

**2-CYCLOHEXENYL HYDROPEROXIDE** $C_6H_{10}O_2$

OOH

Leleu, *Cahiers*, 1973, (71), 226

Accidental contact of a mixture containing the peroxide with the hot top of the freshly sealed glass container led to an explosion.

*See other* ALKYL HYDROPEROXIDES

**† ETHYL CROTONATE** $MeCH{=}CHCO_2Et$ $C_6H_{10}O_2$

**† ETHYL METHACRYLATE** $CH_2{=}CMeCO_2Et$ $C_6H_{10}O_2$

**† VINYL BUTYRATE** $PrCO_2CH{=}CH_2$ $C_6H_{10}O_2$

**DIPROPIONYL PEROXIDE** EtCOOOCOEt $C_6H_{10}O_4$

Swern, 1971, Vol. 2, 815

Pure material explodes on standing at room temperature.

*See other* DIACYL PEROXIDES

**DIALLYL SULPHATE** $C_6H_{10}O_4S$

$(CH_2{=}CHCH_2O)_2SO_2$

von Braun, J. *et al., Ber.*, 1917, **50**, 293

The explosive decomposition during distillation may well have been caused by polymerisation initiated by acidic decomposition products.

*See other* ALLYL COMPOUNDS
POLYMERISATION INCIDENTS

**2-HYDROXY-2-METHYLGLUTARIC ACID** $C_6H_{10}O_5$

$MeCOH(CO_2H)_2$

Preparative hazard

*See* 4-HYDROXY-4-METHYL-1,6-HEPTADIENE, $C_8H_{14}O$: Ozone

**DIETHYL PEROXYDICARBONATE** $EtO_2COOCO_2Et$ $C_6H_{10}O_6$

An oil of extreme instability when impure, and sensitive to heat or impact, exploding more powerfully than dibenzoyl peroxide.

*See* PEROXYCARBONATE ESTERS

### DIALLYL SULPHIDE $(CH_2{=}CHCH_2)_2S$ $C_6H_{10}S$

*N*-Bromosuccinimide

*See* *N*-HALOIMIDES: Alcohols, etc.

*See other* ALLYL COMPOUNDS

### 2-ETHOXY-1-IODO-3-BUTENE $C_6H_{11}IO$

$ICH_2CH(OEt)CH{=}CH_2$

Trent, J. *et al., Chem. Eng. News,* 1966, **44**(43), 7

During a 1g mol-scale preparation by a published method using butadiene, ethanol, iodine and mercury oxide, a violent explosion occurred while ethanol was being distilled off at 35°C under slight vacuum. The cause of the explosion could not be established, and several smaller-scale preparations had been uneventful.

### † DIALLYLAMINE $(CH_2{=}CHCH_2)_2NH$ $C_6H_{11}N$

### 6-HEXANELACTAM $\overline{HN[CH_2]_5CO}$ $C_6H_{11}NO$

Acetic acid,

Dinitrogen trioxide

Huisgen, R. *et al., Ann.*, 1952, **575**, 174–197

During preparation of the *N*-nitroso derivative from 6-hexanelactam in acetic acid solution, treatment with dinitrogen trioxide must be very effectively cooled to prevent explosive decomposition.

### 2-ISOPROPYLACRYLALDEHYDE OXIME $C_6H_{11}NO$

$H_2C{=}C(CHMe_2)CH{=}NOH$

Marvel, C. S. *et al., J. Amer. Chem. Soc.*, 1950, **72**, 5408–5409

Hydroquinone must be added to the oxime before distillation, to prevent formation and subsequent violent reaction of a peroxide.

*See other* OXIMES

### ETHYL 2-FORMYLPROPIONATE OXIME $C_6H_{11}NO_3$

$EtO_2CCHMeCH{=}NOH$

Hydrogen chloride (?)

Loftus, F., private comm., 1972

The formyl ester and hydroxylamine were allowed to react in ethanol, which was then removed by vacuum evaporation at 40°C. The dry residue decomposed violently 5 min later. Analysis of the residue

suggested that the oxime had undergone an exothermic Beckmann rearrangement, possibly catalysed by traces of hydrogen chloride in the reaction residue.
*See also* BUTYRALDOXIME, $C_4H_9NO$

**DIALLYL PHOSPHITE** $C_6H_{11}O_3P$
$(CH_2{=}CH_2CH_2O)_2P(O)H$

Houben-Weyl, 1964, Vol. 12(2), 22
The ester is liable to explode during distillation unless thoroughly dry allyl alcohol is used, and if more than two-thirds of the product is distilled over. Acid-catalysed polymerisation may have been involved.
*See other* ALLYL COMPOUNDS
POLYMERISATION INCIDENTS

† **CYCLOHEXANE** $\overline{CH_2[CH_2]_4CH_2}$ $C_6H_{12}$
*MCA SD-68*, 1957

Dinitrogen tetraoxide
*See* DINITROGEN TETRAOXIDE, $N_2O_4$ : Hydrocarbons

† **ETHYLCYCLOBUTANE** $Et\overline{CH[CH_2]_2CH_2}$ $C_6H_{12}$

† **1-HEXENE** $H_2C{=}CHBu$ $C_6H_{12}$

† **2-HEXENE** $MeCH{=}CHPr$ $C_6H_{12}$

† **METHYLCYCLOPENTANE** $Me\overline{CH[CH_2]_3CH_2}$ $C_6H_{12}$

† **2-METHYL-1-PENTENE** $H_2C{=}CMeBu$ $C_6H_{12}$

† **4-METHYL-1-PENTENE** $H_2C{=}CHCH_2CHMe_2$ $C_6H_{12}$

† ***cis*-4-METHYL-2-PENTENE** $MeCH{=}CHCHMe_2$ $C_6H_{12}$

† ***trans*-4-METHYL-2-PENTENE** $MeCH{=}CHCHMe_2$ $C_6H_{12}$

## 2,2-DICHLORO-3,3-DIMETHYLBUTANE $C_6H_{12}Cl_2$

$MeCCl_2CMe_3$

Sodium hydroxide

1. Bartlett, P. D. *et al., J. Amer. Chem. Soc.*, 1942, **64**, 543
2. Kocienski, P. J., *J. Org. Chem.*, 1974, **39**, 3285–3286

A previous method [1] of preparing 3,3-dimethyl-1-butyne (*tert*-butylacetylene) by dehydrochlorination of the title compound in a sodium hydroxide melt is difficult to control and hazardous on the large scale. Use of potassium *tert*-butoxide as base in dimethyl sulphoxide is a high-yielding, safe and convenient alternative method of preparation of the acetylene [2]. (But see DIMETHYL SULPHOXIDE, $C_2H_6OS$: Metal alkoxides.)

## 3,3-DIMETHYL-1-BUTYNE $HC{\equiv}CCMe_3$ $C_6H_{10}$*

Preparative hazard

*See* 2,2-DICHLORO-3,3-DIMETHYLBUTANE, $C_4H_{12}Cl_2$: Sodium hydroxide

## 1,4-DIAZABICYCLO[2.2.2]OCTANE $C_6H_{12}N_2$

$NC_2H_4NC_2H_4NCH_2CH_2$

Cellulose nitrate

*See* CELLULOSE NITRATE: Amines

## 1,6-DIAZA-3,4,8,9,12,13-HEXAOXABICYCLO[4.4.4]TETRADECANE ('HEXAMETHYLENETRIPEROXYDIAMINE') $C_6H_{12}N_2O_6$

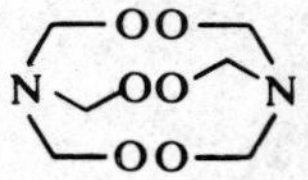

N(CH₂OOCH₂)₃N

Leulier, A., *J. Pharm. Chim.*, 1917, **15**, 222–229

The compound, prepared by interaction of hexamethylenetetramine and acidic hydrogen peroxide, is a heat- and shock-sensitive powerful explosive when dry. It explodes in contact with bromine or sulphuric acid.

*See other* CYCLIC PEROXIDES

*Out of sequence.

**HEXAMETHYLENETETRAMINE** $C_6H_{12}N_4$

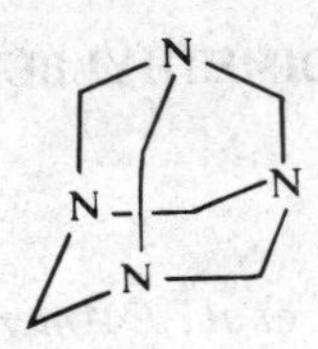


Iodoform

*See* IODOFORM, $CHI_3$ : Hexamethylenetetramine

**† BUTYL VINYL ETHER** $BuOCH{=}CH_2$ $C_6H_{12}O$

**CYCLOHEXANOL** $C_6H_{11}OH$ $C_6H_{12}O$

Oxidants

*See* CHROMIUM TRIOXIDE, $CrO_3$ : Alcohols
NITRIC ACID, $HNO_3$ : Alcohols

**† 2,2-DIMETHYL-3-BUTANONE** $CH_3CMe_2COMe$ $C_6H_{12}O$

**† 2-ETHYLBUTYRALDEHYDE** $CH_3CHEtCH_2CHO$ $C_6H_{12}O$

**† 3-HEXANONE** EtCOPr $C_6H_{12}O$

**† ISOBUTYL VINYL ETHER** $C_6H_{12}O$

$Me_2CHCH_2OCH{=}CH_2$

**† 2-METHYLPENTANAL** $Me_2CH[CH_2]_2CHO$ $C_6H_{12}O$

**† 3-METHYLPENTANAL** $EtCHMeCH_2CHO$ $C_6H_{12}O$

**† 2-METHYL-3-PENTANONE** $Me_2CHCOEt$ $C_6H_{12}O$

**† 3-METHYL-2-PENTANONE** MeCOCHMeEt $C_6H_{12}O$

† **4-METHYL-2-PENTANONE** $MeCOCH_2CHMe_2$ $C_6H_{12}O$

Air

1. Smith, G. H., *Chem. & Ind.*, 1972, 291
2. Bretherick, L., ibid., 363

4-Methyl-2-pentanone had not been considered prone to autoxidation, but an explosion during aerobic evaporation of the solvent [1] was attributed to formation and explosion of a peroxide [2].

Potassium *tert*-butoxide

*See* POTASSIUM *tert*-BUTOXIDE, $C_4H_9KO$: Acids, etc.

† **BUTYL ACETATE** $CH_3CO_2Bu$ $C_6H_{12}O_2$

Potassium *tert*-butoxide

*See* POTASSIUM *tert*-BUTOXIDE, $C_4H_9KO$: Acids, etc.

† **2-BUTYL ACETATE** $CH_3CO_2CHMeEt$ $C_6H_{12}O_2$

† **2,6-DIMETHYL-1,4-DIOXANE** $C_6H_{12}O_2$

$\overline{OCH_2CHMeOCHMeCH_2}$

† **ETHYL ISOBUTYRATE** $Me_2CHCO_2Et$ $C_6H_{12}O_2$

† **2-ETHYL-2-METHYL-1,3-DIOXOLANE** $\overline{OCMeEtOCH_2CH_2}$ $C_6H_{12}O_2$

† **4-HYDROXY-4-METHYL-2-PENTANONE** $C_6H_{12}O_2$

$MeCOCH_2C(OH)Me_2$

† **ISOBUTYL ACETATE** $C_6H_{12}O_2$

$CH_3CO_2CH_2CHMe_2$

† **ISOPROPYL PROPIONATE** $EtCO_2CHMe_2$ $C_6H_{12}O_2$

† **METHYL ISOVALERATE** $C_6H_{12}O_2$

$Me_2CHCH_2CO_2Me$

† **METHYL PIVALATE** $Me_3CCO_2Me$ $C_6H_{12}O_2$

† **METHYL VALERATE** $BuCO_2Me$ $C_6H_{12}O_2$

**TETRAMETHYL-1,2-DIOXETANE** $Me_2\overline{COOC}Me_2$ $C_6H_{12}O_2$

Kopecky, K. R. *et al., Can. J. Chem.*, 1975, **53**, 1107

Several explosions occurred in vacuum-sealed samples kept at ambient temperature. Storage under air at −20°C, or as solutions up to 2 M appear safe.

*See* DIOXETANES

*See other* CYCLIC PEROXIDES

† **2,5-DIMETHOXYTETRAHYDROFURAN** $C_6H_{12}O_3$

$\overline{OCH(OMe)C_2H_4CHO}Me$

***tert*-BUTYL PEROXYACETATE** $MeCO_2OCMe_3$ $C_6H_{12}O_3$

1. Martin, J. H., *Ind. Eng. Chem.*, 1960, **52**(4), 49A
2. Castrantas, 1965, 16

The perester explodes with great violence when rapidly heated to a critical temperature. Previous standard explosivity tests had not shown this behaviour. The presence of benzene as solvent prevents the explosive decomposition, but if the solvent evaporates, the residue is dangerous [1]. The pure ester is also shock-sensitive and detonable, but commercial 75% solutions are not [2].

*See other* PEROXYESTERS

**2-ETHOXYETHYL ACETATE** $MeCO_2C_2H_4OEt$ $C_6H_{12}O_3$

Anon., private comm., 1975

Mild explosions have occurred at the end of technical-scale distillations of 'ethyl glycol acetate' in a copper batch still about 20 min after the kettle heater was shut off. It is possible that air was drawn into the distillation column as it cooled, creating a flammable mixture, but no mechanism of initiation could be established. Oxidation studies did not indicate the likely formation of high peroxide concentrations in the liquid phase of this ether-ester.

† **ISOBUTYL PEROXYACETATE** $MeCO_2OCH_2CHMe_2$ $C_6H_{12}O_3$

*trans*-2-HEXENE OZONIDE $C_6H_{12}O_3$

**MeCHCH(Pr)O–O⁺–O⁻**

Loan, L. D. *et al., J. Amer. Chem. Soc.*, 1965, **87**, 741
Attempts to get carbon and hydrogen analyses by combustion of the ozonides of 2-butene, 2-pentene and 2-hexene caused violent explosions, though oxygen analyses were uneventful.
*See other* OZONIDES

**PEROXYHEXANOIC ACID** $BuCH_2CO_2OH$ $C_6H_{12}O_3$

Swern, D., *Chem. Rev.*, 1949, **45**, 10
Fairly stable at ambient temperature, it explodes and ignites on rapid heating.
*See other* PEROXYACIDS

**† 2,4,6-TRIMETHYLTRIOXANE (PARALDEHYDE)** $C_6H_{12}O_3$

**OCHMeOCHMeOCHMe**

Nitric acid
*See* NITRIC ACID, $HNO_3$: 2,4,6-Trimethyltrioxane

**TETRAMETHOXYETHYLENE** $C_6H_{12}O_4$

$(MeO)_2C{=}C(OMe)_2$

Preparative hazard
*See* 1,2,3,4-TETRACHLORO-7,7-DIMETHOXY-5-PHENYLBICYCLO [2.2.1]-2,5-HEPTADIENE, $C_{15}H_{12}Cl_4O_2$

**3,3,6,6-TETRAMETHYL-1,2,4,5-TETRAOXANE** $C_6H_{12}O_4$

$OOCMe_2OOCMe_2$

Baeyer, A. *et al., Ber.*, 1900, **33**, 858
Action of hydrogen peroxide or permonosulphuric acid on acetone produces this dimeric acetone peroxide, which explodes violently on impact, friction or rapid heating.
*See other* CYCLIC PEROXIDES

**BIS(DIMETHYLTHALLIUM) ACETYLIDE** $C_6H_{12}Tl_2$

$Me_2TlC{\equiv}CTlMe_2$

Nast, R. *et al., J. Organomet. Chem.*, 1966, **6**, 461
Extremely explosive, heat- and friction-sensitive.
*See other* ALKYLMETALS
METAL ACETYLIDES

**† CYCLOHEXYLAMINE** $C_6H_{11}NH_2$ $C_6H_{13}N$

Nitric acid
*See* NITRIC ACID, $HNO_3$: Cyclohexylamine

**† 1-METHYLPIPERIDINE** $MeN[CH_2]_4CH_2$ $C_6H_{13}N$

**† 2-METHYLPIPERIDINE** $CHMe[CH_2]_4NH$ $C_6H_{13}N$

**† 3-METHYLPIPERIDINE** $C_6H_{13}N$

$CH_2CHMe[CH_2]_3NH$

**† 4-METHYLPIPERIDINE** $C_2H_4CHMeC_2H_4NH$ $C_6H_{13}N$

**† PERHYDROAZEPINE (HEXAMETHYLENEIMINE)** $C_6H_{13}N$

$NH[CH_2]_5CH_2$

**† 2,2-DIMETHYLBUTANE** $CH_3CMe_2Et$ $C_6H_{14}$

**† 2,3-DIMETHYLBUTANE** $CH_3CHMeCHMe_2$ $C_6H_{14}$

**† HEXANE** $CH_3[CH_2]_4CH_3$ $C_6H_{14}$

Dinitrogen tetraoxide
*See* DINITROGEN TETRAOXIDE, $N_2O_4$: Hydrocarbons

**† ISOHEXANE** $Me_2CH[CH_2]_2CH_3$ $C_6H_{14}$

**† 3-METHYLPENTANE** $C_6H_{14}$

$CH_3CH_2CHMeCH_2CH_3$

**DIISOPROPYLBERYLLIUM** $Me_2CHBeCHMe_2$ $C_6H_{14}Be$

Coates, G. E. *et al., J. Chem. Soc.*, 1954, 22

Uncontrolled reaction with water is explosive.

*See other* ALKYLMETALS

**DIPROPYLMERCURY** $Pr_2Hg$ $C_6H_{14}Hg$

Iodine

Whitmore, 1921, 100

Interaction is violent.

*See other* MERCURY COMPOUNDS

**DIPROPYL HYPONITRITE** PrON=NOPr $C_6H_{14}N_2O_2$

*See* DIALKYL HYPONITRITES

**† BUTYL ETHYL ETHER** $BuOC_2H_5$ $C_6H_{14}O$

**† DIISOPROPYL ETHER** $Me_2CHOCHMe_2$ $C_6H_{14}O$

1. Anon., *Chem. Eng. News,* 1942, **20**, 1458
2. *MCA Case Histories Nos. 603, 1607*
3. Douglas, I.B., *J. Chem. Educ.*, 1963, **40**, 469
4. Hamstead, A. C. *et al., J. Chem. Eng. Data*, 1960, **5**, 383; *Ind. Eng. Chem.*, 1961, **53**(2), 63A; Rieche, A. *et al., Ber.*, 1942, **75**, 1016

There is a long history of violent explosions involving peroxidised diisopropyl ether, with initiation by disturbing a drum [1], unscrewing a bottle cap [2], or accidental impact [3]. This ether, with two tertiary hydrogen atoms adjacent to the oxygen link, is extremely readily peroxidised after a few hours' exposure to air, initially to a di-hydroperoxide, which disproportionates to dimeric and trimeric acetone peroxides which are highly explosive [4]. It has been found that the ether may be inhibited completely against peroxide formation for a few years by addition of *N*-benzyl-*p*-aminophenol at 16 p.p.m., or by diethylenetriamine, triethylenetetramine or tetraethylenepentamine at 50 p.p.m. [4].

*See* 3,3,6,6-TETRAMETHYL-1,2,4,5-TETRAOXANE, $C_6H_{12}O_4$
3,3,6,6,9,9-HEXAMETHYL-1,2,4,5,7,8-HEXAOXAONANE, $C_9H_{18}O_6$

*See other* PEROXIDISABLE COMPOUNDS

**† DIPROPYL ETHER** **PrOPr** **$C_6H_{14}O$**

**† 2-METHYL-2-PENTANOL** **$CH_3C(Me)OHPr$** **$C_6H_{14}O$**

**† 3-METHYL-3-PENTANOL** **$Et_2C(Me)OH$** **$C_6H_{14}O$**

**† 1,1-DIETHOXYETHANE (DIETHYLACETAL)** **$C_6H_{14}O_2$**
**$(EtO)_2CHCH_3$**

Mee, A. J., *School Sci Rev*., 1940, 22(86), 95
A peroxidised sample exploded violently during distillation.
*See other* PEROXIDISABLE COMPOUNDS

**† 1,2-DIETHOXYETHANE** **$EtOC_2H_4OEt$** **$C_6H_{14}O_2$**

**DIPROPYL PEROXIDE** **PrOOPr** **$C_6H_{14}O_2$**

Swern, 1972, Vol. 3, 21
Unexpected explosions have occurred with dipropyl peroxides.
*See other* DIALKYL PEROXIDES

**BIS(2-METHOXYETHYL) ETHER** **$(MeOC_2H_4)_2O$** **$C_6H_{14}O_3$**

'Diglyme Properties and Uses', Brochure, London, ICI, 1967
Details of properties, applications and safe handling. Light, heat and oxygen (air) promote formation of potentially explosive peroxides. These may be removed by stirring with a suspension of iron oxide in aqueous alkali.

Metal hydrides

*See* ALUMINIUM HYDRIDE, $AlH_3$ : Carbon dioxide, etc.
LITHIUM TETRAHYDROALUMINATE, $AlH_4Li$ : Bis(2-methoxyethyl) ether

**DIPROPYLALUMINIUM HYDRIDE** **$Pr_2AlH$** **$C_6H_{15}Al$**

*See other* ALKYLALUMINIUM ALKOXIDES, etc.

**† TRIETHYLALUMINIUM** **$Et_3Al$** **$C_6H_{15}Al$**

It ignites in air.

Alcohols

Mirviss, S. B. *et al., Ind. Eng. Chem*., 1961, **53**(1), 54A

Interaction with methanol, ethanol and 2-propanol with the undiluted trialkyl is explosive, while *tert*-butanol reacts vigorously.
*See* ALKYLALUMINIUM DERIVATIVES

*N,N*-Dimethylformamide
'DMF Brochure', Billingham, ICI, 1965
A mixture of the amide and triethylaluminium explodes when heated.
*See other* ALKYLMETALS

**ETHOXYDIETHYLALUMINIUM** **$EtOAl(Et)_2$** **$C_6H_{15}AlO$**
*Dangerous Loads,* 1972
The pure material, or solutions concentrated above 20%, ignite in air.
*See other* ALKYLALUMINIUM ALKOXIDES AND HYDRIDES

**TRIETHOXYDIALUMINIUM TRIBROMIDE** **$C_6H_{15}Al_2Br_3O_3$**
**$(EtO)_3Al \cdot AlBr_3$**
Anon., *Chem. Eng. Progr.*, 1966, **62**(9), 128
It ignites in air and explodes with ethanol or water.
*See related* ALKYLALUMINIUM ALKOXIDES

**TRIETHYLDIALUMINIUM TRICHLORIDE** **$C_6H_{15}Al_2Cl_3$**
**$Et_3Al \cdot AlCl_3$**

Carbon tetrachloride
Reineckel, H., *Angew. Chem. (Intern. Ed.)*, 1964, **3**, 65
A mixture exploded when warmed to room temperature.
*See other* ALKYLALUMINIUM HALIDES

**TRIETHYLARSINE** **$Et_3As$** **$C_6H_{15}As$**
Sidgwick, 1950, 762
Inflames in air.
*See other* ALKYLNON-METALS

**TRIETHYLPHOSPHINEGOLD(I) NITRATE** **$C_6H_{15}AuNO_3P$**
**$Et_3P \cdot AuNO_3$**
Coates, G. E. *et al.*, *Austral. J. Chem.*, 1966, **19**, 541
After thorough desiccation, the crystalline solid spontaneously exploded.
*See other* GOLD COMPOUNDS

**TRIETHYLBORANE** **$Et_3B$** **$C_6H_{15}B$**

Meerwein, H., *J. Prakt. Chem.*, 1937, **147**, 240

Hurd, D. T., *J. Amer. Chem. Soc.*, 1948, **70**, 2053

Ignites on exposure to air. Mixtures with triethylaluminium have been used as hypergolic igniters for rocket propulsion systems.

*See other* ALKYLNON-METALS

**† TRIETHYLBORATE** **$(EtO)_3B$** **$C_6H_{15}BO_3$**

**TRIETHYLBISMUTH** **$Et_3Bi$** **$C_6H_{15}Bi$**

Coates, 1967, Vol. 1, 537

Ignites in air, and explodes at about 150°C before distillation begins.

*See other* TRIALKYLBISMUTHS

**TRIETHYLSILICON PERCHLORATE** **$Et_3SiClO_4$** **$C_6H_{15}ClO_4Si$**

*See* ORGANOSILICON PERCHLORATES

**TRIETHYLGALLIUM** **$Et_3Ga$** **$C_6H_{15}Ga$**

Dennis, L. M. *et al., J. Amer. Chem. Soc.*, 1932, **54**, 182

The compound spontaneously inflames in air. Breaking a bulb containing 0.2 g under cold water shattered the container. The monoetherate behaved similarly under 6 M nitric acid.

*See other* ALKYLMETALS

**† BUTYLETHYLAMINE** **BuNHEt** **$C_6H_{15}N$**

**† DIISOPROPYLAMINE** **$(Me_2CH)_2NH$** **$C_6H_{15}N$**

**† 1,3-DIMETHYLBUTYLAMINE** **$C_6H_{15}N$**

**$Me_2CHCH_2CHMeNH_2$**

**† DIPROPYLAMINE** **$Pr_2NH$** **$C_6H_{15}N$**

**† TRIETHYLAMINE** **$Et_3N$** **$C_6H_{15}N$**

Dinitrogen tetraoxide

*See* DINITROGEN TETRAOXIDE, $N_2O_4$ : Triethylamine

Maleic anhydride

*See* MALEIC ANHYDRIDE, $C_4H_2O_3$: Cations, etc.

**TRIETHYLAMINE HYDROGEN PEROXIDATE** $C_6H_{15}N \cdot 4H_2O_2$

$Et_3N \cdot 4H_2O_2$

*See* CRYSTALLINE HYDROGEN PEROXIDATES

**2-DIETHYLAMMONIOETHYL NITRATE NITRATE** $C_6H_{15}N_3O_6$

$Et_2N^+HC_2H_4ONO_2\ NO_3^-$

1. Barbière, J., *Bull. Soc. Chim. Fr.*, 1944, **11**, 470–480
2. Fakstorp, J. *et al., Acta Chem. Scand.*, 1951, **5**, 968–969

Repetition of the original preparation [1] involving interaction of diethylaminoethanol with fuming nitric acid, followed by vacuum distillation of excess acid, invariably caused explosions during this operation. A modified procedure without evaporation is described [2].

*See related* ALKYL NITRATES
OXOSALTS OF NITROGENOUS BASES

**DIETHYL ETHANEPHOSPHONITE** $EtP(OEt)_2$ $C_6H_{15}O_2P$

Arbuzov, B. A. *et al., Chem. Abs.*, 1953, **47**, 3226a

Absorbed onto filter paper it ignites in air.

*See related* ALKYLNON-METALS

**TRIETHYLPHOSPHINE** $Et_3P$ $C_6H_{15}P$

Oxygen

Engler, C. *et al., Ber.*, 1901, **34**, 2935

Action of oxygen at low temperature on the phosphine produces an explosive product.

*See other* ALKYLNON-METALS

**TRIETHYLANTIMONY** $Et_3Sb$ $C_6H_{15}Sb$

von Schwartz, 1918, 322

Inflames in air.

*See other* ALKYLMETALS

**AMMONIUM HEXACYANOFERRATE(4–)** $C_6H_{16}FeN_{10}$

$[NH_4]_4[Fe(CN)_6]$

Metal nitrates

*See* COBALT(II) NITRATE, $CoN_2O_6$: Ammonium hexacyanoferrate(4–)
COPPER(II) NITRATE, $CuN_2O_6$: Ammonium hexacyanoferrate(4–)

**LITHIUM TRIETHYLSILYLAMIDE** **$LiNHSiEt_3$** **$C_6H_{16}LiNSi$**

Oxidants

It is hypergolic with fluorine or fuming nitric acid, and explodes with ozone.

*See* SILYLHYDRAZINES: Oxidants

*See other* *N*-METAL DERIVATIVES

**TRIETHYLAMMONIUM NITRATE** **$Et_3N^+H\ NO_3^-$** **$C_6H_{16}NO_3$**

Dinitrogen tetraoxide

*See* DINITROGEN TETRAOXIDE, $N_2O_4$: Triethylammonium nitrate

**† 1,2-BIS(DIMETHYLAMINO)ETHANE** **$Me_2NC_2H_4NMe_2$** **$C_6H_{16}N_2$**

**† DIETHOXYDIMETHYLSILANE** **$(EtO)_2SiMe_2$** **$C_6H_{16}O_2Si$**

**TRIETHYLTIN HYDROPEROXIDE** **$Et_3SnOOH$** **$C_6H_{16}O_2Sn$**

Aleksandrov, Y. A. *et al., Zh. Obshch. Khim.*, 1965, **35**, 115

While the hydroperoxide is stable at room temperature, the addition compound which it readily forms with excess hydrogen peroxide is not, decomposing violently.

*See* ORGANOMINERAL PEROXIDES

**ETHYLENEBIS(DIMETHYLPHOSPHINE)** **$C_6H_{16}P_2$**

**$Me_2PC_2H_4PMe_2$**

Butter, S. A. *et al., Inorg. Synth.*, 1974, **15**, 190

It ignites in air.

*See other* ALKYLNON-METALS

**1,2-BIS(ETHYLAMMONIO)ETHANE PERCHLORATE** **$C_6H_{18}Cl_2N_2O_8$**

**$EtN^+H_2C_2H_4N^+H_2Et\ (ClO_4^-)_2$**

Sunderlin, K. G. R., *Chem. Eng. News*, 1974, **52**(31), 3

It exploded mildly under an impact of 11.5 kg m.

*See other* OXOSALTS OF NITROGENOUS BASES

### LITHIUM HEXAMETHYLCHROMATE(3–) $C_6H_{18}CrLi_3$

$Li_3[CrMe_6]$

Kurras, E. *et al., J. Organomet. Chem.*, 1965, **4**, 114–118

Isolated as a dioxane complex, it ignites in air.

*See related* ALKYLMETALS

### BIS-TRIMETHYLSILYL CHROMATE $C_6H_{18}CrO_4Si_2$

$(Me_3SiO)_2CrO_2$

1. *ABCM Quart. Safety Summ.*, 1960, **31**, 16
2. Schmidt, M. *et al., Angew. Chem.*, 1958, **70**, 704

Small quantities can be distilled at *ca.* 75°C/1.3 mbar, but larger quantities are liable to explode violently owing to local overheating [1]. An attempt to prepare an analogous poly(dimethylsilyl) chromate by heating a dimethylpolysiloxane with chromium trioxide at 140°C exploded violently after 20 min at this temperature [2].

### LITHIUM BIS(TRIMETHYLSILYL)AMIDE $C_6H_{18}LiNSi_2$

$LiN(SiMe_3)_2$

Amonoo-Neizer, E. H. *et al., Inorg. Synth.*, 1966, **8**, 21

It is unstable in air and ignites when compressed.

### DILITHIUM 1,1-BIS(TRIMETHYLSILYL)HYDRAZIDE $C_6H_{18}Li_2N_2Si_2$

$Li_2NN(SiMe_3)_2$

It ignites in air.

*See* SILYLHYDRAZINES

### BIS(DIMETHYLAMINO)DIMETHYLSTANNANE $C_6H_{18}N_2Sn$

$(Me_2N)_2SnMe_2$

Chloroform

Randall, E. W. *et al., Inorg. Nucl. Chem. Lett.*, 1965, **1**, 105

Gentle heating of a 1:1 chloroform solution of the stannane led to a mild explosion. Similar explosions had been experienced at the conclusion of fractional distillation of other bis(dialkylamino)stannanes where the still-pot temperature had risen to *ca.* 200°C.

*See other* *N*-METAL DERIVATIVES

### TRIS(DIMETHYLAMINO)ANTIMONY $(Me_2N)_3Sb$ $C_6H_{18}N_3Sb$

Ethyl diazoacetate

*See* ETHYL DIAZOACETATE, $C_4H_6N_2O_2$: Tris(dimethylamino)antimony

**$N,N'$-BIS(2-AMINOETHYL)1,2-DIAMINOETHANE (TRIETHYLENETETRAMINE)** $(H_2NC_2H_4NHCH_2)_2$ $C_6H_{18}N_4$

Cellulose nitrate

*See* CELLULOSE NITRATE: Amines

**2,4,6,8,9,10-HEXAMETHYLHEXAAZA-1,3,5,7-TETRAPHOSPHAADAMANTANE** $C_6H_{18}N_6P_4$

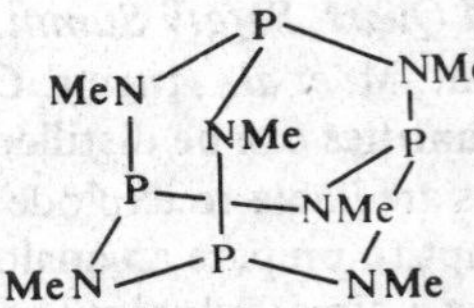

Oxidants

Verdier, F., *Rech. Aerosp.*, 1970, (137), 181–189

Among other solid P–N compounds examined, the title compound ignited immediately on contact with nitric acid, hydrogen peroxide or dinitrogen trioxide.

**† BIS-TRIMETHYLSILYL OXIDE** $(Me_3Si)_2O$ $C_6H_{18}OSi_2$

**BIS-TRIMETHYLSILYL PEROXOMONOSULPHATE** $C_6H_{18}O_5SSi_2$

$Me_3SiOSO_2OOSi(Me)_3$

Blaschette, A. *et al., Angew. Chem. (Intern. Ed.)*, 1969, **8**, 450

It is stable at $-30^\circ C$, but on warming to ambient temperature decomposes violently evolving sulphur trioxide.

*See other* PEROXYESTERS

**(DIMETHYLSILYLMETHYL)TRIMETHYLLEAD** $C_6H_{18}PbSi$

$Me_2SiHCH_2PbMe_3$

Schmidbaur, H., *Chem. Ber.*, 1964, **97**, 270

It decomposes above $100^\circ C$, and with explosive violence in presence of oxygen.

*See related* ALKYLMETALS
ALKYLNON-METALS

**HEXAMETHYLDIPLATINUM** $Me_3PtPtMe_3$ $C_6H_{18}Pt_2$

Gilman, H. *et al., J. Amer. Chem. Soc.*, 1953, **75**, 2065

It explodes sharply in a shower of sparks on heating.

*See other* PLATINUM COMPOUNDS

**HEXAMETHYLRHENIUM** $Me_6Re$ $C_6H_{18}Re$

Mertis, K. *et al.*, *J. Organomet. Chem.*, 1975, **97**, C65;
*J. Chem. Soc., Dalton Trans.*, 1976, 1489

It should be handled with extreme caution, even in absence of air, because dangerous explosions have occurred on warming frozen samples, or during transfer operations. It is unstable above −20°C, and on one occasion admission of nitrogen to the solid under vacuum caused a violent explosion.

*See other* ALKYLMETALS

**HEXAMETHYLTUNGSTEN** $Me_6W$ $C_6H_{18}W$

Galyer, L. *et al.*, *J. Organomet. Chem.*, 1975, **85**, C37

Several unexplained and violent explosions during preparation and handling indicate that the compound must be handled with great care and as potentially explosive, particularly during vacuum sublimation.

*See other* ALKYLMETALS

**HEXAMETHYLDISILAZANE** $Me_3SiNHSiMe_3$ $C_6H_{19}NSi_2$

**BIS-1,2-DIAMINOPROPANE-*cis*-DICHLOROCHROMIUM(III) PERCHLORATE** $C_6H_{20}Cl_3CrN_4O_4$

$[Cr(C_3H_{10}N_2)_2Cl_2]ClO_4$

Perchloric acid

Anon., *Chem. Eng. News*, 1963, **41**(27), 47

A mixture of the complex and 20 volumes of *ca* 70% acid was being stirred at 22°C when it exploded violently. Traces of ether or ethanol may have been present. Extreme caution must be exercised when concentrated perchloric acid is contacted with organic materials with agitation or without cooling.

*See other* AMMINEMETAL OXOSALTS

**1,2-BIS(TRIMETHYLSILYL)HYDRAZINE** $C_6H_{20}N_2Si_2$

$Me_3SiNHNHSiMe_3$

Oxidants

Hypergolic with fluorine or fuming nitric acid, it explodes with ozone.

*See* SILYLHYDRAZINES: Oxidants

**HEXAUREAGALLIUM(III) PERCHLORATE** $C_6H_{24}Cl_3GaN_{12}O_{18}$

$[Ga(NH_2CONH_2)_6][ClO_4]_3$

Lloyd, D. J. *et al.*, *J. Chem. Soc.*, 1943, 76

Decomposes violently when heated strongly above its m.p., 179°C.
*See other* AMMINEMETAL OXOSALTS

**TRIS-1,2-DIAMINOETHANECOBALT(III) NITRATE** $C_6H_{24}CoN_9O_9$
$[Co(NH_2C_2H_4NH_2)_3][NO_3]_3$

No explosion on heating, but medium impact-sensitivity.
*See* AMMINEMETAL OXOSALTS

**HEXAUREACHROMIUM(III) NITRATE** $C_6H_{24}CrN_{15}O_{15}$
$[Cr(NH_2CONH_2)_6][NO_3]_3$

Explodes at 265°C, medium impact-sensitivity.
*See* AMMINEMETAL OXOSALTS

**TRIPOTASSIUM CYCLOHEXANEHEXONE-1,3,5-TRIOXIMATE** $C_6K_3N_3O_6$

See p. 582.

**POTASSIUM BENZENEHEXOLATE ('POTASSIUM CARBONYL')** $C_6K_6O_6$

OK
KO OK
KO OK
OK

1. Sorbe, 1968, 68, 69
2. Mellor, 1963, Vol. 2, Suppl. 2.2, 1567–1568; 1946, Vol. 5, 951

The compound 'potassium carbonyl' is now known to be the hexameric potassium salt of hexahydroxybenzene, and reacts with moist air to give a very explosive product [1]. It reacts violently with oxygen, and explodes on heating in air or in contact with water [2].

*See* POTASSIUM, K: Non-metal oxides (reference 4)
*See other* CARBONYLMETALS

**LITHIUM BENZENEHEXOLATE ('CARBONYLLITHIUM')** $C_6Li_6O_6$

By analogy with the sodium and potassium compounds, 'carbonyllithium' should probably be reformulated as the hexameric benzene derivative.

**HEXACARBONYLMOLYBDENUM** $Mo(CO)_6$ $C_6MoO_6$

Diethyl ether
Owen, B. B. *et al., Inorg. Synth.*, 1950, 3, 158
Solutions of hexacarbonylmolybdenum have exploded after extended storage.
*See other* CARBONYLMETALS

## 3(3-CYANO-1,2,4-OXADIAZOL-5-YL)-4-CYANOFURAZAN 2- (or 5) OXIDE $C_6N_6O_3$

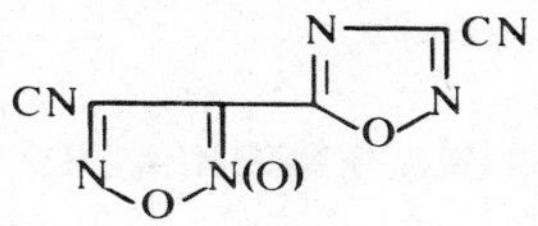

Hydrazine derivatives

Lupton, E. C. *et al., J. Chem. Eng. Data*, 1975, **20**, 136

Interaction is explosive.

*See also* DICYANOFURAZAN, $C_4N_4O$: Nitrogenous bases

## TETRAAZIDO-*p*-BENZOQUINONE $O{=}C_6(N_3)_4{=}O$ $C_6N_{12}O_2$

Friess, K. *et al., Ber.*, 1923, **56**, 1304

Extremely explosive, sensitive to heat, impact or friction. A single crystal heated on a spatula exploded and bent it badly.

*See other* 2-AZIDOCARBONYL COMPOUNDS
HIGH-NITROGEN COMPOUNDS

## SODIUM BENZENEHEXOLATE $C_6(ONa)_6$ $C_6Na_6O_6$

*See* SODIUM, Na: Non-metal oxides (reference 4)

## HEXACARBONYLVANADIUM $(CO)_6V$ $C_6O_6V$

1. Pruett, R. L. *et al., Chem. & Ind.*, 1960, 119
2. Bailar, 1973, Vol. 3, 529

The pyrophoric compound previously regarded as 'dodecacarbonyl divanadium' [1] is now known to be the monomeric hexacarbonylvanadium [2].

*See other* CARBONYLMETALS

## HEXACARBONYLTUNGSTEN $W(CO)_6$ $C_6O_6W$

Hurd, D. T. *et al., Inorg. Chem.*, 1957, **5**, 136

It is dangerous to attempt the preparation from tungsten hexachloride, aluminium powder and carbon monoxide in an autoclave of greater than 0.3 litre capacity.

*See other* CARBONYLMETALS

## 4-IODOBENZENEDIAZONIUM-2-CARBOXYLATE $C_7H_3IN_2O_2$

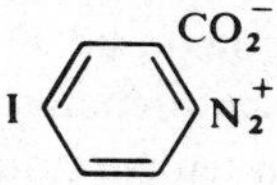

$CO_2^-$ I $N_2^+$

Gommper, R. *et al., Chem. Ber.*, 1968, **101**, 2348

It is a highly explosive solid.
*See other* DIAZONIUM CARBOXYLATES

**2,4,6-TRINITROBENZOIC ACID** $C_7H_3N_3O_8$

$(NO_2)_3C_6H_2CO_2H$

Preparative hazard

*See* SODIUM DICHROMATE, $CrNa_2O_7$ : Sulphuric acid, TNT

Kranz, A. *et al., Chem. Age (London)*, 1925, **13**, 392

All heavy metal salts prepared were explosive on heating or impact.

*See other* POLYNITROARYL COMPOUNDS

**HEPTA-1,3,5-TRIYNE** $H[C{\equiv}C]_3Me$ $C_7H_4$

Cook, C. L. *et al., J. Chem. Soc.*, 1952, 2890

It explodes very readily (in absence of air) above 0°C, and the distillation residues also exploded on admission of air.

*See other* ACETYLENIC COMPOUNDS

**SILVER 3,5-DINITROANTHRANILATE** $C_7H_4AgN_3O_6$

$(NO_2)_2NH_2C_6H_2CO_2Ag$

Gupta, S. P. *et al., Def. Sci. J.*, 1975, **25**, 101–106

It explodes at 394°C, and the copper(II) salt at 371°C, each after a 10 s delay. They are not impact-sensitive.

*See other* HEAVY METAL DERIVATIVES
POLYNITROARYL COMPOUNDS

***o*-, *m*- or *p*-TRIFLUOROMETHYLPHENYLMAGNESIUM BROMIDE**

$CF_3C_6H_4MgBr$ $C_7H_4BrF_3Mg$

1. Appleby, J. C., *Chem. & Ind.*, 1971, 120
2. 'Benzotrifluorides Catalogue 6/15', West Chester, Pa., Marshallton Res. Labs., 1971
3. Taylor, R. *et al.*, private comm., 1973

During a 2 kg preparation under controlled conditions, the *m*-isomer exploded violently [1]. The same had occurred previously in a 1.25 kg preparation 30 min after all magnesium had dissolved. A small preparation of the *o*- isomer had also exploded. It was found that *m*- preparations in ether decomposed violently at 75°C and in benzene–tetrahydrofuran at 90°C. Molar preparations or less have been done several times uneventfully when temperatures below 40°C were maintained, but caution is urged. The *m*- and *p*- reagents prepared in ether exploded when the temperature was raised by adding benzene and distilling the ether out [3].

*See other* HALO-ARYLMETALS

**p-BROMOBENZOYL AZIDE** $BrC_6H_4CON_3$ $C_7H_4BrN_3O$

Curtiss, T. *et al., J. Prakt. Chem.*, 1898, **58**, 201

Explodes violently above its m.p., 46°C.

*See other* ACYL AZIDES

**p-CHLOROTRIFLUOROMETHYLBENZENE** $C_7H_4ClF_3$

$ClC_6H_4CF_3$

Sodium dimethylsulphinate

Meschino, J. A. *et al., J. Org. Chem.*, 1971, **36**, 3637, footnote 9

Interaction with dimethylsulphinate carbanion (from the sulphoxide and sodium hydride) at −5°C is very exothermic, and addition of the chloro compound must be slow to avoid violent eruption.

**2-CHLORO-5-TRIFLUOROMETHYLBENZENEDIAZONIUM HYDROGENSULPHATE** $C_7H_4ClF_3N_2O_4S$

$ClCF_3C_6H_3N_2^+ HSO_4^-$

*See* DIAZONIUM SULPHATES

**p-CHLOROPHENYLISOCYANATE** $ClC_6H_4NCO$ $C_7H_4ClNO$

Preparative hazard

*See* p-CHLOROBENZOYL AZIDE, $C_7H_4ClN_3O$

**o-NITROBENZOYL CHLORIDE** $NO_2C_6H_4COCl$ $C_7H_4ClNO_3$

1. Cook, N. C. *et al., Chem. Eng. News*, 1945, **23**, 2394
2. Bonner, W. D. *et al., J. Amer. Chem. Soc.*, 1946, **68**, 344; *Chem. & Ind.*, 1948, 89
3. *MCA Case History No. 1915*
4. Lockemann, G. *et al., Ber.*, 1947, **80**, 488

The hot material remaining after vacuum stripping of solvent up to 130°C decomposed with evolution of gas and then exploded violently 50 min after heating had ceased. Further attempts to distil the acid chloride even in small quantities at below 1.3 mbar caused exothermic decomposition at 110°C. It was, however, possible to flash-distil the chloride in special equipment [1]. Two later similar publications recommend use in solution of the unisolated material [2].

Smaller-scale distillation of the chloride at 94–95°C/0.03 mbar had been uneventful, but a 1.2 mol scale preparation exploded during distillation at 128°C/1.2 mbar [3], even in presence of phosphorus pentachloride, previously recommended to reduce the danger of explosion during distillation [4]. Many previous explosions had been reported, and the need for adequate purity of intermediates was stressed.

*See related* POLYNITROARYL COMPOUNDS
*See other* NITRO-ACYL HALIDES

**p-CHLOROBENZOYL AZIDE** **$ClC_6H_4CON_3$** **$C_7H_4ClN_3O$**

1. Cobern, D. *et al., Chem. & Ind.*, 1965, 1625; ibid., 1966, 375
2. Tyabji, M. M., *Chem. & Ind.*, 1965, 2070

A violent explosion occurred during vacuum distillation of *p*-chlorophenylisocyanate, prepared by Curtius reaction on *p*-chlorobenzoyl azide. It was found by IR spectroscopy that this isocyanate (as well as others prepared analogously) contained some unchanged azide, to which the explosion was attributed. The use of IR spectroscopy to check for absence of azides in isocyanates is recommended before distillation [1]. Subsequently, the explosion was attributed to free hydrogen azide, produced by hydrolysis of unchanged azide [2].

*o-, m-* **or** *p-* **TRIFLUOROMETHYLPHENYLLITHIUM** **$C_7H_4F_3Li$**
**$CF_3C_6H_4Li$**

(Preparative hazard for *o-*)
*See* ORGANOLITHIUM REAGENTS
*See other* HALO-ARYLMETALS

**SODIUM 2-BENZOTHIAZOLETHIOLATE** **$C_7H_4NNaS_2$**

**$C_6H_4SC(SNa){=}N$** (ring closed between $C_6H_4$ and N)

Maleic anhydride
*See* MALEIC ANHYDRIDE, $C_4H_2O_3$: Cations, etc.

**BENZENEDIAZONIUM-2-CARBOXYLATE** **$C_7H_4N_2O_2$**

(benzene ring bearing $N_2^+$ and $CO_2^-$ on adjacent carbons)

Alone,
or Aniline,
or Isocyanides

1. Yaroslavsky, S., *Chem. & Ind.*, 1965, 765

2. Huisgen, R. *et al., Ber.*, 1965, **98**, 4104
3. Sullivan, J. M., *Chem. Eng. News*, 1971, **49**(16), 5
4. Embree, H. D., *Chem. Eng. News*, 1971, **49**(30), 3
5. Mich, T. K. *et al., J. Chem. Educ.*, 1968, **45**, 272
6. Matuszak, C. A., *Chem. Eng. News*, 1971, **49**(24), 39
7. Logullo, F. M. *et al., Org. Synth.*, 1973, Coll. Vol. 5, 54–59
8. Stiles, R. M. *et al., J. Amer. Chem. Soc.*, 1963, **85**, 1795, footnote 30a

The isolated internal salt is explosive, and should only be handled in small amounts [1]. It reacts explosively with aniline, and violently with aryl isocyanides [2]. The explosive nature of the salt has been amply confirmed [3,4], and it may be precipitated during the generation of benzyne by diazotisation of anthranilic acid at ambient temperature; some heating is essential for complete decomposition [5]. During decomposition of the isolated salt, too-rapid addition to hot solvent caused a violent explosion [6]. The published procedure [7] must be closely followed for safe working. Contrary to earlier beliefs, diazonium carboxylate hydrohalide salts are also shock-sensitive explosives [8].

1-Pyrrolidinylcyclohexene
Kuehne, M. E., *J. Amer. Chem. Soc.*, 1962, **84**, 841
Decomposition of benzenediazonium-2-carboxylate in the enamine at 40°C caused a violent explosion.
*See other* DIAZONIUM CARBOXYLATES

## BENZENEDIAZONIUM-3 or 4-CARBOXYLATE $C_7H_4N_2O_2$

$N_2^+C_6H_4CO_2^-$

Potassium *O,O*-diphenyl dithiophosphate
*See* DIAZONIUM SULPHIDES AND DERIVATIVES (reference 10)

## *o*-NITROBENZONITRILE $NO_2C_6H_4CN$ $C_7H_4N_2O_2$

1. Miller, C. S., *Org. Synth.*, Coll. Vol. 3, 646
2. Partridge, M. W., private comm., 1968

When the published method [1] for preparing the *p*- isomer is used to prepare *o*-nitrobenzonitrile, a moderate explosion often occurs towards the end of the reaction period. (This may be due to formation of nitrogen trichloride as a by-product.) Using an alternative procedure, involving heating *o*-chloronitrobenzene with copper(I) cyanide in pyridine for 7 h at 160°C, explosions occurred towards the end of the heating period in about one in five preparations [2].
*See other* CYANO COMPOUNDS

## 4-HYDROXYBENZENEDIAZONIUM-3-CARBOXYLATE $C_7H_4N_2O_3$

HO–$C_6H_3$(–$N_2^+$)–$CO_2^-$

1. Auden, W., *Chem. News*, 1899, **80**, 302
2. Puxeddu, E., *Gazz. Chim. Ital.*, 1929, **59**, 14

It explodes at either 155°C [1] or 162°C [2], probably depending on the heating rate.

*See other* DIAZONIUM CARBOXYLATES

## *m*-NITROBENZOYL NITRATE $NO_2C_6H_4CO_2NO_2$ $C_7H_4N_2O_6$

Francis, F. E., *J. Chem. Soc.*, 1906, **89**, 1

The nitrate explodes if heated rapidly, like benzoyl nitrate.

*See other* ACYL NITRATES

## 2-CYANO-4-NITROBENZENEDIAZONIUM HYDROGENSULPHATE

$C_7H_4N_4O_6S$

$CN(NO_2)C_6H_3N_2^+ HSO_4^-$

Sulphuric acid

Grewer, T., *Chem. Ing. Tech.*, 1975, **47**, 233–234

DTA examination of a *ca.* 35% solution of the diazonium salt in sulphuric acid showed three exotherms, corresponding to hydrolysis of the nitrile group (peak at 95°C), decomposition of the diazonium salt (peak at 160°C) and loss of the nitro group (large peak at 240°C). Adiabatic decomposition of the solution from 50°C also showed three steps, with induction periods of *ca.* 30, 340 and 380 min, respectively.

*See* THERMAL STABILITY OF REACTION MIXTURES

*See other* DIAZONIUM SULPHATES

## 2,6-DINITROBENZYL BROMIDE $C_7H_5BrN_2O_4$

$(NO_2)_2C_6H_3CH_2Br$

Potassium phthalimide

Reich, S. *et al.*, *Bull. Soc. Chim. Fr.*[4], 1917, **21**, 119

Reaction temperature in absence of solvent must be below 130–135°C to avoid explosive decomposition.

*See other* POLYNITROARYL COMPOUNDS

**PHENYLCHLORODIAZIRINE** N=N–C(Cl)Ph $C_7H_5ClN_2$

Wheeler, J. J., *Chem. Eng. News*, 1970, **48**(30), 10
Manzara, A. P., *Chem. Eng. News*, 1974, **52**(42), 5
The neat material is about three times as shock-sensitive as glyceryl nitrate, and should not be handled undiluted.

It exploded during vacuum distillation at 3.3 mbar from a bath at 140°C. Impact- and spark-sensitivities were determined, and autoignition occurred after 30 s at 107°C.

*See other* DIAZIRINES

**BENZOYL CHLORIDE** $C_6H_5COCl$ $C_7H_5ClO$

Aluminium trichloride,
Naphthalene

*See* ALUMINIUM TRICHLORIDE, $AlCl_3$: Benzoyl chloride, etc.

Dimethyl sulphoxide

*See* DIMETHYL SULPHOXIDE, $C_2H_6OS$: Acyl halides

**† TRIFLUOROMETHYLBENZENE** $CF_3C_6H_5$ $C_7H_5F_3$

**POTASSIUM *aci*-1-NITRO-1-PHENYLNITROMETHANE** $C_7H_5KN_2O_4$

**$PhC(NO_2)=N(O)OK$**

Rohman, A. *et al., J. Org. Chem.*, 1976, **41**, 124
It is unwise to use more than 0.02 mol of potassium salts of dinitroalkanes as they are explosive.

*See other* *aci*-NITRO SALTS

**POTASSIUM *O–O*-BENZOYLMONOPEROXOSULPHATE** $C_7H_5KO_6S$

**$PhCOOOSO_2OK$**

Willstatter, R. *et al., Ber.*, 1909, **42**, 1839
The anhydrous salt explodes on grinding, and the monohydrate on heating at 70–80°C or in contact with sulphuric acid.

*See other* DIACYL PEROXIDES
PEROXOACID SALTS

**DO NOT ASSUME THAT LACK OF INFORMATION MEANS THAT NO HAZARD EXISTS. LOOK FURTHER AT RELATED STRUCTURES.**

### *m*-NITROBENZALDEHYDE $NO_2C_6H_4CHO$ $C_7H_5NO_3$

Large, J. *et al.*, *Chem. & Ind.*, 1967, 1424

Delays in working up the crude product caused violent explosions during attempted vacuum distillation. An alternative method of crystallisation is described.

*See other* C-NITRO COMPOUNDS

### BENZOYL NITRATE $PhCO_2NO_2$ $C_7H_5NO_4$

1. Francis, F. E., *J. Chem. Soc.*, 1906, **89**, 1
2. Ferrario, E., *Gazz. Chim. Ital.*, 1901, [1], **40**, 99

An unstable liquid which is capable of distillation under reduced pressure, but which explodes violently on rapid heating at normal pressure [1]. It may also explode on exposure to light [2].

Water

Francis, F. E., *J. Chem. Soc.*, 1906, **89**, 1

Reactivity of benzoyl nitrate towards moisture is so great that attempted filtration through an undried filter paper causes explosive decomposition (possibly involving cellulose nitrate?).

*See other* ACYL NITRATES

### 2,5-PYRIDINEDICARBOXYLIC ACID $C_7H_5NO_4$

$\overline{N{=}CC(CO_2H){=}CHCH{=}C}CO_2H$

Preparative hazard

*See* NITRIC ACID, $HNO_3$ : 5-Ethyl-2-methylpyridine

### BENZOYL AZIDE $PhCON_3$ $C_7H_5N_3O$

Bergel, F., *Angew. Chem.*, 1927, **40**, 974

The sensitivity towards heat of this explosive compound is increased by previous compression, confinement and presence of impurities. Crude material exploded violently between 120 and 165°C.

*See other* ACYL AZIDES

### 4(?)-CARBOXYBENZENESULPHONYL AZIDE $C_7H_5N_3O_4S$

$HO_2CC_6H_4SO_2N_3$

Anon., *Res. Discl.*, 1975, **134**, 44 (*Chem. Abs.*, 1975, **83**, 180144v)

The azide (structure not stated) is not impact-sensitive, but explosively decomposes at 120°C. Blending with 75% of polymer as diluent eliminated the explosive decomposition.

*See other* ACYL AZIDES

## *o*-NITROPHENYLSULPHONYLDIAZOMETHANE $C_7H_5N_3O_4S$

$NO_2C_6H_4SO_2CHN_2$

Wagenaar, A. *et al., J. Org. Chem.*, 1974, **39**, 411–413

Severe explosions occurred when the cold crystalline solid was allowed to warm to ambient temperature. Normal manipulation of dilute solutions appears feasible, but the solid must be handled at lowest possible temperatures with full safety precautions. The *p*- isomer appears to be much more stable.

*See other* DIAZO COMPOUNDS

## 2,4,6-TRINITROTOLUENE $(NO_2)_3C_6H_2CH_3$ $C_7H_5N_3O_6$

Added impurities

Haüptli, H., private comm., 1972

During investigation of the effect of presence of 1% of added impurities on the thermal explosion temperature of TNT (297°), it was found that fresh red lead, sodium carbonate and potassium hydroxide reduced the explosion temperatures to 192, 218 and 192°C, respectively.

*See also* POTASSIUM, K: Nitrogen-containing explosives

Metals,
Nitric acid

Kovache, A. *et al., Mém. Poudres*, 1952, **34**, 369–378

Trinitrotoluene in contact with nitric acid and lead or iron produces explosive substances which may be readily ignited by shock, friction or contact with nitric or sulphuric acids. Such materials have been involved in industrial explosions.

*See other* HEAVY METAL SALTS

Potassium hydroxide

Copisarow, M., *Chem. News*, 1915, **112**, 283–284

TNT and potassium hydroxide in methanol will interact even at –65°C to give explosive *aci*-nitro salts (presumably *o*-quinonoid). The explosion temperature of TNT is lowered from 240°C to 160°C by the presence of a little potassium hydroxide.

*See other* *aci*-NITROQUINONOID COMPOUNDS
POLYNITROARYL COMPOUNDS

Sodium dichromate,
Sulphuric acid

*See* SODIUM DICHROMATE, $CrNa_2O_7$: Sulphuric acid, etc.

## 3-METHYL-2,4,6-TRINITROPHENOL $Me(NO_2)_3C_6HOH$ $C_7H_5N_3O_7$

Sorbe, 1968, 153

Explodes above 150°C.

*See other* POLYNITROALKYL COMPOUNDS

**BENZIMIDAZOLIUM 1-NITROIMIDATE** $C_7H_5N_4O_2$

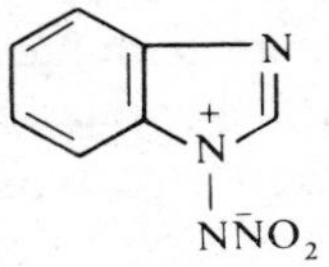

It explodes at the m.p., 169°C.

*See* *N*-AZOLIUM NITROIMIDATES

**1-NITRO-3(2,4-DINITROPHENYL)UREA** $C_7H_5N_5O_7$

$NO_2NHCONHC_6H_3(NO_2)_2$

McVeigh, J. L. *et al., J. Chem. Soc.*, 1945, 621

It is an explosive of lower power, but of greater impact- or frictional sensitivity than picric acid, which should not be stored in bottles with ground-in stoppers.

*See other N*-NITRO COMPOUNDS

***N*,2,4,6-TETRANITRO-*N*-METHYLANILINE (TETRYL)** $C_7H_5N_5O_8$

$(NO_2)_3C_6H_2N(Me)NO_2$

Hydrazine

*See* HYDRAZINE, $H_4N_2$: Oxidants

Trioxygen difluoride

*See* TRIOXYGEN DIFLUORIDE, $F_2O_3$

**1-HEPTEN-4,6-DIYNE** $H_2C{=}CHCH_2[C{\equiv}C]_2H$ $C_7H_6$

Armitage, J. B. *et al., J. Chem. Soc.*, 1952, 1994

It decomposes explosively with great ease.

*See other* ACETYLENIC COMPOUNDS

***o*-AZIDOMETHYLBENZENEDIAZONIUM TETRAFLUOROBORATE**

$C_7H_6BF_4N_5$

$N_3CH_2C_6H_4N_2^+ BF_4^-$

Trichloroacetonitrile

Kreher, R. *et al., Tetrahedron Lett.*, 1976, 4260

Thermal decomposition of the salt in nitriles produces 2-substituted quinazolines. In trichloroacetonitrile it proceeded explosively.

*See* DIAZONIUM TETRAHALOBORATES

**4-CHLORO-3-NITROTOLUENE** $C_7H_6ClNO_2$

$Cl(NO_2)C_6H_3CH_3$

Copper(I) cyanide,
Pyridine

1. Hub, L. *Proc. Chem. Process Haz. Sympos. VI* (Manchester, 1977), No. 49, 39–46, Rugby, I.Ch.E., 1977
2. Hub, L., private comm., 1977

Industrial preparation of 4-cyano-3-nitrotoluene by heating the reaction components at around 170°C for 6 h led to an explosion in 1976. Subsequent investigation by DSC showed that the cyano compound in presence of the starting materials exhibited an exotherm at 180°C. After 6 h reaction, this threshold temperature fell to 170°. Isothermal use of a safety calorimeter showed that a large exotherm occurred during the first hour of reaction and that, in absence of strong cooling, the reaction accelerated and the vessel contents were ejected by the vigour of the decomposition.

*See* REACTION SAFETY CALORIMETRY

**2-CHLORO-4-NITROTOLUENE** $Cl(NO_2)C_6H_3CH_3$ $C_7H_6ClNO_2$

Sodium hydroxide

1. *ABCM Quart. Safety Summ.*, 1962, **33**, 20
2. *MCA Case History No. 907*

The residue from vacuum distillation of crude material (contaminated with sodium hydroxide) exploded after showing signs of decomposition. Experiment showed that 2-chloro-4-nitrotoluene decomposes violently when heated at 170–200°C in presence of alkali. Thorough water washing is therefore essential before distillation is attempted [1]. A similar incident was reported with mixed chloronitrotoluenes [2].

*See other* NITROAROMATIC–ALKALI HAZARDS

**4-CHLORO-2-METHYLBENZENEDIAZONIUM SALTS** $C_7H_6ClN_2^+ Z^-$

$Cl(Me)C_6H_3N_2^+ Z^-$

Sodium hydrogensulphide,
or Disodium sulphide,
or Disodium polysulphide

*See* DIAZONIUM SULPHIDES

## POTASSIUM 4-METHOXY-1-*aci*-NITRO-3,5-DINITRO-2,5-CYCLOHEXADIENE $C_7H_6KN_3O_7$

$O_2N$

MeO ⟨ ⟩ =N(O)OK

$O_2N$

1. Meisenheimer, J., *Ann.*, 1902, **323**, 221
2. Millor, H. D. *et al., Sidgwick's Organic Chemistry of Nitrogen*, 396, Oxford, Clarendon Press, 3rd edn, 1966

The product of interaction of trinitrobenzene and concentrated aqueous potassium hydroxide in methanol is explosive, and analyses as the hemihydrate of a hemi-acetal of the *aci-p*-nitroquinonoid form of picric acid [1], and/or the mesomeric *o*- forms [2].

*See other* *aci*-NITROQUINONOID COMPOUNDS

## THALLIUM *aci*-PHENYLNITROMETHANE $C_7H_6NO_2Tl$

**PhCH=N(O)OTl**

McKillop, A. *et al., Tetrahedron*, 1974, **30**, 1369

No attempt should be made to isolate or dry this compound, as it is treacherously explosive.

*See other* *aci*-NITRO SALTS

## PHENYLDIAZOMETHANE **$PhCHN_2$** $C_7H_6N_2$

Schneider, M. *et al., Tetrahedron*, 1976, **32**, 621

The need for care in distillation of phenyldiazomethane is stressed. Previously it has been used as prepared in solution.

*See other* DIAZO COMPOUNDS

## 2,4-DINITROTOLUENE **$(NO_2)_2C_6H_3CH_3$** $C_7H_6N_2O_4$

*MCA SD-93*, 1966

The commercial material, containing some 20% of the 2,6- isomer, decomposes at 250°, but at 280°C decomposition becomes self-sustaining. Prolonged heating below these temperatures may also cause some decomposition, and the presence of impurities may decrease the decomposition temperatures.

Alkali

Bateman, T. L. *et al., Loss Prevention*, 1974, **8**, 117–122

Dinitrotoluene held at 210°C (rather than the 125°C intended) for 10 days in a 50 mm steam-heated transfer pipeline exploded. Subsequent tests showed decomposition at 210°C (producing a significant pressure rise) in 1 day, and presence of sodium carbonate (but not rust) reduced the induction period. A maximum handling temperature of 150°C was recommended, when the induction period was 32 days, or 14 days for alkali-contaminated material.

*See* NITROAROMATIC–ALKALI HAZARDS

Nitric acid

*See* NITRIC ACID, $HNO_3$: Nitroaromatics
*See other* POLYNITROARYL COMPOUNDS

Sodium oxide

*Rept. of Expl. Div.*, Ottawa, Dept. of Mines, 1929 (*Chem. Abs.*, 1929, **23**, $3809_6$)

Admixture of the two solids caused a rapid reaction and fire.

*See other* POLYNITROARYL COMPOUNDS

**5-PHENYLTETRAZOLE** **N=NNHN=CPh** $C_7H_6N_4$

Wedekind, E., *Ber.*, 1898, **31**, 948

Explodes on attempted distillation.

*See other* TETRAZOLES

***N*-AZIDOCARBONYLAZEPINE** $C_7H_6N_4O$

**$N(CON_3)[CH=CH]_2CH=CH$**

Chapman, L. E. *et al.*, *Chem. & Ind.*, 1966, 1266

Explodes on attempted distillation.

*See other* ACYL AZIDES

**BENZALDEHYDE** **PhCHO** $C_7H_6O$

Peroxyformic acid

*See* PEROXYFORMIC ACID, $CH_2O_3$: Organic materials

**BENZOIC ACID** **$PhCO_2H$** $C_7H_6O_2$

Oxygen

*See* OXYGEN (Gas), $O_2$: Benzoic acid

**PEROXYBENZOIC ACID** **$PhCO_2OH$** $C_7H_6O_3$

1. Baeyer, A. *et al.*, *Ber.*, 1900, **33**, 1577
2. Silbert, L. S. *et al.*, *Org. Synth.*, 1963, **43**, 95

Explodes weakly on heating [1]. A chloroform solution of peroxybenzoic acid exploded during evaporation of solvent. Evaporation should be avoided if possible, or conducted with full precautions [2].

*See other* PEROXYACIDS

**METHYL 4-BROMOBENZENEDIAZOATE** **$C_7H_7BrN_2O$**

**$BrC_6H_4N{=}NOMe$**

Bamberger, E., *Ber.*, 1895, **28**, 233

Explodes on heating.

*See other* ARENEDIAZOATES

**SODIUM *N*-CHLORO-*p*-TOLUENESULPHONAMIDE** **$C_7H_7ClNNaO_2S$**

**$MeC_6H_4SO_2NClNa$**

Klundt, I.L., *Chem. Eng. News*, 1977, **55** (49), 56

The anhydrous salt ('Chloramine-T') explodes at 175°C, but a small quantity exploded after storage in a bottle at ambient temperature.

Calcium carbonate,
Isonitriles

Sorbe, 1968, 137

Mixtures explode when warmed on the steam-bath in presence of calcium carbonate.

*See other* *N*-HALOGEN COMPOUNDS

***o*-TOLUENEDIAZONIUM PERCHLORATE** **$C_7H_7ClN_2O_4$**

**$CH_3C_6H_4N_2^+ ClO_4^-$**

Hofmann, K. A. *et al.*, *Ber.*, 1906, **39**, 3146

The solid salt is explosive even when wet.

*See other* DIAZONIUM PERCHLORATES

**4-CHLORO-2-METHYLPHENOL** **$Cl(Me)C_6H_3OH$** **$C_7H_7ClO$**

Sodium hydroxide

*ABCM Quart. Safety Summ.*, 1957, **28**, 39

A large quantity (700 kg) of the chlorophenol, left in contact with concentrated sodium hydroxide solution for 3 days, decomposed, reaching red heat and evolving fumes which ignited explosively. Although this could not be reproduced under laboratory conditions, it is believed that exothermic hydrolysis to the hydroquinone occurred, the high viscosity of the liquid preventing dissipation of heat.

*See other* HALOARYL COMPOUNDS

**TROPYLIUM PERCHLORATE** $C_7H_7ClO_4$

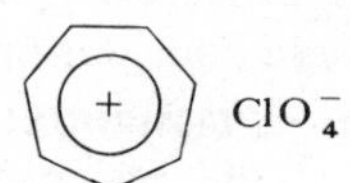


Ferrini, P. G. *et al., Angew. Chem.*, 1962, **74**, 488
Pressing the salt through a funnel with a glass rod caused a violent explosion.
*See other* NON-METAL PERCHLORATES

**TOLYLCOPPER** $CH_3C_6H_4Cu$ $C_7H_7Cu$

Camus, A. *et al., J. Organomet. Chem.*, 1968, **14**, 442–443
The solid *o*-, *m*- and *p*- isomers usually exploded strongly on exposure to oxygen at 0°C, or weakly on heating above 100°C *in vacuo.*
*See other* ARYLMETALS

† *o*-, *m*- or *p*-**FLUOROTOLUENE** $CH_3C_6H_4F$ $C_7H_7F$

*p*-**IODOTOLUENE** $MeC_6H_4I$ $C_7H_7I$

Sorbe, 1968, 66
It explodes above 200°C.
*See other* IODINE COMPOUNDS

*o*- or *m*-**TOLUENEDIAZONIUM IODIDE** $C_7H_7IN_2$

$CH_3C_6H_4N_2^+ I^-$

*See* *m*-TOLUENEDIAZONIUM SALTS, $C_7H_7N_2^+ Z^-$ : Potassium iodide
*See other* DIAZONIUM TRIIODIDES

*p*-**IODOSYLTOLUENE** $MeC_6H_4IO$ $C_7H_7IO$

Sorbe, 1968, 66
It explodes at above 175°C.
*See other* IODINE COMPOUNDS

*p*-**IODYLTOLUENE** $MeC_6H_4IO_2$ $C_7H_7IO_2$

Sorbe, 1968, 66
It explodes at above 200°C.
*See other* IODINE COMPOUNDS

*p*-**IODYLANISOLE** $MeOC_6H_4IO_2$ $C_7H_7IO_3$

Leymann, –, *Chem. Fabrik.*, 1928, 361

A 1.5 kg quantity of the dry material, 95–98% pure, exploded spontaneously as a door was opened.

*See other* IODINE COMPOUNDS

**p-TOLUENEDIAZONIUM TRIIODIDE** $CH_3C_6H_4N_2^+ I_3^-$ $C_7H_7I_3N_2$

*See* DIAZONIUM TRIIODIDES

**o- or p-METHOXYBENZENEDIAZONIUM TRIIODIDE** $C_7H_7I_3N_2O$

$MeOC_6H_4N_2^+ I_3^-$

*See* DIAZONIUM TRIIODIDES

**2(?)-VINYLPYRIDINE** $C_2H_3C_5H_4N$ $C_7H_7N$

Aminium nitrites

Karakuleva, G. I. *et al.*, Russ. Pat. 396 349, 1973 (*Chem. Abs.*, 1975, **82**, 58448q)

Nitrite salts of amines inhibit the sometimes explosive spontaneous polymerisation of '*N*-vinylpyridine' derivatives.

*See other* POLYMERISATION INCIDENTS

***mixo*-NITROTOLUENE** $MeC_6H_4NO_2$ $C_7H_7NO_2$

Staudinger, H., *Z. Angew. Chem.*, 1922, **35**, 657–659

The combined residues (300–400 kg) from several vacuum distillations at up to 180°C/40 mbar of a mixture of the *o*- and *p*- isomers exploded several hours after distillation had been completed. It appeared probable that admission of air to cool the hot residue initiated autoxidative heating, possibly involving nitrogen oxide and catalysis by iron, which eventually led to explosion of the residue (which probably contained poly-nitro compounds).

*See* POLYNITROARYL COMPOUNDS

***o*-NITROTOLUENE** $CH_3C_6H_4NO_2$ $C_7H_7NO_2$

Sodium hydroxide

*ABCM Quart. Safety Summ.*, 1945, **16**, 2

Crude *o*-nitrotoluene, containing some hydrochloric and acetic acids, was charged into a vacuum still with flake sodium hydroxide to effect neutralisation prior to distillation. An explosion occurred later. Similar treatments had been uneventfully used previously, using sodium carbonate or lime for neutralisation. It seems likely that the explosion involved formation and violent decomposition of the sodium salt of an *aci*-nitro species, possibly of quinonoid type.

*See* NITROAROMATIC–ALKALI HAZARDS
*aci*-NITROQUINONOID COMPOUNDS

## *p*-NITROTOLUENE $CH_3C_6H_4NO_2$ $C_7H_7NO_2$

*ABCM Quart. Safety Summ.*, 1938, **9**, 65; 1939, **10**, 2

The residue from large-scale vacuum distillation of *p*-nitrotoluene was left to cool after turning off heat and vacuum, and 8 h later a violent explosion occurred. This was variously attributed to: presence of an excessive proportion of dinitrotoluenes in the residue, which would decompose on prolonged heating; presence of traces of alkali in the crude material, which would form unstable *aci*-nitro salts during distillation; ingress of air to the hot residue on shutting off the vacuum supply, and subsequent accelerating oxidation of the residue. All these factors may have contributed, and several such incidents have occurred.

Sodium

Schmidt, J., *Ber.*, 1899, **32**, 2920

One of the products of treating *p*-nitrotoluene in ether with sodium is a dark brown sodium derivative which ignites in air.

*See also* NITROAROMATIC–ALKALI HAZARDS

Sulphuric acid

1. Hunt, J. K., *Chem. Eng. News*, 1949, **27**, 2504
2. *Chem. Abs.*, 1949, **43**, 8681a
3. McKeand, G., private comm., 1974
4. McKeand, G., *Chem. & Ind.*, 1974, 425

Solutions of *p*-nitrotoluene in 93% sulphuric acid decompose very violently if heated to 160°C. This happened on plant-scale when automatic temperature control failed [1]. The temperature was erroneously reported as 135°C [2]. The explosion temperature of 160°C for the mixture (presumably containing a high proportion of 4-nitrotoluene-2-sulphonic acid) is 22°C lower than that observed for onset of decomposition when *p*-nitrotoluene and 93% sulphuric acid are heated at a rate of 100°C/h [3]. Mixtures of *p*-nitrotoluene with 98% acid or 20% oleum begin to decompose at 180 and 190°C, respectively [3,4]. Thereafter, decomposition accelerates (190–224° in 14 min, 224–270° in 1.5 min) until eruption occurs with evolution of much gas [4].

*See* below

*See also* *m*-NITROBENZENESULPHONIC ACID, $C_6H_5NO_5S$: Sulphuric acid

Sulphuric acid,
Sulphur trioxide

Vervalin, C. H., *Hydrocarbon Process.*, 1976, **55**(9), 323

During sulphonation of *p*-nitrotoluene at 32°C with 24% oleum in a 2000 l vessel, a runaway decomposition reaction set in and ejected the contents as a carbonaceous mass. The thermal decomposition temperature was subsequently estimated as 52°C (but see above).

*See other* SULPHONATION INCIDENTS

Tetranitromethane

*See* TETRANITROMETHANE, $CN_4O_8$ : Aromatic nitrocompounds

*See other* *C*-NITRO COMPOUNDS

**BENZYL NITRATE** $PhCH_2ONO_2$ $C_7H_7NO_3$

Nef, J. U., *Ann.*, 1899, **309**, 172

Decomposed explosively at 180–200°C.

Lewis acids

*See* ALKYL NITRATES

**3-METHYL-4-NITROPHENOL** $MeNO_2C_6H_3OH$ $C_7H_7NO_3$

Dartnell, R. C. *et al.*, *Loss Prevention*, 1971, **5**, 53–56 (*MCA Case History No. 1649*)

A batch of 8 t of material accumulated in storage at 154°C during 72 h decomposed explosively. Stability tests showed that thermal instability developed when 3-methyl-4-nitrophenol is stored molten at above 140°C. Decomposition set in after 14 h at 185°C or 45 h at 165°C, with peak temperatures of 593 and 521°C, respectively. In a closed vessel, a peak pressure of 750 bar was attained, with a maximum rate of increase of 40 kbar/s. Thermal degradation involves an initially slow exothermic free-radical polymerisation process, followed by a rapid and violently exothermic decomposition at take-off.

*See other* *C*-NITRO COMPOUNDS

**4-METHYL-2-NITROPHENOL** $C_7H_7NO_3$

Me OH $NO_2$

Sodium hydroxide,
Sodium carbonate,
Methanol

*MCA Case History No. 701*

Failure to agitate a large-scale mixture of the reagents caused an eruption due to action of the exotherm when mixing did occur.

*See* NITROAROMATIC–ALKALI HAZARDS

## *o*-NITROANISOLE $MeOC_6H_4NO_2$ $C_7H_7NO_3$

Hydrogen

1. Carswell, T. S., *J. Amer. Chem. Soc.*, 1931, **53**, 2417
2. Adkins, H., ibid., 2808

Catalytic hydrogenation on 400 g scale at 34 bar under excessively vigorous conditions (250°C, 12% catalyst, no solvent) caused the thin autoclave (without bursting disc) to rupture [1]. Under more appropriate conditions the hydrogenation is safe [2].

Sodium hydroxide,
Zinc

Anon., *Angew. Chem. (Nachr.)*, 1955, **3**, 186

In preparation of 2,2′-dimethoxyazoxybenzene, solvent ethanol was distilled out of the mixture of *o*-nitroanisole, zinc and sodium hydroxide, before reaction was complete. The exothermic reaction continued unmoderated, and finally exploded.

*See* NITROAROMATIC–ALKALI HAZARDS

*See other* *C*-NITRO COMPOUNDS

## 2-METHYL-5-NITROBENZENESULPHONIC ACID $C_7H_7NO_5S$

$O_2N$ — benzene ring — $SO_3H$, Me

Preparative hazard

*See* *p*-NITROTOLUENE, $C_7H_7NO_2$ : Sulphuric acid
: Sulphuric acid, Sulphur trioxide

## *m*-TOLUENEDIAZONIUM SALTS $CH_3C_6H_4N_2^+ Z^-$ $C_7H_7N_2^+ Z^-$

Ammonium sulphide,
or Hydrogen sulphide

*See* DIAZONIUM SULPHIDES

Potassium *O*-ethyl dithiocarbonate

1. Tarbell, D. S. *et al., Org. Synth.*, 1955, Coll. Vol. 3, 810
2. Parham, W. E. *et al., Org. Synth*, 1967, **47**, 107

During interaction of the diazonium chloride and the *O*-ethyl dithiocarbonate ('xanthate') solutions, care must be taken to ensure that the intermediate diazonium dithiocarbonate decomposes to *m*-thiocresol as fast as it is formed [1]. This can be assured by presence

of a trace of nickel in the solution to effect immediate catalytic decomposition [2]. When the two solutions were mixed cold and then heated to effect decomposition, a violent explosion occurred [2].
*See other* DIAZONIUM SULPHIDES, etc.

Potassium iodide
Trumbull, E. R., *J. Chem. Educ.*, 1971, **48**, 640
Addition of potassium iodide solution to diazotised *m*-toluidine was accompanied on three occasions by an eruption of the beaker contents. This was not observed with the *o*- and *p*- isomers or other aniline derivatives.
*See* DIAZONIUM TRIIODIDES

**o- or p- TOLUENEDIAZONIUM SALTS** $CH_3C_6H_4N_2^+ Z^-$ $C_7H_7N_2^+ Z^-$

Ammonium sulphide,
or Hydrogen sulphide
*See* DIAZONIUM SULPHIDES

**BENZYL AZIDE** $PhCH_2N_3$ $C_7H_7N_3$

Sorbe, 1968, 138
A particularly heat-sensitive explosive oil.
*See other* ORGANIC AZIDES

**p-TOLUENESULPHINYL AZIDE** $MeC_6H_4SON_3$ $C_7H_7N_3OS$

*See* SULPHINYL AZIDES

**p-TOLUENESULPHONYL AZIDE** $C_7H_7N_3O_2S$

$CH_3C_6H_4SO_2N_3$

Rewicki, D. *et al., Angew. Chem. (Intern. Ed.)*, 1972, **11**, 44
Roush, W. R. *et al., Tetrahedron Lett.*, 1974, 1391–1392
The impure compound, present as a majority in the distillation residue from preparation of 1-diazoindene, will explode if the bath temperature exceeds 120°C.
A polymer-bound sulphonyl azide reagent has been described as safer in use than the title azide.
*See other* ACYL AZIDES

**METHYL 2-NITROBENZENEDIAZOATE** $C_7H_7N_3O_3$

$NO_2C_6H_4N{=}NOMe$

Bamberger, E., *Ber.*, 1895, **28**, 237

Explodes violently on heating, or on disturbing after 24 h confinement in a sealed tube at ambient temperature. The 4-nitro analogue is more stable.
*See other* ARENEDIAZOATES

**2-METHANESULPHONYL-4-NITROBENZENEDIAZONIUM HYDROGENSULPHATE** $C_7H_7N_3O_8S_2$

$CH_3SO_2NO_2C_6H_3N_2^+ HSO_4^-$

*See* DIAZONIUM SULPHATES

**TRIS(2,2,2-TRINITROETHYL)ORTHOFORMATE** $C_7H_7N_9O_{27}$

$CH[CH_2C(NO_3)_3]_3$

*See* TRINITROETHYL ORTHOESTERS

**SODIUM *p*-CRESOLATE** $CH_3C_6H_4ONa$ $C_7H_7NaO$

May and Baker Ltd, private comm., 1968

Sodium *p*-cresolate solution was dehydrated azeotropically with boiling chlorobenzene and the filtered solid was dried in an oven, where it soon ignited and glowed locally. This continued for half an hour after it was removed from the oven. A substituted potassium phenolate, prepared differently, also ignited on heating. Finely divided and moist alkali phenolates may be prone to vigorous oxidation when heated in air.

**† BICYCLO-[2.2.1]-2,5-HEPTADIENE** $C_7H_8$

$CH_2CHCH{=}CHCHCH{=}CH$

**† 1,3,5-CYCLOHEPTATRIENE** $CH{=}CH[CH{=}CH]_2CH_2$ $C_7H_8$

**† TOLUENE** $CH_3C_6H_5$ $C_7H_8$

*MCA SD-63,* 1956

Bromine trifluoride

*See* BROMINE TRIFLUORIDE, $BrF_3$: Solvents

1,3-Dichloro-5,5-dimethyl-2,4-imidazolidindione

*See* 1,3-DICHLORO-5,5-DIMETHYL-2,4-IMIDAZOLIDINDIONE, $C_5H_6Cl_2N_2O_2$: Toluene

Dinitrogen tetraoxide
*See* DINITROGEN TETRAOXIDE, $N_2O_4$: Hydrocarbons

Nitric acid
*See* NITRIC ACID, $HNO_3$: Hydrocarbons

Tetranitromethane
*See* TETRANITROMETHANE, $CN_4O_8$: Hydrocarbons

Uranium hexafluoride
*See* URANIUM HEXAFLUORIDE, $F_6U$: Aromatic hydrocarbons

**2-CHLORO-5-METHYLANILINE** **$Cl(Me)C_6H_3NH_2$** **$C_7H_8ClN$**

Preparative hazard
*See* 2-CHLORO-5-METHYLPHENYLHYDROXYLAMINE, $C_7H_8ClNO$

**4-CHLORO-2-METHYLANILINE** **$Cl(Me)C_6H_3NH_2$** **$C_7H_8ClN$**
*See* 4-CHLORO-2-METHYLBENZENEDIAZONIUM SALTS, $C_7H_6ClN_2^+ Z^-$

Copper(II) chloride
Kotogori, T. *et al., J. Haz. Mat.*, 1976, **1**, 252–262
At the end of vacuum distillation of 4-chloro-2-methylaniline, the 1300 l still burst owing to internal gas pressure from an unexpected decomposition reaction. Subsequent investigation showed that copper(I) oxide present in the still-charge would react with hydrogen chloride and leaking air to produce copper(II) chloride. The latter is an effective catalyst for decomposition of the halo compound at 239°C to produce hydrogen chloride. The physical conditions of the installation were compatible with this mechanism, and enough chloroaniline was present to exceed the bursting pressure of the vessel by a factor of 3. It is concluded that halogenated aromatic amines should be distilled under alkaline conditions in absence of air.
*See* CATALYSIS BY IMPURITIES

**2-CHLORO-5-METHYLPHENYLHYDROXYLAMINE** **$C_7H_8ClNO$**
**$Cl(Me)C_6H_3NHOH$**

Rondevstedt, C. S., *Chem. Eng. News*, 1977, **55**(27), 38; *Synthesis*, 1977, 851–852
In an attempt to reduce 2-chloro-5-methylnitrobenzene to the aniline by treatment with excess hydrazine and Pd/C catalyst, the hydroxylamine was unexpectedly produced as major product. During isolation of the product, after removal of solvent it was heated to 120°C under vacuum

and exploded fairly violently. Many arylhydroxylamines decompose violently when heated above 90–100°C, especially in presence of acids. GLC is not suitable as an analytical diagnostic for aryl hydroxylamines because of this thermal decomposition.

**METHYL BENZENEDIAZOATE** $C_6H_5N{=}NOMe$ $C_7H_8N_2O$

Bamberger, E., *Ber.*, 1895, **28**, 228

Explodes on heating or after 1 h storage in a sealed tube at ambient temperature.

*See other* ARENEDIAZOATES

***N*-METHYL-*p*-NITROANILINE** $O_2NC_6H_4NHMe$ $C_7H_8N_2O_2$

2,4-Dithia-1,3-dioxane-2,2,4,4-tetraoxide ('Carbyl sulphate')

*See* 2,4-DITHIA-1,3-DIOXANE-2,2,4,4-TETRAOXIDE ('CARBYL SULPHATE'), $C_2H_4O_6S_2$ : *N*-Methyl-*p*-nitroaniline

**2-METHOXY-5-NITROANILINE** $C_7H_8N_2O_3$

$MeO(O_2N)C_6H_3NH_2$

Preparative hazard

*See* 2-METHOXYANILINIUM NITRATE, $C_7H_{10}N_2O_4$ : Sulphuric acid

**3-PHENYL-1-TETRAZOLYL-1-TETRAZENE** $C_7H_8N_8$

$N{=}NNHN{=}CN{=}NN(Ph)NH_2$ (ring bond from first N to C)

Thiele, J., *Ann.*, 1892, **270**, 60

It explodes on warming, as do *C*-substituted homologues.

*See other* HIGH-NITROGEN COMPOUNDS

**BENZYL ALCOHOL** $PhCH_2OH$ $C_7H_8O$

Sulphuric acid

*See* SULPHURIC ACID, $H_2O_4S$: Benzyl alcohol

***p*-TOLUENESULPHONIC ACID** $MeC_6H_4SO_3H$ $C_7H_8O_3S$

Acetic anhydride,
Water

*See* ACETIC ANHYDRIDE, $C_4H_6O_3$ : *p*-Toluenesulphonic acid, etc.

**$m$-THIOCRESOL** **$CH_3C_6H_4SH$** **$C_7H_8S$**

Preparative hazard

*See* $m$-TOLUENEDIAZONIUM SALTS, $C_7H_7N_2^+ Z^-$. Potassium $O$-ethyl dithiocarbonate

**METHYLPHENYLCHLOROSILANE** **MePhSiHCl** **$C_7H_9ClSi$**

4-Bromobutene,
Chloroplatinic acid

Brook, A. G. *et al., J. Organomet. Chem.*, 1975, **87**, 265
Interaction of methylphenylchlorosilane and 4-bromobutene, catalysed by chloroplatinic acid at 100°C, is extremely exothermic (to 165°C) and is accompanied by vigorous frothing.

*See other* ALKYLNON-METAL HALIDES

**BENZYLAMINE** **$PhCH_2NH_2$** **$C_7H_9N$**

$N$-Chlorosuccinimide

*See* $N$-HALOIMIDES: Alcohols, etc.

**$o$-TOLUIDINE** **$MeC_6H_4NH_2$** **$C_7H_9N$**

*See* $o$-TOLUENEDIAZONIUM SALTS, $C_7H_7N_2^+ Z^-$

Nitric acid

*See* NITRIC ACID, $HNO_3$: Aromatic amines

**$m$-TOLUIDINE** **$MeC_6H_4NH_2$** **$C_7H_9N$**

*See* $m$-TOLUENEDIAZONIUM SALTS, $C_7H_7N_2^+ Z^-$

**$p$-TOLUIDINE** **$MeC_6H_4NH_2$** **$C_7H_9N$**

*See* $p$-TOLUENEDIAZONIUM SALTS, $C_7H_7N_2^+ Z^-$

**2,6-DIMETHYLPYRIDINE-1-OXIDE** **$C_7H_9NO$**

O←N=CMeCH=CHCH=CMe (ring closure from N to the final CMe)

Phosphoryl chloride

*See* PHOSPHORYL CHLORIDE, $Cl_3OP$: 2,6-Dimethylpyridine-1-oxide

## 4(1-HYDROXYETHYL)PYRIDINE *N*-OXIDE $C_7H_9NO_2$

**$MeCHOHC_5H_4N(O)$**

Taylor, R., *J. Chem. Soc., Perkin Trans. 2*, 1975, 279

The compound exploded during vacuum distillation, possibly owing to the presence of traces of peracetic acid.

*See* *N*-OXIDES

## *O-p-*TOLUENESULPHONYLHYDROXYLAMINE $C_7H_9NO_3S$

**$MeC_6H_4SO_2ONH_2$**

Carpino, L. A., *J. Amer. Chem. Soc.*, 1960, **82**, 3134

It deflagrated spontaneously on attempted isolation and drying at ambient temperature.

*See also* *O*-MESITYLENESULPHONYLHYDROXYLAMINE, $C_9H_{13}NO_3S$

## *p*-TOLUENESULPHONYLHYDRAZIDE $C_7H_{10}N_2O_2S$

**$MeC_6H_4SO_2NHNH_2$**

Hansell, D. P. *et al., Chem. & Ind.*, 1975, 464

While drying in a tray drier at 60°C, a batch decomposed fairly exothermically but without fire. Autodecomposition seems to have been involved.

*See also* *O-p-*TOLUENESULPHONYLHYDROXYLAMINE, $C_7H_9NO_3S$

## 2-METHOXYANILINIUM NITRATE $C_7H_{10}N_2O_4$

**$MeOC_6H_4N^+H_3\ NO_3^-$**

Vervalin, C. H., *Hydrocarbon Process.*, 1976, **55**(9), 321

Use of a mechanical screw feeder to charge a reactor with the *o*-anisidine salt led to ignition of the latter. It was known that the salt would exothermically decompose above 140°C, but later investigation showed that lower-quality material could develop an exotherm above 46°C under certain conditions.

Sulphuric acid

Vervalin, C. H., *Hydrocarbon Process.*, 1976, **55**(9), 321

In a process for the preparation of 2-methoxy-5-nitroaniline the *o*-anisidine salt was added to stirred sulphuric acid. An accidental deficiency of the latter prevented proper mixing and dissipation of the heat of solution, and local decomposition spread through the whole contents of the 2000 l vessel, attaining red heat.

*See other* OXOSALTS OF NITROGENOUS BASES

## METHYL 3-METHOXYCARBONYLAZOCROTONATE $C_7H_{10}N_2O_4$

**$MeO_2CN{=}NC(Me){=}CHCO_2Me$**

Sommer, S., *Tetrahedron Lett.*, 1977, 117

Attempted high-vacuum fractional distillation of an isomeric mixture (90% *trans*, 10% *cis*) led to explosive decomposition. The 'diene' polymerises if stored at 20°C.

*See other* AZO COMPOUNDS

**3-METHYL-4-NITRO-1-BUTEN-3-YL ACETATE**
**3-METHYL-4-NITRO-2-BUTEN-1-YL ACETATE** $C_7H_{11}NO_4$

$CH_3CO_2C(C_2H_3)MeCH_2NO_2$; $CH_3CO_2CH_2CH{=}CMeCH_2NO_2$

Wehrli, P. A. *et al., J. Org. Chem.*, 1977, **42**, 2940

The isomeric nitroesters, produced by nitroacetoxylation of isoprene, are of limited thermal stability and it is recommended that neither be heated above 100°C either neat or in solution. Vacuum distillation of the nitroesters should be limited to 1g portions, as decomposition 'fume-offs' have been observed.

*See* 1-NITRO-3-BUTENE, $C_4H_7NO_2$

*See other* *C*-NITRO COMPOUNDS

**† CYCLOHEPTENE** $\overline{CH{=}CH[CH_2]_4CH_2}$ $C_7H_{12}$

**† 1-HEPTYNE** $HC{\equiv}CC_5H_{11}$ $C_7H_{12}$

**† 3-HEPTYNE** $EtC{\equiv}CPr$ $C_7H_{12}$

**† 4-METHYLCYCLOHEXENE** $C_7H_{12}$

$\overline{CH{=}CHCH_2CMeCH_2CH_2}$

**2-HEPTYN-1-OL** $HOCH_2C{\equiv}CBu$ $C_7H_{12}O$

Brandsma, 1971, 69

Dilution of the alcohol with white mineral oil before vacuum distillation is recommended to avoid the possibility of explosion of the undiluted distillation residue.

*See other* ACETYLENIC COMPOUNDS

**3-METHYLCYCLOHEXANONE** $\overline{OC[CH_2]_3CHMeCH_2}$ $C_7H_{12}O$

Hydrogen peroxide,
Nitric acid

*See* HYDROGEN PEROXIDE, $H_2O_2$: Ketones, etc.

**4-METHYLCYCLOHEXANONE** $C_7H_{12}O$

$\overline{OCC_2H_4CHMeCH_2CH_2}$

Nitric acid

*See* NITRIC ACID, $HNO_3$ : 4-Methylcyclohexanone

**1-ETHOXY-3-METHYL-1-BUTYN-3-OL** $EtOC{\equiv}CCMe_2OH$ $C_7H_{12}O_2$

Brandsma, 1971, 78

Traces of acids adhering to glassware are sufficient to induce explosive decomposition of the alcohol during distillation, and must be neutralised by pre-treatment with ammonia gas. Low pressures and temperatures are essential during distillation. Explosions during distillation using a water pump and bath temperatures above 115°C were frequent.

*See* ETHOXYETHYNYL ALCOHOLS
*See other* ACETYLENIC COMPOUNDS

**1,2-DIMETHYLCYCLOPENTENE OZONIDE** $C_7H_{12}O_3$

$CH_2C_2H_4C(Me)C(Me)O{-}O^+{-}O^-$

Criegee, R. *et al., Ber.*, 1953, **86**, 3

Though the ozonide appeared stable to small-scale high-vacuum distillation, the distillation residue exploded violently at 130°C.

*See other* OZONIDES

***N*-CYANO-2-BROMOETHYLBUTYLAMINE** $C_7H_{13}BrN_2$

$BuN(CN)C_2H_4Br$

*BCISC Quart. Safety Summ.*, 1964, **35**, 23

A small sample heated to 160°C decomposed exothermically, reaching 250°C. It had been previously distilled at 95°C/0.65 mbar without decomposition.

*See also* *N*-CYANO-2-BROMOETHYLCYCLOHEXYLAMINE, $C_9H_{15}BrN_2$
*See other* CYANO COMPOUNDS

**† CYCLOHEPTANE** $\overline{CH_2[CH_2]_5CH_2}$ $C_7H_{14}$

**† ETHYLCYCLOPENTANE** $EtCH[CH_2]_3CH_2$ $C_7H_{14}$

**† 1-HEPTENE** $H_2C{=}CHC_5H_{11}$ $C_7H_{14}$

† 2-HEPTENE $MeCH{=}CHBu$ $C_7H_{14}$

† 3-HEPTENE $EtCH{=}CHPr$ $C_7H_{14}$

† METHYLCYCLOHEXANE $Me\overline{CH[CH_2]_4CH_2}$ $C_7H_{14}$

† 2,3,3-TRIMETHYLBUTENE $H_2C{=}CMeCMe_3$ $C_7H_{14}$

† 2,4-DIMETHYL-3-PENTANONE $Me_2CHCOCHMe_2$ $C_7H_{14}O$

† ISOPENTYL ACETATE $C_7H_{14}O_2$

$CH_3CO_2C_2H_4CHMe_2$

† PENTYL ACETATE $CH_3CO_2C_5H_{11}$ $C_7H_{14}O_2$

† 2-PENTYL ACETATE $CH_3CO_2CH(Me)Pr$ $C_7H_{14}O_2$

† 3,3-DIETHOXYPROPENE $H_2C{=}CHCH(OEt)_2$ $C_7H_{14}O_2$

† ETHYL ISOVALERATE $Me_2CHCH_2CO_2Et$ $C_7H_{14}O_2$

† ISOBUTYL PROPIONATE $EtCO_2CH_2CH(Me)_2$ $C_7H_{14}O_2$

† ISOPROPYL BUTYRATE $PrCO_2CH(Me)_2$ $C_7H_{14}O_2$

† ISOPROPYL ISOBUTYRATE $Me_2CHCO_2CHMe_2$ $C_7H_{14}O_2$

1,4-BUTANEDIOL MONO-2,3-EPOXYPROPYL ETHER $C_7H_{14}O_3$

$HOC_4H_8OCH_2\overline{CHOCH_2}$

Trichloroethylene

*See* TRICHLOROETHYLENE, $C_2HCl_3$: Epoxides

**† 2,6-DIMETHYLPIPERIDINE** $C_7H_{15}N$

$\overline{\text{MeCHNHCHMeCH}_2\text{CH}_2}$

**† 1-ETHYLPIPERIDINE** $\overline{CH_2NEt[CH_2]_3CH_2}$ $C_7H_{15}N$

**† 2-ETHYLPIPERIDINE** $\overline{EtCHNH[CH_2]_3CH_2}$ $C_7H_{15}N$

**† 2,3-DIMETHYLPENTANE** $Me_2CHCHMeEt$ $C_7H_{16}$

**† 2,4-DIMETHYLPENTANE** $Me_2CHCH_2CHMe_2$ $C_7H_{16}$

**† HEPTANE** $CH_3[CH_2]_5CH_3$ $C_7H_{16}$

**† 2-METHYLHEXANE** $Me_2CHBu$ $C_7H_{16}$

**† 3-METHYLHEXANE** $EtCHMePr$ $C_7H_{16}$

**† 2,2,3-TRIMETHYLBUTANE** $Me_3CCHMe_2$ $C_7H_{16}$

**3-DIETHYLAMINOPROPYLAMINE** $Et_2N[CH_2]_3NH_2$ $C_7H_{18}N_2$

Cellulose nitrate
*See* CELLULOSE NITRATE: Amines

**TETRA(CHLOROETHYNYL)SILANE** $(ClC{\equiv}C)_4Si$ $C_8Cl_4Si$

Viehe, H. G., *Chem. Ber.*, 1959, **92**, 3075
Though thermally stable, it exploded violently on grinding (with potassium bromide) or on impact.
*See other* HALOACETYLENE DERIVATIVES

**POTASSIUM OCTACYANODICOBALTATE(8–)** $C_8Co_2K_8N_8$

$K_8[Co_2(CN)_8]$

Bailar, 1973, Vol. 4, 175
A very unstable, pyrophoric material.

*See other* CYANO COMPOUNDS
PYROPHORIC MATERIALS

## OCTACARBONYLDICOBALT

$C_8Co_2O_8$

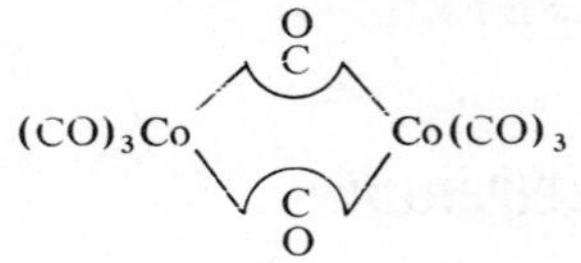

$(CO)_3Co(\mu\text{-}CO)_2Co(CO)_3$

Wender, I. *et al., Inorg. Synth.*, 1957, **5**, 191
When the carbonyl is prepared (by rapid cooling) as a fine powder, it decomposes in air to give pyrophoric dodecacarbonyltetracobalt.
*See other* CARBONYLMETALS

## CAESIUM GRAPHITE

$CsC_8$ $C_8Cs$

*See* CARBON, C: Alkali metals

## 3-BROMO-2,7-DINITRO-5-BENZO[*b*]-THIOPHENEDIAZONIUM-4-OLATE

$C_8HBrN_4O_5S$

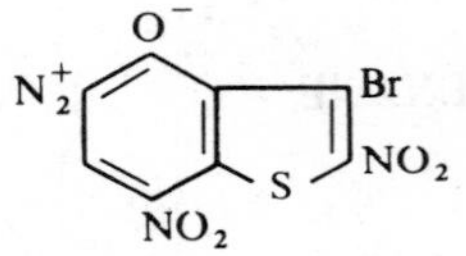

Brown, I. *et al., Chem. & Ind.* 1962, 982
It is an explosive compound.
*See* NITRIC ACID, $HNO_3$: Benzo[*b*]thiophene derivatives

## *N*-BROMO-3-NITROPHTHALIMIDE

$C_8H_3BrN_2O_4$

$NO_2C_6H_4CONBrCO$

Tetrahydrofurfuryl alcohol
*See* *N*-HALOIMIDES: Alcohols

## SILVER ISOPHTHALATE

$C_8H_4Ag_2O_4$

$1,3\text{-}(AgO_2C)_2C_6H_4$

Fields, E. K., *J. Org. Chem.*, 1976, **41**, 918–919
Decomposition of the salt at 375°C under nitrogen, with subsequent cooling under hydrogen, gives a black carbon-like polymer containing metallic silver which ignites at 25°C on exposure to air.

*See other* PYROPHORIC MATERIALS
SILVER COMPOUNDS

**ISOPHTHALOYL DICHLORIDE** **1,3-$C_6H_4(COCl)_2$** **$C_8H_4Cl_2O_2$**

Preparative hazard

*See* 1,3-BIS(TRICHLOROMETHYL)BENZENE, $C_8H_4Cl_6$ : Oxidants

Methanol

Morrell, S. H., private comm., 1968

Violent reaction occurred between isophthaloyl dichloride and methanol when they were accidentally added in succession to the same waste solvent bottle.

**2,4-HEXADIYNYLENE BISCHLOROFORMATE** **$C_8H_4Cl_2O_4$**

**$(ClCO_2CH_2C{\equiv}C)_2$**

Driedger, P. E. *et al., Chem. Eng. News*, 1972, **50**(12), 51

One sample exploded with extreme violence at 15°C/0.2 mbar, while another smaller sample had been distilled uneventfully at 114–115°C/0.2 mbar.

*See other* HALOACETYLENE DERIVATIVES

**1,3-BIS(TRICHLOROMETHYL)BENZENE** **$C_8H_4Cl_6$**

**$(CCl_3)_2C_6H_4$**

Oxidants

Rondestvedt, C. S., *J. Org. Chem.*, 1976, **41**, 3574–3577

Heating the bis(trichloromethyl)benzene with potassium nitrate, selenium dioxide or sodium chlorate to effect conversion to the bis-acyl chloride led to eruptions at higher temperatures, and was too dangerous to pursue.

**TETRAETHYNYLGERMANIUM** **$(HC{\equiv}C)_4Ge$** **$C_8H_4Ge$**

1. Chokiewicz, W. *et al., Compt. Rend.*, 1960, **250**, 866
2. Shikkiev, I. A. *et al., Dokl. Akad. Nauk SSSR*, 1961, **139**, 1138 (Eng. transl. 830)

It explodes on rapid heating [1] or friction [2].

*See other* METAL ACETYLIDES

## POTASSIUM TETRAKIS(ETHYNYL)NICCOLATE(2–) $C_8H_4K_2Ni$

$K_2[Ni(C{\equiv}CH)_4]$

Vigorously pyrophoric in air, it may explode in contact with drops of water.

*See* COMPLEX ACETYLIDES

## POTASSIUM TETRAKIS(ETHYNYL)NICCOLATE(4–) $C_8H_4K_4Ni$

$K_4[Ni(C{\equiv}CH)_4]$

The diammoniate (and that of the sodium salt) explodes on exposure to friction, impact or flame.

*See* COMPLEX ACETYLIDES

## PHTHALOYL DIAZIDE $C_8H_4N_6O_2$

$CON_3$ $CON_3$ CO O C $(N_3)_2$

Lindemann, H. *et al.*, *Ann.*, 1928, **464**, 237

The *symm* diazide is extremely explosive, while the *asymm* isomer is less so, melting at 56°C, but exploding on rapid heating.

*See other* ACYL AZIDES

## PHTHALIC ANHYDRIDE $\overline{C_6H_4COOCO}$ $C_8H_4O_3$

Preparative hazards

1. Kratochvil, V., *Chem. Abs.*, 1974, **80**, 3215m
2. English translation by D. S. Rosenberg, 10.2.69 of German translation by S. Vedrilla (Chemiebau) of 'Low Temperature Pyrophoric Compounds in Process for Preparation of Phthalic Anhydride', a Japanese paper in *Anzen Kogyo Kogaku*, 1968
3. Schwab, R. F. *et al.*, *Chem. Eng. Progr.*, 1970, **66**(9), 49–53

Investigation of an explosion in a phthalic anhydride plant showed that naphthoquinones (by-products from naphthalene oxidation) reacted with iron(III) salts of phthalic, maleic or other acids (corrosion products) to form labile materials. The latter were found to undergo exothermic oxidation and self-ignition at *ca.* 200°C. Process conditions to minimise hazards are discussed, with 27 references [1]. Fires and explosions in the condenser sections of phthalic anhydride plants were traced to the presence of deposits of thermally unstable basic iron salts of maleic or phthalic acids and iron sulphides derived from sulphur in the naphthalene feedstock. A small-scale laboratory test procedure was developed to measure the exotherm point and maximum temperature rise in various mixtures of materials found to be present in the condensing section.

A basic iron(III) maleate hydrate showed an exotherm point of 165°C and temperature rise of 369°C; changed by pretreatment with hydrogen sulphide to 164 and 392°C, respectively. The corresponding figures for basic iron (III) phthalate were 236 and 477°C, both unchanged by pre-sulphiding treatment. The part played by acid-corrosion of the mild steel condensers in formation of the unstable pyrophoric compounds was also investigated.

It was concluded that the origins of the many mishaps which have occurred soon after plant start-up were explicable in terms of these experimental findings [2]. Other hazards in phthalic anhydride production units have also been discussed [3].

*See* PYROPHORIC IRON–SULPHUR COMPOUNDS

Copper oxide

*See* COPPER OXIDE, CuO: Phthalic acid, etc

Nitric acid

*See* NITRIC ACID, $HNO_3$: Phthalic anhydride

Sodium nitrite

*See* SODIUM NITRITE, $NNaO_2$: Phthalic acid, etc.

## PHTHALOYL PEROXIDE $(COC_6H_4COOO)_n$ $C_8H_4O_4$

Jones, M. *et al.*, *J. Org. Chem.*, 1971, **36**, 1536

Detonable by impact or by melting at 123°C. It is probably polymeric.

*See other* DIACYL PEROXIDES

## TETRAETHYNYLTIN $(HC{\equiv}C)_4Sn$ $C_8H_4Sn$

Jenkner, H., Ger. Pat. 115 736, 1961

It explodes on rapid heating.

*See other* METAL ACETYLIDES

## 2,4-DINITROPHENYLACETYL CHLORIDE $C_8H_5ClN_2O_5$

$(NO_2)_2C_6H_3CH_2COCl$

Bonner, T. G., *J. R. Inst. Chem.*, 1957, **81**, 596

Explosive decomposition occurred during attempted vacuum distillation, attributed to either presence of some trinitro compound in the unpurified dinitrophenylacetic acid used, or to the known instability of *o*-nitro acid chlorides. A previous publication (ibid., 407) erroneously described the decomposition of 2,4-dinitrobenzoyl chloride.

*See other* NITRO-ACYL HALIDES

## 3-KALIOBENZOCYCLOBUTENE $C_8H_5K$

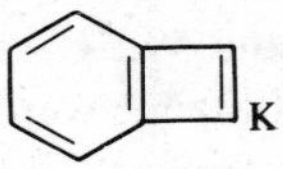

Perchloryl fluoride

*See* PERCHLORYL FLUORIDE, $ClFO_3$: Benzocyclobutene, etc.

## NITROPHENYLACETYLENE $O_2NC_6H_4C{\equiv}CH$ $C_8H_5NO_2$

Sorbe, 1968, 152

The compound (isomer not stated) and its metal derivatives are explosive.

*See other* ACETYLENIC COMPOUNDS
METAL ACETYLIDES

## 3-NITROPHTHALIC ACID $NO_2C_6H_3(CO_2H)_2$ $C_8H_5NO_6$

Preparative hazard

*See* NITRIC ACID, $HNO_3$: Phthalic anhydride, etc.

## NITROTEREPHTHALIC ACID $C_8H_5NO_6$

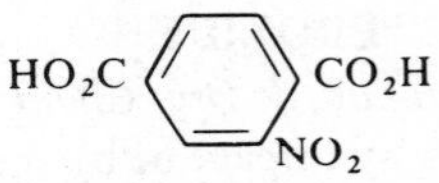

$HO_2C$, $CO_2H$, $NO_2$

Preparative hazard

*See* NITRIC ACID, $HNO_3$: Sulphuric acid, Terephthalic acid

## 3-SODIO-1-(5-NITRO-2-FURFURYLIDENEAMINO)-IMIDAZOLIDIN-2,4-DIONE $C_8H_5N_4NaO_5$

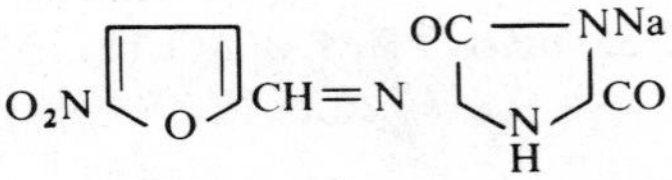

OC —NNa, $O_2N$, O, CH=N, N, H, CO

Boros, L. *et al.*, Ger. Offen., 2 328 927, 1974

The crystalline *N*-sodium salt is explosive.

*See other,* *N*-METAL DERIVATIVES

## SODIUM PHENYLACETYLIDE $NaC{\equiv}CPh$ $C_8H_5Na$

Nef, J. U., *Ann.,* 1899, **308**, 264

The ether-moist powder ignites in air.

*See* ALKALI-METAL DERIVATIVES OF HYDROCARBONS

*See other* METAL ACETYLIDES

## 3-METHYL-2-NITROBENZOYL CHLORIDE $C_8H_6ClNO_3$

**$Me(NO_2)C_6H_3COCl$**

Dahlbom, R., *Acta Chem. Scand.*, 1960, **14**, 2049

Attempted distillation caused an explosion; other *o*-nitro acid chlorides behave similarly.

*See other* NITRO-ACYL HALIDES

## *o*-NITROPHENYLACETYL CHLORIDE $C_8H_6ClNO_3$

**$NO_2C_6H_4CH_2COCl$**

Hayao, S., *Chem. Eng. News*, 1964, **42**(13), 39

Distillation of solvent chloroform from the preparation caused the residue to explode violently on two occasions. Previous publications had recommended use of solutions of the chloride, in preference to isolated material. Other *o*-nitro acid chlorides behave similarly.

*See other* NITRO-ACYL HALIDES

## 4-METHOXY-3-NITROBENZOYL CHLORIDE $C_8H_6ClNO_4$

**$MeO(NO_2)C_6H_3COCl$**

Preparative hazard

Grewer, T. *et al.*, *Chem. Ing. Tech.*, 1977, **49**, 562–563

$$MeO(NO_2)C_6H_3COCl + HCl \rightarrow HO(NO_2)C_6H_3COCl + MeCl \longrightarrow [O(NO_2)C_6H_3CO]_n + nHCl$$

Preparation of the acid chloride by an established procedure involved heating the acid with sulphinyl chloride, finally at 100°C for 4 h, when evolution of hydrogen chloride and sulphur dioxide was complete (and the vessel appears then to have been isolated from the gas absorption system). An apparently normally completed batch led to a double explosion with rupture of the process vessel. Subsequent thermoanalytical investigation revealed that the nitroacyl halide is demethylated by hydrogen chloride produced in the primary reaction, methyl chloride being formed as well as the hydroxyacyl chloride. The latter condenses to form a poly(nitrophenylcarboxylic ester) and hydrogen chloride, which can lead to further demethylation. This sequence of exothermic reactions leads to accelerating decomposition, which becomes violent at about 350–380°C, with generation of pressure up to 40 bar by the volatiles produced. The double explosion was attributed to ignition of chloromethane, followed by that of the decomposing poly(nitro ester). Preventive measures are discussed.

*See other* NITROACYL HALIDES

**4-CYANO-3-NITROTOLUENE** $C_8H_6N_2O_2$

$CN(NO_2)C_6H_3CH_3$

Preparative hazard

*See* 4-CHLORO-3-NITROTOLUENE, $C_7H_6ClNO_2$ : Copper(I) cyanide, Pyridine

**PHENOXYACETYLENE** $PhOC{\equiv}CH$ $C_8H_6O$

Jacobs, T. L. *et al., J. Amer. Chem. Soc.*, 1942, **64**, 223

Small samples rapidly heated in sealed tubes to around 100°C exploded.

*See other* ACETYLENIC COMPOUNDS

† **PHENYLGLYOXAL** PhCOCHO $C_8H_6O_2$

**PHTHALIC ACID** $C_6H_4(CO_2H)_2$ $C_8H_6O_4$

Sodium nitrite

*See* SODIUM NITRITE, $NNaO_2$ : Phthalic acid

**TEREPHTHALIC ACID** $C_6H_4(CO_2H)_2$ $C_8H_6O_4$

Preparative hazard

*See* NITRIC ACID, $HNO_3$ : Hydrocarbons (reference 6)

**DIPEROXYTEREPHTHALIC ACID** $HOO_2CC_6H_4CO_2OH$ $C_8H_6O_6$

Baeyer, A. *et al., Ber.*, 1901, **34**, 762

It explodes when heated or struck.

*See other* PEROXY ACIDS

*p*-**(BROMOMETHYL)BENZOIC ACID** $BrCH_2C_6H_4CO_2H$ $C_8H_7BrO_2$

Preparative hazard

*See* *N*-BROMOSUCCINIMIDE, $C_4H_4BrNO_2$ : Dibenzoyl peroxide, etc.

**BENZYL CHLOROFORMATE** $ClCO_2CH_2Ph$ $C_8H_7ClO_2$

Water

*See* ARYL CHLOROFORMATES: Water

**COPPER 1,3,5-OCTATRIEN-7-YNIDE** $H[CH{=}CH]_3C{\equiv}CCu$ $C_8H_7Cu$

Georgieff, K. K. *et al., J. Amer. Chem. Soc.*, 1954, **76**, 5495

It deflagrates on heating in air.

*See other* METAL ACETYLIDES

**1-FLUORO-1,1-DINITRO-2-PHENYLETHANE** $C_8H_7FN_2O_4$

$FC(NO_2)_2CH_2Ph$

*See* FLUORODINITROMETHYL COMPOUNDS (reference 4)

**PHENYLACETONITRILE** $PhCH_2CN$ $C_8H_7N$

Sodium hypochlorite

*See* SODIUM HYPOCHLORITE, ClNaO : Phenylacetonitrile

*See other* CYANO COMPOUNDS

***o*-NITROACETOPHENONE** $NO_2C_6H_4Ac$ $C_8H_7NO_3$

Potassium methylselenide

Leitem, L. *et al., Compt. Rend., C*, 1974, **278**, 276

Interaction in dimethylformamide was explosive.

**2-METHYL-5-NITROBENZIMIDAZOLE** $C_8H_7N_3O_2$

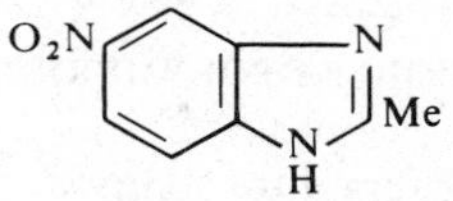

Safe preparation

*See* NITRIC ACID, $HNO_3$ : 2-Methylbenzimidazole, etc.

**3,5-DINITRO-2-TOLUAMIDE** $C_8H_7N_3O_5$

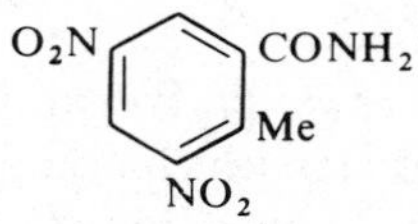

Anon., *Chem. Age*, 1976, **113**(2974), 2

A batch stored at 130°C in a pressurised vessel for 24 h under adiabatic conditions exploded. Accelerated rate calorimetry tests showed that these conditions eventually lead to total decomposition.

*See other* POLYNITROARYL COMPOUNDS

**† 1,3,5,7-CYCLOOCTATETRAENE** **$C_8H_8$**

$\overline{\text{CH=CH(CH=CH)}_2\text{CH=CH}}$

**1,5,7-OCTATRIEN-3-YNE** **$H[CH{=}CH]_3C{\equiv}CH$** **$C_8H_8$**

1. Nieuwland, J. A. *et al., J. Amer. Chem. Soc.*, 1931, **53**, 4201
2. Dolgopolskii, I. M. *et al., Chem. Abs.*, 1948, **42**, 4517e
3. Georgieff, K. K. *et al., J. Amer. Chem. Soc.*, 1954, **76**, 5495

This tetramer of acetylene decomposes violently on distillation at 156°C, but not at reduced pressure [1]. It polymerises on standing to a solid detonable by shock [2]. Explosions occurred during attempted analytical combustion [3].

*See other* ACETYLENIC COMPOUNDS
*See related* DIENES

**STYRENE** **$PhCH{=}CH_2$** **$C_8H_8$**

1. Unpublished information
2. *MCA SD-37,* 1971
3. Harmon, 1974, 2.17

The autocatalytic exothermic polymerisation reaction exhibited by styrene was involved in a plant-scale incident where accidental heating caused violent ejection of liquid and vapour from a storage tank [1]. Polymerisation becomes self-sustaining above 65°C [2].

The monomer has been involved in several plant-scale explosions, and must be stored at below 32°C and for less than 3 months [3].

*See* VIOLENT POLYMERISATION

Alkali metal–graphite compounds

Williams, N. E., private comm., 1968

Interlaminar compounds of sodium or potassium in graphite will ionically polymerise styrene (and other monomers) smoothly. The occasional explosions experienced were probably due to rapid collapse of the layer structure and liberation of very finely divided metal.

Butyllithium

Roper, A. N. *et al., Br. Polym. J.*, 1975, **7**, 195–203

Thermal explosion which occurred during fast anionic polymerisation of styrene catalysed by butyllithium was prevented by addition of low molecular weight polystyrene before the catalyst.

Chlorine,
Iron(III) chloride

*See* CHLORINE, $Cl_2$: Iron(III) chloride, Monomers

Dibenzoyl peroxide

Sebastian, D. H. *et al., Polym. Eng. Sci.*, 1976, **16**, 117–123
The conditions were determined for runaway/non-runaway polymerisation of styrene in an oil-heated batch reactor at 3 bar, using dibenzoyl peroxide as initiator at 3 concentrations. Results are presented diagrammatically.

Initiators

Biesenberger, J. A. *et al., Polym. Eng. Sci.*, 1976, **16**, 101–116
The parameters were determined in a batch reactor for thermal runaway polymerisation of styrene, initiated by azoisobutyronitrile, dibenzoyl peroxide or di-*tert*-butyl peroxide .

*See other* POLYMERISATION INCIDENTS

Oxygen

Barnes, C. E. *et al., J. Amer. Chem. Soc.*, 1950, **72**, 210
Exposure of unstabilised styrene to oxygen at 40–60°C generated a styrene–oxygen interpolymeric peroxide which, when isolated, exploded violently on gentle heating

*See other* POLYPEROXIDES

**DIPOTASSIUM CYCLOOCTATETRAENE** $C_8H_8K_2$

$\overline{CH{=}CH[CH{=}CH]_2CH{=}CH}{\cdot}K_2$

Starks, D. F. *et al., Inorg. Chem.*, 1974, **13**, 1307
The dry solid is violently pyrophoric, exploding on contact with oxygen.

*See other* ORGANOMETALLICS

***N*-NITROSOACETANILIDE** **PhN(NO)Ac** $C_8H_8N_2O_2$

Piperidine

Huisgen, R., *Ann.*, 1951, **573**, 163–181
The dry anilide exploded on contact with a drop of piperidine.

Thiophene

Bamberger, E., *Ber.*, 1887, **30**, 367
A mixture prepared at 0°C exploded when removed from the cooling bath.

*See other* NITROSO COMPOUNDS

## SODIUM 4,4-DIMETHOXY-1-*aci*-NITRO-3,5-DINITRO-2,5-CYCLOHEXADIENE $C_8H_8N_3NaO_8$

$(MeO)_2$ [ring with two $NO_2$ groups] $=N(O)ONa$

Jackson, C. L. *et al., Amer. Chem. J.*, 1898, **20**, 449

It is produced from interaction of trinitroanisole and sodium methoxide and explodes with great violence on heating in a free flame. It should probably be formulated as the title compound, a dimethyl acetal of the *aci-p*-nitroquinonoid form of picric acid.

*See other* *aci*-NITROQUINONOID COMPOUNDS

## *p*-NITROACETANILIDE $NO_2C_6H_4NHAc$ $C_8H_8N_2O_3$

Sulphuric acid

*See* SULPHURIC ACID, $H_2O_4S$ : Nitroaryl bases, etc.

## 5-AMINO-3-PHENYL-1,2,4-TRIAZOLE $N{=}CPhNHN{=}CNH_2$ $C_8H_8N_4$

Polya, J. B., *Chem. & Ind.*, 1965, 812

The solution, obtained by conventional diazotisation of the above, contained some solid which was removed by filtration through sintered glass. The solid, thought to be precipitated diazonium salt (possibly an internal salt), exploded violently on being disturbed with a metal spatula.

*See* 5-AMINOTETRAZOLE, $CH_3N_5$

*See other* DIAZONIUM SALTS

## POLY(STYRENE PEROXIDE) $(C_8H_8O_2)_n$

$(PhCH{=}CHOOH)_n$

*See* STYRENE, $C_8H_8$ : Oxygen

*See other* POLYPEROXIDES

## *p*-METHOXYBENZYL CHLORIDE $MeOC_6H_4CH_2Cl$ $C_8H_9ClO$

1. Carroll, D. W., *Chem. Eng. News*, 1960, 38(34), 40
2. Ager, J. H. *et al.*, ibid., (43), 5

Two incidents involving explosion of bottles of chloride stored at room temperature are described [1,2]. Safe preparation with storage at 5°C is detailed [2].

**1-IODOOCTA-1,3-DIYNE** **IC≡CC≡CBu** $C_8H_9I$

Armitage, J. B. *et al., J. Chem. Soc.*, 1952, 1993

One specimen exploded violently during distillation at 74°C/~13 μbar.

*See other* HALOACETYLENE DERIVATIVES

**2,4-DIMETHYLBENZENEDIAZONIUM TRIIODIDE** $C_8H_9I_3N_2$

$Me_2C_6H_3N_2^+\ I_3^-$

*See* DIAZONIUM TRIIODIDES

**5(1,1-DINITROETHYL)-2-METHYLPYRIDINE** $C_8H_9N_3O_4$

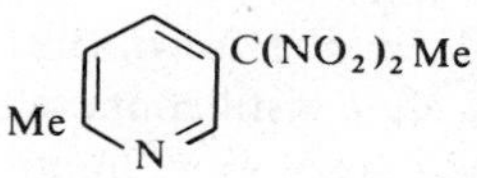

[Structure: pyridine ring with Me at 2-position and $C(NO_2)_2Me$ at 5-position]

*See* NITRIC ACID, $HNO_3$: 5-Ethyl-2-methylpyridine

*See other* POLYNITROALKYL COMPOUNDS

**1,3,5-CYCLOOCTATRIENE** $CH_2(CH{=}CH)_3CH_2$ (cyclic) $C_8H_{10}$

*See* DIETHYL ACETYLENEDICARBOXYLATE, $C_8H_{10}O_4$: Cyclooctatriene

*See also* DIENES

**6,6-DIMETHYLFULVENE** $C_8H_{10}$

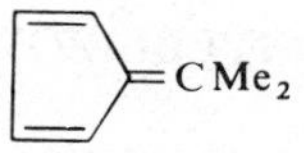

[Structure: cyclopentadiene ring =CMe₂]

Air

Engler, C. *et al., Ber.*, 1901, **34**, 2935

Dimethylfulvene readily peroxidised, giving an insoluble (polymeric?) peroxide which exploded violently on heating to 130°C, and also caused ether used to triturate it to ignite.

*See other* POLYPEROXIDES

**† ETHYLBENZENE** $EtC_6H_5$ $C_8H_{10}$

**† *o*-, *m*- or *p*-XYLENE** $MeC_6H_4Me$ $C_8H_{10}$

Acetic acid,

Air

Shraer, B. I., *Khim. Prom.*, 1970, **46**(10), 747–750

In liquid phase aerobic oxidation of *p*-xylene in acetic acid to terephthalic acid, it is important to eliminate the inherent hazards of this fuel–air mixture. Effects of temperature, pressure and presence of steam on explosive limits of the mixture have been investigated.

1,3-Dichloro-5,5-dimethyl-2,4-imidazolidindione

*See* 1,3-DICHLORO-5,5-DIMETHYL-2,4-IMIDAZOLIDINDIONE, $C_5H_6Cl_2N_2O_2$ : Xylene

Nitric acid

*See* NITRIC ACID, $HNO_3$ : Hydrocarbons

**$p$-BROMO-$N,N$-DIMETHYLANILINE** **$BrC_6H_4NMe_2$** **$C_8H_{10}BrN$**

Anon., *Chem. Eng. News,* 1961, **39**(13), 37

During vacuum distillation, self-heating could not be controlled and proceeded to explosion. Internal dehydrohalogenation to a benzyne seems a possibility.

**2-(2-AMINOETHYLAMINO)-5-CHLORONITROBENZENE** **$C_8H_{10}ClN_3O_2$**

**$(H_2NC_2H_4NH)ClC_6H_3NO_2$**

*See* 2-(2-AMINOETHYLAMINO)-NITROBENZENE, $C_8H_{11}N_3O_2$

**BIS(ETHOXYCARBONYLDIAZOMETHYL)MERCURY** **$C_8H_{10}HgN_4O_4$**

**$(EtO_2CCN_2)_2Hg$**

Buchner, E., *Ber.*, 1895, **28**, 217

It decomposes with foaming on melting at 104°C, and will explode under a hammer-blow.

*See other* DIAZO COMPOUNDS
MERCURY COMPOUNDS

**$N,N$-DIMETHYL-$p$-NITROSOANILINE** **$Me_2NC_6H_4NO$** **$C_8H_{10}N_2O$**

Acetic anhydride,
Acetic acid

Bain, P. J. S. *et al., Chem. Brit.*, 1971, **7**, 81

An exothermic reaction, sufficiently violent to expel the flask contents, occurs after an induction period of 15–30 s when acetic anhydride is added to a solution of the nitroso compound in acetic acid. 4,4′-Azobis-($N,N$-dimethylaniline), isolated from the reaction tar, may have been formed in a redox reaction, possibly involving an oxime derived from the nitroso compound.

**1,4-EPIDIOXY-1,4-DIHYDRO-6,6-DIMETHYLFULVENE** $C_8H_{10}O_2$

Harada, N. *et al.*, *Chem. Lett. (Japan)*, 1973, 1173–1176
When free of solvent it decomposes explosively above −10°C.
*See other* CYCLIC PEROXIDES

**DICROTONOYL PEROXIDE** $(MeCH{=}CHCO_2)_2$ $C_8H_{10}O_4$

Guillet, J. E. *et al.*, Ger. Pat. 1 131 407, 1962
This diacyl peroxide and its higher homologues, though of relatively high thermal stability, are very shock-sensitive. Replacement of the hydrogen atoms adjacent to the carbonyl groups with alkyl groups renders the group non-shock-sensitive.
*See other* DIACYL PEROXIDES

**DIETHYL ACETYLENEDICARBOXYLATE** $C_8H_{10}O_4$

$EtO_2CC{\equiv}CCO_2Et$

1,3,5-Cyclooctatriene
Foote, C. S., private comm., 1965
A mixture of the reactants being heated at 60°C to effect the Diels–Alder addition exploded. Onset of this vigorously exothermic reaction was probably delayed by an induction period, and presence of a solvent and/or cooling would have moderated it.
*See other* ACETYLENIC COMPOUNDS

**DIALLYL PEROXYDICARBONATE** $C_8H_{10}O_6$

$(CH_2{=}CHCH_2OCO_2)_2$

The distilled oil exploded on storage at ambient temperature.
*See* PEROXYCARBONATE ESTERS
*See other* ALLYL COMPOUNDS

**BIS(3-CARBOXYPROPIONYL)PEROXIDE** $C_8H_{10}O_8$

$(HO_2CC_2H_4CO_2)_2$

Lombard, R. *et al.*, *Bull. Soc. Chim. Fr.*, 1963, **12**, 2800
Explodes on contact with flame. The commercial dry 95% material ('succinic acid peroxide') is highly hazard-rated.
*See* COMMERCIAL ORGANIC PEROXIDES
DIACYL PEROXIDES

## 2,4,6-TRIMETHYLPYRILIUM PERCHLORATE $C_8H_{11}ClO_5$

(Structure: 2,4,6-trimethylpyrylium ring, Me at 2, 4 and 6 positions, $\overset{+}{O}$ in ring; $ClO_4^-$)

Hafner, K. *et al., Org. Synth.*, 1964, **44**, 102
The crystalline solid is an impact- and friction-sensitive explosive and must be handled with precautions. These include use of solvent-moist material and storage in a corked rather than glass-stoppered vessel.
*See other* NON-METAL PERCHLORATES

## *N,N*-DIMETHYLANILINE $C_6H_5NMe_2$ $C_8H_{11}N$

Dibenzoyl peroxide
*See* DIBENZOYL PEROXIDE, $C_{14}H_{10}O_4$ : *N,N*-Dimethylaniline

Diisopropyl peroxydicarbonate
*See* DIISOPROPYL PEROXYDICARBONATE, $C_8H_{14}O_6$

## *N*-ETHYLANILINE $C_6H_5NHEt$ $C_8H_{11}N$

Nitric acid
*See* NITRIC ACID, $HNO_3$ : Aromatic amines

## 5-ETHYL-2-METHYLPYRIDINE $C_8H_{11}N$

N=CMeCH=CHC(Et)=CH (ring closed between N and terminal CH)

Nitric acid
*See* NITRIC ACID, $HNO_3$ : 5-Ethyl-2-methylpyridine

## 3,3-DIMETHYL-1-PHENYLTRIAZENE PhN=$NNMe_2$ $C_8H_{11}N_3$

Baeyer, O. *et al., Ber.*, 1875, **8**, 149
Heusler, F., *Ann.*, 1890, **260**, 249
It decomposes explosively on attempted distillation at atmospheric pressure, but may be distilled uneventfully at reduced pressure.
*See other* TRIAZENES

## 2-(2-AMINOETHYLAMINO)NITROBENZENE $C_8H_{11}N_3O_2$

$$C_6H_4(NO_2)NHCH_2CH_2NH_2 \rightleftharpoons C_6H_4(=N(\rightarrow O)O^-)(=NCH_2CH_2\overset{+}{N}H_3)$$

Doleschall, G. *et al., Tetrahedron*, 1976, **32**, 59
This, and the corresponding 5-chloro- and 5-methoxy derivatives, tend to explode during vacuum distillation, and minimal distillation pressures and temperatures are recommended. This instability (which was not noted for the homologues in which the hydrogen atom of the secondary amino group was replaced by alkyl) may be connected with possible isomerisation to the *aci*-nitro iminoquinone internal salt species, 1-(2-ammonioethylimino)-4-R-2-*aci*-nitro-3,5-cyclohexadiene.

*See* NITROAROMATIC–ALKALI HAZARDS
*aci*-NITROQUINONOID COMPOUNDS

## † 4-VINYLCYCLOHEXENE $C_8H_{12}$

$CH_2CH{=}CHC_2H_4CHCH{=}CH_2$ (ring)

*See other* PEROXIDISABLE COMPOUNDS

## 1-*p*-CHLOROPHENYLBIGUANIDIUM HYDROGENDICHROMATE

$C_8H_{12}ClCr_2N_5O_7$

$ClC_6H_4NHC(NH)NHC(NH)N^+H_3\ HCr_2O_7^-$

*See* DICHROMATE SALTS OF NITROGENOUS BASES
*See other* OXOSALTS OF NITROGENOUS BASES

## TETRAMETHYLSUCCINODINITRILE $NCCMe_2CMe_2CN$ $C_8H_{12}N_2$

Preparative hazard

*See* AZOISOBUTYRONITRILE, $C_8H_{12}N_4$: Heptane

## *tert*-BUTYL-2-DIAZOACETOACETATE $C_8H_{12}N_2O_3$

$AcCN_2CO_2CMe_3$

Regitz, M. *et al., Org. Synth.*, 1968, **48**, 38
During low-temperature crystallisation, scratching to induce seeding of the solution must be discontinued as soon as the sensitive solid separates.

*See other* DIAZO COMPOUNDS

## 6-AMINOPENICILLANIC ACID *S*-OXIDE $C_8H_{12}N_2O_4S$

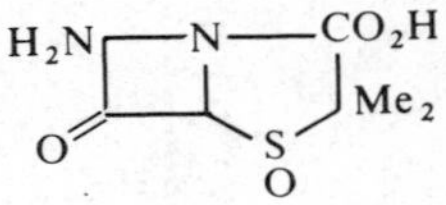

Preparative hazard

1. Micetich, R. G., *Synthesis*, 1976, **4**, 264
2. Noponen, A., *Chem. Eng. News*, 1977, **55**(8), 5
3. Micetich, R. G., *Chem. Brit.*, 1977, **13**, 163

According to a published procedure, the sulphur-containing acid is oxidised to the sulphoxide with hydrogen peroxide and precipitated as the *p*-toluenesulphonate salt in the presence of acetone [1]. During subsequent purification, trimeric acetone peroxide was precipitated and exploded violently after filtration [2]. The use of acetone in this and similar preparations involving hydrogen peroxide is not now recommended, and is highly dangerous [3].

*See* HYDROGEN PEROXIDE, $H_2O_2$: Acetone, etc.
3,3,6,6,9,9-HEXAMETHYL-1,2,4,5,7,8-HEXAOXAONANE, $C_9H_{18}O_6$

## AZOISOBUTYRONITRILE $C_8H_{12}N_4$

**CNCMe$_2$N=NCMe$_2$CN**

Acetone

Carlisle, P. J., *Chem. Eng. News*, 1949, **27**, 150

During recrystallisation of technical material from acetone, explosive decomposition occurred. Non-explosive decomposition occurred when the nitrile was heated alone, or in presence of methanol.

Heptane

Scheffold, R. *et al., Helv. Chim. Acta*, 1975, **58**, 60

During preparation of tetramethylsuccinodinitrile by thermal decomposition of 100 g of the azonitrile by slow warming in unstirred heptane, an explosion occurred. Successive addition of small quantities of the nitrile to heptane at 90–92°C is a safer, preferred method.

*See other* AZO COMPOUNDS
CYANO COMPOUNDS

## *O–O-tert*-BUTYL HYDROGENMONOPEROXYMALEATE $C_8H_{12}O_5$

**Me$_3$COO$_2$CCH=CHCO$_2$H**

Castrantas, 1965, 17

Slightly shock-sensitive, the commercial dry 95% material is highly hazard-rated.

*See other* COMMERCIAL ORGANIC PEROXIDES
PEROXYESTERS

**TETRAVINYLLEAD** $(CH_2{=}CH)_4Pb$ $C_8H_{12}Pb$

Holliday, A. K. *et al., Chem. & Ind.*, 1968, 1699

In the preparation from lead (II) chloride and vinylmagnesium bromide in tetrahydrofuran–hexane, violent explosions occurred during isolation of the product by distillation of solvent. This could be avoided by a procedure involving steam distillation of the tetravinyl-lead, no significant loss of yield (80%) by hydrolysis being noted.

Diborane

*See* DIBORANE, $B_2H_6$ : Tetravinyllead

Phosphorus trichloride

Houben-Weyl, 1975, Vol. 13.3, 244

In absence of solvent, interaction may be explosive.

*See related* ALKYLMETALS

**3-BUTEN-1-YNYLDIETHYLALUMINIUM** $C_8H_{13}Al$

$CH_2{=}CHC{\equiv}CAlEt_2$

Petrov, A. A. *et al., Zh. Obsch. Khim.*, 1962, **32**, 1349

Ignites in air.

*See other* METAL ACETYLIDES
TRIALKYLALUMINIUMS

**1-PHENYLBIGUANIDIUM HYDROGENDICHROMATE** $C_8H_{13}Cr_2N_5O_7$

$PhNHC(NH)NHC(NH)N^+H_3\ HCr_2O_7^-$

*See* DICHROMATE SALTS OF NITROGENOUS BASES

*See other* OXOSALTS OF NITROGENOUS BASES

**1-DIETHYLAMINO-1-BUTEN-3-YNE** $Et_2NCH{=}CHC{\equiv}CH$ $C_8H_{13}N$

Brandsma, 1971, 163

Dilution of the amine with white mineral oil is recommended before vacuum distillation, to avoid explosive decomposition of the residue.

*See other* ACETYLENIC COMPOUNDS

**† 1,7-OCTADIENE** $C_8H_{14}$

$CH_2{=}CH[CH_2]_4CH{=}CH_2$

**† 1-, 2-, 3- or 4-OCTYNE** $HC{\equiv}CC_6H_{13}$, etc. $C_8H_{14}$

**† VINYL CYCLOHEXANE** $H_2C{=}CH\overline{CH[CH_2]_4CH_2}$ $C_8H_{14}$

## DI-2-BUTENYLCADMIUM $C_8H_{14}Cd$

$(MeCH{=}CHCH_2)_2Cd$

Houben-Weyl, 1973, Vol. 13.2a, 881

When rapidly warmed from −5°C to ambient temperature, it decomposes explosively.

*See related* ALKYLMETALS
ALLYL COMPOUNDS

## μ-CYCLOPENTADIENYL(METHYL)-BIS(*N*-METHYL-*N*-NITROSOHYDROXYLAMINO)-TITANIUM $C_8H_{14}N_4O_4Ti$

$C_5H_5MeTi(ON(Me)NO)_2$

Clark, R. J. H. *et al., J. Chem. Soc., Dalton Trans.*, 1974, 122

The complex was stable under nitrogen or vacuum, but detonated quite readily and decomposed in air.

*See other* NITROSO COMPOUNDS
ORGANOMETALLICS

## 4-HYDROXY-4-METHYL-1,6-HEPTADIENE $C_8H_{14}O$

$CH_2{=}CHCH_2C(Me)OHCH_2CH{=}CH_2$

Ozone

1. Rabinowitz, J. L., *Biochem Prep.*, 1958, **6**, 25
2. Miller, F. W., *Chem. Eng. News*, 1973, **51**(6), 29

Following a published procedure [1] but using 75 g of diene instead of the 12.6 g specified, the ozonisation product from the diene exploded during vacuum desiccation. Eight previous preparations on the specified scale had been uneventful.

*See other* OZONIDES

## DIISOBUTYRYL PEROXIDE $(Me_2CHCO_2)_2$ $C_8H_{14}O_4$

1. Kharasch, S. *et al., J. Amer. Chem. Soc.*, 1941, **63**, 526
2. *MCA Case History No. 579*

A sample of the peroxide in ether (prepared according to a published procedure [1]) was being evaporated to dryness with a stream of air when it exploded violently. Handling the peroxide as a dilute solution at low temperature is recommended.

*See other* DIACYL PEROXIDES

## ACETYL CYCLOHEXANESULPHONYL PEROXIDE $C_8H_{14}O_5S$

$AcOOSO_2C_6H_{11}$

'Code of Practice for Storage of Organic Peroxides', GC14, Widnes, Interox Chemicals Ltd, 1970

While the commercial material damped with 30% water is not shock- or friction-sensitive, if it dries out it may become a high-hazard material. As a typical low-melting diacyl peroxide, it may be expected to decompose vigorously or explosively on slight heating or on mechanical initiation.

*See other* DIACYL PEROXIDES

**DIISOPROPYL PEROXYDICARBONATE** $C_8H_{14}O_6$

$(Me_2CHOCO_2)_2$

Alone,
or Amines,
or Potassium iodide

1. Strong, W. A., *Ind. Eng. Chem.*, 1964, **56**(12), 33
2. Strain, F., *J. Amer. Chem. Soc.*, 1950, **72**, 1254
3. 'Bulletin T. S. 350', Pittsburgh, Pittsburgh Plate Glass Co., 1963

When warmed slightly above its m.p. (10°C) the ester undergoes slow self-accelerating decomposition which may become dangerously violent under confinement [1]; at 25–30°C decomposition occurred in 10–30 min [2]. Addition of 1% of aniline, 1,2-diaminoethane or potassium iodide caused instant decomposition, and of *N,N*-dimethylaniline, instant explosion [2]. Bulk solutions of the ester (45%) in benzene-cyclohexane stored at 5°C developed sufficient heat to decompose explosively after a day, and 50–90% solutions were found to be impact-sensitive [1]. The solid is normally stored and transported at below −18°C in loose-topped trays [3].

*See other* DIACYL PEROXIDES
PEROXYCARBONATE ESTERS

**DIPROPYL PEROXYDICARBONATE** $PrO_2COOCO_2Pr$ $C_8H_{14}O_6$

Barter, J. A., US Pat. 3 775 341, 1973

Methylcyclohexane is a suitable solvent to reduce the hazardous properties of the ester.

*See other* PEROXYCARBONATE ESTERS

**4-HYDROXY-4-METHYL-1,6-HEPTADIENE DIOZONIDE** $C_8H_{14}O_7$

$MeCOH(CH_2\underline{CHCH_2O}–O^+–O^-)_2$

*See* 4-HYDROXY-4-METHYL-1,6-HEPTADIENE, $C_8H_{14}O$: Ozone

*See other* OZONIDES

**DI(2-METHOXYETHYL) PEROXYDICARBONATE** $C_8H_{14}O_8$

**$(MeOC_2H_4OCO_2)_2$**

Exploded at 34°C.

*See other* PEROXYCARBONATE ESTERS

**$\mu$-CYCLOPENTADIENYLTRIMETHYLTITANIUM** $C_8H_{14}Ti$

**$C_5H_5TiMe_3$**

Bailar, 1973, Vol. 3, 395

Pyrophoric at ambient temperature.

*See other* ORGANOMETALLICS

**† DIMETHYLCYCLOHEXANES (5 isomers)** **$MeC_6H_{10}Me$** $C_8H_{16}$

**† 1- or 2-OCTENE** **$CH_2{=}CHC_6H_{13}$, etc.** $C_8H_{16}$

**† 2,3,4-TRIMETHYL-1-PENTENE** **$CH_2{=}CMeCHMeCHMe_2$** $C_8H_{16}$

**† 2,4,4-TRIMETHYL-1-PENTENE** **$CH_2{=}CMeCH_2CMe_3$** $C_8H_{16}$

**† 2,3,4-TRIMETHYL-2-PENTENE** **$Me_2C{=}CMeCHMe_2$** $C_8H_{16}$

**† 2,4,4-TRIMETHYL-2-PENTENE** **$Me_2C{=}CHCMe_3$** $C_8H_{16}$

**† 3,4,4-TRIMETHYL-2-PENTENE** **$MeCH{=}CMeCMe_3$** $C_8H_{16}$

**4,4′-DITHIODIMORPHOLINE** $C_8H_{16}N_2O_2S_2$

O NSSN O

Chlorine

*See* 4-MORPHOLINESULPHENYL CHLORIDE, $C_4H_8ClNOS$

**3-AZONIABICYCLO[3.2.2] NONANE NITRATE** $C_8H_{16}N_2O_3$

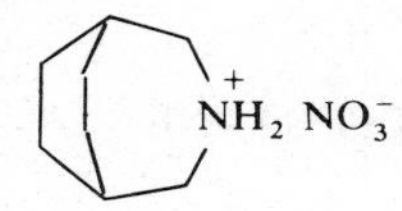


Violent decomposition occurred at 258°C.
*See* DIFFERENTIAL THERMAL ANALYSIS
*See other* OXOSALTS OF NITROGENOUS BASES

**† 2-ETHYLHEXANAL** $C_4H_9CHEtCHO$ $C_8H_{16}O$

**3,6-DIETHYL-3,6-DIMETHYL-1,2,4,5-TETRAOXANE** $C_8H_{16}O_4$

**EtMeCOOCMe(Et)OO**

Castrantas, 1965, 18
Swern, 1970, Vol. 1, 37
By analogy, this dimeric 2-butanone peroxide and the corresponding trimer probably contribute largely to the high shock-sensitivity of the commercial 'MEK peroxide' mixture, which contains these and other peroxides.
*See other* CYCLIC PEROXIDES
KETONE PEROXIDES

**OCTYLSODIUM** $C_8H_{17}Na$
Houben-Weyl, 1970, Vol. 13.1, 384
The powder ignites in air.
*See other* ALKYLMETALS

**† 2,3-DIMETHYLHEXANE** $Me_2CHCHMePr$ $C_8H_{18}$

**† 2,4-DIMETHYLHEXANE** $Me_2CHCH_2CHMeEt$ $C_8H_{18}$

**† 3-ETHYL-2-METHYLPENTANE** $CH_3CHMeCHEt_2$ $C_8H_{18}$

**† 2- or 3-METHYLHEPTANE** $C_8H_{18}$

$Me_2CHC_5H_{11}$, $EtCHMeC_4H_9$

† **OCTANE** $CH_3[CH_2]_6CH_3$ $C_8H_{18}$

† **2,2,3-TRIMETHYLPENTANE** $Me_3CCHMeEt$ $C_8H_{18}$

† **2,2,4-TRIMETHYLPENTANE** $Me_3CCH_2CHMe_2$ $C_8H_{18}$

† **2,3,3-TRIMETHYLPENTANE** $Me_2CHCMe_2Et$ $C_8H_{18}$

† **2,3,4-TRIMETHYLPENTANE** $Me_2CHCHMeCHMe_2$ $C_8H_{18}$

**DIISOBUTYLALUMINIUM CHLORIDE** $C_8H_{18}AlCl$

$(Me_2CHCH_2)_2AlCl$

*See* ALKYLALUMINIUM HALIDES

**DI-*tert*-BUTYL CHROMATE** $CrO_2(OCMe_3)_2$ $C_8H_{18}CrO_4$

*ABCM Quart. Safety Summ.*, 1953, **24**, 2

A preparation by addition of *tert*-butanol to chromium trioxide on large scale in full unstirred flask with poor cooling detonated owing to local overheating. Effective cooling and stirring essential.

**DIBUTYL HYPONITRITE** $BuON{=}NOBu$ $C_8H_{18}N_2O_2$

*See* DIALKYL HYPONITRITES

† **DIBUTYL ETHER** $BuOBu$ $C_8H_{18}O$

Dasler, W. *et al., Ind. Eng. Chem. (Anal. Ed.)*, 1946, **18**, 52

Peroxides formed in storage are effectively removed by percolation through alumina.

*See* ETHERS

*See other* PEROXIDISABLE COMPOUNDS

† **DI-*tert*-BUTYL PEROXIDE** $Me_3COOCMe_3$ $C_8H_{18}O_2$

Preparative hazard

*See* HYDROGEN PEROXIDE, $H_2O_2$: *tert*-Butanol, etc.

*See other* DIALKYL PEROXIDES

† **BIS(2-ETHOXYETHYL)ETHER** $(EtOC_2H_4)_2O$ $C_8H_{18}O_3$

**DI(2-HYDROPEROXYBUTYL-2) PEROXIDE** $C_8H_{18}O_6$

**HOOC(EtMe)OOC(EtMe)OOH**

Leleu, *Cahiers*, 1973, (71), 238

The triperoxide, the main component of 'MEK peroxide', is explosive in the pure state, but insensitive to shock as the commercial 50% solution in dimethyl phthalate. The solution will explode at about 85°C, and slowly liberates oxygen at ambient temperature.

*See* KETONE PEROXIDES

*See related* ALKYL HYDROPEROXIDES

**DIISOBUTYLALUMINIUM HYDRIDE** **$(Me_2CHCH_2)_2AlH$** $C_8H_{19}Al$

1. Mirviss, S. B. *et al.*, *Ind. Eng. Chem.*, 1961, **53**(1), 54A
2. 'Specialty Reducing Agents', Brochure TA-2002/1, New York, Texas Alkyls, 1971

The higher thermal stability of dialkylaluminium hydrides over the corresponding trialkylaluminiums is particularly marked in this case with 2-branched alkyl groups [1]. Used industrially as a powerful reducant, it is supplied as a solution in hydrocarbon solvents. The undiluted material ignites in air unless diluted to below 25% concentration.

*See other* ALKYLALUMINIUM ALKOXIDES AND HYDRIDES

**DIBUTYLAMINE** **$Bu_2NH$** $C_8H_{19}N$

Cellulose nitrate

*See* CELLULOSE NITRATE: Amines

**† DI-2-BUTYLAMINE** **$(EtMeCH)_2NH$** $C_8H_{19}N$

**† DIISOBUTYLAMINE** **$(Me_2CHCH_2)_2NH$** $C_8H_{19}N$

**DIBUTYL HYDROGENPHOSPHITE** **$(BuO)_2P(O)H$** $C_8H_{19}O_3P$

Air

Morrell, S. H., private comm., 1968

The phosphite was being distilled under reduced pressure. At the end of distillation the air bleed was opened more fully, when spontaneous combustion took place inside the flask. It was suggested that phosphine had been formed.

**TETRAETHYLDIARSANE** **$Et_2AsAsEt_2$** **$C_8H_{20}As_2$**

Sidgwick, 1950, 770

Inflames in air.

*See other* ALKYLNON-METALS

**1,3,6,8-TETRAAZONIATRICYCLO[6.2.1.1$^{3,6}$]DODECANE TETRANITRATE** **$C_8H_{20}N_8O_{12}$**

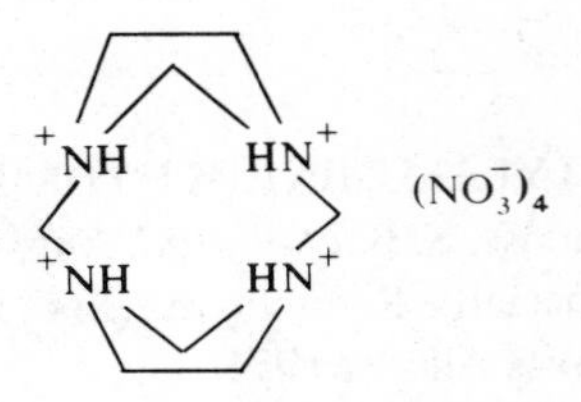


Violent decomposition occurred at 260°C.

*See* DIFFERENTIAL THERMAL ANALYSIS

*See other* OXOSALTS OF NITROGENOUS BASES

**TETRAETHYLLEAD** **$Et_4Pb$** **$C_8H_{20}Pb$**

Doyle, W. H., *Loss Prev*., 1969, **3**, 16

Failure to cover the residue with water after emptying a tank of the compound caused explosive decomposition after several days.

*See other* ALKYLMETALS

**TETRAKIS(ETHYLTHIO)URANIUM** **$(EtS)_4U$** **$C_8H_{20}S_4U$**

Bailar, 1973, Vol. 5, 416

It ignites in air.

*See other* PYROPHORIC MATERIALS

**TETRAETHYLTIN** **$Et_4Sn$** **$C_8H_{20}Sn$**

Sorbe, 1968, 160

It tends to ignite in air.

*See other* ALKYLMETALS

**'TETRAETHYLDIBORANE'** **$C_8H_{22}B_2$**

Köster, R. *et al., Inorg. Synth*., 1974, **15**, 141–146

The material (an equilibrium mixture of diborane and highly ethylated homologues) ignites in air. The propyl analogue behaves similarly.

*See other* BORANES

## 3,6,9-TRIAZA-11-AMINOUNDECANOL ('HYDROXYETHYLTRIETHYLENETETRAMINE') $C_8H_{22}N_4O$

$HOC_2H_4(NHC_2H_4)_3NH_2$

Cellulose nitrate

*See* CELLULOSE NITRATE: Amines

## 1,11-DIAMINO-3,6,9-TRIAZAUNDECANE ('TETRAETHYLENEPENTAMINE') $C_8H_{23}N_5$

$H_2NC_2H_4(NHC_2H_4)NH_2$

Carbon tetrachloride

*See* CARBON TETRACHLORIDE, $CCl_4$: 1,11-Diamino-3,6,9-triazaundecane

Cellulose nitrate

*See* CELLULOSE NITRATE: Amines

## TETRAMETHYLAMMONIUM PENTAPEROXODICHROMATE(2–) $C_8H_{24}Cr_2N_2O_{12}$

$[Me_4N]_2[O(O_2)_2CrOOCr(O_2)_2O]$

Mellor, 1943, Vol. 11, 358

It explodes on moderate heating or in contact with sulphuric acid.

*See other* OXOSALTS OF NITROGENOUS BASES

## TETRAKIS(DIMETHYLAMINO)TITANIUM $C_8H_{24}N_4Ti$

$(Me_2N)_4Ti$

Hydrazine

*See* HYDRAZINE, $H_4N_2$: Titanium compounds

## 1,2-DIAMINOETHANEBIS-TRIMETHYLGOLD $C_8H_{26}Au_2N_2$

$Me_3Au \leftarrow NH_2C_2H_4NH_2 \rightarrow AuMe_3$

Gilman, H. *et al.*, *J. Amer. Chem. Soc.*, 1948, **70**, 550

The solid complex is very sensitive to light, and explodes violently on heating in an open crucible. A drop of concentrated nitric acid added to the dry material causes detonation.

*See other* GOLD COMPOUNDS

**BIS-DIETHYLENETRIAMINECOBALT(III) PERCHLORATE**

$C_8H_{26}Cl_3CoN_6O_{12}$

$[Co2(H_2NC_2H_4)_2NH]^{3+}\ [ClO_4]_3^-$

Explodes at 325°C; high impact-sensitivity.

*See other* AMMINEMETAL OXOSALTS

**POTASSIUM GRAPHITE** $KC_8$ $C_8K$

Bailar, 1973, Vol. 1, 1276

Potassium graphite may explode with water.

*See* GRAPHITE, C: Alkali metals

*See other* METAL NON-METALLIDES

**LITHIUM OCTACARBONYLTRINICCOLATE(2–)** $C_8Li_2Ni_3O_8$

$Li_2[Ni_3(CO)_8]$

Bailar, 1973, Vol. 3, 1117

A pyrophoric salt.

*See related* CARBONYLMETALS

**RUBIDIUM GRAPHITE (corrected name)** $RbC_8$ $C_8Rb$

*See* GRAPHITE, C: Alkali metals

*See other* METAL NON-METALLIDES

**NONACARBONYLDIIRON** $C_9Fe_2O_9$

$(CO)_3Fe(CO)_3Fe(CO)_3$

*ABCM Quart. Safety Summ.*, 1943, **14**, 18

Commercial iron carbonyl (fl.p., 35°C) has an auto-ignition temperature in contact with brass of 93°C, lower than that of carbon disulphide.

*See other* CARBONYLMETALS

**2-NONEN-4,6,8-TRIYN-1-AL** $C_9H_4O$

$H(C{\equiv}C)_3CH{=}CHCHO$

Bohlman, F. *et al., Chem. Ber.*, 1963, **96**, 2586

Extremely unstable, explodes after a few minutes at room temperature and ignites at 110°C.

*See other* ACETYLENIC COMPOUNDS

***o*-NITROPHENYLPROPIOLIC ACID** $C_9H_5NO_4$

$O_2NC_6H_4C{\equiv}CCO_2H$

Sorbe, 1968, 152

It explodes above 150°C.
*See other* ACETYLENIC COMPOUNDS

**3,5-DIMETHYL-4-[BIS(TRIFLUOROACETOXY)IODO]ISOXAZOLE**

$C_9H_6F_6INO_5$

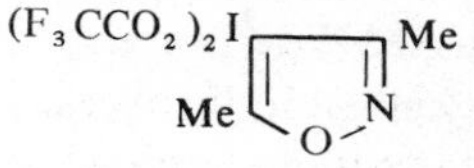

$(F_3CCO_2)_2I$ — Me, Me, N, O (isoxazole ring)

Preparative hazard
*See* PEROXYTRIFLUOROACETIC ACID, $C_2HF_3O_3$: 4-Iodo-3,5-dimethylisoxazole

**1-DIAZOINDENE** $\overline{C_6H_4CN_2CH{=}CH}$ $C_9H_6N_2$

Preparative hazard
*See* *p*-TOLUENESULPHONYL AZIDE, $C_7H_7N_3O_2S$

**2,4-DIISOCYANATOTOLUENE** $MeC_6H_3(NCO)_2$ $C_9H_6N_2O_2$

Acyl chlorides,
or Bases
*MCA SD-73*, 1971
The diisocyanate may undergo exothermic polymerisation in contact with bases or more than traces of acyl chlorides, sometimes used as stabilisers.

**5-(*p*-DIAZONIOBENZENESULPHONAMIDO)THIAZOLE TETRAFLUOROBORATE**

$C_9H_7BF_4N_4S$

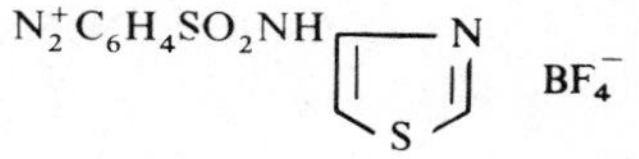

$N_2^+C_6H_4SO_2NH$ — thiazole ring (N, S) $BF_4^-$

Zuber, F. *et al.*, *Helv. Chim. Acta*, 1950, **33**, 1269–1271
It explodes at 135–137°C.
*See* DIAZONIUM TETRAHALOBORATES

**1-IODO-3-PHENYL-2-PROPYNE** $ICH_2C{\equiv}CPh$ $C_9H_7I$

Whiting, M. C., *Chem. Eng. News*, 1972, **50**(23), 86
It detonated on distillation at *ca.* 180°C.
*See other* HALOACETYLENE DERIVATIVES

## QUINOLINE $C_6H_4N{=}CHCH{=}CH$ $C_9H_7N$

1. Blumann, A., *Proc. R. Aust. Chem. Inst.*, 1964, **31**, 286
2. *MCA Case History No. 1008*

The traditional unpredictably violent nature of the Skraup reaction (the preparation of quinoline and its derivatives by treating aniline, etc., with glycerol, sulphuric acid and an oxidant, usually nitrobenzene) is attributed to lack of stirring and adequate temperature control in many published descriptions [1].

A large-scale (450 litre) reaction, in which sulphuric acid was added to a stirred mixture of aniline, glycerol, nitrobenzene, ferrous sulphate and water, went out of control soon after the addition. A 150 mm rupture disc blew out first, followed by the manhole cover of the vessel. The violent reaction is attributed to doubling the scale of the reaction, an unusually high ambient temperature (reaction contents at 32°C) and the accidental addition of excess sulphuric acid. Experiment showed that a critical temperature of 120°C was reached immediately on addition of excess acid under these conditions [2].

Dinitrogen tetraoxide

*See* DINITROGEN TETRAOXIDE, $N_2O_4$: Heterocyclic bases

Linseed oil,
Thionyl chloride

*See* SULPHINYL CHLORIDE, $Cl_2OS$: Quinoline, etc.

Maleic anhydride

*See* MALEIC ANHYDRIDE, $C_4H_2O_3$: Cations, etc.

## DICARBONYL-π-CYCLOHEPTATRIENYLTUNGSTEN AZIDE

$(CO)_2WC_7H_7N_3$ $C_9H_7N_3O_2W$

Hoch, G. *et al., Z. Naturforsch. B*, 1976, **31(B)**, 295

It was insensitive to shock, but decomposed explosively at 130°C.

*See related* METAL AZIDES
ORGANOMETALLICS

## 3-METHOXY-2-NITROBENZOYLDIAZOMETHANE $C_9H_7N_3O_4$

$MeONO_2C_6H_3COCHN_2$

Alone,
or Sulphuric acid

Musajo, L. *et al., Gazz. Chim. Ital.*, 1950, **80**, 171–176

The diazoketone explodes at 138–140°C, or on treatment with concentrated sulphuric acid.

*See other* DIAZO COMPOUNDS

**N-CHLOROCINNAMALDIMINE** **PhCH=CHCH=NCl** **$C_9H_8ClN$**

Hauser, C. R. *et al., J. Amer. Chem. Soc.*, 1935, **57**, 570

It decomposes vigorously after storage at ambient temperature for 30 min.

*See other* N-HALOGEN COMPOUNDS

**3,6-DIMETHYLBENZENEDIAZONIUM-2-CARBOXYLATE** **$C_9H_8N_2O_2$**

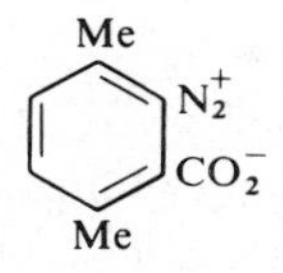


Hart, H. *et al., J. Org. Chem.*, 1972, **37**, 4272

The hydrochloride appears to be stable for considerable periods at ambient temperature, but explodes on melting at 88°C.

*See* BENZENEDIAZONIUM-2-CARBOXYLATE, $C_7H_4N_2O_2$

*See other* DIAZONIUM CARBOXYLATES

**4,6-DIMETHYLBENZENEDIAZONIUM-2-CARBOXYLATE** **$C_9H_8N_2O_2$**

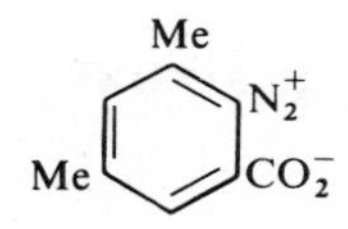


Gommper, R. *et al., Chem. Ber.*, 1968, **101**, 2348

It is a highly explosive solid.

**2,3-EPOXYPROPIONALDEHYDE 2,4-DINITROPHENYLHYDRAZONE**

**$C_9H_8N_4O_5$**

$\overline{OCH_2CH}CH=NNHC_6H_3(NO_2)_2$

Payne, G. B., *J. Amer. Chem. Soc.*, 1959, **81**, 4903

Although it melts at 96–98°C in a capillary, an 0.5 g sample in a test tube decomposed explosively at 100°C.

*See also* 2,3-EPOXYPROPIONALDEHYDE OXIME, $C_3H_5NO_2$

*See other* 1,2-EPOXIDES

**TETRAKIS(2,2,2-TRINITROETHYL) ORTHOCARBONATE** **$C_9H_8N_{12}O_{40}$**

**$C[OCH_2C(NO_3)_3]_4$**

*See* TRINITROETHYL ORTHOESTERS

**BENZYLOXYACETYLENE** $PhCH_2OC{\equiv}CH$ $C_9H_8O$

Olsman, H. *et al., Rec. Trav. Chim.*, 1964, **83**, 305

If heated above 60°C during vacuum distillation, explosive rearrangement occurs.

*See other* ACETYLENIC COMPOUNDS

**CINNAMALDEHYDE** $PhCH{=}CHCHO$ $C_9H_8O$

Sodium hydroxide

Morrell, S. H., private comm., 1968

Rags soaked in sodium hydroxide and in the aldehyde overheated and ignited when they came into contact in a waste bin.

*See other* PEROXIDISABLE COMPOUNDS

**2-METHYL-3,5,7-OCTATRIYN-2-OL** $Me_2C(OH)[C{\equiv}C]_3H$ $C_9H_8O$

Cook, C. J. *et al., J. Chem. Soc.*, 1952, 2885, 2890

The crude material invariably deflagrated at ambient temperature, and once during drying at 0°C/0.013 mbar.

*See other* ACETYLENIC COMPOUNDS

† **1-PHENYL-1,2-PROPANEDIONE** $PhCOCOCH_3$ $C_9H_8O_2$

**4-HYDROXY-*trans*-CINNAMIC ACID** $C_9H_8O_3$

$HOC_6H_4CH{=}CHCO_2H$

Tanaka, Y. *et al., Proc. 4th Int. Conf. High Pressure (1974)*, 704, 1975 (*Chem. Abs.*, 1975, **83**, 114988x)

During spontaneous solid-state polymerisation at 325°C/8–10 kbar, the acid exploded violently.

*See other* POLYMERISATION INCIDENTS

**NITROINDANE** $NO_2C_6H_3C_2H_4CH_2$ $C_9H_9NO_2$

1. Lindner, J. *et al., Ber.*, 1927, **60**, 435
2. Gribble, G. W., *Chem. Eng. News*, 1973, **51**(6), 30, 39

The crude mixture of 4- and 5-nitroindanes produced by mixed acid nitration of indane following a literature method [1] is hazardous to purify by distillation. The warm residue from distillation of 0.015 g mol at 80°C/1.3 mbar exploded on admission of air, and a 1.3 g mol batch exploded as distillation under the same conditions began. Removal of higher-boiling poly-nitrated material before distillation is recommended [2].

*See other C*-NITRO COMPOUNDS

## 2-ISOCYANOETHYL BENZENESULPHONATE $C_9H_9NO_3S$

$PhSO_2OC_2H_4NC$:

Matteson, D. S. *et al., J. Amer. Chem. Soc.*, 1968, **90**, 3761

Heating the product under vacuum to remove pyridine solvent caused a moderately forceful explosion.

*See other* CYANO COMPOUNDS

## 3-PHENYLPROPIONYL AZIDE $PhC_2H_4CON_3$ $C_9H_9N_3O$

Curtiss, T. *et al., J. Prakt. Chem.*, 1901, **64**, 297

A sample exploded on a hot water bath.

*See other* ACYL AZIDES

## 1,3,5-TRIS(NITROMETHYL)BENZENE $(NO_2CH_2)_3C_6H_3$ $C_9H_9N_3O_6$

*See* NITRIC ACID, $HNO_3$ : Mesitylene

## 2-CHLORO-1-NITROSO-2-PHENYLPROPANE $C_9H_{10}ClNO$

$CH_3CClPhCH_2NO$

*MCA Case History No. 747*

A sample of the air-dried material decomposed vigorously on keeping in a closed bottle at ambient temperature overnight.

*See other* NITROSO COMPOUNDS

## 3,5-DIMETHYLBENZOIC ACID $Me_2C_6H_3CO_2H$ $C_9H_{10}O_2$

Preparative hazard

*See* NITRIC ACID, $HNO_3$: Hydrocarbons

## ALLYL BENZENESULPHONATE $C_9H_{10}O_3S$

$C_6H_5SO_3CH_2CH{=}CH_2$

Dye, W. T. *et al., Chem. Eng. News*, 1950, **28**, 3452

The residue, from vacuum distillation at 92–135°C/2.6 mbar, darkened, thickened, then exploded after removal of the heat source. For precautions:

*See* TRIALLYL PHOSPHATE, $C_9H_{15}O_4P_2$

*See other* ALLYL COMPOUNDS

## NITROMESITYLENE $C_9H_{11}NO_2$

Me Me NO$_2$ Me

Preparative hazard

*See* NITRIC ACID, $HNO_3$ : Hydrocarbons (references 7,8)

## 5(4-DIMETHYLAMINOBENZENEAZO)TETRAZOLE $C_9H_{11}N_7$

$\overline{N{=}NNHN{=}C}N{=}NC_6H_4NMe_2$

Thiele, J., *Ann.*, 1892, **270**, 54
Explodes at 155°C.
*See other* TETRAZOLES

## 1,2,3,4-TETRAHYDROISOQUINOLINIUM NITRATE $C_9H_{12}N_2O_3$
## 1,2,3,4-TETRAHYDROQUINOLINIUM NITRATE

$\overline{C_6H_4CH_2N^+H_2[CH_2]_2CH_2}\ NO_3^-$ ; $\overline{C_6H_4N^+H_2[CH_2]_2CH_2}\ NO_3^-$

Violent decomposition at 268 and 236°C, respectively.
*See* DIFFERENTIAL THERMAL ANALYSIS
*See other* OXOSALTS OF NITROGENOUS BASES

## PHENYLPROPANOL $Ph[CH_2]_3OH$ $C_9H_{12}O$

Phosphorus tribromide
*See* PHOSPHORUS TRIBROMIDE, $Br_3P$: Phenylpropanol

## 2-PHENYL-2-PROPYL HYDROPEROXIDE (CUMYL HYDROPEROXIDE)
$PhCMe_2OOH$ $C_9H_{12}O_2$


1. Leroux, A., *Mém. Poudres*, 1955, **37**, 49
2. Simon, A. H. *et al., Chem. Ber.*, 1957, **90**, 1024
3. Hulanicki, A. *Chem. Anal. (Warsaw)*, 1973, **18**, 723–726
4. *MCA Case History No. 906*
5. Leleu, *Cahiers*, 1973, (71), 226
6. Fleming, J. B. *et al., Hydrocarbon Proc.*, 1976, **55**(1), 185–196
7. Redoshkin, B. A. *et al., Chem. Abs.*, 1964, **60**, 14359h; Antonovskii, V. L. *et al.*, ibid., 14360c


The explosibility of this unusually stable compound, 'cumene hydroperoxide', has been investigated [1]. It is difficult, but not impossible, to induce explosive decomposition [2]. A colorimetric method of analysing mixtures of cumene and its hydroperoxide which involves alkaline decomposition of the latter is safer than the use of acidic conditions, or of reducants, which may cause explosive reactions [3].

During vacuum concentration of the hydroperoxide by evaporation of cumene, a 6.5 $m^3$ quantity of 36% material decomposed explosively after storage at 109°C for 5 h. Catalytic decomposition under near-adiabatic conditions may have been involved [4]. Contact with copper, copper or lead alloys, mineral acids or reducants may lead to violent decomposition [5]. Cobalt also catalyses the decompositon reaction, which is autocatalytic. The acid cleavage reaction used industrially is highly exothermic, and safety aspects of commercial phenol/acetone processes are discussed fully [6]. Much detailed kinetic work has also been published [7].

1,2-Dibromo-1,2-diisocyanatoethane polymers
Lapshin, N. M. *et al., Chem. Abs.*, 1974, **81**, 64013m
A mixture of the dimeric and cyclic trimeric dibromo compounds with the hydroperoxide in benzene may react vigorously or explode if the solution is heated to concentrate it.
*See other* ALKYL HYDROPEROXIDES

## BENZYLDIMETHYLAMINE $PhCH_2NMe_2$ $C_9H_{13}N$

Cellulose nitrate
*See* CELLULOSE NITRATE: Amines

## 1,3,5-TRIMETHYLANILINE $Me_3C_6H_2NH_2$ $C_9H_{13}N$

Nitrosyl perchlorate
*See* NITROSYL PERCHLORATE, $ClNO_5$: Organic materials

## *O*-MESITYLENESULPHONYLHYDROXYLAMINE $C_9H_{13}NO_3S$

$Me_3C_6H_2SO_2ONH_2$

1. Tamura, Y. *et al., Tetrahedron Lett.*, 1972, (40), 4133
2. Ning, R. Y., *Chem. Eng. News*, 1973, **51**(51), 36–37
3. Carpino, L. A., *J. Amer. Chem. Soc.*, 1960, **82**, 3134
4. Ning, R. Y., private comm., 1974
5. Johnson, C. R. *et al., J. Org. Chem.*, 1974, **39**, 2459, footnote 15
6. Scopes, D. I. C. *et al., J. Org. Chem.*, 1977, **42**, 376

The title compound was prepared following a published procedure [1] and a dried sample decomposed soon after putting it into an amber bottle for storage, the screw cap being shattered [2]. Although the instability of the compound had been mentioned [3], no suggestion of violent decomposition had previously been made. It seems likely that traces of surface alkali in the soda-glass bottle had catalysed the formation of the highly reactive imidogen diradical ($H\dot{N}\cdot$) from the base-labile compound, and subsequent exothermic decomposition. Storage in dichloromethane solution appears safe [4]. The small crystals produced by a modified method appear to be safe in storage at 0°C, or in use [5]. Attempted vacuum drying of the compound at ambient temperature led to a mild explosion. Subsequently, a solution of the wet solid in dimethoxyethane was dried over molecular sieve and used in further work [6].

*See* *O-p*-TOLUENESULPHONYLHYDROXYLAMINE, $C_7H_9NO_3S$

**2-(2-AMINOETHYLAMINO)-5-METHOXYNITROBENZENE** **$C_9H_{13}N_3O_3$**

**$(H_2NC_2H_4NH)MeOC_6H_3NO_2$**

*See* 2-(2-AMINOETHYLAMINO)NITROBENZENE, $C_8H_{11}N_3O_2$

**3,3,6,6-TETRAKIS(BROMOMETHYL)-9,9-DIMETHYL-1,2,4,5,7,8-HEXAOXAONANE** **$C_9H_{14}Br_4O_6$**

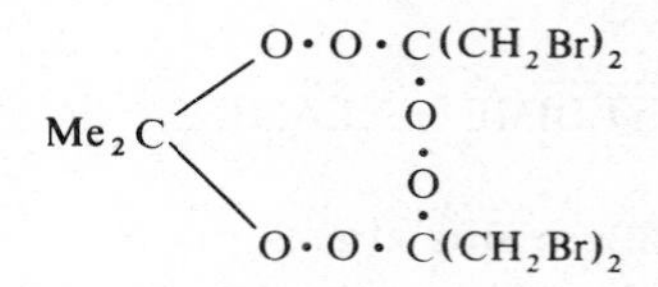

Schulz, M. *et al., Chem. Ber.*, 1967, **100**, 2245

Explodes on impact or friction, as do the tetrachloro and 9-ethyl-9-methyl analogues.

*See other* CYCLIC PEROXIDES

**1-(4-METHYL-1,3-DISELENOLYLIDENE)PIPERIDINIUM PERCHLORATE** **$C_9H_{14}ClNO_4Se_2$**

*See* 1-(1,3-DISELENOLYLIDENE)PIPERIDINIUM PERCHLORATES

**2,6-DIMETHYL-2,5-HEPTADIEN-4-ONE DIOZONIDE** **$C_9H_{14}O_7$**

**$CO(\underline{CHCMe_2O{-}O}^+{-}O^-)_2$**

Harries, G. *et al., Ann.*, 1910, **374**, 338

'Phorone' diozonide ignites at room temperature.

*See other* OZONIDES

***N*-CYANO-2-BROMOETHYLCYCLOHEXYLAMINE** **$C_9H_{15}BrN_2$**

**$C_6H_{11}N(CN)C_2H_4Br$**

*BCISC Quart. Safety Summ.*, 1964, **35**, 23

The reaction product from *N*-cyclohexylaziridine and cyanogen bromide (believed to be the title compound) exploded violently on attempted distillation at 160°C/0.5 mbar.

*See also* *N*-CYANO-2-BROMOETHYLBUTYLAMINE, $C_7H_{13}BrN_2$

*See other* CYANO COMPOUNDS

## TRIS(ETHYLTHIO)CYCLOPROPENYLIUM PERCHLORATE $C_9H_{15}ClO_4S_3$

$EtS\overline{CC(EtS)\overset{+}{=}C}SEt\ ClO_4^-$

Sunderlin, K. G. R., *Chem. Eng. News*, 1974, **52**(31), 3

Some 50 g of the compound had been prepared by a method used for analogous compounds, and the solvent-free oil had been left to crystallise. Some hours later it exploded with considerable violence. Preparation of this and related compounds by other workers had been uneventful.

*See* NON-METAL PERCHLORATES

## TRIALLYLCHROMIUM $(H_2C{=}CHCH_2)_3Cr$ $C_9H_{15}Cr$

O'Brien, S. *et al.*, *Inorg. Chem.*, 1972, **13**, 79

Triallylchromium and its thermal decomposition products are pyrophoric in air.

*See related* ALKYLMETALS
*See other* ALLYL COMPOUNDS

## *p*-TOLYLBIGUANIDIUM HYDROGENDICHROMATE $C_9H_{15}Cr_2N_5O_7$

$MeC_6H_4NHC(NH)NHC(NH)N^+H_3\ HCr_2O_7^-$

*See* DICHROMATE SALTS OF NITROGENOUS BASES
*See other* OXOSALTS OF NITROGENOUS BASES

## TRIALLYL PHOSPHATE $C_9H_{15}O_4P$

$(CH_2{=}CHCH_2O)_3PO$

1. Dye, W. T. *et al.*, *Chem. Eng. News*, 1950, **28**, 3452
2. Steinberg, G. M., ibid., 3755

Alkali-washed material, stabilised with 0.25% of pyrogallol, was distilled at 103°C/4 mbar until slight decomposition began. The heating mantle was removed and the still-pot temperature had fallen below its maximum value of 135°C when the residue exploded violently [1]. The presence of solid alkali [2] or 5% of phenolic inhibitor is recommended, together with low-temperature high-vacuum distillation, to avoid formation of acidic decomposition products, which catalyse rapid polymerisation.

*See other* ALLYL COMPOUNDS

## † 2,6-DIMETHYL-3-HEPTENE $C_9H_{18}$

$Me_2CHCH{=}CHCH_2CHMe_2$

### † 1,3,5-TRIMETHYLCYCLOHEXANE $C_9H_{18}$

$\overline{CH_2CHMeCH_2CHMeCH_2CHMe}$

### DI-*tert*-BUTYL DIPEROXYCARBONATE $C_9H_{18}O_5$

$(Me_3COO)_2CO$

A partially decomposed sample exploded violently at 135° C.

*See* PEROXYCARBONATE ESTERS

### 3,3,6,6,9,9-HEXAMETHYL-1,2,4,5,7,8-HEXAOXAONANE $C_9H_{18}O_6$

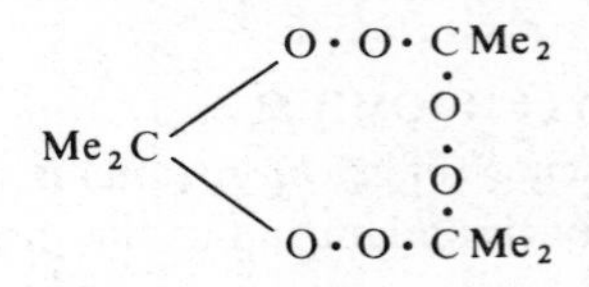


Dilthey, W. *et al., J. Prakt. Chem.*, 1940, **154**, 219

Rohrlich, M. *et al., Z. Ges. Schiess u. Sprengstoffw.*, 1943, **38**, 97–99

This trimeric acetone peroxide is powerfully explosive and will perforate a steel plate when heated on it.

*See also* 6-AMINOPENICILLANIC ACID *S*-OXIDE, $C_8H_{12}N_2O_4S$
OZONE, $O_3$: Citronellic acid

*See other* CYCLIC PEROXIDES

### 3,6,9-TRIETHYL-1,2,4,5,7,8-HEXAOXAONANE $C_9H_{18}O_6$

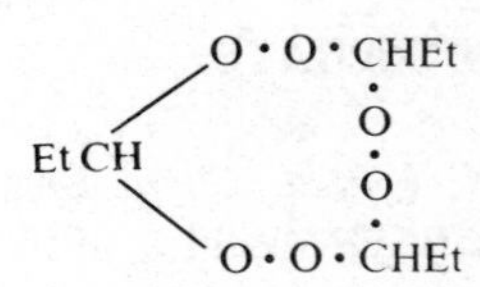


Rieche, A. *et al., Ber.*, 1939, **72**, 1938

This trimeric 'propylidene peroxide' is an extremely explosive and friction-sensitive oil.

*See other* CYCLIC PEROXIDES

### 2,2,4,4,6,6 -HEXAMETHYLTRITHIANE $C_9H_{18}S_3$

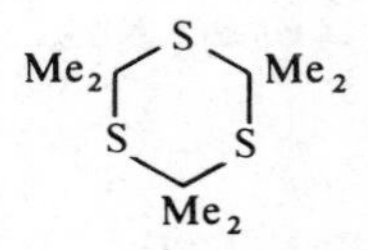


Nitric acid

*See* NITRIC ACID, $HNO_3$ : Hexamethyltrithiane

| | | |
|---|---|---|
| † **3,3-DIETHYLPENTANE** | **$Et_4C$** | **$C_9H_{20}$** |
| † **2,5-DIMETHYLHEPTANE** | **$Me_2HC_2H_4CHMeEt$** | **$C_9H_{20}$** |
| † **3,5-DIMETHYLHEPTANE** | **$EtMeCHCH_2CHMeEt$** | **$C_9H_{20}$** |
| † **4,4-DIMETHYLHEPTANE** | **$Me_2CPr_2$** | **$C_9H_{20}$** |
| † **3-ETHYL-2,3-DIMETHYLPENTANE** | **$Me_2CHC(Me)Et_2$** | **$C_9H_{20}$** |
| † **3-ETHYL-4-METHYLHEXANE** | **$Et_2CHCHMeEt$** | **$C_9H_{20}$** |
| † **4-ETHYL-2-METHYLHEXANE** | **$Me_2CHCH_2CHEt_2$** | **$C_9H_{20}$** |
| † **2-, 3- or 4-METHYLOCTANE** | **$Me_2CH[CH_2]_5Me$, etc.** | **$C_9H_{20}$** |
| † **NONANE** | **$CH_3[CH_2]_7CH_3$** | **$C_9H_{20}$** |
| † **2,2,5-TRIMETHYLHEXANE** | **$Me_3CC_2H_4CHMe_2$** | **$C_9H_{20}$** |
| † **2,2,3,3-TETRAMETHYLPENTANE** | **$Me_3CCMe_2Et$** | **$C_9H_{20}$** |
| † **2,2,3,4-TETRAMETHYLPENTANE** | **$Me_3CCHMeCHMe_2$** | **$C_9H_{20}$** |

**TRIISOPROPYLALUMINIUM** — **$(Me_2CH)_3Al$** — **$C_9H_{21}Al$**

*See* TRIALKYLALUMINIUMS

**TRIPROPYLALUMINIUM** — **$Pr_3Al$** — **$C_9H_{21}Al$**

*See* TRIALKYLALUMINIUMS

**TRIPROPYLSILICON PERCHLORATE** — **$Pr_3SiClO_4$** — **$C_9H_{21}ClO_4Si$**

*See* ORGANOSILICON PERCHLORATES

## TRIISOPROPYLPHOSPHINE $(Me_2CH)_3P$ $C_9H_{21}P$

Chloroform,
or Oxidants

Catalogue entry, Frankfurt, Deutsche Advance Produktion, 1968
Particularly this phosphine reacts, when undiluted, rather vigorously with most peroxides, ozonides, *N*-oxides, and also chloroform. It may be safely destroyed by pouring into a solution of bromine in carbon tetrachloride.

*See other* ALKYLNON-METALS

## TRIS(2-FLUORO-2-METHYL-2-SILAPROPYL)BORANE $C_9H_{24}BF_3Si_3$

$(FSiMe_2CH_2)_3B$

Hopper, S. P. *et al.*, *Synth. React. Inorg. Met.-Org. Chem.*, 1976, **6**, 378
The crude or pure material will ignite in air, especially if exposed on filter paper or a similar extended surface.

*See related* ALKYLBORANES
HALOSILANES

## TRIS(TRIMETHYLSILYL)ALUMINIUM $C_9H_{27}AlSi_3$

$(Me_3Si)_3Al$

Rösch, L., *Angew. Chem. (Intern. Ed.)*, 1977, **16**, 480
The crystalline solid ignites in air.

*See related* ALKYLMETALS
ALKYLNON-METALS

## POTASSIUM DINITROGENTRIS(TRIMETHYLPHOSPHINE)-COBALTATE(1–) $C_9H_{27}CoKN_2P_3$

$K[Co(Me_3P)_3N_2]$

Hammer, R. *et al., Angew. Chem. (Intern. Ed.)*, 1977, **16**, 485
The crystalline solid ignites immediately on exposure to air.

*See other* PYROPHORIC MATERIALS

## TRIS(TRIMETHYLSILYL)HYDRAZINE $C_9H_{28}N_2Si_3$

$Me_3SiNHN(SiMe_3)_2$

Oxidants

It is hypergolic with fluorine or fuming nitric acid and explodes with ozone.

*See* SILYLHYDRAZINES: Oxidants

### *N,N,N'*-TRIS(TRIMETHYLSILYL)DIAMINOPHOSPHANE $C_9H_{29}N_2PSi_3$

$(Me_3Si)_2NPHNHSiMe_3$

Niecke, E. *et al., Angew. Chem. (Intern. Ed.)*, 1977, **16**, 487

The finely divided solid is pyrophoric in air.

*See related* ALKYLNON-METALS
NON-METAL HYDRIDES

### OCTATETRAYNE-1,8-DICARBOXYLIC ACID $C_{10}H_2O_4$

$HO_2C[C{\equiv}C]_4CO_2H$

Sorbe, 1968, 109

Like most polyacetylenic acids, it is extremely explosive on warming.

*See other* ACETYLENIC COMPOUNDS

### TETRANITRONAPHTHALENE $C_{10}H_4(NO_2)_4$ $C_{10}H_4N_4O_8$

Sorbe, 1968, 151

An explosive solid, most sensitive to heating.

*See other* POLYNITROARYL COMPOUNDS

### 2,2'-AZO-3,5-DINITROPYRIDINE $C_{10}H_4N_8O_8$

$O_2N$ $NO_2$ $O_2N$ $NO_2$ N N=N N

Explosive

*See* POLYNITROAZOPYRIDINES

### 1,5-DINITRONAPHTHALENE $(NO_2)_2C_{10}H_6$ $C_{10}H_6N_2O_4$

Sulphur,
Sulphuric acid

Hub, L., *Proc. Chem. Process Haz. Sympos. VI* (Manchester, 1977), No. 49, 43, Rugby, I.Ch.E., 1977

For industrial conversion to 5-aminonaphthoquinone derivatives, dinitronaphthalene had been mixed cold with sulphuric acid and sulphur, then heated to 120°C on over 100 occasions without incident. When dinitronaphthalene from a different supplier was used the unheated mixture exploded violently. Investigation in the safety calorimeter showed that an exothermic reaction begins at only 30°C, and that the onset and intensity of the exotherm markedly depends upon quality of the dinitronaphthalene.

*See* REACTION SAFETY CALORIMETRY

## 6-QUINOLINECARBONYL AZIDE $C_{10}H_6N_4O$

$N_3CO$–[quinoline ring structure]

Houben-Weyl, 1952, Vol. 8, 682

A sample heated above its m.p. (88°C) exploded violently.

*See other* ACYL AZIDES

## 2,4-DIETHYNYLPHENOL $(HC{\equiv}C)_2C_6H_3OH$ $C_{10}H_6O$

Kotlyarevskii, I. L., *Chem. Abs.*, 1971, **74**, 53198e

Oxidation of the methyl or propyl ethers gave insoluble polymeric solids which exploded on heating.

*See other* ACETYLENIC COMPOUNDS
POLYPEROXIDES

## DI-2-FUROYL PEROXIDE $C_{10}H_6O_6$

$(C_4H_3O{\cdot}COO)_2$

Castrantas, 1965, 17

Explodes violently on friction and heating.

*See other* DIACYL PEROXIDES

## 1- or 2-NAPHTHALENEDIAZONIUM PERCHLORATE $C_{10}H_7ClN_2O_4$

$C_{10}H_7N_2^+\ ClO_4^-$

Hofmann, K. A. *et al., Ber.*, 1906, **39**, 3146

Both salts explode under light friction or pressure when dry.

*See other* DIAZONIUM PERCHLORATES

## MERCURY 2-NAPHTHALENEDIAZONIUM TRICHLORIDE

$C_{10}H_7Cl_3HgN_2$

$C_{10}H_7N_2^+\ Cl^-\cdot HgCl_2$

Nesmeyanow, A. M., *Org. Synth.*, 1943, Coll. Vol. 2, 433

The isolated double salt explodes violently if heated during drying.

*See other* DIAZONIUM SALTS

## 1-NITRONAPHTHALENE $C_{10}H_7NO_2$

Tetranitromethane

*See* TETRANITROMETHANE, $CN_4O_8$ : Aromatic nitrocompounds

## 1-NAPHTHALENEDIAZONIUM SALTS $C_{10}H_7N_2^+ Z^-$

Ammonium sulphide,
or Hydrogen sulphide
*See* DIAZONIUM SULPHIDES

## 2-NAPHTHALENEDIAZONIUM SALTS $C_{10}H_7N_2^+ Z^-$

Ammonium sulphide,
or Hydrogen sulphide,
or Sodium sulphides
*See* DIAZONIUM SULPHIDES

## NAPHTHYLSODIUM $C_{10}H_7Na$

Chlorinated diphenyl
*MCA Case History No. 565*
To help extinguish a burning batch of naphthylsodium, a chlorinated diphenyl heat-transfer liquid was added. An exothermic reaction, followed by an explosion, occurred.

Sodium is known to react violently with many halogenated materials.
*See* SODIUM, Na: Halocarbons
*See other* ARYLMETALS

## NAPHTHALENE $C_{10}H_8$

*MCA SD-58*, 1956

Aluminium trichloride,
Benzoyl chloride
*See* ALUMINIUM TRICHLORIDE, $AlCl_3$: Benzoyl chloride, etc.

Dinitrogen pentaoxide
*See* DINITROGEN PENTAOXIDE, $N_2O_5$: Naphthalene

## 2-FURALDEHYDE AZINE $C_{10}H_8N_2O_2$

$(\overline{OCH=CHCH=C}CH=N)_2$

Nitric acid
*See* NITRIC ACID, $HNO_3$: Aromatic amines (reference 5)
: Hydrazine and derivatives

## 2,2′-OXYDI[(IMINOMETHYL)FURAN] MONO-*N*-OXIDE ('DEHYDROFURFURAL OXIME') $C_{10}H_8N_2O_4$

Ponzio, G. *et al., Gazz. Chim. Ital.*, 1906, **36**, 338–344
It explodes at 130°C.
*See related* OXIMES

## 1- or 2-NAPHTHYLAMINE $C_{10}H_7NH_2$ $C_{10}H_9N$

*See* 1- or 2-NAPHTHALENEDIAZONIUM SALTS, $C_{10}H_7N_2^+Z^-$

## DIPYRIDINESILVER(I) PERCHLORATE $C_{10}H_{10}AgClN_2O_4$

$[Ag2C_5H_5N]ClO_4$

Acids

Kauffman, G. B. *et al., Inorg. Synth.*, 1960, **6**, 7, 8
Contact with acids, especially hot, must be avoided to prevent the possibility of violent explosion.
*See other* AMMINEMETAL OXOSALTS

## FERROCENIUM PERCHLORATE $C_{10}H_{10}ClFeO_4$

$C_5H_5Fe^+C_5H_5\ ClO_4^-$

*See* 1,3-BIS(DI-η-CYCLOPENTADIENYLIRON)-2-PROPEN-1-ONE, $C_{23}H_{22}Fe_2O$: Perchloric acid, etc.

## OXODIPEROXODIPYRIDINECHROMIUM(VI) $C_{10}H_{10}CrN_2O_5$

$[CrO(O_2)_2 2C_5H_5N]$

1. Caldwell, S. H. *et al., Inorg. Chem.*, 1969, **8**, 151
2. Adams, D. M. *et al., J. Chem. Educ.*, 1966, **43**, 94
3. Collins, J. C. *et al., Org. Synth.*, 1972, **52**, 5–8
4. Wiede, O. F., *Ber.*, 1897, **30**, 2186

This complex, formerly called 'pyridine perchromate' and now finding wide application as a powerful and selective oxidant, is violently explosive when dry [1]. Use while moist on the day of preparation and destroy any surplus with dilute alkali [2]. Preparation and use of the reagent have been detailed further [3]. The corresponding complexes of aniline, piperidine and quinoline may be similarly hazardous [2]. Dipyridinium dichromate is a much safer similarly powerful oxidant.

The damage caused by a 1 g sample of the pyridine complex exploding during vacuum desiccation on a warm day was extensive. Desiccation of the aniline complex had to be at ice temperature to avoid violent explosion [4].

*See also* CHROMIUM TRIOXIDE, $CrO_3$ : Pyridine
DIPYRIDINIUM DICHROMATE, $C_{10}H_{12}Cr_2NO_7$
*See other* AMMINECHROMIUM PEROXOCOMPLEXES

**BIS(η-CYCLOPENTADIENYLDINITROSYLCHROMIUM)** $C_{10}H_{10}Cr_2N_4O_4$
$[C_5H_5(NO)_2CrCr(NO)_2C_5H_5]$

Flitcroft, N. *et al.*, *Chem. & Ind.*, 1969, 201
A small sample exploded violently upon laser irradiation for Raman spectroscopy.
*See other* ORGANOMETALLICS

**BIS(η-CYCLOPENTADIENYL)MAGNESIUM** $C_{10}H_{10}Mg$
$C_5H_5MgC_5H_5$

Barber, W. A., *Inorg. Synth.*, 1960, **6**, 15
It may ignite on exposure to air.
*See other* ORGANOMETALLICS

**DIPYRIDINESODIUM** $2C_5H_5N{\cdot}Na$ $C_{10}H_{10}N_2Na$

Sidgwick, 1950, 89
The addition product of sodium and pyridine ignites in air.
*See also* DI(2-METHYLPYRIDINE)SODIUM, $C_{12}H_{14}N_2Na$

**BIS(π-CYCLOPENTADIENYL)TUNGSTEN DIAZIDE OXIDE** $C_{10}H_{10}N_6OW$
$(C_5H_5)_2W(N_3)_2O$

Anand, S. P., *J. Inorg. Nucl. Chem.*, 1974, **36**, 926
It is highly explosive.
*See related* METAL AZIDES

**1-KETO-1,2,3,4-TETRAHYDRONAPHTHALENE** $C_{10}H_{10}O$

$\overline{C_6H_4COC_2H_4CH_2}$

Preparative hazards
*See* CHROMIUM TRIOXIDE, $CrO_3$ : Acetic anhydride, etc.
HYDROGEN PEROXIDE, $H_2O_2$ : Acetone, etc.

**BIS-η-CYCLOPENTADIENYLLEAD** $(C_5H_5)_2Pb$ $C_{10}H_{10}Pb$

Preparative hazard

*See* LEAD(II) NITRATE, $N_2O_6Pb$: Cyclopentadienylsodium

**BIS-η-CYCLOPENTADIENYLTITANIUM** $C_{10}H_{10}Ti$

$(C_5H_5)_2Ti$

Bailar, 1973, Vol. 3, 389

Pyrophoric crystals (but see Polar solvents, below)

Polar solvents

Watt, G. W. *et al., J. Amer. Chem. Soc.*, 1966, **88**, 1139–1140

Though not pyrophoric, it reacts very vigorously with oxygen-free water and other polar solvents.

*See other* ORGANOMETALLICS

**BIS-η-CYCLOPENTADIENYLZIRCONIUM** $C_{10}H_{10}Zr$

$(C_5H_5)_2Zr$

Watt, G. W. *et al., J. Amer. Chem. Soc.*, 1966, **88**, 5926

A pyrophoric solid.

*See other* ORGANOMETALLICS

**3-*p*-CHLOROPHENYLBUTANOIC ACID** $C_{10}H_{11}ClO_2$

$ClC_6H_4CHMeCH_2CO_2H$

Preparative hazard

*See* PEROXYACETIC ACID, $C_2H_4O_3$: 5-*p*-Chlorophenyl-2,2-dimethyl-3-hexanone

**DIMETHYL-PHENYLETHYNYLTHALLIUM** $C_{10}H_{11}Tl$

$Me_2TlC{\equiv}CPh$

Nast, R. *et al., J. Organomet. Chem.*, 1966, **6**, 461

May explode on heating, stirring or impact.

*See other* ALKYLMETALS
METAL ACETYLIDES

**† DICYCLOPENTADIENE** $C_{10}H_{12}$

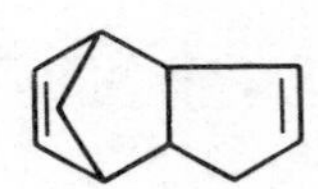

**ETHYLPHENYLTHALLIC ACETATE PERCHLORATE** $C_{10}H_{12}ClO_6Tl$

$EtC_6H_4Tl(II)OAc, ClO_4$

*See* PERCHLORIC ACID, $ClHO_4$ : Ethylbenzene

**6,6′-DIHYDRAZINO-2,2′-BIPYRIDYLNICKEL(II) PERCHLORATE** $C_{10}H_{12}Cl_2N_6NiO_8$

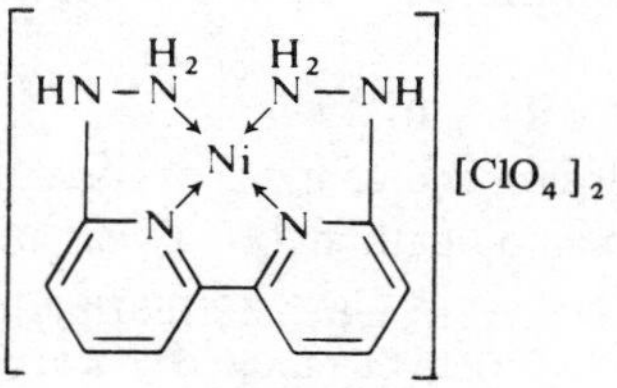

Lewis, J. *et al., J. Chem. Soc., Dalton Trans.*, 1977, 738

It detonates violently on heating, but is stable to shock treatment.

*See other* AMMINEMETAL OXOSALTS

**DIPYRIDINIUM DICHROMATE** $C_{10}H_{12}Cr_2N_2O_7$

$(C_5H_5N^+H)_2\ Cr_2O_7^{2-}$

Coates, W. M. *et al., Chem. & Ind.*, 1969, 1594

Though an oxidant of comparable power to 'pyridine perchromate', the dichromate is free of the explosive properties of the former.

*See* OXODIPEROXODIPYRIDINECHROMIUM(VI), $C_{10}H_{10}CrNO_5$

*See other* DICHROMATE SALTS OF NITROGENOUS BASES

**3,4-DIMETHYL-4-(3,4-DIMETHYL-5-ISOXAZOLYLAZO)-ISOXAZOLIN-5-ONE** $C_{10}H_{12}N_4O_3$

OCON=CMeC(Me)N=NC=CMeC(Me)=NO

Boulton, A. J. *et al., J. Chem. Soc.*, 1965, 5415

It invariably decomposed explosively if heated rapidly to 100°C but was stable to impact or friction.

*See other* AZO COMPOUNDS

**1,2,3,4-TETRAHYDRO-1-NAPHTHYL HYDROPEROXIDE** $C_{10}H_{12}O_2$

$C_6H_4CH(OOH)C_2H_4CH_2$

Hock, H. *et al., Ber.*, 1933, **66**, 61

Explodes on superheating the liquid.

Lithium tetrahydroaluminate

Sutton, D. A., *Chem. & Ind.*, 1951, 272

Interaction in ether is vigorously exothermic.

*See* DIBENZOYL PEROXIDE, $C_{14}H_{10}O_4$: Lithium tetrahydroaluminate

*See other* ALKYL HYDROPEROXIDES

**ALLYL *p*-TOLUENESULPHONATE** **$C_{10}H_{12}O_3S$**

**$MeC_6H_4SO_2OCH_2CH{=}CH_2$**

Rüst, 1948, 302

Explosion and charring of the ester during high-vacuum distillation from an oil-bath at 110°C was ascribed to exothermic polymerisation.

*See* ALLYL BENZENESULPHONATE, $C_9H_{10}O_3S$
TRIALLYL PHOSPHATE, $C_9H_{15}O_4P$
2-BUTEN-1-YL BENZENESULPHONATE, $C_{10}H_{12}O_3S$

*See other* ALLYL COMPOUNDS
POLYMERISATION INCIDENTS

**2-BUTEN-1-YL BENZENESULPHONATE** **$C_{10}H_{12}O_3S$**

**$PhSO_2OCMe{=}CHMe$**

Sorbe, 1968, 122

After evaporation of solvent, the 'crotyl' ester may explode in contact with air (possibly owing to exothermic polymerisation).

*See* ALLYL BENZENESULPHONATE, $C_9H_{10}O_3S$
ALLYL *p*-TOLUENESULPHONATE, $C_{10}H_{12}O_3S$

*See other* POLYMERISATION INCIDENTS

**2-FORMAMIDO-1-PHENYL-1,3-PROPANEDIOL** **$C_{10}H_{13}NO_3$**

**$PhCH(OH)CH(NHCHO)CH_2OH$**

Nitric acid

*See* NITRIC ACID, $HNO_3$: 2-Formamido-1-phenyl-1,3-propanediol

**BUTYLBENZENE** **BuPh** **$C_{10}H_{14}$**

Preparative hazard

*See* SODIUM, Na: Halocarbons (reference 7)

**BIS($\eta$-CYCLOPENTADIENYL)TETRAHYDROBORATONIOBIUM(III)**

**$(C_5H_5)_2NbBH_4$** **$C_{10}H_{14}BNb$**

Lucas, C. R., *Inorg. Synth.*, 1976, **16**, 109–110

It is pyrophoric in air, particularly after sublimation.

*See other* ORGANOMETALLICS

**BIS(2,4-PENTANEDIONATO)CHROMIUM** $C_{10}H_{14}CrO_4$

$Cr(C_5H_7O_2)_2$

Ocone, L. R. *et al., Inorg. Synth.*, 1966, **8**, 131
It ignites in air.
*See also* CHROMIUM DIACETATE, $C_4H_6CrO_4$
*See other* ORGANOMETALLICS

**2-METHYL-1-PHENYL-2-PROPYL HYDROPEROXIDE** $C_{10}H_{14}O_2$

$PhCH_2CMe_2OOH$

Preparative hazard
*See* HYDROGEN PEROXIDE, $H_2O_2$ : 2-Methyl-1-phenyl-2-propanol, etc.

**1,5-*p*-MENTHADIENE** $C_{10}H_{16}$

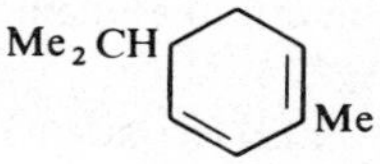

$Me_2CH$ [structure] Me

Air
Bodendorf, K., *Arch. Pharm.*, 1933, **271**, 28
The terpene readily peroxidises with air, and the (polymeric) peroxidic residue exploded violently on attempted distillation at 100°C/0.4 mbar
*See other* POLYPEROXIDES

**2-PINENE** $C_{10}H_{16}$

Me [structure] $Me_2$

Nitrosyl perchlorate
*See* NITROSYL PERCHLORATE, $ClNO_5$ : Organic materials

**1-(4,5-DIMETHYL-1,3-DISELENOLYLIDENE)PIPERIDINIUM PERCHLORATE** $C_{10}H_{16}ClNO_4Se_2$

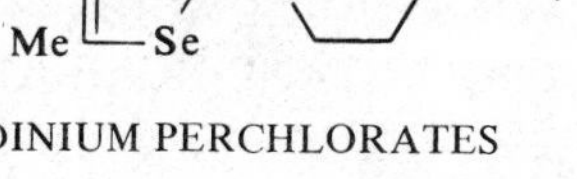

Me, Se, Se, Me, $N^+$ [structure] $ClO_4^-$

*See* 1-(1,3-DISELENOLYLIDENE)PIPERIDINIUM PERCHLORATES

**SEBACOYL DICHLORIDE** $C_{10}H_{16}Cl_2O_2$

$ClCO[CH_2]_8COCl$

Hüning, S. *et al., Org. Synth.*, 1963, **43**, 37

During vacuum distillation of the chloride at 173°C/20 mbar, the residue frequently decomposes spontaneously, producing a voluminous black foam.

### DISODIUM 1,3-DIHYDROXY-1,3-BIS-(*aci*-NITROMETHYL)-2,2,4,4-TETRAMETHYLCYCLOBUTANE $C_{10}H_{16}N_2Na_2O_6$

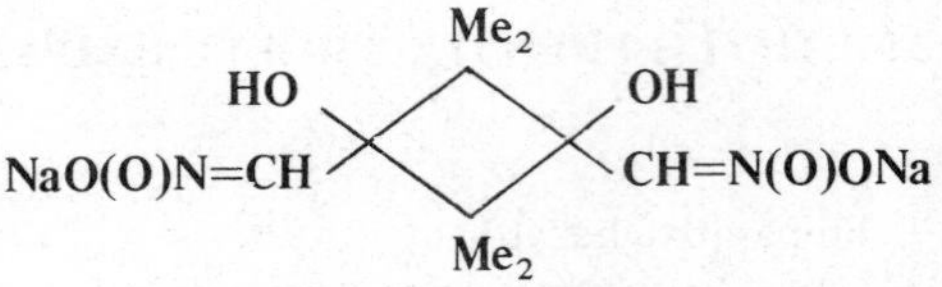

$Me_2$
HO OH
NaO(O)N=CH CH=N(O)ONa
$Me_2$

Dauben, H. J., Jr., *Org. Synth.*, 1963, Coll. Vol. 4, 223
The dry powdered condensation product of sodium *aci*-nitromethane (2 mol) with dimethylketene dimer exploded violently when added to crushed ice.
*See* SODIUM *aci*-NITROMETHANE, $CH_2NNaO_2$ : Water
*See other* *aci*-NITRO SALTS

### 3,7-DIMETHYL-2,6-OCTADIENAL (CITRAL) $C_{10}H_{16}O$

$Me_2C{=}CH[CH_2]_2CMe{=}CHCHO$

Potential preparative hazard
*See* 2-METHYL-1,3-BUTADIENE (ISOPRENE), $C_5H_8$ : Acetone

### 1,4-EPIDIOXY-2-*p*-MENTHENE (ASCARIDOLE) $C_{10}H_{16}O_2$

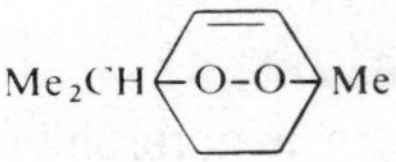

$Me_2CH$–O–O–Me

Castrantas, 1965, 15
Explosive decomposition on heating from 130 to 150°C.
*See other* CYCLIC PEROXIDES

### 3-PEROXYCAMPHORIC ACID (3-PEROXY-1,2,2-TRIMETHYL-1,3-CYCLOPENTANEDICARBOXYLIC ACID) $C_{10}H_{16}O_5$

$Me_2CC(Me)CO_2H[CH_2]_2CHCO_2OH$


1. Milas, N.A. *et al.*, *J. Amer. Chem. Soc.*, 1933, **55**, 350
2. Pirkle, W.H. *et al.*, *J. Org. Chem.*, 1977, **42**, 2080–2082

Explosive decomposition occurs at 80–100°C [1], but the 1-peroxyacid may also have been present in the sample [2].

*See other* PEROXYACIDS

**1-PYRROLIDINYLCYCLOHEXENE** 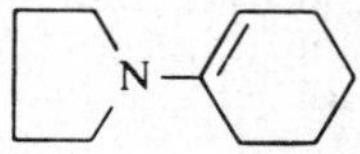 $C_{10}H_{17}N$

Benzenediazonium-2-carboxylate

*See* BENZENEDIAZONIUM-2-CARBOXYLATE, $C_7H_4N_2O_2$: 1-Pyrrolidinylcyclohexene

***cis*-CYCLODECENE** $HC{=}CH[CH_2]_7CH_2$ $C_{10}H_{18}$

Preparative hazard

*See* HYDRAZINE, $H_4N_2$: Metal salts

**CITRONELLIC ACID** $C_{10}H_{18}O_2$

$Me_2C{=}CHC_2H_4CHMeCH_2CO_2H$

Ozone

*See* OZONE, $O_3$: Citronellic acid

**1,4-BUTANEDIOL DI-2,3-EPOXYPROPYL ETHER** $C_{10}H_{18}O_4$

$OCH_2CHCH_2O[CH_2]_4OCH_2CHCH_2O$

Trichloroethylene

*See* TRICHLOROETHYLENE, $C_2HCl_3$: Epoxides

**DI-2-METHYLBUTYRYL PEROXIDE** $(EtCHMeCO_2)_2$ $C_{10}H_{18}O_4$

Swern, 1971, Vol. 2, 815

Pure material explodes on standing at room temperature.

*See other* DIACYL PEROXIDES

**DI-*tert*-BUTYL DIPEROXYOXALATE** $(Me_3COOCO)_2$ $C_{10}H_{18}O_6$

Castrantas, 1965, 17

Exploded on removing from a freezing mixture.

**3-BUTEN-1-YNYLTRIETHYLLEAD** $C_{10}H_{18}Pb$

$H_2C{=}CHC{\equiv}CPbEt_3$

Zavagorodnii, S. V. *et al., Dokl. Akad. Nauk SSSR*, 1962, **143**, 855 (Eng. transl. 268)

It explodes on rapid heating.

*See other* METAL ACETYLIDES

**MANGANESE(II) *N,N*-DIETHYLDITHIOCARBAMATE** $C_{10}H_{20}MnN_2S_4$

$Mn(SCSNEt_2)_2$

Bailar, 1973, Vol. 3, 872

A pyrophoric solid.

*See other* PYROPHORIC MATERIALS

**2,2′-DI-*tert*-BUTYL-3,3′-BIOXAZIRANE** $C_{10}H_{20}N_2O_2$

$Me_3CNOCHCHONCMe_3$

Putnam, S. J. *et al., Chem. Eng. News*, 1958, **36**(23), 46

A sample prepared by an established method and stored overnight at 2°C exploded when disturbed with a metal spatula.

*See* 2,2′-ETHYLENEBIS-(3-PHENYLOXAZIRANE), $C_{16}H_{16}N_2O_2$

**† 2-ETHYLHEXYL VINYL ETHER** $C_{10}H_{20}O$

$BuCHEtCH_2OCH{=}CH_2$

**† ISOPENTYL ISOVALERATE** $C_{10}H_{20}O_2$

$Me_2CHCH_2CO_2C_2H_4CHMe_2$

**TRIPROPYLLEAD FULMINATE** $Pr_3PbC{\equiv}N{\rightarrow}O$ $C_{10}H_{21}NOPb$

Houben-Weyl, 1975, Vol. 13.3, 101

An extremely explosive salt.

*See related* METAL FULMINATES

**† DECANE** $CH_3[CH_2]_8CH_3$ $C_{10}H_{22}$

**† 2-METHYLNONANE** $Me_2CH[CH_2]_6Me$ $C_{10}H_{22}$

**OXODIPEROXODIPIPERIDINECHROMIUM(VI)** $C_{10}H_{22}CrN_2O_5$

$[CrO(O_2)_2(C_5H_{11}N)_2]$

*See* OXODIPEROXODIPYRIDINECHROMIUM(VI), $C_{10}H_{10}CrN_2O_5$

**DIISOPENTYLMERCURY** $(Me_2CHC_2H_4)_2Hg$ $C_{10}H_{22}Hg$

Iodine

Whitmore, 1921, 103

Interaction is violent, accompanied by hissing.

*See other* MERCURY COMPOUNDS

**4-ETHOXYBUTYLDIETHYLALUMINIUM** $C_{10}H_{23}AlO$

$EtOC_4H_8AlEt_2$

Bahr, G. *et al.*, *Chem. Ber.*, 1955, **88**, 256

It ignites in air.

*See related* ALKYLALUMINIUM ALKOXIDES

**ETHOXYDIISOBUTYLALUMINIUM** $C_{10}H_{23}AlO$

$EtOAl(CH_2CHMe_2)_2$

May ignite in air.

*See other* ALKYLALUMINIUM ALKOXIDES

**DI(*O–O-tert*-BUTYL) ETHYL DIPEROXOPHOSPHATE** $C_{10}H_{23}O_6P$

$(Me_3COO)_2P(O)OEt$

Rieche, A. *et al.*, *Chem. Ber.*, 1962, **95**, 385

The liquid deflagrated soon after isolation.

*See other* *tert*-BUTYL PEROXOPHOSPHATE DERIVATIVES

**1,4,8,11-TETRAAZACYCLOTETRADECANENICKEL(II) PERCHLORATE** $C_{10}H_{24}Cl_2N_4NiO_8$

$[Ni\ \overline{CH_2NH[CH_2]_3NHC_2H_4NH[CH_2]_3NHCH_2}][ClO_4]_2$

Barefield, E. K., *Inorg. Chem.*, 1972, **11**, 2274

It exploded violently during analytical combustion.

*See other* AMMINEMETAL OXOSALTS

## 2,4-DI-*tert*-BUTYL-2,2,4,4-TETRAFLUORO-1,3-DIMETHYL 1,3,2,4-DIAZAPHOSPHETIDINE $C_{10}H_{24}F_4N_2P_2$

$Me_3CPF_2N(Me)Me_3CPF_2NMe$

Preparative hazard

*See* *tert*-BUTYLLITHIUM, $C_4H_9Li$: 2,2,2,4,4,4-Hexafluoro-1,3-dimethyl-1,3,2,4-diazaphosphetidine

## ETHYLENEBIS(DIETHYLPHOSPHINE) $C_{10}H_{24}P_2$

$Et_2PC_2H_4PEt_2$

Mays, M. J. *et al., Inorg. Synth.*, 1974, **15**, 22

It ignites in air.

*See other* ALKYLNON-METALS

## TETRAMETHYLBIS(TRIMETHYLSILOXO)DIGOLD $C_{10}H_{30}Au_2O_2Si_2$

$Me_3SiOAu(Me_2)Au(Me_2)OSiMe_3$

Schmidbaur, H. *et al., Inorg. Chem.*, 1966, **5**, 2069

Sublimed crystals decomposed explosively at 120°C.

*See other* GOLD COMPOUNDS

## DECACARBONYLDIRHENIUM $Re_2(CO)_{10}$ $C_{10}O_{10}Re_2$

Bailar, 1973, Vol. 3, 953

It tends to ignite in air above 140°C.

*See other* CARBONYLMETALS

## 3,5-DINITRO-2-(PICRYLAZO)PYRIDINE $C_{11}H_4N_8O_{10}$

$O_2N$, $NO_2$, $N{=}NC_6H_2(NO_2)_3$ (pyridine ring, N)

Explosive.

*See* POLYNITROAZOPYRIDINES

## 1-(3,5-DINITRO-2-PYRIDYL)-2-PICRYLHYDRAZINE $C_{11}H_6N_8O_{10}$

$O_2N$, $NO_2$, $NHNHC_6H_2(NO_2)_3$ (pyridine ring, N)

Explosive.

*See* POLYNITROAZOPYRIDINES

**1-(4-DIAZONIOPHENYL)-1,2-DIHYDROPYRIDINE-2-IMINOSULPHINATE** $C_{11}H_8N_4O_2S$

$N_2^+C_6H_4N$ (1,2-dihydropyridine ring) $=NS(O)O^-$

Hoffmann, H. *et al.*, US Pat. 3 985 724, 1977
Diazotisation of 2-*p*-aminophenylsulphonamidopyridine with subsequent adjustment of pH to 3–6 gives the internal diazonium salt, which decomposes violently at 230°C.
*See other* DIAZONIUM SALTS

**2,4-DIETHYNYL-5-METHYLPHENOL** $(HC{\equiv}C)_2MeC_6H_2OH$ $C_{11}H_8O$
Myasnikova, R. N. *et al.*, *Izv. Akad. Nauk SSSR*, 1970, **11**, 2637
Oxidation of the phenol or its methyl or propyl ethers in pyridine gave insoluble polymers. Those of the ethers exploded on heating.
*See other* ACETYLENIC COMPOUNDS
POLYPEROXIDES

**1(2-NAPHTHYL)-3(5-TETRAZOLYL)TRIAZENE** $C_{11}H_9N_7$
$C_{10}H_7N{=}NNHCHN_4$
Thiele, J., *Ann.*, 1892, **270**, 54; 1893, **273**, 144
Explodes at 184°C.
*See other* TETRAZOLES

**$\eta$-BENZENE-$\eta$-CYCLOPENTADIENYLIRON(II) PERCHLORATE**
$C_{11}H_{11}ClFeO_4$
$[C_5H_5FeC_6H_6][ClO_4]$
Denning, R. G. *et al.*, *J. Organomet. Chem.*, 1966, **5**, 292
The dry material is shock-sensitive and detonated on touching with a spatula.
*See other* ORGANOMETALLICS

**ETHYL $\alpha$-AZIDO-*N*-CYANOPHENYLACETIMIDATE** $C_{11}H_{11}N_5O$
$PhCH(N_3)C(NCN)OEt$
Petersen, H. J., *J. Med. Chem.*, 1974, **17**(1), 104
The crude imidoester should be used as a concentrated solution in ether. A small solvent-free sample exploded violently.
*See related* 2-AZIDOCARBONYL COMPOUNDS

## 3,3-DIMETHYL-1(3-QUINOLYL)TRIAZENE $C_{11}H_{12}N_4$

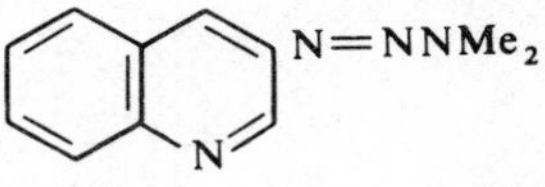

[Structure: quinoline ring with N=NNMe$_2$ at the 3-position]

Rondestvedt, C. S. *et al.*, *J. Org. Chem.*, 1957, **22**, 201
The crude material decomposes violently if allowed to dry, and purified material explodes at 131.5°C or during analytical combustion.
*See other* TRIAZENES

## 3-BUTYN-1-YL *p*-TOLUENESULPHONATE $C_{11}H_{12}O_3S$

$MeC_6H_4SO_3C_2H_4C{\equiv}CH$

Eglington, G. *et al.*, *J. Chem. Soc.*, 1950, 3653
The material could be distilled in small amounts at below 0.01 mbar, but exploded on attempted distillation at 0.65 mbar.
*See other* ACETYLENIC COMPOUNDS

## *tert*-BUTYL *p*-NITROPEROXYBENZOATE $C_{11}H_{13}NO_5$

$NO_2C_6H_4CO_2OCMe_3$

Criegee, R. *et al.*, *Ann.*, 1948, **560**, 135
This and *p*-nitrobenzoates of homologous *tert*-alkyl hydroperoxides explode in a flame.

## † *tert*-BUTYL PEROXYBENZOATE $PhCO_2OCMe_3$ $C_{11}H_{14}O_3$

Criegee, R. *Angew. Chem.*, 1953, **65**, 398–399
Shortly after interruption of vacuum distillation from an oil-bath at 115°C, to change a thermometer, the ester exploded violently. This was attributed to overheating.

Copper(I) bromide,
Limonene

Wilson, C. W. *et al.*, *J. Agric. Food Chem.*, 1975, **23**, 636
Addition of all the perester to limonene and catalytic amounts of copper(I) bromide before oxidation had begun (blue–green coloration) led to an explosion.
*See other* PEROXYESTERS

**METHYL 2-BROMO-5,5-ETHYLENEDIOXY[2.2.1]BICYCLOHEPTANE-7-CARBOXYLATE** $C_{11}H_{15}BrO_4$

Perchloryl fluoride

*See* PERCHLORYL FLUORIDE, ClFO: Methyl 2-bromo-5,5- ...

**2-*tert*-BUTYL-3-PHENYLOXAZIRANE** $Me_3\overline{CNOC}HPh$ $C_{11}H_{15}NO$

Emmons, W. D. *et al., Org. Synth.*, 1969, **49**, 13

Vacuum distillation of this active oxygen compound is potentially hazardous and precautions are necessary.

*See related* 1,2-EPOXIDES

***N*-PHENYLAZOPIPERIDINE** $C_{11}H_{15}N_3$

$PhN{=}N\overline{N[CH_2]_4CH_2}$

Hydrofluoric acid

Wallach, O., *Ann.*, 1886, **235**, 258; 1888, **243**, 219

Interaction to give fluorobenzene is violent and is not suitable for above 10 g quantities

*See other* HIGH-NITROGEN COMPOUNDS

**2-(1,3-DISELENA-4,5,6,7-TETRAHYDROINDANYLIDENE)-PIPERIDINIUM PERCHLORATE** $C_{11}H_{16}ClNO_4Se$

$ClO_4^-$

*See* 1-(1,3-DISELENOLYLIDENE)PIPERIDINIUM PERCHLORATES

**3-IODO-4-METHOXY-4,7,7-TRIMETHYLBICYCLO[4.1.0]HEXANE** $C_{11}H_{19}IO$

*See* *vic*-IODOALKOXY COMPOUNDS

**3-METHYL-3-BUTEN-1-YNYLTRIETHYLLEAD** $C_{11}H_{20}Pb$

$H_2C{=}CMeC{\equiv}CPbEt_3$

Zavgorodnii, S. V. *et al.*, *Dokl. Akad. Nauk SSSR*, 1962, **143**, 855 (Eng. transl. 268)

It explodes on rapid heating.

*See other* METAL ACETYLIDES

**3-DIBUTYLAMINOPROPYLAMINE** $Bu_2N[CH_2]_3NH_2$ $C_{11}H_{26}N_2$

Cellulose nitrate

*See* CELLULOSE NITRATE: Amines

**BIS(PENTAFLUOROPHENYL)ALUMINIUM BROMIDE** $C_{12}AlBrF_{10}$

$(C_6F_5)_2AlBr$

1. Chambers, R. D. *et al.*, *J. Chem. Soc.* (*C*), 1967, 2185; *Tetrahedron Lett.*, 1965, 2389
2. Cohen, S. C. *et al.*, *Advan. Fluorine Chem.*, 1970, **6**, 156

Ignites in air, explodes during uncontrolled hydrolysis and chars during controlled hydrolysis [1]. When isolated as the etherate, attempts to remove solvent ether caused violent decompositions [2].

*See other* HALO-ARYLMETALS

**BIS(HEXAFLUOROBENZENE)COBALT(0)** $C_{12}CoF_{12}$

$Co(C_6F_6)_2$

It decomposes explosively at 10°C.

*See* HALOARENEMETAL π-COMPLEXES

**DODECACARBONYLTETRACOBALT** $Co_4(CO)_{12}$ $C_{12}Co_4O_{12}$

Blake, E. J., private comm., 1974

Cobalt catalysts discharged from 'oxo'-process reactors are frequently pyrophoric, owing to the presence of the carbonylcobalt.

*See* OCTACARBONYLDICOBALT, $C_8Co_2O_8$

*See other* CARBONYLMETALS

**BIS(HEXAFLUOROBENZENE)CHROMIUM(0)** $C_{12}CrF_{12}$

$Cr(C_6F_6)_2$

It decomposes explosively at 40°C.

*See* HALOARENEMETAL π-COMPLEXES

## BIS(HEXAFLUOROBENZENE)IRON(0) $C_{12}F_{12}Fe$

$Fe(C_6F_6)_2$

It decomposes explosively at $-40°C$.

*See* HALOARENEMETAL π-COMPLEXES

## BIS(HEXAFLUOROBENZENE)NICKEL(0) $C_{12}F_{12}Ni$

$Ni(C_6F_6)_2$

Alone,
or Various reagents

Klabunde, K. J. *et al., J. Fluorine Chem.*, 1974, **4**, 114–115
The complex decomposes explosively on slight provocation (flakes falling into the reactor, static charge or uneven warming) and at 70°C with careful heating. At ambient temperature, interaction with air, hydrogen, carbon monoxide, allyl bromide and trifluoromethyl iodide is explosive, but can be controlled by preliminary cooling to $-196°C$, followed by slow warming.

*See* HALOARENEMETAL π-COMPLEXES

## BIS(HEXAFLUOROBENZENE)TITANIUM(0) $C_{12}F_{12}Ti$

$Ti(C_6F_6)_2$

It decomposes explosively at $-50°C$.

*See* HALOARENEMETAL π-COMPLEXES

## BIS(HEXAFLUOROBENZENE)VANADIUM(0) $C_{12}F_{12}V$

$V(C_6F_6)_2$

It decomposes explosively at 100°C.

*See* HALOARENEMETAL π-COMPLEXES

## DODECACARBONYLTRIIRON $C_{12}Fe_3O_{12}$

$(CO)_3Fe(CO)_3Fe(CO)_3Fe(CO)_3$

King, R. B. *et al., Inorg. Synth.*, 1963, **7**, 195
On prolonged storage, pyrophoric decomposition products are formed.

*See other* CARBONYLMETALS

## SILVER HEXANITRODIPHENYLAMIDE $C_{12}H_4AgN_7O_{12}$

$AgN[C_6H_2(NO_2)_3]_2$

Taylor, C. A. *et al., Army Ordnance*, 1926, **7**, 68–69
It has been evaluated as a detonator.

*See other* *N*-METAL DERIVATIVES

## CALCIUM PICRATE

$C_{12}H_4CaN_6O_{14}$

$Ca[OC_6H_2(NO_2)_3]_2$

*See* PICRIC ACID, $C_6H_3N_3O_7$

## BIS-2,4,5-TRICHLOROBENZENEDIAZO OXIDE

$C_{12}H_4Cl_6N_4O$

$(Cl_3C_6H_2N_2)_2O$

Alone,
or Benzene

Kaufmann, T. *et al.*, *Ann.*, 1960, **634**, 77

The dry solid explodes under a hammer blow, or on moistening with benzene.

*See other* BIS-ARENEDIAZO OXIDES

## COPPER DIPICRATE

$C_{12}H_4CuN_6O_{14}$

$Cu[OC_6H_2(NO_2)_3]_2$

*See* PICRIC ACID, $C_6H_3N_3O_7$
*See other* HEAVY METAL DERIVATIVES

## MERCURY DIPICRATE

$C_{12}H_4HgN_6O_{14}$

$Hg[OC_6H_2(NO_2)_3]_2$

*See* PICRIC ACID, $C_6H_3N_3O_7$
*See other* MERCURY COMPOUNDS

## LEAD DIPICRATE

$C_{12}H_4N_6O_{14}Pb$

$Pb[OC_6H_2(NO_2)_3]_2$

Belcher, R., *J. R. Inst. Chem.*, 1960, **84**, 377

During the usual qualitative inorganic analytical procedure, samples containing the lead and salicylate radicals can lead to the formation and possible detonation of lead dipicrate. This arises during evaporation of the filtrate with nitric acid, after precipitation of the copper–tin group metals with hydrogen sulphide. Salicylic acid is converted under these conditions to picric acid, which, in presence of lead, gives explosive lead dipicrate.

An alternative (MAQA) scheme is described which avoids this possibility.

*See* PICRIC ACID, $C_6H_3N_3O_7$
*See other* HEAVY METAL DERIVATIVES

## ZINC DIPICRATE

$C_{12}H_4N_6O_{14}Zn$

$Zn[OC_6H_2(NO_2)_3]_2$

*See* PICRIC ACID, $C_6H_3N_3O_7$
*See other* HEAVY METAL DERIVATIVES

## DINITRO-1-PICRYLBENZOTRIAZOLES

$C_{12}H_4N_8O_{10}$

$(NO_2)_2C_6H_2N{=}NNC_6H_2(NO_2)_3$

Coburn, M. D., *J. Heterocycl. Chem.*, 1973, **10**, 743–746
The isomeric 4,5-, 5,6- and 5,7-dinitro derivatives, and some of the mononitro derivatives, are impact-sensitive explosives.
*See other* POLYNITROARYL COMPOUNDS

## 1,3,6,8-TETRANITROCARBAZOLE

$C_{12}H_5N_5O_8$

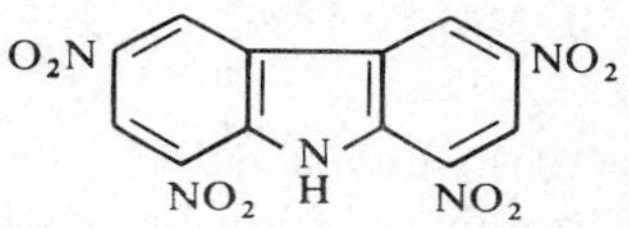

$O_2N$ $NO_2$
N
$NO_2$ H $NO_2$

Sorbe, 1968, 153
It may readily explode when dry, so it should be stored wet with water (10% is normally added to commercial material).
*See other* POLYNITROARYL COMPOUNDS

## 1,3,5-TRIETHYNYLBENZENE

$(HC{\equiv}C)_3C_6H_3$ $C_{12}H_6$

Shvartsberg, M. S. *et al.*, *Izv. Akad. Nauk SSSR, Ser. Khim.*, 1963, **110**, 1836
Polymerised explosively on rapid heating and compression.
*See other* ACETYLENIC COMPOUNDS

## 2,6-DIPERCHLORYL-4,4′-DIPHENOQUINONE

$C_{12}H_6Cl_2O_8$

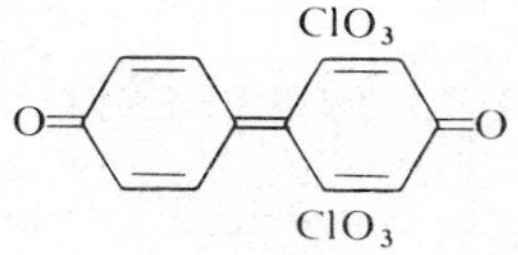

$ClO_3$
O O
$ClO_3$

Gardner, D. M. *et al.*, *J. Org. Chem.*, 1963, **28**, 2650
A shock-sensitive explosive.
*See other* PERCHLORYL COMPOUNDS

## BIS-3,4-DICHLOROBENZENESULPHONYL PEROXIDE $C_{12}H_6Cl_4O_6S_2$

$Cl_2C_6H_3SO_2OOSO_2C_6H_3Cl_2$

Dannley, R. *et al., J. Org. Chem.*, 1966, **31**, 154

The peroxide was too unstable to dry thoroughly; such samples often exploded spontaneously.

*See other* DIACYL PEROXIDES

## POTASSIUM HEXAKIS(ETHYNYL)COBALTATE(4–) $C_{12}H_6CoK_4$

$K_4[Co(C{\equiv}CH)_6]$

Nast, R. *et al., Z. Anorg. Chem.*, 1955, **282**, 210

It is moderately stable at below –30°C, very shock- and friction-sensitive, and explodes violently on contact with water. At ambient temperature, it rapidly forms explosive decomposition products. Its addition compound with ammonia behaves similarly, exploding on contact with air.

*See other* COMPLEX ACETYLIDES

## IRON(III) MALEATE $C_{12}H_6Fe_2O_{12}$

$Fe_2(O_2CCH{=}CHCO_2)_3$

Iron(III) hydroxide

Thermally unstable basic iron(III) maleates produced from mixtures of the above may be pyrophoric.

*See* PHTHALIC ANHYDRIDE, $C_8H_4O_3$: Preparative hazards (reference 2)

Sulphur compounds

*See* PHTHALIC ANHYDRIDE, $C_8H_4O_3$: Preparative hazards

## POTASSIUM HEXAETHYNYLMANGANATE(3–) $C_{12}H_6K_3Mn$

$K_3[Mn(C{\equiv}CH)_6]$

Bailar, 1973, Vol. 3, 855

A highly explosive solid.

*See other* COMPLEX ACETYLIDES

## (2,2-DICHLORO-1-FLUOROVINYL) FERROCENE $C_{12}H_7Cl_2FFe$

See p. 707.

## 2-IODO-3,5-DINITROBIPHENYL $C_{12}H_7IN_2O_4$

$I(NO_2)_2C_6H_2C_6H_5$

Sodium salt of ethyl acetoacetate

Zaheer, S. H. *et al., J. Indian Chem. Soc.*, 1955, **32**, 491

Interaction of 2-halo-3,5-dinitrobiphenyls with the sodium salt should be limited to 5–6 g of the title compound to avoid explosions observed with larger quantities.

*See other* HALOARYL COMPOUNDS

## POTASSIUM 6-*aci*-NITRO-2,4-DINITRO-1-PHENYLIMINO-2,4-CYCLOHEXADIENE

$C_{12}H_7KN_4O_6$

$NO_2$
NPh
$O_2N$
N(O)OK

Sudborough, J. J. *et al., J. Chem. Soc.*, 1906, **89T**, 586

This salt of the mono-*aci-o*-quinonoid form of 2,4,6-trinitro-*N*-phenyl aniline is explosive, like the *N*-1- and -2-naphthyl analogues.

*See* *aci*-NITROQUINONOID COMPOUNDS

## 1,4-DIHYDRODICYCLOPROPA[*b,g*]NAPHTHALENE

$C_{12}H_8$

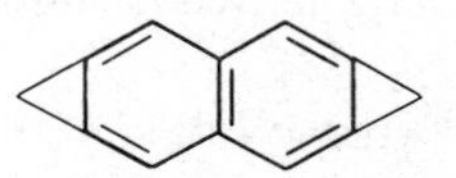

Ippen, J. *et al., Angew. Chem. (Intern. Ed.)*, 1974, **13**, 736

It undergoes explosive decomposition at 132–133°C, and is a shock-sensitive solid which requires handling with caution.

*See other* STRAINED-RING COMPOUNDS

## BIS-*p*-BROMOBENZENESULPHONYL PEROXIDE

$C_{12}H_8Br_2O_6S_2$

$BrC_6H_4SO_2OOSO_2C_6H_4Br$

1. Bolte, J. *et al., Tetrahedron Lett.*, 1965, 1529
2. Dannley, R. L. *et al., J. Org. Chem.*, 1966, **31**, 154

The peroxide was too unstable to dry thoroughly; such samples often exploded spontaneously.

*See other* DIACYL PEROXIDES

## THIANTHRENIUM PERCHLORATE

$C_{12}H_8ClO_4S_2$

$\overline{C_6H_4SC_6H_4S}^+ ClO_4^-$

Shine, H. J. *et al., Chem. & Ind.*, 1969, 782; *J. Org. Chem.*, 1971, **36**, 2925

A small portion (1–2 g) of the freshly prepared suction-dried material exploded violently during transfer from a sintered filter. Initiation may have been caused by friction of transfer or rubbing with a glass rod. Preparation of only 50–100 mg quantities is recommended.

*See other* NON-METAL PERCHLORATES

### BIS-*p*-CHLOROBENZENEDIAZO OXIDE $C_{12}H_8Cl_2N_4O$

$(ClC_6H_4N_2)_2O$

Alone,

or Benzene

Bamberger, E., *Ber.*, 1896, **29**, 464

More stable than unsubstituted analogues, it may be desiccated at 0°C, but is then extremely sensitive and violently explosive. Contact with benzene (even at 0°C) is violent and the reaction may become explosive.

*See other* BIS-ARENEDIAZO OXIDES

### 4,4′-BIPHENYLENEBISDIAZONIUM PERCHLORATE $C_{12}H_8Cl_2N_4O_8$

$N_2^+C_6H_4C_6H_4N_2^+\ 2ClO_4^-$

Vorlander, D., *Ber.*, 1906, **39**, 2713–2715

The perchlorate derived from tetrazotised benzidine is unstable and explosive.

*See other* DIAZONIUM PERCHLORATES

### BIS-*p*-CHLOROBENZENESULPHONYL PEROXIDE $C_{12}H_8Cl_2O_6S_2$

$ClC_6H_4SO_2OOSO_2C_6H_4Cl$

Dannley, R. L. *et al., J. Org. Chem.*, 1966, **31**, 154

The peroxide was too unstable to dry thoroughly; such samples often exploded spontaneously.

*See other* DIACYL PEROXIDES

### BIS(1,4(?)-DIFLUOROBENZENE)CHROMIUM(0) $C_{12}H_8CrF_4$

$Cr(C_6H_4F_2)_2$

Unstable, explosive.

*See* HALOARENEMETAL π-COMPLEXES

### 2-*trans*-1-AZIDO-1,2-DIHYDROACENAPHTHYL NITRATE $C_{12}H_8N_4O_3$

$C_{10}H_6CH(N_3)CHONO_2$

Trahanovsky, W. S. *et al., J. Amer. Chem. Soc.*, 1971, **93**, 5257

Although several other 1-azido-2-nitrato-alkanes appeared thermally stable, the acenaphthane derivative exploded violently on heating (probably during analytical combustion).

*See other* ALKYL NITRATES

ORGANIC AZIDES

**BIS-*p*-NITROBENZENEDIAZO SULPHIDE** $C_{12}H_8N_6O_4S$

$(NO_2C_6H_4N_2)_2S$

Tomlinson, W. R., *Chem. Eng. News*, 1951, **29**, 5473

The dry material is extremely sensitive and can be exploded by very light friction. The material is too sensitive to handle other than as a solution, or dilute slurry in excess solvent, and then only on 1 g scale.

*See* DIAZONIUM SULPHIDES

**(2,2-DICHLORO-1-FLUOROVINYL)FERROCENE** $C_{12}H_7Cl_2FFe$*

$(Cl_2C{=}CF)C_5H_2FeC_5H_5$

Okohura, K., *J. Org. Chem.*, 1976, **41**, 1493, footnote 14

Two attempts to purify the crude material by vacuum distillation led to sudden exothermic decomposition of the still contents. Distillation in steam was satisfactory.

*See other* ORGANOMETALLICS

**POTASSIUM TRICARBONYLTRIS(PROPYNYL)MOLYBDATE(3–)**

$C_{12}H_9KMoO_3$

$K[Mo(CO)_3(C{\equiv}CMe)_3]$

Houben-Weyl, 1975, Vol. 13.3, 470

It is pyrophoric.

*See related* CARBONYLMETALS
COMPLEX ACETYLIDES

**DI(BENZENEDIAZONIUM) ZINC TETRACHLORIDE** $C_{12}H_{10}Cl_4N_4Zn$

$2PhN_2Cl{\cdot}ZnCl_2$

Muir, G. D., *Chem. & Ind.*, 1956, 58

A batch of the double salt exploded, either spontaneously or from slight vibration, after thorough drying under vacuum at ambient temperature overnight. Although dry diazonium salts are known to be light-, heat- and shock-sensitive when dry, the double salts with zinc chloride were considered to be more stable. Presence of traces of solvent reduces the risk of frictional heating and deterioration.

*See other* DIAZONIUM SALTS

**BIS(FLUOROBENZENE)CHROMIUM(0)** $Cr(C_6H_5F)_2$ $C_{12}H_{10}CrF_2$

Unstable, potentially explosive.

*See* HALOARENEMETAL $\pi$-COMPLEXES

**BIS(FLUOROBENZENE)VANADIUM(0)** $V(C_6H_5F)_2$ $C_{12}H_{10}F_2V$

Unstable, potentially explosive.

*See* HALOARENEMETAL $\pi$-COMPLEXES

*Out of sequence.

### FERROCENE-1,1′-DICARBOXYLIC ACID $C_{12}H_{10}FeO_4$

$HO_2CC_5H_4FeC_5H_4CO_2H$

Phosphoryl chloride

*See* PHOSPHORYL CHLORIDE, $Cl_3OP$: Ferrocene-1,1′-dicarboxylic acid

### DIPHENYLMERCURY $Ph_2Hg$ $C_{12}H_{10}Hg$

Non-metal oxides

1. Dreher, E. *et al., Ann.*, 1870, **154**, 127
2. Otto, R., *J. Prakt. Chem.* [2], 1870, **1**, 183

Chlorine monoxide reacts violently [1] and sulphur trioxide very violently [2] with diphenylmercury.

*See other* ARYLMETALS

### POTASSIUM *O,O*-DIPHENYL DITHIOPHOSPHATE $C_{12}H_{10}KO_2PS_2$

$KSP(S)(OPh)_2$

Arenediazonium salts

*See* DIAZONIUM SULPHIDES AND DERIVATIVES (reference 10)

### DIPHENYLMAGNESIUM $Ph_2Mg$ $C_{12}H_{10}Mg$

Air,
Water

Sidgwick, 1950, 234

It ignites in moist (but not dry) air, and reacts violently with water, reaching incandescence.

*See other* ARYLMETALS

### DIPHENYLPHOSPHORUS(III) AZIDE $Ph_2PN_3$ $C_{12}H_{10}N_3P$

Allcock, H. R. *et al., Macromolecules*, 1975, **8**, 380

The azide monomer, prepared in solution from the chlorophosphine and sodium azide, should not be isolated, because it is potentially explosive.

*See other* NON-METAL AZIDES

### BIS-BENZENEDIAZO OXIDE $(PhN_2)_2O$ $C_{12}H_{10}N_4O$

Bamberger, E., *Ber.*, 1896, **29**, 460

Extremely unstable, it explodes on attempted isolation from the liquor, or on allowing the latter to warm to $-18^\circ C$.

*See other* BIS-ARENEDIAZO OXIDES

**DI(BENZENEDIAZO) SULPHIDE** **$(PhN_2)S$** **$C_{12}H_{10}N_4S$**

Tomlinson, W. R., *Chem. Eng. News*, 1951, **29**, 5473

The wet solid can be exploded by impact or heating, and explodes while drying in air at ambient temperature. The material is too sensitive to handle other than as a solution or dilute slurry in excess solvent, and then only on 1 g scale.

*See* DIAZONIUM SULPHIDES

**DIPHENYL ETHER** **$Ph_2O$** **$C_{12}H_{10}O$**

Chlorosulphuric acid

*See* CHLOROSULPHURIC ACID, $ClHO_3S$: Diphenyl ether

**DIPHENYLSELENONE** **$Ph_2SeO_2$** **$C_{12}H_{10}O_2Se$**

Krafft, F. *et al.*, *Ber.*, 1896, **29**, 424

Explodes feebly on heating in a test-tube.

**BENZENESULPHONIC ANHYDRIDE** **$C_{12}H_{10}O_5S_2$**

**$C_6H_5SO_2OSO_2C_6H_5$**

Hydrogen peroxide

*See* HYDROGEN PEROXIDE, $H_2O_2$: Benzenesulphonic anhydride

**DIBENZENESULPHONYL PEROXIDE** **$C_{12}H_{10}O_6S_2$**

**$C_6H_5SO_2OOSO_2C_6H_5$**

1. Davies, 1961, 65
2. Crovatt, L. W. *et al.*, *J. Org. Chem.*, 1959, **24**, 2032
3. Weinland, R. F. *et al.*, *Ber.*, 1903, **36**, 2702

It explodes at 53–54°C [1], and decomposes somewhat violently after storage overnight at ambient temperature, but may be stored unchanged for several weeks at –20°C [2]. It is also shock-sensitive, and explodes with fuming nitric acid, or on addition to boiling water [3].

*See other* DIACYL PEROXIDES

**DIPHENYLDISTIBENE (STIBOBENZENE)** **$C_{12}H_{10}Sb_2$**

**PhSb=SbPh**

Air,
or Nitric acid

Schmidt, H., *Ann.*, 1920, **421**, 235

This antimony analogue of azobenzene ignites in air and is oxidised explosively by nitric acid.

**DIPHENYLTIN** $Ph_2Sn$ $C_{12}H_{10}Sn$

Nitric acid

*See* NITRIC ACID, $HNO_3$: Diphenyltin

**1,3-DIPHENYLTRIAZENE** PhN=NNHPh $C_{12}H_{11}N_3$

Müller, E. *et al., Chem. Ber.*, 1962, **95**, 1257

It decomposes explosively at the m.p., 98°C.

Acetic anhydride

Heusler, F., *Ber.,* 1891, **24**, 4160

A mixture exploded with extraordinary violence on warming.

*See other* TRIAZENES

**1,5-DIPHENYL-1,4-PENTAAZADIENE** PhN=NNHN=NPh $C_{12}H_{11}N_5$

Griess, P., *Ann.*, 1866, **137**, 81

The dry solid explodes violently on warming, impact or friction.

*C*-homologues behave similarly.

*See other* HIGH-NITROGEN COMPOUNDS

**DIPHENYLPHOSPHINE** $(C_6H_5)_2PH$ $C_{12}H_{11}P$

Ireland, R. F. *et al., Org. Synth.*, 1977, **56**, 47

If the phosphine is spilled onto a paper towel, it may ignite in air.

*See related* ALKYLNON-METAL HYDRIDES

**1,2-DI(3-BUTEN-1-YNYL)CYCLOBUTANE** $C_{12}H_{12}$

$H_2C{=}CHC{\equiv}C\overline{CHC_2H_4CH}C{\equiv}CCH{=}CH_2$

Cupery, M. E. *et al., J. Amer. Chem. Soc.*, 1934, **56**, 1167

During isolation of the title product from polymerised 1,5-hexadien-3-yne, high-vacuum distillation must be carried only to a limited extent to prevent sudden explosive decomposition of the more highly polymerised residue.

*See other* ACETYLENIC COMPOUNDS
POLYMERISATION INCIDENTS

**TETRAACRYLONITRILECOPPER(I) PERCHLORATE** $C_{12}H_{12}ClCuN_4O_4$

$[Cu(C_3H_3N)_4]\ [ClO_4]$

Ondrejovic, G., *Chem. Zvesti*, 1964, **18**, 281

Decomposes explosively on heating.

*See* AMMINEMETAL OXOSALTS
*See other* CYANO COMPOUNDS

**TETRAACRYLONITRILECOPPER(II) PERCHLORATE** $C_{12}H_{12}Cl_2CuN_4O_8$
$[Cu(C_3H_3N)_4][ClO_4]_2$

Ondrejovic, G., *Chem. Zvesti*, 1964, **18**, 281
Decomposes explosively on heating.
*See* AMMINEMETAL OXOSALTS
*See other* CYANO COMPOUNDS

**BIS(η-BENZENE)CHROMIUM(0)** $Cr(C_6H_6)_2$ $C_{12}H_{12}Cr$

Oxygen
Anon., *Chem. Eng. News*, 1964, **42**(38), 55
The orange-red complex formed with oxygen in benzene decomposes vigorously on friction or heating in air.
*See other* ORGANOMETALLICS

**BIS(η-BENZENE)IRON(0)** $Fe(C_6H_6)_2$ $C_{12}H_{12}Fe$

Timms, P. L., *Chem. Eng. News*, 1969, **47**(18), 43
Prepared in the vapour phase at low temperature, the solid explodes at −40°C.
*See other* ORGANOMETALLICS

**POTASSIUM TETRAKIS(PROPYNYL)NICCOLATE(4−)** $C_{12}H_{12}K_4Ni$
$K_4[Ni(C{\equiv}CMe)_4]$

The diammoniate ignites on friction, impact or flame contact.
*See* COMPLEX ACETYLIDES

**1,2-DIPHENYLHYDRAZINE** PhNHNHPh $C_{12}H_{12}N_2$

Perchloryl fluoride
*See* PERCHLORYL FLUORIDE, $ClFO_3$: Nitrogenous bases

**DIANILINEOXODIPEROXOCHROMIUM(VI)** $C_{12}H_{14}CrN_2O_5$
$[CrO(O_2)_2 2C_6H_7N]$

*See* OXODIPEROXODIPYRIDINECHROMIUM(VI), $C_{10}H_{10}CrN_2O_5$

**BIS(2-METHYLPYRIDINE)SODIUM** $Na(C_6H_7N)_2$ $C_{12}H_{14}N_2Na$

Sidgwick, 1950, 89

The addition product of sodium and 2-methylpyridine ignites in air.

*See also* DIPYRIDINESODIUM, $C_{10}H_{10}N_2Na$

**2,4,6-TRI(ALLYLOXY)-1,3,5-TRIAZINE ('TRIALLYL CYANURATE')** $C_{12}H_{15}N_3O_3$

$\overline{N=C(OC_3H_5)N=C(OC_3H_5)N=C}OCH_2CH=CH_2$

Preparative hazard

*See* 2,4,6-TRICHLORO-1,3,5-TRIAZINE (CYANURIC CHLORIDE), $C_3Cl_3N_3$ : Allyl alcohol, etc.

**DIANILINIUM DICHROMATE** $C_{12}H_{16}Cr_2N_2O_7$

$(PhN^+H_3)_2\ Cr_2O_7^{2-}$

Gibson, G. M., *Chem. & Ind.*, 1966, 553

It is unstable on storage.

*See other* DICHROMATE SALTS OF NITROGENOUS BASES

**TETRAKIS(PYRAZOLE)MANGANESE(II) SULPHATE** $C_{12}H_{16}MnN_8O_4S$

$[Mn(C_3H_4N_2)_4]SO_4$

*See* AMMINEMETAL OXOSALTS (reference 9)

**TRIS(3-METHYLPYRAZOLE)ZINC SULPHATE** $C_{12}H_{18}N_6O_4SZn$

$[Zn(C_4H_6N_2)_3]SO_4$

*See* AMMINEMETAL OXOSALTS

**1,2- or 1,4-BIS(2-HYDROPEROXY-2-PROPYL)BENZENE** $C_{12}H_{18}O_4$

$(HOOCMe_2)_2C_6H_4$

Velenskii, M. S. *et al.*, *Chem. Abs.*, 1974, **81**, 108066n

These difunctional analogues of 'cumyl hydroperoxide' appear to be no more hazardous than the latter. Though impact-sensitive, the decomposition was mild and incomplete.

*See other* ALKYL HYDROPEROXIDES

## 3-ACETOXY-4-IODO-3,7,7-TRIMETHYLBICYCLO[4.1.0]HEXANE

$C_{12}H_{19}IO_2$

OAc
Me₂ Me
I

*See* *vic*-IODOALKOXY COMPOUNDS

## TETRAALLYL-2-TETRAZENE

$C_{12}H_{20}N_4$

$(H_2C{=}CHCH_2)_2NN{=}NN(CH_2CH{=}CH_2)_2$

Houben-Weyl, 1967, Vol. 10.2, 831

It explodes with great violence when heated above its b.p., 113°C/1 bar.

*See other* ALLYL COMPOUNDS
HIGH-NITROGEN COMPOUNDS

## 3,6-DI(SPIROCYCLOHEXANE)TETRAOXANE

$C_{12}H_{20}O_4$

O—O
O—O

Dilthey, W. *et al.*, *J. Prakt. Chem.*, 1940, **154**, 219

This dimeric cyclohexanone peroxide explodes on impact.

*See other* CYCLIC PEROXIDES

## 1-ACETOXY-6-OXOCYCLODECYL HYDROPEROXIDE

$C_{12}H_{20}O_5$

$HOOC(OAc)[CH_2]_4CO[CH_2]_3CH_2$

Criegee, R. *et al.*, *Ann.*, 1949, **564**, 15

The crystalline compound prepared at −70°C explodes mildly after a few minutes at ambient temperature.

*See other* ALKYL HYDROPEROXIDES

## TETRAALLYLURANIUM

$(H_2C{=}CHCH_2)_4U$ $C_{12}H_{20}U$

Bailar, 1973, Vol. 5, 405

Only stable below −20°C, it ignites in air.

*See other* ALLYL COMPOUNDS
*See related* ALKYLMETALS

## 3-BUTEN-1-YNYLDIISOBUTYLALUMINIUM

$C_{12}H_{21}Al$

$H_2C{=}CHC{\equiv}CAl(CH_2CHMe_2)_2$

Petrov, A. A. *et al.*, *Zh. Obsch. Khim.*, 1962, **32**, 1349

Ignites in air.

*See other* METAL ACETYLIDES
TRIALKYLALUMINIUMS

**1-BUTYLOXYETHYL 3-TRIMETHYLPLUMBYLPROPIOLATE** $C_{12}H_{22}O_3Pb$

$Me_3PbC \equiv CCO_2C(Me)HOBu$

Houben-Weyl, 1975, Vol. 13.3, 80
Explosive.
*See related* METAL ACETYLIDES

**BIS(1-HYDROXYCYCLOHEXYL) PEROXIDE** $C_{12}H_{22}O_4$

$\overline{CH_2[CH_2]_4C}(OH)OO\overline{C(OH)[CH_2]_4CH_2}$

Stoll, M. *et al., Helv. Chim. Acta*, 1930, **13**, 142
Normally stable, it explodes on attempted vacuum distillation.
*See other* 1-OXYPEROXY COMPOUNDS

**DIHEXANOYL PEROXIDE** $C_{12}H_{22}O_4$

$C_5H_{11}COOOCOC_5H_{11}$

Castrantas, 1965, 17
Explodes at 85 °C.
*See other* DIACYL PEROXIDES

**1(1′-HYDROPEROXY-1′-CYCLOHEXYLPEROXY)-CYCLOHEXANOL** $C_{12}H_{22}O_5$

$\overline{CH_2[CH_2]_4C}(OH)OO\overline{C(OOH)[CH_2]_4CH_2}$

Davies, 1961, 74
This appears to be a main constituent of commercial 'cyclohexanone peroxide' together with the symmetrical bis-hydroperoxy peroxide (below), known to be hazardous.
*See* COMMERCIAL ORGANIC PEROXIDES
BIS(1-HYDROPEROXYCYCLOHEXYL) PEROXIDE, $C_{12}H_{22}O_6$
*See other* 1-OXYPEROXY COMPOUNDS

**BIS(1-HYDROPEROXYCYCLOHEXYL) PEROXIDE** $C_{12}H_{22}O_6$

$\overline{CH_2[CH_2]_4C}(OOH)OO\overline{C(OOH)[CH_2]_4CH_2}$

Criegee, R. *et al., Ann.*, 1949, **565**, 17–18
One of the components in 'cyclohexanone peroxide', it explodes violently at elevated temperatures (during vacuum distillation or on exposure to flame).
*See* COMMERCIAL ORGANIC PEROXIDES
*See other* 1-OXYPEROXY COMPOUNDS

### 1,4,7,10,13,16-HEXAOXACYCLOOCTADECANE ('18-CROWN-6')

$C_{12}H_{24}O_6$

$\overline{OCH_2CH_2[OCH_2CH_2]_4OCH_2CH_2}$

Preparative hazard


1. Cram, D. J. *et al.*, *J. Org. Chem.*, 1974, **39**, 2445
2. Stott, P. E., *Chem. Eng. News*, 1976, **54**(37), 5
3. Gouw, T. H., ibid., (44), 5
4. Stott, P. E., ibid., (51), 5


There is a potential explosion hazard during larger-scale operation of the published (small batch) procedure [1]. This arose during thermal decomposition of the crown ether–potassium chloride complex under reduced pressure, when the crude ether distilled out. The larger batch size involved more extensive heating to complete decomposition of the complex, and a considerable amount of 1,4-dioxane was produced by cracking and blocked the vacuum pump trap. Admission of air to the overheated residue led to a violent explosion, attributed to autoignition of the dioxane–air mixture. Dioxane has a relatively low AIT (180°C) and rather wide explosive limits. Practical precautions are detailed [2].

Subsequently the observed hazards were attributed to poor distillation procedures [3], but further more detailed information on the experimental procedure was published to refute this view [4].

### 3,6,9-TRIETHYL-3,6,9-TRIMETHYL-1,2,4,5,7,8-HEXAOXAONANE

$C_{12}H_{24}O_6$

```
          O·O·CEtMe
         /      :
        /       O
EtMeC           :
        \       O
         \      :
          O·O·CEtMe
```


*See* 3,6-DIETHYL-3,6-DIMETHYL-1,2,4,5-TETRAOXANE, $C_8H_{16}O_4$
*See other* CYCLIC PEROXIDES


### † DODECANE

$CH_3[CH_2]_{10}CH_3$ $C_{12}H_{26}$

### † DIHEXYL ETHER

$C_6H_{13}OC_6H_{13}$ $C_{12}H_{26}O$

### 2,2-DI(*tert*-BUTYLPEROXY)BUTANE

$C_{12}H_{26}O_4$

$Me_3COOC_4H_8OOCMe_3$

Dickey, F. H. *et al.*, *Ind. Eng. Chem.*, 1949, **41**, 1673

The pure material explodes on heating to about 130°C, on sparking or on impact.

*See other* DIALKYLPEROXIDES

**DI(2-HYDROPEROXY-4-METHYL-2-PENTYL) PEROXIDE** $C_{12}H_{26}O_6$

$Me_2CHCH_2C(OOH)MeOOC(OOH)MeCH_2CHMe_2$

Leleu, *Cahiers*, 1973, (71), 238

The triperoxide, main component in 'MIBK peroxide', is explosive in the pure state, but insensitive to shock as the commercial 60% solution in dimethyl phthalate. The solution will explode at about 75°C, and slowly liberates oxygen at ambient temperatures.

*See related* ALKYL HYDROPEROXIDES

*See* KETONE PEROXIDES

**† TRIISOBUTYLALUMINIUM** $(Me_2CHCH_2)_3Al$ $C_{12}H_{27}Al$

'Specialty Reducing Agents', Brochure TA-2002/1, New York, Texas Alkyls, 1971

Used industrially as a powerful reducant, it is supplied as a solution in hydrocarbon solvents. The undiluted material is of relatively low thermal stability (decomposing above 50°C) and ignites in air unless diluted to below 25% concentration.

*See other* TRIALKYLALUMINIUMS

***mixo*-TRIBUTYLBORANE** $(Bu)_3B$ $C_{12}H_{27}B$

Hurd, D. T., *J. Amer. Chem. Soc.*, 1948, **70**, 2053

A mixture of the *n*- and iso-isomers ignited on exposure to air.

*See* ALKYLBORANES

*See other* ALKYLNON-METALS

**TRIBUTYLBISMUTH** $(Bu)_3Bi$ $C_{12}H_{27}Bi$

Gilman, H. *et al.*, *J. Amer. Chem. Soc.*, 1939, **61**, 1170

It explodes violently in oxygen and ignites in air.

*See other* TRIALKYLBISMUTHS

**TRIBUTYLINDIUM** $Bu_3In$ $C_{12}H_{27}In$

Houben-Weyl, 1970, Vol. 13.4, 347

It ignites in air.

*See other* ALKYLMETALS

**† TRIBUTYLPHOSPHINE** **$(Bu)_3P$** **$C_{12}H_{27}P$**

**TETRAISOPROPYLCHROMIUM** **$(Me_2CH)_4Cr$** **$C_{12}H_{28}Cr$**

Müller, J. *et al., Angew. Chem. (Intern. Ed.)*, 1975, **14**, 761

It is pyrophoric in finely crystalline form.

*See other* ALKYLMETALS

**TITANIUM TETRAISOPROPOXIDE** **$Ti(OCHMe_2)_4$** **$C_{12}H_{28}O_4Ti$**

Hydrazine

*See* HYDRAZINE, $H_4N_2$ : Titanium compounds

**† TITANIUM TETRAPROPOXIDE** **$Ti(OPr)_4$** **$C_{12}H_{28}O_4Ti$**

**'TETRAPROPYLDIBORANE'** **$C_{12}H_{30}B_2$**

*See* 'TETRAETHYLDIBORANE', $C_8H_{22}B_2$

**BIS(TRIETHYLTIN) PEROXIDE** **$Et_3SnOOSnEt_3$** **$C_{12}H_{30}O_2Sn_2$**

Sorbe, 1968, 160

Readily decomposed, it explodes at 50°C.

*See other* ORGANOMINERAL PEROXIDES

**1,2-BIS(TRIETHYLSILYL)HYDRAZINE** **$C_{12}H_{32}N_2Si_2$**

**$Et_3SiNHNHSiEt_3$**

*See* SILYLHYDRAZINES: Oxidants

**TRIETHYLSILYL-1,2-BIS(TRIMETHYLSILYL)HYDRAZINE** **$C_{12}H_{34}N_2Si_3$**

**$Et_3SiNSiMe_3NHSiMe_3$**

*See* SILYLHYDRAZINES: Oxidants

**HEXA(DIMETHYL SULPHOXIDE)CHROMIUM(III) PERCHLORATE**

**$C_{12}H_{36}CrCl_3O_{18}S_6$**

**$[Cr(Me_2SO)_6][ClO_4]_3$**

Langford, C.H. *et al., J. Chem. Soc., Chem. Comm.*, 1977, 139

The complex salt may be explosive.

*See* DIMETHYL SULPHOXIDE, $C_2H_6OS$: Metal oxosalts

*See related* AMMINEMETAL OXOSALTS

## TRIMETHYLPLATINUM(IV) AZIDE TETRAMER $C_{12}H_{36}N_{12}Pt_4$

$(Me_3PtN_3)_4$

Trimethylphosphite

1. von Dahlen, K. H. *et al., J. Organomet. Chem.*, 1974, **65**, 267
2. Neruda, B. *et al.*, ibid., 1976, **111**, 241–248

The azide is not shock-sensitive, but detonates violently on rapid heating or exposure to flame [1]. In the exothermic reaction with trimethyl phosphite to give *cis*-dimethyl-bis(trimethyl phosphito)-platinum, the azide must be added in small portions with stirring. Addition of the phosphite to the solid azide led to a violent explosion, probably involving the transitory by-product methyl azide [2].

*See related* METAL AZIDES
*See other* PLATINUM COMPOUNDS

## μ-PEROXO-BIS[AMMINE(2,2′,2″-TRIAMINOTRIETHYLAMINE)COBALT-(III)](4+) PERCHLORATE $C_{12}H_{42}Cl_4Co_2N_{10}O_{18}$

$[NH_3(H_2NC_2H_4)_3NCoOOCoN(C_2H_4NH_2)_3NH_3][ClO_4]_4$

1. Yang, C.-H. *et al., Inorg. Chem.*, 1973, **12**, 666
2. Mori, M. *et al., J. Amer. Chem. Soc.*, 1968, **90**, 619

The salt exploded at 220°C [1]. Other salts of this general type (permanganates, possibly nitrates) are explosive [2].

*See related* AMMINEMETAL OXOSALTS

## DI[TRIS-1,2-DIAMINOETHANECOBALT(III)] TRIPEROXODISULPHATE $C_{12}H_{48}Co_2N_{12}O_{24}S_6$

$[Co(C_2H_8N_2)_3]_2[OSO_2OOSO_2O]_3$

Beacom, S. E., *Nature*, 1959, **183**, 38

It explodes upon ignition, or after application of UV irradiation and heating to 120°C.

*See other* AMMINEMETAL OXOSALTS

## DI[TRIS-1,2-DIAMINOETHANECHROMIUM(III)] TRIPEROXO-DISULPHATE $C_{12}H_{48}Cr_2N_{12}O_{24}S_6$

$[Cr(C_2H_8N_2)_3]_2[OSO_2OOSO_2O]_3$

Beacom, S. E., *Nature*, 1959, **183**, 38

It explodes upon ignition, or after application of UV irradiation and then heating to 115°C.

*See other* AMMINEMETAL OXOSALTS

### LEAD(II) TRINITROSOPHLOROGLUCINOLATE $C_{12}N_6O_{12}Pb_3$

$$\left( {}^{-}O, ON, NO, O^{-}, NO, O^{-} \text{ benzene} \right)_2 Pb_3$$

Freund, H. E., *Angew. Chem.*, 1961, 73, 433

An air-dried sample exploded when disturbed, possibly owing to aerobic oxidation to the trinitro compound.

*See other* NITROSO COMPOUNDS

### *N,N'*-DICHLOROBIS(2,4,6-TRICHLOROPHENYL)UREA $C_{13}H_4Cl_8N_2O$

$CO(NClC_6H_2Cl_3)_2$

Ammonia

Pytlewski, L. L., *Rep. AD-A028841*, 14, Richmond (Va.), USNTIS, 1976

Contact of gaseous ammonia with the *N*-chlorourea, either alone or mixed with zinc oxide, leads to ignition. The same could happen in contact with concentrated aqueous ammonia, solid ammonium carbonate, or organic amines.

Dimethyl sulphoxide

Pytlewski, L. L., *Rep. AD-A028841*, 13, Richmond (Va.), USNTIS, 1976

Violent ignition occurs on mixing.

1-*p*-Nitrophenylazo-2-naphthol,
Zinc oxide

Pytlewski, L. L., *Rep. AD-A028841*, Richmond (Va.), USNTIS, 1976

Spontaneous combustion in storage (occasionally at high ambient temperatures) of clothing impregnation kits containing the three title compounds was investigated. The *N*-chlorourea when heated evolves chlorine to give the isocyanate and a nitrene. Chlorine and the azo-dye react violently and serve as an initiation source of heat. Zinc oxide is converted to the chloride, which catalyses violently exothermic polymerisation of the isocyanate, the main contribution to the total exotherm (~2 MJ/mol), which leads to vigorous smouldering decomposition of the whole mass. Temperatures in excess of 315°C are attained. In absence of the dye, heating a mixture of the urea and oxide at 5°C/min leads to ignition at 130°C.

*See other* *N*-HALOGEN COMPOUNDS

### 1-BROMO-1,2-CYCLOTRIDECADIEN-4,8,10-TRIYNE $C_{13}H_9Br$

$BrC{=}C{=}CHC{\equiv}CC_2H_4[C{\equiv}C]_2CH_2CH_2$ (ring closed between first C and last $CH_2$)

Leznoff, C. C. *et al.*, *J. Amer. Chem. Soc.*, 1968, **90**, 731

It explodes at 65°C and slowly decomposes in the dark at 0°C.
*See other* HALOACETYLENE DERIVATIVES

**p-NITROPHENYL-2-CARBOXYBENZENEDIAZOATE** $C_{13}H_9N_3O_5$

$NO_2C_6H_4N{=}NOC_6H_4CO_2H$

Griess, P., *Ber.*, 1884, **17**, 338
It explodes on heating.
*See other* ARENEDIAZO ARYL OXIDES

**SILVER 1-BENZENEAZOTHIOCARBONYL-2-PHENYLHYDRAZINE**

$C_{13}H_{11}AgN_4S$

PhN=NCSNHN(Ag)Ph

Sorbe, 1968, 126
The silver derivative of dithizone decomposes explosively at higher temperatures.
*See other* HEAVY METAL DERIVATIVES

**2-PHENYLAMINO-3-PHENYLOXAZIRANE (BENZALDEHYDE PHENYLHYDRAZONE OXIDE)** $C_{13}H_{12}N_2O$

PhCHONNHPh

Phenylhydrazine
Bergman, M. *et al., Ber.*, 1923, **56**, 681
The 'oxide' reacts violently after a few minutes' warming with phenylhydrazine.

**α-BENZENEDIAZOBENZYL HYDROPEROXIDE** $C_{13}H_{12}N_2O_2$

$C_6H_5N{=}NCH(Ph)OOH$

Busch, M. *et al., Ber.*, 1914, **47**, 3277
Swern, 1971, Vol. 2, 19
The phenylhydrazones of benzaldehyde and its homologues, or of acetone, are readily autoxidised in solution and rearrange to give the diazo-hydroperoxides, isolable as solids which may explode after a short time on standing, though not on friction or impact. Contact with a flame or with concentrated sulphuric or nitric acids also initiates explosion.
*See other* ALKYL HYDROPEROXIDES

## 3-ETHOXYCARBONYL-4,4,5,5-TETRACYANO-3-TRIMETHYL-PLUMBYL-4,5-DIHYDRO-3-*H*-PYRAZOLE $C_{13}H_{14}N_6O_2Pb$

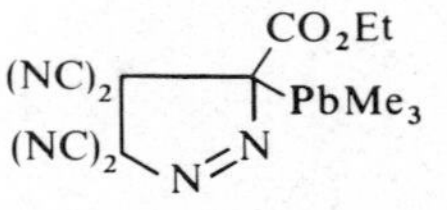

$CO_2Et$, $(NC)_2$, $(NC)_2$, $PbMe_3$, N=N

Houben-Weyl, 1975, Vol. 13.3, 221

It tends to explode on heating.

*See related* CYANO COMPOUNDS

## 2-(DIMETHYLAMINOMETHYL)FLUOROFERROCENE $C_{13}H_{16}FFeN$

$Me_2NCH_2C_5H_3FFeC_5H_5$

Preparative hazard

*See* PERCHLORYL FLUORIDE, $ClFO_3$: 2-Lithio(dimethylaminomethyl)-ferrocene

## DIMETHYLAMINOMETHYLFERROCENE $C_{13}H_{17}FeN$

$Me_2NCH_2C_5H_4FeC_5H_5$

Nitric acid,
Water

*See* NITRIC ACID, $HNO_3$: (Dimethylaminomethyl)ferrocene, etc.

## IRON(II) CHELATE of BIS-*N,N′*(2-PENTANON-4-YLIDENE)-1,3-DIAMINO-2-HYDROXYPROPANE $C_{13}H_{20}FeN_2O_3$

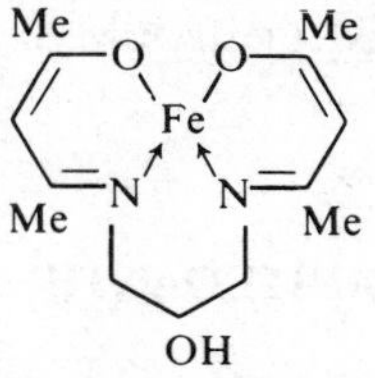

Me, Me, O, O, Fe, N, N, Me, Me, OH

Berenbaum, M. B., US Pat. 3 388 141, 1968 (*Chem. Abs.*, 1969, **70**, 59415n)

The solid chelate is pyrophoric in air, burning to iron(III) oxide.

*See other* PYROPHORIC MATERIALS

## DIBUTYL-3-METHYL-3-BUTEN-1-YNYLBORANE $C_{13}H_{23}B$

$CH_2{=}CMeC{\equiv}CB(Bu)_2$

Davidsohn, W. E., *Chem. Rev.*, 1967, **67**, 75

It ignites in air.

*See other* ACETYLENIC COMPOUNDS
ALKYLNON-METALS

## 1,8-DIHYDROXY-2,4,5,7-TETRANITROANTHRAQUINONE $C_{14}H_4N_4O_{12}$

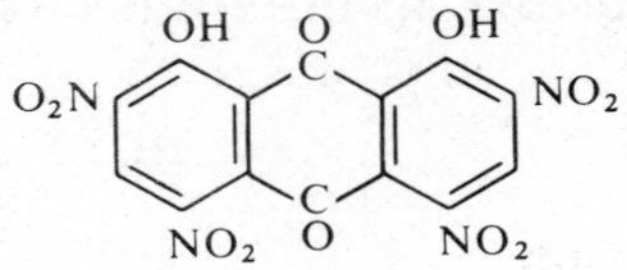

Sorbe, 1968, 153

It is explosive.

*See other* POLYNITROARYL COMPOUNDS

## BIS-2,4-DICHLOROBENZOYL PEROXIDE $C_{14}H_6Cl_4O_4$

$Cl_2C_6H_3COOOCOC_6H_3Cl_2$

'Lucidol Data Sheet', Buffalo, Wallace and Tiernan, 1963

Whereas the pure compound is extremely shock-sensitive and decomposes rapidly at 80°C, the commercial 50% dispersion in plasticiser is not shock-sensitive.

*See* COMMERCIAL ORGANIC PEROXIDES

*See other* DIACYL PEROXIDES

## CALCIUM BIS-*p*-IODYLBENZOATE $C_{14}H_8CaI_2O_8$

$Ca(O_2CC_6H_4IO_2)_2$

Unpublished information, 1948

Formulated granules accidentally dried to below normal moisture content exploded violently.

*See other* IODINE COMPOUNDS

## COPPER(II) 3,5-DINITROANTHRANILATE $C_{14}H_8CuN_6O_{12}$

$Cu[(NO_2)_2NH_2C_6H_2CO_2]_2$

*See* SILVER 3,5-DINITROANTHRANILATE, $C_7H_4AgN_3O_6$

## BIS-*o*-AZIDOBENZOYL PEROXIDE $C_{14}H_8N_6O_4$

$N_3C_6H_4COOOCOC_6H_4N_3$

1. Leffler, J. E., *Chem. Eng. News,* 1963, **41**(48), 45
2. Hoffman, J., *Chem. Eng. News,* **41**(52), 5

A small sample of crystalline material on a sintered glass funnel detonated with extreme violence when touched with a metal spatula [1]. Static electrical initiation may have been involved [2].

*See other* DIACYL PEROXIDES

## 9,10-EPIDIOXYANTHRACENE

$C_{14}H_8O_2$

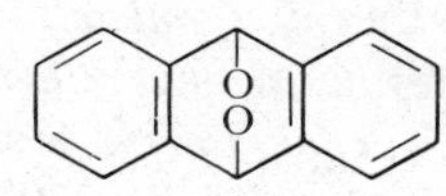

Dufraisse, C. *et al., Compt. Rend.*, 1935, **201**, 428
Decomposes explosively at 120°C.
*See other* CYCLIC PEROXIDES

## 2,2′-BIPHENYLDICARBONYL PEROXIDE

$C_{14}H_8O_4$

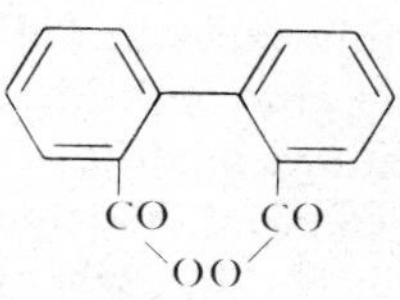


Ramirez, F., *J. Amer. Chem. Soc.*, 1964, **86**, 4394
It explodes violently on heating to 70°C, or on impact, but can be preserved at low temperature.
*See other* DIACYL PEROXIDES

## ANTHRACENE

$C_6H_4{:}(CH)_2{:}C_6H_4$ $C_{14}H_{10}$

Fluorine
*See* FLUORINE, $F_2$ : Hydrocarbons

## MERCURY(II) PEROXYBENZOATE

$Hg(OO_2CPh)_2$ $C_{14}H_{10}HgO_6$

Castrantas, 1965, 19
Explodes if heated above its normal decomposition temperature of 100–110°C.
*See other* PEROXOACID SALTS

## 1,1-BENZOYLPHENYLDIAZOMETHANE

$C_{14}H_{10}N_2O$

$PhCOC(Ph)N_2$

Nenitzescu, C. D. *et al., Org. Synth.*, 1943, Coll. Vol. 2, 497
The material may explode if heated to above 40°C.
*See other* DIAZO COMPOUNDS

## DIBENZOYL PEROXIDE

PhCOOOCOPh $C_{14}H_{10}O_4$

1. *MCA SD-81,* 1960
2. Lappin, G. R., *Chem. Eng. News,* 1948, **26**, 3518; Taub, D., *Chem. Eng. News,* 1949, **27**, 46

3. Nozaki, K. *et al., J. Amer. Chem. Soc.*, 1946, **68**, 1692
4. Fine, D. J. *et al., Combust. Flame*, 1967, **11**, 71–78
5. McCloskey, C. M. *et al., Chem. Abs.*, 1967, **66**, 12613c
6. Uetake, K. *et al., Chem. Abs.*, 1974, **81**, 5175t
7. Sekida, O., Jap. Pat. 40 220, 1974 (*Chem. Abs.*, 1975, **82**, 140814r)
8. Anon., *Sichere Chemiearb.*, 1976, **28**, 49

The dry material is readily ignited, burns very rapidly and is moderately sensitive to heat, shock, friction or contact with combustible materials. When heated above m.p. (103–105°C), instantaneous and explosive decomposition occurs without flame, but the decomposition products are flammable. If under confinement (or in large bulk), decomposition may be violently explosive [1]. An explosion which occurred when a screw-capped bottle of the peroxide was opened was attributed to friction initiating a mixture of peroxide and organic dust in the cap-threads. Waxed paper tubs are recommended to store this and other sensitive solids [2]. Crystallisation of dibenzoyl peroxide from hot chloroform solution involves a high risk of explosion. Precipitation from cold chloroform solution by methanol is safer [3]. Water- or plasticiser-containing pastes of dibenzoyl peroxide are much safer for industrial use.

The explosive decomposition of the solid has been studied in detail [4]. The effect of moisture content upon ignitibility and explosive behaviour under confinement was studied. A moisture content of 3% allowed slow burning only, and at 5% ignition did not occur [5]. Thermal instability was studied using a pressure vessel test, ignition delay time, thermogravimetry and differential scanning calorimetry, and decomposition products were identified [6]. The presence of acyl chlorides renders dibenzoyl peroxide impact-insensitive [7]. There is a further recent report of a violent explosion during purification of the peroxide by Soxhlet extraction with hot chloroform [8].

Aniline

Bailey, P. S. *et al., J. Chem. Educ.*, 1975, **52**, 525

Addition of a drop of aniline to 1 g of the peroxide leads to mildly explosive decomposition after a few seconds' delay.

*N*-Bromosuccinimide,
*p*-Toluic acid

*See* *N*-BROMOSUCCINIMIDE, $C_4H_4BrNO_2$ : Dibenzoyl peroxide, etc.

Carbon tetrachloride,
Ethylene

Bolt, R. O. *et al., Chem. Eng. News*, 1947, **25**, 1866

Interaction of ethylene and carbon tetrachloride at elevated temperatures and pressures, initiated with benzoyl peroxide as radical source, caused violent explosions on several occasions. Precautions recommended include use of minimum pressure and quantity of initiator, maximum agitation, and presence of water as an inert moderator of high specific heat.

*See also* WAX FIRE

*N,N*-Dimethylaniline

Anon., *Angew. Chem.(Nachr.)*, 1954, **2**, 83

The solid peroxide exploded on contact with a drop of dimethylaniline.

Dimethyl sulphide

Pyror, W. A. *et al.*, *J. Org. Chem.*, 1972, **37**, 2885

The rapid decomposition of benzoyl peroxide by dimethyl sulphide is explosive in absence of solvent.

Lithium tetrahydroaluminate

Sutton, D. A., *Chem. & Ind.*, 1951, 272

One of two attempts to reduce the diacyl peroxide in ether led to a moderately violent explosion.

Methyl methacrylate

*MCA Case History No. 996*

Local overheating and ignition occurred when solid benzoyl peroxide was put into a beaker which had been rinsed out with methyl methacrylate. Contact between the peroxide, a powerful oxidising agent and potential source of free radicals, and oxidisable or polymerisable materials should be under controlled conditions.

Vinyl acetate

*See* VINYL ACETATE, $C_4H_6O_2$: Dibenzoyl peroxide, etc.

*See other* DIACYL PEROXIDES

## BIS-3-(2-FURYL)ACRYLOYL PEROXIDE $C_{14}H_{10}O_6$

$C_4H_3OCH{=}CHCOOOCOCH{=}CHC_4H_3O$

Milas, N. A. *et al.*, *J. Amer. Chem. Soc.*, 1934, **56**, 1219

It explodes violently on heating.

*See other* DIACYL PEROXIDES

## *N-m*-TOLYL-*o*-NITROBENZIMIDYL CHLORIDE $C_{14}H_{11}ClN_2O_2$

$NO_2C_6H_4C(Cl){=}NC_6H_4CH_3$

Preparative hazard

*See* PHOSPHORUS PENTACHLORIDE, $Cl_5P$: *o*-Nitrobenzoyl-*m*-toluidide

**1,1-DIPHENYLETHYLENE** $Ph_2C{=}CH_2$ $C_{14}H_{12}$

Oxygen

Staudinger, H., *Ber.*, 1925, **58**, 1075

Exposure of the alkene to oxygen at ambient temperature and pressure produces an alkene–oxygen interpolymeric peroxide, which explodes lightly on heating. An attempt to react the alkene with oxygen at 100 bar and 40–50°C caused a violent explosion in the autoclave.

*See other* POLYPEROXIDES

**BIS-5-CHLOROTOLUENEDIAZONIUM ZINC TETRACHLORIDE** $C_{14}H_{12}Cl_6N_4Zn$

$2ClMeC_6H_3N_2Cl{\cdot}ZnCl_2$

*ABCM Quart. Safety Summ.*, 1953, **24**, 42

A batch containing only half the normal water content (60%) exploded violently during ball-milling. Tests later showed the dry material to be shock-sensitive.

*See other* DIAZONIUM SALTS

**3′-METHYL-2-NITROBENZANILIDE** $C_{14}H_{12}N_2O_3$

$NO_2C_6H_3CONHC_6H_4Me$

Phosphorus pentachloride

*See* PHOSPHORUS PENTACHLORIDE, $Cl_5P$: 3′-Methyl-2-nitrobenzanilide

**POLY(1,1-DIPHENYLETHYLENE PEROXIDE)** $(C_{14}H_{12}O_2)_n$

$(Ph_2C{=}CHOOH)_n$

*See* 1,1-DIPHENYLETHYLENE, $C_{14}H_{12}$: Oxygen

*See other* POLYPEROXIDES

**2-AZOXYANISOLE** $C_{14}H_{14}N_2O_3$

$MeOC_6H_4N{=}N(O)C_6H_4OMe$

Preparative hazard

*See* 2-NITROANISOLE, $C_7H_6NO_3$ : Sodium hydroxide, etc.

**BIS-TOLUENEDIAZO OXIDE** $(CH_3C_6H_4N_2)O$ $C_{14}H_{14}N_4O$

Alone,

or Toluene

Bamberger, E., *Ber.*, 1896, **29**, 452, 458

Extremely unstable, it explodes under its reaction liquor at above −4°C. Very shock- and friction-sensitive, a small sample exploded when dried on a porous tile and set off the moist material some distance away. Contact with toluene, even at −5°C, causes an explosive reaction with flame.

*See other* BIS(ARENEDIAZO) OXIDES

**DIBENZYL ETHER** **$(PhCH_2)_2O$** **$C_{14}H_{14}O$**

Aluminium dichloride hydride

*See* ALUMINIUM DICHLORIDE HYDRIDE ETHERATE, $AlCl_2H \cdot C_4H_{10}O$

*See other* PEROXIDISABLE COMPOUNDS

**DI-*p*-TOLUENESULPHONYL PEROXIDE** **$C_{14}H_{14}O_6S_2$**

**$MeC_6H_4SO_2OOSO_2C_6H_4Me$**

1. Bolte, J. *et al.*, *Tetrahedron Lett.*, 1965, (21), 1529
2. Dannley, R. L. *et al.*, *J. Org. Chem.*, 1966, **31**, 154

The peroxide was too unstable to dry thoroughly; such samples often exploded spontaneously.

*See other* DIACYL PEROXIDES

**DIBENZYL PHOSPHITE** **$(PhCH_2O)_2P(O)H$** **$C_{14}H_{15}O_3P$**

Atherton, F. R. *et al.*, *J. Chem. Soc.*, 1945, 382; 1948, 1106

It decomposes at 160°C, but prolonged heating at 120°C may have the same effect. Not more than 50 g should be distilled at once, using high-vacuum conditions (b.p., 100–120°C/0.001 mbar) unless a preliminary treatment to remove acidic impurities has been used.

**1-(4-PHENYL-1,3-DISELENOLYLIDENE)PIPERIDINIUM PERCHLORATE** **$C_{14}H_{16}ClNO_4Se_2$**

Ph Se
=N$^+$ $ClO_4^-$
Se

*See* 1-(1,3-DISELENOLYLIDENE)PIPERIDINIUM PERCHLORATES

**α-PENTYLCINNAMALDEHYDE** **$C_{14}H_{18}O$**

**$PhCH{=}C(CHO)C_5H_{11}$**

Anon., *Chem. Trade J.*, 1937, **100**, 362

This is very prone to spontaneous oxidative heating. A mixture with absorbent cotton reached 230°C 4 min after exposure to air.
*See other* PEROXIDISABLE COMPOUNDS

**5-*p*-CHLOROPHENYL-2,2-DIMETHYL-3-HEXANONE** $C_{14}H_{19}ClO$

$Me_3CCOCH_2CH(Me)C_6H_4Cl$

Peroxyacetic acid
*See* PEROXYACETIC ACID, $C_2H_4O_3$: 5-*p*-Chlorophenyl-2,2-dimethylhexanone

**2,6-DI-*tert*-BUTYL-4-NITROPHENOL** $C_{14}H_{21}NO_3$

$(Me_3C)_2NO_2C_6H_2OH$

1. *ASESB Expl. Incid. Report 1961*, **24**
2. Barnes, T. J. *et al., J. Chem. Soc.*, 1961, 953

A sample of the compound exploded violently after short heating to 100°C. Although this was attributed to presence of polynitro derivatives [1], the thermal decomposition of this type of nitro compound is known [2].
*See other* *C*-NITRO COMPOUNDS

**DICYCLOHEXYLCARBONYL PEROXIDE** $C_{14}H_{22}O_4$

$C_6H_{11}COOOCOC_6H_{11}$

Castrantas, 1965, 17
Larger quantities may explode without apparent reason.
*See other* DIACYL PEROXIDES

**1-ACETOXY-1-HYDROPEROXY-6-CYCLODODECANONE** $C_{14}H_{24}O_5$

$AcOC(OOH)[CH_2]_4CO[CH_2]_5CH_2$ (ring)

Criegee, R. *et al., Ann.*, 1949, **564**, 9
Explodes on removal from a freezing mixture.
*See other* 1-OXYPEROXY COMPOUNDS

**BIS(DIPROPYLBORINO)ACETYLENE** $C_{14}H_{28}B_2$

$Pr_2BC{\equiv}CBPr_2$

Hartmann, H. *et al., Z. Anorg. Allg. Chem.*, 1959, **299**, 179
Both *n*- and iso-propyl derivatives ignite in air.
*See related* BORANES

**† TETRADECANE** $CH_3[CH_2]_{12}CH_3$ $C_{14}H_{30}$

**ACETYLENEBIS-TRIETHYLLEAD** $Et_3PbC{\equiv}CPbEt_3$ $C_{14}H_{30}Pb_2$

Beerman, C. *et al., Z. Anorg. Allg. Chem.*, 1954, **276**, 20

It is very sensitive to heat, oxygen or light, and should not be distilled.

*See related* METAL ACETYLIDES

**BIS(TRIETHYLTIN)ACETYLENE** $Et_3SnC{\equiv}CSnEt_3$ $C_{14}H_{30}Sn_2$

Stannic chloride

Beerman, C. *et al., Z. Anorg. Chem.*, 1954, **276**, 20

The product of interaction is highly explosive.

*See other* METAL ACETYLIDES

**HEPTAKIS(DIMETHYLAMINO)TRIALUMINIUM TRIBORON PENTAHYDRIDE** $C_{14}H_{47}Al_3B_3N_7$

```
Me2N   H H   NMe2
    \  /   \  /
Me2N-Al\ B /Al-NMe2
   |     |     |
  HB     Al    BH
    \ N /|\ N /
   Me2 / \ Me2
      H  NMe2
```

Hall, R. E. *et al., Inorg. Chem.*, 1969, **8**, 270

The crystalline solid is spontaneously flammable in air.

*See other* COMPLEX HYDRIDES

**1,2,3,4-TETRACHLORO-7,7-DIMETHOXY-5-PHENYLBICYCLO-[2.2.1]-2,5-HEPTADIENE** $C_{15}H_{12}Cl_4O_2$

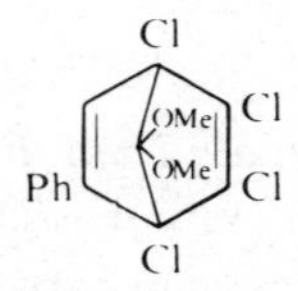


Hoffmann, R. W. *et al., Tetrahedron*, 1965, **21**, 900

Pyrolysis of the material at 130°C under nitrogen at low pressure to give tetramethoxyethylene may be explosive if more than 25g is used.

**TRIS-η-CYCLOPENTADIENYLPLUTONIUM** $C_{15}H_{15}Pu$

$(C_5H_5)_3Pu$

Bailar, 1973, Vol. 5, 407

It ignites in air.

*See other* ORGANOMETALLICS
PYROPHORIC COMPOUNDS

**TRIS-η-CYCLOPENTADIENYLURANIUM** $C_{15}H_{15}U$

$(C_5H_5)_3U$

Bailar, 1973, Vol. 5, 407

It ignites in air.

*See other* ORGANOMETALLICS
PYROPHORIC MATERIALS

**TRIS(2,4-PENTANEDIONATO)MOLYBDENUM(III)** $C_{15}H_{21}MoO_6$

$Mo(C_5H_7O_2)_3$

Larson, M. L. *et al., Inorg. Synth.*, 1966, **8**, 153

Rapidly oxidises in air, sometimes igniting.

*See other* ORGANOMETALLICS

***tert*-BUTYL 1-ADAMANTANEPEROXYCARBOXYLATE** $C_{15}H_{24}O_3$

$C_{10}H_{15}CO_2OCMe_3$

Razuvajev, G. A. *et al., Tetrahedron*, 1969, **25**, 4925

Explodes on heating to 90–100°C.

*See other* PEROXYESTERS

**TRIS(SPIROCYCLOPENTANE)-1,1,4,4,7,7-HEXAOXAONANE** $C_{15}H_{24}O_6$

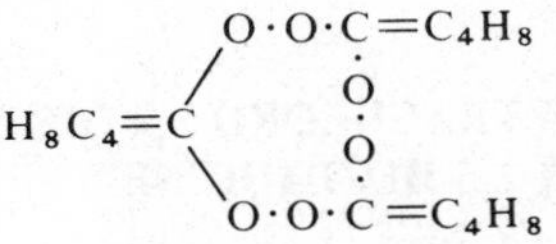

```
          O·O·C=C4H8
         /    O
H8C4=C        O
         \    
          O·O·C=C4H8
```

Bjorklund, G. H. *et al., Trans. R. Soc. Can.(Sect.III)*, 1950, **44**, 25

A violent explosive, very sensitive to shock, friction and rapid heating.

*See* HYDROGEN PEROXIDE, $H_2O_2$ : Ketones, etc.

*See other* CYCLIC PEROXIDES

**TRIS-2,4,6(DIMETHYLAMINOMETHYL)PHENOL** $C_{15}H_{27}N_3O$

$(Me_2NCH_2)_3C_6H_2OH$

Cellulose nitrate

*See* CELLULOSE NITRATE: Amines

**HEXAETHYLTRIALUMINIUM TRITHIOCYANATE** $C_{15}H_{30}Al_3N_3S_3$

$Et_2AlS(CN)AlEt_2S(CN)AlEt_2SCN$

Dehnicke, K., *Angew. Chem. (Intern. Ed.)*, 1967, **6**, 947
On heating at 210°C *in vacuo* it disproportionates explosively, but smoothly at 180°C.
*See related* ALKYLALUMINIUM HALIDES

**1,2-BIS(TRIETHYLSILYL)TRIMETHYLSILYLHYDRAZINE** $C_{15}H_{40}N_2Si_3$

$Et_3SiNHN(SiMe_3)SiEt_3$

*See* SILYLHYDRAZINES: Oxidants

**POTASSIUM BIS(PHENYLETHYNYL)PALLADATE(2–)** $C_{16}H_{10}K_2Pd$

$K_2[Pd(C{\equiv}CPh)_2]$

Immediately pyrophoric in air, and explosive decomposition with aqueous reagents; the sodium salt is simila'
*See* COMPLEX ACETYLIDES

**POTASSIUM BIS(PHENYLETHYNYL)PLATINATE(2–)** $C_{16}H_{10}K_2Pt$

$K_2[Pt(C{\equiv}CPh)_2]$

Pyrophoric in air, and explosive decomposition with water; the sodium salt is similar.
*See* COMPLEX ACETYLIDES

*mixo*-**DIMETHOXYDINITROANTHRAQUINONE** $C_{16}H_{10}N_2O_8$

Sulphuric acid
Hildreth, J. D., *Chem. & Ind.*, 1970, 1592
During hydrolysis of crude dimethoxydinitroanthraquinone by heating in sulphuric acid, a runaway exothermic decomposition occurred causing vessel failure. Experiment showed a threshold decomposition temperature of 150–155°C, and oxidising effect of nitro groups, yielding CO and $CO_2$ above 162°C.
*See other* POLYNITROARYL COMPOUNDS

**BIS($\eta$-CYCLOPENTADIENYL)PENTAFLUOROPHENYLZIRCONIUM HYDROXIDE** $C_{16}H_{11}F_5OZr$

$(C_5H_5)_2C_6F_5ZrOH$

Chaudhari, M. A. *et al., J. Chem. Soc. A*, 1966, 840
Decomposition on heating in air above 260°C is sometimes explosive.

*See* HALOARYLMETALS
*See other* ORGANOMETALLICS

**2,2′-ETHYLENEBIS(3-PHENYLOXAZIRANE)** $C_{16}H_{16}N_2O_2$

$PhCHONC_2H_4NOCHPh$

*MCA Case History No. 2175*
The washed crude product from oxidation of dibenzylidene 1,2-diaminoethane in methylene chloride solution exploded violently during vacuum evaporation at a relatively low temperature.

*See* 2,2′-DI-*tert*-BUTYL-3,3′-BIOXAZIRANE, $C_{10}H_{20}N_2O_2$

**BIS($\eta$-CYCLOOCTATETRAENE)URANIUM(0)** $C_{16}H_{16}U$

$U(C_8H_8)_2$

Streitweiser, A. *et al., J. Amer. Chem. Soc.*, 1968, **90**, 7364
Inflames in air.
*See other* ORGANOMETALLICS

***O–O-tert*-BUTYL DIPHENYL MONOPEROXOPHOSPHATE** $C_{16}H_{19}O_5P$

$Me_3COOPO(OPh)_2$

Rieche, A. *et al., Chem. Ber.*, 1962, **95**, 385
The material deflagrated as a solid, or decomposed exothermically in its reaction mixture soon after preparation.

*See other* *tert*-BUTYLPEROXOPHOSPHATE DERIVATIVES

***N*-METHYL-*p*-NITROANILINIUM 2(*N*-METHYL-*N*-*p*-NITROPHENYLAMINOSULPHONYL)ETHYLSULPHATE** $C_{16}H_{20}N_4O_{10}S_2$

$NO_2C_6H_4N^+H_2Me\ NO_2C_6H_4N(Me)SO_2C_2H_4OSO_2O^-$

There is a preparative hazard for this reaction product of *N*-methyl-*p*-nitroaniline and 'carbyl sulphate'.

*See* 2,4-DITHIA-1,3-DIOXANE-2,2,4,4-TETRAOXIDE, $C_2H_4O_6S_2$ :
*N*-Methyl-*p*-nitroaniline, etc.

**DIBUTYL PHTHALATE** $BuO_2CC_6H_4CO_2Bu$ $C_{16}H_{22}O_4$

Chlorine

*See* CHLORINE, $Cl_2$ : Dibutyl phthalate

**DI-*tert*-BUTYL DIPEROXYPHTHALATE** $C_{16}H_{22}O_6$

$Me_3COO_2CC_6H_4CO_2OCMe_3$

Castrantas, 1965, 17

Shock-sensitive.

*See other* PEROXYESTERS

**TETRAKIS(3-METHYLPYRAZOLE)CADMIUM SULPHATE**

$C_{16}H_{24}CdN_8O_4S$

$[Cd(C_4H_6N_2)_4]SO_4$

*See* AMMINEMETAL OXOSALTS (reference 9)

**TETRAKIS(3-METHYLPYRAZOLE)MANGANESE(II) SULPHATE**

$C_{16}H_{24}MnN_8O_4S$

$[Mn(C_4H_6N_2)_4]SO_4$

*See* AMMINEMETAL OXOSALTS (reference 9)

**BROMO-5,7,7,12,14,14-HEXAMETHYL-1,4,8,11-TETRAAZA-4,11- CYCLOTETRADECADIENEIRON(II) PERCHLORATE**

$C_{16}H_{32}BrClFeN_4O_4$

*See* [14] DIENE-$N_4$ IRON COMPLEXES

**IODO-5,7,7,12,14,14-HEXAMETHYL-1,4,8,11-TETRAAZA-4,11-CYCLOTETRADECADIENEIRON(II) PERCHLORATE**

$C_{16}H_{32}ClFeIN_4O_4$

*See* [14] DIENE-$N_4$ IRON COMPLEXES

**CHLORO-5,7,7,12,14,14-HEXAMETHYL-1,4,8,11-TETRAAZA-4,11-CYCLOTETRADECADIENEIRON(II) PERCHLORATE**

$C_{16}H_{32}Cl_2FeN_4O_4$

*See* [14] DIENE-$N_4$ IRON COMPLEXES

**DICHLORO-5,7,7,12,14,14-HEXAMETHYL-1,4,8,11-TETRAAZA-4,11-CYCLOTETRADECADIENEIRON(III) PERCHLORATE**

*See* [14] DIENE-$N_4$ IRON COMPLEXES $C_{16}H_{32}Cl_3FeN_4O_4$

† **HEXADECANE** $CH_3[CH_2]_{14}CH_3$ $C_{16}H_{34}$

† **DIOCTYL ETHER** $C_8H_{17}OC_8H_{17}$ $C_{16}H_{34}O$

**TETRAKIS(BUTYLTHIO)URANIUM** $(BuS)_4U$ $C_{16}H_{36}S_4U$

Bailar, 1973, Vol. 5, 416

It ignites in air.

*See other* PYROPHORIC MATERIALS

**2,6-BIS(2-PICRYLAZO)-3,5-DINITROPYRIDINE** $C_{17}H_5N_{13}O_{16}$

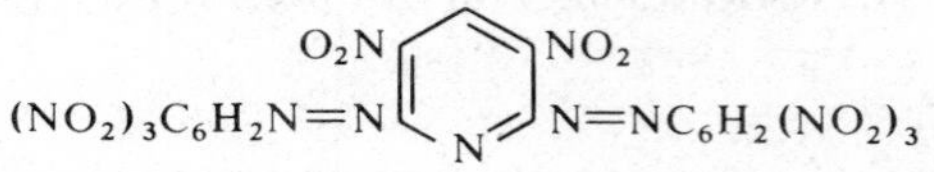

$O_2N$ / $NO_2$

$(NO_2)_3C_6H_2N{=}N$ — N — $N{=}NC_6H_2(NO_2)_3$

Explosive.

*See* POLYNITROAZOPYRIDINES

**2,6-BIS(2-PICRYLHYDRAZINO)-3,5-DINITROPYRIDINE** $C_{17}H_9N_{13}O_{16}$

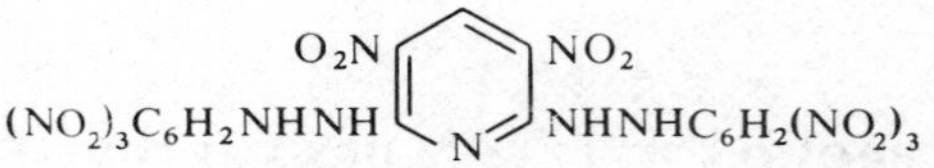

$O_2N$ / $NO_2$

$(NO_2)_3C_6H_2NHNH$ — N — $NHNHC_6H_2(NO_2)_3$

Explosive.

*See* POLYNITROAZOPYRIDINES

**BENZANTHRONE** $C_{17}H_{10}O$

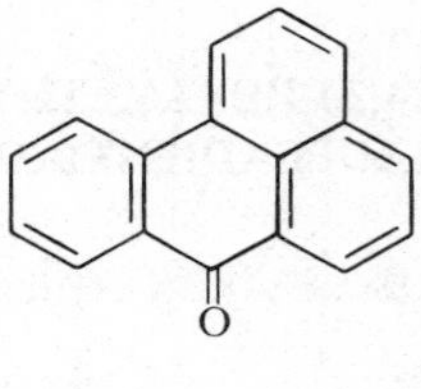

O

Nitrobenzene,

Potassium hydroxide

*See* NITROBENZENE, $C_6H_5NO_2$ : Alkali (reference 4)

***S*-7-METHYLNONYLTHIOURONIUM PICRATE** $C_{17}H_{27}N_5O_7S$

$[C_{10}H_{21}SC(NH_2)_2]^+ (NO_2)_3C_6H_2O^-$

Leese, C. L. *et al., J. Chem. Soc.*, 1950, 2739

The salt explodes if heated rapidly.

*See also* PICRATES

## IRON(III) HEXACYANOFERRATE(4–) $C_{18}Fe_7N_{18}$

$Fe_4[Fe(CN)_6]_3$

Blown castor oil,
Turkey red oil

Gaertner, K., *Farben-Ztg.*, 1938, **43**, 1118
A colour mixture containing the three components ignited spontaneously. Oxidation products in the blown castor oil may have reacted exothermically with the complex cyanide, a reducant.

Lead chromate

*See* LEAD CHROMATE, $CrO_4Pb$: Iron hexacyanoferrate
*See other* METAL CYANIDES

## LANTHANUM PICRATE $C_{18}H_6LaN_9O_{21}$

$La[(NO_2)_3C_6H_2O]_3$

Tucholskii, T., *Rep. AD 633414*, Springfield (Va.), USNTIS, 1966
In a study of the thermal stability of the picrates of group III metals, that of lanthanum occasionally exploded prematurely on heating.
*See also* PICRATES

## 4,4′-DIPHENYL-2,2′-BI(1,3-DITHIOL)-2′-YL-2-YLIUM PERCHLORATE $C_{18}H_{12}ClO_4S_4$

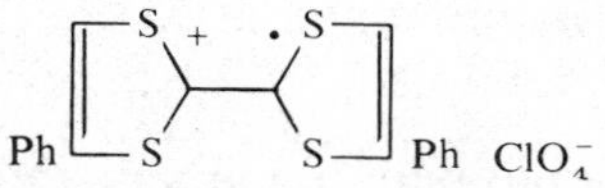

S + • S
Ph S S Ph $ClO_4^-$

Sunderlin, K. G. R., *Chem. Eng. News*, 1974, **52**(31), 3
A few mg exploded violently during determination of m.p., shattering the apparatus.
*See other* NON-METAL PERCHLORATES

## 9-PHENYL-9-IODAFLUORENE $C_{18}H_{13}I$

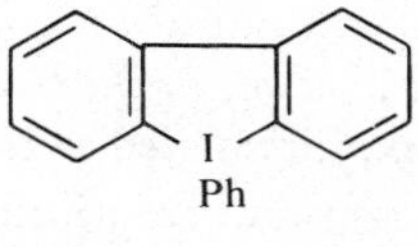

I
Ph

Banks, D. F., *Chem. Rev.*, 1966, **66**, 248
It explodes at 105°C.
*See other* IODINE COMPOUNDS

## TRIPHENYLALUMINIUM $(Ph)_3Al$ $C_{18}H_{15}Al$

Neely, T. A. *et al.*, *Org. Synth.*, 1965, **45**, 107

Triphenylaluminium and its etherate evolved heat and sparks on contact with water.
*See other* ARYLMETALS

## TRIPHENYLSILICON PERCHLORATE $Ph_3SiClO_4$ $C_{18}H_{15}ClO_4Si$

*See* ORGANOSILICON PERCHLORATES

## TRIPHENYLCHROMIUM TETRAHYDROFURANATE $C_{18}H_{15}Cr \cdot 3C_4H_8O$

$Ph_3Cr \cdot 3C_4H_8O$

Diethyl ether

Bailar, 1973, Vol. 4, 974

On warming or on treatment with ether, the solvated triphenylchromium gives a black pyrophoric material of unknown constitution.

*See other* ARYLMETALS

## TRIPHENYLLEAD NITRATE $Ph_3PbNO_3$ $C_{18}H_{15}NO_3Pb$

Sulphuric acid

Gilman, H. *et al.*, *J. Org. Chem.*, 1951, **16**, 466

It ignites in contact with concentrated acid.

*See related* ARYLMETALS

## 1,3,5-TRIPHENYL-1,4-PENTAAZADIENE PhN=NNPhN=NPh $C_{18}H_{15}N_5$

Sorbe, 1968, 141

It explodes at 80°C and is shock-sensitive.

*See other* HIGH-NITROGEN COMPOUNDS

## TRIPHENYLPHOSPHINE OXIDE–HYDROGEN PEROXIDE $C_{18}H_{15}OP \cdot H_2O_2$

$Ph_3PO \cdot H_2O_2$

Bradley, D. C. *et al.*, *J. Chem. Soc., Chem. Comm.*, 1974, 4

The complex, formulated as $2(C_6H_5)_3PO \cdot H_2O_2$, may explode.

*See related* REDOX COMPOUNDS

## 1,3-BIS(PHENYLTRIAZENO)BENZENE $C_{18}H_{16}N_6$

$PhNHN{=}NC_6H_4N{=}NNHPh$

Kleinfeller, H., *J. Prakt. Chem.* [2], 1928, **119**, 61

It explodes if rapidly heated.

*See other* TRIAZENES

**TRIPHENYLTIN HYDROPEROXIDE** **$(Ph)_3SnOOH$** **$C_{18}H_{16}O_2Sn$**

Dannley, R. L. *et al., J. Org. Chem.*, 1965, **30**, 3845

It explodes reproducibly at 75°C.

*See* ORGANOMINERAL PEROXIDES

**SODIUM HEXAKIS(PROPYNYL)FERRATE(4–)** **$C_{18}H_{18}FeNa_4$**

**$Na_4[Fe(C\equiv CMe)_6]$**

Bailar, 1973, Vol. 3, 1025

An explosive complex salt.

*See other* COMPLEX ACETYLIDES

***O–O-tert*-BUTYL DI(*p*-TOLYL) MONOPEROXOPHOSPHATE** **$C_{18}H_{23}O_5P$**

**$Me_3COOPO(OC_6H_4CH_3)_2$**

Rieche, A. *et al., Chem. Ber.*, 1962, **95**, 385

The material deflagrated as a solid, or decomposed exothermically in its reaction mixture soon after preparation.

*See other* *tert*-BUTYL PEROXOPHOSPHATE DERIVATIVES

**2,2,4-TRIMETHYLDECAHYDROQUINOLINE PICRATE** **$C_{18}H_{26}N_4O_7$**

Me

$\overset{+}{N}H_2$ $Me_2$ $(NO_2)_3C_6H_2O^-$

2-(2-Butoxyethoxy)ethanol

Franklin, N. C., private comm., 1967

Evaporation of a solution of the picrate in the diether caused a violent explosion. The solvent had probably peroxidised during open storage and residual mixture of peroxide and picrate had exploded during evaporation. Use of peroxide-free solvent and lower evaporation temperature appeared to be safe.

*See other* POLYNITROARYL COMPOUNDS

**1,4-OCTADECANOLACTONE** **$\overline{OCOC_2H_4CH}C_{14}H_{23}$** **$C_{18}H_{34}O_2$**

Preparative hazard

*See* PERCHLORIC ACID, $ClHO_4$ : Oleic acid

**BIS(DIBUTYLBORINO)ACETYLENE** **$Bu_2BC\equiv CBBu_2$** **$C_{18}H_{36}B_2$**

Hartmann, H. *et al., Z. Anorg. Chem.*, 1959, **299**, 174

Both the *n*- and iso-derivatives ignite in air.

*See other* ACETYLENIC COMPOUNDS
ALKYLNON-METALS

## OLEIC ACID

$C_{18}H_{34}O_2$

*cis*-$CH_3[CH_2]_7CH{=}CH[CH_2]_7CO_2H$

Aluminium

*See* ALUMINIUM, Al: Oleic acid

Perchloric acid

*See* PERCHLORIC ACID, $ClHO_4$ : Oleic acid

## 1,2-BIS(TRIPROPYLSILYL)HYDRAZINE

$C_{18}H_{44}N_2Si_2$

$Pr_3SiNHNHSiPr_3$

*See* SILYLHYDRAZINES: Oxidants

## HEXAMETHYLENETETRAMMONIUM TETRAPEROXOCHROMATE(V) (?)

$C_{18}H_{48}Cr_4N_{12}O_{32}$

$[(CH_2)_6(NH)_4]_3^{4+}[Cr(O_2)_4]_4^{3-}$

House, D. A. *et al., Inorg. Chem.*, 1966, **6**, 1078, footnote 8

Material recrystallised from water, and washed with methanol to dry by suction, ignited and then exploded on the filter funnel.

## 2,7-DINITRO-9-PHENYLPHENANTHRIDINE

$C_{19}H_{11}N_3O_4$

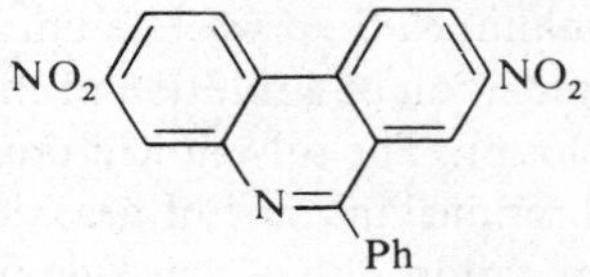

$NO_2$ $NO_2$ N= Ph

Diethyl sulphate

*See* DIETHYL SULPHATE, $C_4H_{10}O_4S$: 2,7-Dinitro-9-phenylphenanthridine

## TRIPHENYLMETHYLPOTASSIUM

$Ph_3CK$ $C_{19}H_{15}K$

Houben-Weyl, 1970, Vol.13.1, 269

The dry powder ignites in air.

*See related* ARYLMETALS

## TRIPHENYLMETHYL NITRATE

$Ph_3C^+ NO_3^-$ $C_{19}H_{15}NO_3$

Lewis acids

*See* ALKYL NITRATES

**TRIPHENYLMETHYL AZIDE** **$Ph_3CN_3$** **$C_{19}H_{15}N_3$**

Robillard, J.J., Ger. Offen. 2 345 787, 1974

The unstable explosive azide may be stabilised by adsorption for reprographic purposes.

*See other* ORGANIC AZIDES

***p*-CHLORO(BIS-*p*-NITROBENZOYLDIOXYIODO)BENZENE**

**$C_{20}H_{12}ClIN_2O_8$**

**$ClC_6H_4I(OCOC_6H_4NO_2)_2$**

*See* (DIBENZOYLDIOXYIODO)BENZENES

***p*-CHLORO(BIS-*m*-CHLOROBENZOYLDIOXYIODO)BENZENE**

**$C_{20}H_{12}Cl_3IO_4$**

**$ClC_6H_4I(OCOC_6H_4Cl)_2$**

*See* (DIBENZOYLDIOXYIODO)BENZENES

**(BIS-*m*-CHLOROBENZOYLDIOXYIODO)BENZENE** **$C_{20}H_{13}Cl_2IO_4$**

**$PhI(OCOC_6H_4Cl)_2$**

*See* (DIBENZOYLDIOXYIODO)BENZENES

**(BIS-*p*-NITROBENZOYLDIOXYIODO)BENZENE** **$C_{20}H_{13}IN_2O_8$**

**$PhI(OCOC_6H_4NO_2)_2$**

*See* (DIBENZOYLDIOXYIODO)BENZENES

**OXODIPEROXODIQUINOLINECHROMIUM(VI)** **$C_{20}H_{14}CrN_2O_5$**

**$[CrO(O_2)_2(C_{10}H_7N)_2]$**

*See* OXODIPEROXODIPYRIDINECHROMIUM(VI), $C_{10}H_{10}CrN_2O_5$

**1,3-DIPHENYL-1,3-EPIDIOXY-1,3-DIHYDROISOBENZOFURAN**

**$C_{20}H_{14}O_3$**

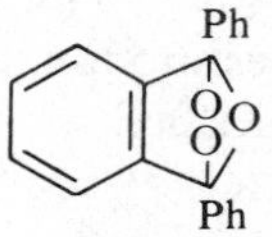

Ph
O
O
O
Ph

Dufraisse, C. *et al., Compt. Rend.*, 1946, **223**, 735

Formally an ozonide, this photo-peroxide explodes at 18°C.

*See other* OZONIDES

**_cis_-DICHLOROBIS(2,2′-BIPYRIDYL)COBALT(III) CHLORIDE** $C_{20}H_{16}Cl_3CoN_4$

$[CoCl_2(C_{10}H_8N_2)_2]Cl$

Preparative hazard

*See* CHLORINE, $Cl_2$ : Cobalt(II) chloride, Methanol

**1,1-BIS(_p_-NITROBENZOYLPEROXY)CYCLOHEXANE** $C_{20}H_{18}N_2O_{10}$

$(NO_2C_6H_4CO_2O)_2C_6H_{10}$

Criegee, R. *et al., Ann.*, 1948, **560**, 135

Explodes at 120°C.

*See other* PEROXYESTERS

**1,1-BIS(BENZOYLPEROXY)CYCLOHEXANE** $C_{20}H_{20}O_6$

$(PhCO_2O)_2C_6H_{10}$

Criegee, R. *et al., Ann.*, 1948, **560**, 135

Explodes sharply in a flame.

*See other* PEROXYESTERS

**SODIUM ABIETATE** $C_{20}H_{29}NaO_2$

*See* METAL ABIETATES

**DI-3-CAMPHOROYL PEROXIDE** $C_{20}H_{30}O_8$

$HO_2CC_8H_{14}COOOCOC_8H_{14}CO_2H$

Milas, N. A. *et al., J. Amer. Chem. Soc.*, 1933, **55**, 350–351

Explosive decomposition occurs at the m.p., 142°C, or on exposure to flame.

*See other* DIACYL PEROXIDES

**1,1,6,6-TETRAKIS(ACETYLPEROXY)CYCLODODECANE** $C_{20}H_{32}O_{12}$

$C(O_2Ac)_2[CH_2]_5C(O_2Ac)_2[CH_2]_4CH_2$ (ring)

Criegee, R. *et al., Ann.*, 1948, **560**, 135

Weak friction causes strong explosion.

*See other* PEROXYESTERS

**TRIBUTYL(PHENYLETHYNYL)LEAD** $C_{20}H_{32}Pb$

$Bu_3PbC{\equiv}CPh$

Houben-Weyl, 1975, Vol. 13.3, 80

Violent decomposition occurred during attempted distillation.

*See related* METAL ACETYLIDES

**DIACETONITRILE-5,7,7,12,14,14-HEXAMETHYL-1,4,8,11-TETRA-AZA-4,11-CYCLOTETRADECADIENEIRON(II) PERCHLORATE**

$C_{20}H_{38}Cl_2FeN_6O_8$

*See* [14]DIENE-$N_4$ IRON COMPLEXES

**TETRACYANOOCTAETHYLTETRAGOLD** $C_{20}H_{40}Au_4N_4$

$$\begin{array}{ccc} Et_2AuN & \equiv & CAuEt_2 \\ \cdot & & \cdot \\ C & & N \\ ||| & & ||| \\ N & & C \\ \cdot & & \cdot \\ Et_2AuC & \equiv & NAuEt_2 \end{array}$$

Burawoy, A. *et al., J. Chem. Soc.*, 1935, 1026

This tetramer of cyanodiethylgold is friction-sensitive and also decomposes explosively above 80°C. The propyl homologue is not friction-sensitive, but also decomposes explosively on heating in bulk.

*See other* CYANO COMPOUNDS
GOLD COMPOUNDS

**1,3,5-TRIS(*p*-AZIDOSULPHONYLPHENYL)-1,3,5-TRIAZINETRIONE**

$C_{21}H_{12}N_{12}O_9S_3$

$N_3SO_2C_6H_4CON(C_6H_4SO_2N_3)CON(C_6H_4SO_2N_3)CO$

Ulrich, H. *et al., J. Org. Chem.*, 1975, **40**, 804

This trimer of 4-azidosulphonylphenyl isocyanate melts at 200°C with violent decomposition.

*See other* ACYL AZIDES

***o*-METHOXY(BIS-*p*-NITROBENZOYLDIOXYIODO)BENZENE**

$C_{21}H_{15}IN_2O_9$

$MeOC_6H_4I(OCOC_6H_4NO_2)_2$

*See* (DIBENZOYLDIOXYIODO)BENZENES

**TRIBENZYLARSINE** $(PhCH_2)_3As$ $C_{21}H_{21}As$

Dondorov, J. *et al., Ber.*, 1935, **68**, 1255

The pure material oxidises slowly at first in air at ambient temperature, but reaction becomes violent through autocatalysis.

*See other* ALKYLNON-METALS

**TRI-*p*-TOLYLSILICON PERCHLORATE** $C_{21}H_{21}ClO_4Si$

$(MeC_6H_4)_3SiClO_4$

*See* ORGANOSILICON PERCHLORATES

**TRI-*p*-TOLYLAMMONIUM PERCHLORATE** $C_{21}H_{22}ClNO_4$

$(CH_3C_6H_4)_3N^+H\ ClO_4^-$

Weitz, E. *et al., Ber.*, 1926, **59**, 2307

Explodes violently when heated above its m.p., 123°C.

*See other* OXOSALTS OF NITROGENOUS BASES

**2,2-BIS[4(2′,3′-EPOXYPROPOXY)PHENYL]PROPANE** $C_{21}H_{24}O_4$

$(\overline{OCH_2CH}CH_2OC_6H_4)_2CMe_2$

Trichloroethylene

*See* TRICHLOROETHYLENE, $C_2HCl_3$: Epoxides

**ACETONITRILEIMIDAZOLE-5,7,7,12,14,14-HEXAMETHYL-1,4,8,11-TETRAAZA-4,11-CYCLOTETRADECADIENEIRON(II) PERCHLORATE** $C_{21}H_{39}Cl_2FeN_7O_8$

*See* [14]DIENE-$N_4$IRON COMPLEXES

**BIS(*η*-CYCLOPENTADIENYL)BIS(PENTAFLUOROPHENYL)-ZIRCONIUM** $C_{22}H_{10}F_{10}Zr$

$(C_5H_5)_2Zr(C_6F_5)_2$

Chaudhari, M. A. *et al., J. Chem. Soc. (A)*, 1966, 838

Explodes in air (but not nitrogen) above its m.p., 219°C.

*See other* HALO-ARYLMETALS

**DI-1-NAPHTHOYL PEROXIDE** $C_{22}H_{14}O_4$

$C_{10}H_7COOOCOC_{10}H_7$

Castrantas, 1965, 17

Explodes on friction.

*See other* DIACYL PEROXIDES

**BIS(DICARBONYL-*η*-CYCLOPENTADIENYLIRON)–BIS(TETRAHYDROFURAN)MAGNESIUM** $C_{22}H_{26}Fe_2MgO_6$

$[C_5H_5Fe(CO)_2]_2Mg \cdot 2C_4H_8O$

McVicker, G. B., *Inorg. Synth.*, 1976, **16**, 56–58

This, like other transition metal carbonyl derivatives of magnesium, is pyrophoric.

*See other* ORGANOMETALLICS
CARBONYLMETALS

**4-[2-(4-HYDRAZINO-1-PHTHALAZINYL)HYDRAZINO]-4-METHYL-2-PENTANONE (4-HYDRAZINO-1-PHTHALAZINYL) HYDRAZONE DINICKEL(II) TETRAPERCHLORATE**

$C_{22}H_{28}N_{12}Ni_2Cl_4O_{16}$

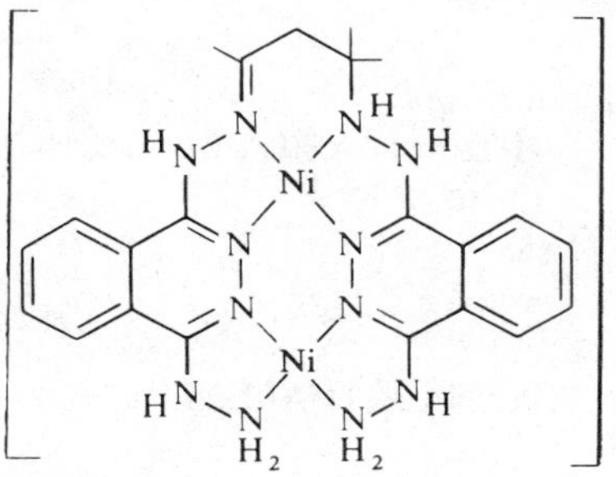

$[ClO_4]_4$

Rosen, W., *Inorg. Chem.*, 1971, **10**, 1833

The green complex isolated directly from the reaction mixture explodes fairly violently at elevated temperatures. The tetrahydrate produced by recrystallisation from aqueous methanol is also moderately explosive when heated rapidly.

*See other* AMMINEMETAL OXOSALTS

**1,3-BIS(DI-η-CYCLOPENTADIENYLIRON)-2-PROPEN-1-ONE** $C_{23}H_{20}Fe_2O$

$C_5H_5FeC_5H_4COCH{=}CHC_5H_4FeC_5H_5$

Perchloric acid,
Acetic anhydride,
Ether,
Methanol

Anon., *Chem. Eng. News*, 1966, **44**(49), 50

Condensation of the iron complex with cyclopentanone in perchloric acid–acetic anhydride–ether medium had been attempted. The non-crystalline residue, after methanol-washing and drying in air for several weeks, exploded on being disturbed. This was attributed to possible presence of a derivative of ferrocenium perchlorate, a powerful explosive and detonator. However, methyl or ethyl perchlorates may have been involved.

*See* ALKYL PERCHLORATES
PERCHLORIC ACID, $ClHO_4$ : Diethyl ether

*See other* ORGANOMETALLICS

## TETRAKIS(PENTAFLUOROPHENYL)TITANIUM $C_{24}F_{20}Ti$

$(C_6F_5)_4Ti$

Houben-Weyl, 1975, Vol. 13.3, 300
It explodes at 120–130°C.

*See* HALOARYLMETALS
*See related* ARYLMETALS

## IRON(III) PHTHALATE $C_{24}H_{12}Fe_2O_{12}$

$Fe_2(O_2CC_6H_4CO_2)_3$

Sulphur compounds
*See* PHTHALIC ANHYDRIDE, $C_8H_4O_3$ : Preparative hazards (reference 2)

## 1,5-DIBENZOYLNAPHTHALENE $PhCOC_{10}H_6COPh$ $C_{24}H_{16}O_2$

Preparative hazard
*See* ALUMINIUM TRICHLORIDE, $AlCl_3$ : Benzoyl chloride, etc.

## 1-(2′-, 3′- or 4′-DIAZONIOPHENYL)-2-METHYL-4,6-DIPHENYLPYRIDINIUM DIPERCHLORATE $C_{24}H_{19}Cl_2N_3O_8$

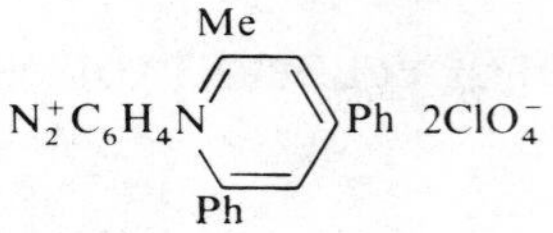

Me
$N_2^+C_6H_4N$ [pyridinium ring] Ph $2ClO_4^-$
Ph

Dorofenko, G. N. *et al.*, *Chem. Abs.*, 1975, **82**, 139908i
The three diazonium perchlorate isomers, and their 2,4,6-triphenyl analogues, exploded on heating.

*See other* DIAZONIUM PERCHLORATES

## 1,3,6,8-TETRAPHENYLOCTAAZATRIENE $C_{24}H_{20}N_8$

PhN=NN(Ph)N=NN(Ph)N=N(Ph)

Wohl, A. *et al.*, *Ber.*, 1900, **33**, 2741
This compound (and several *C*-homologues) is unstable and explodes sharply on heating, impact or friction. In a sealed tube, the explosion is violent.

*See other* HIGH-NITROGEN COMPOUNDS

## TETRAPHENYLLEAD $Ph_4Pb$ $C_{24}H_{20}Pb$

Potassium amide

*See* POTASSIUM AMIDE, $H_2KN$: Tetraphenyllead

Sulphur

Houben-Weyl, 1975, Vol. 13.3, 236

Interaction may be explosive.

*See other* ARYLMETALS

## BIS(DI-$\eta$-BENZENECHROMIUM) DICHROMATE $C_{24}H_{24}Cr_4O_7$

$[Cr(C_6H_6)_2]_2Cr_2O_7$

Anon., *Chem. Eng. News*, 1964, **42**(38), 55

This catalyst exists as explosive orange-red crystals.

*See other* ORGANOMETALLICS

## 3,3,6,6-TETRAPHENYLPERHYDRO-3,6-DISILATETRAZINE $C_{24}H_{24}N_4Si_2$

$Ph_2\overline{SiNHNHSi(Ph)_2NHNH}$

*See* SILYLHYDRAZINES: Oxidants

## TRIS(BIS-2-METHOXYETHYL ETHER)POTASSIUM HEXACARBONYLNIOBATE(1–) $C_{24}H_{42}KNbO_{15}$

$K[Nb(CO)_6]\cdot3(MeOC_2H_4)_2O$

Solvents

Ellis, J. E. *et al.*, *Inorg. Synth.*, 1976, **16**, 70–71

Attempts to recover the complex salt from solutions in other than diglyme invariably led to the formation of pyrophoric solids.

*See related* CARBONYLMETALS

*See other* PYROPHORIC MATERIALS

## DIDODECANOYL PEROXIDE (DILAUROYL PEROXIDE) $C_{24}H_{46}O_4$

$C_{11}H_{23}COOOCOC_{11}H_{23}$

*Haz. Chem. Data*, 1975, 137

Though regarded as one of the more stable peroxides, it becomes shock-sensitive on heating, and self-accelerating decomposition sets in at 49°C.

*See other* DIACYL PEROXIDES

**4,4-DIFERROCENYLPENTANOYL CHLORIDE** $C_{25}H_{27}ClFe_2O$

$MeC(C_{10}H_{10}Fe)_2[CH_2]_2COCl$

Nielsen, A. T. *et al., J. Org. Chem.*, 1976, **41**, 657

The acid chloride is thermally unstable much above 25°C. At 80–100°C it rapidly evolves hydrogen chloride and forms a black solid.

*See other* ACYL HALIDES

**COPPER BIS(1-BENZENEAZOTHIOCARBONYL-2-PHENYLHYDRAZINE)**

$C_{26}H_{22}CuN_8S_2$

PhN=NCSNHN(Ph)CuN(Ph)NHCSN=NPh

Sorbe, 1968, 80

The copper derivative of dithizone explodes at 150°C.

**LEAD BIS(1-BENZENEAZOTHIOCARBONYL-2-PHENYLHYDRAZINE)**

$C_{26}H_{22}N_8PbS_2$

PhN=NCSNHN(Ph)PbN(Ph)NHCSN=NPh

Sorbe, 1968, 36

The lead derivative of dithizone explodes at 215°C.

**ZINC BIS(1-BENZENEAZOTHIOCARBONYL-2-PHENYLHYDRAZINE)**

$C_{26}H_{22}N_8S_2Zn$

PhN=NCSNHN(Ph)ZnN(Ph)NHCSN=NPh

Sorbe, 1968, 159

The zinc derivative of dithizone decomposes explosively at high temperature.

*See other* HEAVY METAL DERIVATIVES

**DIDODECYL PEROXYDICARBONATE** $C_{26}H_{50}O_6$

$C_{12}H_{25}OCOOOCOOC_{12}H_{25}$

D'Angelo, A. J., US Pat. 3 821 273, 1974

It decomposes violently after 90 min at 40°C, while the hexadecyl homologue is stable for a week at 40°C.

*See other* PEROXYCARBONATE ESTERS

**TETRAKIS(PYRIDINE)BIS(TETRACARBONYLCOBALT)MAGNESIUM**

$C_{28}H_{20}Co_2MgN_4O_8$

$[(C_5H_5N)_2Co(CO)_4Mg(CO)_4Co(C_5H_5N)_2]$

McVicker, G. B., *Inorg. Synth.*, 1976, **16**, 58–59

This, like other transition metal carbonyl derivatives of magnesium, is pyrophoric.

*See related* CARBONYLMETALS

## 5,7,7,12,14,14-HEXAMETHYL-1,4,8,11-TETRAAZA-4,11-CYCLOTETRADECADIENE-1,10-PHENANTHROLINEIRON(II) PERCHLORATE

$C_{28}H_{40}Cl_2FeN_6O_8$

*See* [14] DIENE-$N_4$ IRON COMPLEXES

## 1(2′-, 3′- or 4′-DIAZONIOPHENYL)-2,4,6-TRIPHENYLPYRIDINIUM DIPERCHLORATE

$C_{29}H_{21}Cl_2N_3O_8$

$N_2^+C_6H_4N^+$ (2,4,6-triphenylpyridinium: Ph at 2, 4, 6) $2ClO_4^-$

Dorofenko, G. N. *et al., Chem. Abs.*, 1975, **82**, 139908d

The three diazonium perchlorate isomers, and their 2-methyl-4,6-diphenyl analogues, exploded on heating.

*See other* DIAZONIUM PERCHLORATES

## TRIS-2,2′-BIPYRIDINESILVER(II) PERCHLORATE

$C_{30}H_{24}AgCl_2N_6O_8$

$[Ag(C_{10}H_8N_2)_3]^{2+}[ClO_4]_2^-$

Morgan, G. T. *et al., J. Chem. Soc.*, 1930, 2594

Explodes on heating.

*See other* AMMINEMETAL OXOSALTS

## TRIS-2,2′-BIPYRIDINECHROMIUM(II) PERCHLORATE

$C_{30}H_{24}Cl_2CrN_6O_8$

$[Cr(C_{10}H_8N_2)_3]^{2+}[ClO_4]_2^-$

Holah, D. G. *et al., Inorg. Synth.*, 1967, **10**, 34

Explodes violently on slow heating to 250°C and can be initiated by static sparks, but not apparently by impact.

*See other* AMMINEMETAL OXOSALTS

## D-(+)-TRIS(*o*-PHENANTHROLINE)RUTHENIUM(II) PERCHLORATE

$C_{30}H_{24}Cl_2N_6O_8Ru$

$[Ru(C_{10}H_8N_2)_3][ClO_4]_2$

Sulphuric acid

Gillard, R. D. *et al., J. Chem. Soc., Dalton Trans.*, 1974, 1235

Dissolution of the salt in the concentrated acid must be very slow with ice-cooling to prevent an explosive reaction.

*See other* AMMINEMETAL OXOSALTS

**TRIS-2,2'-BIPYRIDINECHROMIUM(0)** $C_{30}H_{24}CrN_6$

$[Cr(C_{10}H_8N_2)_3]$

Herzog, S. *et al.*, *Z. Naturforsch.*, 1957, **12**, 809
Ignites in air.
*See related* AMMINEMETAL OXOSALTS

**POTASSIUM TETRAKIS(PHENYLETHYNYL)NICCOLATE(4–)**

$C_{32}H_{20}K_4Ni$

$K_4[Ni(C{\equiv}CPh)_4]$

It ignites on flame contact.
*See* COMPLEX ACETYLIDES

**BIS(5,7,7,12,14,14-HEXAMETHYL-1,4,8,11-TETRAAZA-4,11-CYCLOTETRADECADIENE) HYDROXODIIRON(II) TRIPERCHLORATE**

$C_{32}H_{65}Cl_3Fe_2N_8O_{13}$

*See* [14]DIENE-$N_4$ IRON COMPLEXES

**BIS[AQUA-5,7,7,12,14,14-HEXAMETHYL-1,4,8,11-TETRAAZA-4,11-CYCLOTETRADECADIENEIRON(II)] OXIDE TETRAPERCHLORATE**

$C_{32}H_{68}Cl_4Fe_2N_8O_{19}$

*See* [14]DIENE-$N_4$ IRON COMPLEXES

**HEXAPHENYLHEXAARSANE** $PhAs[AsPh]_4AsPh$ $C_{36}H_{30}As_6$

Oxygen

1. Maschmann, E., *Ber.*, 1926, **59**, 1143
2. Blicke, F. F. *et al., J. Amer. Chem. Soc.*, 1930, **52**, 2946–2950

The explosively violent interaction of 'arsenobenzene' with oxygen [1] was later ascribed to the presence of various catalytic impurities because the pure compound is stable towards oxygen [2].

**LITHIUM HEXAPHENYLTUNGSTATE(2–)** $C_{36}H_{30}Li_2W$

$Li_2[WPh_6]$

Sarry, B. *et al., Z. Anorg. Allg. Chem.*, 1964, **329**, 218
Pyrophoric in air.
*See related* ARYLMETALS

**HEXACYCLOHEXYLDILEAD** $C_{36}H_{66}Pb_2$

$(C_6H_{11})_3PbPb(C_6H_{11})_3$

Halocarbons

Houben-Weyl, 1975, Vol. 13.3, 168

In absence of solvent and presence of air, interaction with carbon tetrabromide is explosive, and with hexabromoethane, more so.

*See related* ALKYLMETALS

**LEAD OLEATE** $C_{36}H_{66}O_4Pb$

$Pb(O_2C[CH_2]_7HC{=}CH[CH_2]_7CH_3)_2$

Mineral oil

Williams, C. G., *Mech. Eng.*, 1932, **54**, 128–129

Lead oleate greases tended to cause violent explosions when used on hot-running bearings. The cause is not immediately apparent, but peroxidation may have been involved.

*See other* HEAVY METAL DERIVATIVES

**CALCIUM STEARATE** $Ca(O_2CC_{17}H_{35})_2$ $C_{36}H_{70}CaO_4$

Schmutzler, G. *et al., Plaste Kaut.*, 1967, **14**, 827–829

Dust explosion and spontaneous ignition hazards for calcium and related stearate plastics additives are detailed and discussed.

*See* DUST EXPLOSIONS

**CARBONYL-BIS-TRIPHENYLPHOSPHINEIRIDIUM–SILVER DIPERCHLORATE** $C_{37}H_{30}AgCl_2IrO_9P_2$

$[(CO)AgIr(PPh_3)_2][ClO_4]_2$

Kuyper, J. *et al., J. Organomet. Chem.*, 1976, **107**, 130

Highly explosive, detonated by heat or shock.

*See related* AMMINEMETAL OXOSALTS

**CALCIUM ABIETATE** $C_{40}H_{58}CaO_4$

*See* METAL ABIETATES

**MANGANESE ABIETATE** $C_{40}H_{58}MnO_4$

*See* METAL ABIETATES

**LEAD ABIETATE** $C_{40}H_{58}O_4Pb_4$

*See* METAL ABIETATES

**ZINC ABIETATE** $C_{40}H_{58}O_4Zn$

*See* METAL ABIETATES

**HEXAPYRIDINEIRON(II) TRIDECACARBONYL-TETRAFERRATE(2−)** $C_{43}H_{30}Fe_5N_6O_{13}$

$[Fe(C_5H_5N)_6]^{2+}[Fe_4(CO)_{13}]^{2-}$

Brauer, 1965, Vol. 2, 1758

Extremely pyrophoric.

*See other* AMMINEMETAL OXOSALTS

**TETRAKIS(4-*N*-METHYLPYRIDINIO)PORPHINECOBALT(II)(4+) PERCHLORATE** $C_{44}H_{36}Cl_4CoN_8O_{16}$

Pasternack, R. F. *et al., J. Inorg. Nucl. Chem.*, 1974, **36**, 600

A sample obtained by precipitation exploded.

*See other* AMMINEMETAL OXOSALTS

**OXYBIS(*N,N*-DIMETHYLACETAMIDETRIPHENYLSTIBONIUM) DIPERCHLORATE** $C_{44}H_{50}Cl_2N_2O_{11}Sb_2$

$[(Ph_3HSb^+ \leftarrow N(Ac)Me_2)_2O][ClO_4]_2$

Goel, R. G. *et al., Inorg. Chem.*, 1972, **11**, 2143

It exploded on several occasions during handling and attempted analysis.

*See other* AMMINEMETAL OXOSALTS

**BIS[DICARBONYL-η-CYCLOPENTADIENYL(TRIBUTYLPHOSHINE)-MOLYBDENUM]TETRAKIS(TETRAHYDROFURAN)MAGNESIUM** $C_{54}H_{96}MgMo_2O_8P_2$

$[Mg\{(CO)_2C_5H_5MoPBu_3\}_2 4C_4H_8O]$

McVicker, G. B., *Inorg. Synth.*, **16**, 59–61

This and other transition metal carbonyl derivatives of magnesium is pyrophoric.

*See related* CARBONYLMETALS

*See other* ORGANOMETALLICS

**ALUMINIUM ABIETATE** $C_{60}H_{87}AlO_6$

*See* METAL ABIETATES

**CALCIUM** Ca

Air

'Product Information Sheet No. 212', Sandwich, Pfizer Chemicals, 1969

Calcium is pyrophoric when finely divided.

Alkalies
Merck, 1968, 189
Interaction with alkali-metal hydroxides or carbonates may be explosive.

Ammonia
1. Partington, 1967, 369
2. Gibson, 1969, 369

At ambient temperature, ammonia gas reacts exothermally with calcium, but if warmed, the latter becomes incandescent [1]. The metal dissolves unchanged in liquid ammonia, but if the latter evaporates, the finely divided residual metal is pyrophoric [2].
*See other* PYROPHORIC METALS

Asbestos cement
Scott, P. J., *School Sci. Rev.,* 1967, **49**(167), 252
Drops of molten calcium falling on to hard asbestos cement sheeting caused a violent explosion which perforated the sheet. Interaction with sorbed water in the cement seems likely to have occurred.
*See* Water, below

Carbon dioxide,
Nitrogen
*See* CARBON DIOXIDE, $CO_2$: Metals, Nitrogen

Chlorine tri- or penta-fluorides
*See* CHLORINE TRIFLUORIDE, $ClF_3$: Metals

Dinitrogen tetraoxide
*See* DINITROGEN TETRAOXIDE (NITROGEN DIOXIDE), $N_2O_4$: Metals

Halogens
Mellor, 1941, Vol. 3, 638
Massive calcium ignites in fluorine at ambient temperature, and finely divided (but not massive) calcium ignites in chlorine similarly.

Lead dichloride
Mellor, 1941, Vol. 3, 639
Interaction is explosive on warming.

Mercury
Pascal, 1958, Vol. 4, 290
Amalgam formation at 340°C is violent.

Phosphorus(V) oxide

*See* PHOSPHORUS(V) OXIDE, $O_5P_2$ : Metals

Silicon

Mellor, 1940, Vol. 6, 176–177

Interaction is violently incandescent above 1050°C after a short delay.

Sodium,
Mixed oxides

*BCISC Quart. Safety Summ.*, 1966, **37**(145), 6

An operator working above the charging hole of a sludge reactor was severely burned when a quantity of a burning sludge containing calcium and sodium metals and their oxides was ejected. This very reactive mixture is believed to have been ignited by drops of perspiration falling from the operator.

*See* Water, below

Sulphur

Mellor, 1941, Vol. 3, 639

A mixture reacts explosively when ignited.

Water

'Product Information Sheet No. 212', Sandwich, Pfizer Chemicals, 1969

Calcium or its alloys react violently with water (or dilute acids) and the heat of reaction may ignite evolved hydrogen under appropriate contact conditions.

*See* Asbestos cement, above

*See other* METALS

## CALCIUM CHLORIDE $CaCl_2$

Methyl vinyl ether

*See* METHYL VINYL ETHER, $C_3H_6O$: Acids

Water

*MCA Case History No. 69*

The exotherm produced by adding solid calcium chloride to hot water caused violent boiling.

Zinc

*ABCM Quart. Safety Summ.*, 1932, **3**, 35

Prolonged action of calcium chloride solution upon the zinc coating of galvanised iron caused slow evolution of hydrogen, which became ignited and exploded.

*See other* METAL HALIDES

## CALCIUM HYPOCHLORITE $Ca(OCl)_2$ $CaCl_2O_2$

Sidgwick, 1950, 1217

Clancey, V. J., *J. Haz. Mat.*, 1975, **1**, 83–94

This powerful oxidant is technically of great importance for bleaching and sterilisation applications, and contact with reducants or combustible materials must be under controlled conditions. It is present in diluted state in bleaching powder, which is a less powerful oxidant with lower available chlorine content.

There have been recently several instances of mild explosions and/or intense fires on ships carrying cargoes of commercial hypochlorite, usually packed in lacquered steel drums and of Japanese origin. Investigational work is still in progress, but preliminary indications suggest that presence of magnesium oxide in the lime used to prepare the hypochlorite may have led to the presence of magnesium hypochlorite, which is known to be of very limited stability. Unlike many other oxidants, the hypochlorite constitutes a hazard in the absence of other combustible materials, because, after initiation, local rapid thermal decomposition will spread through the contained mass of hypochlorite as a vigorous fire which *evolves* oxygen.

*See* BLEACHING POWDER

Acetylene

Rüst, 1948, 338

Contact of acetylene with bleaching powder, etc., may lead (as with chlorine itself) to formation of explosive chloroacetylenes. Application of hot water to free a partially frozen acetylene purification system containing bleaching powder caused a violent explosion to occur.

*See* HALOACETYLENE DERIVATIVES

Ammonium chloride

1. Morris, D. L., *The Science Teacher,* 1968, **35**(6), 4
2. Anderson, M. B., *The Science Teacher,* 1968, **35**(9), 4

Report of an explosion through unintentional use of calcium hypochlorite instead of calcium hydroxide in the preparation of ammonia gas. Nitrogen trichloride was produced [1, 2].

*See* Nitrogenous bases, below

Carbon

Mellor, 1941, Vol. 2, 262

A confined intimate mixture of hypochlorite and finely divided charcoal exploded on heating.

Contaminants

*MCA Case History No. 666*

The contents of a drum erupted and ignited during intermittent use. This was attributed to contamination of the soldered metal scoop (normally kept in the drum) by oil, grease or water, or all three, and a subsequent exothermic reaction with the hypochlorite.

*See* Organic matter, below

*N,N*-Dichloromethylamine

*See* *N,N*-DICHLOROMETHYLAMINE, $CH_3Cl_2N$

Hydroxy compounds

Fawcett, H. H., *Ind. Eng. Chem.*, 1959, **51**(4), 90A

Contact of the solid hypochlorite with glycerol, digol monomethyl ether or phenol causes ignition within a few minutes, accompanied by irritant smoke, particularly with phenol. Ethanol may cause an explosion, as may methanol.

Nitrogenous bases

Kirk-Othmer, 1963, Vol. 2, 105

Primary aliphatic or aromatic amines react with calcium (or sodium) hypochlorite to form *N*-mono- or di-chloroamines which are explosive, but less so than nitrogen trichloride.

*See* Ammonium chloride, above

Nitromethane

Fawcett, H. H., *Trans. Nat. Safety Cong. Chem. Fertilizer Ind.*, Vol. 5, 32, Chicago, NSC, 1963

They interact, after a delay, with extreme vigour.

Organic matter

'Halane' Information Sheet, Wyandotte Chem. Co., Michigan, 1958

Mixtures of the solid hypochlorite with 1% of admixed organic contaminants are sensitive to heat in varying degree. Wood caused ignition at 176°C, while oil caused violent explosion at 135°C.

*See* Contaminants, above

Organic matter,
Water

Tatara, S. *et al.*, Ger. Offen., 2 450 816, 1975 (*Chem. Abs.*, 1975, **83**, 100239s)

Hypochlorite containing over 60% active chlorine normally ignites in contact with lubricating oils, but addition of 16–22% water will prevent this. Examples show the effect of 18% (but not 15%) of added water in preventing ignition when glycerol was dripped onto a hypochlorite (79% active chlorine) containing 2% of oil.

Organic sulphur compounds

1. Stephenson, F. G., *Chem. Eng. News*, 1973, **51**(26), 14

2. *Laboratory Waste Disposal Manual*, 142, Washington, MCA, 1969–1972 edns
3. *MCA Guide for Safety in the Chemical Laboratory*, 464, New York, Van Nostrand Reinhold, 1972

Contact of the solid hypochlorite with organic thiols or sulphides may cause a violent reaction and flash fire [1]. This procedure was recommended formerly for treating spills of sulphur compounds [2,3] but is now withdrawn as potentially hazardous. Use of an aqueous solution of up to 15% concentration, or of 5% sodium hypochlorite solution, is now recommended.

*See* Sulphur, below

Sodium hydrogensulphate,
Starch,
Sodium carbonate

Anon., *Ind. Eng. Chem. (News Ed.),* 1937, **15**, 282

Shortly after a mixture of the four ingredients had been compressed into tablets, incandescence and an explosion occurred. This may have been due to interaction of the hypochlorite and starch, accelerated by the acid sulphate, and may also have involved dichlorine monoxide.

Sulphur

Katz, S. A. *et al., Chem. Eng. News*, 1965, **46**(29), 6

Admixture of damp sulphur and solid 'swimming pool chlorine' caused a violently exothermic reaction, and ejection of molten sulphur.

*See other* METAL OXOHALOGENATES

## CALCIUM CHLORITE $Ca(ClO_2)_2$ $CaCl_2O_4$

Potassium thiocyanate

Pascal, 1960, Vol. 16, 264

Mixtures may ignite spontaneously.

*See* CHLORITE SALTS
*See other* METAL OXOHALOGENATES

## CALCIUM DIHYDRIDE $CaH_2$

Halogens

Mellor, 1941, Vol. 3, 651

Heating the hydride strongly with chlorine, bromine or iodine leads to incandescence.

Manganese dioxide

*See* MANGANESE DIOXIDE, $MnO_2$: Calcium hydride

Metal halogenates

Mellor, 1946, Vol. 3, 651

Mixtures of the hydride with various bromates, chlorates or perchlorates explode on grinding.

*See* METAL HALOGENATES

Silver halides

Mellor, 1941, Vol. 3, 389, 651

A mixture of silver fluoride and the hydride incandesces on grinding, and the iodide reacts vigorously on heating.

Tetrahydrofuran

*See* LITHIUM TETRAHYDROALUMINATE, $AlH_4Li$: Tetrahydrofuran

**CALCIUM HYDROXIDE *O*-HYDROXYLAMIDE** **$CaH_3NO_2$**

**$Ca(OH)ONH_2$**

Sorbe, 1968, 43

It explodes on heating (and presumably would be formed if distillation of hydroxylamine from calcium oxide were attempted).

*See* HYDROXYLAMINE, $H_3NO$: Metals

**CALCIUM BIS-HYDROXYLAMIDE** **$Ca(NHOH)_2$** **$CaH_4N_2O_2$**

*See* HYDROXYLAMINE, $H_3NO$: Metals

**CALCIUM PHOSPHINATE ('HYPOPHOSPHITE')** **$CaH_4O_4P_2$**

**$Ca(H_2PO_2)_2$**

Potassium chlorate

*See* POTASSIUM CHLORATE, $ClKO_3$ : Reducants

**CALCIUM PERMANGANATE** **$Ca(MnO_4)_2$** **$CaMn_2O_8$**

Acetic acid,
or Acetic anhydride

*See* POTASSIUM PERMANGANATE, $KMnO_4$ : Acetic acid

Hydrogen peroxide

*See* HYDROGEN PEROXIDE, $H_2O_2$ : Metals, etc.

## CALCIUM NITRATE $Ca(NO_3)_2$ $CaN_2O_6$

Ammonium nitrate,
Formamide,
Water

Wilson, J. F. *et al.*, S. Afr. Pat. 74 03 305, 1974 (*Chem. Abs.*, 1976, **84**, 76546n)

A mixture containing 51% of calcium nitrate and 12% of ammonium nitrate with 27% of formamide and 10% of water is detonable at –12°C.

Ammonium nitrate,
Hydrocarbon oil

Haid, A. *et al., Jahresber. Chem. Tech. Reichsanstalt*, 1930, **8**, 108–115

The explosibility of the double salt, calcium ammonium nitrate, is increased by the presence of oil.

*See* AMMONIUM NITRATE, $H_4N_2O_3$ : Organic fuels

Ammonium nitrate,
Water-soluble fuels

Clark, W. F. *et al.*, US Pat. 3 839 107, 1970

Up to, or over, 40% of the ammonium nitrate content of explosive mixtures with water-soluble organic fuels may be replaced with advantage by calcium nitrate.

Organic materials

Sorbe, 1968, 42

Mixtures of the nitrate and organic materials may be explosive.

*See other* METAL NITRATES

## CALCIUM DIAZIDE $Ca(N_3)_2$ $CaN_6$

Mellor, 1940, Vol. 8, 349

Calcium, strontium and barium diazides are not shock-sensitive, but explode on heating at about 150, 170 and 225 (or 152)°C, respectively. In sealed tubes the explosion temperatures are greater.

*See other* METAL AZIDES

## CALCIUM OXIDE $CaO$

Ethanol

Keusler, V., *Apparatebau*, 1928, **40**, 88–89

The lime–alcohol residue from preparation of anhydrous alcohol ignited on discharge from the autoclave and caused a vapour explosion. The finely divided and reactive lime may have heated on exposure to atmospheric moisture and caused ignition.

*See* Water, below

Hydrogen fluoride

*See* HYDROGEN FLUORIDE, FH: Oxides

Interhalogens

*See* BROMINE PENTAFLUORIDE, $BrF_5$ : Acids, etc.
CHLORINE TRIFLUORIDE, $ClF_3$ : Metals, etc.

Phosphorus(V) oxide

*See* PHOSPHORUS(V) OXIDE, $O_5P_2$ : Inorganic bases

Water

1. Anon., *Fire*, 1935, **28**, 30
2. Amos, T., *Zentr. Zuckerind.*, 1923, **32**, 103
3. *BCISC Quart. Safety Summ.*, 1967, **38**, 15
4. *BCISC Quart. Safety Summ.*, 1971, **42**(168), 4
5. Anon., *Engrg. Digest*, 1908, **4**, 661

Quicklime, when mixed with $\frac{1}{3}$ of its weight of water, will reach 150–300°C (depending on quantity) and may ignite combustible material. Occasionally 800–900°C has been attained [1]. Moisture present in wooden storage bins caused ignition of the latter [2]. A water jet was used unsuccessfully to try to clear a pump hose blocked with quicklime. On standing, the exothermic reaction proceeded far enough to generate enough steam to clear explosively the blocked pipe [3].

Two glass bottles of calcium oxide burst while in a laboratory, owing to the considerable increase in bulk which occurs on hydration. Storage in plastics bottles under desiccation is recommended. Granular oxide should cause fewer problems in this respect than the powder [4]. The powerful expansion effect was formerly used in coal-mining operations [5].

*See other* METAL OXIDES

## CALCIUM PEROXIDE $CaO_2$

Oxidisable materials

Castrantas, 1965, 4
Sorbe, 1968, 43

Grinding the peroxide with oxidisable materials may cause fire, and the octahydrated solid explodes at high temperature.

*See other* METAL PEROXIDES

## CALCIUM SULPHATE $CaSO_4$ $CaO_4S$

Aluminium

*See* ALUMINIUM, Al: Metal oxosalts

Diazomethane

*See* DIAZOMETHANE, $CH_2N_2$ : Calcium sulphate

Phosphorus

*See* PHOSPHORUS, P: Metal sulphates

**CALCIUM PEROXODISULPHATE** **$caOSO_2OOSO_2Oca$** **$CaO_8S_2$**

Castrantas, 1965, 6

Shock-sensitive; explodes violently.

*See other* PEROXOACID SALTS

**CALCIUM SULPHIDE** **CaS**

Oxidants

Mellor, 1941, Vol. 3, 745

Alkaline-earth sulphides react vigorously with chromyl chloride, lead dioxide, potassium chlorate (explodes lightly) and potassium nitrate (explodes violently).

*See other* METAL SULPHIDES

**CALCIUM POLYSULPHIDE** **$CaS_x$**

Edwards, P. W., *Chem. Met. Eng.*, 1922, 27, 986

The powdered sulphide, admixed with small amounts of calcium thiosulphate and sulphur, has been involved in several fires and explosions, some involving initiation by static discharges.

*See other* METAL SULPHIDES

**CALCIUM SILICIDE** **CaSi**

Acids

Mellor, 1940, Vol. 6, 177

Interaction is vigorous, and the silanes evolved ignite in air.

**CALCIUM DISILICIDE** **$CaSi_2$**

Carbon tetrachloride

*Zirconium Fire and Explosion Incidents,* TID-5365, Washington, USAEC, 1956

Calcium disilicide exploded when milled in the solvent.

Diiron trioxide

*See* DIIRON TRIOXIDE, $Fe_2O_3$ : Calcium disilicide

Metal fluorides

Berger, E., *Compt. Rend.*, 1920, **170**, 29

Calcium silicide ignites in close contact with alkali metal fluorides (forming silicon tetrafluoride).

Potassium nitrate

*See* POTASSIUM NITRATE, $KNO_3$ : Calcium disilicide

*See other* METAL NON-METALLIDES

**CALCIUM PEROXOCHROMATE(3–)** **$Ca_3Cr_2O_{12}$**

**$Ca_3[Cr(O_2)_2O_2]_2^{3-}$**

Raynolds, J. H. *et al., J. Amer. Chem. Soc.*, 1930, **52**, 1851

It explodes at 100°C.

*See other* PEROXOACID SALTS

**TRICALCIUM DINITRIDE** **$Ca_3N_2$**

von Schwartz, 1918, 322

Spontaneously flammable in air (probably when finely divided in moist air).

Halogens

Mellor, 1940, Vol. 8, 103

Reaction with incandescence in chlorine gas or bromine vapour.

*See other N*-METAL DERIVATIVES

**TRICALCIUM DIPHOSPHIDE** **$Ca_3P_2$**

Dichlorine oxide

*See* DICHLORINE OXIDE, $Cl_2O$: Oxidisable materials

Oxygen

Van Wazer, 1958, Vol. 1, 145

Calcium and other alkaline earth phosphides incandesce in oxygen at about 300°C.

Water

Mellor, 1940, Vol. 8, 841

Calcium and other phosphides on contact with water liberate phosphine, which is usually spontaneously flammable in air, owing to the diphosphane content.

*See other* METAL NON-METALLIDES

## CADMIUM $Cd$

The finely divided metal is pyrophoric.

*See* CADMIUM HYDRIDE, $CdH_2$

*See other* PYROPHORIC METALS

Oxidants

*See* NITRYL FLUORIDE, $FNO_2$ : Metals
AMMONIUM NITRATE, $H_4N_2O_3$ : Metals

Selenium,
or Tellurium

Mellor, 1940, Vol. 4, 480

Reaction on warming powdered cadmium with selenium or tellurium is exothermic, but less vigorous than that of zinc.

*See other* METALS

## CADMIUM CHLORATE $Cd(ClO_3)_3$ $CdCl_2O_6$

Sulphides

Mellor, 1956, Vol. 2, Suppl. 1, 584

Interaction with the sulphide of copper(II) is explosive, and with those of antimony(III), arsenic(III), tin(II) and (IV), incandescent.

*See* METAL CHLORATES: Phosphorus, etc.
CHLORIC ACID, $ClHO_3$: Copper sulphide

*See other* METAL OXOHALOGENATES

## CADMIUM HYDRIDE $CdH_2$

Barbaras, G. D. *et al., J. Amer. Chem. Soc.*, 1951, **73**, 4585

The hydride, prepared at $-78.5°C$, suddenly decomposes during slow warming at $2°C$, leaving a residue of pyrophoric cadmium.

*See other* METAL HYDRIDES

## CADMIUM DIAMIDE $Cd(NH_2)_2$ $CdH_4N_2$

Alone,
or Water

Mellor, 1940, Vol. 8, 261

When heated rapidly, the amide may explode. Reaction with water is violent.

*See other* *N*-METAL DERIVATIVES

**TETRAAMMINECADMIUM(II) PERMANGANATE** **$CdH_{12}Mn_2N_4O_8$**

**$[Cd(NH_3)_4][MnO_4]_2$**

Mellor, 1942, Vol. 12, 335

Explodes on impact.

*See other* AMMINEMETAL OXOSALTS

**CADMIUM DIAZIDE** **$Cd(N_3)_2$** **$CdN_6$**

1. Turney, T. A., *Chem. & Ind.,* 1965, 1295
2. Mellor, 1967, Vol. 8, Suppl. 2.2, 25, 50

A solution, prepared by mixing saturated solutions of cadmium sulphate and sodium azide in a 10 ml glass tube, exploded violently several hours after preparation [1]. The dry solid is extremely hazardous, exploding on heating or light friction. A violent explosion occurred with cadmium rods in contact with aqueous hydrogen azide [2].

*See other* METAL AZIDES

**CADMIUM OXIDE** **CdO**

Magnesium

*See* MAGNESIUM, Mg: Metal oxides

**CADMIUM SELENIDE** **CdSe**

Preparative hazard

*See* SELENIUM, Se: Metals (reference 6)

**TRICADMIUM DINITRIDE** **$Cd_3N_2$**

Fischer, F. *et al., Ber.,* 1910, 43, 1469

The shock of the violent explosion caused by heating a sample of the nitride caused an unheated adjacent sample to explode.

Acids,
or Bases

Mellor, 1964, Vol. 8, Suppl. 2.1, 161

It reacts explosively with dilute acids or bases.

Water

Mellor, 1940, Vol. 8, 261

Explodes on contact.

*See other* *N*-METAL DERIVATIVES

## TRICADMIUM DIPHOSPHIDE $Cd_3P_2$

Nitric acid

*See* NITRIC ACID, $HNO_3$ : Tricadmium diphosphide

## CERIUM Ce

Alone,
or Metals

Mellor, 1945, Vol. 5, 602–603

Cerium, or its alloys, readily give incendive sparks (pyrophoric particles) on frictional contact, this effect of iron alloys being widely used in various forms of 'flint' lighters. The massive metal ignites and burns brightly at 160°C, and cerium wire burns in a Bunsen flame more brilliantly than magnesium.

Of its alloys with aluminium, antimony, arsenic, bismuth, cadmium, calcium, copper, magnesium, mercury, sodium and zinc, those containing major proportions of cerium are often extremely pyrophoric. The mercury amalgams ignite spontaneously in air without the necessity for frictional generation of small particles. The interaction of cerium with zinc is explosively violent, and with antimony or bismuth very exothermic.

Carbon dioxide,
Nitrogen

*See* CARBON DIOXIDE, $CO_2$ : Metals, Nitrogen

Halogens

Mellor, 1946, Vol. 5, 603

Cerium filings ignite in chlorine or bromine vapour at about 215°C.

Phosphorus

*See* PHOSPHORUS, P: Metals

Silicon

Mellor, 1946, Vol. 5, 605

Interaction at 1400°C to form cerium silicide is violently exothermic, often destroying the containing vessel.

*See other* METALS

## CERIUM DIHYDRIDE $CeH_2$

Libowitz, G. G. *et al., Inorg. Synth.*, 1973, **14**, 189–192

The polycrystalline hydride is frequently pyrophoric, particularly at higher hydrogen contents (up to $CeH_{2.85}$), while monocrystalline material appears to be less reactive to air.

*See other* METAL HYDRIDES

**CERIUM TRIHYDRIDE** $CeH_3$

Muthmann, W. *et al., Ann.*, 1902, **325**, 261

The hydride is stable in dry air, but may ignite in moist air.

*See other* METAL HYDRIDES

**CERIUM NITRIDE** CeN

Water

Mellor, 1940, Vol. 8, 121

Contact with water vapour slowly causes incandescence, while a limited amount of water or dilute acid causes rapid incandescence with ignition of evolved ammonia and hydrogen.

*See other* *N*-METAL DERIVATIVES

**DICERIUM TRISULPHIDE** $Ce_2S_3$

Mellor, 1946, Vol. 5, 649

Pyrophoric in ambient air when finely divided.

*See other* METAL SULPHIDES

**CHROMYL AZIDE CHLORIDE** $CrO_2(Cl)N_3$ $ClCrN_3O_2$

Explosive solid.

*See other* METAL AZIDE HALIDES

**CAESIUM TETRAFLUOROCHLORATE(1−)** $ClCsF_4$

$Cs[ClF_4]$

*See* METAL POLYHALOHALOGENATES

**CAESIUM CHLOROXENATE** $Cs[ClXeO_3]$ $ClCsO_3Xe$

Jaselskis, B. *et al., J. Amer. Chem. Soc.*, 1967, **89**, 2770

It explodes at 205°C *in vacuo*.

*See other* XENON COMPOUNDS

**CHLORINE FLUORIDE** ClF

Sidgwick, 1950, 1149

'Product Information Sheet ClF', Tulsa, Ozark-Mahoning Co., 1970

This powerful oxidant reacts with other materials similarly to chlorine trifluoride or fluorine, but more readily than the latter.

*See* CHLORINE TRIFLUORIDE, $ClF_3$
FLUORINE, $F_2$

Aluminium
Mellor, 1956, Vol. 2, Suppl. 1, 63
Aluminium burns more readily in chlorine fluoride than in fluorine.

*tert*-Butanol
Young, D. F. *et al., J. Amer. Chem. Soc.*, 1970, **92**, 2314
A mixture prepared at −196°C exploded at −100°C during slow warming.

*N*-Chlorosulphinylamine
Kuta, G. S. *et al., Int. J. Sulphur Chem.*, 1973, **8**, 335–340
Interaction of the reagents in equimolar proportions produced a highly explosive and strongly oxidising material (possibly an N–F compound).

Fluorocarbon polymers
'Product Information Sheet ClF', Tulsa, Ozark-Mahoning Co., 1970
Chlorine fluoride can probably ignite Teflon and Kel-F at high temperatures or under friction or flow conditions.

Phosphorus trifluoride
1. Fox, W. B. *et al., Inorg. Nucl. Chem. Lett.*, 1971, **7**, 861
2. Neilson, R. H. *et al., Inorg. Chem.*, 1975, **14**, 2019

Interaction of the fluorides to produce chlorotetrafluorophosphorane [1] is uncontrollably violent even at −196°C [2]. An improved method of making the phosphorane from phosphorus pentafluoride and boron trichloride is detailed [2].

Tellurium
Mellor, 1943, Vol. 11, 26
Interaction is incandescent.

Water
Pascal, 1960, Vol. 16.1, 189
Interaction with liquid water or that in hydrates is violent.
*See other* INTERHALOGENS

**FLUORONIUM PERCHLORATE** $F^+H_2\ ClO_4^-$ $ClFH_2O_4$

Water
Hantzsch, A., *Ber.*, 1930, **63**, 97

This hydrogen fluoride–perchloric acid complex reacts explosively with water.

**CHLORYL HYPOFLUORITE** **$ClO_2OF$** **$ClFO_3$**

Hoffman, C. J., *Chem., Rev.*, 1964, **64**, 97

Not then completely purified or characterised, its explosive nature was in contrast to the stability of the isomeric perchloryl fluoride.

*See other* HALOGEN OXIDES
HYPOHALITES

**PERCHLORYL FLUORIDE** **$ClO_3F$** **$ClFO_3$**

1. 'Booklet DC-1819', Philadelphia, Pennsalt Chem. Corp., 1957
2. Anon., *Chem. Eng. News*, 1960, **38**, 62
3. Barton, D. H. R., *Pure Appl. Chem.*, 1970, **21**, 285
4. Sharts, C. M. *et al., Org. React.*, 1974, **21**, 232–234

Procedures relevant to safe handling and use are discussed. Perchloryl fluoride is stable to heat, shock and moisture, but is a powerful oxidiser comparable with liquid oxygen. It forms flammable and/or explosive mixtures with combustible gases and vapours [1,2]. It only reacts with strongly nucleophilic centres, and the by-product, chloric acid, is dangerously explosive in admixture with organic compounds [3]. Safety aspects of practical use of perchloryl fluoride have been reviewed recently [4].

Benzocyclobutene,
Butyllithium,
Potassium *tert*-butoxide

Adcock, W. *et al., J. Organomet. Chem.*, 1975, **91**, C20

An attempt to convert 3-kaliobenzocyclobutene (prepared from the three reagents above) to the fluoro-derivative at −70°C with excess perchloryl fluoride led to a violent explosion when the reaction mixture was removed from the cooling bath after stirring for an hour.

Calcium acetylide,
or Potassium cyanide,
or Potassium thiocyanate,
or Sodium iodide

Kirk-Othmer, 1966, Vol. 9, 602

Unreactive at 25°C, these solids react explosively in the gas at 100–300°C.

Charcoal

Inman, C. E. *et al., Friedel–Craft and Related Reactions* (Olah, G. A., Ed.), Vol. 3, 1508, New York, Interscience, 1964

Adsorption of perchloryl fluoride on charcoal can, like liquid oxygen, produce a powerful 'Sprengel' explosive.

*See* OXYGEN (Liquid), $O_2$ : Carbon

Ethyl *p*-fluorobenzoylacetate

Fuqua, S. A. *et al., J. Org. Chem.*, 1964, **29**, 395, footnote 2

Fluorination of the ester with perchloryl fluoride by an established method led to a violent explosion.

Finely divided solids

McCoy, G., *Chem. Eng. News*, 1960, **38**(4), 62

Oxidisable organic materials of high surface to volume ratio (carbon powder, foamed elastomers, lampblack, sawdust) may react very violently, even at $-78^\circ C$, with perchloryl fluoride, which should be handled with the same precautions as liquid oxygen.

Hydrocarbons,
or Hydrogen sulphide,
or Nitrogen oxide,
or Sulphur dichloride
or Vinylidene chloride

Braker, 1971, 459

At ambient temperature, perchloryl fluoride is unreactive with the above compounds, but reaction is explosive at $100-300^\circ C$, or if the mixtures are ignited.

3α-Hydroxy-5β-androstane-11, 17-dione 17-hydrazone

·Nomine, G. *et al., J. Chem. Educ.*, 1969, **46**, 329

A reaction mixture in aqueous methanol exploded violently at $-65^\circ C$. (Hydrazine–perchloryl fluoride redox reaction?) A previous reaction at $20^\circ C$ had been uneventful, and the low-temperature explosion could not be reproduced.

Laboratory materials

Schlosser, M. *et al., Ber.*, 1969, **102**, 1944

Contact of perchloryl fluoride with laboratory greases or rubber tubing, etc., has led to several explosions.

Lithiated compounds

Houben-Weyl, 1970, Vol. 13.1, 223

There is a danger of explosion during replacement of a lithium substituent by fluorine using perchloryl fluoride.

*See* Benzocyclobutene, etc., above

2-Lithio(dimethylaminomethyl)ferrocene

Peet, J. H. J. *et al., J. Organomet. Chem.*, 1974, **82**, C57

Preparation of 2-(dimethylamino)fluoroferrocene by interaction of the lithio compound in tetrahydrofuran at −70°C with perchloryl fluoride diluted with helium unexpectedly exploded violently.

Methyl 2-bromo-5,5-(1,2-ethylidenedioxy)-[2.2.1] bicycloheptane-7-carboxylate

Wang, C.-L. J. *et al*., *J. Chem. Soc*., *Chem. Comm*., 1976, 468

In an attempt to improve isomer distribution from the fluorination reaction, one run was cooled to below −40°C but on quenching with water, a violent explosion occurred.

Nitrogenous bases

Scott, F. L. *et al., Chem. & Ind*., 1960, 528

Gardiner, D. M. *et al., J. Org. Chem*., 1967, **32**, 1115

Interaction, in presence of diluent below 0°C, with isopropylamine or isobutylamine caused separation of explosive liquids, and, with aniline, phenylhydrazine and 1,2-diphenylhydrazine, explosive solids.

In absence of diluents, contact with most aliphatic or non-aromatic heterocyclic amines often leads to violent uncontrolled oxidation and/or explosions.

Sodium methoxide,
Methanol

Papesch, V., *Chem. Eng. News,* 1959, **36**, 60

Addition of solid methoxide to a reaction vessel containing methanol vapour and gaseous perchloryl fluoride caused ignition and explosion. This could be avoided by adding all the methoxide first, or by nitrogen purging before addition of methoxide.

*See other* HALOGEN OXIDES
PERCHLORYL COMPOUNDS

**CHLORINE FLUOROSULPHATE** **$ClSO_3F$** **$ClFO_3S$**

Bailar, 1973, Vol. 2, 1469

Hydrolysis is violent, producing oxygen.

*See related* HYPOHALITES

**FLUORINE PERCHLORATE (PERCHLORYL HYPOFLUORITE)** **$ClFO_4$**

**$ClO_3OF$**

Alone,
or Hydrogen,
or Potassium iodide

1. Rohrback, G. H. *et al., J. Amer. Chem. Soc*., 1949, **69**, 677
2. Hoffman, C. J., *Chem. Rev*., 1964, **64**, 97

The pure liquid explodes on freezing at −167°C, and the gas is readily initiated by sparks, flame or contact with grease, dust or rubber tube. Contact of the gas with aqueous potassium iodide also caused an explosion [1] and ignition occurs in excess hydrogen gas [2].
*See* HYPOHALITES
*See other* HALOGEN OXIDES

**XENON(II) FLUORIDE PERCHLORATE** **$ClFO_4Xe$**

**$XeFClO_4$**

Bartlett, N. *et al.*, *Chem. Comm.*, 1969, 703
Thermodynamically unstable, it explodes readily and sometimes with violence.
*See other* XENON COMPOUNDS

**NITROGEN CHLORIDE DIFLUORIDE** **$NClF_2$** **$ClF_2N$**

Petry, R. C., *J. Amer. Chem. Soc.*, 1960, 82, 2401
Caution in handling is recommended for this *N*-halogen compound.
*See other* *N*-HALOGEN COMPOUNDS

**PHOSPHORUS CHLORIDE DIFLUORIDE** **$ClF_2P$**

**$PClF_2$**

Hexafluoroisopropylideneaminolithium
*See* HEXAFLUOROISOPROPYLIDENEAMINOLITHIUM, $C_3F_6LiN$: Non-metal halides

**THIOPHOSPHORYL CHLORIDE DIFLUORIDE** **$ClF_2PS$**

**$P(S)ClF_2$**

Mellor, 1971, Vol. 8, Suppl. 3, 536
Mixtures with air explode spontaneously at certain concentrations.
*See related* NON-METAL HALIDES

**CHLORINE TRIFLUORIDE** **$ClF_3$**

1. Anon., *J. Chem. Educ.*, 1967, **44**, A1057–1062
2. O'Connor, D. J. *et al.*, *Chem. & Ind.*, 1957, 1155

Handling procedures for this highly reactive oxidant gas have been detailed [1]. Surplus gas is best burnt with town or natural gas, followed by absorption in alkali [2].

Acids
Mellor, 1956, Vol. 2, Suppl. 1, 157
Strong nitric and sulphuric acids reacted violently.

Ammonium fluoride
Gardner, D. M. *et al., Inorg. Chem.*, 1963, **2**, 413
The reaction gases (containing chlorodifluoramine) must be handled below −5°C to avoid explosion. Ammonium hydrogenfluoride reacts similarly.

Boron-containing materials
Bryant, J. T. *et al., J. Spacecr. Rockets*, 1971, **8**, 192–193
Finely divided boron, tetraboron carbide and boron–aluminium mixtures will ignite on exposure to the gas.
*See* Metals, etc., below

Carbon tetrachloride
Mellor, 1956, Vol. 2, Suppl. 1, 156
Chlorine trifluoride will dissolve in carbon tetrachloride at low temperatures without interaction. Such solutions are dangerous, being capable of detonation. If it is used as a solvent for fluorination with the trifluoride, it is therefore important to prevent build-up of high concentrations.

Chromium trioxide
Mellor, 1943, Vol. 11, 181
Interaction of the two oxidants is incandescent.

Deuterium,
or Hydrogen
Haberland, H. *et al., Chem. Phys.*, 1975, **10**, 36
Studies of the interaction of chlorine trifluoride with deuterium or hydrogen atoms in a scattering chamber were accompanied by frequent flashes or explosions, within 3 h if the reactor had been vented, or after 8 h if it had not.

Fluorinated polymers
'Chlorine Trifluoride', Tech. Bull. TA 8522-3, Morristown, N.J., Baker & Adamson Div. of Allied Chemicals Corp., 1968
Bulk surfaces of polytetrafluoroethylene or polychlorotrifluoroethylene are resistant to the liquid or vapour under static conditions, but breakdown and ignition may occur under flow conditions.
*See* Polychlorotrifluoroethylene, below

Fuels
*See* ROCKET PROPELLANTS

Hydrogen-containing materials
Mellor, 1956, Vol. 2, Suppl. 1, 157
Explosive reactions occur with ammonia, coal-gas, hydrogen or hydrogen sulphide.

Iodine
Mellor, 1956, Vol. 2, Suppl.1, 157
Ignites on contact.

Metals,
or Metal oxides,
or Metal salts
or Non-metals,
or Non-metal oxides
Mellor, 1956, Vol. 2, Suppl.1, 155–157
Sidgwick, 1950, 1156
Rhein, R. A., *J. Spacecr. Rockets*, 1969, **6**, 1328–1329
'Chlorine Trifluoride', Tech. Bull. TA 8532-3, Morristown, N.J., Baker & Adamson Div. of Allied Chem Corp., 1968
Chlorine trifluoride is a hypergolic oxidiser with recognised fuels, and contact with the following at ambient or slightly elevated temperatures is violent, ignition often occurring. The state of subdivision may affect the results.

Antimony, arsenic, selenium, silicon, tellurium; iridium, iron, molybdenum, osmium, potassium, rhodium, tungsten; (and when primed with charcoal, aluminium, copper, lead, magnesium, silver, tin, zinc). Interaction of lithium or calcium with chlorine tri- or penta-fluorides is hypergolic and particularly energetic.

Aluminium oxide, arsenic trioxide, bismuth trioxide, calcium oxide, chromic oxide, lanthanum oxide, lead dioxide, magnesium oxide, manganese dioxide, molybdenum trioxide, phosphorus pentoxide, stannic oxide, sulphur dioxide (explodes), tantalum pentoxide, tungsten trioxide, vanadium pentoxide.

Red phosphorus, sulphur; but with carbon, the observed ignition has been attributed to presence of impurities.

Iodides of mercury, potassium; silver nitrate, potassium carbonate.

Methane
Baddiel, C. B. *et al., Proc. 8th Combust. Symp.*, 1960, 1089–1095
The explosive interaction of chlorine trifluoride with methane and its homologues has been studied in detail.

Nitro compounds
Mellor, 1956, Vol. 2, Suppl.1, 156
Several nitro compounds are soluble in chlorine trifluoride, but the solutions are extremely shock-sensitive. These include trinitrotoluene, hexanitrodiphenyl, hexanitrodiphenyl-amine, -sulphide or -ether. Highly chlorinated compounds behave similarly.

Organic materials
Mellor, 1956, Vol. 2, Suppl.1, 155

Violence of the reaction, sometimes explosive, with, e.g., acetic acid, benzene, ether, is associated with both their carbon and hydrogen contents. If nitrogen is also present, explosive fluoroamino compounds may also be involved. Fibrous materials–cotton, paper, wood – invariably ignite.

Polychlorotrifluoroethylene

Anon., *Chem. Eng. News*, 1965, **43**(20), 41

An explosion occurred while chlorine trifluoride was being bubbled through the fluorocarbon oil at –4°C. Moisture (snow) may have fallen into the mixture, reacted exothermically with the trifluoride and initiated the mixture.

Refractory materials

Cloyd 1965, 58

Mellor 1956, Vol. 2, Suppl.1, 157

Fibrous or finely divided refractory materials, asbestos, glass wool, sand or tungsten carbide, may ignite with the liquid and continue to burn in the gas. The presence of adsorbed or lattice-water seems necessary for attack on the siliceous materials to occur.

Selenium tetrafluoride

*See* SELENIUM TETRAFLUORIDE, $F_4Se$: Chlorine trifluoride

Water

1. Sidgwick, 1950, 1156
2. Mellor, 1956, Vol. 2, Suppl.1, 156, 158
3. Ruff, O. *et al., Z. Anorg. Chem.*, 1930, **190**, 270

Interaction is violent and may be explosive, even with ice, oxygen being evolved [1]. Part of the water dropped into a flask of the gas was expelled by the violent reaction ensuing [2]. An analytical procedure, involving absorption of chlorine trifluoride into 10% sodium hydroxide solution from the open capillary neck of a quartz ampoule to avoid explosion, was described [3].

*See other* INTERHALOGENS

## CHLORINE TRIFLUORIDE OXIDE $ClF_3O$

Bougon, R. *et al.*, Fr. Pat. 2 110 555, 1972

It is a powerful oxidant potentially useful in rocketry.

*See other* HALOGEN OXIDES

## CHLORINE DIOXYGEN TRIFLUORIDE $ClF_3O_2$

**$ClO_2F_3$**

Streng, A. G., *Chem. Rev.*, 1963, **63**, 607

Christe, K. O. *et al., Inorg. Chem.*, 1973, **12**, 1357
It is a very powerful oxidant, but of low stability, which reacts explosively with organic materials; such combinations should be avoided.
*See other* HALOGEN OXIDES

**POTASSIUM TETRAFLUOROCHLORATE(1–)** **$ClF_4K$**

**$K[ClF_4]$**

*See* METAL POLYHALOHALOGENATES

**NITROSYL TETRAFLUOROCHLORATE** **$ClF_4NO$**

**$NO[ClF_4]$**

Organic compounds
Sorbe, 1968, 133
The powerful oxidant reacts explosively with many organic compounds.
*See other* OXIDANTS

**CHLOROTETRAFLUOROPHOSPHORANE** **$ClF_4P$**

**$PClF_4$**

Preparative hazard
*See* CHLORINE FLUORIDE, ClF: Phosphorus trifluoride

**RUBIDIUM TETRAFLUOROCHLORATE(1–)** **$ClF_4Rb$**

**$Rb[ClF_4]$**

*See* METAL POLYHALOHALOGENATES

**CHLORINE PENTAFLUORIDE** **$ClF_5$**

Metals
*See* CHLORINE TRIFLUORIDE, $ClF_3$: Metals

Nitric acid
Christe, K. O., *Inorg. Chem.*, 1972, **11**, 1220
Interaction of anhydrous nitric acid with chlorine pentafluoride vapour at –40°C, or with the liquid at above –100°C, is very vigorous.

Water
1. Pilipovich, D. *et al., Inorg. Chem.*, 1967, **6**, 1918
2. Christe, K. O., *Inorg. Chem.*, 1972, **11**, 1220
Interaction of liquid chlorine pentafluoride with ice at –100°C [1], or

of the vapour with water vapour above 0°C [2], is extremely vigorous.
*See other* INTERHALOGENS

## PENTAFLUOROSULPHUR PEROXYHYPOCHLORITE $ClF_5O_2S$

$F_5SOOCl$

Haloalkenes
Hopkinson, M. J. *et al., J. Org. Chem.*, 1976, **41**, 1408
The peroxyhypochlorite is especially reactive, and the fluoroperoxy compounds produced by its interaction with haloalkenes can detonate when subjected to thermal or mechanical shock. However, no explosions were experienced during this work.
*See related* HYPOHALITES

## DIFLUOROPERCHLORYL HEXAFLUOROPLATINATE $ClF_8O_2Pt$

$F_2ClO_2[PtF_6]$

*See* DIFLUOROPERCHLORYL SALTS

## CHLOROGERMANE $ClGeH_3$

Ammonia
Johnson, O. H., *Chem. Rev.*, 1951, **48**, 274
Both mono- and di-chlorogermanes react with ammonia to give involatile products which explode on heating.
*See related N*-METAL DERIVATIVES

## HYDROGEN CHLORIDE ClH

Preparative hazard
*See* Sulphuric acid, below

Alcohols,
Hydrogen cyanide
1. Pinner, *Die Imidoäther*, Berlin, Openheim, 1892
2. Erickson, J. G., *J. Org. Chem.*, 1955, **20**, 1573

Preparation of alkyliminioformate chlorides by passing hydrogen chloride rapidly into alcoholic hydrogen cyanide proceeds explosively (probably owing to a sudden exotherm), even with strong cooling [1]. Alternative procedures involving very slow addition of hydrogen chloride into a well-stirred mixture kept cooled to ambient temperature, or rapid addition of cold alcoholic hydrogen cyanide to cold alcoholic hydrogen chloride, are free of this hazard [2].

Aluminium

*See* ALUMINIUM, Al: Hydrogen chloride
*See also* ALUMINIUM–TITANIUM ALLOYS, Al–Ti: Oxidants

Fluorine

*See* FLUORINE, $F_2$ : Hydrogen halides

Hexalithium disilicide

*See* HEXALITHIUM DISILICIDE, $Li_6Si_2$: Acids

Metal acetylides or carbides

*See* DICAESIUM ACETYLIDE, $C_2Cs_2$ : Acids
DIRUBIDIUM ACETYLIDE, $C_2Rb_2$ : Acidic materials
URANIUM DICARBIDE, $C_2U$: Hydrogen chloride

Potassium permanganate

*See* POTASSIUM PERMANGANATE, $KMnO_4$ : Hydrochloric acid

Silicon dioxide

*See* SILICON DIOXIDE, $O_2Si$: Hydrochloric acid

Sodium

*See* SODIUM, Na: Acids

Sulphuric acid

1. *MCA Case History No. 1785*
2. Libman, D. D., *Chem. & Ind.*, 1948, 728
3. Smith, G. B. L. *et al., Inorg. Synth.*, 1950, **3**, 132

Accidental addition of 6500 litres of concentrated hydrochloric acid to a bulk sulphuric acid storage tank released sufficient hydrogen chloride by dehydration to cause the tank to explode violently [1]. Complete dehydration of hydrochloric acid solution releases some 250 volumes of gas. A laboratory apparatus for effecting this safely has been described [2], which avoids the possibility of layer formation in unstirred flask generators [3].

Tetraselenium tetranitride

*See* TETRASELENIUM TETRANITRIDE, $N_4Se_4$

## MONOPOTASSIUM PERCHLORYLAMIDE $ClHKNO_3$

**$KNHClO_3$**

Sorbe, 1968, 67

Like the dipotassium salt, it will explode on impact or exposure to flame.

*See* PERCHLORYLAMIDE SALTS
*See other* *N*-METAL DERIVATIVES

Acetic anhydride

Rüst, 1948, 341

A mixture of the anhydride and hypochlorous acid exploded violently while being poured. Some acetyl hypochlorite may have been produced.

*See* HYPOHALITES

Alcohols

Mellor, 1956, Vol. 2, Suppl.1, 560

Contact of these, or of chlorine and alcohols, readily forms unstable alkyl hypochlorites.

*See* HYPOHALITES

Ammonia

Mellor, 1940, Vol. 8, 217

The violent explosion occurring on contact with ammonia gas is due to formation of nitrogen trichloride and its probable initiation by the heat of solution of ammonia.

Arsenic

Mellor, 1941, Vol. 2, 254

Ignition on contact.

*See other* HYPOHALITES
OXOHALOGEN ACIDS

## CHLORIC ACID $HClO_3$ $ClHO_3$

Muir, G. D., private comm., 1968

Aqueous chloric acid solutions decompose explosively if evaporative concentration is carried too far.

Cellulose

Mellor, 1946, Vol. 2, 310

Filter paper ignites after soaking in chloric acid.

*See* Oxidisable materials, below

Copper sulphide

Mellor, 1956, Vol. 2, Suppl. 1, 584

Copper sulphide explodes with concentrated chloric acid solution, or cadmium, magnesium or zinc chlorates.

*See also* METAL HALOGENATES

Metals,
or Organic substances

Majer, V., *Chemie (Prague)*, 1948, **3**, 90–91

The recorded explosions of chloric acid have been attributed to the formation of explosive compounds with antimony, bismuth and iron (including hydrogen in the latter case). Organic substances (and ammonia) are violently oxidised.

Oxidisable materials

In contact with oxidisable substances, reactions are similar to those of the metal chlorates.

*See* Cellulose, above
METAL HALOGENATES

*See other* OXOHALOGEN ACIDS

**CHLOROSULPHURIC ACID** **$ClSO_2OH$** **$ClHO_3S$**

Diphenyl ether

Ehama, T. *et al.*, Jap. Pats. 74 45 034; 74 45 035, 1974 (*Chem. Abs.*, 1974, **81**, 100509k, 100510d)

Presence of various nitrogen-containing compounds, or of fatty acids or their derivatives, controls the vigorous interaction of the ether and chlorosulphuric acid at above 40°C, producing higher yields of *p,p'*-oxybisbenzenesulphonyl chloride.

Phosphorus,
or Water

1. Heumann, K. *et al.*, *Ber.*, 1882, **15**, 417
2. *MCA SD-33*, 1968

The acid is a strong oxidising agent, and above 25–30°C, interaction with yellow phosphorus is vigorous and accelerates to explosion. A higher temperature is needed to start reaction with red phosphorus [1]. Reaction with water is highly exothermic and violent, owing to combined heat of hydrolysis to sulphuric and hydrochloric acids and their heats of dilution. Handling precautions are detailed [2].

Silver nitrate

*See* SILVER NITRATE, $AgNO_3$ : Chlorosulphuric acid

**PERCHLORIC ACID** **$HClO_4$** **$ClHO_4$**

1. *MCA SD-11*, 1965
2. Lazerte, D., *Chem. Eng. News*, 1971, **49**(3), 33
3. Muse, L. A., *J. Chem. Educ.*, 1972, **49**, A463
4. Graf, F. A., *Chem. Eng. Progr.*, 1966, **62**(10), 109
5. Friedmann, W., *Chem. Eng. News*, 1947, **25**, 3458
6. *ABCM Quart. Safety Summ.*, 1936, **7**, 51

Most of the numerous and frequent hazards experienced with perchloric acid have been associated with either its exceptional oxidising power or the inherent instability of its covalent compounds, some of which form readily. Although the 70–72% acid of commerce behaves as a very strong but non-oxidising acid, it becomes an extreme oxidant and powerful dehydrator at elevated temperatures (160°C) or when anhydrous [1].

*See* Dehydrating agents, below

Where an equally strong but non-oxidising acid can be used, trifluoromethanesulphonic acid is recommended [2]. Safe laboratory handling procedures have been detailed [3] and an account of safe handling of perchloric acid in the large-scale preparation of hydrazinium diperchlorate, with recommendations for materials of construction, has been published [4].

The lack of oxidising power of the unheated 70–72% aqueous acid was confirmed during analytical treatment of oil sludges on over 100 occasions [5]. A report of two explosions in a method for estimating potassium during evaporation of aqueous perchloric acid infers that the latter may be unstable. It is more likely that oxidisable material (probably ethanol) was present during the evaporation [6].

*See* ALKYL PERCHLORATES

Acetic anhydride,
Acetic acid,
Organic materials


1. Burton, H. *et al., Analyst*, 1955, **80**, 4
2. Schumacher, 1960, 187, 193
3. Kuney, J. H., *Chem. Eng. News,* 1947, **25**, 1659
4. Tech. Survey No. 2, *Fire Hazards and Safeguards for Metalworking Industries,* US Board of Fire Underwriters, 1954


Mixtures of hydrated perchloric acid with enough acetic anhydride produce a solution of anhydrous perchloric acid in acetic acid/anhydride, which is of high catastrophic potential [1]. Sensitivity to shock and heat depends on composition of the mixture, and vapour evolved on heating is flammable [2]. Such solutions have been used for electropolishing operations, and during modifications to an electropolishing process, a cellulose acetate rack was introduced into a large volume of an uncooled mixture of perchloric acid and acetic anhydride. Dissolution of the rack introduced organic material into the virtually anhydrous acid and caused it to explode disastrously. This cause was confirmed experimentally [3,4].

*See* Alcohols, below

*See also* ACETIC ANHYDRIDE, $C_4H_6O_3$: Perchloric acid

Acetic anhydride,
Carbon tetrachloride,
2-Methylcyclohexanone


Gall, M. *et al., Org. Synth.*, 1972, **52**, 40

During acetylation of the enolised ketone, the 70% perchloric acid must be added last to the reaction mixture to provide maximum dilution and cooling effect.

Acetic anhydride,
Organic materials,
Transition metals

Hikita, T. *et al., J. Chem. Soc. Japan, Ind. Chem. Sect.*, 1951, **54**, 253–255

The stability ranges of mixtures of the acid, anhydride and organic materials (ethanol, gelatine) used in electropolishing were studied. Presence of transition metals (chromium, iron, nickel) increases the possibilities of explosion.

Acetonitrile

Andrussow, L., *Chim. Ind.*, 1961, **86**, 542–545

The latent hazards in storing and handling the explosive mixtures of perchloric acid with acetonitrile or dimethyl ether are discussed.

Alcohols

1. Sweasey, D., *Lab. Practice*, 1968, **17**, 915
2. Moureu, H. *et al., Arch. Mal. Prof. Med.*, 1951, **12**, 157–159
3. Michael, A. *et al., Amer. Chem. J.*, 1900, **23**, 444
4. Bailar, 1973, Vol. 2, 1451

During digestion of a lipid extract (1 mg) with 72% acid (1.5 ml), a violent explosion occurred. This was attributed to residual traces of the extraction solvent (methanol–chloroform) having formed explosive methyl perchlorate [1]. In the analytical determination of potassium as perchlorate, heating the solid containing traces of ethanol and perchloric acid caused a violent explosion [2]. Contact of drops of anhydrous perchloric acid and ethanol caused immediate violent explosion [3]. Partial esterification of polyfunctional alcohols (ethylene glycol, glycerol, pentaerythritol) with the anhydrous acid gives extremely explosive liquids which may explode on pouring from one vessel to another [4].

*See* Glycols and their ethers, below
Methanol, Triglycerides, below
ALKYL PERCHLORATES

Aniline,
Formaldehyde

*Aniline*, 86, New York, Allied Chem. Corp., 1964

Aniline reacts with perchloric acid and then formaldehyde to give an explosively combustible condensed resin.

Antimony(III) compounds

Burton, H. *et al., Analyst*, 1955, **80**, 4

Treatment of tervalent compounds of antimony with perchloric acid can be very hazardous.
*See also* Bismuth, below

Azo-pigment,
Orthoperiodic acid
1. *ABCM Quart. Safety Summ.*, 1961, **32**, 125
2. Smith, F. G. *et al., Talanta,* 1960, **4**, 185

During the later stages of the wet oxidation of an azo-pigment with mixed perchloric and orthoperiodic acids, a violent reaction, accompanied by flashes of light, set in, and terminated in an explosion [1]. The general method upon which this oxidation was based is described as hazard-free [2].

Bis-1,2-diaminopropane-*cis*-dichlorochromium(III) perchlorate
*See* BIS-1,2-DIAMINOPROPANE-*cis*-DICHLOROCHROMIUM(III) PERCHLORATE, $C_6H_{20}Cl_3CrN_4O_4$ : Perchloric acid

1,3-Bis(di-$\eta$-cyclopentadienyliron)-2-propen-1-one
*See* 1,3-BIS(DI-$\eta$-CYCLOPENTADIENYLIRON)-2-PROPEN-1-ONE, $C_{23}H_{20}Fe_2O$: Perchloric acid

Bismuth
Nicholson, D. G. *et al., J. Amer. Chem. Soc.*, 1935, **57**, 817
Attempts to dissolve bismuth and its alloys in hot perchloric acid carry a very high risk of explosion. At 110°C a dark brown coating is formed, and if left in contact with the acid (hot or cold), explosion occurs sooner or later. The same is true of antimony and its tervalent compounds.

Carbon
Mellor, 1946, Vol. 2, 380
Pascal, 1960, Vol. 16, 300–301
Contact of a drop of the anhydrous acid with wood charcoal causes a very violent explosion, and carbon black also reacts violently.

Charcoal,
Chromium trioxide
Randall, W. R., private comm., 1977
A wet-ashing technique used for dissolution of graphite in perchloric acid involved boiling a mixture of 70% perchloric acid and 1% of chromium trioxide as an aqueous solution. This was later applied to 6–14 mesh charcoal, and after boiling for 30 min the reaction rate increased (foaming increased) and accelerated to explosion. The charcoal contained traces of extractable tar.

Cellulose and derivatives

1. Schumacher, 1960, 187, 195
2. Harris, E. M., *Chem. Eng.*, 1949, **56**, 116
3. Sutcliffe, G. R., *J. Textile Ind.*, 1950, **41**, 196T

Contact of the hot concentrated acid or of the cold anhydrous acid with cellulose (as paper, wood fibre or sawdust, etc.) is very dangerous and may cause a violent explosion. Many fires have been caused by long-term contact of diluted acid with wood with subsequent evaporation and ignition [1,2]. Contact of cellulose acetate with 1200 litres of uncooled anhydrous perchloric acid in acetic anhydride caused an extremely violent explosion [1] and interaction of benzyl cellulose with boiling 72% acid was also explosive [3]. Perchlorate esters of cellulose may have been involved in all these incidents.

Combustible materials

Elliot, M. A. *et al., Rep. Invest. No. 4169*, Washington, US Bur. Mines, 1948

Tests of sensitivity to initiation by heat, impact, shock or ignition sources were made on mixtures of a variety of absorbent materials containing a stoicheiometric amount of 40–70% perchloric acid. Wood meal with 70% acid ignited at 155°C and a mixture of coal and 60% acid which did not ignite below 200°C ignited at 90°C when metallic iron was added. Many of the mixtures were more sensitive and dangerous than common explosives.

*See* Acetic anhydride, Organic materials, etc., above

Copper dichromium tetraoxide

Solymosi, F. *et al., Proc. 14th Combust. Symp.*, 1309–1316, 1973

The oxide was the most effective of several catalysts for the vapour phase decomposition of perchloric acid vapour, decomposition occurring above 120°C.

Dehydrating agents

1. Mellor, 1941, Vol. 2, 373, 380; 1956, Vol. 2, Suppl. 2.1, 598, 603
2. Kuney, J. H., *Chem. Eng. News,* 1947, **25**, 1659
3. Burton, H. *et al., Analyst,* 1955, **80**, 4
4. Schumacher, 1960, 71, 187, 193
5. Wirth, C. M. P., *Lab. Practice*, 1966, **15**, 675
6. Plesch, P. H. *et al., Chem.& Ind.,* 1971, 1043
7. Musso, H. *et al., Angew Chem.*, 1970, **82**, 46

Although commercial 70–72% perchloric acid (approximating to the dihydrate) itself is stable, incapable of detonation and readily stored, it may be fairly readily dehydrated by contact with dehydrating agents to anhydrous perchloric acid. This is not safe when stored at room temperature, since it slowly decomposes, even in the dark, with accumulation of chlorine dioxide in the solution, which darkens and finally explodes after about 30 days [1,2,5]. The 72% acid (or perchlorate salts) may be converted to the anhydrous acid by heating with sulphuric acid, phosphorus pentoxide or phosphoric acid, or by distillation under reduced pressure [1,4,5]. In contact with cold acetic anhydride, mixtures of the anhydrous acid with excess anhydride and acetic acid are formed, which are particularly dangerous, being sensitive to mechanical shock, heating or the introduction of organic contaminants [2–4].

*See* Acetic anhydride, etc., above

A solution of the monohydrate in chloroform exploded in contact with phosphorus pentoxide [1]. A safer method of preparing anhydrous solutions of perchloric acid in methylene chloride, which largely avoids the risk of explosion, has been described [6]. Further precautions are detailed in an account of an explosion during a similar preparation [7]. Solutions of the anhydrous acid of less than 55% concentration in acetic acid or anhydride are relatively stable [8].

Deoxyribonucleic acid

Cochrane, A. R. G. *et al., School Sci. Rev.*, 1977, **58**, 706–708

Hazards of using perchloric acid to hydrolyse DNA are stressed. Perchloric acid can cause ignition of any organic material, even a considerable time after contact. Other acids to effect hydrolysis are suggested.

*See* Nitric acid, etc., below

Diethyl ether

Michael, A. T. *et al., Amer. Chem. J.*, 1900, **23**, 444

The explosions sometimes observed on contact of the anhydrous acid with ether are probably due to formation of ethyl perchlorate by scission of the ether, or possibly to formation of diethyloxonium perchlorate.

Dimethyl ether

*See* Acetonitrile, above

Ethylbenzene,
Thallium triacetate

Uemura, S. *et al., Bull. Chem. Soc. Japan,* 1971, **44**, 2571

Application of a published method of thallation to ethylbenzene caused a violent explosion. A reaction mixture of thallium triacetate, acetic acid, perchloric acid and ethylbenzene was stirred at 65°C for 5 h, then filtered from precipitated thallous salts. Vacuum evaporation of the filtrate at 60°C gave a pasty residue, which exploded. This preparation of ethylphenylthallic acetate perchlorate monohydrate had been done twice previously uneventfully, as had been analogous preparations involving thallation of benzene, toluene, three isomeric xylenes and anisole in a total of 150 runs, where excessive evaporation had been avoided.

Fluorine

Rohrbock, G. H. *et al., J. Amer. Chem. Soc.*, 1947, **69**, 677–678

Contact of fluorine and 72% perchloric acid at ambient temperature produces a high yield of the explosive gas, fluorine perchlorate.

Glycerol,

Lead oxide

*MCA Case History No. 799*

During maintenance work on casings of fans used to extract perchloric acid fumes, seven violent explosions occurred when flanges sealed with lead oxide–glycerol cement were disturbed. Explosions, attributed to formation of explosive compounds by interaction of the cement with perchloric acid, may have involved perchlorate esters and/or lead salts. Use of an alternative silicate–hexafluorosilicate cement is recommended.

Glycols and their ethers

Schumacher, 1960, 195, 214

Comas, S. M. *et al., Metallography*, 1974, **7**, 45–47

Glycols and their ethers undergo violent decomposition in contact with ~70% perchloric acid. This seems likely to involve formation of the glycol perchlorate esters (after scission of ethers) which are explosive, those of ethylene glycol and 1-chloro-2,3-propanediol being more powerful than glyceryl nitrate, and the former so sensitive that is explodes on addition of water.

Investigation of the hazards associated with use of 2-butoxyethanol mixtures for alloy electropolishing showed that mixtures with 50–95% of acid at 20°C, or 40–90% at 75°C, were explosive and initiable by sparks. Sparking caused mixtures with 40–50% of acid to become explosive, but 30% solutions appeared safe under static conditions of temperature and concentration.

*See* Alcohols, above

Hydrogen

Schumacher, 1960, 189

Dietz, W., *Angew Chem.*, 1939, **52**, 616

Occasional explosions experienced during use of hot perchloric acid to dissolve steel samples for analysis is attributed to formation of hydrogen –perchloric acid mixtures and their ignition by steel particles at temperatures as low as 215°C.

Hydrogen halides

*See* FLUORONIUM PERCHLORATE, $ClFH_2O_4$
CHLORONIUM PERCHLORATE, $Cl_2H_2O_4$

Iodides

Michael, A. *et al.*, *Amer.Chem.J.*, 1900, **23**, 444

The anhydrous acid ignites in contact with sodium iodide or hydriodic acid.

Iron(II) sulphate

Tod, H., private comm., 1968

During preparation of iron diperchlorate, a mixture of iron sulphate and perchloric acid was being strongly heated when a most violent explosion occurred. Heating should be gentle to avoid initiating this redox system.

Ketones

Schumacher, 1960, 195

Ketones may undergo violent decomposition in contact with ~70% acid.

Methanol,
Triglycerides

1. Mavrikos, P. J. *et al.*, *J. Amer. Oil Chem. Soc.*, 1973, **50**(5), 174
2. Wharton, H. W., ibid., 1974, **51**(2), 35–36

Attention is drawn to the hazards involved in the use of perchloric acid in a published method [1] for transesterification of triglycerides with methanol. Alternative acidic catalysts and safety precautions are suggested [2].

2-Methylpropene,
Metal oxides

Lesnikovich, L. I. *et al.*, *Chem. Abs.*, 1975, **83**, 149760v; 1976, **85**, 7929y

Mixtures of the alkene and perchloric acid vapour (5:1 molar) in nitrogen ignite spontaneously at 250°C. Some metal oxides of high specific surface reduced the ignition temperature to below 178°C.

Nitric acid,
Organic matter

1. Anon., *Ind. Eng. Chem. (News Ed.)*, 1937, **15**, 214
2. Lambie, D. A., *Chem. & Ind.*, 1962, 1421
3. Mercer, E. R., private comm., 1967
4. Muse, L. A., *Chem. Eng. News*, 1973, **51**(6), 29–30
5. Cooke, G. W., private comm., 1967
6. Kuney, J. H., *Chem. Eng. News*, 1947, **25**, 1659
7. Balks, R. *et al.*, *Chem. Abs.*, 1939, **33**, $2438_9$

8. Rooney, R. C., *Analyst*, 1975, **100**, 471–475
9. Martinie, G. D. *et al.*, *Anal. Chem.*, 1976, **48**, 70–74
10. *MCA Case History No. 2145*

The mixed acids have been used to digest organic material prior to analysis, but several explosions have been reported, including those with vegetable oil [1], milk [2] and calcium oxalate precipitates from plants [3]. To avoid trace metal contamination by homogenising rat carcases in a blender, the carcases were dissolved in nitric acid. After separation of fat and addition of perchloric acid (125 ml), evaporation of samples to near-dryness caused a violent explosion [4].

Finely ground plant material in contact with perchloric/nitric acid mixture on a heated sand bath became hot before all the plant material was saturated with the acid and exploded. Subsequent digests left overnight in contact with cold acids proceeded smoothly [5]. Cellulose nitrate and/or perchlorate may have been involved.

Treatment with nitric acid before adding perchloric acid was a previously used and well-tried safe procedure [6]. Application to animal tissues of a degradation procedure previously found satisfactory for plant material caused a violent explosion [7]. A method which involves gradually increasing the liquid temperature by controlled distillation during the digestion of organic matter with the hot mixed acids had been described as safe provided that strict control of the distillation process were observed [4]. During oxidation of large, high-fat samples of animal matter by the usual technique, the existence of a layer of separated fat on the perchloric acid mixture created a highly hazardous situation. A modified procedure is free of explosion risk [8]. The effectiveness of mixed perchloric and nitric acids in wet oxidation of a wide range of organic materials has been studied. Violent oxidation occasionally occurred, but addition of sulphuric acid prevented any explosions or ignition during digestions [9]. The case history describes an explosion during wet-ashing of a phosphorus-containing polymer. One of the flasks may not have been topped up with nitric acid during the digestion phase [10].

Nitric acid,
Pyridine,
Sulphuric acid

*Safety in Handling Hazardous Chemicals*, UCID-16610, Randall, W. L., 8–10, Lawrence Livermore Laboratory, Univ. Calif., 1974

Some rare earth fluoride samples had been wet-ashed incompletely with the three mixed acids and some gave low results. These samples, now containing pyridine, were reprocessed by addition of more acids and slow evaporation on a hot-plate. One of the samples frothed up and then exploded violently. Pyridinium perchlorate seems likely to have been involved.

Nitrogenous epoxides

Harrison, G. E., private comm., 1966

Traces of perchloric acid used as hydration catalyst for ring opening of nitrogenous epoxides caused precipitation of organic perchlorate which was highly explosive. Concentration of acid was less than 1% by volume.

*See other* OXOSALTS OF NITROGENOUS BASES

Oleic acid

1. Swern, D. *et al.*, US Pat. 3 054 804; *Chem. Eng. News,* 1963, **41**(12), 39
2. Anon., *Chem. Eng. News,* 1963, **41**(27), 47

The improved preparation of 1,4-octadecanolactone [1] involves heating oleic acid (or other $C_{18}$ acids) with *ca.* 70% perchloric acid to 115°C. This is considered to be a potentially dangerous method [2].

Phosphine

*See* PHOSPHONIUM PERCHLORATE, $ClH_4O_4P$

Pyridine

*See* PYRIDINIUM PERCHLORATE, $C_5H_6ClNO_4$

Sodium phosphinate ('hypophosphite')

Smith, F. G., *Analyst,* 1955, **80**, 16

Though no interaction occurs in the cold, these powerful reducing and oxidising agents violently explode on heating.

Sulphinyl chloride

Bailar, 1973, Vol. 2, 1442

The anhydrous acid ignites the chloride.

Sulphoxides

1. Therésa, J. de B., *Anales Soc. Españ. Fis. y Quim.*, 1949, **45B**, 235
2. Graf. F. A., *Chem. Eng. Prog.*, 1966, **62**(10), 109
3. Uemura, S. *et al.*, *Bull. Chem. Soc. Japan*, 1971, **44**, 2571
4. Eigenmann, K. *et al.*, *Angew. Chem. (Intern. Ed.)*, 1975, **14**, 647–648

Lower members of the series of salts formed between organic sulphoxides and perchloric acid are unstable and explosive when dry. The salt formed from dibenzyl sulphoxide explodes at 125°C [1]. Dimethyl sulphoxide explodes on contact with 70% perchloric acid solution [2]; one drop of acid added to 10 ml of sulphoxide at 20°C caused a violent explosion [3].

Aryl sulphoxides condense uneventfully with phenols in 70% perchloric acid, but application of these conditions to the alkyl sulphoxide (without the addition of essential phosphoryl chloride) led to a violent explosion. Subsequent investigation showed that mixtures of phenol and perchloric acid (ester formation?) are thermally unstable, and may decompose violently, the temperature range depending upon composition. DSC measurements showed that sulphoxides alone may decompose violently at elevated temperatures; e.g. dimethyl sulphoxide, 270–355°, cyclohexyl methyl sulphoxide, 181–255°, or methyl phenyl sulphoxide, 233–286°C, respectively [4].

*See also* DIMETHYL SULPHOXIDE, $C_2H_6OS$: Metal oxosalts

Sulphuric acid,
Organic materials

Young, E. G. *et al., Science, N.Y.*, 1946, **104**, 353

Precautions are necessary to prevent explosions when using the mixed acids to oxidise organic materials for subsequent analysis. The sulphuric acid probably tends to dehydrate the 70% perchloric acid to produce the hazardous anhydrous acid,

*See* Nitric acid, etc., above

Sulphur trioxide

Pascal, 1960, Vol. 16, 301

Interaction of the anhydrous acid and sulphur trioxide is violent and highly exothermic, even in presence of chloroform as diluent, and explosions are frequent.

*See* Dehydrating agents, above

Trichloroethylene

Prieto, M. A. *et al., Research on the Stabilization and Characterization of Highly Concentrated Perchloric Acid*, 36, Whittier, Calif., American Potash and Chem. Corp., 1962

The solvent reacts violently with the anhydrous acid.

Trizinc diphosphide

Muir, G. D., private comm., 1968

Use of perchloric acid to assist solution of a sample for analysis caused a violent reaction.

*See other* OXIDANTS
OXOHALOGEN ACIDS

**MERCURY(II) AMIDE CHLORIDE** **$HgNH_2Cl$** **$ClH_2HgN$**

Schwarzenbach, V., *Ber.*, 1875, **8**, 1231–1234

Several minutes after addition of ethanol to a mixture of the amide chloride ('fusible white precipitate') and iodine, an explosion occurs. Addition of the compound to chlorine gas or bromine vapour leads to a delayed violent or explosive reaction. Amminemetal salts behave similarly. Formation of *N*-halogen compounds is involved in all cases.

*See other* *N*-METAL DERIVATIVES

**CHLOROAMINE** $ClNH_2$ $ClH_2N$

1. Marckwald, W. *et al., Ber.*, 1923, **56**, 1323
2. Coleman, G. H., *et al., Inorg. Synth.*, 1939, **1**, 59
3. Walek, W. *et al., Tetrahedron*, 1976, **32**, 627

The solvent-free material, isolated at −70°C, violently disproportionates (sometimes explosively) at −50°C to ammonium chloride and nitrogen trichloride [1]. Ethereal solutions of chloroamine are readily handled [2]. In the preparation of chloroamine by reaction of sodium hypochlorite with ammonia, care is necessary to avoid excess chlorine in the preparation of the hypochlorite from sodium hydroxide, because nitrogen trichloride may be formed in the subsequent reaction with ammonia [3].

*See other N*-HALOGEN COMPOUNDS

**AMMONIUM CHLORIDE** $NH_4Cl$ $ClH_4N$

Interhalogens

*See* BROMINE TRIFLUORIDE, $BrF_3$: Ammonium halides
BROMINE PENTAFLUORIDE, $BrF_5$: Acids, etc.

Potassium chlorate

*See* POTASSIUM CHLORATE, $ClKO_3$: Ammonium chloride

**HYDROXYLAMINIUM CHLORIDE** $NH_3OHCl$ $ClH_4NO$

Manganese dioxide

Chatterjee, B. P. *et al., Talanta*, 1977, **24**, 180–181

In a method for titrimetric determination of manganese in pyrolusite ore, addition of the powdered ore to a cold 20% solution of the salt causes vigorous decomposition to occur.

*See* HYDROXYLAMINIUM SALTS

**AMMONIUM CHLORATE** $NH_4ClO_3$ $ClH_4NO_3$

1. Brauer, 1963, Vol. 1, 314
2. Urbanski, 1965, Vol. 2, 476

It occasionally explodes spontaneously, and invariably above 100°C [1]. It will explode after 11 h at 40°C, and after 45 min at 70°C. Ammonium and chlorate salts should not be mixed together [2].

Water

Mellor, 1941, Vol. 2, 339; 1956, Vol. 2, Suppl.1, 591

A cold saturated solution may decompose explosively after a few days

if much excess salt is present. Hot aqueous solutions have exploded during evaporation in steam-heated vessels.
*See other* OXOSALTS OF NITROGENOUS BASES

**AMMONIUM PERCHLORATE** $NH_4ClO_4$ $ClH_4NO_4$

*MCA Case History No. 1002*
Barret, P., *Cah. Therm.*, 1974, **4**, 13–22 (*Chem. Abs.*, 1975, 82, 158330k)

Materials for a batch of ammonium perchlorate castable propellant were charged into a mechanical mixer. A metal spatula was left in accidentally, and the contents ignited when the mixer was started, owing to local friction caused by the spatula. A tool-listing safety procedure has been instituted.

The literature on the kinetics of thermal decomposition of the salt and other solid propellant constituents has been critically reviewed.

*See* Impurities, below

Aluminium
Loftus, H. J. *et al., Rep. AD-769283/3GA*, 1–98, Springfield (Va.), USNTIS, 1973
The powdered solid materials have been evaluated as a practical propellant pair.

Carbon
Galwey, A. K. *et al., Trans. Faraday. Soc.*, 1960, **56**, 581
Below 240°C intimate mixtures with sugar charcoal undergo exothermic decomposition, while mild explosions occur above 240°C.

Catalysts
1. Solymosi, F., *Acta Phys. Chem.*, 1974, **20**, 83–103
2. Shadman-Yazdi, F., *Proc. 1973 Iran Congr. Chem. Eng.*, 1974, **1**, 353–356
3. Glazkova, A. P., *Chem. Abs.*, 1976, **85**, 162846n

In a review of the course and mechanism of the catalytic decomposition of ammonium perchlorate, the considerable effects of metal oxides in reducing the explosion temperature of the salt are described [1]. Solymosi's previous work had shown reductions from 440°C to about 270° by dichromium trioxide, to 260° by 10 mol% of cadmium oxide and to 200° by 0.2% of zinc oxide. The effect of various concentrations of 'copper chromite', copper oxide, iron oxide and potassium permanganate on the catalysed combustion of the propellant salt was studied [2]. Similar studies on the effects of compounds of 11 metals, and potassium dichromate in particular, have been reported [3].

Copper

Anon., *Chem. Eng.*, 1955, **62**(12), 335

Crystalline ammonium perchlorate ignited in contact with hot copper pipes.

Ethylene dinitrate

*MCA Case History No. 1768*

Samples of mixtures of ammonium perchlorate and the highly explosive liquid nitrate kept at 60°C ignited after 7 days. Many adverse criticisms of the general planning and execution of the experiments were made.

Impurities

Jacobs, P. W. M. *et al., Chem. Rev.*, 1969, **69**, 590

The medium impact-sensitivity of this solid propellant component is greatly increased by co-crystallisation of certain impurities, notably nitryl perchlorate, potassium periodate and potassium permanganate.

Metal perchlorates

Solymosi, F., *Combust. Flame*, 1966, **10**, 398–399; *Z. Phys. Chem.*, 1969, **67**, 76–85

Admixture of lithium or zinc perchlorates leads to decomposition with explosion at 290°C or ignition at 240°C, respectively.

Metals,
or Organic materials,
or Sulphur

*Haz. Chem. Data*, 1969, 41

This powerful oxidant functions as an explosive when mixed with finely divided metals, organic materials or sulphur, which increase the shock-sensitivity up to that of picric acid.

*See other* OXOSALTS OF NITROGENOUS BASES

**HYDROXYLAMINIUM PERCHLORATE** **$ClH_4NO_5$**

$N^+H_3OH\ ClO_4^-$

Triaminotrinitrobenzene

Quong, R., *Chem. Abs.*, 1975, **82**, 113753j

A saturated aqueous solution of the perchlorate, and the solid fuel, are individually non-explosive, but form a viable explosive composition on admixture.

*See other* OXOSALTS OF NITROGENOUS BASES
REDOX COMPOUNDS

**PHOSPHONIUM PERCHLORATE** $PH_4ClO_4$ $ClH_4O_4P$

Fichter, F. *et al.*, *Helv. Chim. Acta*, 1934, **17**, 222

The crystalline salt obtained by action of phosphine on 68% perchloric acid at −20°C is dangerously explosive, and sensitive to contact with moist air, increase in temperature, or friction.

*See related* OXOSALTS OF NITROGENOUS BASES

**HYDRAZINIUM CHLORITE** $H_2NNH_3ClO_2$ $ClH_5N_2O_2$

Levi, G. R., *Gazz. Chim. Ital.*, 1923, [2], **53**, 105–108

It is spontaneously flammable when dry.

*See other* CHLORITE SALTS

**AMMONIUM PERCHLORYLAMIDE** $NH_4NHClO_3$ $ClH_5N_2O_3$

*See* PERCHLORYLAMIDE SALTS

**HYDRAZINIUM CHLORATE** $H_2NNH_3ClO_3$ $ClH_5N_2O_3$

Salvadori, J., *Gazz. Chim. Ital.*, 1907, [2], **37**, 32–40

It explodes violently at its m.p., 80°C.

*See other* OXOSALTS OF NITROGENOUS BASES

**HYDRAZINIUM PERCHLORATE** $H_2NNH_3ClO_4$ $ClH_5N_2O_4$

Alone,
or Copper dichloride

1. Levy, J. B. *et al.*, *ACS 54*, 1966, 55; Grelecki, C. J. *et al.*, *ibid.*, 73
2. Shidlovskii, A. A. *et al.*, *Zh. Prikkad. Chim.*, 1962, **35**, 756
3. Salvadori, R., *Gazz. Chim. Ital.*, 1907, **37**(2), 32–40

The deflagration and thermal decomposition of the salt, a component of solid rocket propellants, have been studied [1]. Presence of 5% of copper dichloride caused explosion to occur at 170°C [2].
It also explodes on impact [3].

*See other* OXOSALTS OF NITROGENOUS BASES

**POLY(DIMERCURYIMMONIUM PERCHLORATE)** $(ClHg_2NO_4)_n$
$(Hg{=}N^+{=}Hg\ ClO_4^-)_n$

Sorbe, 1968, 97

Highly explosive.

*See other* POLY(DIMERCURYIMMONIUM) COMPOUNDS

### IODINE CHLORIDE $ICl$ $ClI$

Metals

Mellor, 1940, Vol. 2, 119; 1956, Vol. 2, Suppl. 1, 452; 1963, Vol. 2, Suppl. 2.2, 1563

Mixtures containing sodium explode only on impact, while potassium explodes on contact with the chloride. Aluminium foil ignites after prolonged contact.

*See other* INTERHALOGENS

### INDIUM(I) PERCHLORATE $InClO_4$ $ClInO_4$

Ashraf, M. *et al., J. Chem. Soc., Dalton Trans.*, 1977, 172

The solvent-free material detonated when crushed with a glass rod. The nitrate (also a redox compound) probably would behave similarly.

*See other* METAL PERCHLORATES
REDOX COMPOUNDS

### POTASSIUM CHLORITE $KClO_2$ $ClKO_2$

Sulphur

Leleu, *Cahiers*, 1974, (74), 137

Interaction is violent.

*See other* CHLORITE SALTS

### POTASSIUM CHLORATE $KClO_3$ $ClKO_3$

Rüst, 1948, 294

Although most explosive incidents have involved mixtures of the chlorate with combustible materials, the exothermic decomposition of the chlorate (to chloride and oxygen) can accelerate to explosion if a sufficient quantity and/or powerful heating are involved. A case history of a fire-initiated explosion of a store of 80 tonnes of chlorate is given. The more stable sodium chlorate will also explode under similar conditions.

Agricultural materials

*See* SODIUM CHLORATE, $ClNaO_3$: Agricultural materials

Aluminium,
Diantimony trisulphide

Crozier, T. H. *et al., 52nd Ann. Rep. HM Insp. Explos.*, London, Home Office, 1928

A pyrotechnic mixture containing the powdered ingredients was found dangerously sensitive to frictional initiation and highly explosive.

*See* Metals, below
Metal sulphides, below

Ammonia,
or Ammonium sulphate

Mellor, 1941, Vol. 2, 702; 1940, Vol. 8, 217

High concentrations of ammonia in air react so vigorously with potassium chlorate as to be dangerous. Mixtures with ammonium sulphate when heated decompose with incandescence.

Ammonium chloride

Potjewijd, T., *Pharm. Weekbl.*, 1935, 72, 68–69

Addition of ammonium chloride to a drum of weed-killer was suspected as the cause of a violent explosion (involving ammonium chlorate).

Aqua regia,
Ruthenium

Sidgwick, 1950, 1459

Ruthenium is insoluble in aqua regia, but addition of potassium chlorate causes explosive oxidation.

Carbon

Read, C. W. W., *School Sci. Rev.*, 1941, 22(87), 341

Accidental substitution of powdered carbon for manganese dioxide in 'oxygen mixture' caused a violent explosion when the mixture was heated.

Cyanides

*See* METAL CYANIDES: Oxidants

Cyanoguanidine

*See* CYANOGUANIDINE, $C_2H_4N_4$: Oxidants

Diarsenic trisulphide

Ganguly, A., *J. Indian Acad. Forensic Sci.*, 1973, **12**, 29–30

Dry powdered mixtures containing over 30% oxidant exploded under a hammer blow.

*See* Metal sulphides, below

Dinickel trioxide

Mellor 1942, Vol. 15, 395

Interaction at 300°C is violently exothermic, red-heat being attained.

Fabric

Anon., *Accidents*, 1968, **74**, 24

Fabric gloves (wrongly used in place of impervious plastics gloves), became impregnated during handling operations and subsequently ignited from cigarette ash.

Fluorine

Pascal, 1960, Vol. 16, 316

Interaction at low pressure leads to the formation of the explosive gas, fluorine perchlorate.

*See also* FLUORINE, $F_2$: Sodium bromate

Hydrocarbons

1. Bjorkman, P. O., *Chem. Abs*., 1934, **28**, 6311a
2. Feilitzen, G. von, ibid., 6311b

Mixtures of powdered chlorate and hydrocarbons explode as violently as nitro compound explosives [1]. Porous masses of chlorate impregnated with hydrocarbons are friction-sensitive explosives [2].

Hydrogen iodide

Mellor, 1941, Vol. 2, 310

Molten potassium chlorate ignites hydrogen iodide gas.

Manganese dioxide

Mellor MIC, 1961, 333

When oxygen is generated in the laboratory by heating potassium chlorate with manganese dioxide as catalyst, the latter must be free of organic matter or an explosion will occur.

Manganese dioxide,
Potassium hydroxide

Molinari, E. *et al., Inorg. Chem.,* 1964, **3**, 898

The oxidation of manganese dioxide to manganate by solid alkali–chlorate mixtures becomes explosive above 80–90°C at pressures above 19 kbar.

Metal,
Wood

Davis, G. C., *J. Ind. Eng. Chem*., 1909, **1**, 118

Explosions were caused during transportation of metal castings in wooden kegs previously used to store potassium chlorate, impact or friction of the metal causing initiation of the chlorate-impregnated wood.

Metal phosphides

Mellor, 1940, Vol. 8, 839, 844

Tricopper diphosphide and trimercury tetraphosphide form impact-sensitive mixtures with potassium chlorate. By analogy, the phosphides of aluminium, magnesium, silver and zinc, etc., would be expected to form similar mixtures with metal halogenates.

Metal phosphinates ('hypophosphites')

Mellor, 1940, Vol. 8, 881, 883

Dry mixtures of barium phosphinate and potassium chlorate burn rapidly with a feeble report if unconfined, but even under the slight confinement of enclosing in paper, a sharp explosion occurs. The mixture is readily initiated by sparks, impact or friction. A mixture of calcium phosphinate, potassium chlorate and quartz exploded during mixing. Mixtures of various phosphinates and chlorates have been proposed as explosives, but they are very sensitive to initiation by sparks, friction or shock. Admixture of powdered magnesium causes a brilliant flash on initiation of the mixture.

*See also* SODIUM PHOSPHINATE, $H_2NaO_2P$: Oxidants

Metals

1. Mellor, 1941, Vol. 2, 310; Anon., *Chem. Eng. News,* 1936, **14**, 451
2. Mellor, 1940, Vol. 4, 480
3. Mellor, 1943, Vol. 11, 163
4. Mellor, 1941, Vol. 7, 20, 116, 260
5. Jackson, H., *Spectrum*, 1969, 7, 82
6. Ganguly, A., *J. Indian Acad. Forensic Sci.*, 1973, **12**, 29–30

Mixtures of finely divided aluminium, copper, magnesium [1] and zinc [2] with potassium chlorate (or other metal halogenates) are explosives and may be initiated by heat, impact or light friction. Chromium incandesces in the molten salt [3] and germanium explodes on heating with potassium chlorate. Titanium explodes on heating, while zirconium gives mild explosions on heating, and ignites when the mixture is impacted [4].

Contact of molten chlorate with steel wool causes violent combustion of the latter [5]. Qualitative experiments showed that hammer impact would explode mixtures with aluminium powder containing over 10% of chlorate [6].

*See also* METAL HALOGENATES

Metal sulphides

1. Mellor, 1941, Vol. 2, 310
2. Mellor, 1939, Vol. 9, 523
3. Mellor, 1941, Vol. 3, 447
4. Rüst, 1948, 335

Many metal sulphides when mixed intimately with metal halogenates form heat-, impact- or friction-sensitive explosive mixtures [1]. That with diantimony trisulphide can be initiated by a spark [2] and with disilver sulphide a violent reaction occurs on heating [3].

For the preparation of 'oxygen mixture' (potassium chlorate–manganese dioxide), diantimony trisulphide was mistakenly used instead of the dioxide, and during grinding, the mixture exploded very violently [4].

*See also* METAL HALOGENATES

Metal thiocyanates

1. von Schwartz, 1918, 299–300, 328
2. Anon., *Chem. Age,* 1936, **35**, 42

Mixtures of thiocyanates with chlorates (or nitrates) are friction- and heat-sensitive, and explode on rubbing, heating to 400°C, or initiation by spark or flame [1]. A violent explosion occurred when a little chlorate was ground in a mortar contaminated with ammonium thiocyanate. A similar larger-scale explosion involving traces of barium thiocyanate is also described [2].

Nitric acid,
Organic matter

Asthana, S. S. *et al., Chem. & Ind.*, 1976, 953–954

In a new method for destruction of organic matter prior to analysis, small portions (0.3 g only) of chlorate are added to a hot suspension of the organic matter in concentrated nitric acid. Use of larger portions of chlorate may lead to explosive oxidation of the organic matter.

Non-metals

1. Mellor, 1941, Vol. 2, 310
2. Mellor, 1940, Vol. 8, 785–786
3. Mellor, 1946, Vol. 5, 15
4. Ganguly, A., *J. Indian Acad. Forensic Sci.*, 1973, 12, 29–30

Potassium chlorate (or other metal halogenates) intimately mixed with arsenic, carbon, phosphorus, sulphur or other readily oxidised materials give friction-, impact- and heat-sensitive mixtures which may explode violently [1]. When potassium chlorate is moistened with a solution of phosphorus in carbon disulphide, it eventually explodes as the solvent evaporates and oxidation proceeds [2]. Boron burns in molten potassium chlorate with dazzling brilliance [3]. Mixtures of the chlorate and finely powdered sulphur containing over 20% of the latter will explode under a hammer blow [4].

*See* Sulphur, below
METAL CHLORATES
*See also* METAL HALOGENATES

Reducants

Mellor, 1941, Vol. 3; 651; 1940, Vol. 8, 881, 883

Mixtures with calcium or strontium hydrides may explode readily, and interaction with the molten chlorate is, of course, violent. A mixture of syrupy sodium phosphinate ('hypophosphite') and the powdered chlorate on heating eventually explodes as powerfully as glyceryl trinitrate. Calcium phosphinate mixed with the chlorate and quartz detonates (the latter producing friction to initiate the mixture). Dried mixtures of barium phosphinate and the chlorate are very sensitive and highly explosive under the lightest confinement (screwed up in paper).

Sugars

1. Zaehringer, A. J., *Rocketscience*, 1949, **3**, 64–68
2. Ganguly, A., *J. Indian Acad. Forensic Sci.*, 1973, **12**, 29–30
3. Scanes, F. S., *Combust. Flame*, 1974, **23**, 363–371

A stoicheiometric mixture with sucrose ignites at 159°C, and has been evaluated as a rocket propellant [1]. Dry powdered mixtures with glucose containing above 50% of oxidant explode under a hammer blow [2]. Pyrotechnic 1:1 mixtures with lactose begin to react exothermically at about 200°C, when the lactose melts, and carbon is formed. This is then oxidised by the chlorate at about 340°C. The mechanism was studied by thermal analysis [3].

*See* METAL CHLORATES: Phosphorus, etc.

Sugar,
Sulphuric acid

Hanson, R. M., *J. Chem. Educ.*, 1976, **53**, 578

Addition of a drop of sulphuric acid to a mound of the chlorate–sugar mixture leads to ignition.

Sulphur

Tanner, H. G., *J. Chem. Educ.*, 1959, **36**, 59

A review of the chemistry involved in this explosively unstable system.

Sulphur dioxide

Mellor, 1947, Vol. 10, 217; 1941, Vol. 2, 311

Contact at temperatures above 60°C causes flashing of the evolved chlorine dioxide. Solutions of sulphur dioxide in ethanol or ether cause an explosion on contact at ambient temperature.

Sulphuric acid

Mellor, 1947, Vol. 10, 435

Bailar, 1973, Vol. 2, 1362

Addition of potassium chlorate in portions to sulphuric acid maintained at below 60°C or above 200°C causes brisk effervescence. At intermediate temperatures, explosions are caused by the chlorine dioxide produced, and these reach maximum intensity at 120–130°C. Uncontrolled contact of any chlorate with sulphuric acid may be explosive. The need for great caution in this system was stressed by Davy in 1815.

*See* METAL CHLORATES: Acids

Sodium amide

Mellor, 1940, Vol. 8, 258

A mixture explodes.

Tannic acid

Rüst, 1948, 336

A mixture exploded during grinding in a mortar.

*See* METAL CHLORATES

Thorium dicarbide

*See* THORIUM DICARBIDE, $C_2Th$: Non-metals, etc.

*See other* METAL OXOHALOGENATES

**POTASSIUM PERCHLORATE** **$KClO_4$** **$ClKO_4$**

Burton, H. *et al., Analyst,* 1955, **80**, 16

Many explosions have been experienced during the gravimetric determination of either perchlorates or potassium as potassium perchlorate by a standard method involving an ethanol extraction. During subsequent heating, formation and explosion of ethyl perchlorate is very probable.

Aluminium,
Aluminium fluoride

Freeman, E. S. *et al., Combust. Flame*, 1966, **10**, 337–340

Presence of aluminium fluoride increases the ease of ignition of aluminium–perchlorate mixtures, owing to complex fluoride formation.

*See* Metal powders, below

Aluminium,
Barium nitrate,
Potassium nitrate,
Water

*See* ALUMINIUM, Al: Metal nitrates, etc.

Aluminium powder,
Titanium dioxide

*Fire Prot. Assoc. J.,* 1957, (36), 9

A mixture of the three compounds exploded violently during mixing. Previously the mixture had been accidentally ignited by a spark. Aluminium powder is incompatible with oxidants.

*See* ALUMINIUM, Al: Metal oxides

Barium chromate,
Tungsten

Carrazza, J. A. *et al., Chem. Abs.*, 1975, 82, 113751g

Mixtures containing 65% of tungsten with organic binders were developed as priming charges for surface flares.

*See* Metal powders, below

Combustible materials

Grodzinski, J., *J. Appl. Chem.*, 1958, 8, 523–528

Explosion temperatures were determined for mixtures of a wide range of combustible liquid and solid organic materials with potassium perchlorate sealed into glass tubes. The lowest temperatures were shown by mixtures with ethylene glycol (240°C), cotton linters (245°C) and furfural (270°C).

Ferrocenium diamminetetrakis(thiocyanato-*N*)chromate(1–)
(Ferrocenium reineckate)

Guslev, V. G. *et al., Chem. Abs.*, 1974, **86**, 57615j

Presence of 25% of the organometallic salt considerably increases the rate of thermal decomposition of the chlorate, involving hydrogen cyanide arising from the thiocyanato groups.

Metal powders

Schumacher, 1960, 210

Riffault, M. L., *Chem. Abs.*, 1975, 82, 127130d

The mixture of aluminium and/or magnesium powders with potassium perchlorate (a photo-flash composition) is very readily ignited, and three industrial explosions have occurred. Mixtures of nickel and titanium powders with the perchlorate and infusorial earth are very friction-sensitive, causing severe explosions, and easily ignited by very small (static) sparks.

A mixture containing 70% of molybdenum powder ignites at 330°C.

*See* Aluminium, etc., above
Barium chromate, etc., above
METAL PERCHLORATES

*See other* METAL OXOHALOGENATES

Potassium hexacyanocobaltate(3–)

Massis, T. M. *et al., Chem. Abs.*, 1977, **87**, 8142q

Mixtures serve as gasless pyrotechnic compositions.

Reducants

*See* Titanium hydride, below
PERCHLORATES: Reducants

Sulphur

Schumacher, 1960, 211–212

Mixtures of sulphur and potassium perchlorate, used in pyrotechnic devices, can be exploded by moderate impact. All other inorganic perchlorates form such impact-sensitive mixtures.

Titanium hydride

Massis, T. M. *et al., Rep. SAND-75-5889*, Richmond (Va.), USNTIS, 1976

The stability of the pyrotechnic mixture has been studied.

*See other* METAL OXOHALOGENATES

**DIPOTASSIUM PERCHLORYLAMIDE** **$K_2NClO_3$** **$ClK_2NO_3$**

Sorbe, 1968, 67

Like the monopotassium salt, it will explode on impact or exposure to flame.

*See* PERCHLORYLAMIDE SALTS

**LITHIUM PERCHLORATE** **$LiClO_4$** **$ClLiO_4$**

Hydrazine

Rosolovskii, V. Ya., *Chem. Abs.*, 1969, **70**, 53524a

Interaction gives only the complex $LiClO_4 \cdot 2N_2H_4$, which explodes on grinding, but loses hydrazine uneventfully on heating at atmospheric or reduced pressure. The sodium salt is similar.

Nitromethane

*See* NITROMETHANE, $CH_3NO_2$: Lithium perchlorate

**MANGANESE CHLORIDE TRIOXIDE** **$ClMnO_3$**

Briggs, T. S., *J. Inorg. Nucl. Chem.*, 1968, **30**, 2867–2868

Explosively unstable if isolated as liquid at ambient temperature, it may be handled safely in carbon tetrachloride.

*See also* MANGANESE DICHLORIDE DIOXIDE, $Cl_2MnO_2$
MANGANESE TRICHLORIDE OXIDE, $Cl_3MnO$

*See related* METAL OXIDES

**NITROSYL CHLORIDE** **NOCl** **ClNO**

Acetone,
Platinum

Kaufmann, G. B., *Chem. Eng. News,* 1957, **35**(43), 60

A cold sealed tube containing nitrosyl chloride, platinum wire and traces of acetone exploded violently on being allowed to warm up.

Hydrogen,
Oxygen

*See* NITROGEN OXIDE, NO: Hydrogen, etc.
*See other* *N*-HALOGEN COMPOUNDS

## *N*-CHLOROSULPHINYLIMIDE — O=S=NCl — ClNOS

Anon., *Angew. Chem. (Nachr.),* 1970, **18**, 318
The ampouled solid exploded violently on melting. Distillation at normal pressure and impact tests had not previously indicated instability.

Chlorine fluoride

*See* CHLORINE FLUORIDE, ClF: *N*-Chlorosulphinylimide
*See other* *N*-HALOGEN COMPOUNDS

## NITRYL CHLORIDE — $NO_2Cl$ — $ClNO_2$

Inorganic materials,
or Organic matter

1. Batey, H. H. *et al., J. Amer. Chem. Soc.,* 1952, **74**, 3408
2. Kaplan, R. *et al., Inorg. Synth.,* 1954, **4**, 54

Interaction of the chloride with ammonia or sulphur trioxide is very violent, even at −75°C, and is vigorous with stannic bromide or iodide [1]. It attacks organic matter rapidly, sometimes explosively [2].
*See other* *N*-HALOGEN COMPOUNDS

(*B*)-Trimethylborazine

Hirata, T., *Rep. AD 729339*, Richmond (Va.), USNTIS, 1971
Interaction is violent in absence of a diluent.
*See other* OXIDANTS

## NITRYL HYPOCHLORITE ('CHLORINE NITRATE') — $NO_2OCl$ — $ClNO_3$

Metals,
or Metal chlorides

Bailar, 1973, Vol. 2, 379
Interaction of the hypochlorite with most metal chlorides is explosive at ambient temperature, but controllable at temperatures between −40° and −70°C.

Organic materials

Schmeisser, M., *Inorg. Synth.,* 1967, **9**, 129

Not inherently explosive, but it reacts explosively with alcohols, ethers and most organic materials.
*See also* NITRYL HYPOFLUORITE, $FNO_3$
*See other* OXIDANTS

**NITROSYL PERCHLORATE** $NOClO_4$ $ClNO_5$

Gerding, H. *et al., Chem. Weekbl.*, 1956, **52**, 282–283
Although stable at room temperature, it begins to decompose below 100°C, and at 115–120°C the decomposition becomes a low-order explosion.
*See related* NON-METAL PERCHLORATES

Pentaammineazidocobalt(III) perchlorate,
Phenylisocyanate
Burmeister, J. L. *et al., Chem. Eng. News*, 1968, **46**(8), 39
During an attempt to introduce phenylisocyanate into the Co co-ordination sphere, a mixture of the three components exploded when stirring was stopped.

Organic materials
Hoffman, K. A. *et al., Ber.*, 1909, **42**, 2031
As the anhydride of nitrous and perchloric acids, it is a very powerful oxidant. Pinene explodes sharply; acetone and ethanol ignite, then explode. Ether evolves gas, then explodes after a few seconds' delay. Small amounts of primary aromatic amines–aniline, toluidines, xylidines, mesidine–ignite on contact, while larger quantities explode dangerously, probably owing to rapid formation of diazonium perchlorates. Urea ignites on stirring with the perchlorate.
*See other* OXIDANTS

**NITRONIUM PERCHLORATE** $NO_2^+ \ ClO_4^-$ $ClNO_6$

Albright, Hanson, 1976, 2
The explosively unstable behaviour of stored nitronium perchlorate is attributed to the formation of small equilibrium concentrations of the isomeric covalent nitryl perchlorate ester (below).

**NITRYL PERCHLORATE** $NO_2ClO_4$ $ClNO_6$

Ammonium perchlorate
*See* AMMONIUM PERCHLORATE, $ClH_4NO_4$: Impurities

Organic solvents
1. Spinks, J. W. T., *Chem. Eng. News,* 1960, **38**(15), 5

2. Gordon, W. E. *et al.*, *Can. J. Res.*, 1940, **18B**, 358

Interaction with benzene gave a slight explosion and flash [1], while sharp explosions with ignition were observed with acetone and ether [2].

*See other* OXIDANTS

### CHLORINE AZIDE $ClN_3$

Alone,
or Ammonia,
or Phosphorus,
or Silver azide,
or Sodium

1. Frierson, W. J. *et al.*, *J. Amer. Chem. Soc.*, 1943, **65**, 1696, 1698
2. Rice, W. J. *et al.*, *J. Chem. Educ.*, 1971, **48**, 659

The undiluted material is extremely unstable, usually exploding violently without cause at any temperature, even as solid at –100°C. It gives an explosive yellow liquid with liquid ammonia; when condensed on to yellow phosphorus at –78°C an extremely violent explosion soon occurs. Addition of phosphorus to a solution of the azide in carbon tetrachloride at 0°C causes a series of mild explosions if the mixture is stirred or a violent explosion without stirring. Contact of the liquid or gaseous azide with silver azide at –78°C gave a blue colour, soon followed by explosion, and sodium reacted similarly under the same conditions [1].

When chlorine azide (25 mol %) is used as a thermally activated explosive initiator in a chemical gas laser tube, the partial pressure of the azide should never exceed 16 mbar [2].

*See* HALOGEN AZIDES

### SULPHURYL AZIDE CHLORIDE $SO_2N_3Cl$ $ClN_3O_2S$

Shozda, R. J. *et al.*, *J. Org. Chem.*, 1967, **32**, 2876

During the preparation of this explosive liquid by interaction of sulphuryl chloride fluoride and sodium azide, traces of chlorine must be eliminated from the former to avoid detonation. The product is nearly as shock-sensitive as glyceryl trinitrate and may explode on rapid heating. Solutions (25% wt.) in solvents may be handled safely. The corresponding fluoride is believed to behave similarly.

*See other* NON-METAL AZIDES

### THIOTRITHIAZYL PERCHLORATE $\overline{SNSNSNS}\ ClO_4$ $ClN_3O_4S_4$

Organic solvents

Goehring, 1957, 74

The precipitated perchlorate salt exploded on washing with acetone or ether.

**THIOTRITHIAZYL CHLORIDE** **SNSNSNS Cl** **$ClN_3S_4$**

Ammonia

Mellor, 1940, Vol. 8, 631–2.

Bailar, 1973, Vol. 2, 903

The dry chloride, which explodes on heating in air, will rapidly absorb ammonia gas and then explode. The structure of the cation is now known to be a seven-membered ring with only two adjacent sulphur atoms. Thiotrithiazyl salts other than the chloride are also explosive.

*See also* THIOTRITHIAZYL NITRATE, $N_4O_3S_4$
THIOTRITHIAZYL PERCHLORATE, $ClN_3O_4S_4$

**TRIAZIDOCHLOROSILANE** **$(N_3)_3SiCl$** **$ClN_9Si$**

*See* TETRAAZIDOSILANE, $N_{12}Si$

**SODIUM CHLORIDE** **NaCl** **ClNa**

Lithium

*See* LITHIUM, Li: Sodium carbonate, etc.

**SODIUM HYPOCHLORITE** **NaOCl** **ClNaO**

Alone

Brauer, 1963, Vol. 1, 311

Sorbe, 1968, 85

The anhydrous solid obtained by desiccation of the pentahydrate will decompose violently on heating or friction.

Amines

Kirk-Othmer, 1963, Vol. 2, 105

Primary aliphatic or aromatic amines react with sodium (or calcium) hypochlorite to form *N*-mono- or di-chloroamines which are explosive but less so than nitrogen trichloride.

Aziridine

Graefe, A. F. *et al., J. Amer. Chem. Soc.,* 1958, **80**, 3939

Interaction with sodium (or other) hypochlorite gives the explosive *N*-chloro compound.

*See* 1-CHLOROAZIRIDINE, $C_2H_4ClN$

Methanol

ICI Mond Div., private comm., 1968

Several explosions involving methanol and sodium hypochlorite were attributed to formation of methyl hypochlorite, especially in presence of acid or other esterification catalyst.

*See* HYPOHALITES

Phenylacetonitrile

Libman, D. D., private comm., 1968

Use of sodium hypochlorite solution to destroy acidified benzyl cyanide residues caused a violent explosion, thought to have been due to formation of nitrogen trichloride.

Water

*Hazard Note HN(76)189*, Dept. of Health and Social Security, London, 1976

Two 2.5 l bottles of strong sodium hypochlorite solution (10–14% available chlorine) burst in storage, owing to failure of the cap designed to vent oxygen slowly evolved during storage. This normal tendency may have been accelerated by the unusually hot summer. Vent caps should be checked with full personal protection, and material should be stored at 15–18°C and out of direct sunlight, which accelerates decomposition.

*See other* METAL OXOHALOGENATES

**SODIUM CHLORITE** **$NaClO_2$** **$ClNaO_2$**

Alone

Brauer, 1963, Vol. 1, 312

Leleu, *Cahiers*, 1974, (74), 137

The anhydrous salt explodes on impact, and decomposes violently at 200°C.

Acids

Bailar, 1973, Vol. 2, 1413

Under normal conditions, solutions of sodium (and other) chlorites when acidified do not evolve chlorine dioxide in dangerous amounts. However, explosive concentrations may result if acid is dropped onto solid chlorites.

*See* Oxalic acid, below

Organic matter

'The Diox Process', Newark, N. J., Wallace and Tiernan, 1949

Intimate mixtures of the solid chlorite with finely divided or fibrous organic matter may be very sensitive to heat, impact or friction.

Oxalic acid

*MCA Case History No. 839*

A bleach solution was being prepared by mixing solid sodium chlorite, oxalic acid and water, in that order. As soon as water was added, chlorine dioxide was evolved, and later exploded. The lower explosive limit of the latter is 10%, and the mixture is photo- and heat-sensitive.

Sodium dithionite

Anon., *Chem. Trade J.*, 1953, **132**, 564

Use of a scoop contaminated with sodium dithionite for sodium chlorite caused ignition of the latter. Materials containing sulphur (dithionite, natural rubber gloves) cause decomposition of sodium chlorite and contact should be avoided.

Sulphur-containing materials

Leleu, *Cahiers*, 1974, (74), 138

Sodium chlorite reacts very violently with organic compounds of divalent sulphur, or with free sulphur (which may ignite), even in presence of water. Contact of the chlorite with rubber vulcanised with sulphur or a divalent sulphur compound must therefore be avoided.

*See* Sodium dithionite, above

*See other* CHLORITE SALTS

**SODIUM CHLORATE** **$NaClO_3$** **$ClNaO_3$**

Normally thermally stable, it may explode under confinement and intense heating.

*See* POTASSIUM CHLORATE, $ClKO_3$

Agricultural materials

Reimer, B. *et al., Chem. Technik*, 1974, **26**, 446–447 (condensed paper)

The potential for explosive combustion of mixtures of sodium chlorate-based herbicides with other combustible agricultural materials was determined. Initiation temperatures and maximum temperatures were measured for mixtures of sodium (or potassium) chlorate with peat, powdered sulphur, sawdust, urotropine (hexamethylenetetramine), thiuram and other formulated materials. With many combinations, maximum temperature increases of 500–1000°C at rates of 400–1200°/s were recorded for 2 g samples.

Alkenes,

Potassium osmate

Lloyd, W. D. *et al., Synthesis*, 1972, 610

The hydroxylation of alkenes to diols with potassium osmate–oxidant mixtures has been described, with either hydrogen peroxide or sodium chlorate as the oxidant. The sodium chlorate method is not applicable where the diol is to be distilled from the reaction mixture, because of the danger of explosive oxidation of the diol by the chlorate.

Aluminium,
Rubber

Olson, C. M., *J. Electrochem. Soc.*, 1949, **116**, 34C

The rubber belt of a bucket elevator, fitted with aluminium buckets and used for transporting solid chlorate, jammed during use. Friction from the rotating drive pulley heated and powdered the jammed belt. A violent explosion consumed all the rubber belt and most of the 90 aluminium buckets. Bronze and steel equipment is now installed.

Ammonium salts,
or Metals,
or Non-metals,
or Sulphides

*MCA SD-42,* 1952
*MCA Case History No. 2019*

Mixtures of the chlorate with ammonium salts, powdered metals, phosphorus, silicon, sulphur or sulphides are readily ignited and potentially explosive.

Residues of ammonium thiosulphate in a bulk road tanker contaminated the load of dry sodium chlorate subsequently loaded. An exothermic reaction occurred with gas evolution during several hours. Laboratory tests showed that such a mixture could be made to decompose explosively. A reaction mechanism is suggested.

*See* METAL HALOGENATES

1,3-Bis(trichloromethyl)benzene

*See* 1,3-BIS(TRICHLOROMETHYL)BENZENE, $C_8H_4Cl_6$ : Oxidants

Diarsenic trioxide

Ellern, 1968, 51

Ignition may occur on contact.

Diols

*See* Alkenes, etc., above

Grease

Olson, C. M., *J. Electrochem. Soc.*, 1969, **116**, 34C

The greased bearing of a small grinder exposed to chlorate dust exploded violently during cleaning. Fluorocarbon-based greases and armoured bearings are recommended for chlorate service, with full operator protection during cleaning operations.

*See* Organic matter, below

Leather

*MCA Case History No. 1979*

Shoes became contaminated with a weed-killer solution which dried out in wear. A welding spark later fell into the shoe and the front was blown off.

Nitrobenzene

Hodgson, J. F., private comm., 1973

The combination is powerfully explosive and has been widely used in recent guerrilla activities.

Organic matter

*MCA SD-42,* 1952

Cook, W. H., *Can. J. Res.,* 1933, **8**, 509

Mixtures of sodium (and other) chlorates with fibrous or absorbent organic materials (charcoal, flour, shellac, sawdust, sugar) are hazardous. If the chlorate concentration is high, the mixtures may be ignited or caused to explode by static, friction or shock. Even at 10–15% concentration, low relative humidity may allow easy ignition and rapid combustion to occur.

*See* Paper, Static electricity; Wood, below
*See also* OXIDANTS AS HERBICIDES

Osmium

Rogers, D. B. *et al., Inorg. Synth.,* 1972, **13**, 141

During the preparation of osmium dioxide, the ampoule containing the reaction mixture must be cooled during sealing operations to prevent violent reaction occurring. Subsequent heating to 300 and then 600°C must also be effected slowly.

Paper,
Static electricity

Ewing, O. R., *Chem. Eng. News,* 1952, **30**, 3210

Grimmett, R. E. R., *New Zealand J. Agric.,* 1938, **57**, 224–225

Paper impregnated with sodium chlorate and dried can be ignited by static sparks, but not by friction or impact. Paper bags or card cartons are unsuitable packing materials.

A previous incident involving paper sacks which had formerly contained mixed sodium and calcium chlorates had been noted.

Phosphorus

Anon., *Angew. Chem. (Nachr.),* 1957, **5**, 78

A mixture of red phosphorus and sodium chlorate exploded violently.

Sodium phosphinate

*See* SODIUM PHOSPHINATE, $H_2NaO_2P$: Oxidants

Sulphuric acid

1. *ABCM Quart. Safety Summ.*, 1944, **15**, 3
2. *MCA Case History No. 282*

Erroneous addition of concentrated sulphuric acid to solid sodium chlorate instead of sodium chloride caused an explosion due to formation of chlorine dioxide [1]. Accidental contact of 93% acid on clothing previously splashed with sodium chlorate caused immediate ignition [2].

Wood

1. *ABCM Quart. Safety Summ.*, 1947, **18**, 25
2. *BCISC Quart. Safety Summ.*, 1967, **38**, 42

Various fires and explosions caused by use of wooden containers with chlorates, and precautions necessary during handling and storage, are discussed [1,2].

*See* Organic matter, above

**SODIUM PERCHLORATE** **$NaClO_4$** **$ClNaO_4$**

Hydrazine

Rosolovskii, V. Ya., *Chem. Abs.*, 1969, **70**, 53524a

Interaction gives only an equimolecular complex which explodes on grinding but dissociates uneventfully on heating at atmospheric or reduced pressure. The lithium salt is similar.

*See other* METAL PERCHLORATES

**ANTIMONY(III) CHLORIDE OXIDE** **Sb(O)Cl** **ClOSb**

Bromine trifluoride

*See* BROMINE TRIFLUORIDE, $BrF_3$: Antimony(III) chloride oxide

**CHLORINE DIOXIDE** **$ClO_2$**

1. Sidgwick, 1950, 1203
2. Mellor, 1941, Vol. 2, 288
3. Stedman, R. F., *Chem. Eng. News,* 1951, **29**, 5030
4. Cameron, A. E., ibid., 3196
5. Fawcett, H. H., ibid., 4459
6. McHale, E. T. *et al., J. Phys. Chem.,* 1968, **72**, 1849
7. *Hazards of Chlorine Dioxide,* New York, National Board of Fire Underwriters, 1950; *Chem. Eng. News,* 1950, **28**, 611
8. Derby, R. I. *et al., Inorg. Synth.,* 1954, **4**, 152

This is a powerful oxidant and explodes violently on the slightest provocation as gas or liquid [1]. It is initiated by contact with several materials (below), on heating rapidly to 100°C or on sparking [2], or by impact as solid at −100°C [3]. A small sample exploded during vacuum distillation at below −50°C [4], and it was stated that decomposition by sparking begins to become hazardous at concentrations of 7–8% in air [3], and that at 10% concentrations in air or at 0.1 bar partial pressure explosion may occur from any source of initiation such as sunlight, heat or electrostatic discharge [5].

A kinetic study of the decomposition shows that it is explosive above 45°C even in absence of light, and subject to long induction periods due to formation of intermediate dichlorine trioxide. UV irradiation greatly sensitises the dioxide to explosion [6].

A guide on fire and explosion hazards in industrial use of chlorine dioxide is available [7], and preparative precautions have been detailed [8].

Carbon monoxide

Mellor, 1941, Vol. 2, 288

Explosion on mixing.

Difluoroamine

Lawless, 1968, 171

Interaction in the gas phase is explosive.

Hydrogen

Mellor, 1941, Vol. 2, 288

Near-stoicheiometric mixtures detonate on sparking, or on contact with platinum sponge.

Mercury

Mellor, 1941, Vol. 2, 288

Chlorine dioxide explodes on shaking with mercury.

Non-metals

Mellor, 1941, Vol. 2, 289; 1956, Vol. 2, Suppl. 1, 532

Phosphorus, sulphur, sugar or combustible materials ignite on contact and may cause explosion.

Phosphorus pentachloride

*See* PHOSPHORUS PENTACHLORIDE, $Cl_5P$: Chlorine dioxide

Potassium hydroxide

Mellor, 1941, Vol. 2, 289

The liquid or gaseous oxide will explode in contact with solid potassium hydroxide or its concentrated solution.

*See other* HALOGEN OXIDES

**THALLIUM(I) CHLORITE** $TlClO_2$ $ClO_2Tl$

Bailar, 1973, Vol. 1, 1167

Detonable by shock.

*See other* CHLORITE SALTS

Acetylene

*See* ACETYLENE (ETHYNE), $C_2H_2$ : Halogens

Air,

Ethylene

*See* ETHYLENE, $C_2H_4$ : Air, Chlorine

**CHLORINE TRIOXIDE** $ClO_3$

Schmeisser, M., *Angew. Chem.*, 1955, **67**, 498

The dimeric form is formulated as $ClO_2^+ ClO_4^-$.

*See* CHLORYL PERCHLORATE, $Cl_2O_6$

**THALLIUM(I) CHLORIDE** $TlCl$ $ClTl$

Fluorine

*See* FLUORINE, $F_2$ : Metal salts

**CHLORINE** $Cl_2$

*MCA SD-80*, 1970

Alcohols

*See* *tert*-Butanol, below

HYPOHALITES

Aluminium

*Ann. Rep. Chief. Insp. Factories*, 1953 (Cmd. 9330), 171

Corrosive failure of a vaporiser used in manufacture of aluminium chloride caused liquid chlorine to contact molten aluminium. A series of explosions occurred.

*See* ALUMINIUM, Al: Halogens

: Oxidants

*See other* Metals, below

Antimony trichloride,

Tetramethylsilane

Bush, R. P., Br. Pat. 1 388 991, 1975 (*Chem. Abs.*, 1975, **83**, 59026e)

Antimony trichloride-catalysed chlorination of the silane to chloro-trimethylsilane in absence of diluent was explosive at 100°C, but controllable at below 30°C.

*See other* HALOGENATION INCIDENTS

Bromine pentafluoride

*See* BROMINE PENTAFLUORIDE, $BrF_5$ : Acids, etc.

*tert*-Butanol

1. Bradshaw, C. P. C. *et al., Proc. Chem. Soc.*, 1963, 213
2. Mintz, M. J. *et al., Org. Synth.*, 1969, **49**, 9

Rate of admission of chlorine into the alcohol during preparation of *tert*-butyl hypochlorite must be regulated to keep temperature below 20°C to prevent explosion [1]. A safer and simpler preparation uses hypochlorite solution in place of chlorine [2].

*See also tert*-BUTYL HYPOCHLORITE, $C_4H_9ClO$

Carbon disulphide

*MCA Case History No. 971*

When liquid chlorine was added to carbon disulphide in an iron cylinder, an explosion occurred, due to the iron-catalysed chlorination of carbon disulphide to carbon tetrachloride. The operation had been done previously in glassware without incident.

1-Chloro-2-propyne

Anon., *Loss Prev. Bull.*, 1974, **001**, 10

A vigorous explosion during chlorination of 1-chloro-2-propyne in benzene at 0°C over 4 h was attributed to presence of excess chlorine arising from the slow rate of reaction at low temperature.

*See also* ACETYLENE (ETHYNE), $C_2H_2$ : Halogens
Hydrocarbons, below (reference 10)

*See other* HALOGENATION INCIDENTS

Cobalt(II) chloride,
Methanol

1. Vlcek, A. A., *Inorg. Chem.*, 1967, **6**, 1425–1427
2. Gillard, R. D. *et al., Chem. & Ind.*, 1973, 777
3. Gillard, R. D., private comm., 1973

During the preparation of *cis*-dichlorobis(2,2′-bipyridyl)cobalt(III) chloride repeating a published procedure [1], passage of chlorine into an ice-cold solution of cobalt chloride, bipyridyl and lithium chloride in methanol soon caused an explosion followed by ignition of the methanol inside the reaction vessel. The fire could not be extinguished with carbon dioxide but went out when the flow of chlorine stopped [2]. This is consistent with the formation and spontaneous ignition of methyl hypochlorite in the system at a slightly elevated temperature.

An alternative route to the cobalt complex not involving chlorine is available [3].

*See* Methanol, below
METHYL HYPOCHLORITE, $CH_3ClO$
*See other* HALOGENATION INCIDENTS

Dibutyl phthalate
Statesir, W. A., *Chem. Eng. Progr.*, 1973, **69**(4), 54
A mixture of the ester and liquid chlorine confined in a stainless steel bomb reacted explosively at 118°C.

Dicaesium oxide
*See* DICAESIUM OXIDE, $Cs_2O$: Halogens

Dichloro(methyl)arsine
Dillon, K. B. *et al., J. Chem. Soc., Dalton Trans.*, 1976, 1479
In the attempted preparation of tetrachloromethylarsine by interaction of chlorine with the arsine while slowly warming from −196°C, several sealed ampoules exploded at well below 0°C, probably owing to liberation of chloromethane.

Diethyl ether
1. Harrison, G. E., private comm., 1966
2. Unpublished observation, 1949

Chlorine caused ignition of ether on contact [1]. Exposure of an ethereal solution of chlorine to daylight caused a mild, photocatalysed, explosion [2].
*See also* BROMINE, $Br_2$ : Diethyl ether

Diethylzinc
Weast, 1974, C-715
Ignition occurs on contact.

Dioxygen difluoride
*See* DIOXYGEN DIFLUORIDE, $F_2O_2$

Disilyl oxide
Bailar, 1973, Vol. 1, 1377
Interaction is explosive.

4,4′-Dithiomorpholine
*See* 4-MORPHOLINESULPHENYL CHLORIDE, $C_4H_8ClNOS$

Fluorine
*See* FLUORINE, $F_2$ : Halogens

Glycerol

Statesir, W. A., *Chem. Eng. Progr.*, 1973, **69**(4), 54

A mixture of glycerol and liquid chlorine confined in a stainless steel bomb exploded at 70–80°C

Hexachlorodisilane

Martin, G., *J. Chem. Soc.*, 1914, **105**, 2859

Hexachlorodisilane vapour ignited in chlorine above 300°C; violent explosions sometimes occurred.

Hydrocarbons

1. Mellor, 1956, Vol. 2, Suppl. 1, 380
2. Eisenlohr, D. H., US Pat. 2 989 571, 1961
3. von Schwartz, 1918, 142, 321
4. von Schwartz, 1918, 324
5. Mamadaliev, Y. G. *et al., Chem. Abs.*, 1937, **31**, 8502[5]
6. Brooks, B. T., *Ind. Eng. Chem.*, 1924, **17**, 752
7. Johnson, J. H. *et al., Hydrocarbon Proc. Petr. Ref.*, 1963, **42**(2), 174
8. Statesir, W. A., *Chem. Eng. Prog.*, 1973, **69**(4), 52–54
9. de Oliveria, D. B., *Hydrocarbon Proc.*, 1973, **53**(3), 112–126
10. Kilby, J. L., *Chem. Eng. Progr.*, 1968, **64**(6), 52
11. Lawrence, W. W. *et al., Ind. Eng. Chem., Proc. Des. Develop.*, 1970, **9**, 47–49
12. Kokochashvili, V. I. *et al., Chem. Abs.*, 1976, **84**, 20030x
13. Rozlovskii, A. I. *et al., Chem. Abs.*, 1976, **84**, 49418f
14. Jenkins, P. A., unpublished information, 1974

Interaction of chlorine with methane is explosive at room temperature over yellow mercury oxide [1], and mixtures containing above 20 vol. % of chlorine are explosive [2]. Mixtures of acetylene and chlorine may explode on initiation by sunlight, other UV source, or high temperatures, sometimes very violently [3]. Mixtures with ethylene explode on initiation by sunlight, etc., or over mercury, dimercury or disilver oxides at ambient temperature, or over lead oxide at 100°C [1,4]. Interaction with ethane over activated carbon at 350°C has caused explosions, but added carbon dioxide reduces the risk [5]. Accidental introduction of gasoline into a cylinder of liquid chlorine caused a slow exothermic reaction which accelerated to detonation. This effect was verified [6]. Injection of liquid chlorine into a naphtha–sodium hydroxide mixture (to generate hypochlorite *in situ*) caused a violent explosion. Several other incidents involving violent reactions of saturated hydrocarbons with chlorine were noted [7].

In a review of incidents involving explosive reactivity of liquid chlorine with various organic auxiliary materials, two involved hydrocarbons. A polypropylene filter element fabricated with zinc oxide filler reacted explosively, rupturing the steel case previously tested to over 300 bar. Zinc chloride derived from the oxide may have initiated the runaway reaction. Hydrocarbon-based diaphragm pump oils or metal-drawing waxes were violently or explosively reactive [8]. A violent explosion in a wax chlorination plant may have involved unplanned contact of liquid chlorine with wax or chlorinated wax residues in a steel trap. Corrosion products in the trap may have catalysed the runaway reaction, but hydrogen (also liberated by corrosion in the trap) may also have been involved [9].

During maintenance work, simultaneous release of chlorine and acetylene from two plants into a common vent line leading to a flare caused an explosion in the line [10]. The violent interaction of liquid chlorine injected into ethane at 80°C/10 bar becomes very violent if ethylene is also present [11]. The relationship between critical pressure for self-ignition of chlorine–propane mixtures at 300°C was studied, and the tendency is minimal for 60 : 40 mixtures. Combustion is explosive under some conditions [12]. Precautions to prevent explosions during chlorination of solid paraffin hydrocarbons are detailed [13]. In the continuous chlorination of polyisobutene at below 100°C in absence of air, changes in conditions (increase in chlorine flow, decrease in polymer feed) leading to over-chlorination caused an exotherm to 130° and ignition [14].

*See* TURPENTINE: Halogens

Hydrocarbons,
Lewis acids

Howard, W., *Loss Prev.*, 1973, **7**, 78

During chlorination of hydrocarbons with Lewis acid catalysis, the catalyst must be premixed with the hydrocarbon before admission of chlorine. Addition of catalyst to the chlorine–hydrocarbon mixture is very hazardous, causing instantaneous release of large volumes of hydrogen chloride.

*See* Iron(III) chloride, Monomers, below
*See other* HALOGENATION INCIDENTS

Hydrogen

1. Mellor, 1956, Vol. 2, Suppl. 1, 373–375
2. Weissweiler, A., *Z. Elektrochem.*, 1936, **42**, 499
3. Eichelberger, W. C. *et al., Chem. Eng. Prog.*, 1961, **57**(8), 94
4. Wood, J. L., *Loss Prevention,* 1969, **3**, 45–47
5. Stephens, T. J. R. *et al.*, Paper 62B, *65th A.I. Chem. Eng. Meeting, New York, 1972*
6. de Oliveria, D. B., *Hydrocarbon Proc.*, 1973, **52**(3), 112–126
7. Antonov, V. N. *et al., Chem. Abs.*, 1974, **81**, 172184b

Combination of the elements may be explosive over a wide range of physical conditions, with initiation by sparks, radiant energy or catalysis, e.g. by yellow mercuric oxide at ambient temperature [1]. There is a narrow range of concentrations in which the mixture is supersensitive to initiation [2]. Explosion–detonation phenomena in chlorine production cells have been investigated [3,4]. The explosive limits of the mixture vary with the container shape and method of initiation, but are usually within the range 5–89% of hydrogen by volume [1]. After an explosion in a chlorine distillate receiver where hydrogen had been produced by corrosion, no initiation source could be identified or reasonably postulated following a thorough investigation [5]. Several other hydrogen–chlorine explosions without identifiable ignition sources are also mentioned. Hydrogen may have been involved in a severe wax chlorinator explosion [6].

Pressure increase during explosive combustion of mixtures within the critical concentration region of 5–15% vol of hydrogen was studied, and found less than anticipated [7].

Hydrogen,
Nitrogen trichloride

*See* NITROGEN TRICHLORIDE, $Cl_3N$: Chlorine, Hydrogen

Hydrogen,
Other gases

Munke, *Chem. Technik*, 1974, **26**, 292–295

Available data on explosibility of chlorine and hydrogen in admixture with air, hydrogen chloride, oxygen or inert gases is discussed and presented as triangular or rectangular diagrams.

Iron(III) chloride,
Monomers

*MCA Case History Nos. 2115, 2147*

During chlorination of an aromatic monomer (styrene) in carbon tetrachloride at 50°C, a violent reaction occurred when some 10% of the chlorine gas had been fed in. Laboratory examination showed that the eruption was caused by a rapid decomposition reaction catalysed by ferric chloride. Various aromatic monomers decomposed in this way when treated with gaseous chlorine or hydrogen chloride (either neat, or in a solvent) in the presence of steel or iron(III) chloride. Exotherms of 90°C (in 50% solvent) to 200°C (no solvent) were observed, and much gas and polymeric residue was forcibly ejected.

*See* Hydrocarbons, Lewis acids, above

Metal acetylides and carbides

The mono- and di-alkali metal acetylides, copper acetylides, iron, uranium and zirconium carbides all ignite in chlorine, the former often at ambient temperature.

*See* TRIIRON CARBIDE, $CFe_3$: Halogens
DICAESIUM ACETYLIDE, $C_2Cs_2$: Halogens
COPPER ACETYLIDE, $C_2Cu$: Halogens

DICOPPER(I) ACETYLIDE, $C_2Cu_2$: Halogens
CAESIUM ACETYLIDE, $C_2HCs$
RUBIDIUM ACETYLIDE, $C_2HRb$
DILITHIUM ACETYLIDE, $C_2Li_2$: Halogens
STRONTIUM ACETYLIDE, $C_2Sr$: Halogens
URANIUM DICARBIDE, $C_2U$: Halogens
ZIRCONIUM DICARBIDE, $C_2Zr$: Halogens
*See also* LITHIUM ACETYLIDE–AMMONIA, $C_2HLi{\cdot}H_3N$: Gases

Metal hydrides
Mellor, 1941, Vol. 2, 483; Vol. 3, 73
Potassium, sodium and copper hydrides all ignite in chlorine at ambient temperatures.

Metal phosphides
*See* TRICOPPER DIPHOSPHIDE, $Cu_3P_2$: Oxidants

Metals
1. Mellor, 1941, Vol. 2, 92, 95; 1956, Vol. 2, Suppl. 1, 380, 469; 1941, Vol. 3, 638; 1940, Vol. 4, 267, 480; 1941, Vol. 7, 208, 260, 436; Vol. 9, 379, 626, 849; 1942, Vol. 12, 312; Vol. 15, 146
2. Hanson, B. H., *Process Eng.*, 1975, (2), 77
3. Davies, D. J., ibid.,
4. Anon., *Loss Prev. Bull.*, 1975, **006**, 1

Tin ignites in liquid chlorine at –34°C, aluminium powder in the gas at –20°C, while vanadium powder explodes on contact at 0°C with the pressurised liquid. A solution in heptane ignites in contact with powdered copper well below 0°C. Aluminium, brass foil, calcium powder, copper foil, iron wire, manganese powder and potassium all ignite in the dry gas at ambient temperature, as do powdered antimony, bismuth and germanium sprinkled into the gas, while magnesium, sodium and zinc ignite in the moist gas. Thorium, tin and uranium ignite and incandesce on warming (uranium to 150°C), and powdered nickel burns at 600°C. Aluminium–titanium alloys also ignite on heating. in chlorine. Niobium ignites in chlorine after gentle warming [1]. Unlike most other metals, titanium is not suitable for components in contact with *dry* gaseous or liquid chlorine, as ignition may occur [2]. Minimum water content to prevent attack may be from 0.015 to 1.5%, depending upon conditions [3]. A statement [4] on ignition of titanium in wet chlorine is incorrect.

*See* Aluminium, above
Steel, below
BERYLLIUM, Be: Halogens

Methanol
*Trans. National Safety Congr.*, 1937, 273
Passage of chlorine through cold recovered methanol (but not fresh commercial methanol) led to a mild explosion and ignition.

*See* Cobalt(II) chloride, Methanol, above
METHYL HYPOCHLORITE, $CH_3ClO$

*See also* CATALYSIS BY IMPURITIES
BROMINE, $Br_2$ : Methanol

Nitrogen compounds

1. Mellor, 1941, Vol. 2, 95; 1940, Vol. 8, 99, 288, 313, 607
2. Bowman, W. R. *et al., Chem. & Ind.,* 1963, 979
3. Bainbridge, E. G., ibid., 1350
4. Folkers, K. H. *et al., J. Amer. Chem. Soc.,* 1941, **63**, 3530
5. Schierwater, F.-W., *Major Loss Prevention*, 1971, **49**
6. Sorbe, 1968, 120
7. Kirk-Othmer, 1964, Vol. 5, 2

Ammonia–chlorine mixtures are explosive if warmed or if chlorine is in excess, owing to formation of nitrogen trichloride. Hydrazine, hydroxylamine and calcium nitride ignite in chlorine, and nitrogen triiodide may explode on contact with chlorine [1]. During chlorination of impure biuret in water at 20°C, a violent explosion occurred [2]. This was attributed to conversion of the cyanuric acid impurity (3%) to nitrogen trichloride and spontaneous explosion of the latter [3]. During interaction of chlorine and alkylthiouronium salts to give alkanesulphonyl chlorides, the dangerously explosive nitrogen trichloride may be produced if excess chlorine or slow chlorination is used. General precautions are discussed [4]. Aziridine readily forms the explosive *N*-chloro derivative.

*See* 1-CHLOROAZIRIDINE, $C_2H_4ClN$

During chlorination of 2,4,6-triketo-1,3,5-triazine (cyanuric acid), presence of the diaminoketo and aminodiketo analogues as impurities, or of an unusually low pH value, may lead to formation of nitrogen trichloride. UV irradiation may be used to destroy this in a continuous process circulation reactor and prevent build-up of dangerous concentrations [5]. Chlorination of amidosulphuric acid [6] or acidic ammonium chloride solutions [7] gives the powerfully explosive oil, nitrogen trichloride.

*See* *N*-HALOGEN COMPOUNDS

Non-metal hydrides

Mellor, 1939, Vol. 9, 55, 396; 1939, Vol. 8, 65; 1940, Vol. 6, 219; 1941, Vol. 5. 37

Arsine, phosphine and silane all ignite in contact with chlorine at ambient temperature, while diborane and stibine react explosively, the latter also with chlorine water.

*See also* ETHYLPHOSPHINE, $C_2H_7P$: Halogens

Non-metals

Mellor, 1941, Vol. 2, 92; 1956, Vol. 2, Suppl. 1, 380; 1943, Vol. 11, 26

Liquid chlorine at –34°C explodes with white phosphorus, and a solution in heptane at 0°C ignites red phosphorus. Boron, active carbon, silicon and phosphorus all ignite in contact with gaseous chlorine at ambient temperature. Arsenic incandesces on contact with liquid chlorine at –34°C, and the powder ignites when sprinkled into the gas at ambient temperature. Tellurium must be warmed slightly before incandescence occurs.

Oxygen difluoride

*See* OXYGEN DIFLUORIDE, $F_2O$: Halogens

Phenylmagnesium bromide

1. Datta, R. L. *et al., J. Amer. Chem. Soc.*, 1919, **41**, 287
2. Zakharkin, L. I. *et al., J. Organomet. Chem.*, 1970, **21**, 271

After treatment of the Grignard reagent with chlorine, the solid which separated exploded when shaken [1]. This solid was not seen later, using either diethyl ether or tetrahydrofuran as solvent [2]. The explosive solid could have been magnesium hypochlorite if moisture had been present in the chlorine.

*See* MAGNESIUM HYPOCHLORITE, $Cl_2MgO_2$

Phosphorus compounds

1. Mellor, 1940, Vol. 8, 812, 842, 844–845, 897
2. von Schwartz, 1918, 324

Boron diiodophosphide, phosphine, phosphorus trioxide and trimercury tetraphosphide all ignite on contact with chlorine at ambient temperature. Trimagnesium diphosphide and trimanganese diphosphide ignite in warm chlorine [1], while ethylphosphine explodes with chlorine [2]. Unheated boron phosphide incandesces in chlorine.

Polychlorobiphenyl

Statesir, W. A., *Chem. Eng. Prog.*, 1973, **69**(4), 53–54

A mixture of a polychlorobiphenyl process oil and liquid chlorine confined in a stainless steel bomb reacted exothermically between 25 and 81°C.

Silicones

Statesir, W. A., *Chem. Eng. Prog.*, 1973, **69**(4), 53–54

Silicone process oils mixed with liquid chlorine confined in a stainless steel bomb reacted explosively on heating: polydimethylsiloxane at 88–118°C, and polymethyltrifluoropropylsiloxane at 68–114°C. Previously, leakage of a silicone pump oil into a liquid chlorine feed system had caused rupture of a stainless steel ball valve under a pressure surge of about 2 kbar.

Sodium hydroxide

*MCA Case History No. 1880*

Attempted disposal of a small amount of liquid chlorine by pouring it into 20% sodium hydroxide solution caused a violent reaction leading to personal contamination.

Steel

1. *MCA Case History No. 608*
2. Stephens, T. J. R. *et al.*, Paper 62B, 5, *65th A. I. Chem. Eng. Meeting, New York, 1972*
3. Vervalin, 1973, 85

Chlorine leaking into a steam-heated mild steel pipe caused ignition of the latter at *ca.* 250°C [1]. Sheet steel in contact with chlorine usually ignites at 200–250°C, but the presence of soot, rust, carbon or other catalysts may reduce the ignition temperature to 100°C. Dry steel wool ignites in chlorine at only 50°C [2]. Use of a carbon steel inlet pipe in a paraffin chlorination system led to rupture of the pipe and a fire [3].

Sulphides

Mellor, 1940, Vol. 4, 952; 1946, Vol. 5, 144; 1939, Vol. 9, 270

Diarsenic disulphide, diboron trisulphide and mercuric sulphide all ignite in chlorine at ambient temperature, the first only in a rapid stream.

Synthetic rubber

Murray, R. L., *Chem. Eng. News,* 1948, **26**, 3369

During interaction of synthetic rubber and liquid chlorine, a violent explosion occurred. It is known that natural and synthetic rubbers will burn in liquid chlorine.

Tetraselenium tetranitride

*See* TETRASELENIUM TETRANITRIDE, $N_4Se_4$

Trialkylboranes

Coates, 1967, Vol. 1, 199

The lower homologues tend to ignite in chlorine or bromine.

Tungsten dioxide

Mellor, 1943, Vol. 11, 851

Incandescence on warming.

Water

Kosharov, P., *Chem. Abs.*, 1940, **34**, 3917q

Mixtures of chlorine and water at certain concentrations are capable of explosion by spark ignition. Other hazards in production of chlorine are detailed.

*See other* HALOGENS
OXIDANTS

**COBALT(II) CHLORIDE** **$CoCl_2$** **$Cl_2Co$**

Metals

*See* POTASSIUM, K: Metal halides
SODIUM, Na: Metal halides

**DIHYDRAZINECOBALT(II) CHLORATE** **$Cl_2CoH_8N_4O_6$**

**$[Co(N_2H_4)_2][ClO_3]_2$**

Salvadori, R., *Gazz. Chim. Ital.*, 1910, [2], **40**, 9

Explodes powerfully on slightest impact or friction, or on heating to 90°C.

*See other* AMMINEMETAL OXOSALTS

**HEXAAQUACOBALT(II) PERCHLORATE** **$Cl_2CoH_{12}O_{14}$**

**$[Co(H_2O)_6][ClO_4]_2$**

Salvadori, R., *Gazz. Chim. Ital.*, 1910, [2], **40**, 9

Explodes on impact; deflagrates on rapid heating.

*See other* AMMINEMETAL OXOSALTS
REDOX COMPOUNDS

**PENTAAMMINEPHOSPHINATOCOBALT(III) PERCHLORATE** **$Cl_2CoH_{17}N_5O_{10}P$**

**$[Co(NH_3)_5H_2PO_2][ClO_4]_2$**

*BCISC Quart. Safety Summ.*, 1965, **36**, 58

When a platinum wire (which may have been hot) was dipped for a flame test into a sintered funnel containing the air-dried complex, detonation occurred. This may have been due to heat and/or friction on a compound containing both strongly oxidising and strongly reducing radicals. Avoid dipping platinum wire into bulk samples of materials of unknown potential.

*See other* AMMINEMETAL OXOSALTS
REDOX COMPOUNDS

**CHROMIUM(II) CHLORIDE** **$CrCl_2$** **$Cl_2Cr$**

1. *MCA Case History No. 1660*
2. Bretherick, L., *Chem. Brit.*, 1976, **12**, 204

An unopened bottle of chromous chloride solution exploded after prolonged storage [1]. This was almost certainly caused by internal pressure of hydrogen developed by slow reduction of the solvent water by the powerfully reducant $Cr^{2+}$ ion [2].

*See* CHROMIUM(II) SULPHATE, $CrO_4S$

**CHROMYL CHLORIDE** **$CrO_2Cl_2$** $Cl_2CrO_2$

1. Sidgwick, 1950, 1004
2. Pitwell, L. R., private comm., 1964

Though a powerful and often violent oxidant of inorganic and organic materials in absence of a diluent, it has found use as a solution in preparative organic chemistry for the controlled oxidation of alkyl aromatics [1]. In such a reaction, failure of a stirrer during addition of the chloride caused a build-up of unreacted material, followed by a violent explosion [2].

Ammonia
Mellor, 1943, Vol. 11, 394
Contact with ammonia causes incandescence.

Disulphur dichloride
Mellor, 1943, Vol. 11, 394
A jet of chromyl chloride vapour ignites in the vapour of disulphur dichloride.

Non-metal hydrides
Pascal, 1959, Vol. 14, 153
Hydrogen sulphide and phosphine may ignite in contact with the chloride.

Organic solvents
Mellor, 1943, Vol. 11, 396
Acetone, ethanol and ether ignite on contact with the chloride. Turpentine behaves similarly.

Phosphorus,
or Phosphorus trichloride
Mellor, 1943, Vol. 11, 395
Pascal, 1959, Vol. 14, 153
Moist phosphorus explodes in contact with the liquid chloride. Addition of drops of chromyl dichloride to cooled phosphorus trichloride causes incandescence, and sometimes explosion. Phosphorus tribromide may also ignite with chromyl chloride.

Sodium azide
Mellor, 1967, Vol. 8, Suppl. 2.2, 36

Interaction of chromyl chloride and sodium azide to form chromyl azide is explosive in absence of a diluent.

Sulphur

Mellor, 1943, Vol. 11, 394

Contact of the liquid chloride with flowers of sulphur causes ignition.

Urea

Pascal, 1959, Vol. 14, 153

Urea ignites in contact with the chloride.

*See other* OXIDANTS

**CHROMYL PERCHLORATE** $CrO_2(ClO_4)_2$ $Cl_2CrO_{10}$

Alone,

or Organic solvents

Schmeisser, M., *Angew. Chem.*, 1955, 67, 499

A powerful oxidant, which causes organic solvents to ignite on contact and which explodes violently above 80°C.

*See other* OXIDANTS

**DICHLOROFLUOROAMINE** $Cl_2NF$ $Cl_2FN$

1. Sukornick, B. *et al.*, *Inorg. Chem.*, 1963, 2, 875
2. Batty, W. E., *Chem. & Ind.*, 1969, 1232

In the liquid phase, it is an extremely friction- and shock-sensitive explosive [1], like the trichloro analogue [2].

*See other* *N*-HALOGEN COMPOUNDS

**IRON(II) CHLORIDE** $FeCl_2$ $Cl_2Fe$

Ozonides

*See* OZONIDES: Metals, etc.

**IRON(II) PERCHLORATE** $Fe(ClO_4)_2$ $Cl_2FeO_8$

Preparative hazard

*See* PERCHLORIC ACID, $ClHO_4$: Iron sulphate

*See other* REDOX COMPOUNDS

## DICHLOROGERMANE $Cl_2GeH_2$

Ammonia

*See* CHLOROGERMANE, $ClGeH_3$: Ammonia

## DICHLOROAMINE $Cl_2NH$ $Cl_2HN$

Eckert, P. *et al.*, *Chem. Abs.*, 1951, **45**, 7527g

Dissolving *N,N,N',N'*-tetrachloroadipamide in boiling water gives a highly explosive oil, probably dichloroamine.

*See other* *N*-HALOGEN COMPOUNDS

## *N,N'*-BIS(CHLOROMERCURIO)HYDRAZINE $Cl_2H_2Hg_2N_2$

**ClHgNHNHHgCl**

Mellor, Vol. 4, 874, 881; Vol. 8, 318

It explodes when heated or struck, and the bromo and other analogues are similar.

*See other* *N*-METAL DERIVATIVES

## CHLORONIUM PERCHLORATE $Cl^+H_2\ ClO_4^-$ $Cl_2H_2O_4$

Hantzsch, A., *Ber.*, 1930, **63**, 1789

This hydrogen chloride–perchloric acid complex spontaneously dissociates with explosive violence.

## BISHYDROXYLAMINEZINC(II) CHLORIDE $Cl_2H_6N_2O_2Zn$

$[Zn(H_2NOH)_2]Cl_2$

Walker, J. E. *et al.*, *Inorg. Synth.*, 1967, **9**, 2

It explodes at 170°C.

*See other* AMMINEMETAL OXOSALTS

## HYDRAZINIUM DIPERCHLORATE $Cl_2H_6N_2O_8$

$H_3NNH_3(ClO_4)_2$

Grelecki, C. J. *et al.*, *ACS 54*, 1966, 73

The thermal decomposition of this solid rocket propellant component has been studied.

*See other* OXOSALTS OF NITROGENOUS BASES

## BISHYDRAZINENICKEL(II) PERCHLORATE $Cl_2H_8N_4NiO_8$

$[Ni(N_2H_4)_2][ClO_4]_2$

Maissen, B. *et al.*, *Helv. Chim. Acta*, 1951, **34**, 2084–2085

The salt known to be explosive when heated dry (but not under a

hammer blow), exploded violently when stirred as a dilute aqueous suspension.
*See other* AMMINEMETAL OXOSALTS

**BISHYDRAZINETIN(II) CHLORIDE** **$[Sn(N_2H_4)_2]Cl_2$** **$Cl_2H_8N_4Sn$**

Mellor, 1941, Vol. 7, 430
Explodes on heating.
*See related* AMMINEMETAL OXOSALTS

**TETRAAMINEBIS-DINITROGENOSMIUM DIPERCHLORATE** **$Cl_2H_{12}N_8O_8Os$**
**$[Os(NH_3)_4(N_2)_2][ClO_4]_2$**

Creutz, C. A., private comm., 1969
The dry salt may explode on touching.
*See other* AMMINEMETAL OXOSALTS

**PENTAAMMINEAZIDORUTHENIUM(III) CHLORIDE** **$Cl_2H_{15}N_8Ru$**
**$[Ru(NH_3)_5N_3]Cl_2$**

*See* PENTAAMMINECHLORORUTHENIUM(III) CHLORIDE, $Cl_2H_{15}N_5Ru$: Sodium azide

**MERCURY(II) CHLORIDE** **$HgCl_2$** **$Cl_2Hg$**

Sodium *aci*-nitromethane
*See* SODIUM *aci*-NITROMETHANE, $CH_2NNaO_2$: Mercury(II) chloride

**MERCURY(II) CHLORITE** **$Hg(ClO_2)_2$** **$Cl_2HgO_4$**

Levi, G. R., *Gazz. Chim. Ital.*, 1915, [2], **45**, 161
It is extremely unstable when dry, exploding spontaneously.
*See other* CHLORITE SALTS

**MERCURY(I) CHLORITE** **$Hg_2(ClO_2)_2$** **$Cl_2Hg_2O_4$**

Levi, G. R., *Gazz. Chim. Ital.*, 1915, [2], **45**, 161
It is extremely unstable when dry, exploding spontaneously.

**MAGNESIUM CHLORIDE** **$MgCl_2$** **$Cl_2Mg$**

Jute,
Sodium nitrate
*See* SODIUM NITRATE, $NNaO_3$: Jute, Magnesium chloride

## MAGNESIUM HYPOCHLORITE — $Mg(OCl)_2$ — $Cl_2MgO_2$

Clancey, V. J., *J. Haz. Mat.*, 1975, **1**, 83–94

The compound is very unstable, and its presence may be one of the causes of the observed explosive and apparently spontaneous decomposition of calcium hypochlorite, if produced from magnesia-containing lime.

*See also* CHLORINE, $Cl_2$: Phenylmagnesium bromide

*See other* HYPOHALITES

## MAGNESIUM CHLORATE — $Mg(ClO_3)_2$ — $Cl_2MgO_6$

Sulphides

Mellor, 1956, Vol. 2, Suppl. 1, 584

Interaction with the sulphide of copper(I) is explosive, and with those of antimony(III), arsenic(III), tin(II) and (IV), incandescent.

*See other* METAL CHLORATES

## MAGNESIUM PERCHLORATE — $Mg(ClO_4)_2$ — $Cl_2MgO_8$

Marusch, H., *Chem. Tech. (Berlin)*, 1956, **8**, 482–485

The drying agent may contain traces of perchloric acid remaining from manufacturing operations, and owing to the great desiccating power of the salt, the acid will be anhydrous. Many of the explosions experienced with magnesium perchlorate may have their origins in contact of the anhydrous acid with oxidisable materials, or materials able to form unstable perchlorate esters or salts.

Alkenes

Heertjes, P. M. *et al.*, *Chem. Weekbl.*, 1941, **38**, 85

The anhydrous salt which had been used for drying unsaturated hydrocarbons exploded on heating to 220°C for reactivation. The need to avoid contact with acidic materials is stressed.

*See* Organic materials, below

Argon

Dam, J. W., *Chem. Weekbl.*, 1958, **54**, 277

An explosion which occurred when argon was being dried must have involved some (unstated) impurity in the system.

Arylhydrazine

Belcher, R., private comm., 1968

Anhydrous magnesium perchlorate was used to thoroughly dry an ethereal solution of an arylhydrazine. During evaporation of the filtered solution it exploded completely and violently. Magnesium perchlorate is rather soluble in ether and may contain traces of free perchloric acid (probably in the anhydrous form, as the magnesium salt is a powerful dehydrator). It is entirely unsuitable for drying organic solvents.

Cellulose,
Dinitrogen tetraoxide,
Oxygen

*ABCM Quart. Safety Summ.*, 1961, **32**, 6

Magnesium perchlorate contained in a glass tube between wads of cotton wool was used to dry a mixture of oxygen and dinitrogen tetraoxide. After several days, the drying tube exploded violently. It seems probable that the acidic fumes and cotton produced cellulose nitrate, aided by the dehydrating action of the perchlorate.

Dimethyl sulphoxide

1. Tobe, M. L. *et al., J. Chem. Soc.*, 1964, 2991
2. Anon., *Chem. Eng. News*, 1965, **43**(47), 62
3. Dessy, R. E. *et al., J. Amer. Chem. Soc.*, 1964, **86**, 28

In the preparation of anhydrous dimethyl sulphoxide by a literature method [1] an explosion occurred during distillation from anhydrous magnesium perchlorate [2]. This may have been due to the presence of some free methanesulphonic acid as an impurity in the solvent, which could liberate perchloric acid. It is known that sulphoxides react explosively with 70% perchloric acid. The alternative procedure for drying dimethylsulphoxide with calcium hydride [3] seems preferable, as this would also remove any acidic impurities.

*See* DIMETHYL SULPHOXIDE, $C_2H_6OS$: Metal oxosalts

Hydrogen fluoride

Anon., *Ind. Eng. Chem. (News Ed.)*, 1939, **17**, 70

A violent explosion followed the use of magnesium perchlorate to dry wet fluorobutane. The latter had hydrolysed to give hydrogen fluoride which liberated anhydrous, explosive perchloric acid.* Magnesium perchlorate is unsuitable for drying acid or inflammable materials; calcium sulphate would be suitable.

Organic materials

1. Hodson, R. J., *Chem. & Ind.*, 1965, 1873
2. *MCA Case History No. 243*

The use of the perchlorate as desiccant in a drybag where contamination with organic compounds is possible is considered dangerous [1]. Magnesium perchlorate ('Anhydrone') was inadvertently used instead of calcium sulphate (anhydrite) to dry a reaction product (unstated)

* This seems an unlikely explanation because of the large difference in dissociation constants of the acids.

before vacuum distillation. The error was realised and all solid was filtered off. Towards the end of the distillation, decomposition and a violent explosion occurred, possibly due to the presence of dissolved magnesium perchlorate, or more probably to perchloric acid present as impurity in the magnesium perchlorate [2].

Phosphorus

*See* PHOSPHORUS, P: Magnesium perchlorate

Trimethyl phosphite

Jercinovic, L. M. *et al., J. Chem. Educ.*, 1968, **45**, 751

Contact between the components caused violent explosions on several occasions.

*See other* METAL OXOHALOGENATES

**MANGANESE(II) CHLORIDE** **$MnCl_2$** **$Cl_2Mn$**

Zinc

*See* ZINC, Zn: Manganese dichloride

**MANGANESE DICHLORIDE DIOXIDE** **$MnCl_2O_2$** **$Cl_2MnO_2$**

Briggs, T. S., *J. Inorg. Nucl. Chem.*, 1968, **30**, 2867–2868

Explosively unstable if isolated as the liquid at ambient temperature, it may be handled safely in carbon tetrachloride solution.

*See also* MANGANESE CHLORIDE TRIOXIDE, $ClMnO_3$
MANGANESE TRICHLORIDE OXIDE, $Cl_3MnO$

*See related* METAL HALIDES

**MANGANESE CHLORATE** **$Mn(ClO_3)_2$** **$Cl_2MnO_6$**

Bailar, 1973, Vol. 3, 835

The hexahydrated salt decomposes explosively above 6°C, producing chlorine dioxide.

*See other* METAL CHLORATES

**MANGANESE DIPERCHLORATE** **$Mn(ClO_4)_2$** **$Cl_2MnO_8$**

Sidgwick, 1950, 1285

Explodes at 195°C

2,2-Dimethoxypropane

*See* 2,2-DIMETHOXYPROPANE, $C_5H_{12}O_2$: Metal perchlorates

*See other* METAL OXOHALOGENATES

## VANADYL AZIDE DICHLORIDE $VO(N_3)Cl_2$ $Cl_2N_3OV$

Explosive solid.

*See* METAL AZIDE HALIDES

## DIAZIDODICHLOROSILANE $(N_3)_2SiCl_2$ $Cl_2N_6Si$

*See* TETRAAZIDOSILANE, $N_{12}Si$

## NICKEL DIPERCHLORATE $Ni(ClO_4)_2$ $Cl_2NiO_8$

2,2-Dimethoxypropane

*See* 2,2-DIMETHOXYPROPANE, $C_5H_{12}O_2$: Metal perchlorates

## DICHLORINE OXIDE $Cl_2O$

Alone

Sidgwick, 1950, 1202

Gray, P. *et al., Combust. Flame*, 1972, **18**, 361–371

The liquid may explode on pouring or on boiling at 2°C, and the gas readily explodes on heating or sparking. The spontaneously explosive decomposition of the gas was studied at 42–86°C, and induction periods up to several hours long were noted.

Alcohols

Gallais, 1957, 677

Alcohols are oxidised explosively.

Carbon

Mellor, 1956, Vol. 5, 824

Addition of charcoal to the gas causes an immediate explosion, probably due to initiation by the heat of adsorption on the solid.

Carbon disulphide

Mellor, 1940, Vol. 6, 110

The vapours explode on contact.

Dicyanogen

Brotherton, T. K. *et al., Chem. Rev.,* 1959, **59**, 843

Contact causes ignition or explosion.

Diphenylmercury

*See* DIPHENYLMERCURY, $C_{12}H_{10}Hg$: Non-metal oxides

Ethers

Gallais, 1957, 677

Ethers are oxidised explosively.

Hydrocarbons

Ip, J. K. K. *et al., Combust. Flame*, 1972, **19**, 117–129

The spontaneously explosive interaction of the oxide with methane, ethane, propane, ethylene or butadiene was investigated at 50–150°C. Self-heating occurs with ethylene, ethane and propane mixtures.

Nitrogen oxide

Mellor, 1940, Vol. 8, 433

Interaction is explosive.

Oxidisable materials

1. Mellor, 1941, Vol. 2, 241–242; 1946, Vol. 5, 824
2. Jacobs, P. W. M. *et al., Chem. Rev.*, 1969, **69**, 559
3. Pilipovich, D. *et al., Inorg. Chem.*, 1972, **11**, 2190
4. Cady, G. H., *Inorg. Synth.*, 1957, **5**, 156

The heat-sensitivity (above) may explain the explosions which occur on contact of many readily oxidisable materials with this powerful oxidant. Such materials include ammonia, potassium; arsenic, antimony; sulphur, charcoal (adsorptive heating may also contribute); calcium phosphide, phosphine, phosphorus; hydrogen sulphide and antimony, barium, mercury and tin sulphides. Various organic materials (paper, cork, rubber, turpentine, etc.) behave similarly [4]. Mixtures with hydrogen detonate on ignition [1]. The oxide explodes if heated rapidly or overheated locally (sparking, or adiabatic compression in a U-tube [2, 3]), or often towards the end of slow thermal decomposition. Kinetic data are summarised [2]. Preparative precautions have been detailed [4].

Potassium

*See* POTASSIUM, K: Non-metal oxides
*See other* HALOGEN OXIDES

## SULPHINYL CHLORIDE (THIONYL CHLORIDE) $Cl_2OS$

**$SOCl_2$**

Ammonia

Foote, C. S., private comm., 1965

Addition of a solution of *p*-nitrobenzoyl chloride in excess thionyl chloride to ice-cold concentrated ammonia solution caused a violent explosion. This may have been due to formation of nitrogen trichloride.

Bis(dimethylamino)sulphoxide

Armitage, D. A. *et al., J. Inorg. Nucl. Chem.*, 1974, **36**, 993

Interaction of the chloride with the sulphoxide or its higher homologues causes extensive decomposition, possibly explosive above 80°C.

Chloryl perchlorate

*See* CHLORYL PERCHLORATE, $Cl_2O_6$: Organic matter, etc.

1,2,3-Cyclohexanetrione trioxime,
Sulphur dioxide

1. Tokura, N. *et al., Bull. Chem. Soc. Japan*, 1962, **35**, 723
2. Lewis, J. J., *J. Heterocycl. Chem.*, 1975, **12**, 601

A previous method of making 4-oximino-4,5,6,7-tetrahydrobenzofurazan by cyclisation of the oxime with sulphinyl chloride in liquid sulphur dioxide sometimes led to explosive reactions [1]. A new procedure involving aqueous calcium carbonate is quite safe [2].

*See* OXIMES

*N,N*-Dimethylformamide

Spitulnik, M. J., *Chem. Eng. News*, 1977, **55**(31), 31

Some 200 kg of a mixture of sulphinyl chloride and DMF decomposed vigorously after several hours' storage at ambient temperature. This was attributed to the presence of 90 ppm each of iron and zinc in the chloride used for the preparation. Mixtures of the pure components remained unchanged for 48 h at ambient temperature, but addition of 200 ppm of iron powder led to exothermic decomposition after 22 h stirring.

*See* CATALYSIS BY IMPURITIES
DIMETHYLFORMAMIDE, $C_3H_7NO$: Halogenated compounds, Iron

Dimethyl sulphoxide

*See* DIMETHYL SULPHOXIDE, $C_2H_6OS$: Acyl halides

Hexafluoroisopropylideneaminolithium

*See* HEXAFLUOROISOPROPYLIDENEAMINOLITHIUM, $C_3F_6LiN$: Non-metal halides

Linseed oil
Quinoline,

Laws, G. F., private comm., 1966

It is important to add the quinoline and linseed oil, used in purifying thionyl chloride, to the chloride. If reversed addition is used, a vigorous decomposition may occur.

Sodium

*See* SODIUM, Na: Non-metal halides

Water

*MCA Case History No. 1808*

Passage of thionyl chloride through a flexible metal transfer hose which was contaminated with water or sodium hydroxide solution caused the hose to burst. Interaction with water violently decomposes the chloride to hydrogen chloride (2 mol) and sulphur dioxide (1 mol), the total expansion ratio from liquid to gas being 3700 : 1, so very high pressures may be generated.

*See related* NON-METAL HALIDES

**SELENINYL CHLORIDE** **$SeOCl_2$** **$Cl_2OSe$**

Antimony

Mellor, 1947, Vol. 10, 906

Powdered antimony ignites on contact.

Metal oxides

Mellor, 1947, Vol. 10, 909

In contact with disilver oxide, light is evolved and sufficient heat to decompose some silver oxide. Similar effects were observed with the three lead oxides.

Potassium

Mellor, 1947, Vol. 10, 908

Potassium explodes violently in contact with the liquid.

Phosphorus

Mellor, 1947, Vol. 10, 906

Red phosphorus evolves light and heat in contact with the chloride, while white phosphorus explodes.

*See other* NON-METAL HALIDES

**LEAD HYPOCHLORITE** **$Pb(OCl)_2$** **$Cl_2O_2Pb$**

Hydrogen sulphide

Pascal, 1960, Vol. 13.1, 1004

Ignition occurs on contact.

*See other* HYPOHALITES

**SULPHONYL DICHLORIDE (SULPHURYL CHLORIDE)** **$Cl_2O_2S$**

**$SO_2Cl_2$**

Alkalies

Brauer, 1963, Vol. 1, 385

Reaction with alkalies may be explosively violent.

Diethyl ether

Dunstan, I. *et al., Chem. & Ind.*, 1966, 73

A solution of sulphuryl chloride in ether vigorously decomposed, evolving hydrogen chloride. This was shown to be accelerated by the presence of peroxides. Peroxide-free ether should be used, and with care.

Dimethyl sulphoxide

*See* DIMETHYL SULPHOXIDE, $C_2H_6OS$: Acyl halides

Dinitrogen pentaoxide

*See* DINITROGEN PENTAOXIDE, $N_2O_5$: Sulphur dichloride, etc.

Lead dioxide

*See* LEAD DIOXIDE, $O_2Pb$: Non-metal halides

Phosphorus

*See* PHOSPHORUS, P: Non-metal halides

*See other* NON-METAL HALIDES

**DICHLORINE TRIOXIDE** $Cl_2O_3$

McHale, E. T. *et al., J. Amer. Chem. Soc.*, 1967, **89**, 2796; *J. Phys. Chem.*, 1968, **72**, 1849–1856

It is very unstable and the vapour explodes at about 2 mbar and well below 0°C. It has been identified as the intermediate involved in the delayed explosion of chlorine dioxide.

*See other* HALOGEN OXIDES

**CHLORINE PERCHLORATE** $ClOClO_3$ $Cl_2O_4$

Schack, C. J. *et al., Inorg. Chem.*, 1970, **9**, 1387; 1971, **10**, 1078

A shock-sensitive compound; bromine perchlorate is also unstable.

Chlorotrifluoroethylene

Schack, C. J. *et al., Inorg. Chem.*, 1973, **12**, 897

Interaction of the reactants (pre-mixed at −196°C) during warming to −78°C, then ambient temperature, was explosive. Progressive addition of the alkene to the perchlorate at −78°C was uneventful.

Perfluoroalkyl iodides

Schack, C. J. *et al., Inorg. Chem.*, 1975, **14**, 145–151

Reaction mixtures of chlorine perchlorate with perfluoromethyl iodide, 1,2-diiodoperfluoroethane and 1,3-diiodoperfluoropropane occasionally deflagrated at or below ambient temperature.

*See other* HALOGEN OXIDES

## LEAD DICHLORITE $Pb(ClO_2)_2$ $Cl_2O_4Pb$

Alone,
or Antimony sulphide,
or Sulphur

Mellor, 1941, Vol. 2, 283

It explodes on heating above 100°C or on rubbing with antimony sulphide or fine sulphur.

Non-metals

Pascal, 1960, Vol. 16, 264

Carbon, red phosphorus and sulphur are oxidised violently by the chlorite.

*See* BARIUM CHLORITE, $BaCl_2O_4$

*See other* CHLORITE SALTS
METAL OXOHALOGENATES

## DISULPHURYL DICHLORIDE $ClSO_2OSO_2Cl$ $Cl_2O_5S_2$

Phosphorus

*See* PHOSPHORUS, P: Non-metal halides

Water

Sveda, M., *Inorg. Synth.*, 1950, **3**, 127

The pure chloride reacts slowly with water but, if more than a few % of chlorosulphuric acid (a usual impurity) are present, the reaction is rapid and could become violent with large quantities.

*See other* NON-METAL HALIDES

## CHLORYL PERCHLORATE $ClO_2^+ ClO_4^-$ $Cl_2O_6$

Organic matter,
or Thionyl chloride,
or Water

1. Mellor, 1956, Vol. 2, Suppl. 1, 539
2. Schmeisser, M., *Angew. Chem.*, 1955, **67**, 499
3. Wechsberg, M. *et al.*, *Inorg. Chem.*, 1972, **11**, 3066
4. Sorbe, 1968, 45

Though the least explosive of the chlorine oxides, being insensitive to heat or shock, it is a very powerful oxidant and needs careful handling. It violently or explosively oxidises ethanol, stopcock grease, wood and organic matter generally [1]; it is liable to explode on contact with thionyl chloride in absence of solvent [2], or with water [1]. Recently its explosion on heating has been noted [3]. Explosions in contact

with organic matter below −70°C have also been recorded [4].
*See other* HALOGEN OXIDES

**LEAD(II) CHLORATE** **$Pb(ClO_3)_2$** **$Cl_2O_6Pb$**

Bailar, 1973, Vol. 2, 130
Thermal decomposition may be explosive.
*See other* METAL CHLORATES

**ZINC DICHLORATE** **$Zn(ClO_3)_2$** **$Cl_2O_6Zn$**

Sorbe, 1968, 158
The tetrahydrated salt decomposes explosively at 60°C.

Sulphides
Mellor, 1956, Vol. 2, Suppl. 1, 584
Interaction with the sulphide of copper(II) is explosive, and with those of antimony(III), arsenic(III), tin(II) and (IV), incandescent.
*See other* METAL CHLORATES

**PERCHLORYL PERCHLORATE (DICHLORINE HEPTAOXIDE)** **$ClO_3^+ ClO_4^-$** $Cl_2O_7$

1. Mellor, 1956, Vol. 2, Suppl. 1, 542
2. Schmeisser, M., *Angew. Chem.*, 1955, **67**, 498

It explodes violently on impact or rapid heating, but is a less powerful oxidant than other chlorine oxides [1]. It is now formulated as perchloryl perchlorate [2].

Bromine pentafluoride
Lawless, 1968, 173
Mixtures are shock-sensitive explosives.

Iodine
Kirk-Othmer, 1964, Vol. 5, 5
Explosion on contact.
*See other* HALOGEN OXIDES

**LEAD DIPERCHLORATE** **$Pb(ClO_4)_2$** **$Cl_2O_8Pb$**

Methanol
Willard, H. H. *et al.*, *J. Amer. Chem. Soc.*, 1930, **52**, 2396
A saturated solution of anhydrous lead diperchlorate in dry methanol exploded violently when disturbed. Methyl perchlorate may have been involved.

*See* ALKYL PERCHLORATES
*See other* METAL OXOHALOGENATES

**TIN(II) PERCHLORATE** $Sn(ClO_4)_2$ $Cl_2O_8Sn$

Bailar, 1973, Vol. 2, 76
The trihydrated salt decomposes explosively at 250°C.
*See other* METAL PERCHLORATES
REDOX COMPOUNDS

**XENON(II) PERCHLORATE** $Xe(ClO_4)_2$ $Cl_2O_8Xe$

Wechsberg, M. *et al., Inorg. Chem.,* 1972, **11**, 3066
During preparation from perchloric acid and xenon difluoride at −50°C, violent explosions occurred if the reaction mixture was allowed to warm up rapidly.
*See other* XENON COMPOUNDS

**URANYL DIPERCHLORATE** $UO_2(ClO_4)_2$ $Cl_2O_{10}U$

Ethanol
Erametsa, O., *Suomen Kemist,* 1942, **15B**, 1
Attempted recrystallisation of the salt from ethanol caused an explosion.
*See* ALKYL PERCHLORATES
*See other* METAL OXOHALOGENATES

**LEAD DICHLORIDE** $PbCl_2$ $Cl_2Pb$

Calcium
*See* CALCIUM, Ca: Lead dichloride

**SULPHUR DICHLORIDE** $SCl_2$ $Cl_2S$

Acetone
Fawcett, F. S. *et al., Inorg. Synth.,* 1963, **7**, 121
Acetone is an effective solvent for cleaning traces of sulphur chlorides from reaction vessels, but care is necessary, as the reaction is vigorous if more than traces are present.

Dimethyl sulphoxide
*See* DIMETHYL SULPHOXIDE, $C_2H_6OS$: Acyl halides, etc.

Hexafluoroisopropylideneaminolithium

*See* HEXAFLUOROISOPROPYLIDENEAMINOLITHIUM, $C_3F_6LiN$: Non-metal halides

Metals

*See* POTASSIUM, K: Non-metal halides
SODIUM, Na: Non-metal halides

Oxidants

*See* PERCHLORYL FLUORIDE, $ClFO_3$: Hydrocarbons, etc.
NITRIC ACID, $HNO_3$: Sulphur halides
DINITROGEN PENTAOXIDE, $N_2O_5$: Sulphur dichloride

Water

*MCA SD-77,* 1960
Exothermic reaction with water or steam.
*See other* NON-METAL HALIDES

**DISULPHUR DICHLORIDE** **ClSSCl** $Cl_2S_2$

Aluminium

*See* ALUMINIUM, Al: Non-metal halides

Antimony,
or Antimony or arsenic sulphides

Mellor, 1947, Vol. 10, 641
Interaction at ambient temperature is surprisingly energetic.

Chromyl chloride

*See* CHROMYL CHLORIDE, $Cl_2CrO_2$: Disulphur dichloride

Dimethyl sulphoxide

*See* DIMETHYL SULPHOXIDE, $C_2H_6OS$: Acyl halides, etc.

Diphosphorus trioxide

*See* PHOSPHORUS(III) OXIDE, $O_3P_2$: Disulphur dichloride

Mercury oxide

Mellor, 1947, Vol. 10, 643
Interaction is rapid and very exothermic.

Potassium

Mellor, 1947, Vol. 10, 642
A mixture of potassium and the liquid chloride is shock-sensitive and explodes violently on heating.

Sodium peroxide

*See* SODIUM PEROXIDE, $Na_2O_2$: Non-metal halides

Unsaturated materials

1. Mellor, 1947, Vol. 10, 641–642
2. Huestis, B. L., *Safety Eng.*, 1927, **54**, 95

Alkenes, terpenes and unsaturated glycerides react exothermically, some vigorously [1]. Ignition may occur with some organic materials [2].

Water

*MCA SD-77*, 1960

As with sulphinyl chloride, the exothermic reaction with limited amounts of water may be dangerously violent under confinement because of rapid gas evolution.

*See other* NON-METAL HALIDES

## DISELENIUM DICHLORIDE ClSeSeCl $Cl_2Se_2$

Alkali metals or oxides

*See* POTASSIUM, K: Non-metal halides
POTASSIUM DIOXIDE, $KO_2$: Diselenium dichloride
SODIUM, Na: Non-metal halides
SODIUM PEROXIDE, $Na_2O_2$: Non-metal halides

Aluminium

*See* ALUMINIUM, Al: Non-metal halides

## TIN DICHLORIDE $SnCl_2$ $Cl_2Sn$

Bromine trifluoride

*See* BROMINE TRIFLUORIDE, $BrF_3$: Tin dichloride

Calcium acetylide

*See* CALCIUM ACETYLIDE, $C_2Ca$: Tin dichloride

Hydrogen peroxide

*See* HYDROGEN PEROXIDE, $H_2O_2$: Tin(II) chloride

Metal nitrates

*See* METAL NITRATES: Esters, etc.

## TITANIUM DICHLORIDE $TiCl_2$ $Cl_2Ti$

1. Mellor, 1941, Vol. 7, 75; Brauer, 1965, Vol. 2, 1187
2. *NSC Data Sheet 485*, 1966

Readily ignites in air, particularly if moist [1]. The dichloride on heating under inert atmospheres disproportionates into the tetrachloride and pyrophoric titanium [2].

Water

Gallais, 1957, 431

Interaction at ambient temperature is violent and accompanied by evolution of hydrogen.

*See other* METAL HALIDES

**ZIRCONIUM DICHLORIDE** **$ZrCl_2$** **$Cl_2Zr$**

Sidgwick, 1950, 652

When warm it ignites in air.

*See other* METAL HALIDES

**COBALT(III) CHLORIDE** **$CoCl_3$** **$Cl_3Co$**

Pentacarbonyliron,
Zinc

*See* PENTACARBONYLIRON, $C_5FeO_5$: Transition metal halides, etc.

**PENTAAMMINECHLOROCOBALT(III) PERCHLORATE** **$Cl_3CoH_{15}N_5O_8$**

**$[Co(NH_3)_5Cl][ClO_4]_2$**

Explodes at 320°C; high impact-sensitivity.

*See* AMMINEMETAL OXOSALTS

**PENTAAMMINEAQUACOBALT(III) CHLORATE** **$Cl_3CoH_{17}N_5O_{10}$**

**$[Co(NH_3)_5H_2O][ClO_3]_3$**

Salvadori, R., *Gazz. Chim. Ital.*, 1910[II], **40**, 9

Explodes on impact, or heating to 130°C.

*See other* AMMINEMETAL OXOSALTS

**HEXAAMMINECOBALT(III) CHLORITE** **$Cl_3CoH_{18}N_6O_6$**

**$[Co(NH_3)_6][ClO_2]_3$**

Levi, G. R., *Atti Acad. Lincei*, 1923[V], 32(1), 623

It is impact-sensitive and explosive.

*See other* AMMINEMETAL OXOSALTS

## HEXAAMMINECOBALT(III) CHLORATE $Cl_3CoH_{18}N_6O_9$

$[Co(NH_3)_6][ClO_3]_3$

Friederich, W. *et al.*, *Z. ges. Schiess-Sprengstoffw.*, 1926, 21, 49

It is explosive.

*See other* AMMINEMETAL OXOSALTS

## HEXAAMMINECOBALT(III) PERCHLORATE $Cl_3CoH_{18}N_6O_{12}$

$[Co(NH_3)_6][ClO_4]_3$

Explodes at 360°C, highly impact-sensitive.

*See* AMMINEMETAL OXOSALTS

## CHROMIUM(III) CHLORIDE $CrCl_3$ $Cl_3Cr$

Lithium,
Nitrogen

*See* LITHIUM, Li: Metal chlorides, etc.

## DIDYMIUM PERCHLORATE $Dy(ClO_4)_3$ $Cl_3DyO_{12}$

Birnbaum, E. R. *et al.*, *Inorg. Chem.*, 1973, **12**, 379

The anhydrous salt was stable to 300°C or to impact of a steel hammer, but a mild explosion occurred when a grease-contaminated sample was disturbed with a metal spatula.

*See other* METAL PERCHLORATES

## ERBIUM PERCHLORATE $Er(ClO_4)_3$ $Cl_3ErO_{12}$

Acetonitrile

Wolsey, W. C., *J. Chem. Educ.*, 1973, **50**(6), A336–337

The shock-sensitive glassy residue (containing traces of acetonitrile), left after heating tetraacetonitrileerbium perchlorate to above 150°C on a vacuum line, exploded violently when scraped with a spatula.

*See other* METAL PERCHLORATES

## IRON(III) CHLORIDE $FeCl_3$ $Cl_3Fe$

Chlorine,
Monomers

*See* CHLORINE, $Cl_2$: Iron(III) chloride, Monomers

Ethylene oxide

*See* ETHYLENE OXIDE, $C_2H_6O$: Contaminants

Metals

*See* POTASSIUM, K: Metal halides
SODIUM, Na: Metal halides

## IRON(III) PERCHLORATE $Fe(ClO_4)_3$ $Cl_3FeO_{12}$

Acetonitrile

Bancroft, G. M. *et al., Can. J. Chem.*, 1974, **52**, 783
The violent reaction which occurred on dissolution of the anhydrous salt in acetonitrile did not occur with the hydrated salt.

Dimethyl sulphoxide

*See* DIMETHYL SULPHOXIDE, $C_2H_6OS$: Metal oxosalts

## GALLIUM TRIPERCHLORATE $Ga(ClO_4)_3$ $Cl_3GaO_{12}$

Foster, L. S., *J. Amer. Chem. Soc.*, 1933, **61**, 3123; *Inorg. Synth.*, 1946, **2**, 28
During preparation by dissolving the metal in 72% perchloric acid, the hexahydrate separates as a crystalline solid. After filtration, the damp crystals must not contact any organic material (filter paper, horn spatula) since the adherent perchloric acid liquor is above 72% concentration owing to the hexahydrate formation.

*See* PERCHLORIC ACID, $ClHO_4$: Dehydrating agents
*See other* METAL OXOHALOGENATES

## † TRICHLOROSILANE $Cl_3SiH$ $Cl_3HSi$

Mueller, R. *et al., J. Prakt. Chem.*, 1966, **31**, 1–6
The pure material is not impact-ignitible in absence of electrostatic charges, but technical material (possibly containing dichlorosilane) is.

*See related* NON-METAL HALIDES
*See other* HALOSILANES

## PENTAAMMINECHLORORUTHENIUM(III) CHLORIDE $Cl_3H_{15}N_5Ru$

$[Ru(NH_3)_5Cl]Cl_2$

Sodium azide

Allen, A. D. *et al., Inorg. Synth.*, 1970, **12**, 5–6
During treatment with sodium azide of an intermediate (believed to be pentaammine aquoruthenium(III)) derived from the title compound to produce pentaamminedinitrogenruthenium(II) solutions, a dangerously

explosive red solid may be produced. The solid, pentaammineazido-ruthenium(III), will, however, decompose on standing to the desired dinitrogen species.

*See also* HYDRAZINE, $H_4N_2$ : Ruthenium(III) chloride

**HEXAAMMINETITANIUM(III) CHLORIDE** **$Cl_3H_{18}N_6Ti$**

**$[Ti(NH_3)_6]Cl_3$**

Schumb, W. C. *et al., J. Amer. Chem. Soc.,* 1938, **55**, 599
Reacts violently with water.
*See related* METAL HALIDES

**IODINE TRICHLORIDE** **$ICl_3$** **$Cl_3I$**

Phosphorus
Pascal, 1960, Vol. 16.1, 578
Phosphorus ignites on contact.
*See other* INTERHALOGENS

**IODINE(III) PERCHLORATE** **$I(ClO_4)_3$** **$Cl_3IO_{12}$**

Christe, K. O. *et al., Inorg. Chem.,* 1972, **11**, 1683
The solid was stable at $-45^\circ$C but exploded on laser irradiation at low temperature.
*See* INORGANIC PERCHLORATES
*See other* IODINE COMPOUNDS
NON-METAL PERCHLORATES

**MANGANESE TRICHLORIDE OXIDE** **$MnCl_3O$** **$Cl_3MnO$**

Briggs, T. S., *J. Inorg. Nucl. Chem.*, 1968, **30**, 2867–2868
Explosively unstable if isolated as a liquid at ambient temperature, it may be handled safely in carbon tetrachloride solution.
*See also* MANGANESE CHLORIDE TRIOXIDE, $ClMnO_3$
MANGANESE DICHLORIDE DIOXIDE, $Cl_2MnO_2$
*See related* METAL HALIDES

**NITROGEN TRICHLORIDE** **$NCl_3$** **$Cl_3N$**

1. Mellor, 1940, Vol. 8, 598–604; 1967, Vol. 8, Suppl. 2.2, 411
2. Sidgwick, 1950, 705
3. Brauer, 1963, Vol. 1, 479
4. *ABCM Quart. Safety Summ.*, 1946, **17**, 17
5. Schlessinger, G. C., *Chem. Eng. News,* 1966, **44**(33), 46
6. Kovacic, P. *et al., Org. Synth.,* 1968, **48**, 4

Contact above 0°C of excess chlorine or a chlorinating agent with aqueous ammonia, ammonium salts or a compound containing a hydrolysable amino group, or electrolysis of ammonium chloride solution produces the endothermic explosive nitrogen trichloride as a water-insoluble yellow oil [1,2,4]. It is usually prepared [3] in solution in a solvent, and such solutions in chloroform are reported as stable up to 18% concentration. However, an 18% solution in dibutyl ether exploded on cooling in a refrigerator, and a 12% solution prepared without cooling had vigorously decomposed [5]. In absence of other materials, explosion of the trichloride may be initiated in a wide variety of ways. The solid frozen *in vacuo* in liquid nitrogen explodes on thawing, and the liquid on heating to 60 or 95°C. Exposure to impact, light or ultrasonic irradiation will cause (or sensitise) detonation [1]. The preparation and synthetic use of solutions of nitrogen trichloride in dichloromethane is described in detail [6].

*See* CHLORINE, $Cl_2$: Nitrogen compounds
PHOSPHORUS PENTACHLORIDE, $Cl_5$: Urea

Chlorine,
Hydrogen

Ashmore, P. G., *Nature*, 1953, **172**, 449–450

Equimolar mixtures of chlorine and hydrogen containing 0.1–2.0% of the trichloride will explode in absence of light if the pressure is below a limiting value dependent upon temperature (30 mbar at 20°C, 132 mbar at 57°C)

Initiators

Mellor, 1940, Vol. 8, 601–604; 1967, Vol. 8, Suppl. 2.2, 412
Pascal, 1956, Vol. 10, 264

A wide variety of solids, liquids and gases will initiate the violent and often explosive decomposition of nitrogen trichloride. These include concentrated ammonia, arsenic, dinitrogen tetraoxide (above –40°C, with more than 25% solutions of trichloride in chloroform), hydrogen sulphide, hydrogen trisulphide, nitrogen oxide, organic matter (including grease from the fingers), ozone, phosphine, phosphorus (solid, or as carbon disulphide solution), potassium cyanide (solid or aqueous solution), potassium hydroxide solution and selenium.
All four hydrohalide acids will also initiate explosion of the trichloride.

*See other* *N*-HALOGEN COMPOUNDS

## NITROSYLRUTHENIUM TRICHLORIDE **$Cl_3NORu$**

**$NORuCl_3$**

Sidgwick, 1950, 1486

It decomposes violently at 440°C.

*See other* NITROSO COMPOUNDS

## 1,3,5-TRICHLOROTRITHIA-1,3,5-TRIAZINE (THIAZYL CHLORIDE) $Cl_3N_3S_3$

SN(Cl)SN(Cl)SNCl

Bailar, 1973, Vol. 2, 908
It explodes violently on sudden heating.

Ammonia,
Silver nitrate
Goehring, 1957, 67
Thiazyl chloride, treated with aqueous ammonia and then silver nitrate, gives a compound $AgN_5S_3$ (unknown structure) which is shock-sensitive and explodes violently.
*See other* *N*-HALOGEN COMPOUNDS

## TIN AZIDE TRICHLORIDE $Sn(N_3)Cl_3$ $Cl_3N_3Sn$

Explosive solid.
*See* METAL AZIDE HALIDES

## TITANIUM AZIDE TRICHLORIDE $Ti(N_3)Cl_3$ $Cl_3N_3Ti$

Explosive solid.
*See* METAL AZIDE HALIDES

## PHOSPHORYL CHLORIDE $POCl_3$ $Cl_3OP$

Carbon disulphide
*ABCM Quart. Safety Summ.*, 1964, **35**, 24
Disposal of a benzene solution of phosphoryl chloride into a waste drum containing carbon disulphide (and other solvents) caused an instantaneous reaction, with evolution of (probably) hydrogen chloride.

*N,N*-Dimethylformamide,
2,5-Dimethylpyrrole
*MCA Case History No. 1460*
Poor stirring during formylation of 2,5-dimethylpyrrole with the preformed complex of dimethylformamide with phosphoryl chloride caused eruption of flask contents. Reaction of the complex with local excess of the pyrrole may have been involved.

2,6-Dimethylpyridine *N*-oxide
1. Kato, T. *et al., J. Pharm. Soc. Japan,* 1951, **71**, 217
2. Anon., *Chem. Eng. News,* 1965, **43**(47), 40

3. Evans, R. F. *et al., J. Org. Chem.*, 1962, **27**, 1333
Interaction of the reagents in absence of diluent, according to a published procedure [1], caused an explosion [2]. The use of a chlorinated solvent as diluent prevented explosion, confirming an earlier report [3].

Dimethyl sulphoxide

*See* DIMETHYL SULPHOXIDE, $C_2H_6OS$: Acyl halides

Ferrocene-1,1′-dicarboxylic acid

Marvel, C. S., *Chem. Eng. News*, 1962, **40**(3), 55

An explosion occurred immediately after pouring and capping of the chloride recovered from preparation of ferrocene-1,1′-dicarbonyl chloride. The storage bottle contained phosphoryl chloride recovered from similar preparations and which had been stored for about 3 months. No explanation was offered.

Sodium

*See* SODIUM, Na: Non-metal halides (reference 7)

Water

*MCA SD-26*, 1968

*MCA Case History No. 1274*

Unpublished observations

The hazards arising from interaction of phosphoryl chloride and water are due to there often being a considerable delay in onset of the exothermic hydrolysis reaction, which may proceed with enough vigour to generate steam and liberate hydrogen chloride gas. Conditions tending to favour delayed or violent reaction include limited quantities of water and/or ice for hydrolysis, lack of stirring, cold or frozen phosphoryl chloride, and reaction in closed or virtually closed containers. A layer of the dense and cold liquid may survive for several minutes under water before violent, almost instantaneous hydrolysis occurs, particularly when disturbed. The Case History describes a violent explosion which occurred when water was added to a drum containing some phosphoryl chloride which had been stored below its freezing point, 2°C.

Zinc

Mellor, 1940, Vol. 8, 1025

Zinc dust ignites in contact with a little phosphoryl chloride, and subsequent addition of water liberates phosphine which ignites.

*See other* NON-METAL HALIDES

**ANTIMONY TRICHLORIDE OXIDE** **$SbOCl_3$** **$Cl_3OSb$**

Bromine trifluoride

*See* BROMINE TRIFLUORIDE, $BrF_3$: Antimony trichloride oxide

## VANADIUM TRICHLORIDE OXIDE $VOCl_3$ $Cl_3OV$

Bailar, 1973, Vol. 3, 508

The chloride (and corresponding bromide) is violently hygroscopic.

Rubidium

*See* RUBIDIUM, Rb: Vanadium trichloride oxide

## VANADYL TRIPERCHLORATE $VO(ClO_4)_3$ $Cl_3O_{13}V$

Organic solvents

Schmeisser, M., *Angew. Chem.*, 1955, **67**, 499

A powerful oxidant, igniting organic solvents on contact, which explodes violently above 80°C.

*See other* METAL OXOHALOGENATES
OXIDANTS

## PHOSPHORUS TRICHLORIDE $PCl_3$ $Cl_3P$

Acetic acid

1. Coghill, R. D., *J. Amer. Chem. Soc.*, 1938, **60**, 488
2. Peacocke, T. A., *School Sci. Rev.*, 1962, **44**(152), 217

Use of a free flame instead of a heating bath to distil acetyl chloride produced from the two reactants caused the residual phosphonic acid to decompose violently to give spontaneously flammable phosphine [1]. Two later explosions after reflux but before distillation from a water-bath may have been due to ingress of air into the cooling flask and combustion of traces of phosphine [2].

Carboxylic acids

1. Scrimgeour, C. M., *Chem. Brit.*, 1975, **11**, 267
2. Taylor, D. A. H., ibid., 1976, **12**, 105
3. Bretherick, L., ibid., 26

During preparation of furoyl chloride, the excess phosphorus trichloride was distilled off at atmospheric pressure, then vacuum was applied prior to intended distillation at 100°C/120 mbar, and an explosion occurred shortly after [1]. A similar incident occurred with propionic acid and phosphorus trichloride [2]. These incidents were attributed to thermal decomposition of the by-product phosphonic ('phosphorous') acid to give a phosphine–diphosphane mixture and leakage of air into the evacuated system to produce a spontaneously explosive mixture. Sulphinyl chloride is a safer chlorinating agent for carboxylic acids [3].

*See* Acetic acid, above

Dimethyl sulphoxide

*See* DIMETHYL SULPHOXIDE, $C_2H_6OS$: Acyl halides

Hexafluoroisopropylideneaminolithium

*See* HEXAFLUOROISOPROPYLIDENEAMINOLITHIUM, $C_3F_6LiN$: Non-metal halides

Hydroxylamine

*See* HYDROXYLAMINE, $H_3NO$: Phosphorus chlorides

Metals

Mellor, 1940, Vol. 8, 1006; 1941, Vol. 2, 470

Potassium ignites in phosphorus trichloride, while molten sodium explodes on contact.

*See* ALUMINIUM, Al: Non-metal halides

Oxidants

*See* CHROMYL CHLORIDE, $Cl_2CrO_2$: Phosphorus, etc.
CHROMIUM PENTAFLUORIDE, $CrF_5$: Phosphorus trichloride
FLUORINE, $F_2$: Phosphorus halides
NITRIC ACID, $HNO_3$: Phosphorus halides
SODIUM PEROXIDE, $Na_2O_2$: Non-metal halides
LEAD(II) OXIDE, $O_2Pb$: Non-metal halides
SELENIUM DIOXIDE, $O_2Se$: Phosphorus trichloride

Oxygen

Vorob'ev, N. I. *et al.*, *Izv. Akad. Nauk SSSR, Neorg. Mater.*, 1974, **10**(11), 2039–2042

Interaction to form phosphoryl chloride at 100–700°C was investigated, and the limiting concentrations to prevent explosions were determined.

Tetravinyllead

*See* TETRAVINYLLEAD, $C_8H_{12}Pb$: Phosphorus trichloride

Water

1. *MCA SD-27*, 1972
2. Coghill, R. D., *J. Amer. Chem. Soc.*, 1938, **60**, 488
3. *MCA Case History No. 520*

Reaction with water is exothermic and immediately violent (unlike phosphoryl chloride), and is accompanied by liberation of some diphosphane which ignites [1]. Evaporation of the trichloride from an open beaker on a steam-bath led to ignition, which did not occur if a hot-plate was used as a dry heat source [2]. Interaction of the trichloride and water in a virtually sealed container caused the latter to burst under the pressure of hydrogen chloride generated [3].

*See other* NON-METAL HALIDES

**THIOPHOSPHORYL CHLORIDE** **$PSCl_3$** **$Cl_3PS$**

Moeller, T. *et al.*, *Inorg. Synth.*, 1954, **4**, 72

During the preparation from phosphorus trichloride and sulphur, the quantity and quality of the aluminium chloride catalyst is critical to prevent the exothermic reaction going out of control.

Pentaerythritol

*MCA Case History No. 1315*

On two occasions violent explosions occurred after heating of equimolar proportions of the reagents for 4 h at 160°C according to a literature method had been discontinued.

*See other* NON-METAL HALIDES

**RHODIUM(III) CHLORIDE** **$RhCl_3$** **$Cl_3Rh$**

Pentacarbonyliron,
Zinc

*See* PENTACARBONYLIRON, $C_5FeO_5$: Transition metal halides

**RUTHENIUM(III) CHLORIDE** **$RuCl_3$** **$Cl_3Ru$**

Hydrazine

*See* HYDRAZINE, $H_4N_2$: Ruthenium(III) chloride

Pentacarbonyliron,
Zinc

*See* PENTACARBONYLIRON, $C_5FeO_5$: Transition metal halides

**ANTIMONY TRICHLORIDE** **$SbCl_3$** **$Cl_3Sb$**

*MCA SD-66*, 1957

Aluminium

*See* ALUMINIUM, Al: Antimony trichloride

**TITANIUM TRICHLORIDE** **$TiCl_3$** **$Cl_3Ti$**

Air,
Water

1. Lerner, R. W., *Ind. Eng. Chem.*, 1961, **53**(12), 56A
2. Ingraham, T. K. *et al.*, *Inorg. Synth.*, 1960, **6**, 56

It reacts vigorously with air and/or water (vapour or liquid) and adequate

handling precautions are necessary [1]. The finely divided powder is pyrophoric in air [2].
*See other* METAL HALIDES

**VANADIUM TRICHLORIDE** **$VCl_3$** **$Cl_3V$**

Methylmagnesium iodide
Cotton, F. A., *Chem. Rev.*, 1955, **55**, 560
Reaction of vanadium trichloride and other halides with Grignard reagents is almost explosively violent under a variety of conditions.
*See other* METAL HALIDES

**ZIRCONIUM TRICHLORIDE** **$ZrCl_3$** **$Cl_3Zr$**

Water
Pascal, 1963, Vol. 9, 540
Interaction is very violent, hydrogen being evolved.
*See other* METAL HALIDES

**CAESIUM TETRAPERCHLORATOIODATE** **$Cl_4CsIO_{16}$**
**$Cs[I(ClO_4)_4]$**

Christe, K. O. *et al., Inorg. Chem.*, 1972, **11**, 683
Though stable at ambient temperature, samples exploded under laser irradiation at low temperatures.
*See other* NON-METAL PERCHLORATES

**GERMANIUM TETRACHLORIDE** **$GeCl_4$** **$Cl_4Ge$**

Water
Mellor, 1941, Vol. 7, 270
Interaction is very exothermic, accompanied by crackling if the chloride is dropped into water.
*See other* METAL HALIDES

**MOLYBDENUM DIAZIDE TETRACHLORIDE** **$Cl_4MoN_6$**
**$Mo(N_3)_2Cl_4$**

Highly explosive.
*See* METAL AZIDE HALIDES

### CHLOROIMINOVANADIUM TRICHLORIDE $Cl_4NV$

$V(NCl)Cl_3$

1. Strähle, J. *et al.* *Angew. Chem.*, 1966, **78**, 450
2. Strähle, J. *et al.*, *Z. Anorg. Allg. Chem.*, 1965, **338**, 287

A new method of preparation from vanadium nitride and chlorine [1] is free of the explosion hazards of chlorine azide and vanadium azide tetrachloride present in an earlier method [2].

### VANADIUM AZIDE TETRACHLORIDE $V(N_3)Cl_4$ $Cl_4N_3V$

Explosive solid.

*See* METAL AZIDE HALIDES

### RHENIUM TETRACHLORIDE OXIDE $ReOCl_4$ $Cl_4ORe$

Ammonia

Sidgwick, 1950, 1302

Interaction with gaseous or liquid ammonia is violent.

*See related* METAL HALIDES

### SILICON TETRAPERCHLORATE $Si(ClO_4)_4$ $Cl_4O_{16}Si$

Sorbe, 1968, 127

A highly explosive liquid.

*See other* NON-METAL PERCHLORATES

### TITANIUM TETRAPERCHLORATE $Ti(ClO_4)_4$ $Cl_4O_{16}Ti$

Diethyl ether,
or Formamide,
or *N,N*-Dimethylformamide

Laran, R. J., US Pat. 3 157 464, 1964

Insensitive to heat or shock, this powerful oxidant explodes on contact with diethyl ether, and ignites with formamide or its dimethyl derivative.

*See other* METAL OXOHALOGENATES

### TETRACHLORODIPHOSPHANE $Cl_2PPCl_2$ $Cl_4P_2$

Besson, A. *et al.*, *Compt. Rend.*, 1910, **150**, 102

Oxidises rapidly in air, sometimes igniting.

*See other* NON-METAL HALIDES

**LEAD TETRACHLORIDE** **$PbCl_4$** **$Cl_4Pb$**

Bailar, 1973, Vol. 2, 136, 1066

It may explosively decompose to lead(II) chloride and chlorine above 100°C.

Potassium

*See* Potassium, K: Metal halides

Sulphuric acid

Friedrich, H., *Ber.*, 1893, **26**, 1434

It explodes on warming with diluted sulphuric acid or on attempted distillation at 105°C from the concentrated acid in a stream of chlorine.

*See other* METAL HALIDES

**TETRACHLOROSILANE** **$SiCl_4$** **$Cl_4Si$**

Dimethyl sulphoxide

*See* DIMETHYL SULPHOXIDE, $C_2H_6OS$: Acyl halides, etc.

Sodium

*See* SODIUM, Na: Non-metal halides (reference 8)

**TIN TETRACHLORIDE** **$SnCl_4$** **$Cl_4Sn$**

Alkyl nitrates

*See* ALKYL NITRATES: Lewis acids

Ethylene oxide

*See* ETHYLENE OXIDE, $C_2H_4O$: Contaminants

Turpentine

Mellor, 1941, Vol. 7, 446

Interaction is strongly exothermic and may lead to ignition.

*See other* METAL HALIDES

**TELLURIUM TETRACHLORIDE** **$TeCl_4$** **$Cl_4Te$**

Ammonia

Mellor, 1943, Vol. 11, 58

Interaction with liquid ammonia at −15°C forms tellurium nitride (?), which explodes at 200°C.

*See other* NON-METAL HALIDES

## TITANIUM TETRACHLORIDE $TiCl_4$ $Cl_4Ti$

Hydrogen fluoride

*See* HYDROGEN FLUORIDE, FH: Metal chlorides

Urea

Ionova, E. A. *et al., Chem. Abs.*, 1966, **64**, 9219b

The liquid hexaurea complex formed during 6 weeks at 80°C decomposed violently at above 90°C. *N*-halogen compounds may have been formed.

*See other* METAL HALIDES

## ZIRCONIUM TETRACHLORIDE $ZrCl_4$ $Cl_4Zr$

Ethanol,
or Water

Rosenheim, A. *et al., Ber.*, 1907, **40**, 811

Interaction is violent.

Hydrogen fluoride

*See* HYDROGEN FLUORIDE, FH: Metal chlorides

Lithium,
Nitrogen

*See* LITHIUM, Li: Metal chlorides, etc.

*See other* METAL HALIDES

## DIAMMINEDICHLOROAMINOTRICHLOROPLATINUM(IV) $Cl_5H_7N_3Pt$

$[Pt(NH_3)_2(Cl_2NH)Cl_3]$

Kukushkin, Yu. N., *Chem. Abs.*, 1958, **52**, 13509i

It tended to decompose violently.

*See other* *N*-HALOGEN COMPOUNDS
PLATINUM COMPOUNDS

## MOLYBDENUM AZIDE PENTACHLORIDE $Cl_5MoN_3$

$Mo(N_3)Cl_5$

Extremely explosive.

*See* METAL AZIDE HALIDES

## URANIUM AZIDE PENTACHLORIDE $Cl_5N_3U$

$U(N_3)Cl_5$

Explosive solid.

*See* METAL AZIDE HALIDES

## TUNGSTEN AZIDE PENTACHLORIDE $W(N_3)Cl_5$ $Cl_5N_3W$

Extremely explosive.

*See* METAL AZIDE HALIDES

## PHOSPHORUS PENTACHLORIDE $PCl_5$ $Cl_5P$

Aluminium

*See* ALUMINIUM, Al: Phosphorus pentachloride

Carbamates

*See* *N*-CARBOMETHOXYIMINOPHOSPHORYL CHLORIDE, $C_2H_3Cl_3NO_2P$

Chlorine dioxide,
Chlorine

Mellor, 1941, Vol. 2, 281; 1940, Vol. 8, 1013

Contact between phosphorus pentachloride and a mixture of chlorine and chlorine dioxide (previously considered to be dichlorine trioxide) usually causes explosion, possibly due to formation of the more sensitive chlorine monoxide.

Diphosphorus trioxide

Mellor, 1940, Vol. 8, 898

Interaction is rather violent at ambient temperature.

Fluorine

*See* FLUORINE, $F_2$: Phosphorus halides

Hydroxylamine

*See* HYDROXYLAMINE, $H_3NO$: Phosphorus chlorides

Magnesium oxide

Mellor, 1940, Vol. 8, 1016

A heated mixture incandesces brilliantly.

3′-Methyl-2-nitrobenzanilide

Partridge, M. W. private comm., 1968

The residue from interaction of the chloride and anilide in benzene and removal of solvent and phosphoryl chloride *in vacuo* exploded violently on admission of air.

Nitrobenzene

Unpublished observations

A solution of phosphorus pentachloride in nitrobenzene is stable at

110°C but begins to decompose with accelerating violence above 120°C, with evolution of nitrous fumes.

Sodium

*See* SODIUM, Na: Non-metal halides

Urea

Anon., *Angew. Chem. (Nachr.)*, 1960, 8, 33

A dry mixture exploded after heating, probably owing to formation of nitrogen trichloride. Other chlorinating agents will react similarly with nitrogenous materials.

*See* CHLORINE, $Cl_2$: Nitrogen compounds
NITROGEN TRICHLORIDE, $Cl_3N$

Water

Mellor, 1940, Vol. 8, 1012

Interaction with water in limited quantities is violent, and the hydrolysis products may themselves react violently with more water.

*See* PHOSPHORYL CHLORIDE, $Cl_3OP$: Water
*See other* NON-METAL HALIDES

## ANTIMONY PENTACHLORIDE $SbCl_5$ $Cl_5Sb$

Oxygen difluoride

*See* OXYGEN DIFLUORIDE, $F_2O$: Halogens, etc.

Phosphonium iodide

*See* PHOSPHONIUM IODIDE, $H_4IP$: Oxidants

## AMMONIUM HEXACHLOROPLATINATE(2–) $Cl_6H_8N_2Pt$

$[NH_4]_2[PtCl_6]$

Potassium hydroxide,
Combustible material

Mellor, 1942, Vol. 16, 336

Boiling ammonium chloroplatinate with alkali gives a product which, after drying, will explode violently on heating alone to 205°C or with combustible materials.

*See other* PLATINUM COMPOUNDS

## POTASSIUM HEXACHLOROPLATINATE(2–) $Cl_6K_2Pt$

$K_2[PtCl_6]$

Bromine trifluoride

*See* BROMINE TRIFLUORIDE, $BrF_3$: Potassium hexachloroplatinate(2–)

**BIS(TRICHLOROPHOSPHORANYLIDENE)SULPHAMIDE** $Cl_6N_2O_2P_2S$

$SO_2(N{=}PCl_3)_2$

Cellulose,
or Water

Vandi, A. *et al., Inorg. Synth.*, 1966, 8, 119
It is extremely reactive with water or alcohol and causes filter paper to ignite.

**BIS-TRIPERCHLORATOSILICON OXIDE** $Cl_6O_{25}Si_2$

$[(ClO_4)_3Si]_2O$

Schmeisser, M., *Angew. Chem.*, 1955, **67**, 499
This solid decomposition product of silicon tetraperchlorate was so explosive, even in small amounts, that work was discontinued.
*See other* NON-METAL PERCHLORATES

**HEXACHLORODISILANE** $Cl_3SiSiCl_3$ $Cl_6Si_2$

*See* HALOSILANES

Chlorine

*See* CHLORINE, $Cl_2$: Hexachlorodisilane

**URANIUM HEXACHLORIDE** $UCl_6$ $Cl_6U$

Water

Bailar, 1973, Vol. 5, 189
Interaction is violent.
*See other* METAL HALIDES

**OCTACHLOROTRISILANE** $Cl_3SiSiCl_2SiCl_3$ $Cl_8Si_3$

*See* HALOSILANES

**TETRAZIRCONIUM TETRAOXIDE HYDROGEN-NONAPERCHLORATE** $Cl_9HO_{40}Zr_4$

Mellor, 1946, Vol. 2, 403
The salt, 'zirconyl perchlorate', formulated as $4ZrO(ClO_4)_2 \cdot HClO_4$, explodes if heated rapidly.
*See related* METAL OXOHALOGENATES

**DECACHLOROTETRASILANE** $Si_4Cl_{10}$ $Cl_{10}Si_4$

*See* HALOSILANES

**DODECACHLOROPENTASILANE** $Si_5Cl_{12}$ **$Cl_{12}Si_5$**

*See* HALOSILANES

**COBALT** **Co**

1. Bailar, 1973, Vol. 3, 1056
2. Chadwell, A. J. *et al., J. Phys. Chem.*, 1956, **60**, 1340

Finely divided cobalt is pyrophoric in air [1]. Raney cobalt catalyst appears to be less hazardous than Raney nickel [2].

Acetylene

*See* ACETYLENE, $C_2H_2$: Cobalt

Hydrazinium nitrate

*See* HYDRAZINIUM NITRATE, $H_5N_3O_3$

Oxidants

*See* BROMINE PENTAFLUORIDE, $BrF_5$: Acids, etc.
NITRYL FLUORIDE, $FNO_2$: Metals
AMMONIUM NITRATE, $H_4N_2O_3$: Metals

**COBALT TRIFLUORIDE** **$CoF_3$**

Hydrocarbons,
or Water

Priest, H. F., *Inorg. Synth.*, 1950, **3**, 176

It reacts violently with hydrocarbons or water, and finds use as a fluorinating agent.

Silicon

Mellor, 1956, Vol. 2, Suppl. 1, 64

A gently warmed mixture reacts exothermically, attaining red heat.

*See other* METAL HALIDES

**COBALT(III) AMIDE** **$Co(NH_2)_3$** **$CoH_6N_3$**

Schmitz-Dumont, O. *et al., Z. Anorg. Chem.*, 1941, **284**, 175

Powdered material is sometimes pyrophoric.

*See other N-*METAL DERIVATIVES

**DIAMMINENITRATOCOBALT(II) NITRATE** **$CoH_6N_4O_6$**

**$[Co(NH_3)_2NO_3]NO_3$**

McPherson, G. L. *et al., Inorg. Chem.*, 1971, **10**, 1574

A sample of the molten salt exploded at 200°C.

*See other* AMMINEMETAL OXOSALTS
REDOX COMPOUNDS

**COBALT TRIS(DIHYDROGENPHOSPHIDE)** **$CoH_6P_3$**

**$Co(PH_2)_3$**

Mellor, 1971, Vol. 8, Suppl. 3, 331

It ignites in air.

*See other* PYROPHORIC MATERIALS

**TRIAMMINETRINITROCOBALT(III)** **$CoH_9N_6O_6$**

**$[Co(NH_3)_3(NO_2)_3]$**

Explodes at 305°C, medium impact-sensitivity.

*See* AMMINEMETAL OXOSALTS

**TRIHYDRAZINECOBALT(II) NITRATE** **$CoH_{12}N_8O_6$**

**$[Co(N_2H_4)_3][NO_3]_2$**

Franzen, H. *et al., Z. Anorg. Chem,* 1908, **60**, 247, 274

It is explosive.

*See* AMMINEMETAL OXOSALTS

**AMMONIUM HEXANITROCOBALTATE(3−)** **$CoH_{12}N_9O_{12}$**

**$[NH_4]_3[Co(NO_2)_6]$**

Explodes at 230°C, medium impact-sensitivity.

*See* AMMINEMETAL OXOSALTS

**PENTAAMMINENITRATOCOBALT(III) NITRITE** **$CoH_{15}N_8O_7$**

**$[Co(NH_3)_5NO_3][NO_2]_2$**

Explodes at 310°C, medium impact-sensitivity.

*See* AMMINEMETAL OXOSALTS

**HEXAAMMINECOBALT(III) IODATE** **$CoH_{18}I_3N_6O_9$**

**$[Co(NH_3)_6][IO_3]_3$**

Explodes at 360°C, low impact-sensitivity.

*See* AMMINEMETAL OXOSALTS

**HEXAAMMINECOBALT(III) PERMANGANATE** **$CoH_{18}Mn_3N_6O_{12}$**

**$[Co(NH_3)_6][MnO_4]_3$**

Mellor, 1942, Vol. 12, 336

Joyner, T. B., *Can. J. Chem.*, 1969, **47**, 2730

It explodes on heating, and is of treacherously high impact-sensitivity.

*See other* AMMINEMETAL OXOSALTS

**HEXAAMMINECOBALT(III) NITRATE** $CoH_{18}N_9O_9$

$[Co(NH_3)_6][NO_3]_3$

Explodes at 295°C, medium impact-sensitivity.

*See* AMMINEMETAL OXOSALTS

**HEXAHYDROXYLAMINECOBALT(III) NITRATE** $CoH_{18}N_9O_{15}$

$[Co(NH_2OH)_6][NO_3]_3$

Werner, A. *et al., Ber.,* 1905, **38**, 897

It usually exploded during preparation or handling.

*See other* AMMINEMETAL OXOSALTS

**POTASSIUM TRIAZIDOCOBALTATE(1−)** $CoKN_9$

$K[Co(N_3)_3]$

Fritzer, H. P. *et al., Angew. Chem. (Intern. Ed.),* 1971, **10**, 829

Fritzer, H. P. *et al., Inorg. Nucl. Chem. Lett.*, 1974, **10**, 247–252

The complex azide is highly explosive and must be handled with extreme care. The analogous potassium and caesium derivatives of zinc and nickel azides deflagrate strongly in a flame and some are shock-sensitive. The potassium salt alone out of eight metal azides exploded during X-irradiation in an ESCA study.

*See other* METAL AZIDES

**POTASSIUM HEXANITROCOBALTATE(3−)** $CoK_3N_6O_{12}$

$K_3[Co(NO_2)_6]$

1. Broughton, D. B. *et al., Anal. Chem.*, 1947, **19**, 72
2. Horowitz, O., *Anal. Chem.*, 1948, **20**, 89
3. Tomlinson, W. R. *et al., J. Amer. Chem. Soc.,* 1949, **71**, 375

Evaporation by heating a filtrate from precipitation of potassium cobaltinitrite caused it to turn purple and explode violently [1]. This was attributed to interaction of nitrite, nitrate, acetic acid and residual cobalt with formation of fulminic and methylnitrolic acids or their cobalt salts, all of which are explosive [2]. Mixtures containing nitrates, nitrites and organic material are potentially dangerous, especially in presence of acidic materials and heavy metals. A later publication confirms the suggestion of formation of nitro- or nitrito-cobaltate(III) [3].

**COBALT NITRIDE** CoN

Schmitz-Dumont, O., *Angew. Chem.*, 1955, **67**, 231

Pyrophoric powder.
*See other N*-METAL DERIVATIVES

**COBALT(II) NITRATE** $Co(NO_3)_2$ $CoN_2O_6$

Ammonium hexacyanoferrate(4–)
Wolski, W. *et al., Acta Chim. (Budapest)*, 1972, 72, 25–32
Interaction is explosive at 220°C.
*See also* COPPER(II) NITRATE, $CuN_2O_6$ : Ammonium hexacyanoferrate(4–)
: Potassium hexacyanoferrate(4–)

Carbon
Crowther, J. R., private comm., 1970
Charcoal impregnated with the nitrate exploded lightly during sieving. Possibly a dust or black powder explosion.
*See other* METAL OXONON-METALLATES

**COBALT(II) OXIDE** **CoO**

Hydrogen peroxide
*See* HYDROGEN PEROXIDE, $H_2O_2$: Metals, etc.

**COBALT(II) SULPHIDE** **CoS**
Brauer, 1965, Vol. 2, 1523
Material dried at 300°C is pyrophoric.
*See other* METAL SULPHIDES

**HEXAAMMINECOBALT(3+) HEXANITROCOBALTATE(3–)** $Co_2H_{18}N_{12}O_{12}$
$[Co(NH_3)_6][Co(NO_2)_6]$
Unstable compound, low impact-sensitivity.
*See* AMMINEMETAL OXOSALTS

***trans*-TETRAAMMINEDIAZIDOCOBALT(III) *trans*-DIAMMINE-TETRAAZIDOCOBALTATE(1–)** $Co_2H_{18}N_{24}$
$[Co(NH_3)_4(N_3)_2][Co(NH_3)_2(N_3)_4]$
Druding, L. F. *et al., Inorg. Chem.*, 1975, **14**, 1365
During its preparation, solutions of the salt, which is a dangerous detonator, must not be evaporated to dryness. Surprisingly, mixtures with potassium bromide could be compressed (for IR examination) to 815 bar without decomposition.

*See related* METAL AZIDES
*See other* HIGH-NITROGEN COMPOUNDS

**DICOBALT TRIOXIDE** $Co_2O_3$

Hydrogen peroxide
*See* HYDROGEN PEROXIDE, $H_2O_2$: Metals, etc.

Nitroalkanes
*See* NITROALKANES: Metal oxides

**CHROMIUM** Cr

Sidgwick, 1950, 1013
Evaporation of mercury from chromium amalgam leaves pyrophoric chromium.

Carbon dioxide
*See* CARBON DIOXIDE, $CO_2$: Metals

Oxidants
*See* BROMINE PENTAFLUORIDE, $BrF_5$: Acids, etc.
AMMONIUM NITRATE, $H_4N_2O_3$: Metals
NITROGEN OXIDE, NO: Metals
SULPHUR DIOXIDE, $O_2S$: Metals
*See other* METALS

**CHROMIUM–COPPER CATALYST** Cr–Cu

Alcohols
Budniak, H., *Chem. Abs.*, 1975, **83**, 134493k
Chromium–copper hydrogenation catalysts used to hydrogenate fatty acids to alcohols are pyrophoric. Separation of the catalyst from the alcohols at 130°C in a centrifuge led to a rapid exotherm and auto-ignition at 263°C.
*See other* PYROPHORIC CATALYSTS

**COPPER CHROMATE** $CuCrO_4$ $CrCuO_4$

Hydrogen sulphide
Pascal, 1960, Vol. 13.1, 1025
The gas may ignite on contact with the chromate.
*See other* METAL OXOMETALLATES

## AMMONIUM FLUOROCHROMATE $NH_4FCrO_3$ $CrFH_4NO_3$

Ribas Bernat, J. G., *Ion (Madrid)*, 1975, **35**(409), 573–576
It decomposes violently at 220°C.

*See related* OXOSALTS OF NITROGENOUS BASES

## CHROMYL FLUOROSULPHATE $CrO_2(FSO_3)_2$ $CrF_2O_8S_2$

Water
Rochat, W. V. *et al., Inorg. Chem.*, 1969, **8**, 158
Interaction is violent.

*See related* METAL OXONON-METALLATES

## CHROMIUM PENTAFLUORIDE $CrF_5$

Bailar, 1973, Vol. 2, 1086
It reacts violently in halogen-exchange and oxidation–reduction reactions.

Phosphorus trichloride
O'Donnell, T. A. *et al., Inorg. Chem.*, 1966, **5**, 1435, 1437
Interaction is violent after slight warming.

*See other* METAL HALIDES
OXIDANTS

## POTASSIUM HYDROXOOXODIPEROXOCHROMATE(1−) $CrHKO_6$

$K[CrOH(O)(O_2)_2]$

*See* HYDROXOOXODIPEROXOCHROMATE SALTS

## THALLIUM HYDROXOOXODIPEROXOCHROMATE $CrHO_6Tl$

$Tl[CrOH(O)(O_2)_2]$

*See* HYDROXOOXODIPEROXOCHROMATE SALTS

## CHROMIC ACID $H_2CrO_4$ $CrH_2O_4$

1. Bryson, W. R., *Chem. Brit.*, 1975, **11**, 377
2. Pitt, M. J., ibid., 456
3. Downing, S., ibid.
4. Baker, P. B., ibid.
5. Kipling, B., *Chem. Brit.*, 1976, **12**, 169

A closed bottle of unused potassium dichromate–sulphuric acid mixture exploded after several months in storage [1]. Previous similar incidents were summarised, and the possibility of the bottle having burst from internal pressure of carbon dioxide arising from trace contamination by carbon compounds was advanced [1,2]. Two further reports of incidents within 1 or 2 days of preparation of the mixture were reported [3,4], the latter involving exothermic precipitation of chromium trioxide. Presence of traces of chloride in the dichromate leads to formation of chromyl chloride, which may be unstable [2,5]. The use of other glass-cleaning agents, and non-storage of chromic acid mixtures is again recommended [2].

Acetone

*MCA Case History No. 1583*

During glass cleaning operations, acetone splashed into a beaker containing traces of potassium dichromate–sulphuric acid mixture and the solvent ignited. Alcohols behave similarly.

Oxidisable material

*MCA Case History No. 1919*

A waste plating solution (containing 22% sulphuric acid and 40% w/v of chromium) was being sucked into an acid disposal tanker. When 500 litres had been transferred, a mild explosion in the tanker blew acid back through the transfer pump and hose. The oxidisable component in the tanker was not identified.

*See other* OXIDANTS

## AMMONIUM HYDROXOOXODIPEROXOCHROMATE(1–) $CrH_5NO_6$

$NH_4[CrOH(O)(O_2)_2]$

*See* HYDROXOOXODIPEROXOCHROMATE SALTS

## TRIAMMINEDIPEROXOCHROMIUM(IV) $CrH_9N_3O_4$

$[Cr(NH_3)_3(O_2)_2]$

1. Hughes, R. G. *et al., Inorg. Chem.*, 1968, **7**, 882
2. Kauffman, G. B., *Inorg. Synth.*, 1966, 8, 133

It must be handled with care because it may explode or become incandescent on sudden heating or shock. Heating at 20°C/min caused a violent explosion at *ca.* 120°C [1]. Preparative and handling precautions have been detailed [2].

*See other* AMMINECHROMIUM PEROXOCOMPLEXES

## AMMONIUM TETRAPEROXOCHROMATE(3–) $CrH_{12}N_3O_8$

$[NH_4]_3[Cr(O_2)_4]$

Mellor, 1943, Vol. 11, 356

Explodes at 50°C, on impact, or in contact with sulphuric acid.

*See other* PEROXOACID SALTS

**HEXAAMMINECHROMIUM(III) NITRATE** $CrH_{18}N_9O_9$

$[Cr(NH_3)_6][NO_3]_3$

Moderately impact-sensitive, explodes at 265°C.

*See* AMMINEMETAL OXOSALTS

**POTASSIUM TETRAPEROXOCHROMATE(3–)** $CrK_3O_8$

$K_3[Cr(O_2)_4]$

Mellor, 1943, Vol. 11, 356

Not sensitive to impact, but explodes at 178°C, or in contact with sulphuric acid. The impure salt is less stable, and explosive.

*See other* PEROXOACID SALTS

**LITHIUM CHROMATE** $LiCrO_4$ $CrLiO_4$

Zirconium

de Boer, J. H. *et al., Z. Anorg. Chem.*, 1930, **191**, 113

During reduction of the chromate to lithium at 450–600°C, a considerable excess of zirconium must be used to avoid explosions.

*See other* METAL OXOSALTS

**CHROMIUM NITRIDE** $CrN$

Potassium nitrate

*See* POTASSIUM NITRATE, $KNO_3$: Chromium nitride

**CHROMYL NITRATE** $CrO_2(NO_3)_2$ $CrN_2O_8$

Organic materials

1. Schmeisser, M., *Angew. Chem.*, 1955, **67**, 495
2. Harris, A. D. *et al., Inorg. Synth.*, 1967, **9**, 87

Many hydrocarbons and organic solvents ignite on contact with this powerful oxidant and nitrating agent which reacts like fuming nitric acid in contact with paper, rubber or wood.

*See other* METAL OXONON-METALLATES

**CHROMYL AZIDE** $CrO_2(N_3)_2$ $CrN_6O_2$

Preparative hazard

*See* CHROMYL CHLORIDE, $Cl_2CrO_2$: Sodium azide

*See other* METAL AZIDES

## SODIUM TETRAPEROXOCHROMATE(3−) $CrNa_3O_8$

$Na_3[Cr(O_2)_4]$

Mellor, 1943, Vol. 11, 356

Explodes at 115°C.

*See other* PEROXOACID SALTS

## CHROMIUM(II) OXIDE $CrO$

1. Férée, J., *Bull. Soc. Chim. Fr.*, 1901, **25**, 620
2. Ellern, 1961, 33

The black powder ignites if ground or heated in air [1]. The oxide obtained by oxidation of chromium amalgam is pyrophoric [2].

*See other* METAL OXIDES
PYROPHORIC MATERIALS

## CHROMIUM TRIOXIDE $CrO_3$

Baker, W., *Chem. & Ind.*, 1965, 280

Goertz, A., *Arbeitschutz*, 1935, 323

Presence of nitrates in chromium trioxide may cause oxidation reactions to accelerate out of control, possibly owing to formation of chromyl nitrate. Samples of the oxide should be tested by melting before use, and those evolving oxides of nitrogen should be discarded. A closed container of 50 kg of the pure oxide exploded violently when laid down. This was attributed to unsuspected contamination of the container.

*See* CHROMYL NITRATE, $CrN_2O_8$

Acetic acid

*BCISC Quart. Safety Summ.*, 1966, **37**, 30

An explosion occurred during initial heating up of a large volume of glacial acetic acid being treated with chromium troxide. This was attributed to violent interaction of solid chromium trioxide and liquid acetic acid on a hot, exposed steam coil, and subsequent initiation of an explosive mixture of acetic acid vapour and air. The risk has been obviated by using a solution of dichromate in sulphuric acid as oxidant, in place of chromium trioxide. The sulphuric acid is essential, as the solid dichromate moist with acetic acid, obtained by evaporating an acetic acid solution to near-dryness, will explode.

*See* Butyric acid, below

Acetic anhydride

1. Dawber, J. G., *Chem. & Ind.*, 1964, 973
2. Bretherick, L., ibid., 1196
3. Baker, W., *Chem. & Ind.*, 1956, 280
4. Eck, C. R. *et al.*, *J. Chem. Soc., Chem. Comm.*, 1974, 865

A literature method for preparation of chromyl acetate by interaction of chromium trioxide and acetic anhydride was modified by omission of cooling and agitation. The warm mixture exploded violently on being moved. [1]. A later publication emphasised the need for cooling, and summarised several such previous occurrences [2]. An earlier reference attributes the cause of chromium trioxide–acetic anhydride oxidation mixtures going out of control to presence of nitric acid or nitrates in the chromium trioxide. The latter can readily be checked for freedom from nitrate by melting. If oxides of nitrogen are evolved, the sample should be discarded [3]. Mixtures used as a reagent for remote oxidation of carboxylic esters are potentially explosive, and must be made up and used at below 25°C under controlled conditions [4].

Acetic anhydride,
Tetrahydronaphthalene
Peak, D. A., *Chem. & Ind.*, 1949, 14
Use of an anhydride solution of the trioxide to prepare tetralone caused a vigorous fire. This was attributed to use of the more hygroscopic granular trioxide, which is less preferable than the flake type.

Acetone
Delhez, R., *Chem. & Ind.*, 1956, 931
The use of chromium trioxide to purify acetone is hazardous, ignition on contact occurring at ambient temperature. Methanol behaves similarly when used to reduce the trioxide in preparing hexaaquachromium(III) sulphate.

Acetylene
Grignard, 1935, Vol. 3, 167
Acetylene is oxidised violently.

Alcohols
1. Newth, F. H. *et al., Chem. & Ind.,* 1964, 1482
2. Neumann, H., *Chem. Eng. News,* 1970, **48**(28), 4
When methanol was used to rinse a pestle and mortar which had been used to grind coarse chromium trioxide, immediate ignition occurred, due to vigorous oxidation of the solvent. The same occurred with ethanol, 2-propanol, butanol and cyclohexanol. Water is a suitable cleaning agent [1]. For oxidation of *sec*-alcohols in dimethylformamide, the oxide must be finely divided, as lumps cause violent local reaction on addition to the solution [2].
*See* *N,N*-Dimethylformamide, below

Alkali metals
Mellor, 1943, Vol. 11, 237
Sodium or potassium reacts with incandescence.

Ammonia

Mellor, 1943, Vol. 11, 233

Pascal, 1959, Vol. 14, 215

Gaseous ammonia leads to incandescence, and the aqueous solution is oxidised very exothermically.

Arsenic

Mellor, 1943, Vol. 11, 234

Reaction incandesces.

Bromine pentafluoride

*See* BROMINE PENTAFLUORIDE, $BrF_5$: Acids, etc.

Butyric acid

Wilson, R. D., *Chem. & Ind.*, 1957, 758

A mixture of chromium trioxide and butyric acid became incandescent on heating to 100°C.

*See* Acetic acid, above

Chlorine trifluoride

*See* CHLORINE TRIFLUORIDE, $ClF_3$ : Chromium trioxide

Chromium(II) sulphide

Mellor, 1943, Vol. 11, 430

Interaction causes ignition.

*N,N*-Dimethylformamide

Neumann, H., *Chem. Eng. News,* 1970, **48**(28), 4

During oxidation of a *sec*-alcohol to ketone in cold DMF solution, addition of solid trioxide caused ignition. Addition of lumps of trioxide was later found to cause local ignition on addition to ice-cooled DMF under nitrogen.

Glycerol

Pieters, 1957, 30

Interaction is violent; the mixture may ignite owing to oxidation of the trihydric alcohol, which is viscous and unable to dissipate the heat of oxidation.

*See* GLYCEROL, $C_3H_3O_3$: Solid oxidants

Hexamethylphosphoric triamide

Cardillo, G. *et al., Synthesis*, 1976, **6**, 394–396

Stirring chromium trioxide (added in small portions) with the unheated triamide leads to the formation of a complex useful for oxidising alcohols to carbonyl derivatives. The trioxide must not be crushed before being added to the solvent, because violent decomposition may then occur.

*See* *N,N*-Dimethylformamide, above
Organic materials and solvents, below

Hydrogen sulphide

Mellor, 1943, Vol. 11, 232

Contact with the heated oxide causes incandescence.

Organic materials and solvents

1. Leleu, *Cahiers*, 1976, (83), 286–287
2. Fawcett, H. H., *Ind. Eng. Chem.*, 1959, **51**(4), 90A
3. Mikhailov, V., *Chem. Abs.*, 1960, **54**, 23331f

Combustible materials may ignite or explode on contact with the oxide. A few drops of oil which fell into a container of the oxide led to an explosion which produced fatal burns [1]. If a few drops of organic solvent (acetone, 2-butanone, ethanol) contact solid chromium trioxide, a few seconds' delay ensues while some of the oxidant attains the critical temperature of 330°C. Then combustion occurs with enough vigour to raise a fire-ball several feet, and spattering also occurs [2]. Possible ignition hazards were studied for a range of 60 organic liquids and solids in contact with the solid oxide. Hot liquids were added to the oxide; solids were covered with a layer of the oxide. The most dangerous materials were methanol, ethanol, butanol, isobutanol, acetaldehyde, propionaldehyde, butyraldehyde, benzaldehyde, acetic acid, pelargonic acid, ethyl acetate, isopropyl acetate, pentyl acetate, diethyl ether, methyldioxane, dimethyldioxane, acetone and benzylethylaniline. Other materials evolved heat, especially in presence of water. Segregation in storage or transport is essential [3].

*See* *N,N*-Dimethylformamide, above
Hexamethylphosphoric triamide, above

Peroxyformic acid

*See* PEROXYFORMIC ACID, $CH_2O_3$: Metals, etc.

Phosphorus

Moissan, H., *Ann. Chim. Phys.* [6], 1885, **5**, 435

Phosphorus and the molten trioxide react explosively.

Potassium hexacyanoferrate(3−)

*BCISC Quart. Safety Summ.*, 1965, **36**(144), 55

Mixtures of the ferrate ('ferricyanide') and chromium trioxide explode and inflame when heated above 196°C. Friction alone is sufficient to ignite violently the mixture when ground with silver sand.

Pyridine

1. Dauben, W. G. *et al., J. Org. Chem.*, 1969, **34**, 3587
2. Poos, G. I. *et al., J. Amer. Chem. Soc.*, 1953, **75**, 427
3. Collins, J. C. *et al., Org. Synth.*, 1972, **52**, 7
4. Ratcliffe, R. *et al., J. Org. Chem.*, 1970, **35**, 4001
5. Stensio, K. E., *Acta Chem. Scand.*, 1971, **25**, 1125
6. *MCA Case History No. 1284*

During preparation of the trioxide–pyridine complex (a powerful oxidant) lack of really efficient stirring led to violent flash fires as the oxide was added to the pyridine at −15 to −18°C. Reversed addition of pyridine to the oxide is extremely dangerous [1], ignition usually occurring [2]. A more recent preparation specifies temperature limits of 10–20°C to avoid an excess of unreacted trioxide [3]. A safe method of preparing the complex in solution has been described [4], and preparation and use of solutions of the isolated complex in dichloromethane [3] or acetic acid [5] have been detailed. The Case History gives further information on preparation of the complex. Solution of the oxide is not smooth; it first swells, then suddenly dissolves in pyridine with heat evolution. This hazard may be eliminated without loss of yield by dissolving chromium trioxide in an equal volume of water before adding it to 10 volumes of pyridine. Pulverising chromium trioxide before use is not recommended, as this increases its rate of reaction with organic compounds to a hazardous level.

*See* OXODIPEROXODIPYRIDINECHROMIUM(VI), $C_{10}H_{10}CrN_2O_5$

Selenium

Mellor, 1943, Vol. 11, 233

Interaction is violent.

Sodium

*See* SODIUM, Na: Metal oxides

Sulphur

Mellor, 1943, Vol. 11, 232

Ignition on warming.

*See other* OXIDANTS

**LEAD CHROMATE** **$PbCrO_4$** **$CrO_4Pb$**

Aluminium,
Dinitronaphthalene

Nagaishi, T. *et al., Chem. Abs.*, 1977, **86**, 59602b

The considerable energy released by the mixture derives from chromate-catalysed exothermic decomposition of the nitro compound, coupled with a thermite-type reaction of the aluminium and chromate. It is useful for cracking concrete.

Iron(3+) hexacyanoferrate(4–)
Anon., *Chem. Processing*, 1967, **30**, 118
Creevey, J., *Chem. Age (London)*, 1942, **46**, 99
During grinding operations, the intimate mixture was ignited by a spark and burned fiercely. A similar incident had been reported previously.

Sulphur
Jackson, H., *Spectrum*, 1969, 7(2), 82
The mixture is pyrophoric
*See other* METAL OXONON-METALLATES
OXIDANTS

**CHROMIUM(II) SULPHATE** **$CrSO_4$** **$CrO_4S$**

Water
1. van Bemmelen, J. M., *Rec. Trav. Chim.*, 1887, **6**, 202–204
2. Bailar, 1973, Vol. 3, 657–658

Crystals of the heptahydrate, damp with surplus water, were sealed into a glass tube and stored in darkness. After a year the tube exploded. This was attributed to the pressure of hydrogen liberated by reduction of the water by the chromium(II) salt [1]. More recent information [2] confirms this hypothesis.
*See* CHROMIUM(II) CHLORIDE, $Cl_2Cr$
*See other* REDUCANTS

**CHROMIUM(II) SULPHIDE** **CrS**

Chromium trioxide
*See* CHROMIUM TRIOXIDE, $CrO_3$: Chromium(II) sulphide

Fluorine
*See* FLUORINE, $F_2$: Sulphides

**AMMONIUM DICHROMATE** **$Cr_2H_8N_2O_7$**
**$[NH_4]_2[O_3CrOCrO_3]$**

1. Mellor MIC, 1961, 872
2. *MCA SD-45, 1952*

Thermal decomposition of the salt is initiated by locally heating to

190°C, and flame spreads rapidly through the mass which, if confined, will explode [1].
*See other* OXOSALTS OF NITROGENOUS BASES

**AMMONIUM PENTAPEROXODICHROMATE(2−)** $Cr_2H_8N_2O_{12}$
$[NH_4]_2[(O_2)_2OCrOOCr(O_2)_2O]$

Brauer, 1965, Vol. 2, 1392
Explodes at 50°C.
*See other* PEROXOACID SALTS

**POTASSIUM PENTAPEROXODICHROMATE(2−)** $Cr_2K_2O_{12}$
$K_2[(O_2)_2OCrOOCr(O_2)_2O]$

Mellor, 1943, Vol. 11, 357
Sidgwick, 1950, 1007
The powdered salt explodes above 0°C.
*See other* PEROXOACID SALTS

**POTASSIUM DICHROMATE** $K_2[O_3CrOCrO_3]$ $Cr_2K_2O_7$

Boron,
Silicon
*See* BORON, B: Dichromates, etc.

Hydroxylamine
*See* HYDROXYLAMINE, $H_3NO$: Oxidants

Tungsten
Boddington, T. *et al., Combust. Flame*, 1975, **24**, 137–138
Combustion of a pyrotechnic mixture of the two materials (studied by DTA and temperature analysis) attains a temperature of about 1700°C in 0.1–0.2 s.
*See other* OXIDANTS

**SODIUM DICHROMATE** $Na_2[O_3CrOCrO_3]$ $Cr_2Na_2O_7$

Acetic anhydride
Marszalek, G., private comm., 1973
Addition of the dihydrated salt to acetic anhydride caused an exothermic reaction which accelerated to explosion. Presence of acetic acid (including that produced by hydrolysis of the anhydride by the hydrate water) has a delaying effect on the onset of violent reaction, which occurs when the proportion of anhydride to acid (after

hydrolysis) exceeds 0.37:1, with an initial temperature above 35°C. Mixtures of dichromate (30 g) with mixtures of anhydride–acid (70 g, to give ratios of 2:1, 1:1, 0.37:1) originally at 40°C accelerated out of control after 18, 43 and 120 min, to 160, 155 and 115°C, respectively.

Boron,
Silicon
*See* BORON, B: Dichromates, etc.

Ethanol,
Sulphuric acid
Annable, E. H., *School Sci. Rev.*, 1951, **32**(117), 249
During preparation of acetic acid by acid dichromate oxidation of ethanol according to a published procedure, minor explosions occurred on two occasions after refluxing had been discontinued.

Hydroxylamine
*See* HYDROXYLAMINE, $H_3NO$: Oxidants

Sulphuric acid
Bradshaw, J. R., *Process Biochem.*, 1970, **5**(11), 19
The well-known 'chromic acid' mixture for cleaning laboratory glassware is a powerful oxidant by design, and contact with large amounts of tarry residues in vessels should be avoided as it may lead to a violent reaction. Further, if solvents are first used roughly to clean glassware before acid treatment, traces of readily oxidisable solvents must be removed before adding the acid mixture. In many cases treatment with a properly formulated detergent will ensure adequate cleanliness and avoid possible hazard.
*See* CHROMIC ACID, $CrH_2O_4$ : Acetone
*See also* CHROMIUM TRIOXIDE, $CrO_3$ : Acetic acid

Sulphuric acid,
Trinitrotoluene
Clarke, H. T. *et al., Org. Synth.*, 1941, Coll. Vol. 1, 543–544
During oxidation of TNT in sulphuric acid to trinitrobenzoic acid, stirring of the viscous reaction mixture must be very effective to prevent added portions of solid dichromate causing local ignition.
*See other* METAL OXOMETALLATES

## DICHROMIUM TRIOXIDE $Cr_2O_3$

Chlorine trifluoride
*See* CHLORINE TRIFLUORIDE, $ClF_3$: Metals, etc.

Dirubidium acetylide

*See* DIRUBIDIUM ACETYLIDE, $C_2Rb_2$: Metal oxides

Lithium

*See* LITHIUM Li: Metal oxides

Nitroalkanes

*See* NITROALKANES: Metal oxides

## CAESIUM Cs

Acids

Pascal, 1957, Vol. 3, 94

Most acids react violently, even when anhydrous.

Air,
or Oxygen,
or Water

Mellor, 1941, Vol. 2, 468; 1963, Vol. 2, Suppl. 2.2, 2291, 2328

Markowitz, M. M., *J. Chem. Educ.*, 1963, **40**, 633–636

It ignites immediately in air or oxygen, and in contact with cold water the evolved hydrogen ignites. The reactivity of caesium and other alkali metals with water has been discussed in detail.

Halogens

Gibson, 1969, 8

Interaction at ambient temperature is violent with all halogens.

Non-metals

Pascal, 1957, Vol. 3, 94

Interaction with sulphur or phosphorus attains incandescence.

*See other* METALS

## CAESIUM FLUORIDE CsF

Benzenediazonium tetrafluoroborate,
Difluoroamine

*See* DIFLUOROAMINE, $F_2HN$: Benzenediazonium tetrafluoroborate, etc.

## CAESIUM AMIDOPENTAFLUOROTELLURATE $CsF_5H_2NTe$

$Cs[TeNH_2F_5]$

Seppelt, K., *Inorg. Chem.*, 1973, **12**, 2838

Heating must be avoided during the preparation or subsequent drying, as occasional explosions occurred. It exploded immediately upon laser irradiation for Raman spectroscopy.

*See related* AMMINEMETAL OXOSALTS

**CAESIUM HYDRIDE** **CsH**

Gibson, 1969, 76

The unheated hydride ignites in oxygen.

*See other* METAL HYDRIDES

**CAESIUM HYDROGENXENATE** **$CsXeO_3OH$** **$CsHO_4Xe$**

Jaselskis, B. *et al., J. Amer. Chem. Soc.*, 1966, **88**, 2150; 1969, **91**, 1874

It is unstable to friction, thermal or mechanical shock and may explode on contact with alcohols.

*See other* XENON COMPOUNDS

**CAESIUM AMIDE** **$CsNH_2$** **$CsH_2N$**

Water

Mellor, 1940, Vol. 8, 256

Interaction is incandescent in presence of air.

*See other N*-METAL DERIVATIVES

**CAESIUM TRIOXIDE ('OZONATE')** **$CsO_3$**

Water

Whaley, T. P. *et al., J. Amer. Chem. Soc.*, 1951, **73**, 79

Reaction of caesium or potassium 'ozonates' with water or aqueous acids is violent, producing oxygen and flashes of light.

*See other* METAL OXIDES

**DICAESIUM OXIDE** **$Cs_2O$**

Ethanol

Pascal, 1957, Vol. 3, 104

Contact of a little alcohol with the oxide may ignite the solvent.

Halogens,
or Non-metal oxides

Mellor, 1941, Vol. 2, 487

Above 150–200°C, incandescence occurs with fluorine, chlorine or iodine. In presence of moisture, contact at ambient temperature with carbon monoxide or dioxide causes ignition, while dry sulphur dioxide causes incandescence on heating. Iodine also reacts incandescently above 150–200°C.

Water

Pascal, 1957, Vol. 3, 104

Incandescence on contact.

*See other* METAL OXIDES

**DICAESIUM SELENIDE** **$Cs_2Se$**

Sidgwick, 1950, 92

When warm, it ignites in air.

*See related* METAL SULPHIDES

**TRICAESIUM NITRIDE** **$Cs_3N$**

Mellor, 1940, Vol. 8, 99

Burns in air, and is readily attacked by chlorine, phosphorus or sulphur.

*See other N*-METAL DERIVATIVES

**COPPER** **Cu**

Acetylenic compounds

*See* ACETYLENIC COMPOUNDS: Metals

Aluminium,
Sulphur

*See* ALUMINIUM, Al: Copper, Sulphur

1-Bromo-2-propyne

*See* 1-BROMO-2-PROPYNE, $C_3H_3Br$: Metals

Ethylene oxide

*See* ETHYLENE OXIDE, $C_2H_4O$

Lead azide

*See* LEAD(II) AZIDE, $N_6Pb$: Copper

Oxidants

*See* CHLORINE TRIFLUORIDE, $ClF_3$: Metals, etc.
CHLORINE, $Cl_2$: Metals
AMMONIUM PERCHLORATE, $ClH_4NO_4$: Copper
FLUORINE, $F_2$: Metals

SULPHURIC ACID, $H_2O_4S$: Copper
HYDROGEN SULPHIDE, $H_2S$: Metals
AMMONIUM NITRATE, $H_4N_2O_3$: Metals
HYDRAZINIUM NITRATE, $H_5N_3O_3$
POTASSIUM DIOXIDE, $KO_2$: Metals

Water

Zyszkowski, W., *Int. J. Heat Mass Tranf.*, 1976, **19**, 849–868
The vapour explosion which occurs when liquid copper is dumped into water has been studied.

*See* MOLTEN METALS

## COPPER–ZINC ALLOYS **Cu–Zn**

Alkyl halides

*See* DIALKYLZINCS

Diiodomethane,
Ether

Foote, C. S., private comm., 1965
Lack of cooling during preparation of the Simmonds–Smith organozinc reagent caused the reaction to erupt. Possible pyrophoric nature of organozinc compounds and the presence of ether presents a severe fire hazard.

*See other* ALLOYS

## COPPER IRON(II) SULPHIDE **$CuFeS_2$**

Ammonium nitrate

*See* AMMONIUM NITRATE, $H_4N_2O_3$: Copper iron(II) sulphide

Water

Gribin, A. A., *Chem. Abs.*, 1943, 37, $1272_8$
A large dump of copper pyrites ore ignited after heavy rain. The thick layer (6–7 m) and absence of ventilation were contributory factors to the accelerating aerial oxidation which finally led to ignition.

*See other* METAL SULPHIDES
PYROPHORIC MATERIALS

## COPPER(I) HYDRIDE **CuH**

Neunhöffer, O. *et al.*, *J. Prakt. Chem.* [2], 1935, **144**, 63
This impure, unstable material may decompose explosively on heating.

Halogens

Mellor, 1941, Vol. 3, 73

Ignition on contact with fluorine, chlorine or bromine.

*See other* METAL HYDRIDES

**COPPER(II) AZIDE HYDROXIDE** $CuN_3OH$ $CuHN_3O$

Cirulis, A., *Z. Anorg. Allg. Chem.*, 1943, **251**, 332–334

It explodes at 203°C or on impact. The dimeric double salt is similarly sensitive.

*See related* METAL AZIDES

**LITHIUM DIHYDROCUPRATE** $Li[CuH_2]$ $CuH_2Li$

Ashby, E. C. *et al., J. Chem. Soc., Chem. Comm.*, 1974, 157–158

The solid is highly pyrophoric, but stable as a slurry in ether at ambient temperature.

*See other* COMPLEX HYDRIDES

**COPPER(II) PHOSPHINATE** $Cu(H_2PO_2)_2$ $CuH_4O_4P_2$

Mellor, 1940, Vol. 8, 883

The solid suddenly explodes at about 90°C.

*See other* METAL OXONON-METALLATES

**TETRAAMMINECOPPER(II) SULPHATE** $CuH_{12}N_4O_4S$

$[Cu(NH_3)_4][SO_4]$

Iodine

*See* IODINE, $I_2$: Tetraamminecopper(II) sulphate

**TETRAAMMINECOPPER(II) NITRITE** $CuH_{12}N_6O_4$

$[Cu(NH_3)_4][NO_2]_2$

Mellor, 1940, Vol. 8, 480

The salt is nearly as shock-sensitive as picric acid. When pure it does not explode on heating but traces of nitrate cause explosive decomposition.

*See other* AMMINEMETAL OXOSALTS

**TETRAAMMINECOPPER(II) NITRATE** $CuH_{12}N_6O_6$

$[Cu(NH_3)_4][NO_3]_2$

Explodes at 330°C, high impact-sensitivity.

*See* AMMINEMETAL OXOSALTS

**TETRAAMMINECOPPER(II) AZIDE** $[Cu(NH_3)_4][N_3]_2$ **$CuH_{12}N_{10}$**

Mellor, 1940, Vol. 8, 348; 1967, Vol. 8, Suppl. 2, 26

Explosive on heating or impact.

*See other* METAL AZIDES

**LITHIUM HEXAAZIDOCUPRATE(4–)** $Li_4[Cu(N_3)_6]$ **$CuLi_4N_{18}$**

Urbanski, 1967, Vol. 3, 185

It is, like copper(II) azide, an exceptionally powerful initiating detonator.

*See other* METAL AZIDES

**COPPER(II) NITRATE** $Cu(NO_3)_2$ **$CuN_2O_6$**

Acetic anhydride

*See* ACETIC ANHYDRIDE, $C_4H_6O_3$: Metal salts

Ammonia,
Potassium amide

Sorbe, 1968, 80

Interaction gives an explosive precipitate containing the Cu(I) derivatives: $Cu_3N$, $Cu_3N \cdot nNH_3$, $Cu_2NH$, $CuNK_2 \cdot NH_3$.

Ammonium hexacyanoferrate(4–)

Wolski, J. *et al.*, *Z. Chem.*, 1973, **13**, 95–97; *J. Therm. Anal.*, 1973, 5, 67

At 220°C interaction is explosive in wet mixtures, or in dry mixtures if the nitrate is in excess.

Potassium hexacyanoferrate(4–)

Wolski, J. *et al.*, *Explosivstoffe*, 1969, **17**(5), 103–110

Interaction is explosive at 220°C.

*See other* METAL NITRATES

**COPPER(I) AZIDE** **$CuN_3$**

1. Mellor, 1940, Vol. 8, 348; 1967, Vol. 8, Suppl. 2, 42–50
2. Urbanski, 1967, Vol. 3, 185

One of the more explosive metal azides, it decomposes at 205°C [1] and is very highly impact-sensitive.

*See other* METAL AZIDES

**COPPER(II) AZIDE** $Cu(N_3)_2$ **$CuN_6$**

1. Mellor, 1940, Vol. 8, 348; 1967, Vol. 8, Suppl. 2.2, 42–50

2. Urbanski, 1967, Vol. 3, 185
Very explosive, even when moist. Loosening the solid from filter paper caused frictional initiation. Explosion initiated by impact is very violent, and spontaneous explosion has also been recorded [1]. It is also an exceptionally powerful initiator.

*See* SODIUM AZIDE, $N_3Na$: Heavy metals
*See other* METAL AZIDES

**COPPER(II) OXIDE** **CuO**

Boron
Mellor, 1946, Vol. 5, 17
On heating a mixture, the exothermic reaction melted the glass container.

Dichloro(methyl)silane
*See* DICHLORO(METHYL)SILANE, $CH_4Cl_2Si$: Oxidants

Dirubidium acetylide
*See* DIRUBIDIUM ACETYLIDE, $C_2Rb_2$: Metal oxides

Hydrogen
Read, C. W. W., *School Sci. Rev.*, 1941, **22**(87), 340
Reduction of the heated oxide in a combustion tube by passage of hydrogen caused a violent explosion. (The hydrogen may have been contaminated with air.)

Hydrogen sulphide
*See* HYDROGEN SULPHIDE, $H_2S$: Metal oxides
HYDROGEN TRISULPHIDE, $H_2S_3$: Metal oxides

Metals
Browne, T. E. W., *School Sci. Rev.*, 1967, **48**(166), 921
An attempted thermite reaction with aluminium powder and copper(II) oxide in place of diiron trioxide caused a violent explosion. An anonymous comment suggests that a greater reaction rate and exothermic effect were involved and adds that attempted use of disilver oxide would be even more violent.

*See* ALUMINIUM, Al: Metal oxides, etc.
POTASSIUM, K: Metal oxides
MAGNESIUM, Mg: Metal oxides
SODIUM, Na: Metal oxides

Phospham
*See* PHOSPHAM, $HN_2P$: Oxidants

Phthalic anhydride
'Leaflet No.5', Inst. of Chem., London, 1940
A mixture of the anhydride and anhydrous oxide exploded violently on heating.

Reducants
Mellor, 1941, Vol. 3, 137
Interaction with hydroxylamine or hydrazine is vigorous.
*See other* METAL OXIDES

**COPPER(II) SULPHATE** $CuSO_4$ $CuO_4S$

Hydroxylamine
*See* HYDROXYLAMINE, $H_3NO$: Copper(II) sulphate

**COPPER MONOPHOSPHIDE** $CuP$
*See* TRICOPPER DIPHOSPHIDE, $Cu_3P_2$

**COPPER DIPHOSPHIDE** $CuP_2$
*See* TRICOPPER DIPHOSPHIDE, $Cu_3P_2$

**COPPER(II) SULPHIDE** $CuS$

Chlorates
Mellor, 1956, Vol. 2, Suppl. 1, 584
The copper sulphide explodes in contact with magnesium, zinc, or cadmium chlorates, or with a concentrated solution of chloric acid.
*See* METAL HALOGENATES
*See other* METAL SULPHIDES

**COPPER(I) HYDRIDE** $Cu_2H_2$
Gibson, 1969, 77
The dry hydride ignites in air.
*See other* METAL HYDRIDES

**COPPER(I) OXIDE** $Cu_2O$

Peroxyformic acid
*See* PEROXYFORMIC ACID, $CH_2O_3$: Metals, etc.

**COPPER(I) NITRIDE** $Cu_3N$

Mellor, 1940, Vol. 8, 100

May explode on heating in air.

Nitric acid

*See* NITRIC ACID, $HNO_3$ : Copper(I) nitride

*See other* *N*-METAL DERIVATIVES

**TRICOPPER DIPHOSPHIDE** $Cu_3P_2$

Oxidants

Mellor, 1940, Vol. 8, 839

The powdered phosphide burns vigorously in chlorine. Mixtures with potassium chlorate explode on impact, and with potassium nitrate, on heating. The monophosphide, CuP, and diphosphide, $CuP_2$, behave similarly.

*See also* POTASSIUM CHLORATE, $ClKO_3$ : Metal phosphides

*See other* METAL NON-METALLIDES

**† DEUTERIUM** $D_2$

**DEUTERIUM OXIDE** $D_2O$

Pentafluorophenyllithium

*See* PENTAFLUOROPHENYLLITHIUM, $C_6F_5Li$: Deuterium oxide

**EUROPIUM** Eu

Bailar, 1974, Vol. 4, 69

Europium is the most reactive lanthanide metal, and may ignite on exposure to air if finely divided.

*See* LANTHANIDE METALS: Oxidants

*See other* METALS

PYROPHORIC METALS

**EUROPIUM(II) SULPHIDE** EuS

Sidgwick, 1950, 454

Pyrophoric in air.

*See other* METAL SULPHIDES

1. Keen, M. J. *et al., Chem. & Ind.*, 1957, 805
2. Braker, 1971, 305

Handling precautions for the gas or anhydrous liquid are detailed [1]. A polythene condenser for disposal of hydrogen fluoride is described [2].

Bismuthic acid

*See* BISMUTHIC ACID, $BiHO_3$: Hydrofluoric acid

Cyanogen fluoride

*See* CYANOGEN FLUORIDE, CFN: Hydrogen fluoride

Mercury(II) oxide,
Organic materials

Ormston, J., *School Sci. Rev.,* 1944, **26**(98), 32

During fluorination of organic materials by passing hydrogen fluoride into a vigorously stirred suspension of the oxide (to form transiently mercury difluoride, a powerful fluorinator), it is essential to use adequate and effective cooling below 0°C to prevent loss of control of the reaction system.

Oxides

Mellor, 1956, Vol. 2, Suppl. 1, 122; 1939, Vol. 9, 101

Arsenic trioxide and calcium oxide incandesce in contact with liquid hydrogen fluoride.

*N*-Phenylazopiperidine

*See* *N*-PHENYLAZOPIPERIDINE, $C_{11}H_{15}N_3$: Hydrofluoric acid

Phosphorus(V) oxide

*See* PHOSPHORUS(V) OXIDE, $O_5P_2$: Hydrogen fluoride

Potassium permanganate

*See* POTASSIUM PERMANGANATE, $KMnO_4$: Hydrofluoric acid

Potassium tetrafluorosilicate(2−)

Mellor, 1956, Vol. 2, Suppl. 1, 121

Contact with liquid hydrogen fluoride causes violent evolution of silicon tetrafluoride. (The same is probably true of metal silicides and other silicon compounds generally.)

*See* DIALUMINIUM OCTAVANADIUM TRIDECASILICIDE, $Al_2Si_{13}V_8$

Sodium

*See* SODIUM, Na: Acids

## FLUOROSELENIC ACID $FSeO_2OH$ $FHO_3Se$

Cellulose

Bartels, H. *et al., Helv. Chim. Acta*, 1962, **45**, 179

A strong oxidant, reacts violently with filter paper or similar organic matter, igniting it if dry.

*See other* OXIDANTS

## FLUOROAMINE $FNH_2$ $FH_2N$

Hoffman, C. J. *et al., Chem. Rev.*, 1962, **62**, 7

The impure material is very explosive.

*See other N*-HALOGEN COMPOUNDS

## FLUOROPHOSPHORIC ACID $FPO(OH)_2$ $FH_2O_3P$

Sodium tetrahydroborate

*See* SODIUM TETRAHYDROBORATE, $BH_4Na$: Acids

## FLUOROSILANE $FSiH_3$ $FH_3Si$

Azidogermane

*See* AZIDOGERMANE, $GeH_3N_3$: Fluorosilane

## AMMONIUM FLUORIDE $NH_4F$ $FH_4N$

Chlorine trifluoride

*See* CHLORINE TRIFLUORIDE, $ClF_3$: Ammonium fluoride

## POTASSIUM FLUORIDE HYDROGEN PEROXIDATE $FK \cdot H_2O_2$ $KF \cdot H_2O_2$

*See* CRYSTALLINE HYDROGEN PEROXIDATES

## MANGANESE FLUORIDE TRIOXIDE $MnFO_3$ $FMnO_3$

Alone,
or Organic compounds,
or Water

Engelbrecht, A. *et al., J. Amer. Chem. Soc.*, 1954, **76**, 2042–2044

It decomposes, usually explosively, above 0°C or in contact with moisture. It is a powerful oxidant and reacts violently with organic compounds.

*See related* METAL OXIDES

*See other* OXIDANTS

**NITROSYL FLUORIDE** **NOF** **FNO**

Haloalkene (unspecified)

*MCA Case History No. 928*

Interaction of a mixture in a pressure vessel at −78°C caused it to rupture when moved from the cooling bath.

Metals,
or Non-metals

Schmutzler, R., *Angew. Chem. (Intern. Ed.)*, 1968, **7**, 442

Pascal, 1956, Vol. 10, 346

Antimony, bismuth, arsenic, boron, red phosphorus, silicon and tin all react with incandescence.

Oxygen difluoride

Ruff, O. *et al., Z. Anorg. Chem.*, 1932, **208**, 293

Explosion on mixing, even at low temperatures.

Sodium

Mellor, 1940, Vol. 8, 612

Interaction is incandescent.

*See other* *N*-HALOGEN COMPOUNDS
OXIDANTS

**NITRYL FLUORIDE** **$NO_2F$** **$FNO_2$**

Metals

Aynsley, E. E. *et al., J. Chem. Soc.*, 1954, 1122

When nitryl fluoride is passed at ambient temperature over molybdenum, potassium, sodium, thorium, uranium or zirconium, glowing or white incandescence occurs. Mild warming is needed to initiate similar reactions of aluminium, cadmium, cobalt, iron, nickel, titanium, tungsten, vanadium or zinc, and 200–300°C for lithium or manganese.

Non-metals

Aynsley, E. E. *et al., J. Chem. Soc.*, 1954, 1122

Boron and red phosphorus glow in the fluoride at ambient temperature, while hydrogen explodes at 200–300°C. Carbon and sulphur are also attacked.

*See other* OXIDANTS

## NITRYL HYPOFLUORITE ('FLUORINE NITRATE') $FNO_3$

$NO_2OF$

1. Schmutzler, R., *Angew. Chem. (Intern. Ed.)*, 1968, 7, 454
2. Engelbrecht, A., *Monatsh.*, 1964, **95**, 633

It is a toxic colourless gas which is dangerously explosive in the gaseous, liquid and solid states [1]. It is produced during electrolysis of nitrogenous compounds in hydrogen fluoride [2].

*See* FLUORINE, $F_2$ : Nitric acid
*See other* HYPOHALITES

Gases
Hoffman, C. J., *Chem. Rev.*, 1964, **64**, 94
Immediate ignition in the gas phase occurs with ammonia, dinitrogen oxide or hydrogen sulphide.

Organic materials
Brauer, 1963, 189
The very powerful liquid oxidant explodes when vigorously shaken, or immediately on contact with alcohol, ether, aniline or grease. It is also sensitive in the vapour or solid state.
*See also* NITRYL HYPOCHLORITE, $ClNO_3$
*See other* OXIDANTS

## SULPHUR OXIDE (*N*-FLUOROSULPHONYL)IMIDE $FNO_3S_2$

$O{=}S{=}NSO_2F$

Roesky, H. W., *Angew. Chem. (Intern. Ed.)*, 1967, **6**, 711
It reacts explosively with water at ambient temperature, but smoothly at –20°C.

## FLUORINE AZIDE $FN_3$

Bauer, S. H., *J. Amer. Chem. Soc.*, 1947, **69**, 3104
This unstable material usually explodes on vaporisation (at –82°C)
*See* HALOGEN AZIDES

## FLUOROTHIOPHOSPHORYL DIAZIDE $FN_6PS$

$FPS(N_3)_2$

O'Neill, S. R. *et al.*, *Inorg. Chem.*, 1972, **11**, 1630
The explosion of the glassy material at –183°C was attributed to crystallisation of the glass.
*See other* ACYL AZIDES

**FLUORINE**

1. Kirk-Othmer, 1966, Vol. 9, 506
2. Gall, J.F. *et al., Ind. Eng. Chem.*, 1947, **39**, 262
3. Landau, R. *et al.*, ibid., 262; Turnbull, S. G. *et al.*, ibid., 286
4. Long, G., 'Apparatus for the Disposal of Fluorine on a Laboratory Scale', Harwell, AERE, 1956
5. Gordon, J. *et al., Ind. Eng. Chem.*, 1960, **52**(5), 63A
6. Ring, R. J. *et al., Review of Fluorine Cells and Production*, Lucas Heights, Austral. At. En. Comm., 1973
7. Bailar, 1973, Vol. 2, 1015–1019

Fluorine is the most electronegative and reactive element known, reacting, often violently, with most of the other elements and their compounds. Handling hazards and disposal of fluorine on laboratory and plant scales are adequately described [1–5].

Safety practices associated with the use of laboratory- and industrial-scale fluorine cells and facilities have been reviewed [6]. Equipment and procedures for the laboratory use of fluorine and volatile fluorides have been detailed [7].

Acetylene

*See* ACETYLENE (ETHYNE), $C_2H_2$: Halogens

Alkanes,
Oxygen

Von Elbe, G., US Pat. 3 957 883, 1976 (*Chem. Abs.*, 1976, **85**, 45974s)

Interaction of propane, butane or 2-methylpropane with fluorine and oxygen produces peroxides. Appropriate reaction conditions are necessary to prevent explosions.

Ammonia

1. Mellor, 1940, Vol. 8, 216
2. Mueller, W., *Umsch. Wiss. Tech.*, 1974, **74**(15), 485–486

Ammonia ignites in contact with fluorine [1], and the anhydrous liquids have been used as a propellant pair [2]. Aqueous ammonia ignites or explodes on contact [1].

Boron nitride

Moissan, 1900, 232

Unheated interaction leads to incandescence.

Caesium heptafluoropropoxide

Gumprecht, W. H., *Chem. Eng. News,* 1965, **43**(9), 36

*MCA Case History No. 1045*

Fluorination of caesium heptafluoropropoxide at –40°C with nitrogen-diluted fluorine exploded violently after 10 h. This may have been caused by ingress of moisture, formation of some pentafluoropropionyl derivative, and conversion of this to pentafluoropropionyl hypofluorite, known to be explosive if suitably initiated. Other possible explosive intermediates are peroxides or peresters.

*See* PENTAFLUOROPROPIONYL HYPOFLUORITE, $C_3F_6O_2$

Covalent halides

Mellor, 1940, Vol. 2, 12; 1956, Vol. 2, Suppl. 1, 64; 1940, Vol. 8, 995, 1003, 1013

Leleu, *Cahiers*, 1973, (73), 509; 1974, (74), 427

Chromyl chloride at high concentration ignites in fluorine, while phosphorus pentachloride, trichloride and trifluoride ignite on contact. Boron trichloride ignites in cold fluorine, and silicon tetrachloride on warming.

*See also* Hydrogen fluoride, etc., below

Cyanoguanidine

Rausch, D. A. *et al., J. Org. Chem.,* 1968, **33**, 2522

The products, perfluoro-1-aminomethylguanidine and perfluoro-*N*-aminomethyltriaminomethane, and by-products of the reaction of fluorine and cyanoguanidine, are extremely explosive in gas, liquid and solid states.

*See also* Sodium dicyanamide, below

Graphite

1. Lagow, R. J. *et al., J. Chem. Soc., Dalton Trans.*, 1974, 1270
2. Ruff, O. *et al., Z. Anorg. Chem.*, 1934, **217**, 1
3. Simons, J. H. *et al., J. Amer. Chem. Soc.*, 1939, **61**, 2962–2966

During interaction at ambient temperature in a bomb to produce poly(carbon monofluoride), admission of fluorine beyond a pressure of 13.6 bar must be extremely slow and carefully controlled to avoid a violently exothermic explosion [1]. Previously it had been shown that explosive interaction of carbon and fluorine was due to formation and thermal decomposition of the graphite intercalation compound, poly(carbon monofluoride) [2]. Presence of mercury compounds prevents explosion during interaction of charcoal and fluorine [3].

*See* Non-metals, below

Halocarbons

Mellor, 1956, Vol. 2, Suppl. 1

Schmidt, 1967, 82

Fletcher, E. A. *et al., Combust. Inst. Paper WSS/CI-67-23*, 1967

Moissan, 1900, 241

The violent or explosive reactions which carbon tetrachloride, chloroform, etc., exhibit on direct local contact with gaseous fluorine, can be moderated by suitable dilution, catalysis, and diffused contact.

Combustion of perfluorocyclobutane–fluorine mixtures was detonative between 9.04 and 57.9 vol.% of the hydrocarbon. Iodoform reacts very violently with fluorine owing to its high iodine content.

*See* Polytetrafluoroethylene, etc., below

Halogens,
or Dicyanogen

Mellor, 1940, Vol. 2, 12
Sidgwick, 1950, 1148

While bromine, iodine and dicyanogen all ignite in fluorine at ambient temperature, a mixture of chlorine and fluorine (containing essential moisture) needs sparking before ignition occurs, though an explosion immediately follows.

Hexalithium disilicide

Mellor, 1940, Vol. 6, 169

Incandesces on warming in fluorine.

Hydrocarbons

Mellor, 1940, Vol. 2, 11; 1956, Vol. 2, Suppl. 1, 198
Sidgwick, 1950, 1117
Schmidt, 1967, 82
Moissan, 1900, 240–241

Violent explosions occur when fluorine directly contacts liquid hydrocarbons, even at −210°C with anthracene or turpentine, or solid methane at −190°C with liquid fluorine. Many lubricants ignite in fluorine. Contact under carefully controlled conditions with dilution and catalysis can now be effected smoothly.

Gaseous hydrocarbons (towns gas, methane) ignite in contact with fluorine, and mixtures with unsaturated hydrocarbons (ethylene, acetylene) may explode on exposure to sunlight. Each bubble of fluorine passed through benzene causes ignition, but a rapid stream may lead to explosion.

Hydrofluoric acid

*See* Hydrogen halides, below

Hydrogen

1. Mellor, 1940, Vol. 2, 11; 1956, Vol. 2, Suppl. 1, 55
2. Sidgwick, 1950, 1102
3. Kirshenbaum, 1956, 46

The violently explosive reactions which sometimes occur when the two elements come into contact under conditions ranging from solid fluorine and liquid hydrogen at −252°C to the mixed gases at ambient temperature [1] are caused by the catalytic effects of impurities or the physical nature of the walls of the containing vessel [2]. Even in absence of such impurities, spontaneous explosions still occur in the range of 75–90 mol % fluorine in the gas phase at –183°C [3].

Hydrogen,
Oxygen
Getzringer, R. W., *Rept. LA-5659*, Los Alamos Sci. Lab., 1974 (*Chem. Abs.*, 1975, **82**, 160801h)
The conditions under which mixtures of the gases at 1 bar and ambient temperature will react non-explosively have been studied.

Hydrogen fluoride,
Seleninyl fluoride
Seppelt, K., *Angew. Chem. (Intern. Ed.)*, 1972, **11**, 630
Preparation of pentafluoroorthoselenic acid from the above reagents in an autoclave above ambient temperature caused occasional explosions. A safer alternative preparation is described.
*See also* Covalent halides, above

Hydrogen halides
Mellor, 1940, Vol. 2, 12
Hydrogen chloride, bromide and iodide ignite in contact with fluorine, and the concentrated aqueous solutions, including that of hydrogen fluoride, also produce flame.

Hydrogen sulphide
Mellor, 1947, Vol. 10, 133
Ignition on contact.

Ice
*UK Scientific Mission Report 68/79*, Washington, UKSM, 1968
English, W. D. *et al., Cryog. Technol.*, 1965, **1**, 260
Mixtures of liquid fluorine and ice are highly impact-sensitive, with a power comparable to that of TNT. Contact of moist air or water with liquid fluorine can thus be very hazardous.
*See* Water, below

Metal acetylides and carbides
Mellor, 1946, Vol. 5, 849, 885, 890–891
Mono- and di-caesium, -lithium and -rubidium acetylides, both tungsten carbides, and zirconium dicarbide, all ignite in cold fluorine, while uranium dicarbide ignites in warm fluorine.

Metal borides
Bailar, 1973, Vol. 1, 729
Interaction frequently attains incandescence.

Metal cyanocomplexes

Moissan, 1900, 228

Potassium and lead hexacyanoferrates(3–) and potassium hexacyano-ferrate(4–) become incandescent in fluorine, and the liberated dicyanogen also ignites.

Metal hydrides

Mellor, 1940, Vol. 2, 12, 483; 1956, Vol. 2, Suppl. 1, 56; 1941, Vol. 3, 73

Copper, potassium and sodium hydrides all ignite on contact at ambient temperature.

Metal iodides

Moissan, 1900, 227

Fluorine decomposes the iodides of calcium, lead, mercury and potassium at ambient temperature, and the liberated iodine ignites, evolving much heat.

Metal oxides

Mellor, 1940, Vol. 2, 11–13, 487; 1941, Vol. 3, 663; 1942, Vol. 13, 715; 1936, Vol. 15, 380, 399

Oxides of the alkali and alkaline earth metals and nickel monoxide incandesce in cold fluorine, and iron monoxide when warmed. Nickel dioxide also burns in fluorine.

*See also* Miscellaneous materials, below

Metals

Mellor, 1940, Vol. 2, 13, 469; 1941, Vol. 3, 638; 1940, Vol, 4, 267, 476; 1946, Vol. 5, 421; 1939, Vol. 9, 379, 891; 1943, Vol. 11, 513, 730; 1942, Vol. 12, 344; 1942, Vol. 15, 675

Schmidt, 1967, 78–80

Kirk-Othmer, 1966, Vol. 9, 507–508

Vigour of reaction is greatly influenced by state of subdivision of the metals involved. Massive calcium, moist magnesium, manganese powder, molybenum powder, potassium, sodium, rubidium and antimony all ignite in cold fluorine gas. Warm tantalum powder or cold thallium ignites on contact with fluorine. Fine copper wire (as wool) ignites at 121°C, and osmium and tin begin to burn at 100°C, while iron powder (100-mesh, but not 20-mesh) ignites in liquid fluorine. Titanium will ignite if impacted under the liquid at –188°C and has ignited in presence of catalysts in the gas at –80°C, but in all cases the fluoride film prevents further propagation. Tungsten, and uranium powders, ignite in the gas without heating, while zinc ignites at about 100°C. Molybdenum, tungsten and Monel wires ignited in atmospheric fluorine at 205, 283 and 396°C (averaged values), respectively. Generally, strongly electropositive metals, or those forming volatile fluorides, are attacked the most vigorously.

*See* BERYLLIUM, Be: Halogens

Metal salts

Mellor, 1940, Vol. 2, 13; 1956, Vol. 2, Suppl. 1, 63
Schmidt, 1967, 83–84
Moissan, 1900, 228–239
Pascal, 1960, Vol. 16, 57–60
Fichter, F. *et al., Helv. Chim. Acta*, 1930, **13**, 99–102

Calcium, lead, basic lead, lithium and sodium carbonates all ignite and burn fiercely in contact with fluorine. Chlorides and cyanides are vigorously attacked by cold fluorine, including lead difluoride and thallium(I) chloride, both of which become molten. Mercury dicyanide ignites in fluorine when warmed gently, and silver cyanide reacts explosively when cold. Sodium metasilicate ignites in fluorine.

Unheated calcium phosphate, sodium thiosulphate and diphosphate all incandesce in contact with fluorine, and barium or mercury thiocyanates ignite. On warming, chromium trichloride, calcium arsenate or copper borate incandesces, and sodium arsenate ignites.

Introduction of fluorine into solutions of silver fluoride, nitrate, perchlorate or sulphate causes violent exothermic reactions to occur, with liberation of ozone-rich oxygen.

Metal silicides

Mellor, 1940, Vol. 6, 169, 178

Calcium disilicide readily ignites, and dilithium hexasilicide becomes incandescent, when warmed in fluorine.

Miscellaneous materials

1. Mellor, 1940, Vol. 2, 13
2. Schmidt, 1967, 84, 110
3. Lafferty, R. H. *et al., Chem. Eng. News*, 1948, **26**, 3336

Town gas ignites in contact with gaseous fluorine as does a mixture of lead oxide and glycerol (used as a jointing compound) [1]. Spillage tests involving action of liquid fluorine alone or as a 30% solution in liquid oxygen caused asphalt and crushed limestone to ignite, and coke and charcoal to burn, the latter brilliantly, while liquid JP4 fuel produced violent explosions and a large fireball. Rich soil also burned with a bright flame [2]. Immersion of various glove materials in liquid fluorine was examined. Cotton exploded violently and Neoprene slightly with ignition, and leather charred but did not ignite [3].

Nitric acid

1. Cady, G. H., *J. Amer. Chem. Soc.*, 1934, **56**, 2635

2. Schmutzler, R., *Angew. Chem. (Intern. Ed.)*, 1968, 8, 453

Interaction of fluorine with either the concentrated or very dilute acid caused explosions, while use of 4N acid did not [1]. Later and safer methods of preparing nitryl hypofluorite are summarised [2].

Nitrogenous bases

Hoffman, C. J., *Chem. Rev.*, 1962, **62**, 12

Aniline, dimethylamine and pyridine incandesce on contact with fluorine.

*See* Ammonia, above
Cyanoguanidine, above

Non-metal oxides

1. Mellor, 1940, Vol. 2, 11; 1939, Vol. 9, 101
2. Hoffman, C. J. *et al., Chem. Rev.*, 1962, **62**, 10
3. Pascal, 1960, Vol. 16, 53, 62
4. Moissan, 1900, 138

Diarsenic trioxide reacts violently and nitrogen oxide ignites in excess fluorine. Bubbles of sulphur dioxide explode separately on contacting fluorine, while addition of the latter to sulphur dioxide causes an explosion at a certain concentration [1]. Reaction of fluorine with dinitrogen tetraoxide usually causes ignition [2]. Interaction with carbon monoxide and oxygen may be explosive.

Anhydrous silica incandesces in the gas, and interaction with liquid fluorine at −80°C is explosive [3,4]. Boron trioxide also incandesces in the gas [3].

*See* BISFLUOROFORMYL PEROXIDE, $C_2F_2O_4$

Non-metals

Mellor, 1940, Vol. 2, 11, 12; 1956, Vol. 2, Suppl. 1, 60; 1946, Vol. 5, 785, 822; 1940, Vol. 6, 161; 1939, Vol. 9, 34; 1943, Vol. 11, 26
Schmidt, 1967, 52, 107
Pascal, 1960, Vol. 16, 58
Moissan, 1900, 125–128

Boron, phosphorus (yellow or red), selenium, tellurium and sulphur all ignite in contact with fluorine at ambient temperature, silicon attaining a temperature above 1400°C. The reactivity shown by various forms of carbon (charcoal, lampblack, soot), all of which ignite and burn vigorously in fluorine, has been reported to be due to presence of various impurities, moisture and hydrocarbons. Carefully purified carbon (massive graphite) is inert to fluorine at ambient or slightly elevated temperatures for a short period, but may then react explosively. Phosphorus and sulphur incandesce in liquid fluorine, and sulphur ignites, even at −188°C.

*See* Graphite, above

Oxygenated organic compounds

1. Pascal, 1960, Vol. 16, 65

2. Moissan, 1900, 242–245
3. Mellor, 1940, Vol. 2, 13

Methanol, ethanol and 4-methylbutanol [1], acetaldehyde, trichloroacetaldehyde [2] and acetone [3] all ignites in contact with gaseous fluorine. Lactic, benzoic and salicylic acids ignite, while gallic acid becomes incandescent. Ethyl acetate and methyl borate ignite in fluorine [2].

Perchloric acid

*See* PERCHLORIC ACID, $ClHO_4$ : Fluorine

Phosphorus halides

*See* Covalent halides, above

Polymeric materials,
Oxygen

Schmidt, 1967, 87

Various polymeric materials were tested statically with both gaseous and liquefied mixtures of fluorine and oxygen containing from 50 to 100% of the former. The materials which burned or reacted violently were: phenol-formaldehyde resins (Bakelite); polyacrylonitrilebutadiene (Buna N); polyamides (Nylon); polychloroprene (Neoprene); polyethylene; poly-trifluoropropylmethyl siloxane (LS 63); polyvinylchloride-vinyl acetate (Tygan); polyvinylidene fluoride-hexafluoropropylene (Viton); polyurethane foam. Under dynamic conditions of flow and pressure, the more resistant materials which burned were: chlorinated polyethylenes; polymethylmethacrylate (Perspex); polytetrafluoro-ethylene (Teflon).

Polytetrafluoroethylene

Appelman, E. H., *Inorg. Chem.*, 1969, **8**, 223

Teflon tubing, when used to conduct fluorine into a reaction mixture, sometimes catches fire. Combustion stops when the fluorine flow is shut off.

Polytetrafluoroethylene,
Trichloroethylene

Lawler, A. E., *Advan. Cryog. Eng.*, 1967, **12**, 780–783

Tests showed that Teflon gaskets containing more than 0.35% wt. of sorbed trichloroethylene were potentially hazardous in contact with liquid fluorine.

*See* Halocarbons, above

Potassium chlorate

*See* POTASSIUM CHLORATE, $ClKO_3$ : Fluorine

Potassium hydroxide

1. Fichter, F. *et al., Helv. Chim. Acta*, 1927, **10**, 551
2. Kacmarek, A. J. *et al., Inorg. Chem.*, 1962, **1**, 659–661
3. Bailar, 1973, Vol. 2, 789–791

Interaction at –20°C produces potassium ozonate, a spontaneously explosive solid [1]. Later references [2,3] suggest that ozonates are not spontaneously explosive.

Sodium acetate

Mellor, 1956, Vol. 2, Suppl. 1, 56

Application of fluorine to aqueous sodium acetate solution causes an explosion, involving formation of diacetyl peroxide.

*See* DIACETYL PEROXIDE, $C_4H_6O_4$

Sodium bromate

Appelman, E. H., *Inorg. Chem.*, 1969, **8**, 223

The oxidation of alkaline bromate to perbromate is not smooth and small explosions may occur in the vapour above the solution. The reaction should not be run unattended.

Sodium dicyanamide

Rausch, D. A. *et al., J. Org. Chem.*, 1968, **33**, 2522

The product, perfluoro-*N*-cyanodiaminomethane, and many of the by-products from interaction of fluorine and sodium dicyanamide, are extremely explosive in gaseous, liquid and solid states.

Stainless steel

Stewart, J. W., *Proc. 7th Intern. Conf. Low Temp. Phys., Toronto, 1960,* 671

During the study of phase transitions of solidified gases at high pressures, solid fluorine reacted explosively with apparatus made out of stainless steel.

Sulphides

Mellor, 1940, Vol. 2, 11, 13; Vol. 6, 110; 1939, Vol. 9, 522; 1947, Vol. 10, 133; 1943, Vol. 11, 430

Moissan, 1900, 231–232

Diantimony trisulphide, carbon disulphide vapour, chromium(II) sulphide and hydrogen sulphide all ignite in contact with fluorine at ambient temperature, the solids becoming incandescent.

Iron(II) sulphide reacts violently on mild warming, and barium, potassium or zinc sulphides incandesce in the gas, as does dimolybdenum trisulphide at 200°C.

Trinitromethane

Smith, W. L. *et al., Chem. Abs.*, 1976, **84**, 46818k

During preparation of fluorotrinitromethane, an instrumental method can be used to avoid occurrence of dangerous over-fluorination.

Water

Mellor, 1940, Vol. 2, 11

Schmidt, 1967, 53, 119

Treatment of liquid air (containing condensed atmospheric moisture) with fluorine gives a potentially explosive precipitate, thought to be fluorine hydrate. Contact of liquid fluorine with a bulk of water caused violent explosions. Ice tends to react explosively with fluorine gas after an indeterminate induction period.

Xenon

Levec, J. *et al.*, *J. Inorg. Nucl. Chem.*, 1974, **36**, 997–1001

Interaction may be explosive in the presence of finely divided nickel or silver difluorides, or nickel(III) or silver(I) oxides, or if initiated by local heating. The mechanism is discussed.

*See other* OXIDANTS

**DIFLUOROAMINE** **$F_2NH$** **$F_2HN$**

1. Stevens, T. E., *J. Org. Chem.*, 1968, **33**, 2664, 2671
2. Petry, R. C. *et al.*, *J. Org. Chem.*, 1967, **32**, 4034
3. Rosenfeld, D. D. *et al.*, *J. Org. Chem.*, 1968, **33**, 2521
4. Martin, K. J., *J. Amer. Chem. Soc.*, 1965, **87**, 395
5. *MCA Case History No. 768*
6. Parker, C. O. *et al.*, *Inorg. Synth.*, 1970, **12**, 310–311
7. Christe, K. O., *Inorg. Chem.*, 1975, **14**, 2821

It is a dangerous explosive and must be handled with skill and care and appropriate precautions [1,2]. Explosions have occurred when it was condensed at −196°C or allowed to melt [5], and a glass bulb containing the gas exploded violently when accidentally dropped [5]. Although difluoroamine may be safely condensed at −78° or −130° [6], or at −142° [7], liquid nitrogen should not be used to give a trapping temperature of −196°C, as explosions are very likely to occur [6].

Benzenediazonium tetrafluoroborate,

Caesium fluoride

Baum, K., *J. Org. Chem.*, 1968, **33**, 4333

Use of caesium fluoride as base to effect condensation caused an explosion in absence of solvent. Pyridine or potassium fluoride and use of methylene chloride gave satisfactory results.

*See other N*-HALOGEN COMPOUNDS

## DIFLUORODIAZENE   FN=NF   $F_2N_2$

Hydrogen

Kuhn, L. P. *et al., Inorg. Chem.*, 1970, **9**, 60

Explosive interaction occurs above 90°C.

*See other* *N*-HALOGEN COMPOUNDS

## PHOSPHORUS AZIDE DIFLUORIDE   $P(N_3)F_2$   $F_2N_3P$

1. O'Neill, S. R., *et al., Inorg. Chem.*, 1972, **11**, 1630
2. Lines, E. L. *et al.*, ibid., 2270

It is photolytically and thermally unstable and has exploded at 25°C. It is also explosively sensitive to sudden changes in pressure, as in expansion into a vacuum or surging during boiling [1]. It also ignites in air [2].

*See other* NON-METAL AZIDES

## PHOSPHORUS AZIDE DIFLUORIDE–BORANE   $F_2N_3P{\cdot}BH_3$

$P(N_3)F_2 \cdot BH_3$

Lines, E. L. *et al., Inorg. Chem.*, 1972, **11**, 2270

The liquid exploded violently during transfer operations.

*See other* NON-METAL AZIDES

## OXYGEN DIFLUORIDE   $OF_2$   $F_2O$

Adsorbents

1. Metz, F. I., *Chem. Eng. News*, 1965, **43**(7), 41
2. Streng, A. G., *Chem. Eng. News*, 1965, **43**(12), 5
3. Streng, A. G., *Chem. Rev.*, 1963, **63**, 611

Mixtures of silica gel and the liquid difluoride sealed in tubes at 334 mbar exploded above −196°C, presence of moisture rendering the mixture shock-sensitive at this temperature [1]. Reaction of oxygen difluoride with silica, alumina, molecular sieve or similar surface-active solids is exothermic, and under appropriate conditions may be explosive [2]. A quartz fibre can be ignited in the difluoride [3].

Combustible gases

Streng, A. G., *Chem. Rev.*, 1963, **63**, 612

Mixtures with carbon monoxide, hydrogen and methane are stable at ambient temperatures, but explode violently on sparking. Hydrogen sulphide explodes with oxygen difluoride at ambient temperature and, though interaction is smooth at −78°C under reduced pressure, the white solid produced exploded violently when cooling was stopped.

Diborane

Rhein, R. A., *Combust. Inst. Paper WSC1-67-10*, 73–80, 1967

Although it reacts slowly at ambient temperature, a mixture of the components which is stable at –195°C could explode during warming to ambient conditions.

Diboron tetrafluoride

Holliday, A. K. *et al., J. Chem. Soc.*, 1964, 2732

Ignition occurred on mixing at, or on warming mixtures to, –80°C.

Halogens
or Metal halides

Streng, A. G., *Chem. Rev.*, 1963, **63**, 611

Mixtures with chlorine, bromine or iodine explode on warming. A mixture with chlorine passed through a copper tube at 300°C exploded with variable intensity. Aluminium trichloride explodes in the difluoride, and antimony pentachloride lightly at 150°C.

Metals

Streng. A. G., *Chem. Rev.,* 1963, **63**, 611

Finely divided platinum-group metals react on gentle warming and coarser materials at higher temperatures; aluminium, barium, cadmium, magnesium, strontium, zinc and zirconium evolving light. Lithium, potassium and sodium incandesce brilliantly at 400°C, while tungsten explodes.

Nitrogen oxide

Streng, A. G., *Chem. Rev.,* 1963, **63**, 612

Gaseous mixtures may explode on sparking. The mixed gases slowly react to give a mixture (NO, NOF), which, if liquefied by cooling, will explode on warming.

Nitrosyl fluoride

Hoffman, C. J. *et al., Chem. Rev.*, 1962, **62**, 9

A solid mixture explodes on melting, and the gaseous components ignite on mixing.

*See* Nitrogen oxide, above

Non-metals

1. Sidgwick, 1950, 1136
2. Streng, A. G., *Chem. Rev.,* 1963, **63**, 611

Pressure of the gas must be limited during concentration by contact with cooled charcoal to avoid violent explosions [1]. Red phosphorus ignites when gently warmed, and powdered boron and silicon generate sparks on heating in the difluoride [2].

Phosphorus(V) oxide
Streng, A. G., *Chem. Rev.*, 1963, **63**, 611
Ignition occurs spontaneously on contact.

Water
Streng, A. G., *Chem. Rev.*, 1963, **63**, 610
Presence of water or water vapour in oxygen difluoride is dangerous, the mixture (even when diluted with oxygen) exploding violently on spark initiation, especially at 100°C (i.e. with steam).
*See* Adsorbents, above
OXYGEN FLUORIDES

**SULPHINYL FLUORIDE** **$SOF_2$** **$F_2OS$**

Sodium
*See* SODIUM, Na: Non-metal halides (reference 8)

**XENON DIFLUORIDE OXIDE** **$XeF_2O$** **$F_2OXe$**
Jacob, E. *et al., Angew. Chem. (Intern. Ed.)*, 1976, **15**, 158–159
Although stable at below –40°C in absence of moisture, it will explode if warmed rapidly (>20°C/h). Explosive decomposition of the solid difluoride oxide at –196°C occurs on contact with mercury, or with antimony or arsenic pentafluorides.
*See other* XENON COMPOUNDS

**DIOXYGEN DIFLUORIDE** **$O_2F_2$** **$F_2O_2$**

Sulphur trioxide
Solomon, I. J. *et al., J. Chem. Eng. Data*, 1968, **13**, 530
Interaction of the endothermic fluoride with the trioxide is very vigorous, and explosive in absence of solvent.

Various materials
Streng, A. G., *Chem. Rev.*, 1963, **63**, 615
Though not shock-sensitive, it is of limited thermal stability, decomposing below its b.p., –57°C, and explosively in contact with fluorided platinum at –113°C. It is a very powerful oxidant and reacts vigorously or violently with many materials at cryogenic temperatures. It explodes with methane at –194°C, with ice at –140°C, with solid ethanol at below –130°C and with acetone–solid carbon dioxide at –78°C. Ignition or explosion may occur with chlorine, phosphorus trifluoride, sulphur tetrafluoride or

tetrafluoroethylene, in the range −130 to −190°C. Even a 2% solution in hydrogen fluoride ignites solid benzene at −78°C.
*See* OXYGEN FLUORIDES

## SELENIUM DIFLUORIDE DIOXIDE $SeO_2F_2$ $F_2O_2Se$

Ammonia
1. Bailar, 1973, Vol. 2, 966
2. Engelbrecht, A. *et al., Monatsh.*, 1962, **92**, 555, 581

Interaction is violent [1] and many of the products and derivatives are both shock- and heat-sensitive explosives [2]. These include the ammonium, potassium, silver and thallium salts of the 'triselenimidate' ion, systematically 2,4,6-tris(dioxoselena)perhydrotriazine-1,3,5-triide.
*See related* NON-METAL HALIDES
*See other* *N*-METAL DERIVATIVES

## 'TRIOXYGEN DIFLUORIDE' $F_2O_3$

Bailar, 1973, Vol. 2, 761–763
This material is now considered to be a mixture of dioxygen difluoride and tetraoxygen difluoride, rather than the title compound.

Various materials
Streng, A. G., *Chem. Rev.*, 1963, **63**, 619
Though thermally rather unstable, decomposing above its m.p., −190°C, it appears not to be inherently explosive. It is, however, an extremely potent oxidiser and contact with oxidisable materials causes ignition or explosions, even at −183°C. At this temperature, single drops added to solid hydrazine or liquid methane cause violent explosions, while solid ammonia, bromine, charcoal, iodine, red phosphorus and sulphur react with ignition and/or mild explosion. It is also extremely effective at initiating ignition of combustible materials in liquid oxygen, even at 0.1% concentration, whereas mixtures of ozone and fluorine in oxygen are ineffective. This effect has been examined for use in hypergolic rocket propellant systems. Tetryl detonates spontaneously on contact with the difluoride.
*See* OXYGEN FLUORIDES

## FLUORINE FLUOROSULPHATE $FSO_3F$ $F_2O_3S$

Cady, G. H., *Chem. Eng. News*, 1966, **44**(8), 40; *Inorg. Synth.*, 1968, **11**, 155
The crude fluorosulphate, produced as by-product in preparation of peroxodisulphuryl difluoride, was distilled into a cooled steel

cylinder and, on warming to room temperature, the cylinder exploded. It decomposes at 200°C, but not explosively.
*See related* HYPOHALITES

**DISULPHURYL DIFLUORIDE** **$FSO_2OSO_2F$** **$F_2O_5S_2$**

Ethanol
Hayek, E., *Monatsh.*, 1951, 82, 942
Violent reaction on mixing at room temperature.
*See other* NON-METAL HALIDES

**HEXAOXYGEN DIFLUORIDE** **$O_6F_2$** **$F_2O_6$**
Bailar, 1973, Vol. 2, 764
Flashlight illumination or rapid warming of the solid at 60K to 90K may lead to explosion.
*See other* HALOGEN OXIDES

**PEROXODISULPHURYL DIFLUORIDE** **$F_2O_6S_2$**
**$FSO_2OOSO_2F$**
Shreeve, J. M. *et al., Inorg. Synth.*, 1963, 7, 124
It ignites organic materials immediately on contact.

Carbon monoxide
Gatti, R. *et al., Z. Phys. Chem.*, 1965, 47, 323–336
Interaction proceeds explosively at above 20°C.

Dichloromethane
Kirchmeier, R. L. *et al., Inorg. Chem.*, 1973, 12, 2889
Equimolar amounts exploded while warming to ambient temperature during 30 min after initial contact at −183°C. Dilution of the dichloromethane with 2 vols of trichlorofluoromethane prevented explosion at −20°C.
*See also* FLUORINE FLUOROSULPHATE, $F_2O_3S$
*See other* OXIDANTS

**LEAD DIFLUORIDE** **$PbF_2$** **$F_2Pb$**

Fluorine
*See* FLUORINE, $F_2$: Metal salts

**POLY(DIFLUOROSILYLENE)** $(SiF_2)_n$ $(F_2Si)_n$

1. Bailar, 1973, Vol. 1, 1352
2. Perry, D. L. *et al., J. Chem. Educ.*, 1976, **53**, 696–699

Produced by condensation at low temperature, the rubbery polymer ignites in air [1], and preparation, handling and reactions have been detailed [2].

*See other* NON-METAL HALIDES
PYROPHORIC MATERIALS

**XENON DIFLUORIDE** $XeF_2$ $F_2Xe$

*See* XENON TRIOXIDE, $O_3Xe$

Silicon–nitrogen compounds

Gibson, J. A. *et al., Can. J. Chem.*, 1975, **53**, 3050

Interaction of xenon difluoride and dimethylaminotrimethylsilane in presence or absence of solvent became explosive at subzero temperatures.

*See also* XENON TETRAFLUORIDE, $F_4Xe$

*See other* XENON COMPOUNDS

† **TRIFLUOROSILANE** $F_3SiH$ $F_3HSi$

**IODINE DIOXYGEN TRIFLUORIDE** $IO_2F_3$ $F_3IO_2$

Engelbrecht, A. *et al., Angew. Chem. (Intern. Ed.)*, 1969, **8**, 769

It ignites on contact with flammable organic materials.

*See other* HALOGEN OXIDES

**MANGANESE TRIFLUORIDE** $MnF_3$ $F_3Mn$

Glass

Mellor, 1942, Vol. 12, 344

When heated in contact it attacks glass violently, silicon tetrafluoride being evolved.

*See other* METAL HALIDES

**NITROGEN TRIFLUORIDE** $NF_3$ $F_3N$

Charcoal

Massonne, J. *et al., Angew. Chem. (Intern. Ed.)*, 1966, **5**, 317

Adsorption of the fluoride on to activated granular charcoal at $-100^\circ C$ caused an explosion, attributed to the heat of adsorption not being dissipated and causing decomposition to nitrogen and carbon tetrafluoride. No reaction occurs at +100°C in a flow system, but incandescence occurs at 150°C.

Hydrogen-containing compounds

1. Ruff, O., *Z. Angew. Chem.*, 1929, **42**, 807
2. Hoffman, C. J. *et al.*, *Chem. Rev.*, 1962, **62**, 4

Sparking of mixtures with ammonia or hydrogen causes violent explosions, and with steam, feeble ones [1]. Mixtures with ethylene, methane and hydrogen sulphide (also carbon monoxide) explode on sparking [2].

Tetrafluorohydrazine

*MCA Case History No. 683*

A crude mixture of the two compounds, kept for 3 days in a stainless steel cylinder, exploded violently during valve manipulation.

*See other N*-HALOGEN COMPOUNDS

**PHOSPHORUS TRIFLUORIDE** **$PF_3$** **$F_3P$**

Borane

*See* BORANE–PHOSPHORUS TRIFLUORIDE, $BH_3 \cdot F_3P$

Dioxygen difluoride

*See* DIOXYGEN DIFLUORIDE, $F_2O_2$

Fluorine

*See* FLUORINE, $F_2$ : Covalent halides

Hexafluoroisopropylideneaminolithium

*See* HEXAFLUOROISOPROPYLIDENEAMINOLITHIUM, $C_3F_6LiN$: Non-metal halides

**THIOPHOSPHORYL FLUORIDE** **$PSF_3$** **$F_3PS$**

Air,
or Sodium

Mellor, 1940, Vol. 8, 1072–1073

In contact with air, the fluoride ignites or explodes, depending on contact conditions. Heated sodium ignites in the gas.

**PALLADIUM TRIFLUORIDE** **$PdF_3$** **$F_3Pd$**

Hydrogen

Sidgwick, 1950, 1574

Contact with hydrogen causes the unheated fluoride to be reduced incandescently.
*See other* METAL HALIDES

**MANGANESE TETRAFLUORIDE** **$MnF_4$** **$F_4Mn$**

Petroleum oil
Sorbe, 1968, 84
Interaction leads to fire.
*See other* METAL HALIDES

**† TETRAFLUOROHYDRAZINE** **$F_2NNF_2$** **$F_4N_2$**

Logothetis, A. L., *J. Org. Chem.*, 1966, **31**, 3686, 3689
Modica, A. P. *et al.*, *Rep. No. 357-275*, Princeton Univ., 1963
General precautions for use of the explosive gas tetrafluorohydrazine and derived reaction products include: reactions on as small a scale as possible and behind a barricade; adequate shielding during work-up of products because explosions may occur; storage of tetrafluorohydrazine at –80°C under 1–2 bar pressure in previously fluorinated Monel or stainless steel cylinders with Monel valves; distillation of volatile (difluoroamino) products in presence of an inert halocarbon oil to prevent explosions in dry distilling vessels. Light-initiated explosion of the gas has been reported.

Air
Martin, K. J., *J. Amer. Soc.*, 1965, **87**, 394
This explodes on contact with air or combustible vapours.

Alkenyl nitrates
Reed, S. F. *et al.*, *J. Org. Chem.*, 1972, **37**, 3329
The products of interaction of tetrafluorohydrazine and alkenyl nitrates, bis(difluoroamino)alkyl nitrates, are heat- and impact-sensitive explosives.
*See* DIFLUOROAMINO COMPOUNDS

Hydrocarbons
Petry, R. C. *et al.*, *J. Org. Chem.*, 1967, **32**, 4034
Mixtures are potentially highly explosive, approaching the energy of hydrocarbon–oxygen systems.

Hydrogen
Kuhn, L. P. *et al.*, *Inorg. Chem.*, 1970, **9**, 602
Explosive interaction is rather unpredictable, the initiation temperature required (20–80°C) depending on the condition of the vessel walls.

Nitrogen trifluoride

*See* NITROGEN TRIFLUORIDE, $F_3N$: Tetrafluorohydrazine

Organic materials

Reed, S. F., *J. Org. Chem.*, 1968, **33**, 2634

Mixtures with organic materials in presence of air constitute explosion hazards. Appropriate precautions are essential.

Ozone

Sessa, P. A. *et al., Inorg. Chem.*, 1971, **10**, 2067

When tetrafluorohydrazine was pyrolysed at 310°C to generate $NF_2$ radicals and the mixture contacted liquid ozone at −196°C, a violent explosion occurred.

*See other N*-HALOGEN COMPOUNDS

## XENON TETRAFLUORIDE OXIDE $XeF_4O$ $F_4OXe$

Preparative hazard

*See* XENON HEXAFLUORIDE, $F_6Xe$: Silicon dioxide

Graphite,
Potassium iodide

Selig, H. *et al., Inorg. Nucl. Chem. Lett.*, 1975, **11**(1), 75–77

The graphite–xenon tetrafluoride oxide intercalation compound exploded in contact with potassium iodide solution.

*See other* XENON COMPOUNDS

## PALLADIUM TETRAFLUORIDE $PdF_4$ $F_4Pd$

Water

Bailar, 1973, Vol. 3, 1278

Interaction is violent.

*See other* METAL HALIDES

## PLATINUM TETRAFLUORIDE $PtF_4$ $F_4Pt$

Water

Sidgwick, 1950, 1614

Interaction is violent.

*See other* METAL HALIDES

## RHODIUM TETRAFLUORIDE $RhF_4$ $F_4Rh$

Water
Bailar, 1973, Vol. 3, 1235
Interaction is violent.
*See other* METAL HALIDES

## SULPHUR TETRAFLUORIDE $SF_4$ $F_4S$

2-Methyl-3-butyn-2-ol
Boswell, G. A. *et al., Org. React.*, 1974, **21**, 8
Interaction at −78°C is explosively vigorous.

Dioxygen difluoride
*See* DIOXYGEN DIFLUORIDE, $F_2O_2$

## SELENIUM TETRAFLUORIDE $SeF_4$ $F_4Se$

Chlorine trifluoride
Olah, G. A. *et al., J. Amer. Chem. Soc.*, 1974, **96**, 927
The tetrafluoride is prepared by interaction of chlorine trifluoride and selenium in selenium tetrafluoride as solvent. The crude tetrafluoride must be substantially free from excess chlorine trifluoride to avoid danger during subsequent distillation at 106°C/l bar.

Water
Aynsley, E. E. *et al., J. Chem. Soc.*, 1952, 1231
Violent interaction.
*See other* NON-METAL HALIDES

## SILICON TETRAFLUORIDE $SiF_4$ $F_4Si$

Sodium
*See* SODIUM, Na: Non-metal halides (reference 8)

## XENON TETRAFLUORIDE $XeF_4$ $F_4Xe$

1. Shieh, J. C. *et al., J. Org. Chem.*, 1970, **35**, 4022
2. Malm, J. G. *et al., Inorg. Synth.*, 1966, **8**, 254
3. Chernick, C. L., *J. Chem. Educ.*, 1966, **43**, 619
4. Holloway, J. H., *Talanta,* 1967, **14**, 871
5. Falconer, E. E. *et al., J. Inorg. Nucl. Chem.*, 1967, **29**, 1380

Moisture converts it to highly shock-sensitive xenon oxides [1].
Necessary precautions are given [2–5].
*See* XENON TRIOXIDE, $O_3Xe$

Flammable materials
Klimov, B. D. *et al., Chem. Abs.*, 1970, 72, 85784t
Xenon tetrafluoride or difluoride or their mixtures could not be caused to detonate by impact, although after exposure to moisture the tetrafluoride becomes explosive owing to the formation of xenon trioxide. Both the di- and tetra-fluorides may cause explosion in contact with acetone, aluminium, pentacarbonyliron, styrene, polyethylene, lubricants, paper, sawdust, wool or other combustible materials. Their vigorous reactions with ethanol, potassium iodate or permanganate are not explosive, however.
*See other* XENON COMPOUNDS

**PENTAFLUOROORTHOSELENIC ACID** **$F_5HOSe$**

**$F_5SeOH$**

Preparative hazard
*See* FLUORINE, $F_2$: Hydrogen fluoride
: Seleninyl fluoride

**IODINE PENTAFLUORIDE** **$IF_5$** **$F_5I$**

Benzene
Ruff, O. *et al., Z. Anorg. Chem.*, 1931, **201**, 245
Interaction becomes violent above 50°C.

Calcium carbide,
or Potassium hydride
Booth, H. S. *et al., Chem. Rev.*, 1947, **41**, 425
Both incandesce on contact (the carbide when warmed).

Diethylaminotrimethylsilane
Oates, G. *et al., J. Chem. Soc., Dalton Trans.*, 1974, 1383
A mixture exploded at *ca.* −80°C. Reactions with other silanes were very exothermic.

Dimethyl sulphoxide
Lawless, E. M., *Chem. Eng. News,* 1969, **47**(13), 8, 109
Unmoderated reaction with the sulphoxide is violent, and in the presence of diluents reaction may be delayed and become explosively violent. Although small-scale reactions were uneventful,

reactions involving about 0.15 g mol of the pentafluoride and sulphoxide in presence of trichlorofluoromethane or tetrahydrothiophene-1,1-dioxide as diluents caused delayed and violent explosions. Silver difluoride and other fluorinating agents also react violently with the sulphoxide.

*See* DIMETHYL SULPHOXIDE, $C_2H_6OS$: Acyl halides, etc.

Limonene,
Tetrafluoroethylene

*See* TETRAFLUOROETHYLENE, $C_2F_4$: Iodine pentafluoride, etc.

Metals,
or Non-metals

1. Sidgwick, 1950, 1159
2. Mellor, 1940, Vol. 2, 114
3. Booth, H. S. *et al., Chem. Rev.*, 1947, **41**, 424–425

Contact with boron, silicon, red phosphorus, sulphur, or arsenic, antimony or bismuth usually causes incandescence [1]. Solid potassium or molten sodium explodes with the pentafluoride, and aluminium foil ignites on prolonged contact [2]. Molybdenum and tungsten incandesce when warmed [3].

Organic materials,
or Potassium hydroxide

Pascal, 1960, Vol. 16.1, 582

The pentafluoride chars and usually ignites organic materials, and interaction with potassium hydroxide is violently exothermic.

Potassium hydride

*See* Calcium carbide, etc., above

Water

Ruff, O. *et al., Z. Anorg. Chem.*, 1931, **201**, 245

Reaction with water or water-containing materials is violent.

*See other* INTERHALOGENS

## IRIDIUM HEXAFLUORIDE $IrF_6$ $F_6Ir$

Silicon

*See* SILICON, Si: Metal hexafluorides

## DIPOTASSIUM HEXAFLUOROMANGANATE(IV) $F_6K_2Mn$

$K_2[MnF_6]$

Preparative hazard

*See* POTASSIUM PERMANGANATE, $KMnO_4$: Hydrofluoric acid

## POTASSIUM HEXAFLUOROSILICATE(2–) $K_2[SiF_6]$ $F_6K_2Si$

Hydrogen fluoride

*See* HYDROGEN FLUORIDE, FH: Potassium tetrafluorosilicate(2–)

## NEPTUNIUM HEXAFLUORIDE $NpF_6$ $F_6Np$

Water

Bailar, 1973, Vol. 5, 168

Interaction with water at ambient temperature is violent.

*See other* METAL HALIDES

## PENTAFLUOROSULPHUR HYPOFLUORITE $F_5SOF$ $F_6OS$

Ruff, J. K., *Inorg. Synth.*, 1968, **11**, 137

Considered to be potentially explosive.

*See other* HYPOHALITES

## PENTAFLUOROSELENIUM HYPOFLUORITE $F_5SeOF$ $F_6OSe$

1. Mitra, G. *et al., J. Amer. Chem. Soc.*, 1959, **81**, 2646
2. Smith, J. E. *et al., Inorg. Chem.*, 1970, **9**, 1442

A cooled sample exploded when allowed to warm rapidly [1], but this may have been due to impurities, as it did not happen in the later work [2].

*See other* HYPOHALITES

## OSMIUM HEXAFLUORIDE $OsF_6$ $F_6Os$

Paraffin oil

Sorbe, 1968, 91

It causes ignition of paraffin oil and other organic materials.

Silicon

*See* SILICON, Si: Metal hexafluorides

*See other* METAL HALIDES
OXIDANTS

### PLATINUM HEXAFLUORIDE $PtF_6$ $F_6Pt$

Bistrifluoromethyl nitroxide
Christe, K. O. *et al., J. Fluorine Chem.*, 1974, 4, 425
Interaction of the nitroxide radical and this powerful oxidant was very violent during warming from −196°C.
*See other* METAL HALIDES
OXIDANTS

### PLUTONIUM HEXAFLUORIDE $PuF_6$ $F_6Pu$

Water
Bailar, 1973, Vol. 5, 168
Interaction with water at ambient temperature is violent.
*See other* METAL HALIDES

### RHENIUM HEXAFLUORIDE $ReF_6$ $F_6Re$

Silicon
*See* SILICON, Si: Metal hexafluorides

### SULPHUR HEXAFLUORIDE $SF_6$ $F_6S$

Disilane
*See* DISILANE, $H_6Si_2$: Sulphur hexafluoride

### URANIUM HEXAFLUORIDE $UF_6$ $F_6U$

Aromatic hydrocarbons,
or Hydroxy compounds
Sidgwick, 1950, 1072
Interaction with benzene, toluene or xylene is very vigorous, with separation of carbon, and violent with ethanol or water.
*See other* METAL HALIDES

### XENON HEXAFLUORIDE $XeF_6$ $F_6Xe$

Fluoride donors,
Water
Bailar, 1973, Vol. 1, 317
Adducts of the hexafluoride with sodium, potassium, rubidium, caesium or nitrosyl fluorides react violently with water.

Hydrogen

Malm, J. G. *et al., J. Amer. Chem. Soc.*, 1963, **85**, 110

Interaction is violent.

Silicon dioxide

1. McKee, D. E. *et al., Inorg. Chem.*, 1973, **12**, 1722
2. Aleinikov, N. N. *et al., J. Chromatog.*, 1974, **89**, 367

Interaction of the yellow hexafluoride with silica to give xenon tetrafluoride oxide must be interrupted before completion (disappearance of colour) to avoid the possibility of formation and detonation of xenon trioxide [1]. An attempt to collect the hexafluoride in quartz traps at −20°C after separation by preparative gas chromatography failed because of reaction with the quartz and subsequent explosion of the oxygen compounds of xenon so produced [2].

*See* XENON TRIOXIDE, $O_3Xe$

Water

Gillespie, R. J. *et al., Inorg. Chem.*, 1974, **13**, 2370–2374

Although uncontrolled reaction of xenon hexafluoride and moisture produces explosive xenon trioxide, controlled action by progressive addition of limited amounts of water vapour with agitation to a frozen solution of the hexafluoride in anhydrous hydrogen fluoride at −196°C to give xenon oxide tetrafluoride or xenon dioxide difluoride is safe.

*See other* XENON COMPOUNDS

## IODINE HEPTAFLUORIDE $IF_7$ $F_7I$

Carbon,
or Combustible gases

Booth, H. S. *et al., Chem. Rev.*, 1947, **41**, 428

Activated carbon ignites immediately in the gas, mixtures with methane ignite, and with carbon monoxide ignite on warming, while those with hydrogen explode on heating or sparking.

Metals

Booth, H. S. *et al., Chem. Rev.*, 1947, **41**, 427

Interaction with barium, potassium and sodium is immediate, accompanied by evolution of heat and light. Aluminium, magnesium and tin are passivated on contact, but on heating react similarly to the former metals.

Organic solvents

Booth, H. S. *et al., Chem. Rev.*, 1947, **41**, 428

Benzene, light petroleum, ethanol and ether ignite in contact with

the gas, while the exotherm with acetic acid, acetone or ethyl acetate caused rapid boiling. General organic materials (cellulose, grease, oils) ignite if excess heptafluoride is present.
*See other* INTERHALOGENS

**DIFLUOROAMMONIUM HEXAFLUOROANTIMONATE** **$F_8H_2NSb$**

**$F_2N^+H_2$ $[SbF_6]^-$**

*See* DIFLUOROAMMONIUM HEXAFLUOROARSENATE, $AsF_8H_2N$

**BIS(*S*,*S*-DIFLUORO-*N*-SULPHIMIDO)SULPHUR TETRAFLUORIDE** **$F_8N_2S_3$**

**$F_2S{=}NSF_4N{=}SF_2$**

Water
Hofer, R. *et al., Angew. Chem. (Intern. Ed.)*, 1973, **12**, 1000
Either the liquid tetrafluoride or its viscous polymer decomposes explosively in contact with water.
*See related* NON-METAL HALIDES

**XENON(II)PENTAFLUOROORTHOSELENATE** **$F_{10}O_2Se_2Xe$**

**$Xe[F_5SeO]_2$**

Oxidisable materials
Seppelt, K., *Angew. Chem. (Intern. Ed.),* 1972, **11**, 724
Interaction is explosive.
*See other* XENON COMPOUNDS

**XENON(II) PENTAFLUOROORTHOTELLURATE** **$F_{10}O_2Te_2Xe$**

**$Xe[F_5TeO]_2$**

Organic solvents
Sladky, F., *Angew. Chem. (Intern. Ed.),* 1969, 8, 523
Explosive or very vigorous reactions occur on contact with acetone, benzene or ethanol.
*See other* XENON COMPOUNDS

**OCTAKIS(TRIFLUOROPHOSPHINE)DIRHODIUM** **$F_{24}P_8Rh_2$**

**$[(F_3P)_4RhRh(PF_3)_4]$**

Acetylenic esters
Bennett, M. A. *et al., Inorg. Chem.*, 1976, **15**, 107–108

Formation of complexes with excess methyl propiolate or dimethyl acetylenedicarboxylate must not be allowed to proceed at above +20°C, or violently explosive polymerisation of the acetylenic esters will occur.

*See other* POLYMERISATION INCIDENTS

**IRON** **Fe**

1. Bailar, 1973, Vol. 3, 985
2. Gusein, M. A. *et al., Chem. Abs.*, 1974, **81**, 124774x

The known pyrophoric [1] and explosive properties of ultrafine iron powders were examined in detail [2].

*See other* PYROPHORIC METALS

Acetaldehyde

*See* ACETALDEHYDE, $C_2H_4O$: Metals

Air,
Oil

Glaser, A., *Arbeitschutz*, 1941, 134–135

Oxidative heating of oily iron dust in a collecting vessel caused vaporisation of oil and subsequent ignition, causing an explosion.

Air,
Water

1. Brimelow, H. C., private comm., 1972
2. Unpublished observations, 1949

*o*-Nitrophenylpyruvic acid was reduced to oxindole using iron pindust–ferrous sulphate in water. The iron oxide residues, after filtering and washing with chloroform, rapidly heated on exposure to air and shattered the Buchner funnel [1]. Previously, rapid heating effects had been observed on sucking air through the metal oxide residue from hot filtration of aqueous liquor from reduction of a nitro compound with reduced iron powder [2].

Chloric acid

*See* CHLORIC ACID, $ClHO_3$: Metals, etc.

Chloroformamidinium nitrate

*See* CHLOROFORMAMIDINIUM NITRATE, $CH_4ClN_3O_3$: Alone, or Metals

Disodium acetylide

*See* DISODIUM ACETYLIDE, $C_2Na_2$: Metals

Halogens or Interhalogens

*See* BROMINE PENTAFLUORIDE, $BrF_5$: Acids, etc.

CHLORINE TRIFLUORIDE, $ClF_3$: Metals
CHLORINE, $Cl_2$: Metals
FLUORINE, $F_2$: Metals

Oxidants

*See* Halogens or Interhalogens, above
PEROXYFORMIC ACID, $CH_2O_3$: Metals
NITRYL FLUORIDE, $FNO_2$: Metals
HYDROGEN PEROXIDE, $H_2O_2$: Metals
AMMONIUM NITRATE, $H_4N_2O_3$: Metals
AMMONIUM PEROXODISULPHATE, $H_8N_2O_8S_2$: Iron
DINITROGEN TETRAOXIDE, $N_2O_4$: Metals
SODIUM PEROXIDE, $Na_2O_2$: Metals

Polystyrene

Unpublished observation, 1971

Iron flake powder and polystyrene beads had been blended in a high-speed mixer. The mixture ignited and burned rapidly when discharged into a polythene bag. Rapid oxidation of the finely divided metal and/or static discharge may have initiated the fire. No ignition occurred when the iron powder was surface-coated with stearic acid.

*See other* METALS

**IRON–SILICON** **Fe–Si**

Anon., *Chem. Trade J.*, 1956, **139**, 1180

Ferrosilicon containing from 30 to 75% of silicon is hazardous, particularly when finely divided, and must be kept in a moisture-tight drum. In contact with water, the impurities present (arsenide, carbide, phosphide) evolve poisonous arsine, combustible acetylene, and spontaneously flammable phosphine.

**IRON(II) HYDROXIDE** **$Fe(OH)_2$** **$FeH_2O_2$**

Gibson, 1969, 121

Prepared under nitrogen, it is pyrophoric in air, producing sparks.

*See other* PYROPHORIC MATERIALS

**AMMONIUM IRON(III) SULPHATE** **$NH_4Fe(SO_4)_2$** **$FeH_4NO_8S_2$**

Sulphuric acid

*See* SULPHURIC ACID, $H_2O_4S$: Ammonium iron(III) sulphate

## IRON(II) IODIDE $FeI_2$

Alkali metals

*See* POTASSIUM, K: Metal halides
SODIUM, Na: Metal halides

## POTASSIUM PEROXOFERRATE(2–) $K_2[Fe(O_2)O_3]$ $FeK_2O_5$

Alone,
or Non-metals,
or Sulphuric acid

Goralevich, D. K., *J. Russ. Phys. Chem. Soc.*, 1926, **58**, 1155
Explodes on heating or impact, or in contact with charcoal, phosphorus, sulphur or sulphuric acid.
*See other* PEROXOACID SALTS

## IRON(III) NITRATE $Fe(NO_3)_3$ $FeN_3O_9$

Dimethyl sulphoxide

*See* DIMETHYL SULPHOXIDE, $C_2H_6OS$: Metal oxosalts

## IRON(II) OXIDE FeO

Air,
or Sulphur dioxide

Mellor, 1941, Vol. 13, 715
Bailar, 1973, Vol. 3, 1008–1009
The oxide (prepared at 300°C) incandesces when heated in sulphur dioxide, and burns in air above 200°C. The finely divided oxide prepared by reduction may be pyrophoric in air at ambient temperature. The oxide produced by thermal decomposition under vacuum of iron(II) oxalate is also pyrophoric.
*See other* PYROPHORIC MATERIALS

Oxidants

*See* NITRIC ACID, $HNO_3$: Iron(II) oxide
HYDROGEN PEROXIDE, $H_2O_2$: Metals, etc.
*See other* METAL OXIDES

## IRON(II) SULPHATE $FeSO_4$ $FeO_4S$

Diarsenic trioxide,
Sodium nitrate

*See* SODIUM NITRATE, $NNaO_3$: Diarsenic trioxide, Iron(II) sulphate

## IRON(II) SULPHIDE $FeS$

1. Mellor, 1942, Vol. 14, 157
2. Anon., *Chem. Age*, 1939, **40**, 267

The moist sulphide readily oxidises in air exothermically, and may reach incandescence. Grinding in a mortar hastens this [1]. The impure sulphide formed when steel equipment is used with materials containing hydrogen sulphide or volatile sulphur compounds is pyrophoric, and has caused many fires and explosions when such equipment is opened without effective purging. Various methods of purging are discussed [2].

Lithium

*See* LITHIUM, Li: Metal oxides, etc.
*See other* METAL SULPHIDES

## IRON DISULPHIDE $FeS_2$

Anon., *Angew. Chem. (Nachr.)*, 1954, **2**, 219
Bowes, P. C., *Ind. Chemist*, 1954, **30**, 12–14
Ruiss, I. G. *et al., J. Chem. Ind. (Moscow)*, 1935, **12**, 692–696

Finely powdered pyrites, especially in presence of water, will rapidly heat spontaneously and ignite, particularly in contact with combustible materials. Inert gas blanketing will prevent this.

Precautions to reduce the self-ignition hazards of powdered pyrites, and the explosion hazards of pyrites–air mixtures in the burning ovens of sulphuric acid plants have been detailed and discussed.

Carbon

Ruiss, I. G. *et al., J. Chem. Ind. (Moscow)*, 1935, **12**, 696

The presence of carbon in pyrites lowers the ignition temperature to 228–242°C and increases the explosivity of dust suspensions in air.
*See other* METAL SULPHIDES

## DIIRON TRIOXIDE $Fe_2O_3$

Aluminium

Mellor, 1946, Vol. 5, 217

An intimately powdered mixture, usually ignited by magnesium ribbon, reacts with an intense exotherm to give molten iron and has been used commercially as 'thermite' for welding purposes. Incendive particles have been produced by this reaction on impact between aluminium and rusty iron.

*See* Calcium disilicide, below
LIGHT ALLOYS
ALUMINIUM, Al: Metal oxides, etc.

Aluminium,
Propene

Batty, G. F., private comm., 1972

Use of a rusty iron tool on an aluminium compressor piston caused incendive sparks which ignited residual propene–air mixture in the cylinder.

*See* Aluminium, above

Aluminium–magnesium alloy,
Water

*See* ALUMINIUM–MAGNESIUM ALLOY, Al–Mg: Iron(III) oxide, etc.

Aluminium–magnesium–zinc alloys

*See* ALUMINIUM–MAGNESIUM–ZINC ALLOYS, Al–Mg–Zn: Rusted steel

Calcium disilicide

Berger, E., *Compt. Rend.*, 1920, **170**, 29

The mixture ('silicon thermite') attains a very high temperature when heated, producing molten iron, similar to the normal thermite.

*See* Aluminium, above

Carbon monoxide

Othen, C. W., *School Sci. Rev.*, 1964, **45**(156), 459

The reason for a previously reported explosion during reduction of iron oxide with carbon monoxide is given as the formation of pentacarbonyliron at temperatures between 0 and 150°C. Suitable heating arrangements and precautions will eliminate this hazard.

*See* PENTACARBONYLIRON, $C_5FeO_5$

Ethylene oxide

*See* ETHYLENE OXIDE, $C_2H_4O$: Contaminants

Guanidinium perchlorate

*See* GUANIDINIUM PERCHLORATE, $CH_6ClN_3O_4$: Diiron trioxide

Hydrogen peroxide

*See* HYDROGEN PEROXIDE, $H_2O_2$: Metals, etc.

Magnesium

*See* MAGNESIUM, Mg: Metal oxides

Metal acetylides

*See* CALCIUM ACETYLIDE, $C_2Ca$: Iron(III) chloride, etc.

DICAESIUM ACETYLIDE, $C_2Cs_2$: Diiron trioxide
DIRUBIDIUM ACETYLIDE, $C_2Rb_2$: Metal oxides
*See other* METAL OXIDES

**DIIRON TRISULPHIDE** **$Fe_2S_3$**

Moore, F. M., *Proc. Gas. Cond. Conf.*, 1976, **26**, I
Hydrogen sulphide is removed from natural gas by passage over iron sponge, when flammable iron sulphide is produced. Handling precautions during regeneration of the reactor beds are detailed.
*See other* METAL SULPHIDES
PYROPHORIC IRON–SULPHUR COMPOUNDS

**TRIIRON TETRAOXIDE** **$FeO \cdot Fe_2O_3$** **$Fe_3O_4$**

Aluminium,
Calcium silicide,
Sodium nitrate
Schierwater, F.-W., *Sichere Chemiearb.*, 1976, **28**, 30–31
During the preparation of a foundry mixture of the finely divided oxide with aluminium powder and small amounts of calcium fluoride, calcium silicide and sodium nitrate, a violent explosion occurred in the conical mixer. It had been established previously that the mixture could not be ignited by impact or friction and that if ignited by a very high energy source (magnesium ribbon), it burned rather slowly to a glowing liquid. The possibility of ignition by silane produced from water or acid acting on the silicide content was discounted, but a dust explosion may have been involved.

Aluminium,
Sulphur
Crozier, T. H., *HM Insp. Expl. Spec. Rep. 237*, London, HMSO, 1920
A 20 tonne quantity of an incendiary bomb mixture of the finely powdered oxide, aluminium and sulphur became accidentally ignited and burned with almost explosive violence. It is similar to thermite mixture.
*See* DIIRON TRIOXIDE, $Fe_2O_3$: Aluminium

Hydrogen trisulphide
*See* HYDROGEN TRISULPHIDE, $H_2S_3$: Metal oxides

**GALLIUM** **Ga**

Halogens
Walker, H. L., *School Sci. Rev.*, 1956, **37**(132), 196

Bailar, 1973, Vol. 1, 1084
The metal reacts with cold chlorine strongly exothermically, and the compact metal with bromine even at −33°C, reaction being violent at ambient temperature.

Interaction of gallium with liquid bromine at 0°C proceeds with a flash, resembling reaction of an alkali metal with water.
*See other* METALS

**LITHIUM TETRAHYDROGALLATE** **$Li[GaH_4]$** **$GaH_4Li$**

Gaylord, 1956, 26
Though of lower stability than the analogous aluminate, its reactivity is generally similar to that of the latter.
*See other* COMPLEX HYDRIDES

**SODIUM TETRAHYDROGALLATE** **$Na[GaH_4]$** **$GaH_4Na$**

Mackay, 1966, 169
It is explosively hydrolysed by water.
*See other* COMPLEX HYDRIDES

**DIGALLANE** **$H_3GaGaH_3$** **$Ga_2H_6$**

Leleu, *Cahiers*, 1976, (85), 585
According to some authors it ignites in air.
*See other* METAL HYDRIDES

**DIGALLIUM OXIDE** **$Ga_2O$**

Bromine
Bailar, 1973, Vol. 1, 1091
It is a strong reducant, reacting violently with bromine.
*See other* METAL OXIDES
REDUCANTS

**GERMANIUM** **Ge**

Halogens
Mellor, 1941, Vol. 7, 260
The powdered metal ignites in chlorine, and lumps will ignite on heating in chlorine or bromine.

Oxidants
Mellor, 1941, Vol. 7, 260–261

The powdered metal reacts violently with nitric acid, and mixtures with potassium chlorate or nitrate explode on heating. Heated germanium burns with incandescence in oxygen.

*See* POTASSIUM HYDROXIDE, HKO: Germanium

*See other* METALS

**POLY(GERMANIUM MONOHYDRIDE)** **$(GeH)_n$**

Jolly, W. L. *et al., Inorg. Synth.*, 1963, **7**, 39

The solid polymeric hydride sometimes decomposes explosively into its elements on exposure to air.

*See other* METAL HYDRIDES

**GERMANIUM(II) IMIDE** **Ge=NH** **GeHN**

Oxygen

Johnson, O. H., *Chem. Rev.*, 1952, **51**, 449

On exposure to air it reacts violently, and in oxygen incandescence occurs.

*See other* *N*-METAL DERIVATIVES

**POLY(GERMANIUM DIHYDRIDE)** **$(GeH_2)_n$**

Bailar, 1973, Vol. 2, 13

Impact may cause explosive decomposition to the elements, with ignition of the liberated hydrogen.

*See other* METAL HYDRIDES

**AZIDOGERMANE** **$N_3GeH_3$** **$GeH_3N_3$**

Fluorosilane

Anon., *Angew. Chem. (Nachr.)*, 1970, **18**, 27

An attempt to prepare azidosilane by interaction of azidogermane and fluorosilane exploded.

*See related* METAL AZIDES

**† GERMANE** **$GeH_4$**

Brauer, 1963, Vol. 1, 715

Germane and its higher homologues decompose in air, often igniting.

Bromine

*See* BROMINE, $Br_2$: Germane

*See other* METAL HYDRIDES

**SODIUM GERMANIDE** NaGe **GeNa**

Air,
or Water
Johnson, O. H., *Chem. Rev.*, 1952, **51**, 452
The binary alloy is pyrophoric and may ignite in contact with water, as do other alkali metal germanides.
*See other* ALLOYS

**GERMANIUM(II) SULPHIDE** **GeS**

Potassium nitrate
*See* POTASSIUM NITRATE, $KNO_3$: Metal sulphides

**DIGERMANE** $H_3GeGeH_3$ $Ge_2H_6$

1. Mellor, 1941, Vol. 7, 264
2. Brauer, 1963, 715
3. MacKay, K. M., *Inorg. Synth.*, 1974, **15**, 170

It may ignite in air [1], particularly if air is admitted suddenly into the gas at reduced pressure [2]. Although digermane and its homologues do not usually ignite on exposure to air, their autoignition temperatures appear to be about 50°C, and combustion is rapid or explosive [3].
*See other* METAL HYDRIDES

**TRIGERMANE** $H_3GeGeH_2GeH_3$ $Ge_3H_8$

Brauer, 1963, 715
Air-sensitive, may ignite.
*See other* METAL HYDRIDES

**POLY(DIMERCURYIMMONIUM HYDROXIDE)** $(HHg_2NO)_n$
**('MILLON'S BASE ANHYDRIDE')** $(Hg{=}N^+{=}Hg\ OH^-)_n$

Sidgwick, 1950, 318
The anhydride of Millon's base explodes if touched or heated to 130°C.
*See* POLY(DIMERCURYIMMONIUM COMPOUNDS)
MERCURY, Hg: Ammonia
*See other* *N*-METAL DERIVATIVES

**HYDRIODIC ACID** **HI**

Muir, G. D., private comm., 1968
During preparation of hydriodic acid by distillation of

phosphorus and wet iodine, the condenser became blocked with by-product phosphonium iodide, and an explosion, possibly also involving phosphine, occurred. There is also a purification hazard.

*See* PHOSPHORUS, P: Hydriodic acid

Metals

*See* MAGNESIUM, Mg: Hydrogen iodide
POTASSIUM, K: Hydrogen iodide

Oxidants

Leleu, *Cahiers*, 1974, (75), 271

Hydrogen iodide ignites in contact with fluorine, dinitrogen tri- and tetra-oxides, and fuming nitric acid.

*See* ETHYL HYDROPEROXIDE, $C_2H_6O_2$: Hydriodic acid
PERCHLORIC ACID, $ClHO_4$: Iodides
POTASSIUM CHLORATE, $ClKO_3$: Hydrogen iodide

Phosphorus

*See* PHOSPHORUS, P: Hydriodic acid

## IODIC ACID $IO_2OH$ $HIO_3$

Non-metals

1. Mellor, 1946, Vol. 5, 15
2. Partington, 1967, 813

Interaction with boron below 40°C is vigorous, attaining incandescence [1]. Charcoal, phosphorus and sulphur deflagrate on heating [2].

*See* METAL HALOGENATES
*See other* OXIDANTS
OXOHALOGEN ACIDS

## PERIODIC ACID $IO_3OH$ $HIO_4$

Dimethyl sulphoxide

Rowe, J. J. M. *et al.*, *J. Amer. Chem. Soc.*, 1968, **90**, 1924

Although 1.5 M solutions of periodic acid in dimethyl sulphoxide explode after a few minutes, 0.15 M solutions appear stable.

*See* DIMETHYL SULPHOXIDE, $C_2H_6OS$: Metal oxosalts
: Perchloric acid

*See other* OXIDANTS
OXOHALOGEN ACIDS

**DIIODOAMINE** $I_2NH$ $HI_2N$

Mellor, 1940, Vol. 8, 607

Explosive, formed on prolonged contact of nitrogen triiodide with water.

*See* NITROGEN TRIIODIDE–AMMONIA, $I_3N{\cdot}H_3N$
*See other* *N*-HALOGEN COMPOUNDS

**POTASSIUM HYDRIDE** KH HK

1. Sorbe, 1968, 67
2. Brown, C. A., *J. Org. Chem.*, 1974, **39**, 3913–3918

It ignites on exposure to air [1], and the hydride dispersed in oil is much more highly reactive than sodium hydride dispersions, and rather more careful handling is necessary for safe working. Such precautions are detailed. Contact of even traces of the dispersion in flammable solvents with water will lead to ignition [2].

Air

*See* POTASSIUM HEXAHYDROALUMINATE(3–), $AlH_6K_3$

Fluoroalkene

*MCA Case History No. 2134*

After a few minutes' reflux at 12°C, a mixture of the hydride (0.01 mol) and a fluoroalkene (0.02 mol) exploded violently. This was attributed to possible presence of metallic potassium in the hydride causing polymerisation or formation of a fluoroacetylene.

Oxidants

*See* FLUORINE, $F_2$: Metal hydrides
OXYGEN (Gas), $O_2$: Metal hydrides
*See other* METAL HYDRIDES

**POTASSIUM HYDROXIDE** KOH HKO

Acids

*MCA Case History No. 920*

Incautious addition of acetic acid to a vessel contaminated with potassium hydroxide caused eruption of the acid.

Ammonium hexachloroplatinate(2–)

*See* AMMONIUM HEXACHLOROPLATINATE (2–), $Cl_6H_8N_2Pt$: Potassium hydroxide

Chlorine dioxide

*See* CHLORINE DIOXIDE, $ClO_2$: Potassium hydroxide

Germanium

Partington, 1967, 181

Germanium is oxidised by the fused hydroxide with incandescence.

Hyponitrous acid

*See* HYPONITROUS ACID, $H_2N_2O_2$

Maleic anhydride

*See* MALEIC ANHYDRIDE, $C_4H_2O_3$: Cations, etc.

Nitroalkanes

*See* NITROALKANES: Inorganic bases
NITROMETHANE, $CH_3NO_2$: Acids, etc.

Nitrobenzene

*See* NITROBENZENE, $C_6H_5NO_2$: Alkali

Nitrogen trichloride

*See* NITROGEN TRICHLORIDE, $Cl_3N$: Initiators

Potassium peroxodisulphate

*See* POTASSIUM PEROXODISULPHATE, $K_2O_8S_2$: Potassium hydroxide

2,2,3,3-Tetrafluoropropanol

*See* 2,2,3,3-TETRAFLUOROPROPANOL, $C_3H_4F_4O$: Potassium hydroxide

Tetrahydrofuran

*See* TETRAHYDROFURAN, $C_4H_8O$: Alkalies

Thorium dicarbide

*See* THORIUM DICARBIDE, $C_2Th$: Non-metals, etc.

2,4,6-Trinitrotoluene

*See* 2,4,6-TRINITROTOLUENE, $C_7H_5N_3O_6$: Added impurities

Water

*CISHC Chem. Safety Summ.*, 1976, **46**, 8–9

A mixture of flake potassium and sodium hydroxides was added to a reaction mixture without the agitator running. When this was started the batch erupted, owing to the sudden solution exotherm. Although this is a physical rather than chemical hazard, similar incidents have occurred frequently.

*See* SODIUM HYDROXIDE, HNaO: Water

**POTASSIUM PEROXOMONOSULPHATE KOSO$_2$OOH HKO$_5$S**

Castrantas, 1965, 5

Melts with decomposition at 100°C; forms explosive mixtures with as little as 1% of organic matter.
*See other* PEROXOACID SALTS

**DIPOTASSIUM PHOSPHINATE** **$K_2HPO_2$** **$HK_2O_2P$**

Mellor, 1971, Vol. 8, Suppl. 3, 623

The salt, probably KHPO·OK, ignites in contact with a little water.
*See other* REDUCANTS

**LITHIUM HYDRIDE** **LiH** **HLi**

Bailar, 1973, Vol. 1, 344

The powdered material burns readily on exposure to air.

Oxygen

*See* OXYGEN (Gas), $O_2$: Metal hydrides
OXYGEN (Liquid), $O_2$: Lithium hydride
*See other* METAL HYDRIDES

**PERMANGANIC ACID** **$HMnO_4$**

Organic materials

1. Frigerio, N. A., *J. Amer. Chem. Soc.*, 1969, **91**, 6201
2. von Schwartz, 1918, 327

The crystalline acid and its dihydrate are very unstable, often exploding at about 3 and 18°C, respectively, but they may be stored virtually unchanged at −75°C. The anhydrous solid ignited explosively every organic compound with which it came into contact except mono-, di- or tri-chloromethanes [1]. The solution of permanganic acid (or its explosive anhydride, dimanganese heptoxide) produced by interaction of permanganates and sulphuric acid, will explode on contact with benzene, carbon disulphide, diethyl ether, ethanol, flammable gases, petroleum or other organic substances [2].
*See other* OXIDANTS

**NITROUS ACID** **$HNO_2$**

Ammonium decahydrodecaborate(2−)

*See* AMMONIUM DECAHYDRODECABORATE(2−), $B_{10}H_{18}N_2$: Nitrous acid

A semicarbazone,
Silver nitrate

Mitchell, J. J., *Chem. Eng. News,* 1956, **34**, 4704

Use of nitrous acid to liberate a free keto-acid from its semicarbazone caused formation of hydrogen azide which was co-extracted into ether with the product. Addition of silver nitrate to precipitate the silver salt of the acid also precipitated silver azide, which later exploded on scraping from a sintered disc. The possibility of formation of free hydrogen azide from interaction of nitrous acid and hydrazine or hydroxylamine derivatives is stressed.

Phosphine

*See* PHOSPHINE, $H_3P$: Air

Phosphorus trichloride

Mellor, 1940, Vol. 8, 1004

The trichloride explodes with nitrous (or nitric) acid.

*See other* OXIDANTS

**NITRIC ACID** **$HNO_3$**

*MCA SD-5*, 1961

The oxidising power (and hazard potential) of nitric acid increases progressively with increase in strength from the concentrated acid (70% wt. $HNO_3$) through fuming acids (above 85% wt.) to the anhydrous 100% acid. The presence of dissolved oxides of nitrogen in the red fuming grades of acid further enhances the potency of the oxidant.

Winchesters of fuming nitric acid may develop pressure on storage, owing to attack on the plastics screw cap if the inert protective cap liner fails. Regular pressure release is recommended by one laboratory supplier.

*See* Polyalkenes, below

Acetic acid,

Sodium hexahydroxyplatinate(IV)

1. Davidson, J. M. *et al., Chem. & Ind.*, 1966, 306
2. Malerbi, B. W., *Chem. & Ind.*, 1970, 796

During preparation of diacetatoplatinum(II) by alternative procedures, the hexahydroxyplatinate in mixed nitric–acetic acids was evaporated to a syrup and several explosions were experienced [1], possibly due to formation of acetyl nitrate. On one occasion a brown solid was isolated and dried, but subsequently exploded with great violence when touched with a glass rod. The material was thought to be a mixture of platinum(IV) acetate–nitrate species [2].

Acetic anhydride

1. Brown, T. A. *et al., Chem. Brit.*, 1967, **3**, 504
2. Dubar, J. *et al., Compt. Rend.*, 1968C, **266**, 1114
3. Dingle, L. E. *et al., Chem. Brit.*, 1968, **4**, 136

4. *MCA Case History No. 103*

Mixtures containing between 50 and 85% wt. of fuming nitric acid are detonable and very sensitive to initiation by friction or shock (possibly owing to formation of acetyl nitrate or tetranitromethane). For preparation of mixtures outside these limits, the order of mixing is important (below 50%, acid into anhydride; above 85%, vice versa; and all below 10°C) [1]. Similar information is also presented diagrammatically [2].

Mixtures containing less than 50% of nitric acid are also dangerous in that addition of small amounts of water (or water-containing mineral acids) readily initiates an uncontrollable exothermic fume-off, which will evaporate most of the liquid present. Equimolar mixtures of 38% nitric acid with acetic anhydride can be detonated at room temperature after a few hours' ageing [3]. Accidental contact of the two materials caused a violent explosion [4].

Acetone

Kennedy, R., private comm., 1975

A winchester of fuming nitric acid with a plastics cap burst, probably owing to pressure build-up and uneven wall thickness. The explosion fractured an adjacent bottle of acetone which ignited on contact with the powerful oxidant. Segregation of oxidants and fuels in storage is essential to prevent such incidents.

*See* Polyalkenes, below

Acetone,
Acetic acid

1. Frant, M. S., *Chem. Eng. News*, 1960, **38**(43), 56
2. Secunda, W. J., *Chem. Eng. News*, 1960, **38**(46), 5

A mixture of equal parts of nitric acid, acetone and 75% acetic acid, used to etch nickel, will explode 1½–6 h after mixing if kept in a closed bottle. The presence of the diluted acetic acid would probably slow the known violent oxidation of acetone by nitric acid [1]. Alternatively, the formation of tetranitromethane and subsequent oxidation of acetone is suspected [2].

Acetone,
Sulphuric acid

Fawcett, H. H., *Ind. Eng. Chem.*, 1959, **51**, 89A

Acetone is oxidised violently by mixed nitric–sulphuric acids, and if the mixture is confined in a narrow-mouthed vessel, it may be ejected or explode.

Acetonitrile

Andrussow, L., *Chim. Ind.*, 1961, **86**, 542

Mixtures of fuming nitric acid and acetonitrile are high explosives.

Acetylene,
Mercury(II) salts,
(Sulphuric acid)

Orton, K. J. P. *et al., J. Chem. Soc.*, 1920, **117**, 29

Contact of acetylene with the concentrated acid in presence of mercury salts forms trinitromethane, explosive above its m.p., 15°C. Subsequent addition of sulphuric acid produces tetranitromethane in high yield.

*See* TRINITROMETHANE ('NITROFORM'), $CHN_3O_6$
TETRANITROMETHANE, $CN_4O_8$

Acrylonitrile

*See* ACRYLONITRILE, $C_3H_3N$: Acids

Alcohols

1. Fawcett, H. H., *Chem. Eng. News*, 1949, **27**, 1396;
   Fromm, F. *et al.*, ibid., 1958
2. Unpublished observations, 1956
3. *MCA Case History No. 1152*
4. Potter, C. R., *Chem. & Ind.*, 1971, 501
5. Spengler, G. *et al., Brennst. Chem.*, 1965, **46**, 117
6. Long, L. A. private comm., 1972
7. Kurbangalina, R. K., *Zh. Prikl. Khim.*, 1959, **32**, 1467–1470
8. Albright, Hanson, 1976, 341–343

A 15% solution of nitric acid in ethanol was used to etch a bismuth crystal. After removing the metal, the mixture decomposed vigorously. Mixtures or nitric acid and alcohols ('Nital') are quite unstable when the concentration of acid is above 10%, and mixtures containing over 5% of acid should not be stored [1]. The use of a little alcohol and excess nitric acid to clean sintered glassware (by 'nitric acid fizzing') is not recommended. At best it is a completely unpredictable approximation to a nitric acid–alcohol rocket propulsion system. At worst, if heavy metals are present, fulminates capable of detonating the mixture may be formed [2]. Chromic acid mixture is less hazardous for such cleaning operations. The Case History describes a violent explosion caused by addition of concentrated acid to a tank car contaminated with a little alcohol [3]. During oxidation of cyclohexanol to the 1,2-dione by an established process, a violent explosion occurred. Two intermediates are possible suspects [4]. Furfuryl alcohol is hypergolic with high-strength nitric acid [5] and methanol has been used as a propellant fuel. It also readily forms the explosive ester, methyl nitrate [6].

Mixtures with ethylene glycol are easily detonated by heat, impact or friction [7]. The injector process for safe continuous nitration of glycerol and other alcohols is reviewed [8].

*See* NITRATION INCIDENTS

Alcohols,
Disulphuric acid

Rastogi, R. P. *et al., Amer. Inst. Aero. Astronaut. J.*, 1966, **4**, 1083–1085
Ignition delays were determined for contact of various aliphatic alcohols with mixtures of disulphuric acid (*d*, 1.9) and red fuming nitric acid (*d*, 1.5). With ethanol, a minimum of 30% wt. of disulphuric acid was required for ignition.

Alcohols,
Potassium permanganate

*See* POTASSIUM PERMANGANATE, $KMnO_4$: Alcohols, Nitric acid

Alkanethiols

Filby, W. G., *Lab. Practice*, 1974, **23**, 355
Oxidation of thiols to sulphonic acids with excess concentrated acid as usually described is potentially hazardous, the exotherm usually causing ignition of the thiol. A modified method involving oxidation under nitrogen and at temperatures 1–2°C above the melting point of the thiol is safer and gives purer products.

*See* Butanethiol, below

Aliphatic amines

1. Schalla, R. L. *et al., Amer. Rocket Soc. J.*, 1959, **29**, 33–39
2. Schneebeli, P. *et al., Rech. Aeron.*, 1962, **87**, 33–46

In contact with white fuming nitric acid, ignition delays for tri-butyl-, propyl- and ethyl-amines were less than those of the corresponding *sec*-amines, and, under good mixing conditions, n-butylamine did not ignite [1]. Triethylamine ignites on contact with the concentrated acid [2].

*See* Aromatic amines, below (reference 2)
Cyclohexylamine, below
Dinitrogen tetraoxide, Nitrogenous fuels, below

2-Alkoxy-1,3-dithia-2-phospholane ('*O*-alkyl ethylene dithiophosphate')

Arbuzov, A. E. *et al., Chem. Abs.*, 1953, **47**, 4833e,f
The methoxy and ethoxy derivatives ignite with the concentrated acid.

*See* Phosphorus compounds, below

2-Aminothiazole,
Sulphuric acid

Silver, L., *Chem. Eng. Progr.*, 1967, **63**(8), 43
Nitration of 2-aminothiazole with nitric/sulphuric acids was normally effected by mixing the reactants at low temperature, heating to 90°C during 30 min and then positively cooling. When positive cooling was omitted, a violent explosion occurred. Experiment showed that this was due to a slow exothermic reaction accelerating out of control under the adiabatic conditions. *N*-Nitroamines were not involved.

Ammonia

Mellor, 1940, Vol. 8, 219

A jet of ammonia will ignite in nitric acid vapour.

Anilinium nitrate

Andrussow, L., *Chim. Ind.*, 1961, **86**, 542

Although aniline may be hypergolic with nitric acid (below), anilinium nitrate dissolves unchanged in 98% nitric acid and can be stored for long periods, though the solution has high-explosive properties.

*See* Cyclohexylamine, below

Anilinium nitrate,
Inorganic additives

Munjal, N. L. *et al., Indian J. Chem.*, 1967, **5**, 320–322

The effect of inorganic additives on ignition delay in anilinium nitrate–red fuming nitric systems was examined. The insoluble compounds copper(I) chloride, potassium permanganate, sodium pentacyanonitrosylferrate and vanadium(V) oxide were moderately effective promoters, while the soluble ammonium or sodium metavanadates were very effective, producing vigorous ignition.

Aromatic amines

1. Kit and Evered, 1960, 239, 242
2. *Aniline,* 85, Allied Chemical Corp., New York, 1964
3. Miller, R. O., *Tech. Note No. 3884,* 1–32, Washington, Nat. Advisory Comm. Aeronaut., 1956
4. Spengler, G. *et al., Brennst. Chem.*, 1965, **46**, 117
5. Bernard, M. L. *et al., Compt. Rend. C*, 1966, **263**, 24–26
6. Schalla, R. L. *et al., Amer. Rocket Soc. J.*, 1959, **29**, 38

Many aromatic amines (aniline, *N*-ethylaniline, *o*-toluidine, xylidine, etc., and their mixtures) are hypergolic with red fuming nitric acid [1]. When the amines are dissolved in triethylamine, ignition occurs at −60°C and below [2]. Addition of a mixture of aniline, dimethylaniline, xylidine and pentacarbonyliron renders hydrocarbons hypergolic with concentrated nitric acid [3]. Although aniline is not hypergolic with 96% nitric acid, presence of sulphuric acid (5% or above) renders it so. Presence of dinitrogen tetraoxide further reduces ignition delay [4]. The mechanism of ignition of *p*-phenylenediamine (or furaldehyde azine) on contact with red fuming nitric acid is quantified [5]. Traces of sulphuric acid may be essential for ignition to occur [6].

Aromatic amines,
Metal compounds

Rastogi, R.P. *et al., Indian J. Chem.*, 1964, **2**, 301–307

The effects of various metal oxides and salts which promote ignition of amine–red fuming nitric acid systems were examined. Among soluble catalysts, copper(I) oxide, ammonium or sodium metavanadates, iron(III) chloride (and potassium hexacyanoferrate(4−) with *o*-toluidine) are most effective. Of the insoluble materials, copper(II) oxide, iron(III) oxide, vanadium(V) oxide, potassium chromate, potassium dichromate, potassium hexacyanoferrate(3−) and sodium pentacyanonitrosylferrate(2−) were effective.

*See* Anilinium nitrate, etc., above

Arsine–borontribromide

*See* ARSINE–BORONTRIBROMIDE, $AsH_3 \cdot BBr_3$: Oxidants

Benzo[*b*] thiophene derivatives

Brown, I. *et al., Chem. & Ind.*, 1962, 982

During nitration of several derivatives, diazotisation and oxidation occurred to produce internal diazonium phenolate derivatives. 5-Acetyl-amino-3-bromo-benzo[*b*]thiophene unexpectedly underwent hydrolysis, diazotisation and oxidation to the explosive compound below.

*See* 4-Chloro-2-nitroaniline, below
3-BROMO-2,7-DINITRO-5-BENZO[*b*]THIOPHENEDIAZONIUM-4-OLATE, $C_8HBrN_4O_5S$

1,4-Bis(methoxymethyl)-2,3,5,6-tetramethylbenzene

Anon., *Jahresber.*, 1972, 84, Berufsgenossenschaft der Chem. Ind.

Oxidation of the durene derivative to benzenehexacarboxylic acid (mellitic acid) in an autoclave is normally effected in stages, initially by heating at 80–104°C with the vent open to allow escape of the evolved gases. Subsequent heating to a higher temperature with the vent closed completes the reaction. On one occasion omission of the first vented heating phase led to violently explosive rupture of the autoclave at 80°C.

*See* Hydrocarbons, below

Bromine pentafluoride

*See* BROMINE PENTAFLUORIDE, $BrF_5$: Acids, etc.

Butanethiol

McCullough, F. *et al., Proc. Fifth Combustion Symp.*, 181, New York, Reinhold, 1955

Technical material (containing 28% of propane- and 7% of pentane-thiols) is hypergolic with 96% nitric acid.

Cellulose

*MCA SD-5,* 1961

Cellulose may be converted to the highly flammable nitrate ester on contact with the vapour of nitric acid, as well as the liquid itself.

Chlorobenzene

Anon., *Jahresber.*, 1974, 86, Berufsgenossenschaft der Chem. Ind.

In a plant for the continuous nitration of chlorobenzene, maloperation during start-up caused the addition of substantial amounts of the reagents into the reactor before effective agitation and mixing had been established. The normal reaction temperature of 60°C was rapidly exceeded by at least 60°C and an explosion occurred. Subsequent investigation showed that at 80°C an explosive gas–air mixture was formed above the reaction mixture, and that the adiabatic vapour-phase nitration would attain a temperature of 700°C and ignite the explosive atmosphere in the reactor.

*See other* NITRATION INCIDENTS

4-Chloro-2-nitroaniline

1. Elderfield, R. C. *et al.*, *J. Org. Chem.*, 1946, **11**, 820
2. *MCA Case History No. 1489*

The literature procedure for preparation of 4-chloro-2,6-dinitroaniline [1], involving direct nitration in 65% nitric acid, was modified by increasing the reaction temperature to 60°C 1 h after holding at 30–35°C as originally specified. This procedure was satisfactory on the bench scale, and was scaled up into a 900 litre reactor. After the temperature had reached 30°C, heating was discontinued, but the temperature continued to rise to 100–110°C and decomposition set in with copious evolution of nitrous fumes and production of a very shock-sensitive explosive solid. This was identified as 4-chloro-2,5-dinitrobenzenediazonium-6-olate produced by hydrolysis of a nitro group in the expected product by the diluted nitric acid at high temperature, diazotisation of the free amino group by the nitrous acid produced in the hydrolysis (or by the nitrous fumes), and introduction of a further nitro group under the prevailing reaction conditions. It is recommended that primary aromatic amines should be protected by acetylation before nitration, to avoid the possibility of accidental diazotisation [2].

*See* Benzo[*b*]thiophene derivatives, above

*See other* ARENEDIAZONIUMOLATES

Copper(I) nitride

Mellor, 1940, Vol. 8, 100

Interaction with concentrated acid is very violent.

Crotonaldehyde

Andrussow, L., *Chim. Ind.*, 1961, **86**, 542

Crotonaldehyde is hypergolic with concentrated nitric acid, ignition delay being 1 ms.

*See* ROCKET PROPELLANTS

Cyclohexylamine

Andrussow, L., *Chim. Ind.*, 1961, **86**, 542

Although cyclohexylamine has been used as a fuel with nitric acid in rocket motors, cyclohexylammonium nitrate dissolves unchanged in fuming nitric acid to give a solution stable for long periods.

*See* Aromatic amines, above
ROCKET PROPELLANTS

1,2-Diaminoethanebis-trimethylgold

*See* 1,2-DIAMINOETHANEBIS-TRIMETHYLGOLD, $C_8H_{26}Au_2N_2$

1,2-Dichloroethane

Kurbangalina, R. K., *Zh. Prikl. Khim.*, 1959, **32**, 1467

Mixtures are easily detonated by heat, impact or friction.

Dichloromethane

Andrussow, L., *Chim. Ind.*, 1961, **86**, 542

Dichloromethane dissolves endothermically in concentrated nitric acid to give a detonable solution.

Dichromates,
Organic fuels

Munjal, N. L. *et al.*, *Amer. Inst. Aero. Astronaut. J.*, 1972, **10**, 1722–1723

The effects of dichromates in promoting ignition of non-hypergolic mixtures of red fuming nitric acid with cyclohexanol, *m*- or *o*-cresol and furfural were studied. Ammonium dichromate was most effective in all cases, and the only effective catalyst for cyclohexanol. Potassium chromate and dichromate were also examined.

Diethyl ether

Foote, C. S., private comm., 1965

Addition of ether to a nitration mixture (*o*-bromotoluene and concentrated nitric acid) diluted with an equal volume of water in a separating funnel caused a low-order explosion. This was attributed to oxidation of the ether (possibly containing alcohol) by the acid. Addition of more water before adding ether was recommended.

Diethyl ether,
(Sulphuric acid)

1. Kirk-Othmer, 1965, Vol. 8, 479
2. Van Alphen, J., *Rec. Trav. Chim.*, 1930, 492–500

Interaction of ether with anhydrous nitric acid (to ethyl nitrate) may proceed explosively [1], and in presence of concentrated sulphuric acid, ether and nitric acid explode violently [2].

*See* Dimethyl ether, below

3,6-Dihydro-1,2,2*H*-oxazine

Wichterle, O. *et al., Coll. Czech. Chem. Comm.*, 1950, **15**, 309–321
The product of interaction (possibly an *N*-nitro and/or ring-opened nitrate ester derivative) is explosive.

Dimagnesium silicide

Mayes, R. B. *et al., School Sci. Rev.*, 1975, **56**(197), 819–820
When the residue from combustion of magnesium in air was removed from the porcelain crucible, a grey stain remained. Addition of cold concentrated nitric acid to remove the stain led to a violent reaction. This was found to be caused by the presence of dimagnesium silicide in the stain.

(Dimethylaminomethyl)ferrocene,
Water

Koch-Light Laboratories Ltd., private comm., 1976
In an assay procedure the amine was heated with diluted nitric acid, and a near-explosive decomposition reaction occurred.

Dimethyl ether

Andrussow, L., *Chim. Ind.*, 1961, **86**, 542–545
The latent hazards in storing and handling the explosive mixtures of the concentrated acid with dimethyl ether are discussed.

*See* Diethyl ether, above

1,1-Dimethylhydrazine

*See* 1,1-DIMETHYLHYDRAZINE, $C_2H_8N_2$: Oxidants

1,1-Dimethylhydrazine,
Organic compounds

Spengler, G. *et al., Brennst. Chem.*, 1965, **46**, 117
Contact of nitric acid (or dinitrogen tetraoxide) with dimethylhydrazine is hypergolic and well described in rocket technology. While hydrocarbons and several other classes of organic compounds are not hypergolic with these oxidisers, addition of a proportion of dimethylhydrazine to a wide range of hydrocarbons, alcohols, amines, esters and heterocyclic compounds renders them hypergolic in contact with nitric acid or dinitrogen tetraoxide.

Dimethyl sulphide,
*p*-Dioxane

Rudakov, E. S. *et al., Chem. Abs.*, 1972, **76**, 13515r
The mechanism leading to the delayed explosion of the system (even when cooled in liquid nitrogen) was investigated.

Dinitrogen tetraoxide,
Nitrogenous fuels

1. Durgapal, U. C. *et al., Amer. Inst. Aero. Astronaut. J.*, 1974, **12**, 1611
2. Munjal, N. L. *et al., J. Spacecr. Rockets*, 1974, **11**, 428–430
3. Durgapal, U. C. *et al.*, ibid., 447–448

The effect on decrease in hypergolic ignition delay of increasing concentrations of dinitrogen tetraoxide in red fuming nitric acid was studied with triethylamine, dimethylhydrazine or *mixo*-xylidene as fuels [1]. The effect of various catalysts on ignition delay after contact of red fuming nitric acid with various arylamine–formaldehyde condensation products was also studied [2,3].

Diphenyldistibene
*See* DIPHENYLDISTIBENE (STIBOBENZENE), $C_{12}H_{10}Sb_2$ : Air, etc.

Diphenylmercury
Whitmore, 1921, 43, 168
Interaction in carbon disulphide, even at −15° C, is violent.

Diphenyltin
Krause, E. *et al., Ber.*, 1920, **53**, 177
Ignition occurs on contact with fuming nitric acid.

Disodium phenylorthophosphate
Muir, G. D., private comm., 1968
Concentration of the nitration product of the phosphate ester caused a violent explosion. Picric acid derivatives may have been involved.

Divinyl ether
Andrussow, L., *Chim. Ind.*, 1961, **86**, 542
Divinyl ether is hypergolic with concentrated nitric acid, ignition delay being 1 ms.
*See* ROCKET PROPELLANTS

5-Ethyl-2-methylpyridine

1. Frank, R. L., *Chem. Eng. News*, 1952, **30**, 3348
2. Rubinstein, H. *et al., J. Chem. Eng. Data*, 1967, **12**, 149
3. Elam, E. V. *et al., Chem. Eng. News*, 1973, **51**(34), 42

Following a patented procedure, the two reactants were being heated together at 145°C/14.5 bar to produce 2,5-pyridinedicarboxylic acid. The temperature and pressure rose to 160°C/43.5 bar, and the autoclave was vented and cooled, but 90 s later, a violent explosion occurred, although both rupture discs (105 and 411 bar) had gone. General precautions are discussed [1]. Up to 20% of 5(1,1-dinitroethyl-2-methylpyridine, probably an explosive compound, is produced in this reaction.

However, in presence of added water, no instability was seen in a series of reactions at temperatures up to 160° C and pressure to 102 bar [2].

A further incident in an electrically heated autoclave involved heating the pyridine with excess 70% acid to 140° C during 1 h. When heating was then discontinued, the temperature continued to rise in spite of air cooling, and an explosion which bulged the autoclave occurred at 172° C [3]. It is remotely possible that interaction of the titanium liner with the pressurised hot acid may have contributed to the incident.

*See* TITANIUM, Ti: Nitric acid

Fat,
Sulphuric acid

1. Tyler, L. J., *Chem. Eng. News*, 1973, **51**(31), 32
2. Engan, W. L., ibid., (51), 37

A wet-ashing procedure for analysis of fatty animal tissue was modified by using Teflon-lined bombs rated for use at 340 bar instead of open crucibles. Bombs cooled to well below 0° C were charged with fuming nitric and fuming sulphuric acid (1 ml of each) and adipose tissue (0.5 g), removed from the cooling bath and sealed. After 10 min delay, the bombs exploded, probably owing to development of high internal temperature and pressure (calculated as 4000K and 1000 bar max.) from complete oxidation of all the organic material [1]. Formation of glyceryl nitrate from the lipid content may have contributed to the violence of the explosion. The presence of 2–5% of water in the mixed acids is recommended to reduce the nitrating potential when in contact with organic fuels [2].

*See* Organic matter, below
Organic matter, Sulphuric acid, below

2-Formamido-1-phenyl-1,3-propanediol

Biasutti, G. S. *et al.*, *Loss Prevention*, 1974, 8,123

Large-scale nitration to the *p*-nitro derivative, previously effected without incident for 15 years, exploded with great violence. Subsequent investigation showed that with adequate priming energy, the reaction mixture could detonate (potential energy 3.77 MJ or 875 kcal/kg).

*See* NITRATION INCIDENTS

Fluorine

*See* FLUORINE, $F_2$: Nitric acid

Furfurylidene ketones

Panda, S. P. *et al.*, *J. Armam. Stud.*, 1974, **10**(2), 44–45; 1976, **12**, 138–139

In 10 out of 12 cases, furfurylidene derivatives of ketones were hypergolic with red fuming nitric acid. Difurfurylidenehydrazine behaves similarly and a mechanism involving polymerisation is proposed.



Glycerol,
Hydrofluoric acid

Buck, R. H., *J. Electrochem. Soc.*, 1966, **113**, 1352–1353

A chemical polishing solution consisting of nitric and hydrofluoric acids (1 vol. of each) and glycerol (2 vols) generated enough pressure during 4 h storage to rupture the closed plastics container. This was attributed to oxidation (with gas evolution) of the glycerol by the strongly oxidising medium.

*See* Lactic acid, etc., below

Glycerol,
Sulphuric acid

Dunn, B. W., *Bur. Expl. Accid. Bull.*, 35, Pennsylvania, Amer. Rail Assoc., 1916

Charging mixed nitrating acid into an insufficiently cleaned glycerol drum led to a violent explosion, attributed to formation of glyceryl trinitrate.

Hexalithium disilicide

*See* HEXALITHIUM DISILICIDE, $Li_6Si_2$: Acids

2,2,4,4,6,6-Hexamethyltrithiane

Baumann, E. *et al., Ber.*, 1889, **22**, 2596

Interaction of 'tri(thioacetone)' with the concentrated acid is explosively violent.

*See* Thioaldehydes, etc., below

Hexenal

Lobanov, V. I., *Chem. Abs.*, 1966, **64**, 14031c

Hexenal is determined by photocolorimetry after oxidation with nitric acid. The yellow–orange oxidation product explodes on heating.

Hydrazine and derivatives

1. Andrussow, *Chim. Ind.*, 1961, **86**, 542
2. Jain, S. R., *Combust. Flame.*, 1977, **28**, 101–103

Hydrazine is hypergolic with concentrated nitric acid [1]. Of a series of hydrazones and azines derived from aldehydes and ketones, only those which decomposed when heated alone were hypergolic with the acid when heated at 12.5°C/min [2].

*See* ROCKET PROPELLANTS

Hydrocarbons

1. Andrussow, L., *Chim. Ind.*, 1961, **86**, 542

2. Wilson, P. J. *et al., Chem. Rev.*, 1944, **34**, 8
3. Sykes, W. G. *et al., Chem. Eng. Prog.*, 1963, **59**(1), 70–71
4. Mason, C. M. *et al., J. Chem. Eng. Data*, 1965, **10**, 173
5. Urbanski, 1961, Vol. 1, 140
6. *Proc. Symp. Chem. Process Hazards Plant Design, Manchester, I.Ch.E., 1960,* 37–41
7. Powell, G. *et al., Org. Synth.*, 1943, Coll. Vol. 2, 450
8. Wilms, H. *et al., Angew. Chem.*, 1962, **74**, 465
9. Trent, C. H. *et al., J. Amer. Rocket Soc.*, 1951, **21**, 129–131
10. Chaigneau, M., *Compt. Rend.*, 1951, **223**, 657–659
11. Anon., *Chemisches Ind.*, 1914, **37**, 337–342
12. Bryant, J. T. *et al., Rep. AD-B000447*, Richmond (Va.), USNTIS, 1974

Dienes and acetylene derivatives are hypergolic in contact with concentrated nitric acid, ignition delay being 1 ms [1].

*See* Phenylacetylene, etc., below
ROCKET PROPELLANTS

Cyclopentadiene reacts explosively with fuming nitric acid [2].

*See also* SULPHURIC ACID, $H_2O_4S$: Cyclopentadiene

Burning fuel oil and other petroleum products detonate immediately on contact with concentrated nitric acid [3]. Very high sensitivity to detonation is shown by mixtures with benzene close to the stoicheiometric proportions of *ca.* 84% acid [4]. Lack of proper control in nitration of toluene with mixed acids may lead to runaway or explosive reaction. A contributory factor is the oxidative formation, and subsequent nitration and decomposition, of nitrocresols [5]. Oxidation of *p*-xylene with nitric acid under pressure in manufacture of terephthalic acid carries explosion hazards in the autoclaves and condensing systems [6]. During nitration of mesitylene in acetic acid–anhydride solution, fuming nitric acid must be added slowly to the cooled mixture to prevent temperature exceeding 20°C, when an explosive reaction may occur [7]. During oxidation of mesitylene with nitric acid in an autoclave at 115°C to give 3,5-dimethylbenzoic acid, a violent explosion occurred. This was attributed to local overheating, formation of 1,3,5-tri(nitromethyl) benzene and violent decomposition of the latter. Smaller-scale preparations with better temperature control were uneventful [8].

Under nitrogen, cyclopentadiene ignites in contact with white fuming nitric acid [9]. The explosive hazards involved in the preparation of mellitic acid by fuming nitric acid oxidation of hexamethylbenzene make the procedure only suitable for small-scale operation [10]. Large-scale addition of too-cold nitrating acid to benzene without agitation later caused an uncontrollably violent reaction to occur when stirring was started. The vapour–air mixture produced was ignited by interaction of benzene and nitric acid at 100–170° C and caused an extremely violent explosion [11]. In the impact-initiated combustion of hydrocarbons in acid at high pressure, the effects of impact, hydrocarbon structure and stoicheiometry upon ignition and rate of combustion were investigated [12].

Hydrogen iodide,
or Hydrogen selenide,
or Hydrogen sulphide
Hofmann, A. W., *Ber.*, 1870, **3**, 660
Ignition occurs on contact of fuming nitric acid with excess hydrogen iodide, and on contact of the sulphide or selenide with the acid.
*See* Non-metal hydrides, below

Hydrogen peroxide,
Ketones
*See* HYDROGEN PEROXIDE, $H_2O_2$: Ketones, etc.

Hydrogen peroxide,
Mercury(II) oxide
*See* HYDROGEN PEROXIDE, $H_2O_2$: Mercury(II) oxide, etc.

Hydrogen peroxide,
Soils
Krishnamurty, K. V. *et al., At. Abs. Newslett.*, 1976, **15**, 68–70
When preparing soil and sediment samples for atomic absorption spectral analysis for trace metals, pre-oxidation with nitric acid before addition of hydrogen peroxide eliminates the danger of explosion.

Ion exchange resins
1. Barghusen, J. *et al., Reactor Fuel Processing*, 1964, **7**, 297
2. McBride, J. A. *et al., 3rd Geneva Conf. on Peaceful Uses of Atomic Energy,* A/CONF. 28 28/P/278, 1965

Several cases of interaction between anion exchange resins and nitric acid causing rapid release of energy or explosion have occurred [1]. The cause has been attributed to oxidative degradation of the organic resin matrix and/or nitration of the latter. Suggested precautions include control of temperature, acid concentration and contact time. Presence of heavy ions (Pu) or oxidising agents (dichromates) tends to accelerate the decomposition [2].

Iron(II) oxide
Mellor, 1941, Vol. 13, 716
The finely divided (pyrophoric) oxide incandesces with nitric acid.

Lactic acid,
Hydrofluoric acid
Bubar, S. F. *et al., J. Chem. Educ.*, 1966, **43**, A956

Mixtures of the three acids, used as metal polishing solutions, are unstable and should not be stored. Lactic and nitric acids react autocatalytically after a quiescent period, producing a temperature of about 90°C and vigorous gas evolution after 12 h. Prepare freshly, discard after use and handle carefully.

Lead-containing rubber

Johnson, T. C. *et al., Rep. RFP-1354*, 1–8, Washington, US At. En. Comm., 1969

Lead-containing dry-box gloves may ignite in nitric acid environments.

Metal acetylides

Mellor, 1946, Vol. 5, 848

Dicaesium and dirubidium acetylides explode in contact with nitric acid, and the sodium and potassium analogues probably react violently.

Metal hexacyanoferrates (3–) or (4–)

Sidgwick, 1950, 1344

The action of 30% nitric acid on hexacyanoferrates (3–) or (4–) to produce pentacyanonitrosoferrates(2–) ('nitroprussides') is violent.

Metals

1. Mellor, 1940, Vol. 2, 470; Vol. 4, 270, 483; 1941, Vol. 7, 260; 1940, Vol. 9, 627; 1942, Vol. 12, 32, 188
2. Pascal, 1956, Vol. 10, 504
3. Condike, G. F., *Chemistry*, 1974, **47**(4), 27–28
4. McGarman, A. R., ibid., (11), 27–28

Bismuth powder glows red hot in contact with fuming nitric acid, while the molten metal (271°C) explodes in contact with the concentrated acid. Powdered germanium reacts violently with concentrated acid, and lithium ignites. Manganese powder incandesces and explodes feebly with nitric acid, and sodium ignites in contact with nitric acid of density above 1·056. Titanium alloys form an explosive deposit with fuming nitric acid. Although uranium powder reacts vigorously with red fuming nitric acid, under some conditions explosive deposits may be formed. Addition of concentrated nitric acid to molten zinc (419°C) causes it to incandesce. Magnesium burns brilliantly in nitric acid vapour [1]. Antimony may be attacked violently by fuming nitric acid [2]. The experimental quantities of tin and nitric acid specified for a laboratory demonstration [3] are likely to lead to formation of the explosive nitrate oxide [4].

*See* TIN(II) NITRATE OXIDE, $N_2O_7Sn_2$

Metal salicylates

Belcher, R., private comm., 1968

Metal salicylates are occasionally incorporated in mixtures of 'unknowns' for qualitative inorganic analysis. During the conventional group separation, organic radicals are removed by evaporation with nitric acid. When salicylates are present, this can lead to the formation of trinitrophenol through nitration and decarboxylation. This may react with any heavy metal ions present to form unstable explosive picrates, if the evaporation is taken to dryness. An alternative scheme of analysis obviates this danger.

Metal thiocyanate

*MCA Case History No. 853*

When the (unspecified) thiocyanate solution was pumped through a 80 mm pipeline containing nitric acid, a violent explosion occurred. This was later confirmed experimentally and attributed to oxidation of the thiocyanate solution by nitric acid.

2-Methylbenzimidazole,
Sulphuric acid

Zmojdzin, A. *et al.*, Ger. Offen. 2 310 414, 1974 (*Chem. Abs.*, 1975, **82**, 16839t)

A safe method of preparing 5-nitro-2-methylbenzimidazole involves preliminary addition of the heterocycle to nitric acid (*d*, 1.40) with subsequent addition of sulphuric acid, keeping the temperature below 110°C.

*See other* NITRATION INCIDENTS

4-Methylcyclohexanone

Dye, W. T., *Chem. Eng. News,* 1959, **37**, 48

Oxidation of 4-methylcyclohexanone by addition to nitric acid at about 75°C caused a detonation to occur. These conditions had been used previously to oxidise the corresponding alcohol, but, although the ketone is apparently an intermediate in the oxidation of the alcohol, the former requires a much higher temperature to start and maintain the reaction. An OTS report, PB73591, mentions a similar violent reaction with cyclohexanone.

*See* Acetone, etc., above

Methylthiophene

Rust, 1948, 318

During the nitration of methylthiophene, direct liquid contact causes ignition to occur, so air saturated with the organic vapour was passed into the cooled concentrated acid. After a while, the whole apparatus exploded violently, probably owing to ignition of the air–fuel mixture being passed in.

*See other* NITRATION INCIDENTS

Nitroaromatics

Urbanski, 1967, Vol. 3, 290

A series of mixtures of nitric acid with one or more of mono- and dinitrobenzenes, di- and tri-nitrotoluenes have been shown to possess high explosive properties.

*See* Nitrobenzene, etc., below

Nitrobenzene,
Nitric acid,
Water

1. Anon., *J. R. Inst. Chem.*, 1960, **84**, 451
2. Van Dolah, R. W., *Loss. Prev.*, 1969(3), 32

A plant explosion involved a mixture of nitrobenzene, nitric acid and a substantial quantity of water. Detonation occurred with a speed and power comparable to TNT. This was unexpected in view of the presence of water in the mixture [1]. The later reference deals with a detailed practical and theoretical study of this system and determination of the detonability limits and shock sensitivity. The limits of detonability coincided with the limits of miscibility over a wide portion of the ternary composition diagram. In absence of water, very high sensitivity (similar to that of glyceryl trinitrate) occurred between 50 and 80% nitric acid, the stoicheiometric proportion being 73%.

*See* Hydrocarbons, above
Nitroaromatics, above

Nitrobenzene,
Sulphuric acid

Fritz, E. J., *Loss Prevention*, 1969, **3**, 41–44

Failure of the agitator during addition of nitrating acid allowed unreacted reagents to build up in the reaction system, which owing to absence of the usual reaction exotherm became undercooled. Application of heat and agitation caused a runaway reaction, terminating in explosion, to occur. Laboratory-scale repetition of this sequence showed an exotherm to 200°C during 0.1 s. Operating improvements are detailed.

*See other* NITRATION INCIDENTS

Nitromethane

Olah, G. A. *et al., Org. Synth.*, 1967, **47**, 60

Mixtures are extremely explosive.

*See* NITROMETHANE, $CH_3NO_2$: Acids

Non-metal hydrides

1. Mellor, 1946, Vol. 5, 36; 1940, Vol. 6, 814; 1939, Vol. 9, 56, 397
2. Hofmann, A. W., *Ber.*, 1870, **3**, 658
3. Pascal, 1956, Vol. 10, 505

Arsine, phosphine and tetraborane(10) are all oxidised explosively by fuming nitric acid, while stibine behaves similarly with the concentrated acid [1]. Phosphine, hydrogen sulphide and selenide all ignite when the fuming acid is dripped into the gas [2]. Hydrogen telluride ignites with the cold concentrated acid, sometimes exploding [3].

*See* Phosphine derivatives, below

Non-metals

1. Mellor, 1946, Vol. 5, 16; 1940, Vol. 8, 787, 845
2. Bailar, 1973, Vol. 1, 1337
3. Pascal, 1956, Vol. 10, 504
4. *MCA Case History No. 1969*

Boron (finely divided forms) reacts violently with concentrated acid and may attain incandescence. The vapour of phosphorus, heated in nitric acid in presence of air, may ignite. Boron phosphide ignites with the concentrated acid [1]. Silicon crystallised from its eutectic with aluminium reacts violently with the concentrated acid [2]. Arsenic may react violently with the fuming acid, and finely divided carbon similarly with the concentrated acid [3]. Use of concentrated acid to clean a stainless steel hose contaminated with phosphorus led to an explosion [4].

Organic matter

Bowen, H. J. M., private comm., 1968

When 16M (70%) nitric acid was poured down a sink without diluting water, interaction with (unspecified) organic matter in the trap caused a delayed explosion.

Organic matter,
Perchloric acid

*See* PERCHLORIC ACID, $ClHO_4$: Nitric acid, etc.

Organic matter,
Potassium chlorate

*See* POTASSIUM CHLORATE, $ClKO_3$: Nitric acid, Organic matter

Organic matter,
Sulphuric acid

*ABCM Quart. Safety Summ.*, 1934, **5**, 17

Use of the mixed concentrated acids to dissolve an organic residue caused a violent explosion. Nitric acid is a very powerful and rapid oxidant and may form unstable fulminic acid or polynitro compounds under these conditions.

Phenylacetylene,
1,1-Dimethylhydrazine

Spengler, G. *et al.*, *Sci. Tech. Aerospace Rep.*, **2**(17), 2392, Washington NASA, 1964

Phenylacetylene does not itself ignite on contact with nitric acid but addition of 1,1-dimethylhydrazine causes it to become hypergolic.

*See* Hydrocarbons, above

Phosphine derivatives

von Schwartz, 1918, 325

Mellor, 1947, Vol. 8, 827, 1041

Graham, T., *Trans. R. Soc. Edinburgh*, 1835, **13**, 88

Phosphine ignites in concentrated nitric acid and addition of warm fuming nitric acid to phosphine causes explosion. Phosphonium iodide ignites with nitric acid, and ethylphosphine explodes with fuming acid. Tris(iodomercuri)phosphine is violently decomposed by nitric acid or aqua regia.

Phosphorus compounds

Mellor, 1971, Vol. 8, Suppl. 3, 335, 373

Nickel tetraphosphide ignites with the fuming acid, and tetraphosphorus diiodide triselenide reacts explosively with nitric acid.

*See* 2-Alkoxy-1,3-dithia-2-phospholane ('*O*-alkyl ethylene dithiophosphate')
Phosphine derivatives, above

Phosphorus halides

Mellor, 1947, Vol. 8, 827, 1004, 1038

Tetraphosphorus iodide ignites in contact with cold concentrated nitric acid. Phosphorus trichloride explodes with nitric (or nitrous) acid.

Phthalic anhydride,
Sulphuric acid

1. Tyman, J. H. P. *et al., Chem. & Ind.*, 1972, 664
2. Bentley, R. K., ibid., 767
3. Bretherick, L., ibid., 790

Attempts to follow a published method for nitration of phthalic anhydride in sulphuric acid at 80–100°C with fuming nitric acid caused an eruptive decomposition to occur after 2 h delay [1]. The hazard can be eliminated by use of a smaller excess of nitrating acid at 55–65°C [2]. Possible causes of the delayed eruption are suggested as acyl nitrates [3].

*See* Sulphuric acid, Terephthalic acid, below

Polyalkenes

Marsh, J. R., *Chem. & Ind.*, 1968, 1718

Dabeka, R. W. *et al., Anal. Chem.*, 1976, **48**, 2048

Fuming nitric acid had seeped past the protective polytetrafluoroethylene liner inside the polyethylene or polypropylene screw cap and attacked the latter, causing pressure build-up in the glass bottle.

Polypropylene bottles are unsuitable for long-term storage of nitric acid, because slow embrittlement and cracking occur.

Polydibromosilane

*See* POLYDIBROMOSILANE, $(Br_2Si)_n$: Oxidants

Poly(ethylene oxide) derivatives

1. Corby, M. P., *Chem. Brit.*, 1975, **11**, 334; private comm., 1975
2. Corby, M. P., ibid., 456

A mixture of nitric and phosphoric acids (50, 17%, respectively) with a primary alcohol ethoxylate surfactant (0.1%) and water exploded after 7 months' storage in a glass bottle [1]. Progressive hydrolysis under these conditions would be expected to lead to production of the readily oxidised ethylene glycol, and the formation of ethylene dinitrate may also have contributed to the violence observed. A general warning against mixing surfactants and oxidising acids is given [2].

Polysilylene

*See* POLYSILYLENE, $(H_2Si)_n$; Oxidants

Pyrocatechol

Andrussow, L., *Chim. Ind.*, 1961, **86**, 542

The phenol is hypergolic with concentrated nitric acid, with a 1 ms ignition delay.

*See* ROCKET PROPELLANTS

Reducants

A variety of reducants ignite or explode with nitric acid.

*See* HYDROGEN IODIDE, HI: Nitric acid
HYDROGEN SULPHIDE, $H_2S$: Nitric acid
POTASSIUM PHOSPHINATE, $H_2KO_2P$: Nitric acid
HYDRAZINE, $H_4N_2$: Nitric acid
Sulphur dioxide, below

Silver buten-3-ynide

*See* SILVER BUTEN-3-YNIDE, $C_4H_3Ag$: Nitric acid

Steel gas cylinder

*CISHC Chem. Safety Summ.*, 1974–5, **45–46**, 3–4

Concentrated nitric acid leaking from a faulty road tanker became partially diluted by water and was prevented from running away along the roadside gully by a full oxygen cylinder in the horizontal position. After several hours, the cylinder was sufficiently weakened by corrosion to split open under the internal pressure.

Sulphur dioxide

Coleman, G. H. *et al.*, *Inorg. Synth.*, 1939, **1**, 55

Presence of dinitrogen tetraoxide appears to be essential to catalyse smooth formation of nitrosylsulphuric acid. In its absence, reaction may be delayed and then proceed explosively.

Sulphur halides

Mellor, 1947, Vol. 10, 646, 651–652

Interaction with sulphur dichloride, dibromide or disulphur dibromide is violent, the hydrogen halide being liberated.

Sulphuric acid

*BCISC Quart. Safety Summ.*, 1964, **35**, 3

The gland of a centrifugal pump being used to pump nitrating acid (nitric–sulphuric acids, 1:3) exploded after 10 min use. This was attributed to nitration of the gland packing, followed by frictional detonation. Inert shaft sealing material is advocated.

Sulphuric acid,
Terephthalic acid

Withers, C. V., *Chem. & Ind.*, 1972, 821; private comm., 1972

During nitration of the acid with fuming nitric acid in oleum, a delayed exotherm increased the temperature after 2 h from 100°C to 160°C, causing eruption of the contents. At 120°C the delay was 1 h and at 130°C 30 min.

*See* Phthalic anhydride, etc., above

Thioaldehydes,
or Thioketones

Campaigne, E., *Chem. Rev.*, 1946, **39**, 57

Nitric acid generally reacts too violently with thials or thiones for the reactions to be of synthetic interest.

*See* 2,2,4,4,6,6-Hexamethyltrithiane, above

Thiophene

1. Meyer, V., *Ber.*, 1883, **16**, 1472
2. Babasinian, V. S., *Org. Synth.*, Coll. Vol. 2, 467

Interaction of thiophene with fuming nitric acid is very violent if uncontrolled, extensive oxidation occurring [1]. Use of a diluent and close control of temperature is necessary for preparation of nitrothiophene [2].

Tricadmium diphosphide

Juza, R. *et al.*, *Z. Anorg. Chem.*, 1956, **283**, 230

Reaction with concentrated acid is explosive.

Triethylgallium monoetherate

*See* TRIETHYLGALLIUM, $C_6H_{15}Ga$

Trimagnesium diphosphide
Mellor, 1940, Vol. 8, 842
Oxidation proceeds with incandescence.

2,4,6-Trimethyltrioxane
Muir, G. D., private comm., 1968
Oxidation of 'paraldehyde' to glyoxal by action of nitric acid is subject to an induction period, and the reaction may become violent if addition of trioxane is too fast. Presence of nitrous acid eliminates the induction period.

Turpentine
Hermoni, A. *et al., J. Appl. Chem.*, 1958, **8**, 670–672
Turpentine and fuming nitric acid do not ignite unless catalysts (fuming sulphuric acid, iron(III) chloride, ammonium metavanadate or copper(II) chloride) are added.

Uranium disulphide
Sidgwick, 1950, 1081
Interaction is violent.

Wood
1. Personal experience, 1974
2. Anon., *Chem. Eng. News*, 1975, **53**(11), 14
3. *MCA Case History No. 1797*

A cracked winchester of concentrated acid leaked into sawdust packing and caused a fire [1]. A similar incident was involved in a freight plane crash [2]. Fuming acid, leaking from a cracked bottle, ignited a wooden truck [3].
*See other* OXIDANTS

## PEROXONITRIC ACID $NO_2OOH$ $HNO_4$

Schwarz, R., *Z. Anorg. Chem.*, 1948, **256**, 3
The pure material, prepared at −80°C, decomposes explosively at −30°C. Solutions in acetic acid or water of below the limiting concentration (corresponding to a stoicheiometric mixture of 70% aqueous nitric acid and 100% hydrogen peroxide) are stable, while those above the limit decompose autocatalytically, eventually exploding.
*See other* PEROXOACIDS

## NITROSYLSULPHURIC ACID $O{=}NOSO_2OH$ $HNO_5S$

Preparative hazard
*See* NITRIC ACID, $HNO_3$: Sulphur dioxide

6-Chloro-2,4-dinitroaniline

*See* 6-CHLORO-2,4-DINITROANILINE, $C_6H_4ClN_3O_4$: Nitrosylsulphuric acid

Dinitroaniline

*BCISC Quart. Safety Summ.*, 1970, **41**, 28

During plant-scale diazotisation of a dinitroaniline hydrochloride, local increase in temperature, due to high concentration of reaction mixture, caused a violent explosion.

*See also* NITRIC ACID, $HNO_3$: 4-Chloro-2-nitroaniline

**LEAD IMIDE** Pb=NH **HNPb**

Mellor, 1940, Vol. 8, 265

It explodes on heating or in contact with water or dilute acids.

*See other* *N*-METAL DERIVATIVES

**PHOSPHAM** $(N{=}PNH)_n$ **$HN_2P$**

Hydrogen sulphide

Mellor, 1940, Vol. 8, 270

The solid produced by interaction of phospham and hydrogen sulphide at red heat is probably a trimeric triphosphatriazine derivative such as phospham. The solid ignites in slightly warm air or in dinitrogen tetraoxide, and is violently oxidised by nitric acid.

Oxidants

Mellor, 1940, Vol. 8, 269–270

Interaction with copper(II) or mercury(II) oxides proceeds incandescently. Mixtures with a chlorate or nitrate explode on heating. Phospham ignites in dinitrogen tetraoxide.

**HYDROGEN AZIDE (HYDRAZOIC ACID)** **$HN_3$**

1. Smith, 1966, Vol. 2, 214
2. Bowden, F. P. *et al., Endeavour*, 1962, **21**, 121
3. Audreith, L. F. *et al., Inorg. Synth.*, 1939, **1**, 77
4. Kemp, M. D., *J. Chem. Educ.*, 1960, **37**, 142
5. Birkofer, L. *et al., Org. Synth.*, 1970, **50**, 109
6. Shapiro, E. L., *Chem. Eng. News*, 1974, **52**(2), 5
7. Chandross, E. A.; Gunderson, H., ibid., (24), 5

Hydrogen azide is quite safe in dilute solution, but is violently explosive and of variable sensitivity in the concentrated (17–50%) or pure states. Wherever possible a low-boiling solvent (ether, pentane) should be added to its solutions to prevent inadvertent concentration by evaporation and recondensation. If this is not possible, no unwetted part of apparatus containing its solutions should be kept appreciably below the boiling point (35°C) to prevent condensation of the pure acid. The pure acid has often been isolated by distillation, but appears to undergo rapid sensitisation on standing, so that, after an hour, faint vibrations or speech are enough to initiate detonation [1]. The solid acid is also very unstable [2]. Preparative procedures have been detailed [3]. It is readily formed on contact of hydrazine or its salts with nitrous acid or its salts. A safe procedure for the preparation of virtually anhydrous hydrogen azide is described [4]. Trimethylsilyl azide serves as a safe and stable substitute for hydrogen azide in many cases [5].

A laboratory explosion [6] seems likely to have been caused by the use of a huge (90-fold) excess of azide in too-concentrated solution and at too low an ambient temperature, leading to condensation of highly concentrated hydrogen azide. This penetrated into a ground joint, and explosion was initiated on removing the stopper [7].

*See* METHYL AZIDE, $CH_3N_3$
*p*-CHLOROPHENYL ISOCYANATE, $C_7H_4ClNO$
NITROUS ACID, $HNO_2$: A semicarbazone, etc.

Heavy metals

1. Napier, D. H., private comm., 1972
2. Cowley, B. R. *et al., Chem. & Ind.*, 1973, 444

Great care is necessary to prevent formation of explosive heavy metal azides from unsuspected contact of hydrogen azide with heavy metals. Interaction of hydrazine and nitrite salts in a copper drainage system caused formation and explosion of copper azide [1]. Use of a brass water-pump and vacuum gauge during removal of excess hydrogen azide under vacuum formed deposits which exploded when the pump and gauge were handled later [2].

*See* SODIUM AZIDE, $N_3Na$: Heavy metals
*See other* NON-METAL AZIDES

**PENTAZOLE** N=NNHN=N (ring) **$HN_5$**

Sorbe, 1968, 140

Pentazole (98.7% nitrogen) and its compounds are explosive.

*See other* HIGH-NITROGEN COMPOUNDS

**SODIUM HYDRIDE** **NaH** **HNa**

Gibb, T. R. P., US Pat. 2 702 281, 1955

A safe form of sodium hydride (as a solid solution in a halide) for large-scale industrial use has been described.

Acetylene

Mellor, 1941, Vol. 2, 483

Dry acetylene does not react with sodium hydride below 42°C, but in presence of moisture reaction is vigorous even at –60°C.

Air

Plesek, J. *et al., Sodium Hydride* (Eng. Transl., Jones, G.) 5,8, London, Iliffe Books, 1968

Commercial sodium hydride may contain traces of sodium which render it spontaneously flammable in moist air, or air enriched by carbon dioxide. The very finely divided dry powder ignites in dry air. Dispersions of the hydride in oil are safe to handle. All normal extinguishers are unsuitable for solid sodium hydride fires. Powdered sand, ashes or sodium chloride are suitable.

Dimethyl sulphoxide

*See* DIMETHYL SULPHOXIDE, $C_2H_6OS$ : Sodium hydride

Ethyl 2,2,3-trifluoropropionate

Bagnall, R. D., private comm., 1972

The ester decomposes violently in presence of sodium hydride, probably owing to hydride-induced elimination of hydrogen fluoride and subsequent exothermic polymerisation.

*See* METHYL TRICHLOROACETATE, $C_3H_3Cl_3O_2$ : Trimethylamine

Glycerol

Unpublished observations, 1956

Exothermic interaction of granular hydride with undiluted (viscous) glycerol with inadequate stirring caused charring to occur. Dilution with tetrahydrofuran to reduce viscosity and improve mixing prevented local overheating during formation of monosodium glyceroxide.

Halogens

Mellor, 1940, Vol. 2, 483

Interaction with chlorine or fluorine is incandescent at ambient temperature, and with iodine at 100°C.

Oxygen

*See* OXYGEN, $O_2$ : Metal hydrides

Sulphur

*See* SULPHUR, S: Sodium hydride

Sulphur dioxide

Moissan, H., *Compt. Rend.*, 1902, **135**, 647

Sulphur dioxide reacts explosively in contact with sodium hydride unless diluted with hydrogen.

Water
1. Anon., *J. Chem. Educ.*, 1967, **44**, 321
2. *MCA Case History No. 1587*

Addition of sodium hydride to a damp reactor which had not been purged with an inert gas caused evolution of hydrogen, and a violent explosion. Solid dispersions of the hydride in mineral oil are more easily and safely handled [1]. When an unprotected polythene bag containing the hydride was moved, some of the powder leaked from a hole, contacted moisture and immediately ignited. Such materials should be kept in tightly closed metal containers in an isolated, dry location [2].
*See other* METAL HYDRIDES

**SODIUM HYDROXIDE** **NaOH** **HNaO**

Aluminium,
Arsenical materials
*See* ALUMINIUM, Al: Arsenic trioxide, etc.

Bromine
*See* BROMINE, $Br_2$: Sodium hydroxide

Chloroform,
Methanol
*See* CHLOROFORM, $CHCl_3$: Sodium hydroxide, etc.

4-Chloro-2-methylphenol
*See* 4-CHLORO-2-METHYLPHENOL, $C_7H_7ClO$: Sodium hydroxide

Cinnamaldehyde
*See* CINNAMALDEHYDE, $C_9H_8O$: Sodium hydroxide

Cyanogen azide
*See* CYANOGEN AZIDE, $CN_4$: Sodium hydroxide

Diborane
*See* DIBORANE, $B_2H_6$: Octanal oxime, etc.

Maleic anhydride
*See* MALEIC ANHYDRIDE, $C_4H_2O_3$: Cations, etc.

4-Methyl-2-nitrophenol
*See* 4-METHYL-2-NITROPHENOL, $C_7H_7NO_3$: Sodium hydroxide

3-Methyl-2-penten-4-yn-1-ol
*See* 3-METHYL-2-PENTEN-4-YN-1-OL, $C_6H_8O$: Sodium hydroxide

Nitrobenzene

*See* NITROBENZENE, $C_6H_5NO_2$ : Alkali

Sodium tetrahydroborate

*See* SODIUM TETRAHYDROBORATE, $BH_4Na$: Alkali

1,2,4,5-Tetrachlorobenzene

*See* 1,2,4,5-TETRACHLOROBENZENE, $C_6H_2Cl_4$ : Sodium hydroxide

1,1,1-Trichloroethanol

*See* 1,1,1-TRICHLOROETHANOL, $C_2H_3Cl_3O$: Sodium hydroxide

Trichloronitromethane

*See* TRICHLORONITROMETHANE, $CCl_3NO_3$ : Sodium hydroxide

Water

*MCA SD-9, 1968*

*Haz. Chem. Data,* 1969, 199

*MCA Case History No. 2166*

The heat of solution is very high, and, with limited amounts of water, violent boiling or even ignition of adjacent combustibles may occur. The case history describes the disposal of open-ended drums of waste caustic soda by putting into an empty hopper and sluicing out the alkali with a hot water jet. When the outlet became blocked, the hopper partly filled with hot water. Addition of a further drum caused a violent eruption to occur.

*See* POTASSIUM HYDROXIDE, HKO, Water

Zinc

*See* ZINC, Zn: Sodium hydroxide

Zirconium

*See* ZIRCONIUM, Zr: Oxygen-containing compounds

**SODIUM HYDROGENSULPHATE** **$NaOSO_2OH$** **$HNaO_4S$**

Acetic anhydride,

Ethanol

*See* ACETIC ANHYDRIDE, $C_4H_6O_3$ : Ethanol, etc.

Calcium hypochlorite

*See* CALCIUM HYPOCHLORITE, $CaCl_2O_2$ : Sodium hydrogensulphate

**'SOLID PHOSPHORUS HYDRIDE'** **$HP_2$**

Mellor, 1940, Vol. 8, 851

Sidgwick, 1950, 730
This material (possibly phosphine adsorbed on phosphorus and produced by decomposition of diphosphane in light) ignites in air, on impact, or on sudden heating to 100°C.
*See other* NON-METAL HYDRIDES

**RUBIDIUM HYDRIDE** **RbH** **HRb**

Acetylene
Mellor, 1941, Vol. 2, 483
In presence of moisture interaction of the hydride and acetylene is vigorous at –60°C. In dry acetylene reaction only occurs above 42°C.

Oxygen
*See* OXYGEN (Gas), $O_2$ : Metal hydrides

Water
Mellor, 1963, Vol. 2, Suppl. 2.2, 2187
Interaction with water is too violent to permit of safe use of the hydride as a drying agent. When dispersed as a solid solution in a metal halide, it can be used as a drying or reducing agent.
*See other* METAL HYDRIDES

**SILICON MONOHYDRIDE** **$(SiH)_n$** **$(HSi)_n$**

Alkali
Stock, G. *et al., Angew, Chem.*, 1956, **68**, 213
The polymeric hydride is relatively stable to water, but reacts violently with aqueous alkali, evolving hydrogen.
*See other* NON-METAL HYDRIDES

**'UNSATURATED' SILICON HYDRIDE** **$(SiH_{1.5})_n$** **$(H_{1.5}Si)_n$**
Bailar, 1973, Vol. 1, 1350
The polymeric material (composition varies from $SiH_{1.42}$ to $SiH_{1.56}$) burns with a shower of sparks if heated in air.
*See* NON-METAL HYDRIDES

**† HYDROGEN (Gas)** **$H_2$**

Acetylene,
Ethylene
Anon., *BCISC Quart. Safety Summ.*, 1974, **45**, 2–3
In a plant producing 200 ktonnes per annum of ethylene from cracked

naphtha, acetylene in the product was hydrogenated to ethylene in a catalytic unit operating under conditions mild enough not to hydrogenate ethylene. During a temporary shutdown and probably owing to operating errors, the internal temperature in the catalytic unit rose to about 400°C, although the external wall temperature was recorded as excessive at 120°C. Attempts to reduce the temperature by passing in additional ethylene were unsuccessful, as the conditions were now severe enough to hydrogenate ethylene and the exothermic reaction increased the temperature finally to 950°C, and the extensive cracking to methane, carbon and hydrogen now occurring was accompanied by further pressure increase. Plant failure was followed by an explosion and fire which took 4 days to extinguish. Damage totalled £6M.

*See other* HYDROGENATION INCIDENTS

Air

1. Pignot, A., *Chaleur Ind.*, 1939, **20**, 251–259
2. Fenning, R. W. *et al., Engineering*, 1930, **130**, 252
3. Messelt, W., *Engineering*, 1922, **113**, 502
4. Neer, M. E., *Diss. Abstr. Int. B*, 1972, **33**, 686
5. Carhart, H. W. *et al., AD Rep. No. 781862/1GA*, 1–39, USNTIS (*Chem. Abs.*, 1975, **82**, 113719c)
6. Neer, M. E., *Amer. Inst. Aero. Astron. Paper*, 1974, 74–1159, 1–7

Sudden release of hydrogen into the atmosphere from storage at above 79 bar will cause spontaneous ignition, owing to the inverse Joule–Thompson effect [1] (but this does not always occur in practice). Release of hydrogen at 47.5 bar into a vented 17.5 litre chromium-plated sphere had caused explosive ignition [2]. Earlier, it had been found that release of hydrogen used for filling balloons would ignite under certain circumstances when the aperture was rusted, a brush discharge then being visible [3]. Spontaneous ignition of flowing hydrogen–air mixtures has been studied [4]. All aspects of hazards involved in the production, storage, handling and use of hydrogen as a Navy fuel are discussed [5]. The cause of auto-ignition of fast-flowing hydrogen–air mixtures was studied. At speeds above 750 m/s ignition occurred at temperatures 400°C below those established for static mixtures [6].

Air,
Catalysts

Mellor, 1942, Vol. 1, 325; 1937, Vol. 16, 146

Catalytically active platinum and similar metals containing adsorbed oxygen or hydrogen will heat and cause ignition in contact with hydrogen or air, respectively. Nitrogen-purging before exposure to atmosphere will eliminate the possibility.

Air,
Various vapours

Schumacher, H. J., *Angew. Chem.*, 1951, **63**, 560–561

The effects of the presence of 44 gaseous or volatile materials upon the upper explosion limits of hydrogen–air mixtures have been tabulated.

Calcium carbonate,
Magnesium

*See* MAGNESIUM, Mg: Calcium carbonate, Hydrogen

Catalysts,
Vegetable oils

Smirnov, V. M., *Chem. Abs.*, 1938, **32**, $4368_6$

Flash fires and explosions which frequently occurred on the discharge of the hot products of catalytic hydrogenation of vegetable oils were attributed to formation of phosphine from the phosphatides present to a considerable extent in, for example, rape-seed and linseed oils.

*See other* HYDROGENATION INCIDENTS

Halogens,
or Interhalogens

*See* BROMINE FLUORIDE, BrF: Hydrogen
BROMINE TRIFLUORIDE, $BrF_3$ : Hydrogen-containing materials
CHLORINE TRIFLUORIDE, $ClF_3$ : Hydrogen-containing materials
BROMINE, $Br_2$ : Hydrogen
CHLORINE TRIFLUORIDE, $ClF_3$ : Hydrogen-containing materials
CHLORINE, $Cl_2$ : Hydrogen
FLUORINE, $F_2$ : Hydrogen
IODINE HEPTAFLUORIDE, $F_7I$: Carbon, etc.

Liquid nitrogen

1. Mel'nik, B. D., *Chem. Abs.*, 1963, **59**, 7309g
2. Bohlken, S. F., *Chem. Eng. Prog.*, 1961, **57**(4), 49–52

During the purification of washed hydrogen from cracking of natural gas, cooling with liquid nitrogen can lead to formation of explosive products from interaction of alkenes in the gas with oxides of nitrogen arising from biologically derived ammonium nitrate or nitrite in the scrubbing water. Various measures to prevent this are discussed. Similar effects may be observed when alkenes are oxidised in the presence of nitrogen [1]. Analysis of a similar incident involving hydrogen derived from fuel oil implicated resins derived from interaction of nitrogen oxide (and possibly dinitrogen oxide) in a low-temperature heat exchanger section operating at –130 to –145°C [2].

*See* NITROGEN OXIDE ('NITRIC OXIDE'), NO: Dienes, Oxygen

Nickel,
Oxygen

*See* NICKEL, Ni: Hydrogen, Oxygen

*o*-Nitroanisole

*See* *o*-NITROANISOLE, $C_7H_7NO_2$: Hydrogen

Oxidants

*See* FLUORINE PERCHLORATE, $ClFO_4$: Hydrogen
CHLORINE DIOXIDE, $ClO_2$: Hydrogen
DICHLORINE OXIDE, $Cl_2O$: Oxidisable materials
COPPER(II) OXIDE, CuO: Hydrogen
NITRYL FLUORIDE, $FNO_2$: Non-metals
FLUORINE, $F_2$: Hydrogen, Oxygen
DIFLUORODIAZENE, $F_2N_2$: Hydrogen
XENON HEXAFLUORIDE, $F_6Xe$: Hydrogen
NITROGEN OXIDE, NO: Hydrogen, etc.
DINITROGEN OXIDE ('NITROUS OXIDE'), $N_2O$: Hydrogen, etc.
DINITROGEN TETRAOXIDE, $N_2O_4$: Hydrogen, etc.
PALLADIUM(II) OXIDE, OPd: Hydrogen
OXYGEN (Gas), $O_2$: Hydrogen

Palladium trifluoride

*See* PALLADIUM TRIFLUORIDE, $F_3Pd$: Hydrogen

Poly(carbon monofluoride)

*See* FLUORINE, $F_2$: Graphite

1,1,1-Tris(azidomethyl)ethane,
Palladium catalyst

*See* 1,1,1-TRIS(AZIDOMETHYL)ETHANE, $C_5H_9N_9$: Hydrogen, etc.

1,1,1-Tris(hydroxymethyl)nitromethane,
Nickel

*See* 1,1,1-TRIS(HYDROXYMETHYL)NITROMETHANE, $C_4H_9NO_5$: Hydrogen, etc.

## HYDROGEN (Liquid) $H_2$

Air

1. Kit and Evered, 1960, 123
2. Report *UCRL-3072,* Univ. of California, Berkeley, 1955
3. Weintraub, A. A. *et al., Health Phys.,* 1962, **8**, 11
4. Edutsky, F. J., *Progr. Refrig. Sci. Technol.: Proc. 1967 (12th) Int. Congr. Refrig.*, 1969, **1**, 283–300

The main precaution necessary for use of liquid hydrogen is to prevent air leaking into the system, where it will be condensed and solidified. Fracture of a crystal of solid air or oxygen could produce a spark to initiate explosion [1]. Procedures for the safe handling of liquid hydrogen in the laboratory [2] and in liquid hydrogen bubble chambers [3] have been detailed.

Safety problems in large-scale handling and transport of liquid hydrogen have been discussed, and safety codes described [4].
*See other* CRYOGENIC LIQUIDS

Ozone
*See* OZONE, $O_3$ : Hydrogen (liquid)

**POLY(DIMERCURYIMMONIUM IODIDE HYDRATE)** **$(H_2Hg_2INO)_n$**
**$(Hg{=}N^+{=}HgI^- H_2O)_n$**

Sorbe, 1968, 97
It explodes on heating.
*See other* POLY(DIMERCURYIMMONIUM) COMPOUNDS

**POTASSIUM AMIDE** **$KNH_2$** $H_2KN$

Brandsma, 1971, 20–21
Sorbe, 1968, 68
It has similar properties to the much more widely investigated sodium amide, but may be expected on general grounds to be more violently reactive than the former. The frequent fires or explosions observed during work-up of reaction mixtures involving the amide were attributed to presence of unreacted (oxide-coated) particles of potassium in the amide solution in ammonia. A safe filtration technique is described.
It also ignites on heating or friction in air.

Ammonia,
Copper(II) nitrate
*See* COPPER(II) NITRATE, $CuN_2O_6$

Potassium nitrite
*See* POTASSIUM NITRITE, $KNO_2$ : Potassium amide

Tetraphenyllead
Houben-Weyl, 1975, Vol. 13.3, 241
One of the by-products of interaction is a highly explosive tetravalent lead compound.

Water
Mellor, 1940, Vol. 8, 255
Interaction is violent and ignition may occur, even in contact with humid air. Old samples may explode with a considerable delay after contact with liquid water.
*See other* *N*-METAL DERIVATIVES

**POTASSIUM AMIDOSULPHATE** **$KOSO_2NH_2$** **$H_2KNO_3S$**

Metal nitrates or nitrites
*See* METAL AMIDOSULPHATES

**POTASSIUM PHOSPHINATE ('HYPOPHOSPHITE')** **$H_2KO_2P$**
**KOP(H)OH**

Nitric acid
Mellor, 1940, Vol. 8, 882
The salt burns (owing to evolution of phosphine) when heated in air, and explodes when evaporated with nitric acid.
*See other* REDUCANTS

**POTASSIUM DIHYDROGENPHOSPHIDE** **$H_2KP$**
**$KPH_2$**
Mellor, 1971, Vol. 8, Suppl. 3, 283
The solid ignites in air.
*See related* NON-METAL HYDRIDES

**LANTHANUM DIHYDRIDE** **$LaH_2$** **$H_2La$**
Kirk-Othmer, 1966, Vol. 11, 207
The very reactive hydride ignites in air.
*See other* METAL HYDRIDES

**MAGNESIUM HYDRIDE** **$MgH_2$** **$H_2Mg$**
Bailar, 1973, Vol. 1, 34–35
The finely divided hydride produced by pyrolysis is pyrophoric in air, while synthesis from the elements produces a substantially air-stable product.
*See other* PYROPHORIC MATERIALS

Oxygen
*See* OXYGEN (Gas), $O_2$: Metal hydrides
*See other* METAL HYDRIDES

**SODIUM AMIDE** **$NaNH_2$** **$H_2NNa$**

Air
1. Bergstrom, F. W. *et al.*, *Chem. Rev.*, 1933, **12**, 61; Brauer, 1963, 467

2. Krüger, G. R. *et al., Inorg. Synth.*, 1966, **8**, 15
3. Shreve, R. N. *et al., Ind. Eng. Chem.*, 1940, **32**, 173
4. Sandor, S., *Munkavédelem*, 1960, **6**(4–6), 20
5. Rüst, 1948, 283

It frequently ignites or explodes on heating or grinding in air, particularly if previously exposed to air or moisture to produce degradation products (possibly peroxidic) [1,2]. Only one explosion not involving exposure to air has been recorded, during pulverisation [3]. The following oxidation products, all liable to explode, have been identified [4]:

sodium hyponitrite, $N_2Na_2O_2$;
sodium trioxodinitrate(2–), $N_2Na_2O_3$;
sodium tetraoxodinitrate(2–), $N_2Na_2O_4$;
sodium pentaoxodinitrate(2–), $N_2Na_2O_5$;
sodium hexaoxodinitrate(2–), $N_2Na_2O_6$.

Several cases of explosive incidents during packing or use of air-exposed material are described [5].

Halocarbons

Sorbe, 1968, 85

Interaction is explosively violent.

Oxidants

*See* POTASSIUM CHLORATE, $ClKO_3$: Sodium amide
SODIUM NITRITE, $NNaO_2$: Sodium amide
DINITROGEN TETRAOXIDE, $N_2O_4$: Sodium amide

Water

1. Mellor, 1941, Vol. 2, 255
2. Personal experience

Fresh material behaves like sodium, hissing, forming a diminishing mobile globule, and often finally exploding [1]. Old, degraded (yellow) samples may be immersed in water for appreciable periods with little action, and then explode very violently. Disposal by controlled burning is safer [2].

*See other N*-METAL DERIVATIVES

**SODIUM HYDROXYLAMIDE** NaNHOH $H_2NNaO$

*See* HYDROXYLAMINE, $H_3NO$: Metals

**SODIUM AMIDOSULPHATE** $NaOSO_2NH_2$ $H_2NNaO_3S$

Metal nitrates,
or Metal nitrites

*See* METAL AMIDOSULPHATES

**LEAD(II) NITRATE PHOSPHINATE** $PbNO_3H_2PO_2$ $H_2NO_5PPb$

Bailar, 1973, Vol. 2, 131

It is powerfully explosive.

*See other* REDOX COMPOUNDS

**HYPONITROUS ACID** HON=NOH $H_2N_2O_2$

Sidgwick, 1950, 693

Mellor, 1940, Vol. 8, 407

An extraordinarily explosive solid, of which the sodium salt also explodes on heating to 260°C. An attempt to prepare the acid by treating its silver salt with hydrogen sulphide caused explosive decomposition. Contact with solid potassium hydroxide caused ignition.

*See* LEAD HYPONITRITE, $N_2O_2Pb$

*See other* REDUCANTS

**NITRIC AMIDE (NITRAMIDE)** $NO_2NH_2$ $H_2N_2O_2$

Canis, C., *Rev. Chim. Minerale,* 1964, **1**, 521

Nitramide is quite unstable and various reactions in which it is formed are violent. Attempts to prepare it by interaction of various nitrates and sulphamates showed that the reactions became explosive at specific temperatures.

Alkalies

Thiele, J. *et al., Ber.,* 1894, **27**, 1909

A drop of concentrated alkali added to solid nitramide causes a flame and explosive decomposition.

Sulphuric acid

Urbanski, 1967, Vol. 3, 16

Nitramide decomposes explosively on contact with concentrated sulphuric acid.

*See other* *N*-NITRO COMPOUNDS

**ZINC HYDRAZIDE** znNHNHzn $H_2N_2Zn$

Sorbe, 1968, 159

It ignites in air.

*See other* *N*-METAL DERIVATIVES
PYROPHORIC MATERIALS

**AMIDOSULPHURYL AZIDE** $H_2NSO_2N_3$ $H_2N_4O_2S$

Shozda, R. J. *et al., J. Org. Chem.*, 1967, **32**, 2876

It is a low-melting explosive solid, as shock-sensitive as glyceryl nitrate.

*See other* ACYL AZIDES

## SODIUM PHOSPHINATE ('HYPOPHOSPHITE') $H_2NaO_2P$

**NaOP(H)OH**

Mellor, 1940, Vol. 8, 881

Evaporation of aqueous solutions by heating may cause an explosion, phosphine being evolved.

Oxidants

1. Costa, R. L., *Chem. Eng. News,* 1947, **25**, 3177
2. Mellor, 1940, Vol. 8, 881
3. Mellor, 1971, Vol. 8, Suppl. 3, 624

Evaporation of a moist mixture of sodium phosphinate and a trace of sodium chlorate by slow heating caused a violent explosion. It was concluded that, once started, the decomposition of the phosphinate proceeds spontaneously [1]. Similar reactions have been reported with nitrates instead of chlorates. Such mixtures had previously been proposed as explosives [2].

Interaction of iodine with the anhydrous salt is violently exothermic, causing ignition [3].

*See* PERCHLORIC ACID, $ClHO_4$: Sodium phosphinate

*See other* REDUCANTS

## SODIUM DIHYDROGENPHOSPHIDE **$NaPH_2$** $H_2NaP$

Albers, H. *et al., Ber.*, 1943, **76**, 23

It ignites in air.

*See related* NON-METAL HYDRIDES

## OXOSILANE **$H_2Si{=}O$** $H_2OSi$

Kautsky, K., *Z. Anorg. Chem.*, 1921, **117**, 209

It ignites in air.

*See related* NON-METAL HYDRIDES

## HYDROGEN PEROXIDE **HOOH** $H_2O_2$

1. Shanley, E. S. *et al., Ind. Eng. Chem.*, 1947, **39**, 1536
2. Naistat, S. S. *et al., Chem. Eng. Prog.*, 1961, **57**(8), 76
3. Kirk-Othmer, 1966, Vol. 11, 407; *MCA SD-53,* 1969
4. Anon., *Fire Prot. Ass. J.*, 1954, 215
5. Smith, I. C. P., private comm., 1973
6. *MCA Case History No. 1121*
7. Campbell, G. A. *et al., Proc. 4th I.Ch.E. Symp. on Chem. Proc. Hazards*, 1971, 37–43
8. Clark, M. C. *et al., Chem. & Ind.*, 1974, 113
9. Cookson, P. G. *et al., J. Organomet. Chem.*, 1975, **99**, C31–32

The hazards attendant upon use of concentrated hydrogen peroxide solutions have been reviewed [1–3]. Salient points include:

Release of enough energy during catalytic decomposition of 65% peroxide to evaporate all water present and formed, and subsequent liability of ignition of combustible materials.

Most cellulosic materials contain enough catalyst to cause spontaneous ignition with 90% peroxide.

Contamination of concentrated peroxide causes possibility of explosion. Readily oxidisable materials, or alkaline substances containing heavy metals, may react violently.

Soluble fuels (acetone, ethanol, glycerol) will detonate on admixture with peroxide of over 30% concentration, the violence increasing with concentration.

Handling systems must exclude fittings of iron, brass, copper, Monel, and screwed joints caulked with red lead.

Concentrated peroxide may decompose violently in contact with iron, copper, chromium, and most other metals or their salts, and dust (which frequentiy contains rust). Absolute cleanliness, suitable equipment (PVC, butyl or Neoprene rubber) and personal protection are essential for safe handling [4]. During concentration under vacuum of aqueous [5] or of aqueous–alcoholic [6] solutions of hydrogen peroxide, violent explosions occurred when the concentration was sufficiently high (probably above 90%) [3]. Detonation of hydrogen peroxide vapour has been studied experimentally [7].

Explosion of a screw-capped winchester of 35% peroxide solution after 2 years owing to internal pressure of liberated oxygen emphasises the need to date-label materials of limited stability, and to vent the container automatically by fitting a Bunsen valve or similar device [8]. Possible hazards in use and handling of concentrated hydrogen peroxide can be avoided by using the solid 2:1 complex of hydrogen peroxide with 1,4-diazabicyclo[2.2.2]octane. This is hygroscopic but stable for at least several months in storage, although it has been reported to decompose above 60°C [9].

Acetal,
Acetic acid

Ashley, J. N. *et al., Chem. & Ind.*, 1957, 702

An organic sulphide containing an acetal group in the molecule had been oxidised to the sulphone with 30% hydrogen peroxide in acetic acid. After the residue had been concentrated by vacuum distillation at 50–60°C, the residue exploded during handling. This was attributed to formation of the peroxide of the acetal (formally a *gem*-diether) or of the aldehyde formed by hydrolysis, but peracetic acid may also have been involved.

Acetaldehyde,
Desiccants

Karnojitzky, V. J., *Chim. Ind. (Paris)*, 1962, 88, 235

Interaction gives the extremely explosive poly(ethylidene) peroxide, also formed on warming peroxidised diethyl ether.

Acetic acid

Grundmann, C. *et al.*, *Ber.*, 1939, **69**, 1755

During preparation of peracetic acid, the temperature should not be too low to prevent reaction as the reagents are mixed, because reaction may begin later with explosive violence.

*See* Oxygenated compounds, etc., below
PEROXYACETIC ACID, $C_2H_4O_3$

Acetic acid,
*N*-Heterocycles

1. Wommack, J. B., *Chem. Eng. News*, 1977, **55**(50), 5
2. Dholakia, S. *et al.*, *Chem. & Ind.*, 1977, 963

During isolation of the di-*N*-oxide of 2,5-dimethylpyrazine [1] and of the mono-*N*-oxide of 2,2′-bipyridyl [2], prepared from action of hydrogen peroxide in acetic acid on the heterocycles, violent explosions occurred during vacuum evaporation of the excess peroxyacetic acid. For a method not involving this considerable explosion risk,

*See* *N*-OXIDES

Acetic acid,
3-Hydroxythietane

1. Dittmer, D. C. *et al.*, *J. Org. Chem.*, 1971, **36**, 1324
2. Lamm, B. *et al.*, *Acta Chem. Scand. Ser. B*, 1974, **28**, 701–703

During preparation of the 1,1-dioxide by oxidation of the precursory 3-hydroxythietane with hydrogen peroxide in acetic acid, evaporation led to a peracetic acid explosion [1]. An alternative oxidation with 30% peroxide solution catalysed by sodium tungstate avoids this hazard [2].

Acetic anhydride

1. Prett, K., *Textilveredlung*, 1966, **1**, 288–290
2. Swern, D., *Chem. Rev.*, 1945, **45**, 5

During preparation of peracetic acid solutions for textile bleaching operations, the reaction mixture must be kept acid. Under alkaline conditions, highly explosive diacetyl peroxide separates from solution [1]. An excess of the anhydride has the same effect [2].

*See* DIACETYL PEROXIDE, $C_4H_6O_4$

Acetone,
Other reagents

1. Anon., *Angew. Chem. (Nachr.)*, 1970, **18**, 3
2. *MCA Case History No. 233*
3. *MCA Case History No. 223*

4. Stirling, C. J. M., *Chem. Brit.*, 1969, **5**, 36
5. Seidl, H., *Angew. Chem.* (*Intern. Ed.*), 1964, **3**, 640
6. Treibs, W., ibid., 802
7. Brewer, A. D., *Chem. Brit.*, 1975, **11**, 335

Acetone and hydrogen peroxide readily form explosive dimeric and trimeric peroxides, particularly during evaporation of the mixture. Many explosions have occurred during work-up of peroxide reactions run in acetone, including partial hydrolysis of a nitrile [1] and oxidation of 2,2′-thiodiethanol [2] and of an unspecified material [3]. The reaction mixture from oxidation of a sulphide with hydrogen peroxide in acetone exploded violently during vacuum evaporation at 90°C. On another occasion oxidation of a sulphide in acetone in presence of molybdate catalyst proceeded with explosive violence. A general warning about use of acetone as solvent for peroxide oxidations is given [4]. During the isolation of 1-tetralone, produced by oxidation of tetralin with hydrogen peroxide in acetone, a violent explosion occurred which was attributed to acetone peroxide [5]. The originator of the method later gave detailed instructions for a safe procedure, which include the exclusion of mineral acids, even in traces [6].

During oxidation of a sulphur heterocycle (unspecified) in acetone with excess 35% hydrogen peroxide, a white solid separated during 3 days standing in a cool place. The solid (20 g) appears to have been acetone peroxide, because it exploded with great violence during drying in a vacuum oven. A previous warning [4] on the incompatibility of acetone and hydrogen peroxide is repeated [7].

*See* Oxygenated compounds, etc., below
KETONE PEROXIDES
*See also* 3,3,6,6,9,9-HEXAMETHYL-1,2,4,5,7,8-HEXAOXAONANE, $C_9H_{18}O_6$

Alcohols

1. *MCA SD-53,* 1969
2. Spengler, G. *et al., Brennst. Chem.*, 1965, **46**, 117

Homogeneous mixtures of concentrated peroxide and alcohols or other peroxide-miscible organic liquids are capable of detonation by shock or heat [1]. Furfuryl alcohol ignites in contact with 85% peroxide within 1 s [2].

*See* Oxygenated compounds, etc., below

Alcohols,
Sulphuric acid

Hedaya, E. *et al., Chem. Eng. News,* 1967, **45**(43), 73

During conversion of alcohols to hydroperoxides, the order of mixing the reagents is important. Addition of concentrated acid to mixtures of an alcohol and concentrated peroxide almost inevitably leads to explosion, particularly if the mixture is inhomogeneous and the alcohol is a solid.

*See* *tert*-Butanol, etc., below
2-Methyl-1-phenyl-2-propanol, below

Aluminium isopropoxide,
Heavy metal salts

Ward, D. S., unpublished information, 1974

During preparation of an alumina catalyst, the isopropoxide was stirred with 6% hydrogen peroxide solution to generate a slurry of pseudo-boehemite alumina, to which was added a solution of heavy metal salts. The heating and foaming of the mixture was excessive and foam overflowed into a safety tray, then ignited. The incident was attributed to either spontaneous or static-induced ignition of the foam consisting of oxygen-rich bubbles in the isopropanol-containing liquid medium. The heavy metal salts would catalyse decomposition of the hydrogen peroxide and may have been involved in ignition.

*See* Metals, etc., below

Benzenesulphonic anhydride

Fichter, F., *Helv. Chim. Acta*, 1924, 7, 1072

Attempts to prepare benzeneperoxysulphonic acid by interaction of the anhydride and 90–95% peroxide led to explosively violent decomposition within a few seconds. When diluted with acetic acid, the reaction mixture soon began to decompose, leading to violent boiling. It was concluded that the peroxyacid is too unstable for more than transitory existence.

*tert*-Butanol,
Sulphuric acid

Schenach, T. A., *Chem. Eng. News*, 1973, **51**(6), 39

Preparation of di-*tert*-butyl peroxide by addition of *tert*-butanol to 50% hydrogen peroxide: 78% sulphuric acid mixtures (1:2 by wt.) is a dangerously deceptive procedure. On the small scale and with adequate cooling capacity it may be possible to prevent the initial stage (exothermic formation of *tert*-butyl hydroperoxide) getting out of control and initiating violent or explosive decomposition of the peroxide–peroxomonosulphuric acid mixture. This hazard diminishes as the reaction proceeds with consumption of hydrogen peroxide and dilution by the water of reaction. On the plant scale several severe explosions have occurred, preceded only by a gradual temperature increase, during attempted process development work.

*See* Alcohols, above

Carbon

1. Mellor, 1939, Vol. 1, 936–938

2. Schumb, 1955, 402, 478

The violent decomposition observed on adding charcoal to concentrated hydrogen peroxide is mainly due to catalysis by metallic impurities present and the surface of the charcoal, rather than to direct oxidation of the carbon [1]. Charcoal mixed with a trace of manganese dioxide ignites immediately on contact with concentrated peroxide [2].

Carboxylic acids

Kuchta, J. M. *et al., Rep. Invest. No. 5877*, Washington, US Bur. Mines, 1961

Admixture produces peroxyacids, some of which are unstable and explosive. Aqueous peroxide solutions containing formic, acetic or tartaric acids above certain concentrations can be caused to detonate by a severe explosive shock.

*See* Oxygenated compounds, etc., below
PEROXYACIDS

Diethyl ether

Bruhl, J. W., *Ber.*, 1895, **28**, 2856–2857

Evaporation of an ethereal solution of hydrogen peroxide gave a residue of which a drop on a platinum spatula exploded weakly on exposure to flame. When the sample (1–2 g) was stirred with a glass rod (not fire-polished), an extremely violent detonation occurred.

*See* ETHERS

Formic acid,
Metaboric acid

*See* PEROXYFORMIC ACID, $CH_2O_3$

Gadolinium hydroxide

Bogdanov, G. A. *et al., Chem. Abs.*, 1976, **85**, 52253j

Interaction gives a hydrated peroxide, $Gd(OH)_3 \cdot 0.26\ O_2 \cdot 1.5\ H_2O$, which decomposed explosively at 80–90° C.

Hydroboration product of 2-methyl-2-propenyl tetrahydropyranyl ether

*See* TETRAHYDROPYRANYL ETHER DERIVATIVES

Iron(II) sulphate,
2-Methylpyridine,
Sulphuric acid

Mond Div., ICI, private comm., 1969

Addition of 30% peroxide and sulphuric acid to 2-methylpyridine and iron(II) sulphate caused a sudden exotherm, followed by a vapour phase explosion and ignition. Lack of stirring is thought to have caused localised overheating, vaporisation of the base, and ignition in the possibly oxygen-enriched atmosphere.

Ketene

Swern, D., *Chem. Rev.*, 1949, **45**, 7

Interaction with excess ketene rapidly forms explosive diacetyl peroxide.

Ketones,
Nitric acid

Bjorklund, G. H. *et al., Trans. R. Soc. Can.*, 1950(Sect. III), **44**, 25

Unless the temperature and concentrations of reagents were carefully controlled, mixtures of hydrogen peroxide, nitric acid and acetone overheated and exploded violently. Under controlled conditions, the explosive dimeric or trimeric acetone peroxides were produced. Butanone-2 and pentanone-3 gave shock- and heat-sensitive oily peroxides. Cyclopentanone reacts vigorously, producing a solid which soon produces a series of explosions if left in contact with the undiluted reaction liquor. The isolated trimeric peroxide is very sensitive to shock, slight friction or rapid heating, and explodes very violently. Cyclohexanone and 3-methylcyclohexanone gave oily, rather explosive peroxides.

*See* KETONE PEROXIDES

Lead,
Trioxane

Bamberger, M. *et al., Z. Ges. Schiess- u. Sprengstoffw.*, 1927, **22**, 125–128

Mixtures of trioxane with 60% hydrogen peroxide solution are detonable by heat or shock, or spontaneously after contact with metallic lead. The latter may be due to the heat of oxidation of lead.

*See* Metals, below
Oxygenated compounds, etc., below

Lithium tetrahydroaluminate

Osmon, R. V., *Chem. Eng. Prog. Symp. Ser.*, 1966, **62**(61), 92–102

This oxidiser–fuel combination showed promise in rocketry.

*See* ROCKET PROPELLANTS

Mercury(II) oxide,
Nitric acid

Mellor, 1940, Vol. 4, 781

Although red mercury oxide usually vigorously decomposes hydrogen peroxide, the presence of traces of nitric acid inhibits decomposition and promotes the formation of red mercury(II) peroxide. This explodes on impact or friction, even when wet, if the mercury oxide was finely divided.

Metals,
or Metal oxides,
or Metal salts

1. Mellor, 1939, Vol. 1, 936–944
2. Schumb, 1955, 480
3. Kit and Evered, 1960, 136
4. 'Hydrogen Peroxide Data Manual', Laporte Chemicals Ltd., Luton, 1960
5. Hardin, M. C. *et al., Chem. Abs.*, 1963, **59**, 2583h
6. Winnacker, 1970, Vol. 2, 552

The noble metals are all very active catalysts, particularly when finely divided, for the decomposition of hydrogen peroxide, silver being used for this purpose in peroxide-powered rocket motors. Gold and the platinum group metals behave similarly [2,4]. Addition of platinum black to concentrated peroxide solution may cause an explosion, and powdered magnesium and iron, promoted by traces of manganese dioxide, ignite on contact [1].

Oxides of cobalt, iron (especially rust), lead (also the hydroxide), manganese, mercury and nickel are also very active and the parent metals and their alloys must be rigorously excluded from peroxide handling systems [1,4]. Soluble derivatives of many other metals, particularly under alkaline conditions, will also catalyse the exothermic decomposition, even at low concentrations [4]. Calcium permanganate has been used, either as a solid or in concentrated solution, to ignite peroxide rocket motors [3]. Comprehensive data on all aspects of handling and use of concentrated peroxide are available [4].

Potassium and sodium are oxidised violently by concentrated peroxide solutions [1], and interaction of lithium and hydrogen peroxide is hypergolic and controllable under rocket motor conditions [5]. The effectiveness of metals for catalytic decomposition of peroxide solutions decreases in the order: osmium, palladium, platinum, iridium, gold, silver, manganese, cobalt, copper, lead [6].

*See also* Carbon, above

Methanol,
Phosphoric acid

Anon., *J. Electrochem. Soc.*, 1965, **112**, 251C
King, C. V. (Ed.), ibid., 1057; 1966, **113**, 519

In mixtures of the three components intended for use as chemical polishing or etching solutions, concentrated peroxide must not be used.

*See* Oxygenated compounds, etc., below

2-Methyl-1-phenyl-2-propanol,
Sulphuric acid

1. Winstein, S. *et al., J. Amer. Chem. Soc.,* 1967, **89**, 1661

2. Hedaya, E. *et al., Chem. Eng. News,* 1967, **45**(43), 73
3. Hiatt, R. R. *et al., J. Org. Chem.,* 1963, 28, 1893

Directions given [1] for preparation of 2-methyl-1-phenyl-2-propyl hydroperoxide by adding sulphuric acid to a mixture of the alcohol and 90% hydrogen peroxide are wrong [2] and will lead to explosion [3]. The acidified peroxide (30–50% solution is strong enough) is preferably added to the alcohol, with suitable cooling and precautions [2].

*See* Alcohols, above
Oxygenated compounds, etc., below

Nitric acid,
Soils

*See* NITRIC ACID, $HNO_3$: Hydrogen peroxide, Soils

Nitric acid,
Thiourea

Bjorklund, G. H. *et al., Trans. R. Soc. Can.,* 1950 (Sect.III), **44**, 28

The solid peroxide produced by action of hydrogen peroxide and nitric acid on thiourea in acetic acid solution decomposed violently on drying in air with evolution of sulphur dioxide and free sulphur. (The solid may have been a peroxidate of thiourea dioxide.)

*See* CRYSTALLINE HYDROGEN PEROXIDATES

Nitrogenous bases

1. Stone, F. S. *et al., J. Chem. Phys.,* 1952, **20**, 1339
2. Urbanski, 1967, Vol. 3, 306
3. *Haz. Chem. Data,* 1969, 220

Ammonia dissolved in 99.6% peroxide gave an unstable solution which exploded violently [1]. In the absence of catalysts, concentrated peroxide does not react immediately with hydrazine hydrate. This induction period has caused a number of explosions and accidents due to sudden reaction of accumulated materials [2]. 1,1-Dimethylhydrazine is hypergolic with high-test peroxide [3].

Organic compounds

Hutton, E., *Chem. Brit.,* 1969, **5**, 287

Although under certain circumstances mixtures of hydrogen peroxide and organic compounds are capable of developing more explosive power than an equivalent weight of TNT, in many cases interaction can be effected safely and well under control by applying well-established procedures, to which several references are given.

Organic matter,
Sulphuric acid

Analytical Methods Committee, *Analyst,* 1976, **101**, 62–66

Advantages and potential hazards in the use of mixtures of 50% hydrogen peroxide solution and concentrated sulphuric acid to destroy various types of organic matter prior to analysis are discussed in detail. The method is appreciably safer than those using perchloric and/or nitric acids, but the use of an adequate proportion of sulphuric acid with a minimum of peroxide is necessary to avoid the risk of explosive decomposition. The method is not suitable for use in pressure-digestion vessels (PTFE-lined steel bombs), in which an explosion occurred at 80° C.

*See* PERCHLORIC ACID, $ClHO_4$: Nitric acid, Organic matter

Oxygenated compounds,
Water

Monger, J. M. *et al., J. Chem. Eng. Data,* 1961, **6**(1), 23

The explosion limits have been determined for liquid systems containing hydrogen peroxide, water and acetaldehyde, acetic acid, acetone, ethanol, formaldehyde, formic acid, methanol, 2-propanol or propionaldehyde, under various types of initiation.

*See* Alcohols, above

α-Phenylselenoketones

Reich, H. J. *et al., J. Amer. Chem. Soc.*, 1975, **97**, 5442

During conversion of the selenoketone to the nor-phenylselenoenones with hydrogen peroxide, care is necessary to control the strongly exothermic reaction. Under no circumstances should oxidations of amounts above 5 mmol be effected by adding the full amount of peroxide before oxidation has started.

Phosphorus

Anon., *J. R. Inst. Chem.*, 1957, **81**, 473

If yellow or red phosphorus is incompletely immersed while undergoing oxidation in hydrogen peroxide solutions, heating at the air–solution interface can ignite the phosphorus and lead to a violent reaction. Such behaviour has been observed with hydrogen peroxide solutions above 30% (110 volume) concentration.

Phosphorus(V) oxide

Toennies, G., *J. Amer. Chem. Soc.,* 1937, **59**, 555

The extremely violent interaction of phosphorus(V) oxide and concentrated hydrogen peroxide to give permonophosphoric acid may be moderated by using acetonitrile as a diluent.

Sulphuric acid

Analytical Methods Committee Rep., *Analyst*, 1967, **92**, 404

Evaporation of mixtures of excess 50% hydrogen peroxide solution with sulphuric acid (10:1) leads to loud but non-shattering explosions of the peroxomonosulphuric acid formed.

*See* Organic matter, Sulphuric acid, above
PEROXOMONOSULPHURIC ACID, $H_2O_5S$

Tetrahydrothiophene

Koppel, H. C., *Chem. Eng. News*, 1974, **52**(39), 3

Preparative reactions involving oxidation of tetrahydrothiophene to the sulphoxide by slow addition of 37% peroxide solution exploded violently on three occasions. No explanation is apparent, and similar reactions had been run uneventfully over a period of 10 years.

Tin(II) chloride

Vickery, R. C. *et al.*, *Chem. & Ind.*, 1949, 657

Interaction is strongly exothermic, even in solution. Addition of peroxide solutions of greater than 3% wt. strength causes a violent reaction.

Unsaturated compounds

Swern, 1971, Vol. 2, 432

In commercial-scale epoxidation of unsaturated organic compounds, 50–70% hydrogen peroxide is added during 2–8 h to the stirred mixture of unsaturated compound, aliphatic acid and acid catalyst at 50–80° C, the rate of addition being dependent on heat-transfer capacity of the reaction system. When 70% peroxide is being used, care must be taken to avoid the dangerous detonable region by ensuring that an exotherm has occurred before more than 25% of the hydrogen peroxide is added.

*See* Carboxylic acids, above
Organic compounds, above
Oxygenated compounds, etc., above

Vinyl acetate

*ABCM Quart. Safety Summ.*, 1948, **19**, 18

Vinyl acetate had been hydroxylated by treatment with excess hydrogen peroxide in presence of osmium tetraoxide catalyst. An explosion occurred while excess vinyl acetate and solvent were being removed by vacuum distillation. This was attributed to the presence of peracetic acid, formed by interaction of excess hydrogen peroxide and acetic acid produced by hydrolysis of the vinyl acetate.

Wood

1. *MCA Case History No. 1626*
2. *MCA Case History No. 1648*

Leakage from drums of 35% hydrogen peroxide onto a wooden pallet caused ignition of the latter when it was moved. Combustion, though limited in area, was fierce and took some time to extinguish [1]. Leakage of 50% peroxide onto supporting pallets under polythene sheeting led to spontaneous ignition and a fierce fire. Contact of 50% peroxide with wood does not usually lead to spontaneous ignition, but

hot weather, dry wood (possibly contaminated) and the thermal insulation of the cover may have contributed to occurrence of ignition [2].
*See other* OXIDANTS

**ZINC HYDROXIDE** $Zn(OH)_2$ $H_2O_2Zn$

Chlorinated rubber
*See* CHLORINATED RUBBER: Metal oxides, hydroxides

**SULPHURIC ACID** $HOSO_2OH$ $H_2O_4S$

*MCA SD-20,* 1963
Safe handling procedures for sulphuric acid and fuming sulphuric acid (oleum) are detailed.

Acetaldehyde
*See* ACETALDEHYDE, $C_2H_4O$: Sulphuric acid

Acetone,
Nitric acid
*See* NITRIC ACID, $HNO_3$: Acetone, etc.

Acetonitrile,
Sulphur trioxide
*See* ACETONITRILE, $C_2H_3N$: Sulphuric acid

Acrylonitrile
*See* ACRYLONITRILE, $C_3H_3N$: Acids

Alkyl nitrates
*See* ALKYL NITRATES: Lewis acids

Ammonium iron(III) sulphate dodecahydrate
Clark, R. E. D., private comm., 1973
A few dense crystals heated with sulphuric acid exploded, owing to the exotherm in contact with water liberated as the crystals disintegrated.

Aniline,
Glycerol,
Nitrobenzene
*See* QUINOLINE, $C_9H_7N$

Benzyl alcohol
Grignard, 1960, Vol. 5, 1005

A mixture of the alcohol with 58% sulphuric acid decomposes explosively at about 180° C.

Bromine pentafluoride

*See* BROMINE PENTAFLUORIDE, $BrF_5$: Acids, etc.

1-Chloro-2,3-epoxypropane

*See* 1-CHLORO-2,3-EPOXYPROPANE, $C_3H_5ClO$: Sulphuric acid

*p*-Chloronitrobenzene,
Sulphur trioxide

1. Grewer, T., *Chem. Ing. Techn.*, 1975, **47**(6), 233
2. Vervalin, C. H., *Hydrocarbon Process.*, 1976, **55**(9), 323

DTA shows that the reaction mixture from sulphonation of the nitro-compound in 20% oleum, containing 35% wt. of 2-chloro-5-nitrobenzene-sulphonic acid, shows two exothermic stages at 100 and 220° C, respectively, the latter being violently rapid. The adiabatic reaction mixture, initially at 89° C, attained 285° C with boiling after 17 h. At 180° C the induction period was about 20 min [1]. Sulphonation of *p*-chloronitrobenzene with 65% oleum at 46° C led to a runaway decomposition reaction in a 2000 litre vessel. The original process using 20% oleum was less sensitive to heating rate and temperature. Knowledge that the reaction could be dangerous above 50° C had not been applied [2].

*See* THERMAL STABILITY OF REACTION MIXTURES
*See other* SULPHONATION INCIDENTS

Copper,

Campbell, D. A., *School Sci. Rev.*, 1939, **20**(80), 631

The generation of sulphur dioxide by reduction of sulphuric acid with copper is considered too dangerous for a school experiment.

2-Cyano-4-nitrobenzenediazonium hydrogensulphate

*See* 2-CYANO-4-NITROBENZENEDIAZONIUM HYDROGENSULPHATE, $C_7H_4N_4O_6S$: Sulphuric acid

2-Cyano-2-propanol

1. *Occupancy Fire Record,* FR 57-5, 5, Boston, NFPA, 1957
2. Kirk-Othmer, 1967, Vol. 13, 333

Addition of sulphuric acid to the cyano-alcohol caused a vigorous reaction which pressure-ruptured the vessel [1]. This seems likely to have been due to insufficient cooling to prevent dehydration of the alcohol to methacrylonitrile and lack of inhibitors to prevent exothermic polymerisation of the nitrile [2].

Cyclopentadiene

Wilson, P. J. *et al., Chem. Rev.*, 1944, **34**, 8

It reacts violently with charring, or explodes in contact with concentrated sulphuric acid.

*See also* NITRIC ACID, $HNO_3$ : Hydrocarbons

Cyclopentanone oxime

Brookes, F. R., private comm., 1968

Heating the oxime with 85% sulphuric acid to effect the Beckmann rearrangement caused eruption of the stirred flask contents. Benzenesulphonyl chloride in alkali was a less vigorous reagent.

*See* BUTYRALDOXIME, $C_4H_9NO$
ETHYL FORMYLPROPIONATE OXIME, $C_6H_{11}NO_3$

1,3-Diazidobenzene

*See* 1,3-DIAZIDOBENZENE, $C_6H_4N_6$ : Sulphuric acid

Dimethoxydinitroanthraquinone

*See* *mixo*-DIMETHOXYDINITROANTHRAQUINONE, $C_{16}H_{10}N_2O_8$: Sulphuric acid

*p*-Dimethylaminobenzaldehyde

*MCA Case History No. 2101*

During preparation of a solution of the aldehyde in dilute sulphuric acid, the latter should be prepared before addition of the aldehyde. An attempt to prepare the solution in concentrated acid from a slurry of the aldehyde in a little water caused the stoppered flask to explode. This was attributed to the exotherm caused by addition of a little water and the basic aldehyde to the concentrated acid.

1,5-Dinitronaphthalene,
Sulphur

*See* 1,5-DINITRONAPHTHALENE, $C_{10}H_6N_2O_4$ : Sulphur, etc.

Hexalithium disilicide

*See* HEXALITHIUM DISILICIDE, $Li_6Si_2$ : Acids

Hydrogen peroxide

*See* HYDROGEN PEROXIDE, $H_2O_2$ : Sulphuric acid

Metal acetylides or carbides

1. Mellor, 1946, Vol. 5, 849
2. *MCA SD-20*, 1963

Monocaesium and monorubidium acetylides ignite with concentrated sulphuric acid [1]. Other carbides are hazardous in contact [2].

Metal chlorates

*See* METAL CHLORATES: Acids
METAL HALOGENATES: Metals and oxidisable derivatives, etc.

Metal perchlorates

*See* METAL PERCHLORATES: Sulphuric acid

Nitramide

*See* NITRIC AMIDE, $H_2N_2O_2$: Sulphuric acid

Nitric acid,
Organic matter

*See* NITRIC ACID, $HNO_3$: Organic matter, etc.

Nitric acid,
Toluene

*See* NITRIC ACID, $HNO_3$: Hydrocarbons (reference 5)

Nitroaryl bases and derivatives

1. Hodgson, J. F., *Chem. & Ind.*, 1968, 1399; private comm., 1973
2. Poshkus, A. C. *et al.*, *J. Appl. Polymer Sci.*, 1970, **14**, 2049–2052

A series of *o*- and *p*-nitroaniline derivatives and analogues when heated with sulphuric acid to above 200°C undergo, after an induction period, a vigorous reaction. This is accompanied by gas evolution which produces up to a 150-fold increase in volume of a solid foam, and is rapid enough to be potentially hazardous if confined. *o*-Nitroaniline reacts almost explosively [1] and *p*-nitroaniline, *p*-nitroacetanilide, aminonitrodiphenyls, -naphthalenes and various derivatives [2], as well as some nitro-*N*-heterocycles [1,2], also react vigorously. *p*-Nitroanilinium sulphate and 4-nitroaniline-2-sulphonic acid and its salts also generate foams when heated without sulphuric acid. The mechanism is not clear, but involves generation of a polymeric matrix foamed by sulphur dioxide and water eliminated during the reaction [1].

*See also* DIETHYL SULPHATE, $C_4H_{10}O_4S$: 2,7-Dinitro-9-phenylphenanthridine

Nitrobenzene

*See* NITROBENZENE, $C_6H_5NO_2$: Sulphuric acid

*m*-Nitrobenzenesulphonic acid

*See* *m*-NITROBENZENESULPHONIC ACID, $C_6H_5NO_5S$: Sulphuric acid

Nitromethane

*See* NITROMETHANE, $CH_3NO_2$: Acids

*N*-Nitromethylamine

*See* *N*-NITROMETHYLAMINE, $CH_4N_2O_2$: Sulphuric acid

*p*-Nitrotoluene

*See* *p*-NITROTOLUENE, $C_7H_7NO_2$: Sulphuric acid

Permanganates

Interaction produces the powerful oxidant, permanganic acid.

*See* PERMANGANIC ACID, $HMnO_4$: Organic materials
POTASSIUM PERMANGANATE, $KMnO_4$: Sulphuric acid

Phosphorus

Mellor, 1940, Vol. 8, 786

White phosphorus ignites in contact with boiling sulphuric acid or its vapour.

Phosphorus trioxide

*See* PHOSPHORUS(III) OXIDE, $O_3P_2$: Sulphuric acid

Polysilylene

*See* POLYSILYLENE, $(H_2Si)_n$

Potassium

*See* POTASSIUM, K: Sulphuric acid

Potassium *tert*-butoxide

*See* POTASSIUM *tert*-BUTOXIDE, $C_4H_9KO$: Acids

Silver peroxochromate

*See* SILVER PEROXOCHROMATE, $AgCrO_5$: Sulphuric acid

Sodium

*See* SODIUM, Na: Acids

Sodium carbonate

*See* SODIUM CARBONATE, $CNa_2O_3$: Sulphuric acid

Sodium tetrahydroborate

*See* SODIUM TETRAHYDROBORATE, $BH_4Na$: Acids

Tetramethylbenzenes

1. Birch, S. F. *et al., J. Amer. Chem. Soc.*, 1949, **71**, 1364
2. Smith, L. I. in *Organic Reactions* (Ed. Adams, R. *et al.*), Vol. 1, 382, New York, Wiley, 1942

Sulphonation of the mixed isomers of 1,2,3,5- and 1,2,4,5-tetramethylbenzenes was too violent [1] for shaking in a closed glass vessel as originally described [2].

*See other* SULPHONATION REACTIONS

1,2,4,5-Tetrazine

*See* 1,2,4,5-TETRAZINE, $C_2H_2N_4$: Sulphuric acid

Thallium(I) azidodithiocarbonate

*See* THALLIUM AZIDODITHIOCARBONATE, $CN_3S_2Tl$

Trimercury dinitride

*See* TRIMERCURY DINITRIDE, $Hg_3N_2$

1,3,5-Trinitrosohexahydro-1,3,5-triazine

*See* 1,3,5-TRINITROSOHEXAHYDROTRIAZINE, $C_3H_6N_6O_3$: Sulphuric acid

Water

Mellor, 1947, Vol. 10, 405–408

Dilution of sulphuric acid with water is vigorously exothermic, and must be effected by adding acid to water to avoid local boiling. Mixtures of sulphuric acid and excess snow form powerful freezing mixtures. Fuming sulphuric acid (containing sulphur trioxide) reacts violently with water.

*See also* *m*-NITROBENZENESULPHONIC ACID, $C_6H_5NO_5S$: Sulphuric acid

Zinc iodide

Pascal, 1962, Vol. 5, 168

Interaction with the concentrated acid is violent.

*See other* OXIDANTS

**POLY(DIHYDROXODIOXODISILANE)** **$(H_2O_4Si_2)_n$**

**$(HOSiOSiOOH)_n$**

1. Sorbe, 1968, 127
2. Mellor, 1940, Vol. 6, 230

Dry ‘silico-oxalic acid’ explodes on impact or heating [1]. This polymeric hydrolysis product of hexahalo- or hexaethoxy-disilane decomposes with more or less violence if heated in air (when it ignites) or in a test-tube (when it explodes) [2].

*See related* NON-METAL OXIDES

**PEROXOMONOSULPHURIC ACID** **$HOOSO_2OH$** **$H_2O_5S$**

1. Edwards, J. O., *Chem. Eng. News,* 1955, **33**, 3336
2. Brauer, 1963, Vol. 1, 389

A small sample had been prepared from chlorosulphuric acid and 90% hydrogen peroxide; the required acid phase was separated and stored at 0°C overnight. After warming slightly, it exploded [1]. The handling of large amounts is dangerous owing to possibility of local overheating (e.g. contact with moisture) and explosive decomposition [2].

Acetone

1. *MCA Case History No. 662*
2. Bayer, A., *Ber.*, 1900, **33**, 858

Accidental addition of a little acetone to the residue from wet ashing a polymer with mixed nitric–sulphuric acids and hydrogen peroxide caused a violent explosion [1]. The peroxo acid would be produced under these conditions, and is known to react with acetone to produce the explosive acetone peroxide [2].

Alcohols

Toennies, G., *J. Amer. Chem. Soc.*, 1937, **59**, 552

Contact of the acid with secondary and tertiary alcohols, even with cooling, may lead to violent explosions.

Aromatic compounds

Sidgwick, 1950, 939

Mixtures with aniline, benzene, phenol, etc., explode.

Catalysts

Mellor, 1946, Vol. 2, 483–484

The 92% acid is decomposed explosively on contact with massive or finely divided platinum, manganese dioxide or silver. Neutralised solutions of the acid also froth violently on treatment with silver nitrate, lead or manganese dioxides.

Fibres

Ahrle, H., *Z. Angew. Chem.*, 1909, **23**, 1713

Wool and cellulose are rapidly carbonised by the 92% acid, while cotton ignites after a short delay.

*See other* OXIDANTS

**PEROXODISULPHURIC ACID** **$HOSO_2OOSO_2OH$** **$H_2O_8S_2$**

Organic liquids

1. D'Ans, J. *et al.*, *Ber.*, 1910, **43**, 1880; *Z. Anorg. Chem.*, 1911, **73**, 1911
2. Sidgwick, 1950, 938

A very powerful oxidant; uncontrolled contact with aniline, benzene, ethanol, ether, nitrobenzene or phenol may cause explosion [1]. Alkanes are slowly carbonised [2].

*See other* OXIDANTS

**† HYDROGEN SULPHIDE** **$H_2S$**

*MCA SD-36*, 1968

*p*-Bromobenzenediazonium chloride

*See* DIAZONIUM SULPHIDES

Copper,

Oxygen

Merz, V. *et al., Ber.*, 1880, **13**, 722

Addition of powdered copper to a 1:2 mixture of hydrogen sulphide and oxygen causes the metal to become incandescent and ignite the explosive mixture.

*See* Metals, below

Metal oxides

Mellor, 1947, Vol. 10, 129, 141

Hydrogen sulphide is rapidly oxidised, and may ignite, in contact with a range of metal oxides, including barium dioxide, chromium trioxide, copper oxide, lead dioxide, manganese dioxide, nickel oxide, silver mono- and di-oxides, sodium peroxide, thallium trioxide. In the presence of air, contact with mixtures of calcium or barium oxide with mercury or nickel oxide may cause vivid incandescence or explosion.

*See* Rust, below

Metals

Mellor, 1947, Vol. 10, 140; 1943, Vol. 11, 731

A mixture with air passed over copper powder may attain red heat. Finely divided tungsten glows red-hot in a stream of hydrogen sulphide.

*See* Copper, Oxygen, above
SODIUM, Na: Sulphides

Nitrogen trichloride

*See* NITROGEN TRICHLORIDE, $Cl_3N$: Initiators

Oxidants

Interaction of hydrogen sulphide with a variety of oxidants may be violent if uncontrolled.

*See* Metal oxides, above
Oxygen, below
Rust, below
SILVER BROMATE, $AgBrO_3$: Sulphur compounds
HEPTASILVER NITRATE OCTAOXIDE, $Ag_7NO_{11}$
DIBISMUTH DICHROMIUM NONAOXIDE ('BISMUTH CHROMATE'), $Bi_2Cr_2O_9$: Hydrogen sulphide
BROMINE PENTAFLUORIDE, $BrF_5$: Hydrogen-containing materials
MERCURY(I) BROMATE, $Br_2Hg_2O_6$: Hydrogen sulphide
PERCHLORYL FLUORIDE, $ClFO_3$: Hydrocarbons, etc.
CHLORINE TRIFLUORIDE, $ClF_3$: Hydrogen-containing materials
DICHLORINE OXIDE, $Cl_2O$: Oxidisable materials
LEAD HYPOCHLORITE, $Cl_2O_2Pb$: Hydrogen sulphide
COPPER CHROMATE, $CrCuO_4$: Hydrogen sulphide
CHROMIUM TRIOXIDE, $CrO_3$: Hydrogen sulphide

FLUORINE, $F_2$: Hydrogen sulphide
OXYGEN DIFLUORIDE, $F_2O$: Combustible gases
NITRIC ACID, $HNO_3$: Hydrogen iodide, etc.
SODIUM PEROXIDE, $Na_2O_2$: Hydrogen sulphide
LEAD(IV) OXIDE, $O_2Pb$: Hydrogen sulphides

Oxygen

Gray, P. *et al., J. Chem. Soc., Faraday Trans. I*, 1974, **70**, 2338–2350
During an investigation of the spontaneously explosive oxidation of near-stoicheiometric gas mixtures in the range 280–360°C, extensive self-heating was observed before ignition occurred. The second and third ignition limits were also investigated,

*See* Oxidants, above

Rust

Mee, A. J., *School Sci. Rev.*, 1940, 22(85), 95
Hydrogen sulphide may ignite if passed through rusty iron pipes.

*See* Metal oxides, above

Silver fulminate

*See* SILVER FULMINATE, CAgNO, Hydrogen sulphide

Soda-lime

1. Mellor, 1947, Vol. 10, 140
2. Bretherick, L., *Chem. & Ind.*, 1971, 1042

Interaction is exothermic, and if air is present, incandescence may occur with freshly prepared granular material. Admixture with oxygen causes a violent explosion [1]. Soda-lime, used to absorb hydrogen sulphide, will subsequently react with atmospheric oxygen and carbon dioxide with a sufficient exotherm in contact with moist paper (in a laboratory waste bin) to cause ignition [2]. Spent material should be saturated with water before separate disposal.

Mixtures analogous to soda-lime, such as barium hydroxide with sodium or potassium hydroxides, also behave similarly [1].

*See other* NON-METAL HYDRIDES

**† HYDROGEN DISULPHIDE** **HSSH** **$H_2S_2$**

Bloch, I. *et al., Ber.*, 1908, **41**, 1977
Rather more flammable than hydrogen trisulphide, it is decomposed violently by alkalies.

*See other* NON-METAL SULPHIDES

**HYDROGEN TRISULPHIDE** **HSSSH** **$H_2S_3$**

Benzenediazonium chloride

King, W. B. *et al., J. Amer. Chem. Soc.*, 1932, **54**, 3073
Addition of the dry diazonium salt to the crude liquid sulphide causes

explosively violent interaction. Slow addition of the sulphide to the cooled salt, or dilution with a solvent, moderates the reaction.
*See* DIAZONIUM SULPHIDES AND DERIVATIVES

Metal oxides
Mellor, 1947, Vol. 10, 159
Contact with copper oxide, lead mono- or di-oxides, mercury(II) oxide, tin dioxide and triiron tetraoxide causes violent decomposition and ignition. Dry powdered silver oxide causes an explosion.

Nitrogen trichloride
*See* NITROGEN TRICHLORIDE, $Cl_3N$: Initiators

Pentyl alcohol
Mellor, 1947, Vol. 10, 158
Interaction is explosively violent.

Potassium permanganate
*See* POTASSIUM PERMANGANATE, $KMnO_4$: Hydrogen trisulphide
*See other* NON-METAL SULPHIDES

### † HYDROGEN SELENIDE $H_2Se$

### POLYSILYLENE $(H_2Si)_n$

Brauer, 1963, Vol, 1, 682
The dry solid ignites in air.

Oxidants
Bailar, 1973, Vol. 1, 1352
It ignites in contact with concentrated nitric acid and explodes with sulphuric acid.
*See other* NON-METAL HYDRIDES

### THORIUM DIHYDRIDE $ThH_2$ $H_2Th$

Hartmann, I. *et al., Rep. Invest. No. 4835*, Washington, US Bur. Mines, 1951
Extremely pyrophoric when powdered.
*See other* METAL HYDRIDES

### TITANIUM DIHYDRIDE $TiH_2$ $H_2Ti$

Alekseev, A. G. *et al., Chem. Abs.*, 1974, **81**, 123688k
Pyrophoricity and detonation behaviour of titanium hydride powders

of various particle sizes were studied in comparison with titanium metal powders.

*See also* ZIRCONIUM, Zr (reference 6)

*See other* METAL HYDRIDES

**TITANIUM–ZIRCONIUM HYDRIDE** $TiH_2$–$ZrH_2$ $H_2Ti$–$H_2Zr$

*See* ZIRCONIUM, Zr (reference 6)

**ZINC HYDRIDE** $ZnH_2$ $H_2Zn$

Barbaras, G. D. *et al., J. Amer. Chem. Soc.*, 1951, **73**, 4587

Fresh samples reacted slowly with air, but aged and partly decomposed samples (containing finely divided zinc) may ignite in air.

*See other* METAL HYDRIDES

**'ZIRCONIUM HYDRIDE'** $ZrH_2$ $H_2Zr$

Mellor, 1941, Vol. 7, 114

Sidgwick, 1950, 633

The product of sorbing hydrogen on to hot zirconium powder when heated in air burns with incandescence and mild explosions.

*See also* ZIRCONIUM, Zr (reference 6)

*See other* METAL HYDRIDES

**POTASSIUM TRIHYDROMAGNESATE** $H_3KMg$

$K[MgH_3]$

Ashby, E. C. *et al., Inorg. Chem.*, 1970, **92**, 2182

The solid reacted violently on exposure to atmosphere.

*See other* COMPLEX HYDRIDES

**LANTHANUM TRIHYDRIDE** $LaH_3$ $H_3La$

Mackay, 1966, 66

Ignites in air.

*See other* METAL HYDRIDES

† **AMMONIA** $NH_3$ $H_3N$

1. *MCA SD-8, 1960*
2. *ABCM Quart. Safety Summ.*, 1950, **21**, 1
3. *BCISC Quart. Safety Summ.*, 1974, **45**, 14
4. Unpublished observations, 1968–1974
5. Henderson, D., in *Ammonia Plant Safety*, a CEP Technical Manual, 1975, **17**, 132–134, Amer. Inst. Chem. Engrs.

6. Krivulin, V. N. *et al., Chem. Abs.*, 1976, **84**, 124133z
Although there is a high lower explosive limit in air and ignition is not easy, there is a long history of violent explosions in refrigeration practice, where ammonia was previously widely used [1].
Mixtures of ammonia and air lying within the explosive limits can occur above aqueous solutions of certain strengths. Welding operations on a vessel containing aqueous ammonia caused a violent explosion [2].

Several incidents involving sudden 'boiling' of concentrated ammonia solution (*d*, 0.880, *ca.* 35% wt.) have occurred when screw-capped Winchesters are opened [3,4]. These are attributable to supersaturation of the solution with ammonia, caused by increase in its temperature subsequent to preparation and bottling. The effect is particularly noticeable with Winchesters filled in winter and opened in summer. Usually 'boiling-off' begins immediately the cap is loosened, and if this is done carefully using gloves and eye-protection in a fume cupboard or under local ventilation, no great problems arise [4]. However, a bottle which had successfully been opened in this way without 'boiling-off' subsequently erupted violently when a pipette disturbed the surface of the liquid [3]. This hazard may be avoided by using the slightly less concentrated solution (*d*, 0.990, *ca.* 25% wt.) where this is technically acceptable.

The recent publication in the series previously entitled 'Safety in Air and Ammonia Plants', though mainly devoted to engineering aspects, contains an account of an incident involving combustion/explosion in the free space of a weak ammonia-liquor tank. Detailed examination revealed no evidence for an ignition source [5]. The existence of ammonia–air mixtures able to burn only in sufficiently large enclosed volumes was established [6].

Boron halides

Sidgwick, 1950, 380

The boron halides react violently with ammonia.

Calcium

*See* CALCIUM, Ca: Ammonia

1-Chloro-2,4-dinitrobenzene

*See* 1-CHLORO-2,4-DINITROBENZENE, $C_6H_3ClN_2O_4$: Ammonia

Chlorine azide

*See* CHLORINE AZIDE, $ClN_3$: Alone, or Ammonia

Chloroformamidinium nitrate

*See* CHLOROFORMAMIDINIUM NITRATE, $CH_4ClN_3O_3$: Amines, etc.

*o*- or *p*-Chloronitrobenzene

*See* *o*- or *p*-CHLORONITROBENZENE, $C_6H_4ClNO_2$ : Ammonia

Ethanol,
Silver nitrate

*See* SILVER NITRATE, $AgNO_3$ : Ammonia, Ethanol

Ethylene oxide

*See* ETHYLENE OXIDE, $C_2H_4O$: Ammonia

Germanium derivatives

Paschenko, I. S. *et al., Chem. Abs.*, 1976, **84**, 8568w
'Causes of the Explosiveness of Gas–Liquid Mixtures during Neutralisation of Germanium-containing Solutions by Ammonia' (title only translated)

*See also* GERMANIUM(II) IMIDE, GeHN

Gold(III) chloride

*See* GOLD(III) CHLORIDE, $AuCl_3$ : Ammonia, etc.

Halogens,
or Interhalogens

Ammonia either reacts violently, or produces explosive products, with all four halogens and some of the interhalogens.

*See* BROMINE, $Br_2$ : Ammonia
CHLORINE, $Cl_2$ : Nitrogen compounds
FLUORINE, $F_2$ : Hydrides
IODINE, $I_2$ : Ammonia
BROMINE PENTAFLUORIDE, $BrF_5$ : Hydrogen-containing materials
CHLORINE TRIFLUORIDE, $ClF_3$ : Hydrogen-containing materials

Iodine,
Potassium

Staley, S. W. *et al., Tetrahedron*, 1975, **31**, 1133
During the reductive cleavage of cyclopolyenes with potassium in liquid ammonia, the intermediate anionic species are quenched with iodine–pentane mixtures. The possibility of formation of the highly explosive nitrogen triiodide and the need for precautions are stressed.

Mercury

*See* MERCURY, Hg: Ammonia

Nitrogen trichloride

*See* NITROGEN TRICHLORIDE, $Cl_3N$: Initiators

Oxidants

*See* POTASSIUM CHLORATE, $ClKO_3$ : Ammonia

NITRYL CHLORIDE, $ClNO_2$ : Inorganic materials
CHROMYL CHLORIDE, $Cl_2CrO_2$ : Ammonia
DICHLORINE OXIDE, $Cl_2O$: Oxidisable materials
CHROMIUM TRIOXIDE, $CrO_3$ : Ammonia
TRIOXYGEN DIFLUORIDE, $F_2O_3$
NITRIC ACID, $HNO_3$ : Ammonia
HYDROGEN PEROXIDE, $H_2O_2$ : Nitrogenous bases
AMMONIUM PEROXODISULPHATE, $H_8N_2O_8S_2$ : Ammonia, etc.
NITROGEN OXIDE, NO: Phosphine, etc.
DINITROGEN TETRAOXIDE, $N_2O_4$ : Ammonia
OXYGEN (Liquid), $O_2$ : Ammonia, etc.

Oxygen,
Platinum

Nealy, D., *School Sci. Rev.*, 1935, **16**(63), 410; Williams, E. J. *et al.*, ibid.
In school demonstrations of oxidation of ammonia to nitric acid over platinum catalysts, substitution of oxygen for air causes fairly vigorous explosions to occur.

Pentaborane(9)

*See* PENTABORANE(9), $B_5H_9$ : Ammonia

Silver compounds

*See* SILVER CHLORIDE, AgCl: Ammonia
SILVER NITRATE, $AgNO_3$ : Ammonia
SILVER OXIDE, $Ag_2O$: Ammonia, etc.

Stibine

*See* STIBINE, $H_3Sb$: Ammonia

Tellurium halides

*See* TELLURIUM TETRABROMIDE, $Br_4Te$: Ammonia
TELLURIUM TETRACHLORIDE, $Cl_4Te$: Ammonia

Tetramethylammonium amide

*See* TETRAMETHYLAMMONIUM AMIDE, $C_4H_{14}N_2$ : Ammonia

Thiocarbonyl azide thiocyanate

*See* THIOCARBONYL AZIDE THIOCYANATE, $C_2N_4S_2$ : Ammonia, etc.

Sulphinyl chloride

*See* SULPHINYL CHLORIDE, $Cl_2OS$: Ammonia

Thiotrithiazyl chloride

*See* THIOTRITHIAZYL CHLORIDE, $ClN_3S_4$ : Ammonia
*See other* NON-METAL HYDRIDES

1. Ashford, J. S., private comm., 1967
2. Brauer, 1963, Vol. 1, 502
3. Vervalin, C. H., *Hydrocarbon Proc. Petr. Ref.*, 1963, **42**(2), 174
4. Rüst, 1948, 302
5. Bailar, 1973, Vol. 2, 272

The material was being prepared by distilling a mixture of hydroxylamine hydrochloride and sodium hydroxide in methanol under reduced pressure A violent explosion occurred towards the end of the distillation [1], probably due to an increase in pressure to above 53 mbar. It explodes when heated under atmospheric pressure [2]. Traces of hydroxylamine remaining after reaction with acetonitrile to form acetamide oxime caused an explosion during evaporation of solvent. Traces can be removed by treatment with diacetyl monoxime and ammoniacal nickel sulphate, forming nickel dimethylglyoxime [3].

An account of an extremely violent explosion towards the end of vacuum distillation had been published previously [4]. Anhydrous hydroxylamine is usually stored at 0°C to prevent internal oxidation–reduction reactions which occur at ambient temperature [5].

Barium oxide

*See* Oxidants, below

Carbonyl compound,
Pyridine

Anon., *Chem. Processing (Chicago),* 1963, **26**(24), 30

During preparation of an unspecified oxime the carbonyl compound, pyridine, hydroxylamine hydrochloride and sodium acetate were heated in a stainless steel autoclave. At 90°C sudden reaction caused a pressure increase to 340 bar, when the bursting disc failed. The reaction had been previously run uneventfully on one-tenth scale in a glass-lined autoclave.

Copper(II) sulphate

Mellor, 1940, Vol. 8, 292

Hydroxylamine ignites with the anhydrous salt.

Metals

Mellor, 1940, Vol. 8, 290

Sorbe, 1968, 158

Sodium ignites in contact with hydroxylamine alone, but reacts smoothly in ether solution to give sodium hydroxylamide which may be pyrophoric in air. Calcium reacts to give the bis-hydroxylamide which explodes at 180°C. Finely divided zinc either ignites or explodes when warmed in contact with hydroxylamine.

Zinc and hydroxylamine form the bishydroxylamide solvated with hydroxylamine (3 mol), an explosive compound.
*See also* OXIMES

Oxidants
Mellor, 1941, Vol. 3, 670; 1940, Vol. 8, 287–294
Brauer, 1963, Vol. 1, 502
Hydroxylamine is a powerful reducant, particularly when anhydrous, and if exposed to air on a fibrous extended surface (filter paper), it rapidly heats by aerial oxidation. It explodes in contact with air above 70°C. Barium dioxide will ignite aqueous hydroxylamine, while the solid ignites in contact with dry barium mono- and dioxides, lead dioxide, potassium permanganate, but with chlorates, bromates and perchlorates only when moistened with sulphuric acid. Contact of the anhydrous base with potassium or sodium dichromate is violently explosive, but less so with ammonium dichromate or chromium trioxide. Ignition occurs in gaseous chlorine, and vigorous oxidation occurs with hypochlorites.

Phosphorus chlorides
Mellor, 1940, Vol. 8, 290
The base ignites in contact with phosphorus tri- or pentachloride.
*See* HYDROXYLAMINIUM SALTS
*See other* REDUCANTS

**AMIDOSULPHURIC ACID** **$H_2NSO_2OH$** **$H_3NO_3S$**

Metal nitrates or nitrites.
*See* METAL AMIDOSULPHATES

**SODIUM HYDRAZIDE** **$NaNHNH_2$** **$H_3N_2Na$**

Air,
or Alcohol,
or Water
1. Mellor, 1940, Vol. 8, 317
2. Kauffmann, T., *Angew. Chem. (Intern. Ed.)*, 1964, **3**, 342

Contact with traces of air, alcohol or water causes a violent explosion [1 as does heating to 100°C [2].
*See other N*-METAL DERIVATIVES

**AZIDOSILANE** $N_3SiH_3$ $H_3N_3Si$

Preparative hazard

*See* AZIDOGERMANE, $GeH_3N_3$: Fluorosilane

*See other* NON-METAL AZIDES

**PHOSPHINIC ('HYPOPHOSPHOROUS') ACID** $H_3O_2P$

**$HP(OH)_2$**

Mellor, 1971, Vol. 8, Suppl. 3, 614

The explosion hazard associated with the usual laboratory preparation from white phosphorus and alkali may be avoided by an alternative method involving oxidation of phosphine with an aqueous iodine solution.

Mercury(II) nitrate

*See* MERCURY(II) NITRATE, $HgN_2O_6$: Phosphinic acid

Mercury(II) oxide

Mellor, 1940, Vol. 4, 778

The redox reaction is explosive.

*See other* REDUCANTS

**PHOSPHOROUS ACID** $H_3O_3P$

**$P(OH)_3$ or $HP(O)(OH)_2$**

*See* PHOSPHORUS TRICHLORIDE, $Cl_3P$: Acetic acid

**ORTHOPHOSPHORIC ACID** $P(O)(OH)_3$ $H_3O_4P$

*MCA SD-70*, 1958

Although it is not an oxidant, it is a strong acid which, because of its availability in high concentration (90% soln. contains 50 geq./1), can develop a large exotherm on neutralisation (or dilution). Handling precautions are detailed.

Chlorides,
Stainless steel

Piekarz, J., *Chem. Can.*, 1961, **13**(4), 40–41

Presence of traces of chloride ion in technical 75% phosphoric acid in a closed stainless storage tank caused corrosive liberation of hydrogen which later exploded at a sparking contact.

Nitromethane

*See* NITROMETHANE, $CH_3NO_2$: Acids, etc.

Sodium tetrahydroborate

*See* SODIUM TETRAHYDROBORATE, $BH_4Na$: Acids

## PEROXOMONOPHOSPHORIC ACID $HOOP(O)(OH_2)$ $H_3O_5P$

Preparative hazard

*See* HYDROGEN PEROXIDE, $H_2O_2$: Phosphorus(V) oxide

Coal,
Potassium permanganate

*See* POTASSIUM PERMANGANATE, $KMnO_4$: Coal, etc.

Organic matter

Castrantas, 1965, 5

The 80% solution causes ignition when dropped into organic matter.

*See other* PEROXOACIDS

## † PHOSPHINE $PH_3$ $H_3P$

Rüst, 1948, 301

Liquefied phosphine (an endothermic compound) can be detonated by powerful initiation.

Air

1. Sidgwick, 1950, 729
2. Mellor, 1940, Vol. 8, 811, 814
3. McKay, H. A. C., *Chem. & Ind.*, 1964, 1978

Pure phosphine does not spontaneously ignite in air below 150°C unless it is thoroughly dried, when it ignites in cold air. The presence of traces (0.2%) of diphosphane in phosphine as normally prepared causes it to ignite spontaneously in air, even at below −15°C [1]. Pure phosphine is rendered flammable by a trace of dinitrogen trioxide, nitrous acid, or similar oxidant [2].

Phosphine, generated by action of water on calcium phosphide, was dried by passage through towers packed with the latter. Soon after refilling the generator (but not the towers) and starting purging with argon, a violent explosion occurred. This was attributed to the air, displaced from the generator by argon, reacting explosively with dry phosphine present in the drying towers, possibly catalysed by the orange-yellow polyphosphine which forms on the surface of calcium phosphide. Fresh calcium phosphide in generator and drying towers, with separate purging, is recommended [3].

*See* Oxygen, below

Dichlorine oxide

*See* DICHLORINE OXIDE, $Cl_2O$: Oxidisable materials

Halogens

Mellor, 1940, Vol. 8, 812

Ignition occurs on contact with chlorine or bromine or their aqueous solutions.

Metal nitrates

Mellor, 1941, Vol. 3, 471; 1940, Vol. 4, 993

Passage of phosphine into concentrated silver nitrate solution causes ignition or explosion, depending on the gas rate. Mercury(II) nitrate solution gives a complex phosphide, explosive when dry.

Nitric acid

*See* NITRIC ACID, $HNO_3$: Non-metal hydrides

Nitrogen trichloride

*See* NITROGEN TRICHLORIDE, $Cl_3N$: Initiators

Oxygen

Fischer, E. O. *et al., Angew. Chem. (Intern. Ed.),* 1968, **7**, 136

Mellor, 1971, Vol. 8, Suppl. 3, 281–282

Even small amount of oxygen present in phosphine gives explosive mixture, in which autoignition occurs at low pressures.

The effects of other materials on the explosive interaction have been reviewed.

*See* Air, above

*See other* NON-METAL HYDRIDES

**PLUTONIUM(III) HYDRIDE** **$PuH_3$** **$H_3Pu$**

Stout, E. L., *Chem. Eng. News,* 1958, **36**(8), 64

The hydride, and the metal, when finely divided are spontaneously flammable, and burning causes a specially dangerous contamination problem, in view of the radioactivity and toxic hazards.

Water

Bailar, 1973, Vol. 5, 150

Rapid addition of water often causes ignition of the hydride.

*See other* METAL HYDRIDES

† **STIBINE** **$SbH_3$** **$H_3Sb$**

Mellor, 1939, Vol. 9, 394

During evaporation of liquid stibine at −17°C, a relatively weak and isothermal explosive decomposition may occur. Gaseous stibine at ambient temperature may propagate an explosion from a hot spot on the retaining vessel, and autocatalytically decomposes, sometimes explosively, at 200°C.

Ammonia

Mellor, 1939, Vol. 9, 397

A heated mixture explodes.

Oxidants

*See* CHLORINE, $Cl_2$ : Non-metal hydrides
NITRIC ACID, $HNO_3$ : Non-metal hydrides
OZONE, $O_3$ : Stibine

*See other* NON-METAL HYDRIDES

**URANIUM(III) HYDRIDE** **$UH_3$** **$H_3U$**

Stout, E. L., *Chem. Eng. News*, 1958, **36**(8), 64

The dry powdered hydride readily ignites in air.

Halocarbons

Bailar, 1973, Vol. 5, 150

Contact of the hydride with halogen-containing solvents may lead to explosive interaction.

Water

Bailar, 1973, Vol. 5, 150

Rapid addition of water often causes ignition of the hydride.

*See other* METAL HYDRIDES

**AMMONIUM IODIDE** **$NH_4I$** **$H_4IN$**

Bromine trifluoride

*See* BROMINE TRIFLUORIDE, $BrF_3$ : Ammonium halides

**AMMONIUM IODATE** **$NH_4IO_3$** **$H_4INO_3$**

*ABCM Quart. Safety Summ.*, 1955, **26**, 24

Violent decomposition occurred on touching with a scoop. A similar batch of material contained less than 100 p.p.m. of periodate.

*See related* METAL OXOHALOGENATES

**AMMONIUM PERIODATE** **$NH_4IO_4$** **$H_4INO_4$**

1. Anon., *Chem. Eng. News*, 1951, **29**, 1770

2. Mellor, 1946, Vol. 2, 408

It exploded while being transferred by scooping [1]. The sensitivity towards heat, but not to abrasive impact, was known previously [2].

*See related* METAL OXOMETALLATES

**PHOSPHONIUM IODIDE** $PH_4I$ $H_4IP$

Oxidants

Mellor, 1947, Vol. 8, 827

Bromates, chlorates or iodates ignite in contact with phosphonium iodide at ambient temperature if dry, or in presence of acid to generate bromic acid, etc. Ignition also occurs with nitric acid, and reaction with dry silver nitrate is very exothermic. Interaction with antimony pentachloride at ambient temperature proceeds explosively.

Water

*See* PHOSPHORUS, P: Hydriodic acid

*See other* REDUCANTS

**POTASSIUM TETRAHYDROZINCATE(2–)** $H_4K_2Zn$

$K_2[ZnH_4]$

Ashby, E. C. *et al., Inorg. Chem.*, 1971, **10**, 2486

Extremely reactive, pyrophoric in air.

*See other* COMPLEX HYDRIDES

**AMMONIUM PERMANGANATE** $NH_4MnO_4$ $H_4MnNO_4$

Mellor, 1942, Vol. 12, 302

Pavlychenko, M. M. *et al., Chem. Abs.*, 1962, **57**, 2897e

The dry material is friction-sensitive and explodes at 60°C in air. Under vacuum, decomposition becomes explosive above 100°C, and sparks will also initiate explosion.

*See* OXOSALTS OF NITROGENOUS BASES (reference 1)

*See related* METAL OXOMETALLATES

**AMMONIUM HYDROXIDE** $NH_4OH$ $H_4NO$

Nitromethane

*See* NITROMETHANE, $CH_3NO_2$: Acids, etc.

Alone

Troyan, J. E., *Ind. Eng. Chem.*, 1953, **45**, 2608

Hydrazine vapour is exceptionally hazardous in that once it is ignited it will continue to burn by exothermic decomposition in complete absence of air or other oxidant.

Air

Day, A. C. *et al.*, *Org. Synth.*, 1970, **50**, 4

Distillation of anhydrous hydrazine (prepared by dehydrating hydrazine hydrate with solid sodium hydroxide) must be carried out under nitrogen to avoid the possibility of an explosion if air is present.

*See* Barium oxide, below

Barium oxide,
or Calcium oxide

1. Gmelin, 1935, Syst. 4, 318
2. *ABCM Quart. Safety Summ.*, 1950, **21**, 18

The residue from dehydrating hydrazine with barium oxide slowly decomposes exothermically in daylight and finally explodes [1]. Dehydration of 95% hydrazine by boiling with calcium oxide under nitrogen caused violent explosions on two occasions. It is concluded that use of calcium or barium oxides for this purpose is potentially dangerous, and that boiling under reduced pressure may be advisable to lower the liquid temperature. General precautions in handling hydrazine are discussed [2].

2-Chloro-5-methylnitrobenzene,
Palladium–charcoal catalyst

*See* 2-CHLORO-5-METHYLPHENYLHYDROXYLAMINE, $C_7H_8ClNO$

Dicyanofurazan

*See* DICYANOFURAZAN, $C_4N_4O$: Nitrogenous bases

Metal catalysts

1. Mellor, 1940, Vol. 8, 317; 1967, Vol. 8, Suppl. 2.2, 83
2. Audrieth, L. F. *et al.*, *Ind. Eng. Chem.*, 1951, **43**, 1774
3. Troyan, J. E., *Ind. Eng. Chem.*, 1953, **45**, 2608
4. Blumenthal, J. L. *et al.*, US Pat. 3 846 339, 1974 (*Chem. Abs.*, 1975, **82**, 113788z)
5. Rao, K. R. *et al.*, *Trans. Powder Metall. Assoc. India*, 1975, **2**, 64–66 (*Chem. Abs.*, 1976, **85**, 145385u)

In contact with metallic catalysts (platinum black, Raney nickel) hydrazine is catalytically decomposed, yielding ammonia, nitrogen and hydrogen. With concentrated hydrazine the reaction may be violent, and the ammonia and hydrogen evolved could be ignited by particles of dry catalyst [1]. Measures to prevent autoxidation of hydrazine catalysed by traces of copper are discussed, including displacement of air from transportation containers with nitrogen [2]. Catalytic decomposition is effected by iron oxide, molybdenum-stabilised stainless steel, molybdenum and its oxides, and finely divided solids. Handling procedures are discussed [3].

A rhenium–alumina catalyst causes immediate ignition of undiluted hydrazine [4]. The effect upon ignition delay of thermal treatment and particle size of powdered transition metal catalysts has been studied [5].

*See* Rust, below
Ruthenium trichloride, below

Metal salts

1. Mellor, 1941, Vol. 7, 430; 1940, Vol. 8, 318; 1967, Vol. 8, Suppl. 2.2, 88
2. Traynham, J. G. *et al., J. Org. Chem.*, 1967, **32**, 3285
3. McCoy, P. O., *Chem. Eng. News*, 1970, **48**(48), 9

Several of its complexes with metallic salts are unstable, including those of basic cadmium perchlorate (highly explosive), cadmium nitrite (explosive), copper chlorate (explodes on drying without heat), manganese nitrate (ignites at 150°C), mercury(I) and (II) chlorides and nitrates (all are explosive) and tin(II) chloride (explodes on heating). Other examples are given [1].

There is a published method for reducing *cis-trans*-cyclodecadiene to the *cis*-monoene with diazene ('diimide') generated *in situ* from hydrazine, copper(II) salts and air [2]. During a modified reaction sequence using crude diene, much sludge containing polymer, copper(II) acetate and hydrazine was produced. When filtered off in air, the sludge heated and eventually glowed. Use of purified diene and filtration under nitrogen avoids the problem [3].

*See also* AMMINEMETAL OXOSALTS

Methanol,
Nitromethane

*See* NITROMETHANE, $CH_3NO_2$: Hydrazine, Methanol

Oxidants

1. *Haz. Chem. Data*, 1969, 127
2. Mellor, 1941, Vol. 3, 137; 1940, Vol. 8, 313–319; 1943, Vol. 11, 234
3. *ASESB Op. Incid. Rep. No. 105*
4. Troyan, J. E., *Ind. Eng. Chem.*, 1953, **45**, 2608
5. Wannagat, U. *et al., Monats. für Chem.*, 1969, **97**, 1157–1162

Hydrazine is a powerful, endothermic reducing agent, and its interaction with oxidants may be expected to be violent if unmoderated, as in rocket propulsion systems (often hypergolic). Mixtures of hydrazine vapour and air are explosive over a very wide range (4.7–100%) and ignition temperatures can be very low (24°C on a rusty iron surface). When mixed with, or spilt on, highly porous materials (asbestos, cloth, earth, wood), autoxidation in air may proceed fast enough to cause ignition. Contact with hydrogen peroxide or nitric acid may cause ignition with concentrated reactants, and dinitrogen tetraoxide is hypergolic [1]. Contact with chromate salts or chromium trioxide causes explosive decomposition. Copper or lead oxides cause a vigorous decomposition, while dropping hydrazine hydrate on to mercury oxide can cause an explosion. This may be due to known oxidation of hydrazine to hydrogen azide by two-electron transfer, and formation of explosive mercury azide [2]. Contact of hydrazine with (explosive) *N*,2,4,6-tetranitro-*N*-methylaniline caused ignition [3]. Contact with iron oxide, chlorates or peroxides may lead to a violent reaction [4].

The ignition delay with fuming nitric acid was 8 ms, explosion also occurring [5].

*See* *N*-HALOIMIDES
ROCKET PROPELLANTS
SILVER OXIDE, $Ag_2O$: Ammonia, etc.
LITHIUM PERCHLORATE, $ClLiO_4$: Hydrazine
SODIUM PERCHLORATE, $ClNaO_4$: Hydrazine
TRIOXYGEN DIFLUORIDE, $F_2O_3$
POTASSIUM PEROXODISULPHATE, $K_2O_8S_2$: Hydrazine, etc.

*See also* SILYLHYDRAZINES: Oxidants

Potassium

Gmelin, 1935, Syst. 4, 318

Explosive interaction.

*See also* SODIUM HYDRAZIDE, $H_3N_2Na$

Rust

*MCA Case History No. 1893*

Use of rusty tweezers, rather than the glass rod specified to handle specimens being immersed in hot 64% hydrazine, caused the hydrazine vapour to ignite.

*See* Metal catalysts, above

Ruthenium(III) chloride

Allen, A. D. *et al.*, *Inorg. Synth.*, 1970, **12**, 3

During the preparation of pentaamminedinitrogenruthenium(II) solutions, the initial gas-producing reaction is so vigorous that increase in scale above that described (9 mmol) is not advised.

*See* Metal catalysts, above

*See also* PENTAAMMINECHLORORUTHENIUM(III) CHLORIDE, $Cl_3H_{15}N_5Ru$: Sodium azide

Sodium

*See* SODIUM, Na: Hydrazine

Thiocarbonyl azide thiocyanate

*See* THIOCARBONYL AZIDE THIOCYANATE, $C_2N_4S_2$: Ammonia, etc.

Titanium compounds

Bains, M. S., *Can. J. Chem.*, 1966, **44**, 534–538

Interaction of anhydrous hydrazine and titanium tetraisopropoxide is explosive at 130°C in absence of solvent. Evaporation of solvent ether from the reaction product of tetrakisdimethylaminotitanium and anhydrous hydrazine caused an explosion, attributed to formation of dimethylamine. *N*-metal derivatives may also have been involved.

*See other* NON-METAL HYDRIDES
REDUCANTS

**AMMONIUM NITRITE** $NH_4NO_2$ $H_4N_2O_2$

*See* NITRITE SALTS OF NITROGENOUS BASES

**AMMONIUM NITRATE** $NH_4NO_3$ $H_4N_2O_3$


1. Federoff, 1960, 35
2. Urbanski, 1965, Vol. 2, 460
3. Popper, H., *Chem. Eng.*, 1962, **70**, 91
4. Sykes, W. G. *et al., Chem. Eng. Progr.*, 1963, **59**(1), 66
5. Mellor, 1964, Vol. 8, Suppl. 1, 543
6. Croysdale, L. G. *et al., Chem. Eng. Progr.*, 1965, **61**(1), 76
7. *MCA Case History No. 873*
8. Bailar, 1973, Vol. 2, 317
9. Vakhrushev, Yu. A. *et al., Chem. Abs.*, 1975, **82**, 113745h
10. Rozman, B. Yu., *Chem. Abs.*, 1960, **54**, 12587c
11. Lee, P. R., *J. Appl. Chem.*, 1969, **19**, 345–351
12. Freeman, R., *Chem. Eng. Prog.*, 1975, **71**(11), 71–74


The decomposition, fire and explosion hazards of this salt of negative oxygen balance have been extensively reviewed [1–4]. In the absence of impurities, it is difficult, but not impossible, to cause ammonium nitrate to detonate. Use of explosives to break up the caked double salt, ammonium nitrate: sulphate (2:1), caused a 4.5 Mkg dump to detonate [4] (even though some 45% of inert ammonium sulphate was present in the salt). The effect of various impurities and additions on the thermal stability of solid ammonium nitrate has been widely studied [1,2,5].

A few incidents involving explosive decomposition of aqueous solutions of the salt during transfer operations (or in presence of oil) have been recorded [6]. Flame-cutting a mild steel pipe blocked by the solid (impure) salt caused it to explode [7].

During preparation of dinitrogen oxide by the exothermic thermal decomposition of the nitrate at 170°C, the temperature must not be allowed to exceed 250°C, as explosion may occur [8]. The thermal stability of ammonium nitrate solutions is decreased by presence of disodium 1,1′-methylenebis(6-naphthalenesulphonate) at 0.15%, but zinc oxide or zinc sulphate–apatite mixtures act as stabilisers [9]. The parameters involved in the self-ignition of the nitrate during storage or transport are described, and a mechanism for instantaneous decomposition is proposed [10]. The factors affecting potentially explosive decomposition of bulk storage of the salt are discussed and steady state thermal explosion theory is applied to the prediction of critical masses in relation to temperature in bulk storage. Application of the results to current storage practice are also discussed [11].

A detailed account of the investigation of a fire and explosion (basically unexplained) involving some 5 tonnes of nitrate of a 14 ktonne warehouse stock has been published [12].

*See* Metals, below
Organic fuels, below

Acetic acid

von Schwartz, 1918, 322

Concentrated mixtures ignite on warming.

Ammonium chloride,
(Barium nitrate),
Water,
Zinc

1. Bailey, P. S. *et al.*, *J. Chem. Educ.*, 1975, **52**, 525
2. Jackson, H., *Spectrum*, 1969, 7(2), 82

Addition of water to an intimate mixture of zinc powder and the metal salts causes spontaneous ignition [1] and a mixture of the nitrate and chloride (9:1) sprinkled with zinc dust ignites vigorously when moistened [2].

*See* Metals, below

Ammonium sulphate,
Potassium

Staudinger, H., *Z. Elektrochem.*, 1925, **31**, 549

Ammonium nitrate containing the sulphate readily explodes on contact with potassium or its alloy with sodium.

Alkali metals

Mellor, 1964, Vol. 8, Suppl. 1, 546

Sodium progressively reduces the nitrate, eventually forming a yellow explosive solid, probably sodium hyponitrite.

Charcoal,
Metal oxides

Herbst, H., *Chem. Ztg.*, 1935, **59**, 744–745

The pelleted explosive ('ammonpulver', containing 10% charcoal) normally ignites at 160–165°C, but presence of rust, or copper or zinc oxides, lowers the temperature to 80–120°C.

*See* Non-metals, below

Chloride salts

1. Pascal, 1956, Vol. 10, 216
2. Pany, V. *et al.*, *Chem. Abs.*, 1976, **85**, 56018j

The nitrate containing 0.1% of ammonium chloride decomposes vigorously below 175°C [1]. Presence of 0.1% of calcium or iron(III) chlorides in the nitrate lowers its initiation temperature sufficiently to give violent or explosive decomposition. Thermal analysis plots for aluminium, calcium and iron(III) chlorides are given [2].

Copper iron(II) sulphide

Kuznetsov, G. V. *et al.*, *Chem. Abs.*, 1975, **82**, 75133x

During preparations for blasting the sulphide mineral (copper pyrites), ammonium nitrate-based cartridges exploded prematurely in the blast holes. This was attributed to exothermic interaction of acid groundwater with the sulphide–oxidant combination.

Cyanoguanidine

*See* CYANOGUANIDINE ('DICYANDIAMIDE'), $C_2H_4N_4$: Oxidants

Fertiliser materials

Davies, R. O. E. *et al.*, *Ind. Eng. Chem.*, 1945, **37**, 59–63

Mixtures of ammonium nitrate, superphosphate and organic materials stored in bulk may ignite if the internal temperature exceeds 90°C. This is owing to the free acid in the superphosphate, and may be prevented by neutralisation with ammonia.

Hydrocarbon oils

Goffart, P., *Chem. Abs.*, 1975, **82**, 61420x

Detonability of several ammonium nitrate-based fertilisers, with or without the addition of fuel oil, was studied.

*See* Organic fuels, below

Metals

Mellor, 1964, Vol. 8, Suppl. 1, 543–546

Soda, N. *et al.*, *Chem. Abs.*, 1969, **70**, 49151j

Many of the following powdered metals reacted violently or explosively with fused ammonium nitrate below 200°C: aluminium, antimony, bismuth, cadmium, chromium, cobalt, copper, iron, lead, magnesium, manganese, nickel, tin, zinc; also brass and stainless steel. Sodium reacts to form the yellow explosive compound sodium hyponitrite and presence of potassium sensitises the nitrate to shock.

Shock-sensitivity of mixtures of ammonium nitrate and powdered metals decreases in the order titanium, tin, aluminium, magnesium, zinc, lead, iron, antimony, copper.

*See* Ammonium sulphate, etc., above
ALUMINIUM, Al: Ammonium nitrate
POTASSIUM, K: Nitrogen-containing explosives

Metal salts

1. Glazkova, A. P., *Chem. Abs.*, 1970, **73**, 132646g
2. Rosser, W. A. *et al.*, *Trans. Faraday Soc.*, 1964, **60**, 1618–1625

Catalytic effects on the thermal decomposition and burning under nitrogen of the nitrate were determined for ammonium and potassium dichromates, potassium chromate, barium and sodium chlorides and potassium nitrate. Chromium(VI) salts are most effective in decomposition, and the halide salts during burning of the nitrate [1]. The effect of chromium compounds soluble in the molten nitrate, all of which promote decomposition of the latter, was studied (especially using ammonium dichromate) in kinetic experiments [2].

Non-metals

Mellor, 1946, Vol. 2, 841–842; Vol. 5, 830

The salt in admixture with charcoal explodes at 170°C, or the solid on contact with glowing charcoal. Phosphorus ignites on the fused salt.

*See* Sulphur, below
PHOSPHORUS, P: Nitrates

Organic fuels

1. Sykes, W. G. *et al.*, *Chem. Eng. Progr.*, 1963, **59**(1), 66
2. Urbanski, 1965, Vol. 2, 461
3. Bernoff, R. A. *et al.*, US Pat. 3 232 940, 1969

In a review of ammonium nitrate explosion hazards several incidents involving mixtures of this negative oxygen balance material with various organic materials were analysed [1]. In general, fire incidents involving ammonium nitrate admixed with 1% of wax, oil or stearates (as anti-caking additives) tended more towards explosion than the pure salt, although the degree of confinement is also important [1,2]. The ease of detonation of mixtures is much greater when 2–4% of oil is present, and such mixtures have been used as explosives [2].

The preparation of ammonium nitrate–fuel systems at a molecular level is covered by the patent, which describes several alkylammonium nitrates as high-energy materials [3].

Potassium nitrite

Mellor, Vol. 2, 1946, 842

Contact of solid nitrate with fused nitrite causes incandescence.

*See* NITRITE SALTS OF NITROGENOUS BASES

Potassium permanganate

*See* POTASSIUM PERMANGANATE, $KMnO_4$: Ammonium nitrate

Sulphur

1. Mason, C. M. *et al., J. Agric. Food Chem.*, 1967, **15**, 954
2. Prugh, R. W., *Chem. Eng. Progr.*, 1967, **63**(11), 53–55

The fire risks of nitrate–sulphur mixtures have been discussed [1] and explosion risks assessed by differential thermal analysis [2].

Urea

Croysdale, L. C. *et al., Chem. Eng. Progr.*, 1965, **61**(1), 72

Concentrated solutions of ammonium nitrate and urea exploded during large-scale mixing operations. Although the cause was not established, hazards associated with these operations are discussed in relation to the circumstances.

*See other* OXIDANTS
OXOSALTS OF NITROGENOUS BASES

**HYDROXYLAMINIUM NITRATE** **$NH_3OHNO_3$** **$H_4N_2O_4$**

*See* HYDROXYLAMINIUM SALTS
OXOSALTS OF NITROGENOUS BASES

**TETRAHYDROXOTIN(2+) NITRATE** **$H_4N_2O_{10}Sn$**

**$[Sn(OH)_4][NO_3]_2$**

1. Anon., *J. Soc. Chem. Ind.*, 1922, **41**, 423–434R
2. Donaldson, J. D. *et al., J. Chem. Soc.*, 1961, 1996

An explosion in flour-bleaching operations was attributed to violent decomposition of the basic nitrate [1], which is an impact-, friction- and heat-sensitive explosive [2]. The instability is associated with the presence of reducant and oxidant functions in the same molecule. The previous formulation as tin(II) nitrate oxide is revised to that above.

*See other* REDOX COMPOUNDS

**AMMONIUM AZIDE** **$NH_4N_3$** **$H_4N_4$**

1. Mellor, 1940, Vol. 8, 344
2. Obenland, C. O. *et al., Inorg. Synth.*, 1966, **8**, 53
3. Sorbe, 1968, 129

It explodes on rapid heating [1] and contains ~93% of nitrogen.

Preparative precautions have been detailed [2].

It is also friction- and impact-sensitive [3].

*See other* NON-METAL AZIDES

**TETRAIMIDE** **HNNHNHNH** $H_4N_4$

Rice, F. O. *et al., J. Amer. Chem. Soc.*, 1957, **79**, 1880–1881

This blue solid, obtained by freezing out at −195°C the pyrolysis products of hydrogen azide, is extremely explosive above this temperature. An explosion at −125°C destroyed the apparatus.

*See other* HIGH-NITROGEN COMPOUNDS

**'SODIUM PERPYROPHOSPHATE'** $H_4Na_4O_{11}P_2$

*See* SODIUM PYROPHOSPHATE HYDROGEN PEROXIDATE, $Na_4O_7P_2 \cdot 2H_2O_2$

**OXODISILANE** $H_3SiSi(O)H$ $H_4OSi_2$

Kautsky, K., *Z. Anorg. Chem.*, 1921, **117**, 209

Ignites in air.

*See related* NON-METAL HYDRIDES

**XENON TETRAHYDROXIDE** $Xe(OH)_4$ $H_4O_4Xe$

Sorbe, 1968, 158

Very unstable, it explodes violently on heating.

*See other* XENON COMPOUNDS

**ZIRCONIUM HYDROXIDE** $Zr(OH)_4$ $H_4O_4Zr$

Zirconium

*See* ZIRCONIUM, Zr: Oxygen-containing compounds

**TETRAHYDROXODIOXOTRISILANE ('TRISILICIC ACID')** $H_4O_6Si_3$

$HOSi(O)Si(OH)_2Si(O)OH$

1. Mellor, 1940, Vol. 6, 230
2. Mueller, R. *et al., J. Prakt. Chem.*, 1966, **31**, 1–6

The hydrolysis product ('silicomesoxalic acid', probably polymeric) of trisilicon octachloride, it decomposes violently or explosively when heated in air, or sometimes spontaneously [1], and ignites when subjected to friction [2].

*See also* POLY(DIHYDROXODIOXODISILANE), $(H_2O_4Si_2)_n$

*See related* NON-METAL OXIDES

**DIPHOSPHANE** $H_2PPH_2$ $H_4P_2$

Mellor, 1940, Vol. 8, 829

Ignites in air, and will cause other flammable gases to ignite in air when present at 0.2% v/v concentration.

*See also* 'PHOSPHORUS HYDRIDE', $HP_2$
PHOSPHINE, $H_3P$

*See other* NON-METAL HYDRIDES

**SILANE** $SiH_4$ $H_4Si$

Alone,
or Covalent halides,
or Halogens

1. Mellor, 1940, Vol. 6, 220–221
2. Braker, 1971, 505
3. Bailar, 1973, Vol. 1, 1366

Pure material is said not to ignite in air unless the temperature be increased or the pressure reduced. Presence of other hydrides as impurities causes ignition always to occur on air contact [1]. However, 99.95% pure material, even at concentrations down to 1% in hydrogen and/or nitrogen, ignites in contact with air unless emerging at very high gas velocity, when mixtures with up to 10% silane content may not ignite [2]. Silane burns in contact with bromine, chlorine or covalent chlorides (carbonyl chloride, antimony pentachloride, tin(IV) chloride, etc.)[1]. Extreme caution is necessary when handling silane in systems with halogenated compounds, as a trace of free halogen may cause violent explosion [2]. Very pure silane does not immediately explode with oxygen, but the decomposition products may ignite after a delay [3].

Air,
*cis*-2-Butene

Urtiew, P. A. *et al., Rep. UCRL-52007*, Richmond (Va.), USNTIS, 1976

Presence of the alkene delays or stops spontaneous ignition of silane–air mixtures.

*See other* NON-METAL HYDRIDES

**THORIUM HYDRIDE** $ThH_4$ $H_4Th$

1. Mellor, 1941, Vol. 7, 207
2. Stout, E. L., *Chem. Eng. News,* 1958, **36**(8), 64
3. Hartmann, I. *et al., Rep. Invest. No. 4835*, Washington, US Bur. Mines, 1951

Thorium hydride explodes on heating in air [1], and the powdered hydride readily ignites on handling in air [2].

Layers of thorium and uranium hydrides ignited spontaneously after exposure to ambient air for a few minutes [3].

*See other* METAL HYDRIDES

### URANIUM(IV) HYDRIDE $UH_4$ $H_4U$

Hartmann, I. *et al., Rep. Invest. No. 4835*, Washington, US Bur. Mines, 1951

Layers of uranium and thorium hydrides ignited after exposure to ambient air for a few minutes.

Oxygen

*See* OXYGEN (Gas), $O_2$: Metal hydrides

*See other* METAL HYDRIDES

### ORTHOPERIODIC ACID $IO(OH)_5$ $H_5IO_6$

Azo-pigment,
Perchloric acid

*See* PERCHLORIC ACID, $ClHO_4$: Azo-pigment, etc.

### IODODISILANE $ISiH_2SiH_3$ $H_5ISi_2$

Bailar, 1973, Vol. 1, 1371

It ignites in air.

*See other* HALOSILANES

### HYDRAZINIUM SALTS $H_2NN^+H_3\ Z^-$ $H_5N_2^+\ Z^-$

Metal nitrites

*See* HYDROGEN AZIDE, $HN_3$
NITRITE SALTS OF NITROGENOUS BASES

### DIAMIDOPHOSPHOROUS ACID $(NH_2)_2POH$ $H_5N_2OP$

Water

Mellor, 1940, Vol. 8, 704

Violent reaction with incandescence.

**HYDRAZINIUM NITRITE** $H_2NNH_3NO_2$ $H_5N_3O_2$

*See* NITRITE SALTS OF NITROGENOUS BASES

**HYDRAZINIUM NITRATE** $H_2NNH_3NO_3$ $H_5N_3O_3$

Alone,
or Metals,
or Metal acetylides, nitrides, oxides or sulphides

Mellor, 1940, Vol. 8, 327; 1967, Vol. 8, Suppl. 2.2, 84, 96

It is an explosive, less stable than ammonium nitrate, and has been studied in detail. Stable on slow heating to 300°C, it decomposes explosively on rapid heating or under confinement. Presence of zinc, copper, most other metals and their nitrides, oxides or sulphides causes flaming decomposition above the melting point (70°C). Commercial cobalt (cubes) causes an explosion also.

2-Hydroxyethylamine

Fujihara, S. *et al., Chem. Abs.*, 1976, **84**, 7212p

Mixtures of the salt (80%) and amine (15%) with water are useful as an impact-insensitive but powerful liquid explosive:

Potassium dichromate

Shidlovskii, A. A. *et al., Chem. Abs.*, 1960, **54**, 22132g

Thermal decomposition becomes explosive above 270°C, or above 100°C in presence of 5% of potassium dichromate.

*See other* OXOSALTS OF NITROGENOUS BASES
REDOX COMPOUNDS

**HYDRAZINIUM AZIDE** $H_2NNH_3N_3$ $H_5N_5$

It contains ~94% of nitrogen.

*See* HYDRAZINIUM SALTS
*See other* NON-METAL AZIDES

**DIOXONIUM HEXAMANGANATO(VII)MANGANATE** $H_6Mn_7O_{26}$

$[H_3O^+]_2[Mn(MnO_4)_6]^{2-}$

Krebs, B. *et al., Angew. Chem. (Intern. Ed.)*, 1974, **13**, 603

It has been shown that 'solid permanganic acid dihydrate' is the undecahydrate of the title compound, formulated as $(H_3O)_2[Mn(MnO_4)_6]\cdot 11H_2O$.

*See* PERMANGANIC ACID, $HMnO_4$
*See other* OXIDANTS

### HYDROXYLAMINIUM PHOSPHINATE $H_6NO_3P$

$HON^+H_3\ H_2PO_2^-$

Mellor, 1940, Vol. 8, 880

The salt detonates above its melting point, 92°C.

*See* HYDROXYLAMINIUM SALTS

### AMMONIUM AMIDOSULPHATE ('SULPHAMATE') $H_6N_2O_3S$

$NH_4OSO_2NH_2$

1. Hunt, J. K., *Chem. Eng. News,* 1952, **30**, 707
2. Rogers, M. G., private comm., 1973

A 60% solution of ammonium sulphamate (pH above 4.5) will not undergo rapid hydrolysis below 200°C. Addition of acid (to pH 2) causes a runaway exothermic hydrolysis to set in at 130°C. Superheating and vigorous boiling can occur under appropriate physical conditions [1]. The use of urea–formaldehyde resins as temporary binders in the firing of refractories and ceramics at high temperatures can lead to the formation of substantial deposits of ammonium sulphamate in the cooler parts of kilns, should a fuel oil containing appreciable quantities of sulphur be used for firing. Ammonia from decomposition of the resin combines with sulphur trioxide to form ammonium sulphamate which accumulates as either a solidified mass in flues or a white deposit on walls, causing corrosion and handling problems [2].

*See* OXOSALTS OF NITROGENOUS BASES

### AMMONIUM AMIDOSELENATE $H_6N_2O_3Se$

$N^+H_4\ H_2NSeO_2O^-$

Dostal, K. *et al., Z. Anorg. Allg. Chem.*, 1958, **296**, 29–35

It explodes if rapidly heated to 120°C.

*See related* OXOSALTS OF NITROGENOUS BASES

### HYDRAZINIUM HYDROGENSELENATE $H_6N_2O_4Se$

$H_2NNH_3OSeO_2OH$

Meyer, J. *et al., Ber.*, 1928, **61**, 1839

The salt explodes in contact with a hot glass rod.

*See other* OXOSALTS OF NITROGENOUS BASES

### *cis*-DIAMMINEPLATINUM(II) NITRITE ('*cis*-DIAMMINEDINITROPLATINUM(II)') $H_6N_4O_4Pt$

$[Pt(NH_3)_2][NO_2]_2$

1. Brauer, 1965, Vol. 2, 1560

2. Holifield, P. J., private comm., 1974

It decomposes explosively at 200°C [1]. Dry material stored in clear bottles in sunlight for several weeks became sensitive and exploded violently on slight mechanical shock. The material is now supplied commercially moistened with water [2].

*See other* AMMINEMETAL OXOSALTS
PLATINUM COMPOUNDS

**HYDRAZINIUM DINITRATE** $H_3N^+N^+H_3(NO_3^-)_2$ $H_6N_4O_6$

Sorbe, 1968, 130

Like the mono-salt, it explodes on heating or impact.

*See other* OXOSALTS OF NITROGENOUS BASES

**DIAMMINEPALLADIUM(II) NITRATE** $H_6N_4O_6Pd$

$[Pd(NH_3)_2][NO_3]_2$

White, J. H., private comm., 1965

There is a danger of explosion if the nitrate is dried.

*See other* AMMINEMETAL OXOSALTS

**ZINC DIHYDRAZIDE** $Zn(NHNH_2)_2$ $H_6N_4Zn$

Mellor, 1940, Vol. 8, 315

It explodes at 70°C.

*See other* *N*-METAL DERIVATIVES

**SODIUM HEXAHYDROXOPLATINATE(2–)** $H_6Na_2O_6Pt$

$Na_2[Pt(OH)_6]$

Acetic acid,
Nitric acid

*See* NITRIC ACID $HNO_3$: Acetic acid, etc.

*See other* PLATINUM COMPOUNDS

**DISILYL OXIDE** $(H_3Si)_2O$ $H_6OSi_2$

Chlorine

*See* CHLORINE, $Cl_2$: Disilyl oxide

**2,4,6-TRISILATRIOXANE ('TRIPROSILOXANE')** $H_6O_3Si_3$

$\overline{H_2SiOSiH_2OSiH_2O}$

Air,
or Chlorine
Mellor, 1940, Vol. 6, 234
The solid polymer (approximating to the trimer) ignites in air or chlorine.
*See related* NON-METAL HYDRIDES

**TETRAPHOSPHORIC ACID** $H_6O_{13}P_4$
$(HO)_2P(O)O[HOP(O)O]_2P(O)(OH)_2$

Water
Mellor, 1971, Vol. 8, Suppl. 3, 736
Dilution of polyphosphoric acids with water in absence of cooling may lead to a large exotherm. Thus, tetraphosphoric acid diluted from 84 to 52% $P_2O_5$ content rapidly attains a temperature of 120–140°C.

**DISILYL SULPHIDE** $(H_3Si)_2S$ $H_6SSi_2$
Sorbe, 1968, 127
It ignites in moist air.
*See related* NON-METAL HYDRIDES

**† DISILANE** $H_3SiSiH_3$ $H_6Si_2$

Bromine
*See* BROMINE, $Br_2$: Non-metal hydrides

Non-metal halides,
or Oxygen
Mellor, 1940, Vol. 6, 220–224
It explodes on contact with carbon tetrachloride or sulphur hexafluoride and contact with chloroform causes incandescence. Disilane ignites spontaneously in air, even when pure, and ingress of air or oxygen into a volume of disilane causes explosion.
*See* SILANES
*See other* NON-METAL HYDRIDES

**DIHYDRAZINEMANGANESE(II) NITRATE** $H_8MnN_6O_6$
$[Mn(N_2H_4)_2][NO_3]_2$

Bailar, 1973, Vol. 3, 827

It ignites at 150°C, but is not shock-sensitive.
*See other* AMMINEMETAL OXOSALTS

**AMMONIUM THIOSULPHATE** $(NH_4)_2S_2O_3$ $H_8N_2O_3S_2$

Sodium chlorate
*See* SODIUM CHLORATE, $ClNaO_3$: Ammonium thiosulphate

**AMMINEPENTAHYDROXOPLATINUM** $H_8NO_5Pt$
$[PtNH_3(OH)_5]$

Jacobsen, I., *Compt. Rend.*, 1909, **149**, 575
Explodes fairly violently above 250°C, as does the pyridine analogue.
*See other* PLATINUM COMPOUNDS

**AMMONIUM SULPHATE** $(NH_4)_2SO_4$ $H_8N_2O_4S$

Potassium chlorate
*See* POTASSIUM CHLORATE, $ClKO_3$: Ammonia, etc.

**HYDROXYLAMINIUM SULPHATE** $(HONH_3)_2SO_4$ $H_8N_2O_6S$
*See* HYDROXYLAMINIUM SALTS

**AMMONIUM PEROXODISULPHATE** $H_8N_2O_8S_2$
$NH_4OSO_2OOSO_2ONH_4$

Aluminium,
Water
Pieters, 1957, 30
A mixture including the powdered metal may explode.

Ammonia,
Silver salts
Mellor, 1947, Vol. 10, 466
In concentrated solutions the silver-catalysed oxidation of ammonia to nitrogen may be very violent.

Iron
Mellor, 1947, Vol. 10, 470
Iron exposed to the action of a slightly acid concentrated solution of ammonium peroxodisulphate dissolves violently.

Sodium peroxide

*See* SODIUM PEROXIDE, $Na_2O_2$: Ammonium peroxodisulphate

Zinc,
Ammonia

*Report of H.M. Inspector of Explosives,* 33, London, HMSO, 1950

The salt exploded during drying, but no cause was determined. However, the reputed explosive character of tetraamminezinc peroxodisulphate, possibly formed from interaction of the ammonium salt, galvanised iron and ammonia, was mentioned as a possible cause.

*See other* PEROXOACID SALTS

**AMMONIUM SULPHIDE** **$(NH_4)_2S$** **$H_8N_2S$**

Zinc

*Chemiearbeit,* 1960, **12**(4), 29

A closed zinc container filled with a concentrated solution of ammonium sulphide exploded, owing to liberation of hydrogen sulphide and hydrogen, accompanying the formation of zinc ammonium sulphide.

**AMMONIUM TETRANITROPLATINATE(II)** **$H_8N_6O_8Pt$**

**$[NH_4]_2[Pt(NO_2)_4]$**

Nilson, L. F., *J. Prakt. Chem.*, [2], 1877, **16**, 249

Decomposes explosively on heating.

*See other* PLATINUM COMPOUNDS

**† TRISILANE** **$H_3SiSiH_2SiH_3$** **$H_8Si_3$**

Air,
or Carbon tetrachloride

Mellor, 1940, Vol. 6, 224

Ignites and explodes in air or oxygen, and reacts vigorously with carbon tetrachloride.

*See* SILANES

*See other* NON-METAL HYDRIDES

**TRISILYLAMINE** **$(H_3Si)_3N$** **$H_9NSi_3$**

Mellor, 1940, Vol. 8, 262

The liquid ignites in air.

*See related* NON-METAL HYDRIDES

**TRIAMMINENITRATOPLATINUM(II) NITRATE** $H_9N_5O_6Pt$

$[Pt(NH_3)_3(NO_3)]NO_3$

Mellor, 1942, Vol. 16, 409
Decomposes violently on heating.
*See other* AMMINEMETAL OXOSALTS
PLATINUM COMPOUNDS

**TRISILYLPHOSPHINE** $(H_3Si)_3P$ $H_9PSi_3$

Amberger, E. *et al., Angew. Chem.*, 1962, **74**, 32–33
The liquid ignites in air.
*See related* NON-METAL HYDRIDES

**TETRASILANE** $Si_4H_{10}$ $H_{10}Si_4$

Air,
or Carbon tetrachloride
Mellor, 1940, Vol. 6, 224
Ignites and explodes in air or oxygen, and reacts vigorously with carbon tetrachloride.
*See* SILANES
*See other* NON-METAL HYDRIDES

**TETRASILYLHYDRAZINE** $(H_3Si)_2NN(SiH_3)_2$ $H_{12}N_2Si_4$

Bailar, 1973, Vol. 1, 1377
It explodes in air.
*See related* NON-METAL HYDRIDES

**TETRAAMMINEZINC PEROXODISULPHATE** $H_{12}N_4O_8S_2Zn$

$[Zn(NH_3)_4]S_2O_8$

Barbieri, G. A. *et al., Z. Anorg. Chem.*, 1911, **71**, 347
Explodes on heating or impact.
*See* AMMONIUM PEROXODISULPHATE, $H_8N_2O_8S_2$: Zinc
*See other* AMMINEMETAL OXOSALTS

**TETRAAMMINEPALLADIUM(II) NITRATE** $H_{12}N_6O_6Pd$

$[Pd(NH_3)_4][NO_3]_2$

White, J. H., private comm., 1965
Evaporation of a solution of the salt used for plating gave a moist residue which ignited and burned violently. Reclamation of palladium from such solutions by direct reduction is recommended.
*See other* AMMINEMETAL OXOSALTS

**AMMONIUM 2,4,6-TRIS(DIOXASELENA)PERHYDROTRIAZINE-1,3,5-TRIIDE ('AMMONIUM TRISELENIMIDATE')** $H_{12}N_6O_6Se_3$

$(N^+H_4)_3\,N^-SeO_2N^-SeO_2N^-SeO_2$

Explosive.

*See* SELENIUM DIFLUORIDE DIOXIDE, $F_2O_2Se$: Ammonia

**TRIHYDRAZINENICKEL(II) NITRATE** $H_{12}N_8NiO_6$

$[Ni(N_2H_4)_3][NO_3]_2$

Ellern, H. *et al., J. Chem. Educ.*, 1955, **32**, 24

A dry sample exploded violently and a moist sample spontaneously deflagrated.

*See other* AMMINEMETAL OXOSALTS

**AMMINEDECAHYDROXODIPLATINUM** $H_{13}NO_{10}Pt_2$

$[Pt_2NH_3(OH)_{10}]$

Jacobsen, J., *Compt. Rend.*, 1909, **149**, 574–577

The compound of unknown structure explodes violently on heating.

*See other* PLATINUM COMPOUNDS

**TETRAAMMINEHYDROXONITRATOPLATINUM(IV) NITRATE**

$H_{13}N_7O_{10}Pt$

$[Pt(NH_3)_4OH(NO_3)][NO_3]_2$

Mellor, 1942, Vol. 16, 411

Explodes violently on heating.

*See other* AMMINEMETAL OXOSALTS
PLATINUM COMPOUNDS

**TETRAAMMINELITHIUM DIHYDROGENPHOSPHIDE** $H_{14}LiN_4P$

$[Li(NH_3)_4]PH_2$

Legoux, C., *Compt. Rend.*, 1938, **207**, 634

Reacts vigorously with water, evolving phosphine and some ammonia, which may ignite.

*See related* NON-METAL HYDRIDES

**PENTAAMMINEDINITROGENRUTHENIUM(II) SALTS** $H_{15}N_7Ru^{2+}$

$[Ru(NH_3)_5N_2]^{2+}$

Preparative hazard

*See* PENTAAMMINECHLORORUTHENIUM(III) CHLORIDE, $Cl_3H_{15}N_5Ru$: Sodium azide
HYDRAZINE, $H_4N_2$: Ruthenium(III) chloride

## UNDECAAMMINETETRARUTHENIUM DODECAOXIDE $H_{33}N_{11}O_{12}Ru_4$

$[Ru_4(NH_3)_{11}O_{12}]$

Preparative hazard

*See* RUTHENIUM TETRAOXIDE, $O_4Ru$: Ammonia

## HAFNIUM Hf

Alone,
or Non-metals,
or Oxidants

*MCA SD-92,* 1966

Although the massive metal is relatively inert, when powdered it becomes very reactive. The dry powder may react explosively at elevated temperatures with nitrogen, phosphorus, oxygen, sulphur and other non-metals. The halogens react similarly, and in contact with hot concentrated nitric acid or other oxidants it may explode (often after a delay with nitric acid). The powder is pyrophoric and readily ignitable by friction, heat or static sparks, and if dry burns fiercely. Presence of water (5–10%) slightly reduces the ease of ignition, but combustion of the damp powder proceeds explosively. A minimum of 25% water is necessary to reduce handling hazards to a minimum. Full handling precautions are detailed.

*See* PYROPHORIC METALS

## MERCURY Hg

Acetylenic compounds

*See* DISODIUM ACETYLIDE, $C_2Na_2$: Metals
1-BROMO-2-PROPYNE, $C_3H_3Br$: Metals
2-BUTYNE-1,4-DIOL, $C_4H_6O_2$
ACETYLENIC COMPOUNDS: Metals

Alkynes,
Silver perchlorate

*See* SILVER PERCHLORATE, $AgClO_4$: Alkynes, etc.

Ammonia

1. Sampey, J. J., *Chem. Eng. News*, 1947, **25**, 2138; *J. Chem. Educ.*, 1967, **44**, A324
2. Thodos, G., *Amer. Inst. Chem. Engrs. J.*, 1964, **10**, 275
3. Henderson, L. M., *Ind. Eng. Chem. (News Ed.)*, 1932, **10**, 73
4. Brunt, C. Van, *Science, N.Y.*, 1927, **65**, 63–64

A mercury manometer used with ammonia became blocked by deposition of a grey-brown solid, which exploded during attempts to remove it mechanically or on heating. The solid appeared to be a dehydration product of Millon's base and was freely soluble in sodium thiosulphate solution. This method of cleaning is probably safer than others, but the use of mercury manometers with ammonia should be avoided as intrinsically unsafe.

Although pure dry ammonia and mercury do not react even under pressure at 340 kbar and 200°C, the presence of traces of water leads to the formation of an explosive compound, which may explode during depressurisation of the system [2]. Explosions in mercury–ammonia systems had been reported previously [3,4].

*See* POLY(DIMERCURYIMMONIUM HYDROXIDE), $(HHg_2NO)_n$

Boron diiodophosphide

*See* BORON DIIODOPHOSPHIDE, $BI_2P$: Metals

Calcium

*See* CALCIUM, Ca: Mercury

Ethylene oxide

*See* ETHYLENE OXIDE, $C_2H_4O$

Metals

Bretherick, L., *Lab. Pract.*, 1973, 22, 533

The high mobility and tendency to dispersion exhibited by mercury, and the ease with which it forms alloys (amalgamates) with many laboratory and electrical contact metals, can cause severe corrosion problems in laboratories. A trap is described to contain mercury accidentally ejected by overpressuring of mercury manometers and similar items.

*See also* ALUMINIUM, Al: Mercury
POTASSIUM, K: Mercury
LITHIUM, Li: Mercury
SODIUM, Na: Mercury
RUBIDIUM, Rb: Mercury

Methyl azide

*See* METHYL AZIDE, $CH_3N_3$: Mercury

Methylsilane,
Oxygen

*See* METHYLSILANE, $CH_6Si$: Mercury, etc.

Oxidants

*See* BROMINE, $Br_2$: Metals
PEROXYFORMIC ACID, $CH_2O_3$: Metals
CHLORINE DIOXIDE, $ClO_2$: Mercury
NITRIC ACID, $HNO_3$: Alcohols

Tetracarbonylnickel,
Oxygen
*See* TETRACARBONYLNICKEL, $C_4NiO_4$: Mercury, etc.
*See other* METALS

**ZINC AMALGAM** **Hg–Zn**

Air
Brimelow, H. C., private comm., 1972
Amalgamated zinc residues isolated from Clemmensen reduction of an alkyl aryl ketone in glacial acetic acid were pyrophoric, and had to be immediately dumped into water after filtration to prevent ignition.
*See other* ALLOYS

**MERCURY(II) IODIDE** **$HgI_2$**

Chlorine trifluoride
*See* CHLORINE TRIFLUORIDE, $ClF_3$: Metals, etc.

**MERCURY(II) NITRATE** **$Hg(NO_3)_2$** **$HgN_2O_6$**

Acetylene
Mellor, 1940, Vol. 4, 993
Contact of acetylene with the nitrate solution gives mercury acetylide, an explosive sensitive to heat, friction or contact with sulphuric acid.

Ethanol
Mellor, 1940, Vol. 4, 993
Addition of aqueous nitrate solution to ethanol gives mercury fulminate, $C_2HgN_2O_2$.

Isobutene
*See* 2-METHYL-2-NITRATOMERCURIOPROPYL(NITRATO)-DIMERCURY(II), $C_4H_8Hg_3N_2O_6$

Petroleum hydrocarbons
1. Mixer, R. Y., *Chem. Eng. News,* 1948, **26**, 2434
2. Ball, J., ibid., 3300
Gas oil was stirred vigorously with the finely divided solid nitrate to complete the removal of sulphur compounds. After the mixture had congealed, preventing further stirring, a violent reaction set in, which reached incandescence, accompanied by vigorous evolution of nitrous fumes. This was attributed to the self-catalysed nitrating action of

mercury nitrate in a semi-solid environment unable to lose heat effectively. A similar occurrence was observed when crushed nitrate was just covered with cracked naphtha [1].

The second publication reveals that this type of hazard was known in cracked distillates containing high proportions of unsaturates and aromatics, when allowed to stand in prolonged contact with mercury(II) nitrate. The hazard may be avoided by using several small portions of the salt sequentially and working on 100 g portions of hydrocarbons [2].

Phosphine
Mellor, 1940, Vol. 4, 993
Phosphine reacts with an aqueous salt solution to give a complex nitrate–phosphide, which explodes when dry on heating or impact.

Phosphinic acid
Mellor, 1940, Vol. 4, 993
Phosphinic ('hypophosphorous') acid violently reduces the salt to the metal.

Potassium cyanide
Rüst, 1948, 337
Mixtures exploded when heated, but only if contained in narrow ignition tubes. Formation of nitrite, a more powerful oxidant than nitrate, may have been involved.

*See other* METAL OXONON-METALLATES
OXIDANTS

**MERCURY(II) AZIDE** **$Hg(N_3)_2$** **$HgN_6$**

1. Mellor, 1940, Vol. 8, 351; Mellor 1967, Vol. 8, Suppl. 2, 43, 50; Barton, A. F. M. *et al., Chem. Rev.,* 1973, 73, 138
2. Heathcock, C. H., *Angew. Chem. (Intern. Ed.),* 1969, 8, 134

The very high friction-sensitivity, particularly of large crystals, and brisance on explosion are to be expected from the thermodynamic properties of the salt. Its great sensitivity, even under water, renders it unsuitable as a practical detonator. Spontaneous explosions during intercrystalline transformations have been observed, or on crystallisation from hot water [1]. A safe method of preparing solutions in aqueous tetrahydrofuran for synthetic purposes is available [2].

*See other* METAL AZIDES

**MERCURY(II) OXIDE** **HgO**

Under appropriate conditions, it can function as a powerful oxidant and/ or catalyst, owing to the tendency to dissociate.

Acetyl nitrate

*See* ACETYL NITRATE, $C_2H_3NO_4$: Mercury(II) oxide

Butadiene,
Ethanol,
Iodine

*See* 2-ETHOXY-1-IODO-3-BUTENE, $C_6H_{11}IO$

Chlorine,
Hydrocarbons

*See* METHANE, $CH_4$: Chlorine
ETHYLENE, $C_2H_4$: Chlorine

Diboron tetrafluoride

*See* DIBORON TETRAFLUORIDE, $B_2F_4$: Metal oxides

Disulphur dichloride

*See* DISULPHUR DICHLORIDE, $Cl_2S_2$: Mercury(II) oxide

Hydrogen peroxide

*See* HYDROGEN PEROXIDE, $H_2O_2$: Mercury(II) oxide
: Metals, etc.

Hydrogen trisulphide

*See* HYDROGEN TRISULPHIDE, $H_2S_3$: Metal oxides

Metals

Staudinger, H., *Z. Elektrochem.*, 1925, **31**, 549
Mellor, 1940, Vol. 4, 272
Mixtures of the red or yellow oxides with sodium–potassium alloy explode violently on impact, the yellow (more finely particulate) oxide giving the more sensitive mixture. Mixtures with magnesium or potassium may explode on heating.

Methanethiol

Klason, P., *Ber.*, 1887, **20**, 3410
Interaction is rather violent in absence of diluent.

Non-metals

Mellor, 1940, Vol. 4, 777–778
Mixtures with phosphorus explode on impact or on boiling with water. A mixture with sulphur exploded on heating.

Phospham

*See* PHOSPHAM, $HN_2P$: Oxidants

Reducants

Mellor, 1940, Vol. 4, 778; Vol. 8, 318

Hydrazine hydrate and phosphinic acid both explosively reduce the oxide when dropped on to it.

*See other* METAL OXIDES

## MERCURY PEROXIDE $HgO_2$

*See* HYDROGEN PEROXIDE, $H_2O_2$: Mercury(II) oxide

## MERCURY(II) SULPHIDE $HgS$

Oxidants

Mellor, 1940, Vol. 2, 242; 1941, Vol. 3, 377; 1940, Vol. 4, 952

The sulphide causes dichlorine oxide to explode and incandesces in chlorine. Grinding with silver oxide ignites the mixture.

*See other* METAL SULPHIDES

## POLY(DIMERCURYIMMONIUM PERMANGANATE) $(Hg_2MnNO_4)_n$

$(Hg{=}N^+{=}Hg\ MnO_4^-)_n$

Sorbe, 1968, 97

Highly explosive.

*See other* POLY(DIMERCURYIMMONIUM) COMPOUNDS

## MERCURY(I) NITRATE $Hg_2(NO_3)_2$ $Hg_2N_2O_6$

Carbon

Mellor, 1940, Vol. 4, 987

Contact with red-hot carbon causes a mild explosion.

Phosphorus

*See* Phosphorus, P: Nitrates

*See other* METAL NITRATES

## MERCURY(I) THIONITROSYLATE $(Hg_2N_2S_2)_n$

$$\left[\begin{array}{c} -N \overset{S}{\frown} N-HgHg- \\ S \qquad\quad S \\ -N \underset{S}{\smile} N-HgHg- \end{array}\right]_n$$

Goehring, M. *et al.*, *Z. Anorg. Chem.*, 1956, **285**, 70
Explodes on heating in a flame. The structure may be a linear polymer of tetrasulphur tetraimidate rings linked by pairs of dimercury bonds.

## POLY(DIMERCURYIMMONIUM AZIDE) $(Hg_2N_4)_n$

$(Hg{=}N^+Hg\ N_3^-)_n$

Sorbe, 1968, 97
Highly explosive.
*See other* POLY(DIMERCURYIMMONIUM) COMPOUNDS

## MERCURY(I) AZIDE $Hg_2(N_3)_2$ $Hg_2N_6$

Mellor, 1940, Vol. 8, 351; 1967, Vol. 8, Suppl. 2, 25, 50
It is less sensitive and a less powerful explosive than silver or lead azides. It explodes on heating in air to above 270°C, or at 140°C *in vacuo*.
*See other* METAL AZIDES

## 'MERCURY(I) OXIDE' $Hg_2O$

Sidgwick, 1950, 292
This material is known to be an intimate mixture of mercury(II) oxide and metallic mercury.

Alkali metals
Mellor, 1940, Vol. 4, 771
Interaction with molten potassium or sodium is accompanied by a brilliant light and light explosion.

Chlorine,
Ethylene
*See* ETHYLENE, $C_2H_4$: Chlorine

Hydrogen peroxide
Antropov, V., *J. Prakt. Chem.*, 1908, 77, 316
Contact causes explosive decomposition.
*See* HYDROGEN PEROXIDE, $H_2O_2$: Metals, etc.

Non-metals
Mellor, 1940, Vol. 4, 771

Mixtures with phosphorus explode on impact and that with sulphur ignites on frictional initiation.

### TRIS(IODOMERCURIO)PHOSPHINE $P(HgI)_3$ $Hg_3I_3P$

Nitric acid

*See* NITRIC ACID, $HNO_3$ : Phosphine derivatives

### TRIMERCURY DINITRIDE $Hg_3N_2$

Alone,

or Sulphuric acid

1. Mellor, 1940, Vol. 8, 108
2. Fischer, F. *et al., Ber.*, 1910, **43**, 1469

It explodes on friction, impact, heating, or in contact with sulphuric acid [1]. A sample at below −40°C exploded when disturbed [2].

*See other N*-METAL DERIVATIVES

### TRIMERCURY TETRAPHOSPHIDE $Hg_3P_4$

Oxidants

Mellor, 1940, Vol. 8, 844

It ignites when warmed in air, or cold on contact with chlorine. A mixture with potassium chlorate explodes on impact.

*See* POTASSIUM CHLORATE, $ClKO_3$ : Metal phosphides

*See other* METAL NON-METALLIDES

### POTASSIUM IODIDE KI IK

Diazonium salts

*See* DIAZONIUM TRIIODIDES

Diisopropyl peroxydicarbonate

*See* DIISOPROPYL PEROXYDICARBONATE, $C_8H_{14}O_6$

Oxidants

*See* BROMINE PENTAFLUORIDE, $BrF_5$ : Acids, etc.
TRIFLUOROACETYL HYPOFLUORITE, $C_2F_4O_2$
FLUORINE PERCHLORATE, $ClFO_4$ : Potassium iodide
CHLORINE TRIFLUORIDE, $ClF_3$ : Metals, etc.

## POTASSIUM IODATE $KIO_3$ $IKO_3$

Metals and oxidisable derivatives

*See* METAL HALOGENATES: Metals and oxidisable derivatives
PHOSPHORUS, P: Metal halogenates

## POTASSIUM PERIODATE $KIO_4$ $IKO_4$

Ammonium perchlorate

*See* AMMONIUM PERCHLORATE, $ClH_4NO_4$: Impurities

## IODINE AZIDE $IN_3$

Hantzsch, A., *Ber.*, 1900, **33**, 525

Dehnicke, K., *Angew. Chem. (Intern. Ed.)*, 1976, **15**, 553

The isolated solid is a very shock- and friction-sensitive explosive, but the preparation and safe handling of dilute solutions in solvents other than ether have been described.

*See* HALOGEN AZIDES

Sulphur-containing alkenes

Hassner, A. *et al.*, *J. Org. Chem.*, 1968, **33**, 2690

Interaction is accompanied by violent decomposition of the azide.

## SODIUM IODIDE NaI INa

Oxidants

*See* PERCHLORYL FLUORIDE, $ClFO_3$: Calcium acetylide, etc.
PERCHLORIC ACID, $ClHO_4$: Iodides

## SODIUM IODATE $NaIO_3$ $INaO_3$

Metals

Webster, H. A. *et al.*, *Rep. AD-782510/2GA*, Springfield (Va.), USNTIS, 1974

The use of mixtures with magnesium in flares is discussed.

*See* POTASSIUM, K: Oxidants
SODIUM, Na: Iodates

*See other* OXIDANTS

Nitric acid

*See* NITRIC ACID, $HNO_3$: Phosphorus halides

## IODINE $I_2$

Acetaldehyde

*MCA SD-43*, 1952

Interaction may be violent.

Ammonia

Mellor, 1940, Vol. 8, 605; 1967, Vol. 8, Suppl. 2.2, 416

Ammonia solutions react with iodine (or potassium iodide) to produce highly explosive addition compounds between nitrogen triiodide and ammonia.

*See* NITROGEN TRIIODIDE–AMMONIA, $I_3N \cdot H_3N$

Ammonia,
Potassium

*See* AMMONIA, $H_3N$: Iodine, Potassium

Butadiene,
Ethanol,
Mercuric oxide

*See* 2-ETHOXY-1-IODO-3-BUTENE, $C_6H_{11}IO$

Dicaesium oxide

*See* DICAESIUM OXIDE, $Cs_2O$: Halogens

Ethanol,
Phosphorus

1. Read, C. W. W., *School Sci. Rev.*, 1940, **21**(83), 967
2. Llowarch, D., *School Sci. Rev.*, 1956, **37**(133), 434

Interaction of ethanol, phosphorus and iodine to form iodoethane is considered too dangerous for a school experiment [1]. A safer modification has now become available [2].

*See* PHOSPHORUS, P: Halogens

Formamide,
Pyridine,
Sulphur trioxide

*See* FORMAMIDE, $CH_3NO$: Iodine, etc.

Halogens,
or Interhalogens

*See* BROMINE TRIFLUORIDE, $BrF_3$: Halogens
BROMINE PENTAFLUORIDE, $BrF_5$: Acids, etc.
CHLORINE TRIFLUORIDE, $ClF_3$: Iodine
FLUORINE, $F_2$: Iodine

Metal acetylides or carbides

Several react very exothermally with iodine.

*See* BARIUM ACETYLIDE, $C_2Ba$: Halogens
CALCIUM ACETYLIDE, $C_2Ca$: Halogens
DICAESIUM ACETYLIDE, $C_2Cs_2$: Halogens
CUPROUS ACETYLIDE, $C_2Cu$: Halogens
DILITHIUM ACETYLIDE, $C_2Li_2$: Halogens
DIRUBIDIUM ACETYLIDE, $C_2Rb_2$: Halogens
STRONTIUM ACETYLIDE, $C_2Sr$: Halogens
ZIRCONIUM DICARBIDE, $C_2Zr$: Halogens

Metals

Mellor, 1941, Vol. 2, 469; 1963, Vol. 2, Suppl. 2.2, 1563
1939, Vol. 9, 379

Iodine and antimony powder may react so violently as to cause ignition or explosion of the bulk of the mixture. A mixture of potassium and iodine explodes weakly on impact, while potassium ignites in contact with molten iodine.

*See* ALUMINIUM, Al: Oxidants
ALUMINIUM–TITANIUM ALLOYS, Al–Ti: Oxidants
HAFNIUM, Hf
SODIUM, Na: Halogens (reference 5)

Metals,
Water

Jackson, H., *Spectrum*, 1969, 7, 82

Flash-ignition occurs when mixtures of iodine with powdered aluminium, magnesium or zinc are moistened with a drop of water.

Non-metals

*See* BORON, B: Halogens
PHOSPHORUS, P: Halogens

Oxygen difluoride

*See* OXYGEN DIFLUORIDE, $F_2O$: Halogens, etc.

Silver azide

*See* SILVER AZIDE, $AgN_3$: Halogens

Sodium phosphinate

*See* SODIUM PHOSPHINATE ('HYPOPHOSPHITE'), $H_2NaO_2P$: Oxidants

Tetraamminecopper(II) sulphate
Schwarzenbach, V., *Ber.*, 1875, 8, 1233
Addition of ethanol to a mixture soon led to explosions of variable intensity, involving formation of *N*-iodine derivatives.
*See* MERCURY(II) AMIDE CHLORIDE, $ClH_2HgN$

Trioxygen difluoride
*See* TRIOXYGEN DIFLUORIDE, $F_2O_3$
*See other* HALOGENS

## IODINE(V) OXIDE $I_2O_5$

Bromine pentafluoride
*See* BROMINE PENTAFLUORIDE, $BrF_5$: Acids, etc.

Non-metals
Mellor, 1941, Vol. 2, 295
Iodine pentaoxide reacts explosively with warm carbon, sulphur, resin, sugar or powdered, easily combustible elements.
*See other* HALOGEN OXIDES

## IODINE(VII) OXIDE $I_2O_7$

Diethyl ether
Mishra, H. C. *et al., J. Chem. Soc.*, 1962, 1195–1196
Washing of the (incompletely characterised) solid with ether occasionally led to explosive decomposition.
*See other* HALOGEN OXIDES

## PHOSPHORUS DIIODIDE TRISELENIDE $I_2P_4Se_3$

$P_4I_2Se_3$

Nitric acid
*See* NITRIC ACID, $HNO_3$: Phosphorus compounds

## TITANIUM DIIODIDE $TiI_2$ $I_2Ti$

Gibson, 1969, 63
It may ignite in moist air.
*See other* PYROPHORIC MATERIALS

## ZINC IODIDE $ZnI_2$ $I_2Zn$

Sulphuric acid

*See* SULPHURIC ACID, $H_2O_4S$: Zinc iodide

## NITROGEN TRIIODIDE–SILVER AMIDE $NI_3 \cdot AgNH_2$ $I_3N \cdot AgH_2N$

Mellor, 1967, Vol. 8, Suppl. 2, 418

The dry complex may explode.

*See other* *N*-HALOGEN COMPOUNDS
*N*-METAL DERIVATIVES

## NITROGEN TRIIODIDE–AMMONIA $NI_3 \cdot NH_3$ $I_3N \cdot H_3N$

Alone,
or Halogens,
or Oxidants,
or Concentrated acids

Mellor, 1940, Vol. 8, 607; 1967, Vol. 8, Suppl. 2, 418

Garner, W. E. *et al.*, *Nature*, 1935, **135**, 832

When dry it is an extremely sensitive, unstable detonator capable of initiation by virtually any form of energy (light, heat, sound, nuclear radiation or mechanical vibration), even at sub-zero temperatures, and occasionally even with moisture present. It may be handled cautiously when wet, but heavy friction will still initiate it. It explodes in boiling water and is decomposed by cold water to explosive diiodoamine. It explodes, possibly owing to heat-initiation, on contact with virtually any concentrated acid, with chlorine or bromine, ozone or hydrogen peroxide solution.

Crystals desiccated in vacuum explode spontaneously when dry.

*See other* *N*-HALOGEN COMPOUNDS

## PHOSPHORUS TRIIODIDE $PI_3$ $I_3P$

Hydroxylic compounds,
or Oxygen

Leleu, *Cahiers*, 1974, (75), 273

Interaction with water or glycerol is violent, and the iodide ignites in oxygen.

*See other* NON-METAL HALIDES

**TETRAIODODIPHOSPHANE** $I_2PPI_2$ $I_4P_2$

Preparative hazard

*See* PHOSPHORUS, P: Halogens (reference 3)

**ZIRCONIUM TETRAIODIDE** $ZrI_4$ $I_4Zr$

Ethanol

Pascal, 1963, Vol. 9, 565

Interaction is very violent.

*See other* METAL HALIDES

**INDIUM** In

Acetonitrile,
Dinitrogen tetraoxide

*See* DINITROGEN TETRAOXIDE, $N_2O_4$: Acetonitrile, etc.

Mercury(II) bromide

*See* MERCURY(II) BROMIDE, $Br_2Hg$: Indium

Sulphur

*See* SULPHUR, S: Metals

**INDIUM(II) OXIDE** InO

Ellern, 1968, 33

The lower oxide prepared by hydrogenation incandesces on exposure to air.

*See other* PYROPHORIC MATERIALS

**IRIDIUM** Ir

The finely divided catalytic metal may be pyrophoric.

*See* ZINC, Zn: Catalytic metals

*See other* HYDROGENATION CATALYSTS

Interhalogens

*See* BROMINE PENTAFLUORIDE, $BrF_5$: Acids, etc.
CHLORINE TRIFLUORIDE, $ClF_3$: Metals

Peroxyformic acid

*See* PEROXYFORMIC ACID, $CH_2O_3$: Metals, etc.

## POTASSIUM K

1. Gilbert, H. N., *Chem. Eng. News,* 1948, **26**, 2604
2. Johnson, W. S. *et al., Org. Synth.*, 1963, Coll. Vol. 4, 134–135
3. *MCA Case History No. 1891*
4. Taylor, D. A. H., *Chem. Brit.*, 1974, **10**, 101
5. Diaper, D. G. M., ibid., 312
6. Houben-Weyl, 1970, Vol. 13.1, 264

A review of the comparative properties of sodium and potassium, the latter invariably being the more hazardous. Potassium readily reacts with carbon monoxide to form an explosive carbonyl, while sodium does so only under exceptional circumstances [1]. Detailed laboratory procedures for safe handling of potassium have been published [2].

A safe method for disposal of potassium residues in bulk storage and processing vessels is detailed. Basically the method involves reaction of the metallic residues with dry (condensate-free) steam under closely controlled conditions. It may be used where small-scale techniques (dissolution in higher alcohols) are inapplicable [3]. Hazards associated with the storage of metallic potassium in aluminium containers are discussed. The severe corrosion leading to perforation of such commercially supplied containers [4] was attributed to the deliquescent and subsequently corrosive nature of potassium carbonate formed from atmospheric carbon dioxide inside the container. Sodium may be safely stored in aluminium cans because the derived carbonate is not deliquescent and causes no corrosion [5]. There is a considerable risk of fire if powdered potassium ('potassium sand') dispersed in benzene is exposed to air [6].

*See* POTASSIUM CARBONATE, $CK_2O_3$: Carbon

Acetylene

*See* ACETYLENE, $C_2H_2$: Potassium

Air

It is convenient to consider interaction of potassium and air under two headings.

*Rapid oxidation*

Mellor, 1941, Vol. 2, 468; 1963, Vol. 2, Suppl. 2.2, 1559

Reaction with air or oxygen in complete absence of moisture does not occur, even on heating to boiling point. However, in contact with normally moist air, oxidation may become so fast that melting or ignition occurs, particularly if pressure is applied locally to cause melting and exposure of a fresh surface, as when potassium is pressed through a die to form potassium wire.

*Slow oxidation*

1. Mellor, 1941, Vol. 2, 493
2. Gilbert, H. N., *Chem. Eng. News*, 1948, **26**, 2604
3. Short, J. F., *Chem. & Ind.*, 1964, 2132; Brazier, A. D., *Chem. & Ind.*, 1965, 220; Balfour, A. E., ibid., 353; Bil, M. S., ibid., 812; Cole, R. J., ibid., 944
4. Brandsma, 1971, 10, 21

Metallic potassium on prolonged exposure to air forms a coating of yellow potassium superoxide ($KO_2$) under which is a layer of potassium monoxide in contact with the metal. Normal contact of potassium superoxide with potassium causes ignition to occur [1], but if the layer of superoxide is driven into the underlying potassium by dry metal-cutting operations or a hammer blow, very violent explosions occur [2,3]. It has also been observed that potassium-cutting operations under oil cause flashes of light or fire to occur [3].
(But *see* POTASSIUM DIOXIDE (SUPEROXIDE), $KO_2$)

Fresh potassium should be stored under dry xylene in airtight containers to prevent oxidation [1]. Old stocks, where the coating is orange or yellow, should not be cut but destroyed by burning on an open coke fire, or by addition of *tert*-butanol to small portions under xylene in a hood. It is dangerous to use methyl or ethyl alcohol (either dry or wet) as a replacement for *tert*-butanol. Recommended procedures include cutting and handling the metal with forceps under dry xylene and disposal of scraps in xylene by addition of *tert*-butanol [2,3]. Even the latter may react violently [4].
*See also* POTASSIUM–SODIUM ALLOY, K–Na

Alcohols

Pratt, E. F. *et al.*, *J. Amer. Chem. Soc.*, 1954, **75**, 54

Interaction of potassium with a range of alcohols (n-propanol to n-octanol, benzyl alcohol, cyclohexanol) to form the alkoxides usually led to explosions unless air in the containing vessel was displaced by nitrogen before addition of potassium in small portions with stirring.
*See* Air (slow oxidation), above

Carbon

1. Mellor, 1963, Vol. 2, Suppl. 2.2, 1566
2. Mellor, D. P., *Chem. & Ind.*, 1965, 723
3. *Alkali Metals,* 1957, 169

Reaction of various forms of carbon (soot, graphite or activated charcoal) is exothermic and vigorous at elevated temperatures, and if the carbon is finely divided and air is present, ignition may occur leading to explosion, possibly due to interaction with the potassium superoxide which would be formed [1]. Explosions caused by attempts to extinguish potassium fires with graphite powder have been so attributed [2]. Potassium cannot be produced by electrolysis of potassium chloride because of interaction of the metal with graphite electrodes and formation of explosive carbonyl-potassium [3].

Carbon disulphide

Rüst, 1948, 339

Mixtures with carbon disulphide or other organosulphur compounds will explode on impact, but not on heating.

*See* POTASSIUM–SODIUM ALLOY, K–Na: Carbon dioxide, etc.

Covalent halides

Leleu, *Cahiers*, 1975, (80), 390

Mixtures of potassium with covalent inorganic halides may explode if subjected to shock.

*See* Non-metal halides, below

Dimethyl sulphoxide

*See* DIMETHYL SULPHOXIDE, $C_2H_6OS$: Potassium

Ethylene oxide

*See* ETHYLENE OXIDE, $C_2H_4O$: Contaminants

Halocarbons

1. Staudinger, H., *Z. Angew. Chem.*, 1922, **35**, 657; *Z. Elektrochem.*, 1925, **31**, 549
2. Lenze, F. *et al.*, *Chem. Ztg.*, 1932, **56**, 921
3. Rampino, L. D., *Chem. Eng. News*, 1958, **36**(32), 62

Although apparently stable on contact, mixtures of potassium (or its alloys) with a wide range of halocarbons are shock-sensitive and may explode with great violence on light impact. Mono-, di-, tri- and pentachloro-ethane, bromoform, dibromo- and diiodo-methane are among those investigated. Sensitivity increases generally with the degree of substitution, and potassium–sodium alloy gives extremely sensitive mixtures. The mixture with carbon tetrachloride is 150–200 times as shock-sensitive as mercury fulminate, and a mixture of potassium with bromoform was exploded by a door slamming nearby. Mixtures with tetra- and penta-chloroethane will often explode spontaneously after a short delay during which a voluminous solid separates out [1,2]. When heated together, potassium and tetra-chloroethylene exploded at 97–99°C, except when the metal had been very recently freed of its usual oxide film. Sodium did not explode under the same conditions [3].

*See* METALS: Halocarbons

Halogens,
or Interhalogens

Mellor, 1941, Vol. 2, 114, 469; 1963, Vol. 2, Suppl. 2.2, 1563
Pascal, 1960, Vol. 16, 578

Potassium ignites in fluorine and in dry chlorine (unlike sodium). In bromine vapour it incandesces, and explodes violently in liquid bromine. Mixtures with iodine incandesce on heating, and explode weakly on impact. Potassium reacts explosively with molten iodine bromide and iodine, and a mixture with the former is shock-sensitive and explodes strongly. Molten potassium reacts explosively with iodine pentafluoride. Contact with iodine trichloride causes ignition.

*See* IODINE CHLORIDE, ClI: Metals
IODINE HEPTAFLUORIDE, $F_7I$: Metals

Hydrazine

*See* HYDRAZINE, $H_4N_2$: Potassium

Hydrogen halides

1. Cueilleron, J., *Bull. Soc. Chim. Fr.*, 1945, **12**, 88
2. Pascal, 1963, Vol. 2.2, 31

Impact causes a mixture of potassium and anhydrous hydrogen iodide to explode very violently [1]. Molten potassium ignites in contact with hydrogen chloride, bromide or iodide [2].

Maleic anhydride

*See* MALEIC ANHYDRIDE, $C_4H_2O_3$: Cations, etc.

Mercury

Mellor, 1941, Vol. 2, 469

Interaction to form amalgams is vigorously exothermic and may become violent if too much potassium is added at once.

Metal halides

Staudinger, H., *Z. Elektrochem.*, 1925, **31**, 549
Cueilleron, J., *Bull. Soc. Chim. Fr.*, 1945, **12**, 88
Mellor, 1963, Vol. 2, Suppl. 2.2, 1571
Rieke, R. D., *Accts Chem. Res.*, 1977, **10**, 301–305

Mixtures of potassium with metal halides are sensitive to mechanical shock, and the ensuing explosions have been classified. Very violent explosions occurred with calcium bromide, iron(III) bromide or chloride, iron(II) bromide or iodide, or cobalt(II) chloride.

Strong explosions occurred with silver fluoride, all four mercury(II) halides, copper(I) chloride, bromide or iodide, copper(II) chloride and bromide, and ammonium tetrachlorocuprate; zinc and cadmium chlorides, bromides, iodides; aluminium fluoride, chloride, bromide, thallium(I) bromide; tin(II) or (IV) chloride, tin(IV) iodide (with sulphur), arsenic trichloride and triiodide, antimony and bismuth trichlorides, tribromides and triiodides; vanadium(V) chloride chromium(IV) chloride, manganese(II) and iron(II) chlorides, and nickel chloride, bromide and iodide. Weak explosions occurred with ammonium bromide or iodide, silver chloride or iodide, strontium iodide, thallium(I) chloride; potassium tetrachlorocuprate, zinc, cadmium and nickel fluorides, chromium(III) fluoride and manganese(II) bromide or iodide. Vanadyl chloride reacts violently with potassium at 100°C, and explosions have occurred with copper and lead chloride oxides or lead(IV) chloride.

The recent reference describes the moderation of such reactions by the use of ether or hydrocarbon solvents. The reactions are usually highly exothermic but controllable, and give finely divided and highly reactive metals, some pyrophoric in air, such as magnesium and aluminium.

Metal oxides

1. Mellor, 1963, Vol. 2, Suppl. 2.2, 1571
2. Mellor, 1941, Vol. 3, 138
3. Mellor, 1940, Vol. 4, 770, 779
4. Mellor, 1941, Vol. 7, 401
5. Mellor, 1943, Vol. 11, 237, 542
6. Mellor, 1941, Vol. 9, 649

Lead peroxide reacts explosively [1] and copper(II) oxide incandescently [2] with warm potassium. Mercury(II) and (I) oxides react with molten potassium with incandescence and explosion [3]. Tin(IV) oxide is reduced incandescently on warming [4] and molybdenum(III) oxide on heating [5]. Warm dibismuth trioxide is reduced with incandescence [6].

*See* 'MERCURY(I) OXIDE', $Hg_2O$

Nitric acid

Pascal, 1963, Vol. 2.2, 31

Interaction with concentrated acid is explosive.

Nitrogen-containing explosives

Staudinger, H., *Z. Elektrochem.*, 1925, **31**, 549

Nitro or nitrate explosives, normally shock-insensitive, are rendered extremely sensitive by addition of traces of potassium or potassium–sodium alloy. Ammonium nitrate, and nitrate–sulphate mixtures, picric acid, and even nitrobenzene respond in this way.

Non-metal halides

1. Mellor, 1963, Vol. 2, Suppl. 2.2, 1564–1568
2. Mellor, 1940, Vol. 8, 1006
3. Mellor, 1947, Vol. 10, 642–646, 908, 912

Diselenium dichloride and seleninyl chloride both explode on addition of potassium [1,3], while the metal ignites in contact with phosphorus trichloride vapour or liquid [2]. Mixtures of potassium with sulphur dichloride or dibromide, phosphorus trichloride or tribromide, and with phosgene are shock-sensitive, usually exploding violently on impact. Potassium also explodes violently on heating with disulphur dichloride, and with sulphur dichloride or seleninyl bromide without heating [3].

Non-metal oxides

1. Gilbert, H. N., *Chem. Eng. News,* 1948, **26**, 2604
2. Mellor, 1941, Vol. 2, 241
3. Mellor, 1940, Vol. 8, 436, 544, 554, 945
4. Pascal, 1963, Vol. 2.2, 31

Mixtures of potassium and solid carbon dioxide are shock-sensitive and explode violently on impact, and carbon monoxide readily reacts to form explosive carbonylpotassium [1]. Dichlorine oxide explodes on contact with potassium [2]. Potassium ignites in dinitrogen tetra- or pentaoxide at ambient temperature, and incandesces when warmed with phosphorus(V) or nitrogen oxides [3].

At −50° C, potassium and carbon monoxide react to give dicarbonylpotassium, $C_2KO_2$, which explodes in contact with air or water, or at 100° C. At 150° C the product is a trimer of this, potassium benzenehexolate. The just-molten metal ignites in sulphur dioxide [4].

Oxalyl dihalides

Staudinger, H., *Z. Angew. Chem.*, 1922, **35**, 657; *Ber.*, 1913, **46**, 1426

In absence of mechanical disturbance, potassium or potassium–sodium alloy appears to be stable in contact with oxalyl dibromide or dichloride, but the mixtures are very shock-sensitive and explode very violently.

*See* Halocarbons, above

Oxidants

Mellor, 1963, Vol. 2, Suppl. 2.2, 1571

The potential for violence of interaction between the powerful reducing agent potassium and oxidant classes has been well described. Other miscellaneous oxygen-containing substances which react violently or explosively include sodium and silver iodates, lead sulphate and boric acid.

*See* Air, above
Metal oxides, above
Non-metal oxides, above
Selenium, etc., below

CHLORINE TRIFLUORIDE, $ClF_3$: Metals
DICHLORINE OXIDE, $Cl_2O$: Oxidisable materials
CHROMIUM TRIOXIDE, $CrO_3$: Alkali metals
NITRYL FLUORIDE, $FNO_2$: Metals
AMMONIUM NITRATE, $H_4N_2O_3$: Ammonium sulphate, etc.

Selenium,
or Tellurium

Mellor, 1947, Vol. 10, 767; 1943, Vol. 11, 40
Interaction of selenium and potassium is exothermic and ignition occurs, or, with excess potassium, a mild explosion. Tellurium and potassium become incandescent when warmed in hydrogen atmosphere to prevent aerial oxidation.

Sulphur

Pascal, 1963, Vol. 2.2, 30
Interaction is violent on warming.

Sulphuric acid

Kirk-Othmer, 1961, Vol. 16, 362
Interaction is explosive.

Water

1. Mellor, 1941, Vol. 2, 469; 1963, Vol. 2, Suppl. 2.2, 1560
2. Angus, L. H., *School Sci. Rev.*, 1950, **31**(115), 402
3. Markowitz, M. M., *J. Chem. Educ.*, 1963, **40**, 633–636

Interaction is violently exothermic, and the heat evolved with water at 20°C is enough to ignite evolved hydrogen. Larger pieces of potassium invariably explode in water and scatter burning potassium particles over a wide area. Aqueous alcohols should not be used for waste metal disposal [1]. Small pieces of potassium will also explode with a restricted amount of water [2].

The reactivity of potassium and other alkali metals with water has been discussed in detail. The vigour of reaction is scarcely reduced by contact with ice–water slurries [3].

*See* Air (slow oxidation), above
ALKALI METALS

## POTASSIUM–SODIUM ALLOY — K–Na

*Health & Safety Information,* **251**, Washington, USAEC, 1967
Ellis, J. E. *et al., Inorg. Synth.*, 1976, **16**, 70

Appropriate precautions in handling potassium–sodium alloy, used as a liquid coolant, are described. It is generally more hazardous than either of the component metals, because alloys in the range 50–85% wt. potassium are liquid at ambient temperature and can therefore come into more intimate contact with reagents than the solid metals.

Most of the entries under Potassium, and some of those under Sodium, may be expected to apply to their alloys, depending on the composition.

For the destruction of residual small amounts of the alloy, treatment with a 1:1 mixture of isopropanol and heptane over a safety tray is recommended.

Air,

Mellor, 1961, Vol. 2, Suppl. 2.1, 562

Anon., *Fire Prot. Assoc. J.*, 1965, **68**, 140

The alloy usually ignites on exposure to air, with which it reacts in any case to form potassium dioxide, a very powerful oxidant. Fires may be extinguished with dried sodium carbonate, calcium carbonate, sand or resin-coated sodium chloride. Graphite is not suitable since violent reaction with the superoxide is possible. Carbon dioxide or halocarbon extinguishers must not be used.

*See* Carbon dioxide, etc., below
Halocarbons, below
POTASSIUM, K: Air (slow oxidation)
POTASSIUM DIOXIDE, $KO_2$: Carbon

Carbon dioxide,
or Carbon disulphide

Staudinger, H., *Z. Elektrochem.*, 1925, **31**, 549; *Z. Angew. Chem.*, 1922, **35**, 657

Mixtures of the alloy and solid carbon dioxide are powerful explosives, some 40 times more sensitive to shock than mercury fulminate. Carbon disulphide behaves similarly.

Halocarbons

1. Staudinger, H., *Z. Elektrochem.*, 1925, **31**, 550
2. *Inform. Exch. Bull.*, Lawrence Radiation Lab., 1961, **1**, 2
3. Kimura, T. *et al.*, *J. Chem. Educ.*, 1973, **50**, A85
4. Schmidt, E. *et al.*, *Chem. Abs.*, 1975, **83**, 163034s

The liquid alloy gives mixtures with halocarbons even more shock-sensitive than those with potassium [1]. Highly chlorinated methane derivatives are more reactive than those of ethane, often exploding spontaneously after a delay [1]. Contact of 1,1,1-trichloroethane with a trapped alloy residue in a valve caused an explosion [2]. It is to be expected that chloro-fluorocarbons will also form hazardous mixtures in view of their reactivity with barium. A Teflon-coated magnetic bar used to stir the alloy under propane atmosphere ignited when the speed was increased and generated enough heat to melt the glass. Triboelectric initiation was postulated [1].

Precautionary measures for demonstration of explosion of the alloy with chloroform are detailed [4].

*See other* METALS: Halocarbons

Metal halides

Rieke, R. D., *Accts Chem. Res.*, 1977, **10**, 302

Use of the alloy to reduce metal halides in solvents to the finely divided and highly reactive metals is not recommended for cases where the halide is highly soluble in the solvent (e.g. zinc and iron(III) chlorides in THF). Explosive reaction may ensue.

*See also* POTASSIUM, K: Metal halides

Metal oxides

Staudinger, H., *Z. Elektrochem.*, 1925, **31**, 551

Mixtures of the alloy with silver and mercury oxides are shock-sensitive powerful explosives. The red form of mercury(I) oxide is 40 times, and the yellow form 120 times, as shock-sensitive as mercury fulminate.

Nitrogen-containing explosives

*See* POTASSIUM, K: Nitrogen-containing explosives

*See other* ALLOYS

## POTASSIUM PERMANGANATE $KMnO_4$

Acetic acid,

or Acetic anhydride

von Schwartz, 1918, 34

Cooling is necessary to prevent possible explosion from contact of potassium permanganate (or the calcium or sodium salts) with acetic acid or its anhydride.

Acetone,

*tert*-Butylamine

1. Kornblum, N. *et al., Org. Synth.*, 1973, Coll. Vol. 5, 847
2. idem., *Org. Synth.*, 1963, **43**, 89
3. Haynes, R. K. *et al., Lab. Equip. Digest*, 1974, **12**(6), 98; private comm., 1974

In the reprinted description [1] of a general method of oxidising *tert*-alkylamines with potassium permanganate to the corresponding nitroalkanes, a superscript reference indicating that *tert*-butylamine was oxidised in water, rather than in acetone containing 20% water, is omitted, although it is present in the original description [2]. This appears to be important, since running the reaction in 20% aqueous acetone led to a violent reaction with eruption of the flask contents. This was attributed to caking of the solid permanganate due to inadequate agitation, and onset of an exothermic reaction between the oxidant and solvent [3].

Alcohols,
Nitric acid

Numjal, N. L. *et al., Amer. Inst. Aero. Astronaut. J.*, 1972, **10**, 1345
Methyl, ethyl, isopropyl, pentyl or isopentyl alcohols do not ignite immediately upon mixing with red fuming nitric acid, but addition of potassium permanganate (20%) to the acid before mixing causes immediate ignition.

*See also* NITRIC ACID, $HNO_3$: Alcohols

Aluminium carbide

Mellor, 1946, Vol. 5, 872
Incandescence on warming.

Ammonia,
Sulphuric acid

Mellor, 1941, Vol. 1, 907
Ammonia is oxidised with incandescence in contact with the permanganic acid formed in the mixture.

Ammonium nitrate

Urbanski, 1965, Vol. 2, 491
Admixture of 0.5% of potassium permanganate with an ammonium nitrate explosive caused an explosion 7 h later. This was due to formation and exothermic decomposition of ammonium permanganate, leading to ignition.

Ammonium perchlorate

*See* AMMONIUM PERCHLORATE, $ClH_4NO_4$: Impurities

Antimony,
or Arsenic

Mellor, 1942, Vol. 12, 322
Antimony ignites on grinding in a mortar with the solid permanganate, while arsenic explodes.

Coal,
Peroxomonophosphoric acid

Rawat, N. S., *Chem. & Ind.*, 1976, 743
In a new method for determination of sulphur in coal, the samples are oxidised with an aqueous mixture of permanganate and the peroxoacid. During the digestion, a reflux condenser is essential to prevent loss of water, which could lead to explosively violent oxidation.

Dichloro(methyl)silane

*See* DICHLORO(METHYL)SILANE, $CH_4Cl_2Si$: Oxidants

Formaldehyde

Piefel, W. *et al., Chem. Abs.*, 1977, **86**, 60508g

Formaldehyde gas for disinfection purposes may be released from the aqueous solution ('formalin') by treatment with potassium permanganate, but the quantities used must be limited to avoid the risk of fire or explosion.

Glycerol

1. 'Leaflet No. 5', London, Inst. of Chem., April, 1940
2. *BCISC Quart. Safety Summ.*, 1974, **45**, 11–12

Contact of glycerol with solid potassium permanganate caused a vigorous fire [1]. During the preparation of a solution for the decontamination of tetramethyllead spills, addition of the solid oxidant to a polythene bucket contaminated with glycerol caused ignition to occur after a few seconds [2]. The trifunctionality and high viscosity of glycerol give high heat-release and poor dissipation of reaction heat, respectively.

*See* Oxygenated organic compounds, below

Hydrochloric acid

Curry, J. C., *School Sci. Rev.*, 1965, **46**(160), 770

During preparation of chlorine by addition of the concentrated acid to solid permanganate, a sharp explosion occurred on one occasion. (Contamination of the acid with sulphuric acid could have produced permanganic acid.)

Hydrofluoric acid

Black, A. M. *et al., J. Chem. Soc., Dalton Trans.*, 1974, 977

During preparation of dipotassium hexafluoromanganate(IV), addition of the solid oxidant to exceptionally concentrated hydrofluoric acid (60–90%, rather than the 40% used previously) caused a violent exotherm with light emission.

Hydrogen peroxide

Anon., *J. Pharm. Chim.*, 1927, **6**, 410

Contact of hydrogen peroxide solution from a broken bottle with pervious packages of permanganate caused a violent reaction and fire.

*See also* HYDROGEN PEROXIDE, $H_2O_2$: Metals, etc. (reference 3)

Hydrogen trisulphide

Mellor, 1947, Vol. 10, 159

Contact with solid permanganate ignites the liquid sulphide.

Hydroxylamine

*See* HYDROXYLAMINE, $H_3NO$: Oxidants

Non-metals

Mellor, 1942, Vol. 12, 319–323

A mixture of carbon and potassium permanganate is not friction-sensitive, but burns vigorously on heating. Mixtures with phosphorus or sulphur react explosively on grinding and heating, respectively.

Organic nitro compounds

Blinov, I., *J. Chem. Ind. USSR*, 1937, **14**, 1151–1153

Mixtures ignite easily on heating, shock or contact with sulphuric acid.

*See also* Oxygenated organic compounds, below

Oxygenated organic compounds

1. Rathsbury, H., *Chem. Ztg.*, 1943, **37**, 3272q
2. Gallais, 1957, 697
3. Partington, 1967, 830–831

Liquid oxygenated compounds, or mixtures of solids with water, ignite in contact with the solid permanganate. Such compounds include ethylene glycol, propane-1,2-diol, erythritol, mannitol, triethanolamine, 3-chloropropane-1,2-diol; acetaldehyde, isobutyraldehyde, benzaldehyde; acetylacetone; esters of ethylene glycol, lactic, acetic or oxalic acids. The necessary presence of hydroxy or keto groups to cause ignition may be connected with solubility requirements [1]. Mixtures of the solid oxidant with solid reducing sugars may react violently or explosively [2]. Powdered oxalic acid and the oxidant ignite soon after mixing [3].

Polypropylene

*MCA Case History No. 1842*

While using a screw-conveyor to move the solid oxidant, ignition of a polypropylene tube in the feed system occurred. This could not subsequently be reproduced, even when likely contaminants were present.

Slag wool

Hallam, B. J. *et al., School Sci. Rev.*, 1975, **56**(197), 820

Generation of oxygen by heating a layer of potassium permanganate retained in a test tube by a plug of 'Rocksil' slag wool led to minor explosions. The wool liberated an organic distillate on heating, and the explosions were attributed to combustion of the distillate in the liberated oxygen. Roasting the wool before use with oxidants is recommended.

Sulphuric acid,
Water

1. Archer, J. R., *Chem. Eng. News*, 1948, **26**, 205
2. *ABCM Quart. Safety Summ.*, 1946, **17**, 2

3. Stevens, G. C., private comm., 1971

Addition of concentrated sulphuric acid to the slightly damp permanganate caused an explosion. This was attributed to formation of permanganic acid, dehydration to dimanganese heptoxide and explosion of the latter, caused by heat liberated from interaction of sulphuric acid and moisture [1]. A similar incident was reported previously, when a solution of potassium permanganate in sulphuric acid, prepared as a cleaning agent, exploded violently [2]. There is, however, a reputedly safe procedure for the final ultra-cleaning of glassware in cases where the adsorbed chromium film left by chromic acid cleaning would be intolerable. This involves treatment of the glassware with concentrated sulphuric acid (10 ml) to which one or two small crystals (not more) of permanganate have been added. If the solution changes to a brown colour, it must immediately be discarded into excess water [3].

Titanium

Mellor, 1941, Vol. 7, 20

A mixture of powdered metal and oxidant explodes on heating.

Wood

Anon., *Chem. Ztg.*, 1927, **51**, 221

Dunn, B. W., Amer. Rail Assoc., Bur. Explosives, *Report No. 10*, 1917

Contact between the solid material and wood, in presence of either moisture or mechanical friction, may cause a fire.

*See other* METAL OXOMETALLATES
OXIDANTS

## POTASSIUM NITRITE $KNO_2$

Ammonium salts

Mellor, 1941, Vol. 2, 702

Addition of ammonium sulphate to the fused nitrate causes effervescence and ignition.

*See* NITRITE SALTS OF NITROGENOUS BASES

Boron

Mellor, 1946, Vol. 5, 16

Addition of boron to the fused nitrite causes violent decomposition.

Potassium amide

Bergstrom, F. W. *et al., Chem. Rev.*, 1933, **12**, 64

Heating a mixture of the solids under vacuum causes a vigorous explosion.

*See other* METAL OXONON-METALLATES

Aluminium,
Barium nitrate,
Potassium perchlorate,
Water
*See* ALUMINIUM, Al: Metal nitrates, etc.

1,3-Bis(trichloromethyl)benzene
*See* 1,3-BIS(TRICHLOROMETHYL)BENZENE, $C_8H_4Cl_6$ : Oxidants

Calcium disilicide
Berger, F., *Compt. Rend.*, 1920, **170**, 1492
A mixture of potassium (or sodium) nitrate and calcium silicide (60:40) is a readily ignited primer which burns at a very high temperature. It is capable of initiating many high-temperature reactions.

Chromium nitride
Partington, 1967, 744
The nitride deflagrates with the molten nitrate.

Metals
Mellor, 1941, Vol. 7, 20, 116, 261; 1939, Vol. 9, 382
Mixtures of potassium nitrate and powdered titanium, antimony or germanium explode on heating, and with zirconium at fusion temperature of the mixture.

Metal sulphides
1. Mellor, 1939, Vol. 9, 270, 524
2. Mellor, 1941, Vol. 3, 745
3. Mellor, 1941, Vol. 7, 91, 274
4. Mellor, 1943, Vol. 11, 647
5. Pascal, 1963, Vol. 8.3, 404

Mixtures of potassium nitrate with antimony trisulphide [1], barium or calcium sulphides [2], germanium monosulphide or titanium disulphide [3], all explode on heating. The mixture with arsenic disulphide is detonable, and addition of sulphur gives a pyrotechnic composition [3]. Mixtures with molybdenum disulphide are also detonable [4]. Interaction with sulphides in molten mixtures is violent [5].

Non-metals
1. *MCA Case History No. 1334*
2. Mellor, 1946, Vol. 5, 16
3. Mellor, 1941, Vol. 2, 820, 825; 1963, Vol. 2, Suppl. 2.2, 1939

4. Mellor, 1940, Vol. 8, 788
5. Mellor, 1939, Vol. 9, 35

A pyrotechnic blend of a finely divided mixture with boron ignited and exploded when the aluminium container was dropped [1]. Boron is not attacked at below 400°C, but is at fusion temperature or at lower temperatures if decomposition products (nitrites) are present [2]. Contact of powdered carbon with the fused nitrate causes vigorous combustion and a mixture explodes on heating. Gunpowder is the oldest explosive known and contains potassium nitrate, charcoal and sulphur, the latter to reduce ignition temperature and increase the speed of combustion [3]. Mixtures of white phosphorus and potassium nitrate explode on percussion and a mixture of red phosphorus and potassium nitrate reacts vigorously on heating [4]. Mixtures of potassium nitrate with arsenic explode vigorously on ignition [5].

Organic materials

Smith, A. J., *Quart. Natl. Fire Prot. Assoc.*, 1930, **24**, 39–44

Potassium nitrate in cloth sacks stowed next to baled peat moss became involved in a ship fire and caused rapid flame spread and explosions.

*See* SODIUM NITRATE, $NaNO_3$: Jute, Magnesium chloride

Phosphides

Mellor, 1940, Vol. 8, 839, 845

Boron phosphide ignites in molten nitrates; mixtures of the nitrate with tricopper diphosphide explode on heating, and that with copper monophosphide on impact.

Reducants

Mellor, 1941, Vol. 2, 820

Mixtures of potassium nitrate with sodium phosphinate and sodium thiosulphate are explosive, the former being rather powerful.

Sodium acetate

Pieters, 1957, 30

Mixtures may be explosive.

Thorium dicarbide

*See* THORIUM DICARBIDE, $C_2Th$: Non-metals, etc.

*See other* METAL OXONON-METALLATES
OXIDANTS

**POTASSIUM NITRIDOOSMATE(1−)** **$K[Os(N)O_3]$** **$KNO_3Os$**

Clifford, A. F. *et al., Inorg. Synth.,* 1960, **6**, 205

Heating above 180°C at atmospheric pressure causes it to explode.

*See other* *N*-METAL DERIVATIVES

## POTASSIUM AZIDE $KN_3$

Mellor, 1940, Vol. 8, 347

Insensitive to shock. On heating progressively it melts, then decomposes evolving nitrogen, and the residue ignites with a feeble explosion.

Carbon disulphide

*See* CARBON DISULPHIDE, $CS_2$: Metal azides

Manganese dioxide

*See* MANGANESE DIOXIDE, $MnO_2$: Potassium azide

Sulphur dioxide

Mellor, 1940, Vol. 8, 347

The salt explodes at 120°C when heated in liquid sulphur dioxide.

*See other* METAL AZIDES

## POTASSIUM AZIDOSULPHATE $KOSO_2N_3$ $KN_3O_3S$

Mellor, 1940, Vol. 8, 314

Explodes on heating.

*See other* ACYL AZIDES

## POTASSIUM AZIDODISULPHATE $KOSO_2OSO_2N_3$ $KN_3O_6S_2$

Alone,
or Water

Mellor, 1967, Vol. 8, Suppl. 2.2, 36

On keeping or slow heating, the salt produces explosive disulphuryl azide. The salt reacts explosively with water.

*See* DISULPHURYL AZIDE, $N_6O_5S_2$

*See other* ACYL AZIDES

## POTASSIUM DIOXIDE (SUPEROXIDE) $KO_2$

1. *Alkali Metals*, 1957, 174
2. Gilbert, H. N., *Chem. Eng. News*, 1948, **26**, 2604
3. Madaus, J. H., private comm., 1976
4. Kirk-Othmer, 1968, Vol. 16, 365–366
5. Commander, J. C., *ERDA Rep. ANCR 1217: TID-4500*, Springfield (Va.), USNTIS, 1975
6. Sloan, S. A., *Proc. Int. Conf. Liquid Met. Technol.*, Champion (Pa.), Int. At. En. Auth., 1976

The earlier references, which state that this powerful oxidant is stable when pure, but explosive when formed as a layer on metallic potassium [1,2], are not wholly correct [3], because the superoxide is manufactured uneventfully by spraying the molten metal into air to effect oxidation [4]. Previous incidents appear to have involved the explosive oxidation of mineral oil or solvents [3].

However, mixtures of the superoxide with liquid or solid potassium–sodium alloys will ignite spontaneously after an induction period of 18 min, but combustion while violent is not explosive [3]. The additional presence of water (which reduces the induction period) or hydrocarbon contaminant did produce explosion hazards under various circumstance [5]. Contact of liquid potassium with the superoxide gives no obvious reaction below 117°C and a controlled reaction between 117 and 177°C, but an explosive reaction occurs above 177°C. Heating at 100°C/min from 77°C caused explosion at 208°C [6].

*See* POTASSIUM, K: Air (slow oxidation)

Carbon

Mellor, D. P., *Chem. & Ind.,* 1965, 723

Residues from extinguishing small sodium–potassium alloy fires with graphite were accumulated in an airtight drum. Later, burning alloy fell into the drum and caused a violent explosion. This was attributed to formation of potassium superoxide during storage and explosive reaction of the latter with graphite, initiated by the burning alloy. Graphite is not a suitable extinguisher for potassium or alloy fires.

Diselenium dichloride

Mellor, 1947, Vol. 10, 897

Interaction is very violent.

Ethanol

*Health & Safety Information,* **251**, Washington, USAEC, 1967

Disposal of a piece of sodium–potassium alloy under argon in a glove box by addition of alcohol caused violent gas evolution which burnt a glove and produced a flame. A piece of highly oxidised potassium exploded when dropped into alcohol. Both incidents were attributed to violent interaction of potassium superoxide and ethanol.

Hydrocarbons

*Health & Safety Information,* **251,** Washington, USAEC, 1967

Residues of sodium–potassium alloy in metal containers were covered with oil prior to later disposal. When a lid was removed later, a violent explosion occurred. This was attributed to frictional initiation of the mixture of potassium superoxide (formed on long standing) and oil.

Metals
Mellor, 1941, Vol. 2, 493
Oxidation of arsenic, antimony, copper, potassium, tin or zinc proceeds with incandescence.
*See* POTASSIUM, K: Air (slow oxidation)

Phosphorus
Mellor, 1939, Vol. 2, 492
Interaction with phosphorus is vigorous.
*See related* METAL PEROXIDES

## POTASSIUM TRIOXIDE (POTASSIUM OZONATE) $KO_3$

*See* Fluorine, $F_2$: Potassium hydroxide
CAESIUM TRIOXIDE ('OZONATE'), $CsO_3$

## POTASSIUM SILICIDE KSi

Bailar, 1973, Vol. 1, 1357
It ignites in air.
*See other* METAL NON-METALLIDES

## POTASSIUM DIPEROXOMOLYBDATE(2–) $K_2MoO_6$

**$K_2[MoO_6]$**

Mellor, 1943, Vol. 11, 607
It explodes on grinding.
*See other* PEROXOACID SALTS

## POTASSIUM TETRAPEROXOMOLYBDATE(2–) $K_2MoO_8$

**$K_2[Mo(O_2)_4]$**

Jahr, K. F., *FIAT Rev. of German Science: Inorg. Chem.*, Part III, 1948, 170
It is explosive.
*See other* PEROXOACID SALTS

## POTASSIUM NITROSODISULPHATE(2–) $K_2NO_7S_2$

**$KOSO_2OSO_2NO$**

1. Zimmer, H. *et al., Chem. Rev.*, 1971, **71**, 230, 243
2. Wehrli, P. A. *et al., Org. Synth.*, 1972, **52**, 86

The observed instability of the solid radical on storage, ranging from slow decomposition to violent explosion (probably depending on the degree of confinement), depends largely upon the degree of contamination by nitrite ion. Storage conditions to enhance

stability are detailed [1]. Synthetic use of solutions of the salt has been preferred, and the storage of any isolated salt as an alkaline slurry is recommended [2].
*See other* NITROSO COMPOUNDS

**POTASSIUM DINITROSOSULPHITE** **$(KO)_2(O)S(NO)_2$** **$K_2N_2O_5S$**

Weitz, E. *et al., Ber.* 1933, **66**, 1718
Explodes on heating.
*See other* NITROSO COMPOUNDS

**POTASSIUM AZODISULPHONATE(2−)** **$K_2N_2O_6S_2$**
**$KOSO_2N{=}NSO_2OK$**

Konrad, E. *et al., Ber.* 1926, **59**, 135
There is a possibility of explosion during vacuum desiccation of the ether-damp solid.
*See related* AZO COMPOUNDS

**POTASSIUM SULPHUR DIIMIDE(2−)** **$KN{=}S{=}NK$** **$K_2N_2S$**

Goehring, 1957, 35, 69
Ignites in air, reacts violently and may explode with traces of water.

**POTASSIUM THALLIUM(I)AMIDE(2−) AMMONIATE** **$K_2NTl{\cdot}NH_3$**
**$K_2[TlN(NH_3)]$**

Alone,
or Acids,
or Water
Mellor, 1946, Vol. 5, 262
The amide solvated with 4, 2 or 1⅓ mol of ammonia explodes violently when subjected to shock, heat, dilute acids or water.
*See other N*-METAL DERIVATIVES

**POTASSIUM PEROXIDE** **$K_2O_2$**

Water
Pascal, 1963, Vol. 2.2, 59
Interaction is violent or explosive.
*See other* METAL PEROXIDES

**POTASSIUM DIPEROXOORTHOVANADATE(2−)** **$K_2O_6V$**
**$K_2[V(O_2)_2O_2]$**

Mellor, 1939, Vol. 9, 795

Emeléus, 1960, 425
Explodes on heating.
*See other* PEROXOACID SALTS

**POTASSIUM PEROXODISULPHATE(2–)** $K_2O_8S_2$

**$KOSO_2OOSO_2OK$**

Hydrazine salts,
Alkali

Mellor, 1947, Vol. 10, 466
Addition of alkali to the mixed salts liberates hydrazine which is vigorously oxidised to nitrogen gas.

Potassium hydroxide

*BCISC Quart. Safety Summ.*, 1965, **36**, 41
Surface contamination of 2 kg of the dry salt with as little as two flakes of moist potassium hydroxide caused a vigorous, self-sustaining fire, which was extinguished with water, but not by carbon dioxide or dry powder extinguishers.

Water

Castrantas, 1965, 5
The salt liberates oxygen rapidly above 100°C when dry, but at only 50°C when wet.
*See other* PEROXOACID SALTS

**POTASSIUM TETRAPEROXOTUNGSTATE(2–)** $K_2O_8W$

**$K_2[W(O_2)_4]$**

Mellor, 1943, Vol. 11, 836
Explodes on friction, or rapid heating to 80°C.
*See other* PEROXOACID SALTS

**POTASSIUM SULPHIDE** $K_2S$

Nitrogen oxide

*See* NITROGEN OXIDE, NO: Potassium sulphide

**POTASSIUM NITRIDE(3–)** $K_3N$

Mellor, 1940, Vol. 8, 99
Usually ignites in air.

Phosphorus

*See* PHOSPHORUS, P: Potassium nitride

Sulphur

*See* SULPHUR, S: Potassium nitride

*See other* *N*-METAL DERIVATIVES

**POTASSIUM 2,4,6-TRIS(DIOXOSELENA)PERHYDROTRIAZINE-1,3,5-TRIIDE ('POTASSIUM TRISELENIMIDATE')** **$K_3N_3O_6Se_3$**

**$KNSeO_2N(K)SeO_2N(K)SeO_2$**

Explosive.

*See* SELENIUM DIFLUORIDE DIOXIDE, $F_2O_2Se$: Ammonia

*See other* *N*-METAL DERIVATIVES

**TRIPOTASSIUM ANTIMONIDE** **$K_3Sb$**

1. Mellor, 1939, Vol. 9, 403
2. Rüst, 1948, 342

It usually ignites when broken in air [1] and explodes on exposure to moisture (breath) [2].

*See other* PYROPHORIC MATERIALS

**POTASSIUM HEXAOXOXENONATE(4−)–XENON TRIOXIDE**

**$K_4O_6Xe \cdot 2O_3Xe$**

**$K_4[XeO_6] \cdot 2XeO_3$**

1. Grosse, A. V., *Chem. Eng. News*, 1964, **42**(29), 42
2. Spittler, T. M. *et al.*, *J. Amer. Chem. Soc.*, 1966, **88**, 2942
3. Appelman, E. H. *et al.*, *J. Amer. Chem. Soc.*, 1964, **86**, 2141

The complex salt is very sensitive to mechanical shock and explodes violently [1,2], even when wet [3].

*See other* XENON COMPOUNDS

**LANTHANUM** **La**

Nitric acid

Mellor, 1946, Vol. 5, 603

Oxidation is violent.

Phosphorus

*See* PHOSPHORUS, P: Metals

*See other* METALS

## DILANTHANUM TRIOXIDE $La_2O_3$

Chlorine trifluoride

*See* CHLORINE TRIFLUORIDE, $ClF_3$: Metals, etc.

Water

Sidgwick, 1950, 441

Interaction is vigorously exothermic, accompanied by hissing, as for calcium oxide.

*See* CALCIUM OXIDE, CaO: Water

*See other* METAL OXIDES

## LITHIUM Li

Atmospheric gases

1. Mellor, 1941, Vol. 2, 468; 1961, Vol. 2, Suppl. 2.1, 71
2. *Haz. Chem. Data,* 1969, 141
3. Bullock, A., *Chem. Brit.*, 1975, **11**, 115
4. Bullock, A., *School Sci. Rev.*, 1975, **57**(119), 311–314; private comm., 1975
5. Lloyd, W. H. and Emley, E. F., *Chem. Brit.*, 1975, **11**, 334
6. Ireland, R. E. *et al., Org. Synth.*, 1977, **56**, 47

Finely divided metal may ignite in air at ambient temperature, and massive metal at temperatures above the m.p. (180°C), especially if oxide or nitride is present. Since lithium will burn in oxygen, nitrogen or carbon dioxide, and then reacts with sand, sodium carbonate, etc., it is difficult to extinguish once alight [1]. Molten lithium is very reactive, and will attack concrete and refractory materials. Use of normal fire extinguishers (containing water, foam, carbon dioxide, halocarbons, dry powders) will either accelerate combustion or cause explosion. Powdered graphite, lithium or potassium chlorides or zirconium silicate are suitable extinguishants [2].

A well-tried and usually uneventful demonstration of atmospheric oxidation of molten lithium led to an explosion [3]. A high degree of correlation of incidence of explosions with high atmospheric humidity was demonstrated, with the intensity of explosion apparently directly related to the purity of the sample of metal [4]. Other possible factors are also postulated [5]. While cleaning lithium wire by washing with hexane, the wire must be dried carefully with a paper towel. Too-vigorous rubbing will cause a fire [6].

*See* Metal chlorides, below
Water, below

Bromine pentafluoride

*See* BROMINE PENTAFLUORIDE, $BrF_5$: Acids, etc.

Bromobenzene

Koch-Light Laboratories Ltd., private comm., 1976

In a modified preparation of phenyllithium, bromobenzene was added to finely powdered lithium (rather than coarse particles) in ether. The reaction appeared to be proceeding normally, but after 30 min it became very vigorous and accelerated to explosion. It is thought that the powder may have been partially coated with oxide which abraded during stirring, exposing a lot of fresh metal surface on the powdered metal.

Chlorine tri- or penta-fluorides

*See* CHLORINE TRIFLUORIDE, $ClF_3$: Metals, etc.

Diazomethane

*See* DIAZOMETHANE, $CH_2N_2$: Alkali metals

Diborane

*See* DIBORANE, $B_2H_6$: Metals

Ethylene

Pascal, 1966, Vol. 2.1, 38

Passage of the gas over heated lithium causes the latter to incandesce, a mixture of lithium hydride and acetylide being produced.

*See* Metal chlorides, etc., below

Halocarbons

1. Staudinger, H., *Z. Elektrochem.*, 1925, **31**, 549
2. *Pot. Incid. Rep.*, 1968, **39**, ASESB, Washington
3. Pittwell, L. R., *J. R. Inst. Chem.*, 1959, **80**, 552
4. BDH Catalogue Safety Note 969DD/14.0/0773, 1973

Mixtures of lithium shavings and several halocarbon derivatives are impact-sensitive and will explode, sometimes violently [1,2]. Such materials include: bromoform, carbon tetra-bromide, -chloride and -iodide, chloroform, dichloro- and diiodo-methane, fluorotrichloromethane, tetrachloroethylene, trichloroethylene and 1,1,2-trichlorotrifluoroethane. In an operational incident, shearing samples off a lithium billet immersed in carbon tetrachloride caused an explosion and continuing combustion of the immersed metal [3].

Lithium which had been washed in carbon tetrachloride to remove traces of oil exploded when cut with a knife. Hexane is recommended as a suitable washing solvent [4].

*See* Poly-1,1-difluoroethylene–hexafluoropropylene (Viton), below

*See other* HALOCARBONS: Metals

Halogens

1. Staudinger, H., *Z. Elektrochem.*, 1925, **31**, 549
2. Mellor, 1961, Vol. 2, Suppl. 2.1, 82

Mixtures of lithium and bromine are unreactive unless subject to heavy impact, when explosion occurs [1]. Lithium and iodine react above 200°C with a large exotherm [2].

Mercury

Smith, G. McP. *et al., J. Amer. Chem. Soc.*, 1909, **31**, 799

Alexander, J. *et al., J. Chem. Educ.*, 1970, **47**, 277

Interaction to form lithium amalgam is violently exothermic and may be explosive if large pieces of lithium are used [1]. An improved technique, using *p*-cymene as inerting diluent, is described in the later reference [2].

Metal chlorides,
Nitrogen

*BCISC Quart. Safety Summ.*, 1969, **40**, 16

Accidental contamination of lithium strip with anhydrous chromium trichloride or zirconium tetrachloride caused it to ignite and burn vigorously in the nitrogen atmosphere in a glove-box.

*See* Atmospheric gases, above

Metal oxides and chalcogenides

1. Mellor, 1961, Vol. 2, Suppl. 2.1, 81–82
2. *Alkali Metals,* 1957, 11

Lithium is used to reduce metallic oxides in metallurgical operations, and the reactions, after initiation at moderate temperatures, are violently exothermic and rapid. Chromium(III) oxide reacts at 180°C, reaching 965°C; similarly molybdenum trioxide (180 to 1400°C), niobium pentoxide (320 to 490°C), titanium dioxide (200–400 to 1400°C), tungsten trioxide (200 to 1030°C), vanadium pentoxide (394 to 768°C); also iron(II) sulphide, (260 to 945°C) and manganese telluride, (230 to 600°C)[1]. Residual mixtures from lithium production cells containing lithium and rust sometimes ignite when left as thin layers exposed to air [2].

Metals

Bailar, 1973, Vol. 1, 343–344

Formation of various intermetallic compounds of lithium by melting with aluminium, bismuth, calcium, lead, mercury, silicon, strontium, thallium or tin may be very vigorous and dangerous to carry out.

*See* Mercury, above
Platinum, below

Nitric acid

*Accident & Fire Prevention Information,* Washington, USAEC, March, 1954

Lithium ignited on contact with nitric acid and the reaction became violent, ejecting burning lithium.

Nitryl fluoride

*See* NITRYL FLUORIDE, $FNO_2$: Metals

Non-metal oxides

Mellor, 1961, Vol. 2, Suppl. 2.1, 74, 84, 88

Although carbon dioxide reacts slowly with lithium at ambient temperature, the molten metal will burn vigorously in the gas, which cannot be used as an extinguisher on lithium fires. Carbon monoxide reacts in liquid ammonia to give the carbonyl, which reacts explosively with water or air. Lithium rapidly attacks silica or glass at 250°C.

Platinum

*See* PLATINUM; Pt: Lithium

Poly-1,1-difluoroethylene–hexafluoropropylene (Viton)

Markowitz, M. M. *et al., Chem. Eng. News,* 1961, **39**(32), 4

Mixtures of finely divided metal and shredded polymer ignited in air on contact with water or on heating to 369°C, or at 354°C under argon.

*See* Halocarbons, above

Sodium carbonate,
or Sodium chloride

Mellor, 1961, Vol. 2, Suppl. 2.1, 25

Sodium carbonate and chloride are unsuitable to use as extinguishants for lithium fires, since burning lithium will liberate the more reactive sodium in contact with them.

Sulphur

Mellor, 1961, Vol. 2, Suppl. 2.1, 75

Interaction when either is molten is very violent and, even in presence of an inert diluent, the reaction begins explosively. Reaction of sulphur with lithium dissolved in liquid ammonia at −33°C is also very vigorous.

Trifluoromethyl hypofluorite

*See* TRIFLUOROMETHYL HYPOFLUORITE, $CF_4O$: Lithium

Water

1. Mellor, 1961, Vol. 2, Suppl. 2.1, 72
2. Bailar, 1973, Vol. 1, 337
3. Markowitz, M. M., *J. Chem. Educ.*, 1963, **40**, 633–636

Reaction with cold water is of moderate vigour, but violent with hot water, and the liberated hydrogen may ignite [1]. The powdered metal reacts explosively with water [2]. The reactivity of lithium and other alkali metals with various forms of water has been discussed in detail. Prolonged contact with steam forms a thermally insulating layer which promotes overheating of the metal and may lead to a subsequent explosion as the layer breaks up [3].

*See* Atmospheric gases, above
ALKALI METALS

*See other* METALS

**LITHIUM–TIN ALLOYS** **Li–Sn**

*Alkali Metals,* 1957, 13

Lithium–tin alloys are pyrophoric.

*See other* ALLOYS

**LITHIUM SODIUM NITROXYLATE** **$LiNaNO_2$** **$LiNNaO_2$**

Mellor, 1961, Vol. 2, Suppl. 2.1, 78

Decomposes violently at 130°C.

*See other* METAL OXONON-METALLATES
*N*-METAL DERIVATIVES

**LITHIUM NITRATE** **$LiNO_3$**

Propene,
Sulphur dioxide

*See* PROPENE, $C_3H_6$: Lithium nitrate, etc.

**LITHIUM AZIDE** **$LiN_3$**

Mellor, 1940, Vol. 8, 345

Insensitive to shock, the moist or dry salt decomposes explosively at 115–298° C, depending on the rate of heating.

Alkyl nitrates,
Dimethylformamide

Polansky, O. E., *Angew. Chem. (Intern. Ed.),* 1971, **10**, 412

Although these reaction mixtures are stable at 25°C during preparation of *tert*-alkyl azides, above 200°C the mixtures are shock-sensitive and highly explosive.

*See other* METAL AZIDES

**LITHIUM NITRIDE** **$Li_3N$**

'Data Sheet TD 121', Exton, Foote Mineral Co., 1965

The finely divided powder may ignite if sprayed into moist air.

*See other* *N*-METAL DERIVATIVES

Acids

Mellor, 1940, Vol. 6, 170

The silicide incandesces in concentrated hydrochloric acid, and with dilute acid evolves silicon hydrides which ignite. It explodes with nitric acid and incandesces when floated on sulphuric acid.

Halogens

Mellor, 1940, Vol. 6, 169

When warmed, it ignites in fluorine, but higher temperatures are required with chlorine, bromine and iodine.

Non-metals

Mellor, 1940, Vol. 6, 169

Interaction with phosphorus, selenium or tellurium causes incandescence. Sulphur also reacts vigorously.

Water

Mellor, 1940, Vol. 6, 170

Reaction with water is very violent, and the silicon hydrides evolved ignite.

*See other* METAL NON-METALLIDES

**MAGNESIUM** **Mg**

1. *Haz. Chem. Data*, 1975, 189–190
2. *Dust Explosion Prevention: Magnesium Powder*, NFPA Standard Codes, 48, 1967; 652, 1968
3. Popov, E. I., *Chem. Abs.*, 1975, 82, 61415d

All aspects of prevention of magnesium dust explosions in storage, handling or processing operations are covered in two US Standard Codes [2]. Effects of various parameters on ignition of magnesium powders were studied [3].

Acetylenic compounds

*See* ACETYLENIC COMPOUNDS: Metals

Air

The rapid and brilliant combustion of magnesium powder, initiated in air at 520°C, was used formerly for photographic flash illumination. Very finely divided magnesium powder dispersed in air is a serious dust explosion hazard.

*See* DUST EXPLOSIONS

Aluminium,
Rusted steel,
Zinc

*See* ALUMINIUM–MAGNESIUM–ZINC ALLOYS, Al–Mg–Zn:
Rusted steel

Atmospheric gases,
Water

Darras, R. *et al., Chem. Abs.*, 1961, **55**, 13235a

The temperature at which massive magnesium and its alloys will ignite in air or carbon dioxide at 1 bar or 15 bar pressure depends on the heating programme and the presence or absence of moisture. Steady progressive heating caused ignition in carbon dioxide at 900° (780°C if moist), or at 800° after 4 h (650° C/40 min if moist), combustion being violent.

Barium carbonate,
Water

*MCA Case History No. 1849*

Fusion of the metal and salt formed barium acetylide, to which water was added without effective cooling. The vigorous evolution of acetylene blew off the reactor lid and the hot acetylene ignited in air.

Beryllium fluoride

Walker, H. L., *School Sci. Rev.*, 1954, **35**(127), 348

Interaction is violently exothermic.

Boron diiodophosphide

*See* BORONDIIODOPHOSPHIDE, $BI_2P$: Metals

Calcium carbonate,
Hydrogen

Mellor, 1940, Vol. 4, 271

Heating an intimate mixture of the powdered metal and carbonate in a stream of hydrogen leads to a violent explosion.

Carbon dioxide

Partington, 1967, 475

A mixture of solid carbon dioxide and magnesium burns very rapidly and brilliantly when ignited.

*See* Atmospheric gases, above
CARBON DIOXIDE, $CO_2$ : Metals

Chloroformamidinium nitrate

*See* CHLOROFORMAMIDINIUM NITRATE, $CH_4ClN_3O_3$

Ethylene oxide

*See* ETHYLENE OXIDE, $C_2H_4O$

Halocarbons

1. Clogston, C. C., *Bull. Res. Underwriters Lab.*, 1945, **34**, 5
2. *Pot. Incid. Rep.*, **39**, ASESB, Washington, 1968
3. Mond Div. ICI, private comm., 1968
4. Hartmann, I., *Ind. Eng. Chem.*, 1948, **40**, 756
5. *Haz. Chem. Data*, 1973, 140

Powdered metal reacts vigorously and may explode on contact with chloromethane, chloroform or carbon tetrachloride, or mixtures of these [1]. Mixtures of powdered magnesium with carbon tetrachloride or trichloroethylene will flash on heavy impact [2]. Violent decomposition with evolution of hydrogen chloride can occur when 1,1,1-trichloroethane comes into contact with magnesium or its alloys with aluminium [3]. Magnesium dust ignited at 400° C in dichlorodifluoromethane (used as an extinguishant for hydrocarbon fires) and suspensions of the dust exploded violently on sparking [4]. Interaction with 1,2-dibromoethane may become violent, and gives air-sensitive Grignard compounds [5].

*See* Polytetrafluoroethylene, below
BROMOMETHANE, $CH_3Br$: Metals

Halogens,
or Interhalogens

Mellor, 1940, Vol. 4, 267

It ignites if moist and burns violently in fluorine, and ignites in moist or warm chlorine. It burns not very readily in bromine vapour and may ignite if finely divided on heating in iodine vapour.

*See* CHLORINE TRIFLUORIDE, $ClF_3$
IODINE HEPTAFLUORIDE, $F_7I$
IODINE, $I_2$: Metals, Water

Hydrogen iodide

Mellor, 1940, Vol. 2, 206

Contact causes momentary ignition.

Metal cyanides

Mellor, 1940, Vol. 4, 271

Magnesium reacts with incandescence on heating with cadmium, cobalt, copper, lead, nickel and zinc cyanides. With gold and mercury cyanide, the cyanogen released by thermal decomposition of these salts reacts explosively with magnesium.

Metal oxides

1. Mellor, 1941, Vol. 3, 138, 378; 1940, Vol. 4, 272; 1941, Vol. 7, 401
2. Stout, E. L., *Chem. Eng. News,* 1958, **3**(8), 64

Magnesium will violently reduce metal oxides on heating, similarly to aluminium powder in the 'thermite' reaction. Beryllium, cadmium, copper, mercury, molybdenum, tin and zinc oxides are all reduced explosively on heating. Silver oxide reacts with explosive violence when heated with magnesium powder in a sealed tube [1]. Interaction of molten magnesium and iron oxide scale is violent[2].

Metal oxosalts

Mellor, 1940, Vol. 4, 272; Vol. 8, Suppl. 2.1, 545

Pieters, 1957, 30

Interaction with fused ammonium nitrate or with metal nitrates, phosphates or sulphates may be explosively violent. Lithium and sodium carbonates also react vigorously.

*See* SILVER NITRATE, $AgNO_3$: Magnesium
POTASSIUM PERCHLORATE, $ClKO_4$: Metal powders

Methanol

Personal experience

Vogel, 1957, 169

Shidlovskii, A. A., *Chem. Abs.*, 1947, **41**, 1105d

The reaction of magnesium and methanol to form magnesium methoxide and used to prepare dry methanol is very vigorous, but often subject to a lengthy induction period. Sufficient methanol must be present to absorb the violent exotherm which sometimes occurs.

Mixtures of powdered magnesium (or aluminium) and methanol are capable of detonation and more powerful than military explosives.

*See also* Water, below

Oxidants

Mellor, 1940, Vol. 2, 90; Vol. 4, 270–272

Magnesium is a powerful reducing agent and undergoes violent or explosive reactions with a variety of oxidants, particularly when powdered.

*See* Carbon dioxide, above
Halogens, or Interhalogens, above
Metal oxides, above
Metal oxosalts, above
Sulphur, etc., below
BARIUM PEROXIDE, $BaO_2$: Metals
POTASSIUM CHLORATE, $ClKO_3$: Metals
POTASSIUM PERCHLORATE, $ClKO_4$: Powdered metals
NITRIC ACID, $HNO_3$: Metals
HYDROGEN PEROXIDE, $H_2O_2$: Metals
AMMONIUM NITRATE, $H_4N_2O_3$: Metals
SODIUM IODATE, $INaO_3$: Metals
SODIUM NITRATE, $NNaO_3$: Magnesium
DINITROGEN TETRAOXIDE, $N_2O_4$: Metals
SODIUM PEROXIDE, $Na_2O_2$: Metals

OXYGEN (Liquid), $O_2$: Metals
LEAD DIOXIDE, $O_2Pb$: Metals

Polytetrafluoroethylene
Anon., *Indust. Res.*, 1968(9),15
Finely divided magnesium and Teflon are described as a hazardous combination of materials.
*See* Halocarbons, above

Potassium carbonate
Winckler, C., *Ber.*, 1890, **23**, 44
The mixture of magnesium and potassium carbonate recommended by Castellana as a safer substitute for molten sodium in the Lassaigne test can itself be hazardous, as an equimolar mixture gives an explosive substance (probably carbonylpotassium) on heating.

Silicon dioxide
Barker, W. B., *School Sci. Rev.*, 1938, **20**(77), 150
Marle, E. R., *Nature*, 1909, **82**, 428
Heating a mixture of powdered magnesium and silica (later found not to be absolutely dry) caused a violent explosion rather than the vigorous interaction expected. A warning had been published previously.

Sulphur,
or Tellurium
Brauer, 1963, Vol. 1, 913
Hutton, K., *School Sci. Rev.*, 1950, **31**(114), 265
Interaction may be very violent or explosive at elevated temperatures.
*See* SULPHUR, S: Metals

Water
Shidlovskii, A. A., *Chem. Abs.*, 1947, **41**, 1105d
Mixtures of magnesium (or aluminium) powder with water can be caused to explode powerfully by initiation with a boosted detonator.
*See also* Methanol, above
*See other* METALS
REDUCANTS

**MAGNESIUM NITRATE** **$Mg(NO_3)_2$** **$MgN_2O_6$**

*N,N*-Dimethylformamide
'DMF Brochure', Billingham, ICI Ltd., 1965
Magnesium nitrate has been reported to undergo spontaneous decomposition in DMF. Although this effect has not been observed with other nitrates, such reactions should be treated with care.
*See other* METAL OXONON-METALLATES

## MAGNESIUM OXIDE $MgO$

Interhalogens

*See* BROMINE PENTAFLUORIDE, $BrF_5$: Acids, etc.
CHLORINE TRIFLUORIDE, $ClF_3$: Metals, etc.

Phosphorus pentachloride

*See* PHOSPHORUS PENTACHLORIDE, $Cl_5P$: Magnesium oxide

## MAGNESIUM SULPHATE $MgSO_4$ $MgO_4S$

Ethoxyethynyl alcohols

*See* ETHOXYETHYNYL ALCOHOLS

## MAGNESIUM SILICIDE $Mg_2Si$

Acids

Sorbe, 1968, 84

Contact with moisture under acidic conditions generates silanes which ignite in air.

*See other* METAL NON-METALLIDES

## TRIMAGNESIUM DIPHOSPHIDE $Mg_3P_2$

Oxidants

Mellor, 1940, Vol. 8, 842

It ignites on heating in chlorine, or in bromine and iodine vapours at higher temperatures. Reaction with nitric acid causes incandescence.

Water

Mellor, 1940, Vol. 8, 842

Phosphine is evolved and may ignite.

*See other* METAL NON-METALLIDES

## MANGANESE $Mn$

Carbon dioxide

*See* CARBON DIOXIDE, $CO_2$: Metals

Oxidants

1. *Occup. Hazards,* 1966, 28(11), 44
2. Mellor, 1942, Vol. 12, 186–188

The finely divided metal is pyrophoric, and a mixture of manganese and aluminium dusts accidentally released into air from a filter bag exploded violently [1]. Powdered manganese ignites and becomes incandescent in fluorine or on warming in chlorine. Contact with concentrated hydrogen peroxide causes violent decomposition and/or ignition, and with nitric acid incandescence and a feeble explosion were observed. Contact with dinitrogen tetraoxide caused ignition [2].

*See* BROMINE PENTAFLUORIDE, $BrF_5$: Acids, etc.
NITRYL FLUORIDE, $FNO_2$: Metals
AMMONIUM NITRATE, $H_4N_2O_3$: Metals

Phosphorus

*See* PHOSPHORUS, P: Metals

Sulphur dioxide

Mellor, 1942, Vol. 12, 187

Pyrophoric manganese burns brilliantly on warming in sulphur dioxide.

*See other* METALS

## MANGANESE(II) NITRATE $Mn(NO_3)_2$ $MnN_2O_6$

Urea

Novikov, A. V. *et al., Chem. Abs.*, 1975, 82, 48151v

The anhydrous complex finally decomposes at 240°C with a light explosion.

*See related* AMMINEMETAL OXOSALTS

## SODIUM PERMANGANATE $NaMnO_4$ $MnNaO_4$

Acetic acid,
or Acetic anhydride

*See* POTASSIUM PERMANGANATE, $KMnO_4$: Acetic acid

## MANGANESE(II) OXIDE $MnO$

Hydrogen peroxide

*See* HYDROGEN PEROXIDE, $H_2O_2$: Metals, etc.

## MANGANESE(IV) OXIDE $MnO_2$

Aluminium

Sidgwick, 1950, 1265

Interaction on heating is very violent.

*See* ALUMINIUM, Metal oxides

Calcium hydride

Pascal, 1958, Vol. 4, 304

Interaction becomes incandescent on warming.

Diboron tetrafluoride

*See* DIBORON TETRAFLUORIDE, $B_2F_4$ : Metal oxides

Dirubidium acetylide

*See* DIRUBIDIUM ACETYLIDE, $C_2Rb_2$ : Metal oxides

Hydrogen sulphide

*See* HYDROGEN SULPHIDE, $H_2S$: Metal oxides

Oxidants

1. Mellor, 1942, Vol. 12, 254
2. Mellor, 1941, Vol. 1, 936
3. Ahrle, H., *Z. Angew. Chem.*, 1909, 22, 1713
4. von Schwartz, 1918, 323
5. Blanco Prieto, J., *Chem. Abs.*, 1976, **84**, 46791

Action of chlorine trifluoride causes incandescence [1]. Manganese dioxide catalytically decomposes powerful oxidising agents, often violently. Dropped into concentrated hydrogen peroxide, the powdered oxide may cause explosion [2]. Either the massive or powdered oxide explosively decomposes 92% peroxomonosulphuric acid [3], and mixtures with chlorates (heated to generate oxygen) may react with explosive violence [4].

Cuban pyrolusite can be used in place of potassium dichromate to promote thermal decomposition of potassium chlorate in match-head formulations.

*See* PEROXYFORMIC ACID, $CH_2O_3$ : Metals, etc.

Potassium azide

Mellor, 1940, Vol. 8, 347

On gentle warming, interaction is violent.

*See other* METAL OXIDES
OXIDANTS

**MANGANESE(II) SULPHIDE** **MnS**

Mellor, 1942, Vol. 12, 394

The vacuum-dried red sulphide becomes red-hot on exposure to air.

*See other* METAL SULPHIDES

**MANGANESE(IV) SULPHIDE** $MnS_2$

Sorbe, 1968, 84

It explodes at 580°C.

*See other* METAL SULPHIDES

**MANGANESE(II) TELLURIDE** MnTe

Lithium

*See* LITHIUM, Li: Metal oxides, etc.

**MANGANESE(VII) OXIDE** $Mn_2O_7$

Mellor, 1942, Vol. 12, 293

Bailar, 1973, Vol. 3, 805

It explodes at between 40 and 70°C, or on friction or impact, sensitivity being as great as that of mercury fulminate.

Detonation occurs at 95°C, and under vacuum explosive decomposition occurs above 10°C.

*See* POTASSIUM PERMANGANATE, $KMnO_4$: Sulphuric acid

Organic material

1. Patterson, A. M., *Chem. Eng. News*, 1948, **26**, 711
2. Mellor, 1940, Vol. 12, 293

A sample of the heptoxide exploded in contact with the grease on a stopcock [1], and explosion or ignition had been noted with various solvents, oils, fats and fibres [2].

*See other* OXIDANTS

**ZINC PERMANGANATE** $Zn(MnO_4)_2$ $Mn_2O_8Zn$

Cellulose

Soergel, U. C., *Pharm. Praxis, Beil. Pharmaz.*, 1960, (2), 30

An explosion involving contact of the permanganate with cellulose is described.

*See other* METAL OXOMETALLATES

**TRIMANGANESE DIPHOSPHIDE** $Mn_3P_2$

Chlorine

*See* CHLORINE, $Cl_2$: Phosphorus compounds

## MOLYBDENUM $Mo$

Oxidants

*See* BROMINE TRIFLUORIDE, $BrF_3$: Halogens, etc.
BROMINE PENTAFLUORIDE, $BrF_5$: Acids, etc.
CHLORINE TRIFLUORIDE, $ClF_3$: Metals
POTASSIUM PERCHLORATE, $ClKO_4$: Metal powders
NITRYL FLUORIDE, $FNO_2$: Metals
FLUORINE, $F_2$: Metals
IODINE PENTAFLUORIDE, $F_5I$: Metals
SODIUM PEROXIDE, $Na_2O_2$: Metals
LEAD DIOXIDE, $O_2Pb$: Metals

## SODIUM TETRAPEROXOMOLYBDATE(2–) $MoNa_2O_8$

$Na_2[Mo(O_2)_4]$

Castrantas, 1965, 5
Decomposes explosively under vacuum.
*See other* PEROXOACID SALTS

## MOLYBDENUM(IV) OXIDE $MoO_2$

Gibson, 1969, 102
It becomes incandescent in contact with air.
*See other* PYROPHORIC MATERIALS

## MOLYBDENUM(VI) OXIDE $MoO_3$

Interhalogens

*See* BROMINE PENTAFLUORIDE, $BrF_5$: Acids, etc.
CHLORINE TRIFLUORIDE, $ClF_3$: Metals, etc.

Metals

Mellor, 1943, Vol. 11, 542
Reduction of the oxide by heated sodium or potassium proceeds with incandescence, and explosion occurs in contact with molten magnesium.
*See* LITHIUM Li: Metal oxides
*See other* METAL OXIDES

## OXODIPEROXOMOLYBDENUM–HEXAMETHYLPHOSPHORAMIDE

$MoO_5 \cdot C_6H_{18}N_3OP$

$MoO(O_2)_2 \cdot PO(NMe_2)_3$

Fieser, 1974, Vol. 4, 203
The monohydrated complex exploded after storage at ambient

temperature for a month. Refrigerated storage is recommended.

*See related* METAL PEROXIDES

## MOLYBDENUM(IV) SULPHIDE $MoS_2$

Potassium nitrate

*See* POTASSIUM NITRATE, $KNO_3$ : Metal sulphides

## SODIUM NITRITE $NaNO_2$ $NNaO_2$

Amidosulphates

Heubel, J. *et al., Compt. Rend.*, 1963, **257**, 684–686

Interaction of nitrites when heated with metal amidosulphates ('sulphamates') may become explosively violent owing to liberation of nitrogen and steam. Mixtures with ammonium sulphamate form ammonium nitrite which decomposes violently around 80°C.

*See also* SODIUM NITRATE, $NNaO_3$ : Metal sulphamates

*See other* OXIDANTS

Aminoguanidine salts

Urbanski, 1967, Vol. 3, 207

Interaction, without addition of acid, produces tetrazolylguanyltriazene ('tetrazene'), a primary explosive of equal sensitivity to mercury(II) azide, but more readily initiated.

Ammonium salts

1. von Schwartz, 1918, 299
2. Mellor, 1967, Vol. 8, Suppl. 2, 388
3. *RoSPA Occup. Safety and Health Suppl.*, 1972, 2(10), 3

Heating a mixture of an ammonium salt with a nitrite salt causes a violent explosion on melting [1], due to formation and decomposition of ammonium nitrite. Salts of other nitrogenous bases behave similarly. Mixtures of ammonium chloride and sodium nitrite are used as commercial explosives [2]. Accidental contact of traces of ammonium nitrate with sodium nitrite residues caused wooden decking on a lorry to ignite [3].

*See* Wood, below
NITRITE SALTS OF NITROGENOUS BASES

1,3-Butadiene

Beer, R. N., private comm., 1972

Keister, R. G. *et al., Loss Prev.*, 1971, **5**, 69

Sodium nitrite solution is used to inhibit 'popcorn' polymerisation of butadiene in processing plants. If concentrated nitrite solutions (5%)

are used, a black sludge is produced which, when dry, will ignite and burn when heated to 150°C, even in absence of air.

The sludge produced from use of nitrite solutions to scavenge oxygen in butadiene distillation systems contained 80% of organic polymer, and a nitrate:nitrite ratio of 2:1. Use of dilute nitrite solution (0.5%) or pH above 8 prevents sludge formation.

*See* BUTEN-3-YNE, $C_4H_4$ : Butadiene

*See also* NITROGEN OXIDE, NO: Dienes

Metal cyanides

1. von Schwartz, 1918, 299; Mellor, 1940, Vol. 8, 478
2. Greenwood, P. H. S., *J. and Proc. R. Inst. Chem.*, 1947, 137
3. Elson, C. H. R., ibid., 19
4. Eiter, K. *et al.*, Austrian Pat. 176 784, 1953

Mixtures of sodium (or other) nitrites and various cyanides explode on heating, including potassium cyanide [1] or hexacyanoferrate (3–) or sodium pentacyanonitrosylferrate(2–), potassium hexacyanoferrate (4–) [3] or mercury(II) cyanide [4]. Such mixtures have been proposed as explosives, initiable by heating or detonator [4].

Phenol

Hatton, J. P., private comm., 1976

A mixture exploded violently on heating in a test tube.

Phthalic acid,
Phthalic anhydride

Hawes, B. V. W., *J. R. Inst. Chem.*, 1955, **79**, 668

Mixtures of sodium nitrite and phthalic acid or anhydride explode violently on heating. A nitrite ester may have been produced.

*See* ACYL NITRITES

Reducants

*See* Sodium disulphite, below
Sodium thiosulphate, below
SODIUM NITROXYLATE(2–), $NNa_2O_2$

Sodium amide

Bergstrom, F. W. *et al., Chem. Rev.*, 1933, **12**, 64

Addition of solid nitrate to the molten amide causes immediate gas evolution, followed by a violent explosion.

Sodium disulphite

*MCA Case History No. 183*

Large-scale addition of solid disulphite to an unstirred and too-concentrated solution of nitrite caused a vigorous exothermic reaction.

Sodium thiocyanate
Mellor, 1940, Vol. 8, 478
A mixture explodes on heating.

Sodium thiosulphate
Mellor, 1940, Vol. 8, 478; 1947, Vol. 10, 501
There is no interaction between solutions, but evaporation of the mixture gives a residue which explodes on heating. The mixed solids behave similarly.

Urea
Bucci, F., *Ann.Chim. Rome,* 1951, **41**, 587
Fusion of urea (2 mol) with sodium (or potassium) nitrite (1 mol) to give high yields of the cyanate must be carried out exactly as described to avoid the risk of explosion.

Wood
*ABCM Quart. Safety Summ.*, 1944, **15**, 30
Wooden staging, which had become impregnated over a number of years with sodium nitrite, became accidentally ignited and burned as fiercely as if impregnated with potassium chlorate. Although the effect of impregnating cellulosic material with sodium nitrate is well known, that due to nitrite was unexpected.
*See* Ammonium salts, above
*See other* METAL OXONON-METALLATES
OXIDANTS

## SODIUM NITRATE $NaNO_3$ $NNaO_3$

Acetic anhydride
*See* ACETIC ANHYDRIDE, $C_4H_6O_3$. Metal salts

Aluminium,
or Aluminium oxide
Anon., *Fire*, 1935, 28, 30
Farnell, P. L. *et al., Chem. Abs.*, 1972, **76**, 129564k
Mixtures of the nitrate with the powdered metal or oxide were reported to be explosive, and the performance characteristics of flares containing compressed mixtures of the metal and nitrate have been described.

Aluminium,
Water
Jackson, B. *et al., Chem. Abs.*, 1976, **84**, 46798d
During investigation of pyrotechnic flare formulations, it was found

that mixtures of the metal powder and oxidant undergo a low-temperature exothermic reaction at 70–135°C in presence of moisture.

Antimony
Mellor, 1939, Vol. 9, 382
Powdered antimony explodes when heated with an alkali metal nitrate.

Barium thiocyanate
Pieters, 1957, 30
Mixtures may explode.

Boron phosphide
Mellor, 1940, Vol. 8, 845
Deflagration occurs in molten alkali nitrates.

Calcium–silicon alloy
Smolin, A. O., *Chem. Abs.*, 1974, **81**, 138219g
Combustion of the alloy in admixture with sodium nitrate is mentioned in an explosives context, but no details are translated.

Diarsenic trioxide,
Iron(II) sulphate
Dunn, B. W., *Bur. Expl. Rept. 13*, Pennsylvania, Amer. Rail Assoc., 1920
A veterinary preparation containing the oxidant and reducant materials ignited spontaneously.
*See related* REDOX COMPOUNDS

Fibrous material
1. Mellor, 1961, Vol. 2, Suppl. 2.1, 1244
2. *ABCM Quart. Safety Summ.*, 1944, **15**, 30
Fibrous organic material (jute storage bags) is oxidised in contact with sodium nitrate above 160°C, and will ignite at or below 220°C[2]. Wood and similar cellulosic material are rendered highly combustible by nitrate impregnation [2].

Jute,
Magnesium chloride
van Hoogstraten, C. W., *Chem. Abs.*, 1947, **41**, 3629b
Solid crude sodium nitrate packed in jute bags sometimes ignited the latter in storage. Normally ignition did not occur below 240°C, but in cases where magnesium chloride was present (up to 16%), ignition occurred at 130°C. This was attributed to formation of magnesium nitrate hexahydrate, which hydrolyses above its m.p. (90°C) liberating nitric acid. The latter was thought to have caused ignition of the jute under unusual conditions of temperature and friction.

Magnesium

Bond, B. D. *et al., Combust. Flame*, 1966, **10**, 349–354

A study of the kinetics in attack of magnesium by the molten nitrate indicates that decomposition of the nitrate releases oxygen atoms which oxidise the metal so exothermically that ignition ensues.

*See* MAGNESIUM, Mg: Metal oxosalts

Metal amidosulphates

Heubel, J. *et al., Compt. Rend.*, 1962, **255**, 708–709

Interaction of nitrates when heated with amidosulphates ('sulphamates') may become explosively violent owing to the liberation of dinitrogen oxide and steam.

*See also* SODIUM NITRITE, $NNaO_2$: Amidosulphates

Metal cyanides

*See* MOLTEN SALT BATHS

Non-metals

Mellor, 1941, Vol. 2, 820

Contact of powdered charcoal with the molten nitrate, or of the solid nitrate with glowing charcoal, causes vigorous combustion of the carbon. Mixtures with charcoal and sulphur have been used as black powder.

*See* POTASSIUM NITRATE, $KNO_3$: Non-metals

Peroxyformic acid

*See* PEROXYFORMIC ACID, $CH_2O_3$: Sodium nitrate

Phenol,
Trifluoroacetic acid

Spitzer, U. A. *et al., J. Org. Chem.*, 1974, **39**, 3936

When the salt–acid nitration mixture was applied to phenol, a potentially hazardous rapid exothermic reaction occurred producing tar.

*See other* NITRATION INCIDENTS

Sodium

Mellor, 1961, Vol. 2, Suppl. 2.1, 518

Interaction of sodium nitrate and sodium alone, or dissolved in liquid ammonia, eventually gives a yellow explosive compound.

*See* SODIUM NITROXYLATE(2–), $NNa_2O_2$

Sodium phosphinate

Mellor, 1941, Vol. 2, 820

Rüst, 1947, 337

A mixture exploded violently on warming.

Sodium thiosulphate

Mellor, 1941, Vol. 2, 820

A mixture is explosive when heated.

Wood

*See* Fibrous material, above

*See other* METAL OXONON-METALLATES

OXIDANTS

**SODIUM NITROXYLATE (2–)** **$Na_2NO_2$** **$NNa_2O_2$**

Mellor, 1963, Vol. 2, Suppl. 2.2, 1566

The solid is very reactive towards air, moisture and carbon dioxide, and tends to explode readily, also decomposing violently on heating. It is produced from sodium nitrite or nitrate by electrolytic reduction, or action of sodium alone or in liquid ammonia solution.

*See other N*-METAL DERIVATIVES

**SODIUM NITRIDE(3–)** **$Na_3N$** **$NNa_3$**

Fischer, F. *et al., Ber.*, 1910, **43**, 1468

It decomposes explosively on gentle warming.

*See other N*-METAL DERIVATIVES

**NITROGEN OXIDE ('NITRIC OXIDE')** **NO**

*UK Sci. Mission Rep. 68/79,* Washington, UKSM, 1968

Miller, R. O., *Ind. Eng. Chem., Proc. Res. Dev.*, 1968, 7, 590–593

Ribovich, J. *et al., J. Haz. Mat.*, 1977, **1**, 275–287

Liquid nitrogen oxide and other cryogenic oxidisers (ozone, fluorine in presence of water) are very sensitive to detonation in absence of fuel and can be initiated as readily as glyceryl nitrate.

Detonation of the liquid oxide close to its b.p. (–152°C) generated a 100 kbar pressure pulse and fragmented the test equipment. It is the simplest molecule that is capable of detonation in all three phases.

Alkenes,
Oxygen

*See* NITROGEN DIOXIDE, $NO_2$ : Alkenes

Carbon,
Potassium hydrogentartrate

Pascal, P. *et al., Mém. Poudres*, 1953, **35**, 335–347

The ignition temperature of 400°C for carbon black in nitrogen oxide was reduced to 100°C by the presence of the tartrate.

Carbon disulphide

Winderlich, R., *J. Chem. Educ.*, 1950, **27**, 669

A demonstration of combustion of carbon disulphide in nitrogen oxide exploded violently.

Dichlorine oxide

*See* DICHLORINE OXIDE, $Cl_2O$: Nitrogen oxide

Dienes,
Oxygen

1. Haseba, S. *et al.*, *Chem. Eng. Prog.*, 1966, **62**(4), 92
2. Schuftan, P. M., *Chem. Eng. Prog.*, **62**(7), 8

Violent explosions which occurred at –100 to –180°C in ammonia synthesis gas units were traced to the formation of explosive addition products between dienes and oxides of nitrogen, produced from interaction of nitrogen oxide and oxygen. Laboratory experiments showed that the addition products from 1,3-butadiene or cyclopentadiene formed rapidly at about –150°C, and ignited or exploded on warming to –35 to –15°C. The unconjugated propadiene, and alkenes, or acetylene reacted slowly and the products did not ignite until +30 to 50°C [1]. This type of derivative ('pseudo-nitrosite') was formerly used (Wallach) to characterise terpene hydrocarbons. Further comments were made later [2].

*See also* SODIUM NITRITE, $NNaO_2$ : Butadiene

Fluorine

*See* FLUORINE, $F_2$: Oxides

Hydrogen,
Oxygen

Chanmugam, J. *et al.*, *Nature*, 1952, **170**, 1067

Pre-addition of nitrogen oxide (or nitrosyl chloride as its precursor) to stoicheiometric hydrogen–oxygen mixtures at 240 mbar/360°C will cause immediate ignition under a variety of circumstances.

Metal acetylides or carbides

Mellor, 1946, Vol. 5, 848, 891

Dirubidium acetylide ignites on heating and uranium dicarbide incandesces in the gas at 370°C.

Metals

Mellor, 1940, Vol. 8, 436; 1943, Vol. 11, 162; 1942, Vol. 12, 32

Pyrophoric chromium attains incandescence in the oxide, while calcium, potassium and uranium need heating before ignition occurs, when combustion is brilliant.

Methanol

Partel, G., *Riv. Ing.*, 1965, **15**, 969–976

This oxidant–fuel combination was evaluated as a rocket propellant system.

Nitrogen trichloride

*See* NITROGEN TRICHLORIDE, $Cl_3N$: Initiators

Non-metals

Mellor, 1940, Vol. 8, 109, 433, 435

Amorphous (not crystalline) boron reacts with brilliant flashes at ambient temperature, and charcoal and phosphorus continue to burn more brilliantly than in air.

Ozone

*See* OZONE, $O_3$: Nitrogen oxide

Pentacarbonyliron

Manchot, W. *et al.*, *Ann.*, 1929, **470**, 275

Rapid heating to above 50°C in an autoclave caused an explosive reaction.

Perchloryl fluoride

*See* PERCHLORYL FLUORIDE, $ClFO_3$: Hydrocarbons, etc.

Phosphine,
Oxygen

Mellor, 1940, Vol. 8, 435

Addition of oxygen to a mixture of phosphine and nitrogen oxide causes ignition.

*See also* PHOSPHINE, $H_3P$: Oxygen

Potassium sulphide

Mellor, 1940, Vol. 8, 434

The pyrophoric sulphide ignites in the oxide.

*See other* NON-METAL OXIDES
OXIDANTS

**NITROGEN DIOXIDE** **$NO_2$**

Alkenes

Rozlovskii, A. I., *Chem. Abs.*, 1975, 83, 118122h

The mechanisms of explosions in solidified gas mixtures containing unsaturated hydrocarbons and oxides of nitrogen is discussed. Fast radical addition of nitrogen dioxide to double bonds is involved, and with dienes it is a fast reaction of very low activation energy. Possibilities of preventing explosions are discussed.

*See* NITROGEN OXIDE ('NITRIC OXIDE'), NO: Dienes, Oxygen

*See other* OXIDANTS

**PLUTONIUM NITRIDE** **PuN** **NPu**

Oxygen,
Water

Bailar, Vol. 5, 343

The ignition temperature of the nitride in oxygen is lowered from 300° to below 100°C by the presence of moisture.

*See other* *N*-METAL DERIVATIVES

**RUBIDIUM NITRIDE(3–)** **$Rb_3N$** **$NRb_3$**

Mellor, 1940, Vol. 8, 99

The alkali nitrides burn in air.

*See other* *N*-METAL DERIVATIVES

**ANTIMONY(III) NITRIDE** **SbN** **NSb**

Alone,
or Water

1. Fischer, F. *et al.*, *Ber.*, 1910, **43**, 1471
2. Franklin, E. C., *J. Amer. Chem. Soc.*, 1905, **27**, 850

Explosive decomposition occurs on warming *in vacuo* [1] and impure material explodes mildly on heating in air, or on contact with water or dilute acids [2].

*See other* *N*-METAL DERIVATIVES

**THALLIUM(I) NITRIDE** **$Tl_3N$** **$NTl_3$**

Alone,
or Acids,
or Water

Mellor, 1946, Vol. 5, 262

The nitride explodes violently on exposure to shock, heat, water or dilute acids.

*See other* *N*-METAL DERIVATIVES

**URANIUM(III) NITRIDE** **UN** **NU**

Bailar, 1973, Vol. 5, 340

Finely powdered material is pyrophoric in air at ambient temperature.

*See other* *N*-METAL DERIVATIVES
PYROPHORIC MATERIALS

**NITROGEN (Gas)** $N_2$

Lithium

*See* LITHIUM, Li: Atmospheric gases
: Metal chlorides, etc.

Ozone

*See* OZONE, $O_3$: Nitrogen

Titanium

*See* TITANIUM, Ti: Nitrogen

**NITROGEN (Liquid)**

*BCISC Quart. Safety Summ.*, 1964, **35**, 25; 1974, **45**, 11

Although liquid nitrogen is inherently safer than liquid oxygen as a coolant, its ability to condense liquid oxygen out of the atmosphere can create hazards. A distillation tube containing a little solvent was cooled in liquid nitrogen while being sealed off in a blowpipe flame. A few minutes later the tube exploded violently, probably owing to high internal pressure caused by evaporation of liquid oxygen which had condensed into the tube during sealing. It is possible that the atmosphere close to the blowpipe was oxygen-enriched.

Open vessels which contain organic materials, or which are to be hermetically sealed, should not be cooled in liquid nitrogen, but in a coolant at a higher temperature. Liquid nitrogen should normally be used only for cooling evacuated or closed vessels where extreme cooling is necessary, and removed from around such vessels before opening them to atmosphere.

In two incidents involving forceful explosion of 75 litre cryogenic containers of liquid nitrogen, the cause of the first appeared to be blockage of the 19 mm neck by an ice plug formed in a very humid atmosphere. The cause of the second was probably 'roll-over' of a surface layer of superheated liquid, and fracture of the vessel by the ensuing violent boil-off of gas. Superheating of 10–15°C is known to occur in undisturbed high-purity liquid nitrogen, but much higher levels of superheating may be possible. Various remedies (stirring, generation of a warm-spot, or submerged conductive gauzes) are discussed.

Hydrogen

*See* HYDROGEN (Gas), $H_2$ : Liquid nitrogen

Oxygen,
Radiation

1. Chen, C. W. *et al.*, *Cryogenics*, 1969, **9**, 131–132
2. Takehisa, M. *et al.*, *Chem. Abs.*, 1977, **87**, 30593j
3. Perdue, P. T., *Health Phys.*, 1972, **23**, 116–117

Liquid nitrogen subject to nuclear radiation (high neutron and gamma fluxes) must be kept free of oxygen to prevent explosions occurring in reactor cryostats. The explosive species generated is not NO or $NO_2$, because solutions of these will not cause explosion on contact with a drop of acrylonitrile, whereas irradiated oxygen-containing nitrogen does. Ozone was thought to be responsible [1], and this has now been confirmed by a detailed experimental investigation [2], though trace organic impurities may also have been involved [1]. The subject has been reviewed, and oxygen must also be eliminated from liquid helium, hydrogen or noble gases before irradiation [3].

*See other* CRYOGENIC LIQUIDS

**SODIUM HYPONITRITE** **NaON=NONa** **$N_2Na_2O_2$**

*See* HYPONITROUS ACID, $H_2N_2O_2$
SODIUM AMIDE, $H_2NNa$: Air

**SODIUM TRIOXODINITRATE(2−)** **NaON(O)=NONa** **$N_2Na_2O_3$**

*See* SODIUM AMIDE, $H_2NNa$: Air

**SODIUM TETRAOXODINITRATE(2−)** **$N_2Na_2O_4$**
**NaON(O)=N(O)ONa**

*See* SODIUM AMIDE, $H_2NNa$: Air

**SODIUM PENTAOXODINITRATE(2−)** **$N_2Na_2O_5$**
**NaON($O_2$)=N(O)ONa**

*See* SODIUM AMIDE, $H_2NNa$: Air

**SODIUM HEXAOXODINITRATE(2−)** **$N_2Na_2O_6$**
**NaON($O_2$)=N($O_2$)ONa**

*See* SODIUM AMIDE, $H_2NNa$: Air

**DINITROSYLNICKEL** **$(NO)_2Ni$** **$N_2NiO_2$**

Addison, C. C. *et al.*, *Quart. Rev.*, 1955, **9**, 138

The blue solid (probably polymeric) decomposes violently above 60°C.
*See other* NITROSO COMPOUNDS

### NITRITONITROSYLNICKEL NONiONO $N_2NiO_3$

Bailar, 1973, Vol. 3, 1119
It is probably polymeric and ignites on contact with water.
*See other* NITROSO COMPOUNDS

### DINITROGEN OXIDE ('NITROUS OXIDE') $N_2O$

Preparative hazard
*See* AMMONIUM NITRATE, $H_4N_2O_3$ (reference 9)

Rüst, 1948, 278
During transfer of the liquefied gas from a stock steel cylinder into smaller cylinders, the effective expansion caused cooling of the stock cylinder, and a fall in pressure to occur. Application of a flame to the stock cylinder of the endothermic oxide led to decomposition and explosive rupture of the cylinder.
*See other* ENDOTHERMIC COMPOUNDS

Boron
Mellor, 1947, Vol. 8, 109
Amorphous boron (but not crystalline) ignites on heating in the dry oxide.

Carbon monoxide
Wilton, C. *et al., Chem. Abs.*, 1975, **82**, 113724a
Possible blast hazards associated with use of the propellant combination liquid dinitrogen oxide–liquid carbon monoxide have been evaluated.

Combustible gases
Sorbe, 1968, 131
The oxide (with an oxygen content 1.7 times that of air) forms explosive mixtures with ammonia, carbon monoxide, hydrogen, hydrogen sulphide and phosphine.

Hydrogen,
Oxygen
Bailar, 1973, Vol. 2, 321–322
The ignition temperature of mixtures of hydrogen with dinitrogen oxide is lower than that of hydrogen with air or oxygen. The oxide also sensitises mixtures of hydrogen and oxygen, so that addition of

oxygen to a hydrogen–oxide mixture will cause instantaneous ignition or explosion. Explosive limits are extremely wide.

*See other* NON-METAL OXIDES

Phosphine

Thénard, J., *Compt. Rend.*, 1844, **18**, 652

A mixture with excess oxide can be exploded by sparking.

Tin oxide

*See* TIN(II) OXIDE, OSn: Non-metal oxides

Tungsten carbides

*See* TUNGSTEN CARBIDE, CW: Nitrogen oxides
DITUNGSTEN CARBIDE, $CW_2$: Oxidants

*See other* NON-METAL OXIDES

**LEAD HYPONITRITE** **pbON=NOpb** **$N_2O_2Pb$**

Partington, J. R. *et al.*, *J. Chem. Soc.*, 1932, **135**, 2589

The lead salt decomposes with explosive violence at 150–160°C and should not be dried by heating. The salt prepared from lead acetate explodes more violently than that from the nitrate.

*See* HYPONITROUS ACID, $H_2N_2O_2$

*See other* METAL OXONON-METALLATES

**DINITROGEN TRIOXIDE** **$N_2O_3$**

Acetic acid,
6-Hexanelactam

*See* 6-HEXANELACTAM, $C_6H_{11}NO$: Acetic acid, etc.

Phosphine

*See* PHOSPHINE, $H_3P$: Air

Phosphorus

*See* PHOSPHORUS, P: Non-metal oxides

**DINITROGEN TETRAOXIDE (NITROGEN DIOXIDE)** **$N_2O_4$**

Acetonitrile,
Indium

Addison, C. C. *et al.*, *Chem. & Ind.*, 1958, 1004

Shaking a slow-reacting mixture caused detonation, attributed to indium-catalysed oxidation of acetonitrile.

Alcohols
Daniels, F., *Chem. Eng. News,* 1955, **33**, 2372
A violent explosion occurred during the ready interaction to produce alkyl nitrates.

Ammonia
Mellor, 1940, Vol. 8, 541
Liquid ammonia reacts explosively with the solid tetraoxide at −80°C, while aqueous ammonia reacts vigorously with the gas at ambient temperature.

Barium oxide
Mellor, 1940, Vol. 8, 545
In contact with the gas at 200°C the oxide suddenly reacts, reaches red heat and fuses.

Carbon disulphide
Mellor, 1940, Vol. 8, 543
Sorbe, 1968, 132
Mixtures proposed for use as explosives are stable up to 200°C, but can be detonated by mercury fulminate, and the vapours by sparking.

Carbonylmetals
Cloyd, 1965, 74
Combination is hypergolic.

Cellulose,
Magnesium perchlorate
*See* MAGNESIUM PERCHLORATE, $Cl_2O_8Mg$: Cellulose, etc.

Cycloalkenes,
Oxygen
Lachowicz, D. R. *et al.,* US Pat. 3 621 050, 1971
Contact of cycloalkenes with a mixture of dinitrogen tetraoxide and excess oxygen at temperatures of 0°C or below produces nitroperoxonitrates of the general formula $-CHNO_2-CH(O_2NO_2)-$ which appear to be unstable at temperatures above 0°C, owing to the presence of the peroxonitrate grouping.
*See* Hydrocarbons, below

Dimethyl sulphoxide
*See* DIMETHYL SULPHOXIDE, $C_2H_6OS$: Dinitrogen tetraoxide

Formaldehyde
Rastogi, R. P. *et al., Chem. Abs.*, 1975, **83**, 12936m
Interaction is explosive even below 180°C.

Halocarbons

1. Turley, R. E., *Chem. Eng. News,* 1964, **42**(47), 53
2. Benson, S. W., *Chem. Eng. News,* 1964, **42**(51), 4
3. Shanley, E. S., *Chem. Eng. News,* 1964, **42**(52), 5
4. Kuchta, J. M. *et al., J. Chem. Eng. Data*, 1968, **13**, 421–428

Mixtures of the tetraoxide with dichloromethane, chloroform, carbon tetrachloride, 1,2-dichloroethane, trichloroethylene and tetrachloroethylene are explosive when subjected to shock of 25 g TNT equivalent or less [1]. Mixtures of the tetraoxide with trichloroethylene react violently on heating to 150°C [2]. Partially fluorinated chloroalkanes were more stable to shock. Theoretical aspects are discussed in the later references [2, 3]. The effect of pressure on flammability limits of mixtures of the oxide, nitrogen and several chlorinated hydrocarbons has been studied [4].

*See also* URANIUM, U: Nitric acid

Heterocyclic bases

Mellor, 1940, Vol. 8, 543

Pyridine and quinoline are violently attacked by the liquid oxide.

Hydrazine derivatives

Cloyd, 1965, 74

Miyajima, H. *et al., Combust. Sci. Technol.*, 1973, **8**, 199–200

Combination with hydrazine, methylhydrazine, 1,1-dimethylhydrazine or mixtures is hypergolic and used in rocketry. The hypergolic gas-phase ignition of hydrazine at 70–160°C and 53–120 mbar has been studied.

*See* ROCKET PROPELLANTS

Hydrogen,
Oxygen

Lewis, B., *Chem. Rev.*, 1932, **10**, 60

The presence of small amounts of the oxide in non-explosive mixtures of hydrogen and oxygen renders them explosive.

Hydrocarbons

1. Mellor, 1967, Vol. 8, Suppl. 2.2, 264
2. Fierz, H. E., *J. Soc. Chem. Ind.*, 1922, **41**, 114R

2a. Raschig, F., *Z. Angew. Chem.*, 1922, **35**, 117–119
Berl, E., ibid., 1923, **36**, 87–91
Schaarschmidt, A., ibid., 533–536
Berl, E., ibid., 1924, **37**, 164–165
Schaarschmidt, A., ibid., 1925, **38**, 537–541

3. *MCA Case History No. 128*
4. Folecki, J. *et al., Chem. & Ind.*, 1967, 1424
5. Cloyd, 1965, 74
6. Urbanski, 1967, Vol. 3, 289

A mixture of tetraoxide and toluene exploded, possibly initiated by unsaturated impurities [1]. During attempted separation by low temperature distillation of an accidental mixture of light petroleum and the tetraoxide, a large bulk of material awaiting distillation became heated by unusual atmospheric conditions to 50°C and exploded violently [2].

Subsequently, discussion of possible alternative causes involving either unsaturated or aromatic components in the hydrocarbon mixture was published [2a].

Erroneous addition of liquid instead of gaseous tetraoxide to hot cyclohexane caused an explosion [3]. During kinetic studies, one sample of a 1:1 molar solution of tetraoxide in hexane exploded during (normally slow) decomposition at 28°C [4]. Cyclopentadiene is hypergolic with the tetraoxide [5]. These incidents are understandable because of their similarity to rocket propellant systems and liquid mixtures previously used as bomb fillings [6].

*See* Cycloalkenes, etc., above
Unsaturated hydrocarbons, below

Metal acetylides or carbides

Mellor, 1946, Vol. 5, 849

Caesium acetylide ignites at 100°C in the gas.

*See* TUNGSTEN CARBIDE, CW: Nitrogen oxides
DITUNGSTEN CARBIDE, $CW_2$: Oxidants

Metals

Mellor, 1940, Vol. 8, 544–545; 1942, Vol. 13, 342

Pascal, 1956, Vol. 10, 382; 1958, Vol. 4, 291

Reduced iron, potassium and pyrophoric manganese all ignite in the gas at ambient temperature. Magnesium filings burn vigorously when heated in the gas.

Slightly warm sodium ignites in contact with the gas, and interaction with calcium is explosive.

*See* ALUMINIUM, Al: Oxidants

Nitrobenzene

Urbanski, 1967, Vol. 3, 288

Mixtures were formerly used as liquid high explosives, with addition of carbon disulphide to reduce the freezing point. High sensitivity to mechanical stimuli was disadvantageous.

Nitrogen trichloride

*See* NITROGEN TRICHLORIDE, $Cl_3N$: Initiators

Organic compounds

Riebsomer, J. L., *Chem. Rev.*, 1945, **36**, 158

In a review of the interaction of the tetraoxide with organic compounds, attention is drawn to the possibility of formation of unstable or explosive products.

Ozone

*See* OZONE, $O_3$: Dinitrogen tetraoxide

Phospham

*See* PHOSPHAM, $HN_2P$: Oxidants

Phosphorus

*See* PHOSPHORUS, P: Non-metal oxides

Sodium amide

Beck, G., *Z. Anorg. Chem.*, 1937, **233**, 158

Interaction with the oxide in carbon tetrachloride is vigorous, producing sparks.

Tetracarbonylnickel

Bailar, 1973, Vol. 3, 1130

Interaction of the liquids is rather violent.

*See* Carbonylmetals, above

Tetramethyltin

Bailar, 1973, Vol. 2, 355

Interaction is explosively violent even at $-80°C$, and dilution with inert solvents is required for moderation.

Triethylamine

Davenport, D. A. *et al., J. Amer. Chem. Soc.*, 1953, **75**, 4175

The complex, containing excess oxide over amine, exploded at below 0°C when free of solvent.

*See* Triethylammonium nitrate, below

Triethylammonium nitrate

Addison, C. C. *et al., Chem. & Ind.*, 1953, 1315

The two ingredients form an addition complex with diethyl ether, which exploded violently after partial desiccation: an ether-free complex is also unstable.

*See* Triethylamine, above

Unsaturated hydrocarbons

Sergeev, G. P. *et al., Chem. Abs.*, 1966, **65**, 3659g

Dinitrogen tetraoxide reacts explosively between $-32°$ and $-90°C$ with propene, 1-butene, isobutene, 1,3-butadiene, cyclopentadiene and 1-hexene, but six other unsaturated hydrocarbons failed to react.

*See other* NON-METAL OXIDES
OXIDANTS

**DINITROGEN PENTAOXIDE** $N_2O_5$

Acetaldehyde
Kacmarek, A. J. *et al., J. Org. Chem.*, 1975, **40**, 1853
Direct combination to produce ethylidene dinitrate at −196°C is violently explosive, but uneventful when the aldehyde is diluted with nitrogen.
*See* ETHYLIDENE DINITRATE, $C_2H_4N_2O_6$

Dichloromethane
*See* NITRATING AGENTS

Disodium acetylide
*See* DISODIUM ACETYLIDE, $C_2Na_2$: Oxidants

Metals
Mellor, 1940, Vol. 8, 554
Potassium and sodium burn brilliantly in the gas, while mercury and arsenic are vigorously oxidised.

Naphthalene
Mellor, 1940, Vol. 8, 554
Naphthalene explodes and other organic materials react vigorously with the pentaoxide.

Sulphur dichloride,
or Sulphuryl chloride
Schmeisser, M., *Angew. Chem.*, 1955, **67**, 495, 499
Interaction is explosively violent.
*See other* NON-METAL OXIDES
OXIDANTS

**LEAD(II) NITRATE** $Pb(NO_3)_2$ $N_2O_6Pb$

Carbon
Mellor, 1941, Vol. 7, 863
Contact with red-hot carbon causes an explosion with showers of sparks.

Calcium–silicon alloy
Smolin, A. O., *Chem. Abs.*, 1975, **85**, 7591z

'Combustion of Calcium–Silicon in a Mixture with Lead Nitrate'
(title only translated)

Cyclopentadienylsodium
Houben-Weyl, 1975, Vol. 13.3, 200
If lead nitrate is used rather than the chloride or acetate as the source of divalent lead, the crude dicyclopentadienyllead may explode violently during purification by high-vacuum sublimation at 100–130°C.

Potassium acetate
Mee, A. J., *School Sci. Rev.*, 1940, 22(86), 95
A heated mixture exploded violently.
*See other* METAL OXONON-METALLATES

**TIN(II) NITRATE** $Sn(NO_3)_2$ $N_2O_6Sn$

Bailar, 1973, Vol. 2, 74
During attempted isolation of the nitrate by evaporation of its aqueous solutions, a number of explosions have occurred.
*See other* REDOX COMPOUNDS

**ZINC NITRATE** $Zn(NO_3)_2$ $N_2O_6Zn$

Carbon
Mellor, 1940, Vol. 4, 655
When the nitrate is sprinkled on to hot carbon, an explosion occurs.
*See other* METAL OXONON-METALLATES

**TIN(II) NITRATE OXIDE** $Sn(NO_3)_2O$ $N_2O_7Sn_2$

Alone,
or Organic dust
1. Mellor, 1941, Vol. 7, 481
2. Anon., *J. Soc. Chem. Ind.*, 1922, **41**, 424R

The basic nitrate burns brilliantly when dry, and explodes on shock, friction or heating to above 100°C [1]. It was also involved, either as oxidant or initiator, in a flour mill explosion [2].
*See* TETRAHYDROXOTIN(2+) NITRATE, $H_4N_2O_{10}Sn$
*See related* METAL OXONON-METALLATES

**URANYL NITRATE** $UO_2(NO_3)_2$ $N_2O_8U$

Cellulose
Clinton, T. G., *J. R. Inst. Chem.*, 1958, **82**, 633

The analytical use of cellulose to absorb uranyl nitrate solution prior to ignition has caused explosions during ignition, due to formation of cellulose nitrate. An alternative method is described.

Diethyl ether

1. Ivanov, W. N., *Chem. Ztg.*, 1912, **36**, 297
2. Andrews, L. W., *J. Amer. Chem. Soc.*, 1912, **34**, 1686–1687
3. Muller, A., *Chem. Ztg.*, 1916, **40**, 38; 1917, **41**, 439

The mild detonations reported when the crystalline salt was disturbed [1] were postulated as probably being caused by the presence of ether in the crystals ($6H_2O$ may be replaced by $2C_4H_{10}O$ [2]). This was later confirmed as correct [3], and the formation of ethyl nitrate or diethyloxonium nitrate may have been involved, as the anhydrous salt functions as a powerful nitrator. Solutions of the nitrate in ether should not be exposed to sunlight to avoid the possibility of explosions occurring.

*See* Organic solvents, below

Organic solvents

Burriss, W. H., *Nucl. Sci. Abs.*, 1976, **33**, 19790

During routine operations to reduce the hexahydrate to uranium trioxide, an excessive amount of organic solution entered the denitrator pots and ignited.

*See* Diethyl ether, above

*See other* METAL NITRATES

## TRILEAD DINITRIDE $Pb_3N_2$ $N_2Pb_3$

Fischer, F. *et al.*, *Ber.*, 1910, **43**, 1470

Very unstable, it decomposes explosively during vacuum degassing.

*See other N*-METAL DERIVATIVES

## DISULPHUR DINITRIDE SN=S=N (cyclic) $N_2S_2$

Brauer, 1963, Vol. 1, 410

Explosive, initiated by shock, friction, pressure, or temperatures above 30°C.

*See other* NON-METAL SULPHIDES

## TETRASULPHUR DINITRIDE S=NSS=NS (cyclic) $N_2S_4$

Brauer, 1963, Vol. 1, 408

Heal, H. G. *et al.*, *J. Inorg. Nucl. Chem.*, 1975, **37**, 286

It explosively decomposes to its elements at 100°C. Previously prepared from the explosive tetranitride, it may now safely be prepared from the stable intermediate, heptasulphur imide.

*See other* NON-METAL SULPHIDES

1. Mellor, 1940, Vol. 8, 345
2. Lambert, B., *School Sci. Rev.*, 1927, 8(31), 218

Insensitive to impact, it decomposes, sometimes explosively, above the m.p. [1], particularly if heated rapidly [2].

Barium carbonate

Henneburg, G. O. *et al., Can. J. Res.*, 1950, **28B**, 345

Interaction to form cyanide ion requires careful control of temperature at 630°C to prevent explosions.

Bromine

*See* BROMINE, $Br_2$: Metal azides

Carbon disulphide

*See* CARBON DISULPHIDE, $CS_2$: Metal azides

Chromyl chloride

*See* CHROMYL CHLORIDE, $Cl_2CrO_2$: Sodium azide

Heavy metals

1 Wear, J. O., *J. Chem. Educ.*, 1975, **52**(1), A23–25
2. Becher, H. H., *Naturwissenschaften*, 1970, **57**, 671
3. Anon., *Chem. Eng. News*, 1976, **54**(36), 6

The effluent from automatic blood-analysers in which 0.01–0.1% sodium azide solutions are used, may lead over several months to formation of explosive heavy metal azides in brass, copper or lead plumbing lines, especially if acids are also present. Several incidents during drain-maintenance are described, and preventative flushing and decontamination procedures are discussed [1]. Brass plates exposed to sodium azide solution during several months in soil percolation tests and then dried caused explosions, due to formation of copper and/or zinc azides [2]. During repairs to a metal thermostat bath in which sodium azide solution had been used as a preservative, a violent explosion occurred [3].

Sulphuric acid

Ross, F. F., *Water & Waste Treatment*, 1964, **9**, 528; private comm., 1966

One of the reagents required for the determination of dissolved oxygen in polluted water is a solution of sodium azide in 50% sulphuric acid. It is important that the diluted acid should be quite cold before adding the azide, since hydrogen azide boils at 36°C and is explosive in the condensed liquid state.

Trifluoroacryloyl fluoride

*See* TRIFLUOROACRYLOYL FLUORIDE, $C_3F_4O$: Sodium azide

Water

Anon., *Angew. Chem.*, 1952, **64**, 169

Addition of water to sodium azide which had been strongly heated caused a violent reaction. This was attributed to formation of metallic sodium or sodium nitride in the azide.

*See other* METAL AZIDES

**SODIUM AZIDOSULPHATE** $NaOSO_2N_3$ $N_3NaO_3S$

Shozda, R. J. *et al., J. Org. Chem.*, 1967, **32**, 2876

It is a weak explosive with variable sensitivity to mechanical shock and heating.

*See other* ACYL AZIDES

**THALLIUM 2,4,6-TRIS(DIOXOSELENA)PERHYDROTRIAZINE-1,3,5-TRIIDE (THALLIUM 'TRISELENIMIDATE')** $N_3O_6Se_3Tl_3$

$\overline{TlNSeO_2N(Tl)SeO_2N(Tl)SeO_2}$

Explosive.

*See* SELENIUM DIFLUORIDE OXIDE, $F_2O_2Se$: Ammonia

*See other* *N*-METAL DERIVATIVES

**VANADIUM TRINITRATE OXIDE** $VO(NO_3)_3$ $N_3O_{10}V$

1. Schmeisser, M., *Angew. Chem.*, 1955, **67**, 495
2. Harris, A. D. *et al., Inorg. Synth.*, 1967, **9**, 87

Many hydrocarbons and organic solvents ignite on contact with this powerful oxidant and nitrating agent [1], which reacts like fuming nitric acid with paper, rubber or wood [2].

*See related* METAL OXONON-METALLATES

*See other* OXIDANTS

**TRIS(THIONITROSYL)THALLIUM** $(SN)_3Tl$ $N_3S_3Tl$

Bailar, 1973, Vol. 1, 1162

The compound and its adduct with ammonia explode very easily with heat or shock.

*See related* NITROSO COMPOUNDS

*See other* *N*-METAL DERIVATIVES

**THALLIUM(I) AZIDE** $TlN_3$ $N_3Tl$

Mellor, 1940, Vol. 8, 352

A relatively stable azide, it can be exploded on fairly heavy impact, or by heating at 350–400°C.

*See other* METAL AZIDES

**NITROSYL AZIDE** $O{=}NN_3$ $N_4O$

Lucien, H. W., *J. Amer. Chem. Soc.*, 1958, **80**, 4458

Explosions were experienced on several occasions during preparation of nitrosyl azide by various methods. It decomposes even at –50°C.

*See other* NON-METAL AZIDES

**THIOTRITHIAZYL NITRATE** SNSNSNS $NO_3$ $N_4O_3S_4$

Goehring, 1957, 74

Explodes on friction or impact.

**PLUTONIUM(IV) NITRATE** $Pu(NO_3)_4$ $N_4O_8Pu$

*MCA Case History No. 1498*

Polythene bottles are not suitable for long-term storage of plutonium nitrate solution as stress cracks appeared in bases of several 10 litre bottles during 6 months' storage. Short-term storage and improved venting are recommended.

*See other* METAL OXONON-METALLATES

**TETRASULPHUR TETRANITRIDE** $N_4S_4$

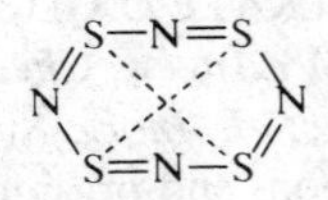

1. Mellor, 1940, Vol. 8, 625; Brauer, 1963, Vol. 1, 406
2. Villema-Blanco, M. *et al., Inorg. Synth.*, 1967, **9**, 101

The endothermic nitride is susceptible to explosive decomposition on friction, shock or heating above 100°C. Explosion is violent if initiated by a detonator [1]. Sensitivity towards heat and shock increases with purity. Preparative precautions have been detailed [2].

Oxidants

Pascal, 1956, Vol. 10, 645–646

Contact with fluorine leads to ignition, and interaction with metal chlorates or oxides is violent.

*See other* NON-METAL SULPHIDES

**TETRASELENIUM TETRANITRIDE** $N_4Se_4$

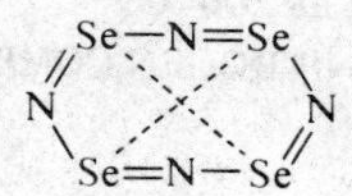

Alone,
or Halogens and derivatives
Mellor, 1947, Vol. 10, 789
The dry material explodes on slight compression, or on heating at 130–230°C. Contact with bromine, chlorine or a little fuming hydrochloric acid also causes explosion.
*See related N*-METAL DERIVATIVES

**TRITELLURIUM TETRANITRIDE** $Te_3N_4$ $N_4Te_3$

Fischer, F. *et al., Ber.,* 1910, **43**, 1472
Two forms are described, one black which explodes on impact, and one yellow which explodes at 200°C.
*See related N*-METAL DERIVATIVES

**TRITHORIUM TETRANITRIDE** $Th_3N_4$ $N_4Th_3$

Air,
or Oxygen
Mellor, 1940, Vol. 8, 122
It burns incandescently on heating in air, and very vividly in oxygen.
*See other N*-METAL DERIVATIVES

**SODIUM TETRASULPHURPENTANITRIDATE** $N_5NaS_4$

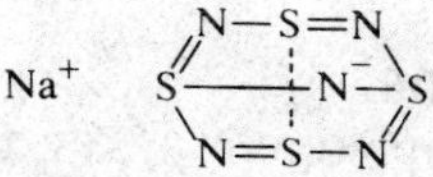

Scherer, O. J. *et al., Angew. Chem. (Intern. Ed.)*, 1975, **14**, 485
It explodes at about 180°C, or under pressure, or by friction of a spatula against a sintered glass filter.
*See other* HIGH-NITROGEN COMPOUNDS

**TRIPHOSPHORUS PENTANITRIDE** $P_3N_5$ $N_5P_3$

Metals
Mellor, 1971, Vol. 8, Suppl. 3, 370–371

Interaction is explosive at the m.p. of magnesium (651°C), and more so with calcium at 200°C.
*See other* HIGH-NITROGEN COMPOUNDS

**SULPHURYL DIAZIDE** $N_3SO_2N_3$ $N_6O_2S$

Curtius, T. *et al., Ber.*, 1922, **55**, 1571
It explodes violently when heated and often spontaneously at ambient temperature.
*See other* ACYL AZIDES

**DISULPHURYL DIAZIDE** $N_3SO_2OSO_2N_3$ $N_6O_5S_2$

Alone,
or Alkali
Lehmann, H. A. *et al., Z. Anorg. Chem.*, 1957, **293**, 314
The azide decomposes explosively below 80°C and should only be stored in 1 g quantities. In contact with dilute alkali at 0°C an explosive deposit is produced.
*See other* ACYL AZIDES

**'TETRAPHOSPHORUS HEXANITRIDE'** '$N_6P_4$'

Mellor, 1971, Vol. 8, Suppl. 3, 371
Apparently mainly $P_3N_5$, it ignites in air.
*See other* HIGH-NITROGEN COMPOUNDS

**LEAD(II) AZIDE** $Pb(N_3)_2$ $N_6Pb$

1. Mellor, 1940, Vol. 8, 353; 1967, Vol. 8, Suppl. 2, 21, 50
2. Taylor, G. W. C. *et al., J. Crystal Growth,* 1968, **3**, 391
3. Barton, A. F. M. *et al., Chem. Rev.*, 1973, **73**, 138
4. *MCA Case History No. 2053*

As a widely used detonator, its properties have been studied in great detail. Although quantitatively inferior to mercury fulminate as a detonator, it has proved to be more reliable in service. The pure compound occurs in two crystalline forms, one of which appears to be much more sensitive to initiation [1]. Factors which suppressed spontaneous explosions of lead azide during crystallisation were vigorous agitation and use of hydrophilic colloids [2]. These aspects have been recently reviewed [3].

A vacuum-desiccated sample of 5 g of the azide exploded violently when touched with a metal spatula [4].
*See also* SODIUM AZIDE, $N_3Na$: Heavy metals

Calcium stearate
*MCA Case History No. 949*
An explosion occurred during blending and screening operations on a mixture of lead azide and 0.5% of calcium stearate. If free stearic acid were present as impurity in the calcium salt, free hydrogen azide may have been involved.

Copper,
or Zinc
Federoff, 1960, A532, 551
Lead azide, on prolonged contact with copper, zinc or their alloys, forms traces of the extremely sensitive copper or zinc azides which may initiate explosion of the whole mass of azide.
*See other* METAL AZIDES

**STRONTIUM AZIDE** **$Sr(N_3)_2$** **$N_6Sr$**
*See* BARIUM AZIDE, $BaN_6$
CALCIUM AZIDE, $CaN_6$

**ZINC AZIDE** **$Zn(N_3)_2$** **$N_6Zn$**
Bailar, 1973, Vol. 3, 217
Zinc azide is said to explode easily.
*See other* METAL AZIDES

**PHOSPHORUS TRIAZIDE OXIDE** **$PO(N_3)_3$** **$N_9OP$**
Buder, W. *et al.*, *Z. Anorg. Allgem. Chem.*, 1975, **415**, 263
The liquid may explode violently on warming from 0°C to ambient temperature, but can be used safely in solution.
*See related* NON-METAL AZIDES

**PHOSPHORUS TRIAZIDE** **$P(N_3)_3$** **$N_9P$**
Lines, E. L. *et al.*, *Inorg. Chem.*, 1972, **11**, 2270
Buder, W. *et al.*, *Z. Anorg. Allgem. Chem.*, 1975, **415**, 263
It is a highly explosive liquid, which may explode on warming from 0°C to ambient temperature, but can be used safely in solution.
*See other* NON-METAL AZIDES

## LEAD(IV) AZIDE $Pb(N_3)_4$ $N_{12}Pb$

1. Mellor, 1967, Vol. 8, Suppl. 2, 22
2. Moeller, H., *Z. Anorg. Chem.*, 1949, **260,** 10
3. Zbirol, E., *Synthesis*, 1972, (6), 285–302

The crystalline product appears less stable than the diazide, spontaneously decomposing, sometimes explosively [1]. It was rated as too unstable for use as a practical explosive or detonator [2]. Lead(IV) acetate azide (probably the triacetate azide) is also rather unstable, evolving nitrogen above 0°C with precipitation of lead(II) azide [3].

## SILICON TETRAAZIDE $Si(N_3)_4$ $N_{12}Si$

1. Mellor, 1967, Vol. 8, Suppl. 2, 20
2. Anon., *Angew. Chem. (Nachr.)*, 1970, **18**, 27

Spontaneous explosions have been observed [1]. A residue containing the tetraazide, the chlorotriazide, and probably the dichlorodiazide, exploded on standing for 2 or 3 days, possibly owing to hydrazoic acid produced by hydrolysis.

*See other* NON-METAL AZIDES

## THALLIUM(I) TETRAAZIDOTHALLATE $N_{12}Tl_2$

$Tl[Tl(N_3)_4]$

Bailar, 1973, Vol. 1, 1152

A heat-, friction- and shock-sensitive explosive.

*See related* METAL AZIDES

## 'PHOSPHORUS PENTAAZIDE' '$N_{15}P$'

*See* SODIUM HEXAAZIDOPHOSPHATE(1–), $N_{18}NaP$

## SODIUM HEXAAZIDOPHOSPHATE(1–) $N_{18}NaP$

$Na[P(N_3)_6]$

1. Volgnandt, P. *et al.*, *Z. Anorg. Allg. Chem.*, 1976, **425**, 189–192
2. Buder, W. *et al.*, ibid., 1975, **415**, 263

The product of interaction of sodium azide and phosphorus trichloride occasionally exploded on warming from 0°C to ambient temperature, but was examined safely in solution. The structure of the explosive product is determined as the title compound [1] rather than phosphorus pentaazide as originally reported [2].

## 1,1,3,3,5,5-HEXAAZIDO-2,4,6-TRIAZA-1,3,5-TRIPHOSPHORINE $N_{21}P_3$

$(N_3)_2P{=}NP(N_3)_2{=}NP(N_3)_2{=}N$ (ring)

Grundmann, C., *Z. Naturforsch.*, 1955, **106**, 116

This viscous liquid (containing 73.5% N) is violently explosive when subjected to shock or friction.
*See other* NON-METAL AZIDES

**SODIUM** **Na**

1. Hawkes, A. S. *et al., J. Chem. Educ.*, 1953, **30**, 467
2. *MCA SD-47,* 1952
3. *Alkali Metals,* 1957
4. Houben-Weyl, 1970, Vol. 13.1, 287–289

In a discussion of safe methods for laboratory use of sodium, disposal of small quantities (up to 5–10 g) by immersion in isopropanol, which may contain up to 2% of water to increase the rate of reaction, is recommended. Quantities up to 50 g should be burned in a heavy metal dish, using a gas flame [1]. Handling techniques and safety precautions for large-scale operations have also been detailed for this reactive metal [2,3]. Techniques for preparation and precautions in use of finely dispersed sodium ('sodium sand') are detailed. It is to some extent pyrophoric [4].

Acids
Mellor, 1941, Vol. 2, 469–70
Anhydrous hydrogen chloride, hydrogen fluoride or sulphuric acids react slowly with sodium, while the aqueous solutions react explosively. Nitric acid of density above 1.056 causes ignition of sodium.

Air
Muir, G. D., private comm., 1968
Dispersions of sodium in volatile solvents become pyrophoric if the solvent evaporates round the neck of a flask or bottle. Serum cap closures are safer.

Calcium,
Mixed oxides
*See* CALCIUM, Ca: Sodium, etc.

Chloroform,
Methanol
*See* CHLOROFORM, $CHCl_3$: Sodium, Methanol

Diazomethane
*See* DIAZOMETHANE, $CH_2N_2$: Alkali metals

Diethyl ether
Hey, P., private comm., 1965
While sodium wire was being pressed into ether, the jet blocked.

Increasing the pressure to free it caused ignition of the ejected sodium and explosion of the flask of ether. Pressing the sodium into xylene or toluene and subsequent transfer to the ether is recommended.

*N, N*-Dimethylformamide

'DMF Brochure', Billingham, ICI, 1965

A vigorous reaction occurs on heating DMF with sodium metal.

Ethanol

Brändstrom, A., *Acta Chem. Scand.*, 1951, **4**, 1608–1610

Air must be excluded during exothermic interaction of ethanol with sodium finely dispersed in hydrocarbons to avoid the possibility of hydrogen–air mixture explosions.

*See also* POTASSIUM, K: Alcohols

Fluorinated compounds

Herring, D. E., private comm., 1964

Fusion of fluorinated compounds with sodium for qualitative analysis requires a high temperature for reaction because of their unreactivity. When reaction does occur, there is often an explosion and suitable precautions must be taken.

Halocarbons

1. Staudinger, H., *Z. Elektrochem*, 1925, **31**, 549
2. Lenze, F. *et al., Chem. Ztg.*, 1932, **56**, 921
3. Ward, E. R., *Proc. Chem. Soc.*, 1963, 15
4. *MCA SD-34*, 1949
5. Rayner, P. N. G., private comm., 1968
6. Anon., *Angew. Chem. (Nachr.)*, 1970, **18**, 397
7. Read, R. R. *et al., Org. Synth.*, 1955, Coll. Vol. 3, 158
8. Houben-Weyl, 1970, Vol. 13.1, 394–395
9. Firat, Y. *et al., Makromol. Chem.*, 1975, **Suppl. 1**, 207

Although apparently stable standing in contact, mixtures of sodium with a range of halogenated alkanes (solvents) are metastable and capable of explosion by shock or impact. Carbon tetrachloride [1–3], chloroform, dichloromethane and chloromethane [1,2], tetrachloroethane [1,2,4] have been investigated, among others.

Generally the sensitivity to initiation and the force of the explosion increase with the degree of halogen substitution; the two former are less than in the corresponding systems with potassium or potassium–sodium alloy. Any aliphatic halocarbon (except fully fluorinated alkanes) may be expected to behave in this way. Small portions of sodium and hexachlorocyclopentadiene were mixed in preparation for a sodium fusion test. On shaking a few minutes later, the tube exploded [5]. On the small scale, no reaction occurred in boiling perfluorohexyl iodide at 114°C in contact with metallic sodium.

With 140 g of iodide and 7 g of sodium an explosion occurred after 30 min [6].

The temperature range for smooth interaction of bromobenzene, 1-bromobutane and sodium in ether to give butylbenzene is critical. Below 15°C reaction is delayed but later becomes vigorous, and above 30°C the reaction becomes violent [7]. Sodium wire and chlorobenzene react exothermically in benzene under nitrogen to give phenylsodium and the reaction must be controlled by cooling. Use of finely divided sodium will lead to an uncontrollable, explosive reaction [8]. The hazards of contacting dichloromethane with a sodium film during an ultra-drying procedure are stressed. Thorough pre-drying and operations under vacuum are essential precautions [9].

*See* Chloroform, etc. above
CHLOROFORM, $CHCl_3$: Metals
IODOMETHANE, $CH_3I$: Sodium
*See other* METALS: Halocarbons

Halogen azides
*See* HALOGEN AZIDES

Halogens,
or Interhalogens

1. Mellor, 1961, Vol. 2, Suppl. 2. 1, 450–452
2. Mellor, 1941, Vol. 2, 114, 469
3. Mellor, 1941, Vol. 2, 92, 469; Staudinger, H., *Z. Elektrochem.*, 1925, **31**, 549
4. Booth, H. S. *et al.*, *Chem. Rev.*, 1947, **41**, 427
5. Pascal, 1966, Vol. 2.1, 221

Sodium ignites in fluorine gas, but is inert in the liquefied gas [1]. Cold sodium ignites in moist chlorine [2] but may be distilled unchanged in the dry gas [1]. Sodium and liquid bromine appear to be unreactive on prolonged contact, but mixtures may be detonated violently by mechanical shock [3]. Finely divided sodium luminesces in bromine vapour [1]. Iodine bromide or chloride react slowly with sodium, but mixtures will explode under a hammer blow [1]. Interaction of iodine pentafluoride with solid sodium is initially vigorous, but soon slows with film-formation, while that with molten sodium is explosively violent [2]. Sodium reacts immediately and incandescently with iodine heptafluoride [4]. Mixtures of solid sodium and iodine explode lightly when initiated by shock [5].

Hydrazine
Mellor, 1940, Vol. 8, 316
Anhydrous hydrazine and sodium react in ether to form sodium hydrazide, which explodes in contact with air. Hydrazine hydrate

and sodium react very exothermically, generating hydrogen and ammonia.

Hydroxylamine

*See* HYDROXYLAMINE, $H_3NO$: Metals

Iodates

Cueilleron, J. *Bull. Soc. Chim. Fr.*, 1945, **12**, 88–89

Mixtures of sodium with silver or sodium iodates explode when initiated by shock.

*See* Oxidants, below

Mercury

Brauer, 1965, Vol. 2, 1802

Interaction of sodium and mercury to form sodium amalgam is violently exothermic, and moderation of the reaction with an inert liquid or by adding mercury slowly to the sodium is necessary. Even so, temperatures of 400°C may be attained.

Metal halides

1. *Alkali Metals,* 1957, 129
2. Mellor, 1961, Vol. 2, Suppl. 2.1, 494–496
3. Staudinger, H., *Z. Elektrochem.*, 1925, **31**, 549
4. Cueilleron, J., *Bull. Soc. Chim. Fr.*, 1945, **12**, 88

Sodium dispersions will reduce many metal halides exothermically. Iron(III) chloride is reduced fairly smoothly at room temperature or below in presence of 1,2-dimethoxyethane. Nickel, cobalt, lead or cadmium chlorides require higher temperatures to initiate the reaction, the exotherm with cobalt increasing the temperature from 325 to 375°C and causing evaporation of most of the dispersing oil [1]. The finely powdered metals produced by reduction of halides of cadmium, chromium, cobalt, copper, iron, manganese, molybdenum, nickel, silver, tin or zinc with sodium, dispersed in ether or toluene, are all pyrophoric [2]. Interaction of sodium and vanadyl chloride at 180°C is violent [2]. Mixtures of sodium with metal halides are sensitive to mechanical shock [3]; the ensuing explosions have been classified [4]. Very violent explosions occurred with iron(III) chloride or bromide, iron(II) bromide or iodide, and cobalt(II) chloride or bromide. Strong explosions occurred with the various halides of aluminium, antimony, arsenic, bismuth, copper(II), mercury, silver and tin (including a mixture of tin(IV) iodide and sulphur), also vanadium pentachloride and ammonium tetrachlorocuprate. Ammonium, copper(I), cadmium and nickel halides generally gave weak explosions. Most of the alkali and alkaline earth halides were insensitive.

Metal oxides

1. Mellor, 1939, Vol. 9, 649
2. Mellor, 1943, Vol. 11, 237, 542
3. Mellor, 1941, Vol. 3, 138
4. Mellor, 1941, Vol. 7, 658, 401
5. Mellor, 1940, Vol. 4, 770
6. Brewer, L. *et al., Rep. UCRL 1864*, 7, Washington, USAEC, 1952
7. Mellor, 1961, Vol. 2, Suppl. 2.1, 474

Bismuth(III) oxide [1], chromium trioxide [2] and copper(II) oxide [3] are reduced with incandescence on heating with sodium. Finely divided sodium ignites on admixture with fine lead oxide without heating, while the coarse material reacts vigorously with molten sodium [4]. The latter reduces mercury(I) oxide [5] or molybdenum trioxide [2] with incandescence, the former producing a light explosion also. Sodium reduces sodium peroxide vigorously at 500°C [6] and tin(IV) oxide incandescently on gentle heating [4]. Finely dispersed sodium reduces metal oxides on heating at temperatures between 100 and 300°C, producing, e.g., pyrophoric iron, nickel and zinc. Tin dioxide is reduced with incandescence [4].

Nitrogen-containing explosives

Leleu, *Cahiers*, 1975, (79), 266

Mixtures of sodium (or its alloy with potassium) and nitromethane, trichloronitromethane, nitrobenzene, dinitrobenzene, dinitronaphthalene, ethyl nitrite, ethyl nitrate or glyceryl trinitrate are shock-sensitive, the sensitivity increasing with the number of nitro groups.

Non-metal halides

1. Mellor, 1940, Vol. 8, 1033
2. Mellor, 1941, Vol. 2, 470; 1940; Vol. 8, 1016
3. Mellor, 1961, Vol. 2, Suppl. 2.1, 455, 460
4. Mellor, 1940, Vol. 8, 1073
5. Cueilleron, J., *Bull. Soc. Chim. Fr.*, 1945, **12**, 88
6. Mellor, 1947, Vol. 10, 912
7. Kirk-Othmer, 1969, Vol. 18, 438
8. Leleu, *Cahiers*, 1975, (79), 270

Sodium floats virtually unchanged on phosphorus tribromide, but added drops of water cause a violent explosion [1]. Molten sodium explodes with phosphorus trichloride, and may ignite or explode with the pentachloride [2]. Diselenium dichloride reacts vigorously with sodium on heating, emitting light and heat [3]. Sodium ignites in sulphinyl chloride vapour at 300°C [3], or in a stream of thiophosphoryl fluoride [4]. The shock-sensitive mixtures of sodium with phosphorus pentachloride, phosphorus tribromide or sulphur dichloride gave very violent explosions on impact, while those with boron tribromide or sulphur dibromide gave strong explosions [5]. Seleninyl bromide reacts explosively [6].

Sodium and phosphoryl chloride interact explosively on heating [7], and mixtures of sodium with sulphinyl fluoride or silicon tetra-chloride or -fluoride are shock-sensitive explosives [8].

Non-metal oxides

1. Mellor, 1940, Vol. 6, 70; 1961, Vol. 2, Suppl. 2.1, 468
2. Gilbert, H. N., *Chem. Eng. News,* 1948, **26**, 2604
3. Mellor, 1940, Vol. 8, 554, 945; 1961, Vol. 2, Suppl. 2.1, 458
4. Leleu, *Cahiers*, 1975, (79), 269

Sodium and carbon dioxide are normally unreactive till red heat is attained [1], but mixtures of the two solids are impact-sensitive and explode violently [2]. Carbon dioxide is unsuitable as an extinguishant for the burning metal alone, as the intensity of combustion is increased by replacing air with carbon dioxide. However, it has been used successfully to extinguish solvent fires where sodium is also present, since it often fails to ignite because of the blanketing effect of solvent vapour. Conversely, addition of kerosene to burning sodium enables the whole to be extinguished with carbon dioxide [1]. Carbon monoxide reacts with sodium in liquid ammonia (but not otherwise) to form sodium carbonyl, which explodes on heating to 90°C [2]. Finely divided silica (sand) will often react with burning sodium, so is not entirely suitable as an extinguishant [1, 2]. (Sodium carbonate and chloride are suitably unreactive for this purpose.) Solid sodium is inert to dry liquid or gaseous sulphur dioxide, but molten sodium reacts violently [2]. The moist gas reacts with sodium almost as vigorously as water [1]. Phosphorus pentaoxide becomes incandescent on warming with sodium, which also ignites in dinitrogen pentaoxide [3]. The product of interaction of carbon monoxide with sodium in liquid ammonia (probably the hexameric sodium benzenehexolate) is also shock-sensitive, and contact with water causes explosion and ignition to occur [4].

Non-metals

1. Mellor, 1961, Vol. 2, Suppl. 2.1, 466–467
2. Ibid., 454–455; Mellor, 1941, Vol. 2, 469
3. Mellor, 1947, Vol. 10, 766

Explosions have occasionally occurred when carbon powder is in contact with evaporating sodium and air [1]. The violent interaction of ground or heated mixtures of sodium and sulphur may be moderated by the presence of sodium chloride or boiling toluene [2]. Selenium reacts incandescently with sodium when heated [3] and molten tellurium reacts vigorously when poured on to solid sodium [2].

Oxidants

*See* Halogens or Interhalogens, above
Metal halides, above
Metal oxides, above
Non-metal oxides, above
NITROSYL FLUORIDE, FNO: Sodium
NITRYL FLUORIDE, $FNO_2$: Metals
AMMONIUM NITRATE, $H_4N_2O_3$: Metals
SODIUM NITRITE, $NNaO_2$: Reducants
SODIUM NITRATE, $NNaO_3$: Sodium

Oxygenated compounds

Leleu, *Cahiers*, 1975, (79), 267

Mixtures of inorganic oxygenated compounds (halide oxides and oxide sulphides) or oxygen-rich organic compounds (alkyl oxalates) with sodium (or its alloy with potassium) are shock-sensitive explosives.

*See also* Non-metal oxides, above

Sulphides

Leleu, *Cahiers*, 1975, (79), 268, 270

Passage of moist hydrogen sulphide over unheated sodium causes melting and then usually ignition of the sodium. Mixtures of sodium (or its alloy with potassium) and carbon disulphide are shock-sensitive explosives.

2,2,3,3-Tetrafluoropropanol

*See* 2,2,3,3-TETRAFLUOROPROPANOL, $C_3H_4F_4O$: Potassium hydroxide, etc.

Water

1. Mellor, 1941, Vol. 2, 469
2. Mellor, 1961, Vol. 2, Suppl. 2.1, 362
3. *MCA Case History No. 456*
4. Markowitz, M.M., *J. Chem. Educ.*, 1963, **40**, 633–636
5. *MCA Case History No. 1653*
6. *MCA Case History No. 2082*

Small pieces of sodium in contact with water react vigorously but do not usually ignite the evolved hydrogen unless the water is at above 40°C, or if rapid dissipation of heat is prevented by immobilising the sodium by use of a viscous solution (starch paste) or wet filter paper [1]. In contact with ice, sodium explodes violently [2]. Small, hot particles of sodium (remaining from dissolution of larger pieces) may finally explode as do large lumps of the metal in water [1]. Sodium residues from a Wurtz reaction were treated with alcohol to destroy sodium, but accidental contact with water caused a fire, showing that the alcohol treatment was incomplete. This may have been due to a crust of alcohol-insoluble halide over the residual sodium [3].

Reactivity of sodium and other alkali metals with water or steam has been discussed in detail [4]. Rolling a drum containing a 12 cm layer of sodium sludge mixed with soda ash led to a mild pressure explosion. Moisture present in the drum contacted the sodium residue when rolled [5]. Preparing to dispose of a few litres of old sodium dispersion by adding a solvent and burning it, an operator opened the container under cover but in an atmosphere rendered humid by very recent rain. The dispersion immediately ignited and exploded [6].

*See* ALKALI METALS
*See other* METALS

**SODIUM–ANTIMONY ALLOY** **Na–Sb**

*See* NITROSYL TRIBROMIDE, $Br_3NO$: Sodium–antimony alloy

**SODIUM–ZINC ALLOY** **Na–Zn**

Houben-Weyl, 1973, Vol. 13.2a, 571
Preparation by addition of sodium to molten zinc in absence of air is rather violent.

**SODIUM OZONATE** **$NaO_3$**

Petrocelli, A. W. *et al., J. Chem. Educ.*, 1962, **39**, 557
This is the least stable of the alkali metal ozonates.
*See related* METAL PEROXIDES

**SODIUM SILICIDE** **NaSi**

Mellor, 1961, Vol. 2, Suppl. 2.1, 564
Bailar, 1973, Vol. 1, 446
Sodium silicide ignites in air, and like its potassium, rubidium and caesium analogues, ignites explosively on contact with water or dilute acids.
*See other* METAL NON-METALLIDES

**SODIUM OXIDE** **$Na_2O$**

2,4-Dinitrotoluene
*See* 2,4-DINITROTOLUENE, $C_7H_6N_2O_4$: Sodium oxide

Phosphorus(V) oxide
*See* PHOSPHORUS(V) OXIDE, $O_5P_2$: Inorganic bases

Water
Interaction is likely to be violently exothermic.

**SODIUM PEROXIDE** **NaOONa** **$Na_2O_2$**

Allen, C. F. H. *et al., J. Chem. Educ.*, 1942, **19**, 72
Rüst, 1948, 297
Hazards attendant upon use of this powerful oxidant may in many cases be eliminated by substitution with sodium perborate. Several boxes of the peroxide (not clearly labelled as such) exploded with great violence during handling operations. It seems likely that contamination with combustible material had occurred.

Acetic acid
von Schwartz, 1918, 321
Admixture causes explosion. This may be due to direct oxidation of the acetic acid, or may involve peracetic acid or its sodium salt.

Aluminium,
Aluminium chloride
Anon., *Chem. Eng. News,* 1954, **32**, 258
A mixture of the three slowly reacted, creating a pressure of 122 bar in 41 days, and the residue reacted spontaneously on exposure to air.
*See* Metals, etc., below

Ammonium peroxodisulphate
Mellor, 1947, Vol. 10, 464
A mixture explodes on being subjected to friction in a mortar, heating above 75°C, or exposure to carbon dioxide or drops of water.

Boron nitride
Mellor, 1940, Vol. 8, 111
Addition of powdered nitride to the molten peroxide causes incandescence.

Calcium acetylide
Mellor, 1941, Vol. 2, 490
A mixture is explosive.

Fibrous materials
von Schwartz, 1918, 328
Kirk-Othmer, 1967, Vol. 14, 749
Betteridge, D., private comm., 1973
Contact of the solid with moist cloth, paper or wood often causes

ignition, and addition of water to intermixed cotton wool and peroxide causes violent ignition.

Hydrogen sulphide

1. Barrs, C. E., *J. R. Inst. Chem.*, 1955, **79**, 43
2. Mellor, 1947, Vol. 10, 129

Solid sodium peroxide causes immediate ignition in contact with gaseous hydrogen sulphide [1]. Barium peroxide and other peroxides behave similarly [2].

Hydroxy compounds

Castrantas, 1965, 4

The exothermic oxidation of ethanol, glycerol, sugar or acetic acid may lead to fire or explosion.

*See* Water, below

Metals

Bunzel, E. G. *et al.*, *Z. Anorg. Allg. Chem.*, 1947, **254**, 20

At 240°C mixtures of finely divided metals (aluminium, iron, tungsten) with the peroxide ignite under high friction, and molybdenum powder reacts explosively.

Metals,
Carbon dioxide

Mellor, 1941, Vol. 2, 490; 1961, Vol. 2, Suppl. 2.1, 633; 1946, Vol. 5, 217

Intimate mixtures with aluminium, magnesium or tin powders ignite on exposure to moist air and become incandescent on heating or on moistening with water (magnesium may explode). Exposure of such mixtures to carbon dioxide causes an explosion. (Interaction of the peroxide and carbon dioxide is highly exothermic.) Sodium is oxidised vigorously at 500°C.

Non-metal halides

1. Mellor, 1947, Vol. 10, 897; 1961, Vol. 2, Suppl. 2.1, 634
2. Comanducci, E., *Chem. Ztg.*, 1911, **15**, 706

Violent interactions occur with diselenium or disulphur dichlorides, the latter emitting light and heat [1]. The very exothermic reaction with phosphorus trichloride may accelerate to explosion [2].

Non-metals

1. von Schwartz, 1918, 328
2. Mellor, 1941, Vol. 2, 490
3. *ABCM Quart. Safety Summ.*, 1948, **19**, 13

Intimate mixtures with carbon or phosphorus ignite or explode [1]. Other readily oxidisable materials (probably antimony, arsenic, boron sulphur, selenium) also form explosive mixtures [2]. Use of finely powdered carbon, rather than the granular carbon specified for a reagent, mixed with sodium peroxide caused an explosion [3].

Organic liquids,
Water

von Schwartz, 1918, 328

Simultaneous contact of sodium peroxide with water and aniline, or benzene, or diethyl ether, or glycerol, causes ignition. (Equivalent to contact with concentrated hydrogen peroxide.)

Organic materials

Rüst, 1948, 337

A 'medicinal' mixture of the peroxide (30%) with liquid paraffin (18%), dried soap (42%) and almond oil (10%) ignited explosively during preparation.

Peroxyformic acid

*See* PEROXYFORMIC ACID, $CH_2O_3$: Metals, etc.

Silver chloride,
Charcoal

Mellor, 1941, Vol. 3, 401

An intimate mixture ignites after a short delay, but the same is probably true if silver chloride is omitted.

*See* Non-metals, above

Water

1. *Haz. Chem. Data*, 1969, 201
2. Friend, J. N. *et al.*, *Nature*, 1934, **134**, 778
3. Cheeseman, G. H. *et al.*, ibid., 971

Reaction with water is vigorous, and with large quantities of peroxide it may be explosive. Contact of the peroxide with combustible materials and traces of water may cause ignition [1]. Violent explosions on two occasions during attempted preparation of oxygen were attributed to presence of sodium metal in the peroxide. The former would liberate hydrogen and ignite the detonable mixture [2,3].

Wood

Dupré, A., *J. Soc. Chem. Ind.*, 1897, **16**, 492

Friction of the peroxide between wooden surfaces causes ignition of the latter.

*See also* Fibrous materials, above

*See other* METAL PEROXIDES

## SODIUM THIOSULPHATE $Na_2[S(S)O_3]$ $Na_2O_3S_2$

Metal nitrates

*See* POTASSIUM NITRATE, $KNO_3$: Sodium thiosulphate
SODIUM NITRATE, $NNaO_3$: Sodium thiosulphate

Sodium nitrite

Stevens, H. P., *J. Proc. R. Inst. Chem.*, 1946, 285

A mixture of these reactants will explode violently when most of the water of crystallisation has been driven off by heating.

*See other* METAL OXONON-METALLATES
REDUCANTS

## SODIUM METASILICATE $(Na_2SiO_3)_n$ $(Na_2O_3Si)_n$

Fluorine

*See* FLUORINE, $F_2$: Metal salts

## SODIUM SULPHATE $Na_2SO_4$ $Na_2O_4S$

Aluminium

ALUMINIUM, Al: Sodium sulphate

## SODIUM DITHIONITE ('HYDROSULPHITE') $Na_2O_4S_2$

NaOS(O)S(O)ONa

*MCA Case History No. 350*

Bailar, 1973, Vol. 2, 882

A batch decomposed violently during drying in a graining bowl. No explanation was offered but contamination with water and/or an oxidant seems likely.

Thermal decomposition occurs violently at 190°C.

Sodium chlorite

*See* SODIUM CHLORITE, $ClNaO_2$: Sodium dithionite

Water

Anon., *Chem. Trade J.*, 1939, **104**, 355

Addition of 10% of water to the solid anhydrous material caused a vigorous exotherm and spontaneous ignition. Bulk material may decompose at 135°C.

*See* METAL OXONON-METALLATES
REDUCANTS

## SODIUM DISULPHITE ('METABISULPHITE') $Na_2O_5S_2$

**NaOS(O)OS(O)ONa**

Sodium nitrite

*See* SODIUM NITRITE, $NNaO_2$: Sodium disulphite

## SODIUM TETRAPEROXOTUNGSTATE(2–) $Na_2O_8W_2$

**$Na_2[W(O_2)_4]$**

Mellor, 1943, Vol. 11, 835

Explodes feebly on warming.

*See other* PEROXOACID SALTS

## SODIUM SULPHIDE $Na_2S$

*ABCM Quart. Safety Summ.*, 1942, **13**, 5

Dunn, B. W., *Bur. Explos. Accid. Bull., Amer. Rail Assoc.*, 1917, **25**, 36

Fused sodium sulphide in small lumps is liable to spontaneous heating, temperatures of up to 120°C being observed after exposure to moisture and oxygen. Packing in hermetically closed containers is essential. Previously, similar material packed in wooden barrels had ignited in transit.

*See also* SULPHUR BLACK

Carbon

Creevey, J., *Chem. Age*, 1941, **44**, 257

Mixtures of sodium sulphide and finely divided carbon react exothermically, probably owing to co-promotion of aerial oxidation.

Diazonium salts

*See* DIAZONIUM SULPHIDES

*N, N*-Dichloromethylamine

*See* *N,N*-DICHLOROMETHYLAMINE, $CH_3Cl_2N$

Sodium carbonate,
Water

*See* SMELT: Water

*See other* METAL SULPHIDES

## SODIUM DISULPHIDE $Na_2S_2$

Diazonium salts

*See* DIAZONIUM SULPHIDES

## SODIUM POLYSULPHIDE $Na_2S_x$

Diazonium salts

*See* DIAZONIUM SULPHIDES

*See also* SULPHUR BLACK

## SODIUM PHOSPHIDE(3–) $Na_3P$

Water

Mellor, 1940, Vol. 8, 834

Sodium (and potassium) phosphide is decomposed by moist air or water, evolving phosphine, which often ignites.

*See other* METAL NON-METALLIDES

## SODIUM PYROPHOSPHATE HYDROGEN PEROXIDATE $Na_4O_7P_2 \cdot 2H_2O_2$

$(NaO)_2P(O)OP(O)(ONa)_2 \cdot 2H_2O_2$

*See* CRYSTALLINE HYDROGEN PEROXIDATES

## NIOBIUM Nb

Halogens,

or Interhalogens

Mellor, 1939, Vol. 9, 849; 1956, Vol. 2, Suppl. 1, 165

Niobium ignites in cold fluorine, and in chlorine at 205°C, and incandesces in contact with bromine trifluoride.

*See other* METALS

## NIOBIUM(V) OXIDE $Nb_2O_5$

Lithium

*See* LITHIUM, Li: Metal oxides

## NEODYMIUM Nd

Phosphorus

*See* PHOSPHORUS, P: Metals

## NICKEL Ni

1. Sasse, W. H. F., *Org. Synth.*, 1966, **46**, 5
2. Whaley, T. P., *Inorg. Synth.*, 1957, 5, 197

3. Brown, C. A. *et al., J. Org. Chem.*, 1973, **38**, 2226
Raney nickel catalysts must not be degassed by heating under vacuum, as large amounts of heat and hydrogen may be evolved suddenly and dangerous explosions may be caused [1]. Nickel powder prepared by several alternative methods may be pyrophoric if particles are fine enough [2]. A non-pyrophoric highly active colloidal hydrogenation catalyst (P-2 nickel) is produced by reduction of nickel acetate in ethanol with sodium tetrahydroborate [3].

Aluminium
*See* ALUMINIUM, Al: Nickel

Aluminium trichloride,
Ethylene
*See* ETHYLENE, $C_2H_4$: Aluminium trichloride

*p*-Dioxane
*See* *p*-DIOXANE, $C_4H_8O_2$: Nickel

Hydrogen
Adkins, H. *et al., J. Amer. Chem. Soc.*, 1948, **70**, 695; *Org. Synth.*, 1955, Coll. Vol. 3, 176
Chadwell, A. J. *et al., J. Phys. Chem.*, 1956, **60**, 1340
During hydrogenation of an unspecified substrate (possibly *p*-nitrotoluene) at high pressure with the highly active W6 type of Raney nickel catalyst at 150°C, a sudden exotherm caused the initial pressure to rapidly double to 680 bar. This does not happen at 100°C or below. Care is necessary with selection of reaction conditions for highly active catalysts. Hydrogen-laden Raney nickel catalyst when heated strongly under vacuum undergoes explosive release of the hydrogen.

Hydrogen,
Oxygen
Lee, W. B., *Ind. Eng. Chem., Prod. Res. Develop.*, 1967, **6**(1), 59–64
Raney nickel powder entrained in cryogenic hydrogen gas causes ignition on contact with liquid or cryogenic gaseous oxygen.
*See* CATALYSIS BY IMPURITIES

Magnesium silicate
Blake, E. J., private comm., 1974
The pyrophoricity of nickel-on-sepiolite catalysts after use in petroleum processing operations may be caused by the presence of finely divided nickel and/or carbon.
*See other* PYROPHORIC CATALYSTS

Methanol
*MCA Case History No. 1225*

Ignition occurred when methanol was poured through the open manhole of a 230 litre reactor containing Raney nickel catalyst. Though the reactor had been thoroughly purged with nitrogen before opening, it was later shown that air was entrained during pouring operations, and this would cause nickel particles round the manhole to glow. A closed system was recommended.

Non-metals

Mellor, 1942, Vol. 15, 148, 151

On heating, mixtures of powdered nickel with sulphur or selenium react incandescently.

*See* Sulphur compounds, below

Organic solvents

Hotta, K. *et al., Chem. Abs.*, 1969, **70**, 81308b

Raney nickel catalyst evaporated with small amounts of methanol, ethanol, isopropanol, pentanol, acetone, benzene, cyclohexane or *p*-dioxane and then heated towards 200°C eventually explodes.

*See* Methanol, above
*p*-DIOXANE, $C_4H_8O_2$: Nickel

Oxidants

*See* BROMINE PENTAFLUORIDE, $BrF_5$: Acids, etc.
PEROXYFORMIC ACID, $CH_2O_3$: Metals
POTASSIUM PERCHLORATE, $ClKO_3$: Metal powders
CHLORINE, $Cl_2$: Metals
NITRYL FLUORIDE, $FNO_2$: Metals
AMMONIUM NITRATE, $H_4N_2O_3$: Metals

Sulphur compounds

Kornfeld, E. C., *J. Org. Chem.*, 1951, **16**, 137

Raney nickel catalyst, containing appreciable amounts of the sulphide (after use to desulphurise thioamides) is rather pyrophoric.

*See* HYDROGENATION CATALYSTS
PYROPHORIC CATALYSTS
*See other* PYROPHORIC METALS

## NICKEL(II) OXIDE NiO

Fluorine

*See* FLUORINE, $F_2$: Metal oxides

Hydrogen peroxide

*See* HYDROGEN PEROXIDE, $H_2O_2$: Metals, etc.

Hydrogen sulphide

*See* HYDROGEN SULPHIDE, $H_2S$: Metal oxides

## NICKEL(IV) OXIDE $NiO_2$

Fluorine
*See* FLUORINE, $F_2$: Metal oxides

## DINICKEL TRIOXIDE $Ni_2O_3$

Hydrogen peroxide
*See* HYDROGEN PEROXIDE, $H_2O_2$: Metals, etc.

Nitroalkanes
*See* NITROALKANES: Metal oxides

## LEAD(II) OXIDE PbO OPb

Aluminium carbide
Mellor, 1946, Vol. 5, 872
The carbide is oxidised with incandescence on warming with lead oxide.
*See* Metal acetylides, below

Chlorinated rubber
*See* CHLORINATED RUBBER: Metal oxides

Chlorine,
Ethylene
*See* ETHYLENE, $C_2H_4$: Chlorine

Dichloro(methyl)silane
*See* DICHLORO(METHYL)SILANE, $CH_4Cl_2Si$: Oxidants

Fluorine,
Glycerol
*See* FLUORINE, $F_2$: Miscellaneous materials

Glycerol,
Perchloric acid
*See* PERCHLORIC ACID, $ClHO_4$: Glycerol, etc.

Hydrogen trisulphide
*See* HYDROGEN TRISULPHIDE, $H_2S_3$: Metal oxides

Linseed oil
Stolyarov, A. A., *Chem. Abs.*, 1935, 29, $3860_3$
Spontaneous ignition occurring during the early stages of mixing and

grinding the components was traced to individual grades of litharge, and the presence of undispersed lumps in the mixture.

Metal acetylides
Mellor, 1946, Vol. 5, 849
Interaction at 200°C with dirubidium acetylide is explosive, and with dilithium acetylide, incandescent.

Metals
Mellor, 1946, Vol. 5, 217; 1941, Vol. 7, 116, 656–658
Mixtures of the oxide with aluminium powder give a violent or explosive 'thermite' reaction on heating. Finely divided sodium ignites on admixture with the oxide, and a mixture of the latter with zirconium explodes on heating. Titanium is also oxidised violently on warming.

Non-metals
Mellor, 1941, Vol. 7, 657
A mixture with boron incandesces on heating, and with silicon the reaction is vigorous. If aluminium is present, the mixture explodes on heating (but the same is true if silicon is absent).

Peroxyformic acid
*See* PEROXYFORMIC ACID, $CH_2O_3$: Metals, etc.

Seleninyl chloride
*See* SELENINYL CHLORIDE, $Cl_2OSe$: Metal oxides
*See other* METAL OXIDES
OXIDANTS

**PALLADIUM(II) OXIDE** **PdO** **OPd**

Hydrogen
Sidgwick, 1950, 1558
It is a strong oxidant and glows in contact with hydrogen at ambient temperature.

**THORIUM OXIDE SULPHIDE** **ThOS** **OSTh**
Mellor, 1941, Vol. 7, 240
Ignites in contact with air.
*See related* METAL SULPHIDES

**ZIRCONIUM OXIDE SULPHIDE** **ZrOS** **OSZr**

Sorbe, 1968, 160

It ignites in air.

*See other* PYROPHORIC MATERIALS

*See related* METAL SULPHIDES

**SILICON OXIDE** **SiO** **OSi**

Zintl, E. *et al.*, *Z. Anorg. Chem.*, 1940, **245**, 1

The freshly prepared material is pyrophoric in air.

*See other* NON-METAL OXIDES

**TIN(II) OXIDE** **SnO** **OSn**

Bailar, 1973, Vol. 2, 64

On heating to 300°C in air, oxidation proceeds incandescently.

Non-metal oxides

Mellor, 1941, Vol. 7, 388

The oxide ignites in nitrous oxide at 400°C, and incandesces when heated in sulphur dioxide.

*See other* METAL OXIDES

**ZINC OXIDE** **ZnO** **OZn**

Chlorinated rubber

*See* CHLORINATED RUBBER: Metal oxides

Linseed oil

Anon., *Chem. Trade J.*, 1933, **92**, 278

Slow addition of zinc white (a voluminous oxide containing much air) to cover the surface of linseed oil varnish caused generation of heat and ignition. Lithopone, a denser oxide, did not cause heating.

Magnesium

*See* MAGNESIUM, Mg: Metal oxides

*See other* METAL OXIDES

**OXYGEN (Gas)** **$O_2$**

Acetaldehyde

*See* ACETALDEHYDE, $C_2H_4O$: Air, or oxygen

Acetone,
Acetylene
*CISHC Chem. Safety Summ.*, 1976, **47**, 33–34
A combination of faulty equipment and careless working led to an extremely violent explosion during oxyacetylene cutting work. The oxygen cylinder was nearly empty and the regulator had a cracked diaphragm. The acetylene cylinder was lying on its side and was feeding a mixture of liquid acetone and acetylene gas to the burner head. When the oxygen ran out, the excess pressure from the acetylene line forced the acetone–acetylene mixture back up the oxygen line and into the cylinder via the cracked diaphragm. The explosion destroyed the whole plant.
*See* ACETYLENE (ETHYNE), $C_2H_2$ : Oxygen

Alcohols
1. Davies, 1961, **80**
2. Redemann, E. G., *J. Amer. Chem. Soc.*, 1942, **64**, 3049
Secondary alcohols are readily autoxidised in contact with oxygen or air, forming ketones and hydrogen peroxide [1]. A partly full bottle of 2-propanol exposed to sunlight for a long period became 0.36M in peroxide and potentially explosive [2].
*See* 2-PROPANOL, $C_3H_8O$: 2-Butanone
HYDROGEN PEROXIDE, $H_2O_2$ : Acetone
: Alcohols

Alkali metals
Mellor, 1941, Vol. 2, 469
Sidgwick, 1950, 65
Reactivity towards air or oxygen increases from lithium to caesium and intensity depends on state of subdivision and on presence or absence of moisture. Lithium normally ignites in air above its melting point, while potassium may ignite after exposure to the atmosphere, unless it is unusually dry. Rubidium and caesium ignite immediately on exposure. It is reported that sodium and potassium may be distilled unchanged under perfectly dried oxygen.

Alkaline earth metals
Mellor, 1941, Vol. 3, 637–638
Finely divided calcium may ignite in air, and the massive metal ignites on heating in air, and burns vigorously at 300°C in oxygen. Strontium and barium behave similarly.

Aluminium–titanium alloys
*See* ALUMINIUM–TITANIUM ALLOYS, Al–Ti: Oxidants

Organic analytical samples

Pack, D. E., *Chem. Eng. News,* 1966, **44**(2), 4
Devereaux, H. D., ibid., (6), 4
Heyssel, R. M. *et al.*, ibid., (10), 4
MacDonald, A. M. G., ibid., (19), 7
Jenkin, J. B., *Chem. & Ind.*, 1974, 585

Hazards involved in use of the oxygen flask combustion technique are discussed and illustrated with seven examples. Some explosions occurred after completion of combustion, when the oxygen concentration in the flask is still *ca.* 75%. Shielding of the flask, minimally with wire gauze, seems advisable.

Explosion during a blank combustion (of a folded filter paper) was attributed possibly to thermal strains. Similar occurrences had been noted previously.

Pentaborane(9)

*See* PENTABORANE(9), $B_5H_9$: Oxygen

Phosphine

*See* PHOSPHINE, $H_3P$: Oxygen

Phosphorus tribromide

Bailar, 1973, Vol. 2, 426

Oxidation of the bromide with gaseous oxygen is not easily controlled and becomes explosive.

Phosphorus trioxide

*See* PHOSPHORUS(III) OXIDE, $O_3P_2$: Oxygen

Polymers

1. *MCA Case History No. 395*
2. *MCA Case History No. 1111*

A foam rubber sample, being tested for oxidation resistance under oxygen at 34 bar at 90°C, exploded with extreme violence after 4 days [1]. Use of Neoprene-lined hose in a high-pressure oxygen manifold caused failure and ignition of the burst hose. Possible ignition sources include adiabatic compressive heating, and friction from vibration of metal fibres in a high-velocity stream of oxygen [2].

Rhenium

Mellor, 1942, Vol. 12, 471

The metal ignites in oxygen at 300°C

Tetrafluoroethylene

1. Pajaczkowski, A., *Chem. & Ind.*, 1964, 659
2. Ger. Pat. 773 900, 1971

Accidental admixture of oxygen gas with unstabilised liquid tetrafluoroethylene produced a polymeric peroxide which was powerfully explosive, and sensitive to heat, impact or friction [1]. Removal of oxygen by treatment with pyrophoric copper to prevent explosion of tetrafluoroethylene has been claimed [2].

*See other* POLYPEROXIDES

Titanium

*ABCM Quart. Safety Summ.*, 1962, **33**, 24

Passage of oxygen through a titanium feed pipe into a titanium autoclave caused a titanium–oxygen fire and explosion at 44 bar. When the surface oxide film is damaged, titanium can ignite at 24 bar under static conditions, and at 3.4 bar under dynamic conditions, with oxygen at ambient temperature.

*See also* ALUMINIUM–TITANIUM ALLOYS, Al–Ti: Oxidants

Titanium trichloride,
Vapours of low ignition temperature

Huneke, K. H., *Draegerheft*, 1976, **304**, 188–22 (*Chem. Abs.*, 1977, **87**, 15458a)

The Draeger oxygen-measuring tube contains titanium trichloride as the indicating substance, and a possible hazard in using these tubes to measure oxygen contents above 25% in gas mixtures is noted. The temperature in the tube can reach 120°C during oxidation of titanium trichloride to titanium dichloride oxide, so the test should not be used if compounds with ignition temperatures below 135°C (e.g. carbon disulphide) are present.

Tricalcium diphosphide

*See* TRICALCIUM DIPHOSPHIDE, $Ca_3P_2$: Oxygen
*See other* NON-METALS
OXIDANTS

**OXYGEN (Liquid)** $O_2$

Although the section below refers to the specific hazards observed with a few organic or oxidisable materials with liquid oxygen, it is probable that most organic materials and inorganic reducing agents are highly hazardous with liquid oxygen under appropriate conditions of contact and initiation.

*See* CRYOGENIC FLUIDS

Acetone

Blau, K., private comm., 1965

Accidental addition of liquid oxygen to vacuum jars containing residual acetone caused a violent explosion. Liquid nitrogen is less hazardous as a trap coolant, but only under controlled conditions.

*See* NITROGEN (Liquid), $N_2$

Acetylene,
Oil

Basyrov, Z. B. *et al., Chem. Abs.*, 1960, **54**, 12587a

The detonation capacity of mixtures of acetylene and liquid oxygen is increased by the presence of organic materials (oils) in the oxygen. Hazards of accumulation of oil in air-liquefaction and -fractionation plants are emphasised.

*See* Hydrocarbons, below

Ammonia,
Ammonium nitrate,
Diphenyl carbonate

Buzard, J. *et al., J. Amer. Chem. Soc.*, 1952, **74**, 2925–2926

During the synthesis of $^{15}$N-labelled urea by interaction of labelled ammonium nitrate, liquid ammonia and diphenyl carbonate in presence of copper powder, a series of explosions of the refrigerated sealed tubes was encountered. This was almost certainly caused by condensation of traces of oxygen in the tubes cooled to −196°C during condensation of ammonia before sealing the tubes. Cooling to −80°C would have been adequate and have avoided the hazard.

*See* NITROGEN (Liquid), $N_2$

Asphalt

Weber, J., *Chem. Ind.*, 1963, **90**, 178

Mechanical impact upon a road surface on to which liquid oxygen had leaked caused a violent explosion.

Carbon

Kirshenbaum, 1965, 4; Stettbacher, A., *Spreng- und Schiess-stoffe*, 71, Zurich, Racher, 1948

Angot, A., *Ann. Mines [13]*, 1934, **5**, 46–51

Mixtures of carbon and liquid oxygen have been used as detonable explosives for some time. Mixtures with carbon black appear unusually sensitive to impact, and a blasting cartridge exploded when dropped.

Carbon,
Iron(II) oxide

Leleu, *Cahiers*, 1975, (78), 121

Carbon containing 3.5% of the oxide explodes on contact with liquid oxygen.

*See* CATALYSIS BY IMPURITIES

Halocarbons

Anon., *Chem. Eng. News,* 1965, **43**(24), 41

Mixture of liquid oxygen with dichloromethane, 1,1,1-trichloroethane, trichloroethylene and 'chlorinated dye penetrants 1 and 2' exploded violently when initiated with a blasting cap. Carbon tetrachloride exploded only mildly, and a partly fluorinated chloroalkane not at all. Trichloroethylene has been used for degreasing metallic parts before use with liquid oxygen, but is not safe.

*See* OXYGEN (Gas): Halocarbons

Hydrocarbons

1. Mellor, 1941, **1**, 379
2. Kirshenbaum, 1956, 17, 22
3. *MCA Case History No. 865*
4. Wright, G. T., *Chem. Eng. Prog.*, 1961, **57**(4), 45–48
5. Rotzler, R. W. *et al.*, ibid., 1960, **56**(6), 68–73
6. Matthews, L. G., ibid., 1961, **57**(4), 48–49
7. Karwat, E., ibid., 1958, **54**(10), 96–101
8. Streng, A. G. *et al., J. Chem. Eng. Data*, 1959, **4**, 127–131
9. Pelman, *AD Rep. 768744/5GA*, Springfield (Va.), USNTIS, 1973

Mixtures of hydrocarbons with liquid oxygen are highly dangerous explosives, not always requiring external initiation. The earliest example must be that of Claudé, who accidentally knocked a lighted candle into a bucket of liquid oxygen in 1903 and suffered from the violent explosion ensuing [1]. Mixtures with liquid methane and benzene are specifically described as explosive [2]. Traces of oil in a liquid oxygen transfer pump caused the explosion described in the Case History [3]. Mixtures with petroleum and absorbent charcoal have been used experimentally as blasting explosives [1]. Addition of a little powdered aluminium to liquid methane–oxygen mixtures increases the explosive power [2].

An explosion in a liquid oxygen evaporator was attributed to the presence of acetylene, arising from unusual plant conditions and higher than usual hydrocarbon concentrations in the atmospheric air taken in for liquefaction [4]. Two similar explosions in main condensers were attributed to presence of acetylene [5] or unspecified hydrocarbons [6]. Suitable precautions are detailed in all cases. Experimental work appeared to implicate presence of ozone as a major contributory factor [7]. The explosive properties of liquid oxygen–methane mixtures were determined [8]. During investigation of an explosion in a portable air liquefaction–separation plant, hydrocarbon oil was found in a silica filtration bed [9].

Liquefied gases

1. Kirshenbaum, 1956, 17, 30, 34
2. Lester, D. H., *Rep. NASA-CR-148711*, Richmond (Va.), USNTIS, 1976

Mixtures with liquefied carbon monoxide, cyanogen (solidified) and methane are highly explosive [1]. Auto-ignition in liquid oxygen–hydrogen propellant systems has been reviewed [2].

Lithium hydride

Kirshenbaum, 1956, 44

Mixtures of lithium hydride powder and liquid oxygen are detonable explosives of greater power than TNT.

Metals

1. Anon., *Chem. Eng. News,* 1957, **35**(24), 90
2. Ibid., (31), 10
3. Austin, C. M. *et al., J. Chem. Educ.,* 1959, **36**, 54, 308
4. *MCA Case History No. 824*
5. Kirshenbaum, 1956, 4, 16
6. *MCA Case History No. 988*
7. Strauss, W. A., *Amer. Inst. Aeronaut. Astronaut. J.*, 1968, **6**, 1753

A demonstration of combustion of aluminium powder in oxygen exploded violently, probably owing to presence of unevaporated liquid at ignition [1]. Stoicheiometric mixtures of the two are explosives much more powerful than TNT [2]. Further details and comments were given later [3]. An aluminium filter in a high-capacity liquid oxygen transfer line exploded violently, possibly owing to frictional or impact ignition of an aluminium component in contact with an abrasive particle [4]. Powdered magnesium, titanium and zirconium mixed with liquid oxygen are detonable [5]. A nitrogen-pressurised liquid oxygen dispenser made of titanium alloy exploded during nitrogen pressurising. The vessel failed because of reaction of the titanium alloy with liquid oxygen. Titanium is more reactive towards oxygen than either stainless steel or aluminium, and should therefore not be used for oxygen service, either gaseous or liquid [6].

Mixtures of liquid oxygen with 48–64% wt. of fine aluminium powder are detonable, and the detonation parameters were investigated [7].

*See* ALUMINIUM, Al: Oxidants
MAGNESIUM, Mg: Oxidants

Organic materials

1. Hauser, R. L. *et al., Adv. Cryog. Eng.*, 1962, **8**, 242–250
2. Clippinger, D. E. *et al., Aircraft Eng.*, 1959, **32**(365), 204–205

Many common polymers, polymeric additives and lubricants oxidise so rapidly after impact in liquid oxygen that they are hazardous. Of those tested, only acrylonitrile–butadiene, poly(cyanoethylsiloxane), poly(dimethylsiloxane) and polystyrene exploded after impact of 6.8–95 J intensity (5–70 ft.lbf). All plasticisers (except dibutyl sebacate) and antioxidants examined were very reactive. A theoretical treatment of rates of energy absorption and transfer is included [1]. Previously, many resins and lubricants had been examined similarly, and 35 were found acceptable for use in liquid oxygen systems [2].

1,3,5-Trioxane
Kirshenbaum, 1956, 31
Mixtures with liquid oxygen are highly explosive.

Wood (+ charcoal)
1. Weber, U., *Chim. Ind.*, 1963, **90**, 178
2. Lang, A., *Chem. Eng. Prog.*, 1962, **58**(2), 71

Several major accidents during handling of tonnage quantities are described [1]. One involved a violent explosion caused by leakage of liquid oxygen on to metal-encased timber (which had previously been charred) and spark ignition from welding in the enriched atmosphere of an air separation plant [2].
*See other* OXIDANTS

## OSMIUM(IV) OXIDE $OsO_2$ $O_2Os$

Preparative hazard
*See* SODIUM CHLORATE, $ClNaO_3$: Osmium

Sidgwick, 1950, 1493
The amorphous form, prepared by dehydration at low temperature, is a pyrophoric powder.
*See other* METAL OXIDES

## LEAD(IV) OXIDE $PbO_2$ $O_2Pb$

Chlorine trifluoride
*See* CHLORINE TRIFLUORIDE, $ClF_3$: Metals, etc.

Hydrogen sulphides
Mellor, 1941, Vol. 7, 689; 1947, Vol. 10, 129, 159
Contact of hydrogen sulphide with dry or moist lead dioxide causes attainment of red heat and ignition. Contact of hydrogen trisulphide with the dioxide causes violent decomposition and ignition.

Metal acetylides or carbides
Mellor, 1946, Vol. 5, 850, 872
Interaction with aluminium carbide is incandescent, and with caesium acetylide at 350°C is explosive. Other related compounds may be expected to be oxidised violently.

Metals

Mellor, 1963, Vol. 2, Suppl. 2.2, 1571; 1940, Vol. 4, 272; 1946, Vol. 5, 217; 1941, Vol. 7, 658, 691

Warm potassium reacts explosively, and sodium probably behaves similarly. Magnesium reacts violently and powdered aluminium probably does also, as the monoxide reacts violently. Mixtures of powdered molybdenum or tungsten with the oxide incandesce on heating.

Metal sulphides

Mellor, 1941, Vol. 3, 745

Calcium, strontium and barium sulphides react vigorously on heating.

Nitroalkanes

*See* NITROALKANES: Metal oxides

Nitrogen compounds

Mellor, 1941, Vol. 7, 637; 1940, Vol. 8, 291

Hydroxylamine ignites in contact with the oxide, while phenylhydrazine immediately reacts vigorously.

Non-metals

Mellor, 1946, Vol. 5, 17; 1941, Vol. 7, 689–690

Boron or yellow phosphorus explodes violently on grinding with the oxide, while red phosphorus ignites. Mixtures with sulphur ignite on grinding or addition of sulphuric acid.

Peroxyformic acid

*See* PEROXYFORMIC ACID, $CH_2O_3$: Metals, etc.

Non-metal halides

Mellor, 1941, Vol. 7, 690; 1947, Vol. 10, 676, 909

Warm phosphorus trichloride reacts with incandescence, and seleninyl chloride vigorously. Sulphonyl dichloride may react explosively.

Potassium

*See* POTASSIUM, K: Metal oxides

Sulphur dioxide

Mellor, 1941, Vol. 7, 689

Interaction is incandescent.

*See other* METAL OXIDES
OXIDANTS

## PALLADIUM OXIDE $PdO_2$ $O_2Pd$

Bailar, 1973, Vol. 3, 1279

The hydrated oxide is a strong oxidant, and slowly evolves oxygen at ambient temperature.

*See other* METAL OXIDES

## PLATINUM(IV) OXIDE $PtO_2$ $O_2Pt$

Acetic acid,
Hydrogen

Adams, R. *et al., Org. Synth.*, 1941, Coll. Vol. 1, 66

Addition of fresh platinum oxide catalyst to a hydrogenation reaction in acetic acid caused an immediate explosion. Several similar incidents, usually involving acetic acid as solvent, are known to the author.

*See other* METAL OXIDES

## SULPHUR DIOXIDE $SO_2$ $O_2S$

*MCA SD-52*, 1953

Preparative hazard

*See* SULPHURIC ACID, $H_2O_4S$: Copper

Barium peroxide

*See* BARIUM PEROXIDE, $BaO_2$: Non-metal oxides

Diethylzinc

*See* DIETHYLZINC, $C_4H_{10}Zn$: Sulphur dioxide

Halogens,
or Interhalogens

*See* BROMINE PENTAFLUORIDE, $BrF_5$: Acids, etc.
CHLORINE TRIFLUORIDE, $ClF_3$: Metals, etc.
FLUORINE, $F_2$: Non-metal oxides

Lithium acetylide–ammonia

*See* LITHIUM ACETYLIDE–AMMONIA, $C_2HLi \cdot H_3N$: Gases

Lithium nitrate,
Propene

*See* PROPENE, $C_3H_6$: Lithium nitrate, etc.

Metal acetylides

Mellor, 1946, Vol. 5, 848–849

Monocaesium or monopotassium acetylides, and the ammoniate of monolithium acetylide, all ignite and incandesce in unheated sulphur dioxide. The di-metal (including sodium) salts appear to be less reactive, needing heat before ignition occurs.

Metal oxides

1. Mellor, 1941, Vol. 2, 487
2. Mellor, 1942, Vol. 13, 715
3. Mellor, 1941, Vol. 7, 388, 689

Dicaesium monoxide [1], iron(II) oxide [2], tin oxide and lead(IV) oxide [3] all ignite and incandesce on heating in the gas.

Metals

1. Mellor, 1943, Vol. 11, 161
2. Mellor, 1942, Vol. 12, 187
3. Gilbert, H. N., *Chem. Eng. News,* 1948, **26**, 2604; Mellor, 1961, Vol. 2, Suppl. 2.1, 468

Finely divided (pyrophoric) chromium incandesces in sulphur dioxide [1], while pyrophoric manganese burns brilliantly on heating in the gas [2]. Molten sodium reacts violently with the dry gas or liquid, while the moist gas reacts as vigorously as water with cold sodium.

Polymeric tubing

*MCA Case History No. 1044*

Plastics tubing normally capable of withstanding an internal pressure of 7 bar failed below 2 bar when used to convey gaseous sulphur dioxide.

Potassium chlorate

*See* POTASSIUM CHLORATE, $ClKO_3$: Sulphur dioxide

Sodium hydride

*See* SODIUM HYDRIDE, HNa: Sulphur dioxide

*See other* NON-METAL OXIDES

**SELENIUM DIOXIDE** **$SeO_2$** **$O_2Se$**

1,3-Bis(trichloromethyl)benzene

*See* 1,3-BIS(TRICHLOROMETHYL)BENZENE, $C_8H_4Cl_6$: Oxidants

Phosphorus trichloride

Mellor, 1940, Vol. 8, 1005

A mixture of the cold components attains red-heat.

*See other* OXIDANTS

## SILICON DIOXIDE $SiO_2$ $O_2Si$

Hydrochloric acid

*MCA Case History No. 1857*

A glass drying trap on a hydrochloric acid storage tank was filled with silica gel instead of the calcium sulphate specified. The glass trap failed, probably caused by thermal shock from the much higher heat of adsorption of water and hydrogen chloride on silica gel.

Metals

*See* MAGNESIUM, Mg: Silicon dioxide
SODIUM, Na: Non-metal oxides

Oxygen difluoride

*See* OXYGEN DIFLUORIDE, $F_2O$: Adsorbents

Ozone

*See* OZONE, $O_3$: Silica gel

Vinyl acetate

*See* VINYL ACETATE, $C_4H_6O_2$: Desiccants

Xenon hexafluoride

*See* XENON HEXAFLUORIDE, $F_6Xe$: Silicon dioxide

*See other* NON-METAL OXIDES

## TIN(IV) OXIDE $SnO_2$ $O_2Sn$

Chlorine trifluoride

*See* CHLORINE TRIFLUORIDE, $ClF_3$: Metals, etc.

Hydrogen trisulphide

*See* HYDROGEN TRISULPHIDE, $H_2S_3$: Metal oxides

Metals

*See* ALUMINIUM, Al: Metal oxides
POTASSIUM, K: Metal oxides
MAGNESIUM, Mg: Metal oxides
SODIUM, Na: Metal oxides

## STRONTIUM PEROXIDE $SrO_2$ $O_2Sr$

Organic materials

*Haz. Chem. Data*, 1975, 272

Mixtures of the peroxide with combustible organic materials very readily ignite on friction or contact with moisture.
*See other* METAL PEROXIDES

**TITANIUM(IV) OXIDE** **$TiO_2$** **$O_2Ti$**

Metals
Mellor, 1941, Vol. 7, 10, 44; 1961, Vol. 2, Suppl.2.1, 81
Reduction of the oxide by aluminium, calcium, magnesium, potassium, sodium or zinc is accompanied by more or less incandescence (lithium magnesium and zinc especially).
*See other* METAL OXIDES

**URANIUM(IV) OXIDE** **$UO_2$** **$O_2U$**
Mellor, 1942, Vol. 12, 42
The finely divided oxide prepared at low temperature is pyrophoric when heated in air, and burns brilliantly.
*See other* METAL OXIDES

**TUNGSTEN(IV) OXIDE** **$WO_2$** **$O_2W$**

Chlorine
*See* CHLORINE, $Cl_2$ : Tungsten dioxide

**ZINC PEROXIDE** **$ZnO_2$** **$O_2Zn$**

Alone,
or Metals
Mellor, 1940, Vol. 4, 530
Sidgwick, 1950, 270
The hydrated peroxide (of indefinite composition) explodes at 212°C, and mixtures with aluminium or zinc powders burn brilliantly.
*See other* OXIDANTS

**OZONE** **$O_3$**

1. Sidgwick, 1950, 860
2. Adley, F. E., *Nucl. Sci. Abs.*, 1962, **17**, 18
3. Clough, P. N. *et al.*, *Chem. & Ind.*, 1966, 1971
4. Gatwood, G. T. *et al.*, *J. Chem. Educ.*, 1969, **46**, A103
5. Streng, A. G., *Explosivstoffe,* 1960, 8, 225
6. Loomis, J. S. *et al.*, *Nucl. Instr. Methods,* 1962, **15**, 243

7. *Ozone Reactions with Organic Compounds,* ACS 112, Washington, ACS, 1972

Pure solid or liquid materials are highly explosive, and evaporation of a solution of ozone in liquid oxygen causes ozone enrichment and ultimately explosion [1]. Organic liquids and oxidisable materials dropped into liquid ozone will also cause explosion of the ozone [2]. Ozone technology and hazards have been reviewed [5], and a safe process to concentrate ozone by selective adsorption on silica gel at low temperature has been described [3], and safe techniques for generation and handling of ozone have been detailed [4]. Ozone is now commercially available as a dissolved gas in cylinders. Explosive hazards involved in use of liquid nitrogen as coolant where ozone may be incidentally produced during operation of van der Graaff generators, etc., have been discussed [6]. The chemistry of interaction with organic compounds has been recently reviewed [7].

Alkenes

Sidgwick, 1950, 862

Davies, 1961, 97

Interaction of alkenes with ozonised oxygen tends to give several types of products or their polymers, some of which show more pronounced explosive tendencies than others. The cyclic *gem*-diperoxides are more explosive than the true ozonides.

*See* CYCLIC PEROXIDES

Alkylmetals

Lee, H. U. *et al., Combust. Flame*, 1975, **24**, 27–34

The spectra of flames from interaction of dimethyl- and diethyl-zinc in a flow system have been studied.

Aromatic compounds

Mellor, 1941, Vol. 1, 911

Benzene, aniline and other aromatic compounds give explosive gelatinous ozonides, among other products, on contact with ozonised oxygen.

Benzene,
Rubber

Morrell, S. H., private comm., 1968

During ozonisation of rubber dissolved in benzene, an explosion occurred. This seems unlikely to have been due to formation of benzene tri-ozonide (which separates as a gelatinous precipitate after prolonged ozonisation), since the solution remained clear. A rubber ozonide may have been involved.

Bromine

Lewis, B. *et al., J. Amer. Chem. Soc.*, 1931, **53**, 2710

Interaction becomes explosive above 20°C and a minimum critical pressure.

*See* Hydrogen bromide, below
BROMINE TRIOXIDE, $BrO_3$

Citronellic acid

Yost, Y., *Chem. Eng. News*, 1977, **55**(21), 4

During work-up of the products of ozonolysis of R- and S-citronellic acids, a substantial quantity of the highly explosive trimeric acetone peroxide (3,3,6,6,9,9-hexamethyl-1,2,4,5,7,8 hexaoxaonane, $C_9H_{18}O_6$) was unwittingly isolated by distillation at 105–135° to give the solid m.p. 95°C. The peroxide appears to have been produced by ozonolysis of the isopropylidene group in citronellic acid, and presumably the same could occur when any isopropylidene group is ozonised. Appropriate care is advised.

Combustible gases

1. Streng, A. G., *Explosivstoffe*, 1960, **8**, 225–235 (*Chem. Abs.*, 1964, **55**, 8862c)
2. Nomura, Y. *et al., Chem. Abs.*, 1952, **46**, 4234a

Carbon monoxide and ethylene admitted into contact with ozone via an aluminium tip ignited and burned smoothly, while normal contact of carbon monoxide, nitrogen oxide, ammonia or phosphine causes immediate explosion at 0° or −78°C [1]. At pressures below 10 mbar, contact with ethylene is explosive at −150°C [2].

*trans*-2,3-Dichloro-2-butene

Griesbaum, K. *et al., J. Amer. Chem. Soc.*, 1976, **98**, 2880

The crude products of ozonolysis at −30°C of the chloroalkene tended to decompose explosively on warming to room temperature, particularly in absence of solvents. The products included the individually explosive compounds:

ACETYL 1,1-DICHLOROETHYL PEROXIDE, $C_4H_6Cl_2O_3$
3,6-DICHLORO-3,6-DIMETHYL-2,3,5,6-TETRAOXANE, $C_4H_6Cl_2O_4$
DIACETYL PEROXIDE, $C_4H_6O_4$

Dicyanogen

Anon., *Chem. Eng. News*, 1957, **35**(21), 22

This hypergolic combination appears to have rocket-propellant capabilities.

*See* ROCKET PROPELLANTS

Diethyl ether

Mellor, 1941, Vol. 1, 911

Contact with ozonised oxygen produces some of the explosive diethyl peroxide.

Dinitrogen tetraoxide

Pinkus, A. *et al.*, *J. Chim. Phys.*, 1920, **18**, 366

Luminescent reaction is often explosive.

Ethylene

Mellor, 1941, Vol. 1, 911

Interaction may be explosive, owing to the very low stability of the ozonide formed.

*See* Combustible gases, above

ETHYLENE OZONIDE, $C_2H_4O_3$

Hydrogen (liquid)

Schwab, G. M., *Umschau*, 1922, 538–539

Liquid hydrogen and solid ozone form very powerfully explosive mixtures.

Hydrogen,

Oxygen difluoride

Boehm, R., *Chem. Abs.*, 1975, **83**, 82254m

The use of liquid oxygen difluoride (40 or 90%) to stabilise liquid ozone as oxidant for gaseous hydrogen in a rocket motor was not entirely successful, explosions occurring at both concentrations.

*See also* OXYGEN DIFLUORIDE, $F_2O$: Combustible gases

Hydrogen bromide

Lewis, B. *et al.*, *J. Amer. Chem. Soc.*, 1931, **53**, 3565

Interaction is rapid and becomes explosive above a total pressure of ~40 mbar, even at −104°C.

*See* Bromine, above

4-Hydroxy-4-methyl-1,6-heptadiene

*See* 4-HYDROXY-4-METHYL-1,6-HEPTADIENE, $C_8H_{14}O$: Ozone

Isopropylidene compounds

*See* Citronellic acid, above

Nitrogen

Strakhov, B. V. *et al.*, *Chem. Abs.*, 1963, **58**, 5069a

The explosive oxidation of gaseous nitrogen in admixture with ozone in metallic vessels has been studied.

Nitrogen oxide

Nomura, Y. *et al.*, *Sci. Reps. Tohoku Univ.*, 1950[A], 2, 229–232

Violent explosions occurred in mixtures at −189°C.

*See also* Combustible gases, above

Nitrogen trichloride

*See* NITROGEN TRICHLORIDE, $Cl_3N$: Initiators

Oxygen fluorides

Streng, A. G., *Chem. Rev.*, 1963, **63**, 613, 619

Mixtures with dioxygen difluoride explode at −148°C, and with 'dioxygen trifluoride' (probably mixed dioxygen di- and tetrafluorides) at −183°C.

Silica gel

Cohen, Z. *et al.*, *J. Org. Chem.*, 1975, **40**, 2142, footnote 6

Silica gel at −78°C adsorbs 4.5% wt. of ozone, and below this temperature the concentration increases rapidly. At below −112°C ozone liquefies and there is a potential explosion hazard at temperatures below −100°C if organic material is present.

*See* reference 3 above

Stibine

Stock, A. *et al.*, *Ber.*, 1905, **38**, 3837

Passage of oxygen containing 2% ozone through stibine at −90°C caused an explosion. On standing, a suspension of solid stibine in a liquid oxygen solution of ozone eventually exploded, as oxygen evaporated and increased the concentration of ozone and the temperature.

Tetrafluorohydrazine

*See* TETRAFLUOROHYDRAZINE, $F_4N_2$: Ozone

Tetramethylammonium hydroxide

Solomon, I. J. *et al.*, *J. Amer. Chem. Soc.*, 1960, **82**, 5640

During interaction to form tetramethylammonium ozonide, constant agitation and slow ozonisation were necessary to prevent ignition.

Unsaturated acetals

Schuster, L., Ger. Offen. 2 514 001, 1976 (*Chem. Abs.*, 1977, **86**, 71908p)

Ozonolysis of unsaturated acetals at −70 to 0°C to give glyoxal monoacetals is uneventful in organic solvents, but leads to explosions in aqueous solutions.

*See other* OXIDANTS

## PHOSPHORUS(III) OXIDE $P_2O_3(P_4O_6)$ $O_3P_2$

Ammonia

Thorpe, T. E. *et al.*, *J. Chem. Soc.*, 1891, **59**, 1019

Interaction of ammonia and the molten oxide under nitrogen is rather violent, and the mixture ignites. Use of a solvent and cooling controls the reaction to produce phosphorus diamide.

Disulphur dichloride
Mellor, 1940, Vol. 8, 898
Interaction is very violent.

Halogens
Thorpe, T. E. *et al., J. Chem. Soc.,* 1891, **59**, 1019
Mellor, 1940, Vol. 8, 897
The oxide ignites in contact with excess chlorine gas, and reacts violently, usually igniting, with liquid bromine.

Organic liquids
1. Thorpe, T. E. *et al., J. Chem. Soc.*, 1890, **57**, 569–573
2. Mellor, 1971, Vol. 8, Suppl. 3, 382

The oxide ignites immediately with ethanol at ambient temperature [1]. The liquid oxide (above 24°C) reacts very violently with methanol, dimethylformamide, dimethyl sulphite or dimethyl sulphoxide (also with arsenic trifluoride) and charring may occur [2].

Oxygen
Mellor, 1940, Vol. 8, 897–898; 1971, Vol. 8, Suppl. 3, 380–381
Interaction with air or oxygen is rapid and, at slightly elevated temperatures in air or at high concentration of oxygen, ignition is very probable, particularly if the oxide is molten (above 22°C) or distributed as a thin layer. The solid in contact with oxygen at 50–60°C ignites and burns very brilliantly. During distillation of phosphorus trioxide, air must be rigorously excluded to avoid the possibility of explosion. The later reference states that the carefully purified oxide does not ignite in oxygen, and that the earlier observations were on material containing traces of white phosphorus.

Phosphorus pentachloride
*See* PHOSPHORUS PENTACHLORIDE, $Cl_5P$: Diphosphorus trioxide

Sulphur
1. Thorpe, T. E. *et al., J. Chem. Soc.,* 1891, **59**, 1019
2. Pernert, J. C. *et al., Chem. Eng. News,* 1949, **27**, 2143

Interaction of a mixture under inert atmosphere above 160°C to form tetraphosphorus tetrathiohexoxide (or phosphorus oxysulphide) is violent and dangerous on scales of working other than small [1]. A safer procedure involving distillation of phosphorus pentaoxide and pentasulphide is described [2].

Sulphuric acid
Mellor, 1940, Vol. 8, 898

Addition of sulphuric acid to the oxide causes violent oxidation, and ignition if more than 1–2 g is used.

Water

Mellor, 1940, Vol. 8, 897

Reaction with cold water is slow, but with hot water, violent, the evolved phosphine igniting. With more than 2 g of oxide, violent explosions occur.

*See other* NON-METAL OXIDES

**DIPALLADIUM TRIOXIDE** $Pd_2O_3$ $O_3Pd_2$

Sidgwick, 1950, 1573

If the hydrated oxide is heated to remove water, it incandesces or explodes, giving the monoxide.

*See other* METAL OXIDES

**SULPHUR TRIOXIDE** $SO_3$ $O_3S$

Acetonitrile,
Sulphuric acid

*See* ACETONITRILE, $C_2H_3N$: Sulphuric acid, etc.

Dimethyl sulphoxide

Seel, F., *Inorg. Synth.*, 1967, **9**, 114

Dissolution of sulphur trioxide in the sulphoxide is very exothermic and must be done slowly with cooling to avoid decomposition.

Dioxane

Sisler, H. H. *et al.*, *Inorg. Synth.*, 1947, **2**, 174

Since the 1:1 addition complex sometimes decomposes violently on storing at ambient temperature, it should only be prepared immediately before use.

Dioxygen difluoride

*See* DIOXYGEN DIFLUORIDE, $F_2O_2$: Sulphur trioxide

Diphenylmercury

*See* DIPHENYLMERCURY, $C_{12}H_{10}Hg$: Non-metal oxides

Formamide,
Iodine,
Pyridine

*See* FORMAMIDE, $CH_3NO$: Iodine, etc.

Metal oxides

Partington, 1967, 706

Mellor, 1941, Vol. 7, 654

The violent interaction causes incandescence with barium and lead oxides.

Nitryl chloride

*See* NITRYL CHLORIDE, $ClNO_2$ : Inorganic materials

Phosphorus

*See* PHOSPHORUS, P: Non-metal oxides

Water

Mellor, 1947, Vol. 10, 344

Interaction is vigorously exothermic, sometimes explosive, with evolution of light and heat. A mixture of oxide and water in ratio of 4:1 completely vaporises, with simultaneous emission of light.

*See other* NON-METAL OXIDES

**SULPHUR TRIOXIDE–DIMETHYLFORMAMIDE** **$O_3S \cdot C_3H_7NO$**

**$SO_3 \cdot Me_2NCHO$**

Koch-Light Laboratories Ltd., private comm., 1976

A bottle of the complex exploded in storage. No cause was established, but diffusive ingress of moisture to form sulphuric acid, and subsequent hydrolysis of the solvent with formation of carbon monoxide appear to be possible contributory factors.

**ANTIMONY(III) OXIDE** **$Sb_2O_3$** **$O_3Sb_2$**

Mellor, 1939, Vol. 9, 425

The powdered oxide ignites on heating in air.

Bromine trifluoride

*See* BROMINE TRIFLUORIDE, $BrF_3$ : Antimony trichloride oxide

Chlorinated rubber

*See* CHLORINATED RUBBER: Metal oxides

**SELENIUM TRIOXIDE** **$SeO_3$** **$O_3Se$**

Organic materials

Sorbe, 1968, 124

A powerful oxidant which forms explosive mixtures with organic materials.

*See other* NON-METAL OXIDES

### TELLURIUM TRIOXIDE $TeO_3$ $O_3Te$

Metals,
or Non-metals
Bailar, 1973, Vol. 2, 971
The yellow $\alpha$-form is a powerful oxidant, reacting violently when heated with a variety of metallic and non-metallic elements.
*See other* NON-METAL OXIDES

### THALLIUM(III) OXIDE $Tl_2O_3$ $O_3Tl_2$

Sulphur,
or Sulphur compounds
Mellor, 1946, Vol. 5, 434
Mixtures of the oxide with sulphur or antimony sulphide explode on grinding in a mortar. Dry hydrogen sulphide ignites, and sometimes feebly explodes, over thallium oxide.
*See other* METAL OXIDES

### VANADIUM(III) OXIDE $V_2O_3$ $O_3V_2$

Sidgwick, 1950, 825
Ignites on heating in air.
*See other* METAL OXIDFS

### TUNGSTEN(VI) OXIDE $WO_3$ $O_3W$

Bromine,
Tungsten
*See* BROMINE, $Br_2$ : Tungsten, etc.

Interhalogens
*See* BROMINE PENTAFLUORIDE, $BrF_5$ : Acids, etc.
CHLORINE TRIFLUORIDE, $ClF_3$ : Metals, etc.

Lithium
*See* LITHIUM, Li: Metal oxides

### XENON TRIOXIDE $XeO_3$ $O_3Xe$

1. Malm, J. G. *et al., Chem. Rev.*, 1965, **65**, 199
2. Chernick, C. L. *et al., J. Chem. Educ.*, 1966, **43**, 619; *Inorg. Synth.*, 1968, **11**, 206, 209
3. Anon., *Chem. Eng. News,* 1963, **41**(13), 45

4. Holloway, J. H., *Talanta,* 1967, **14**, 871–873
5. Shackelford, S. A. *et al., Inorg. Nucl. Chem. Lett.*, 1973, **9**, 609, footnote 9
6. Shackelford, S. A. *et al., J. Org. Chem.*, 1975, **40**, 1869–1875

It is a powerful explosive [1] produced when xenon tetra- or hexafluorides are exposed to moist air and hydrolysed. Some tetrafluoride is usually present in xenon difluoride, so the latter is potentially dangerous. Although safe to handle in small amounts in aqueous solution, great care must be taken to avoid solutions drying out, e.g. around ground stoppers [2]. Full safety precautions have been discussed [2–4]. Precautions necessary for use of aqueous solutions of the trioxide as an epoxidation reagent are detailed [5,6].

*See other* XENON COMPOUNDS

**OSMIUM(VIII) OXIDE** $OsO_4$ $O_4Os$

Hydrogen peroxide

*See* HYDROGEN PEROXIDE, $H_2O_2$: Metals, etc.

**TETRAPHOSPHORUS TETRAOXIDE TRISULPHIDE** $O_4P_4S_3$

$P_4O_4S_3$

Water

Mellor, 1971, Vol. 8, Suppl. 3, 435

It ignites if moistened with water.

*See related* NON-METAL OXIDES
NON-METAL SULPHIDES

**LEAD SULPHATE** $PbSO_4$ $O_4PbS$

Potassium

*See* POTASSIUM, K: Oxidants

**TRILEAD TETRAOXIDE** $2PbO \cdot PbO_2$ $O_4Pb_3$

Dichloro(methyl)silane

*See* DICHLORO(METHYL)SILANE, $CH_4Cl_2Si$: Oxidants

Peroxyformic acid

*See* PEROXYFORMIC ACID, $CH_2O_3$: Metals, etc.

Seleninyl chloride

*See* SELENINYL CHLORIDE, $Cl_2OSe$: Metal oxides

2,4,6-Trinitrotoluene

*See* 2,4,6-TRINITROTOLUENE, $C_7H_5N_3O_6$ : Added impurities

**RUTHENIUM(VIII) OXIDE** **$RuO_4$** **$O_4Ru$**

Mellor, 1942, Vol. 15, 520
Brauer, 1965, Vol. 2, 1600
Bailar, 1973, Vol. 3, 1194
The liquid (which shows no true boiling point) is said to decompose with explosive violence either above 106° or above 180°C.

Ammonia

Watt, J. *et al.*, *J. Inorg. Nucl. Chem.*, 1965, **27**, 262
Rapid interaction at –70°C/1 mbar causes ignition. Slow interaction produced a solid which exploded at 206°C, probably owing to the formation of $RuO(NH_2)$ and RuN.

Organic materials

Mellor, 1942, Vol. 15, 520
Brauer, 1965, Vol. 2, 1600
The solid material, or its concentrated solutions or vapour, tends to oxidise ethanol, cellulose fibres, etc., explosively.
*See other* METAL OXIDES
OXIDANTS

**XENON TETRAOXIDE** **$XeO_4$** **$O_4Xe$**

Selig, H. *et al., Science, N.Y.*, 1964, **143**, 1322
Decomposition may be explosive, even at –40°C.
*See other* XENON COMPOUNDS

**PHOSPHORUS(V) OXIDE** **$P_2O_5(P_4O_{10})$** **$O_5P_2$**

*MCA SD-28,* 1948
Safe handling procedures for this extremely powerful desiccant are detailed.

Barium sulphide

*See* BARIUM SULPHIDE, BaS: Phosphorus(V) oxide

Formic acid

*See* FORMIC ACID, $CH_2O_2$ : Phosphorus pentaoxide

Hydrogen fluoride

Gore, G., *J. Chem. Soc.*, 1869, **22**, 368
Interaction is vigorous below 20°C.

Inorganic bases

Mellor, 1940, Vol. 8, 945

Van Wazer, 1958, Vol. 1, 279

Dry mixtures with sodium or calcium oxides do not react in the cold, but interact violently if warmed or moistened, evolving phosphorus(V) oxide vapour. A mixture of the oxide and sodium carbonate may be initiated by strong local heating, when the whole mass will suddenly become hot.

Iodides

Pascal, 1960, Vol. 16.1, 538

Interaction is violent.

Metals,

Mellor, 1940, Vol. 8, 945

Interaction with warm sodium or potassium is incandescent and explosive with heated calcium.

Methyl hydroperoxide

*See* METHYL HYDROPEROXIDE, $CH_4O_2$: Phosphorus(V) oxide

Oxidants

*See* BROMINE PENTAFLUORIDE, $BrF_5$: Acids, etc.
CHLORINE TRIFLUORIDE, $ClF_3$: Metals, etc.
PERCHLORIC ACID, $ClHO_4$: Dehydrating agents
OXYGEN DIFLUORIDE, $F_2O$: Phosphorus(V) oxide
HYDROGEN PEROXIDE, $H_2O_2$: Phosphorus(V) oxide

Water

Mellor, 1940, Vol. 8, 944

*MCA SD-28*, 1948

Interaction with water is very energetic and highly exothermic. The increase in temperature may be enough to ignite combustible materials if present and in contact.

*See other* NON-METAL OXIDES

**TANTALUM(V) OXIDE** **$Ta_2O_5$** $O_5Ta_2$

Chlorine trifluoride

*See* CHLORINE TRIFLUORIDE, $ClF_3$: Metals, etc.

**VANADIUM(V) OXIDE** **$V_2O_5$** $O_5V_2$

Chlorine trifluoride

*See* CHLORINE TRIFLUORIDE, $ClF_3$: Metals, etc.

Lithium

*See* LITHIUM, Li: Metal oxides

Peroxyformic acid

*See* PEROXYFORMIC ACID, $CH_2O_3$: Metals, etc.

**TETRAPHOSPHORUS HEXAOXIDE–BIS(BORANE)** **$P_4O_6 \cdot 2BH_3$** **$O_6P_4 \cdot B_2H_6$**

Water

Mellor, 1971, Vol. 8, Suppl. 3, 382

The compound ($P_4O_6 \cdot 2BH_3$) ignites in contact with a little water.

*See related* NON-METAL OXIDES

**TETRAPHOSPHORUS HEXAOXIDE TETRASULPHIDE** **$P_4O_6S_4$** **$O_6P_4S_4$**

Preparative hazard

*See* PHOSPHORUS(III) OXIDE, $O_3P_2$: Sulphur

**DISULPHUR HEPTAOXIDE** **$(OSO_2OOSO_2)_n$** **$O_7S_2$**

Meyer, F. *et al.*, *Ber.*, 1922, **55**, 2923

The crystalline material (of unknown structure) soon explodes on exposure to moist air.

*See other* NON-METAL OXIDES

**TRIURANIUM OCTAOXIDE** **$U_3O_8$** **$O_8U_3$**

Barium oxide

Mellor, 1942, Vol. 12, 49

Interaction below 300°C is vigorous and very exothermic.

*See other* METAL OXIDES

**OSMIUM** **Os**

Chlorine trifluoride

*See* CHLORINE TRIFLUORIDE, $ClF_3$: Metals

Fluorine

*See* FLUORINE, $F_2$: Metals

Phosphorus

*See* PHOSPHORUS, P: Metals

## † PHOSPHORUS $P_4$ P

1. MCA SD-16, 1947
2. Anon., *Chem. Trade J.*, 1936, **98**, 522
3. *CISHC Chem. Safety Summ.*, 1974–1975, **45–46**, 3

Handling precautions for the reactive allotrope, white (or yellow) phosphorus, are detailed [1]. Red phosphorus powder being charged into a reaction vessel exploded when the lid of the tin fell in and was struck by the agitator [2]. A further case of ignition of red phosphorus by impact from its metal container is reported [3].

*See* Oxygen, below

Alkalies

Mellor, 1940, Vol. 8, 802

Contact of phosphorus with boiling caustic alkalies or with hot calcium hydroxide evolves phosphine, which usually ignites in air.

Chlorosulphuric acid

Heumann, K. *et al., Ber.,* 1882, **15**, 417

White phosphorus begins to reduce the acid at 25–30°C and the vigorous reaction accelerates to explosion. Red phosphorus is similar at a higher initial temperature.

Cyanogen iodide

Mellor, 1940, Vol. 8, 791

Molten phosphorus reacts incandescently.

Halogen azides

*See* HALOGEN AZIDES

Halogen oxides

*See* CHLORINE DIOXIDE, $ClO_2$: Non-metals
DICHLORINE OXIDE, $Cl_2O$: Oxidisable materials
OXYGEN DIFLUORIDE, $F_2O$: Non-metals
TRIOXYGEN DIFLUORIDE, $F_2O_3$

Halogens,
or Interhalogens

1. Mellor, 1940, Vol. 8, 785; 1956, Vol. 2, Suppl.1, 379
2. Christomanos, A. C., *Z. Anorg. Chem.*, 1904, **41**, 279
3. Kuhn, R. *et al., Helv. Chim. Acta*, 1928, **11**, 107
4. Newkome, G. R. *et al., J. Chem. Soc., Chem. Comm.*, 1975, 885–886

Both yellow and red phosphorus ignite on contact with fluorine and chlorine; red ignites in liquid bromine or in a heptane solution of chlorine at 0°C. Yellow phosphorus explodes in liquid bromine or chlorine, and ignites in contact with bromine vapour or solid iodine [1]. Interaction of bromine and white phosphorus in carbon disulphide gives a slimy by-product which explodes violently on heating [2].

Interaction of phosphorus and iodine in carbon disulphide is rather rapid [3]. A less hazardous preparation of diphosphorus tetraiodide from phosphorus trichloride and potassium iodide in ether is recommended [4].

*See* BROMINE TRIFLUORIDE, $BrF_3$: Halogens, etc.
BROMINE PENTAFLUORIDE, $BrF_5$: Acids, etc.
CHLORINE TRIFLUORIDE, $ClF_3$: Metals, etc.
IODINE TRICHLORIDE, $Cl_3I$: Phosphorus
IODINE PENTAFLUORIDE, $F_5I$: Metals, etc.

Hexalithium disilicide

Mellor, 1940, Vol. 6, 169

Interaction is incandescent.

Hydriodic acid

Blau, K., private comm., 1965

Agranat, I. *et al., J. Chem. Soc., Perkin Trans. I*, 1974, 1159

Villain, F. J. *et al., J. Med. Chem.*, 1964, 7, 457, footnote 9

During removal of free iodine from hydriodic acid by distillation from red phosphorus, phosphine was produced. When air was admixed by changing the receiver, an explosion occurred. Omission of distillation, by boiling the reactants in inert atmosphere, and separating the phosphorus by filtration through a sintered glass funnel containing solid carbon dioxide, gave a colourless product.

Potential hazards incidental to the use of red phosphorus and hydriodic acid as a reducing agent for organic carbonyl compounds have been noted. Accumulation in a reflux condenser of phosphonium iodide which will react violently with water, and two unexplained explosions, possibly due to the same cause, have been reported.

Hydrogen peroxide

*See* HYDROGEN PEROXIDE, $H_2O_2$: Phosphorus

Magnesium perchlorate

*1965 Summary of Serious Accidents,* Washington, USAEC, 1966

Phosphorus and magnesium perchlorate exploded violently while being mixed. It seems likely that this was initiated by interaction of phosphorus and traces of perchloric acid to form magnesium phosphate. The acid may have been present in the magnesium perchlorate, or have been formed by traces of phosphoric acid in the phosphorus. Once formed, the perchloric acid would be rendered anhydrous by the powerful dehyrating action of magnesium perchlorate.

Metal acetylides

Mellor, 1946, Vol. 5, 848–849

Mono-rubidium and -caesium acetylides incandesce with warm phosphorus. Di-lithium and -sodium acetylides burn vigorously in phosphorus vapour; the di-potassium, -rubidium and -caesium derivatives should react with increasing violence.

Metal halogenates

Mellor, 1941, Vol. 2, 310; 1940, Vol. 8, 785–786

Dry finely divided mixtures of red (or white) phosphorus with chlorates, bromates or iodates of barium, calcium, magnesium, potassium, sodium or zinc will readily explode upon initiation by friction, impact or heat. Fires have been caused by accidental contact in the pocket between the red phosphorus in the friction strip on safety match boxes and potassium chlorate tablets. Addition of a little water to a mixture of white or red phosphorus and potassium iodate causes a violent or explosive reaction. Addition of a little of a solution of phosphorus in carbon disulphide to potassium chlorate causes an explosion when the solvent evaporates. The extreme danger of mixtures of red phosphorus (or sulphur) with chlorates was recognised in the UK some 40 years ago when unlicensed preparation of such mixtures was prohibited by Orders in Council.

Metal halides

Mellor, 1939, Vol. 9, 467; 1943, Vol. 11, 234

Phosphorus ignites in contact with antimony pentafluoride and explodes with chromyl chloride in presence of moisture at ambient temperature.

Metal oxides

Mellor, 1941, Vol. 2, 490; Vol. 7, 690; 1940, Vol. 8, 792; 1943, Vol. 11, 234

von Schwartz, 1918, 328

Red phosphorus reacts vigorously on heating with copper oxide or manganese dioxide; and on grinding or warming with lead monoxide or mercury or silver oxides, ignition may occur. Red phosphorus ignites in contact with lead, sodium or potassium peroxides, while white phosphorus explodes, and also in contact with molten chromium trioxide at 200°C.

Metals

1. Mellor, 1941, Vol. 7, 115; 1940, Vol. 8, 842, 847, 853
2. Van Wazer, 1958, Vol. 1, 159
3. Mellor, 1942, Vol. 15, 696; 1937, Vol. 16, 160

Reaction of beryllium, manganese, thorium or zirconium is incandescent when heated with phosphorus [1] and that of cerium, lanthanum, neodymium and praseodymium is violent above 400°C [2]. Osmium incandesces in phosphorus vapour, and platinum burns vividly below red-heat [3].

*See* ALUMINIUM, Al: Non-metals

Metal sulphates

Berger, E., *Compt. Rend.*, 1920, **170**, 1492

Excess red phosphorus will burn admixed with barium or calcium sulphates if primed at a high temperature with potassium nitrate–calcium silicide mixture.

*See* POTASSIUM NITRATE, $KNO_3$ : Calcium silicide

Nitrates

Mellor, 1941, Vol. 3, 470; 1940, Vol. 4, 987; Vol. 8, 788

Anon., *Jahresber.*, 1975, 83–84, Berufsgennossenschaft der Chem. Ind.

Yellow phosphorus ignites in molten ammonium nitrate, and mixtures of phosphorus with ammonium, mercury (I) and silver nitrates explode on impact. Red phosphorus is oxidised vigorously when heated with potassium nitrate.

During development of new refining agents for aluminium manufacture, a mixture containing 16% of red phosphorus and 35% sodium nitrate was being pressed into 400 g tablets. When the die pressure was increased to 70 bar, a violent explosion occurred.

*See* HEPTASILVER NITRATE OCTAOXIDE, $Ag_7NO_{11}$

Nitric acid

*See* NITRIC ACID, $HNO_3$ : Non-metals

Nitrogen halides

*See* NITROGEN TRIBROMIDE HEXAAMMONIATE, $Br_3N \cdot H_{18}N_6$ : Initiators
NITROGEN TRICHLORIDE, $Cl_3N$: Initiators

Nitrosyl fluoride

*See* NITROSYL FLUORIDE, FNO: Metals, etc.

Nitryl fluoride

*See* NITRYL FLUORIDE, $FNO_2$ : Non-metals

Non-metal halides

Pascal, 1960, Vol. 13.2, 1162

Mellor, 1946, Vol. 5, 136; 1940, Vol. 8, 787; 1947, Vol. 10, 906

Either the white or red form incandesces with boron triiodide. Red phosphorus incandesces in seleninyl chloride while white explodes. Red phosphorus reacts vigorously on warming with sulphuryl or disulphuryl chlorides, and violently with disulphur dibromide.

Non-metal oxides

Mellor, 1940, Vol. 8, 786–787

Warm or molten white phosphorus burns vigorously in nitrogen oxide, dinitrogen tetroxide or pentaoxide. White phosphorus ignites after some delay in contact with the vapour of sulphur trioxide, but immediately in contact with the liquid if a large piece is used.

Oxygen
Mellor, 1940, Vol. 8, 771
Van Wazer, 1958, 1, 97
Mellor, 1971, Vol. 8, Suppl. 3, 237–245
The reactivity of phosphorus with air or oxygen depends on the allotrope of phosphorus involved and the conditions of contact, white (yellow) phosphorus being by far the most reactive. White phosphorus readily ignites in air if warmed, finely divided (e.g. from an evaporating solution) or under conditions where the slow oxidative exotherm cannot be dissipated. Contact with finely divided charcoal or lampblack promotes ignition, probably by the adsorbed oxygen. Contact with amalgamated aluminium also promotes ignition.

Oxidation of phosphorus under both explosive and non-explosive conditions, and the effects of other gases, has recently been reviewed in detail.

Peroxyformic acid
*See* PEROXYFORMIC ACID, $CH_2O_3$: Non-metals

Potassium nitride
Mellor, 1940, Vol. 8, 99
Potassium and other alkali metal nitrides react on heating with phosphorus to give a highly inflammable mixture which evolves ammonia and phosphine with water

Potassium permanganate
Mellor, 1942, Vol. 12, 322
Grinding mixtures of phosphorus and potassium permanganate causes explosion, more violent if also heated.

Selenium
*See* SELENIUM, Se: Phosphorus

Sulphur
*See* SULPHUR, S: Non-metals

Sulphuric acid
*See* SULPHURIC ACID, $H_2O_4S$: Phosphorus
*See other* NON-METALS

## PHOSPHORUS(V) SULPHIDE ($(P_4S_{10})$) $P_2S_5$

Air

*MCA SD-71,* 1958

The general reactivity of the sulphide depends markedly on the physical form, and differences of a factor of 10 may be involved. It is ignitable by friction, sparks or flames, and ignites if heated in dry air close to the m.p., 275–280°C. The dust (200-mesh) forms explosive mixtures in air above a concentration of 0.5% w/v.

Alcohols

1. *MCA Case History No. 227*
2. Ashford, J. S., private comm., 1964

A mixture of the sulphide, ethylene glycol and hexane in a mantle-heated flask spontaneously overheated and exploded at an internal temperature of *ca.* 180°C. It had been intended to maintain the reaction at 60°C, but since alcoholysis of phosphorus pentasulphide is exothermic, presence of the heating mantle prevented dissipation of heat, and the reaction accelerated continuously until explosive decomposition occurred [1]. An incident in similar circumstances involving interaction of the sulphide with 4-methyl-2-pentanol also led to violent eruption of the flask contents [2].

Water

*Haz. Chem. Data,* 1971, 180

It heats and may ignite in contact with limited amounts of water, hydrogen sulphide also being evolved.

*See other* NON-METAL SULPHIDES

## TRIZINC DIPHOSPHIDE $Zn_3P_2$ $P_2Zn_3$

Perchloric acid

*See* PERCHLORIC ACID, $ClHO_4$: Trizinc diphosphide

Water

Fehse, W., *Feuerschutz*, 1938, **18,** 17

Contact of one drop of water with a zinc phosphide rodenticide preparation caused ignition.

*See other* METAL NON-METALLIDES

## TETRAPHOSPHORUS TRISULPHIDE $P_4S_3$

Gibson, 1969, 118–119

It ignites in air above 100°C.

*See other* NON-METAL SULPHIDES

**TETRAPHOSPHORUS TRISELENIDE** $Se_3P_4$ $P_4Se_3$

Mellor, 1947, Vol. 10, 790

Ignites when heated in air.

**LEAD PENTAPHOSPHIDE** $PbP_5$ $P_5Pb$

Bailar, 1973, Vol. 2, 117

It ignites in air.

*See other* METAL NON-METALLIDES
PYROPHORIC MATERIALS

**LEAD** Pb

Mézáros, L., *Tetrahedron Lett.*, 1967, 4951

The finely divided lead produced by reduction of the oxide with furfural vapour at 290°C is pyrophoric and chemically reactive.

Disodium acetylide

*See* DISODIUM ACETYLIDE, $C_2Na_2$ : Metals

*See other* METALS
PYROPHORIC METALS

Hydrogen peroxide,
Trioxane

*See* HYDROGEN PEROXIDE, $H_2O_2$ : Lead, Trioxane

Nitric acid,
Rubber

*See* NITRIC ACID, $HNO_3$ : Lead-containing rubber

Oxidants

*See* CHLORINE TRIFLUORIDE, $ClF_3$ : Metals, etc.
HYDROGEN PEROXIDE, $H_2O_2$ : Metals
AMMONIUM NITRATE, $H_4N_2O_3$ : Metals

**ZIRCONIUM–LEAD ALLOYS** Pb–Zr

Alexander, P. P., US Pat., 2 611 316, 1952

Zirconium–lead alloys containing 10–70% of the former will pulverise and ignite on impact.

*See other* ALLOYS

**PALLADIUM** Pd

Arsenic

*See* ARSENIC, As: Metals

Carbon

1. Mozingo, R., *Org. Synth.*, 1955, Coll. Vol. 3, 687
2. Rachlin, A. I. *et al., Org. Synth.*, 1971, **51**, 9
3. Fraser, R., private comm., 1973

Palladium-on-carbon catalysts prepared by formaldehyde reduction are less pyrophoric than those reduced with hydrogen [1]. Palladium-on-carbon catalysts become extremely pyrophoric on thorough vacuum drying [2]. Those prepared on high surface area supports (up to 2000 $m^2/g$) are highly active and will readily cause catalytic ignition of hydrogen/air or solvent/air mixtures. Methanol is notable for easy ignition because of its high volatility [3].

Formic acid

*See* FORMIC ACID, $CH_2O_2$: Palladium–carbon catalyst

Ozonides

*See* OZONIDES

Sodium tetrahydroborate

Augustine, 1968, 75

In preparation for the reduction of a nitro compound, the tetrahydroborate solution is added to an aqueous suspension of palladium-on-charcoal catalyst. Addition of catalyst to the tetrahydroborate solution may cause ignition of liberated hydrogen.

Sulphur

*See* SULPHUR, S: Metals

*See other* HYDROGENATION CATALYSTS
PYROPHORIC METALS

## PLATINUM Pt

1. van Campen, M. G., *Chem. Eng. News,* 1954, **32**, 4698
2. Tilford, C. H., private comm., 1965

Finely divided (catalytic forms) of platinum are hazardous to handle if allowed to dry. Used Adams' hydrogenation catalyst exploded while being sieved in air. It had been well washed, finally with water, and then air-dried [1]. Manipulation of used catalyst on filter paper caused a violent explosion [2]. The explosive oxidation of adsorbed hydrogen is suspected here.

*See also* ZINC, Zn: Catalytic metals

Acetone,
Nitrosyl chloride

*See* NITROSYL CHLORIDE, ClNO: Acetone, etc.

Arsenic

Wöhler, L., *Z. Anorg. Chem.*, 1930, **186**, 324

During interaction to give platinum diarsenide in a sealed tube at 270°C, the highly exothermic reaction may explode.

Dioxygen difluoride

*See* DIOXYGEN DIFLUORIDE, $F_2O_2$

Ethanol

Elkins, J. S., private comm., 1968

Addition of platinum-black catalyst to ethanol caused ignition. Pre-reduction with hydrogen and/or nitrogen purging of air prevented this.

Hydrazine

*See* HYDRAZINE, $H_4N_2$: Catalysts

Hydrogen,
Air

Coleman, J. W., private comm., 1965

A platinised alumina catalyst had been treated with flowing hydrogen at ambient temperature. Purging with air caused an explosion. Inert gas purging is essential after treating catalysts or adsorbents with hydrogen.

Hydrogen peroxide

*See* HYDROGEN PEROXIDE, $H_2O_2$: Metals

Lithium

Nash, C. P., *Chem. Eng. News,* 1961, **39**(5), 42

Interaction to form an intermetallic compound sets in violently at 540°±20°C.

Methyl hydroperoxide

*See* METHYL HYDROPEROXIDE, $CH_4O_2$: Platinum

Ozonides

*See* OZONIDES

Peroxomonosulphuric acid

*See* PEROXOMONOSULPHURIC ACID, $H_2O_5S$: Catalysts

Phosphorus

*See* PHOSPHORUS, P: Metals

Selenium or Tellurium

Mellor, 1937, Vol. 16, 158

Finely divided or spongy platinum reacts incandescently when heated with selenium or tellurium.

*See other* HYDROGENATION CATALYSTS
PYROPHORIC METALS

**PLUTONIUM** **Pu**

Williamson, G. K., *Chem. & Ind.*, 1960, 1384

This pyrophoric metal may be worked under nitrogen, argon or helium (but not carbon dioxide) at temperatures above 600°C.

Air

Bailar, 1973, Vol. 5, 44–45

The reactive metal is pyrophoric in air at elevated temperatures; lathe turnings have ignited at 265°C and 140-mesh powder at 135°C.

Carbon tetrachloride

*Serious Accid. Series,* No. 246, Washington, USAEC, 1965

While draining after degreasing in the solvent, plutonium chips ignited and, when accidentally dropped into the solvent, caused an explosion.

*See other* METALS: Halocarbons

Water

*MCA Case History No. 1212*

Plutonium components, normally kept under argon, were accidentally exposed to air and moisture, probably forming plutonium hydride and hydrated oxides. When the plastics containing bag was disturbed, ignition occurred, causing widespread radiation contamination.

*See* PLUTONIUM HYDRIDE, $H_3Pu$
*See other* PYROPHORIC METALS

**RUBIDIUM** **Rb**

Rubidium is a typical but very reactive member of the series of alkali metals. It is appreciably more reactive than potassium but less so than caesium, and so would be expected to react more violently with those materials which are hazardous with potassium or sodium.

Air,
or Oxygen

Mellor, 1941, Vol. 2, 468; 1963, Vol. 2, Suppl. 2.2, 2172

Rubidium ignites on exposure to air or dry oxygen, largely forming the oxide.

Halogens

Mellor, 1941, Vol. 2, 13; 1963, Vol. 2, Suppl. 2.2, 2174
Ignition occurs in fluorine, chlorine, and bromine or iodine vapours. Contact with liquid bromine would be expected to cause a violent explosion.

Mercury

Mellor, 1941, Vol. 2, 469
Interaction is exothermic and may be violent.

Non-metals

Mellor, 1963, Vol. 2, Suppl. 2.2, 2174–2175
The molten metal ignites in sulphur vapour at 200–300°C, and reacts with various forms of carbon exothermically, one product, rubidium octacarbide, being pyrophoric.

Vanadium trichloride oxide

Mellor, 1963, Vol. 2, Suppl. 2.2, 2176
Interaction is violent at 60°C.

Water

Mellor, 1941, Vol. 2, 469
Markowitz, M. M., *J. Chem. Educ.*, 1963, **40**, 633–636
Contact with cold water is exothermic enough to ignite the hydrogen evolved. Reactivity of rubidium and other alkali metals with water has been discussed in detail.
*See other* ALKALI METALS

## RHENIUM — Re

Oxygen

*See* OXYGEN (Gas), $O_2$: Rhenium

## RHENIUM(VII) SULPHIDE — $Re_2S_7$

Sidgwick, 1950, 1299
Readily oxidised in air, sometimes igniting.
*See other* METAL SULPHIDES

## RHODIUM — Rh

Sidgwick, 1950, 1513
Metallic rhodium prepared by heating its compounds in hydrogen must be allowed to cool in an inert atmosphere to

prevent catalytic ignition of the adsorbed hydrogen on exposure to air.

*See also* ZINC, Zn: Catalytic metals

Interhalogens

*See* BROMINE PENTAFLUORIDE, $BrF_5$ : Acids, etc.
CHLORINE TRIFLUORIDE, $ClF_3$ : Metals

*See other* HYDROGENATION CATALYSTS
PYROPHORIC METALS

## RUTHENIUM

**Ru**

Aqua regia,
Potassium chlorate

*See* POTASSIUM CHLORATE, $ClKO_3$ : Aqua regia, etc.

## RUTHENIUM SALTS

**$Ru^{2+ \text{ or } 3+}$**

Sodium tetrahydroborate

*See* SODIUM TETRAHYDROBORATE, $BH_4Na$: Ruthenium salts

## SULPHUR

**$S_8$** **S**

*MCA SD-74,* 1959

It may become ignited by frictional heat or incendive sparks, particularly when suspended as dust in air, for which the approximate explosive limits are 35–1400 g/m$^3$. Handling precautions for the solid and molten element are detailed. In the hazardous reactions given below, the dual capacity of sulphur to oxidise and be oxidised is apparent.

Aluminium,
Copper

*See* ALUMINIUM, Al: Copper, Sulphur

Diethyl ether

Taylor, H. F., *J. R. Inst. Chem.,* 1955, **79**, 43

Evaporation of an ethereal extract of sulphur caused an explosion of great violence. Experiment showed that evaporation of wet, peroxidised ether gave a mildly explosive residue, which became violently explosive on addition of sulphur.

Fibreglass,
Iron

Anon., *Info. Circ. No. 8272(10)*, Washington, US Bur. Mines, 1964

A mixture of sulphur with iron filings and fibreglass reacts exothermically at 125–145°C.

*See* Metals, below

Fluorine

*See* FLUORINE, $F_2$: Non-metals

Halogen oxides

Mellor, 1941, Vol. 2, 289, 241, 295; 1956, Vol. 2, Suppl. 1, 540, 542

Sulphur ignites in chlorine dioxide gas, and may cause an explosion, which usually happens on contact of sulphur with the very unstable dichlorine monoxide. Iodine(V) oxide reacts explosively on warming with sulphur. Chlorine trioxide would be expected to oxidise sulphur violently but dichlorine heptaoxide does not react.

*See* TRIOXYGEN DIFLUORIDE, $F_2O_3$

Interhalogens

*See* BROMINE TRIFLUORIDE, $BrF_3$: Halogens, etc.
BROMINE PENTAFLUORIDE, $BrF_5$: Acids, etc.
CHLORINE TRIFLUORIDE, $ClF_3$: Metals, etc.
IODINE PENTAFLUORIDE, $F_5I$: Metals, etc.

Metal acetylides or carbides

Mellor, 1946, Vol. 5, 849, 862, 886, 891

Monorubidium acetylide ignites in molten sulphur; barium carbide ignites in sulphur vapour at 150°C and incandesces; while calcium, strontium, thorium and uranium carbides need a temperature around 500°C to ignite.

Metal halogenates

*See* METAL HALOGENATES: Non-metals
METAL CHLORATES: Phosphorus, etc.

Metal oxides

*See* SILVER OXIDE, $Ag_2O$: Non-metals
CHROMIUM TRIOXIDE, $CrO_3$: Sulphur
MERCURY(II) OXIDE, HgO: Non-metals
'MERCURY(I) OXIDE' $Hg_2O$: Non-metals
SODIUM PEROXIDE, $Na_2O_2$: Non-metals
LEAD(IV) OXIDE, $O_2Pb$: Non-metals
THALLIUM(III) OXIDE, $O_3Tl_2$: Sulphur

Metals

1. Mellor, 1940, Vol. 4, 268, 480
2. Mellor, 1941, Vol. 3, 639

3. Mellor, 1946, Vol. 5, 210, 393
4. Mellor, 1942, Vol. 15, 149, 527, 627, 696
5. Mellor, 1941, Vol. 7, 208, 328
6. Mellor, 1942, Vol. 12, 31
7. Rothman, R., *Philad. Astronaut. Soc. Bull.*, 1953, **1**, 7
8. Sleight, A. W. *et al., Inorg. Synth.*, 1973, **14**, 152–153

A mixture of sulphur with powdered zinc explodes on warming, and that with cadmium reacts less vigorously [1]. Mixtures of calcium and sulphur explode on ignition and calcium burns in sulphur vapour at 400°C [2]. Indium [3], palladium and rhodium [4], thorium and tin [5] all ignite and incandesce on heating admixed with sulphur. Uranium [6], osmium and nickel [4] powders ignite and incandesce in boiling sulphur or its vapour at 600°C, although more finely divided (catalytic) forms of nickel should react more readily. Magnesium will react vigorously when red-hot or molten with molten sulphur or its vapour [1] and a mixture of aluminium powder and sulphur will burn violently or explode if ignited with a high-temperature (magnesium) fuse [3].

Intimate mixtures of micrograined zinc and sulphur were used as a rocket propellant [7]. In the reaction of gadolinium and sulphur in a heated quartz ampoule to prepare digadolinium trisulphide, good temperature control is necessary to prevent violent interaction, with attack and possible rupture of the ampoule [8].

*See* Aluminium, Copper, above
POTASSIUM, K: Sulphur
LITHIUM, Li: Sulphur
SODIUM, Na: Non-metals
RUBIDIUM, Rb: Non-metals
*See also* SELENIUM, Se: Cadmium, etc.

Non-metals

1. Mellor, 1946, Vol. 5, 15
2. von Schwartz, 1918, 328
3. Mellor, 1940, Vol. 8, 786
4. Phillips, R., *Org. Synth.*, 1943, Coll. Vol. 2, 579

Boron reacts with sulphur at 600°C, becoming incandescent [1]. Mixtures of sulphur with lamp black or freshly calcined charcoal ignite spontaneously, probably owing to adsorbed oxygen on the catalytic surface [2]. Mixtures of yellow phosphorus and sulphur ignite and/or explode on heating [3]. Ignition of an intimate mixture of red phosphorus and sulphur causes a violent exothermic reaction [4].

Oxidants

*See* Halogen oxides, above
Interhalogens, above
Metal oxides, above

SILVER BROMATE, $AgBrO_3$ : Sulphur compounds
SILVER CHLORITE, $AgClO_2$ : Hydrochloric acid, etc.
SILVER NITRATE, $AgNO_3$ : Non-metals
HEPTASILVER NITRATE OCTAOXIDE, $Ag_7NO_{11}$
BARIUM BROMATE, $BaBr_2O_6$ : Sulphur
POTASSIUM BROMATE, $BrKO_3$ : Non-metals
POTASSIUM CHLORITE, $ClKO_2$ : Sulphur
POTASSIUM CHLORATE, $ClKO_3$ : Non-metals
POTASSIUM PERCHLORATE, $ClKO_4$ : Sulphur
SODIUM CHLORATE, $ClNaO_3$ : Ammonium salts, etc.
CHROMYL CHLORIDE, $Cl_2CrO_2$ : Sulphur
LEAD DICHLORITE, $Cl_2O_4Pb$
LEAD CHROMATE, $CrO_4Pb$: Sulphur
FLUORINE, $F_2$ : Non-metals
AMMONIUM NITRATE, $H_4N_2O_3$ : Sulphur
POTASSIUM PERMANGANATE, $KMnO_4$ : Non-metals

Phosphorus trioxide
*See* PHOSPHORUS(III) OXIDE, $O_3P_2$ : Sulphur

Potassium nitride
Mellor, 1940, Vol. 8, 99
Potassium and other alkali metal nitrides react with sulphur to form a highly flammable mixture, which evolves ammonia and hydrogen sulphide in contact with water.
*See* TRICAESIUM NITRIDE, $Cs_3N$

Sodium hydride
Mellor, 1941, Vol. 2, 483
Interaction with sulphur vapour is vigorous.

Static discharges
Enstad, G., *A Reconsideration of the Concept of Minimum Ignition Energy*, paper at Euro. Fed. Chem. Engrs. Wkg Party meeting, 16th March, 1975
Sulphur dust has a very low minimum ignition energy (3 mJ), so particular care is necessary in formulating protective measures.

Tetraphenyllead
*See* TETRAPHENYLLEAD, $C_{24}H_{20}Pb$: Sulphur
*See other* NON-METALS

**SILICON MONOSULPHIDE** **SiS** **SSi**
Bailar, 1973, Vol. 1, 1355
It reacts rapidly with water and may ignite.
*See other* NON-METAL SULPHIDES

## SAMARIUM SULPHIDE $SmS$ $SSm$

Gadolinium sulphide

Jayaraman, A. *et al., Phys. Rev. Lett.*, 1973, **31**, 700

Cooling of samples of crystalline samarium sulphide containing 22 atom% of gadolinium in the lattice to below −153°C causes a near-explosive rearrangement to a powder form. This is attributed to changes in the lattice constants corresponding to a volume increase of 7.5%

*See other* METAL SULPHIDES

## TIN(II) SULPHIDE $SnS$ $SSn$

Oxidants

*See* CHLORIC ACID, $ClHO_3$: Oxidisable materials
DICHLORINE OXIDE, $Cl_2O$: Oxidisable materials
METAL HALOGENATES: Metals, etc.

## STRONTIUM SULPHIDE $SrS$ $SSr$

Lead dioxide

*See* LEAD(IV) OXIDE, $O_2Pb$: Metal sulphides

## TIN(IV) SULPHIDE $SnS_2$ $S_2Sn$

Oxidants

*See* CHLORIC ACID, $ClHO_3$: Oxidisable materials
DICHLORINE OXIDE, $Cl_2O$: Oxidisable materials
METAL HALOGENATES: Metals, etc.

## TITANIUM(IV) SULPHIDE $TiS_2$ $S_2Ti$

Potassium nitrate

*See* POTASSIUM NITRATE, $KNO_3$: Metal sulphides

## URANIUM(IV) SULPHIDE $US_2$ $S_2U$

Nitric acid

*See* NITRIC ACID, $HNO_3$: Uranium(IV) sulphide

## DIANTIMONY TRISULPHIDE $Sb_2S_3$ $S_3Sb_2$

Oxidants

*See* SILVER OXIDE, $Ag_2O$: Metal sulphides
HEPTASILVER NITRATE OCTAOXIDE, $Ag_7NO_{11}$
POTASSIUM CHLORATE, $ClKO_3$: Metal sulphides
DICHLORINE OXIDE, $Cl_2O$: Oxidisable materials
LEAD DICHLORITE, $Cl_2O_4Pb$
FLUORINE, $F_2$: Sulphides
POTASSIUM NITRATE, $KNO_3$: Metal sulphides
THALLIUM(III) OXIDE, $O_3Tl_2$: Sulphur, etc.

## ANTIMONY Sb

Krebs, H. *et al., Z. Anorg. Chem.*, 1955, **282**, 177
Electrolysis of acidified, stirred antimony halide solutions at low current density produces explosive antimony which contains substantial amounts of halogen.

Alkali nitrates
Mellor, 1947, Vol. 9, 382
Mixtures detonate on heating, forming antimonates.

Aluminium
*See* ALUMINIUM, Al: Antimony, etc.

Disulphur dibromide
*See* DISULPHUR DIBROMIDE, $Br_2S_2$: Metals

Halogens
Mellor, 1947, Vol. 9, 379
Antimony ignites in fluorine, chlorine and bromine. With iodine, ignition or explosion may occur if quantities are large enough.

Oxidants

*See* Halogens, above
BROMINE TRIFLUORIDE, $BrF_3$: Halogens, etc.
BROMINE PENTAFLUORIDE, $BrF_5$: Acids, etc.
CHLORINE TRIFLUORIDE, $ClF_3$: Metals, etc.
CHLORIC ACID, $ClHO_3$: Metals, etc.
PERCHLORIC ACID, $ClHO_4$: Antimony(III) compounds
DICHLORINE OXIDE, $Cl_2O$: Oxidisable materials
SELENINYL CHLORIDE, $Cl_2OSe$: Antimony
NITROSYL FLUORIDE, FNO: Metals
IODINE PENTAFLUORIDE, $F_5I$: Metals
AMMONIUM NITRATE, $H_4N_2O_3$: Metals
IODINE, $I_2$: Metals
POTASSIUM PERMANGANATE, $KMnO_4$: Antimony
POTASSIUM DIOXIDE, $KO_2$: Metals

SODIUM NITRATE, $NNaO_3$: Antimony

*See other* METALS

## SELENIUM

Se

Hexalithium disilicide

*See* HEXALITHIUM DISILICIDE, $Li_6Si_2$: Non-metals

Metal acetylides or carbides

Mellor, 1946, Vol. 5, 862, 886

Metal acetylides incandesce on heating in selenium vapour; those of barium at 150°C, calcium and strontium at 500°C; and thorium carbide at an unstated temperature.

*See* DIRUBIDIUM ACETYLIDE, $C_2Rb_2$: Non-metals

Metal amides

Pascal, 1960, Vol. 16.2, 1724

Interaction with alkali and alkaline earth metal amides gives explosive products.

Metal chlorates

Mellor, 1956, Vol. 2, Suppl.1, 583

A slightly moist mixture of selenium with any chlorate (except of alkali metals) becomes incandescent.

*See* METAL HALOGENATES

Metals

1. Mellor, 1942, Vol. 15, 151
2. Mellor, 1947, Vol. 10, 766–777
3. Mellor, 1942, Vol. 12, 31
4. Mellor, 1940, Vol. 4, 480
5. Mellor, 1942, Vol. 16, 158
6. Reisman, A. *et al., J. Phys. Chem.*, 1963, **67**, 22
7. Bailar, 1973, Vol. 2, 79

Nickel and selenium interact with incandescence on gentle heating [1], as also do sodium and potassium, the latter mildly explosively [2]. Uranium [3] and zinc [4] also incandesce when their mixtures with selenium are heated, and platinum sponge incandesces vividly [5]. The particle size of cadmium and selenium must be below a critical size to prevent explosions during synthesis of cadmium selenide by heating the elements together. Similar considerations apply to interaction of cadmium or zinc with sulphur, selenium or tellurium [6]. Interaction of powdered tin and selenium at 350°C is extremely exothermic [7].

Nitrogen trichloride

*See* NITROGEN TRICHLORIDE, $Cl_3N$: initiators

Oxidants

*See* Metal chlorates, above
Oxygen, etc. below
SILVER(I) OXIDE, $Ag_2O$: Non-metals
BARIUM PEROXIDE, $BaO_2$: Selenium
BROMINE PENTAFLUORIDE, $BrF_5$: Acids, etc.
POTASSIUM BROMATE, $BrKO_3$: Non-metals
CHLORINE TRIFLUORIDE, $ClF_3$: Metals, etc.
CHROMIUM TRIOXIDE, $CrO_3$: Selenium
FLUORINE, $F_2$: Non-metals
SODIUM PEROXIDE, $Na_2O_2$: Non-metals

Oxygen,
Organic matter

Astin, S. *et al., J. Chem. Soc.*, 1933, 391
During attempted conversion of recovered selenium metal to the dioxide by heating in oxygen, a vigorous explosion occurred. This was attributed to selenium-catalysed oxidation of traces of organic impurities in the selenium. Oxidation of recovered selenium with nitric acid is rendered vigorous by presence of organic impurities, but is a safe procedure.

Phosphorus

Pascal, 1956, Vol. 10, 724
Interaction warm attains incandescence.
*See other* NON-METALS

## SILICON Si

Calcium

*See* CALCIUM, Ca: Silicon

Metai acetylides

*See* DICAESIUM ACETYLIDE, $C_2Cs_2$: Non-metals
DIRUBIDIUM ACETYLIDE, $C_2Rb_2$: Non-metals

Metal carbonates

Mellor, 1940, Vol. 6, 164
Amorphous or crystalline silicon both react exothermically when heated with alkali metal carbonates, attaining incandescence and evolving carbon monoxide.

Metal hexafluorides

Paine, R. T. *et al., Inorg. Chem.*, 1975, **14**, 1111

During the reduction of the hexafluorides of iridium, osmium or rhenium with silicon to the pentafluorides, the metal hexafluorides must not be condensed directly onto undiluted silicon powder, or explosions may result.

Oxidants

*See* Metal hexafluorides, above
SILVER FLUORIDE, AgF: Non-metals
BROMINE TRIFLUORIDE, $BrF_3$: Halogens, etc.
PEROXYFORMIC ACID, $CH_2O_3$: Non-metals
CHLORINE TRIFLUORIDE, $ClF_3$: Metals, etc.
CHLORINE, $Cl_2$: Non-metals
COBALT TRIFLUORIDE, $CoF_3$: Silicon
NITROSYL FLUORIDE, FNO: Metals, etc.
FLUORINE, $F_2$: Non-metals
OXYGEN DIFLUORIDE, $F_2O$: Non-metals
IODINE PENTAFLUORIDE, $F_5I$: Metals, etc.
LEAD(II) OXIDE, OPb: Non-metals

Water

*See* Boron, B: Water
*See other* NON-METALS

## SILICON–ZIRCONIUM ALLOYS — Si–Zr

*See* ZIRCONIUM, Zr (reference 6)

## SAMARIUM — Sm

1,1,2-Trichlorotrifluoroethane

Anon., *NSC Newsletter, R & D Sect.*, 1970, (8)

An attempt to mill a slurry of the metal in the solvent caused a violent explosion. Experiments showed violent interaction of the freshly ground metal surfaces with the halocarbon solvent.

*See other* METALS: Halocarbons

## TIN — Sn

Oxidants

*See* CHLORINE TRIFLUORIDE, $ClF_3$: Metals
CHLORINE, $Cl_2$: Metals
FLUORINE, $F_2$: Metals
IODINE HEPTAFLUORIDE, $F_7I$: Metals
AMMONIUM NITRATE, $H_4N_2O_3$: Metals

POTASSIUM DIOXIDE, $KO_2$: Metals
SODIUM PEROXIDE, $Na_2O_2$: Metals
SULPHUR, S: Metals
TELLURIUM, Te: Tin

Water

1. Board, S. J. *et al., Int. J. Heat Mass Transf.*, 1974, **17**, 331–339
2. Board, S. J. *et al., Chem. Abs.*, 1975, **83**, 12950m

Experiments involving explosions of molten tin and water are described [1], and the mechanism of propagation of the thermal explosions was studied [2].

*See* MOLTEN METALS

*See other* METALS

**STRONTIUM** **Sr**

Sorbe, 1968, 154

The finely divided metal may ignite in air.

*See other* PYROPHORIC METALS

Halogens

Pascal, 1958, Vol. 4, 579

At 300°C strontium incandesces in chlorine, and at 400°C it ignites in bromine.

Water

Pascal, 1958, Vol. 4, 579

Interaction is more vigorous than with calcium.

*See other* METALS

**TANTALUM** **Ta**

Although very unreactive in massive forms, the finely divided metal may be pyrophoric.

*See* PYROPHORIC METAL POWDERS

Bromine trifluoride

*See* BROMINE TRIFLUORIDE, $BrF_3$: Halogens, etc.

Fluorine

*See* FLUORINE, $F_2$: Metals

**TECHNETIUM** **Tc**

Bailar, 1973, Vol. 3, 881

In finely divided forms (sponge or powder) it readily burns in air to the heptaoxide.

*See other* METALS

## TELLURIUM

**Te**

Halogens,
or Interhalogens

*See* BROMINE PENTAFLUORIDE, $BrF_5$ : Non-metals
CHLORINE FLUORIDE, ClF: Tellurium
CHLORINE TRIFLUORIDE, $ClF_3$ : Metals, etc.
CHLORINE, $Cl_2$ : Non-metals
FLUORINE, $F_2$ : Non-metals

Hexalithium disilicide

*See* HEXALITHIUM DISILICIDE, $Li_6Si_2$ : Non-metals

Metals

*See* CADMIUM, Cd: Selenium, etc.
POTASSIUM, K: Selenium, etc.
SODIUM, Na: Non-metals
PLATINUM, Pt: Selenium, etc.
ZINC, Zn: Non-metals

Silver iodate

*See* SILVER IODATE, $AgIO_3$ : Tellurium

Tin

Bailar, 1973, Vol. 2, 80
Interaction hot to give SnTe attains incandescence.

*See also* SELENIUM, Se: Metals
*See other* NON-METALS

## THORIUM

**Th**

Air

The finely divided metal is rather pyrophoric in air.

*See* Oxygen, below
THORIUM FURNACE RESIDUES

Carbon dioxide,
Nitrogen

*See* CARBON DIOXIDE, $CO_2$ : Metals, Nitrogen

Halogens

Mellor, 1941, Vol. 7, 207
Thorium incandesces in chlorine, bromine or iodine, and would be expected to ignite readily in fluorine.

Nitryl fluoride

*See* NITRYL FLUORIDE, $FNO_2$ : Metals

Oxygen

1. Mellor, 1941, Vol. 7, 207
2. Stout, E. L., *Chem. Eng. News,* 1958, **36**(8), 64

Massive metal ignites on heating in oxygen, or in air below red heat. Powdered metal ignites when rubbed or crushed [1], or on pouring as a stream in air, owing to intergranular friction [2]. Storage and handling precautions are discussed.

*See other* PYROPHORIC METALS

Peroxyformic acid

*See* PEROXYFORMIC ACID, $CH_2O_3$: Metals

Phosphorus

*See* PHOSPHORUS, P: Metals

Sulphur

*See* SULPHUR, S: Metals
*See other* PYROPHORIC METALS

**THORIUM SALTS** **$ThZ_4$**

'Cupferron'

*See* *N*-NITROSOPHENYLHYDROXYLAMINE *O*-AMMONIUM SALT ('CUPFERRON'), $C_6H_9N_3O_2$: Thorium salts

**TITANIUM** **Ti**

Air

1. Mellor, 1941, Vol. 7, 19
2. Stout, E. L., *Chem. Eng. News,* 1958, **36**(8), 64
3. *NSC Data Sheet 485*
4. Lowry, R. N., *et al., Inorg. Synth.*, 1967, **10**, 3
5. *Standard for Production, Processing, Handling and Storage of Titanium*, NFPA 481, 1961
6. Alekseev, A. G. *et al., Chem. Abs.*, 1975, **83**, 83527q
7. Popov, E. I. *et al.*, ibid., 150922e

While massive titanium ignites in air at 1200° [1], the finely divided metal is very pyrophoric and should be stored damp in metal containers [2]. Titanium fires are difficult to extinguish as it burns brilliantly in nitrogen above 800°C, in carbon dioxide above 550°C [3], and may react explosively with metal carbonates [1]. Residual titanium sponge after vapour phase reaction with iodine at 400–425°C is often pyrophoric [4]. All aspects of preventing hazards in use of titanium are covered in a US Standard [5]. Powdered metal made by calcium hydride reduction shows a greater explosion hazard than electrolytically reduced metal. The effect of inert gases was also examined [6]. The relationship between spontaneous ignition temperatures (250–600°C) and lower explosive limits (40–300 mg/l) of titanium powders was studied [7].

*See other* PYROPHORIC METALS

Aluminium

*See* ALUMINIUM–TITANIUM ALLOYS, Al–Ti

Bromine trifluoride

*See* BROMINE TRIFLUORIDE, $BrF_3$: Halogens, etc.

Carbon dioxide,
or Metal carbonates

1. Stout, E. L., *Chem. Eng. News,* 1958, **36**(8), 64
2. Mellor, 1941, Vol. 7, 20

Titanium powder has ignited as a thin layer under carbon dioxide, and burns in the gas above 550°C [1]. Contact of titanium with fused alkali carbonates causes incandescence [1].

Carbon dioxide,
Nitrogen

*See* CARBON DIOXIDE, $CO_2$: Metals
: Metals, Nitrogen

Halocarbons

*Pot. Incid. Rep. 39,* ASESB, Washington, 1968

Mixtures of powdered titanium and trichlorethylene or 1,1,2-trichloro-trifluoroethane flash or spark under heavy impact.

*See other* METALS: Halocarbons

Halogens

1. Mellor, 1941, Vol. 7, 19
2. Kirk-Othmer, 1969, Vol. 20, 372

Reaction of heated titanium with the halogens usually causes ignition and incandescence (chlorine at 350°C, bromine at 360°C, iodine higher, but weak combustion)[1], especially in absence of moisture[2].

*See* FLUORINE, $F_2$: Metals

Metal oxides

Mellor, 1961, Vol. 7, 20

Interaction with copper(II) oxide or trilead tetraoxide is violent.

Metal oxosalts

Mellor, 1941, Vol. 7, 20

Contact of powdered titanium with molten potassium chlorate, alkali carbonates or mixed potassium carbonate–nitrate causes vivid incandescence. Heating mixtures of the powdered metal with potassium chlorate, nitrate or permanganate causes explosions.

*See* POTASSIUM PERCHLORATE, $ClKO_4$ : Metal powders

Nitric acid

1. Anon., *Occ. Health and Safety*, 1957, **7**, 214
2. Anon., *Chem. Eng. News*, 1953, **31**, 3320

Investigation showed that commercial titanium alloys in contact with acid containing less than 1.34% water and more than 6% dinitrogen tetraoxide may become sensitive to impact, and react explosively with the acid. The possible causes are discussed [1]. The spongy residue formed by prolonged corrosion of titanium–manganese alloys by red fuming nitric acid will explode on exposure to friction or heat [2].

Nitrogen

Bailar, 1973, Vol. 3, 359

Titanium is the only element which will burn in nitrogen.

Nitryl fluoride

*See* NITRYL FLUORIDE, $FNO_2$ : Metals

Oxidants

Mellor, 1941, Vol. 7, 20

Mixtures of the (powdered) metal with copper(II) or lead(II,IV) oxides react violently on heating, and with potassium chlorate, nitrate or permanganate, explosively. Addition of the powdered metal to fused potassium chlorate at 370°C leads to incandescence.

Oxygen

Kirk-Othmer, 1969, Vol. 20, 372

Ignition may occur in liquid or gaseous oxygen and in gas mixtures containing above 46% oxygen, under impact or abrasion. Mixtures of powdered metal and liquid oxygen are detonable.

*See* OXYGEN (Gas), $O_2$ : Titanium
OXYGEN(Liquid), $O_2$ : Metals

Silver fluoride

*See* SILVER FLUORIDE, AgF: Titanium

Water

Mellor, 1941, Vol. 7, 19

*NSC Data Sheet 485*

Finely divided titanium reacts with steam at 700°C, and in presence of air the evolved hydrogen may ignite or explode.

*See other* METALS

## TITANIUM SALTS $TiZ_4$

'Cupferron'

*See* N-NITROSOPHENYLHYDROXYLAMINE O-AMMONIUM SALT ('CUPFERRON'), $C_6H_9N_3O_2$ : Thorium salts

## TITANIUM–ZIRCONIUM ALLOYS Ti–Zr

*See* ZIRCONIUM, Zr (reference 6)

## THALLIUM Tl

Fluorine

*See* FLUORINE, $F_2$ : Metals

## URANIUM U

*MCA Case History No. 1296*

Bailar, 1973, Vol. 5, 40

Storage of uranium foil in closed containers in presence of air and moisture may produce a pyrophoric surface. Uranium must be machined in a fume hood because, apart from the radioactivity hazard, the swarf is easily ignited. The massive metal ignites at 600–700°C in air.

*See other* PYROPHORIC METALS

Ammonia

Mellor, 1942, Vol. 12, 31

At dull red heat the metal incandesces in ammonia.

Bromine trifluoride

*See* BROMINE TRIFLUORIDE, $BrF_3$ : Uranium, etc.

Carbon dioxide

Bailar, 1973, Vol. 5, 42

At 750°C interaction is so rapid that ignition will occur with the finely divided metal, and at 800°C the massive metal will ignite.

Carbon dioxide,
Nitrogen

*See* CARBON DIOXIDE, $CO_2$ : Metals, Nitrogen

Halogens

Mellor, 1942, Vol. 12, 31

Uranium powder ignites in fluorine at ambient temperature, in chlorine at 150–180°C, in bromine vapour at 210–240°C and in iodine vapour at 260°C.

Nitric acid,
Trichloroethylene

*MCA Case History No. 1104*

Uranium scrap containing rather finely divided turnings was being pickled, after degreasing with trichlorethylene, by treatment with 4–6M nitric acid solution. During subsequent rinsing with cold water, an explosion ejected the metal and acid liquor from the beaker. It has been reported in the literature that uranium reacts with nitric acid or dinitrogen tetraoxide either explosively or with the formation of an explosive coating or residue. The latter is inhibited by the presence of fluoride ion during pickling. Preventive measures proposed include separate treatment of finely divided uranium scrap, thorough pre-mixing of the acid solution and addition of ammonium hydrogen fluoride. An alternative, or contributory, cause could have been the violent reaction between dinitrogen tetraoxide and residual trichloroethylene.

*See* DINITROGEN TETRAOXIDE, $N_2O_4$: Halocarbons

Nitrogen oxide

*See* NITROGEN OXIDE, NO: Metals

Nitryl fluoride

*See* NITRYL FLUORIDE, $FNO_2$: Metals

Selenium,
or Sulphur

Mellor, 1942, Vol. 12, 31

Uranium incandesces in sulphur vapour, and with selenium.

*See other* METALS

## VANADIUM — V

Oxidants

*See* BROMINE TRIFLUORIDE, $BrF_3$: Halogens, etc.
CHLORINE, $Cl_2$: Metals
NITRYL FLUORIDE, $FNO_2$: Metals

## TUNGSTEN — W

Air,
or Oxidants

Mellor, 1943, Vol. 11, 729

The finely divided metal may ignite on heating in air, or with a range of oxidants.

*See* BROMINE PENTAFLUORIDE, $BrF_5$ : Acids, etc.
BROMINE, $Br_2$ : Tungsten, Tungsten trioxide
CHLORINE TRIFLUORIDE, $ClF_3$ : Metals, etc.
POTASSIUM DICHROMATE, $Cr_2K_2O_7$ : Tungsten
NITRYL FLUORIDE, $FNO_2$ : Metals
FLUORINE, $F_2$ : Metals
OXYGEN DIFLUORIDE, $F_2O$: Metals
IODINE PENTAFLUORIDE, $F_5I$: Metals
HYDROGEN SULPHIDE, $H_2S$: Metals
SODIUM PEROXIDE, $Na_2O_2$ : Metals
LEAD(IV) OXIDE, $O_2Pb$: Metals

*See other* METALS

## XENON — Xe

Fluorine

*See* FLUORINE, $F_2$ : Xenon

## ZIRCONIUM SALTS — $Z_4Zr$

'Cupferron'

*See* *N*-NITROSOPHENYLHYDROXYLAMINE *O*-AMMONIUM SALT ('CUPFERRON'), $C_6H_9N_3O_2$ : Thorium salts

## ZINC — Zn

Acetic acid

Wehrli, S., *Chem. Fabrik.*, 1940, 362–363

The residues from zinc dust–acetic acid reduction operations may ignite after a long delay if discarded into waste-bins with paper. Small amounts appear to ignite more rapidly than larger portions.

*See other* PYROPHORIC MATERIALS

Air

1. Anon., *Chem. Ztg.*, 1940, 289; *Ind. Chem.*, 1944, **20**, 330
2. Robson, S., *School Sci. Rev.*, 1934, **16**(62), 165

A cloud of zinc dust generated by sieving the hot, dried material, exploded violently, apparently after initiation by a spark from the percussive sieve-shaking mechanism. Precautions recommended include use of cold zinc and total enclosure of the process [1]. The possibility of explosions of zinc dust suspended in air is mentioned as a serious hazard [2].

*See* DUST EXPLOSIONS
METAL DUSTS

Aluminium

*See* ALUMINIUM, Al: Zinc

Aluminium,
Magnesium,
Rusted steel

*See* ALUMINIUM–MAGNESIUM–ZINC ALLOYS, Al–Mg–Zn: Rusted steel

Ammonium nitrate,
(Ammonium chloride),
Water

1. Sorbe, 1968, 158
2. Hanson, R. M., *J. Chem. Educ.*, 1976, **53**, 578

A mixture of zinc dust with the nitrate [1], or mixed chloride and nitrate [2] ignites when moistened.

*See also* AMMONIUM NITRATE, $H_4N_2O_3$: Metals

Ammonium sulphide

*See* AMMONIUM SULPHIDE, $H_8N_2S$: Zinc

Arsenic trioxide

Mellor, 1940, Vol. 4, 486

A mixture with excess zinc filings explodes on heating.

Calcium chloride

*See* CALCIUM CHLORIDE, $CaCl_2$: Zinc

Carbon disulphide

Mellor, 1940, Vol. 4, 4

Carbon disulphide and zinc powder interact with incandescence.

Chlorinated rubber

*See* Halocarbons, below

Catalytic metals,
Acids

Sidgwick, 1950, 1530, 1578

Alloys with iridium, platinum, or rhodium, after extraction with acid, leave residues which explode on warming in air owing to the presence of occluded hydrogen (or oxygen) in the catalytic metals.

Electrolytes,
Silver

*See* SILVER, Ag: Electrolytes, etc.

Halocarbons

1. Berger, E., *Compt. Rend.*, 1920, **170**, 29
2. *ABCM Quart. Safety Summ.*, 1963, **34**, 12
3. Lamouroux, A. *et al., Mém. Poudres*, 1957, **39**, 435–445

A paste of zinc powder and carbon tetrachloride (with kieselguhr as thickener) will readily burn after ignition by a high-temperature primer [1]. Intimate mixtures of chlorinated rubber with powdered zinc (or its oxide) in presence or absence of hydrocarbon or halocarbon solvents react violently or explosively at about 216°C [2]. Powdered zinc initially reacts more violently with hexachloroethane in ethanol than does aluminium [3].

*See* BROMOMETHANE, $CH_3Br$: Metals
CHLOROMETHANE, $CH_3Cl$: Metals
*See also* CHLORINATED RUBBER: Metal oxides
*See other* METALS: Halocarbons

Halogens,
or Interhalogens

Mellor, 1940, Vol. 4, 476

Warm zinc powder incandesces in fluorine, and zinc foil ignites in cold chlorine if traces of moisture are present.

*See* BROMINE PENTAFLUORIDE, $BrF_5$: Acids, etc.
CHLORINE TRIFLUORIDE, $ClF_3$: Metals
IODINE, $I_2$: Metals, Water

Lead azide

*See* LEAD(II) AZIDE, $N_6Pb$: Copper, etc.

Manganese dichloride

Terreil, A., *Bull. Soc. Chim. Fr.* [2], 1874, **21**, 289

Zinc reacts explosively when heated with anhydrous manganese dichloride.

Metal oxides

*See* POTASSIUM DIOXIDE, $KO_2$: Metals
TITANIUM DIOXIDE, $O_2Ti$: Metals
ZINC PEROXIDE, $O_2Zn$

*o*-Nitroanisole,
Sodium hydroxide

*See* *o*-NITROANISOLE, $C_7H_7NO_3$: Sodium hydroxide, etc.

Nitrobenzene

Muir, G. D., private comm., 1968

Zinc residues from reduction of nitrobenzene to *N*-phenylhydroxylamine are often pyrophoric and must be kept wet during disposal.

Nitrogen compounds

*See* HYDRAZINIUM NITRATE, $H_5N_3O_3$: Metals
HYDROXYLAMINE, $H_3NO$: Metals

Non-metals

1. Mellor, 1940, Vol. 4, 480, 485
2. Read, C. W. W., *School Sci. Rev.*, 1940, **21**(83), 976

Arsenic, selenium and tellurium all react with incandescence on heating [1]. Interaction of powdered zinc and sulphur on heating is considered to be too violent for use as a school experiment [2].

Oxidants

*See* Halogens, or Interhalogens, above
Metal oxides, above
PEROXYFORMIC ACID, $CH_2O_3$: Metals
POTASSIUM CHLORATE, $ClKO_3$: Metals
NITRYL FLUORIDE, $FNO_2$: Metals
NITRIC ACID, $HNO_3$: Metals
AMMONIUM NITRATE, $H_4N_2O_3$: Metals

Paint primer base

Muller, K. A., *Spontaneous Combustion of Zinc-rich Paint Mixtures, Sub-committee on Storage & Handling Report,* Washington, API, 1969

Zinc residues from poorly mixed priming paint spontaneously inflamed on prolonged exposure to atmosphere. Aluminium residues may behave similarly.

Pentacarbonyliron,
Transition metal halides

*See* PENTACARBONYLIRON, $C_5FeO_5$: Transition metal halides, etc.

Seleninyl bromide

*See* SELENINYL BROMIDE, $Br_2OSe$: Metals

Sodium hydroxide

*ABCM Quart. Safety Summ.*, 1963, **34**, 14
Jackson, H., *Spectrum*, 1969, **7**(2), 82

Accidental contamination of a metal scoop with flake sodium hydroxide, prior to its use with zinc dust, caused ignition of the latter. A stiff paste prepared from zinc dust and 10% sodium hydroxide solution attains a temperature above 100°C after exposure to air for 15 min.

*See* Water, below

Water

1. Anon., *Metallwirtschaft,* 1941, **20**, 475
2. Anon., *Chem. Fabrik,* 1940, 362
3. Mellor, 1940, Vol. 4, 474

In contact with atmospheric oxygen and limited amounts of water, zinc dust will generate heat, and may become incandescent. Presence of acetic acid and copper shortens the induction period. Store the metal dry, and keep residues thoroughly wet until disposal [1,2]. Hydrogen is evolved, especially in the presence of acid or alkali [3].

*See other* METALS

## ZIRCONIUM **Zr**


1. *MCA SD-92,* 1966
2. *Zirconium Fire and Explosion Incidents,* TID-5365, Washington, USAEC, 1956
3. Gleason, R. P., *et al., J. Amer. Soc. Safety Eng.,* 1964, **9**(3), 15
4. Bulmer, G. H. *et al., Rep. AHSB(S) R59,* London, UKAEA, 1964
5. *MCA Case History No. 1234*
6. Ioffe, V. G., *Chem. Abs.*, 1967, **66**, 12614d
7. Karlowicz, P. *et al., J. Electrochem. Soc.*, 1961, **108**, 659–663
8. *Guide for Fire and Explosion Prevention in Plant Producing and Handling Zirconium*, NFPA 482M, 1961


The pyrophoric hazards of zirconium powder are well documented [1]; there is a collection of 43 abstracts of unusual zirconium fire and explosion incidents, most of which involved finely divided metal, moisture and friction, occasionally accompanied by the possibility of static sparks from polythene bags [2]. Increasing use of zirconium necessitated a review of methods of controlling the pronounced pyrophoric properties of finely divided zirconium metal. Five considerations for the safe use of zirconium include: exclusion of air or oxygen by use of inert gases; exclusion of water, its vapour, and other contaminants or oxidants; control of particle size; limitation of amount of powder handled or exposed; and limitation of exposure of personnel [3]. Storage and disposal of irradiated metal wastes in various forms has also been discussed [4]. Ignition may not be immediate, however, as fine zirconium dust left uncleared from previous grinding operations was ignited violently by a spark from subsequent grinding operations [5].

The lower ignition concentration limits of air suspensions of zirconium powder, of its alloys with silicon and titanium, and of zirconium–titanium hydride have been studied in relation to the history of exposure to air [6]. Treatment of zirconium powder with 1% hydrogen fluoride solution considerably desensitises it to electrostatic ignition [7]. All aspects of hazards in use of zirconium are covered in a US guide [8].

*See* LEAD–ZIRCONIUM ALLOYS, Pb–Zr
METAL DUSTS
PYROPHORIC METALS

*See also* HAFNIUM, Hf

Carbon dioxide,
Nitrogen

*See* CARBON DIOXIDE, $CO_2$: Metals
: Metals, Nitrogen

Carbon tetrachloride

*Zirconium Fire and Explosion Incidents,* TID-5365, Washington, USAEC, 1956

A mixture of powdered zirconium and carbon tetrachloride exploded violently while being heated.

*See other* METALS: Halocarbons

Nitryl fluoride

*See* NITRYL FLUORIDE, $FNO_2$: Metals

Oxygen-containing compounds

1. Mellor, 1941, Vol. 7, 116
2. Ellern, 1968, 249

The affinity of zirconium for oxygen is great, particularly when finely divided, and extends to many oxygen-containing compounds. The vigorous reactions with potassium chlorate or nitrate, or with cupric or lead oxides, might be anticipated but there are also explosive reactions with the combined oxygen when zirconium is heated with alkali metal hydroxides or carbonates, or zirconium hydroxide. Hydrated sodium tetraborate behaves similarly [1]. Several alkali metal oxosalts (chromates, dichromates, molybdates, sulphates, tungstates) react violently or explosively when reduced to the parent metals by heating with zirconium [2].

*See* POTASSIUM CHLORATE, $ClKO_3$: Metals
LITHIUM CHROMATE, $CrLiO_4$: Zirconium
COPPER(II) OXIDE, CuO: Metals
POTASSIUM NITRATE, $KNO_3$: Metals
LEAD(II) OXIDE, OPb: Metals

Phosphorus

*See* PHOSPHORUS, P: Metals

Water

1. *MCA SD-92,* 1966
2. Muir, G. D., private comm., 1968
3. *Haz. Chem. Data,* 1972, 253

Powder damp with 5–10% of water may ignite, and although 25% of water is regarded as a safe concentration [1], ignition of a 50% paste on breakage of a glass container has been observed [2]. Although water is used to prevent ignition, the powder, once ignited, will burn under water more violently than in air [3].

*See other* METALS

Appendix I

# Source Title Abbreviations used in Handbook References

The abbreviations used in the references for titles of journals and periodicals are those used in BP publications practice and conform closely to the recommendations of the *Chemical Abstracts* system. Abbreviations which have been used to indicate textbook and reference book sources of information are set out below with the full titles and publication details.

| | |
|---|---|
| *ABCM Quart. Safety Summ.* | *Quarterly Safety Summaries*, London, Association of British Chemical Manufacturers, 1930–1964 |
| ACS 54, 1966 | *Advanced Propellant Chemistry*, ACS 54, Washington, American Chemical Society, 1966 |
| ACS 88, 1969 | *Propellants Manufacture, Hazards and Testing*, ACS 88, Washington, American Chemical Society, 1969 |
| Albright, Hanson, 1976 | *Industrial and Laboratory Nitrations*, ACS Symposium Series 22, Albright, L. F., Hanson, C. (Editors), Washington, American Chemical Society, 1976 |
| *Alkali Metals*, 1957 | *Alkali Metals*, ACS 19, Washington, American Chemical Society, 1957 |
| Augustine, 1968 | *Reduction, Techniques and Applications in Organic Chemistry*, Augustine, R. L., London, Edward Arnold, 1968 |

Bahme, 1972 — *Fire Officers' Guide to Dangerous Chemicals*, FSP-36, Bahme, C. W., Boston, National Fire Protection Association, 1972

Bailar, 1973 — *Comprehensive Inorganic Chemistry*, Bailar, J. C., Emeléus, H. J., Nyholm, R. S., Trotman-Dickenson, A. F. (Editors), Oxford, Pergamon, 5 vols., 1973

*BCISC Quart. Safety Summ.* — *Quarterly Safety Summaries*, London, British Chemical Industry Safety Council, 1965–1973

*Berufsgenossenschaft* — *Zeitschrift fur Unfallversicherung und Betriebssicherheit*, Heidelberg. Berufsgenossenschaft der Chemische Industrie, 1960 to date

Braker, 1971 — *Matheson Gas Data Book*, Braker, W., Mossmann, A. L., East Rutherford, NJ, Matheson Gas Products, 5th edn, 1971

Brandsma, 1971 — *Preparative Acetylenic Chemistry*, Brandsma, L., Barking, Elsevier, 1971

Brauer, 1963, 1965 — *Handbook of Preparative Inorganic Chemistry*, Brauer, G. (Translation Editor: Riley, R. F.), London, Academic Press, 2nd edn. Vol. 1, 1963; Vol 2, 1965

Castrantas, 1965 — *Fire and Explosion Hazards of Peroxy Compounds*, Special Publication No. 394, Castrantas, H. M., Banerjee, D. K., Noller, D. C., Philadelphia, ASTM, 1965

Castrantas, 1970 — *Laboratory Handling and Storage of Peroxy Compounds*, Special Publication No. 491, Castrantas, H. M., Banerjee, D. K., Philadelphia, ASTM, 1970

*Chemiearbeit* — *Chemiearbeit, Schutzen und Helfen*, Dusseldorf, Berufsgenossenschaft der Chemische Industrie, 1949–62 (supplement to *Chemische Industrie*)

*CISHC Chem. Safety Summ.* — *Chemical Safety Summaries*, London, Chemical Industry Safety and Health Council of the Chemical Industries Association, 1974 to date

Cloyd, 1965 — *Handling Hazardous Materials,* Technical Survey No. SP-5032. Cloyd, D. R., Murphy, W. J., Washington, NASA, 1965

Coates, 1960 — *Organometallic Compounds,* Coates, G. E., London, Methuen, 1960

Coates, 1967, 1968 — *Organometallic Compounds,* Coates, G. E., Green, M. L. H., Wade, K., London, Methuen Vol. 1, 1967; Vol. 2, 1968

*Dangerous Loads,* 1972 — *Dangerous Loads,* London, Institute of Fire Engineers, 1972

*Dangerous Substances*, 1972 — *Dangerous Substances: Guidance on Fires and Spillages* (Section 1: 'Inflammable liquids'), London, HMSO, 1972

Davies, 1961 — *Organic Peroxides,* Davies, A. G., London, Butterworths, 1961

Davis, 1943 — *Chemistry of Powders and Explosives,* Davis, T. L., New York, Wiley, 1943

Dunlop, 1953 — *The Furans,* ACS 119, Dunlop, A. P., Peters, F. N., New York, Reinhold, 1953

Ellern, 1968 — *Military and Civilian Pyrotechnics,* Ellern, H., New York, Chemical Publishing Co., 1968

Emeléus, 1960 — *Modern Aspects of Inorganic Chemistry,* Emeléus, H. J., Anderson, J. S., London, Rutledge and Kegan Paul, 1960

Federoff, 1960 to date — *Encyclopaedia of Explosives and Related Compounds*, Federoff, B. T. (Editor), Dover (N.J.), Piccatinny Arsenal. Vol. 1, 1960, then triennially; Vol. 7, 1975 (available USNTIS)

Fieser, 1967–77 — *Reagents for Organic Synthesis*, Fieser, L. F. and Fieser, M., New York, Wiley. Vol. 1, 1967; Vol. 6, 1977

Freifelder, 1971 — *Practical Catalytic Hydrogenation*, Freifelder, M., New York, Wiley–Interscience, 1971

Gallais, 1957 — *Chimie Minérale Théorique et Expérimental (Chimie Électronique)*, Gallais, F., Paris, Masson, 1957

Gaylord, 1956 — *Reductions with Complex Metal Hydrides*, Gaylord, N. G., New York, Interscience, 1956

Gibson, 1969 — *Handbook of Selected Properties of Air- and Water-Reactive Materials*, Gibson, H., Crane (Ind.), US Naval Ammunition Depot, 1969

Goehring, 1957 — *Ergebnisse und Probleme der Chemie der Schwefelstickstoff-verbindung, Scientia Chemica*, Vol. 9, Goehring, M., Berlin, Akademie Verlag, 1957

Grignard, 1935–54 — *Traité de Chimie Organique*, Grignard, V., and various volume Editors. Paris, Masson. Vol. 1, 1935; Vol. 23, 1954

*Guide for Safety*, 1972 — *Guide for Safety in the Chemical Laboratory*, MCA, New York, Van Nostrand Reinhold, 2nd edn, 1972

Harmon, 1974 — *A Review of Violent Monomer Polymerisation*, Harmon, M. and King, J., Rep. AD-A017443, Springfield (Va.), NTIS, 1974

*Haz. Chem. Data*, 1971 — *Hazardous Chemical Data*, NFPA 49, Boston, National Fire Protection Association, 1971

| | |
|---|---|
| Houben-Weyl, 1953 to date | *Methoden der Organischen Chemie,* Ed. Müller, E., Stuttgart, Georg Thieme, 4th edn, 16 vols, not yet completed, published unsequentially, 1953 to date |
| *Inorg. Synth.*, 1939–76 | *Inorganic Syntheses*, various Editors, London, McGraw-Hill. Vol. 1, 1939; Vol. 17, 1976 |
| Karrer, 1950 | *Organic Chemistry,* Karrer, P., London, Elsevier, 1950 |
| Kharasch and Reinmuth, 1954 | *Grignard Reactions of Non-Metallic Substances,* Kharasch, M. S. and Reinmuth, O., London, Constable, 1954 |
| Kirk-Othmer, 1963–71 | *Encyclopaedia of Chemical Technology,* Kirk, R. E., Othmer, D. F., London, Interscience. Vol. 1, 1963; Vol. 22, 1970; Supplementary Vol. 1971 |
| Kirshenbaum, 1956 | *Final Report on Fundamental Studies of New Explosive Reactions,* Kirshenbaum, A. D., Philadelphia, Research Institute of Temple University, 1956 |
| Kit and Evered, 1960 | *Rocket Propellant Handbook,* Kit, B., Evered, D. S., London, Macmillan, 1960 |
| Lawless, 1968 | *High Energy Oxidisers*, Lawless, E. W. and Smith, I. C., New York, Marcel Dekker, 1968 |
| Leleu, *Cahiers*, 1972 on | *Les Réactions Chimique Dangereuse*, Leleu, J., published serially in *Cahiers de Notes Documentaires*, from 1972, (68), 319 |
| *Loss Prev. Bull.*, 1974 on | *Loss Prevention Bulletin*, published six times annually by Institution of Chemical Engineers, London, since 1974 |

Mackay, 1966 — *Hydrogen Compounds of the Metallic Elements,* Mackay, K. M., London, Spon, 1966

Martin, 1971 — *Dimethylsulphoxid,* Martin, D., Hauthal, H. G., Berlin, Akademie Verlag, 1971

*MCA Case Histories*, 1950–78 — *Case Histories of Accidents in the Chemical Industry,* Washington, Manufacturing Chemists' Association Inc. (Published monthly and republished as indexed collected volumes at 2–3 year intervals)

*MCA SD-Series* — *Safety Data Sheets* (1–101 available), Washington, Manufacturing Chemists' Association Inc. (published and revised as necessary, 1947–1976)

*Major Loss Prevention, 1971* — *Major Loss Prevention in the Process Industries*, Symposium Series No. 34, London, Institution of Chemical Engineers, 1971

Meidl, 1972 — *Hazardous Materials Handbook*, Meidl, J. H., Beverley Hills, Glencoe Press, 1972

Mellor, 1941-71 — *Comprehensive Treatise on Inorganic and Theoretical Chemistry,* Mellor, J. W., London, Longmans Green, Vol. 1, reprinted 1941; Vol. 16 published 1937; and isolated Supplementary volumes up to Vol. 8 Suppl. 3, 1971

Mellor MIC, 1961 — *Mellor's Modern Inorganic Chemistry,* Parkes, G. D. (Revision Editor), London, Longmans Green, 1961

Merck, 1968, 1976 — *The Merck Index*, Rahway (N.J.), Merck and Co. Inc. 8th edn, 1968; 9th edn, 1976

Miller, 1965–6 — *Acetylene, its Properties, Manufacture and Uses,* Miller, S. A., London, Academic Press. Vol. 1, 1965; Vol. 2, 1966

| | |
|---|---|
| Moissan, 1900 | *Le Fluor et ses Composés*, Moissan, H., Paris, Steinheil, 1900 |
| Muir, 1977 | *Hazards in the Chemical Laboratory*, Muir, G. D. (Editor), London, The Chemical Society, 2nd edn, 1977 |
| *NFPA 491M* | *Manual of Hazardous Chemical Reactions*, NFPA 491M, Boston, National Fire Protection Association, 5th edn, 1975 |
| *Org. Synth.*, 1944 to date | *Organic Syntheses*, various Editors, New York, John Wiley. Collective Vol. 1, 1944; Collective Vol. 2, 1944; Collective Vol. 3, 1955; Collective Vol. 4, 1962; Collective Vol. 5, 1973; annual volumes thereafter |
| Partington, 1967 | *General and Inorganic Chemistry*, 4th edn, Partington, J. R., London, Macmillan, 1967 |
| Pascal, 1956–70 | *Nouveau Traité de Chimie Minérale*, Pascal, P. (Editor), Paris, Masson. Vol. 1, 1956; Vol. 20.3, 1964; Vol. 15.5, 1970 |
| Pieters, 1957 | *Safety in the Chemical Laboratory*, Pieters, H. A. J., Creyghton, J. W., London, Academic Press, 2nd edn, 1957 |
| *Pot. Incid. Rep.* | *ASESB Potential Incident Report*, Washington, Armed Services Explosives Safety Board |
| *Proc. Nth Combust. Symp.* | *Proceedings of Combustion Symposia*, various Editors, various Publishers, dated year following event. 1st Symposium, 1928; 16th Symposium, 1976 |
| Rodd, 1951–62 | *The Chemistry of Carbon Compounds*, Rodd, E. H. (Editor), London, Elsevier. Vol. 1A, 1951; Vol. 5, 1962 |

Rüst, 1948 — *Unfälle beim Chemischen Arbeiten,* Rust, E., Ebert, A., Zurich, Rascher Verlag, 2nd edn, 1948

Rutledge, 1968 — *Acetylenic Compounds,* Rutledge, T. F., New York, Reinhold, 1968

Schmidt, 1967 — *Handling and Use of Fluorine and Fluorine–Oxygen Mixtures in Rocket Systems,* SP-3037 Schmidt, H. W., Harper, J. T., Washington, NASA, 1967

Schumacher, 1960 — *Perchlorates, their Properties, Manufacture and Uses,* ACS 146, Schumacher, J. C., New York, Reinhold, 1960

Schumb, 1955 — *Hydrogen Peroxide,* ACS 128, Schumb, W. C., Satterfield, C. N., Wentworth, R. L., New York, Reinhold, 1955

Shriver, 1969 — *Manipulation of Air-sensitive Compounds,* Shriver, D. F., London, McGraw-Hill, 1969

*Sichere Chemiearbeit* — *Sichere Chemiearbeit,* Heidelberg, Berufsgenossenchaft der Chemische Industrie, 1963 to date (supplement to *Chemiker Zeitung)*

Sidgwick, 1950 — *The Chemical Elements and their Compounds,* Sidgwick, N. V., Oxford, Oxford University Press, 1950

Smith, 1965–6 — *Open-chain Nitrogen Compounds,* Smith, P. A. S., New York, Benjamin. Vol. 1, 1965; Vol. 2, 1966

Sorbe, 1968 — *Giftige und Explosive Substanzen,* Sorbe, G., Frankfort, Umschau Verlag, 1968

Steere, 1967 — *Handbook of Laboratory Safety,*

Steere, N. V. (Editor), Cleveland, Chemical Rubber Publishing Co., 1967 (2nd edn, 1971)

Stull, 1977 — *Fundamentals of Fire and Explosion*, Stull, D. R., AIChE Monograph Series No. 10, Vol. 73, New York, American Institute of Chemical Engineers, 1977

Swern, 1970–2 — *Organic Peroxides*, Swern, D. (Editor), London, Wiley-Interscience. Vol. 1, 1970; Vol. 2, 1971; Vol. 3, 1972

Urbanski, 1964–7 — *Chemistry and Technology of Explosives,* Urbanski, T., London, Macmillan. Vol. 1, 1964; Vol. 2, 1965; Vol. 3, 1967

USNTIS — United States National Technical Information Service, Richmond, Virginia

Van Wazer, 1958 — *Phosphorus and its Compounds,* Van Wazer, J. R., New York, Interscience. Vol. 1, 1958; Vol. 2, 1961

Vervalin, 1973 — *Fire Protection Manual for Hydrocarbon Processing Plants*, Vervalin, C. H. (Editor), Houston, Gulf Publishing Co., 2nd edn, 1973

Vogel, 1957 — *A Textbook of Practical Organic Chemistry,* Vogel, A. I., London, Longmans Green, 3rd edn, 1957

von Schwartz, 1918 — *Fire and Explosion Risks,* von Schwartz, E., London, Griffin, 1918

Weast, 1972 — *Handbook of Chemistry and Physics,* Weast, R. C. (Editor), Cleveland, Chemical Rubber Publishing Co., 53rd edn, 1972

Whitmore, 1921 — *Organic Compounds of Mercury*,

Whitmore, F. C., New York,
Chemical Catalog Co. Inc., 1921

Zabetakis, 1965 | *Flammability Characteristics of Combustible Gases and Vapours,* Zabetakis, M. G., Washington, US Bureau of Mines, 1965

## Appendix 2

# Tabulated Fire-related Data

Compounds which are considered to be unusually hazardous in a fire context because of their low flash points (below 25°C) or auto-ignition temperatures (below 225°C) are included in the table. The names used are those titles in the text of Section 2 which are prefixed with a dagger. Synonyms may be found either in Section 2 or in the alphabetical index of names and synonyms in Appendix 5.

Boiling points are given for those flammable materials boiling below 50°C.

The figures for flash points are closed-cup values except where a suffix (o) indicates the (usually higher) open-cup value.

The figures for explosive limits (or flammability limits) are % by volume in air at ambient temperature except where indicated. Where no figure for the upper limit has been found, a query has been inserted.

Figures for auto-ignition temperatures are usually those determined in glass (without catalytic effects) except where stated.

Most of the values are those quoted in the references given in the topic entries AUTOIGNITION TEMPERATURES, FIRE, FLAMMABILITY and FLASH POINTS.

| Name | Formula | B.P./°C | Fl.P./°C | E.L./% | A.I.T./°C |
|---|---|---|---|---|---|
| Arsine | $AsH_3$ | −62 | | flammable | |
| Diborane | $B_2H_6$ | −93 | −90 | 0.9–98 | 38–52 |
| Pentaborane(9) | $B_5H_9$ | | 30 | 0.4–? | 35 |
| Bromosilane | $BrH_3Si$ | 2 | | flammable | ambient |
| Hydrogen cyanide | $CHN$ | 25.7 | −18 | 6.0–41 | |
| Formaldehyde | $CH_2O$ | −19 | −19 | 7.0–73 | |
| Bromomethane | $CH_3Br$ | 3.5 | none | see text | |
| Chloromethane | $CH_3Cl$ | −24 | <0 | 8.1–17 | |
| Methyltrichlorosilane | $CH_3Cl_3Si$ | | 8 | 7.6–? | |
| Fluoromethane | $CH_3F$ | −78 | | flammable | |
| Methane | $CH_4$ | −161 | −187 | 5.3–15 | |
| Dichloro(methyl)silane | $CH_4Cl_2Si$ | | −32 | | ambient |
| Methanol | $CH_4O$ | | 10 | 6–36.5/60°C | |
| Methanethiol | $CH_4S$ | 6 | −18 | 3.9–21.8 | |
| Methylamine | $CH_5N$ | −6 | −18 | 4.5–21 | |
| 35%w/v soln in water | | | 7.5 | | |
| Methylhydrazine | $CH_6N_2$ | | 23<br>17.5(o) | 2.5–97 | 196 |
| Carbon monoxide | $CO$ | −192 | | 12.5–74 | |
| Carbonyl sulphide | $COS$ | −50 | | 12–28.2 | |
| Carbon disulphide | $CS_2$ | 46 | −30 | 1.3–50 | 125 (below 100 if rust present |
| Bromotrifluoroethylene | $C_2BrF_3$ | −3 | | | ambient |
| Chlorotrifluoroethylene | $C_2ClF_3$ | −28 | | 8.4–38.7 | |
| Tetrafluoroethylene | $C_2F_4$ | −76 | | 11–60 | 180 |
| Trichloroethylene | $C_2HCl_3$ | | 32 | 12.5–90 (above 30°C) | |
| Trifluoroethylene | $C_2HF_3$ | | | 15.3–27 | |
| Acetylene | $C_2H_2$ | −83 | | 2.5–82 | |
| 1,1-Dichloroethylene | $C_2H_2Cl_2$ | 32 | −15(o) | 7.3–16 | |
| *cis*-1,2-Dichloroethylene | $C_2H_2Cl_2$ | | 6 | 3.3–15 | |
| *trans*-1,2-Dichloroethylene | $C_2H_2Cl_2$ | 48 | 2 | 9.7--12.8 | |
| 1,1-Difluoroethylene | $C_2H_2F_2$ | −86 | | 5.5–21.3 | |
| Glyoxal | $C_2H_2O_2$ | 50 | | | |
| Bromoethylene | $C_2H_3Br$ | 16 | | 6–15 | |
| Chloroethylene | $C_2H_3Cl$ | −14 | −8 | 4–22 (or 33) | |
| 1-Chloro-1,1-difluoroethane | $C_2H_3ClF_2$ | −9 | | 9–14.8 | |
| Acetyl chloride | $C_2H_3ClO$ | | 4 | 5.0–? | |
| Methyl chloroformate | $C_2H_3ClO_2$ | | 12 | | |
| Trichloro(vinyl)silane | $C_2H_3Cl_3Si$ | | <10 | | |
| Fluoroethylene | $C_2H_3F$ | −73 | | 2.6–22 | |
| 1,1,1-Trifluoroethane | $C_2H_3F_3$ | −47 | | 9.2–18.4 | |
| Acetonitrile | $C_2H_3N$ | | 6(o) | 4–16 | |
| Methyl isocyanate | $C_2H_3NO$ | 39 | <−15 | | |
| Ethylene | $C_2H_4$ | −104 | | 3.0–34 | 492 |
| 1,1-Dichloroethane | $C_2H_4Cl_2$ | | −6 | 5.6–11.4 | |
| 1,2-Dichloroethane | $C_2H_4Cl_2$ | | 13 | 6.2–15.9 | |
| Bis-chloromethyl ether | $C_2H_4Cl_2O$ | | <19 | | |
| 1,1-Difluoroethane | $C_2H_4F_2$ | −25 | | 3.7–18 | |
| Acetaldehyde | $C_2H_4O$ | 20 | −38 | 4.0–57 | 204 (140 by DIN 51794) |

| Name | Formula | B.P./°C | Fl.P./°C | E.L./% | A.I.T./°C |
|---|---|---|---|---|---|
| Ethylene oxide | $C_2H_4O$ | 11 | −20 | 3.0–100 | |
| Thioacetic *S*-acid | $C_2H_4OS$ | | <23 | | |
| Methyl formate | $C_2H_4O_2$ | 32 | −19 | 5.9–20 | |
| Bromoethane | $C_2H_5Br$ | 38 | −20 | 6.7–11.3 | |
| Chloroethane | $C_2H_5Cl$ | 12 | −50 | 3.8–15.4 | |
| Chloromethyl methyl ether | $C_2H_5ClO$ | | <23 | | |
| Trichloro(ethyl)silane | $C_2H_5Cl_3Si$ | | 14 | | |
| Fluoroethane | $C_2H_5F$ | −38 | | | |
| Aziridine | $C_2H_5N$ | | −11 | 3.3–54.8 | |
| Acetaldehyde oxime | $C_2H_5NO$ | | <22 | | |
| *N*-Methylformamide | $C_2H_5NO$ | | <22 | | |
| Ethyl nitrite | $C_2H_5NO_2$ | 17 | −35 | 3.1–>50 | expl.>90 |
| Ethyl nitrate | $C_2H_5NO_3$ | | 10 | 3.8–? | expl.85 |
| Ethane | $C_2H_6$ | −89 | −130 | 3–12.5 | |
| Dichlorodimethylsilane | $C_2H_6Cl_2Si$ | | −9 | 3.4–9.5 | |
| Dichloro(ethyl)silane | $C_2H_6Cl_2Si$ | | <23 | | |
| Dimethyl ether | $C_2H_6O$ | −24 | −41 | 3.4–18 (or 27) | |
| Ethanol | $C_2H_6O$ | | 12 | 3.3–19/60°C | |
| Dimethyl sulphoxide | $C_2H_6OS$ | | | | 215 |
| Dimethyl sulphate | $C_2H_6O_4S$ | | | | 188 |
| Dimethyl sulphide | $C_2H_6S$ | 37 | −34 | 2.2–19.7 | 206 |
| Ethanethiol | $C_2H_6S$ | 36 | <−18 | 2.8–18 | |
| Dimethyl disulphide | $C_2H_6S_2$ | | 7 | 1.1–? | |
| Dimethylamine | $C_2H_7N$ | 7 | −50 | 2.8–14.4 | |
| 40%w/v soln in water | | | −16 | | |
| 25%w/v soln in water | | | 6 | | |
| Ethylamine | $C_2H_7N$ | 17 | <−18 | 3.5–14 | |
| 70%w/v soln in water | | | <22 | | |
| 1,1-Dimethylhydrazine | $C_2H_8N_2$ | | −15 | 2.1–95 | |
| 1,2-Dimethylhydrazine | $C_2H_8N_2$ | | <23 | | |
| 1,1-Dimethyldiborane | $C_2H_{10}B_2$ | −3 | <−10 | | |
| 1,2-Dimethyldiborane | $C_2H_{10}B_2$ | −49 | <−55 | | |
| Dicyanogen | $C_2N_2$ | −21 | | 6–32 | |
| Hexafluoropropene | $C_3F_6$ | | | 15–20 | |
| 2-Chloroacrylonitrile | $C_3H_2ClN$ | | 8 | | |
| 1-Bromo-2-propyne | $C_3H_3Br$ | | 10 | 3.0–? | |
| 1-Chloro-2-propyne | $C_3H_3Cl$ | | <15 | | |
| 3,3,3-Trifluoropropene | $C_3H_3F_3$ | | | 4.7–13.5 | |
| 1,1,1-Trifluoroacetone | $C_3H_3F_3O$ | 22 | <10 | | |
| Acrylonitrile | $C_3H_3N$ | | −1 | 3.0–17 | |
| 5%w/w soln in water | | | <9 | | |
| Propadiene | $C_3H_4$ | −34 | | 1.7–12 | |
| Propyne | $C_3H_4$ | −23 | | 2.15–12.5 | |
| 1-Chloro-3,3,3-trifluoropropane | $C_3H_4ClF_3$ | | | 6–14 | |
| 1,3-Dichloropropene | $C_3H_4Cl_2$ | | 21 | | |
| 2,3-Dichloropropene | $C_3H_4Cl_2$ | | 10 | | |
| Acrylaldehyde | $C_3H_4O$ | | −26 | 2.8–31 | |
| Methoxyacetylene | $C_3H_4O$ | 23 | <−20 | | |
| Vinyl formate | $C_3H_4O_2$ | 46 | <0 | | |
| 1-Bromo-2-propene | $C_3H_5Br$ | | −1 | 4.4–7.3 | |
| 1-Bromo-2,3-epoxypropane | $C_3H_5BrO$ | | <22 | | |
| 1-Chloro-1-propene | $C_3H_5Cl$ | 35 | <−6 | 4.5–16 | |

| Name | Formula | B.P./°C | Fl.P./°C | E.L./% | A.I.T./°C |
|---|---|---|---|---|---|
| 2-Chloropropene | $C_3H_5Cl$ | 22 | <−20 | 4.5–16 | |
| 1-Chloro-2-propene | $C_3H_5Cl$ | 45 | −32 | 3.3–11.2 | |
| 1-Chloro-2,3-epoxypropane | $C_3H_5ClO$ | | 21 | | |
| Propionyl chloride | $C_3H_5ClO$ | | 12 | | |
| Ethyl chloroformate | $C_3H_5ClO_2$ | | 2 | | |
| 1-Iodo-2-propene | $C_3H_5I$ | | <21 | | |
| Propionitrile | $C_3H_5N$ | | 2 | 3.1–? | |
| Cyclopropane | $C_3H_6$ | −33 | | 2.4–10.4 | |
| Propene | $C_3H_6$ | −48 | −108 | 2.4–11.1 | |
| 1,1-Dichloropropane | $C_3H_6Cl_2$ | | 21 | 3.1–? | |
| 1,2-Dichloropropane | $C_3H_6Cl_2$ | | 15 | 3.4–14.5 | |
| Acetone | $C_3H_6O$ | | −17 | 2.6–12.8 | |
| Methyl vinyl ether | $C_3H_6O$ | 5.5 | −56(o) | 2.6–39.0 | |
| 2-Propen-1-ol | $C_3H_6O$ | | 21 | 3.0–18.0 | |
| Propionaldehyde | $C_3H_6O$ | 49 | −9 | 2.9–17 | 207 |
| Propylene oxide | $C_3H_6O$ | 34 | −37 | 3.1–27.5 | |
| 1,3-Dioxolane | $C_3H_6O_2$ | | 2 | | |
| Ethyl formate | $C_3H_6O_2$ | | −20 | 2.7–13.5 | |
| Methyl acetate | $C_3H_6O_2$ | | −9 | 3.1–16 | |
| Dimethyl carbonate | $C_3H_6O_3$ | | 18 | | |
| 2-Propen-1-thiol | $C_3H_6S$ | | −10 | | |
| 1-Bromopropane | $C_3H_7Br$ | | <22 | 4.6–? | |
| 2-Bromopropane | $C_3H_7Br$ | | −14 | | |
| 1-Chloropropane | $C_3H_7Cl$ | 47 | <−18 | 2.6–11.1 | |
| 2-Chloropropane | $C_3H_7Cl$ | 36 | −32 | 2.8–10.7 | |
| Chloromethyl ethyl ether | $C_3H_7ClO$ | | <19 | | |
| 1-Iodopropane | $C_3H_7I$ | | <22 | | |
| 2-Iodopropane | $C_3H_7I$ | | 20 | | |
| Allylamine | $C_3H_7N$ | | −29 | 2.2–22 | |
| Cyclopropylamine | $C_3H_7N$ | | 1 | | |
| 2-Methylaziridine | $C_3H_7N$ | | −10 | | |
| Isopropyl nitrite | $C_3H_7NO_2$ | | <10 | | |
| Isopropyl nitrate | $C_3H_7NO_3$ | | 11 | ?–100 | |
| Propyl nitrate | $C_3H_7NO_3$ | | 20 | 2–100 | |
| Propane | $C_3H_8$ | −45 | −104 | 2.2–9.5 | |
| Ethyl methyl ether | $C_3H_8O$ | 11 | −37 | 2–10.1 | 190 |
| Propanol | $C_3H_8O$ | | 15 | 2.5–13.5 | |
| 2-Propanol | $C_3H_8O$ | | 12 | 2.3–12.7 | |
| Dimethoxymethane | $C_3H_8O_2$ | 44 | −18(o) | | |
| Ethyl methyl sulphide | $C_3H_8S$ | | <21 | | |
| Propanethiol | $C_3H_8S$ | | −20 | | |
| 2-Propanethiol | $C_3H_8S$ | | −35 | | |
| Trimethyl borate | $C_3H_9BO_3$ | | <23 | | |
| Chlorotrimethylsilane | $C_3H_9ClSi$ | | −20 | | |
| Isopropylamine | $C_3H_9N$ | 32 | −37(o) | 2.3–10.4 | |
| Propylamine | $C_3H_9N$ | 49 | −10 | 2.0–10.4 | |
| Trimethylamine | $C_3H_9N$ | 3 | −5 | 2.0–11.6 | 190 |
| 25%w/v soln in water | | | 5 | | |
| 1,2-Diaminopropane | $C_3H_{10}N_2$ | | 24 | | |
| 1,3-Diaminopropane | $C_3H_{10}N_2$ | | 24(o) | | |
| Propadien-1,3-dione | $C_3O_2$ | 7 | | | |
| 1,3-Butadiyne | $C_4H_2$ | 10 | | 1.4–100 | |

| *Name* | *Formula* | *B.P./°C* | *Fl.P./°C* | *E.L./%* | *A.I.T./°C* |
|---|---|---|---|---|---|
| Buten-3-yne | $C_4H_4$ | 5 | <–5 | 1.8–40 (1.7–93% in oxygen) | |
| Furan | $C_4H_4O$ | 32 | –36 | 2.3–14.3 | |
| Thiophene | $C_4H_4S$ | | –6 | | |
| 2-Chloro-1,3-butadiene | $C_4H_5Cl$ | | –20 | 4–20 | |
| Ethyl trifluoroacetate | $C_4H_5F_3O_2$ | | –17 | | |
| 1-Cyanopropene | $C_4H_5N$ | | 16 | | |
| 3-Cyanopropene | $C_4H_5N$ | | 19 | | |
| 1,2-Butadiene | $C_4H_6$ | 19 | <0 | | |
| 1,3-Butadiene | $C_4H_6$ | –4 | <–17 | 2–11.5 | |
| 1-Butyne | $C_4H_6$ | 8 | <–7 | | |
| 2-Butyne | $C_4H_6$ | 28 | <–7 | 1.4–? | |
| Cyclobutene | $C_4H_6$ | 2 | <–10 | | |
| 1-Buten-3-one | $C_4H_6O$ | | –7 | 2.1–15.6 | |
| Crotonaldehyde | $C_4H_6O$ | | 13 | 2.1–15.5/60° | 207 |
| Divinyl ether | $C_4H_6O$ | 29 | <–30 | 1.7–27 (or 36.5) | |
| 3,4-Epoxybutene | $C_4H_6O$ | | <–50 | | |
| Ethoxyacetylene | $C_4H_6O$ | 50 | <–7 | | |
| Methacrylaldehyde | $C_4H_6O$ | | <23 | | |
| Allyl formate | $C_4H_6O_2$ | | <22 | | |
| Butane-2,3-dione | $C_4H_6O_2$ | | <21 | | |
| Methyl acrylate | $C_4H_6O_2$ | | –3(o) | 2.8–25 | |
| Vinyl acetate | $C_4H_6O_2$ | | –8 | 2.6–13.4 | |
| 1-Bromo-2-butene | $C_4H_7Br$ | | 13 | | |
| 4-Bromo-1-butene | $C_4H_7Br$ | | 1 | | |
| 2-Chloro-2-butene | $C_4H_7Cl$ | | –25 | 2.3–9.3 | |
| 3-Chloro-1-butene | $C_4H_7Cl$ | | –27 | | |
| 3-Chloro-2-methyl-1-propene | $C_4H_7Cl$ | | –19 | 2.3–9.3 | |
| Butyryl chloride | $C_4H_7ClO$ | | <21 | | |
| 2-Chloroethyl vinyl ether | $C_4H_7ClO$ | | 25 | | |
| Isobutyryl chloride | $C_4H_7ClO$ | | <21 | | |
| Isopropyl chloroformate | $C_4H_7ClO_2$ | | <23 | | |
| Butyronitrile | $C_4H_7N$ | | 21 | | |
| Isobutyronitrile | $C_4H_7N$ | | 8 | | |
| 1-Butene | $C_4H_8$ | –6 | –80 | 1.6–9.3 | |
| *cis*-2-Butene | $C_4H_8$ | 1 | <–6 | 1.7–9.0 | |
| *trans*-2-Butene | $C_4H_8$ | 3 | <–6 | 1.8–9.7 | |
| Cyclobutane | $C_4H_8$ | 13 | <10 | 1.8–? | |
| Methylcyclopropane | $C_4H_8$ | 4 | <0 | | |
| 2-Methylpropene | $C_4H_8$ | –7 | <–10 | 1.8–8.8 | |
| *mixo*-(1,2- and 1,3-)Dichlorobutane | $C_4H_8Cl_2$ | | 21 | | |
| Dimethylaminoacetonitrile | $C_4H_8N_2$ | | <23 | | |
| 2-Butanone | $C_4H_8O$ | | –7 | 1.8–11.5 | |
| Butyraldehyde | $C_4H_8O$ | | –6 | 2.5–? | |
| Cyclopropyl methyl ether | $C_4H_8O$ | | <10 | | |
| 1,2-Epoxybutane | $C_4H_8O$ | | –15 | 1.5–18.3 | |
| Ethyl vinyl ether | $C_4H_8O$ | 36 | –46 | 1.7–28 | 202 |
| Isobutyraldehyde | $C_4H_8O$ | | –40 | 1.6–10.6 | 223 |
| Tetrahydrofuran | $C_4H_8O$ | | –17 | 1.8–11.8 | 224 |
| *m*-Dioxane | $C_4H_8O_2$ | | 2(o) | 2–22 | |
| *p*-Dioxane | $C_4H_8O_2$ | | 12 | 2–22 | 180 |

| *Name* | *Formula* | *B.P./°C* | *Fl.P./°C* | *E.L./%* | *A.I.T./°C* |
|---|---|---|---|---|---|
| Ethyl acetate | $C_4H_8O_2$ | | −4 | 2.2–11.5 | |
| Isopropyl formate | $C_4H_8O_2$ | | −6 | | |
| Methyl propionate | $C_4H_8O_2$ | | −2 | 2.5–13 | |
| Propyl formate | $C_4H_8O_2$ | | −3 | 2.3–? | |
| Tetrahydrothiophene | $C_4H_8S$ | | 13 | | |
| 1-Bromobutane | $C_4H_9Br$ | | 18 | 2.8–6.6/100° | |
| 2-Bromobutane | $C_4H_9Br$ | | 21 | | |
| 1-Bromo-2-methylpropane | $C_4H_9Br$ | | 22 | | |
| 2-Bromo-2-methylpropane | $C_4H_9Br$ | | −18 | | |
| 2-Bromoethyl ethyl ether | $C_4H_9BrO$ | | 5 | | |
| 1-Chlorobutane | $C_4H_9Cl$ | | −12 | 1.9–10.1 | |
| 2-Chlorobutane | $C_4H_9Cl$ | | −10 | | |
| 1-Chloro-2-methylpropane | $C_4H_9Cl$ | | −6 | 2.0–8.7 | |
| 2-Chloro-2-methylpropane | $C_4H_9Cl$ | | 0 | | |
| 2-Iodobutane | $C_4H_9I$ | | −10 | | |
| 1-Iodo-2-methylpropane | $C_4H_9I$ | | 0 | | |
| 2-Iodo-2-methylpropane | $C_4H_9I$ | | −10 | | |
| Pyrrolidine | $C_4H_9N$ | | 3 | | |
| Butyl nitrite | $C_4H_9NO_2$ | | 10 | | |
| Butane | $C_4H_{10}$ | 1 | −60 | 1.9–8.5 | |
| Isobutane | $C_4H_{10}$ | −12 | −81 | 1.9–8.5 | |
| Boron trifluoride diethyl etherate | $C_4H_{10}BF_3O$ | | <22 | | |
| Dichlorodiethylsilane | $C_4H_{10}Cl_2Si$ | | 24 | | |
| 2-Butanol | $C_4H_{10}O$ | | 14 | 1.7–9.8 (both at 100°) | |
| *tert*-Butanol | $C_4H_{10}O$ | | 10 | 2.4–8 | |
| Diethyl ether | $C_4H_{10}O$ | 36 | −45 | 1.8–36.5 | 180 |
| Methyl propyl ether | $C_4H_{10}O$ | 39 | | 2.0–14.8 | |
| 1,1-Dimethoxyethane | $C_4H_{10}O_2$ | | 1(o) | | |
| 1,2-Dimethoxyethane | $C_4H_{10}O_2$ | | <21 | | |
| Trimethyl orthoformate | $C_4H_{10}O_3$ | | 15 | | |
| 1-Butanethiol | $C_4H_{10}S$ | | 2 | | |
| 2-Butanethiol | $C_4H_{10}S$ | | −23 | | |
| Diethyl sulphide | $C_4H_{10}S$ | | −10 | | |
| 2-Methylpropanethiol | $C_4H_{10}S$ | | −10 | | |
| 2-Methyl-2-propanethiol | $C_4H_{10}S$ | | <−29 | | |
| Butylamine | $C_4H_{11}N$ | | 7 | 1.7–9.8 | |
| 2-Butylamine | $C_4H_{11}N$ | | −9 | | |
| *tert*-Butylamine | $C_4H_{11}N$ | 45 | −7 | 1.7–9.8 (both at 100°) | |
| Diethylamine | $C_4H_{11}N$ | | −18 | 1.8–10.1 | |
| Ethyldimethylamine | $C_4H_{11}N$ | | −36 | | |
| Isobutylamine | $C_4H_{11}N$ | | −9 | | |
| 2-Dimethylaminoethylamine | $C_4H_{12}N_2$ | | 11 | | |
| Tetramethylsilane | $C_4H_{12}Si$ | 26 | <0 | | |
| Tetramethyltin | $C_4H_{12}Sn$ | | <21 | 1.9–? | |
| Tetracarbonylnickel | $C_4NiO_4$ | 43 | <−20 | 2.0–? | |
| Pentacarbonyliron | $C_5FeO_5$ | | −15 | | ambient |
| Pyridine | $C_5H_5N$ | | 20 | 1.8–12.4 | |
| Cyclopentadiene | $C_5H_6$ | 42 | <25 | | |
| 2-Methyl-1-butenyne | $C_5H_6$ | | <−7(o) | | |

| Name | Formula | B.P./°C | Fl.P./°C | E.L./% | A.I.T./°C |
|---|---|---|---|---|---|
| 2-Methylfuran | $C_5H_6O$ | | −30 | | |
| 2-Methylthiophene | $C_5H_6S$ | | 8 | | |
| 1-Methylpyrrole | $C_5H_7N$ | | 16 | | |
| Cyclopentene | $C_5H_8$ | 44 | −29 | | |
| 2-Methyl-1,3-butadiene | $C_5H_8$ | 34 | −53 | 2.0–8.9 | 220 |
| 1,3-Pentadiene | $C_5H_8$ | 42 | −43 | 2–8.3 | |
| 1,4-Pentadiene | $C_5H_8$ | 26 | <0 | | |
| 1-Pentyne | $C_5H_8$ | | −20 | | |
| Allyl vinyl ether | $C_5H_8O$ | | <21(o) | | |
| Cyclopropyl methyl ketone | $C_5H_8O$ | | 13 | | |
| 2,3-Dihydropyran | $C_5H_8O$ | | −16 | | |
| 2-Methyl-3-butyn-2-ol | $C_5H_8O$ | | <21 | | |
| Methyl isopropenyl ketone | $C_5H_8O$ | | 21 | 1.8–9.0 (both at 50°) | |
| Allyl acetate | $C_5H_8O_2$ | | 22 | | |
| Ethyl acrylate | $C_5H_8O_2$ | | 9 | 1.8–? | |
| Methyl crotonate | $C_5H_8O_2$ | | −1 | | |
| Methyl methacrylate | $C_5H_8O_2$ | | 11 | 2.1–12.5 | |
| Isopropenyl acetate | $C_5H_8O_2$ | | 16 | 1.9–? | |
| Vinyl propionate | $C_5H_8O_2$ | | 1 | | |
| Chlorocyclopentane | $C_5H_9Cl$ | | 16 | | |
| Pivaloyl chloride | $C_5H_9ClO$ | | 24 | | |
| Pivalonitrile | $C_5H_9N$ | | 21 | | |
| 1,2,5,6-Tetrahydropyridine | $C_5H_9N$ | | 16 | | |
| Cyclopentane | $C_5H_{10}$ | 49 | −7 | | |
| 1,1-Dimethylcyclopropane | $C_5H_{10}$ | 20 | | | |
| Ethylcyclopropane | $C_5H_{10}$ | 36 | <10 | | |
| 2-Methyl-1-butene | $C_5H_{10}$ | 39 | −20 | | |
| 2-Methyl-2-butene | $C_5H_{10}$ | 38 | −20 | | |
| 3-Methyl-1-butene | $C_5H_{10}$ | 20 | −7 | 1.5–9.1 | |
| Methylcyclobutane | $C_5H_{10}$ | 41 | <10 | | |
| 1-Pentene | $C_5H_{10}$ | 30 | −18 | 1.5–8.7 | |
| 2-Pentene | $C_5H_{10}$ | 30 | −18 | | |
| 3-Dimethylaminopropionitrile | $C_5H_{10}N_2$ | | <22 | | |
| Allyl ethyl ether | $C_5H_{10}O$ | | <24 | | |
| Ethyl propenyl ether | $C_5H_{10}O$ | | <−5 | | |
| Isopropyl vinyl ether | $C_5H_{10}O$ | | −32 | | |
| Isovaleraldehyde | $C_5H_{10}O$ | | −5 | | |
| 3-Methyl-2-butanone | $C_5H_{10}O$ | | <22 | | |
| 3-Methyl-1-buten-3-ol | $C_5H_{10}O$ | | 18 | | |
| 2-Methyltetrahydrofuran | $C_5H_{10}O$ | | −12 | | |
| 2-Pentanone | $C_5H_{10}O$ | | 7 | 1.6–8.2 | |
| 3-Pentanone | $C_5H_{10}O$ | | 13 | 1.6–? | |
| 4-Penten-1-ol | $C_5H_{10}O$ | | <23 | | |
| Tetrahydropyran | $C_5H_{10}O$ | | −20 | | |
| Valeraldehyde | $C_5H_{10}O$ | | 12 | | |
| Butyl formate | $C_5H_{10}O_2$ | | 18 | 1.7–8 | |
| 3,3-Dimethoxypropene | $C_5H_{10}O_2$ | | 19 | | |
| 2,2-Dimethyl-1,3-dioxolane | $C_5H_{10}O_2$ | | −1 | | |
| Ethyl propionate | $C_5H_{10}O_2$ | | 12 | 1.9–11 | |
| Isobutyl formate | $C_5H_{10}O_2$ | | <21 | 2.0–8.9 | |
| Isopropyl acetate | $C_5H_{10}O_2$ | | 4 | 1.7–7.8 | |

| Name | Formula | B.P./°C | Fl.P./°C | E.L./% | A.I.T./°C |
|---|---|---|---|---|---|
| 2-Methoxyethyl vinyl ether | $C_5H_{10}O_2$ | | 18(o) | | |
| Methyl butyrate | $C_5H_{10}O_2$ | | 14 | | |
| 4-Methyl-1,3-dioxane | $C_5H_{10}O_2$ | | 16 | | |
| Methyl isobutyrate | $C_5H_{10}O_2$ | | 13(o) | | |
| Propyl acetate | $C_5H_{10}O_2$ | | 14 | 1.7–8 | |
| Diethyl carbonate | $C_5H_{10}O_3$ | | 25 | | |
| 1-Bromo-3-methylbutane | $C_5H_{11}Br$ | | 21 | | |
| 2-Bromopentane | $C_5H_{11}Br$ | | 20 | | |
| 1-Chloro-3-methylbutane | $C_5H_{11}Cl$ | | <21 | 1.5–7.4 | |
| 2-Chloro-2-methylbutane | $C_5H_{11}Cl$ | | 12 | 1.5–7.4 | |
| 1-Chloropentane | $C_5H_{11}Cl$ | | 13(o) | 1.6–8.6 | |
| 2-Iodopentane | $C_5H_{11}I$ | | <23 | | |
| Cyclopentylamine | $C_5H_{11}N$ | | 13 | | |
| 1-Methylpyrrolidine | $C_5H_{11}N$ | | 3 | | |
| Piperidine | $C_5H_{11}N$ | | 3 | | |
| 4-Methylmorpholine | $C_5H_{11}NO$ | | 24 | | |
| Isopentyl nitrite | $C_5H_{11}NO_2$ | | 10 | 1–? (calc.) | 209 |
| Pentyl nitrite | $C_5H_{11}NO_2$ | | <23 | | |
| 2,2-Dimethylpropane | $C_5H_{12}$ | 9 | <−7 | 1.4–7.5 | |
| 2-Methylbutane | $C_5H_{12}$ | 28 | −51 | 1.4–7.6 | |
| Pentane | $C_5H_{12}$ | 36 | −49 | 1.4–8.0 | |
| Butyl methyl ether | $C_5H_{12}O$ | | <18 | | |
| Ethyl isopropyl ether | $C_5H_{12}O$ | | <−15 | | |
| Ethyl propyl ether | $C_5H_{12}O$ | | <−20 | 1.7–9 | |
| Isopentanol | $C_5H_{12}O$ | | <23 | 1.4–9 (both at 100°) | |
| 2-Pentanol | $C_5H_{12}O$ | | <23 | | |
| *tert*-Pentanol | $C_5H_{12}O$ | | 19 | 1.2–9 | |
| Diethoxymethane | $C_5H_{12}O_2$ | | <21 | | |
| 1,1-Dimethoxypropane | $C_5H_{12}O_2$ | | <10 | | |
| 2,2-Dimethoxypropane | $C_5H_{12}O_2$ | | −7 | | |
| 2-Methyl-2-butanethiol | $C_5H_{12}S$ | | −1 | | |
| 3-Methylbutanethiol | $C_5H_{12}S$ | | <23 | | |
| Pentanethiol | $C_5H_{12}S$ | | 18 | | |
| Isopentylamine | $C_5H_{13}N$ | | −1 | 2.3–22 | |
| *N*-Methylbutylamine | $C_5H_{13}N$ | | 2 | | |
| 1-Pentylamine | $C_5H_{13}N$ | | −1 | | |
| 2-Dimethylamino-*N*-methylethylamine | $C_5H_{14}N_2$ | | 14 | | |
| Hexafluorobenzene | $C_6F_6$ | | 10 | | |
| 1,2,4,5-Tetrafluorobenzene | $C_6H_2F_4$ | | 4(o) | | |
| 1,2,4-Trifluorobenzene | $C_6H_3F_3$ | | −5(o) | | |
| *o*-, *m*- or *p*-Chlorofluorobenzene | $C_6H_4ClF$ | | all 18 | | |
| *m*-Difluorobenzene | $C_6H_4F_2$ | | <0 | | |
| *p*-Difluorobenzene | $C_6H_4F_2$ | | −5(o) | | |
| Fluorobenzene | $C_6H_5F$ | | −15 | | |
| Benzene | $C_6H_6$ | | −11 | 1.4–8.0 | |
| 1,5-Hexadien-3-yne | $C_6H_6$ | | <−20 | 1.5–? | |
| Benzenethiol | $C_6H_6S$ | | <21 | | |
| 1,3-Cyclohexadiene | $C_6H_8$ | | <23 | | |
| 1,4-Cyclohexadiene | $C_6H_8$ | | −11 | | |
| 2,5-Dimethylfuran | $C_6H_8O$ | | 16 | | |

| Name | Formula | B.P./°C | Fl.P./°C | E.L./% | A.I.T./°C |
|---|---|---|---|---|---|
| Cyclohexene | $C_6H_{10}$ | | <−6 | 1.2–?/100° | |
| 1,4-Hexadiene | $C_6H_{10}$ | | −21 | 2.0–6.1 | |
| 1,5-Hexadiene | $C_6H_{10}$ | | −46 | | |
| 1-Hexyne | $C_6H_{10}$ | | <10 | | |
| 2- or 3-Hexyne | $C_6H_{10}$ | | <23 | | |
| 2-Methyl-1,3-pentadiene | $C_6H_{10}$ | | <−20 | | |
| 4-Methyl-1,3-pentadiene | $C_6H_{10}$ | | −34 | | |
| Diallyl ether | $C_6H_{10}O$ | | −6 | | |
| Ethyl crotonate | $C_6H_{10}O_2$ | | 2 | | |
| Ethyl methacrylate | $C_6H_{10}O_2$ | | 21 | 1.8–? | |
| Vinyl butyrate | $C_6H_{10}O_2$ | | 21(o) | 1.4–8.8 | |
| Diallylamine | $C_6H_{11}N$ | | 21 | | |
| Cyclohexane | $C_6H_{12}$ | | −17 | 1.3–8.4 | |
| Ethylcyclobutane | $C_6H_{12}$ | | −15 | 1.2–7.7 | 210 |
| 1-Hexene | $C_6H_{12}$ | | −26 | 1.2–6.9 | |
| 2-Hexene | $C_6H_{12}$ | | −21 | | |
| Methylcyclopentane | $C_6H_{12}$ | | −29 | | |
| 2-Methyl-1-pentene | $C_6H_{12}$ | | −28 | | |
| 4-Methyl-1-pentene | $C_6H_{12}$ | | −7 | | |
| *cis*-4-Methyl-2-pentene | $C_6H_{12}$ | | −32 | | |
| *trans*-4-Methyl-2-pentene | $C_6H_{12\ 2}$ | | −29 | | |
| Butyl vinyl ether | $C_6H_{12}O$ | | −1 | | |
| 2,2-Dimethyl-3-butanone | $C_6H_{12}O$ | | 12 | | |
| 2-Ethylbutyraldehyde | $C_6H_{12}O$ | | 21(o) | | |
| 3-Hexanone | $C_6H_{12}O$ | | 14 | | |
| Isobutyl vinyl ether | $C_6H_{12}O$ | | −9 | | |
| 2- or 3-Methylpentanal | $C_6H_{12}O$ | | <23 | | |
| 2-Methyl-3-pentanone | $C_6H_{12}O$ | | 11 | | |
| 3-Methyl-2-pentanone | $C_6H_{12}O$ | | 15 | | |
| 4-Methyl-2-pentanone | $C_6H_{12}O$ | | 17 | 1.2–8 | |
| Butyl acetate | $C_6H_{12}O_2$ | | 23 | 1.4–7.5 | |
| 2-Butyl acetate | $C_6H_{12}O_2$ | | 18 | 1.3–7.5 | |
| 2,6-Dimethyl-1,4-dioxane | $C_6H_{12}O_2$ | | 24 | | |
| Ethyl isobutyrate | $C_6H_{12}O_2$ | | <18 | | |
| 2-Ethyl-2-methyl-1,3-dioxolane | $C_6H_{12}O_2$ | | 23(o) | | |
| 4-Hydroxy-4-methyl-2-pentanone | $C_6H_{12}O_2$ | | 9 | | |
| Isobutyl acetate | $C_6H_{12}O_2$ | | 17 | 2.4–10.5 | |
| Isopropyl propionate | $C_6H_{12}O_2$ | | <20 | | |
| Methyl isovalerate | $C_6H_{12}O_2$ | | <22 | | |
| Methyl pivalate | $C_6H_{12}O_2$ | | 7 | | |
| Methyl valerate | $C_6H_{12}O_2$ | | <25 | | |
| 2,5-Dimethoxytetrahydrofuran | $C_6H_{12}O_3$ | | <10 | | |
| Isobutyl peroxyacetate | $C_6H_{12}O_3$ | | <25 | | |
| 2,4,6-Trimethyltrioxane | $C_6H_{12}O_3$ | | 17 | 1.3–? | |
| Cyclohexylamine | $C_6H_{13}N$ | | 21 | | |
| 1-Methylpiperidine | $C_6H_{13}N$ | | <23 | | |
| 2-Methylpiperidine | $C_6H_{13}N$ | | 10 | | |
| 3-Methylpiperidine | $C_6H_{13}N$ | | 8 | | |
| 4-Methylpiperidine | $C_6H_{13}N$ | | 9 | | |
| Perhydroazepine | $C_6H_{13}N$ | | 22 | | |

| Name | Formula | B.P./°C | Fl.P./°C | E.L./% | A.I.T./°C |
|---|---|---|---|---|---|
| 2,2-Dimethylbutane | $C_6H_{14}$ | 50 | −48 | 1.2–7.0 at 100°) | |
| 2,3-Dimethylbutane | $C_6H_{14}$ | | −29 | 1.2–7.0 | |
| Hexane | $C_6H_{14}$ | | −23 | 1.1–7.5 | 225 |
| Isohexane | $C_6H_{14}$ | | −7 | 1.2–7.0 | |
| 3-Methylpentane | $C_6H_{14}$ | | −7 | 1.2–7.0 | |
| Butyl ethyl ether | $C_6H_{14}O$ | | 4 | | |
| Diisopropyl ether | $C_6H_{14}O$ | | −28 | 1.4–21 | |
| Dipropyl ether | $C_6H_{14}O$ | | <21 | | 215 |
| 2-Methyl-2-pentanol | $C_6H_{14}O$ | | 21 | | |
| 3-Methyl-3-pentanol | $C_6H_{14}O$ | | 24 | | |
| 1,1-Diethoxyethane | $C_6H_{14}O_2$ | | <4 | 1.7–10.4 | |
| 1,2-Diethoxyethane | $C_6H_{14}O_2$ | | | | 205 |
| Triethylaluminium | $C_6H_{15}Al$ | | <−53 | | <−53 |
| Triethyl borate | $C_6H_{15}BO_3$ | | 11 | | |
| Butylethylamine | $C_6H_{15}N$ | | 18(o) | | |
| Diisopropylamine | $C_6H_{15}N$ | | −7 | | |
| 1,3-Dimethylbutylamine | $C_6H_{15}N$ | | 13 | | |
| Dipropylamine | $C_6H_{15}N$ | | 17 | | |
| Triethylamine | $C_6H_{15}N$ | | −7 | 1.2–8.0 | |
| 1,2-Bis(dimethylamino)ethane | $C_6H_{16}N_2$ | | 18 | | |
| Diethoxydimethylsilane | $C_6H_{16}O_2Si$ | | <23 | | |
| Bis-trimethylsilyl oxide | $C_6H_{18}OSi_2$ | | −1 | | |
| Hexamethyldisilazane | $C_6H_{19}NSi_2$ | | 14 | | |
| Trifluoromethylbenzene | $C_7H_5F_3$ | | <21 | | |
| *o*-Fluorotoluene | $C_7H_7F$ | | 8 | | |
| *m*-Fluorotoluene | $C_7H_7F$ | | 12 | | |
| *p*-Fluorotoluene | $C_7H_7F$ | | 10 | | |
| Bicyclo-[2.2.1]-2,5-heptadiene | $C_7H_8$ | | −21 | | |
| 1,3,5-Cycloheptatriene | $C_7H_8$ | | 4 | | |
| Toluene | $C_7H_8$ | | 4 | 1.3–7.0 | |
| Cycloheptene | $C_7H_{12}$ | | <23 | | |
| 1- or 3-Heptyne | $C_7H_{12}$ | | <23 | | |
| 4-Methylcyclohexene | $C_7H_{12}$ | | −1(o) | | |
| Cycloheptane | $C_7H_{14}$ | | 15 | | |
| Ethylcyclopentane | $C_7H_{14}$ | | <21 | 1.1–6.7 | |
| 1-Heptene | $C_7H_{14}$ | | −1 | | |
| 2-Heptene | $C_7H_{14}$ | | <−1 | | |
| 3-Heptene | $C_7H_{14}$ | | <−7 | | |
| Methylcyclohexane | $C_7H_{14}$ | | −4 | 1.2–6.7 | |
| 2,3,3-Trimethylbutene | $C_7H_{14}$ | | <0 | | |
| 2,4-Dimethyl-3-pentanone | $C_7H_{14}O$ | | 15 | | |
| 3,3-Diethoxypropene | $C_7H_{14}O_2$ | | <23 | | |
| Ethyl isovalerate | $C_7H_{14}O_2$ | | 25 | | |
| Isobutyl propionate | $C_7H_{14}O_2$ | | <22 | | |
| Isopentyl acetate | $C_7H_{14}O_2$ | | <23 | 1.1–7.0 (both at 100°C) | |
| Isopropyl butyrate | $C_7H_{14}O_2$ | | 24 | | |
| Isopropyl isobutyrate | $C_7H_{14}O_2$ | | <21 | | |
| Pentyl acetate | $C_7H_{14}O_2$ | | 25 | 1.1–7.5 | |
| 2-Pentyl acetate | $C_7H_{14}O_2$ | | 23 | 1.1–7.5 | |
| 2,6-Dimethylpiperidine | $C_7H_{15}N$ | | 16 | | |

| Name | Formula | B.P./°C | Fl.P./°C | E.L./% | A.I.T./°C |
|---|---|---|---|---|---|
| 1-Ethylpiperidine | $C_7H_{15}N$ | | 19 | | |
| 2-Ethylpiperidine | $C_7H_{15}N$ | | −12 | | |
| 2,3-Dimethylpentane | $C_7H_{16}$ | | <−6 | 1.1–6.7 | |
| 2,4-Dimethylpentane | $C_7H_{16}$ | | −12 | | |
| Heptane | $C_7H_{16}$ | | −4 | 1.2–6.7 | 223 |
| 2-Methylhexane | $C_7H_{16}$ | | −1 | 1.0–6.0 | |
| 3-Methylhexane | $C_7H_{16}$ | | −1 | 1.0–6.0 | |
| 2,2,3-Trimethylbutane | $C_7H_{16}$ | | <0 | 1.0–? | |
| Phenylglyoxal | $C_8H_6O_2$ | | <23 | | |
| 1,3,5,7-Cyclooctatetraene | $C_8H_8$ | | <22 | | |
| Ethylbenzene | $C_8H_{10}$ | | 15 | 1.2–6.8 | |
| *o*-Xylene | $C_8H_{10}$ | | 17 | 1.0–6.0 | |
| *m*-Xylene | $C_8H_{10}$ | | 25 | 1.1–7.0 | |
| *p*-Xylene | $C_8H_{10}$ | | 25 | 1.1–7.0 | |
| 4-Vinylcyclohexene | $C_8H_{12}$ | | 16 | | |
| 1,7-Octadiene | $C_8H_{14}$ | | 25 | | |
| 1-Octyne | $C_8H_{14}$ | | 16 | | |
| 2-, 3- or 4-Octyne | $C_8H_{14}$ | | all <23 | | |
| Vinylcyclohexane | $C_8H_{14}$ | | 16 | | |
| *cis*-1,2-Dimethylcyclohexane | $C_8H_{16}$ | | 11 | | |
| *trans*-1,2-Dimethylcyclohexane | $C_8H_{16}$ | | 7 | | |
| 1,3-Dimethylcyclohexane | $C_8H_{16}$ | | 6 | | |
| *cis*-1,4-Dimethylcyclohexane | $C_8H_{16}$ | | 10 | | |
| *trans*-1,4-Dimethylcyclohexane | $C_8H_{16}$ | | 16 | | |
| 1-Octene | $C_8H_{16}$ | | 8 | | |
| 2-Octene | $C_8H_{16}$ | | 14 | | |
| 2,3,4-Trimethyl-1-pentene | $C_8H_{16}$ | | <21 | | |
| 2,4,4-Trimethyl-1-pentene | $C_8H_{16}$ | | −29 | | |
| 2,3,4-Trimethyl-2-pentene | $C_8H_{16}$ | | 1 | | |
| 2,4,4-Trimethyl-2-pentene | $C_8H_{16}$ | | 2 | | |
| 3,4,4-Trimethyl-2-pentene | $C_8H_{16}$ | | <21 | | |
| 2-Ethylhexanal | $C_8H_{16}O$ | | | | 197 |
| 2,3-Dimethylhexane | $C_8H_{18}$ | | 7(o) | | |
| 2,4-Dimethylhexane | $C_8H_{18}$ | | 10(o) | | |
| 3-Ethyl-2-methylpentane | $C_8H_{18}$ | | <21 | | |
| 2-Methylheptane | $C_8H_{18}$ | | <10 | | |
| 3-Methylheptane | $C_8H_{18}$ | | −23 | | |
| Octane | $C_8H_{18}$ | | 13 | 1.0–4.7 | 220 |
| 2,2,3-Trimethylpentane | $C_8H_{18}$ | | <21 | | |
| 2,2,4-Trimethylpentane | $C_8H_{18}$ | | −12 | 1.1–6.0 | |
| 2,3,3-Trimethylpentane | $C_8H_{18}$ | | <21 | | |
| 2,3,4-Trimethylpentane | $C_8H_{18}$ | | 4 | | |
| Dibutyl ether | $C_8H_{18}O$ | | 25 | 1.5–7.6 | 194 |
| Di-*tert*-butyl peroxide | $C_8H_{18}O_2$ | | 18 | | |
| Bis(2-ethoxyethyl) ether | | | | | 205 |
| Di-2-butylamine | $C_8H_{19}N$ | | 24 | | |
| Diisobutylamine | $C_8H_{19}N$ | | 21 | | |
| 1-Phenyl-1,2-propanedione | $C_9H_8O_2$ | | <21 | | |
| 2,6-Dimethyl-3-heptene | $C_9H_{18}$ | | 21(o) | | |
| 1,3,5-Trimethylcyclohexane | $C_9H_{18}$ | | 19 | | |
| 3,3-Diethylpentane | $C_9H_{20}$ | | <21 | 0.7–7.7 | |
| 2,5-Dimethylheptane | $C_9H_{20}$ | | 24 | | |

| Name | Formula | B.P./°C | Fl.P./°C | E.L./% | A.I.T./°C |
|---|---|---|---|---|---|
| 3,5-Dimethylheptane | $C_9H_{20}$ | | 23 | | |
| 4,4-Dimethylheptane | $C_9H_{20}$ | | 21 | | |
| 3-Ethyl-2,3-dimethylpentane | $C_9H_{20}$ | | 8 | | |
| 3-Ethyl-4-methylhexane | $C_9H_{20}$ | | 24 | | |
| 4-Ethyl-2-methylhexane | $C_9H_{20}$ | | <21 | | |
| 2-Methyloctane | $C_9H_{20}$ | | | | 220 |
| 3-Methyloctane | $C_9H_{20}$ | | | | 220 |
| 4-Methyloctane | $C_9H_{20}$ | | | | 225 |
| Nonane | $C_9H_{20}$ | | | | 190 |
| 2,2,5-Trimethylhexane | $C_9H_{20}$ | | 13 | | |
| 2,2,3,3-Tetramethylpentane | $C_9H_{20}$ | | <21 | 0.8–4.9 | |
| 2,2,3,4-Tetramethylpentane | $C_9H_{20}$ | | <21 | | |
| Dicyclopentadiene | $C_{10}H_{12}$ | | −7(o) | | |
| 2-Ethylhexyl vinyl ether | $C_{10}H_{20}O$ | | | | 202 |
| Isopentyl isovalerate | $C_{10}H_{20}O_2$ | | 24 | | |
| Decane | $C_{10}H_{22}$ | | | | 210 |
| 2-Methylnonane | $C_{10}H_{22}$ | | | | 210 |
| *tert*-Butyl peroxybenzoate | $C_{11}H_{14}O_3$ | | 19 | | |
| Dodecane | $C_{12}H_{26}$ | | | | 205 |
| Dihexyl ether | $C_{12}H_{26}O$ | | | | 185 |
| Triisobutylaluminium | $C_{12}H_{27}Al$ | | <0 | | <4 |
| Tributylphosphine | $C_{12}H_{27}P$ | | | | 200 |
| Titanium tetrapropoxide | $C_{12}H_{28}O_4Ti$ | | <22 | | |
| Tetradecane | $C_{14}H_{30}$ | | | | 200 |
| Hexadecane | $C_{16}H_{34}$ | | | | 205 |
| Dioctyl ether | $C_{16}H_{34}O$ | | | | 205 |
| Trichlorosilane | $Cl_3HSi$ | 32 | −28(o) | | 104 |
| Deuterium | $D_2$ | −249 | | 5–75 | |
| Trifluorosilane | $F_3HSi$ | −95 | | flammable | |
| Tetrafluorohydrazine | $F_4N_2$ | −73 | | | explosive |
| Germane | $GeH_4$ | −90 | | | ambient |
| Hydrogen | $H_2$ | −253 | | 4.1–74.2 | |
| Hydrogen sulphide | $H_2S$ | −62 | | 4–44 | |
| Hydrogen disulphide | $H_2S_2$ | | <22 | | |
| Hydrogen selenide | $H_2Se$ | −42 | | | |
| Ammonia | $H_3N$ | −33 | | 16–25 | |
| Phosphine | $H_3P$ | −88 | | 1–? | see text |
| Stibine | $H_3Sb$ | −17 | | | |
| Hydrazine | $H_4N_2$ | | 38 | 4.7–100 | 23 on rust<br>132 on iron<br>156 on stainless steel |
| Diphosphane | $H_4P_2$ | | | | ambient |
| Silane | $H_4Si$ | −112 | | | ambient |
| Disilane | $H_6Si_2$ | −14.5 | | | ambient |
| Phosphorus (white) | P | | | | 38.5 |
| Sulphur (dust in air) | S | | | 35–1400 (g/m³) | |

## Appendix 3

# Glossary of Abbreviations and Unusual Words used in Text

| | |
|---|---|
| Analogue | compound of the same structural type |
| Ambient | usual or surrounding |
| A.I.T. | auto-ignition temperature (Appendix 2) |
| Autoxidation | slow reaction with air |
| b.p. | boiling point |
| Congener | compound with a related but not identical structure |
| Desiccate | dry intensively |
| Detonable | capable of detonation |
| Diglyme | diethyleneglycol dimethyl ether |
| DMF | dimethylformamide |
| DMSO | dimethyl sulphoxide |
| DSC | differential scanning calorimetry |
| DTA | differential thermal analysis |
| E.L. | explosive limits, % in air (Appendix 2) |
| Endotherm | absorption of heat |
| Exotherm | liberation of (reaction) heat |
| GLC | gas–liquid chromatography |
| Glyme | diethyleneglycol monomethyl ether |
| Homologue | compound of the same (organic) series |
| Hypergolic | ignites on contact |
| IMS | industrial methylated spirit (ethanol) |
| IR | infrared spectroscopy |
| Initiation | triggering off explosion or decomposition |
| m.p. | melting point |
| Oxidant | oxidising agent |
| Propagation | spread or transmission of decomposition, flame or explosion |
| Pyrophoric | igniting on contact with air (occasionally friction) |

| | |
|---|---|
| Redox compound | compound with reducing and oxidising features |
| Reducant | reducing agent |
| Slurry | pourable mixture of solid and liquid |
| THF | tetrahydrofuran |
| TGA | thermogravimetric analysis |
| USNTIS | US National Technical Information Service |
| UV | ultraviolet spectroscopy |

Appendix 4

# Index of Class, Group and Topic Titles used in Section I

Appendix 5

# Index of Chemical Names and Synonyms used in Section 2

The majority of the names for chemicals in this alphabetically arranged index conform to one of the systematic series permitted under the various sections of the IUPAC Definitive Rules for Nomenclature. Where there is a marked difference between these names and the alternative names recommended in the IUPAC-based BS 2474:1965 or ASE lists, or long-established traditional names, these are given as cross-references back to the IUPAC names used as bold titles in the text.

It should be noted that italicised hyphenated prefixes which indicate structure, such as *cis*-, *o*-, *m*-, *tert*-, *mixo*-, *N*-, etc., have been ignored during alphabetical arrangement of successive names in this index, while the roman structural prefixes iso and neo, and roman multiplying prefixes, such as di, tris, tetra and hexakis, have been included for indexing purposes.

There are several general cross-entries in the index for which no page numbers are given, to facilitate finding the systematic names for some of the less common traditional names. There are, however, a few general entries with page numbers which do appear as items in the text.